Spon's Architects' and Builders' Price Book

Spon's Architects' and Builders' Price Book

Edited by
DAVIS LANGDON & EVEREST
Chartered Quantity Surveyors

1990

One hundred and fifteenth Edition

LONDON
E. & F. N. SPON

First published 1873
One hundred and fifteenth edition 1989
E. & F. N. Spon
11 New Fetter Lane, London, EC4P 4EE
© *1989 E. & F. N. Spon Ltd*

Printed in Great Britain by
Richard Clay Ltd,
Bungay, Suffolk

ISBN 0–419–14980–5
ISSN 0306–3046

British Library Cataloguing in Publication Data

Spon's Architects' and builders' price book.
   –115th ed. (1990)
   I. Buildings – Estimates – Great Britain –
   Periodicals
   692'.5   TH435
   ISBN 0–419–14980–5

Preface

The construction industry may no longer be used to regulate the economy, but the Chancellor's policy of raising interest rates in order to reduce over-heating has had the much needed effect of dampening demand, for in areas of major building activity, annual increases in tender prices of as much as 20% were recorded during 1988.

The rise in interest rates in the first half of 1989 created problems for house builders and businesses that are highly geared, but the overall impact was beneficial, for confidence was looking increasingly edgy as tender price increases started to outstrip projected returns for development. As it is, the fall-off in house building should relieve the pressure on certain trades, such as brick layers, carpenters and plasterers. It should also reduce extended delivery periods on basic materials.

However, there is already enough demand and work committed in the commercial and non-residential sectors to ensure that in the short-term shortages will continue to occur both in skilled labour, such as steel fixers and services sub-contractors, and in materials such as curtain walling, cladding and lifts. As a result the annual rise in tender prices will still be between 10 and 12% over the next year.

Regular readers of Spon's Price Book will see from the tender price index on the front cover that the prices in this edition for Major Work have increased by 12% compared with last year.

The current edition includes the annual NJC wage award agreed on 26th June 1989 and the tender price index on the front cover reflects our view as to current market conditions and tender levels in the industry.

A major advantage of Spons's is that, being geared to building tenders, it is possible for the reader to make such adjustments as may be necessary during the year to reflect trends in the competitive world. This is achieved by comparing the Tender Index with subsequent revisions to that index as circulated in Spon's Price Book Update.

The update is circulated free every three months to those readers who have registered with the publishers; a coloured card is bound in with this volume for this purpose.

Last year the Measured Works sections of Spon's Price Book were restructured in accordance with SMM7, which became operative in July 1988. However, the use of SMM7 has not become as widespread as might have been anticipated so we have retained a number of SMM6 labours/items to help Estimators build up additional rates or to break composite SMM7 rates back to their SMM6 equivalents. We have also retained our SMM6/SMM7 index, which was introduced last year to assist readers during the transition period.

The major change to this year's edition has been the expansion of the Approximate Estimates and Comparative Prices section. Now revised and considerably enlarged, it contains a comprehensive range of specifications and costs for both residential and a number of commercial building types. Readers previously applying average rates per square metre for an estimate can now use this revised section to produce a price which more accurately reflects the shape, size and quality of the scheme under consideration. The section also still includes rates from the previous Comparative prices section, enabling the reader to compare alternative forms of construction, finishes, perimeter details etc.

Additional items included in this years' edition include:-

'Anderson' roofing items   'Jumbo' walls
'Brickforce' stainless reinforcement   'Mandolite P20' fire coatings'
'Cordek' troughed moulds   'Plannja' cladding/deckings
'Glasal' sheeting   'Pyran' fire-resisting glass
'Glulam' laminated beams   'Rockwool' fire stops/barriers
'Holorib' floor decking   Stainless Steel coverings

plus an expansion of cavity insulation slabs, damp-proofing items, land
drainage, safety flooring, wall tiling and waterproofing underlay sections.

As previously, it is essential for a full understanding of the book that the
reader should study the introduction or preamble which appears at the beginning
of each section.

The reader's attention is also drawn to a ruling of the European Court on 21
June 1988. As from 1 April 1989, Value Added Tax will be chargeable at the
standard rate, currently 15%, on supplies of services in the course of:

1.   The construction of a non-domestic building
2.   The construction or demolition of a civil engineering work
3.   The demolition of any building, and
4.   The approved alteration of a non-domestic protected building

**Prices included within this edition do not include for Value Added Tax, for
which provision must be made if appropriate.**

For the benefit of new readers, a brief guide to the book follows:

Part I: Fees and Daywork
   This section contains Fees for Professional Services, Daywork and Prime Cost.

Part II: Rates of Wages
   This section includes authorized wage agreements currently applicable to the
Building and Associated Industries.

Part III: Prices for Measured Work
   This section contains Prices for Measured Work - Major Works, and Prices for
Measured Work - Minor Works (on coloured paper).

Part IV: Approximate Estimating
   This section contains the Building Cost and Tender Price Index, information on
regional price variations, prices per square metre for various types of
buildings, approximate estimates, cost limits and allowances and a procedure for
valuing property for insurance purposes.

Part V: Tables and Memoranda
   This section contains general formulae, weights and quantities of materials,
other design criteria and useful memoranda associated with each trade and a list
of useful Trade Associations.

While every effort is made to ensure the accuracy of the information given in
this publication, neither the Editors nor Publishers in any way accept liability
for loss of any kind resulting from the use made by any person of such
information.

D A V I S   L A N G D O N   &   E V E R E S T
Chartered Quantity Surveyors
5, Golden Square
London   W1R 3AE

Contents

PART I
FEES AND DAYWORK

PART II
RATES OF WAGES

PART III
PRICES FOR MEASURED WORK

PART IV
APPROXIMATE ESTIMATING

PART V
TABLES AND MEMORANDA

Index

References in brackets after page numbers, refer to SMM7 and the New Common Arrangement sections.

THE MANAGEMENT OF QUALITY IN CONSTRUCTION

J Ashford

The quality of a product or service is a measure of its ability to satisfy customer requirements. But quality is also vitally important to company prosperity and establishes the management procedures which will ensure that standards are met. Many clients are now insisting that architects, designers and construction contractors can demonstrate the existence of an effective quality assurance system. An effective quality system brings further benefits; it exposes costly weaknesses in an organization and enables a company to increase its efficiency, reduce delays and maximise profitability.

Developing the right approach to the management of quality is essential. For systems to be effective they must cover the complete process from the design stage to the procurement, construction and maintenance of works. **The Management of Quality in Construction** provides the reader with a knowledge of the principles of quality management and an understanding of how they may successfully be applied in the particular circumstances of the construction industry.

The areas covered range from an historical review of traditional methods of assuring quality in the industry and how contractual arrangements have evolved, to an interpretation of quality system standards in the context of construction. Examples are given which highlight specific areas, and specialist chapters on organization structures and the techniques of quality auditing are included.

John Ashford is a recognised authority on quality assurance, he became Group Quality Assurance Manager of the Wimpey Group in 1984. In **The Management of Quality in Construction** he presents a *proven system* for the implementation of quality assurance in the construction industry and draws upon his wide-ranging experience to illustrate the problems which are likely to arise and the tactics which can be used to overcome them.

Contents: The quality management philosophy. The construction tradition. Standards and terminology. The quality system. Documentation. Design. Procurement. The construction site. Quality audits. Quality assurance and the contract. A strategy for success. Appendices. Bibliography. Index.

1989 Hardback 0 419 14910 4 £25.00 250pp

This book may be obtained from your usual bookshop. In case of difficulty please write to the address below or telephone the Order Department on 0264 332424.

 **E & F N SPON**

11 New Fetter Lane, London EC4P 4EE

SMM6/SMM7 Index

For the benefit of readers, during a year of transition, we have included a brief index of comparative trade/work sections between SMM6 and SMM7, used in producing this years edition of 'Spons'.

SMM6		SMM7	
A	GENERAL RULES		
B	PRELIMINARIES	A	PRELIMINARIES/GENERAL CONDITIONS
	Preliminary Particulars	A10	Project particulars
		A11	Drawings
		A12	The site/Existing buildings
		A13	Description of the work
	Contract	A20	The Contract/Sub-contract
		A30	Employer's requirements: Tendering/Sub-letting/Supply
		A31	Employer's requirements: Provision, content and use of documents
		A32	Employer's requirements: Management of the Works
		A33	Employer's requirements: Quality standards/control
		A34	Employer's requirements: Security/Safety/Protection
		A35	Employer's requirements: Specific limitations on method/ sequence/timing
		A36	Employer's requirements: Facilities/Temporary works/Services
		A37	Employer's requirements: Operation/Maintenance of the finished building
		A50	Work/Materials by the Employer
	Works by Nominated Sub Contractors	A51	Nominated sub-contractors
	Goods and Materials from Nominated	A52	Nominated suppliers
	Suppliers and works by Public Bodies	A53	Work by statutory authorities
	General facilities and obligations	A40	Contractor's general cost items: Management and staff
		A41	Contractor's general cost items: Site accommodation
		A42	Contractor's general cost items: Services and facilities
		A43	Contractor's general cost items: Mechanical plant
		A44	Contractor's general cost items: Temporary works
	Contingencies	A54	Provisional work
		A55	Dayworks

SMM6		SMM7	

N   WOODWORK - cont'd

Sundries		P	BUILDING FABRIC SUNDRIES
plugging		-	deemed included with fixed items (however rates retained under G20)
holes in timber		-	deemed included - except P31 Holes/Chases/Covers/Supports for services
insulating materials		P10	Sundry insulation/proofing work/fire stops
metalwork		G20	Carpentry/Timber framing/First fixing
Ironmongery		N15	Signs/Notices
		P21	Ironmongery
Protection		A42	Contractor's services and facilities

P   STRUCTURAL STEELWORK

Steelwork		G10	Structural steel framing
		G11	Structural aluminium framing
		G12	Isolated structural metal members
Protection		A42	Contractor's services and facilities

Q   METALWORK

Composite items			
curtain walling		H11	Curtain walling
		H12	Plastics glazed vaulting/walling
		H13	Structural glass assemblies
windows		L11	Metal windows/rooflights/screens/louvres
		L12	Plastics windows/rooflights/screens/louvres
doors		L21	Metal doors/shutters/hatches
		L22	Plastics/Rubber doors/Shutters/hatches
rooflights		L11	Metal windows/rooflights/screens/louvres
		L12	Plastics windows/rooflights/screens/louvres
balustrades and staircases		L31	Metal stairs/walkways/balustrades
sundries			
duct covers etc.		P31	Holes/Covers/Chases/Supports for services
gates/shutters/hatches		L21	Metal doors/shutters/hatches
cloakroom fittings etc.		N10	General fixtures/furnishings/equipment
steel lintels		F30	Accessories/Sundry items for brick/block/stone walling
Plates, bars etc.			
floor plates		L31	Metal stairs/walkways/balustrades
matwells		N10	General fixtures/furnishings/equipment
Sheet metal, wiremesh and expanded metal		L21	Metal doors/shutters/hatches

SMM6	SMM7

Holts, bolts, screws and rivets

E42 Accessories cast into in situ concrete

G20 Carpentry/Timber framing/First fixing

L10-L12 Windows
L20-L22 Doors
L31 Metal stairs/walkways/balustrades

Protection

A42 Contractor's services and facilities

R PLUMBING AND MECHANICAL ENGINEERING INSTALLATIONS

R DISPOSAL SYSTEMS

S PIPED SUPPLY SYSTEMS

T MECHANICAL HEATING/COOLING REFRIGERATION SYSTEMS

U VENTILATION/AIR CONDITIONING SYSTEMS

X TRANSPORT SYSTEMS

Y MECHANICAL AND ELECTRICAL SERVICES MEASUREMENT

Classification of work
  a. rainwater installation
  b. sanitary installation (including traps)

R10 Rainwater pipework/gutters
R11 Foul drainage above ground

  c. cold water installation

S10 Cold water
S13 Pressurised water
S14 Irrigation
S15 Fountains/Water features
S20 Treated/Deionised/Distilled water
S21 Swimming pool water treatment

  d. firefighting installation

S60 Fire hose reels
S61 Dry risers
S62 Wet risers
S63 Sprinklers
S64 Deluge
S65 Fire hydrants
S70 Gas fire fighting
S71 Foam fire fighting

  e. heated, hot water installations etc.

S11 Hot water
S12 Hot and cold water (small scale)
S51 Steam

T20 Primary heat distribution
T30 Medium temperature hot water heating
T31 Low temperature hot water heating
T32 Low temperature hot water heating (small scale)
T33 Steam heating

  f. fuel oil installation

S40 Petrol/Oil-lubrication
S41 Fuel oil storage/distribution
S32 Natural gas

  g. fuel gas installation

S33 Liquid petroleum gas

SMM6	**SMM7**

Classification of work - cont'd.

h. refrigeration installation	T61 Primary/Secondary cooling distribution
	T70 Local cooling units
	T71 Cold rooms
	T72 Ice pads
j. compressed air installation	S30 Compressed air
	S31 Instrument air
k. hydraulic installation	-
l. chemical installation	-
m. special gas installation	S34 Medical/Laboratory gas
n. medical suction installation	S50 Vaccuum
p. pneumatic tube installation	S50 Vaccuum
q. vaccuum installation	R30 Centralised vacuum cleaning
	S50 Vacuum
r. refuse disposal installation	R14 Laboratory/Industrial waste drainage
	R20 Sewage pumping
	R21 Sewage treatment/sterilisation
	R31 Refuse chutes
	R32 Compactors/Macerators
	R33 Incineration plant
s. air handling installation	T40 Warm air heating
	T41 Warm air heating (small scale)
	T42 Local heating units
	T50 Heat recovery
	U10 General supply/extract
	U11 Toilet extract
	U12 Kitchen extract
	U13 Car parking extract
	U14 Smoke extract/Smoke control
	U15 Safety cabinet/Fume cupboard extract
	U16 Fume extract
	U17 Anaesthetic gas extract
	U20 Dust collection
	U30 Low velocity air conditioning
	U31 VAV air conditioning
	U32 Dual-duct air conditioning
	U33 Multi-zone air conditioning
	U40 Induction air conditioning
	U41 Fan-coil air conditioning
	U42 Terminal re-heat air conditioning
	U43 Terminal heat pump air conditioning
	U50 Hybrid system air conditioning
	U60 Free standing air conditioning units
	U61 Window/Wall air conditioning units
	U70 Air curtains
	Y24 Trace heating
	Y25 Cleaning and chemical treatment
t. automatic control installation	Y53 Control components-mechanical
u. special equipment eg kitchen equipment	N12 Catering equipment
v. other specialist installations	
	X10 Lifts
	X11 Escalators
	X12 Moving pavements
	X20 Hoists
	X21 Cranes
	X22 Travelling cradles

SMM6	SMM7
Sundries	Y81 Testing and commissioning of electrical services
Builders work	P31 Holes/Chases/Covers/Supports for services
Protection	A42 Contractor's services and facilities

T   FLOOR WALL AND CEILING FINISHINGS   M   SURFACE FINISHES

SMM6	SMM7
In situ finishings/Lathing and base boarding/Beds and backings	J10 Specialist waterproof rendering (including accessories)
	J33 In situ glass reinforced plastics
	M10 Sand cement/Concrete/Granolithic screeds/flooring
	M12 Trowelled bitumen/resin/rubber-latex flooring
	M20 Plastered/Rendered/Roughcast coatings including backings
	M21 Insulation with rendered finish
	M22 Sprayed mineral fibre coatings
	M23 Resin bound mineral coatings
	M30 Metal mesh lathing/Anchored reinforcement for plastered coatings
	M41 Terrazzo tiling/In situ terrazzo
surface sealers	M60 Painting/Clear finishing
Tile, slab and block finishings/Mosaic work	M40 Stone/Concrete/Quarry/Ceramic tiling/Mosaic
	M41 Terrazzo tiling/In situ terrazzo
	M42 Wood block/Composition block/Parquet flooring
Flexible sheet finishings	M50 Rubber/Plastics/Cork/Lino/Carpet tiling/sheeting
Dry linings and partitions	K LININGS/SHEATHING DRY PARTITIONING
	K10 Plasterboard dry lining
	K30 Demountable partitions
	K31 Plasterboard fixed partitions/inner walls/linings
	K32 Framed panel cubicle partitions
	K33 Concrete/Terrazzo partitions
Raised floors	K41 Raised access floors
Suspended ceilings, linings, and support work	K40 Suspended ceilings
Fibrous plaster	M31 Fibrous plaster
Fitted carpeting	M51 Edge fixed carpeting
Protection	A42 Contractor's services and facilities

U   GLAZING

SMM6	SMM7
Glass in openings	L40 General glazing
Leaded lights and copper lights in openings	L41 Lead light glazing
Mirrors	N10 General fixtures/furnishings/equipment
	N20, N21, N22, N23 Special purpose furnishings/equipment
Patent glazing	H10 Patent glazing

	SMM6		SMM7

SMM6

U   GLAZING - cont'd

Domelights

Protection

V   PAINTING AND DECORATING

Painting, polishing and
similar work

Signwriting

Decorative paper, sheet plastic
or fabric backing and lining

Protection

W   DRAINAGE

Pipe trenches
Manholes, soakaways, cesspits
and septic tanks
Connections to sewers
Testing drains

Protection

X   FENCING

Open type fencing
Close type fencing
Gates
Sundries
Protection

EXTERNAL WORKS

SMM7

L12   Plastics windows/rooflights/screens/
       louvres

A42   Contractor's services and facilities

M60   Painting/Clear finishing

N15   Signs/Notices

M52   Decorative papers/fabrics

A42   Contractor's services and facilities

R   DISPOSAL SYSTEMS

R12   Drainage below ground

R13   Land drainage

A42   Contractor's services and facilities

Q40   Fencing

D   GROUNDWORK

F   MASONRY

Q   PAVING/PLANTING/FENCING/SITE
       FURNITURE

Q10   Stone/Concrete/Brick kerbs/edgings/
       channels
Q20   Hardcore/Granular/Cement bound
       bases/sub-bases to roads/pavings
Q21   In situ concrete roads/pavings/bases
Q22   Coated macadam/Asphalt roads/pavings
Q23   Gravel/Hoggin roads/pavings
Q24   Interlocking brick/block roads/
       pavings
Q25   Slab/Brick/Sett/Cobble pavings
Q26   Special surfacings/pavings for sport
Q30   Seeding/Turfing
Q31   Planting
Q50   Site/Street furniture/equipment

Index to Advertisers

Advertising agent:
T. G. Scott & Son Ltd
30-32 Southampton Street
London WC2E 7HR

Acknowledgements

The Editors wish to record their appreciation of the assistance given by many individuals and organisations in the compilation of this edition.

Material Suppliers and Sub-Contractors who have contributed this year include:-

Aggregates
  Redland Aggregates   0992-586600
  lightweight
  Boral Lytag   03752-77181
Aluminium pipes and fittings
  Alumasc   0536-722121
Asbestos pipes and fittings
  Eternit Building Prods   0763-60421
Asphalt work
  Prater Asphalt   0737-772331

Beams,
  laminated
  Moelven (UK)   0703-454944
  lattice
  Naylor Buildings   021-526-3851
Blockwork
  Aerated Concrete Ltd   0375-673344
  ARC Concrete   0235-848808
  Boral Edenhall   0708-862881
  Celcon Ltd   01-242-9766
  Forticrete   0533-320277
  Tarmac   0442-54321
  Thermalite   0675-62081
Brickwork
  Armitage   0532-822141
  Ibstock Bricks   01-402-1227
  London Brick   0525-405858
  Redland Bricks   0293-786688
Building admixtures etc.
  Ardex U.K.   0440-63939
  Aston Building
  Products   0785-57265
  Beton Construct Mats   0256-53146
  Cementone Beaver   0280-823823
  FEB   061-794-7411
  Kerner-Greenwood   0553-772293
  RIW Protective
  Products   0734-792566
  Sealocrete   0703-777331
  Sika   0707-329241
  Vandex (UK) Ltd   01-394-2766
  W Hawley & Son   0332-840294
  Washington Mills
  Electro Minerals   0793-28131
Building papers
  British Sisalcraft   0634-290505
  Davidson Packaging   0602-844022

Cement
  Blue Circle Cement   01-731-7762
Cisterns/tanks/cylinders etc.
  Harvey Fabrication   01-981-7811
  IMI Range   061-338-3353

Cladding,
  aluminium
  British Alcan   0905-754030
  asbestos etc.
  Eternit Building Prods   0763-60421
  steel
  British steel   01-735-7654
  Plannja   0628-37313
  transluscent
  BIP Chemicals   021-353-0814
Column guards
  Huntley & Sparks   0460-72222
Concrete,
  paving
  Marshalls Mono   0422-57155
  pipes/manholes etc.
  Drainage Systems   01-286-5151
  Hume Pipe   0420-80086
  Milton pipes   0795-25191
  RBS Brooklyns   09295-6656
  precast floors etc.
  Bison   0753-652909
  ready-mixed
  Greenham Concrete   01-736-6592
Copper,
  fittings
  IMI York Imp Fittings   0532-701104
  tubes
  IMI Yorkshire
  Copper Tube   051-546-2700

Damp proof courses,closers etc.
  Cavity Trays   0935-74769
  Colas Building
  products   0268-728811
  IMI Rolled Metal   021-356-3344
  Marley Waterproofing   0732-741400
  Products   0268-728811
  Westbrick Plastics   0722-331933
Door sets
  Swedoors   0602-725231
Doors
  Crosby Doors   0793-729555
  Sarek Joinery   0787-60808
Doors, garage
  Catnic Garador   0222-885955
Drainage, stoneware
  Hepworth Iron Co   0226-763561
  Sandell Perkins   01-258-0257

Expansion joint fillers
  Expandite   01-965-8877

Fencing & gates
  Binns Fencing   01-802-5211
Finishes, textured
  Artex   0273-513100
Fire,
  protection
    Morceau-Aaronite   0773-812505
  resisting glass
    Schott - UK   0785-46131
Firecheck boards/channels
  Cape Boards & Panels   0895-37111
Fixings etc
  BAT   0952-680193
  Halfen   0296-20141
  Hilti   061-872-5010
Flashings
  Evode   0785-57755
Flooring,
  access
    Phoenix Floors   0708-851441
  accessories
    Carpet & Flooring   021-550-9131
    Roberts Smoothedge   0403-40721
  hardwood
    Hewetsons   04862-21535
    Junckers   0376-517512
    Viger Floors   037882-3035
  tiles, sheet flooring etc.
    Altro   0454-412992
    Daniel Platt   0782-577187
    Marley Floors   0622-858877
    Ruabon, Dennis   0978-843276
    Wicanders   0293-27700
Flooring Sub-Contractor
  GC Flooring
  & Furnishings   01-991-1000
Flue linings & blocks
  True Flue   0242-862551
Formers/linings etc.
  Dufaylite Dev   0480-215000
  Exxon Chemical
  Geopolymers   04955-57722
Fosalsil bricks,
aggregates etc
  Molar products   0206-73191
French polishing/staining etc
  Mosford Joinery   01-459-6241

Galvanised steel gutters etc.
  WP Metals   0922-743111
Glass & glazing
  Solaglass   01-928-8010
Granolithic Sub-Contractor
  Malcolm MacLeod   01-520-1147
Granular fittings etc.
  Yeoman Aggregates   01-993-6411

Hardcore, Cart away
  Western Foundations   01-684-7700
Hardwood
  see Joinery

Insulation
  Celotex   01-579-0811
  Coolag Purlboard   0524-55611
  DOW Chemicals   021-705-6363
  Erisco Bauder   0473-57671
  Isocrete Group Sales   01-906-1077
  Pilkington Insulation   0533-717202
  Pittsburgh Corning   0734-500655
  Plaschem   061-766-9711
  Vencel Resil   0322-27299

Intumescent strips
  Sealmaster   0223-832851
Iron pipes & fittings
  Stanton Pipeline
  Services   01-459-7801
Ironmongery, etc
  Comyn Ching   01-253-8414

Joinery, purpose-made
  Llewellyns   0323-21300
Joinery, standard hardwood
  Sarek Joinery   0787-60808
Joinery, standard softwood
  Boulton & Paul   0603-660133

Kerbs, channels etc.
  ECC Quarries Ltd   0335-70600
  F.R.Dangerfield   01-435-8044

Lathing, expanded metal
  Expanded Metal Co   0429-266633
Lathing, waterproof
  J Newton   01-629-5752
Lifts, escalators
  Express Lifts   0604-51221
Lightning protection
  R.C.Cutting   01-348-0052
Lime
  Blue Circle
  Industries   0482-633381
Lintols, steel
  Catnic Components   0222-885955

Manhole covers
  Glynwed Foundries   0952-641414
Marble linings
  Whiteheads   01735-1602
Matwell frames
  Nuway Manufacturing   0952-680400
Metal,
  framing
    Unistrut Midland   021-784-4178
  staircases
    Crescent of Cambridge   0480-301522
Mortars, ready-mixed
  Tilcon
  (Midland & South)   0732-453633

Paints, preservatives etc.
  Akzo Coatings   0235-815141
  Cuprinol   0373-65151
  Crown Paints   0254-704951
  J. P. MacDougall   01-749-0111
  Sadolin   0480-496868
  Solignium   0322-526966
  Tretol   0753-24164
Patent glazing
  Mellowes PPG   021-553-4011
Pavement lights, glass block
walling
  Luxcrete   01-965-7292
Paving slabs
  Redland Aggregates   050-981-2601
Piling, diaphragm walling
  Cementation   0923-776666
Plant Hire
  Agent Plant Hire Ltd   0322-22221
  LPH Equipment   0494-21481
Plaster products
  Mineralite Products   05435-71312
  Tilcon Special
  Products   0423-862841

Plaster/plasterboard
  British Gypsum   0602-844844
Plastering Sub-Contractor
  Jonathon James   04027-56921
Plumbers 'sundries'
  McQuire Murray   01-237-4646
Polythene sheeting
  McArthur Steel
  and Metal   0272-656242
Preservatives etc.
  Crown Paints   0254-704951
PVC pipes and fittings
  Caradon Terrain   0622-77811
  Hunter Building
  Products   01-855-9851
  Uponur   0532-701160

Rainwater outlets
  Harmer Holdings   07072-73481
Rawplugs, Rawlbolts etc.
  Rawplug Co.   01-546-2191
Reconstructed stone
walling/roof tiles
  ECC Quarries   0793-28131
Reinforcement bars
  Barfab   01-878-7771
Reinforcement mesh
  BRC   0785-57777
Roof decking
  Briggs Amasco   0306-885933
  Plannja   0628-37313
Roofing/cladding fasteners
  Sela Fasteners   0532-430541
Rooflights/windows
  Coxdome   044-282-4222
  Velux   0438-312570
Roofing products
  Nuralite Roofing
  Systems   0474-82-3451
  Permanite   0992-550511
  Ruberoid Building
  Products   01-805-3434
Roofing slates,
 natural
  Burlington Slate   022-989-661
  Greaves Welsh Slates   0766-830522
Roofing tiles,
 clay
  Goxhill Tileries   0427-872696
  Hinton Perry
  Davenhills   0384-77405
  Keymer Brick   04446-2931
  William Blyth   0652-32175
 concrete
  Marley roof tiles   0732-460055
  Redland Roof Tiles   07372-42488
Roofing Sub-Contractors
  Ruberoid Contracts   0256-461431
  Standard Flat Roofing 01-981-2422
Rubble walling
  ARC Southern   0285-712471

Sand
  Tilcon
  (Midland & South)   0732-452325
Sanitary fittings
  Ideal Standard   0482-499425
  Stelrad Bathroom
  Products   0782-49191
  W & G Sissons   0433-30791

Screed Sub-Contractor
  Alan Milne   01-998-9961
Shelving, adjustable
  Spur Systems   0923-26071
Sliding door gear
  Hillaldam Coburn   01-397-5151
Slots, ties & anchors
  Abbey Building
  Supplies   021-550-7674
  Harris & Edgar   01-686-4891
Softwood, panel products etc.
  C.F.Anderson   01-226-1212
  James Latham   0454-315421
Stainless Steel tube and
fittings
  Lancashire Fittings   0423-522355
Stairs
  Boulton & Paul
  (Joinery)   0603-660133
Steel arch frames
  Truline Building
  Products   0245-450450
Steel tubes and fittings
  British Tube
  Stockholdings   0633-290290
Steelwork
  British Steel   01-735-7654
  Graham Wood   01-586-6094
Stone Cladding Sub-Contractor
  Bath & Portland Stone 0305-820331
Stonework
  Bath & Portland Stone 0225-810456
  Gregory Quarries   0623-23092
Suspended ceilings
  Thermal and Acoustic
  installations   0992-38311

Tarmacadam Sub-Contractor
  Constable Hart   0483-224522
Terrazzo Sub-Contractor
  Marriot & Price   01-521-2821
Timber preservation etc
  Rentokil   0342-833022
Trussed rafters
  Montague L. Meyer   01-594-7111

Wall tiles
  Langley London   01-407-4444
  Pilkington tiles   061-794-2024
Waterbars etc.
  Servicised   0753-692929
Windows uPVC
  Anglian Windows   0603-619471
Windows, aluminium and metal
  Crittall Windows   0376-24106
Windows, hardwood
  Sarek Joinery   0787-60808
Windows, softwood
  Boulton & Paul   0603-660133
Woodwool slabs
  Torvale Building
  Products   05447-262

SPON'S PRICE BOOKS
ORDER FORM

Please send me:—

_____ copy/ies of **Spon's Architects' and Builders' Price Book 1990** *115th Edition* @ £45.00

_____ copy/ies of **Spon's Mechanical and Electrical Services Price Book 1990** *21st Edition* @ £45.00

_____ copy/ies of **Spon's Civil Engineering and Highway Works Price Book 1990** *4th Edition* @ £49.50

_____ copy/ies of **Spon's Landscape and External Works Price Book 1990** *9th Edition* @ £39.50

Please tick as appropriate

☐ I enclose cheque/PO for £ _____ payable to E & FN Spon Ltd

☐ Please charge to ☐ Access ☐ Visa ☐ AmEx ☐ Diners Club

Card Number _____ Expiry Date _____

☐ I enclose an official order, please send invoice with books

Name _____

Address _____

_____ Postcode _____

Signature _____

Return to: The Promotion Department, E & FN Spon, Freepost, KE7748, London EC4B 4JB.

STANDING ORDER PLAN

MAKE SURE you're one of the first to get your copies of Spon's Price Books each year.
Place a Standing Order now.

Please send me:—

_____ copy/ies of **Spon's Architects' and Builders' Price Book**

_____ copy/ies of **Spon's Mechanical and Electrical Services Price Book**

_____ copy/ies of **Spon's Civil Engineering and Highway Works Price Book**

_____ copy/ies of **Spon's Landscape and External Works Price Book**

each year when published until further notice.

Name _____

Address _____

_____ **Postcode** _____

Return to: The Promotion Department, E & FN Spon, Freepost KE7748, London EC4B 4JB.

Fees and Daywork

This part of the book contains the following sections:

Keep your figures up to date, free of charge

This section, and most of the other information in this Price Book, is brought up to date every three months in the *Price Book Update*.

The *Update* is available free to all Price Book purchasers.

To ensure you receive your copy, simply complete the reply card from the centre of the book and return it to us.

Fees for Professional Services

Extracts from the scales of fees for architects, quantity surveyors and
consulting engineers are given together with extracts from the Town and Country
Planning Regulations 1989 and the Building (Prescribed Fees etc.) Regulations
1986. These extracts are reproduced by kind permission of the bodies concerned,
in the case of Building Regulation Fees by kind permission of the Controller HM
Stationary Office. Attention is drawn to the fact that the full scales are not
reproduced here and that the extracts are given for guidance only. The full
authorized scales should be studied before concluding any agreement and the
reader should ensure that the fees quoted here are still current at the time of
reference

ARCHITECTS' FEES

QUANTITY SURVEYORS' FEES

CONSULTING ENGINEERS' FEES

PLANNING REGULATION FEES

BUILDING REGULATION FEES

ARCHITECTS' FEES

EFFECTIVE 1st JULY 1982
including amendments up to November 1988

INTRODUCTION

The RIBA requires of its members that before making an engagement for professional services they shall define the terms of the engagement including the scope of the service, the allocation of resposibilities and any limitation of liability, the method of calculation of remuneration and the provision for termination.

Architect's Appointment consists of four related parts:

Part 1 Architect's Services

Preliminary and Basic services normally provided by the architect. These services progress through work stages based on the RIBA Plan of Work (RIBA Publications Limited). The sequence of work stages may be varied or two or more work stages may be combined to suit the particular circumstances.

Preliminary Services

Work stage A: Inception   Work stage B: Feasibility

Basic Services

Work stage C: Outline proposals

Work stage D: Scheme design

Work stage E: Detail design

Work stage F: Production informatation   Work stage G: Bills of Quantities

Work stage H: Tender action   Work stage J: Project planning

Work stage K: Operation on site   Work stage L: Completion

Preliminary Services are normally charged on a time basis and Basic Services on a percentage basis, as described in Part 4.

Part 2 Other Services

Services which may augment the Preliminary and Basic Services or which may be the subject of a separate appointment. Fees for these services are normally charged on a time or lump sum basis as described in Part 4.

Part 3 Conditions of Appointment

Conditions which apply to an architect's appointment.

Part 4 Recommended Fees and Expenses

Recommended and not mandatory methods of calculating the architect's fees and expenses and of apportioning fees between work stages.

A sample Memorandum of Agreement and Schedule of Services and Fees are included with the document for information. They are published separately.

ARCHITECTS' FEES

The client and the architect should discuss the architect's appointment and agree in writing the services, conditions and fee basis. These should be stated in the Schedule of Services and Fees and referred to in the Memorandum of Agreement between client and architect; alternatively, they should be stated in a letter of appointment.

PART 1 ARCHITECT'S SERVICES

This part describes Preliminary and Basic Services which an architect will normally provide.

Preliminary Services

Word stage A: Inception

1.1. Discuss the client's requirements including timescale and any financial limits; assess these and give general advice on how to proceed; agree the architect's services.
1.2. Obtain from the client information on ownership and any lessors and lessees of the site, any existing buildings on the site, boundary fences and other enclosures, and any known easement, encroachments, underground services, rights of way, rights of support and other relevant matters.
1.3. Visit the site and carry out an initial appraisal.
1.4. Advise on the need for other consultants' services and on the scope of these services.
1.5. Advise on the need for specialist contractors, sub-contractors and suppliers to design and execute part of the works to comply with the architect's requirements.
1.6. Advise on the need for site staff.
1.7. Prepare where required an outline timetable and fee basis for further services for the client's approval.

Work stage B: Feasibility

1.8. Carry out such studies as may be necessary to determine the feasibility of the client's requirements; review with the client alternative design and construction approaches and cost implications; advise on the need to obtain planning permissions, approvals under building acts or regulations, and other similar statutory requirements.

BASIC SERVICES

Work stage C: Outline proposals

1.9. With other consultants where appointed, analyse the client's requirements; prepare outline proposals and an approximation of the construction cost for the client's preliminary approval.

Work stages D: Scheme design

1.10. With other consultants where appointed, develop a scheme design from the outline proposals taking into account amendments requested by the client; prepare a cost estimate; where applicable give an indication of possible start and completion dates for the building contract. The scheme design will illustrate the size and character of the project in sufficient detail to enable the client to agree the spatial arrangements, materials and appearance.

1.11. With other consultants where appointed, advise the client of the implications of any subsequent changes on the cost of the project and on the overall programme.

ARCHITECTS' FEES

1.12. Make where required application for planning permission. The permission
      itself is beyond the architect's control and no guarantee that it will be
      granted can be given.

Work stage E: Detail Design

1.13. With other consultants where appointed, develop the scheme design; obtain
      the client's approval of the type of construction, quality of materials
      and standard of workmanship; co-ordinate any design work done by
      consultants, specialist contractors, sub-contractors and suppliers; obtain
      quotations and other information in connection with specialist work.
1.14. With other consultants where appointed, carry out cost checks as
      necessary; advise the client of the consequences of any subsequent change
      on the cost and programme.
1.15. Make and negotiate where required applications for approvals under
      building acts, regulations or other statutory requirements.

Work stage F and G: Production information and bills of quantities.

1.16. With other consultants where appointed, prepare production information
      including drawings, schedules and specification of material and
      workmanship; provide information for bills of quantities, if any, to be
      prepared: all information complete in sufficient detail to enable a
      contractor to prepare a tender.

Work stage H: Tender action

1.17. Arrange, where relevant, for other contracts to be let prior to the
      contractor commencing work.
1.18. Advise on and obtain the client's approval to a list of tenderers.
1.19. Invite tenders from approved contractors; appraise and advise on tenders
      submitted. Alternatively, arrange for a price to be negotiated with a
      contractor.

Work stage J: Project planning

1.20. Advise the client on the appointment of contractor and on the
      responsibilities of the client, contractor and architect under the terms
      of the building contract; where required prepare the building contract and
      arrange for it to be signed by the client and the contractor; provide
      production information as required by the building contract.

Work stage K: Operations on site

1.21. Administer the terms of the building contract during operations on site.
1.22. Visit the site as appropriate to inspect generally the progress and
      quantity of the work.
1.23. With other consultants where appointed, make where required periodic
      financial reports to the client including the effect of any variations on
      the construction cost.

Work stage L: Completion

1.24. Administer the terms of the building contract relating to the completion
      of the work.
1.25. Give general guidance on maintenance.
1.26. Provide the client with a set of drawings showing the building and the
      main lines drainage; arrange for drawings of the services installations to
      be provided.

ARCHITECTS' FEES

PART 2 OTHER SERVICES

**This part describes services which may be provided by the architect to augment
the Preliminary and Basic Service described in Part 1 or which may be subject of
a separate appointment. The list of services so described is not exhaustive.**

Surveys and investigations

2.1.   Advise on the selection and suitability of sites; conduct negotiations
        concerned with sites and buildings.
2.2.   Make measured surveys, take levels and prepare plans of sites and build-
        ings
2.3.   Provide services in connection with soil and other similar
        investigations.
2.4.   Make inspections, prepare reports or give general advise on the condition
        of premises.
2.5.   Prepare schedules of dilapidations; negotiate them on behalf of landlords
        or tenants.
2.6.   Make structural surveys to ascertain whether there are defects in the
        walls, roof, floors, drains or other parts of a building which may
        materially affect its saftey, life and value.
2.7.   Investigate building failures; arrange and supervise exploratory work by
        contractors or specialists.
2.8.   Take particulars on site; prepare specifications and/or schedules for
        repairs and restoration work, and inspect their execution.
2.9.   Investigate and advise on problems in existing buildings such as fire
        protection, floor loadings, sound insulation, or change of use.
2.10.   Advise on the efficient use of energy in new and existing buildings.
2.11.   Carry out life cycle analyses of buildings to determine their cost in
        use.
2.12.   Make an inspection and valuation for morgage or other purposes.

Development services

2.13.   Prepare special drawings, models or technical information for the use of
        the client or for applications under planning, building act, building
        regulation or other statutory requirements, or for negotiations with
        ground landlords, ajoining owners, public authorities, licensing
        authorities, mortgagors and others; prepare plans for conveyancing, land
        registry and other legal purposes.
2.14.   Prepare development plans for a large building or complex of buildings;
        prepare a layout only, or prepare a layout for a greater area than that
        which is to be developed immediately.
2.15.   Prepare layouts for housing, industrial or other estates showing the
        siting of buildings and other works such as roads and sewers.
2.16.   Prepare drawings and specification of materials and workmanship for the
        construction of housing, industrial or other estate roads and sewers.

2.17.   Provide services in connection with demolitions works.
2.18.   Provide services in connection with environmental studies.

Design services

2.19.   Design or advise on the section of furniture and fittings; inspect the
        making up of such furnishings.
2.20.   Advise on and prepare detailed designs for works of special quality such
        as shop-fitting or exhibition design, either independently or within the
        shell of an existing building.
2.21.   Advise on the commissioning or selection of works of art; supervise their
        installation.
2.22.   Carry out specialist acoustical investigations.

ARCHITECTS' FEES

2.23.   Carry out special constructional research in connection with a scheme
        design, including the design, construction or testing of prototype
        buildings or models.
2.24.   Develop a building system or mass-produced building components; examine
        and advise on existing building systems; monitor the testing of prototype
        buildings and models.

Cost estimating and financial advisory services

2.25.   Carry out planning for a building project, including the cost of
        associated design services, site development, landscaping, furnature and
        equiptment; advise on cash flow requirements for design cost,
        construction cost, and cost in use.
2.26.   Prepare schedules of rates or schedules of quantities for tendering
        purposes; value work excuted where no quantity suveyor is appointed.
        Fees for this work are recommended to be in accordance with the
        Professional Charges of the Royal Institution of Chartered Surveyors.
2.27.   Carry out inspections and surveys; prepare estimates for the replacement
        and reinstatement of buildings and plant; submit and negotiate claims
        following damage by fire or other causes.
2.28.   Provide information; make applications for and conduct negotiations in
        connection with local authority, government or other grants.

Negotiations

2.29.   Conduct exceptional negotiations with a planning authority.
2.30.   Prepare and submit an appeal under planning acts; advise on other works
        in connection with planning appeals.
2.31.   Conduct exceptional negotiations for approvals under building acts or
        regulations; negotiate waivers or relaxations.
2.32.   Make submissions to the Royal Fine Art Commission and other non-statutory
        bodies.
2.33.   Submit plans of proposed building works for approval of landlords,
        mortgagors, Freeholders or others.
2.34.   Advise on the rights and responsibilities of owners or lessees including
        rights of light, rights of support, and rights of way; provide
        information; undertake any negotiations.
2.35.   Provide services in connection with party wall negotiations.
2.36.   Prepare and give evidence; settle proofs; confer with solicitors and
        counsel; attend court and arbitrations; appear before other tribunals;
        act as arbitrator.

Administration and management of building projects

2.37.   Provide site staff for frequent or constant inspection of the works.
2.38.   Provide management from inception to completion: prepare briefs; appoint
        and co-ordinate consultants, construction managers, agents and
        contractors; monitor time, cost and agreed targets; monitor progress of
        the works; hand over the building on completion; equip, commission and
        set up any operational organizations.
2.39.   Provide services to the client, whether employer or contractor, in
        carrying out duties under a design and build contract.
2.40.   Private services in connection with separate trades contracts; agree a
        programme of work; act as co-ordinator for the duration of the contract.
2.41.   Provide services in connection with labour employed directly by the
        client; agree a programme of work; co-ordinate the supply of labour and
        materials; provide general supervision; agree the final account.
2.42.   Provide specially prepared drawings of a building 'as built'.

ARCHITECTS' FEES

2.43.   Compile maintainance and operational manuals; incorporate information
        prepared by other consultants, specialist contractors, sub-contractors
        and suppliers.
2.44.   Prepare a programme for the maintenance of a building; arrange
        maintenance contracts.

Services normally provided by consultants

2.45.   Provide such services as:

        a Quantity surveying
        b Structural engineering
        c Mechanical engineering
        d Electrical engineering
        e Landscape and garden design
        f Civil engineering
        g Town planning
        h Furniture design
        j Graphic design
        k Industrial design
        l Interior design

        Where consultants' services are provides from within the architect's own
        office or by consultants in association with the architect it is
        recommended that fees be in accordance with the scales of charges of the
        relevant professional body.

Consultancy services

2.46.   Provide services as a consultant architect on a regular or
        intermittent basis.

ARCHITECTS' FEES

PART 3 CONDITIONS OF APPOINTMENT

This part describes the conditions which normally apply to an architect's appointement. If different or additional conditions are to apply, they should be set out in the Schedule of Services and Fees or letter of appointment.

3.1.   The architect will exercise reasonable skill and care in conformity with with the normal standards of the architect's profession.

Architect's authority

3.2.   The architect will act on behalf of the client in the matters set out or implied in the architect's appointment; the architect will obtain the authority of the client before initiating any service or work stage.
3.3.   The architect shall not make any material alteration, additional to or omission from the approved design without the knowledge and consent of the client, except if found necessary during construction for constructional reasons in which case the architect shall inform the client without delay.
3.4.   The architect will inform the client if the total authorised expenditure or the building contract period is likely to be materially valued.

Consultants

3.5.   Consultants may be nominated by either the client or the architect, subject to acceptance by each party.
3.6.   Where the client employes the consultants, either directly or through the agency of the architect, the client will hold each consultant, and not the architect, responsible for the competence, general inspection and performance of the work entrusted to that consultant; provided that in relation to the execution of such work under the contract between the client and the contractor nothing in this clause shall affect any responsibility of the architect for issuing instructions or for other functions ascribed to the architect under that contract.
3.7.   The architect will have the authority to co-ordinate and integrate into the overall design the services provided by any consultant, however employed.

Contractors, sub-contractors and suppliers

3.8.   A specialist contractor, sub-contractor or supplier who is to be employed by the client to design any part of the works may be nominated by either the architect or the client, subject to acceptance by each party. The client will hold such contractor, sub-contractor or supplier, and not the architect, responsible for the competence, proper excution and performance of the work thereby entrusted to that contractor, sub-contractor or supplier. The architect will have the authority to co-ordinate and integrate such work into the overall design.
3.9.   The client will employ a contractor under a separate agreement to undertake construction or other works. The client will hold the contractor, and not the architect, responsible for the contractor's operational methods and for the proper execution of the works.

Site inspection

3.10.   The architect will visit the site at intervals appropriate to the stage of construction to inspect the progess and quality of the works and to determine that they are being executed generally in accordance with the contract documents. The architect will not be required to make frequent or constant inspections.

ARCHITECTS' FEES

3.11. Where frequent or constant inspection is required a clerk or clerk of works will be employed. They may be employed either by the client or by the architect and will be either event be under the architect's direction and control.
3.12. Where frequent or constant inspection by the architect is agreed to be necessary a resident architect may be appointed by the architect on a part or full time basis.

Client's instructions

3.13. The client will provide the architect with such information and make such decisions as are necessary for the proper performance of the agreed service.
3.14. The client, if a firm or other body of persons, will, when requested by the architect, nominate a responsible representative through whom all instructions will be given.

Copyright

3.15. Copyright in all documents and drawings prepared by the architect and in any works executed from those documents and drawings shall, otherwise agreed, remain the property of the architect.
3.16. The client, unless otherwise agreed, will be entitled to reproduce the architect's design by proceeding to execute the project provided that:

- the entitlement applies only to the site or part of the site to which the design relates; and
- the architect has completed work stages D or has provided detail design and production information in work stages E, F and G; and
- any fees due to the architect have been paid or tendered.

This entitlement will also apply to the maintainence, repair and renewal of the works.
3.17. Where an architect has not completed work stage D, or where the client and the architect have agreed that clause 3.16 shall not apply, the client may not reproduce the design by proceeding to execute the project without the consent of the architect and payment of any additional fee that may be agreed in exchange for the architect's consent.
3.18. The architect shall not unreasonably withhold his consent under clause 3.17 but where his services are limited to making and negotiating planning applications he may withhold his consent unless otherwise determined by an arbitrator appointed in accordance with clause 3.26.

Assignment

3.19. Neither the architect nor the client may assign the whole or any part of his duties without the other's written consent.

Suspension and termination

3.20. The architect will give immediate notice in writing to the client of any siutation arising from force majeure which makes it impraticable to carry out any of the agreed services, and agree with the client a suitable course of action.
3.21. The client may suspend the performance of any or all of the agrees services by giving reasonable notice in writing to the architect.
3.22. If the architect has not been given instructions to resume any suspended service within six months from the date of suspension the architect will make written request for such instructions which must be given in writing. If these have not been received within 30 days of the date of such a request will have the right to treat the appointment as terminated upon the expiry of the 30 days.

ARCHITECTS' FEES

3.23. The architect's appointment may be terminated by either party on the expiry of reasonable notice given in writing.

3.24. Should the architect through death or incapacity be unable to provide the agreed services, the appointment will thereby be terminated. In such an event the client may, on payment or tender of all outstanding fees and expenses, make full use of reports, drawings or other documents prepared by the architect in accordance with and for use under the agreement, but only for the purpose for which they were prepared.

Settlement of disputes

3.25. A difference or dispute arising on the application of the Architect's Appointment to fees charged by a member may, by agreement between the parties, be referred to the RIBA, RIAS or RSUA for an opinion provided that:

- The member's appointment is based on this document and has been agreed and confirmed in writing; and
- the opinion is sought on a joint statement of undisputed facts; and
- the parties undertake to accept the opinion as final and binding upon them.

3.26. Any other difference or dispute arising out of the appointment and any difference or dispute arising on the fees charged which cannot be resolved in accordance with clause 3.25 shall be referred to the arbitration by a person to be agreed between the parties or, failing agreement within 14 days after either party has given to the other a written request to concur in the appointment of an arbitrator, a person to be nominated at the request of either party by the President of the Chartered Institute of Arbitrators, provided that in a difference or dispute arising out of provisions relating to copyright, cause 3.15 to 3.18 above, the arbitrator shall, unless otherwise agreed be a architect.
or

3.26S In Scotland, any difference or dispute arising out of the appointment which cannot be resolved in accordance with clause 3.25 shall be referred to arbitration by a person to be agreed between the parties or, failing agreement within 14 days after either party has given to the other a written request to concur in the appointment of an arbiter, a person has to be nominated at the request of either party by the Dean of the Faculty of Advocates, provided that in a difference or dispute arising out of provisions relating to copyright, clauses 3.15 to 3.18 above, the arbiter shall, unless otherwise agreed, be an architect.

3.27. Nothing herein shall prevent the parties agreeing to settle any difference or dispute arising out of the appointment without recourse to arbitration.

Governing laws

3.28. The application of these conditions shall be governed by the laws of England and Wales.
or

3.28S The application of these conditions shall be governed by the laws of Scotland
or

3.28NI The application of these conditions shall be governed by the laws of Northern Ireland.

ARCHITECTS' FEES

PART 4 RECOMMENDED FEES AND EXPENSES

**This part describes the recommended and not mandatory methods of calculating
fees for the architect's services and expenses. Fees may be based on a
percentage of the total construction cost or on time expended, or may be a lump
sum. This part should be read in conjunction with Parts 1, 2 and 3.**

Percentage fees

4.1.   The percentage fee scales shown in Figure 1 (graph) are for use where the
       architect's appointment is for the Basic Services described in Part 1 for
       new works having a total construction cost between £20,000 and
       £5,000,000. Where the total construction is less than £20,000 or more
       than £5,000,000 client and architect should agree an appropiate fee basis
       at the time of appointment.
4.2.   Percentage fees are based on the total construction cost of the works; on
       the issue of the final certificate fees should be recalculated on the
       actual total construction cost.
4.3.   Total construction cost is defined as the cost, as certified by the
       architect, of all works including site works executed under the
       architect's direction, subject to the following:

       a   The total construction cost includes the cost of all work designed by
           consultants and co-ordinated by the architect, irrespective of whether
           such work is carried out under separate building contracts for which
           the architect may not be responsible. The architect will be informed
           of the cost of any such separate contracts.
       b   The total construction cost does not include specialist sub-
           contractors' design fees for work on which consultants would otherwise
           have been employed. Where such fees are not known, the architect will
           estimate a reduction from the total construction cost.
       c   For the purpose of calculating the appropriate fee, the total
           construction cost - includes the actual or estimated cost of any work
           executed which is excluded from the contract but otherwise designed
           by the Architect; - is not subject to any deductions made in respect
           of work not in accordance with the building contract.
       d   The total construction cost includes the cost of built-in furniture
           and equipment. Where the the cost of any special equipment is
           excluded from the total construction cost, the architect may charge
           additionally for work in connection with such items.
       e   Where any material, labour or carrage is supplied by a client who is
           not the contractor, the cost will be estimated by the architect as if
           it were supplied by the contractor, and included in the total cost.
       f   Where the client is the contractor, a statement of the acertained
           gross cost of the works may be used in caluclating the total
           construction cost of the works. In the absence of such a statement,
           the architect's own estimate will be used. In both a statement of
           ascertained gross cost and an architect's estimate there will be
           included an allowance for the contractor's profit and overheads.

4.4.   Buildings are divided into five classes for fee calculation purposes.
       For guidance only the building types most likely to fall into each class
       are shown in Figure 3.

Repetition

4.5.   The classification of buildings in Figure 3 takes account of reduced
       design work arising from the nature of the building.
4.6.   Where a building is repeated for the same client the recommended fee for
       the superstructure may be reduced on all except the first three of any
       houses of the same design and on all except the first of all other
       building types of the same design.

ARCHITECTS' FEES

4.7.   Where a single building incorporates a number of identical compartments
        such as floors or complete structural bays the recommended fee may be
        reduce on all identical compartments in excess of ten.
4.8.   Reductions should be made by waiving the fee for work stages E, F and G
        where a complete design can be re-used without modification other than
        the handing of a plan.

Time charge fees

4.9.   Time charges are based on hourly rates for principals and other techical
        staff. In assessing the hourly rate all relevant factors should be
        considered, including the complexity of the work, the qualifications,
        experience and responsibility of the architect, and the character of any
        negotiations. Hourly rates for principals shall be agreed. The hourly
        rate for technical staff should be not less than 18 pence per £100 of
        gross annual income.
4.10.   Technical staff are defined as architectural and other professional and
        technical staff, where the architect is responsible for deducting PAYE
        and National Insurance contributions from those persons' salaries on
        behalf of the Inland Revenue.
4.11.   Gross annual income includes bonus payments plus the employer's share of
        contributions towards National Insurance, pension and private medical
        schemes and other emoluments such as car and accomodation allowances.
4.12.   Where the staff are provided by an agency hourly rates shall be agreed.
4.13.   Where site staff are employed by the architect hourly rates shall be
        agreed.
4.14.   Unless otherwise agreed no separate time charges will be made for
        secretarial staff or staff engaged on general accountancy or
        administrative duties.
4.15.   The architect will maintain records of time spent on services performed
        on a time basis. The architect will make such records available to the
        client on reasonable request.

Lump sum fees

4.16.   The architect may agree with the client to charge a lump sum fee for any
        of the services described in Parts 1 and 2 in appropriate circumstances,
        for example where:

        -   the client's requirements are provided in a form such that the
            architect is not obliged to undertake any additional preparatory work;
        -   the full extent of the service can be determined when the architect is
            appointed;   and
        -   the architect's service can be completed within an agreed period.

Works to existing building

4.17.   The percentage fee scales shown in Figure 2 (graph) are for use where the
        architect's appointment is for the Basic Services described in Part 1 for
        alterations or extensions to an existing building having a total
        construction cost of between £20 000 and £5 000 000. Where the total
        construction cost is less than £20 000 or more than £5 000 000 client and
        architect should agree an appropriate fee basis at the time of the
        appointment.
4.18.   Where extensions to existing buildings are substantially independent,
        percentage fees should be as Figure 1 for new works, but the fee for
        those sections of the works which marry existing buildings to the new
        should be charged separately as Figure 2 applicable to an independent
        commission of similar value.
4.19.   Where the architect's appointment is for repair and restoration work fees
        should be on a time basis: alternatively a percentage fee may be agreed.

ARCHITECTS' FEES

4.20.   Where the architect's appointment is in connection with works to a
        building of architectural or historical interest, or to a building in a
        conservation area, higher fees may be charged.

Compounding of fees

4.21.   By agreement the percentage or lump sum fee may be compounded to cover
        all or any part of the architect's services and expenses.

Interim payments

4.22.   Fees and expenses should be paid in instalments either at regular
        intervals or on completion of work stages of the Basic services (Part 1).
4.23.   Where interim payment of percentage or lump sum fees is related to
        completion of work stages of the Basic Services the recommended
        apportionment is as follows:

Work stage	Proportion of fee	Cumulative total
C	15%	15%
D	20%	35%
E	20%	55%
F G	20%	75%
H J K L	25%	100%

Fees in respect of work stages E to L should be paid in
instalments proportionate to the work completed or the values of
the works certified from time to time.   Interim payment should be   based
on the current estimated cost of the works.
The apportionment of fees is a means of assessing interim payments   and
does not necessarily reflect the amount of work completed in
any work stage.   By agreement an adjustment in the apportionment
may be made.

Interest on "Any sums remaining un-paid at the expiry of 30 days from the
date of submission of the fee account shall bear intrest thereafter, such
interest to accrue from day to day at the rate of 2% per annum above the
current base rate of the architect's principal bank".

Partial services

4.24.   The architect may be required to provide part only of the Basic Services
        (Part 1).   In such cases the architect will be entitled to a commensurate
        fee.
4.25.   Where work is to be done by or on behalf of the client, resulting in the
        omission of part of work stages C to L, or a sponsored constructional
        method is to be used, a commensurate reduction in the recommended
        percentage fee may be agreed.   In accessing the reduction, due account
        should be taken for the need for the architect to become thoroughly
        familiar with the work done by others, and a familiarization fee will be
        charged for this work.
4.26.   All percentage fees for partial services should be based on the
        architect's current estimate of the total construction cost of the works.
        Such estimates may be based on an accepted tender or, subject to the
        following, on the lowest of unaccepted tenders.   Where partial services
        are provided in respect of works for which the executed cost is not known
        and no tender has been accepted, percentage fees should be based either
        on the architect's estimated total construction cost or on the most
        recent cost limit agreed with the client, whichever is the lower.
4.27.   Fees for partial services may alternatively be on a time or lump sum
        basis.

ARCHITECTS' FEES

Suspension, resumption and termination

4.28.   On suspension, or termination of the architect's appointment the
        architect will be entitled to fees for all work complete at that
        time.   Fees will be charged on a partial service basis.
4.29.   During such period of suspension the architect will be reimbursed by the
        client for all expenses and disbursement necessarily incurred under the
        appointment.
4.30.   On the resumption of a suspended service within six months, previous
        payments will be regarded solely as payment on account towards the total
        fee.
4.31.   Where the architect's appointment is terminated by the client the
        architect will be reimbursed by the client for all expenses and
        disbursements necessarily incurred in connection with work then in
        progress and arising as a result of the termination.

Expenses and disbursements

4.32.   In addition to the fees charged the architect will be reimbursed for all
        expenses and disbursements properly incurred in connection with the
        appointment, including the following:

        a   Printing, reproduction or purchase costs of all documents, drawings,
            maps, models, photographs, and other records, including all those used
            in communication between architect, client, consultants and
            contractors, and for enquiries to contractors, sub-contractors and
            suppliers, notwithstanding any obligation on the part of the architect
            to supply such documents to those concerned, except that contractors
            will pay for any prints additional to those to which they are entitled
            under the contract.
        b   Hotel and travelling expenses, including milage allowance for cars at
            rates stated in the Schedule of Services and Fees and other similar
            disbursements.
        c   All payments made on behalf of the client, such as expenses incurred
            in advertising for tenders and resident site staff including the time
            and expense of interviewers and reasonable expenses for interviewees.
        d   Fees and other charges for specialist professional advise, including
            legal advise, which have been incurred by the architect with the
            specific authority of the client.
        e   The cost of postage, telephone charges, telex messages, telegrams,
            cables, facsimilies, air-freight and courier services.
        f   Rental and hire charges for specialised equipment, including
            computers, where required and agreed by the client.
        g   Where work charged on a percentage fee is at such distance that an
            exceptional amount of time is spent travelling, additional charges may
            be made.

4.33.   The architect will maintain records of all such expenses and disbursement
        and will make these records available to the client on reasonable
        request.
4.34.   Expenses and disbursements may by agreement be estimated or standardised
        in whole or in part, or compounded for an increase in the percentage or
        lump sum fee.
4.35.   The client will pay all fees in respect of applications under planning
        and building acts and other statutory requirements.

Variations

4.36.   Where the scope of the architect's services is varied fees may be
        adjusted accordingly.

ARCHITECTS' FEES

4.37. Where the architect is involved in extra work and expense for reasons
       beyond the architect's control and for which the architect would not
       otherwise be renumerated additional fees are due. Any of the following
       is likely to involve the architect in extra work and expense:

   a   The need to revise reports, drawings, specifications or other
       documents due to changes in interpretation or enactment or revisions
       of the laws, statutory or other regulations.
   b   Changes in the client's instructions, or delay by the client in
       providing information.
   c   Consideration of notices, applications or claims by the contractor
       under a building contract; delays resulting from defects or
       deficiencies in the work of the contractor, sub-contractors or
       suppliers; default, bankruptcy or liquidation of the contractor, sub-
       contractors or suppliers.
   d   Any other cause beyond the architect's control.

Value Added Tax

4.38. The amount of any Value Added Tax on the services and expenses of the
       architect arising under the Finance Act 1972 will be chargeable to the
       client in addition to the architect's fees and expenses.

ARCHITECTS' FEES

Figure 1: Recommended percentage fee scales: New works

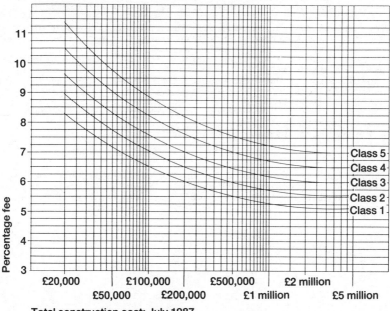

Total construction cost: July 1987

Figure 2: Recommended percentage fee scales: Works to existing buildings

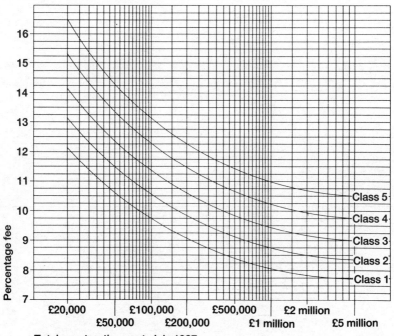

Total construction cost: July 1987

ARCHITECTS' FEES

Figure 3: Classification of buildings, for guidance

Type	Class 1	Class 2	Class 3	Class 4	Class 5
Industrial	Storage sheds	Speculative factories and warehouses Assembly and machine workshops Transport garages	Purpose-built factories and warehouses Garages/ showrooms		
Agricultural	Barns and sheds Stables	Animal breeding units			
Commercial	Speculative shops Single-storey car parks	Speculative offices Multi-storey car parks	Supermarkets Banks Purpose-built offices	Department stores Shopping centres Food processing units Breweries Telecom and computer accommodation	High risk research and production buildings Recording studios
Community		Communal halls	Community centres Branch libraries Ambulance and fire stations Bus stations Police stations Prisons Postal and broadcasting	Civic centres Churches and crematoria Concert halls Specialist libraries Museums Art galleries Magistrates/ county/ sheriff courts	Theatres Opera houses Crown/high courts
Residential		Dormitory hostels	Estate housing Sheltered housing	Parsonages/ manses Hotels	Houses for individual clients
Education			Primary/nursery/ first schools	Other schools including middle and secondary University complexes	University laboratories
Recreation			Sports centres Squash courts Swimming pools		
Medical social services			Clinics Homes for the elderly	Health centres Accommodation for the disabled General hospital complexes Surgeries	Teaching hospitals Hospital laboratories Dental surgeries

QUANTITY SURVEYORS' FEES

Scale 36 inclusive of professional charges for quantity surveying services for building works issued by The Royal Institution of Chartered Surveyors. The scale is recommended and not mandatory.

EFFECTIVE FROM JULY 1988

1.0.   GENERALLY

    1.1. This scale is for use when an inclusive scale of professional charges is considered to be appropriate by mutual agreement between the employer and the quantity surveyor.

    1.2. This scale does not apply to civil engineering works, housing schemes financed by local authorities and the Housing Corporation and housing improvement work for which separate scales of fees have been published.

    1.3. The fees cover quantity surveying services as may be required in connection with a building project irrespective of the type of contract from initial appointment to final certification of the contractor's account such as:

        (a) Budget estimating; cost planning and advice on tendering procedures and contract arrangements.

        (b) Preparing tendering documents for main contract and specialist sub-contracts; examining tenders received and reporting thereon or negotiating tenders and pricing with a selected contractor and or sub-contractors.

        (c) Preparing recommendations for interim payments on account to the contractor; preparing periodic assessments of anticipated final cost and reporting thereon; measuring work and adjusting variations in accordance with the terms of the contract and preparing final account, pricing same and agreeing total with the contractor.

        (d) Providing a reasonable number of copies of bills of quantities and other documents; normal travelling and other expenses. Additional copies of documents, abnormal travelling and other expenses (e.g. in remote areas or overseas) and the provision of checkers on site shall be charged in addition by prior arrangement with the employer

    1.4. If any of the materials used in the works are supplied by the employer or charged at a preferential rate, then the actual or estimated market value thereof shall be included in the amounts upon which fee are to be calculated.

    1.5. If the quantity surveyor incurs additional costs due to exceptional delays in building operations or any other cause beyond the control of the quantity surveyor then the fees may be adjusted by agreement between the employer and the quantity surveyor to cover the reimbursement of these additional costs.

    1.6. The fees and charges are in all cases exclusive of value added tax which will be applied in accordance with legislation.

    1.7. Copyright in bills of quantities and other documents prepared by the quantity surveyor is reserved to the quantity surveyor.

2.0. INCLUSIVE SCALE

    2.1. The fees for the services outlined in para.1.3, subject to the provision of para. 2.2, shall be as follows:

        (a) **Category A:** Relatively complex works and/or works with little or no repetition.

        Examples:
        Ambulance and fire stations; banks; cinemas; clubs; computer buildings; council offices; crematoria; houses; fitting out of existing buildings; homes for the elderly; hospitals and nursing

QUANTITY SURVEYORS' FEES

homes; laboratories; law courts; libraries; 'one off' houses; petrol stations; places of religious worship; police stations; public houses, licensed premises; restaurants; sheltered housing; sports pavillions; theatres;town halls; universities, polytechnics and colleges of further education (other than halls of residence and hostels); and the like.

Value of work £		Category A fee £	£
Up to	150,000	380 + 6.0% (Minimum fee £3,380)	
150,000 -	300,000	9,380 + 5.0% on balance over	150,000
300,000 -	600,000	16,880 + 4.3% on balance over	300,000
600,000 -	1,500,000	29,780 + 3.4% on balance over	600,000
1,500,000 -	3,000,000	60,380 + 3.0% on balance over	1,500,000
3,000,000 -	6,000,000	105,380 + 2.8% on balance over	3,000,000
Over	6,000,000	189,380 + 2.4% on balance over	6,000,000

(a) **Category B**: Less complex works and/or works with some element of repetition.

Examples:
Adult education facilities; canteens; church halls; community centres; departmental stores; enclosed sports stadia and swimming baths; halls of residence; hospitals; hostels; motels; offices other than those included in Categories A and C; railway stations; recreation and leisure centres; residential hotels; schools; self-contained flats and maisonettes; shops and shopping centres; supermarkets and hypermarkets; telephone exchanges; and the like.

Value of Work £		Category B fee £	£
Up to	150,000	360 + 5.8% (Minimum fee £3,260)	
150,000 -	300,000	9,060 + 4.7% on balance over	150,000
300,000 -	600,000	16,110 + 3.9% on balance over	300,000
600,000 -	1,500,000	27,810 + 2.8% on balance over	600,000
1,500,000 -	3,000,000	53,010 + 2.6% on balance over	1,500,000
3,000,000 -	6,000,000	92,010 + 2.4% on balance over	3,000,000
Over	6,000,000	164,010 + 2.0% on balance over	6,000,000

(c) **Category C**: Simple works and/or works with a substantial element of repetition.

Examples:
Factories; garages; multi-storey car parks; open-air sports stadia; structural shell offices not fitted out; warehouses; workshops; and the like,

Value of work £		Category C fee £	£
Up to	150,000	300 + 4.9% (Minimum fee £2,750)	
150,000 -	300,000	7,650 + 4.1% on balance over	150,000
300,000 -	600,000	13,800 + 3.3% on balance over	300,000
600,000 -	1,500,000	23,700 + 2.5% on balance over	600,000
1,500,000 -	3,000,000	46,200 + 2.2% on balance over	1,500,000
3,000,000 -	6,000,000	79,200 + 2.0% on balance over	3,000,000
Over	6,000,000	139,200 + 1.6% on balance over	6,000,000

QUANTITY SURVEYORS' FEES

  (d)   Fees shall be calculated upon the total of the final account for the whole of the work including all nominated sub-contractors' and nominated supplier's accounts. When work normally included in a building contract is the subject of a separate contract for which the quantity surveyor has not been paid fees under any other clause hereof, the value of such work shall be included in the amount upon which fees are charged.

  (e)   When a contract comprises buildings which fall into more than one category, the fee shall be calculated as follows:

      (i)   The amount upon which fees are chargeable shall be allocated to the categories of work applicable and the amounts so allocated expressed as percentages of the total amount upon which fees are chargeable.

      (ii)   Fees shall then be calculated for each category on the total amount upon which fees are chargeable.

      (iii)   The fee chargeable shall then be calculated by applying the percentages of work in each category to the appropriate total fee and adding the resultant amounts.

      (iv)   A consolidated percentage fee applicable to the total value of the work may be charged by prior agreement between the employer and the quantity surveyor. Such a percentage shall be based on this scale and on the estimated cost of the various categories of work and calculated in accordance with the principles stated above.

  (f)   When a project is subject to a number of contracts then, for the purpose of calculating fees, the values of such contracts shall not be aggregated but each contract shall be taken separately and the scale of charges (paras. 2.1 (a) to (e)) applied as appropriate.

2.2. Air conditioning, heating, ventilating and electrical services

  (a)   When the services outlined in para. 1.3 are provided by the quantity surveyor for the air conditioning, heating, ventilating and electrical services there shall be a fee for these services in addition to the fee calculated in accordance with para. 2.1 as follows:

Value of work £	Additional fee £	£
Up to 120,000	5.0%	
120,000 - 240,000	6,000 + 4.7% on balance over	120,000
240,000 - 480,000	11,640 + 4.0% on balance over	240,000
480,000 - 750,000	21,240 + 3.6% on balance over	480,000
750,000 -1,000,000	30,960 + 3.0% on balance over	750,000
1,000,000 -4,000,000	38,460 + 2.7% on balance over	1,000,000
Over 4,000,000	119,460 + 2.4% on balance over	4,000,000

  (b)   The value of such services, whether the subject of separate tenders or not, shall be aggregated and the total value of work so obtained used for the purpose of calculating the additional fee chargeable in accordance with para. (a). (Except that when more than one firm of consulting engineers is engaged on the design of these services, the separate values for which each such firm is responsible shall be aggregated and the additional fees charged shall be calculated independently on each such total value so obtained.)

QUANTITY SURVEYORS' FEES

    (c) Fees shall be calculated upon the basis of the account for the whole of the air conditioning, heating, ventilating and electrical services for which bills of quantities and final accounts have been prepared by the quantity surveyor.

2.3. Works of alteration

On works of alteration or repair, or on those sections of the work which are mainly works of alteration or repair, there shall be a fee of 1.0% in addition to the fee calculated in accordance with paras. 2.1 and 2.2.

2.4. On works of redecoration and associated minor repairs

On works of redecoration and associated minor repairs, there shall be a fee of 1.5% in addition to the fee calculated in accordance with paras. 2.1 and 2.2.

2.5. Generally

If the works are substantially varied at any stage or if the quantity surveyor is involved in an excessive amount of abortive work, then the fees shall be adjusted by agreement between the employer and the quantity surveyor.

3.0. ADDITIONAL SERVICES

3.1.   For additional services not normally necessary, such as those arising as a result of the termination of a contract before completion, liquidation, fire damage to the buildings, services in connection with arbitration, litigation and investigation of the validity of contractors' claims, services in connection with taxation matters and all similar services where the employer specifically instructs the quantity surveyor, the charges shall be in accordance with para. 4.0 below.

4.0. TIME CHARGES

4.1. (a) For consultancy and other services performed by a principal, a fee by arrangement according to the circumstances including the professional status and qualifications of the quantity surveyor.

     (b) When a principal does work which would normally be done by a member of staff, the charge shall be calculated as para. 4.2 below.

4.2. (a) For services by a member of staff, the charges for which are to be based on the time involved, such charges shall be calculated on the hourly cost of the individual involved plus 145%.

     (b) A member of staff shall include a principal doing work normally done by an employee (as para. 4.1 (b) above), technical and supporting staff, but shall exclude secretarial staff or staff engaged upon general administration.

     (c) For the purpose of para. 4.2 (b) above, a principal's time shall be taken at the rate applicable to a senior assistant in the firm.

     (d) The supervisory duties of a principal shall be deemed to be included in the addition of 145% as para. 4.2 (a) above and shall not be charged separately.

     (e) The hourly cost to the employer shall be calculated by taking the sum of the annual cost of the member of staff of:

        (i)   Salary and bonus but excluding expenses;

        (ii) Employer's contributions payable under any Pension and Life Assurance Schemes;

QUANTITY SURVEYORS' FEES

    (iii) Employer's contributions made under the National Insurance A
          Acts, the Redundancy Payments Act and any other payments
          made in respect of the employee by virtue of any statutory
          requirements; and
    (iv)   Any other payments or benefits made or granted by the
          employer in pursuance of the terms of employment of the
          member of staff;

and dividing by 1,650.

5.0. INSTALMENT PAYMENTS

5.1. In the absence of agreement to the contrary, fees shall be paid by
     instalments as follows:

    (a) Upon acceptance by the employer of a tender for the works, one half
        of the fee calculated on the amount of the accepted tender.
    (b) The balance by instalments at intervals to be agreed between the
        date of the first certificate and one month after final
        certification of the contractor's account.

5.2. (a) In the event of no tender being accepted, one half of the fee shall
        be paid within three months of completion of the tender documents.
        The fee shall be calculated upon the basis of the lowest original
        bona fide tender received. In the event of no tender being
        received, the fee shall be calculated upon a reasonable valuation
        of the works based upon the tender documents.
    (b) In the event of the project being abandoned at any stage other than
        those covered by the foregoing, the proportion of fee payable shall
        be by agreement between the employer and the quantity surveyor.

        NOTE: In the foregoing context 'bona fide tender' shall be deemed
        to mean a tender submitted in good faith without major errors of
        computation and not subsequently withdrawn by the tenderer.

QUANTITY SURVEYORS' FEES

Scale 37 itemized scale of professional charges for quantity surveying services
for building work issued by the Royal Institution of Chartered Surveyors.   The
scale is recommended and not mandatory.

EFFECTIVE FROM JULY 1988

1.0. GENERALLY

> 1.1. The fees are in all cases exclusive of travelling and other
> expenses (for which the actual disbursement is recoverable unless there
> is some prior arrangement for such charges) and of the cost of
> reproduction of bills of quantities and other documents, which are
> chargeable in addition at net cost.
> 1.2. The fees are in all cases exclusive of services in connection with
> the allocation of the cost of the works for purposes of calculating
> value added tax for which there shall be an additional fee based on the
> time involved (see paras. 19.1 and 19.2).
> 1.3. If any of the materials used in the works are supplied by the
> employer or charged at a preferential rate, then the actual or
> estimated market value thereof shall be included in the amounts upon
> which fees are to be calculated.
> 1.4. The fees are in all cases exclusive of preparing a specification of
> the materials to be used and the works to be done, but the fees for
> preparing bills of quantities and similar documents do include for
> incorporating preamble clauses describing the materials and workmanship
> (from instructions given by the architect and/or consulting engineer).
> 1.5. If the quantity surveyor incurs additional costs due to
> exceptional delays in building operations or any other cause beyond the
> control of the quantity surveyor then the fees may be adjusted by
> agreement between the employer and the quantity surveyor to cover the
> reimbursement of these additional costs.
> 1.6. The fees and charges are in all cases exclusive of value added tax
> which will be applied in accordance with legislation.
> 1.7. Copyright in bills of quantities and other documents prepared by
> the quantity surveyor is reserved to the quantity surveyor.

CONTRACTS BASED ON BILLS OF QUANTITIES: PRE-CONTRACT SERVICES

2.0. BILLS OF QUANTITIES

   2.1. **Basic scale**

   For preparing bills of quantities and examining tenders received and
   reporting thereon.

   (a)   **Category A:** Relatively complex works and/or works with little or no
         repetition.

         Examples:
         Ambulance and fire stations; banks; cinemas; clubs; computer
         buildings; council offices; crematoria; fitting out of existing
         buildings; homes for the elderly; hospitals and nursing homes;
         laboratories; law courts; libraries; 'one off' houses;
         petrol stations; places of religious worship; police stations;
         public houses, licensed premises; restaurants; sheltered housing;
         sports pavilions; theatres; town halls; universities, polytechnics
         and colleges of further education (other than halls of residence
         and hostels); and the like.

QUANTITY SURVEYORS' FEES

Value of work £		Category A fee £	£
Up to	150,000	230 + 3.0% (Minimum fee £1,730)	
150,000 -	300,000	4,730 + 2.3%on balance over	150,000
300,000 -	600,000	8,180 + 1.8% on balance over	300,000
600,000 -	1,500,000	13,580 + 1.5% on balance over	600,000
1,500,000 -	3,000,000	27,080 + 1.2% on balance over	1,500,000
3,000,000 -	6,000,000	45,080 + 1.1% on balance over	3,000,000
Over	6,000,000	78,080 + 1.0% on balance over	6,000,000

(b) **Category B:** Less complex works and/or works with some element of repetition.

Examples:
Adult education facilities; canteens; church halls; community centres; departmental stores; enclosed sports stadia and swimming baths; halls of residence; hospitals; hostels; motels; offices other than those included in Categories in A and C; railway stations; recreation and leisure centres; residential hotels; schools; self-contained flats and maisonettes; shops and shopping centres; supermarkets and hypermarkets; telephone exchanges; and the like.

Value of work £		Category B fee £	£
Up to	150,000	210 + 2.8% (Minimum fee £1,610)	
150,000 -	300,000	4,410 + 2.0% on balance over	150,000
300,000 -	600,000	7,410 + 1.5% on balance over	300,000
600,000 -	1,500,000	11,910 + 1.1% on balance over	600,000
1,500,000 -	3,000,000	21,810 + 1.0% on balance over	1,500,000
3,000,000 -	6,000,000	36,810 + 0.9% on balance over	3,000,000
Over	6,000,000	63,810 + 0.8% on balance over	6,000,000

(c) **Category C:** Simple works and/or works with a substantial element of repetition

Examples:
Factories; garages; multi-storey car parks; open-air sports stadia; structural shell offices not fitted out; warehouses; workshops and the like.

Value of Work £		Category C fee £	£
Up to	150,000	180 + 2.5% (Minimum fee £1,430)	
150,000 -	300,000	3,930 + 1.8% on balance over	150,000
300,000 -	600,000	6,630 + 1.2% on balance over	300,000
600,000 -	1,500,000	10,230 + 0.9% on balance over	600,000
1,500,000 -	3,000,000	18,330 + 0.8% on balance over	1,500,000
3,000,000 -	6,000,000	30,330 + 0.7% on balance over	3,000,000
Over	6,000,000	51,330 + 0.6% on balance over	6,000,000

(d) The scales of fees for preparing bills of quantities (paras. 2.1 (a) to (c).) are overall scales based upon the inclusion of all provisional and prime cost items, subject to the provision of para. 2.1 (g). When work normally included in a building contract is the subject of a separate contract for which the quantity surveyor has not been paid fees under any other clause hereof, the value of such work shall be included in the amount upon which fees are charged.

QUANTITY SURVEYORS' FEES

(e) Fees shall be calculated upon the accepted tender for the whole of
     the work subject to the provisions of para. 2.6. In the event of
     no tender being accepted, fees shall be calculated upon the basis
     of the lowest original bona fide tender received. In the event of
     no such tender being received, the fees shall be calculated upon a
     reasonable valuation of the works based upon the original bills of
     quantities.
     NOTE: In the foregoing context 'bona fide tender' shall be deemed
     to mean a tender submitted in good faith without major errors of
     computation and not subsequently withdrawn by the tenderer.

(f) In calculating the amount upon which fees are charged the total of
     any credits and the totals of any alternative bills shall be
     aggregated and added to the amount described above. The value of
     any omission or addition forming part of an alternative bill shall
     not be added unless measurement or abstraction from the original
     dimension sheets was necessary.
(g) Where the value of the air conditioning, heating, ventilating and
     electrical services included in the tender documents together
     exceeds 25% of the amount calculated as described in paras. 2.1 (d)
     and (e), then, subject to the provisions of para. 2.2, no fee is
     chargeable on the amount by which the value of these services
     exceeds the said 25%. In this context the term 'value' excludes
     general contractor's profit, attendance, builder's work in
     connection with the services, preliminaries and any similar
     additions.
(h) When a contract comprises buildings which fall into more than one
     category, the fee shall be calculated as follows:

     (i)   The amount upon which fees are chargeable shall be allocated
            to the categories of work applicable and the amounts so
            allocated expressed as percentages of the total amount upon
            which fees are chargeable.
     (ii)   Fees shall then be calculated for each category on the total
            amount upon which fees are chargeable.
     (iii) The fee chargeable shall then be calculated by applying the
            percentages of work in each category to the appropriate
            total fee and adding the resultant amounts.

(j) When a project is the subject of a number of contracts then, for
     the purpose of calculating fees, the values of such contracts shall
     not be aggregated but each contract shall be taken seperately and
     the scale of charges (paras. 2.1 (a) to (h)) applied as
     appropriate.
(k) Where the quantity surveyor is specifically instructed to provide
     cost planning services the fee calculated in accordance with paras.
     2.1 (a) to (j) shall be increased by a sum calculated in accordance
     with the following table and based upon the same value of work as
     that upon which the aforementioned fee has been calculated:

     Categories A & B: (as defined in paras. 2.1 (a) and (b)).

Value of work £	Fee £	£
Up to 600,000	0.7%	
600,000 - 3,000,000	4,200 + 0.4% on balance over 600,000	
3,000,000 - 6,000,000	13,800 + 0.35% on balance over 3,000,000	
Over 6,000,000	24,300 + 0.3% on balance over 6,000,000	

QUANTITY SURVEYORS' FEES

Category C: (as defined in paras. 2.1 (c))

Value of work £	Fee £	£
Up to 600,000	0.5%	
600,000 -3,000,000	3,000 + 0.3% on balance over	600,000
3,000,000 -6,000,000	10,200 + 0.25% on balance over	3,000,000
Over 6,000,000	17,700 + 0.2% on balance over	6,000,000

2.2. **Air conditioning, heating, ventilating and electrical services**

(a) Where bills of quantities are prepared by the quantity surveyor for the air conditioning, heating, ventilating and electrical services there shall be a fee for these services (which shall include examining tenders received and reported thereon), in addition to the fee calculated in accordance with para. 2.1, as follows:

Value of work £	Additional fee £	£
Up to 120,000	2.5%	
120,000 - 240,000	3,000 + 2.25% on balance over	120,000
240,000 - 480,000	5,700 + 2.0% on balance over	240,000
480,000 - 750,000	10,500 + 1.75% on balance over	480,000
750,000 - 1,000,000	15,225 + 1.25% on balance over	750,000
Over - 1,000,000	18,350 + 1.15% on balance over	1,000,000

(b) The values of such services, whether the subject of separate tenders or not, shall be aggregated and the total value of work so obtained used for the purpose of calculating the additional fee chargeable in accordance with para. (a).
(Except that when more than one firm of consulting engineers is engaged on the design of these services, the separate values for which each such firm is responsible shall be aggregated and the additional fees charged shall be calculated independently on each such total value so obtained.)

(c) Fees shall be calculated upon the accepted tender for the whole of the air conditioning, heating, ventilating and electrical services for which bills of quantities have been prepared by the quantity surveyor. In the event of no tender being accepted, fees shall be calculated upon the basis of the lowest original bona fide tender received. In the event of no such tender being received, the fees shall be calculated upon a reasonable valuation of the services based upon the original bills of quantities.

NOTE: In the foregoing context 'bona fide tender' shall be deemed to mean a tender submitted in good faith without major errors of computation and not subsequently withdrawn by the tenderer.

(d) When cost planning services are provided by the quantity surveyor for air conditioning, heating, ventilating and electrical services (or for any part of such services) there shall be an additional fee based on the time involved (see paras. 19.1 and 19.2). Alternatively the fee may be on a lump sum or percentage basis agreed between the employer and the quantity surveyor.

NOTE: The incorporation of figures for air conditioning, heating, ventilating and electrical services provided by the consulting engineer is deemed to be included in the quantity surveyor's services under para. 2.1.

QUANTITY SURVEYORS' FEES

2.3. Works on alteration

On works of alteration or repair, or on those sections of the works hich
are mainly works of alteration or repair, there shall be a fee of 1.0%
in addition to the fee calculated in accordance with paras. 2.1 and 2.2.

2.4. Works of redecoration and associated minor repairs,

On works of redecoration and associated minor repairs, there shall be a
fee of 1.5% in addition to the fee calculated in accordance with paras.
2.1 and 2.2.

2.5. Bills of quantities prepared in special forms

Fees calculated in accordance with paras. 2.1, 2.2, 2.3, and 2.4 include
for the preparation of bills of quantities on a normal trade basis. If
the employer requires additional information to be provided in the bills
of quantities or the bills to be prepared in an elemental, operational
or similar form, then the fee may be adjusted by agreement between the
employer and the quantity surveyor.

2.6. Reduction of tenders

(a) When cost planning services have been provided by the quantity
     surveyor and a tender, when received, is reduced before acceptance
     and if the reduction are not necessitated by amended instructions
     of the employer or by the inclusion in the bills of quantities of
     items which the quantity surveyor has indicated could not be
     contained within the approved estimate, then in such a case no
     charge shall be made by the quantity surveyor for the preparation
     of bills of reductions and the fee for the preparation of the bills
     of quantities shall be based on the amount of the reduced tender.

(b) When cost planning services have not been provided by the quantity
     surveyor and if a tender, when received, is reduced before
     acceptance, fees are to be calculated upon the amount of the
     unreduced tender. When the preparation of bills of reductions is
     required, a fee is chargeable for preparing such bills of
     reductions as follows:

     (i)   2.0% upon the gross amount of all omissions requiring
            measurement or abstraction from original dimensional sheets.
     (ii)   3.0% upon the gross amount of all additions requiring
            measurement.
     (iii) 0.5% upon the gross amount of all remaining additions.

            NOTE: The above scale for the preparation of bills of
            reductions applies to work in all categories.

2.7. Generally

If the works are substantially varied at any stage or if the
quantity surveyor is involved in an excessive amount of abortive
work, then the fees shall be adjusted by agreement between the
employer and the quantity surveyor.

QUANTITY SURVEYORS' FEES

3.0. NEGOTIATING TENDERS

3.1. (a) For negotiating and agreeing prices with a contractor:

Value of work		Fee		
£		£		£
Up to	150,000	0.5%		
150,000 -	600,000	750 + 0.3% on balance over	150,000	
600,000 -	1,200,000	2,100 + 0.2% on balance over	600,000	
Over	1,200,000	3,300 + 0.1% on balance over	1,200,000	

(b) The fee shall be calculated on the total value of the works as defined in paras. 2.1 (d), (e), (f), (g) and (j).

(c) For negotiating and agreeing prices with a contractor for air conditioning, heating, ventilating and electrical services there shall be an additional fee as para. 3.1 (a) calculated on the total value of such services as defined in para. 2.2 (b).

4.0. CONSULTATIVE SERVICES AND PRICING BILLS OF QUANTITIES

4.1. **Consultative services**

Where the quantity surveyor is appointed to prepare approximate estimates, feasibility studies or submissions for the approval of financial grants or similar services, then the fee shall be based on the time involved (see paras. 19.1 and 19.2) or alternatively, on a lump sum or percentage basis agreed between the employer and the quantity surveyor.

4.2. **Pricing bills of quantities**

(a) For pricing bills of quantities, if instructed, to provide an estimate comparable with tenders, the fee shall be one-third (33.33%) of the fee for negotiating and agreeing prices with a contractor, calculated in accordance with paras. 3.1 (a) and (b).

(b) For pricing bills of quantities, if instructed, to provide an estimate comparable with tenders for air conditioning, heating, ventilating and electrical services the fee shall be one-third (33.33%) of the fee calculated in accordance with para. 3.1. (c).

CONTRACTS BASED ON BILLS OF QUANTITIES: POST-CONTRACT SERVICES

Alternative scales (I and II) for post-contract services are set out below to be used at the quantity surveyor's discretion by prior agreement with the employer.

5.0. ALTERNATIVE I: OVERALL SCALE OF CHARGES FOR POST-CONTRACT SERVICES

5.1. If the quantity surveyor appointed to carry out the post-contract services did not prepare the bills of quantities then the fees in paras. 5.2 and 5.3 shall be increased to cover the additional services undertaken by the quantity suveyor.

5.2. **Basic scale**

For taking particulars and reporting valuations for interim certificates for payments on account to the contractor, preparing periodic assessments of anticipated final cost and reporting thereon, measuring and making up bills of variations including pricing and agreeing totals with the contractor, and adjusting fluctuations in the cost of labour and materials if required by the contract.

QUANTITY SURVEYORS' FEES

(a) **Category A**: Relatively complex works and/or works with little or no repetition.

Examples:
Ambulance and fire stations; banks; cinemas; clubs; computer buildings; council offices; crematoria; fitting out existing buildings; homes for the elderly; hospitals and nursing homes; laboratories; law courts; libraries; 'one-off' houses; petrol stations; places of religious worship; police stations; public houses, licensed premises; restaurants; sheltered housing; sports pavilions; theatres; town halls; universities, polytechnics and colleges of further education (other than halls of residence and hostels); and the like.

Value of work £		Category A fee £	£
Up to	150,000	150 + 2.0% (Minimum fee £1,150)	
150,000 -	300,000	3,150 + 1.7% on balance over	150,000
300,000 -	600,000	5,700 + 1.6% on balance over	300,000
600,000 -	1,500,000	10,500 + 1.3% on balance over	600,000
1,500,000 -	3,000,000	22,200 + 1.2% on balance over	1,500,000
3,000,000 -	6,000,000	40,200 + 1.1% on balance over	3,000,000
Over	6,000,000	73,200 + 1.0% on balance over	6,000,000

(b) **Category B**: Less complex works and/or works with some element of repetition.

Examples:
Adult education facilities; canteens; church halls; community centres; departmental stores; enclosed sports stadia and swimming baths; halls of residence; hostels; motels; offices other than those included in Categories A and C; railway stations; recreation and leisure centres; residential hotels; schools; self-contained flats and maisonettes; shops and shopping centres; supermarkets and hypermarkets; telephone exchanges; and the like.

Value of work £		Category B fee £	£
Up to	150,000	150 + 2.0% (Minimum fee £1,150)	
150,000 -	300,000	3,150 + 1.7% on balance over	150,000
300,000 -	600,000	5,700 + 1.5% on balance over	300,000
600,000 -	1,500,000	10,200 + 1.1% on balance over	600,000
1,500,000 -	3,000,000	20,100 + 1.0% on balance over	1,500,000
3,000,000 -	6,000,000	35,100 + 0.9% on balance over	3,000,000
Over	6,000,000	62,100 + 0.8% on balance over	6,000,000

(c) **Category C**: Simple works and/or works with a substantial element of repetition.

Examples:
Factories; garages; multi-storey car parks; open-air sports stadia; structural shell offices not fitted out; warehouses; workshops; and the like.

Value of work £		Category C fee £	£
Up to	150,000	120 + 1.6% (Minimum fee £920)	
150,000 -	300,000	2,520 + 1.5% on balance over	150,000
300,000 -	600,000	4,770 + 1.4% on balance over	300,000
600,000 -	1,500,000	8,970 + 1.1% on balance over	600,000
1,500,000 -	3,000,000	18,870 + 0.9% on balance over	1,500,000
3,000,000 -	6,000,000	32,370 + 0.8% on balance over	3,000,000
Over	6,000,000	56,370 + 0.7% on balance over	6,000,000

QUANTITY SURVEYORS' FEES

(d) The scales of fees for post-contract services (paras. 5.2 (a) to (c)) are overall scales based upon the inclusion of all nominated sub-contractors' and nominated suppliers' accounts, subject to the provision of para. 5.2 (g). When work normally included in a building contract is the subject of a separate contract for which the quantity surveyor has not been paid fees under any other clause hereof, the value of such work shall be included in the amount on which fees are charges.

(e) Fees shall be calculated upon the basis of the account for the whole of the work, subject to the provisions of para. 5.3.

(f) In calculating the amount on which fees are charged the total of any credits is to be added to the amount described above.

(g) Where the value of air conditioning, heating, ventilating and electrical services included in the tender documents together exceeds 25% of the amount calculated as described in paras. 5.2. (d) and (e) above, then, subject to provisions of para. 5.3, no fee is chargeable on the amount by which the value of these services exceeds the said 25%. In this context the term 'value' excludes general contractors' profit, attendance, builders work in connection with the services, preliminaries and other similar additions.

(h) When a contract comprises buildings which fall into more than one category, the fee shall be calculated as follows:

(i) The amount upon which fees are chargeable shall be allocated to the categories of work applicable and the amounts so allocated expressed as percentages of the old total amount upon which fees are chargeable.

(ii) Fees shall then be calculated for each category on the total amount upon which fees are chargeable.

(iii) The fee chargeable shall then be calculated by applying the percentages of work in each category to the appropriate total fee and adding the resultant amounts.

(j) When a project is the subject of a number of contracts then, for the purposes of calculating fees, the values of such contracts shall not be aggregated but each contract shall be taken separately and the scale of charges (paras. 5.2 (a) to (h)), applied as appropriate.

(k) When the quantity surveyor is required to prepare valuations of materials or goods off site, an additional fee shall be charged based on the time involved (see paras. 19.1 and 19.2).

(l) The basic scale for post-contract services includes for a simple routine of periodically estimating final costs. When the employer specifically requests a cost monitoring service which involves the quantity surveyor in additional or abortive measurement an additional fee shall be charged based on the time involved (see paras. 19.1 and 19.2), or alternatively on a lump sum or percentage basis agreed between the employer and the quantity surveyor.

(m) The above overall scales of charges for post-contract services assume normal conditions when the bills of quantities are based on drawing accurately depicting the building work the employer requires. If the works are materially varied to the extent that substantial remeasurements is necessary then the fee for post-contract services shall be adjusted by agreement between the employer and the quantity surveyor.

5.3. **Air conditioning, heating, ventilating and electrical services**

(a) Where final accounts are prepared by the quantity surveyor for the air conditioning, heating, ventilating and electrical services there shall be a fee for these services, in addition to the fee

QUANTITY SURVEYORS' FEES

calculated in accordance with para. 5.2, as follows:

Value of Work £	Additional Fee £	£
Up to 120,000	2.0%	
120,000 - 240,000	2,400 + 1.6% on balance over	120,000
240,000 - 1,000,000	4,320 + 1.25% on balance over	240,000
1,000,000 - 4,000,000	13,820 + 1.0% on balance over	1,000,000
Over 4,000,000	43,820 + 0.9% on balance over	4,000,000

(b) The values of such services, whether the subject of separate tenders or not, shall be aggregated and the total value of work so obtained used for the purpose of calculating the addtional fee chargeable in accordance with para. (a).
(Except that when more than one firm of consulting engineers is engaged on the design of these services the separate values of which each such firm is responsible shall be aggregated and the additional fee charged shall be calculated independently on each such total value obtained.)

(c) The scope of the scale of the services to be provided by the quantity surveyor under para. (a) above shall be deemed to be equivalent to those described for the basic scale for post-contract services.

(d) When the quantity surveyor is required to prepare periodic valuations of materials or goods off site, an additional fee shall be charged based on the time involved (see paras. 19.1 and 19.2).

(e) The basic scale for post-contract services included for a simple routine of periodically estimating final costs. When the employer specifically requests a cost monitoring services which involves the quantity surveyor in addtional or abortive measurement an additional fee shall be based on the time involved (see paras. 19.1 and 19.2), or alternatively on a lump sum or percentage basis agreed between the employer and the quantity surveyor.

(f) Fees shall be calculated upon the basis of the account for the whole of the air conditioning, heating, ventilating and electrical services for which final accounts have been prepared by the quantity surveyor.

6.0. ALTERNATIVE II: SCALE OF CHARGES FOR SEPARATE STAGES OF POST-CONTRACT SERVICES

6.1. If the quantity surveyor appointed to carry out the post-contract services did not prepare the bills of quantities then the fees in paras. 6.2 and 6.3 shall be increased to cover the additional services undertaken by the quantity surveyor.

NOTE: The scales of fees in paras. 6.2 and 6.3 apply to work in all categories (including air conditioning, heating, ventilating and electrical services).

6.2. **Valuations for interim certificates**

(a) For taking particulars and reporting valuations for interim certificates for payments on account to the contractor.

QUANTITY SURVEYORS' FEES

Total of valuations £		Fee £	£
Up to	300,000	0.5%	
300,000 - 1,000,000		1,500 + 0.4% on balance over	300,000
1,000,000 - 6,000,000		4,300 + 0.3% on balance over 1,000,000	
Over	6,000,000	19,300 + 0.2% on balance over 6,000,000	

NOTES:

1 Subject to note 2 below, the fees are to be calculated on the total of all interim valuations (i.e. the amount of the final acount less only the net amount of the final valuation).

2 When consulting engineers are engaged in supervising the installation of air conditioning, heating, ventilating and electrical services and their duties include reporting valuations for inclusion in interim certificates for payments on account in respect of such services, then valuations so reported shall be excluded from any total amount of valuations used for calculating fees.

(b) When the quantity surveyor is required to prepare valuations of materials or goods off site, an additional fee shall be charged based on the time involved (see paras. 19.1 and 19.2).

6.3. **Preparing accounts of variation upon contracts**

For measuring and making up bills of variations including pricing and agreeing totals with the contractor:

(a) An initial lump sum of £600 shall be payable on each contract.

(b) 2.0% upon the gross amount of omissions requiring measurement or abstraction from the original dimension sheets.

(c) 3.0% upon the gross amount of additions requiring measurement and upon dayworks.

(d) 0.5% upon the gross amount of remaining additions which shall be deemed to include all nominated sub-contractors' and nominated suppliers' accounts which do not involve measurement or checking of quantities but only checking against lump sum estimates.

(e) 3.0% upon the aggregate of the amounts of the increases and/or decreases in the cost of labour and materials in accordance with any fluctuations clause in the conditions of contract, except where a price adjustment formula applies.

(f) On contracts where fluctuations are calculated by the use of a price adjustment formula method the following scale shall be applied to the account for the whole of the work:

Value of work £		Fee £	£
Up to	300,000	300 0.5%	
300,000 -1,000,000		1,800 + 0.3% on balance over	300,000
Over	1,000,000	3,900 + 0.1% on balance over 1,000,000	

(g) When consulting engineers are engaged in supervising the installation of air conditioning, heating, ventilating and electrical services and their duties include for the adjustment of accounts and pricing and agreeing totals with the sub-contractors for inclusion in the measured account, then any totals so agreed shall be excluded from any amounts used for calculating fees.

QUANTITY SURVEYORS' FEES

6.4. Cost monitoring services

The fee for providing all approximate estimates of final cost and/or a cost monitoring service shall be based on the time involved (see paras. 19.1 and 19.2), or alternatively on a lump sum or percentage basis agreed between the employer and the quantity surveyor.

7.0. BILLS OF APPROXIMATE QUANTITIES, INTERIM CERTIFICATES AND FINAL ACCOUNTS

7.1. Basic scale

For preparing bills of approximate quantities suitable for obtaining competitive tenders which will provide a schedule of prices and a reasonably close forecast of the cost of the works, but subject to complete remeasurement, examining tenders and reporting thereon, taking particulars and reporting valuations for interim certificates for payments on account to the contractor, preparing periodic assessments of anticipated final cost and reporting thereon, measuring and preparing final account, including pricing and agreeing totals with the contractor and adjusting fluctuations in the cost of labour and materials if required by the contract:

(a) **Category A:** Relatively complex works and/or works with little or no repetition.

Examples: Ambulance and fire stations; banks; cimemas; clubs; computer buildings; council offices; crematoria; fitting out existing buildings; homes for the elderly; hospitals and nursing homes; laboratories; law courts; libraries; 'one-off' houses; petrol stations; places of religious worship; police stations; public houses; licensed premises; resturants; sheltered housing; sports pavilions; theatres; town halls; universities, polytechnics and colleges of further education (other than halls of residence and hostels); and the like.

Value of work £	Category A fee £	£
Up to 150,000	380 + 5.0% (Minimum fee £2,880)	
150,000 - 300,000	7,880 + 4.0% on balance over	150,000
300,000 - 600,000	13,880 + 3.4% on balance over	300,000
600,000 -1,500,000	24,080 + 2.8% on balance over	600,000
1,500,000 -3,000,000	49,280 + 2.4% on balance over	1,500,000
3,000,000 -6,000,000	85,280 + 2.2% on balance over	3,000,000
Over 6,000,000	151,280 + 2.0% on balance over	6,000,000

(b) **Category B:** Less complex works and/or works with some element of repetition

Examples: Adult education facilities; canteens; church halls; comunity centres; departmental stores; enclosed sports stadia and swimming baths; halls of residence; hostels; motels; offices other than those included in Categories A and C; railway stations; recreation and leisure centres; residential hotels; schools; self-contained flats and maisonettes shops and shopping centres; supermarkets and hypermarkets; telephone exchanges; and the like.

QUANTITY SURVEYORS' FEES

Value of work £		Category B fee £		£
Up to	150,000	360	+ 4.8% (Minimum fee £2,760)	
150,000 -	300,000	7,560 +	3.7% on balance over	150,000
300,000 -	600,000	13,110 +	3.0% on balance over	300,000
600,000 -	1,500,000	22,110 +	2.2% on balance over	600,000
1,500,000 -	3,000,000	41,910 +	2.0% on balance over	1,500,000
3,000,000 -	6,000,000	71,910 +	1.8% on balance over	3,000,000
Over	6,000,000	125,910 +	1.6% on balance over	6,000,000

(c)   **Category C**: Simple works and/or works with a substantial element of
repetition.
Examples: Factories; garages; multi-storey car parks; open air
sports stadia; structural shell offices not fitted out;
warehouses; workshops; and the like.

Value of work £		Category C fee £		£
Up to	150,000	300	+4.1% (Minimum fee £2,350)	
150,000 -	300,000	6,450	+3.3% on balance over	150,000
300,000 -	600,000	11,400	+2.6% on balance over	300,000
600,000 -	1,500,000	19,200	+2.0% on balance over	600,000
1,500,000 -	3,000,000	37,200	+1.7% on balance over	1,500,000
3,000,000 -	6,000,000	62,700	+1.5% on balance over	3,000,000
Over	6,000,000	107,700	+1.3% on balance over	6,000,000

(d)   The scales of fees for pre-contract and post-contract services
(paras. 7.1 (a) to (c)) are overall scales based upon the inclusion
of all nominated sub-contractors' and nominated suppliers'
accounts, subject to the provision of para. 7.1. (g). When work
normally included in a building contract is the subject of a
separate contract for which the quantity surveyor has not been paid
fees under any other clause hereof, the value of such work shall be
included in the amount on which fees are charged.

(e)   Fees shall be calculated upon the basis of the account for the
whole of the work, subject to the provisions of para. 7.2.

(f)   In calculating the amount on which fees are charged the total of
any credits is to be added to the amount described above.

(g)   Where the value of air conditioning, heating, ventilating and
electrical services included in tender documents together exceeds
25% of the amount calculated as described in paras. 7.1. (d) and
(e), then, subject to the provisions of para. 7.2 no fee is
chargeable on the amount by which the value of theses services
exceeds the said 25%. In this context the term 'value' excludes
general contractors' profit, attendance, builders' work in
connection with the services, preliminaries and any other similar
additions.

(h)   When a contract comprises buildings which fall into more than one
category, the fee shall be calculated as follows.

(i)   The amount upon which fees are chargeable shall be allocated
to the categories of work applicable and the amount so
allocated expressed as percentages of the total amount upon
which fees are chargeable.

(ii)   Fees shall then be calculated for each category on the total
amount upon which fees are chargeable.

QUANTITY SURVEYORS' FEES

(iii) The fee chargeable shall then be calculated by applying the percentages of work in each category to the appropriate total fee adding the resultant amounts.

(j) When a project is the subject of a number of contracts then, for the purpose of calculating fees, the values of such contracts shall not be aggregated but each contract shall be taken separately and the scale of charges (paras. 7.1 (a) to (h)) applied as appropriate.

(k) Where the quantity surveyor is specifically instructed to provide cost planning services, the fee calculated in accordance with paras. 7.1 (a) to (j) shall be increased by a sum calculated in accordance with the following table and based upon the same value of work as that upon which the aforementioned fee has been calculated:

Categories A & B: (as defined in paras. 7.1 (a) and (b))

Value of work £		Fee £		£
Up to	600,000	0.7%		
600,000	-3,000,000	4,200 + 0.4%	on balance over	600,000
3,000,000	-6,000,000	13,800 + 0.35%	on balance over	3,000,000
Over	6,000,000	24,300 + 0.3%	on balance over	6,000,000

Category C: (as defined in para. 7.1 (c))

Value of work £		Fee £		£
Up to	600,000	0.5%		
600,000	-3,000,000	1,500 + 0.3%	on balance over	600,00
3,000,000	-6,000,000	10,200 + 0.25%	on balance over	3,000,000
Over	6,000,000	17,700 + 0.2%	on balance over	6,000,000

(1) When the quantity surveyor is required to prepare valuations of materials or goods off site, an additional fee shall be charged based on the time involved (see paras. 19.1 and 19.2).

(m) The basic scale for post-contract services includes for a simple routine of periodically estimating final costs. When the employer specifically requests a cost monitering service which involves the quantity surveyor in additional or abortive measurement an additional fee shall be charged based on the time involved (see paras. 19.1 and 19.2), or alternatively on a lump sum or percentage basis agreed between the employer and the quantity surveyor.

7.2. **Air conditioning, heating, ventilating and electrical services**

(a) Where bills of approximate quantities and final accounts are prepared by the quantity surveyor for the air conditioning, heating, ventilating and electrical services there shall be a fee for these services in addition to the fee calculated in accordance with para. 7.1 as follows:

QUANTITY SURVEYORS' FEES

Value of work	Additional fee	
£	£	£
Up to 120,000	4.5%	
120,000 - 240,000	5,400 + 3.85% on balance over	120,000
240,000 - 480,000	10,020 + 3.25% on balance over	240,000
480,000 - 750,000	17,820 + 3.0% on balance over	480,000
750,000 -1,000,000	25,920 + 2.5% on balance over	750,000
1,000,000 -4,000,000	32,170 + 2.15% on balance over	1,000,000
Over 4,000,000	96,670 + 2.05% on balance over	4,000,000

(b) The value of such services, whether the subject of separate tenders or not, shall be aggregated and the value of work so obtained used for the purpose of calculating the additional fee chargeable in accordance with para. (a).
(Except that when more than one firm of consulting engineers is engaged on the design of these services, the separate values for which each such firm is responsible shall be aggregated and the additional fees charged shall be calculated independently on each such total value so obtained.)

(c) The scope of the services to be provided by the quantity surveyor under para. (a) above shall be deemed to be equivalent to those described for the basic scale for pre-contract and post-contract services.

(d) When the quantity surveyor is required to prepare valuations of materials or goods off site, an additional fee shall be charged based on the time involved (see paras. 19.1 and 19.2).

(e) The basic scale for post-contract services includes for a simple routine of periodically estimating final costs. When the employer specifically requests a cost monitoring service, which involves the quantity surveyor in additional or abortive measurement, an additional fee shall be charged based on the time involved (see paras. 19.1 and 19.2), or alternatively on a lump sum or percentage basis agreed between the employer and the quantity surveyor.

(f) Fees shall be calculated upon the basis of the account for the whole of the air conditioning, heating, ventilating and electrical services for which final accounts have been prepared by the quantity surveyor.

(g) When cost planning services are provided by the quantity surveyor for air conditioning, heating, ventilating and electrical services (or for any part of such services) there shall be an additional fee based on the time involved (see paras. 19.1 and 19.2) or alternatively on a lump sum or percentage basis agreed between the employer and quantity surveyor.

NOTE: The incorperation of figures for air conditioning, heating, ventilating and electrical services provided by the consulting engineer is deemed to be included in the quantity surveyor's services under para 7.1.

7.3. Works of alteration

On works of alteration or repair, or on those sections of the work which are mainly works of alteration or repair, there shall be a fee of 1.0% in addition to the fee calculated in accordance with paras. 7.1 and 7.2

7.4. Works of redecoration and associated minor repairs

On works of redecoration and associated minor repairs, there shall be a fee of 1.5% in addition to the fee calculated in accordance with paras. 7.1 and 7.2.

QUANTITY SURVEYORS' FEES

7.5. Bills of quantities and/or final accounts prepared in special forms

Fees calculated in accordance with paras. 7.1, 7.2, 7.3 and 7.4 include
for the preparation of bills of quantities and/or final accounts on a
normal trade basis. If the employer requires additional information to
be provided in the bills of quantities and/or final accounts or the
bills and/or final accounts to be prepared in an elemental, operational
or similar form, then the fee may be adjusted by agreement between the
employer and the quantity surveyor.

7.6. Reduction of tenders

(a) When cost planning services have been provided by the quantity
surveyor and a tender, when received, is reduced before acceptance
and if the reductions are not necessitated by amended instructions
of the employer or by the inclusion in the bills of approximate
quantities of items which the quantity surveyor has indicated could
not be contained within the approved estimate, then in such a case
no charge shall be made by the quantity surveyor for the
preparation of bills of reductions and the fee for the preparation
of bills of approximate quantities shall be based on the amount of
the reduced tender.

(b) When cost planning services have not been provided by the quantity
surveyor and if a tender, when received, is reduced before
acceptance, fees are to be calculated upon the amount of the
unreduced tender. When the preparation of bills of reductions is
required, a fee is chargeable for preparing such bills of
reductions as follows:

(i)   2.0% upon the gross amount of all omissions requiring
measurement or abstraction from original dimension sheets.

(ii) 3.0% upon the gross amount of all additions requiring
measurement.

(iii) 0.5% upon the gross amount of all remaining additions.

NOTE: The above scale for the preparation of bills of reductions
applies to work in all categories.

7.7. Generally

If the works are substantially varied at any stage or if the quantity
surveyor is involved in an excessive amount of abortive work, then the
fees shall be adjusted by agreement between the employer and the
quantity surveyor.

8.0. NEGOTIATING TENDERS

8.1. (a) For negotiating and agreeing prices with a contractor:

Value of Work		Fee		
£		£		£
Up to	150,000	0.5%		
150,000 -	600,000	750 + 0.3% on balance over	150,000
600,000 -	1,200,000	2,100 + 0.2% on balance over	600,000
Over	1,200,000	3,300 + 0.1% on balance over	1,200,000

(b) The fee shall be calculated on the total value of the works as
defined in paras. (d), (e), (f), (g) and (j).

QUANTITY SURVEYORS' FEES

(c)   For negotiating and agreeing prices with a contractor for air
      conditioning, heating, ventilating and electrical services there
      shall be an additional fee as para. 8.1 (a) calculated on the total
      value of such services as defined in para. 7.2 (b).

9.0. CONSULTATIVE SERVICES AND PRICING BILLS OF APPROXIMATE QUANTITIES

9.1. **Consultative services**

Where the quantity surveyor is appointed to prepare approximate,
feasibility studies or submissions for the approval of financial grants
or similar services, then the fee shall be based on the time involved
(see paras. 19.1 and 19.2) or alternatively, on a lump sum or percentage
basis agreed between the employer and the quantity surveyor.

9.2. **Pricing bills of approximate quantities**

For pricing bills of approximate quantities, if instructed, to provide
an estimate comparable with tenders, the fees shall be the same as for
the corresponding services in paras. 4.2 (a) and (b).

10.0. INSTALMENT PAYMENTS

10.1.   For the purpose of instalment payments the fee for preparation of
        bills of approximate quantities only shall be the equivalent of forty
        per cent (40%) of the fees calculated in accordance with the
        appropriate sections of paras. 7.1 to 7.5 and the fee for providing
        cost planning services shall be in accordance with the appropriate
        section of para. 7.1 (k); both fees shall be based on the total value
        of the bills of approximate quantities ascertained in accordance with
        the provisions of para. 2.1 (e).

10.2.   In the absence of agreement to the contary, fees shall be paid by
        installments as follows:

        (a)   Upon acceptance by the employer of a tender for the works the
              above defined fees for the preparation of bills of approximate
              quantities and for providing cost planning services.
        (b)   In the event of no tender being accepted, the aforementioned
              fees shall be paid within three months of completion of the
              bills of approximate quantities.
        (c)   The balance by instalments at intervals to be agreed between
              the date of the first certificate and one month after
              certification of the contractor's account.

10.3.   In the event of the project being abandoned at any stage other than
        those covered by the foregoing, the proportion of fee payable shall be
        by agreement between the employer and the quantity surveyor.

11.0. SCHEDULES OF PRICES

11.1.   The fee for preparing, pricing and agreeing schedules of prices shall
        be based on the time involved (see paras. 19.1 and 19.2).
        Alternatively, the fee may be on a lump sum or percentage basis agreed
        between the employer and the quantity surveyor.

12.0. COST PLANNING AND APPROXIMATE ESTIMATES

12.1   The fee for providing cost planning services or for preparing
        approximate estimates shall be based on the time involved (see paras.
        19.1 and 19.2). Alternatively, the fee may be on a lump sum basis
        agreed between the employer and the quantity surveyor.

QUANTITY SURVEYORS' FEES

CONTRACTS BASED ON SCHEDULES OF PRICES: POST-CONTRACT SERVICES

13.0. FINAL ACCOUNTS

13.1. **Basic Scale**

(a) For taking particulars and reporting valuations for interim certificates for payments on account to the contractor, preparing periodic assessments of anticipated final cost and reporting thereon, measuring and preparing final account includiong pricing and agreeing totals with the contractor, and adjusting fluctuations in the cost of labour and materials if required by the contract, the fee shall be equivalent to sixty per cent (60%) of the fee calculated in accordance with paras. 7.1 (a) to (j).

(b) When the quantity surveyor is required to prepare valuations of materials or goods off site, an additional fee shall be charged on the basis of the time involved (see paras. 19.1 and 19.2).

(c) The basic scale for post-contract services includes for a simple routine of periodically estimating final costs. When the employer specifically requests a cost monitoring service, which involves the quantity surveyor in additional or abortive measurement, an additional fee shall be charged based on the time involved (see paras. 19.1 and 19.2), or alternatively on a lump sum or percentage basis agreed between the employer and the quantity surveyor.

13.2. **Air conditioning, heating, ventilating and electrical services**

Where final accounts are prepared by the quantity surveyor for the air conditioning, heating, ventilating services there shall be a fee for these services, in addition to the fee calculated in accordance with para. 13.1, equivalent to sixty per cent (60%) of the fee calculated in accordance with paras. 7.2 (a) to (f).

13.3. **Works of alterations**

On works of alteration or repair, or on those sections of the work which are mainly works of alteration or repair, there shall be a fee of 1.0% in addition to the fee calculated in accordance with paras. 13.1 and 13.2.

13.4. **Works of redecoration and associated minor repairs**

On works of redecoration and associated minor repairs, there shall be a fee of 1.5% in addition to the fee calculated in accordance with paras. 13.1 and 13.2.

13.5. **Final accounts prepared in special forms**

Fees calculated in accordance with paras. 13.1,13.2,13.2 and 13.4 include for the preparation of final accounts on a normal trade basis. If the employer requires additional information to be provided in the final accounts or the accounts to be prepared in an elemental, operational or similar form, then the fee may be adjusted by agreement between the employer and the quantity surveyor.

14.0. COST PLANNING

14.1. The fee for providing a cost planning service shall be based on the time involved (see paras. 19.1 and 19.2). Alternatively, the fee may be on a lump sum or percentage basis agreed between the employer and the quantity surveyor.

QUANTITY SURVEYORS' FEES

13.3. Works of alterations

On works of alteration or repair, or on those sections of the work which are mainly works of alteration or repair, there shall be a fee of 1.0% in addition to the fee calculated in accordance with paras. 13.1 and 13.2.

13.4. Works of redecoration and associated minor repairs

On works of redecoration and associated minor repairs, there shall be a fee of 1.5% in addition to the fee calculated in accordance with paras. 13.1 and 13.2.

13.5. Final accounts prepared in special forms

Fees calculated in accordance with paras. 38, 39, 40 and 41 include for the preparation of final accounts on a normal trade basis. If the employer requires additional information to be provided in the final accounts or the accounts to be prepared in an elemental, operational or similar form, then the fee may be adjusted by agreement between the employer and the quantity surveyor.

14.0. COST PLANNING

14.1. The fee for providing a cost planning service shall be based on the time involved (see paras. 19.1 and 19.2). Alternatively, the fee may be on a lump sum or percentage basis agreed between the employer and the quantity surveyor.

15.0. ESTIMATES OF COST

15.1. (a) For preparing an approximate estimate, calculated by measurement, of the cost of work, and, if required under the terms of the contract, negotiating, adjusting and agreeing the estimate:

Value of work		Fee		
£		£		£
Up to	30,000	1.25%		
30,000 -	150,000	375 + 1.0% on balance over	30,000
150,000 -	600,000	1,575 + 0.75% on balance over	150,000
Over	600,000	4,950 + 0.5% on balance over	600,000

(b) The fee shall be calculated upon the total of the approved estimates.

16.0. FINAL ACCOUNTS

16.1. (a) For checking prime costs, reporting for interim certificates for payments on account to the contractor and preparing final accounts:

Value of work		Fee		
£		£		£
Up to	30,000	2.25%		
30,000 -	150,000	375 + 2.0% on balance over	30,000
150,000 -	600,000	3,150 + 1.5% on balance over	150,000
Over	600,000	9,900 + 1.25% on balance over	600,000

QUANTITY SURVEYORS' FEES

    (b) The fee shall be calculated upon the total of the final account with the addition of the value of credits received for old materials removed and less the value of any work charged for in accordance with para. 16.1 (c).

    (c) On the value of any work to be paid for on a measured basis, the fee shall be 3%.

    (d) When the quantity surveyor is required to prepare valuations of material or goods off site, an additional fee shall be charged based on the time involved (see paras. 19.1 and 19.2).

    (e) The above charges do not include the provision of checkers on the site. If the quantity surveyor is required to provide such checkers an additional charge shall be made by arrangement.

17.0. COST REPORTING AND MONITORING SERVICES

17.1. The fee for providing cost reporting and/or monitoring services (e.g. preparing periodic assessments of anticipated final costs and reporting thereon) shall be based on the time involved (see paras. 19.1 and 19.2) or alternatively, on a lump sum or percentage basis agreed between the employer and the quantity surveyor.

18.0. ADDITIONAL SERVICES

18.1. For additional services not normally necessary, such as those arising as a result of the termination of a contract before completion, liquidation, fire damage to the buildings, services in connection with arbitration, litigation and investigation of the validity of contractors' claims, services in connection with taxation matters and all similar services where the employer specifically instructs the quantity surveyor, the charges shall be in accordance with paras. 19.1 and 19.2.

19.0. TIME CHARGES

19.1. (a) For consultancy and other services performed by a principle, a fee by arrangment according to the circumstances including the professional status and qualifications of the quantity surveyor.

      (b) When a principal does work which would normally be done by a member of staff, the charge shall be calculated as para. 19.2 below.

19.2. (a) For services by a member of staff, the charges for which are to be based on the time involved, such charges shall be calculated on the hourly cost of the individual involved plus 145%.

      (b) A member of staff shall include a principal doing work normally done by an employee (as para. 19.1 (b) above), technical and supporting staff, but shall exclude secretarial staff or staff engaged upon general administration.

      (c) For the purpose of para. 19.2 (b) above, a principal's time shall be taken at the rate applicable to a senior assistant in the firm.

      (d) The supervisory duties of a principal shall be deemed to be included in the addition of 145% as para. 19.2 (a) above and shall not be charged seperately.

      (e) the hourly cost to the employer shall be calculated by taking the sum of the annual cost of the member of staff of:

         (i)   Salary and bonus but excluding expenses;

         (ii) Empolyer's contributions payable under any Pension and Life Assurance Schemes;

QUANTITY SURVEYORS' FEES

        (iii) Employer's contributions made under the National Insurance
              Acts, the Redundancy Payments Act and any other payments
              made in respect of the employee by virtue of any statutory
              requirements; and
        (iv) Any other payments or benifits made or granted by the
              employer in pursuance of the terms of employment of the
              member of staff;

     and dividing by 1,650.

19.3. The foregoing Time Charges under paras. 19.1 and 19.2 are intended for
use where other paragraphs of the Scale (not related to Time Charges)
form a significant proportion of the overall fee. In all other cases
an increased time charge may be agreed.

20.0. INSTALMENT PAYMENTS

20.1. In the absence of agreement to the contrary, payments to the quantity
surveyor shall be made by instalments by arrangement between the
employer and the quantity surveyor.

QUANTITY SURVEYORS' FEES

Scale 46 professional charges for quantity surveying services in connection with loss assessment of damage to buildings from fire, etc issued by The Royal Institution of Chartered Surveyors. The scale is recommended and not mandatory.

EFFECTIVE FROM JULY 1988

1. This scale of professional charges is for use in assessing loss result-
    ing from damage to buildings by fire etc., under the "building" section
    of an insurance policy and is applicable to all categories of buildings.

2. The fees set out below cover the following quantity surveying services
    as may be required in connection with the particular loss assessment:-

    (a) Examining the insurance policy.
    (b) Visiting the building and taking all necessary site notes.
    (c) Measuring at site and/or from drawings and preparing itemised
         statement of claim and pricing same.
    (d) Negotiating and agreeing claim with the loss adjuster.

3. The fees set out below are exclusive of the following:-

    (a) Travelling and other expenses (for which the actual disbursement is
         recoverable unless there is some special prior arrangement for such
         charge.)
    (b) Cost of reproduction of all documents, which are chargeable in
         addition at net cost.

4. Copyright in all documents prepared by the quantity surveyor is reserved

5. (a) The fees for the services outlined in paragragh 2 shall be
         as follows:-

    Agreed Amount of Damage   Fee
            £   £   £
    Up to   60,000   See note 5(c)below
      60,000 - 180,000   2.5%
     180,000 - 360,000   4,500 + 2.3% on balance over   180,000
     360,000 - 720,000   8,640 + 2.0% on balance over   360,000
    Over   720,000   15,840 + 1.5% on balance over   720,000
    and to the result of that computation shall be added 12.5%

    (b) The sum on which the fees above shall be calculated shall be
         arrived at after having given effect to the following:-
         (i)   The sum shall be based on the amount of damage, including
                such amounts in respect of architects', surveyors and other
                consultants' fees for reinstatement, as admitted by the loss
                adjuster.
         (ii)   When a policy is subject to an average clause, the sum shall
                be the agreed amount before the adjustment for"average".
         (iii) When, in order to apply the average clause, the reinstate-
                ment value of the whole subject is calculated and negotiated
                an additional fee shall be charged commensurate with the
                work involved.
    (c) Subject to 5 (b) above, when the amount of the sum on which fees
         shall be calculated is under £60,000 the fee shall be based on
         time involved as defined in Scale 37 (July 1988) paragraph 19 or
         on a lump sum or percentage basis agreed between the building
         owner and the quantity surveyor.
6. The foregoing scale of charges is exclusive of any services in
    connection with litigation and arbitration.
7. The fees and charges are in all cases exclusive of value added tax
    which shall be applied in accordance with legislation.

QUANTITY SURVEYORS' FEES

Scale 47 professional charges for the assessment of replacement costs of buildings for insurance, current cost accounting and other purposes issued by The Royal Institution of Chartered Surveyors.   The scale is recommended and not mandatory.

EFFECTIVE FROM JULY 1988

1.0. GENERALLY

   1.1. The fees are in all cases exclusive of travelling and other expenses (for which the actual disbursement is recoverable unless there is some prior arrangement for such charges).
   1.2. The fees and charges are in all cases exclusive of value added tax which will be applied in accordance with legislation.

2.0. ASSESSMENT OF REPLACEMENT COSTS OF BUILDINGS FOR
     INSURANCE PURPOSES

   2.1. Assessing the current replacement cost of buildings where adequate drawings for the purpose are available.

Assessed current cost		Fee	
£		£	£
Up to	140,000	0.2%	
140,000 -	700,000	280 + 0.075% on balance over	140,000
700,000 - 4,200,000		700 + 0.025% on balance over	700,000
Over	4,200,000	1,575 + 0.01% on balance over 4,200,000	

   2.2. Fees are to be calculated on the assessed current cost, i.e. base value, for replacement purposes including allowances for demolition and site clearance but excluding inflation allowances and professional fees.
   2.3. Where drawings adequate for the assessed of cost are not available or where other circumstances require that measurements of the whole or part of the buildings are taken, an additional fee shall be charged based on the time involved or alternatively on a lump sum basis agreed between the employer and the surveyor.
   2.4. When the assessment is for buildings of different character or on more than one site, the costs shall not be aggregated for the purpose of calculating fees.
   2.5. For current cost accounting purposes this scale refers only to the assessment of replacement costs of buildings.
   2.6. The scale is appropriate for initial assessments but for annual review or a regular reassessment the fee should be by arangement having regard to the scale and to the amount of work involved and the time taken.
   2.7. The fees are exclusive of services in connection with negotiations with brokers, accountants or insurance companies for which there shall be an additional fee based upon the time involved.

CONSULTING ENGINEERS' FEES

CONDITIONS OF ENGAGEMENT - 1981

INTRODUCTION

The following paragraphs describe the scope of professional services provided by the Consulting Engineer and give general advise about his appointment.
    The Association of Consulting Engineers has drawn up standard Conditions of Engagement to form the basis of the agreement between the Client and the Consulting Engineer, for five different types of appointment as hereafter described.   Each of the standard Conditions of Engagement is accompanied by a recommended Memorandum of Agreement.   These two documents, taken together, constitute the recommended form of Agreement in each case.
    When the standard Conditions of Engagement are not so used, it is in the interests of both the Client and the Consulting Engineer that there should be an exchange of letters defining the duties which the Consulting Engineer is to perform and the terms of payment.

REPORT AND ADVISORY WORK

For reports, and for advisory work, the services required from the Consulting Engineer will usually comprise one or more of the following:

(a)   investigating and advising on a project and submitting a report thereon. The Consulting Engineer may be asked to examine alternatives; review all technical aspects; make an economic appraisal of costs and benifits; draw conclusions and make recommendations.
(b)   inspecting existing works (e.g. a reservoir or a building or an installation) and reporting thereon. If the Client requires continuing advise on maintenance or operation of an existing project, the Consulting Engineer may be appointed to make periodic visits.
(c)   making a special investigation of an engineering problem and reporting thereon.
(d)   making valuations of plant and undertakings.

Payment for services provied under Items (a) to (d) above should normally be on a time basis or, where the duration and extent of the services can be defined clearly, by an agreed lump sum. It is recommended that the conditions governing an appointment for these services should be based on:

Agreement 1 - Conditions of Engagement for Reporting and Advisory Work.

DESIGN AND SUPERVISION OF CONSTRUCTION

When the Client has decided to proceed with the construction of engineering works or the installation of plant, it is normal practice for the Consulting Engineer who prepared or assisted in the preparation of any report to be appointed for the subsequent stages of the work or the relevent part thereof. The Consulting Engineeer will assist the Client in obtaining the requisite approvals, then prepare the designs and tender documents to enable competitive tenders to be obtained or orders to be placed, and will be responsible for the technical control and administration of the construction of he Works.
    It is recommended that the agreement for this type of appointment should be based on the most appropriate of the following:

Agreement 2 -   Conditions of Engagement for Civil, Mechanical and Electrical Work and for Structural Engineering Work where an Architect is not appointed by the Client.
Agreement 3 -   Conditions of Engagement for Structural Engineering Work where an Architect is appointed by the Client.
Agreement 4A -   Conditions of Engagement for Engineering Services in relation to Sub-contract Works.

CONSULTING ENGINEERS' FEES

  Agreement 4B -   Conditions of Engagement for Engineering Services in
                             relation to Direct Contract Works.

TERMS OF PAYMENT

The Association considers that, in normal circumstances, the level of
renumeration represented by the scales of percentage fees and hourly charging
rates set out in the ACE Conditions of Engagement is such as to ensure the
provision by a Consulting Engineer to his Client of a full, competent and
reliable standard of service.

These scales and rates are not mandatory but solely guidelines, and may by
negotiation be adjusted upwards or downwards to take account of the abnormal
complexity or simplicity of design, increase or diminution of the extent of
services to be provided, long-standing client relationships or other
circumstances.

However the Association strongly advises clients to satisfy themselves that the
level and quality of services they want will be covered if they make
appointments based on charges which are appreciably lower than those shown.

DIRECT LABOUR WORKS

The A.C.E. Conditions of Engagement require to be modified and supplemented when
the Client intends to have the Works constructed wholly or partly by direct
labour under the control of he Consulting Engineer.
    The additional services and substantially greater responsibilities undertaken
by the Consulting Engineer in connection with work carried out by direct labour
usually entitle him to a higher level of remuneration than that payable in
respect of Works carried out by contract.

PARTIAL SERVICES

When the Client wishes to appoint the Consulting Engineer for partial services
only, it is important that both parties recognize the limitation which such an
appointment places upon the responsibiliy of the Consulting Engineer who cannot
be held liable for matters that are outside his control.
    The terms of reference for the appointment should be carefully drawn up and
the relevant A.C.E. Conditions of Engagement should be adapted to suit the scope
of services required.
    Professional charges for partial services are usually best calculated on a
time basis, but may, in suitable cases, be a commensurate part of the percentage
fee for normal services shown in the standard Conditions of Engagement.

INSPECTION SERVICES

The inspection of materials and plant during manufacture or on site is usually
required during the construction stage of a project. Consequently, the
arrangements for this service are described in the standard Conditions of
Engagement. If, however the Client wishes to engage the Consulting Engineer to
provide only inspection services, the charges of the Consulting Engineer may be
either on a time basis or a percentage of the cost of the materials to be
inspected.

ACTING AS ARBITRATOR, UMPIRE OR EXPERT WITNESS

A Consulting Engineer may be appointed to act as Arbitrator or Umpire, or be
required to attend as an Expert Witness at Parliamentary Committees, Courts of
Law, Arbitrations or Official Inquiries. Payment for any of these services
should be on a basis of a lump sum retainer plus time charges, not less than
three hours per day being chargeable for attendance, however short, either
before or after a mid-day adjournment.

CONSULTING ENGINEERS' FEES

PAYMENT ON A TIME BASIS

When it is not possible to estimate in advance the duration and extent of the
Consulting Engineer's services, neither a lump sum payment alone nor a
percentage of the estimated construction cost would normally be a fair basis of
remuneration. The most satisfactory and equitable method of payment in these
cases makes allowance for the actual time occupied in providing the services
required, and comprises the following elements, as applicable:

(1) A charge in the form of hourly rate(s) for the services of a self-employed
     Principal or a Consultant of the firm. The hourly rate will depend upon
     his standing, the nature of the work and any special circumstances.
     Alternatively a lump sum fee may be charged instead of the said hourly
     rate(s).

(2) A charge which covers salaried Principals, Directors and technical and
     supporting staff salary and other payroll costs actually incurred by the
     Consulting Engineer, together with a fair proportion of his overhead costs,
     plus an element of profit. This charge is most conveniently calculated by
     applying a multiplier to the salary cost and then adding the net amount of
     other pay roll costs. The major part of the multiplier is attributable to
     the Consulting Engineer's overheads which may include, inter alia, the
     following indirect costs and expenses:

     (a) rent, rates and other expenses of up-keep of his office, its
         furnishings, equipment and supplies;
     (b) insurance premiums other than those recovered in the payroll cost;
     (c) administrative, accounting, secretarial and financing costs;
     (d) the expence of keeping abreast of advances in engineering;
     (e) the expense of preliminary arrangements for new or prospective
         projects;
     (f) loss of productive time of technical staff between assignments.

(3) A charge for use of a computer or other special equipment.

When calculating amounts chargeable on a time basis, a Consulting Engineer is
entitled to include time spent by Partners, Directors,Consultants and technical
and supporting staff in travelling in connection with the performance of the
services. The time spent by secretarial staff or by staff engaged on general
accountancy or administration duties in the Consulting Engineer's office is not
chargeable unless otherwise agreed.

DISBURSEMENTS AND OUT OF POCKET EXPENSES

In addition to the charges and percentage fees referred to in the preceeding
Sections, the Consulting Engineer is entitled to recover from the Client all
disbursements and out of pocket expenses incurred in performing his services.

APPOINTMENTS OUTSIDE THE UNITED KINGDOM

The standard A.C.E. Conditions of Engagement are suitable for appointments in
the United Kingdom. For work overseas, it is impracticable to make definite
recommendations as the conditions vary widely from country to country. There
are added complications in documentation relating to import customs, conditions
of payment, insurance, freight, etc. Furthermore, it is necessary to arrange for
site visits to be undertaken by Partners or senior staff whose absence abroad
during such periods represents a serious reduction of their earning power.
     The additional duties, responsiblities and non-recoverable costs involved, and
the extra work on general co-ordination, therefore justify higher fees in such
cases. Special arrangements are also necessary to cover travelling and other
out of pocket expenses in excess of those normally incurred on similar work in
the United Kingdom - including such matters as local cost-of-living allowances
and the cost of providing home-leave facilities to expatriate staff.

CONSULTING ENGINEERS' FEES

AGREEMENT 1 CONDITIONS OF ENGAGEMENT

1. DEFINITIONS

In construing this Agreement the following expressions shall have the meanings hereby assigned to them except where the context otherwise requires:

'The Consulting Engineer' means the person or firm named in the Memorandum of Agreement and shall include any other person or persons taken into partnership by such person or firm during the currency of this Agreement and the surviving member or members of any such partnership.

'The Task' means the work described in the Memorandum of Agreement in respect of which the Client has engaged the Consulting Engineer to provide professional services.

'Contractor' means any person or persons firm or company under contract to the Client to perform work and/or supply goods in connection with the Task.

'Salary Cost' means the total annual taxable remuneration paid by the Consulting Engineer to any person employed by him, divided by 1600 (being deemed to be the average annual total of effective working hours of an employee) and multiplied by the number of working hours spent by such person in performing any of the services in respect of which payment under this Agreement is to be made to the Consulting Engineer upon the basis of Salary Cost. For the purposes of this definition the annnual remuneration of a person employed by the Consulting Engineer for a period less than a full year shall be calculated pro rata to such person's remuneration for such lesser period.

'Other Payroll Cost' means the annual amount of all contributions and payments made by the Consulting Engineer on behalf of or in respect of a person employed by him for staff pension and life assurance schemes, and also for National Insurance Contributions and for any other tax, charge, levy, impost or payment of any kind whatsoever which the Consulting Engineer at any time during the performance of this Agreement is obliged by law to make on behalf of or in respect of such person, divide by 1600 (being deemed to be the average annual total of effective working hours of an employee) and multiplied by the number of working hours spent by such person in performing any of the services in respect of which payment under this Agreement is to be made to the Consulting Engineer upon the basis of Other Payroll Cost. For the purposes of this definition the annual amount of all contribution and payments made by the Consulting Engineer on behalf of or in respect of a person employed by him for a period less than a full year shall be calculated pro rata to the amount of such contributions and payments for such lesser period.

Words importing the singular include the plural and vise versa where the context requires.

2. DURATION OF ENGAGEMENT

2.1. The appointment of the Consulting Engineer shall commence from the date stated in the Memorandum of Agreement or from the time when the Consulting Engineer shall have begun to perform for the Client any of the services specified in Clauses 6 and 7 of this agreement, whichever is the earlier.

2.2. The Consulting Engineer shall not, without the consent of the Client, assign the benefit or in any way transfer the obligations of this Agreement or any part thereof.

CONSULTING ENGINEERS' FEES

2.3. If at any time the Client decides to postpone or abandon the Task, he may there-upon by notice in writing to the Consulting Engineer terminate the Consulting Engineer's appointment under this Agreement provided that, in any case in which the Consulting Engineer is paid for his services under Clause 6 in accordance with Clause 9.1 or Clause 9.2, the client may, when the Task is postponed, in lieu of so terminating the Consulting Engineer's appointment require the Consulting Engineer in writing to suspend the carrying out of his services under this Agreement for the time being.

2.4. If the Client shall not have required the Consulting Engineer to resume the performance of services in respect of the Task within a period of 12 months from the date of the Client's requirement to the Consulting Engineer to suspend the carrying out of his services, the Task shall be considered to have been abandoned and this Agreement shall terminate.

2.5. In the event of the failure of the Client to comply with any of his obligations under this Agreement, or upon the occurence of any circumstances beyond the control of the Consulting Engineer which are such as to delay for a period of more than 12 months or prevent or unreasonably impede the carrying out by the Consulting Engineer of his services under this Agreement, the Consulting Engineer may upon not less than 60 days' notice in writing to the Client terminate his appointment under this Agreement, provided that, in lieu of so terminating his appointment, the Consulting Engineer may:

(a) forthwith upon any such failure or the occurence of any such circumstances suspend the carrying out of his services hereunder for a period of 60 days (provided that he shall as soon as practicable inform the Client in writing of such suspension and the reasons therefor), and

(b) at the expiry of such period of suspension either continue with the carrying out of his services under this Agreement or else, if any of the reasons for the suspension then remain, forthwith in writing to the Client terminate his appointment under this Agreement.

2.6. The Consulting Engineer shall, upon receipt of any notice or requirement in writing in accordance with Clause 2.3 or the termination by him of his appointment in persuance of Clause 2.5, proceed in an orderly manner but with all reasonable speed and economy to take such steps as are necessary to bring to an end his services under this Agreement.

2.7. Any termination of the Consulting Engineer's appointment under this Agreement shall not prejudice or affect the accrued rights or claims of either party to this Agreement.

3. OWNERSHIP OF DOCUMENTS AND COPYRIGHT

3.1. The copyright in all drawings, reports, calculations and other documents provided by the Consulting Engineer in connection with the Task shall remain vested in the Consulting Engineer, but the Client shall have a licence to use such drawings and other documents for any purpose related to the Task. Save as aforesaid, the Client shall not make copies of such drawings or other documents nor shall he use the same in connection with the making or improvement of any works other than those to which the Task relates without the prior written approval of the Consulting Engineer and upon such terms as may be agreed between the Client and the Consulting Engineer.

3.2. The Consulting Engineer may with the consent of the Client, which consent shall not be unreasonably withheld, publish alone or in conjunction with any other person any articles, photographs or other illustrations relating to the Task.

CONSULTING ENGINEERS' FEES

4. SETTLEMENT OF DISPUTES

Any dispute or difference arising out of this Agreement shall be refered to the arbitration of a person to be agreed upon between the Client and the Consulting Engineer or, failing agreement, nominated by the President for the time being of the Chartered Insitute of Arbitrators.

OBLIGATIONS OF THE CONSULTING ENGINEER

5. CARE AND DILIGENCE

5.1. The Consulting Engineer shall exercise all reasonable skill, care and diligence in the discharge of the services agreed to be performed by him. If in the performance of his services the Consulting Engineer has a descretion exercisable as between the Client and the Contractor, the Consulting Engineer shall exercise his discretion fairly.

5.2. Where any person or persons are engaged, whether by the Client or by the Consulting Engineer on the Client's behalf under Clause 7.4, the Consulting Engineer shall not be liable for acts of negligence, default or omission by such person or persons.

6. NORMAL SERVICES

The services to be provided by the Consulting Engineer shall comprise:

(a)   all or any of the services stated in the Appendix to the Memorandum of Agreement and

(b)   advising the Client as to the need for the Client to be provided with additional services in accordance with Clause 7.

7. ADDITIONAL SERVICES NOT INCLUDED IN NORMAL SERVICES

7.1. As services additional to those specified in Clause 6, the Consulting Engineer shall, if so requested by the Client, provide any of the services specified in Clause 7.2 and provide or take all reasonable steps to arrange for the provision of any of the services specified in Clause 7.3.

7.2. (a)   Carrying out works consequent upon a decision by the Client to seek parlimentary powers.

(b)   Carrying out work in connection with any application by the Client for any order, sanction, licence, permit or other consent, approval or authorization necessary to enable the Task to proceed.

(c)   Carrying out work arising from the failure of the Client to award a contract in due time.

(d)   Carrying out work consequent upon any assignment of a contract by the Contractor or upon the failure of the Contractor properly to perform any contract or upon delay by the Client in fulfilling his obligations under Clause 8 or in taking any other step necessary for the due performance of the Task.

(e)   Advising the Client upon and carrying out work following the taking of any step in or towards and litigation or arbitration relating to the Task.

(f)   Carrying out work in conjunction with others employed to any of the services specified in Clause 7.3.

CONSULTING ENGINEERS' FEES

7.3. (a) Specialist technical advise on any abnormal aspects of the Task.
     (b) Architectual, legal, financial and other professional services.
     (c) Services in connection with the valuation, purchase, sale or leasing of lands and the obtaining of wayleaves.
     (d) The carrying out of marine, air and land surveys, and the making of model tests or special investigations.

7.4. The Consulting Engineer shall obtain the prior agreement of the Client to the arrangements which he proposes to make on the Client's behalf for the provision of any services specified in Clause 7.3. The Client shall be responsible to any person or persons providing such services for the cost thereof.

OBLIGATIONS OF THE CLIENT

8. INFORMATION TO BE SUPPLIED TO THE CONSULTING ENGINEER

8.1. The Client shall supply to the Consulting Engineer without charge and within a reasonable time all necessary and relevant data and information in the possession of the Client and shall give such assistance as shall reasonably be required by the Consulting Engineer in the performance of the Task.

8.2. The Client shall give his decision on all sketches, drawings, reports, recommendations, tender documents and other matters properly referred to him for decision by the Consulting Engineer in such reasonable time as not to delay or disrupt the performance by the Consulting Engineer of the Task.

9. PAYMENT FOR SERVICES - EXPLANATORY NOTE:

Three different methods of payment for services carried out under Clause 6 are detailed in the three succeeding Clauses 9.1, 9.2 and 9.3. The method of payment which the Client and the Consulting Engineer agree to adopt must be specified in the Memorandum of Agreement.

9.1. **Payment at hourly rates**

In respect of services provided by the Consulting Engineer under Clauses 6 and 7 the Client shall pay the Consulting Engineer:

(a) For time spent by Principals and Consultants of the firm, including time spent in travelling in connection with the Task, at the hourly rate or rates specified in Article 3 (a) of the Memorandum of Agreement.

(b) For Directors, salaried Principals and technical and supporting staff working in or based on the Consulting Engineer's Office: Salary Cost times the multiplier stated in Article 3 (b) of the Memorandum of Agreement, plus Other Payroll Cost. Time spent in travelling in connection with the Task shall be chargeable.

(c) For Directors, salaried Principals and technical and supporting staff, working as field staff, in or based on any field office established in pursuance of Clause 10: Salary Cost times the appropriate multiplier specified in Article 3 (c) of the Memorandum of Agreement, plus Other Payroll Cost.

(d) A reasonable charge for the use of a computer or other special equipment which charge shall be agreed between the Client and the Consulting Engineer before the work is put in hand.

(e) For time spent by Directors, salaried Principals and technical and supporting staff in connection with the use of a computer or other special equipment, including the development and writing of programmes and the operation of the computer in trial and

CONSULTING ENGINEERS' FEES

final runs in accordance with Clause 9.1 (b) above. Unless otherwise agreed between the Client and the Consulting Engineer, the Consulting Engineer shall not be entitled to any payment in respect of time spent by secretarial staff or by staff engaged on general accountancy or administration duties in the Consulting Engineer's office.

9.2. **Payment at hourly rates plus a fixed fee**

(a) In respect of services provided by the Consulting Engineer under Clause 6, the Client shall pay the Consulting Engineer in accordance with Clause 9.1, except that in lieu of charging self-employed Principals and Consultants at hourly rates the Client shall pay the Consulting Engineer the fee stated in Article 3 (d) of the Memorandum of Agreement.

(b) In respect of all other services provided by the Consulting Engineer, the Client shall pay the Consulting Engineer on the basis specified in Clause 9.1.

9.3. **Payment of a fixed sum**

(a) The sum payable by the Client to the Consulting Engineer for his services under Clause 6 shall be the sum stated in Article 3 (e) of the Memorandum of Agreement.

(b) In respect of all other services provided by the Consulting Engineer, the Client shall pay the Consulting Engineer on the basis specified in Clause 9.1.

10. PAYMENT FOR FIELD STAFF FACILITIES

The Client shall be responsible for the cost of providing such field office accommodation, furniture, telephones, equipment and transport as shall be necessary for the use of field staff, and for the reasonable running costs of such necessary field office accommodation and other facilities, including those of stationery, telephone calls, telex, telegrams and postage. Unless otherwise agreed between the Client and the Consulting Engineer, the Consulting Engineer shall arrange for the provision of field office accommodation and facilities for the use of field staff.

11. DISBURSEMENTS

The Client shall reimburse the Consulting Engineer in respect of all the Consulting Engineer's disbursements properly made in connection with:

(a) Printing, reproduction and purchase of all documents, drawings, maps, records and photographs.

(b) Telegrams, telex and telephone calls (other than local telephone calls).

(c) Postage and similar delivery charges except in the case of items weighting less than 250 grams sent by ordinary inland post.

(d) Travelling, hotel expenses and other similar disbursements.

(e) Advertising for tenders and for field staff.

(f) The provision of additional services to the Client pursuant to Clause 7.4.

(g) Professional Indemnity Insurance taken out by the Consulting Engineer to accord with the wishes of the Client and as set out in the Memorandum of Agreement.

CONSULTING ENGINEERS' FEES

The Client, by agreement with the Consulting Engineer and in satisfaction of his liability to the Consulting Engineer in respect of payment as specified in Article 3 (f) of the Memorandum of Agreement. these disbursements, may make to the Consulting Engineer a lump sum.

12. **PAYMENT FOLLOWING TERMINATION OR SUSPENSION BY THE CLIENT**

12.1. Upon a termination or suspension by the Client in pursuance of Clause 2.3, the Client shall pay the Consulting Engineer the sums specified in (a), (b) and (c) of this Clause less the amount of payments previously made to the Consulting Engineer under the terms of this Agreement.

(a) All amounts due to the Consulting Engineer at hourly rates in accordance with Clause 9 in respect of services rendered up to the date of termination or suspension together with a sum calculated in accordance with Clause 9 in respect of time worked by the Consulting Engineer's staff in complying with Clause 2.6.

(b) A fair and reasonable proportion of any lump sum specified in Articles 3 (d), 3 (e) and 3 (f) of the Memorandum of Agreement. In the assessment of such proportion, the services carried out by the Consulting Engineer up to the date of termination or suspension and in pursuance of Clause 2.6 shall be compared with a reasonable assessment of the services which the Consulting Engineer would have carried out but for the termination or suspension.

(c) Amounts due to the Consulting Engineer under any other clauses of this Agreement.

12.2. In any case in which the Client has required the Consulting Engineer to suspend the carrying out of the Consulting Engineer's services in pursuance of the power conferred by Clause 2.3, the Client, may at any time within a period of 12 months from the date of his requirement in writing to the Consulting Engineer to suspend the carrying out of the Consulting Engineer's services require the Consulting Engineer in writing to resume the performances of such services. In such event the Consulting Engineer shall within a reasonable time of receipt by him of Client's said requirement in writing resume the performance of his services in accordance with this Agreement. Upon such a resumption, the amount of any payment made to the Consulting Engineer under Clause 12.1 (b) shall rank as a payment made on account of the total sum payable to the Consulting Engineer under this Agreement, but no adjustment shall be made of any sum paid or payable to the Consulting Engineer upon suspension.

12.3. If the Consulting Engineer shall need to perform any additional services in connection with the resumption of his services in accordance with Clause 12.1, th client shall pay the Consulting Engineer in respect of the performance of such additional services in accordance with Clause 9.1 and any appropriate reimbursements in accordance with Clause 11.

13. **PAYMENT FOLLOWING TERMINATION BY THE CONSULTING ENGINEER**

Upon a termination by the Consulting Engineer in pursuance of Clause 2.5, the Client shall pay to the Consulting Engineer the sums specified in Clause 12.1 (a), (b) and (c) less the amount of payments perviously made to the Consulting Engineer under the terms of this Agreement. Upon payment of such sums, the Consulting Engineer shall deliver to the Client such completed drawings and other similar documents relevant to the Task as are in his possession. The Consulting Engineer shall be permitted to retain copies of any documents so delivered to the Client.

CONSULTING ENGINEERS' FEES

The provisions of this Clause are without prejudice to any other rights and remedies which the Consulting Engineer may possess.

14. PAYMENT OF ACCOUNTS

14.1. Unless otherwise agreed between the Client and the Consulting Engineer from time to time

(a) The fee referred to in Clause 9.2 (a) and the sum payable to the Consulting Engineer under Clause 9.3 (a) shall be paid by the Client in the instalments and at the intervals stated in Article 3 (g) of the Memorandum of Agreement. Any lump sum payable under Clause 11 shall be paid as stated in Article 3 (f).

(b) All sums due to the Consulting Engineer, other than those referred to in Clause 14.1 (a), shall be paid by the Client on accounts rendered monthly by the Consulting Engineer.

14.2. The Consulting Engineer shall submit to the Client at the time of submission of the monthly accounts such supporting data as may be agreed between the Client and the Consulting Engineer.

14.3. All sums due from the Client to the Consulting Engineer in accordance with the items of this Agreement shall be paid within 40 days of the submission by the Consulting Engineer of his accounts therefore to the Client, and any sums remaining unpaid at the expiry of such period of 40 days shall bear interest thereafter, such interest to accrue from day to day at the rate of 2% per annum above the Base Rate of the Engineer's Principal Bank as stated in Article 3 (h) of the Memorandum of Agreement.

14.4. If any item or part of an item of an account rendered by the Consulting Engineer is disputed or subject to question by the Client, the payment by the Client of the remainder of that account shall not be withheld on those grounds and the provisions of Clause 14.3 shall apply to such remainder and also to the disputed of questioned item, to the extent that it shall subsequently be agreed or determined to have been due to the Consulting Engineer.

14.5 All fees set out in this Agreement are exclusive of Value Added Tax, the amount of which, at the rate and in the manner prescribed by law, shall be paid by the Client to the Consulting Engineer.

Keep your figures up to date, free of charge

This section, and most of the other information in this Price Book, is brought up to date every three months in the *Price Book Update*.

The *Update* is available free to all Price Book purchasers.

To ensure you receive your copy, simply complete the reply card from the centre of the book and return it to us.

CONSULTING ENGINEERS' FEES

MEMORANDUM OF AGREEMENT

BETWEEN CLIENT AND CONSULTING ENGINEER
FOR REPORT AND ADVISORY WORK

MEMORANDUM OF AGREEMENT   made the ....

day of .... 19......

BETWEEN

....

....(hereinafter called 'the
Client')

of the one part and...

....

(hereinafter called 'the Consulting Engineer') of the other part.

WHEREAS the Client has requested the Consulting Engineer to provide professional
services as described in the Appendix hereto in connection with

....

....

....

...............................(referred to in this Agreement as 'the Task')

NOW IT IS HEREBY AGREED as follows:

1.   The Client agrees to engage the Consulting Engineer subject to and in
     accordance with the Conditions of Engagement attached hereto and the
     Consulting Engineer agrees to provide professional services subject to and
     in accordance with the said Conditions of Engagement.
2.   This memorandum of Agreement and the said Conditions of Engagement shall
     together constitute the Agreement between the Client and the Consulting
     Engineer.
3.   In the said Conditions of Engagement:

     (a)   the rate or rates referred to in Clause 9.1 (a) shall be..............

           ....

           ....

     (b)   the multiplier referred to in Clause 9.1 (b) shall be................

     (c)   the multiplier referred to in Clause 9.1 (c) shall be:
           for field staff who are permanent employees of
           the Consulting Engineer...
           for field staff who are recruited specifically
           for the Task..

     (d)   the fee referred to in Clause 9.2 (a) shall be.......................

     (e)   the sum referred to in Clause 9.3 (a) shall be.......................

CONSULTING ENGINEERS' FEES

    (f)   the lump sum referred to in Clause 11 shall be.................payable in ....equal monthly instalments.

    (g)   the intervals for the payment of instalments under Clause 14.1 (a) and the instalments referred to in the said sub-clause shall be:

    ....

    ....

    (h)   the Engineer's Principal Bank referred to in Clause 14.3 shall be

    ....

4.   The method of payment for services under Clause 6 of the said Conditions of Engagement shall be that described in Clause 9.1*, 9.2*, 9.3*, thereof.

5.   P11

    The amount of professional idemnity insurance referred to in Clause ii........+ of the said Conditions of Engagement shall be............... ....pounds(£   ) for any one occurrence or series of occurrences arising out of this engagement.

    This professional indemnity insurance shall be maintained for a period of .... years from the date of this Memorandum of Agreement, unless such insurance cover ceases to be available in which event the Consulting Engineer will notify the Client immediately.

    The sum payable by the Client to the Consulting Engineer as a contribution to the additional cost of the professional indemnity insurance thus provided shall be ....pounds(£   ) and such amount shall become due to the Consulting Engineer immediately upon acceptance by or on behalf of the Client of any tender in respect of the Works or any part thereof/immediately upon submission by the Consulting Engineer to the Client of his report in relation to the Task.*

6.   LIMITATION OF LIABILITY

    Notwithstanding anything to the contary contained elsewhere in this Agreement, the total liability of the Consulting Engineer under or in connection with this Agreement, whether in contract, in tort, for breach of statutory duty or otherwise, shall not exceed..............pounds(£   ). The Client shall indemnify and keep indemnified the Consulting Engineer from and against all claims, demands, proceedings, damages, costs, charges and expenses arising out of or in connection with this Agreement, the Works and/or the Project in excess thereof.

AS WITNESS the hands of the parties the day and year first above written.

Duly Authorised
Representative   Consulting
of the Client ....   Engineer ....

Witness ....   Witness ....

* Delete as appropriate.

CONSULTING ENGINEERS' FEES

APPENDIX TO THE MEMORANDUM OF
AGREEMENT

The services to be provided by the Consulting Engineer shall be as follows:

BASIS FOR COMPLETING MEMORANDUM OF AGREEMENT- EXPLANATORY NOTE:

1. The Appendix to the Memorandum of Agreement should be a concise but
    comprehensive description of the services to be provided by the Consulting
    Engineer including, for example, reference to

        Objective
        Scope
        Timing
        Surveys
        Geotechnical Investigation
        Field Staff
        Particpation by Client or other parties
        Budget
        Cost Control
        Style of report to suit Client's probable funding agency
        Number of copies of report or method of submitting data to the Client.

2. Articles 3 (a) and 3 (b) must be completed even where payment is to be made
    for normal services under Clause 9.2 or 9.3. In Article 3 (b) the normal
    multiplier is 2.6.
3. In Article 3 (c) the normal multiplier for field staff who are permanent
    employees of the Consulting Engineer is 2.6. The normal multiplier for
    staff recruited specifically for the Task is 1.3.
4. If disbursements are to be reimbursed at cost then Article 3 (f) should be
    deleted in entirety.
5. In Article 3 (g) the intervals for payment of instalments should be monthly
    or quarterly. The amount and number of instalments need to be stated.
6. The amount inserted (in words, figures and time) as P11 shall be based on
    what cover the Client decides appropriate. It should normally be available
    on an each and every occurrence basis. Exceptionally it can be on a
    special basis. It should not be less than the Consulting Engineer provides
    as a general rule.

The sum payable by the Client shall be subject to negotiation. It will have to
be based on the notional premiums which the Consulting Engineer expects to be
having to pay over the whole period involved. Such a figure will need to be
assessed in consultation with the Consulting Engineer's insurance broker
allowing for inflation and based on the anticipated fees to be received during
the whole project. Account may also need to be taken of the level of insurance
taken within the Consulting Engineer's firm itself in the form of an excess or
self-insurance. The proportion of this assessed sum which the Client should be
invited to contribute should take into account whether greater than normal cover
is to be provided and whether the fees being paid for the project are to be
below the guideline level.

The limit of the Consulting Engineer's liability should be the amount of P11 as
completed in the Memorandum of Agreement.

CONSULTING ENGINEERS' FEES

AGREEMENT 3 CONDITIONS OF ENGAGEMENT

1. DEFINITIONS

In construing this Agreement the following expressions shall have the meanings hereby assigned to them except where the context otherwise requires:

'The Consulting Engineer' means the person or firm named in the Memorandum of Agreement and shall include any other person or persons taken into partnership by such person or firm during the currency of this Agreement and the surviving member or members of any such partnership.

'The Architect' means any Architect appointed by the Client to act as the Architect of the project.

'The Project' means the project with which the Client is proceeding and of which the Works form a part.

'The Works' means the Work in connection with which the Client has engaged the Consulting Engineer to perform professional services.

'Contractor' means any person or persons firm or company under contract to the Client to perform work and/or supply goods in connection with the Works.

'Salary Cost' means the total annual taxable remuneration paid by the Consulting Engineer to any person employed by him, divided by 1600 (being deemed to be the average annual total of effective working hours of an employee) and multiplied by the number of working hours spent by such person in performing any of the services in respect of which payment under this Agreement is to be made to the Consulting Engineer upon the basis of Salary Cost. For the purposes of this definition the annual remuneration of a person employed by the Consulting Engineer for a period less than a full year shall be calculated pro rata to such person's remuneration for such lesser period.

'Other Payroll Cost' means the annual amount of all contributions and payment made by the Consulting Engineer on behalf of or in respect of a person employed by him for staff pension and life assurance schemes, and also for National Insurance Contributions and for any other tax, charge, levy, impost or payment of any kind whatsoever which the Consulting Engineer at any time during the performance of this Agreement is obliged by law to make on behalf of or in respect of such person, divided by 1600 (being deemed to be the average annual total of effective working hours of an employee) and multiplied by the number of working hours spent by such person in performing any of the services in respect of which payment under this Agreement is to be made to the Consulting Engineer upon the basis of Other Payroll Cost. For the purpose of this definition the annual amount of all contributions and payments made by the Consulting Engineer on behalf of in respect of a person employed by him for a period less than a full year shall be calculated pro rata to the amount of such contributions and payments for such lesser period.

Words importing the singular include the plural and vise versa where the context requires.

2. DURATION OF ENGAGEMENT

2.1. The appointment of the Consulting Engineer shall commence from the date stated in the Memorandum of Agreement or from the time when the Consulting Engineer shall have begun to perform for the Client any of the services in Clauses 6 and 7 hereof, whichever is the earlier.

2.2. The Consulting Engineer shall not, without the consent of the Client, assign the benefit or in any way transfer the obligations of this Agreement or any part thereof.

2.3. If at any time the Client decides to postpone or abandon the Works, he may thereupon by notice in writing to the Consulting Engineer

CONSULTING ENGINEERS' FEES

forthwith terminate the Consulting Engineer's appointment under this Agreement, provided that the Client may, when the Works or any part thereof are postponed, in lieu of so terminating the Consulting Engineer's appointment require the Consulting Engineer in writing to suspend the carrying out of his services under this Agreement for the time being.

2.4. If at any time the Client decides to postpone or abandon any part of the Works, he may thereupon by notice in writing to the Consulting Engineer seek to vary this Agreement either by excluding the services to be performed by the Consulting Engineer in relation to such part of the Works, or by suspending performance of the same and in such notice the Client shall specify the services affected. The Consulting Engineer shall forthwith comply with the Client's notice and the Client shall pay to the Consulting Engineer a sum caluclated in accordance with the provisions of Clause 18.2 in respect of such compliance.

2.5. If the Client shall not have required the Consulting Engineer to resume the performance of services in respect of the whole or any part of the Works suspended under Clause 2.3 or Clause 2.4 hereof within a period of 12 months from the date of the Client's notice:

(i) In the case of a suspension under Clause 2.3 this Agreement shall forthwith automatically terminate; or

(ii) In the case of a suspension under Clause 2.4 the suspended services shall be deemed to have been excluded from the services to be performed by the Consulting Engineer under this Agreement.

2.6. In the event of the failure of the Client to comply with any of his obligations under this Agreement, or upon the occurrence of any circumstances beyond the control of the Consulting Engineer which are such as to delay for a period of more than 12 months or prevent or unreasonably impede the carrying out by the Consulting Engineer of his services under this Agreement, the Consulting Engineer may upon not less than 60 days' notice in writing to the Client terminate his appointment under this Agreement, provided that, in lieu of so terminating his appointment, the Consulting Engineer may

(a) forthwith upon any such failure or the occurrence of any such circumstances suspend the carrying out of his services hereunder for a period of 60 days (provided that he shall as soon as practicable inform the Client in writing of such suspension and the reasons therefore), and

(b) at the expiry of such period of suspension either continue with the carrying out of his services under this Agreement or else, if any of the reasons for the suspension then remain, forthwith in writing to the Client terminated his appointment under this Agreement.

2.7. The Consulting Engineer shall, upon receipt of any notice in accordance with Clause 2.3 or in the event of the termination by him of his appointment in pursuance of Clause 2.6, proceed in an orderly manner but with all reasonable speed and economy to take such steps as are necessary to bring to an end his services under this Agreement.

2.8. Unless terminated under this Clause the Consulting Engineer's appointment under this Agreement shall terminate when the certificate authorising the final payment to the Contractor is issued.

2.9. Any termination of the Consulting Engineer's appointment under this Agreement shall not prejudice or affect the accrued rights or claims of either party to this Agreement.

CONSULTING ENGINEERS' FEES

3. OWNERSHIP OF DOCUMENTS AND COPYRIGHT

3.1.   The copyright in all drawing, reports, specifications, bills of
        quantities, calculations and other documents provided by the
        Consulting Engineer in connection with the Works shall remain vested
        in the Consulting Engineer, but the Client shall have a licence to use
        such drawings and other documents for any purpose related to the
        Works. Save as aforesaid, the Client shall not make copies of such
        drawings or other documents nor shall he use the same in connection
        with the making or improvement of any works other than those to which
        the Works relate without the prior wrtten approval of the Consulting
        Engineer and upon such terms as may be agreed between the Client and
        the Consulting Engineer.

3.2.   The Consulting Engineer may with the consent of the Client, which
        consent shall not be unreasonable withheld,publish alone or in
        conjunction with any other person any articles, photographs or other
        illustrations relating to the Works.

4. SETTLEMENT OF DISPUTES

Any dispute or difference arising out of this Agreement shall be referred
to the arbitration of a person to be agreed upon between the Client and the
Consulting Engineer or, failing agreement, nominated by the President for
the time being of the Chartered Institute of Arbitrators.

OBLIGATIONS OF THE CONSULTING ENGINEER

5. CARE AND DILIGENCE

5.1. The Consulting Engineer shall exercise all reasonable skill, care and
      diligence in the discharge of the services agreed to be performed by
      him. If in the performance of his services the Consulting Engineer
      has a discretion exercisable as between the Client and the Contractor,
      the Consulting Engineer shall exercise his discretion fairly.

5.2. The Consulting Engineer may recommend that specialist suppliers and/or
      contractors should design and execute certain part or parts of the
      Works in which circumstances the Consulting Engineer shall co-ordinate
      and intergrate the design of such part or parts with the overall
      design of the Works but he shall be relieved of all responsibility for
      the design, manufacture, installation and performance of any such part
      or parts of the works. Where any persons are engaged in accordance
      with Clause 7.4, the Consulting Engineer shall be under no liability
      for any negligence, default or omission of such persons.

6. NORMAL SERVICES

6.1. Preliminary or Sketch Plan Stage.

The services to be provided by the Consulting Engineer at this stage
shall comprise all or any of the following as may be necessary in the
particular case:

(a)   Investigating data and information relating to the Project and
       relevant to the Works which are reasonably accessible to the
       Consulting Engineer and considering any reports relating to the
       Works which have either been prepared by the Consulting Engineer
       or else prepared by others and made available to the Consulting
       Engineer by the Client.

(b)   Advising the Client on the need to carry out any geotechnical
       investigations which may be necessary to supplement the
       geotechnical information already available to the Consulting
       Engineer, arranging for such investigations when authorised by

CONSULTING ENGINEERS' FEES

the Client, certifying the amount of any payments to be made by the Client to the persons or firm carrying out such investigations under the Consulting Engineer's direction, and advising the Client on the results of such investigations.

(c) Advising the client on the need for arrangements to be made, in accordance with Clause 7, for the carrying out of special surveys, special investigations or model tests, and advising the Client of the results of any such surveys, investigations or tests carried out.

(d) Consulting any local or other authorities on matters of principle in connection with the structural design of the Works.

(e) Providing sufficient structural information to enable the Architect to produce his sketch plans.

6.2. **Tender Stage**

The services to be provided by the Consulting Engineer at this stage shall include all or any of the following as may be necessary in the particular case, for the purpose of enabling tenders to be obtained:

(a) Developing the design of the Works in collaboration with the Architect and preparing calculations, drawings and specifications of the Works to enable Bills of Quantities to be prepared.

(b) Advising on conditions of contract relevant to the Works and forms of tender and invitations to tender as they relate to the Works.

(c) Consulting any local or other authorites in connection with the structural design of the Works, and preparing typical details and typical calculations.

(d) Advising the Client as to the suitability for carrying out the Works of persons and firms tendering.

6.3. **Working Drawing Stage**

The services to be provided by the Consulting Engineer at this stage shall include all or any of the following as may be necessary in the particular case:

(a) Advising the Client as to the relative merits of tenders, prices and estimates received for carrying out the Works.

(b) Preparing such calculations and details relating to the Works as may be required for submission to any appropriate authority.

(c) Preparing any further designs, specifications and drawings, including bar bending schedules, necessary for the information of the Contractor to enable him to carry out the Works, but excepting the preparation of any shop details relating to the Works or any part thereof.

(d) Examining shop details for general dimensions and adequancy of members and connections.

6.4 **Construction Stage**

The Consulting Engineer shall not accept any tender in respect of the Works unless the Client gives him instructions in writing to do so, and any acceptance so made by the Consulting Engineer on the instructions of the Client shall be on behalf of the Client. The services to be provided by the Consulting Engineer at this stage shall include all or any of the following as may be necessary in the particular case:

CONSULTING ENGINEERS' FEES

   (a) Advising on the preparation of formal contract documents relating
       to accepted tenders for carrying out the Works or any part
       thereof.
   (b) Advising the Client on the need for special inspections or tests.
   (c) Advising the Client and the Architect on the appointment of site
       staff in accordance with Clause 8.
   (d) Examining the Contractor's proposals.
   (e) Making such visits to site as the Consulting Engineer shall
       consider necessary to satisfy himself as to the performance of
       any site staff appointed pursuant to Clause 8, and to satisfy
       himself that the Works are executed generally according to the
       contract and otherwise in accordance with good engineering
       practice.
   (f) With the prior agreement of the Architect, giving all necessary
       instructions to the contractor, provided that the Consulting
       Engineer shall not without prior approval of the Client give any
       instructions which in the opinion of the Consulting Engineer are
       likely substantially to increase the cost of the Works unless it
       is not in the circumstances practicable for the Consulting
       Engineer to obtain such prior approval.
   (g) Advising on certificates for payment to the Contractor.
   (h) Performing any service which the Consulting Engineer may be
       required to carry out under any contract for the execution of the
       Works including where appropriate the supervision of any
       specified tests, provided that the Consulting Engineer may
       decline to perform any services specified in a contract the terms
       of which have not initially been approved by the Consulting
       Engineer.
   (j) Delivering to the Client on the completion of the Works such
       records as are reasonably necessary to enable the Client to
       operate and maintain the Works.
   (k) Assisting in settling any dispute or difference relating to the
       Works which may arise between the Client and the Contractor
       provided that such assistance shall not relate to the detailed
       examination of any financial claim and shall not extend to
       advising the Client following the taking of any steps in or
       towards any arbitration or litigation in connection with the
       Works.

6.5. **General**

   Without prejudice to the preceding provisions of this clause, the
   Consulting Engineer shall from time to time as may be necessary advise
   the Client as to the need for the Client to be provided with
   additional services in accordance with Clause 7.

7. ADDITIONAL SERVICES NOT INCLUDED IN NORMAL SERVICES

7.1. As services additional to those specified in Clause 6, the Consulting
     Engineer shall, if so requested by the Client, provide any of the
     services specified in Clause 7.2 and provide or take all reasonable
     steps to arrange for the provision by others of any of the services
     specified in Clause 7.3.

7.2. (a) Preparing any report or additional contract documents required
         for consideration of proposals for the carrying out of
         alternative works.
     (b) Carrying out work consequent upon a decision by the Client to
         seek parliamentary powers.

CONSULTING ENGINEERS' FEES

(c) Carrying out work in connection with any application by the Client for any order, sanction, licence, permit or other consent, approval or authorisation necessary to enable the Works to proceed.

(d) Carrying out work arising from the failure of the Client to award a contract in due time.

(e) Preparing preliminary estimates.

(f) Preparing details for shop fabrication of ductwork, metal or plastic frameworks.

(g) Checking and advising upon any part of the Project not designed by the Consulting Engineer.

(h) Preparing intrim or other reports or detailed valuations, including estimates or cost analyses based on measurement or forming an element of a cost planning service.

(j) Carrying out work consequent upon any assignment of a contract by the Contractor or upon the failure of the Contractor to properly perform any contract on or upon delay by the Client in fulfilling his obligations under Clause 9 or in taking any other step necessary for the due performance of the Works.

(k) Advising the Client with regard to any dispute or difference which involves matters excluded from Clause 6.4 (k) above.

(l) Carrying out work in conjunction with others employed to provide any of the services specified in Clause 7.3.

(m) Carrying out such other additional services, if any, as are specified in Article 5 of the Memorandum of Agreement.

7.3. (a) Specialist technical advice on any abnormal aspects of the Works.

(b) Legal, financial and other professional services.

(c) Services in connection with the valuation, purchase, sale or leasing of lands and the obtaining of wayleaves.

(d) The surveying of sites or existing works.

(e) Investigation of the nature and strength works and the making of models tests or special investigations.

(f) The carrying out of special inspections or tests advised by the Consulting Engineer under Clause 6.4 (b).

7.4. The Consulting Engineer shall obtain the prior agreemnet of the Client to the arrangements which he proposes to make on the Client's behalf for the provision of any of the services specified in Clause 7.3. The Client shall be responsible to any person or persons providing such services for the cost thereof.

8. SUPERVISION ON SITE

8.1. If in the opinion of the Consulting Engineer the nature of the Works, including the carrying out of any geotechnical investigation pursuant to Clause 6.1, warrents full-time or part-time engineering supervision on site, the Client shall agree to the appointment of suitably qualified technical and clerical site staff as the Consulting Engineer shall consider reasonably necessary to enable such supervision to be carried out.

8.2. Persons appointed pursuant to Clause 8.1 shall be employed either by the Consulting Engineer or, if the Client and the Consulting Engineer shall so agree, by the Client directly, provided that the Client shall not employ any person as a member of the site staff unless the Consulting Engineer has first selected or approved such person as suitable for employment.

8.3. The terms of service of all site staff to be employed by the Consulting Engineer shall be subject to the approval of the Client, which approval shall not be unreasonably withheld.

CONSULTING ENGINEERS' FEES

8.4. The Client shall procure that the contracts of employment of the site staff employed by the Client shall stipulate that the staff so employed shall in no circumstances take or act upon instructions other than those of the Consulting Engineer.

8.5. Where a clerk of works nominated by the Architect is charged with the supervision of the Works on site, his selection and appointment shall be subject to the approval of the Consulting Engineer. In respect of the Works, the Client shall ensure that such clerk of works shall take instructions solely from the Consulting Engineer, who shall inform the Architect of all such instructions.

8.6. Where any supervision on site is performed by staff employed other than by the Consulting Engineer, the Consulting Engineer shall not be responsible for any failure on the part of such staff properly to comply with any instructions given by the Consulting Engineer.

OBLIGATIONS OF THE CLIENT

9. INFORMATION TO BE SUPPLIED TO THE CONSULTING ENGINEER

9.1. The Client shall supply to the Consulting Engineer, without charge and in such reasonable time as not to delay or disrupt the performance by the Consulting Engineer of his services under this Agreement, all necessary and relevant data and information in the possession of the Client or the Contractors or their respective servants, agents or sub-contractors and the Client shall give and shall procure that such persons give such assistance as shall reasonably be required by the Consulting Engineer in the performance of his services under this Agreement. The information to be provided by the Client to the Consulting Engineer shall include:

(a) All such drawings as may be necessary to make the Client's or the Architect's requirements clear, including plans and sections of all buildings (to a scale of not less than 1 to 100) and essential details (to a scale of not less 1 to 25) together with site plans (to a scale of not less than 1 to 1,250) and levels.

(b) Copies of all contract documents, including priced bills of quanities, relating to those parts of the Project which are relevant to the Works.

(c) Copies of all variation orders and supporting documents relating to the Project.

9.2. The Client shall give his decision on all sketches, drawings, reports, recommendation, tender documents and other matters properly referred to him for decision by the Consulting Engineer in such reasonable time as not to delay or disrupt the performance by the Consulting Engineer of his services under this Agreement.

10. PAYMENT FOR NORMAL SERVICES - EXPLANATORY NOTE:

Alternative methods of payment for services carried out under Clause 6 are detailed in the two succeeding Clauses 10.1 and 10.2. The method of payment which the Client and the Consulting Engineer agree to adopt must be specified in the Memorandum of Agreement. It should be noted that the method of payment specified in Clause 10.2 may also be used for services other than those carried out under Clause 6.

10.1. **Payment depending upon the actual cost of the Works.**

10.1.1. The sum payable to the Client to the Consulting Engineer for his services under Clause 6 shall be:

CONSULTING ENGINEERS' FEES

    (a)   On the total cost of the Works including reinforced concrete and brickwork or blockwork, but excluding unreinforced brickwork or blockwork, a fee calculated in accordance with Article 3(a) of the Memorandum of Agreement.

    (b)   A further fee on the cost of reinforced concrete work and/or reinforced brickwork or blockwork calculated in accordance with the percentage stated in Article 3(b) of the Memorandum of Agreement.

    (c)   A fee on the cost of unreinforced load-bearing brickwork or blockwork calculated in accordance with the percentage stated in Article 3(c) of the Memorandum of Agreement.

10.1.2.   The total cost of the Works shall be calculated in accordance with Clause 20.

10.1.3.   If the Works are to be constructed in more than one phase and as a consequence the services which it may be necessary for the Consulting Engineer to perform under Clause 6 have to be undertake by the Consulting Engineer separately in respect of each phase, then the fee entered in Article 3(a) of the Memorandum of Agreement shall be adjusted for each phase.

10.2.   **Payment on a Time Basis**

10.2.1.   The Client shall pay the Consulting Engineer in accordance with the Scale of Charges set out in Clause 10.2.2.

10.2.2.   Scales of Charges:

    (a)   Self-employed Principals and Consultants: At the hourly rate or rates specified in Article 3 (d) of the Memorandum of Agreement.

    (b)   Directors, salaried Principals and Technical and supporting staff: Salary Cost times the multiplier specified in Article 3 (e) of the Memorandum of Agreement, plus Other Payroll Cost.

    (c)   Time spent by Principals, Directors,Consultants, technical and supporting staff in travelling in connection with the Works shall be chargeable on the above basis.

    (d)   Unless otherwise agreed between the Client and the Consulting Engineer, the Consulting Engineer shall not be entitled to any payment in respect of time spent by secretarial staff or by staff engaged on general accountancy or administration duties in the Consulting Engineer's office.

11. PAYMENT FOR ADDITIONAL SERVICES

In respect of additional services provided by the Consulting Engineer under Clause 7, the Client shall pay the Consulting Engineer in accordance with the Scales of Charges set out in Clause 10.2.2.

12. PAYMENT FOR QUANTITY SURVEYING SERVICES

If the Client requires the Consulting Engineers to carry out quantity surveying services in respect of the Works designed by him, the Client shall pay the Consulting Engineer in respect of such services in accordance with the appropriate scale of professional charges for quantity surveying services published by The Royal Institution of Chartered Surveyors current when such services are carried out.

CONSULTING ENGINEERS' FEES

13. **PAYMENT FOR USE OF COMPUTER OR OTHER SPECIAL EQUIPMENT**

Where the Consulting Engineer decides to use a computer or other special equipment in carrying out any additional services in accordance with Clause 7 or is expressly required by the Client to use a computer or other special equipment in carrying out of his services under Clause 6, the Client shall, unless otherwise agreed between the Client and Consulting Engineer, pay the Consulting Engineer:

(a) for the time spent in connection with the use of a computer or other special equipment including the development and writing of programmes and the operation of the computer in trial and final runs, in accordance with the Scale of charges set out in Clause 10.2.2 and

(b) a reasonable charge for the use of the Computer or other special equipment which charge shall be agreed between the Client and the Consulting Engineer before work is put in hand.

14. **PAYMENT FOR SITE SUPERVISION**

14.1 In addition to any other payment to be made by the Client to the Consulting Engineer under the Agreement, the Client shall

(a) reimburse the Consulting Engineer in respect of all salary and wage payments made by the Consulting Engineer to site staff employed by the Consulting Engineer pursuant to Clause 8 and in respect of all other expenditure incurred by the Consulting Engineer in connection with the selection, engagement and employment of site staff, and

(b) pay to the Consulting Engineer an increment on the amount due under sub-clause 14.1 (a), calculated at the percentage specified in Article 3 (f) of the Memorandum of Agreement in respect of head office overhead costs incurred on site staff administration.

provided that in lieu of payments under (a) and (b) above the Client and the Consulting Engineer may agree upon inclusive monthly or other rates to be paid by the Client to the Consulting Engineer for each member of site staff employed by the Consulting Engineer.

14.2 The Client shall also be responsible for the cost of providing such local office accommodation, furniture, telephones, equipment and transport as shall be reasonably necessary for the use of site staff appointed pursuant to Clause 8, and for the reasonable running costs of such necessary local office accommodation and other facilities, including those of stationery, telephone calls, telex, telegrams and postage. Unless otherwise agreed between the Client and the Consulting Engineer, the Consulting Engineer shall arrange, whether through the Contractor or otherwise, for the provision of local office accommodation and facilities for the use of site staff.

14.3 In cases where the Consulting Engineer has thought it proper that site staff should not be appointed, or where the necessary site staff is not available at site due to sickness or any other cause, the Consulting Engineer shall be paid in accordance with the Scale of Charges set out in Clause 10.2.2 for site visits which would have been unnecessary but for the absence or non-availability of site staff.

15. DISBURSEMENTS

15.1 The Client shall in all cases reimburse the Consulting Engineer in respect of all the Consulting Engineer's disbursements properly made in connection with:

CONSULTING ENGINEERS' FEES

(a) Printing, reproduction and purchase of all documents, drawings, maps, records and photographs.
(b) Telegrams, telex and telephone calls (other than local telephone calls).
(c) Postage and similar delivery charges except in the case of items weighing less than 250 grams sent by ordinary inland post.
(d) Travelling, hotel expenses and other similar disbursements.
(e) Advertising for tenders and for site staff.
(f) The provision of additional services to the Client pursuance to Clause 7.4.
(g) Professional Indemnity Insurance taken out by the Consulting Engineer to accord with the wishes of the Client as set out in the Memorandum of Agreement.

15.2. The Client by agreement with the Consulting Engineer, and in full satisfaction of his liability to the Consulting Engineer in respect of these disbursements, may make to the Consulting Engineer a lump sum payment or sum calculated as a percentage of the fees and charges falling due under Clauses 10, 11, 12 and 13, as specified in Article 3 (g) of the Memorandum of Agreement.

16. PAYMENT FOR ALTERATION OR MODIFICATION TO DESIGN

16.1 Subject to Clause 16.2 if at any time after the commencement of the Consulting Engineer's appointment under Clause 2.1, any design whether completed or in progress or any specification, drawing or other document prepared in whole or in part by the Consulting Engineer shall require to be modified or revised by reason of instructions received by the Consulting Engineer from or on behalf of the Client, or by reason of circumstances which could not reasonably have been foreseen by the Consulting Engineer, then the Client shall make additional payment to the Consulting Engineer for making necessary modifications or revisions and for any consequential reproduction of documents. Unless otherwise agreed between the Client and the Consulting Engineer, the additional sum to be paid to the Consulting Engineer shall be calculated in accordance with the Scale of Charges set out in Clause 10.2, and shall also include any appropriate reimbursements in accordance with Clause 15 provided always that the Consulting Engineer shall give the Client written notice as soon as it is reasonably apparent that additional payments will arise under this Clause.

16.2 Where in the Consulting Engineer's opinion the Client's instruction necessitates a fundamental redesign of the part or parts of the Works affected by the instruction such that designs, specifications, drawings and other documents prepared by the Consulting Engineer cannot be modified or revised to take account thereof or where the modification or revision instructed by the Client results in a reduction in the cost of the part or parts of the Works affected thereby as contained in the Consulting Engineer's most recent estimate under Clause 21.1 (a) by 10% or more, then such part or parts of the Works shall be deemed to have been abandoned and the Consulting Engineer shall be paid therefor in accordance with Clause 18.2. The Consulting Engineer shall carry out such further work and shall produce such further designs, specifications, drawings and other such documents as may be necessary to comply with the Client's instructions and the Consulting Engineer shall be paid therefor in accordance with the provisions of this Agreement.

17. PAYMENT WHEN WORKS ARE DAMAGED OR DESTROYED

If at any time before completion of the Works any part of the Works or any materials, plant or equipment whether incorporated in the Works or not shall be damaged or destroyed, the Client shall make additional payment to the Consulting Engineer in respect of any expenses incurred or additional

CONSULTING ENGINEERS' FEES

work required to be carried out by the Consulting Engineer as a result of such damage or destruction. The amount of such additional payment shall be calculated in accordance with the Scale of Charges set out in Clause 10.2 and shall also include any appropriate reimbursements in accordance with Clause 15.

18. PAYMENT FOLLOWING TERMINATION OR SUSPENSION BY THE CLIENT

18.1. Upon a termination or suspension by the Client in pursuance of Clause 2.3, the Client shall pay to the Consulting Engineer the sums specified in (a), (b) and (c) below less the amount of payments previously made to the Consulting Engineer under the terms of this Agreement.

(a) A fair and reasonable proportion of the sum which would have been payable to the Consulting Engineer under Clauses 10 and 12 if no such termination or suspension had taken place. In the assessment of such proportion, the services carried out by the Consulting Engineer up to the date of termination or suspension and in pursuance of Clause 2.7 shall be compared with a reasonable assessment of the services which the Consulting Engineer would have carried out but for the termination or suspension. In any case in which it is necessary to assess the payment to be made to the Consulting Engineer in accordance with this sub-clause by reference to the cost of the Works, then to the extent that such cost is not known the assessment shall be made upon the basis of the Consulting Engineer's best estimate of the cost.

(b) Amounts due to the Consulting Engineer under any other clauses of this Agreement.

(c) A disruption charge equal to one-sixth of the difference between the sum which would have been payable to the Consulting Engineer under Clauses 10 and 12 but for the termination or suspension, and the sum payable under (a).

18.2. Upon a termination or suspension by the Client in pursuance of Clause 2.4, the Client shall pay to the Consulting Engineer the sums specified in (a) and (b) below.

(a) A fair and reasonable proportion of the sum which would have been payable to the Consulting Engineer under Clauses 10 and 12 in respect of the services affected if no such termination or suspension had taken place. The proportion shall be calculated in accordance with the provisions of Clause 18.1 (a); and

(b) A disruption charge, to be calculated in accordance with the provisions of Clause 18.1 (c)

18.3. In any case in which the Client has required the Consulting Engineer to suspend the carrying out of the Consulting Engineer's services in pursuance of Clauses 2.3 or 2.4 hereof, the Client may, at any time within the period of 12 months from the date of the Client's notice, require the Consulting Engineer in writing to resume the performance of such services. In such event:

(a) the Consulting Engineer shall within a reasonable time of receipt by him of the Client's said requirement in writing resume the performance of his service in accordance with this Agreement, the payment made under Clause 18.1 (a) or 18.2. (a) as the case may be ranking as payment on account towards the total sum payable to the Consulting Engineer under Clauses 10 and 12 but,

CONSULTING ENGINEERS' FEES

      (b) not withstanding such resumption, the Consulting Engineer shall
      be entitled to retain or receive as an additional payment due in
      accordance with this Agreement the disruption charge referred to
      in Clause 18.1 (c) or 18.2 (b) as the case may be.

18.4. If the Consulting Engineer shall need to perform any additional
      services in connection with the resumption of his services in
      accordance with Clause 18.3 the Client shall pay the Consulting
      Engineer in respect of the performance of such additional services
      in accordance with the Scale of Charges set out in Clause 10.2 and
      any appropriate reimbursements in accordance with Clause 15.

19. PAYMENT FOLLOWING TERMINATION BY THE CONSULTING ENGINEER

Upon a termination by the Consulting Engineer in pursuance of Clause 2.6
the Client shall pay to the Consulting Engineer the sums specified in
Clauses 18.1 (a) and (b) less the amount of payments previously made to the
Consulting Engineer under the terms of this Agreement. Upon payment of
such sums, the Consulting Engineer shall deliver to the Client such
completed drawings, specifications and other similar documents relevant to
the Works as are in his possession. The Consulting Engineer shall be
permitted to retain copies of any document so delivered to the Client. The
provisions of this Clause are without prejudice to any other rights and
remedies which the Consulting Engineer may possess.

20. COST OF THE WORKS

20.1. The cost of the Works or any part thereof shall be deemed to
      include:

      (a) The cost to the Client of the Works however incurred, without
      deduction of any liquidated damages or penalties payable by the
      Contractor to the Client but including any payments made by the
      Client to the Contractor by way of bonus, incentive or ex-gratia
      payments, or in settlement of claims, including but not limited
      to:
      (i)   all excavations necessary to enable the Works to be
            carried out and supports thereof, filling, shoring pumping
            and other operations for the control of water
      (ii)   concrete, reinforcement, prestressing tendons and
            anchorages, formwork and inserts
      (iii) load-bearing brickwork or other masonry, including
            facings and damp-proof courses, which has to resist forces
            from vertical loads, wind or other loading or upon which
            the stability of the structure depends.
      (iv)   all labours, sundries and materials associated with the
            Works and the cost of the preliminaries and general items
            in the proportion that the cost of the Works bears to the
            total cost of the Project.

      (b) A fair valuation of any labour, materials, manufactured goods,
      machinery or other facilities provided by the Client, and of the
      full benifit accruing to the Contractor from the use of
      construction plant and equipment belonging to the Client which
      the Client has required to be used in the execution of the
      Works.
      (c) The market value, as if purchased new, of any second-hand
      materials, manufactured goods and machinery incorporated in the
      Works.
      (d) The cost of geotechnical investigation (Clause 6.1 (b)).

CONSULTING ENGINEERS' FEES

20.2. The cost of the Works shall not include:

(a) Administration expenses incurred by the Client.
(b) Costs incurred by the Client under this Agreement.
(c) Interest on capital during construction, and the cost of raising moneys required for carrying out the construction of the Works.
(d) Cost of land and wayleaves.

21. PAYMENT OF ACCOUNTS

21.1. Unless otherwise agreed between the Client and the Consulting Engineer from time to time

(a) The sum payable to the Consulting Engineer under Clause 10.1 shall, until the cost of the Works is known, be paid by the Client to the Consulting Engineer in instalments and shall be calculated by reference to the Consulting Engineer's most recent estimate of the cost of the Works taking into account any acceptable tender or tenders when available and any fluctuations clauses contained therein. Such instalments shall be paid during each of the several stages of the Consulting Engineer's services at the intervals specified in Article 4 of the Memorandum of Agreement so that by the end of each stage the cumulative total of all instalments then paid shall amount to the relevant proportion, as specified in Article 4 of the Memorandum of Agreement, of the estimate total sum payable to the Consulting Engineer under this Agreement.

(b) Any lump sum payable to the Consulting Engineer under Clause 15.2 shall be paid by the Client in the instalments and at the intervals stated in Article 3 (g) (2) of the Memorandum of Agreement.

(c) All other sums due to the Consulting Engineer shall be paid by the Client on accounts rendered monthly by the Consulting Engineer.

21.2. Instalments paid by the Client to the Consulting Engineer in accordance with Clause 21.1 (a) shall constitute no more than payments on account. A statement of the total sum due to the Consulting Engineer shall be prepared when the cost of the Works is fully known. Such statement, after giving credit to the Client for all instalments previously paid, shall state the balance (if any) due from the Client to the Consulting Engineer or from the Consulting Engineer to the Client as the case may be which balance shall be paid to or by the Consulting Engineer as the case may require.

21.3. All sums due from the Client to the Consulting Engineer in accordance with the terms of this agreement shall be paid within 40 days of the submission by the Consulting Engineer of his accounts therefor to the Client, and any sums remaining unpaid at the expiry of such period of 40 days shall bear interest thereafter, such interest to accrue from day to day at the rate of 2% per annum above the Base Rate of the Engineers' Principal Bank as stated in Article 3 (h) of the Memorandum of Agreement.

21.4. If any item or part of an item of an account rendered by the Consulting Engineer is disputed or subject to question by the Client, the payment by the Client of the remainder of that account shall not be withheld on those grounds and the provisions of Clause 21.3 shall apply to such remainder and also to the disputed or questioned item to the extent that it shall subsequently be agreed or determined to have been due to the Consulting Engineer.

21.5. All fees set out in this Agreement are exclusive of Value Added Tax, the amount of which, at the rate and in the manner prescribed by law, shall be paid by the Client to the Consulting Engineer.

CONSULTING ENGINEERS' FEES

MEMORANDUM OF AGREEMENT

BETWEEN CLIENT AND CONSULTING
ENGINEER FOR STRUCTURAL ENGINEERING
WORK WHERE AN ARCHITECT IS APPOINTED
BY THE CLIENT

MEMORANDUM OF AGREEMENT   made the ....

day of........... 19........

BETWEEN ....

....

....(hereinafter called 'the Client')

of the one part and ....

....

(hereinafter called 'the Consulting Engineer') of the other part.

WHEREAS the Client has appointed or proposes to appoint ....

....

to be the Architect for the Project and intends to proceed with ....

....

....

....

and has requested the Consulting Engineer to provide professional services in

connection with ....

....

....

(referred to in this Agreement as 'the Works').

NOW IT IS HEREBY AGREED as follows:

1. The Client agrees to engage the Consulting Engineer subject to and in
   accordance with the Conditions of Engagement attached hereto and the
   Consulting Engineer agrees to provide professional services subject to and in
   accordance with the said Conditions of Engagement.
2. This Memorandum of Agreement and the said Conditions of Engagement shall
   together constitute the Agreement between the Client and the Consulting
   Engineer.
3. The method of payment for services under Clause 6 of the said Conditions of
   Engagement shall be that described in Clause 10.1*/Clause 10.2* thereof.

   (a) The fee referred to in Clause 10.1.1 (a) shall be ......% of the cost of
       the Works.

   (b) The percentage referred to in Clause 10.1.1 (b) shall be ....%

CONSULTING ENGINEERS' FEES

(c)   The percentage referred to in Clause 10.1.1 (c) shall be ....%

(d)   The rate or rates referred to in Clause 10.2.2 (a) shall be ....

      ....

(e)   The multiplier referred to in Clause 10.2.2 (b) shall be ....

      ....

(f)   The percentage to be be applied to the cost of site staff referred to in

      Clause 14.1 (b) shall be ....% for specifically recruited staff and

      ....% for seconded staff.

(g)   The payment for Disbursements referred to in Clause 15 shall be:

      (1) reimbursed at cost*

      (2) a lump sum of ....

      ....(£   ) payable in ....equal monthly instalments*

      (3) ....% of the total fees and charges payable in ....
          equal monthly instalments*.

(h)   The Engineer's Principal Bank referred to in Clause 21.3 shall be ....

      ....

4. The intervals for the payment of instalment under Clause 21.1 (a) of the said
   Conditions of Engagement shall be equal monthly/quarterly* intervals reckoned
   from the commencement of the Consulting Engineer's appointment, and the
   cumulative proportions referred to in the said sub-clause shall be as
   follows:

   On completion of Preliminary or Sketch Plan Stage.......% of the total fee
   On completion of Tender Stage   ....% of the total fee
   On completion of Working Drawing Stage   ....% of the total fee
   On completion of Construction Stage   100 % of the total fee

5. The additional services to be carried out in accordance with Clause 7.2 (m)
   of the said Conditions of Engagement shall be ....

   ....
   ....
   ....
   ....
   ....

6. P11

The amount of Professional Indemnity Insurance referred to in Clause 15.1 (g)
of the said Conditions of Engagement shall be ....
....pounds(£   )
for any one occurence or series of occurences arising out of this engagement.

CONSULTING ENGINEERS' FEES

This Professional Indemnity Insurance shall be maintained for a period of.......
years from the date of this Memorandum of Agreement, unless such insurance cover
ceases to be available in which event the Consulting Engineer will notify the
Client immediately.

The sum payable by the Client to the Consulting Engineer as a contribution to
the additional cost of the Professional Indemnity Insurance thus provided shall
be..pounds(£   )
and such amount shall become due to the Consulting Engineer immediately upon
acceptance by or on behalf of the Client of any tender in respect of the Works
or any part thereof/immediately upon submission by the Consulting Engineer to
the Client of his report in relation to the Works.*

7. LIMITATION OF LIABILITY

Notwithstanding anything to the contrary contained elsewhere in this Agreement,
the total liability of the Consulting Engineer under or in connection with this
Agreement, whether in connection, in tort, for breach of statutory duty or
otherwise, shall not exceed....................................pounds(£   ).
The Client shall indemnify and keep indemnified the Consulting Engineer from and
against all claims, demands, proceedings, damages, costs, charges and expenses
arising out of or in connection with this Agreement, the Works and/or the
Project in excess thereof.

AS WITNESS the hands of the parties the and the year first above written

Duly Authorised
Representative
of the Client ....   Consulting
                                                   Engineer ....

Witness ....   Witness ....

*Delete as appropriate.

CONSULTING ENGINEERS' FEES

BASIS FOR COMPLETING MEMORANDUM OF AGREEMENT - EXPLANATORY NOTES:

1. In Article 3 (a) the percentages for fee calculations are fixed for the
   duration of the Agreement.
2. For works of average complexity the percentage for fee calculation may be
   determined from the graph and notes on page 75. The estimated cost of works
   shall be agreed with the Client and the costs shall be divided by the current
   Output Price Index (1975 = 100) published by the Department of the
   Enviroment.
   The current index is 293 (provisional - Q1 1989). From the figures so
   obtained the percentage fee appropriate for the works can be determined from
   the graph.

   e.g. If the cost of works is estimated to be £400,000 and the Output Price
   Index is 200, the percentage would be 6.1.

3. The normal percentage to be entered in Article 3 (b) is 3%.
4. The normal percentage to be entered in Article 3 (c) is 3.5%.
5. Articles 3 (d) and 3 (e) must be completed in all cases.
6. In Article 3 (e) the normal multiplier is 2.6.
7. In Article 3 (f) the percentage to be added to the direct costs of providing
   site supervisory staff is to cover all overheads and head office charges
   incurred in administering such staff and should include for administration of
   payroll, finance costs and an allowance for potential redundancy payments.
   The normal figure is 15% for staff recruited on a contract basis. For
   seconded staff the terms are to be negotiated between the Client and the
   Consulting Engineer.
8. In Article 4, the following table, provided for guidance only, is typical of
   the proportionate amount of the total services performed at the end of each
   stage:

   On completion of Preliminary or Sketch Plan Stage   15%
   On completion of Tender Stage   60%
   On completion of Working Drawing Stage   85%
   On completion of Construction Stage   100%

9. The amount inserted in Article 6 (in words, figures and time) as P11 shall be
   based on what cover the Client decides appropriate. It should normally be
   available on an each and every occurrence basis. Exceptionally it can be on
   a special basis. It should not be less than the Consulting Engineer provides
   as a general rule.

   The sum payable by the Client shall be subject to negotiation. It will have
   to be based on the notional premiums which the Consulting Engineer expects to
   be having to pay over the whole period involved. Such a figure will need to
   be assessed in consultation with the Consulting Engineer's insurance broker
   allowing for inflation and based on the the anticipated fees to be received
   during the whole of the project. Account may also need to be taken of the
   level of insurance taken within the Consulting Engineer's firm itself in the
   form of an excess or self-insurance. The proportion of this assessed sum
   which the Client should be invited to contribute should take into account
   whether greater than normal cover is to be provided and whether the fees
   being paid for the project are to be below the guideline level. As a guide
   the Client should contribute say half of the P11 normal insurance premiums
   and all of the premium for additional cover.

10. The limit of the Consulting Engineer's liability should be the amount of P11
    as completed in the Memorandum of Agreement.

CONSULTING ENGINEERS' FEES

Graph showing relationship of percentage fee with Cost of Works (excluding unreinforced load bearing brickwork or blockwork)

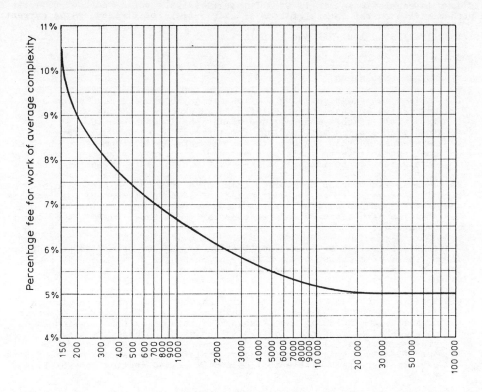

Category of the Works
Cost of Works in £ divided by Output Price Index (1975)=100)

(i)   For small projects, i.e. those where the Cost of the Works (as defined in Clause 20, cost of unreinforced brickwork or blockwork) divided by the Output Price Index is less than 150, time charges should be used.
(ii)   The minimum fee should not be less than 5% of the Cost of the Works (Clause 20, as above).

THE TOWN AND COUNTRY PLANNING (FEES FOR APPLICATIONS
AND DEEMED APPLICATIONS) (AMENDMENTS) REGULATIONS 1989

- operative from 14th March 1989

The following extracts from the Town and Country Planning Fees Regulations
relate only to those applications which meet the "deemed to qualify clauses"
laid down in regulations 1 to 11 of S.I. No. 1989/193.
     Further advice on the interpretation of these regulations can be found in the
1971, 1981 and 1984 Town and County Planning Acts and DOE circulars S.I.
1981/369, 1983/1674, 1987/764 and 1988/1813.

SCHEDULE 1

PART 1: GENERAL PROVISIONS

1.   (1) Subject to paragraphs 3 to 11, the fee payable under regulation 3 or
         regulation 10 shall be calculated in accordance with the table set out
         in Part II of this Schedule and paragraphs 2 and 12 to 16.
     (2) In the case of an applicationn for approval of reserved matters,
         references in this Schedule to the category of development to which an
         application relates shall be construed as references to the category of
         development authorised by the relevant outline planning permission.

2.   Where an application or deemed application relates to the retention of
     buildings or works or to the continuance of a use of land, the fee payable
     shall be calculated as if the application or deemed application, were one
     for planning permission to construct or carry out those buildings or works
     or to institute the use.

3.   Where an application or deemed application is made or deemed to be made by
     or on behalf of a parish council or by or on behalf of a community council,
     the fee payable shall be one-half of the amount as would otherwise be
     payable.

4.   (1) Where and application or deemed application for planning permission is
         made or deemed to be made by or on behalf of a club, society or other
         organisation(including any persons administering a trust) which is not
         established or conducted for profit and whose objects are the
         provision of facilities for sport or recreation, and the conditions
         specified in subparagragh (2) below satisfied, the amount of the fee
         payable in respect of the application or deemed application shall be
         £76.

     (2) The conditions referred to in subparagraph (1) above are:

         (a) that the application or deemed application relates to:-
             (i) the making of a material change in the use of land to use as a
                  playing field; or
             (ii) the carrying out of operations (other than the erection of a
                  building containing floor space) for purposes ancillary to the
                  use of land as a playing field, and to no other development:
                  and
         (b) that the local planning authority with whom the application is
             lodged, or (in the case of a deemed application) the Secretary of
             State, is satisfied that the development is to be carried out on
             land which is, or is intended to be, occupied by the club, society
             or organisation and used wholly or mainly for the carrying out of
             its objects.

THE TOWN AND COUNTRY PLANNING (FEES FOR APPLICATIONS
AND DEEMED APPLICATIONS) (AMENDMENTS) REGULATIONS 1989

5. (1) Where an application for planning permission or an application for
        approval of reserved matters is made not more than 28 days after the
        lodging with the local planning authority of an application for
        planning permission or, as the case may be, an application for
        approval of reserved matters:

        (a) made by or on behalf of the same applicant;
        (b) relating to the same site; and
        (c) relating to the same development or, in the case of an application
            for approval of reserved matters, relating to the same reserved
            matters in respect of the same building or buildings authorised by
            the relevant outline planning permission,

        and a fee of the full amount payable in respect of the category or
        categories of development to which the applications relate has been
        paid in respect of the earlier application, the fee payable in respect
        of the later application shall, subject sub-paragraph (2) below, be
        one-quarter of the full amount paid in respect of the earlier
        application.

    (2) Sub-paragraph (1) apply only in respect of one application made by or
        on behalf of the same applicant in relation to the same development or
        in relation to the same reserved matters (as the case may be).

6. (1) This paragraph applies where:

        (a) an application is made for approval of one or more reserved matters
            ("the current application"); and
        (b) the applicant has previously applied for such approval under the
            same outline planning permission and paid fees in relation to one
            or more such applications; and
        (c) no application has been made under that permission other than by or
            on behalf of the applicant.

    (2) Where the amount as mentioned in sub-paragraph (1) (b) is not less than
        the amount which would be payable if the applicant were by his current
        application seeking approval of all the matters reserved by the outline
        permission (and in relation to the whole of the development authorised
        by the permission), the amount of the fee payable in respect of the
        current application shall be £76.

    (3) Where:
        (a) a fee has been paid as mentioned in sub-paragraph (1) (b) at a rate
            lower than that prevailing at the date of the current application;
            and
        (b) subparagraph (2) would apply if that fee had been paid at the rate
            applying at that date, the fee in respect of the current
            application shall be the amount specified in sub-paragraph (2).

7. Where application is made pursuant to section 31A of the 1971 Act the fee
    payable in respect of the application shall be £38.

8. (1) This paragraph applies where applications are made for planning
        permission or for the approval of reserved matters in respect of the
        development of land lying in the areas of:

        (a) two or more local planning authorities in a Metropolitan County or
            in Greater London; or
        (b) two or more district planning authorities in a Non-metropolitan
            County; or

THE TOWN AND COUNTRY PLANNING (FEES FOR APPLICATIONS
AND DEEMED APPLICATIONS) (AMENDMENTS) REGULATIONS 1989

      (c) one or more such local planning authorities and one or more such
          district planning authorities.

    (2) A fee shall be payable only to the local planning authority or district
       planning authority in whose area the largest part of the relevant land
       is situated: and the amount payable shall not exceed:
      (a) where the applications relate wholly or partly to a county matter
          within the meaning of paragraph 32 of Schedule 16 to the Local
          Government Act 1972 (b), and all the land is situated in a single
          non-metropolitan County, the amount which would have been payable
          if application had fallen to be made to one authority in relation
          to the whole development;
      (b) in any other case, one and a half times the amount which would have
          been payable if application had fallen to be made to a single
          authority.

9.   (1) This paragraph applies where application for planning permission is
       deemed to have been made by virtue of section 88B (3) of the 1971 Act
       in respect of such land as is mentioned in paragraph 8 (1).

    (2)   The fee payable to the Secretary of State shall be a fee of the amount
       which would be payable by virtue of paragraph 8 (2) if application for
       the like permission had been made to the relevant local or district
       planning authority on the date on which notice of appeal was given in
       accordance with section 88 (3) of the 1971 Act.

10. (1) Where:

      (a) application for planning permission is made in respect of two or
          more alternative proposals for the development of the same land; or
      (b) application for approval of reserved matters is made, in respect of
          two or more alternative proposals for the carrying out of the
          development authorised by an outline planning permission, and
          application is made in respect of all of the alternative proposals
          on the same date and by or on behalf of the same applicant, a
          single fee shall be payable in respect of all such alternative
          proposals, calculated as provided in sub-paragraph (2).

    (2) Calculations shall be made, in accordance with this Schedule, of the
       fee appropriate to each of the alternative proposals and the single fee
       payable in respect of all the alternative proposals shall be the sum
       of:
      (i)   an amount equal to the highest of the amounts calculated in
          respect of each of the alternative proposals; and
      (ii) an amount calculated by adding together the amounts
          appropriate to all of the alternative proposals, other than
          the amount referred to in subparagraph (i) above, and dividing
          that total by the figure of 2.

11. In the case of an application for planning permission which is deemed to
    have been made by virtue of section 95 (6) of the 1971 Act, the fee payable
    shall be the sum of £76.

12. Where, in respect of any category of development specified in the table set
    out in Part II of this Schedule, the amount of the fee is to be calculated
    by reference to the site area:-

THE TOWN AND COUNTRY PLANNING (FEES FOR APPLICATIONS
AND DEEMED APPLICATIONS) (AMENDMENTS) REGULATIONS 1989

    (a) that area shall be taken as consisting of the area of land to which
        the application relates or, in the case of an application for
        planning permission which is deemed to have been made by virtue of
        section 88B (3) of the 1971 Act, the area of land to which the
        relevant enforcement notice relates; and

    (b) where the area referred to in sub-paragraph (a) above is not an
        exact multiple of the unit of measurement specified in respect of
        the relevant category of development, the fraction of a unit
        remaining after division of the total area by the unit of
        measurement shall be treated as a complete unit.

13. (1) In relation to development within any of the categories 2 to 4
       specified in the table in Part II of this Schedule, the area of gross
       floor space to be created by the development shall be ascertained by
       external measurement of the floor space, whether or not it is to be
       bounded (wholly or partly) by external walls of a building.

    (2) In relation to development within category 2 specified in the said
       table, where the area of gross floor space to be created by the
       development exceeds 75 sq metres and is not an exact multiple of 75 sq
       metres, the area remaining after division of the total number of square
       metres of gross floor space by the figure of 75 shall be treated as
       being 75 sq metres.

    (3) In relation to development within category 3 specified in the said
       table, where the area of gross floor space exceeds 540 sq metres and
       the amount of the excess is not an exact multiple of 75 sq metres, the
       area remaining after division of the number of square metres of that
       excess area of gross floor space by the figure of 75 shall be treated
       as being 75 sq metres.

14. (1) Where an application (other than an outline application) or a deemed
       application relates to development which is within category 1 in the
       table set out in Part II of this Schedule and in part within category
       2,3, or 4, the following sub-paragraphs shall apply for the purpose of
       calculating the fee payable in respect of the application or deemed
       application.

    (2) An assessment shall be made of the total amount of gross floor space
       which is to be created by that part of the development which is within
       category 2,3 or 4 ("the non-residential floor space"), and the sum
       payable in respect of the non-residential floor space to be created by
       the development shall be added to the sum payable in respect of that
       part of the development which is within category 1, and subject to sub-
       paragraph (4), the sum so calculated shall be the fee payable in
       respect of the application or deemed application.

    (3) For the purpose of calculating the fee under sub-paragraph (2)-

    (a) Where any of the buildings is to contain floor space which it is
        proposed to use for the purposes of providing common access or
        common services or facilities for persons occupying or using that
        building for residential purposes and for persons occupying or
        using it for non-residential purposes ("common floor space"), the
        amount of non-residential floor space shall be assessed, in
        relation to that building, as including such proportion of the
        common floor space as the amount of non-residential floor space in
        the building bears to the total amount of gross floor space in the
        building to be created by the development;

THE TOWN AND COUNTRY PLANNING (FEES FOR APPLICATIONS
AND DEEMED APPLICATIONS) (AMENDMENTS) REGULATIONS 1989

      (b) where the development falls within more than one of categories 2,3,
      and 4 an amount shall be calculated in accordance with each such
      category highest amount so calculated shall be taken as the sum
      payable in respect of all of the non-residental floor space.

    (4) Where an application or deemed application to which this paragraph
    applies relates to development which is also within one one or more
    than one of the categories 5 to 13 in the table set out in Part II of
    this Schedule, an amount shall be calculated in accordance with the
    with each such category and if any of the amounts so calculated exceeds
    the amount calculated in accordance with sub-paragraph (2) that higher
    amount shall be the fee payable in respect of all of the development to
    which the application or deemed application relates.

15. (1) Subject to paragraph 14, and sub-paragraph (2), where an application or
    deemed application relates to development which is within more than one
    of the categories specified in the table set out in Part II of this
    Schedule-

    (a) an amount shall be calculated in accordance with each such
      category; and
    (b) the highest amount so calculated shall be the fee payable in
      respect of the application or deemed application.

    (2) Where an application is for outline planning permission and relates to
    development which is within more than one of the categories specified
    in the said table, the fee payable in respect of the application shall
    be £76 for each 0.1 hectacres of the site area, subject to a maximum of
    £1900.

16. In the case of an application for planning permission which is deemed to
    have been made by virtue of section 88B (3) of the 1971 Act, references in
    this Schedule to the development to which an application relates shall be
    construed as references to the use of land or the operations (as the case
    may be) to which the relevant enforcement notice relates; references to the
    development shall be construed as references to the amount of floor space
    or the number of dwellinghouses to be created by the development shall be
    construed as references to the amount of floor space or the number of
    dwellinghouses to which that enforcement notice relates; and references to
    the purposes for which it is proposed that floor space be used shall be
    construed as references to the purposes for which floor space was stated to
    be used in the enforcement notice.

PART II: SCALE OF FEES

Category of development   Fee payable

I. Operations

1. The erection of dwellinghouses (other than development within category 6 below).	(a) Where the application is for outline planning permission £76 for each 0.1 hectare of the site area, subject to a maximum of £1,900, (b) in other cases, £76 for each dwellinghouse to be created by the development, subject to a maximum of £3,800.

THE TOWN AND COUNTRY PLANNING (FEES FOR APPLICATIONS
AND DEEMED APPLICATIONS) (AMENDMENTS) REGULATIONS 1989

Category of development

Fee payable

2. The erection of buildings (other than
   buildings coming within categories
   1,3, 4, 5 or 7.

(a) Where the application is for
    outline planning permission £76
    for each 0.1 hectare of the
    site area, subject to a maximum
    of £1,900;

(b) in other cases:

    (i)   where no floor space is to
          be created by the develop-
          ment, £38;
    (ii) where the area of gross
          floor space to be created
          by the development does
          not exceed 40 sq metres,
          £38;
    (iii) where the area of gross
          floor space to be created
          by the development exceeds
          40 sq metres but does not
          exceed 75 sq metres, £76;
          and
    (iv) where the area of gross
          floor space to be created
          by the development exceeds
          75 sq metres, £76 for each
          75 sq metres, subject to a
          maximum of £3,800.

3. The erection, on land used for the
   purposes of agriculture, of
   buildings to be used for agricultural
   purposes (other than buildings coming
   within category 4).

(a) Where the application is for
    outline planning permission £76
    for each 0.1 hectare of the
    site area, subject to a maximum
    of £,1900;

(b) in other cases:

    (i)   where the area of gross
          floor space to be created
          by the development does
          not exceed 465 sq metres
          nil;
    (ii) where the area of gross
          floor space to be created
          by the development exceeds
          465 sq metres but does not
          exceed 540 sq metres, £76;
    (iii) where the area of gross
          floor space to be created
          by development exceeds 540
          sq metres, £76 for the
          first 540 sq metres and
          £76 for each 75 sq metres
          in excess of that figure,
          subject to a maximum of
          £3,800.

THE TOWN AND COUNTRY PLANNING (FEES FOR APPLICATIONS
AND DEEMED APPLICATIONS) (AMENDMENTS) REGULATIONS 1989

Category of development	Fee payable
4. The erection of glasshouses on land used for the purposes of agriculture.	(a) Where the area of gross floor space to be created by the development does not exceed 465 sq metres, nil; (b) where the area of gross floor space to be created by the development exceeds 465 sq metres, £450.
5. The erection, alteration or replacement of plant or machinery.	£76 for each 0.1 hectare of the area, subject to a maximum of £3,800.
6. The enlargement, improvement or other alteration of existing dwellinghouses.	(a) Where the applications relates to one dwellinghouse, £38; (b) where the application relates to 2 or more dwellinghouses, £76
7. (a) The carrying out of operations (including the erection of a building) within the curtilage of an existing dwellinghouse, for purposes ancillary to the enjoyment of the dwellinghouse as such, or the erection or construction of gates, fences, walls or other means of enclosure along a boundary of the curtilage of an existing dwellinghouse; or (b) the construction of car parks, service roads and other means of access on land used for the purposes of a single undertaking, where the development is required for a purpose incidental to the existing use of the land.	£38
8. The carrying out of any operations connected with exploratory drilling for oil or natural gas.	£76 for each 0.1 hectare of the site area, subject to a maximum of £5,700.
9. The carrying out of any operations not coming within any of the above categories.	£38 for each 0.1 hectare of the site area, subject to a maximum of: (a) in the case of operations for the winning and working of minerals, £5,700; (b) in other case, £380.

THE TOWN AND COUNTRY PLANNING (FEES FOR APPLICATIONS
AND DEEMED APPLICATIONS) (AMENDMENTS) REGULATIONS 1989

Category of development   Fee payable

II. Uses of Land

10. The change of use of a building to   (a) Where the change is from a
     use as one or more separate   previous use as a single
     dwellinghouses.   dwellinghouse to use as a two
                                                or more single dwellinghouses,
                                                £76 for each additional
                                                dwellinghouse to be created by
                                                the development subject to a
                                                maximum of £3,800.

                                            (b) in other cases, £76 for each
                                                dwellinghouse to be created by
                                                the development, subject to a
                                                maximum of £3,800.

11. (a) The use of land for the disposal   £38 for each 0.1 hectare of the
         of refuse or waste materials or   site area, subject to a maximum of
         for the deposit of material   £5,700.
         remaining after minerals have
         been extracted from land; or

     (b) the use of land for the storage
         of minerals in the open.

12. The making of a material change in   £76.
     the use of a building or land (other
     than a material change of use coming
     within any of the above categories).

13. The continuance of a use of land, or   £38.
     the retention of buildings or works
     on land, without compliance with a
     condition subject to which a previous
     planning permission has been granted
     (including a condition requiring the
     discontinuance of the use or the
     removal of the building or works at
     the end of a specified period).

SCHEDULE 2

SCALE OF FEES IN RESPECT OF APPLICATIONS
FOR CONSENT TO DISPLAY ADVERTISEMENTS

Category of advertisments   Fee payable

1. Advertisements displayed on business premises, on the   £21
   forecourt of business premises or on other land within
   the curtilage of business premises, wholly with
   reference to all or any of the following matters:-

   (a) the nature of the business or other activity
       carried on on the premises;
   (b) the goods sold or the services provided on the
       premises; or
   (c) the name and qualifications of the person
       carrying on such business or activity or
       supplying such goods or services.

THE TOWN AND COUNTRY PLANNING (FEES FOR APPLICATIONS
AND DEEMED APPLICATIONS) (AMENDMENTS) REGULATIONS 1989

Category of advertisements	Fee payable
2. Avertisemnents for the purpose of directing members of the public to, or otherwise drawing attention to the existence of, business premises which are in the same locality as the site on which the advertisement is to be displayed but which are not visible from that site.	£21
3. All other advertisements.	£76

THE BUILDING (PRESCRIBED FEES ETC.)
REGULATIONS 1988

- operative from 13 June 1988

PART I: GENERAL

Interpretation

2 (1)
   'cost" does not include any professional fees paid to an architect, quantity
   surveyor or any other person;

   (2)

      (a the total floor area of a dwelling or extension is the total of the
          floor areas of all the storeys in it; and
      (b) the floor area of -

          (i)   any storey of a dwelling or extension, or
          (ii)   a garage or carport,

          is the total floor area calculated by reference to the finished
          internal faces of the wall enclosing the area, or if at any point there
          is no enclosing wall, by reference to the outermost edge of the floor.

PART II: FEES CHARGED BY LOCAL AUTHORITIES

Prescribed functions

3. The prescribed functions in relation to which local authorities are
   authorised to charge fees are the following -

   (a) the passing or rejection by the local authority, in accordance with
       section 16 of the Act, of plans of proposed work deposited with them
       (including plans of work proposed to be carried out by or on behalf of
       the authority);
   (b) the inspection in connection with the principal regulations of work for
       which such plans have been deposited;
   (c) the inspection in connection with the principal regulations of work for
       which a building notice has been given to the local authority; and
   (d) the consideration of plans of work reverting to local authority
       control, and the inspection of that work.

PART III: FEES FOR DETERMINATION OF QUESTIONS BY THE SECRETARY OF STATE

18.   (1) Where in accordance with section 16 (10) of the Act (determination of
           questions arising under that section) a person has referred a
           question to the Secretary of State for his determination, that
           application shall be accompanied by a fee calculated in accordance
           with paragraph 1 or 2
           of Schedule 4.
       (2) Where in accordance with section 50 (2) of the Act (determination of
           questions arising between approved inspector and developer) the
           person proposing to carry out work has referred a question to the
           Secretary of State for his determination, his application shall be
           accompanied by a fee calculated in accordance with paragraph 3 of
           Schedule 4.

THE BUILDING (PRESCRIBED FEES ETC.)
REGULATIONS 1986

SCHEDULE 1
SMALL DOMESTIC BUILDINGS

Fees for single small domestic buildings

Plan fee

1. Where a plan fee is chargeable in respect of -

    (a) plans for the erection of a single small domestic building, or
    (b) plans for the execution of works of drainage in connection with the
         erection of such a building deposited before plans for the erection
         of the building are deposited,

the plan fee for that building is the amount specified in column 2 of Table 1
below for the number of dwellings in that building.

Inspection fee

2. In relation to the erection of a small domestic building and any work in
    connection with the erection of such a building, the inspection fee is
    the aggregate of -

    (a) £80 multiplied by the number of dwellings in that building; and
    (b) where any of those dwellings has a total floor area (excluding the
         floor area of any garage comprised in the building) exceeding 64 m2,
         the amount specified in column (2) of Table 2 below for the number
         of such dwellings in that building.

Building notice fee

3. In relation to the erection of a single small domestic building and any
    work in connection with the erection of such a building, the building
    notice fee is the aggregate of the plan fee and the inspection fee which
    would have been payable had plans been deposited for the purpose of
    section 16 of the Act.

Fees for small domestic buildings in multiple work scheme

Plan fee for more than one small domestic building

4. (1) Where a plan fee is chargeble in respect of -

    (a)   plans for the erection of a small domestic building, or
    (b)   plans for the execution of works of drainage in connection with
             the erection of such buildings deposited before the plans for
             the erection of those buildings are deposited, and

the works forms part of a multiple work scheme which consists of or includes -

    (i)   the erection of two or more small domestic buildings, or
    (ii)   the execution of works of drainage in connection with such
             buildings,

the plan fee for each small domestic building is determined in accordance with
sub-paragraph (2).
    (2) In a case falling within sub-paragraph (1), the plan fee is determined by
        the formula -

$$\frac{A}{B} \times C$$

THE BUILDING (PRESCRIBED FEES ETC.)
REGULATIONS 1988

where -
A is the amount specified in column 2 of Table 1 below for the number of
   dwellings comprises in all the small domestic buildings in the multiple work
   scheme,
B is the number of dwellings comprised in all the small domestic buildings in
   the multiple work scheme, and
C is the number of dwellings in the small domestic building for which the fee
   is being determined.

Inspection fee for more than one small domestic building

    5. In relation to -

        (a) the erection of small domestic buildings, and
        (b) any work in connection with the erection of small domestic
             buildings.
Where that work forms part of a multiple work scheme which consists of or
includes the erection of two or more small domestic buildings, the inspection
fee for each small domestic building is determined by the formula -

$$\frac{D + E}{B} \times C$$

Where -
D is the amount which results from multiplying £80 by the number of dwellings
   in all the small domestic buildings in the multiple work scheme;
E is the amount specified in column 2 of Table 2 below for the number of
   dwellings (if any) comprised in the small domestic buildings in the multiple
   work scheme with a total floor area (excluding the floor area of any garage)
   exceeding 64m2;
B is the number of dwellings in the small domestic buildings in the multiple
   work scheme; and
C is the number of dwellings in the small domestic building for which the fee
   is being determined.

TABLE 1

Plan fee for erection of small domestic buildings

(1) No. of dwellings	(2) Amount of or for determining plan fee £	(1) No. of dwellings	(2) Amount of or for determining plan fee £
1	48	11	417
2	96	12	433
3	144	13	449
4	192	14	465
5	240	15	481
6	273	16	497
7	305	17	513
8	337	18	529
9	369	19	545
10	401	20 or more	561

THE BUILDING (PRESCRIBED FEES ETC.)
REGULATIONS 1988

Building notice fee for more than one small domestic building

6.   In relation to the erection of small domestic buildings and any work in
     connection with the erection of such buildings, where that work forms
     part of a multiple work scheme which consists of or includes the erection
     of two or more small domestic buildings, the building notice fee for each
     small domestic building is the aggregate of the plan fee and the
     inspection fee   which would have been payable had plans been
     deposited for the purposes of section 16 of the Act.

TABLE 2

Inspection fee for erection of small domestic building

```
-----------------------------------------
        (1)                 (2)
Number of dwellings    Amount for
each having a floor    determining
area exceeding 64m2    inspection fee
-----------------------------------------
                            £
    1                      48
    2                      96
    3                     144
    4                     192
    5                     240
    6                     257
    7                     273
    8                     289
    9                     305
10 or more                321
-----------------------------------------
```


SCHEDULE 2

SMALL GARAGES, CARPORTS AND CERTAIN ALTERATIONS AND EXTENSIONS

Fees for the erection of small garages and carports

1)   In relation to the erection of a detached building which -

     (a)   consists of a garage or carport or both,
     (b)   has a total floor area not exceeding 40 m2, and
     (c)   is intended to be used in common with an existing building, and
           which is not an exempt building,

     (i)   the plan fee is £11;
     (ii)   the inspection fee of £33; and
     (iii)   the building notice fee is £44.

THE BUILDING (PRESCRIBED FEES ETC.)
REGULATIONS 1988

Fees for domestic extensions and alteration

2.   (1)   This paragraph applies to work to or in connection with -

(a)   a small domestic building; or
(b)   a building, other than a small domestic building, which
      consists of flats or maisonettes or both; or
(c)   a building consisting of a garage or carport or both which is
      occupied in common with a building of the kind described in
      paragraphs (a) and (b),

(2)   Where work shown in deposited plan to be carried out to or in
      connection with a buiilding to which this paragraph applies is
      described in column 1 of the Table in Schedule 2 of the Regulation,
      the plan fee is the amount shown in column 2 for that description
      of work.

(3)   In relation to work described in column 1 of the Table in Schedule 2
      of the Regulation to be carried out to or in connection with a
      building to which this paragraph applies, the inspection fee is the
      amount shown in column 3 for that description of work.

(4)   In relation to work described in column (1) of the Table in Schedule
      2 of the Regulation to be carried out to or in connection with a
      building to which this paragraph applies, the building notice fee is
      the aggregate of the amounts shown in column 2 and 3 for that
      description of work.

TABLE

Fees for certain domestic extensions and alterations

(1) Type of work	(2) Amount of plan fee	(3) Amount of inspection fee
1. An extension or alteration consisting of the provision of one or more rooms in roof space, including means of access.	£22	£66
2. Any extension (not falling within paragraph 1 above) the total floor area of which does not exceed 20 m2, including means of access	£11	£33
3. Any extension (not falling within paragraph 1 above) the total floor area of which exceeds 20 m2 but does not exceed 40 m2, including means of access.	£22	£66

References in this table to square metres relate to the total area of the work
in question as it is shown on the plans

SCHEDULE 3

WORK OTHER THAN THAT DESCRIBED IN SCHEDULE 1 OR SCHEDULE 2

Interpretation

1.   In this Schedule the appropriate percentage figure is 70% of the estimate
     of cost for the building in question supplier in accordance with
     regulation 13 (2) (a).

THE BUILDING (PRESCRIBED FEES ETC.)
REGULATIONS 1988

Plan fee for single building

2. Where a plan fee is chargeable in a case described in regulation 5 (3),
    the plan fee is, unless paragraph 3 applies, a fee of the amount shown in
    column 2 of the Table in Schedule 3 of the Regulation for the
    appropriate percentage figure in Column 1.

Plan fee for multiple work scheme

3. Where a plan fee is chargeable in a case described in regulation 5 (3) in
    relation to work which forms part of a multiple work scheme, and the
    scheme includes other such proposed work to or in connection with at
    least one other building, the plan fee is the amount determinded in
    accordance with the formula -

    $$\frac{E}{T} \times A$$

where -

E is the appropriate percentage figure;
T is 70% of the aggregate figure supplied in accordance with regulation 13 (2)
   (b); and
A is the amount shown in column 2 of the Table below for the amount of T.

Inspection fee

4. In relation to work described in regulation 5 (3), the inspection fee is
    a fee of the amount shown in column 3 of the Table in Schedule 3 of the
    Regulation for the appropriate percentage figure in column 1.

Building notice fee

5. In relation to work described in regulation 5 (3), the building notice
    fee is the aggregate of the plan fee and the inspection fee which would
    have been payable had plans been deposited for the purposes of section
    16 of the Act.

SCHEDULE 4

FEES FOR DETERMINATION OF QUESTIONS BY THE SECRETARY OF STATE

Questions about comformity of plans with building regulations

1. Where an application is made for the determination under section 16 (10)
    (a) of the Act of the question whether plans of proposed work are in
    conformity with building regulations, the fee payable to the Secretary of
    State is half the plan fee which is payable in relation to the work shown
    in the plans, subject to a minimum fee of £25 and a maximum of £250.

Questions about rejection of certificated plans

2. Where an application is made for the detemination under section 16 (10)
    (b) of the Act of the question whether the local authority are prohibited
    by virtue of section 16 (9) of the Act (certificates that plans comply
    with certain regulations) from rejecting plans of proposed work, the fee
    payable to the Secretary of State is £25.

THE BUILDING (PRESCRIBED FEES ETC.)
REGULATIONS - 1988

Question arising between approved inspectors and developers

3. Where an application is made for the determination under section 50 (2)
    of the Act of the question whether plans of proposed work are in
    conformity with building regulations, the fee payable to the Secretary of
    State is half the plan fee which would have been payable in relating to
    the work shown in the plans had plans been deposited in accordance with
    section 16 of the Act, subject to a minimum fee of £25 and a maximum of
    £250.

TABLE

Fees for other work

(1) 70% of estimated cost	(2) Amount of plan fee or for determining plan fee	(3) Amount of inspection fee
£	£	£
Under 1,000	6	18
1,000 and under 2,000	10	30
2,000 and under 3,000	12	36
3,000 and under 4,000	16	48
4,000 and under 5,000	20	60
5,000 and under 6,000	24	72
6,000 and under 7,000	28	84
7,000 and under 8,000	32	96
8,000 and under 9,000	34	102
9,000 and under 10,000	36	108
10,000 and under 12,000	40	120
12,000 and under 14,000	46	138
14,000 and under 16,000	52	156
16,000 and under 18,000	58	174
18,000 and under 20,000	64	192
20,000 and under 25,000	75	225
25,000 and under 30,000	85	255
30,000 and under 35,000	95	285
35,000 and under 40,000	110	330
40,000 and under 45,000	120	360
45,000 and under 50,000	130	390
50,000 and under 60,000	145	435
60,000 and under 70,000	170	510
70,000 and under 80,000	195	585
80,000 and under 90,000	210	630
90,000 and under 100,000	230	690
100,000 and under 140,000	255	765
140,000 and under 180,000	330	990
180,000 and under 240,000	410	1230
240,000 and under 300,000	510	1530
300,000 and under 400,000	610	1830
400,000 and under 500,000	775	2325
500,000 and under 700,000	910	2730
700,000 up to and including 1,000,000	1185	3555
Thereafter for each additional 100,000 and part thereof	200	600

COMPETITIVE TENDERING FOR ENGINEERING CONTRACTS

M O'C Horgan

The majority of engineering contracts are negotiated by competitive tendering procedure. All too often shortcomings in the application of these procedures lead to delays, extra costs, disputes and sometimes, legal actions. In seeking to avoid such pitfalls this book provides a comprehensive but compact guide to competitive tendering, drawing on the author's many years of experience as a contracts manager for an international engineering company and subsequently as a contracts and contract management consultant. The book takes the reader step by step through all stages of establishing a reliable contract, including the selection of tenderers, preparation of documentation, submission and appraisal of tenders and finally, negotiation and acceptance or rejection. The emphasis throughout on real life, with check lists, practical advice and guidelines for action makes the book valuable both for experienced engineers looking for quick reference to advice on particular problems and for junior project managers needing more extensive guidance on appropriate, well tried procedures. Finally, a substantial section of appendices gives typical pro-formae, contractural letters, terms of agreement and similar documents essential in professional contract negotiation.

'This is a gem of a book . . . a detailed and totally practical guide to clients, project managers and engineers acting for clients on how to organize the tendering procedure.' — New Civil Engineer

1984 Hardback 0 419 11630 3 £25.00 296pp

CONSTRUCTION TENDERING AND ESTIMATING

J I W Bentley

Accurate estimating is the key to profit in construction contracting. The first step towards accuracy is a clear, logical approach to estimating — an approach which this book will help to teach.

Each chapter, with a set of self-assessment questions, covers a separate topic and can be used on its own as a reference by practising estimators. The book as a whole pays particular attention to the B/Tech Level IV syllabus in 'Tendering and Estimating'. It will also meet the needs of degree course students in building, surveying and civil engineering and those working for CIOB, RICS or other professional examination. The author has concentrated on giving practical, up-to-date advice, with a special chapter on computer-aided estimating systems. A model construction project is used throughout to illustrate tendering procedures and a profile of a medium-sized construction firm helps to explain overhead percentages and profit margins in a realistic way.

1987 Paperback 0 419 14240 1 £9.95 250pp

These books may be obtained from your usual bookshop. In case of difficulty please write to the address below or telephone the Order Department on 0264 332424.

 **E & F N SPON**
11 New Fetter Lane, London EC4P 4EE

Daywork and Prime Cost

When work is carried out which cannot be valued in any other way it is customary to assess the value on a cost basis with an allowance to cover overheads and profit. The basis of costing is a matter for agreement between the parties concerned, but definitions of prime cost for the building industry have been prepared and published jointly by the Royal Institution of Chartered Surveyors and the National Federation of Building Trades Employers (now the Building Employers Confederation) for the convenience of those who wish to use them. These documents are reproduced on the following pages by kind permission of the publishers.

Also reproduced in this section is the daywork schedule published by the Federation of Civil Engineering Contractors.

For larger Prime Cost contracts the reader is referred to the form of contract issued by the Royal Institute of British Architects.

BUILDING INDUSTRY

DEFINITION OF PRIME COST OF DAYWORK CARRIED OUT UNDER A BUILDING CONTRACT (DECEMBER 1975 EDITION)

This definition of Prime Cost is published by the Royal Institution of Chartered Surveyors and the National Federation of Building Trades Employers, for convenience and for use by people who choose to use it. Members of the National Federation of Building Trades Employers are not in any way debarred from defining Prime Cost and rendering their accounts for work carried out on that basis in any way they choose. Building owners are advised to reach agreement with contractors on the Definition of Prime Cost to be used prior to issuing instructions.

SECTION 1 - APPLICATION

1.1. This definition provides a basis for the valuation of daywork executed under such building contracts as provide for its use (e.g. contracts embodying the Standard Forms issued by the Joint Contracts Tribunal).

1.2. It is not applicable in any other circumstances, such as jobbing or other work carried out as a separate or main contract nor in the case of daywork executed during the Defects Liability Period of contracts embodying the above mentioned Standard Forms.

SECTION 2 - COMPOSITION OF TOTAL CHARGES

2.1. The prime cost of daywork comprises the sum of the following costs:

(a) Labour as defined in Section 3.
(b) Material and goods as defined in Section 4.
(c) Plant as defined in Section 5.

BUILDING INDUSTRY

2.2. Incidental costs, overheads and profit as defined in Section 6, as provided in the
building contract and expressed therein as percentage adjustments are applicable to
each of 2.1 (a)-(c).

SECTION 3 - LABOUR

3.1. The standard wage rates, emoluments and expenses referred to below and the standard
working hours referred to in 3.2 are those laid down for the time being in the rules
or decisions of the National Joint Council for the Building Industry and the terms of
the Building and Civil Enginnering Annual and Public Holiday Agreements applicable to
the works, or the rules or decisions or agreements of such body, other than the
National Joint Council for the Building Industry, as may be applicable relating to
the class of labour concerned at the time when and in the area where the daywork is
executed.

3.2. Hourly base rates for labour are computed by dividing the annual prime cost of
labour, based upon standard working hours and as defined in 3.4 (a)-(i), by the
number of standard working hours per annum.

3.3. The hourly rates computed in accordance with 3.2 shall be applied in respect of the
time spent by operatives directly engaged on daywork, including those operating
mechanical plant and transport and erecting and dismantling other plant (unless
otherwise expressly provided in the building contract).

3.4. The annual prime cost of labour comprises the following:

(a) Guaranteed minimum weekly earnings (e.g. Standard Basic Rate of Wages, Joint
Board Supplement and Guaranteed Minimum Bonus Payment in the case of NJCBI
rules).
(b) All other guaranteed minimum payments (unless included in Section 6).
(c) Differentials or extra payments in respect of skill, responsibility, discomfort,
inconvenience or risk (excluding those in respect of supervisory responsibility -
see 3.5).
(d) Payments in respect of public holidays.
(e) Any amounts which may become payable by the Contractor to or in respect of
operatives arising from the operation of the rules referred to in 3.1 which are
not provided for in 3.4 (a)-(d) or in Section 6.
(f) Employer's National Insurance contributions applicable to 3.4 (a)-(e).
(g) Employer's contributions to annual holiday credits.
(h) Employer's contributions to death benefit scheme.
(i) Any contribution, levy or tax imposed by statute, payable by the contractor in
his capacity as an employer.

3.5. Note:

Differentials or extra payments in respect of supervisory responsibility are excluded
from the annual prime cost (see Section 6). The time of principals, foremen,
gangers, leading hands and similar categories, when working manually, is admissible
under this Section at the appropriate rates for the trades concerned.

SECTION 4 - MATERIALS AND GOODS

4.1. The prime cost of materials and goods obtained from stockists or manufacturers is the
invoice cost after deduction of all trade discounts but including cash discounts not
exceeding 5 per cent and includes the cost of delivery to site.

BUILDING INDUSTRY

4.2. The prime cost of materials and goods supplied from the Contractor's stock is based upon the current market prices plus any appropriate handling charges.

4.3. Any Value Added Tax which is treated, or is capable of being treated, as input tax (as defined in the Finance Act, 1972) by the Contractor is excluded.

SECTION 5 - PLANT

5.1. The rates for plant shall be as provided in the building contract.

5.2. The costs included in this Section comprise the following:

  (a) Use of mechanical plant and transport for the time employed on daywork.
  (b) Use of non-mechanical plant (excluding non-mechanical hand tools) for the time employed on daywork.

5.3. **Note:**

The use of non-mechanical hand tools and of erected scaffolding, staging, trestles or the like is excluded (see Section 6).

SECTION 6 - INCIDENTAL COSTS, OVERHEADS AND PROFIT

6.1. The percentage adjustments provided in the building contract, which are applicable to each of the totals of Sections 3, 4 and 5, comprise the following:

  (a) Head Office charges.
  (b) Site staff, including site supervision.
  (c) The additional cost of overtime (other than that referred to in 6.2).
  (d) Time lost due to inclement weather.
  (e) The additional cost of bonuses and all other incentive payments in excess of any guaranteed minimum included in 3.4. (a).
  (f) Apprentices study time.
  (g) Subsistence and periodic allowances.
  (h) Fares and travelling allowances.
  (i) Sick pay or insurance in respect thereof.
  (j) Third-party and employers' liability insurance.
  (k) Liability in respect of redundancy payments to employees.
  (l) Employers' National Insurance contributions not included in Section 3.4.
  (m) Tool allowances.
  (n) Use, repair and sharpening of non-mechanical hand tools.
  (o) Use of erected scaffolding, staging, trestles or the like.
  (p) Use of tarpaulins, protective clothing, artificial lighting, safety and welfare facilities, storage and the like that may be available on the site.
  (q) Any variation to basic rates required by the Contractor in cases where the building contract provides for the use of a specified schedule of basic plant charges (to the extent that no other provision is made for such variation).
  (r) All other liabilities and obligations whatsoever not specifically referred to in this Section nor chargeable under any other Section.
  (s) Profit.

6.2. **Note:**

The additional cost of overtime, where specifically ordered by the Architect/Supervising Officer shall only be chargeable in the terms of prior written agreement between the parties to the building contract.

BUILDING INDUSTRY

Example of calculation of typical standard hourly base rate (as defined in Section 3) for NJCBI building craftsman and labourer in Grade A areas at 1st July, 1975.

		Rate £	Craftsman £	Rate £	Labourer £
Guaranteed minimum weekly earnings					
Standard Basic Rate	49 wks	37.00	1813.00	31.40	1538.60
Joint Board Supplement	49 wks	5.00	245.00	4.20	205.80
Guaranteed Minimum Bonus	49 wks	4.00	196.00	3.60	176.40
			-------		-------
			2254.00		1920.80
Employer's National Insurance Contribution at 8.5%			191.59		163.27
			-------		-------
			2445.59		2084.07
Employer's Contribution to:					
CITB annual levy			15.00		3.00
Annual holiday credits	49 wks	2.80	137.20	2.80	137.20
Public holidays (included in guaranteed minimum weekly earnings above)					
Death benefit scheme	49 wks	0.10	4.90	0.10	4.90
			-------		-------
Annual labour cost as defined in Section 3			£2602.69		£2229.17
Hourly rate of labour as defined in Section 3, Clause 3.2		£2602.69 ------- = 1904	£1.37	£2229.17 ------- = 1904	£1.17

Note:

1. Standard working hours per annum calculated as follows:
   52 weeks @ 40 hours   =   2080
   Less
   3 weeks holiday @ 40 hours   = 120
   7 days public holidays @ 8 hours   = 56   176
                                         ---   ----
                                                    1904

2. It should be noted that all labour costs incurred by the Contractor in his capacity as an employer other than those contained in the hourly rate, are to be taken into account under Section 6.

3. The above example is for the convenience of users only and does not form part of the Definition; all basic costs are subject to re-examination according to the time when and in the area where the daywork is executed.

BUILDING INDUSTRY

NOTE: For the convenience of readers the example which appears on the previous page has been updated by the Editors for London and Liverpool rates as at 31st July 1989.

	Rate £	Craft operative £	Rate £	Labourer £
Guaranteed minimum weekly earnings				
Standard Basic Rate 47.80 wks	120.120	5741.74	102.375	4893.53
Guaranteed Minimum Bonus 47.80 wks	15.015	717.72	12.675	605.87
		-------		-------
		6459.46		5499.40
Employer's National Insurance Contribution 9%		581.35		494.95
		-------		-------
	£	7040.81	£	5994.35
Employer's Contributions to:				
CITB annual levy		75.00		18.00
Annual holiday credits 47 wks	14.70	690.90	14.70	690.90
Public holidays (included in guaranteed minimum weekly earnings above)				
Retirement and death benefit scheme 47 wks	1.40	65.80	1.40	65.80
		-------		-------
Annual labour cost as defined in Section 3		7872.51		6769.05
Hourly rate of labour as defined in Section 3, Clause 3.02	£7872.51 ‾‾‾‾‾‾‾ 1801.80	= £4.37	£6769.05 ‾‾‾‾‾‾‾ 1801.80	= £3.76

Note:

1. Standard working hours per annum calculated as follows:
   52 weeks @ 39 hours   =   2028.0
   Less
       4.20 weeks holiday @ 39 hours   = 163.8
       8 days public holidays @ 7.8 hours = 62.4   226.2
                                            -----   ------
                                                        1801.8

2. It should be noted that all labour costs incurred by the Contractor in his capacity as an employer other than those contained in the hourly rate, are to be taken into account under Section 6.

3. The above example is for the convenience of users only and does not form part of the Definition; all the basic costs are subject to re-examination according to the time when and in the area where the daywork is executed.

4. The rates for annual holiday credits and retirement and death benefits are those which came into force in 31st July 1989 and take into account the discount available.

BUILDING INDUSTRY

**DEFINITION OF PRIME COST OF BUILDING WORKS OF A JOBBING
OR MAINTENANCE CHARACTER (1980 EDITION)**

This Definition of Prime Cost is published by the Royal Institution of Chartered Surveyors
and the National Federation of Building Trades Employers for convenience and for use by
people who choose to use it. Members of the National Federation of Building Trades
Employers are not in any way debarred from defining Prime Cost and rendering their
accounts for work carried out on that basis in any way they choose.
  Building owners are advised to reach agreement with contractors on the Definition of
Prime Cost to be used prior to issuing instructions.

SECTION 1 - APPLICATION

1.1. This definition provides a basis for the valuation of work of a jobbing or
     maintenance character executed under such building contracts as provide for its use.

1.2. It is not appplicable in any other circumstances, such as daywork executed under or
     incidental to a building contract.

SECTION 2 - COMPOSITION OF TOTAL CHARGES

2.1. The prime cost of jobbing work comprises the sum of the following costs:

     (a) Labour as defined in Section 3.
     (b) Materials and goods as defined in Section 4.
     (c) Plant, consumable stores and services as defined in Section 5.
     (d) Sub-contracts as defined in Section 6.

2.2. Incidental costs, overhead and profit as defined in Section 7 and expressed as
     percentage adjustments are applicable to each of 2.1 (a)-(d).

SECTION 3 - LABOUR

3.1. Labour costs comprise all payments made to or in respect of all persons directly
     engaged upon the work, whether on or off the site, except those included in Section
     7.

3.2. Such payments are based upon the standard wage rates, emoluments and expenses as laid
     down for the time being in the rules or decisions of the National Joint Council for
     the Building Industry and the terms of the Building and Civil Engineering Annual and
     Public Holiday Agreements applying to the works, or the rules of decisions or
     agreements of such other body as may relate to the class of labour concerned, at the
     time when and in the area where the work is executed, together with the Contractor's
     statutory obligations, including:

     (a) Guaranteed minimum weekly earnings (e.g. Standard Basic Rate of Wages and
         Guaranteed Minimum Bonus Payment in the case of NJCBI rules).
     (b) All other guaranteed minimum payments (unless included in Section 7).
     (c) Payments in respect of incentive schemes or productivity agreements applicable to
         the works.
     (d) Payments in respect of overtime normally worked; or necessitated by the
         particular circumstances of the work; or as otherwise agreed between the parties.
     (e) Differential or extra payments in respect of skill, responsibility, discomfort,
         inconvenience or risk.
     (f) Tool allowance.

BUILDING INDUSTRY

(g) Subsistence and periodic allowances.
(h) Fares, travelling and lodging allowances.
(j) Employer's contributions to annual holiday credits.
(k) Employer's contributions to death benefit schemes.
(l) Any amounts which may become payable by the Contractor to or in respect of operatives arising from the operation of the rules referred to in 3.2 which are not provided for in 3.2 (a)-(k) or in Section 7.
(m) Employer's National Insurance contributions and any contribution, levy or tax imposed by statute, payable by the Contractor in his capacity as employer.

Note:

Any payments normally made by the Contractor which are of a similar character to those described in 3.2 (a)-(c) but which are not within the terms of the rules and decisions referred to above are applicable subject to the prior agreement of the parties, as an alternative to 3.2 (a)-(c).

3.3. The wages or salaries of supervisory staff, timekeepers, storekeepers, and the like, employed on or regularly visiting site, where the standard wage rates etc. are not applicable, are those normally paid by the Contractor together with any incidental payments of a similar character to 3.2 (c)-(k).

3.4. Where principals are working manually their time is chargeable, in respect of the trades practised, in accordance with 3.2.

SECTION 4 - MATERIALS AND GOODS

4.1. The prime cost of materials and goods obtained by the Contractor from stockists or manufacturers is the invoice cost after deduction of all trade discounts but including cash discounts not exceeding 5 per cent, and includes the cost of delivery to site.

4.2. The prime cost of materials and goods supplied from the Contractor's stock is based upon the current market prices plus any appropriate handling charges.

4.3. The prime cost under 4.1 and 4.2 also includes any costs of:

(a) non-returnable crates or other packaging.
(b) returning crates and other packaging less any credit obtainable.

4.4. Any Value Added Tax which is treated, or is capable of being treated, as input tax (as defined in the Finance Act, 1972 or any re-enactment thereof) by the Contractor is excluded.

SECTION 5 - PLANT, CONSUMABLE STORES AND SERVICES

5.1. The prime cost of plant and consumable stores as listed below is the cost at hire rates agreed between the parties or in the absence of prior agreement at rates not exceeding those normally applied in the locality at the time when the works are carried out, or on a use and waste basis where applicable:

(a) Machinery in workshops.
(b) Mechanical plant and power-operated tools.
(c) Scaffolding and scaffold boards.
(d) Non-mechanical plant excluding hand tools.
(e) Transport including collection and disposal of rubbish.

BUILDING INDUSTRY

    (f) Tarpaulins and dust sheets.
    (g) Temporary roadways, shoring, planking and strutting, hoarding, centering, formwork, temporary fans, partitions or the like.
    (h) Fuel and consumable stores for plant and power-operated tools unless included in 5.1 (a), (b), (d) or (e) above.
    (j) Fuel and equipment for drying out the works and fuel for testing mechanical services.

5.2. The prime cost also includes the net cost incurred by the Contractor of the following services, excluding any such cost included under Sections 3, 4 or 7:

    (a) Charges for temporary water supply including the use of temporary plumbing and storage.
    (b) Charges for temporary electricity or other power and lighting including the use of temporary installations.
    (c) Charges arising from work carried out by local authorities or public undertakings.
    (d) Fees, royalties and similar charges.
    (e) Testing of materials.
    (f) The use of temporary buildings including rates and telephone and including heating and lighting not charged under (b) above.
    (g) The use of canteens, sanitary accommodation, protective clothing and other provision for the welfare of persons engaged in the work in accordance with the current Working Rule Agreement and any Act of Parliament, statutory instrument, rule, order, regulation or bye-law.
    (h) The provision of safety measures necessary to comply with any Act of Parliament.
    (j) Premiums or charges for any performance bonds or insurances which are required by the Building Owner and which are not referred to elsewhere in this Definition.

SECTION 6 - SUB-CONTRACTS

6.1. The prime cost of work executed by sub-contractors, whether nominated by the Building Owner or appointed by the Contractor, is the amount which is due from the Contractor to the sub-contractors in accordance with the terms of the sub-contracts after deduction of all discounts except any cash discount offered by any sub-contractor to the Contractor not exceeding 2.5%.

SECTION 7 - INCIDENTAL COSTS, OVERHEADS AND PROFIT

7.1. The percentage adjustments provided in the building contract, which are applicable to each of the totals of Sections 3-6, provide for the following:
    (a) Head Office charges.
    (b) Off-site staff including supervisory and other administrative staff in the Contractor's workshops and yard.
    (c) Payments in respect of public holidays.
    (d) Payments in respect of apprentices' study time.
    (e) Sick pay or insurance in respect thereof.
    (f) Third party employer's liability insurance.
    (g) Liability in respect of redundancy payments made to employees.
    (h) Use, repair and sharpening of non-mechanical hand tools.
    (j) Any variations to basic rates required by the Contractor in cases where the building contract provides for the use of a specified schedule of basic plant charges (to the extent that no other provision is made for such variation).
    (k) All other liabilities and obligations whatsoever not specifically referred to in this Section nor chargeable under any other section.
    (l) Profit.

BUILDING INDUSTRY

SPECIMEN ACCOUNT FORMAT

If this Definition of Prime Cost is followed the Contractor's account could be in the following format:

£

        Labour (as defined in Section 3)
        Add ... % (see Section 7)
        Materials and goods (as defined in Section 4)
        Add ... % (see Section 7)
        Plant, consumable stores and services
        (as defined in Section 5)
        Add ... % (see Section 7)
        Sub-contracts (as defined in Section 6)
        Add ... % (see Section 7)

£

VAT to be added if applicable.

SCHEDULE OF BASIC PLANT CHARGES (OCTOBER 1980 ISSUE)

This Schedule is published by the Royal Institution of Chartered Surveyors and is for use in connection with Dayworks under a Building Contract.

EXPLANATORY NOTES

1. The rates in the Schedule are intended to apply solely to daywork carried out under and incidental to a Building Contract. They are NOT intended to apply to:
   (i)   jobbing or any other work carried out as a main or separate contract; or
   (ii) work carried out after the date of commencement of the Defects Liability Period.
2. The rates in the Schedule are basic and may be subject to an overall adjustment to be quoted by the Contractor prior to the placing of the Contract.
3. The rates apply to plant and machinery already on site, whether hired or owned by the Contractor.
4. The rates, unless otherwise stated, include the cost of fuel of every description, lubricating oils, grease, maintenance, sharpening of tools, replacement of spare parts, all consumable stores and for licences and insurances applicable to items of plant. They do not include the costs of drivers and attendants.

5. The rates should be applied to the time during which the plant is actually engaged in daywork.
6. Whether or not plant is chargeable on daywork depends on the daywork agreement in use and the inclusion of an item of plant in this schedule does not necessarily indicate that that item is chargeable.
7. Rates for plant not included in the Schedule or which is not on site and is specifically hired for daywork shall be settled at prices which are reasonably related to the rates in the Schedule having regard to any overall adjustment quoted by the Contractor in the Conditions of Contract.

NOTE: All rates in the schedule are expressed per hour and were calculated during the second quarter of 1980.

BUILDING INDUSTRY

MECHANICAL PLANT AND TOOLS

Item of plant	Description	Unit	Rate per hour £
BAR-BENDING AND SHEARING MACHINES			
Bar bending machine	Up to 2 in (50 mm) dia rods	each	1.11
Bar cropper machine		each	1.11
BRICK SAWS			
Brick saw (use of abrasive disc to be charged net and credited)	Power driven (bench type clipper or similar)	each	0.80
COMPRESSORS			
Portable compressors (machine only)	Nominal delivery of free air per min at 100 lb/sq in (7kg/sq cm) pressure.		

Up to and including

	cu ft	(cu m)		
	100	(2.83)	each	1.53
	101-130	(2.86-3.68)	each	1.76
	131-180	(3.71-5.10)	each	1.87
	181-220	(5.13-6.23)	each	2.27
	221-260	(6.26-7.36)	each	2.91
	261-320	(7.39-9.06)	each	3.23
	321-380	(9.09-10.76)	each	3.85
	500-599	(14.16-16.96)	each	6.65
	600-630	(16.90-17.84)	each	7.50
Mobile compressor (machine plus lorry only)	101-150	(2.86-4.24)	each	3.05
Tractor-mounted compressor	101-120	(2.86-3.40)	each	2.78

COMPRESSED AIR EQUIPMENT

(with and including up to 60 ft (18.3 m) of hose)			
Heavy breaker and six steels		each	0.18
Light pneumatic pick and six steels		each	0.14
Pneumatic clay spade and one blade		each	0.16
Hand-held rock drill and rod (bits to be paid for at net cost and credited)	35 lb (15.9 kg) Class	each	0.18
	45 lb (20.5 kg) Class	each	0.19
	55 lb (25.0 kg) Class	each	0.23
Drill, rotary (bit to be paid for at net cost and credited)	Up to 0.75 in (19 mm)	each	0.15
Sander/Grinder		each	0.20
Chipping hammer and 6 steels		each	0.16
Additional hoses	Per 60 ft (18.3 m) length	each	0.05
Muffler, tool silencer		each	0.06

CONCRETE BREAKER

Concrete breaker, portable hydraulic complete with power pack		each	1.14

BUILDING INDUSTRY

Item of plant	Description	Unit	Rate per hour £
CONCRETE OR MORTAR MIXERS			
Concrete mixer	Diesel, electric or petrol		
	cu ft (cu m)		
Open drum without hopper	3/2 (0.09/0.06)	each	0.34
	4/3 (0.12/0.09)	each	0.38
	5/3.5 (0.15/0.10)	each	0.44
Open drum with hopper	7/5 (0.20/0.15)	each	0.58
Reversing drum with hopper weigher and feed shovel	10/7 (0.28/0.20)	each	1.73
Reversing drum with hopper weigher and feed shovel	14/10 (0.40/0.28)	each	1.81
Reversing drum with hopper weigher and feed shovel	18/12 (0.51/0.34)	each	2.08
Reversing drum with hopper weigher and feed shovel	21/14 (0.60/0.40)	each	2.22
CONCRETE PUMP			
Lorry mounted concrete pump (meterage charge to be added net)		each	10.50
CONCRETE EQUIPMENT			
Vibrator poker type	Petrol, diesel or electric	each	0.48
	Air, excluding compressor and hose	each	0.22
Vibrator-tamper	With tamping board	each	0.55
	Double beam screeder	each	0.65
Power float		each	0.68
CONVEYOR BELTS	Power operated up to 25 ft (7.62 m) long, 16 in (400 mm) wide	each	2.27
CRANES			
	Maximum capacity up to		
Mobile, rubber tyred	15 cwt (762 kg)	each	1.61
Lorry mounted, telescopic	6 tons (tonnes)	each	3.17
	7 tons (tonnes)	each	3.52
	10 tons (tonnes)	each	4.02
	12 tons (tonnes)	each	4.95
	15 tons (tonnes)	each	5.92
	18 tons (tonnes)	each	7.46
	Capacity metre/tonnes		
Static tower	40	each	5.80
	42	each	7.65
	80	each	8.10
Track mounted tower	40	each	6.80
	42	each	8.80
	80	each	9.40

BUILDING INDUSTRY

MECHANICAL PLANT AND TOOLS

Item of plant	Description		Unit	Rate per hour £
CRANE EQUIPMENT	cu yd	(cu m)		
	Up to			
Skip, muck	0.5	(0.38)	each	0.08
	0.75	(0.57)	each	0.08
	1.0	(0.76)	each	0.09
	Up to			
Skip, concrete	0.5	(0.38)	each	0.16
	0.75	(0.57)	each	0.18
	1.0	(0.76)	each	0.21
Skip, concrete lay down or roll over	0.5	(0.38)	each	0.16

DEHUMIDIFIERS

Water extraction per 24 hours	15 galls	(68 litres)	each	0.57
	20 galls	(90 litres)	each	0.61

DUMPERS

Dumper (site use only excluding Tax, Insurance and cost of DERV, etc., when operating on highway)	Makers capacity			
2 wheel drive				
Gravity tip	15 cwt	(762 kg)	each	1.00
Gravity tip	23 cwt	(1168 kg)	each	1.05
Hydraulic tip	23 cwt	(1168 kg)	each	1.20
Gravity tip	35 cwt	(1778 kg)	each	1.50
4 wheel drive				
Gravity tip	23 cwt	(1168 kg)	each	1.56
Hydraulic tip	25 cwt	(1270 kg)	each	1.65
Hydraulic tip	35 cwt	(1778 kg)	each	1.90
Hydraulic tip	40 cwt	(2032 kg)	each	2.40
Hydraulic tip	45 cwt	(2286 kg)	each	2.50
Hydraulic tip	50 cwt	(2540 kg)	each	2.60
Hydraulic tip	60 cwt	(3048 kg)	each	3.20

ELECTRIC HAND TOOLS
(drills to be charged net and credited)

Heavy breaker	Kango 1800		each	0.59
	Kango 2500		each	0.63
Rotary hammer	Kango 950		each	0.35
	Kango 627/637		each	0.29
	Hilti TE12/TE17		each	0.23
Pipe drilling tackle	Ordinary type		set	0.27
	Under pressure type		set	0.67

EXCAVATORS

Hydraulic full circle slew excavator	cu yd	(cu m)		
Hymac 580 or similar	0.625	(0.48)	each	5.50
Wheeled tractor type hydraulic excavator, JCB type 3C or similar			each	3.50

FORKLIFTS

	Payload	Max. lift	Unit	Rate
2 wheel drive	20 cwt (1016 kg)	21 ft 4 in (6.50 m)	each	4.40
	20 cwt (1016 kg)	26 ft 0 in (7.92 m)	each	5.05
	30 cwt (1524 kg)	20 ft 0 in (6.09 m)	each	3.75
	36 cwt (1829 kg)	18 ft 0 in (5.48 m)	each	3.75
	50 cwt (2540 kg)	12 ft 0 in (3.66 m)	each	3.55

BUILDING INDUSTRY

Item of plant	Description		Unit	Rate per hour £
	Payload	Max. lift		
4 wheel drive	30 cwt (1524 kg)	20 ft 0 in (6.09 m)	each	4.30
	40 cwt (2032 kg)	20 ft 0 in (6.09 m)	each	5.65
	50 cwt (2540 kg)	12 ft 0 in (3.66 m)	each	4.20
HAMMERS, CARTRIDGE	Excluding cartridges and studs		each	0.16
HEATERS - SPACE	Paraffin/electric Btu/hr			
	50 000-75 000		each	0.64
	80 000-100 000		each	0.68
	150 000		each	1.06
	320 000		each	1.74

HOISTS

		Unit	Rate
Scaffold	Up to 5 cwt (254 kg)	each	0.54
Mobile goods	Up to 10 cwt (508 kg)	each	1.20
Static goods	10 to 15 cwt (508-762 kg)	each	1.53
Rack and pinion goods	16 cwt (813 kg)	each	1.50
	20 cwt (1016 kg)	each	1.72
Rack and pinion goods and passenger	8 person, 1433 lbs (650 kg)	each	2.45
	12 person, 2205 lbs (1000 kg)	each	2.80

LORRIES

	Plated gross vehicle Weight, ton/tonne	Unit	Rate
Fixed body	Up to 5.50	each	3.82
	Up to 7.50	each	4.50
Tipper	Up to 16	each	6.90
	Up to 24	each	8.70
	Up to 30	each	9.45

PIPE WORK EQUIPMENT

		Unit	Rate
Pipe bender, power driven	50-150 mm dia	each	0.64
Pipe cutting machine	Hydraulic	each	0.54
Pipe defrosting equipment	Electrical	set	1.30
Pipe testing equipment	Compressed air	set	0.54
	Hydraulic	set	0.27

PUMPS

Including 20 ft (6 m) length of suction and/or delivery hose, couplings, valves and strainers

		Unit	Rate
'Simplite' 2 in (50 mm) diaphragm	Petrol or electric	each	0.48
'Wickham' 3 in (76 mm) diaphragm	Diesel	each	0.80
Submersible 2 in (50 mm)	Electric	each	0.47
'Spate' 3 in (76 mm) induced flow	Diesel	each	0.76
'Spate' 4 in (102 mm) induced flow	Diesel	each	1.23
'Univac' 2 in (50 mm)	Diesel	each	1.18
'Univac' 4 in (102 mm)	Diesel	each	2.04
'Univac' 6 in (152 mm)	Diesel	each	2.87

BUILDING INDUSTRY

MECHANICAL PLANT AND TOOLS

Item of plant	Description		Unit	Rate per hour £
PUMPING EQUIPMENT	in	(mm)		
Pump hoses, per 20 ft (6.00 m)	2	(51)	each	0.08
flexible, suction or delivery,	3	(76)	each	0.10
including coupling valve and	4	(102)	each	0.12
strainer	6	(152)	each	0.16
RAMMERS AND COMPACTORS				
Power rammer, Pegson or similar			each	0.70
Soil compactor, plate type				
Plate size				
12 x 13 in (305 x 330 mm)	172 lb (78 kg)		each	0.70
20 x 18 in (508 x 457 mm)	264 lb (120 kg)		each	0.77
ROLLERS	cwt	(kg)		
Vibrating rollers	7.25-8.25	(368-420)	each	0.66
Single roller	10.5	(533)	each	1.08
Twin roller	13.75	(698)	each	1.20
	16.75	(851)	each	1.42
Twin roller with seat and steering	21	(1067)	each	1.18
	27.5	(1397)	each	1.98
	ton/tonne			
Pavement rollers dead weight	3-4		each	2.38
	4-6		each	3.34
	6-10		each	4.17
SAWS, MECHANICAL	in	(m)		
Chain saw	21	(0.53)	each	0.58
	30	(0.76)	each	0.62
Bench saw	Up to 20	(0.51) blade	each	0.78
	Up to 24	(0.61) blade	each	0.87
SCREED PUMP	Maximum delivery			
Working volume 7 cu ft (200 litres)	Vertical 300 ft (91 m)			
	Horizontal 600 ft (182 m)		each	6.70
Screed pump hose	50/65 mm dia			
	13.3 m long		each	0.26
SCREWING MACHINES	13- 50 mm dia		each	0.19
	25-100 mm dia		each	0.32
TRACTORS	cu yd	(cu m)		
	Up to			
Shovel, tractor (crawler), any	0.75	(0.57)	each	3.70
type of bucket	1	(0.76)	each	4.33
	1.25	(0.96)	each	4.76
	1.5	(1.15)	each	5.61
	1.75	(1.34)	each	6.46
	2	(1.53)	each	7.30
	Up to			
Shovel, tractor (wheeled)	0.5	(0.38)	each	2.53
	0.75	(0.57)	each	3.25
	1.0	(0.76)	each	3.70
	2.5	(1.91)	each	5.60

BUILDING INDUSTRY

Item of plant	Description	Unit	Rate per hour £
Maker's rated flywheel horsepower			
Tractor (crawler) with dozer	75	each	4.33
	140	each	7.40
Tractor-wheeled (rubber tyred)	Light 48 hp	each	2.50
Agricultural type	Heavy 65 hp	each	2.68

WELDING AND CUTTING AND BURNING SETS

Welding and cutting set (including oxygen and acetylene, excluding underwater equipment and thermic boring)		each	2.25
Electric welding set (excluding electrodes)	Diesel 300 amp, single operator	each	1.00
	Diesel 600 amp, double operator	each	1.75
	Transformer electric 300 amp, single operator	each	0.75

NON-MECHANICAL PLANT

BAR BENDING AND SHEARING MACHINES

Bar bending machine, hand operated	Up to 1 in (25 mm) dia rods	each	0.09
Shearing machine, hand operated	Up to 0.625 (16 mm) dia rods	each	0.09

BROTHER OR SLING CHAINS

	Not exceeding 2 ton/tonne	set	0.10
	Exceeding 2 ton/tonne, not exceeding 5 ton/tonne	set	0.17
	Exceeding 5 ton/tonne, not exceeding 10 ton/tonne	set	0.21

DRAIN TESTING EQUIPMENT

		set	0.27

LIFTING AND JACKING GEAR

	ton/tonne		
Pipe winch - including shear legs	0.5	set	0.29
Pipe winch - including gantry	2	set	0.60
	3	set	0.70
Chain blocks up to 20 ft (6.10 m) lift	1	each	0.15
	2	each	0.18
	3	each	0.21
	4	each	0.24
Pull lift (Trifor type)	0.75	each	0.11
	1.5	each	0.14
	3	each	0.31

BUILDING INDUSTRY

NON-MECHANICAL PLANT

Item of plant	Description	Unit	Rate per hour £
PIPE BENDERS	13- 75 mm dia	each	0.27
	50-100 mm dia	each	0.43
PLUMBER'S FURNACE	Calor gas or similar	each	0.48
ROAD WORKS - EQUIPMENT			
Barrier trestles or similar		10	0.15
Crossing plates (steel sheets)		each	0.06
Danger lamp, including oil		each	0.01
Traffic control lights (not pad operated)	Automatic equipment 150 yds (137.16 m) cable		
Traffic control lights	Mains supply	set	1.04
	With generator	set	1.16
Warning sign		each	0.05
Road cone		10	0.08
Flasher unit (battery to be charged at cost)		each	0.03
Flashing bollard (battery to be charged at cost)		each	0.08
SCAFFOLDING			
Boards		100 ft (30.48m)	0.03
Castor wheels	Steel or rubber tyred	100	0.56
Fall ropes	Up to 200 ft (61 m)	each	0.28
Fittings (including couplers, base plates	Steel or alloy	100	0.02
Ladders, pole	20 rung	each	0.04
	30 rung	each	0.07
	40 rung	each	0.10
Ladders, extension	Extended length ft (m)		
	20 (6.10)	each	0.06
	26 (7.92)	each	0.07
	35 (10.67)	each	0.12
Putlogs	Steel or alloy	100	0.05
Tube	Steel	100 ft (30.48 m)	0.008 / 0.016
Staging, lightweight	Alloy	100 ft (30.48 m)	0.38
Wheeled tower	Working platform up to 20 ft (6 m high. Including castors and boards		
7 ft x 7 ft (2.13 x 2.13 m)		each	0.31
10 ft x 10 ft (3.05 x 3.05 m)		each	0.45
Splithead			
small		10	0.03
medium		10	0.04
large		10	0.04
TARPAULINS		10 m2	0.02
TRENCH STRUTS AND SHEETS			
Adjustable steel trench strut	All sizes from 1 ft (305 mm) closed to 5 ft 6 in (1.8 m) extended	10	0.02
Steel trench sheet	5-14 ft (1.52-4.27 m) lengths	100 ft (30.48 m)	0.08

CIVIL ENGINEERING INDUSTRY

Daywork charges for the civil engineering industry in accordance with the schedule prepared by the Federation of Civil Engineering Contractors

WITH EFFECT FROM 10 AUGUST 1987

SCHEDULES OF DAYWORKS CARRIED OUT INCIDENTAL TO CONTRACT WORK

(For operation throughout England, Scotland and Wales)

1. Labour   2. Materials   3. Plant   4. Supplementary charges

These schedules are the schedules referred to in Clause 52 (3) of the ICE Conditions of Contract 5th Edition and have been prepared for use in connection with Dayworks carried out incidental to contract work where no other rates have been agreed. They are not intended to be applicable for Dayworks ordered to be carried out after the contract works have been substantially completed or to a contract to be carried out wholly on a Daywork basis. The circumstances of such works vary so widely that the rates applicable call for special consideration and agreement between contractor and employing authority.

1. LABOUR

Add to amount of wages paid to workmen   133%

NOTES AND CONDITIONS

(1) 'Amount of wages' means:

Wages, actual bonus paid, daily travelling allowances (fare and/or time), tool allowance and all prescribed payments including those in respect of time lost due to inclement weather paid to the workmen at plain time rates and/or at overtime rates.
All payments shall be in accordance with the Working Rule Agreement of the Civil Engineering Construction Conciliation Board for Great Britain Rule Nos I to XXIII inclusive or that of other appropriate wage-fixing authorities current at the date of executing the work and where there are no prescribed payments or recognized wage-fixing authorities, the actual payments made to the workmen concerned.

(2) The percentage addition provides for all statutory charges at the date of publication and other charges including:

National Insurances and surcharge.
Normal Contract Works, Third Party and Employers' Liability Insurances.
Annual and Public Holidays with Pay and Benefit Scheme.
Non-contributory Sick Pay Scheme.
Industrial Training Levy.
Redundancy Payments Contribution.
Contracts of Employment Act.
Site supervision and staff including foreman and working gangers, but the time of the gangers or charge hands working with their gangs is to be paid for as for workmen.
Small tools - such as picks, shovels, barrows, trowels, hand saws, buckets, trestles, hammers, chisels and all items of a like nature.
Protective clothing.
Head Office charges and profit.

(3) All labour hired plant drivers and only subcontractors' accounts to be charged at full amount of invoice (without deduction of any cash discounts not exceeding 2.5%) plus 64%.

CIVIL ENGINEERING INDUSTRY

> (4) Subsistence or lodging allowances and periodic travel allowances (fare and/or time) paid to or incurred on behalf of workmen are chargeable at cost plus 12.5%.
> (5) The charges referred to above are as at date of publication and will be subject to periodic review.

2. MATERIALS

Add to the cost of materials delivered site   12.5%

NOTES AND CONDITIONS

> (1) The percentage addition provides for Head Office charges and profit.
> (2) The cost of materials means the invoiced price of materials including delivery to site without deduction of any cash discounts not exceeding 2.5%.
> (3) Unloading of materials:
>
> The percentage added to the cost of materials excludes the cost of handling which shall be charged in addition.
> An allowance for unloading into site stock or storage including wastage should be added where materials are taken from existing stock.
> (4) The charges referred to above are as at the date of publication and will be subject to periodic review.

3. PLANT

NOTES AND CONDITIONS

> (1) These rates apply only to plant already on site exclusive of driver and attendants, but inclusive of fuel and consumable stores, unless stated to be charged in addition, repairs and maintenance, insurance of plant, but excluding time spent on general servicing.
>> (1A) Where plant is hired in specifically for dayworks: plant hire (exclusive of drivers and attendants), fuel, oil and grease, insurance, transport etc., to be charged at full amount of invoice (without deduction of any cash discount not exceeding 2.5%) to which should be added consumables where supped by the Contractor, all plus   12.5%.
> (2) Fuel distribution is not included in the rates quoted which shall be an additional charge.
> (3) Metric capacities are adopted and these are not necessarily exact conversions from their imperial equivalents, but cater for the variations arising from comparison of plant manufacturing firms' ratings.
> (4) Minimum hire charge will be for the period quoted.
> (5) Hire rates for mechanical or other special plant not normally classed as small tools and not included below shall be settled at prices reasonably related to the rates quoted.
> (6) The rates for temporary track in Section 17 are for track already in position. Arrangements should be made on site for payment in cases where track has to be relocated or when used exclusively on daywork operations.
> (7) The charges referred to above are as at the date of publication and will be subject to periodic review.
> (8) To differentiate between mechanical and non-mechanical plant, consumables, etc., the latter is indicated by an asterisk (*).
> (9) The rates provide for Head Office charges and profit.

CIVIL ENGINEERING INDUSTRY

Description	Unit	Hire Rate £	Period

ASPHALT EQUIPMENT

Ashpalt and/or coated macadam spreader

	Description	Unit	Hire Rate £	Period
Crawler or wheeled	Up to 37 kW	each	19.09	hour
	38 to 56 kW	each	27.81	hour
	57 to 82 kW	each	40.28	hour
	83 to 150 kW	each	132.86	hour
* Joint matcher(s). Separately or in conjunction with longitudinal beams	Extra over machine	each	1.62	hour
* As above with addition of transverse control devices or fully automatic equipment capable of of control from reference guides	Extra over machine	each	5.09	hour
Self-propelled, metered coated chipping applicator for asphalt work (including trailer)	up to 7.5 kW	each	27.82	hour
	width; m			
	7.6 to 18.7	each	35.70	hour
	18.8 to 56.0	each	48.71	hour
Heater planer	up to 2.5	each	261.92	hour
Cold planer	up to 0.5	each	60.82	hour
	0.51 to 1.00	each	75.91	hour
	1.01 to 2.25	each	178.96	hour
* Asphalt road burner				
portable		each	5.55	hour
self-propelled		each	93.05	hour
* Asphalt kerb machine		each	26.53	hour

BAR BENDING AND
BAR SHEARING MACHINES

For mild steel rods up to mm dia

		Unit	Hire Rate £	Period
* Bar bending machine				
hand operated	25	each	1.00	day
power driven	38	each	1.90	hour
power driven	51	each	2.30	hour
* Bar shearing machine				
hand operated	25	each	1.32	day
power driven	38	each	1.87	hour
power driven	51	each	2.25	hour

For high tensile steel rods up to mm dia

		Unit	Hire Rate £	Period
Bar bending machine				
power driven	55	each	6.01	hour
Bar shearing machine				
power driven	55	each	4.60	hour

BOILERS

Evaporate capacity per hour
8.5 kg per cm2 working pressure

		Unit	Hire Rate £	Period
Oil fired	400 kg	each	9.14	hour
	500 kg	each	12.27	hour
	700 kg	each	16.23	hour
	800 kg	each	19.19	hour
	1000 kg	each	21.91	hour
	1500 kg	each	32.14	hour

CIVIL ENGINEERING INDUSTRY

	Description	Unit	Hire Rate £	Period
COMPRESSORS				

NOTE: Cu m piston displacement is approx. 20% above the nominal free air delivery.

	Description	Unit	Hire Rate £	Period
	Nominal delivery of free air per min at 7 kg per cm2 (for compressors with higher working pressure, rates to be negotiated)			
	Up to and including			
Portable compressor (machine only)	cu m/min			
	2.0	each	3.28	hour
	2.3	each	3.60	hour
	2.6	each	3.79	hour
	3.2	each	4.53	hour
	3.7	each	5.08	hour
	4.0	each	5.17	hour
	4.6	each	5.80	hour
	6.0	each	7.27	hour
	7.4	each	8.46	hour
	9.1	each	10.72	hour
	10.7	each	12.00	hour
	11.4	each	12.82	hour
	12.2	each	13.53	hour
	13.6	each	15.09	hour
	17.0	each	18.84	hour
Portable compressor (silenced) (machine only)	2.6	each	4.13	hour
	3.7	each	5.45	hour
	4.0	each	6.59	hour
	4.6	each	7.32	hour
	6.0	each	9.00	hour
	7.4	each	10.22	hour
	9.7	each	13.01	hour
	10.5	each	14.26	hour
	12.2	each	16.63	hour
	13.6	each	18.26	hour
	17.0	each	22.09	hour
	19.8	each	25.92	hour
	25.5	each	29.00	hour
Lorry mounted compressor (machine, plus lorry only) (site and public highway use - including Tax, Insurance and extra cost of DERV)				
	Range cu m/min			
Portable on lorry	2.86/4.24	each	10.72	hour
	4.27/5.66	each	12.40	hour
Portable silenced	1.44/2.83	each	10.04	hour
	2.86/4.24	each	11.89	hour
	4.27/5.66	each	13.49	hour
Tractor mounted compressor (machine plus rubber tyred tractor only) (site use only - excluding Tax, Insurance and Extra cost of DERV)				
	Range cu m/min			
Portable on tractor	3.43/4.24	each	9.29	hour
	cu m; up to			
* Air receiver	1.14	each	1.80	day
	2.83	each	4.29	day
	8.50	each	7.72	day

CIVIL ENGINEERING INDUSTRY

Description	Unit	Hire Rate £	Period

CONCRETE MIXERS

	Wet capacity litres			
Open drum, mixer only	60	each	0.64	hour
	90	each	0.88	hour
	100	each	1.21	hour
Closed drum, mixer only	150	each	1.71	hour
	185	each	1.98	hour
	200	each	3.22	hour
As above, with swing batch weighing gear	300	each	5.02	hour
	350	each	5.34	hour
	400	each	6.25	hour
Extra for power driven loading shovel	200	each	0.77	hour
	for other sizes	each	0.92	hour
Mixer complete with integral batch weighing gear and power driven loading shovel	litres			
	150	each	3.44	hour
	200	each	4.29	hour
	300	each	6.05	hour
	350	each	6.15	hour
	400	each	7.17	hour
* Extra for aggregate feed apron	single compartment	each	0.17	hour
	2 compartment	each	0.19	hour
	3 compartment	each	0.21	hour
	Wet capacity			
	up to 50 litres	each	3.02	hour
Pan mixer	51- 100	each	3.82	hour
	101- 200	each	5.11	hour
	201- 400	each	7.41	hour
	401- 550	each	8.78	hour
	551- 850	each	11.23	hour
	851-1250	each	15.88	hour
	1251-1500	each	18.78	hour
* Extra for batch weighing gear	401- 500	each	1.75	hour
	501-1250	each	1.84	hour
	1251-1500	each	2.07	hour
Extra for power driven batch loader	101- 200	each	1.98	hour
	201- 400	each	2.05	hour
	401- 550	each	2.48	hour
	551- 850	each	3.25	hour
	851-1250	each	4.96	hour
	1251-1500	each	6.23	hour
Roller pan mortar mixer	up to 200	each	5.30	hour
Extra for power driven batch loader		each	2.30	hour
	tonnes			
Cement silo (low level)	13.00	each	1.53	hour
	26.00	each	2.54	hour
	31.00	each	2.72	hour
	51.00	each	3.50	hour
* Extra for aeration equipment		each	0.15	hour
* Extra for capsule weighing equipment and swinging arm		each	0.41	hour
* Extra for autofeed weighing equipment and screw conveyor		each	1.43	hour
* Cement silo (high level) with weighing equipment and swinging arm (manually operated)	26.00	each	3.39	hour
	31.00	each	3.55	hour
Cement silo (high level) with weighing equipment and screw conveyor (power operated)	25.40	each	4.29	hour
	30.50	each	5.41	hour

CIVIL ENGINEERING INDUSTRY

Description		Unit	Hire Rate £	Period

CONCRETE MIXERS - cont'd

	Maximum output litres/min	Hopper capacity litres			
Grout mixer, single drum	60	120	each	5.82	hour
Grout mixer, double drum	110	120	each	8.17	hour
	170	230	each	12.63	hour
Loading hopper, batch weighing gear and power driven loading shovel		210 kg	each	7.74	hour
Grout mixer, double drum	22	480	each	19.68	hour
Grout mixer, roller	10		each	2.00	hour
	20		each	3.99	hour
	40		each	5.24	hour
Grouting machine including pump and hopper	40	40	each	3.30	hour
	70	40	each	3.84	hour
	110	40	each	5.02	hour

	Maximum output litres/min				
* Grout pump (excl. compressor) ram dia 50 mm	310		each	7.50	hour
ram dia 63 mm	480		each	7.67	hour
* Agitating tank for use with above			each	1.30	hour
	mm				
* Grout pump (hand operated)	76	280	each	0.94	hour
	38	480	each	0.66	hour

	Mixed batch capacity cu m				
Concrete transporter (lorry mounted) (lorry 4 x 2)	2.30		each	11.14	hour
	3.00		each	11.72	hour
Extra for lorry 4 x 4			each	2.82	hour
Concrete truck mixer (agitator) (incl. separate engine with mechanical drive to drum)	5.00		each	34.75	hour
	6.00		each	36.11	hour

CIVIL ENGINEERING INDUSTRY

	Description	Unit	Hire Rate £	Period
CONCRETE EQUIPMENT				
	Output per hour maximum cu m			
Concrete pump (skid mounted) (exclusive of piping)	20/26	each	29.25	hour
	30/45	each	35.98	hour
	46/54	each	37.74	hour
	70/72	each	53.38	hour
	90/110	each	54.30	hour
Concrete pump (lorry mounted) (exclusive of piping) (including boom)	50	each	46.52	hour
	60	each	60.98	hour
	80	each	62.88	hour
	90/110	each	66.52	hour
Concrete pump (mobile trailer) (rubber tyred) (exclusive of piping)	15/26	each	31.20	hour
	30/34	each	32.76	hour
	38/45	each	40.40	hour
	50/75	each	49.42	hour
	90/110	each	57.88	hour
	mm dia			
* Piping per 3 m straight length	102	each	0.43	day
	127	each	0.50	day
	152	each	0.52	day
* Bend	102	each	0.71	day
	127	each	0.79	day
	152	each	1.51	day
* Flexible distributor hose per 4 m length	102	each	3.76	day
	127	each	5.58	day
	152	each	7.79	day
* Shut off pipe section	102	each	1.76	day
	127	each	1.90	day
	152	each	2.42	day
* Pipe cleaning equipment incl. cleaning bend rubber sponge ball and trap basket	102	each	2.65	day
	127	each	2.71	day
	152	each	2.84	day
Vibrator poker				
petrol driven		each	1.47	hour
diesel driven		each	1.84	hour
electric		each	1.10	hour
* air (excl. compressor)		each	0.48	hour
Alternator/frequency convertor	Up to 2.5 kVA	each	1.40	hour
	Up to 4.5 kVA	each	1.81	hour
Vibrator poker (H/F motor in head type)		each	0.52	hour
Vibrator external type clamp on electric	small	each	0.80	hour
	medium	each	1.26	hour
	large	each	1.87	hour
* Vibrator external type clamp on, air (excluding compressor)		each	0.77	hour
Vibrator tamper type including single timber or metal screed board	screed length 3.10 m	each	1.05	hour
* Extra for additional single screed length	additional 0.31 m	each	0.07	hour
Vibrator tamper including double screed board	screed length 3.10 m	each	1.19	hour
* Extra for additional double screed length	additional 0.31 m	each	0.12	hour
Power float				
petrol driven		each	1.60	hour
electric driven		each	1.19	hour

CIVIL ENGINEERING INDUSTRY

	Description	Unit	Hire Rate £	Period
CONCRETE EQUIPMENT - cont'd	Blade diameter			
Concrete saw (exclusive of blades	mm			
and water supply to be charged in	356	each	1.76	hour
addition), manually propelled	kW 457	each	2.32	hour
As above, self-propelled	13 457	each	4.28	hour
As above, self-propelled, with				
hydraulic system for blade movement	17 457	each	5.58	hour
	31 457	each	6.76	hour
Joint former hand propelled	Blade width			
with vibratory blade	Up to			
	2.4 m	each	4.66	hour
	8.2 m	each	6.63	hour
Grinder pedestrian operated		each	2.19	hour
	Struck capacity litre			
* Concrete pram dobbin				
rubber tyred tipping	100	each	1.12	day
	140	each	1.24	day
	200	each	1.69	day
	280	each	2.18	day
Power driven barrow	170	each	1.08	hour

CIVIL ENGINEERING INDUSTRY

	Description	Unit	Hire Rate £	Period
CRANES				
	Maximum working load in accordance with BS 1757 (1964) Clause II - Stability			
Mobile rubber tyred (site use only)	tonnes			
Full circle slew	6.50	each	15.47	hour
	8.50	each	16.03	hour
	12.50	each	20.83	hour
	16.00	each	24.42	hour
	19.00	each	32.55	hour
Mobile rubber tyred rough terrain type (site use only excluding Tax, Insurance and extra cost of DERV)	13.00	each	24.85	hour
	16.50	each	30.19	hour
	18.00	each	37.59	hour
	20.00	each	41.49	hour
	23.00	each	49.72	hour
	40.00	each	68.40	hour
	55.00	each	82.55	hour
Excavator crane tracked (site use only excluding Tax, Insurance and extra cost of DERV) Full circle slew	Up to 18.00	each	18.93	hour
	28.00	each	23.54	hour
	36.00	each	26.99	hour
	56.00	each	42.41	hour
	75.00	each	66.15	hour
	85.00	each	79.25	hour
	115.00	each	86.19	hour
	150.00	each	97.00	hour
	152.40	each	98.99	hour
Lorry mounted crane (site and public highway use including Tax, Insurance and extra cost of DERV)				
Full circle slew	Up to 12.00	each	21.98	hour
	16.50	each	32.14	hour
	20.00	each	38.15	hour
	25.50	each	47.96	hour
	30.00	each	53.33	hour
	45.00	each	67.65	hour
	60.00	each	88.91	hour
	90.00	each	131.27	hour
	110.00	each	152.21	hour
	135.00	each	178.39	hour

CIVIL ENGINEERING INDUSTRY

	Description	Unit	Hire Rate £	Period

CRANES - cont'd

Tower crane - electric - (complete with ballast and/or kentledge). Standard trolley or luffing jib operating on straight/curved rail track. Capacity maximum lift in tonnes x maximum radius in metres at which it can be lifted

(height under hook above ground for trolley jib cranes)

(jib pivot height above ground for luffing jib cranes)

Tonnes; up to

	m	to	m			
10	17.1		22.0	each	6.85	hour
15	16.5		17.4	each	7.85	hour
20	18.3		20.1	each	11.63	hour
25	20.1		22.6	each	12.92	hour
31	22.0		24.7	each	14.50	hour
40	22.0		22.6	each	17.41	hour
50	22.0		22.6	each	21.07	hour
61	22.0		22.6	each	23.11	hour
70	22.0		22.6	each	25.93	hour
81	22.0		22.6	each	27.29	hour
110	22.0		22.6	each	33.50	hour
130	21.4		22.6	each	36.79	hour
150	21.4		22.6	each	39.79	hour
170	21.4		22.6	each	40.72	hour
180	21.4		23.8	each	41.88	hour
210	21.4		23.8	each	45.09	hour
220	21.4		23.8	each	46.34	hour
255	21.4		23.8	each	48.25	hour

Static tower crane to be charged at the following percentage of the above rates — each 84% hour

Climbing or extendable mounted base crane to be charged at the following percentage of the above rates — each 77% hour

* Extra for extended height of tower (incl. accessories) for trolley jib crane

m/t; up to

60	per 1 metre of height	each	0.30	hour
130	per 1 metre of height	each	0.40	hour
170	per 1 metre of height	each	0.48	hour
255	per 1 metre of height	each	0.58	hour

* Extra for extended height of tower (incl. accessories) for luffing crane

m/t; up to

30	per 1 metre of height	each	0.42	hour
50	per 1 metre of height	each	0.46	hour
70	per 1 metre of height	each	0.50	hour

* Tower crane rail track, timbers and fastenings etc.	for tower cranes	per 1 m		
	up to 61 m tonnes	track	0.55	day
	62 to 170	track	0.64	day
	171 upwards	track	0.77	day

CIVIL ENGINEERING INDUSTRY

Description	Unit	Hire Rate £	Period
CRANE EQUIPMENT			
Struck capacity up to and including			
* Crane grab, all types, excavating (normal weight) cu m			
0.55	each	2.29	hour
0.80	each	2.87	hour
1.25	each	3.74	hour
* Crane grab, all types, excavating (heavy weight)			
0.55	each	3.85	hour
0.80	each	4.90	hour
1.25	each	5.61	hour
* Crane grab, all types, rehandling (normal weight)			
0.55	each	1.65	hour
0.80	each	2.02	hour
1.25	each	2.77	hour
* Crane grab, all types, rehandling (heavy duty)			
0.55	each	2.94	hour
0.80	each	3.69	hour
1.25	each	4.93	hour
Up to and including cu m			
* Skip, muck tipping circular			
0.08	each	0.09	hour
0.15	each	0.10	hour
0.20	each	0.11	hour
0.30	each	0.13	hour
0.40	each	0.14	hour
0.60	each	0.16	hour
0.80	each	0.22	hour
1.20	each	0.28	hour
1.60	each	0.34	hour
* Skip, concrete tipping			
0.15	each	0.14	hour
0.20	each	0.15	hour
0.40	each	0.18	hour
0.60	each	0.26	hour
0.80	each	0.27	hour
1.00	each	0.43	hour
* Skip concrete lay down or roll over standard front or bottom discharge			
0.20	each	0.30	hour
0.30	each	0.36	hour
0.40	each	0.40	hour
0.60	each	0.53	hour
0.80	each	0.65	hour
1.20	each	1.00	hour
* Skip concrete roll over geared or hydraulic hand operated clamshell			
0.40	each	0.74	hour
0.60	each	0.91	hour
0.80	each	0.95	hour
1.20	each	1.50	hour
1.60	each	1.80	hour

Description	Chain diameter mm	Safe working load tonnes	Unit	Hire Rate £	Period
* Chain slings or brothers					
1.83 m EWL					
* single	6	1.27	each	0.26	day
double at 90 degrees	6	1.78	each	0.36	day
* single	10	2.19	each	0.31	day
double at 90 degrees	10	3.24	each	0.46	day
* single	13	5.08	each	0.41	day
double at 90 degrees	13	7.36	each	0.65	day
* single	16	7.27	each	0.57	day
double at 90 degrees	16	11.18	each	0.93	day
* single	19	11.68	each	0.78	day
double at 90 degrees	19	15.42	each	1.16	day
* single	22	15.85	each	1.06	day
double at 90 degrees	22	21.46	each	1.91	day

CIVIL ENGINEERING INDUSTRY

Description	Unit	Hire Rate £	Period

CRANE EQUIPMENT - cont'd

cu m
Up to

* Grab tag lines	0.60	each	4.70	day
	1.20	each	5.47	day
	Over 1.20	each	7.70	day

Dead weight up to
and including

kg

* Demolition ball	260	each	1.35	day
	510	each	1.67	day
	770	each	2.06	day
	1020	each	2.48	day
	1300	each	2.88	day
	1550	each	3.47	day
	1800	each	3.82	day
	2050	each	4.18	day

CIVIL ENGINEERING INDUSTRY

	Description	Unit	Hire Rate £	Period
DERRICKS, SCOTCH	tonnes			
* Hand, jib up to 18.3 m	1.60	each	1.39	hour
	2.50	each	1.48	hour
	3.50	each	2.07	hour
* Hand, jib 18.6 m to 21.4 m	3.50	each	2.26	hour
* Extra for hand slew gear		each	0.48	hour
* Hand, jib up to 18.3 m	6.00	each	2.67	hour
* Hand, jib 18.6 m to 21.4 m	6.00	each	2.99	hour
* Extra for hand slew gear		each	0.63	hour
* Extra for 3 tonne bogies		set of 3	0.72	hour
* Extra for 5 tonne bogies		set of 3	0.84	hour
All electric (single motor)				
jib 30.5 m	3.50	each	6.18	hour
jib 30.8 m to 36.6 m	3.50	each	6.85	hour
* Extra for plain bogies				
heavy duty 1.44 m gauge		set of 3	1.13	hour
Extra per additional motor		each	1.91	hour
All electric (single motor)				
jib up to 24.4 m	6.00	each	7.05	hour
jib 24.7 m to 30.5 m	6.00	each	7.61	hour
jib 30.8 m to 36.6 m	6.00	each	7.86	hour
* Extra for plain bogies				
heavy duty 1.44 m gauge		set of 3	1.85	hour
Extra per additional motor		each	2.47	hour
All electric (three motor)				
jib 30.8 m to 36.6 m	11.00	each	26.10	hour
jib 36.9 m to 45.8 m	11.00	each	28.52	hour
* Extra for plain bogies				
heavy duty 1.44 m gauge		set of 3	4.24	hour
Extra per additional motor		each	2.80	hour
All electric (three motor)				
jib 30.8 m to 36.6 m	16.00	each	30.41	hour
jib 36.9 m to 45.8 m	16.00	each	33.11	hour
* Extra for plain bogies				
heavy duty 1.44 m gauge		set of 3	5.29	hour
Extra per additional motor		each	3.03	hour
All electric (three motor)				
jib 30.8 m to 36.6 m	21.00	each	33.81	hour
jib 36.9 m to 45.8 m	21.00	each	36.23	hour
* Extra for plain bogies				
heavy duty 1.44 m gauge		set of 3	5.77	hour
Extra per additional motor		each	3.29	hour
All electric (three motor)				
jib 30.8 m to 36.6 m	31.00	each	43.89	hour
jib 36.9 m to 45.8 m	31.00	each	48.34	hour
* Extra for plain bogies				
heavy duty 1.44 m gauge		set of 3	9.60	hour
Extra for additional motor		each	3.34	hour
* Kentledge weights		per tonne	1.30	week

DIVING GEAR   Rates to be negotiated

CIVIL ENGINEERING INDUSTRY

Description	Unit	Hire Rate £	Period

DUMPERS (site use only excluding Tax, Insurance and extra cost of DERV)

Small dumper

Maker's rated payload

		Unit	Hire Rate £	Period
(manual gravity tipping)	kg; Up to			
2 wheel drive	900	each	2.02	hour
	970	each	2.32	hour
	1040	each	2.65	hour
	1100	each	2.93	hour
	1400	each	3.36	hour
	1750	each	3.51	hour
	1900	each	3.83	hour
	2000	each	4.09	hour
4 wheel drive	1100	each	3.70	hour
	1400	each	4.29	hour
	1750	each	4.45	hour
	1900	each	4.60	hour
	2400	each	5.86	hour
(hydraulic tipping)				
2 wheel drive	500	each	2.19	hour
	970	each	2.53	hour
	1040	each	2.83	hour
	1100	each	3.29	hour
	1400	each	3.85	hour
	1750	each	4.04	hour
	1900	each	4.27	hour
	2000	each	4.55	hour
4 wheel drive	1400	each	4.50	hour
	1750	each	4.90	hour
	1900	each	5.09	hour
	2000	each	5.50	hour
	2100	each	5.87	hour
	2250	each	5.92	hour
	2375	each	6.78	hour
	3000	each	7.87	hour
	3125	each	8.12	hour
	3250	each	8.65	hour
	6500	each	12.63	hour
(high discharge)				
2 wheel drive	1100	each	3.98	hour
	1400	each	4.38	hour
	1750	each	4.82	hour
	1900	each	4.96	hour
	2000	each	5.10	hour
	2250	each	5.43	hour
4 wheel drive	1100	each	4.37	hour
	1400	each	5.38	hour
	1900	each	5.86	hour
	2100	each	6.56	hour
	2250	each	6.74	hour
	2375	each	7.40	hour
(turntable side tipping)				
2 wheel drive	1040	each	3.96	hour
	1100	each	4.46	hour
	1400	each	4.92	hour
	1750	each	5.55	hour
	1900	each	6.23	hour
4 wheel drive	1400	each	5.68	hour
	1750	each	6.15	hour
	1900	each	6.91	hour
	2100	each	7.64	hour
	2375	each	8.56	hour

CIVIL ENGINEERING INDUSTRY

	Description	Unit	Hire Rate £	Period
	Maker's rated payload kg			
Dump truck - rear dump	15 500	each	24.54	hour
	17 000	each	30.82	hour
	22 000	each	36.94	hour
	25 000	each	42.43	hour
	28 000	each	46.29	hour
	32 000	each	62.07	hour
	38 000	each	68.61	hour
	45 000	each	76.39	hour
	50 000	each	84.26	hour
Dump truck - articulated or trailer type	12 200	each	22.75	hour
	18 500	each	30.45	hour
	23 000	each	38.89	hour
	30 000	each	48.51	hour
	31 800	each	55.10	hour

Keep your figures up to date, free of charge

This section, and most of the other information in this Price Book, is brought up to date every three months in the *Price Book Update*.

The *Update* is available free to all Price Book purchasers.

To ensure you receive your copy, simply complete the reply card from the centre of the book and return it to us.

CIVIL ENGINEERING INDUSTRY

Description	Unit	Hire Rate £	Period

EXCAVATORS

	Maker's rated dragline capacity cu m			
Rope operated, full circle slew, crawler mounted, with single equipment (dragline)	0.07	each	21.91	hour
	0.80	each	24.77	hour
	1.00	each	25.40	hour
	1.20	each	27.92	hour
	1.40	each	36.04	hour
	1.50	each	39.67	hour
	2.00	each	49.59	hour
	2.30	each	58.67	hour
	2.70	each	69.38	hour
	Maker's rated face shovel capacity cu m			
Rope operated, full circle slew, crawler mounted, with single equipment (face shovel)	0.70	each	25.57	hour
	1.00	each	34.92	hour
	1.40	each	42.44	hour
	1.50	each	51.23	hour
	2.50	each	76.84	hour
	Maker's rated nominal weight of machine			
Hydraulic, full circle slew, crawler or wheel mounted, with single equipment (backacter)	Up to 2.0 tonnes	each	5.21	hour
	2.1 to 3.0	each	7.80	hour
	3.1 to 4.0	each	8.24	hour
	4.1 to 6.0	each	10.63	hour
	6.1 to 11.0	each	14.83	hour
	11.1 to 14.0	each	23.52	hour
	14.1 to 17.0	each	28.33	hour
	17.1 to 21.0	each	33.94	hour
	21.1 to 25.0	each	45.46	hour
	25.1 to 30.0	each	46.53	hour
	30.1 to 38.0	each	56.78	hour
	38.1 to 55.0	each	84.88	hour
As above (shovel)	Up to 17.0 tonnes	each	27.84	hour
	17.1 to 21.0	each	35.22	hour
	21.5 to 25.0	each	46.78	hour
	25.1 to 30.0	each	50.37	hour
	30.1 to 38.0	each	62.35	hour
	38.1 to 55.0	each	70.43	hour
	55.1 to 75.0	each	80.84	hour
	Unit weight less cradle kg			
Extra for machine mounted percussion breaker	Up to 500	each	6.17	hour
	501 to 1000	each	9.94	hour
	1001 to 1500	each	18.57	hour
	Maker's rated loader bucket capacity cu m			
Hydraulic, offset or centre post, half circle slew wheeled, dual purpose (back hoe/loader)	0.60	each	9.95	hour
	0.80	each	11.15	hour
	1.00	each	11.94	hour
	1.30	each	13.88	hour
Excavator mats	sq m			
* Light, thickness 150 mm	1.0	each	0.31	hour
* Heavy, thickness 250 mm	1.0	each	0.43	hour

CIVIL ENGINEERING INDUSTRY

	Description	Unit	Hire Rate £	Period
GENERATING SETS AND TRANSFORMERS	kVA			
Generating set	1	each	1.30	hour
	1.5	each	1.71	hour
	2	each	2.02	hour
	3	each	2.20	hour
	4	each	2.35	hour
	5	each	2.50	hour
	6.25	each	2.82	hour
	10	each	3.83	hour
	12.5	each	4.85	hour
	16.5	each	5.54	hour
	20	each	5.80	hour
	23.5	each	6.37	hour
	31	each	7.08	hour
	40	each	8.37	hour
	50	each	9.53	hour
	70	each	12.76	hour
	80	each	13.81	hour
	90	each	15.41	hour
	100	each	16.25	hour
	120	each	18.87	hour
	156	each	21.39	hour
	210	each	25.46	hour
	280	each	29.65	hour
	330	each	37.16	hour
* Transformer, stationary (air cooled)	1	each	0.14	hour
	2.5	each	0.25	hour
	5	each	0.50	hour
	7.5	each	0.60	hour
	10	each	0.71	hour
	12.5	each	0.92	hour
	15	each	1.00	hour
(oil immersed)	35	each	1.24	hour
	50	each	1.50	hour
	60	each	1.58	hour
	100	each	2.18	hour
	200	each	2.72	hour
	300	each	3.50	hour
	500	each	4.50	hour
Mobile lighting unit, 2 light tungsten halogen incl. 1.5 kVA generator	Tower or mast height m Up to 8.0	each	4.53	hour
As above, 4 light mercury vapour incl. 6.25 kVA generator	16.0	each	7.30	hour
As above, 4 light tungsten halogen incl. 7.5 kVA generator	20.0	each	11.21	hour

CIVIL ENGINEERING INDUSTRY

Description	Unit	Hire Rate £	Period
HAULAGE			
* Trailer flat (towed with knock on brakes) site use only excluding Tax and Insurance	carrying capacity Up to and including tonnes		
2.0	each	5.39	day
2.1 to 3.0	each	8.11	day
3.1 to 4.0	each	8.87	day
	Maximum lifting capacity (forward loading) kg Up to and including		
Fork lift truck (yard type) 1000	each	4.38	hour
1001 to 1250	each	4.48	hour
1251 to 1500	each	5.05	hour
1501 to 2000	each	6.27	hour
2001 to 2500	each	7.21	hour
Fork lift truck (rough terrain type) (2 wheel drive)	Up to and including		
1000	each	5.58	hour
1001 to 1250	each	6.50	hour
1251 to 1500	each	7.07	hour
1501 to 2000	each	8.05	hour
2001 to 2500	each	8.82	hour
2501 to 3000	each	10.16	hour
Fork lift truck (rough terrain type) (4 wheel drive)	Up to and including		
1500	each	7.12	hour
1501 to 2000	each	8.16	hour
2001 to 2500	each	9.83	hour
2501 to 3000	each	10.87	hour
3001 to 3500	each	12.37	hour
3501 to 4000	each	15.67	hour
4001 to 5000	each	19.33	hour
5001 to 6000	each	23.14	hour
6001 to 7000	each	25.36	hour
7001 to 8000	each	32.64	hour
Telescopic Site Handling Machines (2 wheel drive)	Up to and including		
2000	each	8.81	hour
2001 to 4000	each	9.57	hour
Telescopic Site Handling Machines (4 wheel drive)	Up to and including		
2000	each	11.65	hour
2001 to 4000	each	12.35	hour

	Belt width mm	Length m; Up to	Unit	Hire Rate £	Period
Conveyor, mobile (including loading hopper)	400	3.50	each	2.70	hour
	400	5.50	each	2.76	hour
	400	7.50	each	3.06	hour
	400	9.00	each	3.53	hour
	500	6.00	each	3.03	hour
	500	9.00	each	4.11	hour
	500	12.00	each	4.59	hour
	600	9.00	each	5.01	hour
	600	12.00	each	7.21	hour
	600	18.00	each	11.72	hour
Extra for diesel engine			each	25% on above rate hour	
Extra for elevating type			each	25% on above rate hour	

CIVIL ENGINEERING INDUSTRY

	Description	Hire Unit £	Rate	Period

HOISTS

Cantilever platform and centre slung
scaffold tower type (complete with
safety device and overwind limits):
Goods including winch, cage,
platform, mast and guide rails

m; up to	kg			
16.5	508	each	2.55	hour
16.5	672	each	3.31	hour
15.6	1016	each	4.00	hour
15.6	1524	each	5.38	hour
15.6	2032	each	6.68	hour
* Extra for additional mast sections				
1 m	508	each	0.038	hour
	762	each	0.044	hour
	1016	each	0.050	hour
	1524	each	0.065	hour
	2032	each	0.110	hour
* Extra for additional tubular side guides				
1 m	762	each	0.03	hour
	1016	each	0.04	hour
	1524	each	0.07	hour
	2032	each	0.14	hour
Goods (rack and pinion) incl. cage tower and guides top landing height - 50 m	762	each	4.02	hour
* Extra for additional tower section	1.00 m	each	0.04	hour
* Extra for landing locking safety gate		each	0.06	hour
* Extra for bottom gate and side screens		each	0.12	hour
* Extra for landing electric locking safety gate		each	0.16	hour
* Extra for bottom electric locking safety gate and side screens		each	0.28	hour
Passenger (rack and pinion) incl. cage tower and guides, 2 motor electric top landing height	single cage			
60 m	12 men	each	10.02	hour
60 m	20 men	each	11.29	hour
Passenger (rack and pinion) including cages, tower and guides, 2 motor electric top landing height	twin cage			
60 m	2 x 12 men	each	15.53	hour
60 m	2 x 20 men	each	17.02	hour

CIVIL ENGINEERING INDUSTRY

Description	Hire Unit	Rate £	Period

HOISTS - cont'd

Description		Hire Unit	Rate £	Period
* Extra for additional tower section	per m single cage	each	0.12	hour
*	per m twin cage	each	0.14	hour
* Extra for landing gate (electrically and mechanically interlocking)	single cage	each	0.50	hour
*	twin cage	each	1.00	hour
* Mobile hoist with 1 length of mast landing height	kg			
6.4 m	305	each	2.44	hour
7.3 m	508	each	3.70	hour
* Extra for additional mast sections				
1 m	305	each	0.03	hour
* 1 m	508	each	0.04	hour
* Extra for locking safety gate		each	0.05	hour
* Extra for 3 side floorgate enclosure		each	0.13	hour
Scaffold hoist	254	each	1.00	hour

LIFTING AND JACKING GEAR

Description	tonnes; up to	Hire Unit	Rate £	Period
* Shear legs, steel, 5 metres	1	each	1.53	day
* Pipe gantry	3	each	4.40	day
* Chain blocks	1	each	2.74	day
	2	each	3.39	day
	3.5	each	3.64	day
	4.5	each	3.85	day
	5.5	each	4.51	day
	6.5	each	4.98	day
	8	each	6.64	day
	10	each	8.45	day
	15	each	16.13	day
	20	each	18.60	day
* Screw jack	Up to 5.00	each	0.37	day
* Hydraulic jack (hand operated)	Up to 6.50	each	0.72	day
	10.50	each	1.08	day
	16.00	each	1.40	day
	21.00	each	1.69	day
	30.00	each	1.98	day
	35.00	each	2.40	day
	55.00	each	4.48	day
	105.00	each	5.42	day
* Ratchet jack	5.50	each	1.44	day
	10.50	each	2.13	day
	16.00	each	2.38	day
	21.00	each	2.80	day

Description	Safe working load		Hire Unit	Rate £	Period
	Lifting	Pulling			
* Lifting and pulling machine	800 kg	1200 kg	each	1.10	day
	1600 kg	2500 kg	each	1.49	day
	3000 kg	5000 kg	each	2.27	day

CIVIL ENGINEERING INDUSTRY

	Description	Unit	Rate	Period
		Hire £		

LOCOS AND RAILWAY EQUIPMENT

	Description	Unit	Rate	Period
Loco - 0.61 m gauge electric including battery	Draw bar pull 150 kg normal	each	3.05	hour
* 2nd battery		each	1.42	hour
* Battery charger for above		each	0.36	hour
Loco - 0.61 m gauge electric including battery	Draw bar pull 273 kg normal	each	3.99	hour
* 2nd battery		each	1.68	hour
* Battery charger for above		each	0.71	hour
Loco - 0.61 m gauge electric including battery	Draw bar pull 318 kg normal	each	4.36	hour
* 2nd battery		each	1.77	hour
* Battery charger for above		each	0.84	hour
Loco - 0.61 m gauge electric including battery	Draw bar pull 454 kg normal	each	5.94	hour
* 2nd battery		each	1.98	hour
* Battery charger for above		each	1.01	hour
Loco - 0.61 m gauge electric including battery	Draw bar pull 907 kg normal	each	10.13	hour
* 2nd battery		each	4.80	hour
* Battery charger for above		each	1.72	hour
Loco - 0.61 m gauge diesel	16 kW, weight 2.29 tonnes	each	3.83	hour
	22 kW, weight 3.50 tonnes	each	5.92	hour
	kg/m rail	per		
* Track - 0.61 m	7.4	1.0 m	0.05	day
	12.4	1.0 m	0.06	day
	17.4	1.0 m	0.065	day
* Track - 1.44 m gauge	29.6	1.0 m	0.143	day
	52.2	1.0 m	0.173	day
* Extra over track for points or turnout	Up to			
0.61 m gauge	17.4	each	1.54	day
* 1.44 m gauge	29.6	each	7.39	day
	52.2	each	8.21	day
* Jim Crow	17.4	each	0.92	day
	52.2	each	1.26	day
	Capacity cu m			
* Skip side tipping 'U' shaped	0.43	each	1.65	day
	0.57	each	1.82	day
	0.85	each	2.24	day

CIVIL ENGINEERING INDUSTRY

Description	Unit	Hire Rate £	Period
LORRIES, VANS, ETC.			

Lorry, ordinary (site use and public highway use - including Tax, Insurance and extra cost of petrol or DERV)

Plated gross vehicle weight up to and including tonnes			
6.50	each	7.11	hour
6.6 to 7.5	each	8.22	hour
7.6 to 12.0	each	9.24	hour
12.1 to 14.0	each	11.30	hour
14.1 to 17.0	each	13.59	hour
17.1 to 24.5	each	19.09	hour
24.6 to 30.0	each	25.01	hour
30.1 to 33.0	each	25.27	hour
33.1 to 35.0	each	26.04	hour
35.1 to 39.0	each	30.22	hour

Extra for lorry fitted with crane attachment (whether or not used)

Maximum load tonnes up to and including			
Up to 1.5	each	0.64	hour
1.60 to 2.50	each	0.94	hour
2.60 to 3.5	each	1.17	hour

Lorry, tipper (site use and public highway use - including Tax, Insurance and extra cost of petrol or DERV)

Plated gross vehicle weight up to and including tonnes			
11.00	each	10.07	hour
11.1 to 17.0	each	15.85	hour
17.1 to 25.0	each	20.94	hour
25.1 to 31.0	each	27.88	hour

Extra for side tipping

	each	20% on above rates	hour

Van, pick up or similar utility vehicle, (site use and public highway use - including Tax, Insurance and extra cost of petrol or DERV)

Carrying capacity up to and including tonnes			
0.60	each	5.17	hour
0.70 to 1.10	each	5.95	hour
1.20 to 1.30	each	6.42	hour
1.40 to 1.60	each	7.41	hour
1.70 to 2.10	each	8.01	hour
2.20 to 2.60	each	8.53	hour

Passenger/goods, cross country, (as above)

Wheelbase up to			
2.30 m	each	6.83	hour

Station wagon (as above)

Wheelbase 2.40 m and over			
	each	7.63	hour
7 seater	each	7.73	hour
10 seater	each	8.01	hour
12 seater	each	8.05	hour

Personnel carrier/coach/bus, (as above)

9/13 seater	each	7.30	hour
14/17 seater	each	9.08	hour

Road sweeper/cleaner self-propelled, (as above)

Hopper capacity cu m			
1.00	each	7.29	hour
1.50	each	8.75	hour
2.00	each	9.86	hour

MONORAIL   Rates to be negotiated

CIVIL ENGINEERING INDUSTRY

Description	Unit	Hire Rate £	Period

OFFICES, STORES, SHEDS, ETC.

	Description	Unit	Hire Rate £	Period
* Offices on site with usual fittings, i.e. heating equipment, desk, tables, chairs, stools, plan chest, etc. (including heating and lighting)	Timber sectional (excluding insurance)	per 10 sq m floor area	15.00	week
* Messroom on site with usual fittings, heating equipment, counter, tables, forms, etc. (excluding all kitchen equipment) (including heating and lighting)	Timber sectional (excluding insurance)	per 10 sq m floor area	13.88	week
* Stores on site with usual fittings, i.e. counter, desk, chair, racks, shelves, bins (including heating and lighting)	Timber sectional (excluding insurance)	per 10 sq m floor area	13.14	week
* Mobile office (caravan type) (excluding Tax and Insurance) (including heating and lighting)	Up to 4.57 m	each	34.14	week
* Mobile office (caravan type) (excluding Tax and Insurance) (including heating and lighting)	Up to 7.62 m	each	44.50	week
* Watchman's hut		each	3.00	week
* Latrine (including consumables)	single	each	10.31	week
* Men's shelter, including tarpaulins		each	6.75	week
* Proprietary Prefabricated Units	Rates for special types of prefabricated and mobile buildings owned by main contractors to be based on those charged by proprietary firms, plus 12.5%			

PAINT SPRAYING MACHINES

	Description	Unit	Hire Rate £	Period
* Paint spraying machine, 1 gun type with 7.6 m lengths of air and fluid hose and 9.0 litre pressure tank.	(Excluding compressor)	each	0.53	hour
As above, power driven	(Electric or petrol)	each	2.56	hour
* Paint spraying machine, 2 gun type with 15.3 m lengths of air and fluid hose and 22.7 litre pressure tank.	(Excluding compressor)	each	1.08	hour
As above, power driven	(electric, petrol or diesel)	each	3.60	hour

CIVIL ENGINEERING INDUSTRY

	Description	Unit	Hire Rate £	Period
PILING PLANT				
(excluding boiler or compressor)	Hammer weight			
* Piling hammer, double-acting	kg			
(steam or air)	155	each	1.78	hour
	305	each	2.33	hour
	1143	each	3.32	hour
	2132	each	4.77	hour
	3006	each	5.66	hour
	4334	each	11.24	hour
	6350	each	20.88	hour
	Piston weight			
* Single acting hammer (suitable	kg			
for use with channel leaders)	2500	each	5.88	hour
	3000	each	6.70	hour
	4000	each	8.05	hour
	5000	each	9.87	hour
	6000	each	11.65	hour
* Drop hammer (bare) (suitable for use with channel or tubular leaders fitted with leather guides and rubber inserts)	Hammer weight kg			
	1016	each	0.85	hour
	2032	each	1.28	hour
	3048	each	1.79	hour
* Internal drop hammer (for use with cased piles): Internal diameter of cased pile mm	Hammer weight kg			
305	1270	each	0.85	hour
406	3556	each	1.79	hour
508	5588	each	2.30	hour
* Air operated extractor	Unit weight			
(excl. compressor)	kg			
	764	each	4.42	hour
	1705	each	7.83	hour
	3000	each	12.40	hour
	4590	each	13.50	hour
* Flexible reinforced rubber hose	Diameter			
(for compressed air)	mm			
m;				
10	25	each	2.40	day
20	25	each	4.20	day
10	32	each	2.89	day
20	32	each	5.13	day
10	38	each	3.32	day
20	38	each	5.85	day
10	51	each	4.70	day
20	51	each	7.50	day
10	64	each	6.03	day
20	64	each	10.16	day

N.B. Flexible armoured steam hose, 2.5
times above rates.

CIVIL ENGINEERING INDUSTRY

Description	Unit	Hire Rate £	Period

Diesel hammer (including complete set of guiding equipment) (single acting)

Piston weight kg			
500	each	10.90	hour
800	each	12.83	hour
1300	each	14.86	hour
1500	each	15.96	hour
2200	each	20.72	hour
2500	each	21.63	hour
3000	each	24.30	hour
3500	each	34.83	hour
4600	each	42.42	hour

Diesel hammer (including complete set of guiding equipment) (double acting)

800	each	15.90	hour
1850	each	25.28	hour
2500	each	31.97	hour
3500	each	44.59	hour
4500	each	53.03	hour

Hydraulic vibrating hammer/extractor

Centrifugal force; tonnes	Max. pulling force; tonnes			
21	20	each	35.46	hour
38	27	each	48.25	hour
62	40	each	71.44	hour
143	40	each	131.55	hour
207	120	each	211.44	hour

* Pie helmet for pile

203 x 203 mm	each	0.50	hour
254 x 254 mm	each	0.68	hour
305 x 305 mm	each	0.95	hour
356 x 356 mm	each	1.00	hour
406 x 406 mm	each	1.14	hour
457 x 457 mm	each	1.20	hour
508 x 508 mm	each	1.27	hour

(Plastic Dollies or equivalent to be paid for in addition)

* Hanging leaders channel type (for use with drop hammer)

Length of jib m			
12.2	each	4.18	hour
15.3	each	4.77	hour

* Hanging leaders, rectangular section 0.61 x 0.61 m (for use with drop or diesel hammers)

Crane boom m	Nominal length m			
9.2	15.3	each	7.44	hour
15.3	24.4	each	8.59	hour

* Hanging leaders, rectangular section 0.84 x 0.84 m

m	m			
12.2	21.4	each	10.72	hour
18.3	29.9	each	13.05	hour
24.4	36.6	each	14.61	hour

CIVIL ENGINEERING INDUSTRY

	Description	Unit	Hire Rate £	Period

PILING

Temporary steel piling and steel
trench sheeting to be charged as a
material.   The residual value of
recovered steel piling and steel
trench sheeting to be the subject
of special agreement.

PIPE BENDING EQUIPMENT

	tonnes			
* Pipe winch	0.51	each	0.35	hour
	1.02	each	0.39	hour
	2.03	each	0.54	hour
	3.05	each	0.75	hour
* Pipe bending machine (hand	mm; up to			
operated) single stage	51	each	0.19	hour
* two stage	51	each	0.25	hour
* two stage	76	each	0.36	hour
* two stage	152	each	0.74	hour
Pipe bending machine (power operated)				
* two stage	51	each	1.23	hour
* two stage	76	each	1.92	hour
* two stage	102	each	2.91	hour
* two stage	152	each	3.40	hour

			Hire rate per day £	Extra per working hour £
PUMPS, PORTABLE				
(exclusive of all hoses)	mm			
* Semi rotary (hand)	19	each	0.14	
	25	each	0.25	
Single diaphragm	51	each	3.95	0.63
	76	each	4.64	0.67
	102	each	5.68	0.86
Double diaphragm	51	each	4.80	0.64
	76	each	5.78	0.74
	102	each	6.51	0.96
Self priming centrifugal	38	each	1.43	0.47
	51	each	5.29	0.97
	76	each	7.00	1.28
	102	each	9.00	1.64
	152	each	11.93	2.13
Sludge and sewage	76	each	10.27	1.48
	102	each	14.71	2.01
	152	each	23.04	3.22
	204	each	41.80	5.72
* Sump pneumatic (excluding compressor)	51	each	2.82	0.25
	64	each	3.22	0.28
	76	each	3.58	0.34
Electric submersible	38	each	1.51	0.52
	51	each	2.09	0.61
	64	each	3.01	0.82
	76	each	3.72	1.19
	102	each	4.58	1.61

CIVIL ENGINEERING INDUSTRY

Description	Unit	Hire Rate £	Period

PUMPING EQUIPMENT

Pump hoses, flexible, suction or
delivery, including couplings,
valve and strainer

	Description	Unit	Hire Rate £	Period
	Diameter mm			
* suction	38	Per 1.00 m length	0.16	day
	51	Per 1.00 m length	0.17	day
	76	Per 1.00 m length	0.23	day
	102	Per 1.00 m length	0.40	day
	152	Per 1.00 m length	0.67	day
	204	Per 1.00 m length	1.14	day
* delivery	38	Per 1.00 m length	0.11	day
	51	Per 1.00 m length	0.12	day
	76	Per 1.00 m length	0.20	day
	102	Per 1.00 m length	0.30	day
	152	Per 1.00 m length	0.60	day
	204	Per 1.00 m length	0.96	day

Description	Unit	Hire Rate £	Period
* Additional lengths of hose	above sizes	pro rata	day
* Steel pipe suction or delivery, including flanges, bolts and joint rings (excl. valve and strainer)	mm; Up to		
76	Per 1.00 m	0.016	day
152	Per 1.00 m	0.035	day
204	Per 1.00 m	0.060	day

* Bend to be charged as 9.0 m
length
* Valve to be charged as 9.0 m
length

Description	Unit	Hire Rate £	Period
* Steel pipe suction or delivery screwed and socketed joints (excl. valve and strainer)	mm; Up to		
51	Per 1.00 m	0.010	day
102	Per 1.00 m	0.029	day
152	Per 1.00 m	0.050	day

* Bend to be charged as 3.50 m
length
* Valve to be charged as 3.50 m
length

CIVIL ENGINEERING INDUSTRY

	Description	Unit	Hire Rate £	Period
RAMMERS AND COMPACTORS				
	Weight up to kg			
Vibro compactor or vibration rammer	60	each	1.01	hour
	70	each	1.09	hour
	78	each	1.23	hour
	114	each	1.60	hour
	200	each	1.65	hour
Vibrating plate compactor	78	each	1.04	hour
	90	each	1.20	hour
	120	each	1.71	hour
	135	each	1.80	hour
	176	each	2.00	hour
	220	each	2.26	hour
	310	each	2.96	hour
	470	each	3.96	hour
	494	each	4.36	hour
	520	each	4.76	hour
Trench compactor	265	each	2.40	hour
Jumping rammer including trolley	100	each	1.36	hour

CIVIL ENGINEERING INDUSTRY

Description	Unit	Hire Rate £	Period

ROLLERS

* Hand		each	0.43	day

Road deadweight (steel 3 wheel/3 roll) diesel

Unballasted weight up to and including tonnes

2.60	each	4.76	hour
4.10	each	6.10	hour
6.10	each	8.62	hour
8.50	each	9.73	hour
10.50	each	10.35	hour
13.00	each	11.81	hour

Road deadweight (tandem) diesel	8.50 each	10.73	hour

Average weight/wheel tonnes up to and including 1.0

Rubber tyred, self-propelled	tonne/wheel	each	15.75	hour
	1.1 to 2.0	each	21.81	hour
	2.1 to 3.0	each	32.44	hour

Average weight/wheel tonnes up to and including 1.0

Rubber tyred trailer type	tonne/wheel	each	2.54	hour
	7.0	each	26.77	hour

Maker's weight kg;	Roll width mm;			
Up to	Up to			
Vibratory pedestrian operated (single roller)	550	750	each	1.98	hour
	700	900	each	2.21	hour

Vibratory pedestrian operated (twin roller)

650	650	each	3.00	hour
950	800	each	3.32	hour
1300	950	each	3.90	hour
1750	1000	each	4.20	hour

Up to and including kg / mm

Towed, vibratory trailer	1500	1400	each	3.57	hour
	6500	1950	each	6.30	hour
	8800	1950	each	8.63	hour
	11700	1950	each	12.70	hour
	13300	2100	each	16.18	hour

Up to and including tonne / mm

Self-propelled vibratory tandem (seated control)	1.1	800	each	3.55	hour
	2.0	1100	each	6.88	hour
	3.1	1200	each	10.10	hour
	4.5		each	16.43	hour
	7.0		each	18.70	hour
	8.5		each	20.29	hour
	12.0		each	25.60	hour

As above, single roll, rubber tyred driving wheels

8.2	each	23.83	hour
10.5	each	32.07	hour

* Scarifier (working time) extra over roller including sharpening tines

1 tine	each	2.98	hour
2 tine hydraulic	each	4.76	hour
3 tine hydraulic	each	6.68	hour

Compactor sheeps foot self-propelled tamping foot, heavy duty wheeled

Maker's rated flywheel kW

130	each	42.90	hour
224	each	55.25	hour
300	each	68.05	hour

CIVIL ENGINEERING INDUSTRY

Description	Unit	Hire Rate £	Period

SAWS, MECHANICAL

Guide bar
m; up to

Chain saw (1 man operated)	0.31	each	0.78	hour
	0.41	each	1.14	hour
	0.53	each	1.29	hour
	0.64	each	1.44	hour
	0.79	each	1.56	hour
	0.94	each	1.69	hour
	1.12	each	1.76	hour
Chain saw (2 man operated)	1.27	each	2.00	hour
	1.52	each	2.20	hour
	2.29	each	2.50	hour

Saw diameter
m

Portable saw bench	0.26	each	0.74	hour
	0.31	each	0.79	hour
	0.41	each	1.00	hour
	0.46	each	1.27	hour
	0.51	each	1.41	hour
	0.61	each	1.74	hour
	0.66	each	1.96	hour
	0.76	each	2.30	hour
Band saw		each	2.30	hour

SCAFFOLDING

mm dia nominal

* Tubular steel	51 m	Per metre	0.0167	week
* Tubular alloy	51 m	Per metre	0.0341	week

m

* Putlog steel	1.52 long	each	0.036	week
* Putlog alloy	1.52 long	each	0.065	week
* Putlog steel	1.83 long	each	0.047	week
* Putlog alloy	1.83 long	each	0.074	week
* Fitting steel, single, double or swivel coupler, joint pin, fixed base plate		each	0.017	week
* adjustable base plate		each	0.073	week
* hop up bracket		each	0.140	week
* split head trestle folding type/adjustable	0.48 to 0.84	each	0.54	week
	0.76 1.37	each	0.60	week
	1.07 1.83	each	0.72	week
	1.37 2.44	each	0.99	week
* Castor		each	0.41	week
* rubber tyred		each	0.50	week
* nylon tyred		each	1.03	week
* Jenny wheel including 30.5 m rope	254 mm dia	each	1.75	week
	305 mm dia	each	1.94	week
* Board	4.0 m long	each	0.26	week

Rates for special items of
prefabricated scaffold units owned
by the main contractor to be
similar to those charged by
proprietary firms, plus 12.5%.

CIVIL ENGINEERING INDUSTRY

Description	Unit	Hire Rate £	Period

SHORING, PLANKING AND STRUTTING

* Baulk timber, use and waste (excluding nails, dogs, wedges, etc., to be charged in addition as consumables)

	Per 0.01 cu m	0.25	day

(Minimum charge 18 days hire)
* Timber for planking and strutting use and waste (excluding nails, dogs, wedges, etc., to be charged in addition addition as consumables)

	Per 0.10 cu m	0.35	day

(Minimum charge 12 days hire)
* Adjustable steel strut, extending

Description	Unit	Hire Rate £	Period
457-711 mm	each	0.42	week
686 mm-1.09 m	each	0.47	week
1.04-1.70 m	each	0.58	week

CIVIL ENGINEERING INDUSTRY

Description	Unit	Hire Rate £	Period

SHUTTERING

* Steel shutters, all types.
Rates for items of steel
shuttering to be based on
those chargeable by
Proprietary Firms. (Wedges,
keys and consumables to be
added) plus 12.5%

Description	sub	Unit	Hire Rate £	Period
* Steel road forms	102 mm	Per 10 metres	0.43	day
	152 mm	Per 10 metres	0.53	day
	203 mm	Per 10 metres	0.59	day
	203 mm (heavy section)	Per 10 metres	0.79	day
	203 mm (heavy section with rail)	Per 10 metres	1.00	day
	254 mm (heavy section with rail)	Per 10 metres	1.26	day
	305 mm (heavy section with rail)	Per 10 metres	1.67	day
* Telescopic steel floor centre, extending inverted triangular plate type	1.22-1.83 m	each	0.70	week
	1.83-2.74 m	each	0.99	week
	2.44-3.66 m	each	1.33	week
	2.74-4.88 m	each	1.42	week
* Telescopic steel floor centre, extending lattice girder type	Up to 2.77 m	each	0.71	week
	4.17 m	each	0.98	week
	4.93 m	each	1.08	week
	5.56 m	each	1.28	week
* Telescopic steel prop extending	1.04-1.83 m	each	0.60	week
	1.75-3.12 m	each	0.76	week
	1.98-3.35 m	each	0.80	week
	2.44-3.96 m	each	0.86	week
* Column shutter clamps, extending	254-508 mm	per set	0.54	week
	406-813 mm	per set	0.64	week
	0.61-1.22 m	per set	0.82	week
* Beam shutter clamps, clamping width 114 mm to 850 mm	305 mm arm	per set	0.71	week
	457 mm arm	per set	0.80	week
	619 mm arm	per set	0.93	week
* Wall shutter clamps, concrete thickness	102-305 mm	per set	0.71	week
	102-610 mm	per set	0.80	week
	102-915 mm	per set	0.93	week

CIVIL ENGINEERING INDUSTRY

	Description	Unit	Hire Rate £	Period
Timber used for shuttering (excluding wedges, nails, screws, bolts, etc., to be charged in addition as consumables +12.5%)				
* Rough (minimum charge 10 days hire)		Per 0.10 cu m	0.49	day
* Wrot (minimum charge 10 days hire)		Per 0.10 cu m	0.63	day
Plywood sheeting (excluding nails, etc., to be charged in addition as consumables). Excluding timber and steel or timber supports to be charged in addition +12.5%) Douglas Fir				
* Thickness 13 mm (minimum charge 12 days hire)	Good 1 side	Per 3.00 sq m	0.454	day
* Thickness 13 mm (minimum charge per 12 days hire)	Good 2 sides	Per 3.00 sq m	0.624	day
* Thickness 19 mm (minimum charge 12 days hire)	Good 1 side	Per 3.00 sq m	0.531	day
* Thickness 19 mm (minimum charge 12 days hire)	Good 2 sides	Per 3.00 sq m	0.739	day
* Thickness 25 mm (minimum charge 12 days hire)	Good 1 side	Per 3.00 sq m	0.877	day
* Thickness 25 mm (minimum charge 12 days hire)	Good 2 sides	Per 3.00 sq m	0.986	day
Plastic faced				
* Thickness 13 mm (minimum charge 12 days hire)		Per 3.00 sq m	0.653	day
* Thickness 19 mm (minimum charge 12 days hire)		Per 3.00 sq m	0.865	day

SURVEYING INSTRUMENTS

		Unit	Hire Rate £	Period
* Dumpy level and staff		each	1.96	day
* Quickset level and staff		each	2.42	day
* Engineer's automatic and staff		each	3.18	day
* Engineer's precise parallel plate and staff		each	5.17	day
* Theodolite, vernier reading to 20 seconds		each	4.44	day
* Theodolite, microptic reading to 20 seconds		each	8.16	day
* Theodolite, microptic reading to 1 second		each	12.46	day
* Ranging rod or pole;	timber	each	0.055	day
	alloy	each	0.10	day

CIVIL ENGINEERING INDUSTRY

	Description	Unit	Hire Rate £	Period
TAR SPRAYING AND COLD EMULSION PAINT				
	litres			
* Tar boiler and sprayer (excl. firing), hand operated	600	each	0.87	hour
As above, power operated	1200	each	1.73	hour
* Gritter (attached to lorry) extra over lorry		each	0.72	hour
* Cold Emulsion Sprayer, hand operated		each	2.00	hour
As above, power operated	Up to 205	each	5.46	hour
TOOLS, PNEUMATIC				
(excluding compressor) Compressor tool (including sharpening) with up to and including 15.3 m of hose	Consumables to be charged in addition +12.5% except where otherwise stated			
* Breaker including steels		each	1.89	hour
* Light pneumatic pick including steels		each	1.83	hour
* Pneumatic clay spade including blade		each	1.81	hour
* Chipping/scaling/caulking hammer		each	10.18	hour
Hand-held rock drill without drill rods or detachable bits	kg			
* light weight	16-20	each	0.53	hour
* middle weight	20-24	each	0.65	hour
* heavy duty	25-32	each	0.74	hour
(Drill rods and bits for hand-held rock drills to be paid for in addition +12.5%)				
Addition for silencer tool or muffler		each	0.10	hour
* Additional hoses	Per 15 m	each	0.105	hour
* Drill (excluding consumables)*		each	0.29	hour
* Reversible drill (excluding consumables)		each	0.59	hour
* Grinder (excluding consumables)	76 mm light weight	each	0.10	hour
*	177 mm heavy duty	each	0.40	hour
* Sander (excluding pad and disc, and consumables)		each	0.40	hour
* Riveting hammer (excluding consumables)		each	0.16	hour
* Chain saw up to 0.58 m (excluding consumables)		each	0.68	hour
* Pneumatic paint scraper tool		each	0.25	hour
Polisher		each	0.18	hour

CIVIL ENGINEERING INDUSTRY

Description	Unit £	Hire Rate	Period

TOOLS, PORTABLE
ELECTRIC

(excluding generator or power)
(consumables, to be charged in
addition)

	Diameter mm	Unit	Hire Rate	Period
* Drill	10	each	0.14	hour
	13	each	0.23	hour
	19	each	0.39	hour
	32	each	0.52	hour
* Extra for stand		each	0.10	hour
* Extra for magnetic stand		each	0.48	hour
* Bench grinder and pedestal	152	each	0.27	hour
	204	each	0.37	hour
	250	each	0.57	hour
* Angle grinder	180	each	0.25	hour
*	204	each	0.29	hour
*	230	each	0.33	hour
* Extra for stand		each	0.08	hour
* Portable grinder	102	each	0.24	hour
	152	each	0.32	hour
* Sander	180	each	0.40	hour
*	230	each	0.42	hour
* Polisher/sander		each	0.28	hour
* Electric rotary hammer drill		each	0.84	hour
* Electric demolition hammer				
lightweight		each	0.60	hour
* heavyweight		each	0.70	hour
* Electric hammer kit	29	each	1.02	hour
	51	each	1.35	hour

* Extra for power to be added to
above items where applicable.   each   50%   hour

CIVIL ENGINEERING INDUSTRY

Description	Unit	Hire Rate £	Period

TRACTORS, SCRAPERS, ETC.

Description		Unit	Hire Rate £	Period
Tractor (crawler) with bull angle dozer (hydraulically or winch operated) (Ripper attachments to these items subject to negotiation)	Maker's rated flywheel kW up to			
	70.0	each	25.75	hour
	85.0	each	34.57	hour
	100.0	each	37.62	hour
	115.0	each	40.26	hour
	135.0	each	45.58	hour
	185.0	each	58.30	hour
	215.0	each	61.87	hour
	250.0	each	72.16	hour
	350.0	each	129.58	hour
	450.0	each	199.14	hour
Tractor loading shovel (crawler)	SAE rated capacity (cu m) up to			
	0.60	each	12.57	hour
	0.80	each	17.11	hour
	1.00	each	20.17	hour
	1.20	each	24.88	hour
	1.40	each	30.16	hour
	1.80	each	34.64	hour
	2.00	each	36.34	hour
	2.10	each	48.96	hour
	3.50	each	73.17	hour
Tractor loading shovel (crawler) with back hoe equipment	add to above rates	each	15%	hour
Tractor loading shovel (crawler) with 4 in 1 attachment	add to above rates	each	17.5%	hour
Tractor loading shovel (crawler) with additional hydraulic mounted ripper	SAE rated capacity (cu m) up to			
	0.60	each	13.39	hour
	0.80	each	18.42	hour
	1.00	each	21.90	hour
	1.20	each	26.76	hour
	1.40	each	32.62	hour
	1.60	each	35.15	hour
	1.80	each	37.70	hour
Tractor loading shovel (wheeled) with 4 wheel drive				
	1.00	each	15.34	hour
	1.20	each	18.32	hour
	1.30	each	22.31	hour
	1.40	each	23.37	hour
	1.60	each	24.64	hour
	1.80	each	25.95	hour
	2.00	each	27.13	hour
	2.10	each	30.53	hour
	2.30	each	34.14	hour
	2.70	each	38.89	hour
	3.10	each	45.76	hour
	3.50	each	51.05	hour
	4.70	each	68.46	hour
Tractor loading shovel (wheeled) with 4 wheel drive articulated	SAE rated capacity (cu m) up to			
	1.00	each	16.39	hour
	1.20	each	20.38	hour
	1.40	each	23.39	hour
	1.50	each	24.46	hour
	1.60	each	25.52	hour
	1.80	each	26.43	hour
	2.00	each	28.99	hour
	2.30	each	33.68	hour
	2.70	each	38.80	hour
	3.10	each	41.90	hour
	3.50	each	47.98	hour
	3.85	each	59.97	hour
	5.00	each	78.98	hour

CIVIL ENGINEERING INDUSTRY

Description		Unit	Hire Rate £	Period
Tractor loading shovel (wheeled) with 4 wheel drive	SAE rated capacity (cu m) up to			
articulated with 4 in 1 attachment	1.00	each	19.56	hour
	1.40	each	24.50	hour
	1.60	each	28.17	hour
	2.10	each	32.71	hour
	2.70	each	45.20	hour
Tractor loading shovel (wheeled steer skid loader)				
	0.15	each	3.95	hour
	0.30	each	6.55	hour
	0.40	each	8.91	hour
Tractor loading shovel (wheeled) with 2 wheel drive				
	0.50	each	7.64	hour
	0.60	each	7.88	hour
	0.80	each	9.78	hour
	1.00	each	2.15	hour
	1.10	each	13.87	hour
	1.20	each	15.76	hour
	Variable Blade Flywheel kW; up to			
Motor grader	80	each	19.79	hour
	110	each	25.44	hour
	120	each	32.37	hour
	160	each	39.44	hour
	200	each	44.40	hour
Wheeled tractor (rubber tyred)	Up to and including kW			
	40	each	4.72	hour
	50	each	5.85	hour
	60	each	9.48	hour
	70	each	12.32	hour
	75	each	13.07	hour
	90	each	15.24	hour
	95	each	19.21	hour
	105	each	22.45	hour
	Heaped capacity cu m; up to			
* Scraper, tractor drawn (when hired with tractor)	3.0	each	4.42	hour
	5.0	each	5.46	hour
	8.0	each	6.60	hour
	11.0	each	7.50	hour
	17.0	each	8.28	hour
	20.0	each	9.32	hour
Motorized scraper (rubber tyred) (single engine)				
	16.0	each	75.99	hour
	23.0	each	102.20	hour
Motorized scraper (rubber tyred) (twin engine)				
	16.0	each	96.03	hour
	25.0	each	119.51	hour
	35.0	each	189.48	hour
Motorized elevating scraper (rubber tyred) (single engine)				
	9.0	each	58.64	hour
	15.0	each	88.12	hour
	20.0	each	88.59	hour
	30.0	each	130.39	hour
Motorized elevating scraper (rubber tyred) (twin engine)				
	25.0	each	144.78	hour
	30.0	each	181.95	hour

CIVIL ENGINEERING INDUSTRY

Description	Unit	Hire Rate £	Period

TRENCHERS   Flywheel kW

Chain bucket or wheel type	Up to 9	each	4.56	hour
	10 to 15	each	7.51	hour
	16 to 30	each	9.51	hour
	31 to 40	each	13.18	hour
	41 to 45	each	15.41	hour
	46 to 55	each	17.22	hour
	56 to 65	each	19.11	hour
	66 to 85	each	24.68	hour

WATER AND FUEL
SUPPLY   (Exclusive of supporting structure)

	litres			
* Water storage tank	1140	each	1.35	day
	2280	each	1.75	day
	2281 to 4550	each	2.63	day
	4551 to 9100	each	3.83	day
* Fuel storage tank	2280	each	1.33	day
	2281 to 4550	each	2.73	day
	4551 to 9100	each	4.43	day
* Water or fuel storage tank trailer (rubber tyred)	1140	each	5.26	day
	1141 to 2280	each	8.75	day
Water of fuel tanker mobile - self-propelled	4550	each	8.32	hour
	4551 to 6820	each	8.43	hour
* Water dandy or barrow		each	0.71	day

WELDING AND CUTTING
SETS

Oxy-acetylene cutting and welding set inclusive of oxygen and acetylene (excluding underwater equipment)		each	4.23	hour
Welding set, diesel (exclusive of electrodes to be charged in addition)	150 amp Single operator	each	2.00	hour
	300 amp Single operator	each	4.05	hour
	480 amp Double operator	each	9.15	hour
As above, transformer electric	150 amp Single operator	each	2.00	hour
	300 amp Single operator	each	2.25	hour
* Hand screen		each	0.38	day
* Helmet		each	0.40	day
* Standard kit for 300 amp set		each	2.46	day

CIVIL ENGINEERING INDUSTRY

	Description	Unit	Hire Rate £	Period
WINCHES	Line pull tonnes			
Double drum diesel friction winch	1.00	each	5.56	hour
	3.00	each	8.64	hour
	4.00	each	11.03	hour
Single drum winch, diesel hydraulic	10.00	each	34.39	hour
	15.00	each	50.02	hour
Single drum winch, electric	1.00	each	2.18	hour
	2.00	each	3.77	hour
	3.00	each	6.23	hour
	5.00	each	10.67	hour
	10.00	each	17.76	hour
	15.00	each	24.14	hour
* Hand operated winch	1.00	each	0.25	hour
	2.00	each	0.38	hour
	3.00	each	0.56	hour
	5.00	each	0.65	hour
As above mounted on lorry (power ex lorry battery)	2.00	each	1.52	hour

N.B. Special rope requirements to
be charged extra.

CIVIL ENGINEERING INDUSTRY

Description		Unit	Hire Rate £	Period
MISCELLANEOUS PLANT AND CONSUMABLE STORES				
* Air testing machine for drains		each	0.90	day
* Anvil	up to 100 kg/per	each	0.45	day
* Fencing chestnut	1.0 metre	each	0.013	day
* Fencing pole and trestle or similar	per 1.0 metre	each	0.07	day
* Firing for pipe jointer		each	0.68	hour
* Firing for blacksmith's hearth		each	1.04	hour
* Firing for tar boiler		each	2.56	hour
* Firing for watchman		each	1.04	hour
* Fire devil (watchman's)		each	0.12	day
* Fire devil with lowering hook, lead pot and ladle		each	0.29	day
* Forge (blacksmith's)		each	0.45	day
* Road barrier	per 1.0 metre	each	0.20	day
* Tarpaulin		1.0 sq m	0.046	day
* Tilley lamp (flood)		each	2.00	night
Traffic signals	Fixed time portable	each	101.30	week
	Vehicle activated portable with detector loops	each	124.50	week
	Vehicle activated portable with radar detectors	each	143.00	week
* Watchman's lamp including paraffin		each	0.93	night
Space heater (paraffin/electric)	Btu/hr; Up to			
	30 000	each	0.68	hour
	63 000	each	1.15	hour
	100 000	each	1.66	hour
	150 000	each	2.19	hour
	240 000	each	3.19	hour
	325 000	each	4.29	hour
Extra for gas operated		each	add 10%	
Brickwork and masonry saw (excluding abrasive discs)	Blade dia			
	305 mm	each	0.90	hour
	356 mm	each	1.39	hour
	457 mm	each	2.34	hour
* Crossing plates (steel sheets)	thickness	per 1.0		
	20 mm	sq m	0.12	day
* Flashing traffic warning lamp unit (including batteries)		each	1.09	day
* Flashing traffic warning lamp unit (including batteries, bollard or tripod)		each	1.29	day
* Pendant barrier marker	26 m cord length	each	0.84	day
including fencing pins	102-152 mm dia.	per 10.0 m	0.66	day
* Drainage rods incl. accessories	0.91 m length	each	0.05	day
* As above, extra per rod				
* Drain stopper (expanding)	dia mm; Up to			
	102	each	0.18	day
	152	each	0.25	day
	305	each	0.40	day
	457	each	0.92	day
	610	each	1.79	day
	762	each	3.62	day
	914	each	5.15	day
	1220	each	7.33	day

CIVIL ENGINEERING INDUSTRY

Description	Unit	Hire Rate £	Period
Water extraction per 24 hours litres			
Dehumidifier 68	each	5.51	day
91	each	7.22	day
160	each	11.27	day
Dehumidifier with temperature control 160	each	15.48	day
* Ladders - all types and lengths		Rates to be negotiated	

Description		Unit	Hire Rate £	Period
* Traffic warning cone (plastic)	Height up to 500 mm	each	0.09	day
	501-600	each	0.11	day
	601-800	each	0.22	day
	801-1000	each	0.49	day
* Traffic warning sign (circular)	diameter (mm)			
	600	each	0.38	day
	750	each	0.60	day
	900	each	0.81	day
	1200	each	1.29	day
* Traffic warning sign (triangular)	height (mm)			
	600	each	0.30	day
	700	each	0.44	day
	900	each	0.56	day
	1200	each	0.97	day
* Traffic warning board (rectangular)	1050 x 750	each	0.79	day

ACCESS PLATFORM

	Capacity kg	Platform height metres	Unit	Hire Rate £	Period
Scissor lift self-propelled - yard travel	up to 400	up to 5.0	each	4.68	hour
	up to 400	5.1-9.0	each	5.47	hour
	up to 400	over 9.0	each	5.94	hour
	401-800	up to 5.0	each	6.03	hour
	401-800	5.1-9.0	each	6.97	hour
	401-800	over 9.0	each	9.25	hour
	over 800	any height	each	12.28	hour
Scissor lift self-propelled - rough terrain travel	up to 800	up to 9.0	each	9.01	hour
	up to 800	over 9.0	each	14.14	hour
	over 800	up to 9.0	each	16.62	hour
	over 800	over 9.0	each	17.91	hour
Scissor lift - hand-propelled or towed	any capacity	up to 5.0	each	1.68	hour
	any capacity	5.1-7.0	each	3.35	hour
Telescopic self-propelled - yard travel	any capacity	7.1-9.0	each	3.84	hour
	up to 400	up to 10.0	each	12.32	hour
	up to 400	10.1-15.0	each	16.02	hour
	up to 400	15.1-20.0	each	22.73	hour
Telescopic self-propelled - yard travel	up to 400	over 20.0	each	44.42	hour
	over 400	up to 15.0	each	18.76	hour
	over 400	15.1-20.0	each	23.51	hour
Telescopic self-propelled - rough terrain travel				+50% on above	
Vehicle mounted platform - telescopic/articulated	any capacity	up to 10.0	each	8.39	hour
	any capacity	10.1-20.0	each	13.79	hour
	any capacity	20.1-30.0	each	44.68	hour
	any capacity	over 30.0	each	118.44	hour

CIVIL ENGINEERING INDUSTRY

4. SUPPLEMENTARY CHARGES

Notes and Conditions

(1)   Cost of free transport provided by contractors for workmen to and from the site
      to be charged at cost plus 12.5%.
(2)   Sub-contractors
        (i) Ordinary Sub-contractors
      All sub-contractors' accounts, other than labour only sub-contractors or plant
      hire to be charged at full amount of invoice (without deduction of any cash
      discounts not exceeding 2.5%) plus 12.5%.
      (ii) Labour only Sub-contractors
      Cost of labour to be dealt with as in Section 1 Note 3.
(3)   Plant Hired in Specifically for Dayworks
      (a)   Plant hire, fuel, oil and grease, insurance, transport, etc., to be charged
             at full amount of invoice (without deduction of any cash discounts not
             exceeding 2.5%), to which should be added consumables where supplied by the
             contractor, all plus 12.5%.
      (b)   for driver or operative, full amount of invoice plus 64%.
(4)   The cost of internal transport on a site to be charged in addition at the
      appropriate daywork rates for labour, plant hire, etc.
(5)   The cost of operating welfare facilities to be charged by the contractor plus
      12.5%.
(6)   The cost of additional insurance premiums for abnormal contract work or special
      site conditions to be charged at cost plus 12.5%.
(7)   The cost of watching and lighting specially necessitated by daywork is to be paid
      for separately.
(8)   The charges referred to above are as at date of publication and will be subject
      to periodic review.

5. VALUE ADDED TAX

Note and Condition

(1)   Value Added Tax has not been included in any of the rates in this schedule but
      will be chargeable if payable to HM Customs and Excise by the contractor.

6. CONVERSION TABLE

Note:

Plant items whose specifications and data are related to now discarded imperial capacities
may still be in use on contracts.

To allocate them to appropriate current daywork items the following conversions should be
used:-

LINEAR   WEIGHT
0.03937 in -- 1 -- mm   25.4   0.0196 cwt -- 1 -- kg   50.82
3.28084 ft -- 1 -- metre 0.3048   0.9842 ton -- 1 -- tonne 1.016
1.0936   yd -- 1 -- metre 0.9144   2.20463 lb -- 1 -- kg   0.45359

CAPACITY   AREA
1.7598 pint   -- 1 -- litre 0.56826   0.00155 in2   -- 1 -- mm2   645.16
0.21997 gallon -- 1 -- litre 4.54609   10.7643 ft2   -- 1 -- m2   0.0929
                                             1.1960 yd2   -- 1 -- m2   0.83613
                                             2.4711 acre -- 1 -- ha   0.40469
                                             0.3861 mile2 -- 1 -- km2   2.59

VOLUME   POWER
 0.0610 in3 -- 1 -- cm3 16.387   1.310 Hp -- 1 -- kW 0.7457
35.315 ft3 -- 1 -- m3   0.0283
 1.3079 yd3 -- 1 -- m3   0.7645

Rates of Wages

BUILDING INDUSTRY

Authorized rates of wages, etc., in the building industry in England, Wales and Scotland agreed by the National Joint Council for the Building Industry

AND EFFECTIVE FROM 26th JUNE 1989

Subject to the conditions prescribed in the Working Rule Agreement guaranteed minimum weekly earnings shall be as follows:

	Craft operatives £	Labourers £
LONDON AND LIVERPOOL DISTRICT	135.330	115.245
GRADE A AND SCOTLAND	135.135	115.050

These guaranteed minimum weekly earnings shall be made as follows:

	Craft operatives £	Labourers £
LONDON AND LIVERPOOL DISTRICT		
Standard basic rates of wages	120.315	102.570
Guaranteed minimum bonus payment	15.015	12.675
Guaranteed minimum weekly earning	135.330	115.245

	Craft operatives £	Labourers £
GRADE A AND SCOTLAND		
Standard basic rates of wages	120.120	102.375
Guaranteed minimum bonus payment	15.015	12.675
Guaranteed minimum weekly earnings	135.135	115.050

NOTE
The guaranteed minimum bonus payment shall be set off against existing bonus payments of all kinds or against existing extra payments other than those prescribed in the Working Rule Agreement. The entitlement to the guaranteed minimum bonus shall be pro rata to normal working hours for which the operative is available for work.

BUILDING INDUSTRY

Young Labourers

The rates of wages for young labourers shall be the following proportions of the labourers' rates:

At 16 years of age   50%
At 17 years of age   70%
At 18 years of age 100%

Watchmen

The weekly pay for watchmen shall be equivalent to the labourers' guaranteed minimum weekly earnings provided that not less than five shifts (day or night) are worked.

Apprentices/Trainees under 19

Six-month period	Rate per week £	GMB £
First	55.575	-
Second	75.465	-
Third	82.290	10.140
*Thereafter, until skills test is passed	98.865	12.480
On passing skills test and until completion of training period	111.930	14.040

*The NJCBI has agreed to the introduction of skills testing under the National Joint Training Scheme. The first practical skills tests commenced in May 1986.

Apprentices/Trainees over 19

Six-month period	Rate per week £	GMB £
First and Second	96.525	12.090
Third and Fourth	102.180	12.480
Fifth and Sixth	107.640	13.455

BUILDING INDUSTRY

Authorized rates of wages in the building industry in England and Wales agreed by the Building & Allied Trades Joint Industrial Council

AND EFFECTIVE FROM 26th JUNE 1989

Subject to the conditions prescribed in the Working Rule Agreement the standard weekly rates of wages shall be as follows:

	£
Craft operatives	140.01
Adult general operatives	120.51

ROAD HAULAGE WORKERS EMPLOYED IN THE
BUILDING INDUSTRY

Authorized rates of pay for road haulage workers in the building industry agreed between the Builders Employers Confederation and the Transport and General Workers Union.

AND EFFECTIVE FROM 26th JUNE 1989

Employers	Operatives
The Building Employers Confederation,	The Transport and General Workers Union,
82 New Cavendish Street,	Transport House,
London W1M 8AD	Smith Square,
Telephone: 01-580 5588	Westminster,
	London SW1P 3JB
	Telephone: 01-828 7788

DEFINITIONS OF GRADES
London
London Region as defined by the National Joint Council for the Building Industry.

Grade 1
The Grade A districts as defined by the National Joint Council for the Building Industry, including Liverpool and District (i.e., the area covered by the Liverpool Regional Federation of Building Trades Employers).

	LONDON per week £	GRADE 1 per week £	GMB per week £
Drivers of vehicles of gross vehicle weight up to 3.5 tonnes	120.315	120.120	15.015
over 3.5 tonnes and up to and including 7.5 tonnes	120.705	120.510	15.015
over 7.5 tonnes and up to and including 10 tonnes	121.290	121.095	15.015
over 10 tonnes and up to and including 16 tonnes	121.680	121.485	15.015
over 16 tonnes and up to and including 24 tonnes	123.045	122.850	15.015
over 24 tonnes	123.630	123.435	15.015

Subject to certain conditions workers covered by this agreement are entitled to additonal payments corresponding to the NJCBI's Guaranteed Minimum Bonus. The gross vehicle weight in every case is to include, where applicable, the unladen weight of a trailer designed to carry goods.

BUILDING INDUSTRY - ISLE OF MAN

Authorized rates of wages in the building industry in the Isle of Man agreed by
the Isle of Man Joint Industrial Council for the Building Industry

AND EFFECTIVE FROM 26th JUNE 1989

SECRETARIES

Employers	Operatives
A. Hill, FCA, FCIS,	Mr Moffat,
Kensington House,	Transport House,
Rosemount,	Prospect Hill,
Douglas	Douglas,
Isle of Man	Isle of Man
Telephone: 0624 24331	Telephone: 0624 21156

	Standard basic Wage rate per 39-hour week £	Guaranteed minimum bonus payment £	Guaranteed minimum weekly earnings £
Craft operatives	120.315	15.015	135.330
Labourers	102.570	12.675	115.245

BUILDING AND CIVIL ENGINEERING INDUSTRY -
NORTHERN IRELAND

Authorized rates of wages in the building and civil engineering industry in
Northern Ireland as agreed by the Joint Council for the Building and Civil
Engineering Industry (Northern Ireland)

AND EFFECTIVE FROM 31st JULY 1989

Federation of Building & Civil Engineering Contractors (Northern Ireland) Ltd,
    143 Malone Rd,
    Belfast BT9 6SU
    Telephone: 0232 661711

	Standard basic wage rate per 39-hour week £	Guaranteed minimum bonus payment £	Guaranteed minimum weekly earnings £
Craft operatives	132.99	7.000	139.99
Labourers	115.83	7.000	122.83

CIVIL ENGINEERING INDUSTRY

Authorized rates of wages in the civil engineering industry as agreed by the Civil Engineering Construction Conciliation Board

SECRETARIES

Employers panel
D.L.F. Chapman M.I.P.M.,
M.I.T.D., A.M.I.O.S.H
Cowdray House,
6 Portugal Street,
London WC2A 2HH
Telephone: 01-404 4020

Operatives panel
G. P. Henderson
Transport House,
Smith Square,
Westminster,
London SW1P 3JB
Telephone: 01-828 7788

Rates of pay effective from 26th JUNE 1989	Per hour £
General Operatives:	
London Super Grade	2.630
Class I	2.625
Craft Operatives	
London Super Grade	3.085
Liverpool Grade	3.085
Class I	3.080

Guaranteed bonus from 26th JUNE 1989	Per hour £
General Operatives	0.325
General Operatives plus-rated up to but excluding 37p	0.325
General Operatives plus-rated 37p or over	0.385
Craft Operatives	0.385

Watchmen (based on five shifts)	Per week £
London Super Grade	115.100
Class I	115.050
Craftsmen's tool allowance	
Bricklayers	0.77
Carpenters and Joiners	1.50

PLUMBING MECHANICAL ENGINEERING SERVICES

**Authorized rates of wages agreed by the Joint Industry Board for the Plumbing
Mechanical Engineering Services Industry in England and Wales**

The Joint Industry Board for Plumbing Mechanical
    Engineering Services in England and Wales,
  Brook House, Brook Street,
  St Neots,
  Huntingdon, Cambs PE19 2HW
  Telephone: 0480 76925-8

Rates of pay effective from 3rd APRIL 1989

	Per hour £
Technical plumber	4.90
Advanced plumber	4.31
Trained plumber	3.91
Apprentices	
1st year of training	1.11
2nd year of training	1.65
3rd year of training	2.03
4th year of training	2.60

Prices for Measured Work

Keep your figures up to date, free of charge

This section, and most of the other information in this Price Book,
is brought up to date every three months in the *Price Book Update*.

The *Update* is available free to all Price Book purchasers.

To ensure you receive your copy, simply complete the reply card from
the centre of the book and return it to us.

SPON'S PLANT AND EQUIPMENT PRICE GUIDE

Edited by **Plant Assessment (London) Ltd**

Spon's Plant and Equipment Price Guide has for more than 10 years been *THE* one-stop reference for plant professionals requiring reliable price information and specifications for new and used machinery.

Whether you are offering financial services, operating or trading plant and equipment you'll find **Spon's Plant and Equipment Price Guide** an easy to use, valuable source of reliable price data and specifications.

In two volumes of over 1,500 pages, the Guide is divided into 13 sections giving immediate access to more than 30,000 current prices for all types of machinery. Its convenient organisation – presenting equipment classified by type – makes it simple to find the right machine and establish its cost.

Each volume is published in convenient loose-leaf form for easy replacement of revised sections which are issued to subscribers. The strong PVC binders, (7"×5") include an index and details of how to use the Guide.

Spon's Plant and Equipment Price Guide is a continuing service. Revised sections are mailed to subscribers over the subscription year. In addition a Percentage Variation List is issued periodically to correct market fluctuations occuring since a section was issued.

Subject Areas
The categories used in the Guide are:
Volume One: Compressors, Concreting Plant, Cranes, Dumpers and Dumptrucks, Excavators and Graders. Volume Two: Hoists and Winches, Loading Shovels, Pumps, Piling Equipment, Rollers and Compaction Plant, Tractors, and Miscellaneous Equipment.

To subscribe to the Guide for one year costs £110.00 in the UK, $225.00 in the USA and Canada and £140.00 for other overseas subscribers. Subscription price includes:

- The two volumes of the Guide, up to date at commencement of subscription. Includes 2 PVC binders, 13 sections and divider cards.

- Regular revisions, section by section, of the Guide.

- A schedule of Average Price Variations issued periodically.

- Post and packing free.

ISSN 02635038   E & F N Spon

 **E & F N SPON**
11 New Fetter Lane, London EC4P 4EE

Prices for Measured Work
– Major Works

INTRODUCTION

The 'Prices for Measured Work - Major Works' are intended to apply to a project costing (excluding Preliminaries) about £1 800 000 in the outer London area and assume that reasonable quantities of all types of work are required. Similarly it has been necessary to assume that the size of the job warrants the sub-letting of all types of work normally sub-let.

The distinction between builders' work and work normally sub-let is stressed because prices for work which can be sub-let may well be quite inadequate for the contractor who is called upon to carry out relatively small quantities of such work himself

As explained in more detail later the prices are generally based on wage rates which came into force in June 1989, material costs as stated and include an allowance for overheads and profit. They do not allow for preliminary items which are dealt with under a separate heading (see page 167) or for any Value Added Tax which will be payable on non-domestic buildings after 1st April 1989.

The format of this section is so arranged that, in the case of work normally undertaken by the Main Contractor, the constituent parts of the total rate are shown enabling the reader to make such adjustments as may be required in particular circumstances.

Similar details have also been given for work normally sub-let although it has not been possible to provide this in all instances.

As explained in the Preface, there is now a facility available to readers which enables a comparison to be made between the level of prices in this section and current tenders by means of a tender index.

The tender index for this Major Works section of Spon's is 366 (as shown on the front cover) which coincides with our forecast tender price level index for the outer London region applicable to the third quarter of 1989

To adjust prices for other regions/times, the reader is recommended to refer to the explanations and examples on how to apply these tender indexes, given on pages 723-725.

There follow explanations and definitions of the basis of costs in the 'Prices for Measured Work' section under the following headings:

Overhead charges and profit
Labour hours and Labour £ column
Material £ column
Material/Plant £ columns
Total rate £ column

OVERHEAD CHARGES AND PROFIT

For those items where detailed breakdowns have been given an allowance of 9% has been added to labour, plant and material costs for overhead charges and profit.

In other cases 5% has been included for the General Contractor's attendance, overhead charges and profit.

LABOUR HOURS AND LABOUR £ COLUMNS

'Labour rates are based upon typical gang costs divided by the number of primary working operatives for the trade concerned, and for general building work include an allowance for trade supervision (see below). 'Labour hours' multiplied by 'Labour rate' with the appropriate addition for overhead charges and profit gives 'Labour £'. In some instances, due to variations in gangs used, 'Labour rate' figures have not been indicated, but can be calculated by dividing 'Labour £' by 'Labour hours'. For building operatives the rates used are as set out below and allow for a bonus of plus 55% (in lieu of Guaranteed Minimum Bonus), which is considered to be representative of current labour costs in the outer London area.
    Alternative labour rates are also given based upon bonuses of other values.

Building craft operatives and labourers

From 26th June 1989 guaranteed minimum weekly earnings in the London area for craft operatives and labourers are £135.33 and £115.25 respectively; to these rates have been added allowances for the items below in accordance with the recommended procedure of the Institute of Building in its 'Code of Estimating Practice'. The resultant hourly rates on which the 'Prices for Measured Work' have generally been based are £6.61 and £5.72 for craft operatives and labourers, respectively.
    The items referred to above for which allowances have been made are:
55% bonus (in lieu of guaranteed minimum bonus)
Lost time
Non-productive overtime
Extra payments under National Working Rule 3, 17 and 18
Construction Industry Training Board Levy
Holidays with pay
Pension and death benefit scheme
Sick pay
National Insurance
Severance pay and sundry costs
Employers' liability and third party insurance

NOTE: For travelling allowances and site supervision see 'Preliminaries' section.

    The table which follows shows how the 'all-in' hourly rates referred to above have been calculated. Productive time has been based on a total of 2063 hours worked per year.

		Craft operatives		Labourers	
		£	£	£	£
Wages at standard basic rate					
productive time	52.88 wks	120.32	6362.52	102.57	5423.90
Lost time allowance	1.06 wks	120.32	127.53	102.57	108.72
Non-productive overtime	4.43 wks	120.32	533.00	102.57	454.39
			--------		--------
			7023.05		5987.01
Allowance for bonus	Plus	55%	3862.68	55%	3292.86
Extra payments under					
National Working Rule 3	46.80 wks	1.60	74.88	3.10	145.08
Sick pay	1 wk	-	-	-	-
CITB levy	1 year	-	75.00	-	18.00
Public holiday pay	1.60 wks	135.33	216.53	115.25	184.39
Employer's contribution to:					
annual holiday pay scheme	47 wks	14.70	690.90	14.70	690.90
pension and death benefit					
scheme	47 wks	1.40	65.80	1.40	65.80
National Insurance	47.80 wks	24.43	1167.75	21.01	1004.28
			--------		-------
		C/F	13176.59		11388.32

		Craft operatives		Labourers	
		£	£	£	£
	B/F		13176.59		11388.32
Severance pay and sundry costs	Plus	1.5%	197.65	1.5%	170.82
			13374.24		11559.14
Employer's liability and third-party insurance	Plus	2%	267.48	2%	231.18
			13641.72		11790.32
Total cost per annum			6.61		5.72
Total cost per hour					

NOTES:
1. Absence due to sickness has been assumed to be for individual days when no payment would be due.
2. The annual holiday pay pension and death benefit scheme contributions are those which came into force on 31st July 1989.
3. Employers' contribution to the holiday stamp assumes that the Contractor makes use of the special credit arrangements on buying the stamps and does not therefore receive the discount from the Management Company.

From these rates have been calculated the 'labour rate' applicable to each trade which generally include an allowance for supervision by a foreman or ganger and are based on a bonus rate plus 55%.

Gang		Labour(man)rate		Alternative labour rates			
	Gang rate £ (bonus + 55%)	£/hour (bonus + 55%)		£/hour GMB	+40%	+60%	+70%
Groundwork gang							
1 Ganger	5.72 + 5% = 6.01						
6 Labourers	6 x 5.72 = 34.32						
	40.33	/ 6.5 = 6.20		4.57	5.66	6.38	6.74
Concreting gang							
1 Ganger	5.72 + 5% = 6.01						
4 Labourers	4 x 5.72 = 22.88						
	28.89	/ 4.5 = 6.42		4.72	5.86	6.60	6.97
Steelfixing gang							
1 Foreman	6.61 + 5% = 6.94						
4 Steelfixers	4 x 6.61 = 26.44						
	33.38	/ 4.5 = 7.42		5.51	6.77	7.64	8.08
Formwork gang							
1 Foreman	6.61 + 5% = 6.94						
10 Carpenters	10 x 6.61 = 66.10						
1 Labourer	1 x 5.72 = 5.72						
	78.76	/ 10.5 = 7.50		5.57	6.84	7.73	8.17
Bricklaying/Ltwt blockwork gang							
1 Foreman	6.61 + 5% = 6.94						
6 Bricklayers	6 x 6.61 = 39.66						
4 Labourers	4 x 5.72 = 22.88						
	69.48	/ 6.5 = 10.69		7.92	9.75	11.00	11.63

Gang	Gang rate £ (bonus + 55%)	Labour(man)rate £/hour (bonus + 55%)	Alternative labour rates £/hour			
			GMB	+40%	+60%	+70%
Dense blockwork gang						
1 Foreman	6.61 + 5% = 6.94					
6 Bricklayers	6 x 6.61 = 39.66					
6 Labourers	6 x 5.72 = 34.32					

	80.92	/ 6.5 = 12.45	9.21	11.36	12.81	13.54
Carpentry/joinery gang						
1 Foreman	6.61 + 5% = 6.94					
5 Carpenters	5 x 6.61 = 33.05					
1 Labourer	1 x 5.72 = 5.72					

	45.71	/ 5.5 = 8.31	6.17	7.58	8.56	9.05
Craft Operatives only						
1 Craft operative	1 x 6.61 = 6.61	/ 1 = 6.61	4.91	6.03	6.81	7.20
Craft operatives/Labourer pair						
1 Craft operative	1 x 6.61 = 6.61					
1 Labourer	1 x 5.72 = 5.72					

	12.33	/ 1 = 12.33	9.12	11.25	12.69	13.41
Small labouring gang (making good)						
1 Ganger	5.72 + 5% = 6.01					
4 Labourers	4 x 5.72 = 22.88					

	28.89	/ 4.5 = 6.42	4.72	5.86	6.60	6.97
2 Craft Operatives/1 Labourer						
2 Craft Operatives	2 x 6.61 = 13.22					
1 Labourer	1 x 5.72 = 5.72					

	18.94	/ 2 = 9.47	7.02	8.64	9.75	10.31
Drain laying gang/clayware						
2 Labourers	2 x 5.81 = 11.62	/ 2 = 5.81	4.30	5.31	5.97	6.30

Sub-Contractor's operatives
Similar labour rates are shown in respect of sub-let trades where applicable.

Plumbing operatives
From 3rd April 1989 the hourly earnings for technical and trained plumbers are
£4.90 and £3.91, respectively; to these rates have been added allowances similar
to those added for building operatives (see below). The resultant average
hourly rate on which the 'Prices for Measured Work' have been based is £9.49
 The items referred to above for which allowance has been made are:

    55% Bonus
    Non-productive overtime
    Tool allowance
    Plumbers' welding supplement
    Construction Industry Training Board Levy
    Holidays with pay
    Pension and welfare stamp
    National Insurance 'contracted out' contributions
    Severance pay and sundry costs
    Employer's liability and third party insurance

No allowance has been made for supervision as we have assumed the use of a team
of technical or trained plumbers who are able to undertake such relatively
straightforward plumbing works, e.g. on housing schemes, without further
supervision.

The table which follows shows how the average hourly rate referred to above has been calculated. Productive time has been based on a total of 2034 hours worked per year.

			Technical plumber £	Trained plumber £
Wages at standard basic rate productive time	2034 hrs	4.90	9 966.60	3.91	7 952.94
Non-productive overtime	113 hrs	4.90	553.70	3.91	441.83
Plumbers' welding supplement	2034 hrs	0.36	732.24	-	-
			11 252.54		8 394.77
Allowance for bonus	Plus	55%	6 188.90	55%	4 617.12
Tool allowance	45.20 wks	1.70	76.84	1.70	76.84
CITB levy	1 year	-	110.00	-	110.00
Public holiday pay	60 hrs	4.90	294.00	3.91	234.60
Employer's contributions to holiday credit/welfare stamps	47 wks	18.60	874.20	15.20	714.40
pension (6.50% of earnings)	47.8 wks	24.22	1 157.72	18.12	866.14
National Insurance	47.8 wks	28.22	1 349.06	20.17	964.13
			21 303.26		15 978.00
Severance pay and sundry costs	Plus	1.5%	319.55	1.5%	239.67
			21 622.81		16 217.67
Employers' liability and third-party insurance	Plus	2%	432.46	2%	324.35
Total cost per annum			22 055.27		16 542.02
Total cost per hour			10.84		8.13
Average all-in rate per hour			£9.49 per hour		

MATERIAL £ COLUMN

Many items have reference to a 'PC' value. This indicates the prime cost of the major material delivered to site in the outer London area for large quantities, assuming maximum discounts.

When obtaining material prices from other sources, it is important to identify any discount that may apply. Some manufacturers only offer 5 to 10% discount for the largest of orders; or 'firm' orders (as distinct from quotations). For other materials discounts of 30% to 40% may be obtainable, depending on value of order and any preferential position of the purchaser.

The 'Material £' column indicates the total cost of the material, based upon the PC, including delivery, waste, sundry materials and an allowance for overhead charges and profit for the unit of work concerned.

Alternative material prices are given at the beginning of many sections, by means of which alternative 'Total rate £' prices may be calculated.

All material prices quoted are exclusive of Value Added Tax.

MATERIAL PLANT £ COLUMN

Plant costs have been based on current weekly hire charges and estimated weekly cost of oil, grease, ropes (where necessary), site servicing and cartage charges.

The total amount is divided by 30 (assuming 25% idle time) to arrive at a cost per working hour of plant. To this hourly rate is added one hour fuel consumption and one hour for each of the following operators; the rate to be calculated in accordance with the principles set out earlier in this section, i.e. with an allowance for plus rates, etc.

For convenience the all-in rates per hour used in the calculations of 'Prices for Measured Work' are shown below and where included in Material/Plant £ column will include the appropriate addition allowance of 9% for overhead charges and profit for Major Works and 12.5% for overhead charges and profit for Minor Works.

Plant	Labour	'All-in' rate Per hour £
Excavator (4 wheeled - 0.76 m3 shovel, 0.24 m3 bucket)	Driver	22.50
Excavator (JCB 3C - 0.24 m3 bucket)	Driver	19.00
Excavator (JCB 3C off centre - 0.24 m3 bucket)	Driver	21.00
Excavator (Hymac 580 - 0.53 m3 bucket)	Driver	26.00
Dumper (2.30 m3)	Driver	15.00
Two tool portable compressor (125 cfm)		
per breaking tool	*	2.20
per 'jumping jack' rammer	*	2.60
Roller		
0.75 tonnes	Driver	9.00
6-8 tonnes	Driver	13.00
Concrete mixer 10/7	Operator	4.75
Kango heavy duty hammer	-	1.10
Power float	-	2.50
Light percussion drill	-	0.50

* Operation of compressor by tool operator

TOTAL RATE £ COLUMN

'Total rate £' column is the sum of 'Labour £' and 'Material £' columns and therefore includes the allowances for overhead charges and profit previously described.

This column excludes any allowance for 'Preliminaries' which must be taken into account if one is concerned with the total cost of work.

The example of 'Preliminaries' in the following section indicates that in the absence of detailed calculations at least 14% must be added to all prices for measured work to arrive at total cost.

A PRELIMINARIES

The number of items priced in the 'Preliminaries' section of Bills of Quantities
and the manner in which they are priced vary considerably between Contractors.
Some Contractors, by modifying their percentage factor for overheads and profit,
attempt to cover the costs of 'Preliminary' items in their 'Prices for Measured
Work'. However, the cost of 'Preliminaries' will vary widely according to job
size and complexity, site location, accessibility, degree of mechanization
practicable, position of the Contractor's head office and relationships with
local labour/domestic Sub-Contractors. It is therefore usually far safer to
price 'Preliminary' items separately on their merits according to the job.
    In amending the Preliminaries/General Conditions section for SMM7, the Joint
Committee stressed that the preliminaries section of a bill should contain two
types of cost significant item:
    1. Items which are not specific to work sections but which have an
        identifiable cost which is useful to consider separately in
        tendering e.g. contractual requirements for insurances, site
        facilities for the employer's representative and payments to the
        local authority.
    2. Items for fixed and time-related costs which derive from the
        contractor's expected method of carrying out the work, e.g.
        bringing plant to and from site, providing temporary works and
        supervision.
    A fixed charge is for work the cost of which is to be considered as
independent of duration. A time related charge is for work the cost of which is
to be considered as dependent on duration.
    The fixed and time-related subdivision given for a number of preliminaries
items will enable tenderers to price the elements separately should they so
desire. Tenderers also have the facility at their discretion to extend the list
of fixed and time-related cost items to suit their particular methods of
construction.
    The opportunity for Tenderers to price fixed and time-related items in A30-
A37, A40-A44 and A51-A52 have been noted against the following appropriate items
although we have not always provided guidance as costs can only be assessed in
the light of circumstances of a particular job.
    Works of a temporary nature are deemed to include rates, fees and charges
related thereto in Sections A36, A41, A42, and A44, all of which will probably
be dealt with as fixed charges.
    In addition to the cost significant items required by the method, other
preliminaries items which are important from other points of view, e.g. quality
control requirements, administrative procedures, may need to be included to
complete the Preliminaries/General conditions as a comprehensive statement of
the employer's requirements.
    Typical clause descriptions from a 'Preliminaries/General Conditions' section
are given below together with details of those items that are usually priced in
detail in tenders. As SMM7 only becomes operative on 1 July 1988, it may be
that Tenderers may adopt a different approach to pricing these sections;
possibly necessitating further revisions to this example in the light of their
response.
    An example in pricing 'Preliminaries' follows, and this assumes the form of
contract used is the Standard Form of Building Contract 1980 Edition and the
value, excluding 'Preliminaries', is £1 800 000. The contract is estimated to
take 80 weeks to complete and the value is built up as follows:

	£
Labour value	650 000
Material value	550 000
Provisional sums and all Sub-Contractors	600 000
	£1 800 000

    At the end of the section the examples are summarized to give a total value of
'Preliminaries' for the project.

A PRELIMINARIES/GENERAL CONDITIONS

Preliminary particulars

A10 Project particulars - Not priced

A11 Drawings - Not priced

A12 The site/Existing buildings - Generally not priced

The reference to existing buildings relates only to those buildings which could have an influence on cost. This could arise from their close proximity making access difficult, their heights relative to the possible use of tower cranes or the fragility of, for example, an historic building, necessitating special care.

A13 Description of the work - Generally not priced

A20 The Contract/Sub-contract

(The Standard Form of Building Contract 1980 Edition is assumed)

Clause no
1.   **Interpretation, definitions, etc.** - Not priced
2.   **Contractor's obligations** - Not priced
3.   **Contract Sum - adjustment - Interim certificates** - Not priced
4.   **Architect's Instructions** - Not priced
5.   **Contract documents - other documents - issue of certificates**
     The contract conditions may require a master programme to be prepared. This will normally form part of head office overheads and therefore is 'Not priced' separately here.
6.   **Statutory obligations, notices, fees and charges** - Not priced. Unless the Contractor is specifically instructed to allow for these items.
7.   **Level and setting out of the works** - Not priced
8.   **Materials, goods and workmanship to conform to description, testing and inspection** - Not priced
9.   **Royalties and patent rights** - Not priced
10. **Person-in-charge**
     Under this heading are usually priced any staff that will be required on site. The staff required will vary considerably according to the size, layout and complexity of the scheme, from one foreman-in-charge to a site agent, general foreman, assistants, checkers and storemen etc.
         The costs included for such people should include not only their wages, but their total cost to the site including statutory payments, pension, expenses, holiday relief, overtime, etc.

NOTE: The term 'Not priced' or 'Generally not priced' where used throughout this section means either that the cost implication is negligible or that is is usually included elsewhere in the tender.

A PRELIMINARIES

      Part of the foreman's time, together with that of an assistant will be spent on setting out the site. Allow say £2. 00 per day for levels, staff, pegs and strings, plus the assistant's time if not part of the general management team. Most sites usually include for one operative to clean up generally and do odd jobs around the site.

      Cost of other staff, such as buyers, and quantity surveyors, are usually part of head office overhead costs, but alternatively, may now be covered under A40 Management and staff.

A typical build-up of a foreman's costs might be: £

Annual salary	15 000.00
Expenses	1 500.00
Bonus - say	5 000.00
Employer's National Insurance	
contribution on salary and bonus	
(10.75% on £20 000.00)	2 150.00
Training levy	75.00
Pension scheme	
(say 6% of salary and bonus)	1 200.00
Sundries, including Employer's Liability and Third	
Party (say 3.5% of salary and bonus)	700.00

	£25 625.00
Divide by 47 to allow for holidays:	
Per week	545.21
Say	£545.00

      Corresponding costs for other site staff should be calculated in a similar manner.

Example

   Site administration   £

General foreman 80 weeks @ £545	43 600.00
Holiday relief 4 weeks @ £545	2 180.00
Assistant foreman 60 weeks @ £305	18 300.00
Storeman/checker 60 weeks @ £220	13 200.00
	---------
	77 280.00
Add 9% for overheads and profit - say	6 950.00

	£84 230.00
	---------

11.   **Access for Architect to the Works** - Not priced
12.   **Clerk of Works** - Not priced
13.   **Variations and provisional sums** - Not priced
14.   **Contract Sum** -Not priced
15.   **Value Added Tax - supplemental provisions**
     Major changes to the VAT status of supplies of goods and services by Contractors came into effect on 1 April 1989. It is clear that on and from that date the majority of work supplied under contracts on JCT Forms will be chargeable on the Contractor at the standard rate of tax. In April 1989, the JCT issued Amendment 8 and a guidance note dealing with the amendment to VAT provisions. This involves a revision to clause 15 and the Supplemental Provisions (the VAT agreement).
     Although the standard rating of most supplies should reduce the amount of VAT analysing previously undertaken by contractors, he should still allow for any incidental costs and expenses which may be incurred.

A PRELIMINARIES

A20 **The Contract/Sub-contract** - cont'd.

    16. **Materials and goods unfixed or off-site** - Not priced
    17. **Practical completion and Defects Liability**
        Inevitably some defects will arise after practical completion and an
        allowance will often be made to cover this. An allowance of say 0.25
        to 0.5% should be sufficient.
        Example
        Defects after completion   £

        Based on £0.25% of the contract sum
          £1 800 000 @ £0.25% - say   4 500.00
        Add 9% for overheads and profit - say   400.00

                                           £4 900.00
                                              ----

    18. **Partial possession by Employer** - Not priced
    19. **Assignment and Sub-Contracts** - Not priced
    19A. **Fair wages** - Not priced
    20. **Injury to persons and property and Employers indemnity**
        (See Clause no 21)
    21. **Insurance against injury to persons and property**
        The Contractor's Employer's Liability and Public Liability policies
        (which would both be involved under this heading) are often in the
        region of 0.5 to 0.6% on the value of his own contract work (excluding
        provisional sums and work by Sub-Contractors whose prices should allow
        for these insurances). However, this allowance can be included in the
        all-in hourly rate used in the calculation of 'Prices for Measured
        Work' (see page 146).
        Under Clause 21.2 no requirement is made upon the Contractor to
        insure as stated by the clause unless a provisional sum is allowed in
        the Contract Bills.
    22. **Insurance of the works against Clause 22 Perils**
        If at the Contractor's risk the insurance cover must be sufficient to
        include the full cost of reinstatement, all increases in cost,
        professional fees and any consequential costs such as demolition. The
        average provision for fire risk is £0.10% of the value of the work
        after adding for increased costs and professional fees.
        Example   £
        Contractor's Liability - Insurance of works against fire, etc.

        Contract value (including 'Preliminaries'), say   2 065 000.00
        Estimated increased costs during contract period,
          say 8%   165 000.00
                                                ----
                                          2 230 000.00
        Estimated increased costs incurred during period of
          reinstatement, say an average of 7%   156 000.00

        Fees @ 16%- say   382 000.00

                                      £2 768 000.00

        Allow £0.10% on say £2 768 000.00
        plus 9% for overheads and profit - say   £3 025.00

A PRELIMINARIES

NOTE: Insurance premiums are liable to considerable variation, depending on the Contractor, the nature of the work and the market in which the insurance is placed.

23. **Date of possession, completion and postponement** - Not priced
24. **Damages for non-completion** - Not priced
25. **Extension of time** - Not priced
26. **Loss and expense caused by matters materially affecting regular progress of the works** - Not priced
27. **Determination by Employer** - Not priced
28. **Determination by Contractor** -Not priced
29. **Works by Employer or persons employed or engaged by Employer** - Not priced
30. **Certificates and payments** - Not priced
31. **Finance (No.2) Act 1975 - Statutory tax deduction scheme** - Not priced
32. **Outbreak of hostilities** -Not priced
33. **War damage** - Not priced
34. **Antiquities** - Not priced
35. **Nominated Sub-Contractors**
    Not priced here. An amount should be added to the relevant PC sums, if required, for profit and a further sum for special attendance.
36. **Nominated Suppliers**
    Not priced here. An amount should be added to the relevant PC sums, if required, for profit.
37. **Choice of fluctuation provisions - entry in Appendix**
    The amount which the Contractor may recover under the fluctuations clauses
    (Clause nos 38, 39 and 40) will vary depending on whether the Contract is
    'firm', i.e. Clause no 38 is included, or 'fluctuating', whether the traditional method of assessment is used, i.e. Clause no 39 or the formula method, i.e. Clause no 40.
    An allowance should be made for any shortfall in reimbursement under fluctuating contracts.
38. **Contribution Levy and Tax Fluctuations** (see Clause no 37)
39. **Labour and Materials Cost and Tax Fluctuations** (see Clause no 37)
40. **Use of Price Adjustment Formulae** (see Clause no 37)

Details should include special conditions or amendments to standard conditions, the Appendix insertions and the Employer's insurance responsibilities.
Additional obligations may include the provision of a performance bond. If the Contractor is required to provide sureties for the fulfilment of the work the usual method of providing this is by a bond provided by one or more insurance companies. The cost of a performance bond depends largely on the financial standing of the applying Contractor. Figures tend to range from £0.25 to £0.50% of the contract sum.

A30-A37 EMPLOYERS' REQUIREMENTS

These include the following items but costs can only be assessed in the light of circumstances on a particular job.
Details should be given for each item and the opportunity for the Tenderer to separately price items related to fixed charges and time related charges.

A PRELIMINARIES

A30 Tendering/Sub-letting/Supply

A31 Provision, content and use of documents

A32 Management of the works

A33 Quality standards/control

A34 Security/Safety/Protection

This includes noise and pollution control, maintaining adjoining buildings, public and private roads, live services, security and the protection of work in all sections.

(i)   Control of noise, pollution and other obligations
       The Local Authority, Landlord or Management Company may impose restrictions on the timing of certain operations, particularly noisy of dust-producing operations, which may necessitate the carrying out of these works outside normal working hours or using special tools and equipment.
       The situation is most likely to occur in built-up areas such as city centres, shopping malls etc., where the site is likely to be in close proximity to offices, commercial or residential property.
(ii)   Maintenance of public and private roads
       Some additional value or allowance may be required against this item to insure/protect against damage to entrance gates, kerbs or bridges caused by extraordinary traffic in the execution of the works.

A35 Specific limitations on method/sequence/timing

This includes design constraints, method and sequence of work, access, possession and use of the site, use or disposal of materials found, start of work, working hours, employment of labour and sectional possession or partial possession etc.

A36 Facilities/Temporary work/Services

This includes offices, sanitary accommodation, temporary fences, hoardings, screens and roofs, name boards, technical and surveying equipment, temperature and humidity, telephone/facsimile installation and rental/maintenance, special lighting and other general requirements etc.
   The attainment and maintenance of suitable levels necessary for satisfactory completion of the work including the installation of joinery, suspended ceilings, lift machinery etc. is the responsibility of the contractor.
   The following is an example how to price a mobile office for a Clerk of Works

		£
(a)	Fixed charge	
	Haulage to and from site - say	64.00
	Add 9% for overheads and profit - say	6.00

		£70.00

A PRELIMINARIES

    (b)   Time related charge
          Hire charge - 15 m2 x 76 weeks @ £2.42 m2   2 760.00
          Lighting, heating and attendance on office, say 76 weeks
          @ £20.00   1 520.00
          Rates on temporary buildings based on £8.00/m2 per annum   170.00
                                                        ----
                                                       4 450.00
    Add 9% for overheads and profit - say   400.00

                                               £4 850.00

    (c)   Combined charge   £4 920.00

The installation of telephones or facsimiles for the use of the Employer, and
all related charges therewith, shall be given as a provisional sum.

A37 Operation/Maintenance of the finished building

A40-A44 CONTRACTORS GENERAL COST ITEMS

    For items A41-A44 it shall be clearly indicated whether such items are to be
'Provided by the Contractor' or 'Made available (in any part) by the
Employer'.

A40 Management and staff (Provided by the Contractor)

NOTE: The cost of site administrative staff has previously been included against
Clause no 10 of the Conditions of Contract, where Readers will find an example
of management and staff costs.

When required allow for the provision of a watchman or inspection by a security
organization.
    Other general administrative staff costs, e.g. Engineering, Programming and
production and Quantity Surveying could be priced as either fixed or time
related charges, under this section.
    For the purpose of this example - allow say 1% of contract value for other
administrative staff costs.

    (a)   Time related charge   £
          Based on 1% of £1 800 000 - say   18 000.00
          Add 9% for overheads and profit   1 620.00

                                          £19 620.00

A41 Site accommodation (Provided by the Contractor or made available by the
Employer)

    This includes all temporary offices laboratories, cabins, stores, compounds,
canteens, sanitary facilities and the like for the Contractor's and his
domestic sub-contractors' use (temporary office for a Clerk of Works is
covered under obligations and restrictions imposed by the Employer).
    Typical costs for mobile offices are as follows, based upon a twelve months
minimum hire period they exclude furniture which could add a further £12.50 -
£15.00 per week.

A PRELIMINARIES

Size	Rate per week £
12ft x 7ft 6in (8.36 m2)	27.00 (£3.23 m2)
16ft x 7ft 6in (11.15 m2)	32.00 (£2.87 m2)
22ft x 7ft 6in (15.33 m2)	37.00 (£2.42 m2)
32ft x 9ft (26.75 m2)	49.00 (£1.83 m2)

Typical rates for timber huts are as follows:

Size	Rate per week £
6ft x 12ft (6.69 m2)	10.00 (£1.50 m2)
18ft x 12ft (20.07 m2)	17.30 (£0.86 m2)
24ft x 12ft (26.75 m2)	23.00 (£0.86 m2)
30ft x 12ft (33.44 m2)	28.50 (£0.85 m2)

The following example is for one Foreman's office and two storage sheds.

(a) Fixed charge   £
    Foreman's office - haulage to and from site - say   64.00
    Storage sheds - haulage to and from site -
      erection and dismantling - say   386.00

                                                                450.00
    Add 9% for overheads and profit - say   40.00

                                                               £490.00

(b) Time related charge
    Hire charge - Foreman's office -
      15m2 x 76 weeks @ £2.42 m2   2 760.00
    Hire charge - 2 No. storage sheds - 60 m2 x 70 weeks
      @ £0.85 m2   3 570.00
                                                             ----
                                                             6 330.00
    Lighting, heating and attendance on offices say, 76 weeks
    @ £70.00   5 320.00
    Rates on temporary buildings based on £8.00/m2 per annum   880.00
                                                             ----
                                                            12 530.00
    Add 9% for overheads and profit - say   1 130.00

                                                           £13 660.00

(c) Combined charge   £14 150.00

A42 Services and facilities (provided by the Contractor or made available by the Employer)

This generally includes the provision of all of the Contractor's own services, power, lighting, fuels, water, telephone and administration, safety, health and welfare, storage of materials, rubbish disposal, cleaning, drying out, protection of work in all sections, security, maintaining public and private roads, small plant and tools and general attendance on nominated sub-contractors.

However, this section does not cover fuel for testing and commissioning permanent installations which would come under Sections Y51 and Y81.

A PRELIMINARIES

Examples of build-ups/allowances for some of the major items are provided
below:
(i)   Lighting and power for the works
        The Contractor is usually responsible for providing all temporary
        lighting and power for the works and all charges involved. On large
        sites this could be expensive and involve sub-stations and the like,
        but on smaller sites it is often limited to general lighting
        (depending upon time of year), power for power operated tools, a
        small diesel generator and some transformers.
        Typical costs are:
        Low voltage diesel generator   £50.00 - £80.00 per week
        1.5 to 10 kVA transformer   £3.75 - £7.50 per week

        A typical allowance, including charges, installation and fitting
        costs could be 1% of contract value.

        Example
        Lighting and power for the works
        (c) Combined charge
             The fixed charge would normally represent a proportion of the
             following allowance applicable to connection and supply charges.
             The residue would be allocated to time related charges.

        Dependant on the nature of the work, time of year and   £
          incidence of power operated tools, allow say,
        1% on £1 800 000   18 000.00
        Add 9% for overheads and profit   1 620.00

                                                                 £19 620.00

(ii)   Water for the works
        Charges should properly be ascertained from the local Water
        Authority. If these are not readily available, an allowance of
        £0.25% of the value of the contract is probably adequate, providing
        water can be obtained directly from the mains. Failing this, each
        case must be dealt with on its merits. In all cases an allowance
        should also be made for temporary plumbing including site storage of
        water if required.

        Useful rates for temporary plumbing include:

        Piping   £5.00 per metre
        Connection   £130.00
        Standpipe   £55.00
        Plus an allowance for barrels and hoses.

        Example
        Water for the works
        (c) Combined charge
             The fixed charge would normally represent a proportion of the
             following allowance applicable to connection and supply
             charges. The residue would be allocated to time related
             charges.   £

        £0.25% on £1 800 000   4 500.00
        Temporary plumbing - say   400.00
                                                                    ----
                                                                    4 900.00
        Add 9% for overheads and profit   440.00

                                                                   £5 340.00

A PRELIMINARIES

A42 Services and facilities - cont'd.

(iii)   Temporary telephones for the use of the Contractor
Against this item should be included the cost of installation, rental and an assessment of the cost of calls made during the contract.

Installation costs	£115.00 (for more than 12 months rental)
Rental	£26.75 per quarter
Cost of calls	For sites with one telephone allow about £18.00 per week.

Example
Temporary telephones

	£
(a) Fixed charge	
Connection charges	
for telephone	115.00
for outside bell	10.00

	125.00
Add 9% for overheads and profit - say	12.00

	£137.00

(b) Time related charge	
Rental	
for telephone 6 quarters @ £26.75	160.50
for outside bell 6 quarters @ £1.00	6.00
Calls - 76 weeks @ £18.00	1 368.00
	--------
	1 534.50
Add 9% for overheads and profit - say	138.50

	£1 673.00

(c) Combined charge	£1 810.00

(iv)   Safety, health and welfare of workpeople
The Contractor is required to comply with the Code of Welfare Conditions for the Building Industry which sets out welfare requirements as follows:

1. Shelter from inclement weather
2. Accommodation for clothing
3. Accommodation and provision for meals
4. Provision of drinking water
5. Sanitary conveniences
6. Washing facilities
7. First aid
8. Site conditions

A variety of self-contained mobile or jack-type units are available for hire and a selection of rates is given below:

	£
Kitchen with cooker, fridge, sink unit, water heater	
and basin 32ft x 9ft	£57.50 per week
Mess room with water heater, wash basin and seating	
16ft x 7ft 6in	£33.00 per week
16ft x 9ft	£37.50 per week
Welfare unit with drying rack, lockers, tables, seating, cooker,	
heater, sink and basin	
22ft x 7ft 6in	£48.00 per week

A PRELIMINARIES

(iv)   Safety, health and welfare of workpeople - cont'd

Toilets (mains type)
    Single unit   £14.50 per week
    Two unit   £29.50 per week
    Four unit   £44.70 per week
Add for wheels or jack mounting   £7.50 per week

Allowance must be made in addition for transport costs to and from
site, setting up costs, connection to mains, fuel supplies and
attendance

    Site first aid kit   £3.75 per week

A general provision to comply with the above code is often £0.50
to £0.75% of the contract value.
The costs of safety supervisors (required for firms employing
more than 20 people) are usually part of head office overhead
costs.

Example
Safety, health and welfare
(c) Combined charge
    The fixed charge would normally represent a proportion of the
    following allowance, with the majority allocated to time
    related charges.
                                                           £
    Based on £0.75% of £1 800 000 - say   13 500.00

    Add 9% for overheads and profit - say   1 215.00

                                                      £14 715.00

(v)   Removing rubbish, protective casings and coverings and cleaning the
          works on completion
          This includes removing surplus materials and final cleaning of the
          site prior to handover.
          Allow for sufficient 'bins' for the site throughout the contract
          duration and for some operatives time at the end of the contract for
          final clearing and cleaning ready for handover.

          Cost of 'bins' - approx. £26.50 each
          A general allowance of £0.20% of contract value is probably
          sufficient.

          Example
          Removing rubbish, etc., and cleaning
          (c) Combined charge
               The fixed charge would normally represent an allowance for
               final clearing of the works on completion with the residue for
               cleaning
               throughout the contract period   £

               Say £0.2% of contract value of £1 800 000   3 600.00
               Add 9% for overheads and profit - say   325.00

                                                             £3 925.00

A PRELIMINARIES

A42 Services and facilities - cont'd.

(vi)   Drying the works
         Use or otherwise of an installed heating system will probably
         determine the value to be placed against this item.
         Dependant upon the time of year, say allow 0.1% to 0.2% of the
         contract value to cover this item.

         Example
         Drying of the works   £

         (c) Combined charge
              Generally this cost is likely to be related to a fixed
              charge only

                 Say 0.2% of contract value of £1 800 000   3 600.00
                 Add 9% for overheads and profit - say   325.00

                                                                  £3 925.00

(vii)   Protecting the works from inclement weather
         In areas likely to suffer particularly inclement weather, some
         nominal allowance should be included for tarpaulins, polythene
         sheeting, battening, etc., and the effect of any delays in
         concreting or brickwork by such weather.

(viii)   Small plant and tools
         Small plant and hand tools are usually assessed as between 0.5% and
         1.5% of total labour value.

         (c) Combined charge   £

                 Say 1% of labour value of £650 000   6 500.00
                 Add 9% for overheads and profit   600.00

                                                                  £7 100.00

(ix)   General attendance on nominated sub-contractors
         In the past this item was located after each PC sum. Under SMM7 it
         is intended that two composite items (one for fixed charges and the
         other for time related charges) should be provided in the
         preliminaries bill for general attendance on all nominated sub-
         contractors.

A PRELIMINARIES

A43 Mechanical plant

This includes for cranes, hoists, personnel transport, transport, earthmoving plant, concrete plant, piling plant, paving and surfacing plant etc.
SMM6 required that items for protection or for plant be given in each section, whereas SMM7 provides for these items to be covered under A34, A42 and A43, as appropriate.

(i)   Plant
Quite often, the Contractors own plant and plant employed by sub-contractors are included in with measured rates e.g. for earthmoving, concrete or piling plant, and the Editors have adopted this method of pricing where they believe it to be 'appropriate'.

As for other items of plant e.g. cranes, hoists, these tend to be used by a variety of trades. An example of such an item might be:

Example
Tower crane
- static 30 m radius - 4/5 tonne max. load   £

(a)   Fixed charge
Haulage of crane to and from the site
erection, testing, commisioning & dismantling - say   5 000.00
Add 9% for overheads and profit - say   450.00

                                                              £5 450.00

(b)   Time related charge
Crane, on hire say 30 weeks @ 700.00 per week   21 000.00
Electricity, fuel and oil - say   4 000.00
Operator 30 weeks @ 40 hours @ £10.00   12 000.00
                                                      ----
                                                      37 000.00
Add 9% for overheads and profit - say   3 350.00

                                                     £40 350.00

(c)   Combined charge   £45 800.00

(ii)   Personnel transport
The labour rates per hour on which 'Prices for Measured Work' have been based do not cover travel and lodging allowances which must be assessed according to the appropriate working rule agreement.

Example
Personnel transport

Assuming all labour can be found within the London region, the labour value of £650 000 represents approximately 2200 man weeks. Assume each man receives an allowance of £1.30 per day or £6.50 per week of five days.
                                                                        £
(c)   Combined/time related charge
2200 man weeks at £6.50   14 300.00
Add 9% for overheads and profit - say   1 300.00

                                                     £15 600.00

A PRELIMINARIES

A44 Temporary works (Provided by the Contractor or made available by the Employer)

This includes for temporary roads, temporary walkways, access scaffolding, support scaffolding and propping, hoardings, fans, fencing etc., hardstanding and traffic regulations etc.

The Contractor should include maintaining any temporary works in connection with the items, adapting, clearing away and making good, and all notices and fees to Local Authorities and public undertakings. On fluctuating contracts, i.e. where Clause no 39 or no 40 is incorporated there is no allowance for fluctuations in respect of plant and temporary works and in such instances allowances must be made for any increases likely to occur over the contract period.

Examples of build-ups/allowances for some items are provided below:

> (i) Temporary roads, hardstandings, crossings and similar items
> Quite often consolidated bases of eventual site roads are used throughout a contract to facilitate movement of materials around the site. However, during the initial setting up of a site, with drainage works outstanding, this is not always possible and occasionally temporary roadways have to be formed and ground levels later reinstated.
>
> Typical costs are:
>
> Removal of topsoil and provision of 225 mm stone/hardcore base blinded with ashes as a temporary roadway 3.50 m wide and subsequent reinstatement:
> on level ground   £26.00 per metre
> on sloping ground including 1 m of cut
> or fill   £33.00 per metre
> Removal of topsoil and provision of 225 mm stone/hardcore base blinded with ashes as a temporary hardstanding £10.00 per m2
>
> NOTE: Any allowance for special hardcore hardstandings for piling Sub-Contractors is usually priced against the 'special attendance' clause after the relevant Prime Cost sum.

(ii) Scaffolding
The General Contractor's standing scaffolding is usually undertaken by specialist Sub-Contractors who will submit quotations based on the specific requirements of the works. It is not possible to give rates here for the various types of scaffolding that may be required but for the purposes of this section is is assumed that the cost of supplying, erecting, maintaining and subsequently dismantling the scaffolding required would amount to £12 000.00 inclusive of overheads and profit.

(iii) Temporary fencing, hoarding, screens, fans, planked footways, guardrails, gantries, and similar items.
This item must be considered in some detail as it is dependant on site perimeter, phasing of the work, work within existing buildings, etc.

A PRELIMINARIES

Useful rates include:

Hoarding 2.3 m high of 18 mm plywood with 50 x 100 mm sawn
softwood studding, rails and posts including later dismantling
    undecorated   £33.00/m (£14.35/m^2)
    decorated one side   £38.00/m (£16.50/m^2)
Pair of gates for hoarding   extra £160.00 per pair
Cleft Chestnut fencing 1.2 m high including dismantling £7.00 m

Example
Temporary hoarding   £

  (c) Combined/fixed charge
      Plywood decorated hoarding
      100 metres @ £38.00   3 800.00
      Extra for one pair of gates   160.00
                                ----
                                 3 960.00
  Add 9% for overheads and profit - say   360.00

                              £4 320.00

(iv)   Traffic regulations
      Waiting and unloading restrictions can occasionally add
considerably to costs, resulting in forced overtime or additional
weekend working. Any such restrictions must be carefully assessed
for the job in hand.

A50 Work/Materials by the Employer

A description shall be given of works by others directly engaged by the
Employer and any attendance that is required shall be priced in the same way
as works by nominated sub-contractors.

A51 Nominated sub-contractors

This section governs how nominated sub-contractors should be covered in the
bills of quantities for main contracts. Bills of quantities used for
inviting tenders from potential nominated sub-contractors should be drawn up
in accordance with SMM7 as a whole as if the work was main contractor's work.
This means, for example, that bills issued to potential nominated sub-
contractors should include preliminaries and be accompanied by the drawings.
  As much information as possible should be given in respect of nominated
sub-contractors' work in order that tenderers can make due allowance when
assessing the overall programme and establishing the contract period if not
already laid down. A simple list of the component elements of the work might
not be sufficient, but a list describing in addition the extent and possible
value of each element would be more helpful. The location of the main plant
e.g. whether in the basement or on the roof would clearly have a bearing on
tenderers' programmes. It would be good practice to seek programme
information when obtaining estimates from sub-contractors so that this can be
incorporated in the bills of quantities, for the benefit of tenderers.
  A percentage should be added for the main contractor's profit together with
items for fixed and time related charges for special attendances required by
nominated sub-contractors.
  Special attendances to include scaffolding (additional to the Contractor's
standing scaffolding), access roads, hardstandings, positioning (including
unloading, distributing, hoisting or placing in position items of significant
weight or size), storage, power, temperature and humidity etc.

A PRELIMINARIES

A52 Nominated suppliers

Goods and materials which are required to be obtained from a nominated supplier shall be given as a prime cost sum to which should be added, if required, a percentage for profit.

A53 Work by statutory authorities

Works which are to be carried out by a Local Authority or statutory undertakings shall be given as provisional sums.

A54 Provisional work

One of the more significant revisions in SMM7 is the introduction of two types of provisional sum (defined and undefined work).

The new rules require that each sum for defined work should be accompanied in the bills of quantities by a description of the work sufficiently detailed for the tenderer to make allowance for its effect in the pricing of relevant preliminaries. The information should also enable the length of time required for execution of the work to be estimated and its position in the sequence of construction to be determined and incorporated into the programme.
Where Provisional Sums are given for undefined work the Contractor will be deemed not to have made any allowance in programming, planning and pricing preliminaries.
Any provision for Contingencies shall be given as an undefined provisional sum.

A55 Dayworks

To include provisional sums for:

Labour
Materials and goods
Plant

A PRELIMINARIES

SAMPLE SUMMARY

Item		£
A20.10	Site administration	84 230.00
A20.17	Defects after completion	4 900.00
A20.22	Insurance of the works against fire, etc.	3 025.00
A36	Clerk of Work's Office	4 920.00
A40	Additional management and staff	19 620.00
A41	Contractor's accommodation	14 150.00
A42(i)	Lighting and power for the works	19 620.00
A42(ii)	Water for the works	5 340.00
A42(iii)	Temporary telephones	1 810.00
A42(iv)	Safety, health and welfare	14 715.00
A42(v)	Removing rubbish, etc., and cleaning	3 925.00
A42(vi)	Drying the works	3 925.00
A42(viii)	Small plant and tools	7 100.00
A43(i)	Mechanical plant	45 800.00
A43(ii)	Personnel transport	15 600.00
A44(ii)	Scaffolding	12 000.00
A44(iii)	Temporary hoarding	4 320.00

TOTAL £265 000.00

It is emphasized that the above is an example only of the way in which 'Preliminaries' may be priced and it is essential that for any particular contract or project the items set out in 'Preliminaries' should be assessed on their respective values.

'Preliminaries' as a percentage of a total contract will vary considerably according to each scheme and each Contractors' estimating practice. A recent national study undertaken by the authors indicates that the current trend for preliminaries to be priced at, is approximately 12 to 16%.

The value of the 'Preliminaires' in the above example is about a 14% addition to the value of work; this fact should not be forgotten when using the rates given in the trades following this section.

D GROUNDWORK
Including overheads and profit at 9.00%

Prices are applicable to excavation in firm soil. Multiplying factors for other soils
are as follows:-

	Mechanical	Hand
Clay	x 2.00	x 1.20
Compact gravel	x 3.00	x 1.50
Soft chalk	x 4.00	x 2.00
Hard rock	x 5.00	x 6.00
Running sand or silt	x 6.00	x 2.00

	Labour hours	Labour £	Material Plant £	Unit	Total rate £
D20 EXCAVATION AND FILLING					
Site preparation					
Removing trees					
600 mm - 1.50 m girth	20.00	135.16	-	nr	**135.16**
1.50 - 3.00 m girth	35.00	236.53	-	nr	**236.53**
over 3.00 m girth	50.00	337.90	-	nr	**337.90**
Removing tree stumps					
600 mm - 1.50 m girth	1.00	6.76	34.51	nr	**41.27**
1.50 - 3.00 m girth	1.50	10.14	51.77	nr	**61.91**
over 3.00 m girth	2.00	13.52	69.03	nr	**82.55**
Clearing site vegetation	0.04	0.27	-	m2	**0.27**
Lifting turf for preservation; stacking	0.35	2.37	-	m2	**2.37**
Mechanical excavation using a wheeled					
hydraulic excavator with a 0.76 m3 shovel					
Excavating topsoil to be preserved					
average 150 mm deep	0.03	0.20	0.74	m2	**0.94**
Add or deduct for each 25 mm variation					
in depth	0.01	0.03	0.12	m2	**0.15**
Excavating to reduce levels; not exceeding					
0.25 m deep	0.10	0.68	2.45	m3	**3.13**
1 m deep	0.08	0.54	1.96	m3	**2.50**
2 m deep	0.10	0.68	2.45	m3	**3.13**
4 m deep	0.12	0.81	2.94	m3	**3.75**
Mechanical excavation using a tracked					
hydraulic excavator with a 0.53 m3 bucket					
Excavating basements and the like; starting					
from reduced level; not exceeding					
1 m deep	0.10	0.68	2.83	m3	**3.51**
2 m deep	0.08	0.54	2.27	m3	**2.81**
4 m deep	0.10	0.68	2.83	m3	**3.51**
6 m deep	0.12	0.81	3.40	m3	**4.21**
8 m deep	0.14	0.95	3.97	m3	**4.92**
Excavating pits to receive bases; not					
exceeding					
0.25 m deep	0.30	2.03	8.50	m3	**10.53**
1 m deep	0.27	1.82	7.65	m3	**9.47**
2 m deep	0.30	2.03	8.50	m3	**10.53**
4 m deep	0.33	2.23	9.35	m3	**11.58**
6 m deep	0.36	2.43	10.20	m3	**12.63**
Extra over pit excavation for commencing					
below ground or reduced level					
1 m below	0.03	0.20	0.85	m3	**1.05**
2 m below	0.04	0.27	1.13	m3	**1.40**
3 m below	0.05	0.34	1.42	m3	**1.76**
4 m below	0.07	0.47	1.98	m3	**2.45**

D GROUNDWORK Including overheads and profit at 9.00%	Labour hours	Labour £	Material £	Unit	Total rate £
Excavating trenches; not exceeding 0.30 m wide; not exceeding					
0.25 m deep	0.26	1.76	7.37	m3	9.13
1 m deep	0.22	1.49	6.23	m3	7.72
2 m deep	0.26	1.76	7.37	m3	9.13
4 m deep	0.30	2.03	8.50	m3	10.53
6 m deep	0.36	2.43	10.20	m3	12.63
Excavating trenches; over 0.30 m wide; not exceeding					
0.25 m deep	0.24	1.62	6.80	m3	8.42
1 m deep	0.20	1.35	5.67	m3	7.02
2 m deep	0.24	1.62	6.80	m3	8.42
4 m deep	0.27	1.82	7.65	m3	9.47
6 m deep	0.33	2.23	9.35	m3	11.58
Extra over trench excavation for commencing below ground or reduced level					
1 m below	0.03	0.20	0.85	m3	1.05
2 m below	0.04	0.27	1.13	m3	1.40
3 m below	0.05	0.34	1.42	m3	1.76
4 m below	0.07	0.47	1.98	m3	2.45
Extra over trench excavation for curved trench	0.01	0.07	0.28	m3	0.35
Excavating for pile caps and ground beams between piles; not exceeding					
0.25 m deep	0.33	2.23	9.35	m3	11.58
1 m deep	0.30	2.03	8.50	m3	10.53
2 m deep	0.33	2.23	9.35	m3	11.58
Excavating to bench sloping ground to receive filling; not exceeding					
0.25 m deep	0.09	0.61	2.55	m3	3.16
1 m deep	0.07	0.47	1.98	m3	2.45
2 m deep	0.09	0.61	2.55	m3	3.16
Extra over any type of excavation at any depth for excavating					
alongside services	1.95	11.76		m3	11.76
around services crossing excavation	4.50	25.88		m3	25.88
below ground water level	0.10	0.68	2.83	m3	3.51
Extra over excavation for breaking out existing materials					
rock	2.30	15.54	12.15	m3	27.69
concrete	1.95	13.18	9.10	m3	22.28
reinforced concrete	2.80	18.92	13.35	m3	32.27
brickwork; blockwork or stonework	1.40	9.46	6.68	m3	16.14
Extra over excavation for breaking out existing hard pavings					
tarmacadam 75 mm thick	0.15	1.01	0.56	m2	1.57
tarmacadam and hardcore 150 mm thick	0.20	1.35	0.60	m2	1.95
concrete 150 mm thick	0.30	2.03	1.34	m2	3.37
reinforced concrete 150 mm thick	0.45	3.04	1.94	m2	4.98
Working space allowance to excavations					
reduced levels; basements and the like	0.08	0.54	2.27	m2	2.81
pits	0.17	1.15	4.82	m2	5.97
trenches	0.16	1.08	4.53	m2	5.61
pile caps and ground beams between piles	0.18	1.22	5.10	m2	6.32
Extra over excavation for working space for filling in with					
hardcore	0.10	0.68	6.99	m2	7.67
coarse ashes	0.10	0.68	9.32	m2	10.00
sand	0.10	0.68	13.12	m2	13.80
40 - 20 mm gravel	0.10	0.68	15.61	m2	16.29
in situ concrete - 7.50 N/mm2					
- 40 mm aggregate (1:8)	0.72	5.04	29.07	m2	34.11

D GROUNDWORK Including overheads and profit at 9.00%	Labour hours	Labour £	Material £	Unit	Total rate £
D20 EXCAVATION AND FILLING - cont'd					
Hand excavation					
Excavating topsoil to be preserved					
average 150 mm deep	0.25	1.69	-	m2	1.69
Add or deduct for each 25 mm variation					
in depth	0.04	0.27	-	m2	0.27
Excavating to reduce levels; not exceeding					
0.25 m deep	1.80	12.16	-	m3	12.16
1 m deep	2.00	13.52	-	m3	13.52
2 m deep	2.20	14.87	-	m3	14.87
4 m deep	2.40	16.22	-	m3	16.22
Excavating basements and the like; starting					
from reduced level; not exceeding					
1 m deep	2.40	16.22	-	m3	16.22
2 m deep	2.70	18.25	-	m3	18.25
4 m deep	3.50	23.65	-	m3	23.65
6 m deep	4.30	29.06	-	m3	29.06
8 m deep	5.20	35.14	-	m3	35.14
Excavating pits to receive bases; not					
exceeding					
0.25 m deep	2.80	18.92	-	m3	18.92
1 m deep	3.00	20.27	-	m3	20.27
2 m deep	3.50	23.65	-	m3	23.65
4 m deep	4.50	30.41	-	m3	30.41
6 m deep	5.70	38.52	-	m3	38.52
Extra over pit excavation for commencing					
below ground or reduced level					
1 m below	0.50	3.38	-	m3	3.38
2 m below	1.00	6.76	-	m3	6.76
3 m below	1.50	10.14	-	m3	10.14
4 m below	2.00	13.52	-	m3	13.52
Excavating trenches; not exceeding					
0.30 m wide; not exceeding					
0.25 m deep	2.50	16.90	-	m3	16.90
1 m deep	2.75	18.58	-	m3	18.58
2 m deep	3.20	21.63	-	m3	21.63
4 m deep	4.05	27.37	-	m3	27.37
6 m deep	5.15	34.80	-	m3	34.80
Excavating trenches; over 0.30 m wide;					
not exceeding					
0.25 m deep	2.30	15.54	-	m3	15.54
1 m deep	2.50	16.90	-	m3	16.90
2 m deep	2.90	19.60	-	m3	19.60
4 m deep	3.70	25.00	-	m3	25.00
6 m deep	4.70	31.76	-	m3	31.76
Extra over trench excavation for commencing					
below ground or reduced level					
1 m below	0.50	3.38	-	m3	3.38
2 m below	1.00	6.76	-	m3	6.76
3 m below	1.50	10.14	-	m3	10.14
4 m below	2.00	13.52	-	m3	13.52
Extra over trench excavation for curved					
trench	0.50	3.38	-	m3	3.38
Excavating for pile caps and ground beams					
between piles; not exceeding					
0.25 m deep	3.30	22.30	-	m3	22.30
1 m deep	3.60	24.33	-	m3	24.33
2 m deep	4.20	28.38	-	m3	28.38
Excavating to bench sloping ground					
to receive filling; not exceeding					
0.25 m deep	1.50	10.14	-	m3	10.14
1 m deep	1.80	12.16	-	m3	12.16
2 m deep	2.00	13.52	-	m3	13.52

D GROUNDWORK Including overheads and profit at 9.00%	Labour hours	Labour £	Material £	Unit	Total rate £
Extra over any type of excavation at any depth for excavating					
alongside services	1.00	6.76	-	m3	6.76
around services crossing excavation	2.00	13.52	-	m3	13.52
Extra over any type of excavation at any depth for excavating					
below ground water level	0.35	2.37	-	m3	2.37
Extra over excavation for breaking out existing materials					
rock	5.00	33.79	7.19	m3	40.98
concrete	4.50	30.41	6.00	m3	36.41
reinforced concrete	6.00	40.55	8.39	m3	48.94
brickwork; blockwork or stonework	3.00	20.27	3.60	m3	23.87
Extra over excavation for breaking out existing hard pavings					
tarmacadam 75 mm thick	0.40	2.70	0.48	m2	3.18
tarmacadam and hardcore 150 mm thick	0.50	3.38	0.60	m2	3.98
concrete 150 mm thick	0.70	4.73	0.84	m2	5.57
reinforced concrete 150 mm thick	0.90	6.08	1.20	m2	7.28
Extra for taking up precast concrete paving slabs	0.30	2.03	-	m2	2.03
Working space allowance to excavations					
reduced levels; basements and the like	2.60	17.57	-	m2	17.57
pits	2.70	18.25	-	m2	18.25
trenches	2.35	15.88	-	m2	15.88
pile caps and ground beams between piles	2.75	18.58	-	m2	18.58
Extra over excavation for working space for filling in with					
hardcore	0.80	5.41	5.52	m2	10.93
coarse ashes	0.72	4.87	7.85	m2	12.72
sand	0.80	5.41	11.65	m2	17.06
40 - 20 mm gravel	0.80	5.41	14.14	m2	19.55
in situ concrete - 7.50 N/mm2					
- 40 mm aggregate (1:8)	1.10	7.70	27.60	m2	35.30
Earthwork support (average 'risk' prices) Not exceeding 2 m between opposing faces; not exceeding					
1 m deep	0.10	0.68	0.31	m2	0.99
2 m deep	0.13	0.88	0.34	m2	1.22
4 m deep	0.16	1.08	0.40	m2	1.48
6 m deep	0.19	1.28	0.48	m2	1.76
2 - 4 m between opposing faces; not exceeding					
1 m deep	0.11	0.74	0.34	m2	1.08
2 m deep	0.14	0.95	0.40	m2	1.35
4 m deep	0.17	1.15	0.48	m2	1.63
6 m deep	0.21	1.42	0.57	m2	1.99
Over 4 m between opposing faces; not exceeding					
1 m deep	0.12	0.81	0.40	m2	1.21
2 m deep	0.15	1.01	0.48	m2	1.49
4 m deep	0.19	1.28	0.57	m2	1.85
6 m deep	0.24	1.62	0.69	m2	2.31
Earthwork support (open boarded) Not exceeding 2 m between opposing faces; not exceeding					
1 m deep	0.30	2.03	0.95	m2	2.98
2 m deep	0.38	2.57	1.14	m2	3.71
4 m deep	0.48	3.24	1.43	m2	4.67
6 m deep	0.60	4.05	1.64	m2	5.69
2 - 4 m between opposing faces; not exceeding					
1 m deep	0.34	2.30	1.14	m2	3.44
2 m deep	0.42	2.84	1.45	m2	4.29
4 m deep	0.54	3.65	1.72	m2	5.37
6 m deep	0.66	4.46	2.02	m2	6.48

D GROUNDWORK Including overheads and profit at 9.00%	Labour hours	Labour £	Material £	Unit	Total rate £

D20 EXCAVATION AND FILLING - cont'd

Earthwork support (open boarded) - cont'd
Over 4 m between opposing faces;
not exceeding

1 m deep	0.38	2.57	1.34	m2	3.91
2 m deep	0.48	3.24	1.68	m2	4.92
4 m deep	0.60	4.05	2.02	m2	6.07
6 m deep	0.76	5.14	2.33	m2	7.47

Earthwork support (close boarded)
Not exceeding 2 m between opposing faces;
not exceeding

1 m deep	0.80	5.41	1.72	m2	7.13
2 m deep	1.00	6.76	2.10	m2	8.86
4 m deep	1.25	8.45	2.58	m2	11.03
6 m deep	1.56	10.54	3.05	m2	13.59

2 - 4 m between opposing faces; not exceeding

1 m deep	0.88	5.95	1.91	m2	7.86
2 m deep	1.10	7.43	2.38	m2	9.81
4 m deep	1.25	8.45	2.86	m2	11.31
6 m deep	1.65	11.15	3.34	m2	14.49

Over 4 m between opposing faces;
not exceeding

1 m deep	0.97	6.56	2.10	m2	8.66
2 m deep	1.20	8.11	2.67	m2	10.78
4 m deep	1.40	9.46	3.15	m2	12.61
6 m deep	1.90	12.84	3.62	m2	16.46

Extra over earthwork support for

extending into running silt or the like	0.50	3.38	0.57	m2	3.95
extending below ground water level	0.30	2.03	0.29	m2	2.32
support next to roadways	0.40	2.70	0.48	m2	3.18
curved support	0.03	0.20	0.32	m2	0.52
support left in	0.65	4.39	11.75	m2	16.14

Mechanical disposal of excavated materials
Removing from site to tip not exceeding
13 km (using lorries)

	-	-	-	m3	8.31

Depositing on site in spoil heaps

average 25 m distant	-	-	-	m3	0.74
average 50 m distant	-	-	-	m3	1.23
average 100 m distant	-	-	-	m3	1.96
average 200 m distant	-	-	-	m3	2.94

Spreading on site

average 25 m distant	0.20	1.35	0.74	m3	2.09
average 50 m distant	0.20	1.35	1.23	m3	2.58
average 100 m distant	0.20	1.35	1.96	m3	3.31
average 200 m distant	0.20	1.35	2.94	m3	4.29

Hand disposal of excavated materials
Removing from site to tip not exceeding
13 km (using lorries)

	1.00	6.76	11.92	m3	18.68

Depositing on site in spoil heaps

average 25 m distant	1.00	6.76	-	m3	6.76
average 50 m distant	1.30	8.79	-	m3	8.79
average 100 m distant	1.90	12.84	-	m3	12.84
average 200 m distant	2.80	18.92	-	m3	18.92

Spreading on site

average 25 m distant	1.30	8.79	-	m3	8.79
average 50 m distant	1.60	10.81	-	m3	10.81
average 100 m distant	2.20	14.87	-	m3	14.87
average 200 m distant	3.10	20.95	-	m3	20.95

D GROUNDWORK Including overheads and profit at 9.00%	Labour hours	Labour £	Material £	Unit	Total rate £
Mechanical filling with excavated material					
Filling to excavations	0.15	1.01	1.96	m3	2.97
Filling to make up levels over					
250 mm thick; depositing; compacting	0.22	1.49	1.47	m3	2.96
Filling to make up levels not exceeding					
250 mm thick; compacting	0.26	1.76	1.96	m3	3.72
Mechanical filling with imported soil					
Filling to make up levels over 250 mm thick;					
depositing; compacting in layers	0.22	1.49	11.01	m3	12.50
Filling to make up levels not exceeding					
250 mm thick; compacting	0.26	1.76	14.17	m3	15.93
Mechanical filling with hardcore; PC £6.75/m3					
Filling to excavations	0.17	1.15	11.65	m3	12.80
Filling to make up levels over					
250 mm thick; depositing; compacting	0.26	1.76	11.16	m3	12.92
Filling to make up levels not exceeding					
250 mm thick; compacting	0.30	2.03	14.22	m3	16.25
Mechanical filling with granular fill type 1; **PC £9.05/t (PC £15.40/m3)**					
Filling to excavations	0.17	1.15	23.44	m3	24.59
Filling to make up levels over					
250 mm thick; depositing; compacting	0.26	1.76	22.94	m3	24.70
Filling to make up levels not exceeding					
250 mm thick; compacting	0.30	2.03	29.31	m3	31.34
Mechanical filling with granular fill type 2; **PC £8.75/t (PC £15.00/m3)**					
Filling to excavations	0.17	1.15	22.89	m3	24.04
Filling to make up levels over					
250 mm thick; depositing; compacting	0.26	1.76	22.40	m3	24.16
Filling to make up levels not exceeding					
250 mm thick; compacting	0.30	2.03	28.61	m3	30.64
Mechanical filling with coarse ashes; PC £9.00/m3					
Filling to make up levels over					
250 mm thick; depositing; compacting	0.24	1.62	14.76	m3	16.38
Filling to make up levels not exceeding					
250 mm thick; compacting	0.28	1.89	18.88	m3	20.77
Mechanical filling with sand; **PC £8.90/t (PC £14.25/m3)**					
Filling to make up levels over					
250 mm thick; depositing; compacting	0.26	1.76	21.38	m3	23.14
Filling to make up levels not exceeding					
250 mm thick; compacting	0.30	2.03	27.30	m3	29.33
Hand filling with excavated material					
Filling to excavations	1.00	6.76	-	m3	6.76
Filling to make up levels over					
250 mm thick; depositing; compacting	1.10	7.43	-	m3	7.43
Filling to make up levels exceeding 250 mm					
thick; multiple handling via spoil heap					
average - 25 m distant	2.40	16.22	-	m3	16.22
Filling to make up levels not exceeding					
250 mm thick; compacting	1.35	9.12	-	m3	9.12
Hand filling with imported soil					
Filling to make up levels over 250 mm thick;					
depositing; compacting in layers	1.10	7.43	9.54	m3	16.97
Filling to make up levels not exceeding					
250 mm thick; compacting	1.35	9.12	12.21	m3	21.33

D GROUNDWORK Including overheads and profit at 9.00%	Labour hours	Labour £	Material £	Unit	Total rate £
D20 EXCAVATION AND FILLING - cont'd					
Hand filling with hardcore; PC £6.75/m3					
Filling to excavations	1.30	8.79	9.20	m3	**17.99**
Filling to make up levels over					
250 mm thick; depositing; compacting	0.55	3.72	14.59	m3	**18.31**
Filling to make up levels not exceeding					
250 mm thick; compacting	0.66	4.46	18.25	m3	**22.71**
Hand filling with granular fill type 1; **PC £9.05/t (PC £15.40/m3)**					
Filling to excavations	1.30	8.79	20.98	m3	**29.77**
Filling to make up levels over					
250 mm thick; depositing; compacting	0.55	3.72	26.38	m3	**30.10**
Filling to make up levels not exceeding					
250 mm thick; compacting	0.66	4.46	33.33	m3	**37.79**
Hand filling with granular fill type 2; **PC £8.75/t (PC £15.00/m3)**					
Filling to excavations	1.30	8.79	20.44	m3	**29.23**
Filling to make up levels over					
250 mm thick; depositing; compacting	0.55	3.72	25.83	m3	**29.55**
Filling to make up levels not exceeding					
250 mm thick; compacting	0.66	4.46	32.63	m3	**37.09**
Hand filling with coarse ashes; PC £9.00/m3					
Filling to make up levels over					
250 mm thick; depositing; compacting	0.50	3.38	17.95	m3	**21.33**
Filling to make up levels not exceeding					
250 mm thick; compacting	0.60	4.05	22.56	m3	**26.61**
Hand filling with sand; PC £8.90/t **(PC £14.25/m3)**					
Filling to excavations	0.50	3.38	24.32	3	**27.70**
Filling to make up levels over					
250 mm thick; depositing; compacting	0.65	4.39	25.79	m3	**30.18**
Filling to make up levels not exceeding					
250 mm thick; compacting	0.77	5.20	32.41	m3	**37.61**
Surface packing to filling					
To vertical or battered faces	0.18	1.22	0.51	m2	**1.73**
Surface treatments					
Compacting					
surfaces of ashes	0.04	0.27	-	m2	**0.27**
filling; blinding with ashes	0.08	0.54	0.67	m2	**1.21**
filling; blinding with sand	0.08	0.54	1.17	m2	**1.71**
bottoms of excavations	0.05	0.34	-	m2	**0.34**
Trimming					
sloping surfaces	0.18	1.22	-	m2	**1.22**
sloping surfaces; in rock	1.00	6.23	2.40	m2	**8.63**
Filter membrane; one layer; laid on earth to receive granular material					
'Terram 500'	0.05	0.34	0.40	m2	**0.74**
'Terram 700'	0.05	0.34	0.40	m2	**0.74**
'Terram 1000'	0.06	0.41	0.53	m2	**0.94**

D GROUNDWORK Including overheads and profit at 9.00 % & 5.00%	Labour hours	Labour £	Material £	Unit	Total rate £

D30 CAST IN PLACE PILING

The following approximate prices for the
quantities of piling quoted, are for work on
clear open sites with reasonable access.
They are based on 500 mm nominal diameter
piles normal concrete mix 20 N/mm2 reinforced
for loading up to 40 000 kg depending on
ground conditions and include up to 0.16 m
of projecting reinforcement at top of pile.
The prices do not allow for removal of spoil.

	Labour hours	Labour £	Material £	Unit	Total rate £
Tripod bored cast-in-place concrete piles Provision of all plant; including bringing to and removing from site; maintenance, erection and dismantling at each pile position for 100 nr piles	-	-	-	item	8000.00
500 mm diameter piles; reinforced; 10 m long	-	-	-	nr	520.00
Add for additional pile length up to 15 m	-	-	-	m	52.00
Deduct for reduction in pile length	-	-	-	m	19.75
Cutting off heads of piles*	1.30	13.40	-	nr	13.40
Blind boring; 500 mm diameter	-	-	-	m	17.65
Plant; provisional standing time; per rig unit	-	-	-	hour	67.50
Extra for boring through artificial obstructions; per rig unit	-	-	-	hour	93.50
Testing piles; working to 600 kN/t; using tension piles as reaction					
first pile	-	-	-	nr	3650.00
subsequent piles	-	-	-	nr	3115.00
Rotary bored cast-in-place concrete piles Provision of all plant; including bringing to and removing from site; maintenance, erection and dismantling at each pile position for 100 nr piles	-	-	-	item	7785.00
500 mm diameter piles; reinforced; 10 m long	-	-	-	nr	260.00
Add for additional pile length up to 15 m	-	-	-	m	26.00
Deduct for reduction in pile length	-	-	-	m	12.50
Cutting off heads of piles*	1.30	13.40	-	nr	13.40
Blind boring; 500 mm diameter	-	-	-	m	14.55
Plant; provisional standing time; per rig unit	-	-	-	hour	135.00
Extra for boring through artificial obstructions; per rig unit	-	-	-	hour	156.00
Testing piles; working to 600 kN/t; using tension piles as reaction					
first pile	-	-	-	nr	3635.00
subsequent piles	-	-	-	nr	3115.00

D33 STEEL PILING

	Labour hours	Labour £	Material £	Unit	Total rate £
'Frodingham' steel sheet piling; BS 4360; grade 43A; pitched and driven Provision of all plant; including bringing to and removing from site; maintenance, erection and dismantling; assuming one rig for 1500 m2 of piling	-	-	-	item	4600.00
Type 1N; 99.1 kg/m2	2.60	26.84	41.22	m2	68.06
Type 2N; 112.3 kg/m2	2.75	28.39	46.63	m2	75.02
Type 3N; 137.1 kg/m2	3.00	30.97	57.04	m2	88.01
Type 4N; 170.8 kg/m2	3.35	34.58	71.20	m2	105.78
Type 5N; 236.9 kg/m2	4.00	41.29	98.68	m2	139.97
Burn off tops of piles level	1.00	10.32	-	m	10.32

*Work normally carried out by the Main Contractor

D GROUNDWORK Including overheads and profit at 9.00% & 5.00%	Labour hours	Labour £	Material £	Unit	Total rate £
D33 STEEL PILING - cont'd					
'Frodingham' steel sheet piling; extract only					
Provision of all plant; including bringing to and removing from site; maintenance, erection and dismantling; assuming					
one rig as before	-	-	-	item	3545.00
Type 1N; 99.1 kg/m2	1.25	12.90	-	m2	12.90
Type 2N; 112.3 kg/m2	1.25	12.90	-	m2	12.90
Type 3N; 137.1 kg/m2	1.25	12.90	-	m2	12.90
Type 4N; 170.8 kg/m2	1.25	12.90	-	m2	12.90
Type 5N; 236.9 kg/m2	1.40	14.45	-	m2	14.45
D40 DIAPHRAGM WALLING					
Provision of all plant; including bringing to and removing from site; maintenance, erection and dismantling; assuming one rig for 1000 m2 of walling	-	-	-	item	36350.0
Excavation for diaphragm wall; excavated material removed from site; Bentonite slurry supplied and disposed of					
600 mm thick walls	-	-	-	m3	182.00
1000 mm thick walls	-	-	-	m3	145.35
Ready mixed reinforced in situ concrete; normal Portland cement; mix 30.00 N/mm2-10mm aggregate in walls					
over 300 mm thick	-	-	-	m3	94.00
Reinforcement bars; BS 4461 cold rolled deformed square high yield steel bars; straight or bent					
25 - 40 mm	-	-	-	t	814.00
20 mm	-	-	-	t	814.00
16 mm	-	-	-	t	854.60
Formwork 75 mm thick to form chases	-	-	-	m2	46.72
Construct twin guide walls in reinforced concrete; together with reinforcement and formwork along the axis of the diaphragm wall*	-	-	-	m	207.65
Plant; provisional standing time; per rig unit	-	-	-	hour	225.00
Extra for delay through obstructions; per rig unit	-	-	-	hour	275.00
D50 UNDERPINNING					
Mechanical excavation using a wheeled hydraulic off-centre excavator with a 0.24 m3 bucket					
Excavating preliminary trenches; not exceeding					
1 m deep	0.25	1.69	5.72	m3	7.41
2 m deep	0.30	2.03	6.87	m3	8.90
4 m deep	0.35	2.37	8.01	m3	10.38
Extra for breaking up					
concrete 150 mm thick	0.70	4.73	0.84	m2	5.57
Hand excavation					
Excavating preliminary trenches; not exceeding					
1 m deep	2.90	19.60	-	m3	19.60
2 m deep	3.30	22.30	-	m3	22.30
4 m deep	4.25	28.72	-	m3	28.72
Extra for breaking up					
concrete 150 mm thick	0.30	2.03	1.74	m2	3.77

*Work normally carried out by the Main Contractor

D GROUNDWORK Including overheads and profit at 9.00%	Labour hours	Labour £	Material £	Unit	Total rate £
Excavating underpinning pits starting from 1 m below ground level; not exceeding					
0.25 m deep	4.40	29.74	-	m3	29.74
1 m deep	4.80	32.44	-	m3	32.44
2 m deep	5.75	38.86	-	m3	38.86
Excavating underpinning pits starting from 2 m below ground level; not exceeding					
0.25 m deep	5.40	36.49	-	m3	36.49
1 m deep	5.80	39.20	-	m3	39.20
2 m deep	6.75	45.62	-	m3	45.62
Excavating underpinning pits starting from 4 m below ground level; not exceeding					
0.25 m deep	6.40	43.25	-	m3	43.25
1 m deep	6.80	45.95	-	m3	45.95
2 m deep	7.75	52.37	-	m3	52.37
Extra over any type of excavation at any depth for excavating below ground water level	0.35	2.37	-	m3	2.37
Earthwork support (open boarded) **in 3 m lengths** To preliminary trenches; not exceeding 2 m between opposing faces; not exceeding					
1 m deep	0.40	2.70	1.24	m2	3.94
2 m deep	0.50	3.38	1.53	m2	4.91
4 m deep	0.64	4.33	1.91	m2	6.24
To underpinning pits; not exceeding 2 m between opposing faces; not exceeding					
1 m deep	0.44	2.97	1.34	m2	4.31
2 m deep	0.55	3.72	1.72	m2	5.44
4 m deep	0.70	4.73	2.10	m2	6.83
Earthwork support (closed boarded) **in 3 m lengths** To preliminary trenches; not exceeding 2 m between opposing faces; not exceeding					
1 m deep	1.00	6.76	2.10	m2	8.86
2 m deep	1.25	8.45	2.67	m2	11.12
4 m deep	1.55	10.47	3.24	m2	13.71
To underpinning pits; not exceeding 2 m between opposing faces; not exceeding					
1 m deep	1.10	7.43	2.29	m2	9.72
2 m deep	1.38	9.33	2.86	m2	12.19
4 m deep	1.70	11.49	3.62	m2	15.11
Extra over all types of earthwork support for earthwork support left in	0.75	5.07	12.02	m2	17.09
Preparation/cutting away Cutting away projecting plain concrete foundations					
150 x 150 mm	0.16	1.08	0.38	m	1.46
150 x 225 mm	0.24	1.62	0.58	m	2.20
150 x 300 mm	0.32	2.16	0.77	m	2.93
300 x 300 mm	0.63	4.26	1.51	m	5.77
Cutting away masonry					
one course high	0.05	0.34	0.12	m	0.46
two courses high	0.14	0.95	0.34	m	1.29
three courses high	0.27	1.82	0.65	m	2.47
four courses high	0.45	3.04	1.08	m	4.12
Preparing underside of existing work to receive new underpinning					
380 mm wide	0.60	4.05	-	m	4.05
600 mm wide	0.80	5.41	-	m	5.41
900 mm wide	1.00	6.76	-	m	6.76
1200 mm wide	1.20	8.11	-	m	8.11

D GROUNDWORK Including overheads and profit at 9.00%	Labour hours	Labour £	Material £	Unit	Total rate £
D50 UNDERPINNING - cont'd					
Hand disposal of excavated materials					
Removing from site to tip not exceeding					
13 km (using lorries)	1.00	6.76	11.24	m3	18.00
Hand filling with excavated material					
Filling to excavations	1.00	6.76	-	m3	6.76
Surface treatments					
Compacting bottoms of excavations	0.05	0.34	-	m2	0.34
Plain insitu ready mixed concrete;					
11.50 N/mm2 - 40 mm aggregate (1:3:6);					
poured against faces of excavation					
Underpinning					
over 450 mm thick	2.70	18.89	51.59	m3	70.48
150 - 450 mm thick	3.10	21.69	51.59	m3	73.28
not exceeding 150 mm thick	3.70	25.89	51.59	m3	77.48
Plain insitu ready mixed concrete;					
21.00 N/mm2 - 20 mm aggregate (1:2:4);					
poured against faces of excavation					
Underpinning					
over 450 mm thick	2.70	18.89	55.44	m3	74.33
150 - 450 mm thick	3.10	21.69	55.44	m3	77.13
not exceeding 150 mm thick	3.70	25.89	55.44	m3	81.33
Extra for working around reinforcement	0.30	2.10	-	m3	2.10
Sawn formwork					
Sides of foundations in underpinning					
over 1 m high	1.60	13.08	5.24	m2	18.32
not exceeding 250 mm high	0.55	4.50	1.44	m	5.94
250 - 500 mm high	0.85	6.95	2.74	m	9.69
500 mm - 1 m high	1.30	10.63	5.24	m	15.87
Reinforcement					
Reinforcement bars; BS 4449; hot rolled					
plain round mild steel bars; bent					
20 mm PC £374.50	17.00	137.49	466.29	t	603.78
16 mm PC £380.50	20.00	161.76	476.83	t	638.59
12 mm PC £410.50	23.00	186.02	514.84	t	700.86
10 mm PC £434.50	27.00	218.37	545.98	t	764.35
8 mm PC £465.50	31.00	250.72	585.13	t	835.85
6 mm PC £509.50	36.00	291.16	639.16	t	930.32
8 mm; links or the like PC £465.50	42.00	339.69	596.15	t	935.84
6 mm; links or the like PC £509.50	49.00	396.30	653.86	t	1050.16
Reinforcement bars; BS 4461; cold worked					
deformed square high yield steel bars; bent					
20 mm PC £378.50	17.00	137.49	470.87	t	608.36
16 mm PC £384.50	20.00	161.76	481.41	t	643.17
12 mm PC £414.50	23.00	186.02	519.42	t	705.44
10 mm PC £438.50	27.00	218.37	550.56	t	768.93
8 mm PC £469.50	31.00	250.72	589.71	t	840.43
6 mm PC £513.50	36.00	291.16	643.74	t	934.90
Common bricks; PC £117.00/1000;					
in cement mortar (1:3)					
Walls in underpinning					
one brick thick	2.40	27.97	18.76	m2	46.73
one and a half brick thick	3.30	38.45	28.21	m2	66.66
two brick thick	4.10	47.77	37.53	m2	85.30
Add or deduct for variation of £10.00/1000 in					
PC of common bricks					
one brick thick	-	-	-	m2	1.37
one and a half brick thick	-	-	-	m2	2.06
two brick thick	-	-	-	m2	2.75

D GROUNDWORK Including overheads and profit at 9.00%	Labour hours	Labour £	Material £	Unit	Total rate £
Class A engineering bricks; PC £340.00/1000; **in cement mortar (1:3)**					
Walls in underpinning					
one brick thick	2.60	30.30	56.96	m2	87.26
one and a half brick thick	3.55	41.36	85.49	m2	126.85
two brick thick	4.40	51.27	113.90	m2	165.17
Add or deduct for variation of £10.00/1000 in PC of common bricks					
one brick thick	-	-	-	m2	1.37
one and a half brick thick	-	-	-	m2	2.06
two brick thick	-	-	-	m2	2.75
Class B engineering bricks; PC £192.00/1000; **in cement mortar (1:3)**					
Walls in underpinning					
one brick thick	2.60	30.30	29.06	m2	59.36
one and a half brick thick	3.55	41.36	43.66	m2	85.02
two brick thick	4.40	51.27	58.12	m2	109.39
Add or deduct for variation of £10.00/1000 in PC of common bricks					
one brick thick	-	-	-	m2	1.37
one and a half brick thick	-	-	-	m2	2.06
two brick thick	-	-	-	m2	2.75
'Pluvex' (hessian based) damp proof course **or similar; PC £3.43/m2; 200 mm laps;** **in cement mortar (1:3)**					
Horizontal					
over 225 mm wide	0.25	2.91	4.02	m2	6.93
not exceeding 225 mm wide	0.50	5.83	4.11	m2	9.94
'Hyload' (pitch polymer) damp proof course **or similar; PC £4.09/m2; 150 mm laps;** **in cement mortar (1:3)**					
Horizontal					
over 225 mm wide	0.25	2.91	4.80	m2	7.71
not exceeding 225 mm wide	0.50	5.83	4.91	m2	10.74
'Ledkore' grade A (bitumen based lead cored) **damp proof course or similar; PC £12.04/m2;** **200 mm laps; in cement mortar (1:3)**					
Horizontal					
over 225 mm wide	0.33	3.85	14.10	m2	17.95
not exceeding 225 mm wide	0.66	7.69	14.43	m2	22.12
Two courses of slates in cement mortar (1:3)					
Horizontal					
over 225 mm wide	1.50	17.48	20.95	m2	38.43
not exceeding 225 mm wide	2.50	29.13	21.32	m2	50.45
Wedging and pinning					
To underside of existing construction with slates in cement mortar (1:3)					
102 mm wall	1.10	12.82	3.96	m	16.78
215 mm wall	1.30	15.15	5.99	m	21.14
317 mm wall	1.50	17.48	7.96	m	25.44

E IN SITU CONCRETE/LARGE PRECAST CONCRETE
Including overheads and profit at 9.00%

BASIC CONCRETE PRICES

Concrete aggregates (£/tonne)	£		£		£
40 mm all-in	9.14	40 mm shingle	9.14	10 mm shingle	9.24
20 mm all-in	9.24	20 mm shingle	9.24	sharp sand	8.90

Formwork items	£			£
plywood (£/m2)	8.20	timber (£/m3)		175.00

Lightweight aggregates 'Lytag' (£/m3)			
'fines'	20.72	6-12 mm granular	21.40

Portland cement (£/tonne)			
normal - in bags	60.89	normal - in bulk to silos	54.41
sulphate - resisting - plus	9.00	rapid - hardening - plus	3.50

Tying wire for reinforcement - £0.47/kg

MIXED CONCRETE PRICES (£/m3)

	Mix 7.50 N/mm2 - 40mm aggre-gate (1:8)	Mix 11.50 N/mm2 - 40mm aggre-gate (1:3:6)	Mix 15.00 N/mm2 -40mm aggre-gate	Mix 21.00 N/mm2 - 20mm aggre-gate (1:2:4)	Mix 26.00 N/mm2 - 20mm aggre-gate (1:1.5:3)	Mix 31.00 N/mm2 - 20mm aggre-gate	Mix 40.00 N/mm2 - 20mm aggre-gate
The following prices are for ready or site mixed concrete ready for placing including 5% for waste and 9% for overheads and profit							
	£	£	£	£	£	£	£
Ready mixed concrete							
Normal Portland cement	46.00	46.90	48.10	50.40	52.65	55.00	58.40
Sulphate - resistant cement	48.10	49.20	50.35	53.80	56.10	58.40	63.00
Normal Portland cement with water-repellentd additive	48.30	49.20	50.40	52.70	54.95	57.30	60.70
Normal Portland cement; air-entrained	47.00	47.90	49.10	51.40	53.65	56.00	59.40
Lightweight concrete using sintered PFA aggregate	-	-	66.40	68.70	71.00	73.25	-
Site mixed concrete							
Normal Portland cement	-	48.00	50.50	53.00	57.50	60.00	65.00
Sulphate - resistant cement	-	50.50	53.30	56.30	61.70	65.00	70.00

	Labour hours	Labour £	Material £	Unit	Total rate £
E10 IN SITU CONCRETE					
Plain in situ ready mixed concrete; 11.50 N/mm2 - 40 mm aggregate (1:3:6)					
Foundations	1.60	11.20	46.90	m3	**58.10**
Isolated foundations	1.90	13.30	46.90	m3	**60.20**
Beds					
over 450 mm thick'	1.25	8.75	46.90	m3	**55.65**
150 - 450 mm thick	1.65	11.55	46.90	m3	**58.45**
not exceeding 150 mm thick	2.40	16.79	46.90	m3	**63.69**
Filling to hollow walls					
not exceeding 150 mm thick	4.25	29.74	46.90	m3	**76.64**

E IN SITU CONCRETE/LARGE PRECAST CONCRETE Including overheads and profit at 9.00%	Labour hours	Labour £	Material £	Unit	Total rate £
Plain in situ ready mixed concrete; **11.50 N/mm2 - 40 mm aggregate (1:3:6);** **poured on or against earth or** **unblinded hardcore**					
Foundations	1.70	11.90	49.25	m3	61.15
Isolated foundations	2.00	14.00	49.25	m3	63.25
Beds					
over 450 mm thick	1.30	9.10	49.25	m3	58.35
150 - 450 mm thick	1.75	12.25	49.25	m3	61.50
not exceeding 150 mm thick	2.50	17.49	49.25	m3	66.74
Plain in situ ready mixed concrete; **21.00 N/mm2 - 20 mm aggregate (1:2:4)**					
Foundations	1.60	11.20	50.40	m3	61.60
Isolated foundations	1.90	13.30	50.40	m3	63.70
Beds					
over 450 mm thick	1.25	8.75	50.40	m3	59.15
150 - 450 mm thick	1.65	11.55	50.40	m3	61.95
not exceeding 150 mm thick	2.40	16.79	50.40	m3	67.19
Filling to hollow walls					
not exceeding 150 mm thick	4.25	29.74	50.40	m3	80.14
Plain in situ ready mixed concrete; **21.00 N/mm2 - 20 mm aggregate (1:2:4);** **poured on or against earth or** **unblinded hardcore**					
Foundations	1.70	11.90	52.92	m3	64.82
Isolated foundations	2.00	14.00	52.92	m3	66.92
Beds					
over 450 mm thick	1.30	9.10	52.92	m3	62.02
150 - 450 mm thick	1.75	12.25	52.92	m3	65.17
not exceeding 150 mm thick	2.50	17.49	52.92	m3	70.41
Reinforced in situ ready mixed concrete; **21.00 N/mm2 - 20 mm aggregate (1:2:4)**					
Foundations	2.00	14.00	50.40	m3	64.40
Ground beams	3.50	24.49	50.40	m3	74.89
Isolated foundations	2.30	16.09	50.40	m3	66.49
Beds					
over 450 mm thick	1.60	11.20	50.40	m3	61.60
150 - 450 mm thick	2.00	14.00	50.40	m3	64.40
not exceeding 150 mm thick	2.75	19.24	50.40	m3	69.64
Slabs					
over 450 mm thick	3.10	21.69	50.40	m3	72.09
150 - 450 mm thick	3.50	24.49	50.40	m3	74.89
not exceeding 150 mm thick	4.40	30.79	50.40	m3	81.19
Coffered or troughed slabs					
over 450 mm thick	3.50	24.49	50.40	m3	74.89
150 - 450 mm thick	4.00	27.99	50.40	m3	78.39
Extra over for laying to slopes					
not exceeding 15 degrees	0.30	2.10	-	m3	2.10
over 15 degrees	0.60	4.20	-	m3	4.20
Walls					
over 450 mm thick	3.25	22.74	50.40	m3	73.14
150 - 450 mm thick	3.70	25.89	50.40	m3	76.29
not exceeding 150 mm thick	4.60	32.19	50.40	m3	82.59
Isolated beams	5.00	34.99	50.40	m3	85.39
Isolated deep beams	5.50	38.49	50.40	m3	88.89
Attached deep beams	5.00	34.99	50.40	m3	85.39
Isolated beam casings	5.50	38.49	50.40	m3	88.89
Isolated deep beam casings	6.00	41.99	50.40	m3	92.39
Attached deep beam casings	5.50	38.49	50.40	m3	88.89
Columns	6.00	41.99	50.40	m3	92.39
Column casings	6.60	46.19	50.40	m3	96.59
Staircases	7.50	52.48	50.40	m3	102.88
Upstands	4.80	33.59	50.40	m3	83.99

E IN SITU CONCRETE/LARGE PRECAST CONCRETE Including overheads and profit at 9.00%	Labour hours	Labour £	Material £	Unit	Total rate £
E10 IN SITU CONCRETE - cont'd					
Reinforced in situ ready mixed concrete; **26.00 N/mm2 - 20 mm aggregate (1:1.5:3)**					
Foundations	2.00	14.00	52.65	m3	66.65
Ground beams	3.50	24.49	52.65	m3	77.14
Isolated foundations	2.30	16.09	52.65	m3	68.74
Beds					
over 450 mm thick	1.60	11.20	52.65	m3	63.85
150 - 450 mm thick	2.00	14.00	52.65	m3	66.65
not exceeding 150 mm thick	2.75	19.24	52.65	m3	71.89
Slabs					
over 450 mm thick	3.10	21.69	52.65	m3	74.34
150 - 450 mm thick	3.50	24.49	52.65	m3	77.14
not exceeding 150 mm thick	4.40	30.79	52.65	m3	83.44
Coffered or troughed slabs					
over 450 mm thick	3.50	24.49	52.65	m3	77.14
150 - 450 mm thick	4.00	27.99	52.65	m3	80.64
Extra over for laying to slopes					
not exceeding 15 degrees	0.30	2.10	-	m3	2.10
over 15 degrees	0.60	4.20	-	m3	4.20
Walls					
over 450 mm thick	3.25	22.74	52.65	m3	75.39
150 - 450 mm thick	3.70	25.89	52.65	m3	78.54
not exceeding 150 mm thick	4.60	32.19	52.65	m3	84.84
Isolated beams	5.00	34.99	52.65	m3	87.64
Isolated deep beams	5.50	38.49	52.65	m3	91.14
Attached deep beams	5.00	34.99	52.65	m3	87.64
Isolated beam casings	5.50	38.49	52.65	m3	91.14
Isolated deep beam casings	6.00	41.99	52.65	m3	94.64
Attached deep beam casings	5.50	38.49	52.65	m3	91.14
Columns	6.00	41.99	52.65	m3	94.64
Column casings	6.60	46.19	52.65	m3	98.84
Staircases	7.50	52.48	52.65	m3	105.13
Upstands	4.80	33.59	52.65	m3	86.24
Reinforced in situ ready mixed concrete; **31.00 N/mm2 - 20 mm aggregate (1:1:2)**					
Beds					
over 450 mm thick	1.60	11.20	55.00	m3	66.20
150 - 450 mm thick	2.00	14.00	55.00	m3	69.00
not exceeding 150 mm thick	2.75	19.24	55.00	m3	74.24
Slabs					
over 450 mm thick	3.10	21.69	55.00	m3	76.69
150 - 450 mm thick	3.50	24.49	55.00	m3	79.49
not exceeding 150 mm thick	4.40	30.79	55.00	m3	85.79
Coffered or troughed slabs					
over 450 mm thick	3.50	24.49	55.00	m3	79.49
150 - 450 mm thick	4.00	27.99	55.00	m3	82.99
Extra over for laying to slopes					
not exceeding 15 degrees	0.30	2.10	-	m3	2.10
over 15 degrees	0.60	4.20	-	m3	4.20
Walls					
over 450 mm thick	3.25	22.74	55.00	m3	77.74
150 - 450 mm thick	3.70	25.89	55.00	m3	80.89
not exceeding 150 mm thick	4.60	32.19	55.00	m3	87.19
Isolated beams	5.00	34.99	55.00	m3	89.99
Isolated deep beams	5.50	38.49	55.00	m3	93.49
Attached deep beams	5.00	34.99	55.00	m3	89.99
Isolated beam casings	5.50	38.49	55.00	m3	93.49
Isolated deep beam casings	6.00	41.99	55.00	m3	96.99
Attached deep beam casings	5.50	38.49	55.00	m3	93.49
Columns	6.00	41.99	55.00	m3	96.99
Column casings	6.60	46.19	55.00	m3	101.19
Staircases	7.50	52.48	55.00	m3	107.48
Upstands	4.80	33.59	55.00	m3	88.59

E IN SITU CONCRETE/LARGE PRECAST CONCRETE Including overheads and profit at 9.00%	Labour hours	Labour £	Material £	Unit	Total rate £
Grouting with cement mortar (1:1)					
Stanchion bases					
10 mm thick	1.00	7.00	0.37	nr	7.37
25 mm thick	1.25	8.75	0.90	nr	9.65
Grouting with epoxy resin					
Stanchion bases					
10 mm thick	1.25	8.75	0.27	nr	9.02
25 mm thick	1.50	10.50	0.64	nr	11.14
Filling; plain in situ concrete;					
21.00 N/mm2 - 20 mm aggregate (1:2:4)					
Mortices	0.10	0.70	0.28	nr	0.98
Holes	0.25	1.75	2.00	m3	3.75
Chases					
over 0.01 m2	0.20	1.40	1.10	m3	2.50
not exceeding 0.01 m2	0.15	1.05	0.55	m	1.60
Sheeting to prevent moisture loss					
Building paper; lapped joints					
subsoil grade; horizontal on foundations	0.03	0.21	0.46	m2	0.67
standard grade; horizontal on slabs	0.05	0.35	0.73	m2	1.08
Polyethylene sheeting; lapped joints;					
horizontal on slabs					
65 microns; 0.7 mm thick	0.05	0.35	0.07	m2	0.42
125 microns; 0.13 mm thick	0.05	0.35	0.15	m2	0.50
250 microns; 0.25 mm thick	0.05	0.35	0.31	m2	0.66
'Visqueen' sheeting; lapped joints;					
horizontal on slabs					
1000 Grade; 0.25 mm thick	0.05	0.35	0.32	m2	0.67
1200 Super; 0.30 mm thick	0.06	0.42	0.40	m2	0.82

E20 FORMWORK FOR IN SITU CONCRETE

	Labour hours	Labour £	Material £	Unit	Total rate £
Note: Generally all formwork based on four					
uses unless otherwise stated					
Sides of foundations					
over 1 m high	1.40	11.45	4.08	m2	15.53
not exceeding 250 mm high	0.45	3.68	1.26	m	4.94
250 - 500 mm high	0.75	6.13	2.16	m	8.29
500 mm - 1 m high	1.15	9.40	4.08	m	13.48
Sides of foundations; left in					
over 1 m high	1.40	11.45	14.22	m2	25.67
not exceeding 250 mm high	0.45	3.68	3.66	m	7.34
250 - 500 mm high	0.75	6.13	7.18	m	13.31
500 mm - 1 m high	1.15	9.40	14.22	m	23.62
Sides of ground beams and edges of beds					
over 1 m high	1.65	13.49	5.57	m2	19.06
not exceeding 250 mm high	0.50	4.09	1.54	m	5.63
250 - 500 mm high	0.90	7.36	2.86	m	10.22
500 mm - 1 m high	1.25	10.22	5.57	m	15.79
Edges of suspended slabs					
not exceeding 250 mm high	0.75	6.13	1.81	m	7.94
250 - 500 mm high	1.10	8.99	3.62	m	12.61
500 mm - 1 m high	1.75	14.31	7.12	m	21.43
Sides of upstands					
over 1 m high	2.00	16.35	7.32	m2	23.67
not exceeding 250 mm high	0.63	5.15	1.91	m	7.06
250 - 500 mm high	1.00	8.18	3.81	m	11.99
500 mm - 1 m high	1.75	14.31	7.32	m	21.63
Steps in top surfaces					
not exceeding 250 mm high	0.50	4.09	2.00	m	6.09
250 - 500 mm high	0.80	6.54	4.16	m	10.70
Steps in soffits					
not exceeding 250 mm high	0.55	4.50	2.00	m	6.50
250 - 500 mm high	0.88	7.19	4.16	m	11.35

E IN SITU CONCRETE/LARGE PRECAST CONCRETE Including overheads and profit at 9.00%	Labour hours	Labour £	Material £	Unit	Total rate £
E20 FORMWORK FOR IN SITU CONCRETE - cont'd					
Machine bases and plinths					
over 1 m high	1.60	13.08	5.57	m2	18.65
not exceeding 250 mm high	0.50	4.09	1.54	m	5.63
250 - 500 mm high	0.85	6.95	2.86	m	9.81
500 mm - 1 m high	1.25	10.22	5.57	m	15.79
Soffits of slabs; 1.5 - 3 m height to soffit					
not exceeding 200 mm thick	1.70	13.90	5.41	m2	19.31
not exceeding 200 mm thick (5 uses)	1.65	13.49	4.87	m2	18.36
not exceeding 200 mm thick (6 uses)	1.60	13.08	4.49	m2	17.57
200 - 300 mm thick	1.80	14.72	5.92	m2	20.64
300 - 400 mm thick	1.85	15.12	6.08	m2	21.20
400 - 500 mm thick	1.95	15.94	6.33	m2	22.27
500 - 600 mm thick	2.10	17.17	5.41	m2	22.58
Soffits of slabs; not exceeding 200 mm thick					
not exceeding 1.5 m height to soffit	1.80	14.72	6.63	m2	21.35
3 - 4.5 m height to soffit	1.70	13.90	6.49	m2	20.39
4.5 - 6 m height to soffit	1.80	14.72	7.52	m2	22.24
Soffits of landings; 1.5 - 3m height to soffit					
not exceeding 200 mm thick	1.80	14.72	5.64	m2	20.36
200 - 300 mm thick	1.90	15.53	6.15	m2	21.68
300 - 400 mm thick	1.95	15.94	6.32	m2	22.26
Extra over for sloping					
not exceeding 15 degrees	0.20	1.64	-	m2	1.64
over 15 degrees	0.40	3.27	-	m2	3.27
Soffits of coffered or troughed slabs; including 'Cordeck' troughed forms; 300 mm deep; ribs at 600 mm centres and cross ribs at centres of bay; 300 - 400 mm thick					
1.5 - 3 m height to soffit	2.50	20.44	8.58	m2	29.02
3 - 4.5 m height to soffit	2.60	21.26	9.31	m2	30.57
4.5 - 6 m height to soffit	2.70	22.07	10.33	m2	32.40
Soffits of bands and margins to troughed slabs					
horizontal; 300 - 400 mm thick	2.00	16.35	6.08	m2	22.43
Top formwork	1.50	12.26	4.28	m2	16.54
Walls					
vertical	2.00	16.35	6.43	m2	22.78
vertical; interrupted	2.10	17.17	6.71	m2	23.88
vertical; exceeding 3 m high; inside stairwells	2.20	17.99	6.79	m2	24.78
vertical; exceeding 3 m high; inside lift shaft	2.40	19.62	7.32	m2	26.94
battered	2.80	22.89	7.47	m2	30.36
Beams attached to insitu slabs					
square or rectangular; 1.5 - 3 m height to soffit	2.20	17.99	8.08	m2	26.07
square or rectangular; 3 - 4.5 m height to soffit	2.30	18.80	9.13	m2	27.93
square or rectangular; 4.5 - 6 m height to soffit	2.40	19.62	10.19	m2	29.81
Beams attached to walls					
square or rectangular; 1.5 - 3 m height to soffit	2.30	18.80	8.08	m2	26.88
Isolated beams					
square or rectangular; 1.5 - 3 m height to soffit	2.40	19.62	8.08	m2	27.70
square or rectangular; 3 - 4.5 m height to soffit	2.50	20.44	9.13	m2	29.57
square or rectangular; 4.5 - 6 m height to soffit,	2.60	21.26	10.19	m2	31.45
Beam casings attached to insitu slabs					
square or rectangular; 1.5 - 3 m height to soffit	2.30	18.80	8.08	m2	26.88
square or rectangular; 3 - 4.5 m height to soffit	2.40	19.62	9.13	m2	28.75

E IN SITU CONCRETE/LARGE PRECAST CONCRETE Including overheads and profit at 9.00%	Labour hours	Labour £	Material £	Unit	Total rate £

Beam casings attached to walls
  square or rectangular; 1.5 - 3 m
  height to soffit
	2.40	19.62	8.08	m2	27.70

Isolated beam casings
  square or rectangular; 1.5 - 3 m
  height to soffit
	2.50	20.44	8.08	m2	28.52

square or rectangular; 3 - 4.5 m height to soffit	2.60	21.26	9.13	m2	30.39
Extra over for sloping not exceeding 15 degrees	2.40	19.62	8.54	m2	28.16
over 15 degrees	2.50	20.44	9.00	m2	29.44
Columns attached to walls square or rectangular	2.20	17.99	6.43	m2	24.42
Isolated columns square or rectangular	2.30	18.80	6.43	m2	25.23
Column casings attached to walls square or rectangular	2.30	18.80	6.43	m2	25.23
Isolated column casings square or rectangular	2.40	19.62	6.43	m2	26.05
Extra over for Throat	0.05	0.41	0.16	m	0.57
Chamfer 30 mm wide	0.06	0.49	0.22	m	0.71
60 mm wide	0.07	0.57	0.36	m	0.93
90 mm wide	0.08	0.65	0.74	m	1.39
Rebate or horizontal recess 12 x 12 mm	0.07	0.57	0.09	m	0.66
25 x 25 mm	0.07	0.57	0.15	m	0.72
25 x 50 mm	0.07	0.57	0.29	m	0.86
50 x 50 mm	0.07	0.57	0.52	m	1.09
Nibs 50 x 50 mm	0.55	4.50	1.24	m	5.74
100 x 100 mm	0.78	6.38	2.43	m	8.81
100 x 200 mm	1.04	8.50	3.24	m	11.74
Extra over basic formwork for rubbing down, filling and leaving face of concrete smooth general surfaces	0.33	2.70	0.10	m2	2.80
edges	0.50	4.09	0.12	m2	4.21
Add to prices for basic formwork for curved radius 6 m	27.5%				
curved radius 2 m	50%				
coating with retardant agent	0.02	0.16	0.45	m2	0.61
Wall kickers to both sides 150 mm high	0.50	4.09	1.27	m	5.36
225 mm high	0.65	5.31	1.66	m	6.97
150 mm high; one side suspended	0.63	5.15	2.06	m	7.21
Wall ends, soffits and steps in walls over 1 m wide	1.90	15.53	6.76	m	22.29
not exceeding 250 mm wide	0.60	4.91	1.91	m	6.82
250 - 500 mm wide	0.95	7.77	3.69	m	11.46
500 mm - 1 m wide	1.50	12.26	6.76	m	19.02
Openings in walls over 1 m wide	2.10	17.17	8.11	m	25.28
not exceeding 250 mm wide	0.65	5.31	2.05	m	7.36
250 - 500 mm wide	1.10	8.99	4.11	m	13.10
500 mm - 1 m wide	1.70	13.90	8.11	m	22.01

Stair flights
  1 m wide; 150 mm waist;
  150 mm undercut risers
	5.00	40.88	21.03	m	61.91

2 m wide; 200 mm waist; 150 mm vertical risers	9.00	73.57	33.64	m	107.21
Mortices; not exceeding 250 mm deep not exceeding 500 mm girth	0.15	1.23	1.12	nr	2.35
Holes; not exceeding 250 mm deep not exceeding 500 mm girth	0.20	1.64	1.16	nr	2.80
500 mm - 1 m girth	0.25	2.04	2.23	nr	4.27
1 - 2 m girth	0.45	3.68	4.47	nr	8.15
2 - 3 m girth	0.60	4.91	6.70	nr	11.61

E IN SITU CONCRETE/LARGE PRECAST CONCRETE Including overheads and profit at 9.00%	Labour hours	Labour £	Material £	Unit	Total rate £
E20 FORMWORK FOR IN SITU CONCRETE - cont'd					
Holes; 250 - 500 mm deep					
not exceeding 500 mm girth	0.30	2.45	2.23	nr	4.68
500 mm - 1 m girth	0.38	3.11	4.47	nr	7.58
1 - 2 m girth	0.67	5.48	8.94	nr	14.42
2 - 3 m girth	0.90	7.36	13.41	nr	20.77
Permanent shuttering; left in					
Dufaylite 'Clayboard' shuttering; type KN30;					
horizontal; under concrete beds; left in					
50 mm thick	0.15	1.23	6.97	m2	8.20
75 mm thick	0.16	1.31	7.48	m2	8.79
100 mm thick	0.17	1.39	8.10	m2	9.49
150 mm thick	0.20	1.64	8.68	m2	10.32
Dufaylite 'Clayboard' shuttering; type KN30;					
horizontal or vertical (including temporary					
supports); beneath or to sides of					
foundations; left in					
50 mm thick; vertical	0.25	2.04	7.55	m2	9.59
400 x 50 mm thick; horizontal	0.09	0.74	2.86	m	3.60
600 x 50 mm thick; horizontal	0.12	0.98	4.28	m	5.26
800 x 50 mm thick; horizontal	0.15	1.23	5.72	m	6.95
75 mm thick; vertical	0.27	2.21	8.07	m2	10.28
400 x 75 mm thick; horizontal	0.10	0.82	3.07	m	3.89
600 x 75 mm thick; horizontal	0.13	1.06	4.59	m	5.65
800 x 75 mm thick; horizontal	0.16	1.31	6.12	m	7.43
100 mm thick; vertical	0.30	2.45	8.71	m2	11.16
400 x 100 mm thick; horizontal	0.11	0.90	3.31	m	4.21
600 x 100 mm thick; horizontal	0.14	1.14	4.98	m	6.12
800 x 100 mm thick; horizontal	0.17	1.39	6.63	m	8.02
Hyrib permanent shuttering and reinforcement					
ref 2411 to soffits of slabs; left in					
horizontal	1.50	12.26	11.97	m2	24.23
0.9 mm 'Super Holorib' steel deck permanent					
shuttering; to soffit of slabs; left in					
1.5 - 3 m height to soffit	1.50	12.26	8.34	m2	20.60
3 - 4.5 m height to soffit	1.60	13.08	9.06	m2	22.14
4.5 - 6 m height to soffit	1.70	13.90	10.09	m2	23.99
1.2 mm 'Super Holorib' steel deck permanent					
shuttering; to soffit of slabs; left in					
1.5 - 3 m height to soffit	1.65	13.49	10.05	m2	23.54
3 - 4.5 m height to soffit	1.75	14.31	11.19	m2	25.50
3 - 4.5 m height to soffit; deck stud					
welded to steelwork	2.75	25.33	13.53	m2	38.86
4.5 - 6 m height to soffit	1.85	15.12	12.72	m2	27.84
E30 REINFORCEMENT FOR IN SITU CONCRETE					
Reinforcement bars; BS 4449; hot rolled					
plain round mild steel bars; straight					
40 mm PC £365.50	11.00	88.97	444.97	t	533.94
32 mm PC £352.50	11.50	93.01	433.77	t	526.78
25 mm PC £345.50	13.50	109.19	429.43	t	538.62
20 mm PC £343.50	15.50	125.36	430.81	t	556.17
16 mm PC £344.50	18.00	145.58	435.63	t	581.21
12 mm PC £364.50	21.00	169.84	462.19	t	632.03
10 mm PC £373.50	24.00	194.11	476.17	t	670.28
8 mm PC £384.50	27.00	218.37	492.43	t	710.80
6 mm PC £423.50	32.00	258.81	540.74	t	799.55

E IN SITU CONCRETE/LARGE PRECAST CONCRETE Including overheads and profit at 9.00%		Labour hours	Labour £	Material £	Unit	Total rate £
Reinforcement bars; BS 4449; hot rolled **plain round mild steel bars; bent**						
40 mm	PC £390.50	11.00	88.97	473.58	t	562.55
32 mm	PC £378.50	13.00	105.14	463.52	t	568.66
25 mm	PC £376.50	15.00	121.32	464.91	t	586.23
20 mm	PC £374.50	17.00	137.49	466.29	t	603.78
16 mm	PC £380.50	20.00	161.76	476.83	t	638.59
12 mm	PC £410.50	23.00	186.02	514.84	t	700.86
10 mm	PC £434.50	27.00	218.37	545.98	t	764.35
8 mm	PC £465.50	31.00	250.72	585.13	t	835.85
6 mm	PC £509.50	36.00	291.16	639.16	t	930.32
8 mm; links or the like	PC £465.50	42.00	339.69	596.15	t	935.84
6 mm; links or the like	PC £509.50	49.00	396.30	653.86	t	1050.16
Reinforcement bars; BS 4461; cold worked **deformed square high steel bars; straight**						
40 mm	PC £369.50	10.00	80.88	449.55	t	530.43
32 mm	PC £357.50	11.50	93.01	439.49	t	532.50
25 mm	PC £350.50	13.50	109.19	435.15	t	544.34
20 mm	PC £348.50	15.50	125.36	436.53	t	561.89
16 mm	PC £349.50	18.00	145.58	441.35	t	586.93
12 mm	PC £369.50	21.00	169.84	467.92	t	637.76
10 mm	PC £378.50	24.00	194.11	481.89	t	676.00
8 mm	PC £389.50	27.00	218.37	498.15	t	716.52
6 mm	PC £428.50	32.00	258.81	546.46	t	805.27
Reinforcement bars; BS 4461; cold worked **deformed square high steel bars; bent**						
40 mm	PC £394.50	11.00	88.97	478.16	t	567.13
32 mm	PC £382.50	13.00	105.14	468.10	t	573.24
25 mm	PC £380.50	15.00	121.32	469.48	t	590.80
20 mm	PC £378.50	17.00	137.49	470.87	t	608.36
16 mm	PC £384.50	20.00	161.76	481.41	t	643.17
12 mm	PC £414.50	23.00	186.02	519.42	t	705.44
10 mm	PC £438.50	27.00	218.37	550.56	t	768.93
8 mm	PC £469.50	31.00	250.72	589.71	t	840.43
6 mm	PC £513.50	36.00	291.16	643.74	t	934.90
Reinforcement fabric; BS 4483; lapped; in **beds or suspended slabs**						
Ref A98 (1.54 kg/m2)	PC £0.61	0.12	0.97	0.80	m2	1.77
Ref A142 (2.22 kg/m2)	PC £0.77	0.12	0.97	1.00	m2	1.97
Ref A193 (3.02 kg/m2)	PC £1.04	0.12	0.97	1.36	m2	2.33
Ref A252 (3.95 kg/m2)	PC £1.36	0.13	1.05	1.78	m2	2.83
Ref A393 (6.16 kg/m2)	PC £2.16	0.15	1.21	2.82	m2	4.03
Ref B196 (3.05 kg/m2)	PC £1.16	0.12	0.97	1.52	m2	2.49
Ref B283 (3.73 kg/m2)	PC £1.37	0.12	0.97	1.79	m2	2.76
Ref B385 (4.53 kg/m2)	PC £1.64	0.13	1.05	2.15	m2	3.20
Ref B503 (5.93 kg/m2)	PC £2.10	0.15	1.21	2.75	m2	3.96
Ref B785 (8.14 kg/m2)	PC £2.88	0.17	1.37	3.77	m2	5.14
Ref B1131 (10.90 kg/m2)	PC £3.89	0.19	1.54	5.09	m2	6.63
Ref C283 (2.61 kg/m2)	PC £0.99	0.12	0.97	1.30	m2	2.27
Ref C385 (3.41 kg/m2)	PC £1.26	0.12	0.97	1.65	m2	2.62
Ref C503 (4.34 kg/m2)	PC £1.54	0.13	1.05	2.02	m2	3.07
Ref C636 (5.55 kg/m2)	PC £1.99	0.14	1.13	2.61	m2	3.74
Ref C785 (6.72 kg/m2)	PC £2.40	0.15	1.21	3.14	m2	4.35
Reinforcement fabric; BS 4483; lapped; in **casings to steel columns or beams**						
Ref D49 (0.77 kg/m2)	PC £0.57	0.25	2.02	0.74	m2	2.76
Ref D98 (1.54 kg/m2)	PC £0.57	0.25	2.02	0.74	m2	2.76

E IN SITU CONCRETE/LARGE PRECAST CONCRETE Including overheads and profit at 9.00%		Labour hours	Labour £	Material £	Unit	Total rate £
E40 DESIGNED JOINTS IN IN SITU CONCRETE						
Expandite 'Flexcell' joint filler; or similar						
Formed joint; 10 mm thick						
not exceeding 150 mm wide		0.30	2.45	0.97	m	3.42
150 - 300 mm wide		0.40	3.27	1.68	m	4.95
300 - 450 mm wide		0.50	4.09	2.59	m	6.68
Formed joint; 12.5 mm thick						
not exceeding 150 mm wide		0.30	2.45	1.01	m	3.46
150 - 300 mm wide		0.40	3.27	1.76	m	5.03
300 - 450 mm wide		0.50	4.09	2.73	m	6.82
Formed joint; 20 mm thick						
not exceeding 150 mm wide		0.30	2.45	1.40	m	3.85
150 - 300 mm wide		0.40	3.27	2.38	m	5.65
300 - 450 mm wide		0.50	4.09	3.77	m	7.86
Formed joint; 25 mm thick						
not exceeding 150 mm wide		0.30	2.45	1.57	m	4.02
150 - 300 mm wide		0.40	3.27	2.79	m	6.06
300 - 450 mm wide		0.50	4.09	4.24	m	8.33
Sealing top of joint with Expandite 'Pliastic' hot poured rubberized bituminous compound						
10 x 25 mm		0.18	1.47	0.37	m	1.84
12.5 x 25 mm		0.19	1.55	0.45	m	2.00
20 x 25 mm		0.20	1.64	0.62	m	2.26
25 x 25 mm		0.21	1.72	0.87	m	2.59
Sealing top of joint with Expandite 'Thioflex 600' cold poured polysulphide rubberized compound						
10 x 25 mm		0.06	0.49	2.55	m	3.04
12.5 x 25 mm		0.07	0.57	3.07	m	3.64
20 x 25 mm		0.08	0.65	4.25	m	4.90
25 x 25 mm		0.09	0.74	6.13	m	6.87
Serviciced water stops or similar						
Formed joint; PVC water stop; flat dumbell type; heat welded joints						
100 mm wide	PC £37.40/15m	0.22	1.80	2.87	m	4.67
Flat angle	PC £2.38	0.27	2.21	2.72	nr	4.93
Vertical angle	PC £4.13	0.27	2.21	4.73	nr	6.94
Flat three way intersection	PC £4.68	0.37	3.02	5.35	nr	8.37
Vertical three way intersection	PC £5.03	0.37	3.02	5.76	nr	8.78
Four way intersection	PC £5.87	0.47	3.84	6.71	nr	10.55
170 mm wide	PC £52.65/15m	0.25	2.04	4.04	m	6.08
Flat angle	PC £2.39	0.30	2.45	2.74	nr	5.19
Vertical angle	PC £4.13	0.30	2.45	4.73	nr	7.18
Flat three way intersection	PC £4.72	0.40	3.27	5.40	nr	8.67
Vertical three way intersection	PC £5.85	0.40	3.27	6.69	nr	9.96
Four way intersection	PC £6.40	0.50	4.09	7.32	nr	11.41
210 mm wide	PC £62.23/15m	0.28	2.29	4.77	m	7.06
Flat angle	PC £4.01	0.32	2.62	4.58	nr	7.20
Vertical angle	PC £4.58	0.32	2.62	5.24	nr	7.86
Flat three way intersection	PC £5.85	0.42	3.43	6.69	nr	10.12
Vertical three way intersection	PC £7.24	0.42	3.43	8.28	nr	11.71
Four way intersection	PC £7.34	0.52	4.25	8.40	nr	12.65
250 mm wide	PC £90.99/15m	0.30	2.45	6.97	m	9.42
Flat angle	PC £4.88	0.34	2.78	5.58	nr	8.36
Vertical angle	PC £5.16	0.34	2.78	5.90	nr	8.68
Flat three way intersection	PC £6.55	0.44	3.60	7.50	nr	11.10
Vertical three way intersection	PC £8.15	0.44	3.60	9.33	nr	12.93
Four way intersection	PC £8.44	0.54	4.41	9.66	nr	14.07

E IN SITU CONCRETE/LARGE PRECAST CONCRETE Including overheads and profit at 9.00%		Labour hours	Labour £	Material £	Unit	Total rate £
Formed joint; PVC water stop; centre bulb **type; heat welded joints**						
160 mm wide	PC £50.49/15m	0.25	2.04	3.87	m	5.91
Flat angle	PC £3.15	0.30	2.45	3.61	nr	6.06
Vertical angle	PC £5.18	0.30	2.45	5.92	nr	8.37
Flat three way intersection	PC £6.64	0.40	3.27	7.60	nr	10.87
Vertical three way intersection	PC £8.23	0.40	3.27	9.41	nr	12.68
Four way intersection	PC £7.90	0.50	4.09	9.05	nr	13.14
210 mm wide	PC £72.77/15m	0.28	2.29	5.58	m	7.87
Flat angle	PC £4.68	0.32	2.62	5.35	nr	7.97
Vertical angle	PC £6.16	0.32	2.62	7.05	nr	9.67
Flat three way intersection	PC £6.73	0.42	3.43	7.70	nr	11.13
Vertical three way intersection	PC £9.32	0.42	3.43	10.68	nr	14.11
Four way intersection	PC £9.03	0.52	4.25	10.33	nr	14.58
260 mm wide	PC £103.41/15m	0.30	2.45	7.93	m	10.38
Flat angle	PC £6.11	0.34	2.78	7.00	nr	9.78
Vertical angle	PC £6.71	0.34	2.78	7.68	nr	10.46
Flat three way intersection	PC £9.17	0.44	3.60	10.50	nr	14.10
Vertical three way intersection	PC £11.37	0.44	3.60	13.01	nr	16.61
Four way intersection	PC £10.98	0.54	4.41	12.57	nr	16.98
325 mm wide	PC £173.88/15m	0.33	2.70	13.33	m	16.03
Flat angle	PC £8.85	0.36	2.94	10.12	nr	13.06
Vertical angle	PC £9.14	0.36	2.94	10.46	nr	13.40
Flat three way intersection	PC £13.00	0.46	3.76	14.87	nr	18.63
Vertical three way intersection	PC £13.28	0.46	3.76	15.20	nr	18.96
Four way intersection	PC £14.52	0.56	4.58	16.62	nr	21.20
Formed joint; rubber water stop; **flat dumbell type; sleeved joints**						
150 mm wide	PC £90.73/9m	0.20	1.64	12.68	m	14.32
Flat angle	PC £25.82	0.20	1.64	29.55	nr	31.19
Vertical angle	PC £25.82	0.20	1.64	29.55	nr	31.19
Flat three way intersection	PC £28.48	0.25	2.04	32.60	nr	34.64
Vertical three way intersection	PC £28.48	0.25	2.04	32.60	nr	34.64
Four way intersection	PC £31.47	0.30	2.45	36.02	nr	38.47
230 mm wide	PC £137.25/9m	0.25	2.04	18.92	m	20.96
Flat angle	PC £31.25	0.22	1.80	35.76	nr	37.56
Vertical angle	PC £31.25	0.22	1.80	35.76	nr	37.56
Flat three way intersection	PC £34.04	0.27	2.21	38.95	nr	41.16
Vertical three way intersection	PC £34.04	0.27	2.21	38.95	nr	41.16
Four way intersection	PC £36.68	0.33	2.70	41.97	nr	44.67
Formed joint; rubber water stop; centre bulb **type; sleeved joints**						
150 mm wide	PC £104.40/9m	0.20	1.64	14.41	m	16.05
Flat angle	PC £28.29	0.20	1.64	32.37	nr	34.01
Vertical angle	PC £28.29	0.20	1.64	32.37	nr	34.01
Flat three way intersection	PC £31.21	0.25	2.04	35.72	nr	37.76
Vertical three way intersection	PC £31.21	0.25	2.04	35.72	nr	37.76
Four way intersection	PC £34.15	0.30	2.45	39.08	nr	41.53
230 mm wide	PC £155.99/9m	0.25	2.04	21.39	m	23.43
Flat angle	PC £33.23	0.22	1.80	38.03	nr	39.83
Vertical angle	PC £33.23	0.22	1.80	38.03	nr	39.83
Flat three way intersection	PC £34.76	0.27	2.21	39.78	nr	41.99
Vertical three way intersection	PC £34.76	0.27	2.21	39.78	nr	41.99
Four way intersection	PC £40.28	0.33	2.70	46.10	nr	48.80
305 mm wide	PC £257.24/9m	0.30	2.45	35.00	m	37.45
Flat angle	PC £52.21	0.24	1.96	59.75	nr	61.71
Vertical angle	PC £52.21	0.24	1.96	59.75	nr	61.71
Flat three way intersection	PC £63.12	0.30	2.45	72.24	nr	74.69
Vertical three way intersection	PC £63.12	0.30	2.45	72.24	nr	74.69
Four way intersection	PC £79.04	0.36	2.94	90.46	nr	93.40

E IN SITU CONCRETE/LARGE PRECAST CONCRETE Including overheads and profit at 9.00%	Labour hours	Labour £	Material £	Unit	Total rate £
E41 WORKED FINISHES/CUTTING ON IN SITU CONCRETE					
Tamping by mechanical means	0.03	0.21	0.14	m2	0.35
Power floating	0.17	1.19	0.46	m2	1.65
Trowelling	0.33	2.31	-	m2	2.31
Hacking					
by mechanical means	0.33	2.31	0.40	m2	2.71
by hand	0.70	4.90	-	m2	4.90
Wood float finish	0.13	0.91	-	m2	0.91
Tamped finish	0.05	0.35	-	m2	0.35
to falls	0.07	0.49	-	m2	0.49
to crossfalls	0.10	0.70	-	m2	0.70
Spade finish	0.15	1.05	-	m2	1.05
Cutting chases					
not exceeding 50 mm deep; 10 mm wide	0.33	2.31	0.18	m	2.49
not exceeding 50 mm deep; 50 mm wide	0.50	3.50	0.27	m	3.77
not exceeding 50 mm deep; 75 mm wide	0.66	4.62	0.36	m	4.98
50 - 100 mm deep; 75 mm wide	0.90	6.30	0.49	m	6.79
50 - 100 mm deep; 100 mm wide	1.00	7.00	0.55	m	7.55
100 - 150 mm deep; 100 mm wide	1.30	9.10	0.71	m	9.81
100 - 150 mm deep; 150 mm wide	1.60	11.20	0.87	m	12.07
Cutting chases in reinforced concrete					
50 - 100 mm deep; 100 mm wide	1.50	10.50	0.82	m	11.32
100 - 150 mm deep; 100 mm wide	2.00	14.00	1.09	m	15.09
100 - 150 mm deep; 150 mm wide	2.40	16.79	1.31	m	18.10
Cutting rebates					
not exceeding 50 mm deep; 50 mm wide	0.50	3.50	0.27	m	3.77
50 - 100 mm deep; 100 mm wide	1.00	7.00	0.55	m	7.55
Cutting mortices; not exceeding 100 mm deep; making good					
20 mm dia	0.15	1.05	0.07	nr	1.12
50 mm dia	0.17	1.19	0.09	nr	1.28
150 x 150 mm	0.35	2.45	0.20	nr	2.65
300 x 300 mm	0.70	4.90	0.43	nr	5.33
Cutting mortices in reinforced concrete; not exceeding 100 mm deep; making good					
150 x 150 mm	0.55	3.85	0.28	nr	4.13
300 x 300 mm	1.05	7.35	0.59	nr	7.94
Cutting holes; not exceeding 100 mm deep					
50 mm dia	0.35	2.45	0.42	nr	2.87
100 mm dia	0.40	2.80	0.48	nr	3.28
150 x 150 mm	0.45	3.15	0.54	nr	3.69
300 x 300 mm	0.55	3.85	0.66	nr	4.51
Cutting holes; 100 - 200 mm deep					
50 mm dia	0.50	3.50	0.60	nr	4.10
100 mm dia	0.60	4.20	0.72	nr	4.92
150 x 150 mm	0.75	5.25	0.90	nr	6.15
300 x 300 mm	0.95	6.65	1.14	nr	7.79
Cutting holes; 200 - 300 mm deep					
50 mm dia	0.75	5.25	0.90	nr	6.15
100 mm dia	0.90	6.30	1.08	nr	7.38
150 x 150 mm	1.10	7.70	1.32	nr	9.02
300 x 300 mm	1.40	9.80	1.68	nr	11.48
Add for making good fair finish one side					
50 mm dia	0.05	0.35	0.01	nr	0.36
100 mm dia	0.12	0.84	0.02	nr	0.86
150 x 150 mm	0.20	1.40	0.04	nr	1.44
300 x 300 mm	0.40	2.80	0.07	nr	2.87
Add for fixing only sleeve					
50 mm dia	0.10	0.70	-	nr	0.70
100 mm dia	0.22	1.54	-	nr	1.54
150 x 150 mm	0.33	2.31	-	nr	2.31
300 x 300 mm	0.60	4.20	-	nr	4.20

E IN SITU CONCRETE/LARGE PRECAST CONCRETE Including overheads and profit at 9.00%	Labour hours	Labour £	Material £	Unit	Total rate £
Cutting holes in reinforced concrete; not exceeding 100 mm deep					
50 mm dia	0.55	3.85	0.66	nr	4.51
100 mm dia	0.60	4.20	0.72	nr	4.92
150 x 150 mm dia	0.70	4.90	0.84	nr	5.74
300 x 300 mm dia	0.85	5.95	1.02	nr	6.97
Cutting holes in reinforced concrete; 100 - 200 mm deep					
50 mm dia	0.75	5.25	0.90	nr	6.15
100 mm dia	0.90	6.30	1.08	nr	7.38
150 x 150 mm dia	1.15	8.05	1.38	nr	9.43
300 x 300 mm dia	1.45	10.15	1.74	nr	11.89
Cutting holes in reinforced concrete; 200 - 300 mm deep					
50 mm dia	1.15	8.05	1.38	nr	9.43
100 mm dia	1.35	9.45	1.62	nr	11.07
150 x 150 mm dia	1.65	11.55	1.98	nr	13.53
300 x 300 mm dia	2.10	14.70	2.52	nr	17.22
E42 ACCESSORIES CAST INTO IN SITU CONCRETE					
Temporary plywood foundation bolt boxes					
75 x 75 x 150 mm	0.45	3.15	0.45	nr	3.60
75 x 75 x 250 mm	0.50	3.50	0.72	nr	4.22
'Expamet' cylindrical expanded steel foundation boxes					
76 mm dia x 152 mm high	0.30	2.10	2.93	nr	5.03
76 mm dia x 305 mm high	0.20	1.40	0.90	nr	2.30
102 mm dia x 457 mm high	0.25	1.75	1.63	nr	3.38
10 mm dia x 100 mm long	0.25	1.75	1.11	nr	2.86
12 mm dia x 120 mm long	0.25	1.75	1.26	nr	3.01
16 mm dia x 160 mm long	0.30	2.10	2.90	nr	5.00
20 mm dia x 200 mm long	0.30	2.10	2.93	nr	5.03
'Abbey' galvanized steel masonry slots; 18 G (1.22 mm)					
3.048 m lengths	0.35	2.45	0.98	m	3.43
76 mm long	0.08	0.56	0.13	nr	0.69
102 mm long	0.08	0.56	0.16	nr	0.72
152 mm long	0.09	0.63	0.20	nr	0.83
229 mm long	0.10	0.70	0.28	nr	0.98
'Unistrut' galvanized steel slotted metal inserts; 2.5 mm thick; end caps and foam filling					
41 x 41 mm; ref P3270	0.40	2.80	4.08	m	6.88
41 x 41 x 75 mm; ref P3249	0.10	0.70	1.54	nr	2.24
41 x 41 x 100 mm; ref P3250	0.10	0.70	1.63	nr	2.33
41 x 41 x 150 mm; ref P3251	0.10	0.70	1.84	nr	2.54
Butterfly type wall ties; casting one end into concrete; other end built into joint of brickwork					
galvanized steel	0.10	0.70	0.07	nr	0.77
stainless steel	0.10	0.70	0.09	nr	0.79
Mild steel fixing cramp; once bent; one end shot fired into concrete; other end fanged and built into joint of brickwork					
200 mm girth	0.15	1.05	0.56	nr	1.61
Sherardized steel floor clips; pinned to surface of concrete					
50 mm wide; standard type	0.08	0.56	0.10	nr	0.66
50 mm wide; direct fix acoustic type	0.10	0.70	0.67	nr	1.37
Hardwood dovetailed fillets					
50 x 50/40 x 1000 mm	0.10	0.70	1.08	nr	1.78
50 x 50/40 x 100 mm	0.08	0.56	0.16	nr	0.72
50 x 50/40 x 200 mm	0.08	0.56	0.22	nr	0.78

E IN SITU CONCRETE/LARGE PRECAST CONCRETE Including overheads and profit at 9.00% & 5.00%	Labour hours	Labour £	Material £	Unit	Total rate £
E42 ACCESSORIES CAST INTO IN SITU CONCRETE - cont'd					
'Rigifix' galvanized steel plate column guard; 1 m long					
75 mm x 75 mm x 3 mm	0.60	4.20	7.54	nr	11.74
75 mm x 75 mm x 4.5 mm	0.60	4.20	10.07	nr	14.27
'Rigifix' white nylon coated steel plate corner guard; plugged and screwed to concrete with chromium plated domed headed screws					
75 x 75 x 1.5 mm x 1 m long	0.80	5.60	11.05	nr	16.65
E60 PRECAST/COMPOSITE CONCRETE DECKING					
Prestressed precast flooring planks; Bison 'Drycast' or similar; cement and sand (1:3) grout between planks and on prepared bearings 100 mm thick suspended slabs; horizontal					
400 mm wide planks	-	-	-	m2	29.43
1200 mm wide planks	-	-	-	m2	27.25
150 mm thick suspended slabs; horizontal					
400 mm wide planks	-	-	-	m2	29.98
1200 mm wide planks	-	-	-	m2	27.80
Prestressed precast concrete beam and block floor; Bison 'Housefloor' or similar; in situ concrete 30 N/mm2 - 10 mm aggregate in filling at wall abutments; cement and sand (1:6) grout brushed in between beams and blocks 155 mm thick suspended slab at ground level; 440 x 215 x 100 mm blocks; horizontal					
beams at 520 mm centres; up to 3.30 m span with a superimposed load of 5 kN/m2	-	-	-	m2	17.44
beams at 295 mm centres; up to 4.35 m span with a superimposed load of 5 kN/m2	-	-	-	m2	20.17
Composite floor comprising reinforced in situ ready-mixed concrete 31.00 N/mm2; on and including 1.2 mm 'Super Holorib' steel deck permanent shuttering; complete with reinforcment to support imposed loading and A142 anti-crack reinforcement 150 mm thick suspended slab; 5 kN/m2 loading					
1.5 - 3 m height to soffit	2.40	18.84	18.88	m2	37.72
3 - 4.5 m height to soffit	2.50	19.65	19.59	m2	39.24
4.5 - 6 m height to soffit	2.60	20.47	20.62	m2	41.09
200 mm thick suspended slab; 7.5 kN/m2 loading					
1.5 - 3 m height to soffit	2.44	19.12	22.96	m2	42.08
3 - 4.5 m height to soffit	2.54	19.93	24.09	m2	44.02
4.5 - 6 m height to soffit	2.64	20.75	25.63	m2	46.38

F MASONRY
Including overheads and profit at 9.00%

BASIC MORTAR PRICES

£		£		£		£
Coloured mortar materials (£/tonne); (excluding cement)						
light	25.72	medium	26.72	dark	27.72	extra dark 27.72

Mortar materials (£/tonne)
cement   60.89   lime   88.00   sand   8.36   white cement   67.13

Mortar plasticizer - £1.53/Litre

	Labour hours	Labour £	Material £	Unit	Total rate £
F10 BRICK/BLOCK WALLING					
Common bricks; PC £117.00/1000; in cement mortar (1:3)					
Walls					
half brick thick	1.25	14.57	9.05	m2	23.61
one brick thick	2.10	24.47	18.76	m2	43.23
one and a half brick thick	2.85	33.21	28.21	m2	61.42
two brick thick	3.50	40.78	37.52	m2	78.30
Walls; facework one side					
half brick thick	1.40	16.31	9.05	m2	25.36
one brick thick	2.25	26.22	18.76	m2	44.98
one and a half brick thick	3.00	34.96	28.21	m2	63.17
two brick thick	3.65	42.53	37.52	m2	80.05
Walls; facework both sides					
half brick thick	1.50	17.48	9.05	m2	26.53
one brick thick	2.35	27.38	18.76	m2	46.14
one and a half brick thick	3.10	36.12	28.21	m2	64.33
two brick thick	3.75	43.70	37.52	m2	81.22
Walls; built curved mean radius 6 m					
half brick thick	1.65	19.23	9.74	m2	28.96
one brick thick	2.75	32.04	20.15	m2	52.19
Walls; built curved mean radius 1.50 m					
half brick thick	2.10	24.47	10.24	m2	34.71
one brick thick	3.45	40.20	21.15	m2	61.35
Walls; built overhand					
half brick thick	1.55	18.06	9.05	m2	27.11
Walls; building up against concrete including flushing up at back					
half brick thick	1.35	15.73	10.15	m2	25.88
Walls; backing to masonry; cutting and bonding					
one brick thick	2.50	29.13	19.14	m2	48.27
one and a half brick thick	3.35	39.03	28.79	m2	67.82
Honeycomb walls					
half brick thick	1.00	11.65	6.37	m2	18.02
Dwarf support wall					
half brick thick	1.55	18.06	9.05	m2	27.11
one brick thick	2.50	29.13	18.76	m2	47.89
Battering walls					
one and a half brick thick	3.30	38.45	28.79	m2	67.24
two brick thick	4.10	47.77	38.30	m2	86.07
Walls; tapering one side; average					
337 mm thick	3.65	42.53	29.36	m2	71.89
450 mm thick	4.70	54.76	39.06	m2	93.82
Walls; tapering both sides; average					
337 mm thick	4.20	48.94	29.36	m2	78.30
450 mm thick	5.25	61.17	39.06	m2	100.23

F MASONRY Including overheads and profit at 9.00%	Labour hours	Labour £	Material £	Unit	Total rate £
F10 BRICK/BLOCK WALLING - cont'd					
Common bricks; PC £117.00/1000; **in cement mortar (1:3) - cont'd**					
Isolated piers					
one brick thick	3.20	37.29	19.14	m2	56.43
two brick thick	5.00	58.26	38.30	m2	96.56
three brick thick	6.30	73.41	57.44	m2	130.85
Isolated casings to steel columns					
half brick thick	1.60	18.64	9.25	m2	27.89
one brick thick	2.75	32.04	19.14	m2	51.19
Chimney stacks					
one brick thick	3.20	37.29	19.14	m2	56.43
two brick thick	5.00	58.26	38.30	m2	96.56
three brick thick	6.30	73.41	57.44	m2	130.85
Projections; vertical					
225 x 112 mm	0.40	4.66	2.14	m	6.80
225 x 225 mm	0.75	8.74	4.15	m	12.89
337 x 225 mm	1.10	12.82	6.97	m	19.78
440 x 225 mm	1.25	14.57	8.15	m	22.72
Bonding ends to existing					
half brick thick	0.40	4.66	0.66	m	5.32
one brick thick	0.55	6.41	1.31	m	7.72
one and a half brick thick	0.85	9.90	1.97	m	11.87
two brick thick	1.20	13.98	2.62	m	16.61
ADD or DEDUCT to walls for variation of £10.00/1000 in PC of common bricks					
half brick thick	-	-	-	m2	0.71
one brick thick	-	-	-	m2	1.41
one and a half brick thick	-	-	-	m2	2.10
two brick thick	-	-	-	m2	2.75
Extra over walls for sulphate-resisting cement mortar (1:3) in lieu of cement mortar (1:3)					
half brick thick	-	-	-	m2	0.11
one brick thick	-	-	-	m2	0.28
one and a half brick thick	-	-	-	m2	0.43
two brick thick	-	-	-	m2	0.56
Common bricks; PC £117.00/1000; **in gauged mortar (1:1:6)**					
Walls					
half brick thick	1.25	14.57	8.98	m2	23.55
one brick thick	2.10	24.47	18.58	m2	43.05
one and a half brick thick	2.85	33.21	27.93	m2	61.14
two brick thick	3.50	40.78	37.15	m2	77.94
Walls; facework one side					
half brick thick	1.40	16.31	8.98	m2	25.29
one brick thick	2.25	26.22	18.58	m2	44.80
one and a half brick thick	3.00	34.96	27.93	m2	62.89
two brick thick	3.65	42.53	37.15	m2	79.68
Walls; facework both sides					
half brick thick	1.50	17.48	8.98	m2	26.46
one brick thick	2.35	27.38	18.58	m2	45.96
one and a half brick thick	3.10	36.12	27.93	m2	64.05
two brick thick	3.75	43.70	37.15	m2	80.85
Walls; built curved mean radius 6 m					
half brick thick	1.65	19.23	9.66	m2	28.89
one brick thick	2.75	32.05	19.95	m2	52.00
Walls; built curved mean radius 1.50 m					
half brick thick	2.10	24.47	10.16	m2	34.63
one brick thick	3.45	40.20	20.94	m2	61.14
Walls; built overhand					
half brick thick	1.55	18.06	8.98	m2	27.04
Walls; built up against concrete including flushing up at back					
half brick thick	1.35	15.73	10.01	m2	25.74

F MASONRY Including overheads and profit at 9.00%	Labour hours	Labour £	Material £	Unit	Total rate £
Walls; backing to masonry; cutting and bonding					
one brick thick	2.50	29.13	18.96	m2	48.09
one and a half brick thick	3.35	39.03	28.51	m2	67.54
Honeycomb walls					
half brick thick	1.00	11.65	6.30	m2	17.95
Dwarf support wall					
half brick thick	1.55	18.06	8.98	m2	27.04
one brick thick	2.50	29.13	18.58	m2	47.71
Battering walls					
one and a half brick thick	3.30	38.45	28.51	m2	66.96
two brick thick	4.10	47.77	37.93	m2	85.70
Walls; tapering one side; average					
337 mm thick	3.65	42.53	29.08	m2	71.61
450 mm thick	4.70	54.76	38.69	m2	93.46
Walls; tapering both sides; average					
337 mm thick	4.20	48.94	29.08	m2	78.02
450 mm thick	5.25	61.17	38.69	m2	99.86
Isolated piers					
one brick thick	3.20	37.29	18.96	m2	56.25
two brick thick	5.00	58.26	37.93	m2	96.19
three brick thick	6.30	73.41	56.89	m2	130.30
Isolated casings to steel columns					
half brick thick	1.60	18.64	9.18	m2	27.82
one brick thick	2.75	32.04	18.96	m2	51.00
Chimney stacks					
one brick thick	3.20	37.29	18.96	m2	56.25
two brick thick	5.00	58.26	37.93	m2	96.19
three brick thick	6.30	73.41	56.89	m2	130.30
Projections; vertical					
225 x 112 mm	0.40	4.66	2.12	m	6.78
225 x 225 mm	0.75	8.74	4.12	m	12.86
337 x 225 mm	1.10	12.82	6.92	m	19.74
440 x 225 mm	1.25	14.57	8.09	m	22.66
Bonding ends to existing					
half brick thick	0.40	4.66	0.64	m	5.30
one brick thick	0.55	6.41	1.29	m	7.70
one and a half brick thick	0.85	9.90	1.93	m	11.84
two brick thick	1.20	13.98	2.57	m	16.55
ADD or DEDUCT to walls for variation of £10.00/1000 in PC of common bricks					
half brick thick	-	-	-	m2	0.69
one brick thick	-	-	-	m2	1.37
one and a half brick thick	-	-	-	m2	2.06
two brick thick	-	-	-	m2	2.75
Segmental arches; one ring, 102 mm high on face					
102 mm wide on exposed soffit	1.80	17.84	1.19	m	19.03
215 mm wide on exposed soffit	2.25	21.52	2.56	m	24.08
Segmental arches; two ring, 215 mm high on face					
102 mm wide on exposed soffit	2.20	22.51	1.19	m	23.70
215 mm wide on exposed soffit	2.85	28.51	2.56	m	31.07
Semi-circular arches; one ring, 102 mm high on face					
102 mm wide on exposed soffit	2.10	20.30	1.45	m	21.75
215 mm wide on exposed soffit	2.50	23.57	16.64	m	40.21
Semi-circular arches; two ring, 215 mm high on face					
102 mm wide on exposed soffit	2.10	20.30	1.45	m	21.75
215 mm wide on exposed soffit	2.50	23.57	16.64	m	40.21
Labours on brick fairface					
Fair returns					
half brick wide	0.03	0.35	-	m	0.35
one brick wide	0.05	0.58	-	m	0.58

F MASONRY Including overheads and profit at 9.00%	Labour hours	Labour £	Material £	Unit	Total rate £
F10 BRICK/BLOCK WALLING - cont'd					
Class A engineering bricks; PC £340.00/1000; **in cement mortar (1:3)**					
Walls					
half brick thick	1.35	15.73	23.81	m2	39.54
one brick thick	2.25	26.22	48.28	m2	74.50
one and a half brick thick	3.00	34.96	72.48	m2	107.44
two brick thick	3.75	43.70	96.56	m2	140.26
Walls; facework one side					
half brick thick	1.50	17.48	24.36	m2	41.84
one brick thick	2.40	27.97	49.39	m2	77.36
one and a half brick thick	3.15	36.70	74.15	m2	110.85
two brick thick	3.90	45.44	98.78	m2	144.22
Walls; facework both sides					
half brick thick	1.60	18.64	24.36	m2	43.00
one brick thick	2.50	29.13	49.39	m2	78.52
one and a half brick thick	3.25	37.87	74.15	m2	112.02
two brick thick	4.00	46.61	98.78	m2	145.39
Walls; built curved mean radius 6 m					
one brick thick	3.00	34.96	51.90	m2	86.86
Walls; backing to masonry;					
cutting and bonding					
one brick thick	2.70	31.46	50.50	m2	81.96
one and a half brick thick	3.60	41.95	75.82	m2	117.77
Walls; tapering one side; average					
337 mm thick	3.90	45.44	75.82	m2	121.26
450 mm thick	5.00	58.26	101.00	m2	159.27
Walls; tapering both sides; average					
337 mm thick	4.50	52.43	77.49	m2	129.92
450 mm thick	5.70	66.42	103.23	m2	169.65
Isolated piers					
one brick thick	3.50	40.78	49.39	m2	90.17
two brick thick	5.50	64.09	98.78	m2	162.87
three brick thick	6.75	78.65	148.26	m2	226.91
Isolated casings to steel columns					
half brick thick	1.75	20.39	24.92	m2	45.31
one brick thick	3.00	34.96	50.50	m2	85.46
Projections; vertical					
225 x 112 mm	0.45	5.24	5.80	m	11.05
225 x 225 mm	0.80	9.32	11.20	m	20.52
337 x 225 mm	1.20	13.98	18.99	m	32.97
440 x 225 mm	1.35	15.73	22.00	m	37.73
Bonding ends to existing					
half brick thick	0.45	5.24	1.54	m	6.78
one brick thick	0.60	6.99	3.08	m	10.07
one and a half brick thick	0.90	10.49	4.62	m	15.10
two brick thick	1.30	15.15	6.16	m	21.30
ADD or DEDUCT to walls for variation of £10.00/1000 in PC of engineering bricks					
half brick thick	-	-	-	m2	0.69
one brick thick	-	-	-	m2	1.37
one and a half brick thick	-	-	-	m2	2.06
two brick thick	-	-	-	m2	2.75
Class B engineering bricks; PC £192.00/1000; **in cement mortar (1:3)**					
Walls					
half brick thick	1.35	15.73	13.89	m2	29.62
one brick thick	2.25	26.22	28.44	m2	54.66
one and a half brick thick	3.00	34.96	42.72	m2	77.68
two brick thick	3.75	43.70	56.87	m2	100.57
Walls; facework one side					
half brick thick	1.50	17.48	14.21	m2	31.69
one brick thick	2.40	27.97	29.06	m2	57.03
one and a half brick thick	3.15	36.70	43.66	m2	80.36
two brick thick	3.90	45.44	58.12	m2	103.56

F MASONRY Including overheads and profit at 9.00%	Labour hours	Labour £	Material £	Unit	Total rate £
Walls; facework both sides					
half brick thick	1.60	18.64	14.21	m2	32.85
one brick thick	2.50	29.13	29.06	m2	58.19
one and a half brick thick	3.25	37.87	43.66	m2	81.53
two brick thick	4.00	46.61	58.12	m2	104.73
Walls; built curved mean radius 6 m					
one brick thick	3.00	34.96	30.56	m2	65.52
Walls; backing to masonry;					
cutting and bonding					
one brick thick	2.70	31.46	29.69	m2	61.15
one and a half brick thick	3.60	41.95	44.60	m2	86.55
Walls; tapering one side; average					
337 mm thick	3.90	45.44	44.60	m2	90.04
450 mm thick	5.00	58.26	59.39	m2	117.65
Walls; tapering both sides; average					
337 mm thick	4.50	52.43	45.55	m2	97.98
450 mm thick	5.70	66.42	60.64	m2	127.06
Isolated piers					
one brick thick	3.50	40.78	29.06	m2	69.84
two brick thick	5.50	64.09	58.12	m2	122.21
three brick thick	6.75	78.65	87.29	m2	165.94
Isolated casings to steel columns					
half brick thick	1.75	20.39	14.51	m2	34.90
one brick thick	3.00	34.96	29.69	m2	64.65
Projections					
225 x 112 mm	0.45	5.24	3.37	m	8.61
225 x 225 mm	0.80	9.32	6.52	m	15.84
337 x 225 mm	1.20	13.98	11.01	m	24.99
440 x 225 mm	1.35	15.73	12.80	m	28.53
Bonding ends to existing					
half brick thick	0.45	5.24	0.94	m	6.18
one brick thick	0.60	6.99	1.89	m	8.88
one and a half brick thick	0.90	10.49	2.84	m	13.33
two brick thick	1.30	15.15	3.78	m	18.93
ADD or DEDUCT to walls for variation of £10.00/1000 in PC of engineering bricks					
half brick thick	-	-	-	m2	0.69
one brick thick	-	-	-	m2	1.37
one and a half brick thick	-	-	-	m2	2.06
two brick thick	-	-	-	m2	2.75
Refractory bricks; PC £494.00/1000; stretcher bond lining to flue; in fireclay cement mortar (1:4); built 50 mm clear of flues; one header per m2					
Walls; vertical; facework one side					
half brick thick	1.70	19.81	33.71	m2	53.52

ALTERNATIVE FACING BRICK PRICES (£/1000)

	£		£
Ibstock facing bricks; 215 x 102.5 x 65 mm			
Aldridge brown blend	302.00	Leicester Anglican Red Rustic	294.00
Cattybrook Gloucester Golden	300.00	Leicester Red Stock	326.00
Himley Dark Brown Rustic	345.00	Roughdales Red Multi Rustic	330.00
Himley Mixed Russet	297.00	Roughdales Trafford Buff Multi	340.00
London Brick Company facing bricks; 215 x 102.5 x 65 mm			
Brecken Grey	113.76	Orton Multi Buff	122.36
Chiltern	121.08	Regency	115.20
Claydon Red Multi	116.44	Sandfaced	121.03
Delph Autumn	114.10	Saxon Gold	118.39
Edwardian	127.26	Tudor	123.54
Georgian Red Multi	117.05	Victorian	134.50
Heather	119.22	Wansford Multi	121.74
Ironstone	116.16	Windsor	116.63
Milton Buff	115.27		

F MASONRY
Including overheads and profit at 9.00%

Redland facing bricks; 215 x 102.5 x 65 mm

Arun	374.00	Southwater class B	285.00
Beare Green restoration red	522.00	Sheppy 'matured' yellow	396.00
Chailey yellow multicoloured	391.00	Stourbridge Sherbourne range	293.00
Cottage mixed multicoloured	295.00	Stourbridge Henley range	280.00
Crowborough multicoloured	352.00	Stourbridge Pennine range	270.00
Dorking	302.00	Stourbridge Stratford range	241.00
Funton yellow London	284.00	Surrey bronze multicoloured	304.00
Hamsey multicoloured	368.00	Tonbridge handmade	481.00
Holbrook Castle range	291.00	Tonbridge handmade (50 mm deep)	497.00
Holbrook sandfaced textured	306.00	Tudor (53 mm deep)	599.00
Nutbourne sandfaced	269.00	Wealden	374.00
Pevensey red multicoloured	355.00	Wealdmade	418.00
Pluckley multicoloured	372.00		

	Labour hours	Labour £	Material £	Unit	Total rate £
F10 BRICK/BLOCK WALLING - cont'd					
Facing bricks; sand faced; PC £125.00/1000 (unless otherwise stated); in gauged mortar (1:1:6)					
Extra over common bricks; PC £117.00/1000; for facing bricks in					
stretcher bond	0.40	4.66	0.57	m2	5.23
flemish bond with snapped headers	0.50	5.83	3.49	m2	9.32
english bond with snapped headers	0.50	5.83	3.41	m2	9.24
ADD or DEDUCT for variation of £10.00/1000 in PC of facing bricks	-	-	-	m2	0.94
Half brick thick; stretcher bond; facework one side					
walls	1.65	19.23	9.73	m2	28.96
walls; building curved mean radius 6 m	2.40	27.97	10.47	m2	38.44
walls; building curved mean radius 1.50 m	3.00	34.96	11.01	m2	45.97
walls; building overhand	2.00	23.30	9.73	m2	33.03
walls; building up against concrete including flushing up at back	1.75	20.39	10.95	m2	31.34
walls; as formwork; temporary strutting	2.40	27.97	11.83	m2	39.80
walls; panels and aprons; not exceeding 1 m2	2.10	24.47	9.78	m2	34.25
isolated casings to steel columns	2.50	29.13	9.73	m2	38.86
bonding ends to existing	0.65	7.57	1.51	m	9.08
projections; vertical					
225 x 112 mm	0.40	4.66	2.25	m	6.91
337 x 112 mm	0.75	8.74	3.49	m	12.23
440 x 112 mm	1.10	12.82	4.72	m	17.54
Half brick thick; flemish bond with snapped headers; facework one side					
walls	1.90	22.14	10.69	m2	32.83
walls; building curved mean radius 6 m	2.70	31.46	11.22	m2	42.68
walls; building curved mean radius 1.50 m	3.50	40.78	11.76	m2	52.54
walls; building overhand	2.25	26.22	10.69	m2	36.91
walls; building up against concrete including flushing up at back	2.00	23.30	11.91	m2	35.21
walls; as formwork; temporary strutting	2.65	30.88	12.78	m2	43.66
walls; panels and aprons; not exceeding 1 m2	2.35	27.38	10.69	m2	38.07
isolated casings to steel columns	2.75	32.04	9.73	m2	41.77
bonding ends to existing	0.65	7.57	1.51	m	9.08
projections; vertical					
225 x 112 mm	0.50	5.83	2.55	m	8.38
337 x 112 mm	0.85	9.90	3.79	m	13.69
440 x 112 mm	1.20	13.98	5.16	m	19.14

F MASONRY Including overheads and profit at 9.00%	Labour hours	Labour £	Material £	Unit	Total rate £
One brick thick; two stretcher skins tied together; facework both sides					
walls	2.80	32.63	20.28	m2	52.91
walls; building curved mean radius 6 m	3.90	45.44	21.74	m2	67.18
walls; building curved mean radius 1.50 m	4.80	55.93	24.93	m2	80.86
isolated piers	3.30	38.45	22.02	m2	60.47
bonding ends to existing	0.85	9.90	3.01	m	12.91
One brick thick; flemish bond; facework both sides					
walls	2.90	33.79	20.09	m2	53.88
walls; building curved mean radius 6 m	4.00	46.61	21.56	m2	67.18
walls; building curved mean radius 1.50 m	5.00	58.26	24.75	m2	83.01
isolated piers	3.40	39.62	21.85	m2	61.47
bonding ends to existing	0.85	9.90	3.01	m	12.91
projections; vertical					
225 x 225 mm	0.80	9.32	2.96	m	12.28
337 x 225 mm	1.50	17.48	4.32	m	21.80
440 x 225 mm	2.20	25.63	5.77	m	31.40
ADD or DEDUCT for variation of £10.00/1000 in PC of facing bricks; in stretcher bond					
half brick thick	-	-	-	m2	0.71
one brick thick	-	-	-	m2	1.42
ADD or DEDUCT for variation of £10.00/1000 in PC of facing bricks; in flemish bond					
half brick thick	-	-	-	m2	0.88
one brick thick	-	-	-	m2	1.77
Extra over facing bricks for					
recessed joints	0.03	0.35	-	m2	0.35
raking out joints and pointing in black mortar	0.50	5.83	0.15	m2	5.98
bedding and pointing half brick wall in black mortar	-	-	-	m2	0.68
bedding and pointing one brick wall in black mortar	-	-	-	m2	1.81
flush plain bands; 225 mm wide stretcher bond; horizontal; bricks; PC £145.00/1000	0.25	2.91	0.33	m	3.24
flush quoins; average 320 mm girth; block bond vertical; facing bricks; PC £145.00/1000	0.40	4.66	0.31	m	4.97
Flat arches; 215 mm high on face					
102 mm wide exposed soffit	1.15	12.01	1.47	m	13.48
215 mm wide exposed soffit	1.72	17.96	2.95	m	20.91
Flat arches; 215 mm high on face; bullnosed specials; PC £42.90/100					
102 mm wide exposed soffit	1.20	12.59	7.92	m	20.51
215 mm wide exposed soffit	1.80	18.89	15.85	m	34.74
Segmental arches; one ring; 215 mm high on face					
102 mm wide exposed soffit	2.10	20.99	2.07	m	23.06
215 mm wide exposed soffit	3.15	31.49	4.33	m	35.82
Segmental arches; two ring; 215 mm high on face					
102 mm wide exposed soffit	2.70	27.98	2.07	m	30.05
215 mm wide exposed soffit	4.05	41.98	4.33	m	46.31
Segmental arches; 215 mm high on face; cut voussoirs; PC £59.55/100					
102 mm wide exposed soffit	2.20	22.16	10.96	m	33.12
215 mm wide exposed soffit	3.30	33.24	22.10	m	55.34
Segmental arches; one and a half ring; 320 mm high on face; cut voussoirs; PC £59.55/100					
102 mm wide exposed soffit	3.00	31.48	20.72	m2	52.20
215 mm wide exposed soffit	4.50	47.22	41.64	m2	88.86

F MASONRY Including overheads and profit at 9.00%	Labour hours	Labour £	Material £	Unit	Total rate £
F10 BRICK/BLOCK WALLING - cont'd					
Facing bricks; sand faced; PC £125.00/1000 (unless otherwise stated); in gauged mortar (1:1:6) - cont'd					
Semi circular arches; one ring; 215 mm high on face					
102 mm wide exposed soffit	2.60	26.12	2.63	m	**28.75**
215 mm wide exposed soffit	3.70	37.55	18.99	m	**56.54**
Semi circular arches; two ring; 215 mm high on face					
102 mm wide exposed soffit	3.20	33.11	2.63	m	**35.74**
215 mm wide exposed soffit	4.60	48.04	18.99	m	**67.03**
Semi circular arches; one ring; 215 mm high on face; cut voussoirs PC £59.55/100					
102 mm wide exposed soffit	2.70	27.29	11.23	m	**38.52**
215 mm wide exposed soffit	3.85	39.30	36.17	m	**75.47**
Bullseye window 600 mm dia; two rings; 215 mm high on face					
102 mm wide exposed soffit	6.00	66.44	3.14	nr	**69.58**
215 mm wide exposed soffit	9.00	99.65	6.08	nr	**105.73**
Bullseye window 1200 mm dia; two rings; 215 mm high on face					
102 mm wide exposed soffit	10.50	115.39	4.78	nr	**120.17**
215 mm wide exposed soffit	15.75	173.09	9.56	m	**182.65**
Bullseye window 600 mm dia; one ring; 215 mm high on face; cut voussoirs PC £59.55/100					
102 mm wide exposed soffit	5.00	54.78	25.80	nr	**80.58**
215 mm wide exposed soffit	7.50	82.18	51.43	nr	**133.61**
Bullseye window 1200 mm dia; one ring; 215 mm high on face; cut voussoirs					
102 mm wide exposed soffit	9.00	97.91	44.30	nr	**142.21**
215 mm wide exposed soffit	13.50	146.87	88.59	nr	**235.46**
ADD or DEDUCT for variation of £10.00/1000 in PC of facing bricks	-	-	-	m	**0.29**
Sills; horizontal; headers on edge; pointing top and one side; set weathering					
150 x 102 mm	0.70	8.16	2.31	m	**10.47**
150 x 102 mm; cant headers; PC £55.35/100	0.75	8.74	9.33	m	**18.07**
Sills; horizontal; headers on flat; pointing top and one side					
150 x 102 mm; bullnosed specials; PC £42.90/100	0.65	7.57	5.19	m	**12.76**
Coping; horizontal; headers on edge; pointing top and both sides					
215 x 102 mm	0.56	6.53	2.25	m	**8.78**
260 x 102 mm	0.90	10.49	3.49	m	**13.98**
215 x 102 mm; double bullnosed specials; PC £43.80/100	0.60	6.99	7.39	m	**14.38**
260 x 120 mm; single bullnosed specials; PC £42.90/100	0.90	10.49	11.85	m	**22.34**
ADD or DEDUCT for variation of £10.00/1000 in PC of facing bricks	-	-	-	m	**0.29**
Facing bricks; machine made facings; PC £320.00/1000 (unless otherwise stated; in gauged mortar (1:1:6)					
Extra over common bricks; PC £117.00/1000; for facing bricks in					
stretcher bond	0.40	4.66	14.27	m2	**18.93**
flemish bond with snapped headers	0.50	5.83	17.84	m2	**23.67**
english bond with snapped headers	0.50	5.83	20.21	m2	**26.04**
ADD or DEDUCT for variation of £10.00/1000 in PC of facing bricks	-	-	-	m2	**0.88**

F MASONRY Including overheads and profit at 9.00%	Labour hours	Labour £	Material £	Unit	Total rate £
Half brick thick; stretcher bond; facework **one side**					
walls	1.65	19.23	23.44	m2	42.67
walls; building curved mean radius 6 m	2.40	27.97	25.21	m2	53.18
walls; building curved mean radius 1.50 m	3.00	34.96	26.43	m2	61.39
walls; building overhand	2.00	23.30	23.44	m2	46.74
walls; building up against concrete including flushing up at back	1.75	20.39	24.66	m2	45.05
walls; as formwork; temporary strutting	2.40	27.97	25.54	m2	53.51
walls; panels and aprons; not exceeding 1 m2	2.10	24.47	23.97	m2	48.44
isolated casings to steel columns	2.50	29.13	23.44	m2	52.57
bonding ends to existing	0.65	7.57	1.51	m	9.08
projections; vertical					
225 x 112 mm	0.40	4.66	5.46	m	10.12
337 x 112 mm	0.75	8.74	8.28	m	17.02
440 x 112 mm	1.10	12.82	11.11	m	23.93
Half brick thick; flemish bond with **snapped headers; facework one side**					
walls	1.90	22.14	25.89	m2	48.03
walls; building curved mean radius 6 m	2.70	31.46	27.12	m2	58.58
walls; building curved mean radius 1.50 m	3.50	40.78	28.35	m2	69.13
walls; building overhand	2.25	26.22	25.89	m2	52.11
walls; building up against concrete including flushing up at back	2.00	23.30	27.98	m2	51.28
walls; as formwork; temporary strutting	2.65	30.88	27.11	m2	57.99
walls; panels and aprons; not exceeding 1 m2	2.35	27.38	25.89	m2	53.27
isolated casings to steel columns	2.75	32.04	23.44	m2	55.48
bonding ends to existing	0.65	7.57	1.51	m	9.08
projections; vertical					
225 x 112 mm	0.50	5.83	6.20	m	12.03
337 x 112 mm	0.85	9.90	9.03	m	18.93
440 x 112 mm	1.20	13.98	12.24	m	26.22
One brick thick; two stretcher skins tied **together; facework both sides**					
walls	2.80	32.63	47.69	m2	80.32
walls; building curved mean radius 6 m	3.90	45.44	51.02	m2	96.46
walls; building curved mean radius 1.50 m	4.80	55.93	53.66	m2	109.59
isolated piers	3.30	38.45	47.68	m2	86.13
bonding ends to existing	0.85	9.90	3.01	m	12.91
One brick thick; flemish bond; **facework both sides**					
walls	2.90	33.79	47.51	m2	81.30
walls; building curved mean radius 6 m	4.00	46.61	51.02	m2	97.63
walls; building curved mean radius 1.50 m	5.00	58.26	53.48	m2	111.74
isolated piers	3.40	39.62	47.69	m2	87.31
bonding ends to existing	0.85	9.90	3.01	m	12.91
projections; vertical					
225 x 225 mm	0.80	9.32	10.54	m	19.86
337 x 225 mm	1.50	17.48	17.86	m	35.34
440 x 225 mm	2.20	25.63	20.69	m	46.32
ADD or DEDUCT for variation of £10.00/1000 **in PC of facing bricks; in stretcher bond**					
half brick thick	-	-	-	m2	0.71
one brick thick	-	-	-	m2	1.42
ADD or DEDUCT for variation of £10.00/1000 **in PC of facing bricks; in flemish bond**					
half brick thick	-	-	-	m2	0.88
one brick thick	-	-	-	m2	1.77
Extra over facing bricks for					
recessed joints	0.03	0.35	-	m2	0.35
raking out joints and pointing in black mortar	0.50	5.83	0.15	m2	5.98
bedding and pointing half brick wall in black mortar	-	-	-	m2	0.68
bedding and pointing one brick wall in black mortar	-	-	-	m2	1.81

F MASONRY Including overheads and profit at 9.00%	Labour hours	Labour £	Material £	Unit	Total rate £
F10 BRICK/BLOCK WALLING - cont'd					
Facing bricks; machine made facings; **PC £320.00/1000 (unless otherwise stated;** **in gauged mortar (1:1:6) - cont'd**					
Extra over facing bricks for					
flush plain bands; 225 mm wide stretcher					
bond; horizontal; bricks PC £350.00/1000	0.25	2.91	0.49	m	3.40
flush quoins; average 320 mm girth; black					
bond vertical; bricks PC £350.00/1000	0.40	4.66	0.46	m	5.12
Flat arches; 215 mm high on face					
102 mm wide exposed soffit	1.15	12.01	2.39	m	14.40
215 mm wide exposed soffit	1.72	17.96	4.78	m	22.74
Flat arches; 215 mm high on face; bullnosed					
specials; PC £105.00/100					
102 mm wide exposed soffit	1.20	12.59	18.11	m	30.70
215 mm wide exposed soffit	1.80	18.89	36.22	m	55.11
Segmental arches; one ring; 215 mm high					
on face					
102 mm wide exposed soffit	2.10	20.99	3.44	m	24.43
215 mm wide exposed soffit	3.15	31.49	7.09	m	38.58
Segmental arches; two ring; 215 mm high					
on face					
102 mm wide exposed soffit	2.70	27.98	3.44	m	31.42
215 mm wide exposed soffit	4.05	41.98	7.09	m	49.07
Segmental arches; 215 mm high on face; cut					
voussoirs; PC £226.00/100					
102 mm wide exposed soffit	2.20	22.16	38.26	m	60.42
215 mm wide exposed soffit	3.30	33.24	76.72	m	109.96
Segmental arches; one and a half ring;					
320 mm high on face; cut voussoirs					
PC £226.00/100					
102 mm wide exposed soffit	3.00	31.48	75.34	m	106.82
215 mm wide exposed soffit	4.50	47.22	150.86	m	198.08
Semi circular arches; one ring;					
215 mm high on face					
102 mm wide exposed soffit	2.60	26.12	4.46	m	30.58
215 mm wide exposed soffit	3.70	37.55	22.66	m	60.21
Semi circular arches; two ring; 215 mm					
high on face					
102 mm wide exposed soffit	3.20	33.11	4.46	m	37.57
215 mm wide exposed soffit	4.60	48.04	22.66	m	70.70
Semi circular arches; one ring; 215 mm high					
on face; cut voussoirs PC £226.00/100					
102 mm wide exposed soffit	2.70	27.29	38.52	m	65.81
215 mm wide exposed soffit	3.85	39.30	90.79	m	130.09
Bullseye window 600 mm dia; two rings;					
215 mm high on face					
102 mm wide exposed soffit	6.00	66.44	46.59	m	113.03
215 mm wide exposed soffit	9.00	99.65	92.98	m	192.63
Bullseye window 1200 mm dia; two rings;					
215 mm high on face					
102 mm wide exposed soffit	10.50	115.39	8.45	m	123.84
215 mm wide exposed soffit	15.75	173.09	16.91	m	190.00
Bullseye window 600 mm dia; one ring;					
215 mm high on face; cut voussoirs					
PC £226.00/100					
102 mm wide exposed soffit	5.00	54.78	94.07	m	148.85
215 mm wide exposed soffit	7.50	82.18	187.94	m	270.12
Bullseye window 1200 mm dia; one ring;					
215 mm high on face; cut voussoirs					
PC £226.00/100					
102 mm wide exposed soffit	9.00	97.91	162.06	m	259.97
215 mm wide exposed soffit	13.50	146.87	324.11	m	470.98
ADD or DEDUCT for variation of £10.00/1000					
in PC of facing bricks	-	-	-	m2	0.29

F MASONRY Including overheads and profit at 9.00%	Labour hours	Labour £	Material £	Unit	Total rate £
Sills; horizontal; headers on edge; pointing top and one side; set weathering					
150 x 102 mm;	0.70	8.16	5.50	m	13.66
150 x 102 mm; cant headers; PC £105.00/100	0.75	8.74	17.47	m	26.21
Sills; horizontal; headers on flat; pointing top and one side					
150 x 102 mm; bullnosed specials; PC £105.00/100	0.65	7.57	12.53	m	20.10
Coping; horizontal; headers on edge; pointing top and both sides					
215 x 102 mm	0.56	6.53	11.84	m	18.37
260 x 102 mm	0.90	10.49	8.28	m	18.77
215 x 102 mm; double bullnosed specials; PC £105.00/100	0.60	6.99	17.43	m	24.42
260 x 120 mm; single bullnosed specials; PC £105.00/100	0.90	10.49	28.58	m	39.07
ADD or DEDUCT for variation of £10.00/1000 in PC of facing bricks	-	-	-	m	0.29
Facing bricks; hand made; PC £520.00/1000 (unless otherwise stated); in gauged mortar (1:1:6)					
Extra over common bricks; PC £117.00/1000; for facing bricks in					
stretcher bond	0.40	4.66	28.33	m2	32.99
flemish bond with snapped headers	0.50	5.83	35.42	m2	41.25
english bond with snapped headers	0.50	5.83	40.14	m2	45.97
ADD or DEDUCT for variation of £10.00/1000 in PC of facing bricks	-	-	-	m2	0.88
Half brick thick; stretcher bond; facework one side					
walls	1.65	19.23	37.51	m2	56.74
walls; building curved mean radius 6 m	2.40	27.97	40.31	m2	68.28
walls; building curved mean radius 1.50 m	3.00	34.96	42.26	m2	77.22
walls; building overhand	2.00	23.30	37.51	m2	60.81
walls; building up against concrete including flushing up at back	1.75	20.39	38.72	m2	59.11
walls; as formwork; temporary strutting	2.40	27.97	39.60	m2	67.57
walls; panels and aprons; not exceeding 1 m2	2.10	24.47	37.51	m2	61.98
isolated casings to steel columns	2.50	29.13	37.51	m2	66.64
bonding ends to existing	0.65	7.57	1.51	m	9.08
projections; vertical					
225 x 112 mm	0.40	4.66	8.74	m	13.40
337 x 112 mm	0.75	8.74	13.21	m	21.95
440 x 112 mm	1.10	12.82	17.68	m	30.50
Half brick thick; flemish bond with snapped headers; facework one side					
walls	1.90	22.14	41.47	m2	63.61
walls; building curved mean radius 6 m	2.70	31.46	43.43	m2	74.89
walls; building curved mean radius 1.50 m	3.50	40.78	45.38	m2	86.16
walls; building overhand	2.25	26.22	41.47	m2	67.69
walls; building up against concrete including flushing up at back	2.00	23.30	42.70	m2	66.00
walls; as formwork; temporary strutting	2.65	30.88	43.57	m2	74.45
walls; panels and aprons; not exceeding 1 m2	2.35	27.38	41.47	m2	68.85
isolated casings to steel columns	2.75	32.04	37.51	m2	69.55
bonding ends to existing	0.65	7.57	1.51	m	9.08
projections; vertical					
225 x 112 mm	0.50	5.83	9.95	m	15.78
337 x 112 mm	0.85	9.90	14.43	m	24.33
440 x 112 mm	1.20	13.98	19.51	m	33.49
One brick thick; two stretcher skins tied together; facework both sides					
walls	2.80	32.63	76.17	m2	108.80
walls; building curved mean radius 6 m	3.90	45.44	79.99	m2	125.43
walls; building curved mean radius 1.50 m	4.80	55.93	85.15	m2	141.88
isolated piers	3.30	38.45	76.14	m2	114.59
bonding ends to existing	0.85	9.90	3.01	m	12.9

F MASONRY Including overheads and profit at 9.00%	Labour hours	Labour £	Material £	Unit	Total rate £

F10 BRICK/BLOCK WALLING - cont'd

Facing bricks; hand made; PC £520.00/1000
(unless otherwise stated); in gauged
mortar (1:1:6) - cont'd
One brick thick; flemish bond;
facework both sides

	Labour hours	Labour £	Material £	Unit	Total rate £
walls	2.90	33.79	75.63	m2	109.42
walls; building curved mean radius 6 m	4.00	46.61	81.23	m2	127.84
walls; building curved mean radius 1.50 m	5.00	58.26	85.13	m2	143.40
isolated piers	3.40	39.62	75.63	m2	115.25
bonding ends to existing	0.85	9.90	3.01	m	12.91
projections; vertical					
225 x 225 mm	0.80	9.32	16.86	m	26.18
337 x 225 mm	1.50	17.48	28.64	m	46.12
440 x 225 mm	2.20	25.63	33.12	m	58.75
ADD or DEDUCT for variation of £10.00/1000 in PC of facing bricks; in stretcher bond					
half brick thick	-	-	-	m2	0.71
one brick thick	-	-	-	m2	1.42
ADD or DEDUCT for variation of £10.00/1000 in PC of facing bricks; in flemish bond					
half brick thick	-	-	-	m2	0.88
one brick thick	-	-	-	m2	1.77
Extra over facing bricks for					
recessed joints	0.03	0.35	-	m2	0.35
raking out joints and pointing in black mortar	0.50	5.83	0.15	m2	5.98
bedding and pointing half brick wall in black mortar	-	-	-	m2	0.68
bedding and pointing one brick wall in black mortar	-	-	-	m2	1.81
flush plain bands; 225 mm wide stretcher bond; horizontal; bricks PC £560.00/1000	0.25	2.91	0.65	m	3.56
flush quoins; average 320 mm girth; block bond vertical; bricks PC £560.00/1000	0.40	4.66	0.61	m	5.27
Flat arches; 215 mm high on face					
102 mm wide exposed soffit	1.15	12.01	3.34	m	15.35
215 mm wide exposed soffit	1.72	17.96	6.67	m	24.63
Flat arches; 215 mm high on face; bullnosed specials; PC £105.00/100					
102 mm wide exposed soffit	1.20	12.59	18.11	m	30.70
215 mm wide exposed soffit	1.80	18.89	36.22	m	55.11
Segmental arches; one ring; 215 mm high on face					
102 mm wide exposed soffit	2.10	20.99	4.86	m	25.85
215 mm wide exposed soffit	3.15	31.49	9.91	m	41.40
Segmental arches; two ring; 215 mm high on face					
102 mm wide exposed soffit	2.70	27.98	4.86	m	32.84
215 mm wide exposed soffit	4.05	41.98	9.91	m	51.89
Segmental arches; 215 mm high on face; cut voussoirs; PC £226.00/100					
102 mm wide exposed soffit	2.20	22.16	38.26	m	60.42
215 mm wide exposed soffit	3.30	33.24	76.72	m	109.96
Segmental arches; one and a half ring; 320 mm high on face; cut voussoirs PC £226.00/100					
102 mm wide exposed soffit	3.00	31.48	75.34	m	106.82
215 mm wide exposed soffit	4.50	47.22	150.86	m	198.08
Semi circular arches; one ring; 215 mm high on face					
102 mm wide exposed soffit	2.60	26.12	6.35	m	32.47
215 mm wide exposed soffit	3.70	37.55	26.44	m	63.99
Semi circular arches; two ring; 215 mm high on face					
102 mm wide exposed soffit	3.20	33.11	6.35	m	39.46
215 mm wide exposed soffit	4.60	48.04	26.44	m	74.48

F MASONRY Including overheads and profit at 9.00%	Labour hours	Labour £	Material £	Unit	Total rate £
Semi circular arches; one ring; 215 mm high on face; cut voussoirs PC £226.00/100					
102 mm wide exposed soffit	2.70	27.29	38.52	m	65.81
215 mm wide exposed soffit	3.85	39.30	90.79	m	130.09
Bullseye window 600 mm dia; two rings; 215 mm high on face					
102 mm wide exposed soffit	6.00	66.44	74.84	m	141.28
215 mm wide exposed soffit	9.00	99.65	149.49	m	249.14
Bullseye window 1200 mm dia; two rings; 215 mm high on face					
102 mm wide exposed soffit	10.50	115.39	12.23	m2	127.62
215 mm wide exposed soffit	15.75	173.09	24.44	m2	197.53
Bullseye window 600 mm dia; one ring; 215 mm high on face; cut voussoirs; PC £226.00/100					
102 mm wide exposed soffit	5.00	54.78	94.01	m	148.79
215 mm wide exposed soffit	7.50	82.18	187.94	m	270.12
Bullseye window 1200 mm dia; one ring; 215 mm high on face; cut voussoirs; PC £226.00/100					
102 mm wide exposed soffit	9.00	97.91	162.06	m	259.97
215 mm wide exposed soffit	13.50	146.87	324.11	m	470.98
ADD or DEDUCT for variation of £10.00/1000 in PC of facing bricks	-	-	-	m	0.29
Sills; horizontal; headers on edge; pointing top and one side; set weathering					
150 x 102 mm;	0.70	8.16	8.78	m	16.94
150 x 102 mm; cant headers; PC £105.00/100	0.75	8.74	17.47	m	26.21
Sills; horizontal; headers on flat; pointing top and one side					
150 x 102 mm; bullnosed specials; PC £105.00/100	0.65	7.57	12.53	m	20.10
Coping; horizontal; headers on edge; pointing top and both sides					
215 x 102 mm	0.56	6.53	15.12	m	21.65
260 x 102 mm	0.90	10.49	13.21	m	23.70
215 x 102 mm; double bullnosed specials; PC £105.00/100	0.60	6.99	17.43	m	24.42
260 x 120 mm; single bullnosed specials; PC £105.00/100	0.90	10.49	28.58	m	39.07
ADD or DEDUCT for variation of £10.00/1000 in PC of facing bricks	-	-	-	m2	0.29
50 mm facing bricks slips; PC £93.00/100; in gauged mortar (1:1:6) built up against concrete including flushing up at back (ties measured elsewhere)					
Walls	2.50	29.13	67.08	m2	96.21
Edges of suspended slabs 200 mm wide	0.75	8.74	13.52	m	22.26
Columns 400 mm wide	1.50	17.48	26.93	m	44.41
Engineering bricks; PC £340.00/1000; and specials at PC £105.00/100; in cement mortar (1:3)					
Steps; all headers-on-edge; edges set with					
215 x 102 mm; horizontal; set weathering	0.70	8.16	17.04	m	25.20
Returned ends pointed	0.20	2.33	2.55	nr	4.88
430 x 102 mm; horizontal; set weathering	1.00	11.65	22.84	m	34.49
Returned ends pointed	0.25	2.91	4.96	nr	7.87
Labours on brick facework					
Fair cutting					
to curve	0.20	2.33	1.74	m	4.07
Fair returns					
half brick wide	0.05	0.58	-	m	0.58
one brick wide	0.07	0.82	-	m	0.82
one and a half brick wide	0.09	1.05	-	m	1.05
Fair angles formed by cutting					
squint	0.80	9.32	-	m	9.32
birdsmouth	0.70	8.16	-	m	8.16
external; chamfered 25 mm wide	1.00	11.65	-	m	11.65
external; rounded 100 mm radius	1.20	13.98	-	m	13.98

F MASONRY Including overheads and profit at 9.00%	Labour hours	Labour £	Material £	Unit	Total rate £
F10 BRICK/BLOCK WALLING - cont'd					
Labours on brick facework - cont'd					
Fair chases					
100 x 50 mm; horizontal	1.50	17.48	-	m	17.48
100 x 50 mm; vertical	1.50	17.48	-	m	17.48
300 x 50 mm; vertical	2.70	31.46	-	m	31.46
Bonding ends to existing					
half brick thick	0.65	7.57	1.51	m	9.08
one brick thick	0.85	9.90	3.01	m	12.91
Centering to brickwork soffits; **(prices included within arches rates)**					
Flat soffits; not exceeding 2 m span					
over 0.3 m wide	2.00	16.35	5.48	m2	21.83
102 mm wide	0.40	3.27	0.97	m	4.24
215 mm wide	0.60	4.91	1.95	m	6.86
Segmental soffits; 1500 mm span 200 mm rise					
102 mm wide	1.33	10.87	2.02	nr	12.89
215 mm wide	2.00	16.35	4.03	nr	20.38
Semicircular soffits; 1500 mm span					
102 mm wide	1.80	14.72	2.39	nr	17.11
215 mm wide	2.40	19.62	4.78	nr	24.40
Bullseye window; 265 mm wide					
600 mm dia	1.50	12.26	2.84	nr	15.10
1200 mm dia	3.00	24.53	5.34	nr	29.87

ALTERNATIVE BLOCK PRICES (£/m2)

	£		£		£		£
Aerated Concrete Durox 'Supablocs'; 630 x 225 mm							
75 mm	4.19	125 mm	7.87	175 mm	11.02	225 mm	14.17
90 mm	5.39	130 mm	8.19	190 mm	11.38	250 mm	15.75
100 mm	5.99	140 mm	8.82	200 mm	11.98	280 mm	17.64
115 mm	7.24	150 mm	9.45	215 mm	12.87		

	£			£		
ARC Conbloc blocks; 450 x 225 mm						
Cream fair faced						
75 mm solid	5.45	190 mm solid	13.63	190 mm hollow	11.07	
100 mm solid	6.51	100 mm hollow	6.34	215 mm hollow	12.41	
140 mm solid	9.35	140 mm hollow	8.54			
Fenlite						
90 mm solid	4.25	100 mm solid	4.35	140 mm solid	7.35	
Leca thermal						
100 mm Leca 6	6.75	190 mm Leca 6	12.80	125 mm Leca 45	9.10	
Standard facing						
100 mm solid	5.95	190 mm solid	12.20	190 mm hollow	10.05	
140 mm solid	8.80	140 mm hollow	8.20	215 mm hollow	11.45	

	£		£		£		£
Celcon 'Standard' blocks; 450 x 225 mm							
75 mm	4.34	125 mm	7.24	190 mm	11.00	230 mm	13.32
90 mm	5.21	140 mm	8.11	200 mm	11.58	250 mm	14.48
100 mm	5.79	150 mm	8.69	215 mm	12.45	300 mm	17.37

'Solar' blocks are also available at the same price, in a limited range

	£		£		£
Forticrete painting quality blocks; 450 x 225 mm					
100 mm solid	6.22	215 mm solid	13.39	190 mm hollow	10.07
140 mm solid	9.24	100 mm hollow	5.34	215 mm hollow	10.38
190 mm solid	12.35	140 mm hollow	7.41		

F MASONRY
Including overheads and profit at 9.00%

Lytag blocks; 450 x 225 mm
 3.5 N/mm2 Thermal blocks   Insulating blocks
   75 mm solid   3.33   140 mm solid   6.25   100 mm cellular   6.87
   90 mm solid   6.50   190 mm solid   8.47   140 mm cellular   8.84
  100 mm solid   4.43
 7.0 N/mm2 Insulating extra strength blocks
  100 mm solid   4.59   140 mm solid   6.49   190 mm solid   8.79
 10.5 N/mm2 High strength blocks
  100 mm solid   5.07   140 mm solid   7.21   190 mm solid   9.74
 3.5 N/mm2 and 7.0 N/mm2 Close textured blocks
  100 mm solid (3.5)   5.71   140 mm solid (3.5) 8.03   190 mm solid (3.5) 10.85
  100 mm solid (7.0)   5.93   140 mm solid (7.0) 8.35   190 mm solid (7.0) 11.28

Tarmac 'Topblocks'; 450 x 225 mm
 3.5 N/mm2 'Hemelite' blocks
   70/75 mm solid   4.15   140 mm solid   6.40   190 mm solid   8.40
  100 mm solid   4.00   150 mm solid   6.80   215 mm solid   9.75
 7.0 N/mm2 'Hemelite' blocks
  100 mm solid   4.71   190 mm solid   9.08   215 mm solid   10.56
  140 mm solid   7.27

               £   £   £   £
'Toplite' standard blocks
   75 mm   4.09   140 mm   7.63   100 mm   5.45   215 mm   11.72
   90 mm   4.91   150 mm   8.18   200 mm   10.90

'Toplite' GTI (thermal) blocks
  115 mm   6.44   130 mm   7.28   200 mm   11.20
  125 mm   7.00   150 mm   8.40   215 mm   12.04

Discounts of 0 - 7.5% available depending on quantity/status

		Labour hours	Labour £	Material £	Unit	Total rate £
Lightweight aerated concrete blocks;						
Thermalite 'Shield'/'Turbo' blocks or						
similar; in gauged mortar (1:2:9)						
Walls or partitions or skins of hollow walls						
75 mm thick	PC £4.44	0.60	6.99	5.79	m2	12.78
90 mm thick	PC £5.33	0.66	7.69	6.91	m2	14.60
100 mm thick	PC £5.92	0.70	8.16	7.72	m2	15.88
115 mm thick	PC £7.17	0.74	8.62	9.29	m2	17.91
125 mm thick	PC £7.79	0.77	8.97	10.16	m2	19.13
130 mm thick	PC £8.10	0.80	9.32	10.53	m2	19.85
140 mm thick	PC £8.29	0.82	9.55	10.79	m2	20.34
150 mm thick	PC £8.89	0.85	9.90	11.51	m2	21.41
190 mm thick	PC £11.25	1.00	11.65	14.57	m2	26.22
200 mm thick	PC £11.84	1.05	12.23	15.28	m2	27.51
215 mm thick	PC £12.73	1.10	12.82	16.47	m2	29.29
255 mm thick	PC £15.09	1.20	13.98	19.45	m2	33.43
Isolated piers or chimney stacks						
190 mm thick		1.40	16.31	14.57	m2	30.88
215 mm thick		1.50	17.48	16.35	m2	33.83
255 mm thick		1.70	19.81	19.45	m2	39.26

F MASONRY Including overheads and profit at 9.00%	Labour hours	Labour £	Material £	Unit	Total rate £

F10 BRICK/BLOCK WALLING - cont'd

Lightweight aerated concrete blocks;
Thermalite 'Shield'/'Turbo' blocks or
similar; in gauged mortar (1:2:9) - cont'd
Isolated casings

75 mm thick	0.70	8.16	5.79	m2	13.95
90 mm thick	0.76	8.86	6.98	m2	15.84
100 mm thick	0.80	9.32	7.72	m2	17.04
115 mm thick	0.84	9.79	9.29	m2	19.08
125 mm thick	0.87	10.14	10.16	m2	20.30
140 mm thick	0.92	10.72	10.79	m2	21.51
Extra over for fair face; flush pointing					
walls; one side	0.10	1.17	-	m2	1.17
walls; both sides	0.17	1.98	-	m2	1.98
Bonding ends to common brickwork					
75 mm blockwork	0.15	1.75	0.47	m	2.22
90 mm blockwork	0.15	1.75	0.55	m	2.30
100 mm blockwork	0.20	2.33	0.61	m	2.94
115 mm blockwork	0.20	2.33	0.73	m	3.06
125 mm blockwork	0.23	2.68	0.82	m	3.50
130 mm blockwork	0.23	2.68	0.85	m	3.53
140 mm blockwork	0.25	2.91	0.88	m	3.79
150 mm blockwork	0.25	2.91	0.93	m	3.84
190 mm blockwork	0.30	3.50	1.20	m	4.70
200 mm blockwork	0.33	3.85	1.25	m	5.10
215 mm blockwork	0.36	4.19	1.34	m	5.53
255 mm blockwork	0.40	4.66	1.60	m	6.26

Lightweight smooth face aerated concrete
blocks; Thermalite 'Smooth Face' blocks or
similar; in gauged mortar (1:2:9);
flush pointing one side
Walls or partitions or skins of hollow walls

100 mm thick	PC £8.39	0.88	9.88	10.37	m2	20.25
140 mm thick	PC £11.72	1.02	11.42	14.52	m2	25.94
150 mm thick	PC £12.54	1.05	11.77	15.50	m2	27.27
190 mm thick	PC £15.91	1.22	13.66	19.70	m2	33.36
200 mm thick	PC £16.74	1.27	14.24	20.69	m2	34.93
215 mm thick	PC £18.00	1.34	14.96	22.25	m2	37.21
Isolated piers or chimney stacks						
190 mm thick		1.62	18.32	19.70	m2	38.02
200 mm thick		1.72	19.48	20.69	m2	40.17
215 mm thick		1.94	21.95	22.25	m2	44.20
Isolated casings						
100 mm thick		0.98	11.05	10.37	m2	21.42
140 mm thick		1.12	12.58	14.52	m2	27.10
Extra over for flush pointing						
walls; both sides		0.07	0.82	-	m2	0.82
Bonding ends to common brickwork						
100 mm blockwork		0.25	2.91	0.85	m	3.76
140 mm blockwork		0.28	3.26	1.20	m	4.46
150 mm blockwork		0.30	3.50	1.28	m	4.78
190 mm blockwork		0.35	4.08	1.65	m	5.73
200 mm blockwork		0.38	4.43	1.72	m	6.15
215 mm blockwork		0.40	4.66	1.84	m	6.50

Lightweight aerated high strength concrete
blocks (7 N/mm2); Thermalite 'High Strength'
blocks or similar; in cement mortar (1:3)
Walls or partitions or skins of hollow walls

100 mm thick	PC £7.66	0.74	8.44	9.50	m2	17.94
140 mm thick	PC £10.71	0.85	9.67	13.31	m2	22.98
150 mm thick	PC £11.48	0.90	10.25	14.24	m2	24.49
190 mm thick	PC £14.56	1.06	12.07	18.09	m2	30.16
200 mm thick	PC £15.32	1.11	12.65	18.37	m2	31.02
215 mm thick	PC £16.47	1.17	13.31	20.42	m2	33.73

F MASONRY Including overheads and profit at 9.00%		Labour hours	Labour £	Material £	Unit	Total rate £
Isolated piers or chimney stacks						
190 mm thick		1.41	16.15	16.50	m2	32.65
200 mm thick		1.51	17.32	18.99	m2	36.31
215 mm thick		1.72	19.72	20.38	m2	40.10
Isolated casings						
100 mm thick		0.84	9.60	9.50	m2	19.10
140 mm thick		0.95	10.84	13.31	m2	24.15
150 mm thick		1.00	11.42	14.24	m2	25.66
190 mm thick		1.16	13.24	18.09	m2	31.33
200 mm thick		1.26	14.40	18.99	m2	33.39
215 mm thick		1.32	15.05	20.42	m2	35.47
Extra over for fair face; flush pointing						
walls; one side		0.10	1.17	-	m2	1.17
walls; both sides		0.17	1.98	-	m2	1.98
Bonding ends to common brickwork						
100 mm blockwork		0.25	2.91	0.77	m	3.68
140 mm blockwork		0.28	3.26	1.11	m	4.37
150 mm blockwork		0.30	3.50	1.18	m	4.68
190 mm blockwork		0.35	4.08	1.52	m	5.60
200 mm blockwork		0.38	4.43	1.58	m	6.01
215 mm blockwork		0.40	4.66	1.70	m	6.36
Dense aggregate concrete blocks; 'ARC Conbloc' or similar; in gauged mortar (1:2:9)						
Walls or partitions or skins of hollow walls						
75 mm thick; solid	PC £4.05	0.70	9.50	5.10	m2	14.60
100 mm thick; solid	PC £4.25	0.80	10.86	5.42	m2	16.28
140 mm thick; hollow	PC £6.65	0.90	12.21	8.41	m2	20.62
140 mm thick; solid	PC £6.70	1.00	13.57	8.46	m2	22.03
190 mm thick; hollow	PC £8.40	1.15	15.61	10.70	m2	26.31
215 mm thick; hollow	PC £8.90	1.25	16.96	11.34	m2	28.30
Isolated piers or chimney stacks						
140 mm thick; hollow		1.25	16.96	8.41	m2	25.37
190 mm thick; hollow		1.60	21.71	10.70	m2	32.41
215 mm thick; hollow		1.80	24.43	11.34	m2	35.77
Isolated casings						
75 mm thick; solid		0.80	10.86	5.10	m2	15.96
100 mm thick; solid		0.90	12.21	5.42	m2	17.63
140 mm thick; solid		1.10	14.93	8.46	m2	23.39
Extra over for fair face; flush pointing						
walls; one side		0.10	1.36	-	m2	1.36
walls; both sides		0.17	2.31	-	m2	2.31
Bonding ends to common brickwork						
75 mm blockwork		0.20	2.71	0.43	m	3.14
100 mm blockwork		0.26	3.53	0.44	m	3.97
140 mm blockwork		0.32	4.34	0.72	m	5.06
190 mm blockwork		0.38	5.16	0.93	m	6.09
215 mm blockwork		0.44	5.97	0.97	m	6.94
Dense aggregate concrete blocks; (7 N/mm2) Forticrete 'Leicester Common' blocks or similar; in cement mortar (1:3)						
Walls or partitions or skins of hollow walls						
75 mm thick; solid	PC £4.40	0.70	9.50	5.54	m2	15.04
100 mm thick; hollow	PC £4.85	0.80	10.86	6.16	m2	17.02
100 mm thick; solid	PC £5.65	0.80	10.86	7.13	m2	17.99
140 mm thick; hollow	PC £6.71	0.90	12.21	8.53	m2	20.74
140 mm thick; solid	PC £8.41	1.00	13.57	10.57	m2	24.14
190 mm thick; hollow	PC £9.10	1.15	15.61	11.62	m2	27.23
190 mm thick; solid	PC £11.27	1.25	16.96	14.22	m2	31.18
215 mm thick; hollow	PC £9.43	1.25	16.96	12.05	m2	29.01
215 mm thick; solid	PC £12.15	1.35	18.32	15.31	m2	33.63
Dwarf support wall						
140 mm thick; solid		1.40	19.00	10.57	m2	29.57
190 mm thick; solid		1.60	21.71	14.22	m2	35.93
215 mm thick; solid		1.80	24.43	15.31	m2	39.74

F MASONRY Including overheads and profit at 9.00%	Labour hours	Labour £	Material £	Unit	Total rate £

F10 BRICK/BLOCK WALLING - cont'd

Dense aggregate concrete blocks; (7 N/mm2)
Forticrete 'Leicester Common' blocks or
similar; in cement mortar (1:3) - cont'd

Isolated piers or chimney stacks

140 mm thick; hollow	1.25	16.96	8.53	m2	25.49
190 mm thick; hollow	1.60	21.71	11.62	m2	33.33
215 mm thick; hollow	1.80	24.43	12.05	m2	36.48

Isolated casings

75 mm thick; solid	0.80	10.86	5.54	m2	16.40
100 mm thick; solid	0.90	12.21	7.13	m2	19.34
140 mm thick; solid	1.10	14.93	10.57	m2	25.50

Extra over for fair face; flush pointing

walls; one side	0.10	1.36	-	m2	1.36
walls; both sides	0.17	2.31	-	m2	2.31

Bonding ends to common brickwork

75 mm blockwork	0.20	2.71	0.47	m	3.18
100 mm blockwork	0.26	3.53	0.59	m	4.12
140 mm blockwork	0.32	4.34	0.90	m	5.24
190 mm blockwork	0.38	5.16	1.21	m	6.37
215 mm blockwork	0.44	5.97	1.30	m	7.27

Dense aggregate coloured concrete blocks;
Forticrete 'Leicester Bathstone'; in
coloured gauged mortar (1:1:6); flush
pointing one side

Walls or partitions or skins of hollow walls

100 mm thick; hollow	PC £10.78	0.90	12.21	13.50	m2	25.71
100 mm thick; solid	PC £12.95	0.90	12.21	16.09	m2	28.30
140 mm thick; hollow	PC £14.27	1.00	13.57	17.90	m2	31.47
140 mm thick; solid	PC £19.15	1.10	14.93	23.74	m2	38.67
190 mm thick; hollow	PC £17.83	1.25	16.96	22.52	m2	39.48
190 mm thick; solid	PC £26.09	1.35	18.32	32.43	m2	50.75
215 mm thick; hollow	PC £19.26	1.35	18.32	24.31	m2	42.63
215 mm thick; solid	PC £27.85	1.45	19.68	34.61	m2	54.29

Isolated piers or chimney stacks

140 mm thick; solid	1.50	20.36	23.74	m2	44.10
190 mm thick; solid	1.70	23.07	32.43	m2	55.50
215 mm thick; solid	1.90	25.78	34.75	m2	60.53

Extra over for flush pointing

walls; both sides	0.07	0.95	-	m2	0.95

Extra over blocks for

100 mm thick lintol blocks; ref D14	0.25	3.39	7.75	m	11.14
140 mm thick lintol blocks; ref H14	0.30	4.07	8.51	m	12.58
140 mm thick quoin blocks; ref H16	0.40	5.43	17.88	m	23.31
140 mm thick cavity closer blocks; ref H17	0.40	5.43	20.07	m	25.50
140 mm thick cill blocks; ref H21	0.30	4.07	12.47	m	16.54
190 mm thick lintol blocks; ref A14	0.40	5.43	9.94	m	15.37
190 mm thick cill blocks; ref A21	0.35	4.75	13.33	m	18.08

F11 GLASS BLOCK WALLING

NOTE: The following specialist prices for
glass block walling assume standard blocks;
panels of 50 m2; no fire rating; work in
straight walls at ground floor level; and all
necessary ancillary fixing; strengthening;
easy access; pointing and expansion
materials etc.

F MASONRY Including overheads and profit at 9.00% & 5/00%	Labour hours	Labour £	Material £	Unit	Total rate £

Hollow glass block walling; Luxcrete sealed
'Luxblocks' or similar; in cement mortar
'Luxfix' joints; reinforced with 6 mm dia.
stainless steel rods; with 'Luxfibre' at
head and jambs; pointed both sides with
'Luxseal' mastic
Panel walls; facework both sides

115 x 115 x 80 mm flemish blocks	-	-	-	m2	440.09
190 x 190 x 80 mm flemish;					
cross reeded or clear blocks	-	-	-	m2	248.52
240 x 240 x 80 mm flemish;					
cross reeded or clear blocks	-	-	-	m2	217.46
240 x 115 x 80 mm flemish					
or clear blocks	-	-	-	m2	331.36

F20 NATURAL STONE RUBBLE WALLING

Cotswold Guiting limestone; laid dry
Uncoursed random rubble walling

275 mm thick	1.80	24.19	35.72	m2	59.91
350 mm thick	2.10	28.22	45.18	m2	73.40
425 mm thick	2.35	31.58	55.26	m2	86.84
500 mm thick	2.60	34.94	64.72	m2	99.66

Cotswold Guiting limestone;
bedded; jointed and pointed in
cement - lime mortar (1:2:9)
Uncoursed random rubble walling; faced
and pointed; both sides

275 mm thick	1.70	22.85	37.29	m2	60.14
350 mm thick	1.80	24.19	47.14	m2	71.33
425 mm thick	1.90	25.54	57.61	m2	83.15
500 mm thick	2.00	26.88	67.86	m2	94.74

Coursed random rubble walling; rough
dressed; faced and pointed one side

114 mm thick	1.48	17.25	45.92	m2	63.17
150 mm thick	1.74	20.27	52.69	m2	72.96

Fair returns on walling

114 mm wide	0.03	0.35	-	m	0.35
150 mm wide	0.04	0.47	-	m	0.47
275 mm wide	0.07	0.82	-	m	0.82
350 mm wide	0.09	1.05	-	m	1.05
425 mm wide	0.11	1.28	-	m	1.28
500 mm wide	0.13	1.51	-	m	1.51

Fair raking cutting on walling

114 mm thick	0.20	2.33	6.49	m	8.82
150 mm thick	0.24	2.80	7.46	m	10.26

Level uncoursed rubble walling for damp
proof courses and the like

275 mm wide	0.21	2.82	2.19	m	5.01
350 mm wide	0.22	2.96	2.50	m	5.46
425 mm wide	0.23	3.09	2.80	m	5.89
500 mm wide	0.24	3.23	3.40	m	6.63

Copings formed of rough stones; faced
and pointed all round

275 x 200 mm (average) high	0.51	6.85	7.67	m	14.52
350 x 250 mm (average) high	0.67	9.00	11.63	m	20.63
425 x 300 mm (average) high	0.85	11.42	16.30	m	27.72
500 x 350 mm (average) high	1.05	14.11	22.37	m	36.48

F MASONRY Including overheads and profit at 9.00%	Labour hours	Labour £	Material £	Unit	Total rate £

F22 CAST STONE WALLING/DRESSINGS

Reconstructed limestone walling; 'Bradstone'
100 mm bed weathered Cotswold or
North Cerney masonry blocks or similar;
laid to pattern or course recommended;
bedded, jointed and pointed in approved
coloured cement - lime mortar (1:2:9)

Walls; facing and pointing one side

masonry blocks; random uncoursed	1.12	13.05	18.75	m2	31.80
Extra for					
Return ends	0.40	4.66	1.20	m	5.86
Plain 'L' shaped quoins	0.13	1.51	4.14	m	5.65
traditional walling; coursed squared	1.40	16.31	18.11	m2	34.42
squared random rubble	1.40	16.31	18.50	m2	34.81
squared coursed rubble (large module)	1.30	15.15	18.35	m2	33.50
squared coursed rubble (small module)	1.35	15.73	18.64	m2	34.37
squared and pitched rock faced walling;					
coursed	1.45	16.90	18.11	m2	35.01
rough hewn rockfaced walling; random	1.50	17.48	17.95	m2	35.43
Extra for return ends	0.16	1.86	-	m	1.86

Isolated piers or chimney stacks;
facing and pointing one side

masonry blocks; random uncoursed	1.55	18.06	19.64	m2	37.70
traditional walling; coursed squared	1.95	22.72	18.97	m2	41.69
squared random rubble	1.95	22.72	19.38	m2	42.10
squared coursed rubble (large module)	1.80	20.97	19.22	m2	40.19
squared coursed rubble (small module)	1.90	22.14	19.54	m2	41.68
squared and pitched rock faced walling;					
coursed	2.05	23.89	18.97	m2	42.86
rough hewn rockfaced walling; random	2.10	24.47	18.81	m2	43.28

Isolated casings; facing and pointing
one side

masonry blocks; random uncoursed	1.35	15.73	19.64	m2	35.37
traditional walling; coursed squared	1.70	19.81	18.97	m2	38.78
squared random rubble	1.70	19.81	19.38	m2	39.19
squared coursed rubble (large module)	1.55	18.06	19.22	m2	37.28
squared coursed rubble (small module)	1.65	19.23	19.54	m2	38.77
squared and pitched rock faced walling;					
coursed	1.75	20.39	18.97	m2	39.36
rough hewn rockfaced walling; random	1.80	20.97	18.81	m2	39.78

Fair returns 100 mm wide

masonry blocks; random uncoursed	0.12	1.40	-	m	1.40
traditional walling; coursed squared	0.15	1.75	-	m	1.75
squared random rubble	0.15	1.75	-	m	1.75
squared coursed rubble (large module)	0.14	1.63	-	m	1.63
squared coursed rubble (small module)	0.14	1.63	-	m	1.63
squared and pitched rock faced walling;					
coursed	0.15	1.75	-	m	1.75
rough hewn rockfaced walling; random	0.16	1.86	-	m	1.86

Fair raking cutting on masonry blocks

100 mm thick	0.18	2.10	-	m	2.10

Reconstructed limestone dressings;
'Bradstone Architectural' dressings in
weathered Cotswold or North Cerney shades
or similar; bedded, jointed and pointed in
approved coloured cement - lime
mortar (1:2:9)

Copings; twice weathered and throated

152 x 76 mm; type A	0.33	3.85	8.32	m	12.17
178 x 64 mm; type B	0.33	3.85	8.94	m	12.79
305 x 76 mm; type A	0.40	4.66	15.89	m	20.55
Extra for					
Fair end	-	-	-	nr	3.10
Returned mitred fair end	-	-	-	nr	3.10

F MASONRY Including overheads and profit at 9.00%	Labour hours	Labour £	Material £	Unit	Total rate £
Copings; once weathered and throated					
191 x 76 mm	0.33	3.85	9.68	m	13.53
305 x 76 mm	0.40	4.66	15.60	m	20.26
365 x 76 mm	0.40	4.66	16.79	m	21.45
Extra for					
Fair end	-	-	-	nr	3.10
Returned mitred fair end	-	-	-	nr	3.10
Chimney caps; four times weathered and throated; once holed					
553 x 533 x 76 mm	0.40	4.66	21.49	nr	26.15
686 x 686 x 76 mm	0.40	4.66	35.32	nr	39.98
Pier caps; four times weathered and throated					
305 x 305 mm	0.25	2.91	7.13	nr	10.04
381 x 381 mm	0.25	2.91	10.03	nr	12.94
457 x 457 mm	0.30	3.50	13.93	nr	17.43
533 x 533 mm	0.30	3.50	19.34	nr	22.84
Splayed corbels					
457 x 102 x 229 mm	0.15	1.75	8.98	nr	10.73
686 x 102 x 229 mm	0.20	2.33	14.31	nr	16.64
Air bricks					
229 x 142 x 76 mm	0.08	0.93	3.51	nr	4.44
102 x 152 mm lintels; rectangular; reinforced with mild steel bars					
not exceeding 1.22 m long	-	-	-	m	14.60
1.37 - 1.67 m long	-	-	-	m	14.87
1.83 - 1.98 m long	-	-	-	m	16.27
102 x 229 mm lintels; rectangular; reinforced with mild steel bars					
not exceeding 1.67 m long	0.26	3.03	18.22	m	21.25
1.83 - 1.98 m long	0.28	3.26	18.65	m	21.91
2.13 - 2.44 m long	0.30	3.50	18.78	m	22.28
2.59 - 2.90 m long	0.32	3.73	20.48	m	24.21
197 x 67 mm sills to suit standard softwood windows; stooled at ends					
0.56 - 2.50 m long	0.30	3.50	22.95	m	26.45
Window surround; traditional with label moulding; for single light; sill 146 x 133 mm; jambs 146 x 146 mm; head 146 x 105 mm; including all dowels and anchors					
window size 508 x 1479 mm PC £102.43	0.90	10.49	112.08	nr	122.57
Window surround; traditional with label moulding; three light; for windows 508 x 1219 mm; sill 146 x 133 mm; jambs 146 x 146 mm; head 146 x 103 mm; mullions 146 x 108 mm; including all dowels and anchors					
overall size 1975 x 1479 mm PC £223.10	2.35	27.38	243.89	nr	271.27
Door surround; moulded continuous jambs and head with label moulding; including all dowels and anchors					
door 839 x 1981 mm in 102 x 64 mm frame	1.65	19.23	296.13	nr	315.36

F30 ACCESSORIES/SUNDRY ITEMS FOR BRICK/BLOCK/STONE WALLING

Sundries - brick/block walling

Forming cavities in hollow walls; three wall ties per m2					
50 mm cavity; polypropylene ties	0.06	0.70	0.13	m2	0.83
50 mm cavity; galvanized steel butterfly wall ties	0.06	0.70	0.18	m2	0.88
50 mm cavity; galvanized steel twisted wall ties	0.06	0.70	0.24	m2	0.94
50 mm cavity; stainless steel butterfly wall ties	0.06	0.70	0.23	m2	0.93

F MASONRY Including overheads and profit at 9.00%	Labour hours	Labour £	Material £	Unit	Total rate £
F30 ACCESSORIES/SUNDRY ITEMS FOR BRICK/BLOCK/ STONE WALLING - cont'd					
Sundries - brick/block walling - cont'd					
Forming cavities in hollow walls; three wall					
ties per m2					
50 mm cavity; stainless steel twisted					
wall ties	0.06	0.70	0.59	m2	1.29
75 mm cavity; polypropylene ties	0.06	0.70	0.13	m2	0.83
75 mm cavity; galvanised steel butterfly					
wall ties	0.06	0.70	0.18	m2	0.88
75 mm cavity; galvanised steel twisted					
wall ties	0.06	0.70	0.24	m2	0.94
75 mm cavity; stainless steel butterfly					
wall ties	0.06	0.70	0.23	m2	0.93
75 mm cavity; stainless steel twisted					
wall ties	0.06	0.70	0.59	m2	1.29
Closing at jambs with common brickwork					
half brick thick					
50 mm cavity	0.30	3.50	0.90	m	4.40
50 mm cavity; including damp proof course	0.40	4.66	1.39	m	6.05
75 mm cavity	0.30	3.50	1.18	m	4.68
75 mm cavity; including damp proof course	0.40	4.66	1.67	m	6.33
Closing at jambs with blockwork 100 mm thick					
50 mm cavity	0.25	2.91	0.43	m	3.34
50 mm cavity; including damp proof course	0.30	3.50	0.94	m	4.44
75 mm cavity	0.25	2.91	0.61	m	3.52
75 mm cavity; including damp proof course	0.30	3.50	1.11	m	4.61
Closing at sill with one course common					
brickwork					
50 mm cavity	0.30	3.50	2.00	m	5.50
50 mm cavity; including damp proof course	0.35	4.08	2.49	m	6.57
75 mm cavity	0.30	3.50	2.00	m	5.50
75 mm cavity; including damp proof course	0.35	4.08	2.49	m	6.57
Closing at top with					
single course of slates	0.15	1.75	2.07	m	3.82
'Westbrick' cavity closer	0.15	1.75	3.41	m	5.16
Cavity wall insulation; 'Dritherm'					
filling 50 mm cavity	0.15	1.75	1.44	m2	3.19
filling 75 mm cavity	0.17	1.98	1.91	m2	3.89
Cavity wall insulation; fixing with					
insulation retaining ties					
30 mm Celotex RR cavity insulation	0.20	2.33	0.16	m2	2.49
30 mm cavity wall batts	0.20	2.33	1.92	m2	4.25
25 mm Plaschem 'Aerobuild' foil faced					
polyurethene cavity slabs	0.20	2.33	3.45	m2	5.78
25 mm Styrofoam cavity wall insulation	0.20	2.33	2.89	m2	5.22

ALTERNATIVE DAMP PROOF COURSE PRICES (£/m2)

	£		£		£
Asbestos based					
'Astos'	3.77	'Barchester'	3.77		
Fibre based					
'Challenge'	2.29	'Stormax'	2.30		
Hessian based					
'Callendrite'	3.07	'Nubit'	3.41	'Permaseal'	3.40
Leadcored					
'Astos'	9.16	'Ledumite' grade 2	14.01	'Permalead'	10.08
'Ledkore' grade B	16.37	'Ledumite' grade 3	17.63	'Pluvex'	8.17
'Ledumite' grade 1	10.09	'Nuled'	8.09	'Trindos'	9.10
Pitch polymer					
'Aquagard'	4.22	'Permaflex'	4.22		

Further discounts may be available depending on quantity/status

F MASONRY Including overheads and profit at 9.00%	Labour hours	Labour £	Material £	Unit	Total rate £
'Pluvex' (hessian based) damp proof course **or similar; PC £3.43/m2; 200 mm laps;** **in gauged mortar (1:1:6)**					
Horizontal					
over 225 mm wide	0.25	2.91	4.04	m2	6.95
over 225 mm wide; forming cavity gutters					
in hollow walls	0.40	4.66	4.04	m2	8.70
not exceeding 225 mm wide	0.50	5.83	4.04	m2	9.87
Vertical					
not exceeding 225 mm wide	0.75	8.74	4.04	m2	12.78
'Pluvex' (fibre based) damp proof course or **similar; PC £2.30/m2; 200 mm laps;** **in gauged mortar (1:1:6)**					
Horizontal					
over 225 mm wide	0.25	2.91	2.71	m2	5.62
over 225 mm wide; forming cavity gutters					
in hollow walls	0.40	4.66	2.70	m2	7.36
not exceeding 225 mm wide	0.50	5.83	2.70	m2	8.53
Vertical					
not exceeding 225 mm wide	0.75	8.74	2.70	m2	11.44
'Asbex' (asbestos based) damp proof course **or similar; PC £3.77/m2; 200 mm laps;** **in gauged mortar (1:1:6)**					
Horizontal					
over 225 mm wide	0.25	2.91	4.41	m2	7.32
over 225 mm wide; forming cavity gutters					
in hollow walls	0.40	4.66	4.41	m2	9.07
not exceeding 225 mm wide	0.50	5.83	4.41	m2	10.24
Vertical					
not exceeding 225 mm wide	0.75	8.74	4.41	m2	13.15
Polyethelene damp proof course; PC £0.89/m2; **200 mm laps; in gauged mortar (1:1:6)**					
Horizontal					
over 225 mm wide	0.25	2.91	1.04	m2	3.95
over 225 mm wide; forming cavity gutters					
in hollow walls	0.40	4.66	1.04	m2	5.70
not exceeding 225 mm wide	0.50	5.83	1.04	m2	6.87
Vertical					
not exceeding 225 mm wide	0.75	8.74	1.04	m2	9.78
'Permabit' bitumen polymer damp proof course **or similar; PC £4.22/m2; 150 mm laps; in gauged** **mortar (1:1:6)**					
Horizontal					
over 225 mm wide	0.25	2.91	4.97	m2	7.88
over 225 mm wide; forming cavity gutters					
in hollow walls	0.40	4.66	4.97	m2	9.63
not exceeding 225 mm wide	0.50	5.83	4.97	m2	10.80
Vertical					
not exceeding 225 mm wide	0.75	8.74	4.95	m2	13.69
'Hyload' (pitch polymer) damp proof course **or similar; PC £4.09/m2; 150 mm laps;** **in gauged mortar (1:1:6)**					
Horizontal					
over 225 mm wide	0.25	2.91	4.82	m2	7.73
over 225 mm wide; forming cavity gutters					
in hollow walls	0.40	4.66	4.82	m2	9.48
not exceeding 225 mm wide	0.50	5.83	4.82	m2	10.65
Vertical					
not exceeding 225 mm wide	0.75	8.74	4.82	m2	13.56

F MASONRY Including overheads and profit at 9.00%	Labour hours	Labour £	Material £	Unit	Total rate £
F30 ACCESSORIES/SUNDRY ITEMS FOR BRICK/BLOCK/ STONE WALLING - cont'd					
'Ledkore' grade A (bitumen based lead cored); damp proof course or similar; PC £12.04/m2; 200 mm laps; in gauged mortar (1:1:6)					
Horizontal					
over 225 mm wide	0.33	3.85	14.17	m2	18.02
over 225 mm wide; forming cavity gutters in hollow walls	0.53	6.18	14.17	m2	20.35
not exceeding 225 mm wide	0.50	5.83	14.17	m2	20.00
Vertical					
not exceeding 225 mm wide	0.75	8.74	14.10	m2	22.84
Milled lead damp proof course; PC £20.84/m2; BS 1178; 1.80 mm (code 4), 175 mm laps; in cement-lime mortar (1:2:9)					
Horizontal					
over 225 mm wide	2.00	23.30	24.54	m2	47.84
not exceeding 225 mm wide	3.00	34.96	24.54	m2	59.50
Two courses slates in cement mortar (1:3)					
Horizontal					
over 225 mm wide	1.50	17.48	17.76	m2	35.24
Vertical					
over 225 mm wide	2.25	26.22	18.16	m2	44.38
'Synthaprufe' damp proof membrane; PC £27.02/25L; three coats brushed on					
Vertical					
not exceeding 150 mm wide	0.08	0.94	0.46	m2	1.40
150 - 225 mm wide	0.10	1.17	0.71	m2	1.88
225 - 300 mm wide	0.13	1.51	0.93	m2	2.44
over 300 mm wide	0.28	3.26	3.10	m2	6.36
'Type X' polypropylene abutment cavity tray by Cavity Trays Ltd; built into facing brickwork as the work proceeds; complete with Code 4 lead flashing					
Intermediate tray with short leads (requiring soakers); to suit roof of					
17 - 20 degrees pitch	0.06	0.70	6.84	nr	7.54
21 - 25 degrees pitch	0.06	0.70	9.29	nr	9.99
26 - 45 degrees pitch	0.06	0.70	5.27	nr	5.97
Intermediate tray with long leads (suitable only for corrugated roof tiles); to suit roof of					
17 - 20 degrees pitch	0.06	0.70	4.85	nr	5.55
21 - 25 degrees pitch	0.06	0.70	4.48	nr	5.18
26 - 45 degrees pitch	0.06	0.70	4.46	nr	5.16
Extra for					
ridge tray	0.06	0.70	4.13	nr	4.83
catchment end tray	0.06	0.70	3.47	nr	4.17
Servicised 'Bitu-thene' self-adhesive cavity flashing; type 'CA'; well lapped at joints; in gauged mortar (1:1:6)					
Horizontal					
over 225 mm wide	0.85	9.90	0.54	m2	10.44

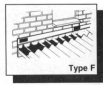
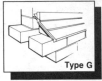
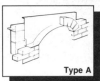
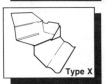

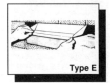
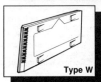
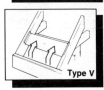
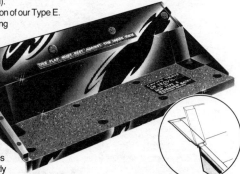

F MASONRY Including overheads and profit at 9.00%	Labour hours	Labour £	Material £	Unit	Total rate £
'Brickforce' galvanised steel joint reinforcement In walls					
40 mm wide; ref. GBF 40	0.02	0.23	0.16	m	0.39
60 mm wide; ref. GBF 60	0.03	0.35	0.17	m	0.52
100 mm wide; ref. GBF 100	0.04	0.47	0.22	m	0.69
160 mm wide; ref. GBF 160	0.05	0.58	0.32	m	0.90
'Brickforce' stainless steel joint reinforcement In walls					
40 mm wide; ref. SBF 40	0.02	0.23	1.65	m	1.88
60 mm wide; ref. SBF 60	0.03	0.35	1.80	m	2.15
100 mm wide; ref. SBF 100	0.04	0.47	1.85	m	2.32
160 mm wide; ref. SBF 160	0.05	0.58	1.92	m	2.50
Fillets/pointing/wedging and pinning etc. Weather or angle fillets in cement mortar (1:3)					
50 mm face width	0.12	1.40	0.07	m	1.47
100 mm face width	0.20	2.33	0.27	m	2.60
Pointing wood frames or sills with mastic					
one side	0.07	0.82	0.40	m	1.22
each side	0.14	1.63	0.80	m	2.43
Pointing wood frames or sills with polysulphide sealant					
one side	0.07	0.82	0.80	m	1.62
each side	0.14	1.63	1.59	m	3.22
Bedding wood plates in cement mortar (1:3)					
100 mm wide	0.06	0.70	0.04	m	0.74
Bedding wood frame in cement mortar (1:3) and point					
one side	0.08	0.93	0.04	m	0.97
each side	0.10	1.17	0.06	m	1.23
one side in mortar; other side in mastic	0.15	1.75	0.44	m	2.19
Wedging and pinning up to underside of existing construction with slates in cement mortar (1:3)					
102 mm wall	0.80	9.32	2.15	m	11.47
215 mm wall	1.00	11.65	3.96	m	15.61
327 mm wall	1.20	13.98	5.77	m	19.75
Raking out joint in brickwork or blockwork for turned-in edge of flashing					
horizontal	0.15	1.75	0.04	m	1.79
stepped	0.20	2.33	0.04	m	2.37
Raking out and enlarging joint in brickwork or blockwork for nib of asphalt					
horizontal	0.20	2.33	-	m	2.33
Cutting grooves in brickwork or blockwork					
for water bars and the like	0.30	3.50	-	m	3.50
for nib of asphalt; horizontal	0.30	3.50	-	m	3.50
Preparing to receive new walls					
top existing 215 mm brick wall	0.20	2.33	-	m	2.33
Sills and tile creasings Sills; two courses of machine made plain roofing tiles; set weathering; bedded and pointed	0.60	6.99	2.87	m	9.86
Extra over brickwork for two courses of machine made tile creasing; bedded and pointed; projecting 25 mm each side; horizontal					
215 mm wide copings	0.50	5.83	4.33	m	10.16
260 mm wide copings	0.70	8.16	5.57	m	13.73
Galvanised steel coping cramp; built in	0.10	1.17	0.24	nr	1.41

F MASONRY Including overheads and profit at 9.00%	Labour hours	Labour £	Material £	Unit	Total rate £

F30 ACCESSORIES/SUNDRY ITEMS FOR BRICK/BLOCK/ STONE WALLING - cont'd

Flue linings etc
Flue linings; True Flue 200 mm refractory
concrete square flue linings; rebated
joints in refractory mortar (1:2.5)

	Labour hours	Labour £	Material £	Unit	Total rate £
linings	0.18	2.10	8.02	m	10.12
offset unit; ref 5u	0.10	1.17	5.99	nr	7.16
offset unit; ref 6u	0.10	1.17	5.39	nr	6.56
off set unit; ref 7u	0.10	1.17	5.04	nr	6.21
pot; ref 8u	0.10	1.17	3.37	nr	4.54
single cap unit; ref 10u	0.25	2.91	13.53	nr	16.44
double cap unit; ref 11u	0.30	3.50	12.33	nr	15.83
U-type lintol with U1 attachment	0.30	3.50	14.29	nr	17.79

Gas flue linings; True Flue 'Typex 115'
refractory concrete blocks, built in; in
refractory mortar (1:2.5); cutting brickwork
or blockwork around

	Labour hours	Labour £	Material £	Unit	Total rate £
recess panel; ref PXY 218	0.10	1.17	2.36	nr	3.53
lintol unit; ref 1Y	0.10	1.17	3.34	nr	4.51
entry and exit unit; ref 2Y	0.10	1.17	4.34	nr	5.51
straight block; ref 3YA	0.10	1.17	2.13	nr	3.30
straight block; ref 3YB	0.10	1.17	2.21	nr	3.38
offset unit; ref 4Y	0.10	1.17	2.64	nr	3.81
reverse rebate unit; ref 5Y	0.10	1.17	2.17	nr	3.34
conversion corbel unit; ref 6Y	0.10	1.17	4.10	nr	5.27
vertical conversion unit; ref 7Y	0.10	1.17	4.32	nr	5.49
back offset unit; ref 8Y	0.10	1.17	6.43	nr	7.60
back offset unit; ref 9Y	0.10	1.17	5.99	nr	7.16

Gas flue linings; True Flue 'Typex 150'
refractory concrete blocks built in; in
refractory mortar (1:2.5); cutting brickwork
and or blockwork around

	Labour hours	Labour £	Material £	Unit	Total rate £
recess panel; ref RPX 305	0.10	1.17	3.29	nr	4.46
offset lintol; ref 1X1	0.10	1.17	3.69	nr	4.86
twin channel flue block; ref 2X1	0.10	1.17	3.43	nr	4.60
side offset block; ref 4X1	0.10	1.17	2.70	nr	3.87
back offset block; ref 4X4	0.10	1.17	2.56	nr	3.73
single flue block; ref 5X	0.10	1.17	1.79	nr	2.96
single flue block; ref 5X1	0.10	1.17	2.14	nr	3.31
single flue block; ref 5X2	0.10	1.17	2.17	nr	3.34
single flue block; ref 5X3/5/8	0.10	1.17	2.16	nr	3.33
separating withe; ref 8X	0.10	1.17	1.36	nr	2.53
separating withe; ref 8X4	0.10	1.17	1.97	nr	3.14
end terminal cap unit; ref 9X	0.25	2.91	4.34	nr	7.25
inter terminal cap unit; ref 9X1	0.25	2.91	3.10	nr	6.01
end terminal cap unit; ref 9X6	0.25	2.91	4.49	nr	7.40
inter terminal cap unit; ref 9X7	0.25	2.91	3.69	nr	6.60
single flue cap unit; ref 9X9	0.25	2.91	7.82	nr	10.73
conversion unit; ref 10X1	0.10	1.17	2.45	nr	3.62
angled conversion unit; ref 10X2	0.10	1.17	3.65	nr	4.82
aluminium louvre; ref MLX	0.10	1.17	4.37	nr	5.54

Parging and coring flues with refractory
mortar (1:2.5); sectional area not exceeding
0.25 m2

	Labour hours	Labour £	Material £	Unit	Total rate £
900 x 900 mm	0.60	6.99	4.96	m	11.95

F MASONRY Including overheads and profit at 9.00%		Labour hours	Labour £	Material £	Unit	Total rate £

Ancillaries

Forming openings; one ring arch over
225 x 225 mm; one brick facing brickwork

making good facings both sides		1.50	17.48	-	nr	17.48

Forming opening through hollow wall; slate
lintel over; sealing 50 cavity with slates
in cement mortar (1:3); making good fair
face or facings one side

225 x 75 mm		0.20	2.33	1.50	nr	3.83
225 x 150 mm		0.25	2.91	2.20	nr	5.11
225 x 225 mm		0.30	3.50	2.90	nr	6.40

Air bricks; red terracotta; building into
prepared openings

215 x 65 mm		0.08	0.93	0.90	nr	1.83
215 x 140 mm		0.08	0.93	1.47	nr	2.40
215 x 215 mm		0.08	0.93	3.11	nr	4.04

Air bricks; cast iron; building into
prepared openings

215 x 65 mm		0.08	0.93	2.63	nr	3.56
215 x 140 mm		0.08	0.93	4.77	nr	5.70
215 x 215 mm		0.08	0.93	6.30	nr	7.23

Bearers; mild steel

51 x 10 mm flat bar		0.25	2.58	2.97	m	5.55
64 x 13 mm flat bar		0.25	2.58	4.75	m	7.33
50 x 50 x 6 mm angle section		0.35	3.61	2.42	m	6.03
Ends fanged for building in		-	-	-		0.78

Proprietary items

Ties in walls; 150 mm long butterfly type;
building into joints of brickwork
or blockwork

galvanized steel or polypropylene		0.03	0.35	0.07	nr	0.42
stainless steel		0.03	0.35	0.09	nr	0.44
copper		0.03	0.35	0.13	nr	0.48

Ties in walls; 20 x 3 x 150 mm long twisted
wall type; building into joints of brickwork
or blockwork

galvanized steel		0.04	0.47	0.10	nr	0.57
stainless steel		0.04	0.47	0.25	nr	0.72
copper		0.04	0.47	0.34	nr	0.81

Anchors in walls; 25 x 3 x 100 mm long; one
end dovetailed; other end building into
joints of brickwork or blockwork

galvanized steel		0.06	0.70	0.22	nr	0.92
stainless steel		0.06	0.70	0.35	nr	1.05
copper		0.06	0.70	0.54	nr	1.24

Fixing cramp 25 x 3 x 250 mm long; once
bent; fixed to back of frame; other end
building into joints of brickwork
or blockwork

galvanized steel		0.06	0.70	0.11	nr	0.81

Chimney pots; red terracotta; plain or
cannon-head; setting and flaunching in
cement mortar (1:3)

185 mm dia x 300 mm long	PC £11.48	1.80	20.97	13.27	nr	34.24
185 mm dia x 600 mm long	PC £19.80	2.00	23.30	22.56	nr	45.86
185 mm dia x 900 mm long	PC £26.33	2.20	25.63	29.86	nr	55.49

F MASONRY		Labour hours	Labour £	Material £	Unit	Total rate £
Including overheads and profit at 9.00% built into brickwork or blockwork						
'CN7' combined lintel; 143 mm high; for standard cavity walls						
750 mm long	PC £10.02	0.25	2.58	11.47	nr	14.05
900 mm long	PC £12.06	0.30	3.10	13.80	nr	16.90
1200 mm long	PC £16.02	0.35	3.61	18.34	nr	21.95
1500 mm long	PC £20.82	0.40	4.13	23.82	nr	27.95
1800 mm long	PC £25.49	0.45	4.65	29.17	nr	33.82
2100 mm long	PC £29.72	0.50	5.16	34.02	nr	39.18
'CN8' combined lintel; 219 mm high; for standard cavity walls						
2400 mm long	PC £40.36	0.60	6.19	46.19	nr	52.38
2700 mm long	PC £45.53	0.70	7.23	52.12	nr	59.35
3000 mm long	PC £55.53	0.80	8.26	63.56	nr	71.82
3300 mm long	PC £61.81	0.90	9.29	70.74	nr	80.03
3600 mm long	PC £67.69	1.00	10.32	77.47	nr	87.79
3900 mm long	PC £90.14	1.10	11.35	103.17	nr	114.52
4200 mm long	PC £97.04	1.20	12.39	111.07	nr	123.46
'CN92' single lintel; for 75 mm internal walls						
900 mm long	PC £1.79	0.30	3.10	2.05	nr	5.15
1050 mm long	PC £2.11	0.30	3.10	2.42	nr	5.52
1200 mm long	PC £2.37	0.35	3.61	2.72	nr	6.33
'CN102' single lintel; for 100 mm internal walls						
900 mm long	PC £2.21	0.30	3.10	2.53	nr	5.63
1050 mm long	PC £2.58	0.30	3.10	2.95	nr	6.05
1200 mm long		0.35	3.61	3.22	nr	6.83

F31 PRECAST CONCRETE SILLS/LINTELS/COPINGS/ FEATURES

Mix 21.00 N/mm2 - 20 mm aggregate (1:2:4)						
Lintels; plate; prestressed; bedded						
100 x 65 x 750 mm	PC £2.70	0.20	2.33	3.38	nr	5.71
100 x 65 x 900 mm	PC £2.88	0.20	2.33	3.58	nr	5.91
100 x 65 x 1050 mm	PC £3.24	0.20	2.33	3.97	nr	6.30
100 x 65 x 1200 mm	PC £3.78	0.20	2.33	4.56	nr	6.89
150 x 65 x 900 mm	PC £3.87	0.25	2.91	4.66	nr	7.57
150 x 65 x 1050 mm	PC £4.23	0.25	2.91	5.05	nr	7.96
150 x 65 x 1200 mm	PC £5.13	0.25	2.91	6.03	nr	8.94
220 x 65 x 900 mm	PC £5.85	0.30	3.50	6.82	nr	10.32
220 x 65 x 1200 mm	PC £7.74	0.30	3.50	8.88	nr	12.38
220 x 65 x 1500 mm	PC £9.90	0.35	4.08	11.23	nr	15.31
265 x 65 x 900 mm	PC £6.75	0.30	3.50	7.80	nr	11.30
265 x 65 x 1200 mm	PC £8.91	0.30	3.50	10.15	nr	13.65
265 x 65 x 1500 mm	PC £11.12	0.35	4.08	12.56	nr	16.64
265 x 65 x 1800 mm	PC £13.35	0.40	4.66	14.99	nr	19.65
Lintels; rectangular; reinforced with mild steel bars; bedded						
100 x 150 x 900 mm	PC £7.20	0.30	3.50	8.29	nr	11.79
100 x 150 x 1050 mm	PC £8.10	0.30	3.50	9.27	nr	12.77
100 x 150 x 1200 mm	PC £9.63	0.30	3.50	10.94	nr	14.44
225 x 150 x 1200 mm	PC £12.15	0.40	4.66	13.69	nr	18.35
225 x 225 x 1800 mm	PC £22.50	0.75	8.74	24.97	nr	33.71
Lintels; boot; reinforced with mild steel bars; bedded						
250 x 225 x 1200 mm	PC £18.00	0.60	6.99	20.06	nr	27.05
275 x 225 x 1800 mm	PC £24.30	0.90	10.49	27.37	nr	37.86
Padstones						
300 x 100 x 75 mm	PC £1.80	0.15	1.75	2.40	nr	4.15
225 x 225 x 150 mm	PC £2.70	0.20	2.33	3.38	nr	5.71
450 x 450 x 150 mm	PC £7.20	0.30	3.50	8.29	nr	11.79

F MASONRY Including overheads and profit at 9.00%		Labour hours	Labour £	Material £	Unit	Total rate £
Mix 31.00 N/mm2 - 20 mm aggregate (1:1:2)						
Copings; once weathered; once throated						
bedded						
and pointed						
152 x 76 mm	PC £8.28	0.33	3.85	9.47	m	13.32
178 x 64 mm	PC £8.28	0.33	3.85	9.47	m	13.32
305 x 76 mm	PC £10.74	0.40	4.66	12.14	m	16.80
Extra for						
fair end	PC £3.03	-	-	3.30	nr	3.30
angles	PC £6.60	-	-	7.19	nr	7.19
Copings; twice weathered; twice throated;						
bedded and pointed						
152 x 76 mm	PC £7.66	0.33	3.85	8.79	m	12.64
178 x 64 mm	PC £7.66	0.33	3.85	8.79	m	12.64
305 x 76 mm	PC £10.11	0.40	4.66	11.46	m	16.12
Extra for						
fair end	PC £3.03	-	-	3.30	nr	3.30
angles		-	-	-	nr	7.19

Keep your figures up to date, free of charge

This section, and most of the other information in this Price Book, is brought up to date every three months in the *Price Book Update*.

The *Update* is available free to all Price Book purchasers.

To ensure you receive your copy, simply complete the reply card from the centre of the book and return it to us.

| G STRUCTURAL/CARCASSING METAL/TIMBER | Labour | Labour | Material | | Total |
| Including overheads and profit at 9.00% | hours | £ | £ | Unit | rate £ |

BASIC STEEL PRICES

NOTE: The following basic prices are based on **Basis** quantities (over 20 tonnes, of one quality, one section, for delivery to one destination) plus the addition of transport charges to the London area.

Prices are based on grade 43A steel. The additional cost of grade 50C steel is £25.00 per tonne.

Figures in brackets are available weights in kg per metre, at the prices indicated.

Universal beams (£/tonne)

	£		£
914x419 mm (343,388)	405.75	356x171 mm (57,67)	351.50
914x305 mm (201,224,253,289)	400.75	356x127 mm (33,39)	356.50
838x292 mm (176,194,226)	395.75	305x165 mm (40)	326.50
762x267 mm (147,173,197)	395.75	305x165 mm (46,54)	341.50
686x254 mm (125,140,152,170)	395.75	305x127 mm (37)	346.50
610x305 mm (149,179,238)	375.75	305x127 mm (42,48)	351.50
610x229 mm (101,113,125,140)	375.75	305x102 mm (25,28,33)	351.50
533x210 mm (82,92,101)	360.75	254x146 mm (31,37)	316.50
533x210 mm (109,122)	375.75	254x146 mm (43)	321.50
457x191 mm (67,74)	330.75	254x102 mm (22,25)	351.50
457x191 mm (82,89,98)	340.75	254x102 mm (28)	346.50
457x152 mm (52,60,67,74,82)	365.75	203x133 mm (25,30)	306.50
406x178 mm (54)	340.75	203x102 mm (33)	301.50
406x178 mm (60,67,74)	345.75	178x102 mm (19)	296.50
406x140 mm (39,46)	351.50	152x 89 mm (16)	336.50
356x171 mm (45,51)	336.50	127x 76 mm (13)	356.15

Universal columns (£/tonne)

	£		£
356x406 mm (235,287,340)	405.75	254x254 mm (107)	340.75
356x406 mm (393,467,551,634)	410.75	254x254 mm (167)	370.75
356x368 mm (129,153,177,202)	405.75	203x203 mm (46)	330.40
305x305 mm (97)	375.75	203x203 mm (52)	340.40
305x305 mm (118)	385.75	203x203 mm (60)	355.40
305x305 mm (137,158)	380.75	203x203 mm (71)	345.40
305x305 mm (240,283)	385.75	203x203 mm (86)	365.40
305x305 mm (198)	365.75	152x152 mm (23)	330.40
254x254 mm (73)	335.75	152x152 mm (30,37)	335.40
254x254 mm (89,132)	350.75		

Joists (£/tonne)

	£		£
254x203 mm (81.85)	355.75	114x114 mm (26.79)	335.75
203x152 mm (52.09)	345.75	102x102 mm (23.06)	335.75
152x127 mm (37.20)	345.75	89x89 mm (19.35)	335.75
127x114 mm (26.79,29.76)	335.75	76x76 mm (12.65)	385.75

G STRUCTURAL/CARCASSING METAL/TIMBER
Including overheads and profit at 9.00%

Channels (£/tonne)

	£		£
432x102 mm (65.54)	385.75	203x89 mm (29.78)	352.15
381x102 mm (55.10)	382.15	203x76 mm (23.82)	302.15
305x102 mm (46.18)	352.15	178x89 mm (26.81)	342.15
305x89 mm (41.69)	352.15	178x76 mm (20.84)	302.15
254x89 mm (35.74)	352.15	152x89 mm (23.84)	342.15
254x76 mm (28.29)	352.15	152x76 mm (17.88)	302.15
229x89 mm (32.76)	352.15	127x64 mm (14.90)	292.15
229x76 mm (26.06)	352.15		

Equal angles (£/tonne)

	£		£		£
200x200x16 mm	310.75	150x150x18 mm	322.15	100x100x12 mm	277.15
200x200x18 mm	315.75	120x120x8 mm	292.15	90x90x6 mm	257.15
200x200x20 mm	315.75	120x120x10 mm	282.15	90x90x8 mm	257.15
200x200x24 mm	315.75	120x120x12 mm	277.15	90x90x10 mm	257.15
150x150x10 mm	307.15	120x120x15 mm	282.15	90x90x12 mm	257.15
150x150x12 mm	307.15	100x100x8 mm	251.15		
150x150x15 mm	312.15	100x100x10 mm	251.15		
150x150x18 mm	322.15	100x100x12 mm	251.15		

Unequal angles (£/tonne)

200x150x12 mm	320.75	150x90x15 mm	315.75	125x75x15 mm	305.75
200x150x15 mm	325.75	150x75x10 mm	305.75	100x75x8 mm	267.15
200x150x18 mm	325.75	150x75x12 mm	305.75	100x75x10 mm	267.15
200x100x10 mm	320.75	150x75x15 mm	305.75	100x75x12 mm	267.15
200x100x12 mm	320.75	125x75x8 mm	270.75	100x65x7 mm	270.75
200x100x15 mm	340.75	125x75x10 mm	270.75	100x65x8 mm	270.75
150x90x10 mm	315.75	125x75x12 mm	270.75	100x65x10 mm	270.75
150x90x12 mm	315.75				

Structural hollow sections to BS 4848: Part 2 (£/100 m)

	BS 4360 grades 43C £	50C £		BS 4360 grades 43C £	50C £
Circular			**Rectangular and square**		
21.3x3.2 mm (1.43)	50.99	-	20x20x2.5 mm (1.35)	49.06	-
26.9x3.2 mm (1.87)	66.68	-	50x50x3.2 mm (4.66)	166.15	182.76
42.4x3.2 mm (3.09)	109.65	-	60x60x4.0 mm (6.97)	260.70	286.77
48.3x4.0 mm (4.37)	155.08	170.59	80x80x5.0 mm (11.70)	437.62	481.37
60.3x4.0 mm (5.55)	204.63	225.10	100x50x4.0 mm (8.86)	331.39	364.53
76.1x5.0 mm (8.77)	323.36	355.69	100x100x6.3 mm (18.40)	688.22	757.03
114.3x6.3 mm (16.80)	619.43	681.37	100x100x10.0 mm (27.90)	1127.66	1240.43
139.7x6.3 mm (20.70)	797.18	876.89	120x80x6.3 mm (18.40)	688.22	757.03
168.3x6.3 mm (25.20)	970.18	1067.52	150x150x6.3 mm (28.30)	1116.21	1227.82
219.1x10.0 mm (51.60)	2490.37	2863.90	200x100x8.0 mm (35.40)	1396.25	1535.86
273x10.0 mm (64.90)	2789.60	3208.02	200x200x10.0 mm (59.30)	2931.79	3371.56
323.9x12.5 mm (96.00)	4890.82	5624.45	300x200x12.5 mm (92.60)	4832.61	5557.48

NOTE: PC prices indicated below relate to an
average price for the sections used, assuming
a clean and level site with easy access..

G10 STRUCTURAL STEEL FRAMING

	Labour hours	Labour £	Material £	Unit	Total rate £
Fabricated steelwork; BS 4360; grade 50					
Single beams; universal beams, joists or channels PC £395.75	-	-	940.00	t	940.00
Single beams; castellated universal beams	-	-	-	t	2267.50
Single beams; rectangular hollow sections	-	-	-	t	1125.00
Single cranked beams; rectangular hollow sections	-	-	-	t	1370.00

G STRUCTURAL/CARCASSING METAL/TIMBER Including overheads and profit at 9.00%	Labour hours	Labour £	Material £	Unit	Total rate £

G10 STRUCTURAL STEEL FRAMING - cont'd

Fabricated steelwork; BS 4360; grade 50 - cont'd

	Labour hours	Labour £	Material £	Unit	Total rate £
Built-up beams; universal beams, joists or channels and plates PC £395.75	-	-	1357.50	t	1357.50
Latticed beams; angle sections PC £292.15	-	-	1082.00	t	1082.00
Latticed beams; circular hollow sections	-	-	-	t	1217.75
Single columns; universal beams, joists or channels PC £395.75	-	-	872.00	t	872.00
Single columns; circular hollow sections PC £955.00	-	-	1690.00	t	1690.00
Single columns; rectangular and square hollow sections PC £850.00	-	-	1570.00	t	1570.00
Built-up columns; universal beams, joists or channels and plates PC £395.75	-	-	1405.00	t	1405.00
Roof trusses; angle sections PC £292.15	-	-	1320.00	t	1320.00
Roof trusses; circular hollow sections PC £466.00	-	-	1460.00	t	1460.00
Bolted fittings; other than in connections; consisting of cleats, brackets etc.	-	-	-	t	1490.00
Welded fittings; other than in connections; consisting of cleats, brackets, etc.	-	-	-	t	1350.00

Erection of fabricated steelwork

	Labour hours	Labour £	Material £	Unit	Total rate £
Erection of steelwork on site	-	-	-	t	110.00
Wedging					
stanchion bases	-	-	-	nr	6.30
Holes for other trades; made on site					
16 mm dia; 6 mm thick metal	-	-	-	nr	2.90
16 mm dia; 10 mm thick metal	-	-	-	nr	4.10
16 mm dia; 20 mm thick metal	-	-	-	nr	5.75
Anchorage for stanchion base; including plate; holding down bolts; nuts and washers; for stanchion size					
152 x 152 mm	-	-	-	nr	9.00
254 x 254 mm	-	-	-	nr	13.50
356 x 406 mm	-	-	-	nr	20.00

Surface treatments off site

	Labour hours	Labour £	Material £	Unit	Total rate £
On steelwork; general surfaces					
galvanizing	-	-	-	t	457.80
shot blasting	-	-	-	m2	0.70
grit blast and one coat zinc chromate primer	-	-	-	m2	1.90
touch up primer and one coat of two pack epoxy zinc phosphate primer	-	-	-	m2	2.00

G12 ISOLATED STRUCTURAL METAL MEMBERS

Unfabricated steelwork; BS 4360; grade 43A

	Labour hours	Labour £	Material £	Unit	Total rate £
Single beams; joists or channels PC £395.75	-	-	780.00	t	780.00
Erection of steelwork on site	-	-	-	t	115.00

Naylor open web steel lattice beams; in single members; one coat red lead primer at works; raised 3.50 m above ground; ends built in

	Labour hours	Labour £	Material £	Unit	Total rate £
Lattice beams (safe distributed loads in brackets)					
200 mm deep; to span 5 m (s.d.l. 444 kg/m): ref N200A	1.20	12.39	148.64	nr	161.03
250 mm deep; to span 6 m (s.d.l. 526 kg/m); ref N250B	1.40	14.45	188.98	nr	203.43
300 mm deep; to span 7 m (s.d.l. 466 kg/m); ref N300B	1.60	16.52	222.13	nr	238.65
350 mm deep; to span 8 m (s.d.l. 611 kg/m); ref N350C	1.80	18.58	276.35	nr	294.93
400 mm deep; to span 9 m (s.d.l. 552 kg/m); ref N400C	2.00	20.64	310.78	nr	331.42
450 mm deep; to span 10 m (s.d.l. 728 kg/m); ref N450D	3.00	30.97	384.41	nr	415.38

G STRUCTURAL/CARCASSING METAL/TIMBER
Including overheads and profit at 9.00%

BASIC TIMBER PRICES

	£		£		£
Hardwood; fair average (£/m3 ex-wharf)					
African Walnut	575.00	Brazil. Mahogany	635.00	Sapele	475.00
Afrormosia	706.00	European Oak	1165.00	Teak	1400.00
Agba	480.00	Iroko	353.00	Utile	560.00
Beech	500.00	Obeche	300.00	W.A.Mahogany	494.00

Softwood
  Carcassing quality (£/m3)

2-4.8 m lengths	180.00	G.S.grade	12.00	
4.8-6 m lengths	171.00	S.S.grade	24.00	
6-9 m lenths	189.00			

  Joinery quality - £310.00/m3
  'Treatment'(£/m3)
  Pre-treatment of timber by vacuum/pressure
  impregnation, excluding transport costs and
  any subsequent seasoning:-

interior work; min. salt ret. 4 kg/m3	30.00
exterior work; min. salt ret. 5.30 kg/m3	35.00

  Pre-treatment of timber including flame
  proofing

all purposes; min. salt ret. 36 kg/m3	104.50

	Labour hours	Labour £	Material £	Unit	Total rate £
G20 CARPENTRY/TIMBER FRAMING/FIRST FIXING					
Sawn softwood; untreated					
Floor members					
38 x 75 mm	0.12	1.09	0.70	m	1.79
38 x 125 mm	0.12	1.09	0.90	m	1.99
38 x 125 mm	0.13	1.18	1.11	m	2.29
38 x 150 mm	0.14	1.27	1.31	m	2.58
50 x 75 mm	0.12	1.09	0.78	m	1.87
50 x 100 mm	0.14	1.27	1.03	m	2.30
50 x 125 mm	0.14	1.27	1.29	m	2.56
50 x 150 mm	0.15	1.36	1.54	m	2.90
50 x 175 mm	0.15	1.36	1.79	m	3.15
50 x 200 mm	0.16	1.45	2.12	m	3.57
50 x 225 mm	0.16	1.45	2.46	m	3.91
50 x 250 mm	0.17	1.54	2.89	m	4.43
75 x 125 mm	0.16	1.45	2.03	m	3.48
75 x 150 mm	0.16	1.45	2.38	m	3.83
75 x 175 mm	0.16	1.45	2.86	m	4.31
75 x 200 mm	0.17	1.54	3.34	m	4.88
75 x 225 mm	0.17	1.54	3.81	m	5.35
75 x 250 mm	0.18	1.63	4.32	m	5.95
100 x 150 mm	0.22	1.99	3.67	m	5.66
100 x 200 mm	0.23	2.08	4.88	m	6.96
100 x 250 mm	0.25	2.26	6.09	m	8.35
100 x 300 mm	0.27	2.45	7.32	m	9.77
Wall or partition members					
25 x 25 mm	0.07	0.63	0.19	m	0.82
25 x 38 mm	0.07	0.63	0.27	m	0.90
25 x 75 mm	0.09	0.82	0.48	m	1.30
38 x 38 mm	0.09	0.82	0.41	m	1.23
38 x 50 mm	0.09	0.82	0.47	m	1.29
38 x 75 mm	0.12	1.09	0.70	m	1.79
38 x 100 mm	0.15	1.36	0.90	m	2.26

G STRUCTURAL/CARCASSING METAL/TIMBER Including overheads and profit at 9.00%	Labour hours	Labour £	Material £	Unit	Total rate £
G20 CARPENTRY/TIMBER FRAMING/FIRST FIXING - cont'd					
Sawn softwood; untreated - cont'd					
Wall or partition members					
50 x 50 mm	0.12	1.09	0.58	m	1.67
50 x 75 mm	0.15	1.36	0.79	m	2.15
50 x 100 mm	0.18	1.63	1.04	m	2.67
50 x 125 mm	0.19	1.72	1.30	m	3.02
75 x 75 mm	0.18	1.63	1.28	m	2.91
75 x 100 mm	0.21	1.90	1.78	m	3.68
100 x 100 mm	0.21	1.90	2.43	m	4.33
Flat roof members					
38 x 75 mm	0.14	1.27	0.72	m	1.99
38 x 100 mm	0.14	1.27	0.90	m	2.17
38 x 125 mm	0.14	1.27	1.11	m	2.38
38 x 150 mm	0.14	1.27	1.31	m	2.58
50 x 100 mm	0.14	1.27	1.03	m	2.30
50 x 125 mm	0.14	1.27	1.29	m	2.56
50 x 150 mm	0.15	1.36	1.54	m	2.90
50 x 175 mm	0.15	1.36	1.79	m	3.15
50 x 200 mm	0.16	1.45	2.12	m	3.57
50 x 225 mm	0.16	1.45	2.46	m	3.91
50 x 250 mm	0.17	1.54	2.89	m	4.43
75 x 150 mm	0.16	1.45	2.37	m	3.82
75 x 175 mm	0.16	1.45	2.85	m	4.30
75 x 200 mm	0.17	1.54	3.34	m	4.88
75 x 225 mm	0.17	1.54	3.81	m	5.35
75 x 250 mm	0.18	1.63	4.32	m	5.95
Pitched roof members					
25 x 100 mm	0.12	1.09	0.58	m	1.67
25 x 125 mm	0.12	1.09	0.72	m	1.81
25 x 150 mm	0.15	1.36	0.87	m	2.23
25 x 150 mm; notching over trussed rafters	0.30	2.72	0.87	m	3.59
25 x 175 mm	0.15	1.36	1.03	m	2.39
25 x 175 mm; notching over trussed rafters	0.30	2.72	1.03	m	3.75
25 x 200 mm	0.18	1.63	1.20	m	2.83
32 x 150 mm; notching over trussed rafters	0.33	2.99	1.21	m	4.20
32 x 175 mm; notching over trussed rafters	0.33	2.99	1.45	m	4.44
32 x 200 mm; notching over trussed rafters	0.33	2.99	1.69	m	4.68
38 x 100 mm	0.15	1.36	0.90	m	2.26
38 x 125 mm	0.15	1.36	1.11	m	2.47
38 x 150 mm	0.15	1.36	1.31	m	2.67
50 x 50 mm	0.12	1.09	0.57	m	1.66
50 x 75 mm	0.15	1.36	0.78	m	2.14
50 x 100 mm	0.18	1.63	1.03	m	2.66
50 x 125 mm	0.18	1.63	1.29	m	2.92
50 x 150 mm	0.21	1.90	1.54	m	3.44
50 x 175 mm	0.21	1.90	1.79	m	3.69
50 x 200 mm	0.21	1.90	2.12	m	4.02
50 x 225 mm	0.21	1.90	2.46	m	4.36
75 x 100 mm	0.25	2.26	1.76	m	4.02
75 x 125 mm	0.25	2.26	2.03	m	4.29
75 x 150 mm	0.25	2.26	2.38	m	4.64
100 x 150 mm	0.30	2.72	3.68	m	6.40
100 x 175 mm	0.30	2.72	4.29	m	7.01
100 x 200 mm	0.30	2.72	4.89	m	7.61
100 x 225 mm	0.33	2.99	5.49	m	8.48
100 x 250 mm	0.33	2.99	6.09	m	9.08
Kerbs, bearers and the like					
19 x 100 mm	0.04	0.36	0.47	m	0.83
19 x 125 mm	0.04	0.36	0.58	m	0.94
19 x 150 mm	0.04	0.36	0.70	m	1.06
25 x 75 mm	0.05	0.45	0.48	m	0.93
25 x 100 mm	0.05	0.45	0.58	m	1.03
38 x 75 mm	0.06	0.54	0.70	m	1.24
38 x 100 mm	0.06	0.54	0.90	m	1.44

G STRUCTURAL/CARCASSING METAL/TIMBER Including overheads and profit at 9.00%	Labour hours	Labour £	Material £	Unit	Total rate £
50 x 75 mm	0.06	0.54	0.78	m	1.32
50 x 100 mm	0.07	0.63	1.03	m	1.66
75 x 100 mm	0.07	0.63	1.75	m	2.38
75 x 125 mm	0.08	0.72	2.02	m	2.74
75 x 150 mm	0.08	0.72	2.37	m	3.09
75 x 150 mm; splayed and rounded	0.10	0.91	4.16	m	5.07
Kerbs, bearers and the like; fixing by bolting					
19 x 100 mm	0.08	0.72	0.47	m	1.19
19 x 125 mm	0.08	0.72	0.58	m	1.30
19 x 150 mm	0.08	0.72	0.70	m	1.42
25 x 75 mm	0.10	0.91	0.48	m	1.39
25 x 100 mm	0.10	0.91	0.58	m	1.49
38 x 75 mm	0.12	1.09	0.70	m	1.79
38 x 100 mm	0.12	1.09	0.90	m	1.99
50 x 75 mm	0.12	1.09	0.78	m	1.87
50 x 100 mm	0.14	1.27	1.03	m	2.30
75 x 100 mm	0.14	1.27	1.76	m	3.03
75 x 125 mm	0.16	1.45	2.03	m	3.48
75 x 150 mm	0.16	1.45	2.37	m	3.82
Herringbone strutting 50 x 50 mm					
to 150 mm deep joists	0.50	4.53	1.36	m	5.89
to 175 mm deep joists	0.50	4.53	1.38	m	5.91
to 200 mm deep joists	0.50	4.53	1.41	m	5.94
to 225 mm deep joists	0.50	4.53	1.44	m	5.97
to 250 mm deep joists	0.50	4.53	1.47	m	6.00
Solid strutting to joists					
50 x 150 mm	0.30	2.72	1.76	m	4.48
50 x 175 mm	0.30	2.72	2.01	m	4.73
50 x 200 mm	0.30	2.72	2.36	m	5.08
50 x 225 mm	0.30	2.72	2.69	m	5.41
50 x 250 mm	0.30	2.72	3.14	m	5.86
Cleats					
225 x 100 x 75 mm	0.20	1.81	0.62	nr	2.43
Sprockets					
50 x 50 x 200 mm	0.15	1.36	0.62	nr	1.98
Extra for stress grading to above timbers					
general structural (GS) grade	-	-	-	m3	13.08
special structural (SS) grade	-	-	-	m3	26.16
Extra for protecting and flameproofing timber with 'Celcure F' protection					
small sections	-	-	-	m3	119.90
large sections	-	-	-	m3	113.91
Wrought faces					
generally	0.30	2.72	-	m2	2.72
50 mm wide	0.04	0.36	-	m	0.36
75 mm wide	0.05	0.45	-	m	0.45
100 mm wide	0.06	0.54	-	m	0.54
Raking cutting					
50 mm thick	0.20	1.81	0.48	m	2.29
75 mm thick	0.25	2.26	0.72	m	2.98
100 mm thick	0.30	2.72	0.95	m	3.67
Curved cutting					
50 mm thick	0.25	2.26	0.67	m	2.93
Scribing					
50 mm thick	0.30	2.72	-	m	2.72
Notching and fitting ends to metal	0.12	1.09	-	nr	1.09
Trimming to openings					
760 x 760 mm; joists 38 x 150 mm	2.20	19.93	0.76	nr	20.69
760 x 760 mm; joists 50 x 175 mm	2.30	20.83	1.34	nr	22.17
1500 x 500 mm; joists 38 x 200 mm	2.50	22.64	1.53	nr	24.17
1500 x 500 mm; joists 50 x 200 mm	2.50	22.64	1.91	nr	24.55
3000 x 1000 mm; joists 50 x 225 mm	4.50	40.76	4.20	nr	44.96
3000 x 1000 mm; joists 75 x 225 mm	5.00	45.29	6.29	nr	51.58

G STRUCTURAL/CARCASSING METAL/TIMBER Including overheads and profit at 9.00%	Labour hours	Labour £	Material £	Unit	Total rate £
G20 CARPENTRY/TIMBER FRAMING/FIRST FIXING - cont'd					
Sawn softwood; 'Tanalised'					
Floor members					
38 x 75 mm	0.12	1.09	0.88	m	**1.97**
38 x 100 mm	0.12	1.09	1.12	m	**2.21**
38 x 125 mm	0.13	1.18	1.38	m	**2.56**
38 x 150 mm	0.14	1.27	1.63	m	**2.90**
50 x 75 mm	0.12	1.09	0.97	m	**2.06**
50 x 100 mm	0.14	1.27	1.29	m	**2.56**
50 x 125 mm	0.14	1.27	1.61	m	**2.88**
50 x 150 mm	0.15	1.36	1.91	m	**3.27**
50 x 175 mm	0.15	1.36	2.24	m	**3.60**
50 x 200 mm	0.16	1.45	2.63	m	**4.08**
50 x 225 mm	0.16	1.45	3.06	m	**4.51**
50 x 250 mm	0.17	1.54	3.61	m	**5.15**
75 x 125 mm	0.16	1.45	2.54	m	**3.99**
75 x 150 mm	0.16	1.45	2.97	m	**4.42**
75 x 175 mm	0.16	1.45	3.56	m	**5.01**
75 x 200 mm	0.17	1.54	4.17	m	**5.71**
75 x 225 mm	0.17	1.54	4.76	m	**6.30**
75 x 250 mm	0.18	1.63	5.39	m	**7.02**
100 x 150 mm	0.22	1.99	4.57	m	**6.56**
100 x 200 mm	0.23	2.08	6.08	m	**8.16**
100 x 250 mm	0.25	2.26	7.59	m	**9.85**
100 x 300 mm	0.27	2.45	9.13	m	**11.58**
Wall or partition members					
25 x 25 mm	0.07	0.63	0.23	m	**0.86**
25 x 38 mm	0.07	0.63	0.32	m	**0.95**
25 x 75 mm	0.09	0.82	0.60	m	**1.42**
38 x 38 mm	0.09	0.82	0.50	m	**1.32**
38 x 50 mm	0.09	0.82	0.58	m	**1.40**
38 x 75 mm	0.12	1.09	0.88	m	**1.97**
38 x 100 mm	0.15	1.36	1.12	m	**2.48**
50 x 50 mm	0.12	1.09	0.73	m	**1.82**
50 x 75 mm	0.15	1.36	0.99	m	**2.35**
50 x 100 mm	0.18	1.63	1.30	m	**2.93**
50 x 125 mm	0.19	1.72	1.62	m	**3.34**
75 x 75 mm	0.18	1.63	1.64	m	**3.27**
75 x 100 mm	0.21	1.90	2.18	m	**4.08**
100 x 100 mm	0.21	1.90	3.09	m	**4.99**
Flat roof members					
38 x 75 mm	0.14	1.27	0.88	m	**2.15**
38 x 100 mm	0.14	1.27	1.12	m	**2.39**
38 x 125 mm	0.14	1.27	1.38	m	**2.65**
38 x 150 mm	0.14	1.27	1.63	m	**2.90**
50 x 100 mm	0.14	1.27	1.29	m	**2.56**
50 x 125 mm	0.14	1.27	1.61	m	**2.88**
50 x 150 mm	0.15	1.36	1.91	m	**3.27**
50 x 175 mm	0.15	1.36	2.24	m	**3.60**
50 x 200 mm	0.16	1.45	2.63	m	**4.08**
50 x 225 mm	0.16	1.45	3.06	m	**4.51**
50 x 250 mm	0.17	1.54	3.61	m	**5.15**
75 x 150 mm	0.16	1.45	2.97	m	**4.42**
75 x 175 mm	0.16	1.45	3.55	m	**5.00**
75 x 200 mm	0.17	1.54	4.17	m	**5.71**
75 x 225 mm	0.17	1.54	4.76	m	**6.30**
75 x 250 mm	0.18	1.63	5.39	m	**7.02**
Pitched roof members					
25 x 100 mm	0.12	1.09	0.72	m	**1.81**
25 x 125 mm	0.12	1.09	0.90	m	**1.99**
25 x 150 mm	0.15	1.36	1.07	m	**2.43**
25 x 150 mm; notching over trussed rafters	0.30	2.72	1.07	m	**3.79**
25 x 175 mm	0.15	1.36	1.29	m	**2.65**
25 x 175 mm; notching over trussed rafters	0.30	2.72	1.29	m	**4.01**

G STRUCTURAL/CARCASSING METAL/TIMBER Including overheads and profit at 9.00%	Labour hours	Labour £	Material £	Unit	Total rate £
25 x 200 mm	0.18	1.63	1.51	m	3.14
32 x 150 mm; notching over trussed rafters	0.33	2.99	1.51	m	4.50
32 x 175 mm; notching over trussed rafters	0.33	2.99	1.81	m	4.80
32 x 200 mm; notching over trussed rafters	0.33	2.99	2.12	m	5.11
38 x 100 mm	0.15	1.36	1.12	m	2.48
38 x 125 mm	0.15	1.36	1.38	m	2.74
38 x 150 mm	0.15	1.36	1.63	m	2.99
50 x 50 mm	0.12	1.09	0.71	m	1.80
50 x 75 mm	0.15	1.36	0.97	m	2.33
50 x 100 mm	0.18	1.63	1.29	m	2.92
50 x 125 mm	0.18	1.63	1.61	m	3.24
50 x 150 mm	0.21	1.90	1.91	m	3.81
50 x 175 mm	0.21	1.90	2.24	m	4.14
50 x 200 mm	0.21	1.90	2.63	m	4.53
50 x 225 mm	0.21	1.90	3.06	m	4.96
75 x 100 mm	0.25	2.26	2.17	m	4.43
75 x 125 mm	0.25	2.26	2.54	m	4.80
75 x 150 mm	0.25	2.26	2.97	m	5.23
100 x 150 mm	0.30	2.72	4.59	m	7.31
100 x 175 mm	0.30	2.72	5.34	m	8.06
100 x 200 mm	0.30	2.72	6.09	m	8.81
100 x 225 mm	0.33	2.99	6.85	m	9.84
100 x 250 mm	0.33	2.99	7.61	m	10.60
Kerbs, bearers and the like					
19 x 100 mm	0.04	0.36	0.58	m	0.94
19 x 125 mm	0.04	0.36	0.72	m	1.08
19 x 150 mm	0.04	0.36	0.88	m	1.24
25 x 75 mm	0.05	0.45	0.60	m	1.05
25 x 100 mm	0.05	0.45	0.72	m	1.17
38 x 75 mm	0.06	0.54	0.88	m	1.42
38 x 100 mm	0.06	0.54	1.12	m	1.66
50 x 75 mm	0.06	0.54	0.97	m	1.51
50 x 100 mm	0.07	0.63	1.29	m	1.92
75 x 100 mm	0.07	0.63	2.15	m	2.78
75 x 125 mm	0.08	0.72	2.52	m	3.24
75 x 150 mm	0.08	0.72	2.96	m	3.68
75 x 150 mm; splayed and rounded	0.10	0.91	5.20	m	6.11
Kerbs, bearers and the like; fixing by bolting					
19 x 100 mm	0.08	0.72	0.58	m	1.30
19 x 125 mm	0.08	0.72	0.72	m	1.44
19 x 150 mm	0.08	0.72	0.88	m	1.60
25 x 75 mm	0.10	0.91	0.61	m	1.52
25 x 100 mm	0.10	0.91	0.72	m	1.63
38 x 75 mm	0.12	1.09	0.88	m	1.97
38 x 100 mm	0.12	1.09	1.12	m	2.21
50 x 75 mm	0.12	1.09	0.97	m	2.06
50 x 100 mm	0.14	1.27	1.29	m	2.56
75 x 100 mm	0.14	1.27	2.17	m	3.44
75 x 125 mm	0.16	1.45	2.54	m	3.99
75 x 150 mm	0.16	1.45	2.96	m	4.41
Herringbone strutting 50 x 50 mm					
to 150 mm deep joists	0.50	4.53	1.64	m	6.17
to 175 mm deep joists	0.50	4.53	1.67	m	6.20
to 200 mm deep joists	0.50	4.53	1.71	m	6.24
to 225 mm deep joists	0.50	4.53	1.75	m	6.28
to 250 mm deep joists	0.50	4.53	1.78	m	6.31
Solid strutting to joists					
50 x 150 mm	0.30	2.72	2.13	m	4.85
50 x 175 mm	0.30	2.72	2.46	m	5.18
50 x 200 mm	0.30	2.72	2.87	m	5.59
50 x 225 mm	0.30	2.72	3.31	m	6.03
50 x 250 mm	0.30	2.72	3.86	m	6.58
Cleats					
225 x 100 x 75 mm	0.20	1.81	0.77	nr	2.58
Sprockets					
50 x 50 x 200 mm	0.15	1.36	0.77	nr	2.13

G STRUCTURAL/CARCASSING METAL/TIMBER Including overheads and profit at 9.00%	Labour hours	Labour £	Material £	Unit	Total rate £
G20 CARPENTRY/TIMBER FRAMING/FIRST FIXING - cont'd					
Sawn softwood; 'Tanalised' - cont'd					
Extra for stress grading to above timbers					
general structural (GS) grade	-	-	-	m3	**13.08**
special structural (SS) grade	-	-	-	m3	**26.16**
Extra for protecting and flameproofing					
timber with 'Celcure F' protection					
small sections	-	-	-	m3	**119.90**
large sections	-	-	-	m3	**113.91**
Wrought faces					
generally	0.30	2.72	-	m2	**2.72**
50 mm wide	0.04	0.36	-	m	**0.36**
75 mm wide	0.05	0.45	-	m	**0.45**
100 mm wide	0.06	0.54	-	m	**0.54**
Raking cutting					
50 mm thick	0.20	1.81	0.60	m	**2.41**
75 mm thick	0.25	2.26	0.91	m	**3.17**
100 mm thick	0.30	2.72	1.20	m	**3.92**
Curved cutting					
50 mm thick	0.25	2.26	0.84	m	**3.10**
Scribing					
50 mm thick	0.30	2.72	-	m	**2.72**
Notching and fitting ends to metal	0.12	1.09	-	nr	**1.09**
Trimming to openings					
760 x 760 mm; joists 38 x 150 mm	2.20	19.93	0.96	nr	**20.89**
760 x 760 mm; joists 50 x 175 mm	2.30	20.83	1.68	nr	**22.51**
1500 x 500 mm; joists 38 x 200 mm	2.50	22.64	1.92	nr	**24.56**
1500 x 500 mm; joists 50 x 200 mm	2.50	22.64	2.40	nr	**25.04**
3000 x 1000 mm; joists 50 x 225 mm	4.50	40.76	5.28	nr	**46.04**
3000 x 1000 mm; joists 75 x 225 mm	5.00	45.29	7.91	nr	**53.20**
Trussed rafters, stress graded sawn softwood pressure impregnated; raised through two storeys and fixed in position					
'W' type truss (Fink); 22.5 degree pitch;					
450 mm eaves overhang					
5.00 m span	1.60	14.49	16.91	nr	**31.40**
7.60 m span	1.75	15.85	25.49	nr	**41.34**
10.00 m span	2.00	18.12	40.22	nr	**58.34**
'W' type truss (Fink); 30 degree pitch;					
450 mm eaves overhang					
5.00 m span	1.60	14.49	17.92	nr	**32.42**
7.60 m span	1.75	15.85	26.27	nr	**42.12**
10.00 m span	2.00	18.12	42.34	nr	**60.45**
'W' type truss (Fink); 45 degree pitch;					
450 mm eaves overhang					
4.60 m span	1.60	14.49	19.83	nr	**34.32**
7.00 m span	1.75	15.85	30.60	nr	**46.45**
'Mono' type truss; 17.5 degree pitch;					
450 mm eaves overhang					
3.30 m span	1.40	12.68	14.68	nr	**27.36**
5.60 m span	1.60	14.49	23.88	nr	**38.37**
7.00 m span	1.85	16.76	34.01	nr	**50.77**
'Mono' type truss; 30 degree pitch;					
450 mm eaves overhang					
3.30 m span	1.40	12.68	16.40	nr	**29.09**
5.60 m span	1.60	14.49	26.94	nr	**41.43**
7.00 m span	1.85	16.76	38.25	nr	**55.01**
Attic type truss; 45 degree pitch; 450 mm					
eaves overhang					
5.00 m span	3.15	28.53	67.47	nr	**96.01**
7.60 m span	3.30	29.89	51.36	nr	**81.25**
9.00 m span	3.50	31.70	57.67	nr	**89.37**

G STRUCTURAL/CARCASSING METAL/TIMBER Including overheads and profit at 9.00%	Labour hours	Labour £	Material £	Unit	Total rate £
Standard 'Toreboda' glulam timber beams; **Moelven (UK) Ltd.; LB grade whitewood;** **pressure impregnated; phenol resorcinal** **adhesive; clean planed finish; fixed** Laminated roof beams					
56 x 255 mm	0.55	4.98	12.83	m	**17.81**
66 x 315 mm	0.70	6.34	17.23	m	**23.57**
90 x 315 mm	0.90	8.15	21.61	m	**29.76**
90 x 405 mm	1.15	10.42	27.05	m	**37.47**
115 x 405 mm	1.45	13.13	32.59	m	**45.72**
115 x 495 mm	1.80	16.30	43.26	m	**59.56**
115 x 630 mm	2.20	19.93	54.09	m	**74.02**

ALTERNATIVE FIRST FIXING MATERIAL PRICES

	£		£		£
Chipboard roofing (£/10 m2)					
12 mm	29.21	18 mm	40.21	25 mm	70.59
Non-asbestos boards (£/10 m2)					
'Masterboard'					
6 mm	39.87	9 mm	72.27	12 mm	95.22
'Masterclad'; sanded finish					
4.5 mm	31.95	6 mm	41.49	9 mm	64.80
Plywood (£/10 m2)					
External quality					
12 mm	67.90	15 mm	83.40	25 mm	125.60
Marine quality					
12 mm	59.90	15 mm	75.90	25 mm	121.75

Discounts of 0 - 10% available depending on quantity/status

	Labour hours	Labour £	Material £	Unit	Total rate £
'Masterboard'; 6 mm thick; PC £39.87/10m2 Boarding to eaves, verges, fascias and the like					
over 300 mm wide	0.70	6.34	4.94	m2	**11.28**
75 mm wide	0.21	1.90	0.45	m	**2.35**
150 mm wide	0.24	2.17	0.83	m	**3.00**
200 mm wide	0.27	2.45	1.09	m	**3.54**
225 mm wide	0.28	2.54	1.22	m	**3.76**
250 mm wide	0.29	2.63	1.37	m	**4.00**
Raking cutting	0.05	0.45	0.23	m	**0.68**
Plywood; external quality; 18 mm thick; PC £99.10/10m2 Boarding to eaves, verges, fascias and the like					
over 300 mm wide	0.82	7.43	11.87	m2	**19.30**
75 mm wide	0.25	2.26	1.07	m	**3.33**
150 mm wide	0.29	2.63	2.02	m	**4.65**
225 mm wide	0.33	2.99	2.98	m	**5.97**
Plywood; marine quality; 18 mm thick; PC £87.65/10m2 Boarding to gutter bottoms or sides; butt joints					
over 300 mm wide	0.93	8.42	10.54	m2	**18.96**
150 mm wide	0.33	2.99	1.79	m	**4.78**
225 mm wide	0.37	3.35	2.74	m	**6.09**
300 mm wide	0.42	3.80	3.48	m	**7.28**

G STRUCTURAL/CARCASSING METAL/TIMBER Including overheads and profit at 9.00%	Labour hours	Labour £	Material £	Unit	Total rate £
G20 CARPENTRY/TIMBER FRAMING/FIRST FIXING - cont'd					
Plywood; marine quality; 18 mm thick; PC £87.65/10m2 - cont'd					
Boarding to eaves, verges, fascias and					
the like					
over 300 mm wide	0.82	7.43	10.54	m2	17.97
75 mm wide	0.25	2.26	0.95	m	3.21
150 mm wide	0.29	2.63	1.79	m	4.42
225 mm wide	0.31	2.81	2.31	m	5.12
Sawn softwood; untreated					
Boarding to gutter bottoms or sides;					
butt joints					
19 mm thick; sloping	1.25	11.32	5.00	m2	16.32
19 mm thick x 75 mm wide	0.35	3.17	0.39	m	3.56
19 mm thick x 150 mm wide	0.40	3.62	0.74	m	4.36
19 mm thick x 225 mm wide	0.45	4.08	1.17	m	5.25
25 mm thick; sloping	1.25	11.32	6.10	m2	17.42
25 mm thick x 75 mm wide	0.35	3.17	0.51	m	3.68
25 mm thick x 150 mm wide	0.40	3.62	0.90	m	4.52
25 mm thick x 225 mm wide	0.45	4.08	1.45	m	5.53
Cesspools with 25 mm thick sides and bottom					
225 x 225 x 150 mm	1.20	10.87	2.89	nr	13.76
300 x 300 x 150 mm	1.40	12.68	4.11	nr	16.79
Firrings					
50 mm wide x 36 mm average depth	0.15	1.36	0.61	m	1.97
50 mm wide x 50 mm average depth	0.15	1.36	0.73	m	2.09
50 mm wide x 75 mm average depth	0.15	1.36	1.09	m	2.45
Bearers					
25 x 50 mm	0.10	0.91	0.37	m	1.28
38 x 50 mm	0.10	0.91	0.51	m	1.42
50 x 50 mm	0.10	0.91	0.62	m	1.53
50 x 75 mm	0.10	0.91	0.83	m	1.74
Angle fillets					
38 x 38 mm	0.10	0.91	0.29	m	1.20
50 x 50 mm	0.10	0.91	0.43	m	1.34
75 x 75 mm	0.12	1.09	0.85	m	1.94
Tilting fillets					
19 x 38 mm	0.10	0.91	0.17	m	1.08
25 x 50 mm	0.10	0.91	0.25	m	1.16
38 x 75 mm	0.10	0.91	0.48	m	1.39
50 x 75 mm	0.10	0.91	0.62	m	1.53
75 x 100 mm	0.15	1.36	0.99	m	2.35
Grounds or battens					
13 x 19 mm	0.05	0.45	0.12	m	0.57
13 x 32 mm	0.05	0.45	0.16	m	0.61
25 x 50 mm	0.05	0.45	0.34	m	0.79
Grounds or battens; plugged and screwed					
13 x 19 mm	0.15	1.36	0.18	m	1.54
13 x 32 mm	0.15	1.36	0.21	m	1.57
25 x 50 mm	0.15	1.36	0.40	m	1.76
Open-spaced grounds or battens; at 300 mm					
centres one way					
25 x 50 mm	0.15	1.36	1.14	m2	2.50
25 x 50 mm; plugged and screwed	0.45	4.08	1.33	m2	5.41
Framework to walls; at 300 mm centres one					
way and 600 mm centres the other					
25 x 50 mm	0.75	6.79	1.88	m2	8.67
38 x 50 mm	0.75	6.79	2.65	m2	9.44
50 x 50 mm	0.75	6.79	3.24	m2	10.03
50 x 75 mm	0.75	6.79	4.41	m2	11.20
75 x 75 mm	0.75	6.79	7.20	m2	13.99

G STRUCTURAL/CARCASSING METAL/TIMBER Including overheads and profit at 9.00%	Labour hours	Labour £	Material £	Unit	Total rate £
Framework to walls; at 300 mm centres one way and 600 mm centres the other way; plugged and screwed					
25 x 50 mm	1.25	11.32	2.16	m2	13.48
38 x 50 mm	1.25	11.32	2.94	m2	14.26
50 x 50 mm	1.25	11.32	3.53	m2	14.85
50 x 75 mm	1.25	11.32	4.69	m2	16.01
75 x 75 mm	1.25	11.32	7.49	m2	18.81
Framework to bath panel; at 500 mm centres both ways					
25 x 50 mm	0.90	8.15	2.06	m2	10.21
Framework as bracketing and cradling around steelwork					
25 x 50 mm	1.40	12.68	2.45	m2	15.13
50 x 50 mm	1.50	13.59	4.22	m2	17.81
50 x 75 mm	1.60	14.49	5.72	m2	20.21
Blockings wedged between flanges of steelwork					
50 x 50 x 150 mm	0.12	1.09	0.28	nr	1.37
50 x 75 x 225 mm	0.13	1.18	0.43	nr	1.61
50 x 100 x 300 mm	0.14	1.27	0.57	nr	1.84
Sawn softwood; 'Tanalised'					
Boarding to gutter bottoms or sides; butt joints					
19 mm thick; sloping	1.25	11.32	6.13	m2	17.45
19 mm thick x 75 mm wide	0.35	3.17	0.49	m	3.66
19 mm thick x 150 mm wide	0.40	3.62	0.92	m	4.54
19 mm thick x 225 mm wide	0.45	4.08	1.45	m	5.53
25 mm thick; sloping	1.25	11.32	7.57	m2	18.89
25 mm thick x 75 mm wide	0.35	3.17	0.63	m	3.80
25 mm thick x 150 mm wide	0.40	3.62	1.12	m	4.74
25 mm thick x 225 mm wide	0.45	4.08	1.80	m	5.88
Firrings					
50 mm wide x 36 mm average depth	0.15	1.36	0.73	m	2.09
50 mm wide x 50 mm average depth	0.15	1.36	0.88	m	2.24
50 mm wide x 75 mm average depth	0.15	1.36	1.33	m	2.69
Bearers					
25 x 50 mm	0.10	0.91	0.46	m	1.37
38 x 50 mm	0.10	0.91	0.63	m	1.54
50 x 50 mm	0.10	0.91	0.76	m	1.67
50 x 75 mm	0.10	0.91	1.02	m	1.93
Cesspools with 25 mm thick sides and bottom					
225 x 225 x 150 mm	1.20	10.87	3.60	nr	14.47
300 x 300 x 150 mm	1.40	12.68	5.12	nr	17.80
Angle fillets					
38 x 38 mm	0.10	0.91	0.35	m	1.26
50 x 50 mm	0.10	0.91	0.53	m	1.44
75 x 75 mm	0.12	1.09	1.04	m	2.13
Tilting fillets					
19 x 38 mm	0.10	0.91	0.21	m	1.12
25 x 50 mm	0.10	0.91	0.30	m	1.21
38 x 75 mm	0.10	0.91	0.58	m	1.49
50 x 75 mm	0.10	0.91	0.76	m	1.67
75 x 100 mm	0.15	1.36	1.23	m	2.59
Grounds or battens					
13 x 19 mm	0.05	0.45	0.15	m	0.60
13 x 32 mm	0.05	0.45	0.19	m	0.64
25 x 50 mm	0.05	0.45	0.43	m	0.88
Grounds or battens; plugged and screwed					
13 x 19 mm	0.15	1.36	0.18	m	1.54
13 x 32 mm	0.15	1.36	0.25	m	1.61
25 x 50 mm	0.15	1.36	0.48	m	1.84
Open-spaced grounds or battens; at 300 mm centres one way					
25 x 50 mm	0.15	1.36	1.40	m2	2.76
25 x 50 mm; plugged and screwed	0.45	4.08	1.59	m2	5.67

G STRUCTURAL/CARCASSING METAL/TIMBER Including overheads and profit at 9.00%	Labour hours	Labour £	Material £	Unit	Total rate £
G20 CARPENTRY/TIMBER FRAMING/FIRST FIXING - cont'd					
Sawn softwood; 'Tanalised' - cont'd					
Framework to walls; at 300 mm centres one					
way and 600 mm centres the other					
25 x 50 mm	0.75	6.79	2.33	m2	9.12
38 x 50 mm	0.75	6.79	3.30	m2	10.09
50 x 50 mm	0.75	6.79	4.02	m2	10.81
50 x 75 mm	0.75	6.79	5.52	m2	12.31
75 x 75 mm	0.75	6.79	9.22	m2	16.01
Framework to walls; at 300 mm centres one					
way and 600 mm centres the other way;					
plugged and screwed					
25 x 50 mm	1.25	11.32	2.62	m2	13.94
38 x 50 mm	1.25	11.32	3.59	m2	14.91
50 x 50 mm	1.25	11.32	4.31	m2	15.63
50 x 75 mm	1.25	11.32	5.81	m2	17.13
75 x 75 mm	1.25	11.32	9.51	m2	20.83
Framework to bath panel; at 500 mm centres					
both ways					
25 x 50 mm	0.90	8.15	2.55	m2	10.70
Framework as bracketing and cradling around					
steelwork					
25 x 50 mm	1.40	12.68	3.04	m2	15.72
50 x 50 mm	1.50	13.59	5.22	m2	18.81
50 x 75 mm	1.60	14.49	7.15	m2	21.64
Blockings wedged between flanges of steelwork					
50 x 50 x 150 mm	0.12	1.09	0.35	nr	1.44
50 x 75 x 225 mm	0.13	1.18	0.52	nr	1.70
50 x 100 x 300 mm	0.14	1.27	0.71	nr	1.98
Floor fillets set in or on concrete					
38 x 50 mm	0.12	1.09	0.58	m	1.67
50 x 50 mm	0.12	1.09	0.71	m	1.80
Floor fillets fixed to floor clips (priced					
elsewhere)					
38 x 50 mm	0.10	0.91	0.60	m	1.51
50 x 50 mm	0.10	0.91	0.73	m	1.64
Wrought softwood					
Boarding to gutter bottoms or sides; tongued					
and grooved joints					
19 mm thick; sloping	1.50	13.59	6.78	m2	20.37
19 mm thick x 75 mm wide	0.40	3.62	0.66	m	4.28
19 mm thick x 150 mm wide	0.45	4.08	1.01	m	5.09
19 mm thick x 225 mm wide	0.50	4.53	1.61	m	6.14
25 mm thick; sloping	1.50	13.59	8.78	m2	22.37
25 mm thick x 75 mm wide	0.40	3.62	0.67	m	4.29
25 mm thick x 150 mm wide	0.45	4.08	1.28	m	5.36
25 mm thick x 225 mm wide	0.50	4.53	2.08	m	6.61
Boarding to eaves, verges, fascias and					
the like					
19 mm thick x over 300 mm wide	1.24	11.23	6.63	m2	17.86
19 mm thick x 150 mm wide; once grooved	0.20	1.81	2.43	m	4.24
25 mm thick x 150 mm wide; once grooved	0.20	1.81	2.98	m	4.79
25 mm thick x 175 mm wide; once grooved	0.22	1.99	3.33	m	5.32
32 mm thick x 225 mm wide; moulded	0.25	2.26	5.02	m	7.28
Mitred angles	0.10	0.91	-	nr	0.91
Rolls					
32 x 44 mm	0.12	1.09	0.58	m	1.67
50 x 50 mm	0.12	1.09	0.94	m	2.03
50 x 75 mm	0.13	1.18	1.46	m	2.64
75 x 75 mm	0.14	1.27	2.16	m	3.43

G STRUCTURAL/CARCASSING METAL/TIMBER Including overheads and profit at 9.00%	Labour hours	Labour £	Material £	Unit	Total rate £
Wrought softwood; 'Tanalised'					
Boarding to gutter bottoms or sides; tongued					
and grooved joints					
19 mm thick; sloping	1.50	13.59	8.36	m2	21.95
19 mm thick x 75 mm wide	0.40	3.62	0.81	m	4.43
19 mm thick x 150 mm wide	0.45	4.08	1.25	m	5.33
19 mm thick x 225 mm wide	0.50	4.53	2.00	m	6.53
25 mm thick; sloping	1.50	13.59	10.84	m2	24.43
25 mm thick x 75 mm wide	0.40	3.62	0.84	m	4.46
25 mm thick x 150 mm wide	0.45	4.08	1.60	m	5.68
25 mm thick x 225 mm wide	0.50	4.53	2.59	m	7.12
Boarding to eaves, verges, fascias and					
the like					
19 mm thick x over 300 mm wide	1.24	11.23	8.22	m2	19.45
19 mm thick x 150 mm wide; once grooved	0.20	1.81	3.03	m	4.84
25 mm thick x 150 mm wide; once grooved	0.20	1.81	3.71	m	5.52
25 mm thick x 175 mm wide; once grooved	0.22	1.99	4.17	m	6.16
32 mm thick x 225 mm wide; moulded	0.25	2.26	6.24	m	8.50
Mitred angles	0.10	0.91	-	nr	0.91
Rolls					
32 x 44 mm	0.12	1.09	0.72	m	1.81
50 x 50 mm	0.12	1.09	1.16	m	2.25
50 x 75 mm	0.13	1.18	1.82	m	3.00
75 x 75 mm	0.14	1.27	2.69	m	3.96
Labours on softwood boarding					
Raking cutting					
12 mm thick	0.06	0.54	0.27	m	0.81
19 mm thick	0.08	0.72	0.27	m	0.99
25 mm thick	0.10	0.91	0.38	m	1.29
Boundary cutting					
19 mm thick	0.09	0.82	0.27	m	1.09
Curved cutting					
19 mm thick	0.12	1.09	0.38	m	1.47
Tongued edges and mitred angle					
19 mm thick	0.14	1.27	0.38	m	1.65
Plugging					
Plugging blockwork					
300 mm centres; both ways	0.12	1.09	0.05	m2	1.14
300 mm centres; one way	0.06	0.54	0.03	m	0.57
isolated	0.03	0.27	0.01	nr	0.28
Plugging brickwork					
300 mm centres; both ways	0.20	1.81	0.05	m2	1.86
300 mm centres; one way	0.10	0.91	0.03	m	0.94
isolated	0.05	0.45	0.01	nr	0.46
Plugging concrete					
300 mm centres; both ways	0.36	3.26	0.05	m2	3.31
300 mm centres; one way	0.18	1.63	0.03	m	1.66
isolated	0.09	0.82	0.01	nr	0.83

G STRUCTURAL/CARCASSING METAL/TIMBER
Including overheads and profit at 9.00%

BASIC BOLT PRICES

	£		£		£		£
Black bolts, nuts and washers (£/100)							
Mild steel hex. hdd. bolts/nuts							
M6x50 mm	5.75	M10x50 mm	13.84	M12x100 mm	28.41	M16x140 mm	76.95
M6x80 mm	9.62	M10x80 mm	19.75	M12x140 mm	43.06	M16x180 mm	112.55
M6x100 mm	10.46	M10x100 mm	24.34	M12x180 mm	81.37	M20x80 mm	78.75
M8x50 mm	9.10	M10x140 mm	33.47	M16x50 mm	34.71	M20x100 mm	91.85
M8x80 mm	12.62	M12x50 mm	19.68	M16x80 mm	44.46	M20x140 mm	125.01
M8x100 mm	15.74	M12x80 mm	24.93	M16x100 mm	49.23	M20x180 mm	169.29
Mild steel cup. hdd. bolts/nuts							
M6x50 mm	5.06	M8x75 mm	9.40	M10x100 mm	21.03	M12x100 mm	30.78
M6x75 mm	6.19	M8x100 mm	15.65	M10x150 mm	31.33	M12x150 mm	42.73
M6x100 mm	9.63	M8x150 mm	21.47	M10x200 mm	65.81	M12x200 mm	104.72
M6x150 mm	14.66	M10x50 mm	12.89	M12x50 mm	19.52		
M8x50 mm	7.76	M10x75 mm	15.39	M12x75 mm	24.44		
Mild steel washers; round							
M6	1.33	M10	1.74	M16	3.68		
M8	1.29	M12	2.74	M20	4.56		
Mild steel washers; square							
38x38 mm	9.06	50x50 mm	5.26				

	Labour hours	Labour £	Material £	Unit	Total rate £
G20 CARPENTRY/TIMBER FRAMING/FIRST FIXING - cont'd					
Metalwork; mild steel; galvanized					
Fix only bolts; 50-200 mm long					
6 mm dia	0.04	0.36	-	nr	0.36
8 mm dia	0.04	0.36	-	nr	0.36
10 mm dia	0.05	0.45	-	nr	0.45
12 mm dia	0.05	0.45	-	nr	0.45
16 mm dia	0.06	0.54	-	nr	0.54
20 mm dia	0.06	0.54	-	nr	0.54
Straps; standard twisted vertical restraint; fixing to softwood and brick or blockwork					
30 x 2.5 x 400 mm girth	0.25	2.26	0.66	nr	2.92
30 x 2.5 x 600 mm girth	0.26	2.36	0.84	nr	3.20
30 x 2.5 x 800 mm girth	0.27	2.45	1.09	nr	3.54
30 x 2.5 x 1000 mm girth	0.30	2.72	1.33	nr	4.05
30 x 2.5 x 1200 mm girth	0.31	2.81	1.55	nr	4.36
Timber connectors; round toothed plate; for 10 mm or 12 mm dia bolts					
38 mm dia; single sided	0.02	0.18	0.13	nr	0.31
38 mm dia; double sided	0.02	0.18	0.15	nr	0.33
50 mm dia; single sided	0.02	0.18	0.15	nr	0.33
50 mm dia; double sided	0.02	0.18	0.16	nr	0.34
63 mm dia; single sided	0.02	0.18	0.20	nr	0.38
63 mm dia; double sided	0.02	0.18	0.22	nr	0.40
75 mm dia; single sided	0.02	0.18	0.28	nr	0.46
75 mm dia; double sided	0.02	0.18	0.32	nr	0.50
Framing anchor	0.15	1.36	0.41	nr	1.77
Joist hangers 1.0 mm thick; for fixing to softwood; joint sizes					
50 x 100 mm PC £0.53	0.12	1.09	0.71	nr	1.80
50 x 125 mm PC £0.53	0.12	1.09	0.71	nr	1.80
50 x 150 mm PC £0.53	0.13	1.18	0.71	nr	1.89
50 x 175 mm PC £0.53	0.13	1.18	0.71	nr	1.89
50 x 200 mm PC £0.53	0.14	1.27	0.71	nr	1.98
50 x 225 mm PC £0.53	0.14	1.27	0.71	nr	1.98
50 x 250 mm PC £0.53	0.15	1.36	0.71	nr	2.07

G STRUCTURAL/CARCASSING METAL/TIMBER Including overheads and profit at 9.00%		Labour hours	Labour £	Material £	Unit	Total rate £
75 x 150 mm	PC £0.57	0.13	1.18	0.75	nr	1.93
75 x 175 mm	PC £0.57	0.13	1.18	0.75	nr	1.93
75 x 200 mm	PC £0.57	0.14	1.27	0.75	nr	2.02
75 x 225 mm	PC £0.57	0.14	1.27	0.75	nr	2.02
75 x 250 mm	PC £0.57	0.15	1.36	0.75	nr	2.11
100 x 200 mm	PC £0.61	0.15	1.36	0.80	nr	2.16
Joist hangers 2.7 mm thick; for building in; joist sizes						
50 x 100 mm	PC £0.99	0.08	0.72	1.18	nr	1.90
50 x 125 mm	PC £01.0	0.08	0.72	1.19	nr	1.91
50 x 150 mm	PC £1.03	0.09	0.82	1.22	nr	2.04
50 x 175 mm	PC £1.04	0.09	0.82	1.24	nr	2.06
50 x 200 mm	PC £1.15	0.10	0.91	1.37	nr	2.28
50 x 225 mm	PC £1.23	0.10	0.91	1.46	nr	2.37
50 x 250 mm	PC £1.60	0.11	1.00	1.88	nr	2.88
75 x 150 mm	PC £1.54	0.09	0.82	1.81	nr	2.63
75 x 175 mm	PC £1.42	0.09	0.82	1.68	nr	2.50
75 x 200 mm	PC £1.56	0.10	0.91	1.83	nr	2.74
75 x 225 mm	PC £1.62	0.10	0.91	1.90	nr	2.81
75 x 250 mm	PC £1.79	0.11	1.00	2.10	nr	3.10
100 x 200 mm		0.10	0.91	2.11	nr	3.02
Herringbone joist struts; to suit joists at						
400 mm centres	PC £18.95/100	0.30	2.72	1.03	m	3.75
450 mm centres	PC £21.39/100	0.27	2.45	1.03	m	3.48
600 mm centres	PC £23.87/100	0.24	2.17	1.04	m	3.21
Expanding bolts; 'Rawlbolt' projecting type; plated; one nut; one washer						
6 mm dia; ref M6 10P		0.08	0.72	0.52	nr	1.24
6 mm dia; ref M6 25P		0.08	0.72	0.59	nr	1.31
6 mm dia; ref M6 60P		0.08	0.72	0.61	nr	1.33
8 mm dia; ref M8 25P		0.08	0.72	0.70	nr	1.42
8 mm dia; ref M8 60P		0.08	0.72	0.74	nr	1.46
10 mm dia; ref M10 15P		0.10	0.91	0.91	nr	1.82
10 mm dia; ref M10 30P		0.10	0.91	0.95	nr	1.86
10 mm dia; ref M10 60P		0.10	0.91	0.99	nr	1.90
12 mm dia; ref M12 15P		0.10	0.91	1.43	nr	2.34
12 mm dia; ref M12 30P		0.10	0.91	1.54	nr	2.45
12 mm dia; ref M12 75P		0.10	0.91	1.92	nr	2.83
16 mm dia; ref M16 35P		0.12	1.09	3.55	nr	4.64
16 mm dia; ref M16 75P		0.12	1.09	3.72	nr	4.81
Expanding bolts; 'Rawlbolt' loose bolt type; plated; one bolt; one washer						
6 mm dia; ref M6 10L		0.10	0.91	0.52	nr	1.43
6 mm dia; ref M6 25L		0.10	0.91	0.55	nr	1.46
6 mm dia; ref M6 40L		0.10	0.91	0.56	nr	1.47
8 mm dia; ref M8 25L		0.10	0.91	0.68	nr	1.59
8 mm dia; ref M8 40L		0.10	0.91	0.72	nr	1.63
10 mm dia; ref M10 10L		0.10	0.91	0.88	nr	1.79
10 mm dia; ref M10 25L		0.10	0.91	0.90	nr	1.81
10 mm dia; ref M10 50L		0.10	0.91	0.95	nr	1.86
10 mm dia; ref M10 75L		0.10	0.91	0.99	nr	1.90
12 mm dia; ref M12 10L		0.10	0.91	1.30	nr	2.21
12 mm dia; ref M12 25L		0.10	0.91	1.43	nr	2.34
12 mm dia; ref M12 40L		0.10	0.91	1.50	nr	2.41
12 mm dia; ref M12 60L		0.10	0.91	1.58	nr	2.49
16 mm dia; ref M16 30L		0.12	1.09	3.65	nr	4.74
16 mm dia; ref M16 60L		0.12	1.09	3.65	nr	4.74
Metalwork; mild steel; galvanised						
Ragbolts; mild steel; one nut; one washer						
M10 x 120 mm long		0.10	1.03	0.97	nr	2.00
M12 x 160 mm long		0.13	1.34	1.21	nr	2.55
M20 x 200 mm long		0.15	1.55	2.64	nr	4.19

G STRUCTURAL/CARCASSING METAL/TIMBER Including overheads and profit at 9.00%	Labour hours	Labour £	Material £	Unit	Total rate £
G32 EDGE SUPPORTED/REINFORCED WOODWOOL SLAB DECKING					
Woodwool interlocking reinforced slabs; Torvale 'Woodcelip' or similar; natural finish; fixing to timber or steel with galvanized nails or clips; flat or sloping					
50 mm slabs; type 503; max. span 2100 mm					
1800 - 2100 mm lengths PC £10.90	0.50	4.53	12.69	m2	17.22
2400 mm lengths PC £11.44	0.50	4.53	13.32	m2	17.85
2700 - 3000 mm lengths PC £11.62	0.50	4.53	13.66	m2	18.19
75 mm slabs; type 751; max. span 2100 mm					
1800 - 2400 mm lengths PC £16.04	0.55	4.98	18.62	m2	23.60
2700 - 3000 mm lengths PC £16.11	0.55	4.98	18.71	m2	23.69
75 mm slabs; type 752; max. span 2100 mm					
1800 - 2400 mm lengths PC £15.98	0.55	4.98	18.56	m2	23.54
2700 - 3000 mm lengths PC £16.03	0.55	4.98	18.77	m2	23.75
75 mm slabs; type 753; max. span 3600 mm					
2400 mm lengths PC £15.96	0.55	4.98	18.54	m2	23.52
2700 - 3000 mm lengths PC £16.64	0.55	4.98	19.32	m2	24.30
3300 - 3900 mm lengths	0.55	4.98	22.80	m2	27.78
Raking cutting; including additional trim	0.22	1.99	4.08	m	6.07
Holes for pipes and the like	0.12	1.09	-	nr	1.09
100 mm slabs; type 1001; max. span 3600 mm					
3000 mm lengths PC £21.25	0.60	5.43	24.63	m2	30.06
3300 - 3600 mm lengths PC £23.22	0.60	5.43	26.88	m2	32.31
100 mm slabs; type 1002; max. span 3600 mm					
3000 mm lengths PC £20.76	0.60	5.43	24.07	m2	29.50
3300 - 3600 mm lengths PC £22.03	0.60	5.43	25.70	m2	31.13
100 mm slabs; type 1003; max. span 4000 mm					
3000 - 3600 mm lengths PC £20.01	0.60	5.43	23.21	m2	28.64
3900 - 4000 mm lengths PC £20.01	0.60	5.43	23.21	m2	28.64
125 mm slabs; type 1252; max. span 3000 mm					
2400 - 3000 mm lengths PC £22.42	0.60	5.43	25.97	m2	31.40
Extra over slabs for					
pre-screeded deck	-	-	-	m2	0.91
pre-screeded soffit	-	-	-	m2	2.09
pre-screeded deck and soffit	-	-	-	m2	2.65
pre-screeded and proofed deck	-	-	-	m2	1.68
pre-screeded and proofed deck plus pre-screeded soffit	-	-	-	m2	3.94
pre-felted deck (glass fibre)	-	-	-	m2	2.22
pre-felted deck plus pre-screeded soffit	-	-	-	m2	4.25
'Weatherdeck'	-	-	-	m2	1.43

H CLADDING/COVERING Including overheads and profit at 9.00% & 5.00%	Labour hours	Labour £	Material £	Unit	Total rate £
H10 PATENT GLAZING					
Patent glazing; aluminium alloy bars 2.44 mm long at 622 mm centres; fixed to supports Roof cladding; glazing with 7 mm thick Georgian wired cast glass	-	-	-	m2	76.25
Extra for associated code 4 lead flashings					
top flashing; 210 mm girth	-	-	-	m	27.28
bottom flashing; 240 mm girth	-	-	-	m	35.48
end flashing; 300 mm girth	-	-	-	m	50.47
Wall cladding; glazing with 7 mm thick Georgian wired cast glass	-	-	-	m2	82.24
Wall cladding; glazing with 6 mm thick plate glass	-	-	-	m2	119.14
Extra for aluminium alloy members					
38 x 38 x 3 mm angle jamb	-	-	-	m	27.66
extruded cill member	-	-	-	m	21.31
extruded channel head and PVC came	-	-	-	m	13.24

H14 CONCRETE ROOFLIGHTS/PAVEMENT LIGHTS

NOTE: The following specialist prices for rooflights/pavement lights assume panels of 50 m2; pavement lights at ground level and rooflights at 2nd floor level; easy access; and all necessary ancillary fixing; strengthening; pointing and expansion materials etc.

	Labour hours	Labour £	Material £	Unit	Total rate £
Reinforced concrete rooflights/pavement lights; 'Luxcrete' or similar; with glass lenses; supplied and fixed complete Rooflights					
2.5 KN/m2 loading; ref. R254/125	-	-	-	m2	326.18
2.5 KN/m2 loading; ref. R200/90 or R254/B191	-	-	-	m2	341.72
2.5 KN/m2 loading; home office; double glazed	-	-	-	m2	455.62
Pavement lights					
20 KN/m2; pedestrian traffic; ref. P150/100	-	-	-	m2	305.47
60 KN; vehicular traffic; ref. P165/165	-	-	-	m2	321.01
Brass terrabond	-	-	-	m2	7.59
150 x 75 mm identification plates	-	-	-	m2	15.53
Escape hatch	-	-	-	m2	2485.20

H20 RIGID SHEET CLADDING

Eternit 2000 'Glasal' sheet; Eternit TAC Ltd; flexible neoprene gasket joints; fixing with stainless steel screws and coloured caps 7.5 mm thick cladding to walls	Labour hours	Labour £	Material £	Unit	Total rate £
over 300 mm wide	2.10	28.22	36.45	m2	64.67
not exceeding 300 mm wide	0.70	9.41	18.30	m	27.71
External angle	0.10	1.34	4.93	m	6.27
7.5 mm thick cladding to eaves; verges fascias or the like					
100 mm wide	0.50	6.72	10.41	m	17.13
150 mm wide	0.55	7.39	11.94	m	19.33
200 mm wide	0.60	8.06	13.48	m	21.54
250 mm wide	0.65	8.74	15.02	m	23.76
300 mm wide	0.70	9.41	17.08	m	26.49

H CLADDING/COVERING Including overheads and profit at 9.00%		Labour hours	Labour £	Material £	Unit	Total rate £
H30 FIBRE CEMENT PROFILED SHEET CLADDING/ **COVERING/SIDING**						
Asbestos-free corrugated sheets; Eternit **'2000' or similar**						
Roof cladding; sloping not exceeding 50 degrees; fixing to timber purlins with drive screws						
'Profile 3'; natural	PC £5.34	0.20	2.69	8.78	m2	11.47
'Profile 3'; coloured	PC £5.87	0.20	2.69	9.35	m2	12.04
'Profile 6'; natural	PC £5.36	0.25	3.36	8.58	m2	11.94
'Profile 6'; coloured		0.25	3.36	9.13	m2	12.49
'Profile 6'; natural; insulated; 60 mm glass fibre infill; lining panel		0.45	6.05	20.62	m2	26.67
Roof cladding; sloping not exceeding 50 degrees; fixing to steel purlins with hook bolts						
'Profile 3'; natural		0.25	3.36	9.24	m2	12.60
'Profile 3'; coloured		0.25	3.36	9.78	m2	13.14
'Profile 6'; natural		0.30	4.03	8.97	m2	13.00
'Profile 6'; coloured		0.30	4.03	9.52	m2	13.55
'Profile 6'; natural; insulated; 60 mm glass fibre infill; lining panel		0.50	6.72	19.03	m2	25.75
Wall cladding; vertical; fixing to steel rails with hook bolts						
'Profile 3'; natural		0.30	4.03	9.24	m2	13.27
'Profile 3'; coloured		0.30	4.03	9.95	m2	13.98
'Profile 6'; natural		0.35	4.70	8.97	m2	13.67
'Profile 6'; coloured		0.35	4.70	9.52	m2	14.22
'Profile 6'; natural; insulated; 60 mm glass fibre infill; lining panel		0.55	7.39	19.03	m2	26.42
Raking cutting		0.15	2.02	1.31	m	3.33
Holes for pipes and the like		0.15	2.02	-	nr	2.02
Accessories; to 'Profile 3' cladding; natural						
eaves filler		0.10	1.34	5.72	m	7.06
vertical corrugation closure		0.12	1.61	5.72	m	7.33
apron flashing		0.12	1.61	6.44	m	8.05
underglazing flashing		0.12	1.61	6.19	m	7.80
plain wing or close fitting two-piece adjustable capping to ridge		0.17	2.28	14.42	m	16.70
ventilating two-piece adjustable capping to ridge		0.17	2.28	14.42	m	16.70
Accessories; to 'Profile 3' cladding; coloured						
eaves filler		0.10	1.34	6.18	m	7.52
vertical corrugation closure		0.12	1.61	6.86	m	8.47
apron flashing		0.12	1.61	8.50	m	10.11
underglazing flashing		0.12	1.61	7.44	m	9.05
plain wing or close fitting two-piece adjustable capping to ridge		0.17	2.28	15.53	m	17.81
ventilating two-piece adjustable capping to ridge		0.17	2.28	17.22	m	19.50
Accessories; to 'Profile 6' cladding; natural						
eaves filler		0.10	1.34	4.48	m	5.82
vertical corrugation closure		0.12	1.61	4.97	m	6.58
apron flashing		0.12	1.61	4.60	m	6.21
plain cranked crown to ridge		0.17	2.28	9.49	m	11.77
plain wing, close fitting or north light two-piece adjustable capping to ridge		0.17	2.28	8.26	m	10.54
ventilating two-piece adjustable capping to ridge		0.17	2.28	12.38	m	14.66
Accessories; to 'Profile 6' cladding; coloured						
eaves filler		0.10	1.34	5.96	m	7.30
vertical corrugation closure		0.12	1.61	6.14	m	7.75

H CLADDING/COVERING Including overheads and profit at 9.00%	Labour hours	Labour £	Material £	Unit	Total rate £
apron flashing	0.12	1.61	4.51	m	6.12
plain cranked crown to ridge	0.17	2.28	11.31	m	13.59
plain wing, close fitting or north light two-piece adjustable capping to ridge	0.17	2.28	10.88	m	13.16
ventilating two-piece adjustable capping to ridge	0.17	2.28	14.77	m	17.05

H31 METAL PROFILED/FLAT SHEET CLADDING/COVERING/SIDING

Galvanised steel strip troughed sheets; BSC Strip Mill Products; colorcoat 'Plastisol' finish
Roof cladding; sloping not exceeding
50 degrees; fixing to steel purlins
with plastic headed self-tapping screws

0.7 mm type 12.5/3 in. corrugated PC £6.49/m	0.35	4.70	12.02	m2	16.72
0.7 mm Long Rib 1000; 35 mm deep PC £6.49/m	0.40	5.38	12.11	m2	17.49

Wall cladding; vertical; fixing to steel
rails with plastic headed self-tapping
screws

0.7 mm type 12.5/3 in. corrugated	0.40	5.38	12.02	m2	17.40
0.7 mm Scan Rib 1000; 19 mm deep	0.45	6.05	11.13	m2	17.18
Raking cutting	0.22	2.96	1.70	m	4.66
Holes for pipes and the like	0.40	5.38	-	nr	5.38

Accessories; colorcoat silicone polyester
finish 0.9 mm standard flashings; bent to
profile

250 mm girth	0.20	2.69	3.64	m	6.33
375 mm girth	0.22	2.96	4.74	m	7.70
500 mm girth	0.24	3.23	5.78	m	9.01
625 mm girth	0.30	4.03	6.93	m	10.96

Galvanised steel profile sheet cladding; Plannja Ltd.; PVF2 coated finish
Roof cladding; sloping not exceeding
50 degrees; fixing to steel purlins
with plastic headed self-tapping screws

0.72 mm 'profile 20B'	-	-	-	m2	15.52
0.72 mm 'profile TOP 40'	-	-	-	m2	14.93
0.72 mm 'profile 45'	-	-	-	m2	16.90
Extra for 80 mm insulation and 0.4 mm coated inner lining sheet	-	-	-	m2	10.05

Wall cladding; vertical; fixing to steel
rails with plastic headed self-tapping
screws

0.60 mm 'profile 20B'; corrugations vertical	-	-	-	m2	16.28
0.60 mm 'profile 30'; corrugations vertical	-	-	-	m2	16.28
0.60 mm 'profile TOP 40'; corrugations vertical	-	-	-	m2	15.19
0.6 mm 'profile 60B'; corrugations vertical	-	-	-	m2	18.97
0.60 mm 'profile 30'; corrugations horizontal	-	-	-	m	16.72
0.60 mm 'profile 60B'; corrugations horizontal	-	-	-	nr	19.40
Extra for 80 mm insulation and 0.4 mm coated inner lining sheet	-	-	-	m	10.09

Accessories for roof/vertical cladding;
PVF2 coated finish; 0.60 mm thick
flashings; once bent

250 mm girth	-	-	-	m	9.37
375 mm girth	-	-	-	m	11.34
500 mm girth	-	-	-	m	12.86
625 mm girth	-	-	-	m	14.44
Extra bends - each	-	-	-	m	0.15
Profile fillers	-	-	-	m	0.49

H CLADDING/COVERING Including overheads and profit at 9.00%		Labour hours	Labour £	Material £	Unit	Total rate £
H31 METAL PROFILED/FLAT SHEET CLADDING/ COVERING/SIDING - cont'd						
Aluminium troughed sheets; British Aluminium 'Rigidal' range; pre-painted finish						
Roof cladding; sloping not exceeding 50 degrees; fixing to steel purlins with plastic headed self-tapping screws						
0.7 mm type WA6	PC £8.16/m	0.45	6.05	14.42	m2	20.47
0.9 mm type A7	PC £10.32/m	0.45	6.05	18.08	m2	24.13
Wall cladding; vertical; fixing to steel rails with plastic headed self-tapping screws						
0.7 mm type WA6	PC £8.16/m	0.50	6.72	14.42	m2	21.14
0.9 mm type A7	PC £10.32/m	0.50	6.72	18.08	m2	24.80
0.9 mm type MM10	PC £10.32/m	0.50	6.72	16.38	m2	23.10
Raking cutting	PC £10.32	0.22	2.96	2.70	m	5.66
Holes for pipes and the like		0.40	5.38	-	nr	5.38
Accessories; 0.9 mm pre-painted standard flashings; bent to profile						
200 mm girth		0.25	3.36	3.25	m	6.61
312 mm girth		0.27	3.63	4.37	m	8.00
380 mm girth		0.30	4.03	5.35	m	9.38
500 mm girth		0.35	4.70	7.39	m	12.09
H41 GLASS REINFORCED PLASTICS CLADDING/ FEATURES						
Glass fibre translucent sheeting grade AB class 3						
Roof cladding; sloping not exceeding 50 degrees; fixing to timber purlins with drive screws; to suit						
'Profile 3'	PC £5.52	0.20	2.69	8.78	m2	11.47
'Profile 6'	PC £6.94	0.25	3.36	10.74	m2	14.10
Roof cladding; sloping not exceeding 50 degrees; fixing to steel purlins with hook bolts; to suit						
'Profile 3'	PC £5.52	0.25	3.36	9.23	m2	12.59
'Profile 6'	PC £6.94	0.30	4.03	11.19	m2	15.22
'Longrib 1000'	PC £7.27	0.30	4.03	11.64	m2	15.67

BASIC NATURAL STONE BLOCK PRICES

Block prices (£/m3)	£			£
Ancaster Hardwhite	312.00	Westwood		250.00
Ancaster Weatherbed	312.00	Portland Whitbed	180.00 -	320.00
Doulting	259.00	Westmorland green slate		330.00
Monks Park	183.00			

H CLADDING/COVERING Including overheads and profit at 9.00%	Labour hours	Labour £	Material £	Unit	Total rate £
H51 NATURAL STONE SLAB CLADDING FEATURES					
Portland Whitbed limestone bedded and jointed in cement - lime - mortar (1:2:9); slurrying with weak lime and stonedust mortar; flush pointing and cleaning on completion (cramps etc measured separately)					
Facework; one face plain and rubbed; bedded against backing					
50 mm thick stones	-	-	-	m2	167.65
63 mm thick stones	-	-	-	m2	186.02
75 mm thick stones	-	-	-	m2	207.33
100 mm thick stones	-	-	-	m2	234.78
Fair returns on facework					
50 mm wide	-	-	-	m	0.50
63 mm wide	-	-	-	m	0.50
75 mm wide	-	-	-	m	0.50
100 mm wide	-	-	-	m	0.59
Fair raking cutting on facework					
50 mm thick	-	-	-	m	4.84
63 mm thick	-	-	-	m	5.19
75 mm thick	-	-	-	m	5.69
100 mm thick	-	-	-	m	5.90
Copings; once weathered; and throated; rubbed; set horizontal or raking					
250 x 50 mm	-	-	-	m	46.69
300 x 50 mm	-	-	-	m	56.01
350 x 75 mm	-	-	-	m	75.13
400 x 100 mm	-	-	-	m	99.42
450 x 100 mm	-	-	-	m	109.08
500 x 125 mm	-	-	-	m	123.96
Extra for angles on copings					
250 x 50 mm	-	-	-	nr	77.93
300 x 50 mm	-	-	-	nr	85.70
375 x 75 mm	-	-	-	nr	91.06
400 x 100 mm	-	-	-	nr	94.80
450 x 100 mm	-	-	-	nr	98.14
500 x 125 mm	-	-	-	nr	103.06
Band courses; plain; rubbed; horizontal					
225 x 112 mm	-	-	-	m	74.67
300 x 112 mm	-	-	-	m	83.98
Extra for					
Ends	-	-	-	nr	21.19
Angles	-	-	-	nr	25.52
Band courses; moulded 100 mm girth on face; rubbed; horizontal					
125 x 75 mm	-	-	-	m	108.40
150 x 75 mm	-	-	-	m	112.94
200 x 100 mm	-	-	-	m	122.45
250 x 150 mm	-	-	-	m	135.46
300 x 250 mm	-	-	-	m	145.18
Extra for ends on band courses					
125 x 75 mm	-	-	-	nr	31.36
150 x 75 mm	-	-	-	nr	41.81
200 x 100 mm	-	-	-	nr	52.28
250 x 150 mm	-	-	-	nr	62.73
300 x 250 mm	-	-	-	nr	83.63
Extra for angles on band courses					
125 x 75 mm	-	-	-	nr	20.90
150 x 75 mm	-	-	-	nr	32.54
200 x 100 mm	-	-	-	nr	41.97
250 x 150 mm	-	-	-	nr	53.71
300 x 250 mm	-	-	-	nr	62.87
Coping apex block; two sunk faces; rubbed					
650 x 450 x 225 mm	-	-	-	nr	154.58

H CLADDING/COVERING Including overheads and profit at 9.00%	Labour hours	Labour £	Material £	Unit	Total rate £
H51 NATURAL STONE SLAB CLADDING FEATURES - cont'd					
Portland Whitbed limestone bedded **and jointed in cement - lime - mortar** **(1:2:9); slurrying with weak lime and** **stonedust mortar; flush pointing and** **cleaning on completion (cramps etc** **measured separately) - cont'd**					
Coping kneeler block; three sunk faces; rubbed					
350 x 350 x 375 mm	-	-	-	nr	112.39
450 x 450 x 375 mm	-	-	-	nr	133.54
Corbel; turned and moulded; rubbed					
225 x 225 x 375 mm	-	-	-	nr	213.94
Slab surrounds to openings; one face splayed; rubbed					
200 x 75 mm	-	-	-	m	64.02
75 x 100 mm	-	-	-	m	58.98
100 x 100 mm	-	-	-	m	64.13
125 x 100 mm	-	-	-	m	70.93
125 x 150 mm	-	-	-	m	78.19
175 x 175 mm	-	-	-	m	84.85
225 x 175 mm	-	-	-	m	93.43
300 x 175 mm	-	-	-	m	100.74
300 x 225 mm	-	-	-	m	109.10
Slab surrounds to openings; one face sunk splayed; rubbed					
200 x 75 mm	-	-	-	m	66.41
75 x 100 mm	-	-	-	m	72.70
100 x 100 mm	-	-	-	m	76.65
125 x 100 mm	-	-	-	m	81.36
125 x 150 mm	-	-	-	m	87.60
175 x 175 mm	-	-	-	m	95.43
225 x 175 mm	-	-	-	m	102.70
300 x 175 mm	-	-	-	m	110.68
300 x 225 mm	-	-	-	m	117.86
Extra for					
Throating	-	-	-	m	6.10
Rebates and grooves	-	-	-	m	7.33
Stooling	-	-	-	nr	34.37
Sundries - stone walling					
Coating backs of stones with brush applied cold bitumen solution; two coats					
limestone facework	0.20	1.40	1.77	m2	3.17
Cutting grooves in limestone masonry for					
water bars or the like	-	-	-	m	4.89
Mortices in limestone masonry for					
metal dowel	-	-	-	nr	0.36
metal cramp	-	-	-	nr	0.83
Cramps and dowels; Harris and Edgar's **'Delta' range or similar; one end built** **into brickwork or set in slot in concrete;** **Stainless Steel**					
Dowel					
8 mm dia x 75 mm long PC £13.02/100	0.05	0.58	0.14	nr	0.72
10 mm dia x 150 mm long PC £34.10/100	0.05	0.58	0.37	nr	0.95
Pattern 'J' tie					
25 x 3 x 100 mm PC £32.21/100	0.07	0.82	0.35	nr	1.17
Pattern 'S' cramp; with two 20 mm turndowns (190 mm girth)					
25 x 3 x 150 mm PC £37.93/100	0.07	0.82	0.41	nr	1.23
Pattern 'B' anchor; with 8 x 75 mm loose dowel					
25 x 3 x 150 mm PC £41.29/100	0.10	1.17	0.45	nr	1.62

H CLADDING/COVERING Including overheads and profit at 9.00%		Labour hours	Labour £	Material £	Unit	Total rate £
Pattern 'Q' tie						
25 x 3 x 200 mm	PC £47.44/100	0.07	0.82	0.52	nr	1.34
38 x 3 x 250 mm (special)	PC £86.51/100	0.07	0.82	0.94	nr	1.76
Pattern 'H1' halftwist tie						
25 x 3 x 200 mm	PC £57.56/100	0.07	0.82	0.63	nr	1.45
38 x 3 x 250 mm (special)	PC £89.75/100	0.07	0.82	0.98	nr	1.80

ALTERNATIVE TILE PRICES (£/1000)

	£		£		£
Clay tiles; plain, interlocking and pantile					
'Langleys' 'Sterreberg' pantiles					
anthracite	1078.00	natural red	762.00	rustic	762.00
deep brown	1203.00				
Sandtoft pantiles					
Bold roll 'Roman'	754.80	'Gaelic'	506.94	'County' i'locking	479.40
William Blyth pantiles					
'Barco' bold roll	408.00	'Celtic' (French)	453.00		
Concrete tiles, plain and interlocking					
Marley roof tiles					
'Anglia'	310.65	plain	179.55	'Roman'	432.25
'Ludlow +'	274.55				
Redland roof tiles					
'49'-granule	285.00	'Grovebury'	564.30	'Stonewold Mk 1	765.70
'50 Roman'	507.30				

Discounts of 2.5 - 15% available depending on quantity/status

---

NOTE: The following items of tile roofing unless otherwise described, include for conventional fixing assuming 'normal exposure' with appropriate nails and/or rivets or clips to pressure impregnated softwood battens fixed with galvanized nails; Prices also include for all bedding and pointing at verges; beneath ridge tiles, etc..	Labour hours	Labour £	Material £	Unit	Total rate £

H60 CLAY/CONCRETE ROOF TILING

Clay interlocking pantiles; Sandtoft Goxhill 'Tudor' red sand faced; PC £747.66/1000; 470 x 285 mm; to 100 mm lap; on 25 x 38 mm battens and type 1F reinforced underlay	Labour hours	Labour £	Material £	Unit	Total rate £
Roof coverings	0.40	5.38	12.40	m2	17.78
Extra over coverings for					
fixing every tile	0.02	0.27	0.27	m2	0.54
eaves course with plastic filler	0.30	4.03	4.49	m	8.52
verges; extra single undercloak course of					
plain tiles	0.30	4.03	1.94	m	5.97
open valleys; cutting both sides	0.18	2.42	4.40	m	6.82
ridge tiles	0.60	8.06	9.76	m	17.82
hip tiles; cutting both sides	0.75	10.08	14.16	m	24.24
Holes for pipes and the like	0.20	2.69	-	nr	2.69

H CLADDING/COVERING Including overheads and profit at 9.00%	Labour hours	Labour £	Material £	Unit	Total rate £
H60 CLAY/CONCRETE ROOF TILING - cont'd					
Clay interlocking pantiles; Langley's 'Sterreberg' black glazed; or similar; PC £1203.00/1000; 355 x 240 mm: to 75 mm lap; on 25 x 38 mm battens and type 1F reinforced underlay					
Roof coverings	0.45	6.05	26.66	m2	32.71
Extra over coverings for					
double course at eaves	0.33	4.44	5.14	m	9.58
verges; extra single undercloak course of					
plain tiles	0.30	4.03	5.97	m	10.00
open valleys; cutting both sides	0.18	2.42	9.97	m	12.39
saddleback ridge tiles	0.60	8.06	17.09	m	25.15
saddleback hip tiles; cutting both sides	0.75	10.08	27.05	m	37.13
Holes for pipes and the like	0.20	2.69	-	nr	2.69
Clay pantiles; Sandtoft Goxhill 'Old English' red sand faced; PC £463.08/1000 342 x 241 mm; to 75 mm lap; on 25 x 38 mm battens and type 1F reinforced underlay					
Roof coverings	0.45	6.05	13.11	m2	19.16
Extra over coverings for					
fixing every tile	0.03	0.40	0.46	m2	0.86
other colours	-	-	-	m2	0.82
double course at eaves	0.33	4.44	3.26	m	7.70
verges; extra single undercloak course of					
plain tiles	0.30	4.03	1.94	m	5.97
open valleys; cutting both sides	0.18	2.42	3.84	m	6.26
ridge tiles; tile slips	0.60	8.06	10.46	m	18.52
hip tiles; tile slips; cutting both sides	0.75	10.08	14.29	m	24.37
Holes for pipes and the like	0.20	2.69	-	nr	2.69
Clay pantiles; William Blyth's 'Lincoln' natural; 343 x 280 mm; to 75 mm lap; PC £516.25/1000; on 19 x 38 mm battens and type 1F reinforced underlay					
Roof coverings	0.45	6.05	14.12	m2	20.17
Extra over coverings for					
fixing every tile	0.03	0.40	0.46	m2	0.86
other colours	-	-	-	m2	1.20
double course at eaves	0.33	4.44	2.95	m	7.39
verges; extra single undercloak course of					
plain tiles	0.30	4.03	5.39	m	9.42
open valleys; cutting both sides	0.18	2.42	4.50	m	6.92
ridge tiles; tile slips	0.60	8.06	10.36	m	18.42
hip tiles; tile slips; cutting both sides	0.75	10.08	14.86	m	24.94
Holes for pipes and the like	0.20	2.69	-	nr	2.69
Clay plain tiles; Hinton, Perry and Davenhill 'Dreadnought' smooth red machine- made; PC £255.00/1000; 265 x 165 mm; on 19 x 38 mm battens and type 1F reinforced underlay					
Roof coverings; to 64 mm lap	0.65	8.74	21.39	m2	30.13
Wall coverings; to 38 mm lap	0.80	10.75	19.10	m2	29.85
Extra over coverings for					
25 x 38 mm battens in lieu	-	-	-	m2	0.51
other colours	-	-	-	m2	3.60
ornamental tiles in lieu	-	-	-	m2	7.72
double course at eaves	0.25	3.36	1.84	m	5.20
verges; extra single undercloak course	0.33	4.44	2.71	m	7.15
valley tiles; cutting both sides	0.65	8.74	29.36	m	38.10
bonnet hip tiles; cutting both sides	0.80	10.75	30.33	m	41.08
external vertical angle tiles; supplementary					
nail fixings	0.40	5.38	21.81	m	27.19
half round ridge tiles	0.50	6.72	10.16	m	16.88
Holes for pipes and the like	0.20	2.69	-	nr	2.69

H CLADDING/COVERING Including overheads and profit at 9.00%	Labour hours	Labour £	Material £	Unit	Total rate £
Clay plain tiles; Keymer best hand-made **sand-faced tiles; PC £534.00/1000;** **265 x 165 mm; on 19 x 38 mm battens** **and type 1F reinforced underlay**					
Roof coverings; to 64 mm lap	0.65	8.74	40.56	m2	49.30
Wall coverings; to 38 mm lap	0.80	10.75	36.03	m2	46.78
Extra over coverings for					
25 x 38 mm battens in lieu	-	-	-	m2	0.51
ornamental tiles in lieu	-	-	-	m2	3.92
double course at eaves	0.25	3.36	3.66	m	7.02
verges; extra single undercloak course	0.33	4.44	5.44	m	9.88
valley tiles; cutting both sides	0.65	8.74	22.05	m	30.79
bonnet hip tiles; cutting both sides	0.80	10.75	32.33	m	43.08
external vertical angle tiles; supplementary					
nail fixings	0.40	5.38	19.10	m	24.48
half round ridge tiles	0.50	6.72	9.66	m	16.38
Holes for pipes and the like	0.20	2.69	-	nr	2.69
Concrete interlocking tiles; Marley 'Bold **Roll' granule finish tiles or similar;** **PC £500.65/1000; 419 x 330 mm;** **to 75 mm lap; on 22 x 38 mm battens** **and type 1F reinforced underlay**					
Roof coverings	0.35	4.70	7.88	m2	12.58
Extra over coverings for					
fixing every tile	0.03	0.40	0.57	m2	0.97
25 x 38 mm battens in lieu	-	-	-	m2	0.07
eaves; eave filler	0.05	0.67	0.75	m	1.42
verges; 150 mm asbestos cement strip					
undercloak	0.23	3.09	1.63	m	4.72
valley trough tiles; cutting both sides	0.55	7.39	12.09	m	19.48
segmental ridge tiles; tile slips	0.55	7.39	9.12	m	16.51
segmental ridge tiles; tile slips; cutting					
both sides	0.70	9.41	11.30	m	20.71
dry ridge tiles; segmental including batten					
sections; unions and filler pieces	0.30	4.03	9.03	m	13.06
segmental monoridge tiles	0.55	7.39	10.47	m	17.86
gas ridge terminal	0.50	6.72	37.15	nr	43.87
Holes for pipes and the like	0.20	2.69	-	nr	2.69
Concrete interlocking tiles; Marley 'Ludlow **Major' granule finish tiles or similar;** **PC £432.25/1000; 413 x 330 mm;** **to 75 mm lap; on 22 x 38 mm battens**					
Roof coverings	0.35	4.70	7.27	m2	11.97
Extra over coverings for					
fixing every tile	0.03	0.40	0.34	m2	0.74
25 x 38 mm battens in lieu	-	-	-	m2	0.08
verges; 150 mm asbestos cement strip					
undercloak	0.23	3.09	1.63	m	4.72
dry verge system; extruded white pvc	0.15	2.02	5.23	m	7.25
Segmental ridge cap	0.03	0.40	1.61	nr	2.01
valley trough tiles; cutting both sides	0.55	7.39	13.37	m	20.76
segmental ridge tiles	0.50	6.72	4.61	m	11.33
segmental hip tiles; cutting both sides	0.65	8.74	6.59	m	15.33
dry ridge tiles; segmental including batten					
sections; unions and filler pieces	0.30	4.03	9.03	m	13.06
segmental monoridge tiles	0.50	6.72	9.26	m	15.98
gas ridge terminal	0.50	6.72	37.15	nr	43.87
Holes for pipes and the like	0.20	2.69	-	nr	2.69

H CLADDING/COVERING Including overheads and profit at 9.00%	Labour hours	Labour £	Material £	Unit	Total rate £
H60 CLAY/CONCRETE ROOF TILING - cont'd					
Concrete interlocking tiles; Marley 'Mendip'					
granule finish double pantiles or similar;					
PC £500.65/1000; 413 x 330 mm;					
to 75 mm lap; on 22 x 38 mm battens					
and type 1F reinforced underlay					
Roof coverings	0.35	4.70	8.07	m2	12.77
Extra over coverings for					
fixing every tile	0.03	0.40	0.34	m2	0.74
25 x 38 mm battens in lieu	-	-	-	m2	0.08
eaves; eave filler	0.03	0.40	0.07	m	0.47
verges; 150 mm asbestos cement strip					
undercloak	0.23	3.09	1.63	m	4.72
dry verge system; extruded white pvc	0.15	2.02	5.23	m	7.25
valley trough tiles; cutting both sides	0.55	7.39	13.69	m	21.08
segmental ridge tiles	0.55	7.39	7.39	m	14.78
segmental hip tiles; cutting both sides	0.70	9.41	9.69	m	19.10
dry ridge tiles; segmental including batten					
sections; unions and filler pieces	0.30	4.03	9.03	m	13.06
segmental monoridge tiles	0.50	6.72	9.26	m	15.98
gas ridge terminal	0.50	6.72	37.15	nr	43.87
Holes for pipes and the like	0.20	2.69	-	nr	2.69
Concrete interlocking tiles; Marley 'Modern'					
smooth finish tiles or similar; PC £495.90/1000;					
413 x 330 mm; to 75 mm lap; on 22 x 38 mm					
battens and type 1F reinforced underlay					
Roof coverings	0.35	4.70	8.11	m2	12.81
Extra over coverings for					
fixing every tile	0.04	0.54	0.30	m2	0.84
25 x 38 mm battens in lieu	-	-	-	m2	0.08
verges; 150 mm asbestos cement strip					
undercloak	0.28	3.76	2.44	m	6.20
dry verge system, extruded white pvc	0.20	2.69	6.19	m	8.88
'Modern' ridge cap	0.03	0.40	1.61	nr	2.01
valley trough tiles; cutting both sides	0.55	7.39	13.67	m	21.06
'Modern' ridge tiles	0.50	6.72	4.86	m	11.58
'Modern' hip tiles; cutting both sides	0.65	8.74	7.13	m	15.87
dry ridge tiles; 'Modern'; including batten					
sections, unions and filler pieces	0.30	4.03	8.43	m	12.46
monoridge tiles	0.50	6.72	9.26	m	15.98
gas ridge terminal	0.50	6.72	37.15	nr	43.87
Holes for pipes and the like	0.20	2.69	-	nr	2.69
Concrete interlocking tiles; Marley 'Wessex'					
smooth finish tiles or similar;					
PC £580.45/1000; 413 x 330 mm;					
to 75 mm lap; on 22 x 38 mm battens					
and type 1F reinforced underlay					
Roof coverings	0.35	4.70	9.09	m2	13.79
Extra over coverings for					
fixing every tile	0.04	0.54	0.30	m2	0.84
25 x 38 mm battens in lieu	-	-	-	m2	0.08
verges; 150 mm asbestos cement strip					
undercloak	0.23	3.09	1.63	m	4.72
dry verge system, extruded white pvc	0.15	2.02	5.23	m	7.25
'Modern' ridge cap	0.03	0.40	1.61	nr	2.01
valley trough tiles; cutting both sides	0.55	7.39	14.05	m	21.44
'Modern' ridge tiles	0.55	7.39	6.24	m	13.63
'Modern' hip tiles; cutting both sides	0.70	9.41	8.89	m	18.30
dry ridge tiles; 'Modern'; including batten					
sections, unions and filler pieces	0.30	4.03	9.29	m	13.32
monoridge tiles	0.50	6.72	9.26	m	15.98
gas ridge terminal	0.50	6.72	37.15	nr	43.87
Holes for pipes and the like	0.20	2.69	-	nr	2.69

H CLADDING/COVERING Including overheads and profit at 9.00%	Labour hours	Labour £	Material £	Unit	Total rate £
Concrete interlocking tiles; Redland **'Delta' smooth finish tiles or similar;** **PC £822.70/1000; 430 x 380 mm;** **to 75 mm lap; on 22 x 38 mm battens** **and type 1F reinforced underlay**					
Roof coverings	0.35	4.70	9.91	m2	14.61
Extra over coverings for					
fixing every tile	0.03	0.40	0.33	m2	0.73
25 x 38 mm battens in lieu	-	-	-	m2	0.07
eaves; eave filler	0.03	0.40	0.10	m	0.50
verges; extra single undercloak course of					
plain tiles	0.25	3.36	3.34	m	6.70
dry verge system; extruded white pvc	0.20	2.69	8.03	m	10.72
Ridge end unit	0.03	0.40	1.86	nr	2.26
valley trough tiles; cutting both sides	0.55	7.39	15.32	m	22.71
universal 'Delta' ridge tiles	0.50	6.72	5.00	m	11.72
universal 'Delta' hip tiles; cutting					
both sides	0.65	8.74	8.23	m	16.97
universal 'Delta' mono-pitch ridge tiles	0.50	6.72	8.62	m	15.34
gas flue terminal; 'Delta' type	0.50	6.72	38.22	nr	44.94
Holes for pipes and the like	0.20	2.69	-	nr	2.69
Concrete interlocking tiles; Redland **'Norfolk' smooth finish pantiles or similar;** **PC £369.55/1000; 381 x 229 mm;** **to 75 mm lap; on 22 x 38 mm battens and** **type 1F reinforced underlay**					
Roof coverings	0.45	6.05	9.15	m2	15.20
Extra over coverings for					
fixing every tile	0.05	0.67	0.28	m2	0.95
25 x 38 mm battens in lieu	-	-	-	m2	0.08
eaves; eave filler	0.05	0.67	0.58	m	1.25
verges; extra single undercloak course of					
plain tiles	0.30	4.03	1.03	m	5.06
valley trough tiles; cutting both sides	0.60	8.06	15.16	m	23.22
segmental ridge tiles	0.60	8.06	7.33	m	15.39
segmental hip tiles; cutting both sides	0.75	10.08	10.39	m	20.47
Holes for pipes and the like	0.20	2.69	-	nr	2.69
Concrete interlocking tiles; Redland **'Regent' granule finish bold roll tiles or** **similar; PC £564.30/1000; 418 x 332 mm;** **to 75 mm lap; on 22 x 38 mm battens and** **type 1F reinforced underlay**					
Roof coverings	0.35	4.70	8.61	m2	13.31
Extra over coverings for					
fixing every tile	0.04	0.54	0.48	m2	1.02
25 x 38 mm battens in lieu	-	-	-	m2	0.07
eaves; eave filler	0.05	0.67	0.71	m	1.38
verges; extra single undercloak course of					
plain tiles	0.25	3.36	2.04	m	5.40
dry verge system; extruded white pvc	0.15	2.02	6.43	m	8.45
Ridge end unit	0.03	0.40	1.86	nr	2.26
cloaked verge system	0.15	2.02	3.60	m	5.62
Blocked end ridge unit	0.03	0.40	3.62	nr	4.02
valley trough tiles; cutting both sides	0.55	7.39	14.56	m	21.95
segmental ridge tiles; tile slips	0.55	7.39	7.33	m	14.72
segmental hip tiles; tile slips; cutting					
both sides	0.70	9.41	9.79	m	19.20
dry ridge system; segmental ridge tiles;					
including fixing straps; 'Nuralite' fillets					
and seals	0.25	3.36	16.28	m	19.64
half round mono-pitch ridge tiles	0.55	7.39	12.54	m	19.93
gas flue terminal; half round type	0.50	6.72	38.22	nr	44.94
Holes for pipes and the like	0.20	2.69	-	nr	2.69

H CLADDING/COVERING Including overheads and profit at 9.00%	Labour hours	Labour £	Material £	Unit	Total rate £
H60 CLAY/CONCRETE ROOF TILING - cont'd					
Concrete interlocking tiles; Redland 'Renown' granule finish tiles or similar; PC £507.30/1000; 418 x 330 mm; to 75 mm lap; on 22 x 38 mm battens and type 1F reinforced underlay					
Roof coverings	0.35	4.70	7.84	m2	12.54
Extra over coverings for					
fixing every tile	0.03	0.40	0.40	m2	0.80
25 x 38 mm battens in lieu	-	-	-	m2	0.08
verges; extra single undercloak course of					
plain tiles	0.25	3.36	1.03	m	4.39
dry verge system; extruded white pvc	0.15	2.02	6:43	m	8.45
Ridge end unit	0.03	0.40	1.86	nr	2.26
cloaked verge system	0.15	2.02	3.80	m	5.82
Blocked end ridge unit	0.03	0.40	3.62	nr	4.02
valley trough tiles; cutting both sides	0.55	7.39	14.31	m	21.70
segmental ridge tiles	0.50	6.72	4.72	m	11.44
segmental hip tiles; cutting both sides	0.65	8.74	6.93	m	15.67
dry ridge system; segmental ridge tiles; including fixing straps; 'Nuralite' fillets and seals	0.25	3.36	16.28	m	19.64
half round mono-pitch ridge tiles	0.50	6.72	8.25	m	14.97
gas flue terminal; half round type	0.50	6.72	38.22	nr	44.94
Holes for pipes and the like	0.20	2.69	-	nr	2.69
Concrete interlocking tiles; Redland 'Stonewold' smooth finish tiles or similar; PC £702.05/1000; 430 x 380 mm; to 75 mm lap; on 22 x 38 mm battens and type 1F reinforced underlay					
Roof coverings	0.35	4.70	8.70	m2	13.40
Extra over coverings for					
fixing every tile	0.03	0.40	0.33	m2	0.73
25 x 38 mm battens in lieu	-	-	-	m2	0.07
verges; extra single undercloak course of					
plain tiles	0.30	4.03	3.14	m	7.17
dry verge system; extruded white pvc	0.20	2.69	7.81	m	10.50
Ridge end unit	0.03	0.40	1.86	nr	2.26
valley trough tiles; cutting both sides	0.55	7.39	14.85	m	22.24
universal 'Stonewold' ridge tiles	0.50	6.72	5.00	m	11.72
universal 'Stonewold' hip tiles; cutting both sides	0.65	8.74	7.76	m	16.50
dry ridge system; universal 'Stonewold' ridge tiles; including fixing straps; 'Nuralite' fillets and seals	0.25	3.36	16.77	m	20.13
universal 'Stonewold' mono-pitch ridge tiles	0.50	6.72	8.62	m	15.34
gas flue terminal; 'Stonewold' type	0.50	6.72	38.22	nr	44.94
Holes for pipes and the like	0.20	2.69	-	nr	2.69
Concrete plain tiles; BS 473 and 550 group A; PC £199.50/1000; 267 x 165 mm; on 19 x 38 mm battens and type 1F reinforced underlay					
Roof coverings; to 64 mm lap	0.65	8.74	17.58	m2	26.32
Wall coverings; to 38 mm lap	0.80	10.75	15.73	m2	26.48
Extra over coverings for					
25 x 38 mm battens in lieu	-	-	-	m2	0.51
ornamental tiles in lieu	-	-	-	m2	6.90
double course at eaves	0.25	3.36	1.47	m	4.83
verges; extra single undercloak course	0.33	4.44	2.16	m	6.60
valley tiles; cutting both sides	0.65	8.74	16.34	m	25.08
bonnet hip tiles; cutting both sides	0.80	10.75	17.31	m	28.06
external vertical angle tiles; supplementary nail fixings	0.40	5.38	10.48	m	15.86

H CLADDING/COVERING Including overheads and profit at 9.00%	Labour hours	Labour £	Material £	Unit	Total rate £
segmental ridge tiles	0.50	6.72	6.87	m	13.59
segmental hip tiles; cutting both sides	0.75	10.08	7.48	m	17.56
Holes for pipes and the like	0.20	2.69	-	nr	2.69

Sundries
Hip irons

galvanized mild steel; fixing with screws	0.10	1.34	1.66	nr	3.00

Fixing

lead soakers (supply included elsewhere)	0.08	1.08	-	nr	1.08

Pressure impregnated softwood counter
battens; 25 x 50 mm

450 mm centres	0.07	0.94	0.80	m2	1.74
600 mm centres	0.05	0.67	0.60	m2	1.27

Underlay; BS 747 type 1B; bitumen felt
weighing 14 kg/10 m2;
PC £13.68/20m2; 75 mm laps

To sloping or vertical surfaces	0.03	0.40	0.80	m2	1.20

Underlay; BS 747 type 1F; reinforced bitumen
felt; weighing 22.5 kg/15 m2; PC £16.15/15m2; 75 mm laps
(prices included within tiling rates)

To sloping or vertical surfaces	0.03	0.40	1.25	m2	1.65

H61 FIBRE CEMENT SLATING

Asbestos-cement slates; Eternit or similar;
to 75 mm lap; on 19 x 50 mm battens and
type 1F reinforced underlay
Coverings; 500 x 250 mm 'blue/black' slates

roof coverings	0.65	8.74	14.81	m2	23.55
wall coverings	0.85	11.42	14.81	m2	26.23

Coverings; 600 x 300 mm 'blue/black' slates

roof coverings	0.50	6.72	14.02	m2	20.74
wall coverings	0.65	8.74	14.02	m2	22.76

Extra over slate coverings for

double course at eaves	0.25	3.36	2.29	m	5.65
verges; extra single undercloak course	0.33	4.44	1.80	m	6.24
open valleys; cutting both sides	0.20	2.69	4.74	m	7.43
valley gutters; cutting both sides	0.55	7.39	16.30	m	23.69
half round ridge tiles	0.50	6.72	11.72	m	18.44
Stop end	0.10	1.34	4.04	nr	5.38
roll top ridge tiles	0.50	6.72	14.65	m	21.37
Stop end	0.10	1.34	6.06	nr	7.40
mono-pitch ridges	0.50	6.72	16.91	m	23.63
Stop end	0.10	1.34	18.37	nr	19.71
duo-pitch ridges	0.50	6.72	14.48	m	21.20
Stop end	0.10	1.34	13.47	nr	14.81
mitred hips; cutting both sides	0.20	2.69	4.74	m	7.43
half round hip tiles; cutting both sides	0.65	8.74	16.46	m	25.20
Holes for pipes and the like	0.20	2.69	-	nr	2.69

Asbestos-free artificial slates; Eternit
'2000' or similar; to 75 mm lap; on
19 x 50 mm battens and type 1F reinforced
underlay
Coverings; 400 x 200 mm 'blue/black' slates

roof coverings	0.80	10.75	17.04	m2	27.79
wall coverings	1.05	14.11	17.04	m2	31.15

Coverings; 500 x 250 mm 'blue/black' slates

roof coverings	0.65	8.74	15.25	m2	23.99
wall coverings	0.85	11.42	15.25	m2	26.67

Coverings; 600 x 300 mm 'blue/black' slates

roof coverings	0.50	6.72	14.53	m2	21.25
wall coverings	0.65	8.74	14.53	m2	23.27

H CLADDING/COVERING Including overheads and profit at 9.00%	Labour hours	Labour £	Material £	Unit	Total rate £
H61 FIBRE CEMENT SLATING - cont'd					
Asbestos-free artificial slates; Eternit '2000' or similar; to 75 mm lap; on 19 x 50 mm battens and type 1F reinforced underlay - cont'd Coverings; 600 x 300 mm 'brown' or 'rose nuit' slates					
roof coverings	0.50	6.72	14.53	m2	21.25
wall coverings	0.65	8.74	14.53	m2	23.27
Extra over slate coverings for					
double course at eaves	0.25	3.36	2.38	m	5.74
verges; extra single undercloak course	0.33	4.44	1.88	m	6.32
open valleys; cutting both sides	0.20	2.69	4.96	m	7.65
valley gutters; cutting both sides	0.55	7.39	16.52	m	23.91
half round ridge tiles	0.50	6.72	11.72	m	18.44
Stop end	0.10	1.34	4.04	nr	5.38
roll top ridge tiles	0.60	8.06	14.65	m	22.71
Stop end	0.10	1.34	6.06	nr	7.40
mono-pitch ridges	0.50	6.72	13.83	m	20.55
Stop end	0.10	1.34	18.37	nr	19.71
duo-pitch ridges	0.50	6.72	13.83	m	20.55
Stop end	0.10	1.34	13.47	nr	14.81
mitred hips; cutting both sides	0.20	2.69	4.96	m	7.65
half round hip tiles; cutting both sides	0.65	8.74	16.68	m	25.42
Holes for pipes and the like	0.20	2.69	-	nr	2.69

ALTERNATIVE SLATE PRICES (£/1000)

	£		£		£		£
Natural slates							
Greaves Portmadoc Welsh blue-grey							
Mediums (Class 1)							
305x255 mm	505.00	405x255 mm	750.00	510x255 mm	1210.00	610x305 mm	1914.25
355x255 mm	595.00	460x255 mm	960.00	560x305 mm	1590.00	610x355 mm	2150.00
Strongs (Class 2)							
305x255 mm	490.00	405x255 mm	700.00	510x255 mm	1155.00	610x305 mm	1885.00
355x255 mm	575.00	460x255 mm	900.00	560x305 mm	1540.00	610x355 mm	2050.00

Discounts of 2.5 - 15% available depending on quantity/status

NOTE: The following items of slate roofing unless otherwise described, include for conventional fixing assuming 'normal exposure' with appropriate nails and/or rivets or clips to pressure impregnated softwood battens fixed with galvanized nails; Prices also include for all bedding and pointing at verges; beneath ridge tiles, etc..	Labour hours	Labour £	Material £	Unit	Total rate £
H62 NATURAL SLATING					
Natural slates; BS 680 Part 2; Welsh blue; uniform size; to 75 mm lap; on 25 x 50 mm battens and type 1F reinforced underlay Coverings; 405 x 255 mm slates					
roof coverings	0.90	12.10	23.40	m2	35.50
wall coverings	1.15	15.46	23.40	m2	38.86
Coverings; 510 x 255 mm slates					
roof coverings	0.75	10.08	27.30	m2	37.38
wall coverings	0.90	12.10	27.30	m2	39.40
Coverings; 610 x 305 mm slates					
roof coverings	0.60	8.06	29.93	m2	37.99
wall coverings	0.75	10.08	29.93	m2	40.01

H CLADDING/COVERING Including overheads and profit at 9.00%	Labour hours	Labour £	Material £	Unit	Total rate £
Extra over coverings for					
double course at eaves	0.30	4.03	5.34	m	9.37
verges; extra single undercloak course	0.42	5.64	4.39	m	10.03
open valleys; cutting both sides	0.22	2.96	11.68	m	14.64
blue/black glazed ware 152 mm half round					
ridge tiles	0.50	6.72	9.37	m	16.09
blue/black glazed ware 125 x 125 mm plain					
angle ridge tiles	0.50	6.72	12.45	m	19.17
mitred hips; cutting both sides	0.22	2.96	11.68	m	14.64
blue/black glazed ware 152 mm half round					
hip tiles; cutting both sides	0.70	9.41	21.06	m	30.47
blue/black glazed ware 125 x 125 mm plain					
angle hip tiles; cutting both sides	0.70	9.41	24.14	m	33.55
Holes for pipes and the like	0.20	2.69	-	nr	2.69
Natural slates; Westmorland green;					
PC £1092.50/t; random lengths;					
457 - 229 mm proportionate widths to					
75 mm lap; in diminishing courses; on					
25 x 50 mm battens and type 1F underlay					
Roof coverings	1.15	15.46	79.76	m2	95.22
Wall coverings	1.45	19.49	79.76	m2	99.25
Extra over coverings for					
double course at eaves	0.66	8.87	13.61	m	22.48
verges; extra single undercloak course					
slates 152 mm wide	0.75	10.08	10.41	m	20.49
Holes for pipes and the like	0.30	4.03	-	nr	4.03

H63 RECONSTRUCTED STONE SLATING/TILING

	Labour hours	Labour £	Material £	Unit	Total rate £
Reconstructed stone slates; Bradstone					
'Cotswold' style or similar; PC £16.69/m2;					
random lengths 550 - 300 mm; proportional					
widths; to 80 mm lap; in diminishing courses;					
on 25 x 50 mm battens and					
type 1F reinforced underlay					
Roof coverings	1.05	14.11	24.15	m2	38.26
Wall coverings	1.35	18.14	25.47	m2	43.61
Extra over coverings for					
double course at eaves	0.50	6.72	3.91	m	10.63
verges; extra single undercloak course	0.66	8.87	3.17	m	12.04
open valleys; cutting both sides	0.45	6.05	8.19	m	14.24
ridge tile	0.66	8.87	9.58	m	18.45
mitred hips; cutting both sides	0.45	6.05	8.19	m	14.24
hip tile; cutting both sides	1.05	14.11	17.34	m	31.45
Holes for pipes and the like	0.30	4.03	-	nr	4.03
Reconstructed stone slates; Bradstone					
'Moordale' style or similar; PC £17.13/m2;					
random lengths 550 - 450 mm; proportional					
widths; to 80 mm lap; in diminishing					
courses; on 25 x 50 mm battens and					
type 1F reinforced underlay					
Roof coverings	0.95	12.77	24.19	m2	36.96
Wall coverings	1.25	16.80	25.74	m2	42.54
Extra over coverings for					
double course at eaves	0.50	6.72	4.01	m	10.73
verges; extra single undercloak course	0.66	8.87	3.24	m	12.11
ridge tile	0.66	8.87	9.43	m	18.30
mitred hips; cutting both sides	0.45	6.05	8.40	m	14.45
Holes for pipes and the like	0.30	4.03	-	nr	4.03

H CLADDING/COVERING Including overheads and profit at 9.00%	Labour hours	Labour £	Material £	Unit	Total rate £
H64 TIMBER SHINGLING					
Red cedar sawn shingles preservative treated; PC £27.12 per bundle (2.11 m2 cover); uniform length 450 mm; varying widths; to 125 mm lap; on 25 x 100 mm battens and type 1F reinforced underlay					
Roof coverings	1.05	14.11	21.76	m2	35.87
Wall coverings	1.35	18.14	21.76	m2	39.90
Extra over coverings for					
double course at eaves; three rows of battens	0.30	4.03	4.25	m	8.28
verges; extra single undercloak course	0.50	6.72	4.39	m	11.11
open valleys; cutting both sides	0.20	2.69	3.25	m	5.94
selected shingles to form hip capping	1.20	16.13	8.24	m	24.37
Double starter course on last	0.20	2.69	1.78	nr	4.47
Holes for pipes and the like	0.15	2.02	-	nr	2.02
H71 LEAD SHEET COVERINGS/FLASHINGS					
Milled lead; BS 1178; PC £1141.92/t					
1.25 mm (code 3) roof coverings					
flat	2.70	27.93	18.55	m2	46.48
sloping 10 - 50 degrees	3.00	31.03	18.55	m2	49.58
vertical or sloping over 50 degrees	3.30	34.14	18.55	m2	52.69
1.80 mm (code 4) roof coverings					
flat	2.90	30.00	26.70	m2	56.70
sloping 10 - 50 degrees	3.20	33.10	26.70	m2	59.80
vertical or sloping over 50 degrees	3.50	36.20	26.70	m2	62.90
1.80 mm (code 4) dormer coverings					
flat	3.40	35.17	27.34	m2	62.51
sloping 10 - 50 degrees	3.90	40.34	27.34	m2	67.68
vertical or sloping over 50 degrees	4.20	43.45	27.34	m2	70.79
2.24 mm (code 5) roof coverings					
flat	3.10	32.07	33.23	m2	65.30
sloping 10 - 50 degrees	3.40	35.17	33.23	m2	68.40
vertical or sloping over 50 degrees	3.70	38.27	33.23	m2	71.50
2.24 mm (code 5) dormer coverings					
flat	3.70	38.27	34.02	m2	72.29
sloping 10 - 50 degrees	4.10	42.41	34.02	m2	76.43
vertical or sloping over 50 degrees	4.50	46.55	34.02	m2	80.57
2.50 mm (code 6) roof coverings					
flat	3.30	34.14	37.10	m2	71.24
sloping 10 - 50 degrees	3.60	37.24	37.10	m2	74.34
vertical or sloping over 50 degrees	3.90	40.34	37.10	m2	77.44
2.50 mm (code 6) dormer coverings					
flat	4.00	41.38	37.98	m2	79.36
sloping 10 - 50 degrees	4.30	44.48	37.98	m2	82.46
vertical or sloping over 50 degrees	4.70	48.62	37.98	m2	86.60
Dressing over glazing bars and glass	0.33	3.41	-	m	3.41
Soldered dot	1.25	12.93	2.22	nr	15.15
Copper nailing 75 mm spacing	0.20	2.07	0.27	m	2.34
1.80 mm (code 4) lead flashings, etc.					
Flashings; wedging into grooves					
150 mm girth	0.80	8.28	4.01	m	12.29
240 mm girth	0.90	9.31	6.45	m	15.76
Stepped flashings; wedging into grooves					
180 mm girth	0.90	9.31	4.81	m	14.12
270 mm girth	1.00	10.34	7.21	m	17.55
Linings to sloping gutters					
390 mm girth	1.20	12.41	10.46	m	22.87
450 mm girth	1.30	13.45	12.02	m	25.47
750 mm girth	1.60	16.55	20.12	m	36.67
Cappings to hips or ridges					
450 mm girth	1.50	15.52	12.02	m	27.54
600 mm girth	1.60	16.55	16.06	m	32.61

H CLADDING/COVERING Including overheads and profit at 9.00%	Labour hours	Labour £	Material £	Unit	Total rate £
Soakers					
200 x 200 mm	0.15	1.55	1.21	nr	2.76
300 x 300 mm	0.20	2.07	2.72	nr	4.79
Saddle flashings; at intersections of hips and ridges; dressing and bossing					
450 x 600 mm	1.80	18.62	8.15	nr	26.77
Slates; with 150 mm high collar					
450 x 450 mm; to suit 50 mm pipe	1.70	17.58	11.45	nr	29.03
450 x 450 mm; to suit 100 mm pipe	2.00	20.69	16.31	nr	37.00
2.24 mm (code 5) lead flashings, etc.					
Flashings; wedging into grooves					
150 mm girth	0.80	8.28	5.00	m	13.28
240 mm girth	0.90	9.31	7.96	m	17.27
Stepped flashings; wedging into grooves					
180 mm girth	0.90	9.31	6.00	m	15.31
270 mm girth	1.00	10.34	8.98	m	19.32
Linings to sloping gutters					
390 mm girth	1.20	12.41	13.00	m	25.41
450 mm girth	1.30	13.45	14.97	m	28.42
750 mm girth	1.60	16.55	25.15	m	41.70
Cappings to hips or ridges					
450 mm girth	1.50	15.52	14.97	m	30.49
600 mm girth	1.60	16.55	20.00	m	36.55
Soakers					
200 x 200 mm	0.15	1.55	1.51	nr	3.06
300 x 300 mm	0.20	2.07	3.33	nr	5.40
Saddle flashings; at intersections of hips and ridges; dressing and bossing					
450 x 600 mm	1.80	18.62	11.31	nr	29.93
Slates; with 150 mm high collar					
450 x 450 mm; to suit 50 mm pipe	1.70	17.58	12.94	nr	30.52
450 x 450 mm; to suit 100 mm pipe	2.00	20.69	17.80	nr	38.49
H72 ALUMINIUM SHEET COVERINGS/FLASHINGS					
Aluminium roofing; commercial grade; PC £2000.00/t					
0.90 mm roof coverings					
flat	3.00	31.03	5.52	m2	36.55
sloping 10 - 50 degrees	3.30	34.14	5.52	m2	39.66
vertical or sloping over 50 degrees	3.60	37.24	5.52	m2	42.76
0.90 mm dormer coverings					
flat	3.60	37.24	5.65	m2	42.89
sloping 10 - 50 degrees	4.00	41.38	5.65	m2	47.03
vertical or sloping over 50 degrees	4.40	45.51	5.65	m2	51.16
Aluminium nailing; 75 mm spacing	0.20	2.07	0.17	m	2.24
0.90 mm commercial grade aluminium flashings, etc.					
Flashings; wedging into grooves					
150 mm girth	0.80	8.28	0.86	m	9.14
240 mm girth	0.90	9.31	1.40	m	10.71
300 mm girth	1.05	10.86	1.72	m	12.58
Stepped flashings; wedging into grooves					
180 mm girth	0.90	9.31	1.04	m	10.35
270 mm girth	1.00	10.34	1.58	m	11.92
H73 COPPER SHEET COVERINGS/FLASHINGS					
Copper roofing; BS 2870					
0.56 mm (24 swg) roof coverings					
flat	3.20	33.10	18.95	m2	52.05
sloping 10 - 50 degrees	3.50	36.20	18.95	m2	55.15
vertical or sloping over 50 degrees	3.80	39.31	18.95	m2	58.26
0.56 mm (24 swg) dormer coverings					
flat	3.80	39.31	19.40	m2	58.71
sloping 10 - 50 degrees	4.20	43.45	19.40	m2	62.85
vertical or sloping over 50 degrees	4.60	47.58	19.40	m2	66.98

H CLADDING/COVERING Including overheads and profit at 9.00%	Labour hours	Labour £	Material £	Unit	Total rate £
H73 COPPER SHEET COVERINGS/FLASHINGS - cont'd					
Copper roofing; BS 2870 - cont'd					
0.61 mm (23 swg) roof coverings					
flat PC £3300.00/t	3.20	33.10	21.91	m2	55.01
sloping 10 - 50 degrees	3.50	36.20	21.91	m2	58.11
vertical or sloping over 50 degrees	3.80	39.31	21.91	m2	61.22
0.61 mm (23 swg) dormer coverings					
flat	3.80	39.31	22.43	m2	61.74
sloping 10 - 50 degrees	4.20	43.45	22.43	m2	65.88
vertical or sloping over 50 degrees	4.60	47.58	22.43	m2	70.01
Copper nailing; 75 mm spacing	0.20	2.07	0.27	m	2.34
0.56 mm copper flashings, etc.					
Flashings; wedging into grooves					
150 mm girth	0.80	8.28	2.98	m	11.26
240 mm girth	0.90	9.31	4.77	m	14.08
300 mm girth	1.05	10.86	5.96	m	16.82
Stepped flashings; wedging into grooves					
180 mm girth	0.90	9.31	3.59	m	12.90
270 mm girth	1.00	10.34	5.38	m	15.72
0.61 mm copper flashings, etc.					
Flashings; wedging into grooves					
150 mm girth	0.80	8.28	3.45	m	11.73
240 mm girth	0.90	9.31	5.54	m	14.85
300 mm girth	1.05	10.86	6.88	m	17.74
Stepped flashings; wedging into grooves					
180 mm girth	0.90	9.31	4.16	m	13.47
270 mm girth	1.00	10.34	6.21	m	16.55
H74 ZINC SHEET COVERINGS/FLASHINGS					
Zinc BS 849; PC £2000.00/t					
0.81 mm roof coverings					
flat	3.20	33.10	13.03	m2	46.13
sloping 10 - 50 degrees	3.50	36.20	13.03	m2	49.23
vertical or sloping over 50 degrees	3.80	39.31	13.03	m2	52.34
0.81 mm dormer coverings					
flat	3.80	39.31	13.33	m2	52.64
sloping 10 - 50 degrees	4.20	43.45	13.33	m2	56.78
vertical or sloping over 50 degrees	4.60	47.58	13.33	m2	60.91
0.81 mm zinc flashings, etc.					
Flashings; wedging into grooves					
150 mm girth	0.80	8.28	2.04	m	10.32
240 mm girth	0.90	9.31	3.28	m	12.59
300 mm girth	1.05	10.86	4.10	m	14.96
Stepped flashings; wedging into grooves					
180 mm girth	0.90	9.31	2.47	m	11.78
270 mm girth	1.00	10.34	3.70	m	14.04
H75 STAINLESS STEEL SHEET COVERINGS/ FLASHINGS					
Terne coated stainless steel roofing					
0.38 mm roof coverings					
flat	3.20	33.10	29.19	m2	62.29
sloping 10 - 50 degrees	3.50	36.20	29.19	m2	65.39
vertical or sloping over 50 degrees	3.80	39.31	29.19	m2	68.50
Flashings; wedging into grooves					
150 mm girth	1.00	10.34	6.11	m	16.45
240 mm girth	1.15	11.90	8.87	m	20.77
300 mm girth	1.30	13.45	10.70	m	24.15
Stepped flashings; wedging into grooves					
180 mm girth	1.15	11.90	7.04	m	18.94
270 mm girth	1.25	12.93	9.79	m	22.72

H CLADDING/COVERING Including overheads and profit at 9.00%	Labour hours	Labour £	Material £	Unit	Total rate £
H76 FIBRE BITUMEN THERMOPLASTIC SHEET **COVERINGS/FLASHINGS**					
Glass fibre reinforced bitumen strip slates; **'Langhome 1000' or similar; PC £19.11/2 m2;** **strip pack; 900 x 300 mm mineral finish;** **fixed to external plywood boarding** **(measured separately)**					
Roof coverings	0.25	3.36	9.15	m2	12.51
Wall coverings	0.40	5.38	9.15	m2	14.53
Extra over coverings for					
double course at eaves; felt soaker	0.20	2.69	1.76	m	4.45
verges; felt soaker	0.25	3.36	1.67	m	5.03
valley slate; cut to shape; felt soaker					
both sides; cutting both sides	0.45	6.05	4.33	m	10.38
ridge slate; cut to shape	0.30	4.03	2.24	m	6.27
hip slate; cut to shape; cutting both sides	0.45	6.05	4.29	m	10.34
Holes for pipes and the like	0.05	0.67	-	nr	0.67
'Evode Flashband' sealing strips and **flashings; special grey finish** Flashings; wedging at top if required; pressure bonded; flashband primer before application; to walls					
100 mm girth	0.25	2.59	1.03	m	3.62
150 mm girth	0.33	3.41	1.54	m	4.95
225 mm girth	0.40	4.14	2.34	m	6.48
300 mm girth	0.45	4.65	3.03	m	7.68
450 mm girth	0.60	6.21	5.89	m	12.10
'Nuralite' semi-rigid bitumen membrane **roofing DC12 jointing strip system** **PC £21.41/3m2**					
Roof coverings					
flat	0.50	5.17	11.34	m2	16.51
sloping 10 - 50 degrees	0.55	5.69	11.34	m2	17.03
vertical or sloping over 50 degrees	0.70	7.24	11.34	m2	18.58
'Nuralite' flashings, etc.					
Flashings; wedging into grooves					
150 mm girth	0.50	5.17	1.44	m	6.61
200 mm girth	0.60	6.21	1.99	m	8.20
250 mm girth	0.60	6.21	2.26	m	8.47
Linings to sloping gutters					
450 mm girth	0.85	8.79	4.90	m	13.69
490 mm girth	0.95	9.83	5.40	m	15.23
Cappings to hips or ridges					
450 mm girth	0.90	9.31	7.24	m	16.55
600 mm girth	1.00	10.34	10.89	m	21.23
Undersoakers					
150 mm girth	0.12	1.24	0.74	nr	1.98
250 mm girth	0.18	1.86	0.77	nr	2.63
Oversoakers					
200 mm girth	0.15	1.55	7.47	nr	9.02
Cavity tray without apron					
450 mm long	0.45	4.65	1.39	nr	6.04
250 mm long	0.25	2.59	1.08	nr	3.67
350 mm long	0.35	3.62	1.30	nr	4.92

J WATERPROOFING Including overheads and profit at 5.00%	Labour hours	Labour £	Material £	Unit	Total rate £
J10 SPECIALIST WATERPROOF RENDERING					
'Sika' waterproof rendering; steel trowelled					
20 mm work to walls; three coat; to concrete base					
over 300 mm wide	-	-	-	m2	32.70
not exceeding 300 mm wide	-	-	-	m2	51.23
25 mm work to walls; three coat; to concrete base					
over 300 mm wide	-	-	-	m2	37.06
not exceeding 300 mm wide	-	-	-	m2	58.86
40 mm work to walls; four coat; to concrete base					
over 300 mm wide	-	-	· -	m2	56.68
not exceeding 300 mm wide	-	-	-	m2	87.20
J20 MASTIC ASPHALT TANKING/DAMP PROOF MEMBRANES					
Mastic asphalt to BS 1097					
13 mm one coat coverings to concrete base; flat; subsequently covered					
over 300 mm wide	-	-	-	m2	6.43
225 - 300 mm wide	-	-	-	m2	12.20
150 - 225 mm wide	-	-	-	m2	12.37
not exceeding 150 mm wide	-	-	-	m2	15.26
20 mm two coat coverings to concrete base; flat; subsequently covered					
over 300 mm wide	-	-	-	m2	8.97
225 - 300 mm wide	-	-	-	m2	18.77
150 - 300 mm wide	-	-	-	m2	19.99
not exceeding 150 mm wide	-	-	-	m2	22.43
30 mm three coat coverings to concrete base; flat; subsequently covered					
over 300 mm wide	-	-	-	m2	12.84
225 - 300 mm wide	-	-	-	m2	28.15
150 - 225 mm wide	-	-	-	m2	29.99
not exceeding 150 mm wide	-	-	-	m2	33.65
13 mm two coat coverings to brickwork base; vertical; subsequently covered					
over 300 mm wide	-	-	-	m2	27.84
225 - 300 mm wide	-	-	-	m2	22.20
150 - 225 mm wide	-	-	-	m2	24.57
not exceeding 150 mm wide	-	-	-	m2	29.30
20 mm three coat coverings to brickwork base; vertical; subsequently covered					
over 300 mm wide	-	-	-	m2	38.03
225 - 300 mm wide	-	-	-	m2	33.31
150 - 225 mm wide	-	-	-	m2	36.85
not exceeding 150 mm wide	-	-	-	m2	43.95
Turning 20 mm into groove	-	-	-	m	0.55
Internal angle fillets; subsequently covered	-	-	-	m	3.61
Mastic asphalt to BS 6577					
13 mm one coat coverings to concrete base; flat; subsequently covered					
over 300 mm wide	-	-	-	m2	8.71
225 - 300 mm wide	-	-	-	m2	14.08
150 - 225 mm wide	-	-	-	m2	15.01
not exceeding 150 mm wide	-	-	-	m2	16.86
20 mm two coat coverings to concrete base; flat; subsequently covered					
over 300 mm wide	-	-	-	m2	12.88
225 - 300 mm wide	-	-	-	m2	21.67
150 - 225 mm wide	-	-	-	m2	23.10
not exceeding 150 mm wide	-	-	-	m2	25.94

J WATERPROOFING Including overheads and profit at 5.00%	Labour hours	Labour £	Material £	Unit	Total rate £
30 mm three coat coverings to concrete base; flat; subsequently covered					
over 300 mm wide	-	-	-	m2	18.72
225 - 300 mm wide	-	-	-	m2	32.51
150 - 225 mm wide	-	-	-	m2	34.65
not exceeding 150 mm wide	-	-	-	m2	38.91
13 mm two coat coverings to brickwork base; vertical; subsequently covered					
over 300 mm wide	-	-	-	m2	30.39
225 - 300 mm wide	-	-	-	m2	24.54
150 - 225 mm wide	-	-	-	m2	26.91
not exceeding 150 mm wide	-	-	-	m2	31.66
20 mm three coat coverings to brickwork base; vertical; subsequently covered					
over 300 mm wide	-	-	-	m2	41.94
225 - 300 mm wide	-	-	-	m2	36.81
150 - 225 mm wide	-	-	-	m2	40.37
not exceeding 150 mm wide	-	-	-	m2	47.50
Turning 20 mm into groove	-	-	-	m	0.55
Internal angle fillets; subsequently covered	-	-	-	m	4.03

J21 MASTIC ASPHALT ROOFING/INSULATION/ FINISHES

Mastic asphalt to BS 988					
20 mm two coat coverings; felt isolating membrane; to concrete (or timber) base; flat or to falls or slopes not exceeding 10 degrees from horizontal					
over 300 mm wide	-	-	-	m2	9.52
225 - 300 mm wide	-	-	-	m2	18.77
150 - 225 mm wide	-	-	-	m2	19.99
not exceeding 150 mm wide	-	-	-	m2	22.45
Add to the above for covering with:					
10 mm limestone chippings in hot bitumen	-	-	-	m2	2.18
coverings with solar reflective paint	-	-	-	m2	2.13
300 x 300 x 8 mm g.r.p. tiles in hot bitumen	-	-	-	m2	32.20
Cutting to line; jointing to old asphalt	-	-	-	m	3.61
13 mm two coat skirtings to brickwork base					
not exceeding 150 mm girth	-	-	-	m	8.00
150 - 225 mm girth	-	-	-	m	9.13
225 - 300 mm girth	-	-	-	m	10.27
13 mm three coat skirtings; expanded metal lathing reinforcement nailed to timber base					
not exceeding 150 mm girth	-	-	-	m	12.58
150 - 225 mm girth	-	-	-	m	14.94
225 - 300 mm girth	-	-	-	m	17.32
13 mm two coat fascias to concrete base					
not exceeding 150 mm girth	-	-	-	m	11.37
150 - 225 mm girth	-	-	-	m	12.50
20 mm two coat linings to channels to concrete base					
not exceeding 150 mm girth	-	-	-	m	19.37
150 - 225 mm girth	-	-	-	m	20.50
225 - 300 mm girth	-	-	-	m	21.64
20 mm two coat lining to cesspools					
250 x 150 x 150 mm deep	-	-	-	nr	27.25
Collars around pipes, standards and like members	-	-	-	nr	9.27

J WATERPROOFING Including overheads and profit at 9.00% & 5.00%	Labour hours	Labour £	Material £	Unit	Total rate £
J21 MASTIC ASPHALT ROOFING/INSULATION/ **FINISHES - cont'd**					
Mastic asphalt to BS 6577 20 mm two coat coverings; felt isolating membrane; to concrete (or timber) base; flat or to falls or slopes not exceeding 10 degrees from horizontal					
over 300 mm wide	-	-	-	m2	12.84
225 - 300 mm wide	-	-	-	m2	21.67
150 - 225 mm wide	-	-	-	m2	23.10
not exceeding 150 mm wide	-	-	-	m2	25.94
Add to the above for covering with:					
10 mm limestone chippings in hot bitumen	-	-	-	m2	2.18
solar reflective paint	-	-	-	m2	2.13
300 x 300 x 8 mm g.r.p. tiles in hot bitumen	-	-	-	m2	32.20
Cutting to line; jointing to old asphalt	-	-	-	m	3.61
13 mm two coat skirtings to brickwork base					
not exceeding 150 mm girth	-	-	-	m	8.77
150 - 225 mm girth	-	-	-	m	10.08
225 - 300 mm girth	-	-	-	m	11.38
13 mm three coat skirtings; expanded metal lathing reinforcement nailed to timber base					
not exceeding 150 mm girth	-	-	-	m	13.54
150 - 225 mm girth	-	-	-	m	16.14
225 - 300 mm girth	-	-	-	m	18.76
13 mm two coat fascias to concrete base					
not exceeding 150 mm girth	-	-	-	m	12.67
150 - 225 mm girth	-	-	-	m	13.97
20 mm two coat linings to channels to concrete base					
not exceeding 150 mm girth	-	-	-	m	21.45
150 - 225 mm girth	-	-	-	m	22.76
225 - 300 mm girth	-	-	-	m	24.06
20 mm two coat lining to cesspools					
250 x 150 x 150 mm deep	-	-	-	nr	27.25
Collars around pipes, standards and like members	-	-	-	nr	9.27
Accessories Eaves trim; extruded aluminium alloy; working asphalt into trim					
'Alutrim'; type A roof edging PC £8.26/2.5m	0.40	5.38	3.85	m	9.23
Angle PC £2.85	0.12	1.61	3.26	nr	4.87
Roof screed ventilator - aluminium alloy Extr-aqua-vent; set on screed over and including dished sinking; working collar					
around ventilator PC £3.71	1.10	14.78	4.25	nr	19.03
J30 LIQUID APPLIED TANKING/DAMP PROOF **MEMBRANES**					
'Synthaprufe'; blinding with sand; horizontal on slabs					
two coats	0.20	1.40	1.82	m2	3.22
three coats	0.28	1.96	2.65	m2	4.61
'Tretolastex 202T'; on vertical surfaces of concrete					
two coats	0.20	1.40	1.59	m2	2.99
three coats	0.28	1.96	2.39	m2	4.35
One coat Vandex 'Super' 0.75/m2 slurry; one consolidating coat of Vandex 'Premix' 1kg/m2 slurry; horizontal on beds					
over 225 mm wide	-	-	-	m2	5.36
'Ventrot' hot applied damp proof membrane; one coat; horizontal on slabs					
over 225 mm wide	-	-	-	m2	3.79

J WATERPROOFING Including overheads and profit at 9.00% & 5.00%	Labour hours	Labour £	Material £	Unit	Total rate £
J40 FLEXIBLE SHEET TANKING/DAMP PROOF MEMBRANES					
'Bituthene' sheeting; lapped joints; horizontal on slabs					
standard 500 grade	0.10	0.70	4.07	m2	4.77
1000 grade	0.11	0.77	4.53	m2	5.30
1200 grade	0.12	0.84	6.20	m2	7.04
heavy duty grade	0.13	0.91	5.14	m2	6.05
'Bituthene' sheeting; lapped joints; dressed up vertical face of concrete					
1000 grade	0.18	1.26	5.04	m2	6.30
'Kork-pak'; 9.5 mm thick joint filler; set vertically against 'Bituthene' as protection					
over 225 mm wide	0.60	4.20	10.57	m2	14.77
'Bituthene' fillet					
40 x 40 mm	0.10	0.70	3.87	m	4.57
'Bituthene' reinforcing strip; 300 mm wide					
1000 grade	0.10	0.70	1.57	m	2.27
Expandite 'Famflex' waterproof tanking; 150 mm laps					
horizontal; over 300 mm wide	0.40	2.80	6.28	m2	9.08
vertical; over 300 mm wide	0.65	4.55	6.28	m2	10.83

J41 BUILT UP FELT ROOF COVERINGS

NOTE: The following items of felt roofing, unless otherwise described, include for conventional lapping, laying and bonding between layers and to base; and laying flat or to falls to cross-falls or to slopes not exceeding 10 degrees - but exclude any insulation etc. (measured separately)

Felt roofing; BS 747; suitable for flat roofs Three layer coverings type 1B (two 18 kg/10 m2 and one 25 kg/10 m2)					
bitumen fibre based felts	-	-	-	m2	10.68
Extra over top layer type 1B for mineral surfaced layer type 1E	-	-	-	m2	1.41
Three layer coverings type 2B bitumen asbestos based felts	-	-	-	m2	14.02
Extra over top layer type 2B for mineral surfaced layer type 2E	-	-	-	m2	1.77
Three layer coverings first layer type 3G; subsequent layers type 3B bitumen glass fibre based felt	-	-	-	m2	11.26
Extra over top layer type 3B for mineral surfaced layer type 3E	-	-	-	m2	1.58

J WATERPROOFING Including overheads and profit at 9.00% & 5.00%	Labour hours	Labour £	Material £	Unit	Total rate £
J41 BUILT UP FELT ROOF COVERINGS - cont'd					
Felt roofing; BS 747; suitable for flat roofs - cont'd					
Extra over felt for covering with and					
bedding in hot bitumen					
13 mm granite chippings	-	-	-	m2	4.29
300 x 300 x 8 mm asbestos tiles	-	-	-	m2	40.08
Working into outlet pipes and the like	-	-	-	nr	7.22
Skirtings; three layer; top layer mineral					
surfaced; dressed over tilting fillet;					
turned into groove					
not exceeding 200 mm girth	-	-	-	m	6.77
200 - 400 mm girth	-	-	-	m	8.27
Coverings to kerbs; three layer					
400 - 600 mm girth	-	-	-	m	10.59
Linings to gutters; three layer					
400 - 600 mm (average) girth	-	-	-	m	18.26
Collars around pipes and the like; three					
layer mineral surfaced; 150 mm high					
not exceeding 55 mm nominal size	-	-	-	nr	5.18
55 - 110 mm nominal size	-	-	-	nr	6.46
Felt roofing; BS 747; suitable for pitched					
timber roofs; sloping not exceeding					
50 degrees					
Two layer coverings; first layer type 2B;					
second layer 2E; mineral surfaced asbestos					
based felts	-	-	-	m2	14.91
Three layer coverings; first two layers type					
2B; top layer type 2E; mineral surfaced					
asbestos based felts	-	-	-	m2	21.06
'Andersons' high performance polyester-based					
roofing system					
Two layer coverings; first layer HT 125					
underlay; second layer HT 350; fully bonded					
to wood; fibre or cork base	0.30	4.03	9.16	m2	13.19
Extra over for					
Top layer mineral surfaced	-	-	-	m2	1.47
13 mm granite chippings	-	-	-	m2	4.29
Third layer of type 3B as underlay for					
concrete or screeded base	0.15	2.02	1.55	m2	3.57
Working into outlet pipes and the like	0.50	6.72	-	nr	6.72
Welted drip; two layer	0.20	2.69	0.80	m	3.49
Skirtings; two layer; top layer mineral					
surfaced; dressed over tilting fillet;					
turned into groove					
not exceeding 200 mm girth	0.15	2.02	2.44	m	4.46
200 - 400 mm girth	0.20	2.69	4.63	m	7.32
Coverings to kerbs; two layer					
400 - 600 mm girth	0.25	3.36	6.34	m	9.70
Linings to gutters; three layer					
400 - 600 mm (average) girth	0.60	8.06	7.88	m	15.94
Collars around pipes and the like; two					
layer; 150 mm high					
not exceeding 55 mm nominal size	0.33	4.44	0.38	nr	4.82
55 - 110 mm nominal size	0.40	5.38	0.63	nr	6.01
'Ruberglas 120 GP' high performance roofing					
Two layer coverings; first and second layers					
'Ruberglas 120 GP'; fully bonded to wood;					
fibre or cork base	-	-	-	m2	9.74
Extra over for					
Top layer mineral surfaced	-	-	-	m2	1.58
13 mm granite chippings	-	-	-	m2	4.29
Third layer of 'Rubervent 3G' as					
underlay for concrete or screeded base	-	-	-	m2	4.10
Working into outlet pipes and the like	-	-	-	nr	7.22

J WATERPROOFING Including overheads and profit at 5.00%	Labour hours	Labour £	Material £	Unit	Total rate £
Welted drip; two layer	-	-	-	m	4.10
Skirtings; two layer; top layer mineral surfaced; dressed over tilting fillet; turned into groove					
not exceeding 200 mm girth	-	-	-	m	5.92
200 - 400 mm girth	-	-	-	m	7.96
Coverings to kerbs; two layer					
400 - 600 mm girth	-	-	-	m	12.75
Linings to gutters; three layer					
400 - 600 mm (average) girth	-	-	-	m	20.14
Collars around pipes and the like; two layer; 150 mm high					
not exceeding 55 mm nominal size	-	-	-	nr	5.17
55 - 110 mm nominal size	-	-	-	nr	6.56
'Ruberfort HP 350' high performance roofing					
Two layer coverings; first layer 'Ruberfort HP 180'; second layer 'Ruberfort HP 350'; fully bonded; to wood; fibre or cork base	-	-	-	m2	14.77
Extra over for					
Top layer mineral surfaced	-	-	-	m2	1.53
13 mm granite chippings	-	-	-	m2	4.29
Third layer of 'Rubervent 3G' as underlay for concrete or screeded base	-	-	-	m2	4.10
Working into outlet pipes and the like	-	-	-	nr	7.22
Welted drip; two layer	-	-	-	m	5.35
Skirtings; two layer; top layer mineral surfaced; dressed over tilting fillet; turned into groove					
not exceeding 200 mm girth	-	-	-	m	7.18
200 - 400 mm girth	-	-	-	m	10.43
Coverings to kerbs; two layer					
400 - 600 mm girth	-	-	-	m	16.93
Linings to gutters; three layer					
400 - 600 mm (average) girth	-	-	-	m	26.98
Collars around pipes and the like; two layer; 150 mm high					
not exceeding 55 mm nominal size	-	-	-	nr	5.17
55 - 110 mm nominal size	-	-	-	nr	6.59
'Polybit 350' elastomeric roofing					
Two layer coverings; first layer 'Polybit 180'; second layer 'Polybit 350'; fully bonded to wood; fibre or cork base	-	-	-	m2	16.32
Extra over for					
Top layer mineral surfaced	-	-	-	m2	1.97
13 mm granite chippings	-	-	-	m2	4.29
Third layer of 'Rubervent 3G' as underlay for concrete or screeded base	-	-	-	m2	4.10
Working into outlet pipes and the like	-	-	-	nr	7.22
Welted drip; two layer	-	-	-	m	6.02
Skirtings; two layer; top layer mineral surfaced; dressed over tilting fillet; turned into groove					
not exceeding 200 mm girth	-	-	-	m	7.51
200 - 400 mm girth	-	-	-	m	11.03
Coverings to kerbs; two layer					
400 - 600 mm girth	-	-	-	m	18.14
Linings to gutters; three layer					
400 - 600 mm (average) girth	-	-	-	m	28.93
Collars around pipes and the like; two layer; 150 mm high					
not exceeding 55 mm nominal size	-	-	-	nr	6.46
55 - 110 mm nominal size	-	-	-	nr	9.81

J WATERPROOFING Including overheads and profit at 5.00%	Labour hours	Labour £	Material £	Unit	Total rate £
J41 BUILT UP FELT ROOF COVERINGS - cont'd					
'Hyload 150 E' elastomeric roofing					
Two layer coverings; first layer 'Ruberglas					
120 GP'; second layer 'Hyload 150 E' fully					
bonded to wood; fibre or cork base	-	-	-	m2	15.64
Extra over for					
13 mm granite chippings	-	-	-	m2	4.29
Third layer of 'Rubervent 3G' as					
underlay for concrete or screeded base	-	-	-	m2	4.10
Working into outlet pipes and the like	-	-	-	nr	7.22
Welted drip; two layer	-	-	-	m	6.14
Skirtings; two layer; dressed over tilting					
fillet; turned into groove					
not exceeding 200 mm girth	-	-	-	m	7.07
200 - 400 mm girth	-	-	-	m	10.25
Coverings to kerbs; two layer					
400 - 600 mm girth	-	-	-	m	16.70
Linings to gutters; three layer					
400 - 600 mm (average) girth	-	-	-	m	26.11
Collars around pipes and the like; two					
layer; 150 mm high					
not exceeding 55 mm nominal size	-	-	-	nr	5.17
55 - 110 mm nominal size	-	-	-	nr	6.46
Felt; 'Paradiene' elastomeric bitumen					
roofing; perforated crepe paper isolating					
membrane; first layer 'Paradiene 20';					
second layer 'Paradiene 30' pre-finished					
surface					
Two layer coverings	-	-	-	m2	16.35
Working into outlet pipes and the like	-	-	-	nr	7.09
Welted drip; three layer	-	-	-	m	5.41
Skirtings; three layer; dressed over tilting					
fillet; turned into groove					
not exceeding 200 mm girth	-	-	-	m	4.90
200 - 400 mm girth	-	-	-	m	8.61
Coverings to kerbs; two layer					
400 - 600 mm girth	-	-	-	m	10.51
Linings to gutters; three layer					
400 - 600 mm (average) girth	-	-	-	m	17.14
Collars around pipes and the like; three					
layer; 150 mm high					
not exceeding 55 mm nominal size	-	-	-	nr	4.91
55 - 110 mm nominal size	-	-	-	nr	5.45
Metal faced 'Veral' glass cloth reinforced					
bitumen roofing; first layer 'Veralvent'					
perforated underlay; second layer					
'Veralglas'; third layer 'Veral' natural					
slate aluminium surfaced					
Three layer coverings	-	-	-	m2	22.24
Working into outlet pipes and the like	-	-	-	nr	7.09
Welted drip; three layer	-	-	-	m	5.31
Skirtings; three layer; dressed over tilting					
fillet; turned into groove					
not exceeding 200 mm girth	-	-	-	m	4.47
200 - 400 mm girth	-	-	-	m	7.68
Coverings to kerbs; three layer					
400 - 600 mm girth	-	-	-	m	11.17
Linings to gutters; three layer					
400 - 600 mm (average) girth	-	-	-	m	15.07
Collars around pipes and the like; three					
layer; 150 mm high					
not exceeding 55 mm nominal size	-	-	-	nr	4.91
55 - 110 mm nominal size	-	-	-	nr	5.45

J WATERPROOFING Including overheads and profit at 9.00% & 5.00%	Labour hours	Labour £	Material £	Unit	Total rate £
Accessories					
Eaves trim; extruded aluminium alloy;					
working felt into trim					
'Alutrim' type F roof edging PC £7.77/2.5m	0.25	3.36	3.63	m	**6.99**
Angle PC £2.85	0.12	1.61	3.26	nr	**4.87**
Roof screed ventilator - aluminium alloy					
'Extr-aqua-vent'; set on screed over and					
including dished sinking; working collar					
around ventilator PC £3.71	0.60	8.06	4.25	nr	**12.31**
Insulation board underlays					
Vapour barrier					
reinforced; metal lined	0.03	0.40	3.32	m2	**3.72**
Cork boards; density 112 - 125 kg/m3					
60 mm thick	0.30	4.03	4.81	m2	**8.84**
Foamed glass boards; density 125 - 135 kg/m2					
60 mm thick	0.30	4.03	12.17	m2	**16.20**
Glass fibre boards; density 120 - 130 kg/m2					
60 mm thick	0.30	4.03	7.55	m2	**11.58**
Perlite boards; density 170 - 180 kg/m3					
60 mm thick	0.30	4.03	7.82	m2	**11.85**
Polyurethene boards; density 32 kg/m3					
30 mm thick	0.20	2.69	3.89	m2	**6.58**
35 mm thick	0.20	2.69	4.38	m2	**7.07**
50 mm thick	0.30	4.03	5.70	m2	**9.73**
Wood fibre boards; impregnated; density					
220 - 350 kg/m3					
12.7 mm thick	0.20	2.69	1.82	m2	**4.51**
Insulation board overlays					
Dow 'Roofmate SL' extruded polystyrene					
foam boards					
50 mm thick	0.30	4.03	7.42	m2	**11.45**
75 mm thick	0.30	4.03	10.94	m2	**14.97**
Dow 'Roofmate LG' extruded polystyrene					
foam boards					
50 mm thick	0.30	4.03	13.39	m2	**17.42**
75 mm thick	0.30	4.03	15.90	m2	**19.93**
100 mm thick	0.30	4.03	18.81	m2	**22.84**
J42 SINGLE LAYER PLASTICS ROOF COVERINGS					
Felt; 'Derbigum' special polyester 4 mm					
roofing; first layer 'Ventilag' (partial					
bond) underlay; second layer 'Derbigum SF';					
glass reinforced weathering surface					
Two layer coverings	-	-	-	m2	**19.94**
Welted drip; two layer	-	-	-	m	**6.38**
Skirtings; two layer; dressed over tilting					
fillet; turned into groove					
not exceeding 200 mm girth	-	-	-	m	**9.51**
200 - 400 mm girth	-	-	-	m	**12.38**
Coverings to kerbs; two layer					
400 - 600 mm girth	-	-	-	m	**21.15**
Collars around pipes and the like; two					
layer; 150 mm high					
not exceeding 55 mm nominal size	-	-	-	nr	**6.46**
55 - 110 mm nominal size	-	-	-	nr	**9.81**

J WATERPROOFING Including overheads and profit at 5.00%	Labour hours	Labour £	Material £	Unit	Total rate £
J43 PROPRIETARY ROOF DECKING WITH FELT FINISH					
'Bitumetal' flat roof construction fixing to timber, steel or concrete; flat or sloping; vapour check; 32 mm polyurethane insulation; 3G perforated felt underlay; two layers of glass fibre base felt roofing; stone chipping finish					
0.7 mm galvanized steel					
35 mm profiled decking; 2.38 m span	-	-	-	m2	**29.63**
46 mm profiled decking; 2.96 m span	-	-	-	m2	**30.52**
60 mm profiled decking; 3.74 m span	-	-	-	m2	**31.07**
100 mm profiled decking; 5.13 m span	-	-	-	m2	**32.57**
0.9 mm aluminium; mill finish					
35 mm profiled decking; 1.79 m span	-	-	-	m2	**37.15**
60 mm profiled decking; 2.34 m span	-	-	-	m2	**37.54**
'Bitumetal' flat roof construction fixing to timber, steel or concrete; flat or sloping; vapour check; 32 mm polyurethane insulation; 3G perforated felt underlay; two layers of polyester based roofing; stone chipping finish					
0.7 mm galvanized steel					
35 mm profiled decking; 2.38 m span	-	-	-	m2	**34.92**
46 mm profiled decking; 2.96 m span	-	-	-	m2	**35.92**
60 mm profiled decking; 3.74 m span	-	-	-	m2	**36.23**
100 mm profiled decking; 5.13 m span	-	-	-	m2	**37.74**
0.9 mm aluminium; mill finish					
35 mm profiled decking; 1.79 m span	-	-	-	m2	**43.29**
60 mm profiled decking; 2.34 m span	-	-	-	m2	**43.69**
'Plannja' flat roof construction; fixing to timber; steel or concrete; flat or sloping; 3B vapour check; 32 mm polyurethene insulation; 3G perforated felt underlay; two layers of glass fibre bitumen felt roofing type 3B; stone chipping finish					
0.72 mm galvanised steel					
45 mm profiles decking; 3.12 m span	-	-	-	m2	**31.89**
70 mm profiled decking; 4.40 m span	-	-	-	m2	**33.35**
'Plannja' flat roof construction; fixing to timber; steel or concrete; flat or sloping; 3B vapour check; 32 mm polyurethene insulation; 3G perforated underlay; one layer of Anderson's HT 125 underlay and HT 350 sanded; stone chipping finish					
0.72 mm galvanised steel					
45 mm profiles decking; 3.12 m span	-	-	-	m2	**35.98**
70 mm profiled decking; 4.40 m span	-	-	-	m2	**37.44**

K LININGS/SHEATHING/DRY PARTITIONING
Including overheads and profit at 9.00%

ALTERNATIVE SHEET LINING MATERIAL PRICES

	£		£			£		£
Asbestos cement flat sheets (£/10 m2)								
Fully compressed								
4.5 mm	49.33	6 mm	61.25	9 mm		92.66	12 mm	115.34
Semi compressed								
4.5 mm	25.06	6 mm	34.12	9 mm		50.79	12 mm	84.62

Blockboard

	£		£			£		£
Gaboon faced (£/10 m2)								
16 mm	79.75	18 mm	80.10	22 mm		90.60	25 mm	101.80
	£				£			£
18 mm Decorative faced (£/10 m2)								
Ash	123.25	Mahogany			93.67	Teak		100.00
Beech	105.51	Oak			93.67			
Edgings; self adhesive (£/25 m roll)								
19 mm Mahogany	2.14	19 mm Oak			3.00	19 mm Ash		3.00
25 mm Mahogany	3.14	25 mm Oak			3.57	19 mm Teak		3.00

	£		£			£		£
Chipboard (£/10 m2)								
Standard grade								
3.2 mm	7.95	9 mm	17.95	16 mm		24.25	22 mm	33.30
4 mm	10.05	12 mm	19.50	18 mm		26.50	25 mm	37.80
6 mm	14.80							
Melamine faced								
12 mm	33.90	18 mm	39.90					

Laminboard; Birch faced (£/10 m2)								
16 mm	107.00	18 mm	111.70	22 mm		130.80	25 mm	145.45

Medium density fibreboard (£/10 m2)								
6.5 mm	22.45	12 mm	35.65	17.5 mm		45.90	25 mm	64.95
9 mm	27.10	16 mm	42.65	19 mm		50.25		

	£		£			£		£
Plasterboard (£/100m2)								
Wallboard plank								
9.5 mm	104.91	12.5 mm	125.12	15 mm		176.94		19
mm	220.45							
Lath (Thistle baseboard)								
9.5 mm	109.94	12.5	131.45					

	£				£			£
Industrial board						**Fireline Industrial board**		
9.5 mm	211.76	12.5 mm			238.28	12.5 mm	(2910)	
Fireline board								
12.5 mm	188.43	15 mm			218.87			

	£				£			£
Plywood (£/10 m2)								
Decorative								
6 mm Afrormosia	36.85							
6 mm Ash	35.91	9 mm Ash			54.94	12 mm Ash		56.82
6 mm Oak	37.40	9 mm Oak			48.40	12 mm Oak		58.37
6 mm Sapele	33.40	9 mm Sapele			44.50	12 mm Sapele		54.30
6 mm Teak	45.20	9 mm Teak			50.60	12 mm Teak		64.30
Prefinished 4 mm random v-grooved decorative								
Afrormosia	53.25	Elm			54.05	Sapele		50.75
Ash	54.30	Oak			57.15	Teak		58.50
Birch	62.00	Knotty Pine			52.55			

Discounts of 0 - 10% available depending on quantity/status

K LININGS/SHEATHING/DRY PARTITIONING Including overheads and profit at 9.00%	Labour hours	Labour £	Material £	Unit	Total rate £
K10 PLASTERBOARD DRY LINING					
Gypsum plasterboard; BS 1230; fixing with **nails; joints left open to receive 'Artex';** **to softwood base** Plain grade tapered edge wallboard 9.5 mm board to ceilings					
over 300 mm wide	0.28	3.76	1.16	m2	4.92
9.5 mm board to beams					
over 300 mm wide	0.35	4.70	0.72	m	5.42
12.5 mm board to ceilings					
total girth not exceeding 600 mm	0.45	6.05	1.39	m	7.44
total girth 600 - 1200 mm	0.30	4.03	1.61	m2	5.64
12.5 mm board to beams					
total girth not exceeding 600 mm	0.36	4.84	1.01	m	5.85
total girth 600 - 1200 mm	0.48	6.45	1.92	m	8.37
Gypsum plasterboard to BS 1230; fixing with **nails; joints filled with joint filler and** **joint tape to receive direct decoration; to** **softwood base** Plain grade tapered edge wallboard 9.5 mm board to walls					
wall height 2.40 - 2.70	1.00	13.44	4.02	m	17.46
wall height 2.70 - 3.00	1.15	15.46	4.48	m	19.94
wall height 3.00 - 3.30	1.30	17.47	4.94	m	22.41
wall height 3.30 - 3.60	1.50	20.16	5.40	m	25.56
9.5 mm board to reveals and soffits of openings and recesses					
not exceeding 300 mm wide	0.20	2.69	0.48	m	3.17
width 300 - 600 mm	0.40	5.38	0.95	m	6.33
9.5 mm board to faces of columns					
total girth not exceeding 600 mm	0.50	6.72	0.92	m	7.64
total girth 600 - 1200 mm	1.00	13.44	1.92	m	15.36
total girth 1200 - 1800 mm	1.30	17.47	2.83	m	20.30
9.5 mm board to ceilings					
over 300 mm wide	0.42	5.64	1.49	m2	7.13
9.5 mm board to faces of beams					
total girth not exceeding 600 mm	0.53	7.12	0.95	m	8.07
total girth 600 - 1200 mm	1.06	14.25	1.92	m	16.17
total girth 1200 - 1800 mm	1.38	18.55	2.83	m	21.38
Add for 'Duplex' insulating grade	-	-	-	m2	0.45
12.5 mm board to walls					
wall height 2.40 - 2.70	1.05	14.11	4.71	m	18.82
wall height 2.70 - 3.00	1.20	16.13	5.23	m	21.36
wall height 3.00 - 3.30	1.35	18.14	5.77	m	23.91
wall height 3.30 - 3.60	1.60	21.50	6.32	m	27.82
12.5 mm board to reveals and soffits of openings and recesses					
not exceeding 300 mm wide	0.21	2.82	0.57	m	3.39
width 300 - 600 mm	0.42	5.64	1.12	m	6.76
12.5 mm board to faces of columns					
total girth not exceeding 600 mm	0.52	6.99	1.12	m	8.11
total girth 600 - 1200 mm	1.04	13.98	2.24	m	16.22
total girth 1200 - 1800 mm	1.35	18.14	3.29	m	21.43
12.5 mm board to ceilings					
over 300 mm wide	0.44	5.91	1.76	m2	7.67
12.5 mm board to faces of beams					
total girth not exceeding 600 mm	0.56	7.53	1.12	m	8.65
total girth 600 - 1200 mm	1.12	15.05	2.24	m	17.29
total girth 1200 - 1800 mm	1.45	19.49	3.29	m	22.78
External angle; with joint tape bedded in joint filler; covered with joint finish	0.12	1.61	0.43	m	2.05
Add for 'Duplex' insulating grade	-	-	-	m2	0.45

K LININGS/SHEATHING/DRY PARTITIONING Including overheads and profit at 9.00%	Labour hours	Labour £	Material £	Unit	Total rate £
Tapered edge plank					
19 mm plank to walls					
wall height 2.40 - 2.70	1.10	14.78	7.76	m	22.54
wall height 2.70 - 3.00	1.25	16.80	8.61	m	25.41
wall height 3.00 - 3.30	1.40	18.82	9.46	m	28.28
wall height 3.30 - 3.60	1.70	22.85	9.88	m	32.72
19 mm plank to reveals and soffits of					
openings and recesses					
not exceeding 300 mm wide	0.22	2.96	0.93	m	3.89
width 300 - 600 mm	0.44	5.91	1.83	m	7.74
19 mm plank to faces of columns					
total girth not exceeding 600 mm	0.54	7.26	1.83	m	9.09
total girth 600 - 1200 mm	1.08	14.51	3.66	m	18.17
total girth 1200 - 1800 mm	1.40	18.82	5.33	m	24.15
19 mm plank to ceilings					
over 300 mm wide	0.46	6.18	2.87	m2	9.05
19 mm plank to faces of beams					
total girth not exceeding 600 mm	0.58	7.80	1.79	m	9.59
total girth 600 - 1200 mm	1.16	15.59	3.66	m	19.25
total girth 1200 - 1800 mm	1.52	20.43	5.33	m	25.76
Thermal board					
25 mm board to walls					
wall height 2.40 - 2.70	1.15	15.46	9.44	m	24.90
wall height 2.70 - 3.00	1.30	17.47	10.45	m	27.92
wall height 3.00 - 3.30	1.45	19.49	11.52	m	31.01
wall height 3.30 - 3.60	1.75	23.52	12.60	m	36.12
25 mm board to reveals and soffits of					
openings and recesses					
not exceeding 300 mm wide	0.23	3.09	1.11	m	4.20
width 300 - 600 mm	0.46	6.18	2.22	m	8.40
25 mm board to faces of columns					
total girth not exceeding 600 mm	0.56	7.53	2.22	m	9.75
total girth 600 - 1200 mm	1.12	15.05	4.44	m	19.49
total girth 1200 - 1800 mm	1.45	19.49	6.51	m	26.00
25 mm board to ceilings					
over 300 mm wide	0.50	6.72	3.48	m2	10.20
25 mm board to faces of beams					
total girth not exceeding 600 mm	0.60	8.06	2.22	m	10.28
total girth 600 - 1200 mm	1.20	16.13	4.44	m	20.57
total girth 1200 - 1800 mm	1.60	21.50	6.50	m	28.00
50 mm board to walls					
wall height 2.40 - 2.70	1.25	16.80	16.13	m	32.93
wall height 2.70 - 3.00	1.40	18.82	17.94	m	36.76
wall height 3.00 - 3.30	1.55	20.83	19.71	m	40.54
wall height 3.30 - 3.60	1.85	24.68	21.50	m	46.38
50 mm board to reveals and soffits of					
openings and recesses					
not exceeding 300 mm wide	0.25	3.36	1.88	m	5.24
width 300 - 600 mm	0.50	6.72	3.77	m	10.49
50 mm board to faces of columns					
total girth not exceeding 600 mm	0.60	8.06	3.77	m	11.83
total girth 600 - 1200 mm	1.20	16.13	7.54	m	23.66
total girth 1200 - 1800 mm	1.55	20.83	11.08	m	31.91
50 mm board to ceilings					
over 300 mm wide	0.53	7.12	5.97	m2	13.09
50 mm board to faces of beams					
total girth not exceeding 600 mm	0.63	8.47	3.77	m	12.24
total girth 600 - 1200 mm	1.27	17.07	7.54	m	24.61
total girth 1200 - 1800 mm	1.70	22.85	11.08	m	33.92

K LININGS/SHEATHING/DRY PARTITIONING Including overheads and profit at 9.00%	Labour hours	Labour £	Material £	Unit	Total rate £
K10 PLASTERBOARD DRY LINING - cont'd					
White plastic faced gypsum plasterboard to **BS 1230; fixing with screws; butt joints; to** **softwood base** Insulating grade square edge wallboard					
9.5 mm board to walls					
wall height 2.40 - 2.70	1.00	13.44	7.80	m	21.24
wall height 2.70 - 3.00	1.15	15.46	8.66	m	24.12
wall height 3.00 - 3.30	1.30	17.47	9.53	m	27.00
wall height 3.30 - 3.60	1.50	20.16	10.39	m	30.55
9.5 mm board to reveals and soffits of openings and recesses					
not exceeding 300 mm wide	0.20	2.69	0.90	m	3.59
width 300 - 600 mm	0.40	5.38	1.80	m	7.18
9.5 mm board to faces of columns					
total girth not exceeding 600 mm	0.50	6.72	1.80	m	8.52
total girth 600 - 1200 mm	1.00	13.44	3.57	m	17.01
total girth 1200 - 1800 mm	1.30	17.47	5.27	m	22.74
9.5 mm board to ceilings					
over 300 mm wide	0.42	5.64	2.89	m2	8.53
9.5 mm board to faces of beams					
total girth not exceeding 600 mm	0.53	7.12	1.80	m	8.92
total girth 600 - 1200 mm	1.06	14.25	3.57	m	17.81
total girth 1200 - 1800 mm	1.38	18.55	5.27	m	23.82
12.5 mm board to walls					
wall height 2.40 - 2.70	1.05	14.11	7.80	m	21.91
wall height 2.70 - 3.00	1.20	16.13	8.66	m	24.79
wall height 3.00 - 3.30	1.35	18.14	9.53	m	27.67
wall height 3.30 - 3.60	1.60	21.50	10.39	m	31.89
12.5 mm board to reveals and soffits of openings and recesses					
not exceeding 300 mm wide	0.21	2.82	0.90	m	3.72
width 300 - 600 mm	0.42	5.64	1.80	m	7.45
12.5 mm board to faces of columns					
total girth not exceeding 600 mm	0.52	6.99	1.80	m	8.79
total girth 600 - 1200 mm	1.04	13.98	3.57	m	17.55
total girth 1200 - 1800 mm	1.35	18.14	5.27	m	23.42
12.5 mm board to ceilings					
over 300 mm wide	0.44	5.91	3.19	m2	9.10
12.5 mm board to faces of beams					
total girth not exceeding 600 mm	0.56	7.53	1.99	m	9.52
total girth 600 - 1200 mm	1.12	15.05	3.96	m	19.01
total girth 1200 - 1800 mm	1.45	19.49	5.83	m	25.32
Two layers of gypsum plasterboard to **BS 1230; fixing with nails; joints filled**					
19 mm two layer board to walls					
wall height 2.40 - 2.70	1.60	21.50	7.64	m	29.14
wall height 2.70 - 3.00	1.80	24.19	8.48	m	32.67
wall height 3.00 - 3.30	2.00	26.88	9.34	m	36.22
wall height 3.30 - 3.60	2.40	32.26	9.73	m	41.99
19 mm two layer board to reveals and soffits of openings and recesses					
not exceeding 300 mm wide	0.30	4.03	0.91	m	4.94
width 300 - 600 mm	0.60	8.06	1.80	m	9.86
19 mm two layer board to faces of columns					
total girth not exceeding 600 mm	0.80	10.75	1.80	m	12.55
total girth 600 - 1200 mm	1.50	20.16	3.61	m	23.77
total girth 1200 - 1800 mm	2.00	26.88	5.25	m	32.13
19 mm two layer board to ceilings					
over 300 mm wide	0.66	8.87	2.83	m2	11.70
19 mm two layer board to faces of beams					
total girth not exceeding 600 mm	0.85	11.42	1.76	m	13.18
total girth 600 - 1200 mm	1.60	21.50	3.61	m	25.11
total girth 1200 - 1800 mm	2.15	28.90	5.25	m	34.15

K LININGS/SHEATHING/DRY PARTITIONING Including overheads and profit at 9.00%	Labour hours	Labour £	Material £	Unit	Total rate £
25 mm two layer board to walls					
wall height 2.40 - 2.70	1.70	22.85	9.13	m	31.98
wall height 2.70 - 3.00	1.90	25.54	10.14	m	35.68
wall height 3.00 - 3.30	2.20	29.57	11.20	m	40.77
wall height 3.30 - 3.60	2.60	34.94	12.27	m	47.21
25 mm two layer board to reveals and soffits of openings and recesses					
not exceeding 300 mm wide	0.33	4.44	1.08	m	5.52
width 300 - 600 mm	0.65	8.74	2.14	m	10.88
25 mm two layer board to faces of columns					
total girth not exceeding 600 mm	0.86	11.56	2.14	m	13.70
total girth 600 - 1200 mm	1.60	21.50	4.29	m	25.79
total girth 1200 - 1800 mm	2.15	28.90	6.31	m	35.20
25 mm two layer board to ceilings					
over 300 mm wide	0.72	9.68	· 3.38	m2	13.06
25 mm two layer board to faces of beams					
total girth not exceeding 600 mm	0.92	12.36	2.14	m	14.50
total girth 600 - 1200 mm	1.75	23.52	4.29	m	27.81
total girth 1200 - 1800 mm	2.33	31.31	6.31	m	37.62
Gypsum plasterboard to BS 1230; 3 mm joints; **fixed by the 'Thistleboard' system of dry** **linings; joints; filled with joint filler** **and joint tape; to receive direct decoration** **Plain grade tapered edge wallboard**					
9.5 mm board to walls					
wall height 2.40 - 2.70	1.05	14.11	3.95	m	18.06
wall height 2.70 - 3.00	1.20	16.13	4.40	m	20.53
wall height 3.00 - 3.30	1.35	18.14	4.85	m	22.99
wall height 3.30 - 3.60	1.60	21.50	5.30	m	26.80
9.5 mm board to reveals and soffits of openings and recesses					
not exceeding 300 mm wide	0.21	2.82	0.45	m	3.27
width 300 - 600 mm	0.42	5.64	0.91	m2	6.55
9.5 mm board to faces of columns					
total girth not exceeding 600 mm	0.52	6.99	0.91	m	7.90
total girth 600 - 1200 mm	1.04	13.98	1.83	m	15.81
total girth 1200 - 1800 mm	1.35	18.14	2.70	m	20.84
Angle; with joint tape bedded in joint filler; covered with joint finish					
internal·	0.06	0.81	0.17	m	0.98
external	0.12	1.61	0.43	m	2.04
Vermiculite gypsum cladding; 'Vicuclad' board **on and including shaped noggins; fixed with** **nails and adhesive; joints pointed in** **adhesive**					
25 mm column casings; 2 hour fire protection rating					
over 1 m girth	1.00	13.44	13.43	m2	26.87
150 mm girth	0.30	4.03	4.07	m	8.10
300 mm girth	0.45	6.05	6.09	m	12.14
600 mm girth	0.75	10.08	9.94	m	20.02
900 mm girth	0.95	12.77	12.77	m	25.54
30 mm beam casings; 2 hour fire protection rating					
over 1 m girth	1.10	14.78	16.50	m2	31.28
150 mm girth	0.33	4.44	4.99	m	9.43
300 mm girth	0.50	6.72	7.48	m	14.20
600 mm girth	0.83	11.15	12.23	m	23.38
900 mm girth	1.05	14.11	15.68	m	29.79
55 mm column casings; 4 hour fire protection rating					
over 1 m girth	1.20	16.13	33.87	m2	50.00
150 mm girth	0.36	4.84	10.20	m	15.04
300 mm girth	0.55	7.39	15.30	m	22.69
600 mm girth	0.90	12.10	25.27	m	37.37
900 mm girth	1.15	15.46	32.19	m	47.65

K LININGS/SHEATHING/DRY PARTITIONING Including overheads and profit at 9.00%	Labour hours	Labour £	Material £	Unit	Total rate £
K10 PLASTERBOARD DRY LINING - cont'd					
Vermiculite gypsum cladding; 'Vicuclad' board **on and including shaped noggins; fixed with** **nails and adhesive; joints pointed in** **adhesive - cont'd** 60 mm beam casings; 4 hour fire protection rating					
over 1 m girth	1.30	17.47	35.72	m2	53.19
150 mm girth	0.39	5.24	10.75	m	15.99
300 mm girth	0.60	8.06	16.13	m	24.19
600 mm girth	0.97	13.04	26.66	m	39.70
900 mm girth	1.25	16.80	33.95	m	50.75
Add to the above for					
Plus 3% for work 3.5 - 5 m high					
Plus 6% for work 5 - 6.5 m high					
Plus 12% for work 6.5 - 8 m high					
Plus 18% for work over 8 m high					
Cutting and fitting around steel joists, angles, trunking, ducting, ventilators, pipes, tubes, etc					
over 2 m girth	0.30	4.03	-	m	4.03
not exceeding 0.30 m girth	0.20	2.69	-	nr	2.69
0.30 - 1 m girth	0.25	3.36	-	nr	3.36
1 - 2 m girth	0.35	4.70	-	nr	4.70
K11 RIGID SHEET FLOORING/SHEATHING/LININGS/ **CASINGS**					
Blockboard (Birch faced)					
12 mm lining to walls					
over 300 mm wide PC £71.05/10m2	0.47	4.26	8.62	m2	12.88
not exceeding 300 mm wide	0.31	2.81	2.83	m	5.64
Raking cutting	0.08	0.72	0.31	m	1.03
Holes for pipes and the like	0.04	0.36	3.10	nr	3.46
18 mm lining to walls					
over 300 mm wide PC £83.30/10m2	0.50	4.53	10.09	m2	14.62
not exceeding 300 mm wide	0.32	2.90	3.31	m	6.21
Raking cutting	0.10	0.91	0.45	m	1.36
Holes for pipes and the like	0.05	0.45	-	nr	0.45
Two-sided 18 mm thick pipe casing; 50 x 50 mm softwood framing; two members plugged to wall					
300 mm girth	1.25	11.32	5.24	m	16.56
450 mm girth	1.35	12.23	6.92	m	19.15
600 mm girth	1.45	13.13	8.60	m	21.73
750 mm girth	1.55	14.04	10.28	m	24.32
Three-sided 18 mm thick pipe casing; 50 x 50 mm softwood framing; two members plugged to wall					
450 mm girth	1.70	15.40	7.56	m	22.96
600 mm girth	1.80	16.30	9.24	m	25.54
750 mm girth	1.90	17.21	10.92	m	28.13
900 mm girth	2.00	18.12	12.60	m	30.72
1050 mm girth	2.10	19.02	14.28	m	33.30
Extra for 400 x 400 mm removable access panel; brass cups and screws; additional framing	1.00	9.06	7.47	nr	16.53
25 mm lining to walls					
over 300 mm wide PC £108.55/10m2	0.54	4.89	13.14	m2	18.03
not exceeding 300 mm wide	0.35	3.17	4.31	m	7.48
Raking cutting	0.12	1.09	0.59	m	1.68
Holes for pipes and the like	0.06	0.54	-	nr	0.54

K LININGS/SHEATHING/DRY PARTITIONING Including overheads and profit at 9.00%	Labour hours	Labour £	Material £	Unit	Total rate £
Chipboard (plain)					
12 mm lining to walls					
over 300 mm wide PC £19.50/10m2	0.38	3.44	2.45	m2	**5.89**
not exceeding 300 mm wide	0.22	1.99	0.81	m	**2.80**
Raking cutting	0.06	0.54	0.11	m	**0.65**
Holes for pipes and the like	0.03	0.27	-	nr	**0.27**
15 mm lining to walls					
over 300 mm wide PC £24.92/10m2	0.40	3.62	3.09	m2	**6.71**
not exceeding 300 mm wide	0.24	2.17	1.02	m	**3.19**
Raking cutting	0.08	0.72	0.14	m	**0.86**
Holes for pipes and the like	0.04	0.36	-	nr	**0.36**
Two-sided 15 mm thick pipe casing;					
50 x 50 mm softwood framing; two members					
plugged to wall					
300 mm girth	1.00	9.06	3.14	m	**12.20**
450 mm girth	1.08	9.78	3.77	m	**13.55**
600 mm girth	1.16	10.51	4.40	m	**14.91**
750 mm girth	1.24	11.23	5.02	m	**16.25**
Three-sided 15 mm thick pipe casing;					
50 x 50 mm softwood framing; two members					
plugged to wall					
450 mm girth	1.49	13.50	4.41	m	**17.91**
600 mm girth	1.58	14.31	5.04	m	**19.35**
750 mm girth	1.70	15.40	5.67	m	**21.07**
900 mm girth	1.79	16.21	6.29	m	**22.50**
1050 mm girth	2.01	18.21	6.92	m	**25.13**
Extra for 400 x 400 mm removable access					
panel; brass cups and screws; additional					
framing	1.00	9.06	7.42	nr	**16.48**
18 mm lining to walls					
over 300 mm wide PC £26.50/10m2	0.42	3.80	3.29	m2	**7.09**
not exceeding 300 mm wide	0.27	2.45	1.08	m	**3.53**
Raking cutting	0.10	0.91	0.14	m	**1.05**
Holes for pipes and the like	0.05	0.45	-	nr	**0.45**
Chipboard (Melamine faced white matt finish)					
15 mm thick; PC £36.50/10m2; laminated masking strips					
Lining to walls					
over 300 mm wide	1.05	9.51	6.10	m2	**15.61**
not exceeding 300 mm wide	0.68	6.16	2.64	m	**8.80**
Raking cutting	0.13	1.18	0.20	m	**1.38**
Holes for pipes and the like	0.07	0.63	-	nr	**0.63**
Chipboard boarding and flooring					
Boarding to floors; butt joints					
18 mm thick PC £29.90/10m2	0.30	2.72	3.56	m2	**6.28**
22 mm thick PC £39.95/10m2	0.33	2.99	4.71	m2	**7.70**
Boarding to floors; tongued and					
grooved joints					
18 mm thick PC £31.40/10m2	0.32	2.90	3.73	m2	**6.63**
22 mm thick PC £41.50/10m2	0.35	3.17	4.88	m2	**8.05**
Boarding to roofs; butt joints					
18 mm thick; pre-felted PC £28.60/10m2	0.33	2.99	3.40	m2	**6.39**
Raking cutting on 18 mm thick					
chipboard	0.08	0.72	1.63	m	**2.35**
Durabella 'Westbourne' flooring system or					
similar; comprising 19 mm thick tongued and					
grooved chipboard panels secret nailed to					
softwood MK 10X-profiled foam backed battens					
at 600 mm centres; on concrete floor					
Flooring tongued and grooved joints					
63 mm thick 36 x 50 mm battens PC £8.65	0.90	8.15	10.58	m2	**18.73**
75 mm thick 43 x 50 mm battens PC £9.30	0.90	8.15	11.36	m2	**19.51**

K LININGS/SHEATHING/DRY PARTITIONING Including overheads and profit at 9.00%		Labour hours	Labour £	Material £	Unit	Total rate £
K11 RIGID SHEET FLOORING/SHEATHING/LININGS/ CASINGS - cont'd						
Plywood flooring Boarding to floors; tongued and grooved joints						
15 mm thick	PC £99.30/10m2	0.40	3.62	11.50	m2	15.12
18 mm thick	PC £115.36/10m2	0.44	3.99	13.34	m2	17.33
Plywood; external quality; 18 mm thick; PC £99.10/10m2 Boarding to roofs; butt joints						
flat to falls		0.40	3.62	11.48	m2	15.10
sloping		0.43	3.89	11.48	m2	15.37
vertical		0.57	5.16	11.48	m2	16.64
Hardboard to BS 1142 3.2 mm lining to walls						
over 300 mm wide	PC £9.30/10m2	0.30	2.72	1.15	m2	3.87
not exceeding 300 mm wide		0.18	1.63	0.38	m	2.01
Raking cutting		0.02	0.18	0.05	m	0.23
Holes for pipes and the like		0.01	0.09	-	nr	0.09
6.4 mm lining to walls						
over 300 mm wide	PC £18.50/10m2	0.33	2.99	2.26	m2	5.25
not exceeding 300 mm wide		0.21	1.90	0.74	m	2.64
Raking cutting		0.03	0.27	0.10	m	0.37
Holes for pipes and the like		0.02	0.18	-	nr	0.18
Glazed hardboard to BS 1142; PC £41.90/10m2; on and including 38 x 38 mm wrought softwood framing 3.2 mm thick panel						
to side of bath		1.80	16.30	7.98	nr	24.28
to end of bath		0.70	6.34	2.84	nr	9.18
Insulation board to BS 1142 12.7 mm lining to walls						
over 300 mm wide	PC £16.61/10m2	0.24	2.17	2.10	m2	4.27
not exceeding 300 mm wide		0.14	1.27	0.69	m	1.96
Raking cutting		0.03	0.27	0.09	m	0.36
Holes for pipes and the like		0.01	0.09	-	nr	0.09
19 mm lining to walls						
over 300 mm wide	PC £27.12/10m2	0.26	2.36	3.36	m2	5.72
not exceeding 300 mm wide		0.16	1.45	1.10	m	2.55
Raking cutting		0.05	0.45	0.15	m	0.60
Holes for pipes and the like		0.02	0.18	-	nr	0.18
25 mm lining to walls						
over 300 mm wide	PC £35.09/10m2	0.32	2.90	4.33	m2	7.23
not exceeding 300 mm wide		0.19	1.72	1.43	m	3.15
Raking cutting		0.06	0.54	0.19	m	0.73
Holes for pipes and the like		0.03	0.27	-	nr	0.27
Laminboard (Birch Faced); 18 mm thick PC £111.70/10m2 Lining to walls						
over 300 mm wide		0.53	4.80	13.50	m2	18.30
not exceeding 300 mm wide		0.34	3.08	4.42	m	7.50
Raking cutting		0.11	1.00	0.61	m	1.61
Holes for pipes and the like		0.06	0.54	-	nr	0.54
Non-asbestos board; 'Supalux'; natural finish 6 mm lining to walls						
over 300 mm wide	PC £48.42/10m2	0.33	2.99	5.88	m2	8.87
not exceeding 300 mm wide		0.20	1.81	1.93	m	3.74
6 mm lining to ceilings						
over 300 mm wide		0.44	3.99	5.88	m2	9.87
not exceeding 300 mm wide		0.27	2.45	1.93	m	4.38
Raking cutting		0.05	0.45	0.26	m	0.71
Holes for pipes and the like		0.03	0.27	-	nr	0.27

K LININGS/SHEATHING/DRY PARTITIONING Including overheads and profit at 9.00%	Labour hours	Labour £	Material £	Unit	Total rate £
9 mm lining to walls					
over 300 mm wide PC £75.42/10m2	0.36	3.26	9.13	m2	12.39
not exceeding 300 mm wide	0.22	1.99	3.00	m	4.99
9 mm lining to ceilings					
over 300 mm wide	0.45	4.08	9.13	m2	13.21
not exceeding 300 mm wide	0.29	2.63	3.00	m	5.63
Raking cutting	0.06	0.54	0.41	m	0.95
Holes for pipes and the like	0.04	0.36	-	nr	0.36
12 mm lining to walls					
over 300 mm wide PC £102.78/10m2	0.40	3.62	12.43	m2	16.05
not exceeding 300 mm wide	0.24	2.17	4.08	m	6.25
12 mm lining to ceilings					
over 300 mm wide	0.53	4.80	12.43	m2	17.23
not exceeding 300 mm wide	0.32	2.90	4.08	m	6.98
Raking cutting	0.07	0.63	0.56	m	1.19
Holes for pipes and the like	0.05	0.45	-	nr	0.45
Non-asbestos board; 'Supalux'; sanded finish					
6 mm lining to walls					
over 300 mm wide PC £50.85/10m2	0.33	2.99	6.18	m2	9.17
not exceeding 300 mm wide	0.20	1.81	2.03	m	3.84
6 mm lining to ceilings					
over 300 mm wide	0.44	3.99	6.18	m2	10.17
not exceeding 300 mm wide	0.27	2.45	2.03	m	4.48
Raking cutting	0.05	0.45	0.28	m	0.73
Holes for pipes and the like	0.03	0.27	-	nr	0.27
9 mm lining to walls					
over 300 mm wide PC £77.58/10m2	0.36	3.26	9.40	m2	12.66
not exceeding 300 mm wide	0.22	1.99	3.09	m	5.08
9 mm lining to ceilings					
over 300 mm wide	0.45	4.08	9.40	m2	13.48
not exceeding 300 mm wide	0.29	2.63	3.09	m	5.72
Raking cutting	0.06	0.54	0.42	m	0.96
Holes for pipes and the like	0.04	0.36	-	nr	0.36
12 mm lining to walls					
over 300 mm wide PC £102.78/10m2	0.40	3.62	12.43	m2	16.05
not exceeding 300 mm wide	0.24	2.17	4.08	m	6.25
12 mm lining to ceilings					
over 300 mm wide	0.53	4.80	12.43	m2	17.23
not exceeding 300 mm wide	0.32	2.90	4.08	m	6.98
Raking cutting	0.07	0.63	0.56	m	1.19
Holes for pipes and the like	0.05	0.45	-	nr	0.45
Plywood (Russian Birch); internal quality					
4 mm lining to walls					
over 300 mm wide PC £21.35/10m2	0.37	3.35	2.66	m2	6.01
not exceeding 300 mm wide	0.24	2.17	0.88	m	3.05
4 mm lining to ceilings					
over 300 mm wide	0.50	4.53	2.66	m2	7.19
not exceeding 300 mm wide	0.32	2.90	0.88	m	3.78
Raking cutting	0.05	0.45	0.12	m	0.57
Holes for pipes and the like	0.03	0.27	-	nr	0.27
6 mm lining to walls					
over 300 mm wide PC £30.40/10m2	0.40	3.62	3.75	m2	7.37
not exceeding 300 mm wide	0.26	2.36	1.23	m	3.59
6 mm lining to ceilings					
over 300 mm wide	0.53	4.80	3.75	m2	8.55
not exceeding 300 mm wide	0.35	3.17	1.23	m	4.40
Raking cutting	0.05	0.45	0.17	m	0.62
Holes for pipes and the like	0.03	0.27	-	nr	0.27
Two-sided 6 mm thick pipe casings; 50 x 50 mm softwood framing; two members plugged to wall					
300 mm girth	1.20	10.87	3.34	m	14.21
450 mm girth	1.30	11.78	4.06	m	15.84
600 mm girth	1.40	12.68	4.79	m	17.47
750 mm girth	1.50	13.59	5.52	m	19.11

K LININGS/SHEATHING/DRY PARTITIONING Including overheads and profit at 9.00%	Labour hours	Labour £	Material £	Unit	Total rate £
K11 RIGID SHEET FLOORING/SHEATHING/LININGS/ **CASINGS - cont'd**					
Plywood (Russian Birch); internal quality - cont'd					
Three-sided 6 mm thick pipe casing;					
50 x 50 mm softwood framing; to members					
plugged to wall					
450 mm girth	1.65	14.95	4.71	m	19.66
600 mm girth	1.75	15.85	5.43	m	21.28
750 mm girth	1.85	16.76	6.16	m	22.92
900 mm girth	1.95	17.66	6.89	m	24.55
1050 mm girth	2.00	18.12	7.62	m	25.74
9 mm lining to walls					
over 300 mm wide PC £42.60/10m2	0.43	3.89	5.22	m2	9.11
not exceeding 300 mm wide	0.28	2.54	1.72	m	4.26
9 mm lining to ceilings					
over 300 mm wide	0.57	5.16	5.22	m2	10.38
not exceeding 300 mm wide	0.37	3.35	1.72	m	5.07
Raking cutting	0.06	0.54	0.23	m	0.77
Holes for pipes and the like	0.04	0.36	-	nr	0.36
12 mm lining to walls					
over 300 mm wide PC £55.80/10m2	0.46	4.17	6.80	m2	10.97
not exceeding 300 mm wide	0.30	2.72	2.22	m	4.94
12 mm lining to ceilings					
over 300 mm wide	0.61	5.53	6.80	m2	12.33
not exceeding 300 mm wide	0.40	3.62	2.22	m	5.84
Raking cutting	0.06	0.54	0.30	m	0.84
Holes for pipes and the like	0.04	0.36	-	nr	0.36
Plywood (Finnish Birch); external quality					
4 mm lining to walls					
over 300 mm wide PC £27.15/10m2	0.37	3.35	3.36	m2	6.71
not exceeding 300 mm wide	0.24	2.17	1.10	m	3.27
4 mm lining to ceilings					
over 300 mm wide	0.50	4.53	3.36	m2	7.89
not exceeding 300 mm wide	0.32	2.90	1.10	m	4.00
Raking cutting	0.05	0.45	0.15	m	0.60
Holes for pipes and the like	0.03	0.27	-	nr	0.27
6 mm lining to walls					
over 300 mm wide PC £40.80/10m2	0.40	3.62	5.00	m2	8.62
not exceeding 300 mm wide	0.26	2.36	1.64	m	4.00
6 mm lining to ceilings					
over 300 mm wide	0.53	4.80	5.00	m2	9.80
not exceeding 300 mm wide	0.35	3.17	1.64	m	4.81
Raking cutting	0.05	0.45	0.22	m	0.67
Holes for pipes and the like	0.03	0.27	-	nr	0.27
Two-sided 6 mm thick pipe casings;					
50 x 50 mm softwood framing; two members					
plugged to wall					
300 mm girth	1.20	10.87	3.72	m	14.59
450 mm girth	1.30	11.78	4.63	m	16.41
600 mm girth	1.40	12.68	5.55	m	18.23
750 mm girth	1.50	13.59	6.46	m	20.05
Three-sided 6 mm thick pipe casing;					
50 x 50 mm softwood framing; to members					
plugged to wall					
450 mm girth	1.65	14.95	5.27	m	20.22
600 mm girth	1.75	15.85	6.18	m	22.03
750 mm girth	1.85	16.76	7.10	m	23.86
900 mm girth	1.95	17.66	8.02	m	25.68
1050 mm girth	2.00	18.12	8.93	m	27.05
9 mm lining to walls					
over 300 mm wide PC £53.30/10m2	0.43	3.89	6.49	m2	10.38
not exceeding 300 mm wide	0.28	2.54	2.13	m	4.67
9 mm lining to ceilings					
over 300 mm wide	0.57	5.16	6.49	m2	11.65
not exceeding 300 mm wide	0.37	3.35	2.13	m	5.48
Raking cutting	0.07	0.63	0.29	m	0.92
Holes for pipes and the like	0.05	0.45	-	nr	0.45

K LININGS/SHEATHING/DRY PARTITIONING Including overheads and profit at 9.00%	Labour hours	Labour £	Material £	Unit	Total rate £
12 mm lining to walls					
over 300 mm wide PC £67.90/10m2	0.46	4.17	8.25	m2	12.42
not exceeding 300 mm wide	0.30	2.72	2.72	m	5.44
12 mm lining to ceilings					
over 300 mm wide	0.61	5.53	8.26	m2	13.79
not exceeding 300 mm wide	0.40	3.62	2.72	m	6.34
Raking cutting	0.10	0.91	0.37	m	1.28
Holes for pipes and the like	0.08	0.72	-	nr	0.72
Extra over wall linings fixed with nails;					
for screwing	0.15	1.36	0.16	m2	1.52
Woodwool unreinforced slabs; Torvale **'Woodcemair' or similar; BS 1105 type SB;** **natural finish; fixing to timber or steel** **with galvanized nails or clips; flat or** **sloping**					
50 mm slabs; type 500; max. span 600 mm					
1800 - 2400 mm lengths PC £4.62	0.40	3.62	5.51	m2	9.13
2700 - 3000 mm lengths PC £4.72	0.40	3.62	5.64	m2	9.26
75 mm slabs; type 750; max. span 900 mm					
2100 mm lengths PC £6.30	0.45	4.08	7.47	m2	11.55
2400 - 2700 mm lengths PC £6.26	0.45	4.08	7.47	m2	11.55
3000 mm lengths	0.45	4.08	7.50	m2	11.58
Raking cutting	0.14	1.27	1.56	m	2.83
Holes for pipes and the like	0.12	1.09	-	nr	1.09
100 mm slabs; type 1000; max. span 1200 mm					
3000 - 3600 mm lengths PC £8.72	0.50	4.53	10.21	m2	14.74
Internal quality West African Mahogany **veneered plywood; 6 mm thick; PC £35.60/10m2; WAM** **cover strips**					
Lining to walls					
over 300 mm wide	0.70	6.34	5.25	m2	11.59
not exceeding 300 mm wide	0.45	4.08	2.07	m	6.15

K13 RIGID SHEET FINE LININGS/PANELLING

	Labour hours	Labour £	Material £	Unit	Total rate £
Formica faced chipboard; 17 mm thick; white **matt finish; balancer; PC £156.66/10m2; aluminium** **cover strips and countersunk screws**					
Lining to walls					
over 300 mm wide	1.90	17.21	20.39	m2	37.60
not exceeding 300 mm wide	1.24	11.23	7.26	m	18.49
Lining to isolated columns or the like					
over 300 mm wide	2.85	25.82	22.72	m2	48.54
not exceeding 300 mm wide	1.65	14.95	7.84	m	22.79

K20 TIMBER BOARD FLOORING/SHEATHING/LININGS/CASINGS

	Labour hours	Labour £	Material £	Unit	Total rate £
Sawn softwood; untreated Boarding to roofs; 150 mm wide boards; butt joints					
19 mm thick; flat; over 300 mm wide					
PC £57.78/100m	0.45	4.08	4.80	m2	8.88
19 mm thick; flat; not exceeding 300 mm wide	0.30	2.72	1.48	m	4.20
19 mm thick; sloping; over 300 mm wide	0.50	4.53	4.80	m2	9.33
19 mm thick; sloping; not exceeding 300 mm wide	0.33	2.99	1.48	m	4.47
19 mm thick; sloping; laid diagonally; over 300 mm wide	0.63	5.71	4.91	m2	10.62
19 mm thick; sloping; laid diagonally; not exceeding 300 mm wide	0.40	3.62	1.50	m	5.12
25 mm thick; flat; over 300 mm wide;					
PC £73.44/100m	0.45	4.08	6.03	m2	10.11
25 mm thick; flat; not exceeding 300 mm	0.30	2.72	1.86	m	4.58
25 mm thick; sloping; over 300 mm wide	0.50	4.53	6.03	m2	10.56
25 mm thick; sloping; not exceeding 300 mm wide	0.33	2.99	1.86	m	4.85

K LININGS/SHEATHING/DRY PARTITIONING Including overheads and profit at 9.00%	Labour hours	Labour £	Material £	Unit	Total rate £
K20 TIMBER BOARD FLOORING/SHEATHING/LININGS/ **CASINGS - cont'd**					
Sawn softwood; untreated - cont'd Boarding to tops or cheeks of dormers; 150 mm wide boards; butt joints					
19 mm thick; diagonally; over 300 mm wide	0.80	7.25	4.91	m2	**12.16**
19 mm thick; diagonally; not exceeding 300 mm wide	0.50	4.53	1.48	m	**6.01**
19 mm thick; diagonally; area not exceeding 1.00 m2; irrespective of width	1.00	9.06	5.01	nr	**14.07**
Sawn softwood; 'Tanalised' Boarding to roofs; 150 mm wide boards; butt joints					
19 mm thick; flat; over 300 mm wide PC £72.22/100m	0.45	4.08	5.93	m2	**10.01**
19 mm thick; flat; not exceeding 300 mm wide	0.30	2.72	1.82	m	**4.54**
19 mm thick; sloping; over 300 mm wide	0.50	4.53	5.93	m2	**10.46**
19 mm thick; sloping; not exceeding 300 mm wide	0.33	2.99	1.82	m	**4.81**
19 mm thick; sloping; laid diagonally; over 300 mm wide	0.63	5.71	6.06	m2	**11.77**
19 mm thick; sloping; laid diagonally; not exceeding 300 mm wide	0.40	3.62	1.86	m	**5.48**
25 mm thick; flat; over 300 mm wide; PC £91.80/100m	0.45	4.08	7.47	m2	**11.55**
25 mm thick; flat; not exceeding 300 mm	0.30	2.72	2.29	m	**5.01**
25 mm thick; sloping; over 300 mm wide	0.50	4.53	7.47	m2	**12.00**
25 mm thick; sloping; not exceeding 300 mm wide	0.33	2.99	2.29	m	**5.28**
Boarding to tops or cheeks of dormers; 150 mm wide boards; butt joints					
19 mm thick; diagonally; over 300 mm wide	0.80	7.25	6.06	m2	**13.31**
19 mm thick; diagonally; not exceeding 300 mm wide	0.50	4.53	1.86	m	**6.39**
19 mm thick; diagonally; area not exceeding 1.00 m2; irrespective of width	1.00	9.06	6.06	nr	**15.12**
Wrought softwood Boarding to floors; butt joints					
19 mm thick x 75 mm wide boards PC £51.97/100m	0.60	5.43	8.58	m2	**14.01**
19 mm thick x 125 mm wide boards PC £61.83/100m	0.50	4.53	6.11	m2	**10.64**
22 mm thick x 150 mm wide boards PC £73.03/100m	0.45	4.08	6.00	m2	**10.08**
25 mm thick x 100 mm wide boards PC £71.55/100m	0.55	4.98	8.76	m2	**13.74**
25 mm thick x 150 mm wide boards PC £104.35/100m	0.45	4.08	8.45	m2	**12.53**
Boarding to floors; tongued and grooved joints					
19 mm thick x 75 mm wide boards PC £51.97/100m	0.70	6.34	9.07	m2	**15.41**
19 mm thick x 125 mm wide boards PC £61.83/100m	0.60	5.43	6.46	m2	**11.89**
22 mm thick x 150 mm wide boards PC £73.03/100m	0.55	4.98	6.35	m2	**11.33**
25 mm thick x 100 mm wide boards PC £71.55/100m	0.65	5.89	9.26	m2	**15.15**
25 mm thick x 150 mm wide boards PC £104.35/100m	0.55	4.98	8.95	m2	**13.93**
Nosings; tongued to edge of flooring					
19 x 75 mm; once rounded	0.25	2.26	1.41	m	**3.67**
25 x 75 mm; once rounded	0.25	2.26	1.68	m	**3.94**

K LININGS/SHEATHING/DRY PARTITIONING Including overheads and profit at 9.00%	Labour hours	Labour £	Material £	Unit	Total rate £
Boarding to internal walls; tongued and grooved and V-jointed 12 mm thick x 100 mm wide boards					
PC £49.68/100m	0.80	7.25	6.55	m2	13.80
16 mm thick x 100 mm wide boards					
PC £58.60/100m	0.80	7.25	7.65	m2	14.90
19 mm thick x 100 mm wide boards					
PC £69.52/100m	0.80	7.25	9.01	m2	16.26
19 mm thick x 125 mm wide boards					
PC £85.99/100m	0.75	6.79	8.86	m2	15.65
19 mm thick x 125 mm wide boards; chevron pattern	1.20	10.87	9.32	m2	20.19
25 mm thick x 125 mm wide boards					
PC £111.78/100m	0.75	6.79	11.43	m2	18.22
12 mm thick x 100 mm wide Knotty Pine boards PC £35.30/100m	0.80	7.25	4.75	m2	12.00
Boarding to internal ceilings; tongued and grooved and V-jointed					
12 mm thick x 100 mm wide boards	1.00	9.06	6.55	m2	15.61
16 mm thick x 100 mm wide boards	1.00	9.06	7.65	m2	16.71
19 mm thick x 100 mm wide boards	1.00	9.06	9.01	m2	18.07
19 mm thick x 125 mm wide boards	0.95	8.61	8.86	m2	17.47
19 mm thick x 125 mm wide boards; chevron pattern	1.40	12.68	9.32	m2	22.00
25 mm thick x 125 mm wide boards	0.95	8.61	11.43	m2	20.04
12 mm thick x 100 mm wide Knotty Pine boards	1.00	9.06	4.75	m2	13.81
Boarding to roofs; tongued and grooved joints					
19 mm thick; flat to falls PC £80.46/100m	0.55	4.98	6.96	m2	11.94
19 mm thick; sloping	0.60	5.43	6.96	m2	12.39
19 mm thick; sloping; laid diagonally	0.75	6.79	7.08	m2	13.87
25 mm thick; flat to falls PC £104.35/100m	0.55	4.98	8.95	m2	13.93
25 mm thick; sloping	0.60	5.43	8.95	m2	14.38
Boarding to tops or cheeks of dormers; tongued and grooved joints					
19 mm thick; laid diagonally	1.00	9.06	7.08	m2	16.14
Wrought softwood; 'Tanalised' Boarding to roofs; tongued and grooved joints					
19 mm thick; flat to falls PC £100.58/100m	0.55	4.98	8.63	m2	13.61
19 mm thick; sloping	0.60	5.43	8.63	m2	14.06
19 mm thick; sloping; laid diagonally	0.75	6.79	8.79	m2	15.58
25 mm thick; flat to falls PC £130.44/100m	0.55	4.98	11.12	m2	16.10
25 mm thick; sloping	0.60	5.43	11.12	m2	16.55
Boarding to tops or cheeks of dormers; tongued and grooved joints					
19 mm thick; laid diagonally	1.00	9.06	8.79	m2	17.85
Wood strip; 22 mm thick; 'Junckers' **pre-treated or similar; tongued and grooved** **joints; pre-finished boards; level fixing to** **resilient battens; to cement and sand base** Strip flooring; over 300 mm wide					
Beech; prime	-	-	-	m2	29.98
Beech; standard	-	-	-	m2	28.32
Beech; sylvia squash	-	-	-	m2	29.98
Oak; quality A	-	-	-	m2	37.54
Wrought hardwood Strip flooring to floors; 25 mm thick x 75 mm wide; tongued and grooved joints; secret fixing; surface sanded after laying					
American Oak	-	-	-	m2	33.79
Canadian Maple	-	-	-	m2	27.25
Gurjun	-	-	-	m2	27.25
Iroko	-	-	-	m2	33.79
Raking cutting	-	-	-	m	2.04
Curved cutting	-	-	-	m	2.88

K LININGS/SHEATHING/DRY PARTITIONING Including overheads and profit at 9.00%		Labour hours	Labour £	Material £	Unit	Total rate £
K29 TIMBER FRAMED AND PANELLED PARTITIONS						
Purpose made screen components; wrought softwood						
Panelled partitions						
32 mm thick frame; 9 mm thick						
plywood panels; over 300 mm wide	PC £28.77	0.90	8.15	32.14	m2	**40.29**
38 mm thick frame; 9 mm thick						
plywood panels over 300 mm wide	PC £30.67	0.95	8.61	34.27	m2	**42.88**
50 mm thick frame; 12 mm thick						
plywood panels; over 300 mm wide	PC £37.32	1.00	9.06	41.70	m2	**50.76**
Panelled partitions; mouldings worked on solid both sides						
32 mm thick frame; 9 mm thick						
plywood panels; over 300 mm wide	PC £33.24	0.90	8.15	37.13	m2	**45.28**
38 mm thick frame; 9 mm thick						
plywood panels over 300 mm wide	PC £34.77	0.95	8.61	38.85	m2	**47.46**
50 mm thick frame; 12 mm thick						
plywood panels; over 300 mm wide	PC £42.20	1.00	9.06	47.14	m2	**56.20**
Panelled partitions; mouldings planted on both sides						
32 mm thick frame; 9 mm thick						
plywood panels; over 300 mm wide	PC £33.64	0.90	8.15	37.59	m2	**45.74**
38 mm thick frame; 9 mm thick						
plywood panels over 300 mm wide	PC £35.08	0.95	8.61	39.19	m2	**47.80**
50 mm thick frame; 12 mm thick						
plywood panels; over 300 mm wide	PC £42.20	1.00	9.06	47.14	m2	**56.20**
Panelled partitions; diminishing stiles over 300 mm wide; upper portion open panels for glass; in medium panes						
32 mm thick frame; 9 mm thick						
plywood panels; over 300 mm wide	PC £47.79	0.90	8.15	53.38	m2	**61.53**
38 mm thick frame; 9 mm thick						
plywood panels over 300 mm wide	PC £53.43	0.95	8.61	59.70	m2	**68.31**
50 mm thick frame; 12 mm thick						
plywood panels; over 300 mm wide	PC £66.52	1.00	9.06	74.31	m2	**83.37**
Purpose made screen components; selected West African Mahogany; PC £494.00/m3						
Panelled partitions						
32 mm thick frame; 9 mm thick						
plywood panels; over 300 mm wide	PC £52.36	1.20	10.87	58.50	m2	**69.37**
38 mm thick frame; 9 mm thick						
plywood panels over 300 mm wide	PC £56.71	1.25	11.32	63.36	m2	**74.68**
50 mm thick frame; 12 mm thick						
plywood panels; over 300 mm wide	PC £71.40	1.35	12.23	79.78	m2	**92.01**
Panelled partitions; mouldings worked on solid both sides						
32 mm thick frame; 9 mm thick						
plywood panels; over 300 mm wide	PC £59.81	1.20	10.87	66.83	m2	**77.70**
38 mm thick frame; 9 mm thick						
plywood panels over 300 mm wide	PC £66.82	1.25	11.32	74.66	m2	**85.98**
50 mm thick frame; 12 mm thick						
plywood panels; over 300 mm wide	PC £74.15	1.35	12.23	82.84	m2	**95.07**
Panelled partitions; mouldings planted on both sides						
32 mm thick frame; 9 mm thick						
plywood panels; over 300 mm wide	PC £75.63	1.20	10.87	84.49	m2	**95.36**
38 mm thick frame; 9 mm thick						
plywood panels over 300 mm wide	PC £76.29	1.25	11.32	85.24	m2	**96.56**
50 mm thick frame; 12 mm thick						
plywood panels; over 300 mm wide	PC £83.07	1.35	12.23	92.82	m2	**105.05**
Panelled partitions; diminishing stiles over 300 mm wide; upper portion open panels for glass; in medium panes						
32 mm thick frame; 9 mm thick						
plywood panels; over 300 mm wide	PC £98.21	1.20	10.87	109.73	m2	**120.60**

K LININGS/SHEATHING/DRY PARTITIONING Including overheads and profit at 9.00%		Labour hours	Labour £	Material £	Unit	Total rate £
38 mm thick frame; 9 mm thick plywood panels over 300 mm wide	PC £105.51	1.25	11.32	117.89	m2	129.21
50 mm thick frame; 12 mm thick plywood panels; over 300 mm wide	PC £111.99	1.35	12.23	125.12	m2	137.35

Purpose made screen components; Afrormosia
PC £706.00/m3
Panelled partitions

32 mm thick frame; 9 mm thick plywood panels; over 300 mm wide	PC £65.20	1.20	10.87	72.85	m2	83.72
38 mm thick frame; 9 mm thick plywood panels over 300 mm wide	PC £70.71	1.25	11.32	79.00	m2	90.32
50 mm thick frame; 12 mm thick plywood panels; over 300 mm wide	PC £89.11	1.35	12.23	99.57	m2	111.80

Panelled partitions; mouldings worked
on solid both sides

32 mm thick frame; 9 mm thick plywood panels; over 300 mm wide	PC £74.55	1.20	10.87	83.28	m2	94.15
38 mm thick frame; 9 mm thick plywood panels over 300 mm wide	PC £83.43	1.25	11.32	93.21	m2	104.53
50 mm thick frame; 12 mm thick plywood panels; over 300 mm wide	PC £92.36	1.35	12.23	103.19	m2	115.42

Panelled partitions; mouldings planted on
both sides

32 mm thick frame; 9 mm thick plywood panels; over 300 mm wide	PC £100.13	1.20	10.87	111.87	m2	122.74
38 mm thick frame; 9 mm thick plywood panels over 300 mm wide	PC £100.93	1.25	11.32	112.76	m2	124.08
50 mm thick frame; 12 mm thick plywood panels; over 300 mm wide	PC £103.61	1.35	12.23	115.76	m2	127.99

Panelled partitions; diminishing stiles
over 300 mm wide; upper portion open panels
for glass; in medium panes

32 mm thick frame; 9 mm thick plywood panels; over 300 mm wide	PC £122.39	1.20	10.87	136.74	m2	147.61
38 mm thick frame; 9 mm thick plywood panels over 300 mm wide	PC £131.38	1.25	11.32	146.78	m2	158.10
50 mm thick frame; 12 mm thick plywood panels; over 300 mm wide	PC £139.76	1.35	12.23	156.14	m2	168.37

K31 PLASTERBOARD FIXED PARTITIONS/
INNER WALLS/ LININGS

'Gyproc' laminated partition; comprising two
skins of gypsum plasterboard bonded to a
centre core of plasterboard square edge
plank 19 mm thick; fixing with nails to
softwood studwork (measured separately);
joints filled with joint filler and joint
tape; to receive direct decoration;
(perimeter studwork measured separately)
50 mm partition; two outer skins of 12.7 mm
tapered edge wallboard

height 2.10 - 2.40 m		2.15	28.90	16.42	m	45.32
height 2.40 - 2.70 m		2.40	32.26	18.54	m	50.80
height 2.70 - 3.00 m		2.70	36.29	20.69	m	56.98
height 3.00 - 3.30 m		3.00	40.32	22.84	m	63.16
height 3.30 - 3.60 m		3.40	45.69	24.99	m	70.68

65 mm partition; two outer skins of 19 mm
tapered edge plank

height 2.10 - 2.40 m		2.35	31.58	21.61	m	53.19
height 2.40 - 2.70 m		2.60	34.94	24.36	m	59.30
height 2.70 - 3.00 m		3.00	40.32	27.13	m	67.45
height 3.00 - 3.30 m		3.35	45.02	29.91	m	74.93
height 3.30 - 3.60 m		3.75	50.40	32.69	m	83.09

K LININGS/SHEATHING/DRY PARTITIONING Including overheads and profit at 9.00%	Labour hours	Labour £	Material £	Unit	Total rate £
K31 PLASTERBOARD FIXED PARTITIONS/ INNER WALLS/ LININGS - cont'd					
Labours and associated additional wrought softwood studwork Floor, wall or ceiling battens					
25 x 38 mm	0.12	1.61	0.45	m	2.06
Forming openings					
25 x 38 mm framing	0.30	4.03	0.71	m	4.74
Fair ends	0.20	2.69	0.45	m	3.14
Angle	0.30	4.03	0.45	m	4.48
Intersection	0.20	2.69	-	m	2.69
Cutting and fitting around steel joists, angles, trunking, ducting, ventilators, pipes, tubes, etc					
over 2 m girth	0.09	1.21	-	m	1.21
not exceeding 0.30 m girth	0.05	0.67	-	nr	0.67
0.30 - 1 m girth	0.07	0.94	-	nr	0.94
1 - 2 m girth	0.11	1.48	-	nr	1.48
'Paramount' dry partition comprising Paramount panels with and including wrought softwood battens at vertical joints; fixing with nails to softwood studwork; (perimeter studwork measured separately) Square edge panels; joints filled with plaster and jute scrim cloth; to receive plaster (measured elsewhere);					
57 mm partition					
height 2.10 - 2.40 m	1.45	19.49	13.68	m	33.17
height 2.40 - 2.70 m	1.60	21.50	15.39	m	36.89
height 2.70 - 3.00 m	1.80	24.19	17.11	m	41.30
height 3.00 - 3.30 m	2.05	27.55	17.05	m	44.60
height 3.30 - 3.60 m	2.35	31.58	20.62	m	52.20
63 mm partition					
height 2.10 - 2.40 m	1.60	21.50	15.74	m	37.24
height 2.40 - 2.70 m	1.75	23.52	17.72	m	41.24
height 2.70 - 3.00 m	1.95	26.20	19.68	m	45.88
height 3.00 - 3.30 m	2.20	29.56	21.70	m	51.26
height 3.30 - 3.60 m	2.50	33.60	23.72	m	57.32
Tapered edge panels; joints filled with joint filler and joint tape to receive direct decoration;					
57 mm partition					
height 2.10 - 2.40 m	2.05	27.55	14.74	m	42.29
height 2.40 - 2.70 m	2.20	29.57	16.60	m	46.17
height 2.70 - 3.00 m	2.40	32.26	18.48	m	50.74
height 3.00 - 3.30 m	2.65	35.61	20.39	m	56.00
height 3.30 - 3.60 m	3.00	40.32	22.30	m	62.62
63 mm partition					
height 2.10 - 2.40 m	2.20	29.57	16.84	m	46.41
height 2.40 - 2.70 m	2.35	31.58	18.96	m	50.54
height 2.70 - 3.00 m	2.55	34.27	21.09	m	55.36
height 3.00 - 3.30 m	2.80	37.63	23.27	m	60.90
height 3.30 - 3.60 m	3.15	42.33	25.43	m	67.76

K LININGS/SHEATHING/DRY PARTITIONING Including overheads and profit at 9.00%	Labour hours	Labour £	Material £	Unit	Total rate £
Labours and associated additional wrought **softwood studwork on 57 mm partition**					
Wall or ceiling battens					
19 x 37 mm	0.12	1.61	0.41	m	2.02
Sole plates; with 19 x 37 x 150 mm long battens spiked on at 600 mm centres					
19 x 57 mm	0.25	3.36	0.77	m	4.13
50 x 57 mm	0.30	4.03	1.21	m	5.24
Forming openings					
37 x 37 mm framing	0.30	4.03	0.41	m	4.44
Fair ends	0.20	2.69	0.41	m	3.10
Angle	0.30	4.03	0.88	m	4.91
Intersection	0.25	3.36	0.59	m	3.95
Cutting and fitting around steel joists, angles, trunking, ducting, ventilators, pipes, tubes, etc					
over 2 m girth	0.10	1.34	-	m	1.34
not exceeding 0.30 m girth	0.06	0.81	-	nr	0.81
0.30 - 1 m girth	0.08	1.08	-	nr	1.08
1 - 2 m girth	0.12	1.61	-	nr	1.61
Plugging; 300 mm centres; one way					
brickwork	0.10	1.34	0.03	m	1.37
concrete	0.18	2.42	0.03	m	2.45
'Gyroc' metal stud partition; comprising **146 mm metal stud frame; with floor channel** **plugged and screwed to concrete through** **38 x 148 mm tanalised softwood sole plate**					
Tapered edge panels; joints filled with joint filler and joint tape to receive direct decoration					
171 mm partition; one hour; one layer of 12.5 mm Fireline board each side					
height 2.10 - 2.40 m	4.20	56.45	20.41	m	76.86
height 2.40 - 2.70 m	4.85	65.18	22.56	m	87.74
height 2.70 - 3.00 m	5.40	72.57	24.85	m	97.42
height 3.00 - 3.30 m	6.25	84.00	27.25	m	111.25
height 3.30 - 3.60 m	6.85	92.06	29.16	m	121.22
height 3.60 - 3.90 m	8.20	110.21	31.91	m	142.12
height 3.90 - 4.20 m	8.80	118.27	34.21	m	152.48
Angles	0.20	2.69	1.19	m	3.88
T-junctions	0.20	2.69	1.12	m	3.81
Fair end	0.30	4.03	1.84	m	5.87
Tapered edge panels; joints filled with joint filler and joint tape to receive direct decoration					
196 mm partition; two hour; two layers of 12.5 mm Fireline board both sides					
height 2.10 - 2.40 m	6.00	80.63	32.24	m	112.87
height 2.40 - 2.70 m	6.75	90.72	35.88	m	126.60
height 2.70 - 3.00 m	7.50	100.80	39.63	m	140.43
height 3.00 - 3.30 m	8.75	117.60	43.28	m	160.88
height 3.30 - 3.60 m	9.60	129.02	46.90	m	175.92
height 3.60 - 3.90 m	11.70	157.25	51.14	m	208.39
height 3.90 - 4.20 m	12.60	169.34	54.91	m	224.25
Angles	0.20	2.69	1.19	m	3.88
T-junctions	0.20	2.69	1.12	m	3.81
Fair end	0.35	4.70	2.33	m	7.03

K LININGS/SHEATHING/DRY PARTITIONING Including overheads and profit at 9.00%	Labour hours	Labour £	Material £	Unit	Total rate £
K33 CONCRETE/TERRAZZO PARTITIONS					
Terrazzo faced partitions; cement and white Sicilian marble aggregate; polished					
Pre-cast reinforced terrazzo faced WC partitions					
38 mm thick; over 300 mm wide	6.75	110.36	8.28	m2	118.64
50 mm thick; over 300 mm wide	7.00	114.45	9.32	m2	123.77
Wall post; once rebated					
64 x 102 mm	3.75	61.31	3.62	m	64.93
64 x 152 mm	4.00	65.40	4.14	m	69.54
Centre post; twice rebated					
64 x 152 mm	4.75	77.66	4.66	m	82.32
64 x 203 mm	5.20	85.02	5.18	m	90.20
Lintel; once rebated					
64 x 102 mm	3.80	62.13	3.62	m	65.75
Brass topped plates or sockets cast into posts for fixings (measured elsewhere)	0.66	10.79	10.36	pr	21.15
K40 SUSPENDED CEILINGS					
Suspended ceilings, Gyproc M/F suspended ceiling system; hangers plugged and screwed to concrete soffite, 900 x 1800 x 12.7 mm tapered edge wallboard infill PC £1.72/m2; joints filled with joint filler and taped to receive direct decoration					
Lining to ceilings; hangers av. 400 mm long					
over 300 mm wide	1.10	14.78	6.87	m2	21.65
not exceeding 300 mm wide in isolated strips	0.55	7.39	2.44	m	9.83
300 - 600 mm wide in isolated strips	0.85	11.42	4.87	m	16.29
Vertical bulkhead; including additional hangers					
over 300 mm wide	1.40	18.82	9.16	m2	27.98
not exceeding 300 mm wide in isolated strips	0.70	9.41	3.04	m	12.45
300 - 600 mm wide in isolated strips	1.10	14.78	6.09	m	20.87
Suspended ceilings; 'Slimline' exposed suspended ceiling system; hangers plugged and screwed to concrete soffits, 600 x 600 x 19 mm Treetex 'Glacier' mineral tile infill; PC £8.29/m2					
Lining to ceilings; hangers av. 400 mm long					
over 300 mm wide	0.50	6.72	18.74	m2	25.46
not exceeding 300 mm wide in isolated strips	0.25	3.36	5.94	m	9.30
300 - 600 mm wide in isolated strips	0.40	5.38	11.89	m	17.27
Extra for					
suspension 800 mm high	0.05	0.67	0.07	m2	0.74
suspension 1200 mm high	0.07	0.94	0.10	m2	1.04
Extra for cutting and fitting around modular downlighting	0.20	2.69	-	nr	2.69
Vertical bulkhead; including additional hangers					
over 300 mm wide	0.60	8.06	23.42	m2	31.48
not exceeding 300 mm wide in isolated strips	0.30	4.03	7.43	m	11.46
300 - 600 mm wide in isolated strips	0.45	6.05	14.85	m	20.90
Edge detail 24 x 19 mm white finished angle edge trim	0.06	0.81	0.71	m	1.52
Cutting and fitting around pipes; not exceeding 0.30 m girth	0.10	1.34	-	nr	1.34
Suspended ceilings; 'Z' demountable suspended ceiling system; hangers plugged and screwed to concrete soffite, 600 x 600 x 19 mm 'Echostop' glass reinforced fibrous plaster lightweight plain bevelled edge tiles; PC £8.74/m2					
Lining to ceilings; hangers av. 400 mm long					
over 300 mm wide	0.65	6.71	18.01	m2	24.72
not exceeding 300 mm wide in isolated strips	0.35	4.70	5.92	m	10.62

K LININGS/SHEATHING/DRY PARTITIONING Including overheads and profit at 9.00%	Labour hours	Labour £	Material £	Unit	Total rate £
Suspended ceilings; concealed galvanised steel suspension system; hangers plugged and screwed to concrete soffite, Burgess white stove enamelled perforated mild steel tiles 600 x 600 mm; PC £9.63/m2					
Lining to ceilings; hangers av. 400 mm long					
over 300 mm wide	0.55	7.39	19.71	m2	**27.10**
not exceeding 300 mm wide in isolated strips	0.27	3.63	6.34	m	**9.97**
Suspended ceilings; concealed galvanised steel 'Trulok' suspension system; hangers plugged and screwed to concrete; Armstrong 'Travertone', 'Sanserra' or 'Highspire' 300 x 300 x 18 mm mineral ceiling tiles; PC £9.40/m2					
Linings to ceilings; hangers av. 700 mm long					
over 300 mm wide	0.65	8.74	18.31	m2	**27.05**
over 300 mm wide; 3.5 - 5 m high	0.85	11.42	18.74	m2	**30.16**
over 300 mm wide; in staircase areas or plant rooms	1.30	17.47	19.90	m2	**37.37**
not exceeding 300 mm wide in isolated strips	0.35	4.70	5.97	m	**10.67**
300 - 600 mm wide in isolated strips	0.70	7.23	11.94	m	**19.17**
Extra for cutting and fitting around modular downlighting	0.25	3.36	-	nr	**3.36**
Extra for					
suspension 800 mm high	0.05	0.67	0.10	m2	**0.77**
suspension 1200 mm high	0.07	0.94	0.13	m2	**1.07**
Vertical bulkhead; including additional hangers					
over 300 mm wide	0.80	10.75	22.12	m2	**32.87**
not exceeding 300 mm wide in isolated strips	0.40	5.38	6.99	m	**12.37**
300 - 600 mm wide in isolated strips	0.60	8.06	13.98	m	**22.04**
Shadow gap edge detail 24 x 19 mm white finished angle edge trim	0.09	1.21	0.84	m	**2.05**
Cutting and fitting around pipes; not exceeding 0.30 m girth	0.10	1.34	-	nr	**1.34**
Suspended ceilings; galvanised steel suspension system; hangers plugged and screwed to concrete soffite, 'Luxalon' stove enamelled aluminium linear panel ceiling, type 80B, complete with mineral insulation; PC £12.10/m2					
Linings to ceilings; hangers av. 700 mm long					
over 300 mm wide	0.65	8.74	23.14	m2	**31.88**
not exceeding 300 mm wide in isolated strips	0.33	4.44	7.41	m	**11.85**

L WINDOWS/DOORS/STAIRS
Including overheads and profit at 9.00%

ALTERNATIVE TIMBER WINDOW PRICES (£/each)

	£		£
Softwood shallow circular bay windows			
3 lightx1200 mm; ref CSB312CV	140.13	4 lightx1350 mm; ref CSB413CV	197.40
3 lightx1350 mm; ref CSB313CV	146.85	4 lightx1500 mm; ref CSB415T	223.74
3 light glass fibre roof unit	61.83	4 light glass fibre roof unit	66.69

	Labour hours	Labour £	Material £	Unit	Total rate £
L10 TIMBER WINDOWS/ROOFLIGHTS/SCREENS/LOUVRES					
Standard windows; 'treated' wrought softwood (refs. refer to Boulton & Paul cat. nos.) Side hung casement windows without glazing bars; with 140 mm wide softwood sills; opening casements and ventilators hung on rustproof hinges; fitted with aluminized laquered finish casement stays and fasteners; knotting and priming by manufacturer before delivery					
500 x 750 mm; ref N07V PC £20.67	0.70	6.34	22.53	nr	28.87
500 x 900 mm; ref N09V PC £24.10	0.80	7.25	26.27	nr	33.52
600 x 750 mm; ref 107V PC £26.16	0.80	7.25	28.51	nr	35.76
600 x 750 mm; ref 107C PC £24.00	0.80	7.25	26.16	nr	33.41
600 x 900 mm; ref 109V PC £23.88	0.90	8.15	26.03	nr	34.18
600 x 900 mm; ref 109C PC £28.47	0.80	7.25	31.03	nr	38.28
600 x 1050 mm; ref 110V PC £27.84	0.80	7.25	30.35	nr	37.60
600 x 1050 mm; ref 110C PC £29.76	1.00	9.06	32.44	nr	41.50
900 x 900 mm; ref 2N09W PC £30.81	1.10	9.96	33.58	nr	43.54
900 x 1050 mm; ref 2N10W PC £31.38	1.15	10.42	34.20	nr	44.62
900 x 1200 mm; ref 2N12W PC £32.28	1.20	10.87	35.19	nr	46.06
900 x 1350 mm; ref 2N13W PC £33.06	1.35	12.23	36.04	nr	48.27
900 x 1500 mm; ref 2N15W PC £33.66	1.40	12.68	36.69	nr	49.37
1200 x 750 mm; ref 207C PC £32.07	1.15	10.42	34.96	nr	45.38
1200 x 750 mm; ref 207CV PC £41.94	1.15	10.42	45.71	nr	56.13
1200 x 900 mm; ref 209C PC £33.87	1.20	10.87	36.92	nr	47.79
1200 x 900 mm; ref 209W PC £35.76	1.20	10.87	38.98	nr	49.85
1200 x 900 mm; ref 209CV PC £43.77	1.20	10.87	47.71	nr	58.58
1200 x 1050 mm; ref 210C PC £35.22	1.35	12.23	38.39	nr	50.62
1200 x 1050 mm; ref 210W PC £36.51	1.35	12.23	39.80	nr	52.03
1200 x 1050 mm; ref 210T PC £43.77	1.35	12.23	47.71	nr	59.94
1200 x 1050 mm; ref 210CV PC £45.24	1.35	12.23	49.31	nr	61.54
1200 x 1200 mm; ref 212C PC £36.90	1.45	13.13	40.22	nr	53.35
1200 x 1200 mm; ref 212W PC £37.35	1.45	13.13	40.71	nr	53.84
1200 x 1200 mm; ref 212T PC £45.27	1.45	13.13	49.34	nr	62.47
1200 x 1200 mm; ref 212CV PC £46.98	1.45	13.13	51.21	nr	64.34
1200 x 1350 mm; ref 213W PC £38.22	1.55	14.04	41.66	nr	55.70
1200 x 1350 mm; ref 213CV PC £49.62	1.55	14.04	54.09	nr	68.13
1200 x 1500 mm; ref 215W PC £39.30	1.70	15.40	42.84	nr	58.24
1770 x 750 mm; ref 307CC PC £50.73	1.40	12.68	55.30	nr	67.98
1770 x 900 mm; ref 309CC PC £53.37	1.70	15.40	58.17	nr	73.57
1770 x 1050 mm; ref 310C PC £42.66	1.75	15.85	46.50	nr	62.35
1770 x 1050 mm; ref 310T PC £52.23	1.70	15.40	56.93	nr	72.33
1770 x 1050 mm; ref 310CC PC £55.68	1.40	12.68	60.69	nr	73.37
1770 x 1050 mm; ref 310WW PC £61.53	1.40	12.68	67.07	nr	79.75
1770 x 1200 mm; ref 312C PC £44.13	1.80	16.30	48.10	nr	64.40

L WINDOWS/DOORS/STAIRS Including overheads and profit at 9.00%		Labour hours	Labour £	Material £	Unit	Total rate £
1770 x 1200 mm; ref 312T	PC £53.88	1.80	16.30	58.73	nr	75.03
1770 x 1200 mm; ref 312CC	PC £58.02	1.80	16.30	63.24	nr	79.54
1770 x 1200 mm; ref 312WW	PC £63.27	1.80	16.30	68.96	nr	85.26
1770 x 1200 mm; ref 312CVC	PC £68.10	1.80	16.30	74.23	nr	90.53
1770 x 1350 mm; ref 313CC	PC £62.43	1.90	17.21	68.05	nr	85.26
1770 x 1350 mm; ref 312CC	PC £64.89	1.90	17.21	70.73	nr	87.94
1770 x 1350 mm; ref 313CVC	PC £72.54	1.90	17.21	79.07	nr	96.28
1770 x 1500 mm; ref 315T	PC £56.82	2.00	18.12	61.93	nr	80.05
2340 x 1050 mm; ref 410CWC	PC £78.03	1.95	17.66	85.05	nr	102.71
2340 x 1200 mm; ref 412CWC	PC £80.76	2.05	18.57	88.03	nr	106.60
2340 x 1350 mm; ref 413CWC	PC £85.59	2.20	19.93	93.29	nr	113.22

Top hung casement windows; with 140 mm wide
softwood sills; opening casements and
ventilators hung on rustproof hinges; fitted
with aluminized laquered finish casement
stays; knotting and priming by manufacturer
before delivery

		Labour hours	Labour £	Material £	Unit	Total rate £
600 x 750 mm; ref 107A	PC £26.22	0.80	7.25	28.58	nr	35.83
600 x 900 mm; ref 109A	PC £27.30	0.90	8.15	29.76	nr	37.91
600 x 1050 mm; ref 110A	PC £28.65	1.00	9.06	31.23	nr	40.29
900 x 750 mm; ref 2N07A	PC £32.73	1.05	9.51	35.68	nr	45.19
900 x 900 mm; ref 2N09A	PC £35.58	1.10	9.96	38.78	nr	48.74
900 x 1050 mm; ref 2N10A	PC £37.11	1.15	10.42	40.45	nr	50.87
900 x 1350 mm; ref 2N13AS	PC £42.96	1.35	12.23	46.83	nr	59.06
900 x 1500 mm; ref 2N15AS	PC £44.34	1.40	12.68	48.33	nr	61.01
1200 x 750 mm; ref 207A	PC £37.65	1.15	10.42	41.04	nr	51.46
1200 x 900 mm; ref 209A	PC £40.23	1.20	10.87	43.85	nr	54.72
1200 x 1050 mm; ref 210A	PC £41.79	1.35	12.23	45.55	nr	57.78
1200 x 1050 mm; ref 210AT	PC £56.01	1.35	12.23	61.05	nr	73.28
1200 x 1200 mm; ref 212A	PC £43.23	1.45	13.13	47.12	nr	60.25
1200 x 1200 mm; ref 212AT	PC £59.10	1.45	13.13	64.42	nr	77.55
1200 x 1350 mm; ref 213AS	PC £47.88	1.55	14.04	52.19	nr	66.23
1200 x 1500 mm; ref 215AS	PC £49.38	1.70	15.40	53.82	nr	69.22
1770 x 1050 mm; ref 310A	PC £50.49	1.70	15.40	55.03	nr	70.43
1770 x 1050 mm; ref 310AV	PC £61.92	1.70	15.40	67.49	nr	82.89
1770 x 1200 mm; ref 312A	PC £52.20	1.80	16.30	56.90	nr	73.20
1770 x 1220 mm; ref 312AV	PC £63.69	1.80	16.30	69.42	nr	85.72
2340 x 1200 mm; ref 412A	PC £58.65	2.05	18.57	63.93	nr	82.50
2340 x 1200 mm; ref 412AW	PC £74.25	2.05	18.57	80.93	nr	99.50
2340 x 1500 mm; ref 415AWS	PC £79.05	2.35	21.29	86.16	nr	107.45

High performance top hung reversible windows;
with 140 mm wide softwood sills; adjustable
ventilators weather stripping; opening
sashes and fanlights hung on rustproof
hinges; fitted with aluminized laquered
espagnolette bolts; knotting and priming
by manufacturer before delivery

		Labour hours	Labour £	Material £	Unit	Total rate £
600 x 900 mm; ref R0609	PC £65.91	0.90	8.15	71.84	nr	79.99
900 x 900 mm; ref R0909	PC £72.09	1.10	9.96	78.58	nr	88.54
900 x 1050 mm; ref R0910	PC £73.50	1.15	10.42	80.12	nr	90.54
900 x 1200 mm; ref R0912	PC £75.54	1.25	11.32	82.34	nr	93.66
900 x 1500 mm; ref R0915	PC £85.59	1.40	12.68	93.29	nr	105.97
1200 x 1050 mm; ref R1210	PC £80.19	1.35	12.23	87.41	nr	99.64
1200 x 1200 mm; ref R1212	PC £82.32	1.45	13.13	89.73	nr	102.86
1200 x 1500 mm; ref R1215	PC £93.30	1.70	15.40	101.70	nr	117.10
1500 x 1200 mm; ref R1512	PC £92.40	1.70	15.40	100.72	nr	116.12
1800 x 1050 mm; ref R1810	PC £96.51	1.70	15.40	105.20	nr	120.60
1800 x 1200 mm; ref R1812	PC £98.22	1.80	16.30	107.06	nr	123.36
1800 x 1500 mm; ref R1815	PC £105.45	1.90	17.21	114.94	nr	132.15
2400 x 1500 mm; ref R2415	PC £117.00	2.35	21.29	127.53	nr	148.82

L WINDOWS/DOORS/STAIRS Including overheads and profit at 9.00%		Labour hours	Labour £	Material £	Unit	Total rate £

L10 TIMBER WINDOWS/ROOFLIGHTS/SCREENS/LOUVRES - cont'd

Standard windows; 'treated' wrought softwood
(refs. refer to Boulton & Paul cat. nos.) - cont'd
High performance double hung sash windows
with glazing bars; solid frames; 63 x 175 mm
softwood sills; standard flush external
linings; spiral spring balances and sash
catch; knotting and priming by manufacturer
before delivery

635 x 1050 mm; ref DH0610B	PC £94.38	2.00	18.12	102.87	nr	120.99
635 x 1350 mm; ref DH0613B	PC £102.66	2.20	19.93	111.90	nr	131.83
635 x 1650 mm; ref DH0616B	PC £110.04	2.45	22.19	119.94	nr	142.13
860 x 1050 mm; ref DH0810B	PC £103.44	2.30	20.83	112.75	nr	133.58
860 x 1350 mm; ref DH0813B	PC £112.38	2.60	23.55	122.49	nr	146.04
860 x 1650 mm; ref DH0816B	PC £121.32	3.00	27.17	132.24	nr	159.41
1085 x 1050 mm; ref DH1010B	PC £112.98	2.60	23.55	123.15	nr	146.70
1085 x 1350 mm; ref DH1013B	PC £124.41	3.00	27.17	135.61	nr	162.78
1085 x 1650 mm; ref DH1016B	PC £134.85	3.70	33.51	146.99	nr	180.50
1699 x 1050 mm; ref DH1710B	PC £205.92	3.70	33.51	224.45	nr	257.96
1699 x 1350 mm; ref DH1713B	PC £225.36	4.60	41.67	245.64	nr	287.31
1699 x 1650 mm; ref DH1716B	PC £242.58	4.70	42.57	264.41	nr	306.98

Purpose made window casements; 'treated'
wrought softwood
Casements; rebated; moulded

38 mm thick	PC £16.53	-	-	18.02	m2	18.02
50 mm thick	PC £19.64	-	-	21.41	m2	21.41

Casements; rebated; moulded; in medium panes

38 mm thick	PC £24.13	-	-	26.31	m2	26.31
50 mm thick	PC £29.86	-	-	32.55	m2	32.55

Casements; rebated; moulded; with
semi-circular head

38 mm thick	PC £30.16	-	-	32.87	m2	32.87
50 mm thick	PC £37.35	-	-	40.71	m2	40.71

Casements; rebated; moulded; to bullseye
window

38 mm thick x 600 mm dia	PC £54.22/nr	-	-	59.10	m2	59.10
38 mm thick x 900 mm dia	PC £80.44/nr	-	-	87.68	m2	87.68
50 mm thick x 600 mm dia	PC £62.37/nr	-	-	67.99	m2	67.99
50 mm thick x 900 mm dia	PC £92.48/nr	-	-	100.80	m2	100.80

Fitting and hanging casements

square or rectangular	0.50	4.53	-	nr	4.53
semicircular	1.25	11.32	-	nr	11.32
bullseye	2.00	18.12	-	nr	18.12

Purpose made window frames; 'treated'
wrought softwood
Frames; rounded; rebated check grooved

25 x 120 mm	0.14	1.27	5.71	m	6.98
50 x 75 mm	0.14	1.27	6.46	m	7.73
50 x 100 mm	0.16	1.45	7.96	m	9.41
50 x 125 mm	0.16	1.45	9.66	m	11.11
63 x 100 mm	0.16	1.45	9.89	m	11.34
75 x 150 mm	0.18	1.63	16.61	m	18.24
90 x 140 mm	0.18	1.63	21.49	m	23.12

Mullions and transoms; twice rounded,
rebated and check grooved

50 x 75 mm	0.10	0.91	7.84	m	8.75
50 x 100 mm	0.12	1.09	9.34	m	10.43
63 x 100 mm	0.12	1.09	11.26	m	12.35
75 x 150 mm	0.14	1.27	17.98	m	19.25

Sill; sunk weathered, rebated and grooved

75 x 100 mm	0.20	1.81	16.30	m	18.11
75 x 150 mm	0.20	1.81	21.45	m	23.26

Add +5% to the above 'Material £ prices'
for 'selected' softwood for staining

L WINDOWS/DOORS/STAIRS Including overheads and profit at 9.00%		Labour hours	Labour £	Material £	Unit	Total rate £
Purpose made double hung sash windows; 'treated' wrought softwood Cased frames of 100 x 25 mm grooved inner linings; 114 x 25 mm grooved outer linings; 125 x 38 mm twice rebated head linings; 125 x 32 mm twice rebated grooved pulley stiles; 150 x 13 mm linings; 50 x 19 mm parting slips; 25 x 13 mm parting beads; 25 x 19 mm inside beads; 150 x 75 mm Oak twice sunk weathered throated sill; 50 mm thick rebated and moulded sashes; moulded horns; over 1.25 m2 each; both sashes in medium panes;						
including spiral spring balances	PC £121.53	2.25	20.38	187.20	m2	207.58
As above but with cased mullions	PC £128.21	2.50	22.64	194.49	m2	217.13
Standard pre-glazed roof windows; 'treated' Nordic Red Pine and aluminium trimmed 'Velux' windows, or equivalent; including type U flashings and soakers (for tiles and pantiles), and sealed double glazing unit; trimming opening measured separately						
550 x 700 mm; ref GGL-9	PC £90.45	5.00	45.29	99.89	nr	145.18
550 x 980 mm; ref GGL-6	PC £101.65	5.00	45.29	112.11	nr	157.40
700 x 1180 mm; ref GGL-5	PC £120.70	6.00	54.35	132.87	nr	187.22
780 x 980 mm; ref GGL-1	PC £116.20	5.50	49.82	127.96	nr	177.78
780 x 1400 mm; ref GGL-2	PC £136.81	6.50	58.88	150.43	nr	209.31
940 x 1600 mm; ref GGL-3	PC £161.08	6.50	58.88	176.89	nr	235.77
1140 x 1180 mm; ref GGL-4	PC £153.87	7.00	63.41	169.03	nr	232.44
1340 x 980 mm; ref GGL-7	PC £156.98	7.00	63.41	172.41	nr	235.82
1340 x 1400 mm; ref GGL-8	PC £181.61	7.50	67.93	199.26	nr	267.19
Standard windows; selected Philippine Mahogany; preservative stain finish Side hung casement windows; with 45 x 140 mm hardwood sills; weather stripping; opening sashes on canopy hinges; fitted with fasteners; aluminized lacquered finish ironmongery						
600 x 600 mm; ref SS0606/L	PC £45.24	0.95	8.61	49.31	nr	57.92
600 x 900 mm; ref SS0609/L	PC £50.16	1.20	10.87	54.67	nr	65.54
600 x 900 mm; ref SV0609/0	PC £68.25	0.95	8.61	74.39	nr	83.00
600 x 1050 mm; ref SS0610/L	PC £61.95	1.30	11.78	67.53	nr	79.31
600 x 1050 mm; ref SV0610/0	PC £70.28	1.30	11.78	76.61	nr	88.39
900 x 900 mm; ref SV0909/0	PC £90.93	1.50	13.59	99.11	nr	112.70
900 x 1050 mm; ref SV0910/0	PC £91.56	1.60	14.49	99.80	nr	114.29
900 x 1200 mm; ref SV0912/0	PC £93.66	1.70	15.40	102.09	nr	117.49
900 x 1350 mm; ref SV0913/0	PC £95.55	1.80	16.30	104.15	nr	120.45
900 x 1500 mm; ref SV0915/0	PC £97.51	1.90	17.21	106.29	nr	123.50
1200 x 900 mm; ref SS1209/L	PC £95.76	1.70	15.40	104.38	nr	119.78
1200 x 900 mm; ref SV1209/0	PC £104.44	1.70	15.40	113.84	nr	129.24
1200 x 1050 mm; ref SS1210/L	PC £86.88	1.80	16.30	94.70	nr	111.00
1200 x 1050 mm; ref SV1210/0	PC £106.47	1.80	16.30	116.05	nr	132.35
1200 x 1200 mm; ref SS1212/L	PC £91.56	1.95	17.66	99.80	nr	117.46
1200 x 1200 mm; ref SV1212/0	PC £108.36	1.95	17.66	118.11	nr	135.77
1200 x 1350 mm; ref SV1213/0	PC £110.11	2.10	19.02	120.02	nr	139.04
1200 x 1500 mm; ref SV1215/0	PC £112.28	2.20	19.93	122.39	nr	142.32
1800 x 900 mm; ref SS189D/0	PC £103.20	2.25	20.38	112.49	nr	132.87
1800 x 1050 mm; ref SS1810/L	PC £103.20	2.25	20.38	112.49	nr	132.87
1800 x 1050 mm; ref SS180D/0	PC £166.46	2.25	20.38	181.44	nr	201.82
1800 x 1200 mm; ref SS1812/L	PC £125.93	2.40	21.74	137.26	nr	159.00
1800 x 1200mm; ref SS182D/0	PC £150.36	2.40	21.74	163.89	nr	185.63
2400 x 1200 mm; ref SS242D/0	PC £166.68	2.80	25.36	181.68	nr	207.04

L WINDOWS/DOORS/STAIRS Including overheads and profit at 9.00%		Labour hours	Labour £	Material £	Unit	Total rate £

L10 TIMBER WINDOWS/ROOFLIGHTS/SCREENS/LOUVRES - cont'd

Standard windows; selected Philippine
Mahogany; preservative stain finish - cont'd
Top hung casement windows; with 45 x 140 mm
hardwood sills; weather stripping; opening
sashes on canopy hinges; fitted with
fasteners; aluminized lacquered finish
ironmongery

600 x 900 mm; ref ST0609/O	PC £59.71	0.95	8.61	65.08	nr	73.69
600 x 1050 mm; ref ST0610/O	PC £64.26	1.30	11.78	70.04	nr	81.82
900 x 900 mm; ref ST0909/O	PC £76.16	1.50	13.59	83.01	nr	96.60
900 x 1050 mm; ref ST0910/O	PC £78.75	1.60	14.49	85.84	nr	100.33
900 x 1200 mm; ref ST0912/O	PC £87.01	1.70	15.40	94.84	nr	110.24
900 x 1350 mm; ref ST0913/O	PC £91.84	1.80	16.30	100.11	nr	116.41
900 x 1500 mm; ref ST0915/O	PC £99.33	1.90	17.21	108.27	nr	125.48
1200 x 900 mm; ref ST1209/O	PC £88.69	1.90	17.21	96.67	nr	113.88
1200 x 1050 mm; ref ST1210/O	PC £82.62	1.80	16.30	90.06	nr	106.36
1200 x 1200 mm; ref ST1212/O	PC £89.76	1.95	17.66	97.84	nr	115.50
1200 x 1350 mm; ref ST1213/O	PC £111.02	2.10	19.02	121.01	nr	140.03
1200 x 1500 mm; ref ST1215/O	PC £96.24	2.15	19.47	104.90	nr	124.37
1500 x 1050 mm; ref ST1510/L	PC £118.02	2.10	19.02	128.64	nr	147.66
1500 x 1200 mm; ref ST1512/L	PC £128.52	2.20	19.93	140.09	nr	160.02
1500 x 1350 mm; ref ST1513/L	PC £136.22	2.30	20.83	148.48	nr	169.31
1500 x 1500 mm; ref ST1515/L	PC £145.81	2.50	22.64	158.93	nr	181.57
1800 x 1050 mm; ref ST1810/L	PC £109.44	2.25	20.38	119.29	nr	139.67
1800 x 1200 mm; ref ST1812/L	PC £118.38	2.40	21.74	129.03	nr	150.77
1800 x 1350 mm; ref ST1813/L	PC £124.92	2.50	22.64	136.16	nr	158.80
1800 x 1500 mm; ref ST1815/L	PC £155.47	2.70	24.46	169.46	nr	193.92

Purpose made window casements; selected
West African Mahogany; PC £494.00/m3
Casements; rebated; moulded

38 mm thick	PC £24.69	-	-	26.91	m2	26.91
50 mm thick	PC £32.17	-	-	35.07	m2	35.07

Casements; rebated; moulded; in medium panes

38 mm thick	PC £29.95	-	-	32.65	m2	32.65
50 mm thick	PC £40.08	-	-	43.69	m2	43.69

Casements; rebated; moulded; with
semi-circular head

38 mm thick	PC £37.44	-	-	40.81	m2	40.81
50 mm thick	PC £50.12	-	-	54.63	m2	54.63

Casements; rebated; moulded; to bullseye
window

38 mm thick x 600 mm dia	PC £81.04/nr	-	-	88.33	m2	88.33
38 mm thick x 900 mm dia	PC £121.29/nr	-	-	132.20	m2	132.20
50 mm thick x 600 mm dia	PC £93.18/nr	-	-	101.57	m2	101.57
50 mm thick x 900 mm dia	PC £138.96/nr	-	-	151.47	m2	151.47

Fitting and hanging casements

square or rectangular		0.70	6.34	-	nr	6.34
semicircular		1.70	15.40	-	nr	15.40
bullseye		2.70	24.46	-	nr	24.46

Purpose made window frames; selected West
African Mahogany; PC £494.00/m3
Frames; rounded; rebated check grooved

25 x 120 mm	PC £7.55	0.18	1.63	8.43	m	10.06
50 x 75 mm	PC £8.56	0.18	1.63	9.56	m	11.19
50 x 100 mm	PC £10.55	0.21	1.90	11.78	m	13.68
50 x 125 mm	PC £12.79	0.21	1.90	14.29	m	16.19
63 x 100 mm	PC £13.09	0.21	1.90	14.63	m	16.53
75 x 150 mm	PC £21.98	0.24	2.17	24.56	m	26.73
90 x 140 mm	PC £28.43	0.24	2.17	31.76	m	33.93

L WINDOWS/DOORS/STAIRS Including overheads and profit at 9.00%		Labour hours	Labour £	Material £	Unit	Total rate £
Mullions and transoms; twice rounded, rebated and check grooved						
50 x 75 mm	PC £10.36	0.14	1.27	11.58	m	12.85
50 x 100 mm	PC £12.34	0.16	1.45	13.79	m	15.24
63 x 100 mm	PC £14.88	0.16	1.45	16.63	m	18.08
75 x 150 mm	PC £23.77	0.18	1.63	26.56	m	28.19
Sill; sunk weathered, rebated and grooved						
75 x 100 mm	PC £21.52	0.27	2.45	24.05	m	26.50
75 x 150 mm	PC £28.37	0.27	2.45	31.70	m	34.15
Purpose made double hung sash windows; selected West African Mahogany; PC £494.00/m3 Cased frames of 100 x 25 mm grooved inner linings; 114 x 25 mm grooved outer linings; 125 x 38 mm twice rebated head linings; 125 x 32 mm twice rebated grooved pulley stiles; 150 x 13 mm linings; 50 x 19 mm parting slips; 25 x 13 mm parting beads; 25 x 19 mm inside beads; 150 x 75 mm Oak twice sunk weathered throated sill; 50 mm thick rebated and moulded sashes; moulded horns; over 1.25 m2 each; both sashes in medium panes;						
including spiral spring balances	PC £167.97	3.00	27.17	237.83	m2	265.00
As above but with cased mullions	PC £177.19	3.33	30.16	247.88	m2	278.04
Purpose made window casements; Afrormosia; **PC £706.00/m3** Casements; rebated; moulded						
38 mm thick	PC £30.01	-	-	32.71	m2	32.71
50 mm thick	PC £36.87	-	-	40.19	m2	40.19
Casements; rebated; moulded; in medium panes						
38 mm thick	PC £38.90	-	-	42.40	m2	42.40
50 mm thick	PC £46.73	-	-	50.94	m2	50.94
Casements; rebated; moulded; with semi-circular head						
38 mm thick	PC £48.63	-	-	53.00	m2	53.00
50 mm thick	PC £58.43	-	-	63.69	m2	63.69
Casements; rebated; moulded; to bullseye window						
38 mm thick x 600 mm dia	PC £98.57/nr	-	-	107.44	m2	107.44
38 mm thick x 900 mm dia	PC £147.28/nr	-	-	160.53	m2	160.53
50 mm thick x 600 mm dia	PC £113.36/nr	-	-	123.56	m2	123.56
50 mm thick x 900 mm dia	PC £169.51/nr	-	-	184.76	m2	184.76
Fitting and hanging casements						
square or rectangular		0.70	6.34	-	nr	6.34
semicircular		1.70	15.40	-	nr	15.40
bullseye		2.70	24.46	-	nr	24.46
Purpose made window frames; Afrormosia; **PC £706.00/m3** Frames; rounded; rebated check grooved						
25 x 120 mm	PC £8.71	0.18	1.63	9.74	m	11.37
50 x 75 mm	PC £9.94	0.18	1.63	11.11	m	12.74
50 x 100 mm	PC £12.40	0.21	1.90	13.85	m	15.75
50 x 125 mm	PC £15.11	0.21	1.90	16.89	m	18.79
63 x 100 mm	PC £15.42	0.21	1.90	17.23	m	19.13
75 x 150 mm	PC £26.15	0.24	2.17	29.21	m	31.38
90 x 140 mm	PC £34.00	0.24	2.17	37.99	m	40.16
Mullions and transoms; twice rounded, rebated and check grooved						
50 x 75 mm	PC £11.75	0.14	1.27	13.12	m	14.39
50 x 100 mm	PC £14.19	0.16	1.45	15.85	m	17.30
63 x 100 mm	PC £17.22	0.16	1.45	19.24	m	20.69
75 x 150 mm	PC £27.94	0.18	1.63	31.22	m	32.85

L WINDOWS/DOORS/STAIRS Including overheads and profit at 9.00%		Labour hours	Labour £	Material £	Unit	Total rate £
L10 TIMBER WINDOWS/ROOFLIGHTS/SCREENS/LOUVRES - cont'd						
Purpose made window frames; Afrormosia; **PC £706.00/m3 - cont'd**						
Sill; sunk weathered, rebated and grooved						
75 x 100 mm	PC £24.29	0.27	2.45	27.14	m	29.59
75 x 150 mm	PC £32.57	0.27	2.45	36.38	m	38.83
Purpose made double hung sash windows; **Afrormosia PC £706.00/m3**						
Cased frames of 100 x 25 mm grooved inner linings; 114 x 25 mm grooved outer linings; 125 x 38 mm twice rebated head linings; 125 x 32 mm twice rebated grooved pulley stiles; 150 x 13 mm linings; 50 x 19 mm parting slips; 25 x 13 mm parting beads; 25 x 19 mm inside beads; 150 x 75 mm Oak twice sunk weathered throated sill; 50 mm thick rebated and moulded sashes; moulded horns; over 1.25 m2 each; both sashes in medium panes;						
including spiral spring balances	PC £188.27	3.00	27.17	237.83	m2	265.00
As above but with cased mullions	PC £198.62	3.33	30.16	247.88	m2	278.04
L11 METAL WINDOWS/ROOFLIGHTS/SCREENS/LOUVRES						
Aluminium fixed and fanlight windows; **Crittall 'Luminaire' stock sliders or** **similar; white acrylic finish; fixed in** **position; including lugs plugged and screwed** **to brickwork or blockwork; or screwed to** **wooden sub-frame (measured elsewhere)**						
Fixed lights; factory glazed with 3, 4 or 5 mm OQ clear float glass						
600 x 900 mm; ref 6FL9A	PC £39.75	2.20	22.71	44.05	nr	66.76
900 x 1500 mm; ref 9FL15A	PC £56.25	3.15	32.52	62.03	nr	94.55
1200 x 900 mm; ref 12FL9A	PC £60.75	3.15	32.52	66.94	nr	99.46
1200 x 1200 mm; ref 12FL12A	PC £62.25	3.15	32.52	68.57	nr	101.09
1500 x 1200 mm; ref 15FL12A	PC £69.00	3.15	32.52	75.93	nr	108.45
1500 x 1500 mm; ref 15FL15A	PC £140.25	4.00	41.29	153.59	nr	194.88
Fixed lights; site double glazed with 11 mm clear float glass						
600 x 900 mm; ref 6FL9A	PC £53.25	3.00	30.97	58.76	nr	89.73
900 x 1500 mm; ref 9FL15A	PC £73.50	4.75	49.03	80.83	nr	129.86
1200 x 900 mm; ref 12FL9A	PC £85.50	5.25	54.19	93.91	nr	148.10
1200 x 1200 mm; ref 12FL12A	PC £84.75	5.45	56.26	93.10	nr	149.36
Fixed lights with fanlight over; factory glazed with 3, 4 or 5 mm OQ clear float glass						
600 x 900 mm; ref 6FV9A	PC £105.75	2.20	22.71	115.99	nr	138.70
600 x 1500 mm; ref 6FV15A	PC £117.00	2.20	22.71	128.25	nr	150.96
900 x 1200 mm; ref 9FV12A	PC £129.00	3.15	32.52	141.33	nr	173.85
900 x 1500 mm; ref 9FV15A	PC £138.00	3.15	32.52	151.14	nr	183.66
Fixed lights with fanlight over; site double glazed with 11 mm clear float glass						
600 x 900 mm; ref 6FV9A	PC £126.75	3.00	30.97	138.88	nr	169.85
600 x 1500 mm; ref 6FV15A	PC £151.50	3.55	36.64	165.85	nr	202.49
900 x 1200 mm; ref 9FV12A	PC £167.25	4.33	44.70	183.02	nr	227.72
900 x 1500 mm; ref 9FV15A	PC £180.75	5.25	54.19	197.74	nr	251.93
Vertical slider; factory glazed with 3; 4 or 5 mm OQ clear float glass						
600 x 900 mm; ref 6VV9A	PC £90.75	2.70	27.87	99.64	nr	127.51
900 x 1100 mm; ref 9VV11A	PC £107.25	2.70	27.87	117.62	nr	145.49
1200 x 1500 mm; ref 12VV15A	PC £130.50	3.75	38.71	142.96	nr	181.67
Vertical slider; factory double glazed with 11 mm clear float glass						
600 x 900 mm; ref 6VV9A	PC £147.75	3.40	35.10	161.77	nr	196.87
900 x 1100 mm; ref 9VV11A	PC £174.00	3.40	35.10	190.38	nr	225.48
1200 x 1500 mm; ref 12VV15A	PC £243.75	4.75	49.03	266.41	nr	315.44

L WINDOWS/DOORS/STAIRS Including overheads and profit at 9.00%		Labour hours	Labour £	Material £	Unit	Total rate £
Horizontal slider; factory glazed with 3, 4 or 5 mm OQ clear float glass						
1200 x 900 mm; ref 12HH9A	PC £116.25	3.75	38.71	127.43	nr	166.14
1200 x 1200 mm; ref 12HH12A	PC £126.75	3.75	38.71	138.88	nr	177.59
1500 x 900 mm; ref 15HH9A	PC £126.75	3.75	38.71	138.88	nr	177.59
1500 x 1300 mm; ref 15HH13A	PC £145.50	3.75	38.71	159.31	nr	198.02
1800 x 1200 mm; ref 18HH12A	PC £153.75	5.25	54.19	168.31	nr	222.50
Horizontal slider; factory double glazed with 11 mm clear float glass						
1200 x 900 mm; ref 12HH9A	PC £158.25	4.75	49.03	173.21	nr	222.24
1200 x 1200 mm; ref 12HH12A	PC £174.00	4.75	49.03	190.38	nr	239.41
1500 x 900 mm; ref 15HH9A	PC £173.25	4.75	49.03	189.56	nr	238.59
1500 x 1300 mm; ref 15HH13A	PC £204.75	4.75	49.03	223.90	nr	272.93
1800 x 1200 mm; ref 18HH12A	PC £216.75	5.75	59.35	236.98	nr	296.33
Galvanized steel fixed light; casement and **fanlight windows; Crittall; 'Homelight'** **range or similar; site glazing measured** **elsewhere; fixed in position; including lugs** **cut and pinned to brickwork or blockwork** Basic fixed lights; including easy-glaze beads						
628 x 292 mm; ref ZNG5	PC £10.31	1.20	12.39	11.54	nr	23.93
628 x 923 mm; ref ZNC5	PC £14.70	1.20	12.39	16.32	nr	28.71
628 x 1513 mm; ref ZNDV5	PC £20.91	1.20	12.39	23.09	nr	35.48
1237 x 292 mm; ref ZNG13	PC £15.75	1.20	12.39	17.47	nr	29.86
1237 x 923 mm; ref ZNC13	PC £20.38	1.75	18.06	22.51	nr	40.57
1237 x 1218 mm; ref ZND13	PC £23.99	1.75	18.06	26.45	nr	44.51
1237 x 1513 mm; ref ZNDV13	PC £26.28	1.75	18.06	28.94	nr	47.00
1846 x 292 mm; ref ZNG14	PC £20.42	1.20	12.39	22.55	nr	34.94
1846 x 923 mm; ref ZNC14	PC £25.16	1.75	18.06	27.73	nr	45.79
1846 x 1513 mm; ref ZNDV14	PC £30.48	2.20	22.71	33.52	nr	56.23
Basic opening lights; including easy-glaze beads and weatherstripping						
628 x 292 mm; ref ZNG1	PC £23.42	1.20	12.39	25.83	nr	38.22
1237 x 292 mm; ref ZNG13G	PC £34.76	1.20	12.39	38.19	nr	50.58
1846 x 292 mm; ref ZNG4	PC £59.45	1.20	12.39	65.10	nr	77.49
One-piece composites; including easy-glaze beads and weatherstripping						
628 x 923 mm; ref ZNC5F	PC £32.75	1.20	12.39	36.00	nr	48.39
628 x 1513 mm; ref ZNDV5F	PC £39.20	1.20	12.39	43.02	nr	55.41
1237 x 923 mm; ref ZNC2F	PC £64.64	1.75	18.06	70.75	nr	88.81
1237 x 1218 mm; ref ZND2F	PC £76.05	1.75	18.06	83.19	nr	101.25
1237 x 1513 mm; ref ZNDV2V	PC £91.29	1.75	18.06	99.81	nr	117.87
1846 x 923 mm; ref NC4F	PC £99.59	1.75	18.06	108.86	nr	126.92
1846 x 1218 mm; ref ZND10F	PC £96.50	2.20	22.71	105.48	nr	128.19
Reversible windows; including easy-glaze beads and weatherstripping						
997 x 923 mm; ref NC13R	PC £87.09	1.55	16.00	95.23	nr	111.23
997 x 1067 mm; ref NCO13R	PC £92.34	1.55	16.00	100.95	nr	116.95
1237 x 923 mm; ref ZNC13R	PC £96.20	2.30	23.74	105.16	nr	128.90
1237 x 1218 mm; ref ZND13R	PC £105.15	2.30	23.74	114.91	nr	138.65
1237 x 1513 mm; ref ZNDV13RS	PC £118.23	2.30	23.74	129.17	nr	152.91
Pressed steel sills; to suit above window widths						
628 mm	PC £4.97	0.35	3.61	6.14	nr	9.75
997 mm	PC £6.98	0.45	4.65	8.33	nr	12.98
1237 mm	PC £7.88	0.55	5.68	9.30	nr	14.98
1486 mm	PC £9.14	0.65	6.71	10.68	nr	17.39
1846 mm	PC £10.38	0.75	7.74	12.03	nr	19.77

L WINDOWS/DOORS/STAIRS Including overheads and profit at 9.00%	Labour hours	Labour £	Material £	Unit	Total rate £

L11 METAL WINDOWS/ROOFLIGHTS/SCREENS/LOUVRES - cont'd

**Factory finished steel fixed light; casement
and fanlight windows; Crittall polyester
powder coated 'Homelight' range or similar;
site glazing measured elsewhere; fixed in
position; including lugs cut and pinned to
brickwork or blockwork**
Basic fixed lights; including easy-glaze
beads

628 x 292 mm; ref ZNG5	PC £12.89	1.20	12.39	14.34	nr	26.73
628 x 923 mm; ref ZNC5	PC £18.37	1.20	12.39	20.32	nr	32.71
628 x 1513 mm; ref ZNDV5	PC £26.13	1.20	12.39	28.78	nr	41.17
1237 x 292 mm; ref ZNG13	PC £19.67	1.20	12.39	21.74	nr	34.13
1237 x 923 mm; ref ZNC13	PC £25.48	1.75	18.06	28.07	nr	46.13
1237 x 1218 mm; ref ZND13	PC £29.99	1.75	18.06	32.99	nr	51.05
1237 x 1513 mm; ref ZNDV13	PC £32.84	1.75	18.06	36.10	nr	54.16
1846 x 292 mm; ref ZNG14	PC £25.52	1.20	12.39	28.11	nr	40.50
1846 x 923 mm; ref ZNC14	PC £31.45	1.75	18.06	34.58	nr	52.64
1846 x 1513 mm; ref ZNDV14	PC £38.10	2.20	22.71	41.83	nr	64.54

Basic opening lights; including easy-glaze
beads and weatherstripping

628 x 292 mm; ref ZNG1	PC £28.49	1.20	12.39	31.35	nr	43.74
1237 x 292 mm; ref ZNG13G	PC £42.36	1.20	12.39	46.47	nr	58.86
1846 x 292 mm; ref ZNG4	PC £72.44	1.20	12.39	79.25	nr	91.64

One-piece composites; including easy-glaze
beads and weatherstripping

628 x 923 mm; ref ZNC5F	PC £40.10	1.20	12.39	44.01	nr	56.40
628 x 1513 mm; ref ZNDV5F	PC £48.16	1.20	12.39	52.79	nr	65.18
1237 x 923 mm; ref ZNC2F	PC £78.49	1.75	18.06	85.85	nr	103.91
1237 x 1218 mm; ref ZND2F	PC £92.45	1.75	18.06	101.06	nr	119.12
1237 x 1513 mm; ref ZNDV2V	PC £110.79	1.75	18.06	121.06	nr	139.12
1846 x 923 mm; ref NC4F	PC £120.67	1.75	18.06	131.83	nr	149.89
1846 x 1218 mm; ref ZND10F	PC £117.52	2.20	22.71	128.39	nr	151.10

Reversible windows; including easy-glaze
beads and weatherstripping

997 x 923 mm; ref NC13R	PC £113.21	1.55	16.00	123.70	nr	139.70
997 x 1067 mm; ref NCO13R	PC £120.05	1.55	16.00	131.15	nr	147.15
1237 x 923 mm; ref ZNC13R	PC £125.06	2.30	23.74	136.62	nr	160.36
1237 x 1218 mm; ref ZND13R	PC £136.69	2.30	23.74	149.29	nr	173.03
1237 x 1513 mm; ref ZNDV13RS	PC £153.70	2.30	23.74	167.83	nr	191.57

Pressed steel sills; to suit above window
widths

628 mm	PC £6.21	0.35	3.61	7.49	nr	11.10
997 mm	PC £8.72	0.45	4.65	10.23	nr	14.88
1237 mm	PC £9.84	0.55	5.68	11.45	nr	17.13
1486 mm	PC £11.43	0.65	6.71	13.18	nr	19.89
1846 mm	PC £12.97	0.75	7.74	14.85	nr	22.59

**L12 PLASTICS WINDOWS/ROOFLIGHTS/SCREENS/
LOUVRES**

**uPVC windows to BS 2782; 'Anglian' or
similar; reinforced where appropriate with
aluminium alloy; including standard
ironmongery; cills and glazing; fixed in
position; including lugs plugged and screwed
to brickwork or blockwork**
Fixed light; including e.p.d.m. glazing
gaskets and weather seals

600 x 900 mm; single glazed	PC £87.87	3.50	36.13	96.08	nr	132.21
600 x 900 mm; double glazed	PC £98.98	3.50	36.13	108.19	nr	144.32

Casement/fixed light; including e.p.d.m.
glazing gaskets and weather seals

600 x 1200 mm; single glazed	PC £126.25	3.75	38.71	137.91	nr	176.62
600 x 1200 mm; double glazed	PC £137.36	3.75	38.71	150.02	nr	188.73

L WINDOWS/DOORS/STAIRS Including overheads and profit at 9.00%		Labour hours	Labour £	Material £	Unit	Total rate £
1200 x 1200 mm; single glazed	PC £226.24	4.50	46.45	246.90	nr	293.35
1200 x 1200 mm; double glazed	PC £248.46	4.50	46.45	271.12	nr	317.57
1800 x 1200 mm; single glazed	PC £323.20	5.00	51.61	352.59	nr	404.20
1800 x 1200 mm; double glazed	PC £355.52	5.00	51.61	387.82	nr	439.43
'Tilt & Turn' light; including e.p.d.m. glazing gaskets and weather seals						
1200 x 1200 mm; single glazed	PC £181.80	4.50	46.45	198.46	nr	244.91
1200 x 1200 mm; double glazed	PC £203.01	4.50	46.45	221.58	nr	268.03
Cox's 'Skydome' or 'Coxdome'; plugged and screwed to concrete or screwed to timber Rooflight; 'Skydome Mark 3'; single skin; square or rectangular dome						
600 x 600 mm	PC £45.05	1.50	10.81	51.55	nr	62.36
900 x 600 mm	PC £93.15	1.65	11.89	106.61	nr	118.50
900 x 900 mm	PC £93.65	1.80	12.97	107.17	nr	120.14
1200 x 900 mm	PC £113.22	1.95	14.05	129.58	nr	143.63
1200 x 1200 mm	PC £117.54	2.10	15.13	134.53	nr	149.66
1800 x 1200 mm	PC £199.35	2.50	18.01	228.16	nr	246.17
Rooflight; 'Coxdome Mark 5' double skin; square or rectangular dome; extruded aluminium plain splayed upstand with 'hit and miss' ventilators two sides						
600 x 600 mm	PC £200.34	3.00	21.61	222.74	nr	244.35
900 x 600 mm	PC £273.01	3.30	23.78	303.54	nr	327.32
900 x 900 mm	PC £287.01	3.60	25.94	319.10	nr	345.04
1200 x 900 mm	PC £318.73	3.90	28.10	354.36	nr	382.46
1200 x 1200 mm	PC £362.57	4.20	30.26	403.10	nr	433.36
1800 x 1200 mm	PC £499.19	5.00	36.02	554.99	nr	591.01

ALTERNATIVE TIMBER DOOR PRICES (£/each)

	£		£		£
Hardwood doors					
Brazilian Mahogany period doors (£/each); 838 x 1981 x 44 mm					
6 panel	119.90	'Carolina'	106.70	'Kentucky'	122.55
8 panel	104.00	'Gothic'	117.20	'Elizabethan'	114.40
Red Meranti period doors					
6 panel					
762x1981x44 mm	105.05	838x1981x44 mm	105.05	807x2000x44 mm	105.05
8 panel					
762x1981x44 mm	104.00	838x1981x44 mm	104.00	807x2000x44 mm	104.00
Half bow					
762x1981x44 mm	122.10	838x1981x44 mm	122.10	807x2000x44 mm	122.10
'Kentucky'					
762x1981x44 mm	102.30	838x1981x44 mm	102.30	807x2000x44 mm	102.30
Softwood doors					
Casement doors					
2 XG; two panel; beaded					
762x1981x44 mm	29.90	838x1981x44 mm	31.20	807x2000x44 mm	31.20
813x2032x44 mm	31.20	726x2040x44 mm	31.20	826x2040x44 mm	31.20
2 XGG; two panel; beaded					
762x1981x44 mm	27.70	838x1981x44 mm	28.95	807x2000x44 mm	28.95
813x2032x44 mm	28.95	726x2040x44 mm	28.95	826x2040x44 mm	28.95
10; one panel; beaded					
762x1981x44 mm	23.40	838x1981x44 mm	24.35	807x2000x44 mm	24.35
SA; 15 panel; beaded					
762x1981x44 mm	37.00	838x1981x44 mm	38.50	807x2000x44 mm	38.50
22; pair of two panel; beaded					
914x1981x44 mm	61.80	1168x1981x44 mm	61.80		
2 SA; pair of 10 panel; beaded					
914x1981x44 mm	81.15	1168x1981x44 mm	81.15		

L WINDOWS/DOORS/STAIRS
Including overheads and profit at 9.00%

Louvre doors

533x1524x28 mm	12.05	610x1524x28 mm	13.20	686x1981x28 mm	17.90
533x1676x28 mm	13.05	610x1676x28 mm	14.45	762x1981x28 mm	19.85
533x1829x28 mm	13.90	610x1829x28 mm	15.35	- pair of 737x	
533x1981x28 mm	14.80	610x1981x28 mm	16.45	1067x28 mm doors	14.60

Period doors
6 panel

762x1981x44 mm	80.30	838x1981x44 mm	80.30	807x2000x44 mm	80.30

8 panel

762x1981x44 mm	78.80	838x1981x44 mm	78.80	807x2000x44 mm	78.80

Half bow

762x1981x44 mm	77.55	838x1981x44 mm	77.55	807x2000x44 mm	77.55

'Kentucky'

762x1981x44 mm	70.95	838x1981x44 mm	70.95	807x2000x44 mm	70.95

	Labour hours	Labour £	Material £	Unit	Total rate £
L20 TIMBER DOORS/SHUTTERS/HATCHES					
Standard matchboarded doors; wrought softwood					
Matchboarded, ledged and braced doors; 25 mm ledges and braces; 19 mm tongued, grooved and V-jointed; one side vertical boarding					
762 x 1981 mm PC £31.61	1.50	13.59	34.46	nr	48.05
838 x 1981 mm PC £33.50	1.50	13.59	36.51	nr	50.10
Matchboarded, framed, ledged and braced doors; 44 mm framing; 25 mm intermediate and bottom rails; 19 mm tongued, grooved and V-jointed; one side vertical boarding					
762 x 1981 x 44 mm PC £41.13	1.80	16.30	44.83	nr	61.13
838 x 1981 x 44 mm PC £42.64	1.80	16.30	46.47	nr	62.77
Standard flush doors; softwood composition					
Flush door; internal quality; skeleton or cellular core; hardboard faced both sides;					
457 x 1981 x 35 mm PC £10.05	1.25	11.32	10.95	nr	22.27
533 x 1981 x 35 mm PC £10.05	1.25	11.32	10.95	nr	22.27
610 x 1981 x 35 mm PC £11.30	1.25	11.32	10.95	nr	22.27
686 x 1981 x 35 mm PC £10.05	1.25	11.32	10.95	nr	22.27
762 x 1981 x 35 mm PC £10.05	1.25	11.32	10.95	nr	22.27
838 x 1981 x 35 mm PC £10.50	1.25	11.32	11.45	nr	22.77
526 x 2040 x 40 mm PC £11.30	1.25	11.32	12.32	nr	23.64
626 x 2040 x 40 mm PC £11.30	1.25	11.32	12.32	nr	23.64
726 x 2040 x 40 mm PC £11.30	1.25	11.32	12.32	nr	23.64
826 x 2040 x 40 mm PC £11.80	1.25	11.32	12.86	nr	24.18
Flush door; internal quality; skeleton or cellular core; plywood faced both sides; lipped on two long edges					
457 x 1981 x 35 mm PC £13.05	1.25	11.32	14.22	nr	25.54
533 x 1981 x 35 mm PC £13.05	1.25	11.32	14.22	nr	25.54
610 x 1981 x 35 mm PC £13.05	1.25	11.32	14.22	nr	25.54
686 x 1981 x 35 mm PC £13.05	1.25	11.32	14.22	nr	25.54
762 x 1981 x 35 mm PC £13.05	1.25	11.32	14.22	nr	25.54
838 x 1981 x 35 mm PC £13.75	1.25	11.32	14.99	nr	26.31
526 x 2040 x 40 mm PC £13.75	1.25	11.32	14.99	nr	26.31
626 x 2040 x 40 mm PC £13.75	1.25	11.32	14.99	nr	26.31
726 x 2040 x 40 mm PC £13.75	1.25	11.32	14.99	nr	26.31
826 x 2040 x 40 mm PC £14.25	1.25	11.32	15.53	nr	26.85
Flush door; internal quality; skeleton or cellular core; Sapele faced both sides; lipped on all four edges					
457 x 1981 x 35 mm PC £14.70	1.35	12.23	16.02	nr	28.25
533 x 1981 x 35 mm PC £14.70	1.35	12.23	16.02	nr	28.25
610 x 1981 x 35 mm PC £14.70	1.35	12.23	16.02	nr	28.25
686 x 1981 x 35 mm PC £14.70	1.35	12.23	16.02	nr	28.25
762 x 1981 x 35 mm PC £14.70	1.35	12.23	16.02	nr	28.25
838 x 1981 x 35 mm PC £14.70	1.35	12.23	16.02	nr	28.25

L WINDOWS/DOORS/STAIRS Including overheads and profit at 9.00%		Labour hours	Labour £	Material £	Unit	Total rate £
Flush door; internal quality; skeleton or cellular core; Teak faced both sides; lipped on all four edges						
457 x 1981 x 35 mm	PC £36.10	1.35	12.23	39.35	nr	51.58
533 x 1981 x 35 mm	PC £31.75	1.35	12.23	34.61	nr	46.84
610 x 1981 x 35 mm	PC £31.75	1.35	12.23	34.61	nr	46.84
686 x 1981 x 35 mm	PC £31.75	1.35	12.23	34.61	nr	46.84
762 x 1981 x 35 mm	PC £37.75	1.35	12.23	41.15	nr	53.38
838 x 1981 x 35 mm	PC £32.70	1.35	12.23	35.64	nr	47.87
526 x 2040 x 40 mm	PC £32.40	1.35	12.23	35.32	nr	47.55
626 x 2040 x 40 mm	PC £32.40	1.35	12.23	35.32	nr	47.55
726 x 2040 x 40 mm	PC £32.40	1.35	12.23	35.32	nr	47.55
826 x 2040 x 40 mm		1.35	12.23	35.64	nr	47.87
Flush door; half hour fire-resisting (30/30); 'Melador'; laminate faced both sides; hardwood lipped on all edges						
610 x 1981 x 47 mm	PC £171.68	2.25	20.38	187.13	nr	207.51
686 x 1981 x 47 mm	PC £171.68	2.25	20.38	187.13	nr	207.51
762 x 1981 x 47 mm	PC £171.68	2.25	20.38	187.13	nr	207.51
838 x 1981 x 47 mm	PC £171.68	2.25	20.38	187.13	nr	207.51
526 x 2040 x 47 mm	PC £171.68	2.25	20.38	187.13	nr	207.51
626 x 2040 x 47 mm	PC £171.68	2.25	20.38	187.13	nr	207.51
726 x 2040 x 47 mm	PC £171.68	2.25	20.38	187.13	nr	207.51
826 x 2040 x 47 mm	PC £171.68	2.25	20.38	187.13	nr	207.51
Flush door; half-hour fire check (30/20); hardboard faced both sides;						
457 x 1981 x 44 mm	PC £26.26	1.75	15.85	28.62	nr	44.47
533 x 1981 x 44 mm	PC £26.26	1.75	15.85	28.62	nr	44.47
610 x 1981 x 44 mm	PC £26.26	1.75	15.85	28.62	nr	44.47
686 x 1981 x 44 mm	PC £26.26	1.75	15.85	28.62	nr	44.47
762 x 1981 x 44 mm	PC £26.26	1.75	15.85	28.62	nr	44.47
838 x 1981 x 44 mm	PC £27.62	1.75	15.85	30.11	nr	45.96
526 x 2040 x 44 mm	PC £27.27	1.75	15.85	29.73	nr	45.58
626 x 2040 x 44 mm	PC £27.27	1.75	15.85	29.73	nr	45.58
726 x 2040 x 44 mm	PC £27.27	1.75	15.85	29.73	nr	45.58
826 x 2040 x 44 mm	PC £28.65	1.75	15.85	31.23	nr	47.08
Flush door; half-hour fire check (30/20); plywood faced both sides; lipped on all four edges						
457 x 1981 x 44 mm	PC £25.32	1.75	15.85	27.60	nr	43.45
533 x 1981 x 44 mm	PC £25.32	1.75	15.85	27.60	nr	43.45
610 x 1981 x 44 mm	PC £25.32	1.75	15.85	27.60	nr	43.45
686 x 1981 x 44 mm	PC £25.32	1.75	15.85	27.60	nr	43.45
762 x 1981 x 44 mm	PC £25.32	1.75	15.85	27.60	nr	43.45
838 x 1981 x 44 mm	PC £26.61	1.75	15.85	29.01	nr	44.86
526 x 2040 x 44 mm	PC £25.32	1.75	15.85	27.60	nr	43.45
626 x 2040 x 44 mm	PC £25.32	1.75	15.85	27.60	nr	43.45
726 x 2040 x 44 mm	PC £26.01	1.75	15.85	28.35	nr	44.20
826 x 2040 x 44 mm	PC £27.34	1.75	15.85	29.80	nr	45.65
Flush door; half-hour fire check (30/20); Sapele faced both sides; lipped on all four edges						
457 x 1981 x 44 mm	PC £35.80	1.85	16.76	39.02	nr	55.78
533 x 1981 x 44 mm	PC £35.80	1.85	16.76	39.02	nr	55.78
610 x 1981 x 44 mm	PC £35.80	1.85	16.76	39.02	nr	55.78
686 x 1981 x 44 mm	PC £35.80	1.85	16.76	39.02	nr	55.78
762 x 1981 x 44 mm	PC £35.80	1.85	16.76	39.02	nr	55.78
526 x 2040 x 44 mm	PC £36.73	1.85	16.76	40.03	nr	56.79
626 x 2040 x 44 mm	PC £36.73	1.85	16.76	40.03	nr	56.79
726 x 2040 x 44 mm	PC £36.73	1.85	16.76	40.03	nr	56.79
826 x 2040 x 44 mm	PC £38.11	1.85	16.76	41.54	nr	58.30

L WINDOWS/DOORS/STAIRS		Labour	Labour	Material		Total
Including overheads and profit at 9.00%		hours	£	£	Unit	rate £

L20 TIMBER DOORS/SHUTTERS/HATCHES - cont'd

Standard flush doors; softwood composition - cont'd
Flush door; half-hour fire resisting (30/30)
Sapele faced both sides; lipped on all four
edges

457 x 1981 x 44 mm	PC £29.58	1.85	16.76	32.24	nr	49.00
533 x 1981 x 44 mm	PC £30.00	1.85	16.76	32.70	nr	49.46
610 x 1981 x 44 mm	PC £30.36	1.85	16.76	33.09	nr	49.85
686 x 1981 x 44 mm	PC £36.09	1.85	16.76	39.33	nr	56.09
762 x 1981 x 44 mm	PC £37.00	1.85	16.76	40.32	nr	57.08
838 x 1981 x 44 mm	PC £38.61	1.85	16.76	42.08	nr	58.84
526 x 2040 x 44 mm	PC £35.49	1.85	16.76	38.68	nr	55.44
626 x 2040 x 44 mm	PC £36.05	1.85	16.76	39.29	nr	56.05
726 x 2040 x 44 mm	PC £37.42	1.85	16.76	40.78	nr	57.54
826 x 2040 x 44 mm	PC £39.31	1.85	16.76	42.84	nr	59.60

Flush door; one hour fire check (60/60);
plywood faced both sides; lipped on all
four edges

610 x 1981 x 54 mm	PC £133.74	2.00	18.12	145.77	nr	163.89
686 x 1981 x 54 mm	PC £133.74	2.00	18.12	145.77	nr	163.89
762 x 1981 x 54 mm	PC £87.11	2.00	18.12	94.95	nr	113.07
838 x 1981 x 54 mm	PC £93.56	2.00	18.12	101.98	nr	120.10
526 x 2040 x 54 mm	PC £133.74	2.00	18.12	145.77	nr	163.89
626 x 2040 x 54 mm	PC £133.74	2.00	18.12	145.77	nr	163.89
726 x 2040 x 54 mm	PC £87.11	2.00	18.12	94.95	nr	113.07
826 x 2040 x 54 mm	PC £93.56	2.00	18.12	101.98	nr	120.10

Flush door; one hour fire check (60/45);
Sapele faced both sides; lipped on all
four edges

610 x 1981 x 54 mm	PC £139.87	2.10	19.02	152.46	nr	171.48
686 x 1981 x 54 mm	PC £139.87	2.10	19.02	152.46	nr	171.48
762 x 2040 x 54 mm	PC £139.87	2.10	19.02	152.46	nr	171.48
838 x 1981 x 54 mm	PC £139.87	2.10	19.02	152.46	nr	171.48
526 x 2040 x 54 mm	PC £139.87	2.10	19.02	152.46	nr	171.48
626 x 2040 x 54 mm	PC £139.87	2.10	19.02	152.46	nr	171.48
762 x 2040 x 54 mm	PC £139.87	2.10	19.02	152.46	nr	171.48
826 x 2040 x 54 mm	PC £139.87	2.10	19.02	152.46	nr	171.48

Flush door; one hour fire resisting (60/60);
Sapele faced both sides; lipped on all
four edges

610 x 1981 x 54 mm	PC £157.19	2.10	19.02	171.33	nr	190.35
686 x 1981 x 54 mm	PC £157.19	2.10	19.02	171.33	nr	190.35
762 x 1981 x 54 mm	PC £157.19	2.10	19.02	171.33	nr	190.35
838 x 1981 x 54 mm	PC £157.19	2.10	19.02	171.33	nr	190.35
526 x 2040 x 54 mm	PC £157.19	2.10	19.02	171.33	nr	190.35
626 x 2040 x 54 mm	PC £157.19	2.10	19.02	171.33	nr	190.35
726 x 2040 x 54 mm	PC £157.19	2.10	19.02	171.33	nr	190.35
826 x 2040 x 54 mm	PC £157.19	2.10	19.02	171.33	nr	190.35

Flush door; one hour fire-resisting (60/60);
'Melador'; laminate faced both sides;
hardwood lipped on all edges

610 x 1981 x 57 mm	PC £238.03	2.75	24.91	259.45	nr	284.36
686 x 1981 x 57 mm	PC £238.03	2.75	24.91	259.45	nr	284.36
762 x 1981 x 57 mm	PC £238.03	2.75	24.91	259.45	nr	284.36
838 x 1981 x 57 mm	PC £238.03	2.75	24.91	259.45	nr	284.36
526 x 2040 x 57 mm	PC £238.03	2.75	24.91	259.45	nr	284.36
626 x 2040 x 57 mm	PC £238.03	2.75	24.91	259.45	nr	284.36
726 x 2040 x 57 mm	PC £238.03	2.75	24.91	259.45	nr	284.36
826 x 2040 x 57 mm	PC £238.03	2.75	24.91	259.45	nr	284.36

Flush door; external quality; skeleton or
cellular core; plywood faced both sides;
lipped on all four edges

762 x 1981 x 44 mm	PC £23.28	1.50	13.59	25.38	nr	38.97
838 x 1981 x 44 mm	PC £24.33	1.50	13.59	26.52	nr	40.11

L WINDOWS/DOORS/STAIRS Including overheads and profit at 9.00%		Labour hours	Labour £	Material £	Unit	Total rate £
Flush door; external quality with standard glass opening; skeleton or cellular core; plywood faced both sides; lipped on all four edges; including glazing beads						
762 x 1981 x 44 mm	PC £28.95	1.75	15.85	31.56	nr	47.41
838 x 1981 x 44 mm	PC £30.03	1.75	15.85	32.73	nr	48.58
Purpose made panelled doors; wrought softwood Panelled doors; one open panel for glass; including glazing beads						
686 x 1981 x 44 mm	PC £40.08	1.75	15.85	43.69	nr	59.54
762 x 1981 x 44 mm	PC £41.34	1.75	15.85	45.06	nr	60.91
838 x 1981 x 44 mm	PC £42.57	1.75	15.85	46.40	nr	62.25
Panelled doors; two open panel for glass; including glazing beads						
686 x 1981 x 44 mm	PC £55.10	1.75	15.85	60.06	nr	75.91
762 x 1981 x 44 mm	PC £58.05	1.75	15.85	63.28	nr	79.13
838 x 1981 x 44 mm	PC £60.93	1.75	15.85	66.41	nr	82.26
Panelled doors; four 19 mm thick plywood panel; mouldings worked on solid both sides						
686 x 1981 x 44 mm	PC £77.45	1.75	15.85	84.42	nr	100.27
762 x 1981 x 44 mm	PC £80.69	1.75	15.85	87.95	nr	103.80
838 x 1981 x 44 mm	PC £84.75	1.75	15.85	92.38	nr	108.23
Panelled doors; six 19 mm thick panels raised and fielded; mouldings worked on solid both sides						
686 x 1981 x 50 mm	PC £166.21	2.10	19.02	181.17	nr	200.19
762 x 1981 x 50 mm	PC £173.11	2.10	19.02	188.69	nr	207.71
838 x 1981 x 50 mm	PC £181.78	2.10	19.02	198.14	nr	217.16
Rebated edges beaded		-	-	-	m	0.95
Rounded edges or heels		-	-	-	m	0.48
Weatherboard; fixed to bottom rail		0.25	2.26	2.04	m	4.30
Stopped groove for weatherboard		-	-	-	m	0.33
Purpose made panelled doors; selected West African Mahogany; PC £494.00/m3 Panelled doors; one open panel for glass; mouldings worked on the solid one side; 19 x 13 mm beads one side; fixing with brass screws and cups						
686 x 1981 x 50 mm	PC £70.63	2.50	22.64	76.99	nr	99.63
762 x 1981 x 50 mm	PC £72.94	2.50	22.64	79.50	nr	102.14
838 x 1981 x 50 mm	PC £75.23	2.50	22.64	82.00	nr	104.64
686 x 1981 x 63 mm	PC £85.60	2.75	24.91	93.30	nr	118.21
762 x 1981 x 63 mm	PC £88.48	2.75	24.91	96.44	nr	121.35
838 x 1981 x 63 mm	PC £91.31	2.75	24.91	99.52	nr	124.43
Panelled doors; 250 mm wide cross-tongued intermediate rail; two open panels for glass mouldings worked on the solid one side; 19 x 13 mm beads one side; fixing with brass screws and cups						
686 x 1981 x 50 mm	PC £96.47	2.50	22.64	105.15	nr	127.79
762 x 1981 x 50 mm	PC £101.15	2.50	22.64	110.25	nr	132.89
838 x 1981 x 50 mm	PC £105.80	2.50	22.64	115.32	nr	137.96
686 x 1981 x 63 mm	PC £115.51	2.75	24.91	125.91	nr	150.82
762 x 1981 x 63 mm	PC £121.18	2.75	24.91	132.08	nr	156.99
838 x 1981 x 63 mm	PC £126.82	2.75	24.91	138.23	nr	163.14
Panelled doors; four panels; (19 mm for 50 mm thick doors and 25 mm for 63 mm thick doors); mouldings worked on solid both sides						
686 x 1981 x 50 mm	PC £123.03	2.50	22.64	134.11	nr	156.75
762 x 1981 x 50 mm	PC £128.15	2.50	22.64	139.68	nr	162.32
838 x 1981 x 50 mm	PC £134.56	2.50	22.64	146.67	nr	169.31
686 x 1981 x 63 mm	PC £144.81	2.75	24.91	157.84	nr	182.75
762 x 1981 x 63 mm	PC £150.86	2.75	24.91	164.44	nr	189.35
838 x 1981 x 63 mm	PC £158.39	2.75	24.91	172.64	nr	197.55

L WINDOWS/DOORS/STAIRS Including overheads and profit at 9.00%		Labour hours	Labour £	Material £	Unit	Total rate £
L20 TIMBER DOORS/SHUTTERS/HATCHES - cont'd						
Purpose made panelled doors; selected West						
African Mahogany; PC £494.00/m3 - cont'd						
Panelled doors; 150 mm wide stiles in one						
width; 430 mm wide cross-tongued bottom						
rail; six panels raised and fielded one side						
(19 mm thick for 50 mm thick doors and 25						
mm thick for 63 mm thick doors); mouldings						
worked on the solid both sides						
686 x 1981 x 50 mm	PC £258.48	2.50	22.64	281.75	nr	304.39
762 x 1981 x 50 mm	PC £269.26	2.50	22.64	293.49	nr	316.13
838 x 1981 x 50 mm	PC £282.71	2.50	22.64	308.15	nr	330.79
686 x 1981 x 63 mm	PC £316.89	2.75	24.91	345.41	nr	370.32
762 x 1981 x 63 mm	PC £330.10	2.75	24.91	359.81	nr	384.72
838 x 1981 x 63 mm	PC £346.61	2.75	24.91	377.80	nr	402.71
Rebated edges beaded		-	-	-	m	1.42
Rounded edges or heels		-	-	-	m	0.71
Weatherboard; fixed to bottom rail		0.33	2.99	3.88	m	6.87
Stopped groove for weatherboard		-	-	-	m	0.64
Purpose made panelled doors; Afrormosia;						
PC £706.00/m3						
Panelled doors; one open panel for glass;						
mouldings worked on the solid one side;						
19 x 13 mm beads one side; fixing with brass						
screws and cups						
686 x 1981 x 50 mm	PC £82.27	2.50	22.64	89.67	nr	112.31
762 x 1981 x 50 mm	PC £85.01	2.50	22.64	92.66	nr	115.30
838 x 1981 x 50 mm	PC £87.75	2.50	22.64	95.64	nr	118.28
686 x 1981 x 63 mm	PC £100.11	2.75	24.91	109.12	nr	134.03
762 x 1981 x 63 mm	PC £103.56	2.75	24.91	112.88	nr	137.79
838 x 1981 x 63 mm	PC £106.96	2.75	24.91	116.58	nr	141.49
Panelled doors; 250 mm wide cross-tongued						
intermediate rail; two open panels for glass						
mouldings worked on the solid one side;						
19 x 13 mm beads one side; fixing with brass						
screws and cups						
686 x 1981 x 50 mm	PC £111.81	2.50	22.64	121.88	nr	144.52
762 x 1981 x 50 mm	PC £117.36	2.50	22.64	127.93	nr	150.57
838 x 1981 x 50 mm	PC £122.87	2.50	22.64	133.93	nr	156.57
686 x 1981 x 63 mm	PC £134.68	2.75	24.91	146.80	nr	171.71
762 x 1981 x 63 mm	PC £141.40	2.75	24.91	154.13	nr	179.04
838 x 1981 x 63 mm	PC £148.11	2.75	24.91	161.44	nr	186.35
Panelled doors; four panels; (19 mm for						
50 mm thick doors and 25 mm for 63 mm thick						
doors); mouldings worked on solid both sides						
686 x 1981 x 50 mm	PC £142.46	2.50	22.64	155.28	nr	177.92
762 x 1981 x 50 mm	PC £148.40	2.50	22.64	161.75	nr	184.39
838 x 1981 x 50 mm	PC £155.80	2.50	22.64	169.82	nr	192.46
686 x 1981 x 63 mm	PC £169.05	2.75	24.91	184.27	nr	209.18
762 x 1981 x 63 mm	PC £176.09	2.75	24.91	191.94	nr	216.85
838 x 1981 x 63 mm	PC £184.90	2.75	24.91	201.54	nr	226.45
Panelled doors; 150 mm wide stiles in one						
width; 430 mm wide cross-tongued bottom						
rail; six panels raised and fielded one side						
(19 mm thick for 50 mm thick doors and 25						
mm thick for 63 mm thick doors); mouldings						
worked on the solid both sides						
686 x 1981 x 50 mm	PC £289.18	2.50	22.64	315.21	nr	337.85
762 x 1981 x 50 mm	PC £301.24	2.50	22.64	328.36	nr	351.00
838 x 1981 x 50 mm	PC £316.30	2.50	22.64	344.77	nr	367.41
686 x 1981 x 63 mm	PC £355.75	2.75	24.91	387.76	nr	412.67
762 x 1981 x 63 mm	PC £370.57	2.75	24.91	403.93	nr	428.84
838 x 1981 x 63 mm	PC £389.11	2.75	24.91	424.13	nr	449.04
Rebated edges beaded		-	-	-	m	1.42
Rounded edges or heels		-	-	-	m	0.71
Weatherboard; fixed to bottom rail		0.33	2.99	4.39	m	7.38
Stopped groove for weatherboard		-	-	-	m	0.64

L WINDOWS/DOORS/STAIRS Including overheads and profit at 9.00%		Labour hours	Labour £	Material £	Unit	Total rate £

Standard joinery sets; wrought softwood
Internal door frame or lining set for 686 x
1981 mm door; all with loose stops unless
rebated; 'finished sizes'

27 x 94 mm lining	PC £12.99	0.80	7.25	14.16	nr	21.41
27 x 107 mm lining	PC £13.65	0.80	7.25	14.88	nr	22.13
35 x 107 mm rebated lining	PC £14.16	0.80	7.25	15.43	nr	22.68
27 x 121 mm lining	PC £14.73	0.80	7.25	16.06	nr	23.31
27 x 121 mm lining with fanlight over						
	PC £20.82	0.95	8.61	22.69	nr	31.30
27 x 133 mm lining	PC £15.87	0.80	7.25	17.30	nr	24.55
35 x 133 mm rebated linings	PC £16.29	0.80	7.25	17.76	nr	25.01
27 x 133 mm lining with fanlight over						
	PC £22.32	0.95	8.61	24.33	nr	32.94
33 x 57 mm frame	PC £10.59	0.80	7.25	11.54	nr	18.79
33 x 57 mm storey height frame	PC £13.02	0.85	7.70	14.19	nr	21.89
33 x 57 mm frame with fanlight over	PC £15.90	0.95	8.61	17.33	nr	25.94
33 x 64 mm frame	PC £11.37	0.80	7.25	12.39	nr	19.64
33 x 64 mm storey height frame	PC £13.86	0.85	7.70	15.11	nr	22.81
33 x 64 mm frame with fanlight over	PC £16.71	0.95	8.61	18.21	nr	26.82
44 x 94 mm frame	PC £17.37	0.92	8.33	18.93	nr	27.26
44 x 94 mm storey height frame	PC £20.67	1.00	9.06	22.53	nr	31.59
44 x 94 mm frame with fanlight over	PC £24.24	1.10	9.96	26.42	nr	36.38
44 x 107 mm frame	PC £19.95	0.92	8.33	21.75	nr	30.08
44 x 107 mm storey height frame	PC £23.37	1.00	9.06	25.47	nr	34.53
44 x 107 mm frame with fanlight over						
	PC £26.76	1.10	9.96	29.17	nr	39.13

Internal door frame or lining set for
762 x 1981 mm door; all with loose stops
unless rebated; 'finished sizes'

27 x 94 mm lining	PC £12.99	0.80	7.25	14.16	nr	21.41
27 x 107 mm lining	PC £13.65	0.80	7.25	14.88	nr	22.13
35 x 107 mm rebated lining	PC £14.16	0.80	7.25	15.43	nr	22.68
27 x 121 mm lining	PC £14.73	0.80	7.25	16.06	nr	23.31
27 x 121 mm lining with fanlight over						
	PC £20.82	0.95	8.61	22.69	nr	31.30
27 x 133 mm lining	PC £15.87	0.80	7.25	17.30	nr	24.55
35 x 133 mm rebated linings	PC £16.29	0.80	7.25	17.76	nr	25.01
27 x 133 mm lining with fanlight over						
	PC £22.32	0.95	8.61	24.33	nr	32.94
33 x 57 mm frame	PC £10.59	0.80	7.25	11.54	nr	18.79
33 x 57 mm storey height frame	PC £13.02	0.85	7.70	14.19	nr	21.89
33 x 57 mm frame with fanlight over	PC £15.90	0.95	8.61	17.33	nr	25.94
33 x 64 mm frame	PC £11.37	0.80	7.25	12.39	nr	19.64
33 x 64 mm storey height frame	PC £13.86	0.85	7.70	15.11	nr	22.81
33 x 64 mm frame with fanlight over	PC £16.71	0.95	8.61	18.21	nr	26.82
44 x 94 mm frame	PC £17.37	0.92	8.33	18.93	nr	27.26
44 x 94 mm storey height frame	PC £20.67	1.00	9.06	22.53	nr	31.59
44 x 94 mm frame with fanlight over	PC £24.24	1.10	9.96	26.42	nr	36.38
44 x 107 mm frame	PC £19.95	0.92	8.33	21.75	nr	30.08
44 x 107 mm storey height frame	PC £23.37	1.00	9.06	25.47	nr	34.53
44 x 107 mm frame with fanlight over						
	PC £26.76	1.10	9.96	29.17	nr	39.13

Internal door frame or lining set for
726 x 2040 mm door; with loose stops

30 x 94 mm lining	PC £15.09	0.80	7.25	16.45	nr	23.70
30 x 94 mm lining with fanlight over						
	PC £21.60	0.95	8.61	23.54	nr	32.15
30 x 107 mm lining	PC £17.76	0.80	7.25	19.36	nr	26.61
30 x 107 mm lining with fanlight over						
	PC £24.36	0.95	8.61	26.55	nr	35.16
30 x 133 mm lining	PC £19.83	0.80	7.25	21.61	nr	28.86
30 x 133 mm lining with fanlight over						
	PC £26.25	0.95	8.61	28.61	nr	37.22

L WINDOWS/DOORS/STAIRS Including overheads and profit at 9.00%		Labour hours	Labour £	Material £	Unit	Total rate £
L20 TIMBER DOORS/SHUTTERS/HATCHES - cont'd						
Standard joinery sets; wrought softwood - cont'd Internal door frame or lining set for 826 x 2040 mm door; with loose stops						
30 x 94 mm lining	PC £15.09	0.80	7.25	16.45	nr	**23.70**
30 x 94 mm lining with fanlight over						
	PC £21.60	0.95	8.61	23.54	nr	**32.15**
30 x 107 mm lining	PC £17.76	0.80	7.25	19.36	nr	**26.61**
30 x 107 mm lining with fanlight over						
	PC £24.36	0.95	8.61	26.55	nr	**35.16**
30 x 133 mm lining	PC £19.83	0.80	7.25	21.61	nr	**28.86**
30 x 133 mm lining with fanlight over		0.95	8.61	28.61		**37.22**
Trap door set; 9 mm plywood in 35 x 107 mm (fin) rebated lining						
762 x 762 mm	PC £16.80	0.80	7.25	18.58	nr	**25.83**
Serving hatch set; plywood faced doors; rebated meeting stiles; hung on nylon hinges; in 35 x 140 mm (fin) rebated lining						
648 x 533 mm	PC £32.01	1.30	11.78	35.22	nr	**47.00**
Purpose made door frames and lining sets; **wrought softwood** Jambs and heads; as linings						
32 x 63 mm		0.17	1.54	3.20	m	**4.74**
32 x 100 mm		0.17	1.54	4.22	m	**5.76**
32 x 140 mm		0.17	1.54	5.48	m	**7.02**
Jambs and heads; as frames; rebated, rounded and grooved						
38 x 75 mm		0.17	1.54	4.15	m	**5.69**
38 x 100 mm		0.17	1.54	4.98	m	**6.52**
38 x 115 mm		0.17	1.54	5.81	m	**7.35**
38 x 140 mm		0.20	1.81	6.78	m	**8.59**
50 x 100 mm		0.20	1.81	6.23	m	**8.04**
50 x 125 mm		0.20	1.81	7.46	m	**9.27**
63 x 88 mm		0.20	1.81	7.31	m	**9.12**
63 x 100 mm		0.20	1.81	7.71	m	**9.52**
63 x 125 mm		0.20	1.81	9.19	m	**11.00**
75 x 100 mm		0.20	1.81	8.82	m	**10.63**
75 x 125 mm		0.22	1.99	10.91	m	**12.90**
75 x 150 mm		0.22	1.99	12.78	m	**14.77**
100 x 100 mm		0.22	1.99	11.49	m	**13.48**
100 x 150 mm		0.22	1.99	16.58	m	**18.57**
Mullions and transoms; in linings						
32 x 63 mm		0.12	1.09	5.19	m	**6.28**
32 x 100 mm		0.12	1.09	6.21	m	**7.30**
32 x 140 mm		0.12	1.09	7.48	m	**8.57**
Mullions and transoms; in frames; twice rebated, rounded and grooved						
38 x 75 mm		0.12	1.09	6.25	m	**7.34**
38 x 100 mm		0.12	1.09	7.08	m	**8.17**
38 x 115 mm		0.12	1.09	7.79	m	**8.88**
38 x 140 mm		0.14	1.27	8.75	m	**10.02**
50 x 100 mm		0.14	1.27	8.23	m	**9.50**
50 x 125 mm		0.14	1.27	9.56	m	**10.83**
63 x 88 mm		0.14	1.27	9.30	m	**10.57**
63 x 100 mm		0.14	1.27	9.69	m	**10.96**
75 x 100 mm		0.14	1.27	10.91	m	**12.18**
Extra for additional labours						
one		0.02	0.18	-	m	**0.18**
two		0.03	0.27	-	m	**0.27**
three		0.04	0.36	-	m	**0.36**
Add +5% to the above 'Material £ prices' for 'selected' softwood for staining						

L WINDOWS/DOORS/STAIRS Including overheads and profit at 9.00%		Labour hours	Labour £	Material £	Unit	Total rate £
Purpose made door frames and lining sets; **selected West African Mahogany; PC £494.00/m3**						
Jambs and heads; as linings						
32 x 63 mm	PC £4.57	0.23	2.08	5.10	m	**7.18**
32 x 100 mm	PC £5.90	0.23	2.08	6.60	m	**8.68**
32 x 140 mm	PC £7.47	0.23	2.08	8.35	m	**10.43**
Jambs and heads; as frames; rebated, rounded and grooved						
38 x 75 mm	PC £5.76	0.23	2.08	6.43	m	**8.51**
38 x 100 mm	PC £6.96	0.23	2.08	7.77	m	**9.85**
38 x 115 mm	PC £7.97	0.23	2.08	8.91	m	**10.99**
38 x 140 mm	PC £9.19	0.27	2.45	10.26	m	**12.71**
50 x 100 mm	PC £8.74	0.27	2.45	9.77	m	**12.22**
50 x 125 mm	PC £10.49	0.27	2.45	11.72	m	**14.17**
63 x 88 mm	PC £9.77	0.27	2.45	10.91	m	**13.36**
63 x 100 mm	PC £10.80	0.27	2.45	12.07	m	**14.52**
63 x 125 mm	PC £12.94	0.27	2.45	14.45	m	**16.90**
75 x 100 mm	PC £12.40	0.27	2.45	13.86	m	**16.31**
75 x 125 mm	PC £15.37	0.30	2.72	17.17	m	**19.89**
75 x 150 mm	PC £18.03	0.30	2.72	20.14	m	**22.86**
100 x 100 mm	PC £16.19	0.30	2.72	18.08	m	**20.80**
100 x 150 mm	PC £23.40	0.30	2.72	26.15	m	**28.87**
Mullions and transoms; in linings						
32 x 63 mm	PC £7.18	0.16	1.45	8.02	m	**9.47**
32 x 100 mm	PC £8.52	0.16	1.45	9.51	m	**10.96**
32 x 140 mm	PC £10.07	0.16	1.45	11.25	m	**12.70**
Mullions and transoms; in frames; twice rebated, rounded and grooved						
38 x 75 mm	PC £8.51	0.16	1.45	9.50	m	**10.95**
38 x 100 mm	PC £9.70	0.16	1.45	10.84	m	**12.29**
38 x 115 mm	PC £10.58	0.16	1.45	11.82	m	**13.27**
38 x 140 mm	PC £11.79	0.18	1.63	13.17	m	**14.80**
50 x 100 mm	PC £11.34	0.18	1.63	12.67	m	**14.30**
50 x 125 mm	PC £13.24	0.18	1.63	14.79	m	**16.42**
63 x 88 mm	PC £12.37	0.18	1.63	13.82	m	**15.45**
63 x 100 mm	PC £13.40	0.18	1.63	14.98	m	**16.61**
75 x 100 mm	PC £15.14	0.18	1.63	16.92	m	**18.55**
Sills; once sunk weathered; once rebated, three times grooved						
63 x 175 mm	PC £24.70	0.33	2.99	27.60	m	**30.59**
75 x 125 mm	PC £22.44	0.33	2.99	25.07	m	**28.06**
75 x 150 mm	PC £25.12	0.33	2.99	28.07	m	**31.06**
Extra for additional labours						
one		0.03	0.27	-	m	**0.27**
two		0.05	0.45	-	m	**0.45**
three		0.07	0.63	-	m	**0.63**
Purpose made door frames and lining sets; **Afrormosia; PC £706.00/m3**						
Jambs and heads; as linings						
32 x 63 mm	PC £5.40	0.23	2.08	6.04	m	**8.12**
32 x 100 mm	PC £7.09	0.23	2.08	7.92	m	**10.00**
32 x 140 mm	PC £9.14	0.23	2.08	10.21	m	**12.29**
Jambs and heads; as frames; rebated, rounded and grooved						
38 x 75 mm	PC £6.81	0.23	2.08	7.61	m	**9.69**
38 x 100 mm	PC £8.36	0.23	2.08	9.34	m	**11.42**
38 x 115 mm	PC £9.68	0.23	2.08	10.82	m	**12.90**
38 x 140 mm	PC £11.14	0.27	2.45	12.44	m	**14.89**
50 x 100 mm	PC £8.74	0.27	2.45	11.83	m	**14.28**
50 x 125 mm	PC £12.81	0.27	2.45	14.31	m	**16.76**
63 x 88 mm	PC £11.86	0.27	2.45	13.25	m	**15.70**
63 x 100 mm	PC £13.14	0.27	2.45	14.69	m	**17.14**
63 x 125 mm	PC £15.86	0.27	2.45	17.73	m	**20.18**
75 x 100 mm	PC £15.18	0.27	2.45	16.96	m	**19.41**
75 x 125 mm	PC £18.85	0.30	2.72	21.06	m	**23.78**
75 x 150 mm	PC £22.20	0.30	2.72	24.81	m	**27.53**
100 x 100 mm	PC £19.89	0.30	2.72	22.22	m	**24.94**
100 x 150 mm	PC £28.97	0.30	2.72	32.36	m	**35.08**

L WINDOWS/DOORS/STAIRS Including overheads and profit at 9.00%		Labour hours	Labour £	Material £	Unit	Total rate £
L20 TIMBER DOORS/SHUTTERS/HATCHES - cont'd						
Purpose made door frames and lining sets; **Afrormosia; PC £706.00/m3 - cont'd**						
Mullions and transoms; in linings						
32 x 63 mm	PC £8.00	0.16	1.45	8.94	m	10.39
32 x 100 mm	PC £9.68	0.16	1.45	10.82	m	12.27
32 x 140 mm	PC £11.75	0.16	1.45	13.12	m	14.57
Mullions and transoms; in frames; twice rebated, rounded and grooved						
38 x 75 mm	PC £9.56	0.16	1.45	10.68	m	12.13
38 x 100 mm	PC £11.11	0.16	1.45	12.41	m	13.86
38 x 115 mm	PC £12.29	0.16	1.45	13.74	m	15.19
38 x 140 mm	PC £12.29	0.18	1.63	13.74	m	15.37
50 x 100 mm	PC £13.20	0.18	1.63	14.75	m	16.38
50 x 125 mm	PC £15.54	0.18	1.63	17.36	m	18.99
63 x 88 mm	PC £14.49	0.18	1.63	16.19	m	17.82
63 x 100 mm	PC £15.74	0.18	1.63	17.59	m	19.22
75 x 100 mm	PC £17.93	0.18	1.63	20.03	m	21.66
Sills; once sunk weathered; once rebated, three times grooved						
63 x 175 mm	PC £28.78	0.33	2.99	32.15	m	35.14
75 x 125 mm	PC £25.91	0.33	2.99	28.95	m	31.94
75 x 150 mm	PC £29.32	0.33	2.99	32.76	m	35.75
Extra for additional labours						
one		0.03	0.27	-	m	0.27
two		0.05	0.45	-	m	0.45
three		0.07	0.63	-	m	0.63
Door sills; European Oak PC £1165.00/m3						
Sills; once sunk weathered; once rebated, three times grooved						
63 x 175 mm	PC £47.92	0.33	2.99	53.55	m	56.54
75 x 125 mm	PC £42.68	0.33	2.99	47.69	m	50.68
75 x 150 mm	PC £48.76	0.33	2.99	54.47	m	57.46
Extra for additional labours						
one		0.03	0.27	-	m	0.27
two		0.06	0.54	-	m	0.54
three		0.09	0.82	-	m	0.82
Bedding and pointing frames						
Pointing wood frames or sills with mastic						
one side		0.07	0.82	0.40	m	1.22
each side		0.14	1.63	0.80	m	2.43
Pointing wood frames or sills with polysulphide sealant						
one side		0.07	0.82	0.80	m	1.62
each side		0.14	1.63	1.59	m	3.22
Bedding wood frame in cement mortar (1:3) and point						
one side		0.08	0.93	0.04	m	0.97
each side		0.10	1.17	0.06	m	1.23
one side in mortar; other side in mastic		0.15	1.75	0.44	m	2.19
L21 METAL DOORS/SHUTTERS/HATCHES						
Aluminium double glazed sliding patio doors; **Crittall 'Luminaire' or similar; white** **acrylic finish; with and including 18 mm** **annealed double glazing; fixed in position;** **including lugs plugged and screwed to** **brickwork or blockwork; or screwed to wooden** **sub-frame (measured elsewhere)**						
Patio doors						
1800 x 2100 mm; ref D18HDL21	PC £405.00	7.50	77.42	444.09	nr	521.51
2400 x 2100 mm; ref D24HDL21	PC £479.63	9.00	92.90	525.43	nr	618.33
3000 x 2100 mm; ref D30HDF21	PC £670.50	10.50	108.38	733.48	nr	841.86

L WINDOWS/DOORS/STAIRS Including overheads and profit at 9.00%	Labour hours	Labour £	Material £	Unit	Total rate £
Galvanized steel 'up and over' type garage doors; Catnic 'Garador' or similar; spring counterbalanced; fixed to timber frame (measured elsewhere) Garage door					
2135 x 1980 mm; ref MK 3C PC £95.63	5.00	51.61	105.67	nr	157.28
2135 x 2135 mm; ref MK 3C PC £105.38	5.25	54.19	116.30	nr	170.49
2400 x 2125 mm; ref MK 3C PC £124.88	6.00	61.93	137.55	nr	199.48
4270 x 2135 mm; ref 'Carlton' PC £417.75	9.00	92.90	456.79	nr	549.69

L30 TIMBER STAIRS/WALKWAYS/BALUSTRADES

Standard staircases; wrought softwood Stairs; 25 mm treads with rounded nosings; 12 mm plywood risers; 32 mm once rounded strings; bullnose bottom tread; 50 x 75 mm hardwood handrail; two 32 x 140 mm balustrade knee rails; 32 x 50 mm stiffeners and 100 x 100 mm newel posts with hardwood newel caps on top					
straight flight; 838 mm wide; 2688 mm going; 2600 mm rise; with two newel posts	13.50	122.28	290.69	nr	412.97
straight flight; 838 mm wide; 2600 mm rise; with two newel posts and three top treads winding	18.00	163.04	359.91	nr	522.95
dogleg staircase; 838 mm wide; 2600 mm rise; quarter space landing third riser from top; with three newel posts	19.00	172.10	368.43	nr	540.53
as last but with half space landing; one 100 x 200 mm newel post; and two 100 x 100 mm newel posts	20.00	181.16	438.94	nr	620.10
Stairs; 25 mm treads with rounded nosings; 12 mm plywood risers; 32 mm once rounded strings; with string cappings; bullnose bottom tread; 50 x 75 mm hardwood handrail; two 32 x 32 mm balusters per tread and 100 x 100 mm newel post with hardwood newel caps on top					
straight flight; 838 mm wide 2688 going 2600 mm rise with two newel posts	15.00	135.87	285.45	nr	421.32
Standard balustrades; wrought softwood Landing balustrade; 50 x 75 mm hardwood handrail; three 32 x 140 mm balustrades knee rails; two 32 x 50 mm stiffeners; one end jointed to newel post; other end built into wall (newel post and mortices both measured separately)					
3 m long	3.00	27.17	92.19	nr	119.36
Landing balustrade; 50 x 75 mm hardwood handrail; 32 x 32 mm balusters; one end of handrail jointed to newel post; other end built into wall; balusters housed in at bottom (newel post and mortices both measured separately)					
3 m long	4.50	40.76	75.14	nr	115.90
Purpose made staircase components; wrought softwood Board landings; cross-tongued joints; 100 x 50 mm sawn softwood bearers					
25 mm thick	1.00	9.06	39.08	m2	48.14
.32 mm thick	1.00	9.06	46.45	m2	55.51

L WINDOWS/DOORS/STAIRS Including overheads and profit at 9.00%	Labour hours	Labour £	Material £	Unit	Total rate £
L30 TIMBER STAIRS/WALKWAYS/BALUSTRADES - cont'd					
Purpose made staircase components; wrought **softwood - cont'd**					
Treads cross-tongued joints and risers; rounded nosings; tongued, grooved, glued and blocked together; one 175 x 50 mm sawn softwood carriage					
25 mm treads; 19 mm risers	1.50	13.59	49.37	m2	62.96
Ends; quadrant	-	-	-	nr	23.92
32 mm treads; 25 mm risers	1.50	13.59	57.90	m2	71.49
Ends; quadrant	-	-	-	nr	28.19
Ends; housed to hardwood	-	-	-	nr	0.42
Winders; cross-tongued joints and risers in one width; rounded nosings; tongued, grooved glued and blocked together; one 175 x 50 mm sawn softwood carriage					
25 mm treads; 19 mm risers	2.50	22.64	53.51	m2	76.15
32 mm treads; 25 mm risers	2.50	22.64	62.90	m2	85.54
Wide ends; housed to hardwood	-	-	-	nr	0.86
Narrow ends; housed to hardwood	-	-	-	nr	0.65
Closed strings; in one width; 230 mm wide; rounded twice					
32 mm thick	0.50	4.53	9.13	m	13.66
38 mm thick	0.50	4.53	11.02	m	15.55
50 mm thick	0.50	4.53	14.46	m	18.99
Closed strings; cross-tongued joints; 280 mm wide; once rounded; fixing with screws; plugging 450 mm centres					
32 mm thick	0.60	5.43	16.54	m	21.97
Extra for short ramp	0.12	1.09	8.27	nr	9.36
38 mm thick	0.60	5.43	18.74	m	24.17
Extra for short ramp	0.12	1.09	9.28	nr	10.37
50 mm thick	0.60	5.43	23.08	m	28.51
Ends; fitted	0.10	0.91	0.28	nr	1.19
Ends; framed	0.12	1.09	2.75	nr	3.84
Extra for tongued heading joint	0.12	1.09	1.42	nr	2.51
Extra for short ramp	0.12	1.09	11.53	nr	12.62
Closed strings; ramped; crossed tongued joints 280 mm wide; once rounded; fixing with screws; plugging 450 mm centres					
32 mm thick	0.60	5.43	18.19	m	23.62
38 mm thick	0.60	5.43	21.62	m	27.05
50 mm thick	0.60	5.43	25.39	m	30.82
Apron linings; in one width 230 mm wide					
19 mm thick	0.33	2.99	3.20	m	6.19
25 mm thick	0.33	2.99	3.92	m	6.91
Handrails; rounded					
44 x 50 mm	0.25	2.26	2.81	m	5.07
50 x 75 mm	0.27	2.45	3.57	m	6.02
63 x 87 mm	0.30	2.72	5.00	m	7.72
75 x 100 mm	0.35	3.17	5.68	m	8.85
Handrails; moulded					
44 x 50 mm	0.25	2.26	3.08	m	5.34
50 x 75 mm	0.27	2.45	3.85	m	6.30
63 x 87 mm	0.30	2.72	5.27	m	7.99
75 x 100 mm	0.35	3.17	5.94	m	9.11
Handrails; rounded; ramped					
44 x 50 mm	0.33	2.99	5.63	m	8.62
50 x 75 mm	0.36	3.26	7.12	m	10.38
63 x 87 mm	0.40	3.62	10.01	m	13.63
75 x 100 mm	0.45	4.08	11.35	m	15.43
Handrails; moulded; ramped					
44 x 50 mm	0.33	2.99	6.16	m	9.15
50 x 75 mm	0.36	3.26	7.68	m	10.94
63 x 87 mm	0.40	3.62	10.53	m	14.15
75 x 100 mm	0.45	4.08	11.89	m	15.97

L WINDOWS/DOORS/STAIRS Including overheads and profit at 9.00%	Labour hours	Labour £	Material £	Unit	Total rate £
Add to above for					
grooved once	-	-	-	m	0.25
ends; framed	0.10	0.91	2.19	nr	3.10
ends; framed on rake	0.15	1.36	2.75	nr	4.11
Heading joints on rake; handrail screws					
44 x 50 mm	0.15	1.36	13.44	nr	14.80
50 x 75 mm	0.15	1.36	13.44	nr	14.80
63 x 87 mm	0.15	1.36	17.01	nr	18.37
75 x 100 mm	0.15	1.36	17.01	nr	18.37
Mitres; handrail screws					
44 x 50 mm	0.20	1.81	13.44	nr	15.25
50 x 75 mm	0.20	1.81	13.44	nr	15.25
63 x 87 mm	0.20	1.81	17.01	nr	18.82
75 x 100 mm	0.20	1.81	17.01	nr	18.82
Balusters; stiffeners					
25 x 25 mm	0.08	0.72	1.13	m	1.85
32 x 32 mm	0.08	0.72	1.34	m	2.06
32 x 50 mm	0.08	0.72	1.66	m	2.38
Ends; housed	0.03	0.27	0.53	nr	0.80
Sub rails					
32 x 63 mm	0.33	2.99	2.29	m	5.28
Ends; housed to newel	0.10	0.91	2.19	nr	3.10
Knee rails					
32 x 140 mm	0.40	3.62	3.81	m	7.43
Ends; housed to newel	0.10	0.91	2.19	nr	3.10
Newel posts					
50 x 100 mm; half	0.40	3.62	3.91	m	7.53
75 x 75 mm	0.40	3.62	4.24	m	7.86
100 x 100 mm	0.50	4.53	6.66	m	11.19
Newel caps; splayed on four sides					
62.5 x 125 x 50 mm; half	0.15	1.36	2.57	nr	3.93
100 x 100 x 50 mm	0.15	1.36	2.57	nr	3.93
125 x 125 x 50 mm	0.15	1.36	2.76	nr	4.12
Purpose made staircase components; selected **West African Mahogany; PC £494.00/m3**					
Board landings; cross-tongued joints; 100 x 50 mm sawn softwood bearers					
25 mm thick PC £51.94	1.50	13.59	57.62	m2	71.21
32 mm thick PC £63.45	1.50	13.59	70.16	m2	83.75
Treads cross-tongued joints and risers; rounded nosings; tongued, grooved, glued and blocked together; one 175 x 50 mm sawn softwood carriage					
25 mm treads; 19 mm risers PC £63.55	2.00	18.12	70.84	m2	88.96
Ends; quadrant PC £37.29	-	-	40.64	nr	40.64
32 mm treads; 25 mm risers PC £76.09	2.00	18.12	84.52	m2	102.64
Ends; quadrant PC £53.26	-	-	58.06	nr	58.06
Ends; housed to hardwood	-	-	-	nr	0.65
Winders; cross-tongued joints and risers in one width; rounded nosings; tongued, grooved glued and blocked together; one 175 x 50 mm sawn softwood carriage					
25 mm treads; 19 mm risers PC £69.89	3.35	30.34	77.11	m2	107.45
32 mm treads; 25 mm risers PC £83.71	3.35	30.34	92.18	m2	122.52
Wide ends; housed to hardwood	-	-	-	nr	1.28
Narrow ends; housed to hardwood	-	-	-	nr	0.98
Closed strings; in one width; 230 mm wide; rounded twice					
32 mm thick PC £13.46/m	0.67	6.07	14.74	m2	20.81
38 mm thick PC £16.12/m	0.67	6.07	17.64	m2	23.71
50 mm thick PC £21.24/m	0.67	6.07	23.22	m2	29.29

L WINDOWS/DOORS/STAIRS Including overheads and profit at 9.00%		Labour hours	Labour £	Material £	Unit	Total rate £
L30 TIMBER STAIRS/WALKWAYS/BALUSTRADES - cont'd						
Purpose made staircase components; selected						
West African Mahogany; PC £494.00/m3 - cont'd						
Closed strings; cross-tongued joints; 280 mm						
wide; once rounded; fixing with screws;						
plugging 450 mm centres						
32 mm thick	PC £23.56	0.80	7.25	25.75	m	33.00
Extra for short ramp	PC £11.78	0.18	1.63	12.85	nr	14.48
38 mm thick	PC £26.70	0.80	7.25	29.17	m	36.42
Extra for short ramp	PC £13.37	0.18	1.63	14.57	nr	16.20
50 mm thick	PC £32.99	0.80	7.25	36.02	m	43.27
Ends; fitted	PC £0.71	0.15	1.36	0.78	nr	2.14
Ends; framed	PC £4.40	0.18	1.63	4.80	nr	6.43
Extra for tongued heading joint	PC £2.17	0.18	1.63	2.37	nr	4.00
Extra for short ramp	PC £16.52	0.18	1.63	18.01	nr	19.64
Closed strings; ramped; crossed tongued						
joints 280 mm wide; once rounded; fixing						
with screws; plugging 450 mm centres						
32 mm thick	PC £25.94	0.80	7.25	28.34	m	35.59
38 mm thick	PC £29.40	0.80	7.25	32.11	m	39.36
50 mm thick	PC £36.29	0.80	7.25	39.62	m	46.87
Apron linings; in one width 230 mm wide						
19 mm thick	PC £5.43	0.44	3.99	5.98	m	9.97
25 mm thick	PC £6.55	0.44	3.99	7.20	m	11.19
Handrails; rounded						
44 x 50 mm	PC £4.59	0.33	2.99	5.25	m	8.24
50 x 75 mm	PC £5.71	0.36	3.26	6.54	m	9.80
63 x 87 mm	PC £7.35	0.40	3.62	8.42	m	12.04
75 x 100 mm	PC £8.91	0.45	4.08	10.20	m	14.28
Handrails; moulded						
44 x 50 mm	PC £4.95	0.33	2.99	5.67	m	8.66
50 x 75 mm	PC £6.08	0.36	3.26	6.95	m	10.21
63 x 87 mm	PC £7.71	0.40	3.62	8.83	m	12.45
75 x 100 mm	PC £9.26	0.45	4.08	10.59	m	14.67
Handrails; rounded; ramped						
44 x 50 mm	PC £9.19	0.44	3.99	10.51	m	14.50
50 x 75 mm	PC £11.43	0.48	4.35	13.08	m	17.43
63 x 87 mm	PC £14.73	0.53	4.80	16.86	m	21.66
75 x 100 mm	PC £17.81	0.60	5.43	20.38	m	25.81
Handrails; moulded; ramped						
44 x 50 mm	PC £9.90	0.44	3.99	11.32	m	15.31
50 x 75 mm	PC £12.15	0.48	4.35	13.90	m	18.25
63 x 87 mm	PC £15.42	0.53	4.80	17.65	m	22.45
75 x 100 mm	PC £18.52	0.60	5.43	21.20	m	26.63
Add to above for						
grooved once		-	-	-	m	0.38
ends; framed	PC £3.34	0.15	1.36	3.64	nr	5.00
ends; framed on rake	PC £4.11	0.22	1.99	4.48	nr	6.47
Heading joints on rake; handrail screws						
44 x 50 mm	PC £15.15	0.22	1.99	16.52	nr	18.51
50 x 75 mm	PC £15.15	0.22	1.99	16.52	nr	18.51
63 x 87 mm	PC £19.42	0.22	1.99	21.16	nr	23.15
75 x 100 mm	PC £19.42	0.22	1.99	21.16	nr	23.15
Mitres; handrail screws						
44 x 50 mm	PC £15.15	0.27	2.45	16.52	nr	18.97
50 x 75 mm	PC £15.15	0.27	2.45	16.52	nr	18.97
63 x 87 mm	PC £19.42	0.27	2.45	21.16	nr	23.61
75 x 100 mm	PC £19.42	0.27	2.45	21.16	nr	23.61
Balusters; stiffeners						
25 x 25 mm	PC £2.24	0.10	0.91	2.44	m	3.35
32 x 32 mm	PC £2.58	0.10	0.91	2.81	m	3.72
32 x 50 mm	PC £3.10	0.10	0.91	3.38	m	4.29
Ends; housed	PC £0.75	0.05	0.45	0.82	nr	1.27
Sub rails						
32 x 63 mm	PC £3.98	0.44	3.99	4.34	m	8.33
Ends; housed to newel	PC £3.02	0.15	1.36	3.29	nr	4.65

L WINDOWS/DOORS/STAIRS Including overheads and profit at 9.00%		Labour hours	Labour £	Material £	Unit	Total rate £
Knee rails						
32 x 140 mm	PC £6.15	0.53	4.80	6.70	m	11.50
Ends; housed to newel	PC £3.02	0.15	1.36	3.29	nr	4.65
Newel posts						
50 x 100 mm; half	PC £6.57	0.53	4.80	7.16	m	11.96
75 x 75 mm	PC £7.09	0.53	4.80	7.72	m	12.52
100 x 100 mm	PC £10.89	0.67	6.07	11.87	m	17.94
Newel caps; splayed on four sides						
62.5 x 125 x 50 mm; half	PC £4.21	0.20	1.81	4.59	nr	6.40
100 x 100 x 50 mm	PC £4.21	0.20	1.81	4.59	nr	6.40
125 x 125 x 50 mm	PC £4.54	0.20	1.81	4.95	nr	6.76
Purpose made staircase components;						
Oak; PC £1165.00/m3						
Board landings; cross-tongued joints;						
100 x 50 mm sawn softwood bearers						
25 mm thick	PC £107.24	1.50	13.59	117.90	m2	131.49
32 mm thick	PC £131.80	1.50	13.59	144.67	m2	158.26
Treads cross-tongued joints and risers;						
rounded nosings; tongued, grooved, glued and						
blocked together; one 175 x 50 mm sawn						
softwood carriage						
25 mm treads; 19 mm risers	PC £116.24	2.00	18.12	128.27	m2	146.39
Ends; quadrant	PC £58.14	-	-	63.37	nr	63.37
32 mm treads; 25 mm risers	PC £142.80	2.00	18.12	157.23	m2	175.35
Ends; quadrant	PC £71.41	-	-	77.84	nr	77.84
Ends; housed to hardwood		-	-	-	nr	0.86
Winders; cross-tongued joints and risers in						
one width; rounded nosings; tongued, grooved						
glued and blocked together; one 175 x 50 mm						
sawn softwood carriage						
25 mm treads; 19 mm risers	PC £127.87	3.35	30.34	140.31	m2	170.65
32 mm treads; 25 mm risers	PC £157.08	3.35	30.34	172.15	m2	202.49
Wide ends; housed to hardwood		-	-	-	nr	1.72
Narrow ends; housed to hardwood		-	-	-	nr	1.28
Closed strings; in one width; 230 mm wide;						
rounded twice						
32 mm thick	PC £28.01	0.67	6.07	30.60	m	36.67
38 mm thick	PC £33.34	0.67	6.07	36.40	m	42.47
50 mm thick	PC £43.97	0.67	6.07	48.00	m	54.07
Closed strings; cross-tongued joints; 280 mm						
wide; once rounded; fixing with screws;						
plugging 450 mm centres						
32 mm thick	PC £44.87	0.80	7.25	48.98	m	56.23
Extra for short ramp	PC £22.43	0.18	1.63	24.45	nr	26.08
38 mm thick	PC £51.38	0.80	7.25	56.07	m	63.32
Extra for short ramp	PC £25.70	0.18	1.63	28.01	nr	29.64
50 mm thick	PC £64.53	0.80	7.25	70.40	m	77.65
Ends; fitted	PC £0.83	0.15	1.36	0.91	nr	2.27
Ends; framed	PC £5.67	0.18	1.63	6.18	nr	7.81
Extra for tongued heading joint	PC £2.84	0.18	1.63	3.10	nr	4.73
Extra for short ramp	PC £32.27	0.18	1.63	35.18	nr	36.81
Closed strings; ramped; crossed tongued						
joints 280 mm wide; once rounded; fixing						
with screws; plugging 450 mm centres						
32 mm thick	PC £49.35	0.80	7.25	53.86	m	61.11
38 mm thick	PC £56.53	0.80	7.25	61.68	m	68.93
50 mm thick	PC £70.99	0.80	7.25	77.45	m	84.70
Apron linings; in one width 230 mm wide						
19 mm thick	PC £12.86	0.44	3.99	14.08	m	18.07
25 mm thick	PC £16.15	0.44	3.99	17.67	m	21.66
Handrails; rounded						
44 x 50 mm	PC £9.30	0.33	2.99	10.63	m	13.62
50 x 75 mm	PC £12.59	0.36	3.26	14.41	m	17.67
63 x 87 mm	PC £17.36	0.40	3.62	19.87	m	23.49
75 x 100 mm	PC £21.89	0.45	4.08	25.04	m	29.12

L WINDOWS/DOORS/STAIRS Including overheads and profit at 9.00%		Labour hours	Labour £	Material £	Unit	Total rate £

L30 TIMBER STAIRS/WALKWAYS/BALUSTRADES - cont'd

Purpose made staircase components;
Oak; PC £1165.00/m3 - cont'd
Handrails; moulded
44 x 50 mm	PC £9.78	0.33	2.99	11.19	m	14.18
50 x 75 mm	PC £13.07	0.36	3.26	14.96	m	18.22
63 x 87 mm	PC £17.84	0.40	3.62	20.42	m	24.04
75 x 100 mm	PC £22.35	0.45	4.08	25.58	m	29.66

Handrails; rounded; ramped
44 x 50 mm	PC £18.61	0.44	3.99	21.30	m	25.29
50 x 75 mm	PC £25.18	0.48	4.35	28.82	m	33.17
63 x 87 mm	PC £34.74	0.53	4.80	39.77	m	44.57
75 x 100 mm	PC £43.77	0.60	5.43	50.10	m	55.53

Handrails; moulded; ramped
44 x 50 mm	PC £19.55	0.44	3.99	22.38	m	26.37
50 x 75 mm	PC £26.12	0.48	4.35	29.90	m	34.25
63 x 87 mm	PC £35.69	0.53	4.80	40.84	m	45.64
75 x 100 mm	PC £44.71	0.60	5.43	51.18	m	56.61

Add to above for
grooved once		-	-	-	m	0.52
ends; framed	PC £4.35	0.15	1.36	4.74	nr	6.10
ends; framed on rake	PC £5.37	0.22	1.99	5.86	nr	7.85

Heading joints on rake; handrail screws
44 x 50 mm	PC £23.02	0.22	1.99	25.10	nr	27.09
50 x 75 mm	PC £23.02	0.22	1.99	25.10	nr	27.09
63 x 87 mm	PC £28.27	0.22	1.99	30.81	nr	32.80
75 x 100 mm	PC £28.27	0.22	1.99	30.81	nr	32.80

Mitres; handrail screws
44 x 50 mm	PC £23.02	0.27	2.45	25.10	nr	27.55
50 x 75 mm	PC £23.02	0.27	2.45	25.10	nr	27.55
63 x 87 mm	PC £28.27	0.27	2.45	30.81	nr	33.26
75 x 100 mm	PC £28.27	0.27	2.45	30.81	nr	33.26

Balusters; stiffeners
25 x 25 mm	PC £3.59	0.10	0.91	3.91	m	4.82
32 x 32 mm	PC £4.59	0.10	0.91	5.00	m	5.91
32 x 50 mm	PC £6.10	0.10	0.91	6.64	m	7.55
Ends; housed	PC £0.98	0.05	0.45	1.07	nr	1.52

Sub rails
| 32 x 63 mm | PC £7.83 | 0.44 | 3.99 | 8.53 | m | 12.52 |
| Ends; housed to newel | PC £4.02 | 0.15 | 1.36 | 4.39 | nr | 5.75 |

Knee rails
| 32 x 140 mm | PC £14.11 | 0.53 | 4.80 | 15.38 | m | 20.18 |
| Ends; housed to newel | PC £4.02 | 0.15 | 1.36 | 4.39 | nr | 5.75 |

Newel posts
50 x 100 mm; half	PC £15.37	0.53	4.80	16.76	m	21.56
75 x 75 mm	PC £16.86	0.53	4.80	18.38	m	23.18
100 x 100 mm	PC £27.92	0.67	6.07	30.43	m	36.50

Newel caps; splayed on four sides
62.5 x 125 x 50 mm; half	PC £6.08	0.20	1.81	6.62	nr	8.43
100 x 100 x 50 mm	PC £6.08	0.20	1.81	6.62	nr	8.43
125 x 125 x 50 mm	PC £7.06	0.20	1.81	7.69	nr	9.50

L31 METAL STAIRS/WALKWAYS/BALUSTRADES

Cat ladders, balustrades and handrails,
etc.; mild steel; BS 4360
Cat ladders; welded construction; 64 x 13 mm
bar strings; 19 mm rungs at 250 mm centres;
| fixing by bolting; 0.46 m wide; 3.05 m high | 3.00 | 30.97 | 109.22 | nr | 140.19 |

Extra for
ends of strings; bent once; holed once for
10 mm dia bolt	-	-	-	nr	1.44
ends of strings; fanged	-	-	-	nr	0.65
ends of strings; bent in plane	-	-	-	nr	6.80

L WINDOWS/DOORS/STAIRS Including overheads and profit at 9.00%	Labour hours	Labour £	Material £	Unit	Total rate £
Balustrades; welded construction; galvanized after manufacture; 1070 mm high; 50 x 50 x 3.2 mm r.h.s. top rail; 38 x 13 mm bottom rail, 50 x 50 x 3.2 mm r.h.s. standards at 1830 mm centres with base plate drilled and bolted to concrete; 13 x 13 mm balusters at 102 mm centres	1.50	15.48	44.96	m	60.44
Balusters; isolated; one end ragged and cemented in; one 76 x 25 x 6 mm flange plate welded on; ground to a smooth finish; countersunk drilled and tap screwed to underside of handrail					
19 x 19 x 914 mm square bar	-	-	-	nr	7.55
Core-rails; joints prepared, welded and ground to a smooth finish; fixing on brackets (measured elsewhere)					
38 x 10 mm flat bar	-	-	-	m	10.79
50 x 8 mm flat bar	-	-	-	m	10.43
Extra for					
ends fanged	-	-	-	nr	0.78
ends scrolled	-	-	-	nr	3.34
ramps in thickness	-	-	-	nr	3.34
wreaths	-	-	-	nr	3.92
Handrails; joints prepared, welded and ground to a smooth finish; fixing on brackets (measured elsewhere)					
38 x 12 mm half oval bar	-	-	-	m	14.39
44 x 13 mm half oval bar	-	-	-	m	15.47
Extra for					
ends fanged	-	-	-	nr	0.78
ends scrolled	-	-	-	nr	3.34
ramps in thickness	-	-	-	nr	3.34
wreaths	-	-	-	nr	3.92
Handrail bracket; comprising 40 x 5 mm plate with mitred and welded angle; one end welded to 100 mm dia x 5 mm backplate; three times holed and plugged and screwed to brickwork; other end scribed and welded to underside of handrail					
140 mm girth	-	-	-	nr	8.44
Holes					
Holes; countersunk; for screws or bolts					
6 mm dia wood screw; 3 mm thick	0.05	0.52	-	nr	0.52
6 mm dia wood screw; 6 mm thick	0.07	0.72	-	nr	0.72
8 mm dia bolt; 6 mm thick	0.07	0.72	-	nr	0.72
10 mm dia bolt; 6 mm thick	0.08	0.83	-	nr	0.83
12 mm dia bolt; 8 mm thick	0.10	1.03	-	nr	1.03

BASIC GLASS PRICES (£/m2)

	£		£		£		£
Trade cut prices - all to limiting sizes							
Ordinary transparent glass							
Float; GG quality							
2 mm	15.61	5 mm	32.40	12 mm	103.57	19 mm	165.58
3 mm	19.70	6 mm	36.22	15 mm	121.52	25 mm	234.46
4 mm	22.33	10 mm	76.97				

Ordinary transluscent/patterned glass
  Obscured ground sheet glass - extra on sheet glass prices £7.29/m2
  Patterned

4 mm tint.	31.88	6 mm tint.	39.50	4 mm white	17.67	6 mm white	30.85

Rough cast

6 mm	26.97	10 mm	39.00

L WINDOWS/DOORS/STAIRS
Including overheads and profit at 9.00%

	£		£		£		£
Ordinary Georgian wired							
7 mm cast	28.48	6 mm polish	62.10				

Polycarbonate standard sheets

	£		£		£		£
2 mm	25.67	4 mm	49.31	6 mm	73.03	10 mm	114.55
3 mm	37.44	5 mm	61.17	8 mm	96.66	12 mm	144.30

Special glasses
'Antisun' float; bronze or grey

	£		£		£		£
4 mm	43.74	6 mm	63.14	10 mm	139.07	12 mm	175.40
6 mm float	103.83	10 mm float	114.49				

'Cetuff' toughened
float

	£		£		£		£
4 mm	33.21	6 mm	40.90	10 mm	81.18	12 mm	123.62
5 mm	37.62						

patterned

	£		£		£		£
6 mm rough	50.00	10 mm rough	75.00	4 mm white	32.08	6 mm white	45.31

solar control; bronze or grey

	£		£		£		£
4 mm	57.50	6 mm	74.11	10 mm	139.81	12 mm	181.94

Clear laminated
'security'

	£		£		£
7.5 mm	94.81	9.5 mm	97.01	11.5 mm	107.70

'safety'

	£		£		£		£
4.4 mm	47.05	6.4 mm	51.29	8.8 mm	79.66	10.8 mm	95.12
5.4 mm	48.14	6.8 mm	63.24				

'Permawal'
'reflective'

	£		£
6 mm	50.00	10 mm	75.00

'Permasol'

	£
6 mm	70.00

'Silvered'

	£		£		£		£
2 mm	31.42	3 mm	31.78	4 mm	36.90	6 mm	47.82

'Silvered tinted'

	£		£		£		£
4 mm bronze	56.73	4 mm grey	56.73	6 mm bronze	77.80	6 mm grey	77.80

'Venetian striped'

	£		£
4 mm	103.83	6 mm	114.49

Trade discounts of 20 - 30% are usually available off the above prices; The following
'measured rates' are provided by a glazing Sub-Contractor and assume quantities in
excess of 500 m2, within 20 miles of the suppliers branch. Therefore deduction of
'trade prices for cut sizes' from 'measured rates' is not a reliable basis for
identifying fixing costs.

EXTRAS:
Panes under 0.10 m2 are charged as 0.10 m2 plus per pane:-

	£
putty or bradded beads	0.26
bradded beads and butyl compound	0.29
screwed beads	0.35
screwed beads and butyl compound	0.42
The following extras are charged per m2	
external nailed or bradded beads	1.06
to preservative stained wood	2.01
to contracts - 100 - 200 m2	2.01
- 50 - 100 m2	2.90

L WINDOWS/DOORS/STAIRS Including overheads and profit at 5.00%	Labour hours	Labour £	Material £	Unit	Total rate £

L40 GENERAL GLAZING

Standard plain glass; BS 952; clear float;
panes area 0.15 - 4.00 m2
3 mm thick; glazed with

putty or bradded beads	-	-	-	m2	13.49
bradded beads and butyl compound	-	-	-	m2	16.78
screwed beads	-	-	-	m2	18.19
screwed beads and butyl compound	-	-	-	m2	21.06

4 mm thick; glazed with

putty or bradded beads	-	-	-	m2	16.20
bradded beads and butyl compound	-	-	-	m2	19.92
screwed beads	-	-	-	m2	21.37
screwed beads and butyl compound	-	-	-	m2	24.69

5 mm thick; glazed with

putty or bradded beads	-	-	-	m2	21.41
bradded beads and butyl compound	-	-	-	m2	23.82
screwed beads	-	-	-	m2	25.27
screwed beads and butyl compound	-	-	-	m2	28.58

6 mm thick; glazed with

putty or bradded beads	-	-	-	m2	22.62
bradded beads and butyl compound	-	-	-	m2	25.02
screwed beads	-	-	-	m2	26.47
screwed beads and butyl compound	-	-	-	m2	29.78

Standard plain glass; BS 952;
white patterned; panes area 0.15 - 4.00 m2
4 mm thick; glazed with

putty or bradded beads	-	-	-	m2	15.36
bradded beads and butyl compound	-	-	-	m2	18.64
screwed beads	-	-	-	m2	20.06
screwed beads and butyl compound	-	-	-	m2	22.93

6 mm thick; glazed with

putty or bradded beads	-	-	-	m2	22.37
bradded beads and butyl compound	-	-	-	m2	25.72
screwed beads	-	-	-	m2	27.03
screwed beads and butyl compound	-	-	-	m2	30.01

Standard plain glass; BS 952; rough cast;
panes area 0.15 - 4.00 m2
6 mm thick; glazed with

putty or bradded beads	-	-	-	m2	18.53
bradded beads and butyl compound	-	-	-	m2	21.88
screwed beads	-	-	-	m2	23.18
screwed beads and butyl compound	-	-	-	m2	26.17

Standard plain glass; BS 952; Georgian wired
cast; panes area 0.15 - 4.00 m2
7 mm thick; glazed with

putty or bradded beads	-	-	-	m2	19.92
bradded beads and butyl compound	-	-	-	m2	23.28
screwed beads	-	-	-	m2	24.58
screwed beads and butyl compound	-	-	-	m2	27.57
Extra for lining up wired glass	-	-	-	m2	2.06

Standard plain glass; BS 952; Georgian wired
polished; panes area 0.15 - 4.00 m2
6 mm thick; glazed with

putty or bradded beads	-	-	-	m2	43.36
bradded beads and butyl compound	-	-	-	m2	45.52
screwed beads	-	-	-	m2	46.83
screwed beads and butyl compound	-	-	-	m2	49.81
Extra for lining up wired glass	-	-	-	m2	2.30

L WINDOWS/DOORS/STAIRS Including overheads and profit at 9.00% & 5.00%	Labour hours	Labour £	Material £	Unit	Total rate £
L40 GENERAL GLAZING - cont'd					
Special glass; BS 952; toughened clear float;					
panes area 0.15 - 4.00 m2					
4 mm thick; glazed with					
putty or bradded beads	-	-	-	m2	41.90
bradded beads and butyl compound	-	-	-	m2	44.06
screwed beads	-	-	-	m2	45.37
screwed beads and butyl compound	-	-	-	m2	48.35
5 mm thick; glazed with					
putty or bradded beads	-	-	-	m2	46.46
bradded beads and butyl compound	-	-	-	m2	48.63
screwed beads	-	-	-	m2	49.93
screwed beads and butyl compound	-	-	-	m2	52.91
6 mm thick; glazed with					
putty or bradded beads	-	-	-	m2	49.86
bradded beads and butyl compound	-	-	-	m2	52.02
screwed beads	-	-	-	m2	53.33
screwed beads and butyl compound	-	-	-	m2	56.31
10 mm thick; glazed with					
putty or bradded beads	-	-	-	m2	91.57
bradded beads and butyl compound	-	-	-	m2	93.73
screwed beads	-	-	-	m2	95.04
screwed beads and butyl compound	-	-	-	m2	98.02
Special glass; BS 952; clear laminated					
safety glass; panes area 0.15 - 4.00 m2					
4.4 mm thick; glazed with					
putty or bradded beads	-	-	-	m2	56.23
bradded beads and butyl compound	-	-	-	m2	58.39
screwed beads	-	-	-	m2	59.70
screwed beads and butyl compound	-	-	-	m2	62.68
5.4 mm thick; glazed with					
putty or bradded beads	-	-	-	m2	57.36
bradded beads and butyl compound	-	-	-	m2	59.52
screwed beads	-	-	-	m2	60.83
screwed beads and butyl compound	-	-	-	m2	63.81
6.4 mm thick; glazed with					
putty or bradded beads	-	-	-	m2	60.62
bradded beads and butyl compound	-	-	-	m2	62.78
screwed beads	-	-	-	m2	64.09
screwed beads and butyl compound	-	-	-	m2	67.07
Special glass; BS 952; 'Antisun' solar					
control float glass infill panels; panes					
area 0.15 - 4.00 m2					
4 mm thick; glazed with					
non-hardening compound to metal	-	-	-	m2	59.25
6 mm thick; glazed with					
non-hardening compound to metal	-	-	-	m2	79.34
10 mm thick; glazed with					
non-hardening compound to metal	-	-	-	m2	157.97
12 mm thick; glazed with					
non-hardening compound to metal	-	-	-	m2	195.59
Special glass; BS 952; 'Pyran' fire-					
resisting glass; Schott Glass Ltd.					
6.5 mm thick rectangular panes; glazed with					
screwed hardwood beads and Interdens					
intumescent strip					
300 x 400 mm pane	1.40	18.82	29.22	nr	48.04
400 x 800 mm pane	2.40	32.26	71.04	nr	103.30
500 x 1400 mm pane	3.80	51.07	147.92	nr	198.99
600 x 1800 mm pane	4.80	64.51	222.75	nr	287.26

L WINDOWS/DOORS/STAIRS Including overheads and profit at 9.00% & 5.00%	Labour hours	Labour £	Material £	Unit	Total rate £
6.5 mm thick irregular panes; glazed with screwed hardwood beads and Intergens intumescent strip					
300 x 400 mm pane	1.75	23.52	44.55	nr	68.07
400 x 800 mm pane	3.00	40.32	111.91	nr	152.23
500 x 1400 mm pane	4.75	63.84	237.34	nr	301.18
600 x 1800 mm pane	6.00	80.64	360.70	nr	441.34
Glass louvres; BS 952; with long edges ground or smooth					
5 mm or 6 mm thick float					
100 mm wide	-	-	-	m	5.34
150 mm wide	-	-	-	m	5.96
7 mm thick Georgian wired cast					
100 mm wide	-	-	-	m	10.79
150 mm wide	-	-	-	m	11.56
6 mm thick Georgian polished wired					
100 mm wide	-	-	-	m	14.59
150 mm wide	-	-	-	m	16.85
Labours on glass/sundries					
Curved cutting on panes					
4 mm thick	-	-	-	m	1.76
6 mm thick	-	-	-	m	1.76
6 mm thick; wired	-	-	-	m	2.69
Imitation washleather or black velvet strip; as bedding to edge of glass	-	-	-	m	0.72
Intumescent paste to fire doors per side of glass	-	-	-	m	4.14
Drill hole exceeding 6 mm and not exceeding 15 mm dia. through panes					
not exceeding 6 mm thick	-	-	-	nr	1.97
not exceeding 10 mm thick	-	-	-	nr	2.54
not exceeding 12 mm thick	-	-	-	nr	3.17
not exceeding 19 mm thick	-	-	-	nr	3.95
not exceeding 25 mm thick	-	-	-	nr	4.93
Drill hole exceeding 16 mm and not exceeding 38 mm dia. through panes					
not exceeding 6 mm thick	-	-	-	nr	2.80
not exceeding 10 mm thick	-	-	-	nr	3.74
not exceeding 12 mm thick	-	-	-	nr	4.46
not exceeding 19 mm thick	-	-	-	nr	5.61
not exceeding 25 mm thick	-	-	-	nr	7.01
Drill hole over 38 mm dia. through panes					
not exceeding 6 mm thick	-	-	-	nr	5.61
not exceeding 10 mm thick	-	-	-	nr	6.80
not exceeding 12 mm thick	-	-	-	nr	8.05
not exceeding 19 mm thick	-	-	-	nr	9.92
not exceeding 25 mm thick	-	-	-	nr	12.35
Add to the above					
Plus 33% for countersunk holes					
Plus 50% for wired or laminated glass					

L WINDOWS/DOORS/STAIRS Including overheads and profit at 5.00%	Labour hours	Labour £	Material £	Unit	Total rate £
L40 GENERAL GLAZING - cont'd					
Factory made double hermetically sealed units; Pilkington's 'Insulight' or similar; to wood or metal with butyl non-setting compound and screwed or clipped beads					
Two panes; BS 952; clear float glass; GG; 3 mm or 4 mm thick; 13 mm air space					
2 - 4 m2	-	-	-	m2	54.33
1 - 2 m2	-	-	-	m2	52.42
0.75 - 1 m2	-	-	-	m2	58.36
0.5 - 0.75 m2	-	-	-	m2	66.64
0.35 - 0.5 m2	-	-	-	m2	74.07
0.25 - 0.35 m2	-	-	-	m2	91.69
not exceeding 0.25 m2	-	-	-	m2	91.69
Two panes; BS 952; clear float glass; GG; 5 mm or 6 mm thick; 13 mm air space					
2 - 4 m2	-	-	-	m2	69.97
1 - 2 m2	-	-	-	m2	66.57
0.75 - 1 m2	-	-	-	m2	73.37
0.5 - 0.75 m2	-	-	-	m2	82.28
0.35 - 0.5 m2	-	-	-	m2	91.20
0.25 - 0.35 m2	-	-	-	m2	106.38
not exceeding 0.25 m2	-	-	-	m2	106.38

Keep your figures up to date, free of charge

This section, and most of the other information in this Price Book, is brought up to date every three months in the *Price Book Update*.

The *Update* is available free to all Price Book purchasers.

To ensure you receive your copy, simply complete the reply card from the centre of the book and return it to us.

M SURFACE FINISHES Including overheads and profit at 9.00% & 5.00%	Labour hours	Labour £	Material £	Unit	Total rate £

M10 SAND CEMENT/CONCRETE/GRANOLITHIC SCREEDS/ FLOORING

Cement and sand (1:3); steel trowelled
Work to floors; one coat; level and to falls
not exceeding 15 degrees from horizontal;
to concrete base; over 300 mm wide

32 mm	0.40	5.38	1.70	m2	7.08
40 mm	0.42	5.64	2.04	m2	7.68
48 mm	0.45	6.05	2.40	m2	8.45
50 mm	0.46	6.18	2.47	m2	8.65
60 mm	0.49	6.59	2.83	m2	9.42
65 mm	0.51	6.85	3.07	m2	9.92
70 mm	0.53	7.12	3.30	m2	10.42
75 mm	0.55	7.39	3.53	m2	10.92
Finishing around pipes; not exceeding 0.30 m girth	0.01	0.10	-	nr	0.10

Add to the above for work

to falls and crossfalls and to slopes not exceeding 15 degrees from horizontal	0.03	0.31	-	m2	0.31
to slopes over 15 degrees from horizontal	0.10	1.03	-	m2	1.03
two coats of surface hardener brushed on	0.04	0.41	0.85	m2	1.26
water-repellent additive incorporated in the mix	-	-	-	m2	0.80
oil-repellent additive incorporated in the mix	-	-	-	m2	2.28

Cement and sand (1:3) beds and backings
Work to floors; one coat; level; to concrete
base; screeded; over 300 mm wide

25 mm thick	-	-	-	m2	5.45
50 mm thick	-	-	-	m2	7.36
75 mm thick	-	-	-	m2	9.54
100 mm thick	-	-	-	m2	11.72

Work to floors; one coat; level; to concrete
base; steel trowelled; over 300 mm wide

25 mm thick	-	-	-	m2	5.45
50 mm thick	-	-	-	m2	7.36
75 mm thick	-	-	-	m2	9.54
100 mm thick	-	-	-	m2	11.72

Granolithic paving; cement and granite chippings 5 mm down (1:2.5); steel trowelled
Work to floors; one coat; level; laid on
concrete while green; over 300 mm wide

20 mm	0.30	4.91	3.51	m2	8.42
25 mm	0.35	5.72	4.24	m2	9.96

Work to floors; two coat; laid on hacked
concrete with slurry; over 300 mm wide

38 mm	0.60	9.81	5.65	m2	15.46
50 mm	0.70	11.45	7.24	m2	18.69
75 mm	0.90	14.72	10.66	m2	25.38

Work to landings; one coat; level; laid
on concrete while green; over 300 mm wide

20 mm	0.45	7.36	3.51	m2	10.87
25 mm	0.45	7.36	4.24	m2	11.60

Work to landings; two coat; laid on hacked
concrete with slurry; over 300 mm wide

38 mm	0.75	12.26	5.65	m2	17.91
50 mm	0.85	13.90	7.24	m2	21.14
75 mm	1.00	16.35	10.66	m2	27.01
Finishing around pipes; not exceeding 0.30 m girth	0.05	0.82	-	nr	0.82

M SURFACE FINISHES Including overheads and profit at 9.00%	Labour hours	Labour £	Material £	Unit	Total rate £
M10 SAND CEMENT/CONCRETE/GRANOLITHIC SCREEDS/ **FLOORING - cont'd**					
Granolithic paving; cement and granite **chippings 5 mm down (1:2.5); steel trowelled - cont'd** Add to the above over 300 mm wide for					
1.35 kg/m2 carborundum grains trowelled in	0.06	0.98	0.92	m2	1.90
two coats of surface hardener brushed on	0.02	0.33	1.02	m2	1.35
liquid hardening additive incorporated in the mix	-	-	-	m2	1.22
oil-repellent additive incorporated in the mix	-	-	-	m2	2.28
25 mm work to treads; one coat; to concrete base					
225 mm wide	0.90	14.72	4.63	m	19.35
275 mm wide	0.90	14.72	4.70	m	19.42
Return end	0.18	2.94	-	nr	2.94
Extra for 38 mm 'Ferodo' nosing	0.20	3.27	6.30	m	9.57
13 mm skirtings; rounded top edge and coved bottom junction; to brickwork or blockwork base					
75 mm wide on face	0.55	8.99	0.66	m	9.65
150 mm wide on face	0.75	12.26	1.00	m	13.26
Ends; fair	0.05	0.82	-	nr	0.82
Angles	0.07	1.14	-	nr	1.14
13 mm outer margin to stairs; to follow profile of and with rounded nosing to treads and risers; fair edge and arris at bottom; to concrete base					
75 mm wide	0.90	14.72	1.33	m	16.05
Angles	0.07	1.14	-	nr	1.14
13 mm wall string to stairs; fair edge and arris on top; coved bottom junction with treads and risers; to brickwork or blockwork base					
275 mm (extreme) wide	0.80	13.08	1.33	m	14.41
Ends	0.05	0.82	-	nr	0.82
Angles	0.07	1.14	-	nr	1.14
Ramps	0.08	1.31	-	nr	1.31
Ramped and wreathed corners	0.10	1.64	-	nr	1.64
13 mm outer string to stairs; rounded nosing on top at junction with treads and risers; fair edge and arris at bottom; to concrete base					
300 mm (extreme) wide	0.80	13.08	2.21	m	15.29
Ends	0.05	0.82	-	nr	0.82
Angles	-	-	-	nr	1.14
Ramps	0.08	1.31	-	nr	1.31
Ramped and wreathed corners	0.10	1.64	-	nr	1.64
19 mm skirtings; rounded top edge and coved bottom junction; to brickwork or blockwork base					
75 mm wide on face	0.55	8.99	1.11	m	10.10
150 mm wide on face	0.75	12.26	1.66	m	13.92
Ends; fair	0.05	0.82	-	nr	0.82
Angles	0.07	1.14	-	nr	1.14
19 mm riser; one rounded nosing; to concrete base					
150 mm high; plain	0.90	14.72	3.75	m	**18.47**
150 mm high; undercut	0.90	14.72	3.75	m	**18.47**
180 mm high; plain	0.90	14.72	3.81	m	**18.53**
180 mm high; undercut	0.90	14.72	3.81	m	**18.53**

M SURFACE FINISHES Including overheads and profit at 5.00%	Labour hours	Labour £	Material £	Unit	Total rate £

M11 MASTIC ASPHALT FLOORING

Mastic asphalt paving to BS 1076; black
20 mm one coat coverings; felt isolating
membrane; to concrete base; flat

over 300 mm wide	-	-	-	m2	9.10
225 - 300 mm wide	-	-	-	m2	17.27
150 - 225 mm wide	-	-	-	m2	17.51
not exceeding 150 mm wide	-	-	-	m2	21.59

25 mm one coat coverings; felt isolating
membrane; to concrete base; flat

over 300 mm wide	-	-	-	m2	10.83
225 - 300 mm wide	-	-	-	m2	20.56
150 - 225 mm wide	-	-	-	m2	20.84
not exceeding 150 mm wide	-	-	-	m2	25.70

20 mm three coat skirtings to brickwork base

not exceeding 150 mm girth	-	-	-	m	10.20
150 - 225 mm girth	-	-	-	m	11.90
225 - 300 mm girth	-	-	-	m	13.59

Mastic asphalt paving; acid-resisting; black
20 mm one coat coverings; felt isolating
membrane; to concrete base; flat

over 300 mm wide	-	-	-	m2	10.74
225 - 300 mm wide	-	-	-	m2	18.90
150 - 225 mm wide	-	-	-	m2	19.14
not exceeding 150 mm wide	-	-	-	m2	23.23

25 mm one coat coverings; felt isolating
membrane; to concrete base; flat

over 300 mm wide	-	-	-	m2	12.47
225 - 300 mm wide	-	-	-	m2	23.66
150 - 225 mm wide	-	-	-	m2	23.98
not exceeding 150 mm wide	-	-	-	m2	29.58

20 mm three coat skirtings to brickwork base

not exceeding 150 mm girth	-	-	-	m	12.03
150 - 225 mm girth	-	-	-	m	14.04
225 - 300 mm girth	-	-	-	m	16.02

Mastic asphalt paving to BS 6577; black
20 mm one coat coverings; felt isolating
membrane; to concrete base; flat

over 300 mm wide	-	-	-	m2	14.50
225 - 300 mm wide	-	-	-	m2	23.45
150 - 225 mm wide	-	-	-	m2	24.98
not exceeding 150 mm wide	-	-	-	m2	30.25

25 mm one coat coverings; felt isolating
membrane; to concrete base; flat

over 300 mm wide	-	-	-	m2	17.57
225 - 300 mm wide	-	-	-	m2	28.42
150 - 225 mm wide	-	-	-	m2	30.28
not exceeding 150 mm wide	-	-	-	m2	34.02

M SURFACE FINISHES Including overheads and profit at 9.00% & 5.00%	Labour hours	Labour £	Material £	Unit	Total rate £
M11 MASTIC ASPHALT FLOORING - cont'd					
Mastic asphalt paving to BS 6577; black - cont'd					
20 mm three coat skirtings to brickwork base					
not exceeding 150 mm girth	-	-	-	m	11.41
150 - 225 mm girth	-	-	-	m	13.49
225 - 300 mm girth	-	-	-	m	15.58
Mastic asphalt paving to BS 1451; red					
15 mm one coat coverings; felt isolating					
membrane; to concrete base; flat					
over 300 mm wide	-	-	-	m2	10.09
225 - 300 mm wide	-	-	-	m2	19.14
150 - 225 mm wide	-	-	-	m2	19.41
not exceeding 150 mm wide	-	-	-	m2	23.95
20 mm three coat skirtings to brickwork base					
not exceeding 150 mm girth	-	-	-	m	10.44
150 - 225 mm girth	-	-	-	m	12.22
M12 TROWELLED BITUMEN/RESIN/RUBBER LATEX FLOORING					
Latex cement floor screeds; steel trowelled					
Work to floors; level; to concrete base;					
over 300 mm wide					
3 mm thick; one coat	0.17	1.75	1.66	m2	3.41
5 mm thick; two coats	0.22	2.27	2.59	m2	4.86
Isocrete K screeds; steel trowelled					
Work to floors; level; to concrete base;					
over 300 mm wide					
35 mm thick; plus polymer bonder coat	-	-	-	m2	8.98
40 mm thick	-	-	-	m2	9.50
45 mm thick	-	-	-	m2	10.02
50 mm thick	-	-	-	m2	10.59
Work to floors; to falls or cross-falls; to					
concrete base; over 300 mm wide					
55 mm (average) thick	-	-	-	m2	6.29
60 mm (average) thick	-	-	-	m2	7.16
65 mm (average) thick	-	-	-	m2	7.68
75 mm (average) thick	-	-	-	m2	8.72
90 mm (average) thick	-	-	-	m2	10.38
'Synthanite' floor screeds; steel trowelled					
Work to floors; one coat; level; paper felt					
underlay; to concrete base; over 300 mm wide					
25 mm thick	0.50	5.16	11.04	m2	16.20
50 mm thick	0.60	6.19	17.53	m2	23.72
75 mm thick	0.70	7.23	21.08	m2	28.31
Bituminous lightweight insulating roof screeds					
'Bit-Ag' or similar roof screed; to falls or					
cross-falls; bitumen felt vapour barrier;					
over 300 mm wide					
75 mm (average) thick	-	-	-	m2	16.16
100 mm (average) thick	-	-	-	m2	18.86

M SURFACE FINISHES
Including overheads and profit at 5.00%

BASIC PLASTER PRICES

	£		£
Plaster prices (£/tonne)			
BS 1191 Part 1; class A			
CB stucco	49.82		
BS 1191 Part 1; class B			
'Thistle' board	56.04	'Thistle' finish	57.28
BS 1191 Part 1; class C			
'Limelite' - backing	128.46	'Limelite' - renovating	171.29
- finishing	103.34		
Pre-mixed lightweight; BS 1191 Part 2			
'Carlite' - bonding	93.73	'Carlite' - finishing	74.81
- browning	91.75	- metal lathing	94.76
- browning HSB	100.31		
Projection 86.64			
Tyrolean 'Cullamix'	163.20		

Plastering sand £8.36/tonne

Sundries

ceramic tile - adhesive (£/5L)	5.64	plasterboard nails (£/25kg)	34.31
- grout (£/kg)	0.74	scrim (£/100 m roll)	3.36
impact adhesive (£/5L)	15.90		

	Labour hours	Labour £	Material £	Unit	Total rate £
M20 PLASTERED/RENDERED/ROUGHCAST COATINGS					
Cement and sand (1:3) beds and backings					
10 mm work to walls; one coat; to brickwork or blockwork base					
over 300 mm wide	-	-	-	m2	6.43
not exceeding 300 mm wide	-	-	-	m	3.87
12 mm work to walls; one coat; to brickwork or blockwork base					
over 300 mm wide	-	-	-	m2	7.03
not exceeding 300 mm wide	-	-	-	m	4.25
13 mm work to walls; one coat; to brickwork or blockwork base; screeded					
over 300 mm wide	-	-	-	m2	8.18
not exceeding 300 mm wide	-	-	-	m	4.63
15 mm work to walls; one coat; to brickwork or blockwork base					
over 300 mm wide	-	-	-	m2	8.28
not exceeding 300 mm wide	-	-	-	m	4.96
Cement and sand (1:3); steel trowelled					
13 mm work to walls; two coats; to brickwork or blockwork base					
over 300 mm wide	-	-	-	m2	9.21
not exceeding 300 mm wide	-	-	-	m	5.45
16 mm work to walls; two coats; to brickwork or blockwork base					
over 300 mm wide	-	-	-	m2	10.03
not exceeding 300 mm wide	-	-	-	m	6.00
19 mm work to walls; two coats; to brickwork or blockwork base					
over 300 mm wide	-	-	-	m2	10.90
not exceeding 300 mm wide	-	-	-	m	6.54
Add to the above over 300 mm wide for first coat in water-repellent cement	-	-	-	m2	1.91
finishing coat in coloured cement	-	-	-	m2	4.63

M SURFACE FINISHES Including overheads and profit at 5.00%	Labour hours	Labour £	Material £	Unit	Total rate £
M20 PLASTERED/RENDERED/ROUGHCAST COATINGS - cont'd					
Cement-lime-sand (1:2:9); steel trowelled					
19 mm work to walls; two coats; to brickwork					
or blockwork base					
over 300 mm wide	-	-	-	m2	10.36
not exceeding 300 mm wide	-	-	-	m	6.21
Cement-lime-sand (1:1:6); steel trowelled					
19 mm work to ceilings; three coats; to					
metal lathing base					
over 300 mm wide	-	-	-	m2	14.72
not exceeding 300 mm wide	-	-	-	m	8.72
Plaster; first and finishing coats of					
'Carlite' pre-mixed lightweight plaster;					
steel trowelled					
13 mm work to walls; two coats; to brickwork					
or blockwork base (or 10 mm work to concrete					
base)					
over 300 mm wide	-	-	-	m2	7.09
over 300 mm wide; in staircase areas or					
plant rooms	-	-	-	m2	8.72
not exceeding 300 mm wide	-	-	-	m	4.36
13 mm work to isolated piers or columns; two					
coats over 300 mm wide	-	-	-	m2	11.45
not exceeding 300 mm wide	-	-	-	m	6.81
10 mm work to ceilings; two coats; to					
concrete base					
over 300 mm wide	-	-	-	m2	7.36
over 300 mm wide; 3.5 - 5 m high	-	-	-	m2	7.52
over 300 mm wide; in staircase areas or					
plant rooms	-	-	-	m2	8.72
not exceeding 300 mm wide	-	-	-	m	4.58
10 mm work to isolated beams; two coats; to					
concrete base					
over 300 mm wide	-	-	-	m2	10.90
over 300 mm wide; 3.5 - 5 m high	-	-	-	m2	11.72
not exceeding 300 mm wide	-	-	-	m	6.54
Plaster; one coat 'Snowplast' plaster; steel					
trowelled					
13 mm work to walls; one coat; to brickwork					
or blockwork base					
over 300 mm wide	-	-	-	m2	8.18
over 300 mm wide; in staircase areas or					
plant rooms	-	-	-	m2	9.27
not exceeding 300 mm wide	-	-	-	m	4.69
13 mm work to isolated columns; one coat					
over 300 mm wide	-	-	-	m2	11.45
not exceeding 300 mm wide	-	-	-	m	6.76
Plaster; first coat of cement and sand					
(1:3); finishing coat of 'Thistle' class B					
plaster; steel trowelled					
13 mm work to walls; two coats; to brickwork					
or blockwork base					
over 300 mm wide	-	-	-	m2	9.54
over 300 mm wide; in staircase areas or					
plant rooms	-	-	-	m2	10.90
not exceeding 300 mm wide	-	-	-	m	5.78
13 mm work to isolated columns; two coats					
over 300 mm wide	-	-	-	m2	12.81
not exceeding 300 mm wide	-	-	-	m	7.85

M SURFACE FINISHES Including overheads and profit at 5.00%	Labour hours	Labour £	Material £	Unit	Total rate £
Plaster; first coat of cement-lime-sand (1:1:6); finishing coat of 'Sirapite' class B plaster; steel trowelled					
13 mm work to walls; two coats; to brickwork or blockwork base					
over 300 mm wide	-	-	-	m2	9.54
not exceeding 300 mm wide	-	-	-	m	5.72
over 300 mm wide; in staircase areas or plant rooms	-	-	-	m2	11.01
not exceeding 300 mm wide	-	-	-	m	5.72
13 mm work to isolated columns; two coats					
over 300 mm wide	-	-	-	m2	13.08
not exceeding 300 mm wide	-	-	-	m	7.85
Plaster; first coat of 'Limelite' renovating plaster; finishing coat of 'Limelite' finishing plaster; steel trowelled					
13 mm work to walls; two coats; to brickwork or blockwork base					
over 300 mm wide	-	-	-	m2	9.79
over 300 mm wide; in staricase areas or compartments under 4 m2	-	-	-	m2	11.74
not exceeding 300 mm wide	-	-	-	m	5.87
Dubbing out existing walls with undercoat plaster; average 6 mm thick					
over 300 mm wide	-	-	-	m2	3.88
not exceeding 300 mm wide	-	-	-	m	2.32
Dubbing out existing walls with undercoat plaster; average 12 mm thick					
over 300 mm wide	-	-	-	m2	5.25
not exceeding 300 mm wide	-	-	-	m	3.13
Plaster; one coat 'Thistle' projection plaster; steel trowelled					
13 mm work to walls; one coat; to brickwork or blockwork base					
over 300 mm wide	-	-	-	m2	10.36
over 300 mm wide; in staircase areas or plant rooms	-	-	-	m2	11.72
not exceeding 300 mm wide	-	-	-	m	6.27
6 mm work to isolated columns; one coat					
over 300 mm wide	-	-	-	m2	11.17
not exceeding 300 mm wide	-	-	-	m	6.54
Plaster; first, second and finishing coats of 'Carlite' pre-mixed lightweight plaster; steel trowelled					
13 mm work to ceilings; three coats to metal lathing base					
over 300 mm wide	-	-	-	m2	11.17
over 300 mm wide; in staircase areas or plant rooms	-	-	-	m2	12.54
not exceeding 300 mm wide	-	-	-	m	6.76
13 mm work to swept soffit of metal lathing arch former					
not exceeding 300 mm wide	-	-	-	m	16.90
300 - 400 mm wide	-	-	-	m	17.99
13 mm work to vertical face of metal lathing arch former					
not exceeding 0.5 m2 per side	-	-	-	nr	32.97
0.5 m2 - 1 m2 per side	-	-	-	nr	39.79

M SURFACE FINISHES Including overheads and profit at 9.00% & 5.00%	Labour hours	Labour £	Material £	Unit	Total rate £

M20 PLASTERED/RENDERED/ROUGHCAST COATINGS - cont'd

Tyrolean decorative rendering; 13 mm first coat of cement-lime-sand (1:1:6); finishing three coats of 'Cullamix' applied with approved hand operated machine external
To walls; four coats; to brickwork or blockwork base

over 300 mm wide	-	-	-	m2	23.44
not exceeding 300 mm wide	-	-	-	m	14.17

'Mineralite' decorative rendering; first coat of cement-lime-sand (1:0.5:4.5); finishing coat of 'Mineralite'; applied by Specialist Subcontractor
17 mm work to walls; to brickwork or blockwork base

over 300 mm wide	-	-	-	m2	37.01
not exceeding 300 mm wide	-	-	-	m	18.51
form 9 x 6 mm expansion joint and point in one part polysulphide mastic	-	-	-	m	6.38

Plaster; one coat 'Thistle' board finish; steel trowelled (prices included within plasterboard rates)
3 mm work to walls or ceilings; one coat; to plasterboard base

over 300 mm wide	-	-	-	m2	4.91
over 300 mm wide; in staircase areas or plant rooms	-	-	-	m2	6.27
not exceeding 300 mm wide	-	-	-	m	3.00

Plaster; one coat 'Thistle' board finish; steel trowelled 3 mm work to walls or ceilings; one coat on and including gypsum plasterboard; BS 1230; fixing with nails; 3 mm joints filled with plaster and jute scrim cloth; to softwood base; plain grade baseboard or lath with rounded edges
9.5 mm board to walls

over 300 mm wide	0.50	6.72	2.44	m2	9.16
not exceeding 300 mm wide	0.25	3.36	0.77	m	4.13
9.5 mm board to walls; in staircase areas or plant rooms					
over 300 mm wide	0.60	8.06	2.44	m2	10.50
not exceeding 300 mm wide	0.30	4.03	0.77	m	4.80
9.5 mm board to isolated columns					
over 300 mm wide	0.66	8.87	2.46	m2	11.33
not exceeding 300 mm wide	0.33	4.44	0.78	m	5.22
9.5 mm board to ceilings					
over 300 mm wide	0.56	7.53	2.44	m2	9.97
over 300 mm wide; 3.5 - 5 m high	0.66	8.87	2.44	m2	11.31
not exceeding 300 mm wide	0.28	3.76	0.77	m	4.53
9.5 mm board to ceilings; in staircase areas or plant rooms					
over 300 mm wide	0.66	8.87	2.44	m2	11.31
not exceeding 300 mm wide	0.33	4.44	0.77	m	5.21
9.5 mm board to isolated beams					
over 300 mm wide	0.70	9.41	2.46	m2	11.87
not exceeding 300 mm wide	0.35	4.70	0.74	m	5.45
Add for 'Duplex' insulating grade	-	-	-	m2	0.45
12.5 mm board to walls					
over 300 mm wide	0.56	7.53	2.89	m2	10.42
not exceeding 300 mm wide	0.28	3.76	0.91	m	4.68
12.5 mm board to walls; in staircase areas or plant rooms					
over 300 mm wide	0.66	8.87	2.89	m2	11.76
not exceeding 300 mm wide	0.33	4.44	0.91	m	5.35

M SURFACE FINISHES Including overheads and profit at 9.00% & 5.00%	Labour hours	Labour £	Material £	Unit	Total rate £
12.5 mm board to isolated columns					
over 300 mm wide	0.70	9.41	2.91	m2	12.32
not exceeding 300 mm wide	0.35	4.70	0.92	m	5.62
12.5 mm board to ceilings					
over 300 mm wide	0.60	8.06	2.89	m2	10.96
over 300 mm wide; 3.5 - 5 m high	0.70	9.41	2.89	m2	12.30
not exceeding 300 mm wide	0.30	4.03	0.91	m	4.94
12.5 mm board to ceilings; in staircase areas or plant rooms					
over 300 mm wide	0.70	9.41	2.89	m2	12.30
· not exceeding 300 mm wide	0.35	4.70	0.91	m	5.62
12.5 mm board to isolated beams					
over 300 mm wide	0.76	10.21	2.91	m2	13.12
not exceeding 300 mm wide	0.38	5.11	0.92	m	6.03
Add for 'Duplex' insulating grade	-	-	-	m2	0.45
Accessories					
'Expamet' render beads; white PVC nosings; to brickwork or blockwork base					
external stop bead; ref 1222	0.06	0.81	1.53	m	2.34
'Expamet' plaster beads; to brickwork or blockwork base					
angle bead; ref 550	0.07	0.94	0.42	m	1.36
architrave bead; ref 579	0.09	1.21	0.66	m	1.87
stop bead; ref 563	0.06	0.81	0.50	m	1.31

M22 SPRAYED MINERAL FIBRE COATINGS

Prepare and apply by spray Mandolite P20 fire protection on structural steel/metalwork					
16 mm (one hour) fire protection					
to walls and columns	-	-	-	m2	7.36
to ceilings and beams	-	-	-	m2	8.09
to isolated metalwork	-	-	-	m2	18.39
22 mm (one and a half hour) fire protection					
to walls and columns	-	-	-	m2	9.32
to ceilings and beams	-	-	-	m2	10.25
to isolated metalwork	-	-	-	m2	21.43
28 mm (two hour) fire protection					
to walls and columns	-	-	-	m2	10.30
to ceilings and beams	-	-	-	m2	11.33
to isolated metalwork	-	-	-	m2	22.66
52 mm (four hour) fire protection					
to walls and columns	-	-	-	m2	21.09
to ceilings and beams	-	-	-	m2	23.20
to isolated metalwork	-	-	-	m2	42.18

M30 METAL MESH LATHING/ANCHORED REINFORCEMENT FOR PLASTERED COATING

Accessories

Pre-formed galvanised expanded steel arch-frames; 'Truline' or similar; semi-circular; to suit walls up to 230 mm thick					
375 mm radius; for 800 mm opening; ref SC 750	PC £13.91	0.25	3.36	15.16	nr	18.52
425 mm radius; for 850 mm opening; ref SC 850	PC £14.77	0.25	3.36	16.10	nr	19.46
450 mm radius; for 900 mm opening; ref SC 900	PC £17.14	0.25	3.36	18.68	nr	22.04
600 mm radius; for 1200 mm opening; ref SC 1200	PC £21.33	0.25	3.36	23.25	nr	26.61

M SURFACE FINISHES Including overheads and profit at 9.00% & 5.00%	Labour hours	Labour £	Material £	Unit	Total rate £

M30 METAL MESH LATHING/ANCHORED REINFORCEMENT
FOR PLASTERED COATING - cont'd

Lathing; Expamet 'BB' expanded metal
lathing or similar; BS 1369; 50 mm laps
6 mm mesh linings to ceilings; fixing with
staples; to softwood base; over 300 mm wide

ref BB263; 0.500 mm thick PC £2.10	0.30	4.03	2.40	m2	6.43
ref BB264; 0.675 mm thick PC £2.46	0.30	4.03	2.82	m2	6.85

6 mm mesh linings to ceilings; fixing with
wire; to steelwork; over 300 mm wide

ref BB263; 0.500 mm thick	0.32	4.30	2.47	m2	6.77
ref BB264; 0.675 mm thick	0.32	4.30	2.89	m2	7.19

6 mm mesh linings to ceilings; fixings with
wire; to steelwork; not exceeding 300 mm
wide

ref BB263; 0.500 mm thick	0.20	2.69	0.78	m	3.47
ref BB264; 0.675 mm thick	0.20	2.69	0.91	m	3.60
Raking cutting	0.20	2.69	0.78	m	3.47
Cutting and fitting around pipes;					
not exceeding 0.30 m girth	0.20	2.69	0.91	nr	3.60

Lathing; Expamet 'Riblath' or 'Spraylath'
stiffened expanded metal lathing or similar;
50 mm laps
10 mm mesh lining to walls; fixing with
nails; to softwood base; over 300 mm wide

Riblath ref 269; 0.30 mm thick PC £2.62	0.25	3.36	3.01	m2	6.37
Riblath ref 271; 0.50 mm thick PC £3.04	0.25	3.36	3.49	m2	6.85
Spraylath ref 273; 0.50 mm thick	0.25	3.36	4.27	m2	7.63

10 mm mesh lining to walls; fixing with
nails; to softwood base; not exceeding 300
mm wide

Riblath ref 269; 0.30 mm thick	0.15	2.02	0.95	m	2.97
Riblath ref 271; 0.50 mm thick	0.15	2.02	1.10	m	3.12
Spraylath ref 273; 0.50 mm thick	0.15	2.02	1.34	m	3.36

10 mm mesh lining to walls; fixing to brick
or blockwork; over 300 mm wide

Red-rib ref 274; 0.50 mm thick PC £3.36	0.20	2.69	5.34	m2	8.03
Stainless steel Riblath ref 267; 0.30 mm					
thick PC £7.37	0.20	2.69	9.94	m2	12.63

10 mm mesh lining to ceilings; fixing with
wire; to steelwork; over 300 mm wide

Riblath ref 269; 0.30 mm thick	0.32	4.30	3.10	m2	7.40
Riblath ref 271; 0.50 mm thick	0.32	4.30	3.58	m2	7.88
Spraylath ref 273; 0.50 mm thick	0.32	4.30	4.36	m2	8.66
Raking cutting	0.15	2.02	0.61	m	2.63
Cutting and fitting around pipes;					
not exceeding 0.30 m girth	0.05	0.67	-	nr	0.67

M31 FIBROUS PLASTER

Fibrous plaster; fixing with screws;
plugging; countersinking; stopping; filling
and pointing joints with plaster
16 mm plain slab coverings to ceilings

over 300 mm wide	-	-	-	m2	90.91
not exceeding 300 mm wide	-	-	-	m	29.30
Coves; not exceeding 150 mm girth					
per 25 mm girth	-	-	-	m	4.25
Coves; 150 - 300 mm girth					
per 25 mm girth	-	-	-	m	5.23
Cornices					
per 25 mm girth	-	-	-	m	5.23
Cornice enrichments					
per 25 mm girth; depending on degree of					
enrichments	-	-	-	m	6.41

M SURFACE FINISHES Including overheads and profit at 9.00% & 5.00%	Labour hours	Labour £	Material £	Unit	Total rate £
Fibrous plaster; fixing with plaster wadding **filling and pointing joints with plaster; to** **steel base**					
16 mm plain slab coverings to ceilings					
over 300 mm wide	-	-	-	m2	**90.91**
not exceeding 300 mm wide	-	-	-	m	**29.30**
16 mm plain casings to stanchions					
per 25 mm girth	-	-	-	m	**2.55**
16 mm plain casings to beams					
per 25 mm girth	-	-	-	m	**2.55**
Gyproc cove; fixing with adhesive; filling **and pointing joints with plaster**					
Cove					
125 mm girth	0.20	2.06	0.70	m	**2.76**
Angles	0.04	0.41	0.22	nr	**0.63**

ALTERNATIVE TILE MATERIALS

	£		£
Dennis Ruabon clay floor quarries (£/1000)			
Heather brown			
150x72x12.5 mm	149.40	194x94x12.5 mm	197.50
150x150x12.5 mm	196.70	194x194x12.5 mm	381.40
Red			
150x150x12.5 mm; square	138.90	150x150x12.5 mm; hexagonal	219.30
Rustic			
150x150x12.5 mm	204.50		
Daniel Platt heavy duty floor tiles (£/m2)			
'Ferrolite' flat; 150x150x9 mm			
black	9.38	steel	9.16
cream	9.16	red	7.88
'Ferrolite' flat; 150x150x12 mm			
black	12.44	chocolate	11.99
cream	10.99	red	11.33
'Ferrundum' anti-slip; 150x150x12 mm			
black	19.70	red	18.73
chocolate	19.28		
Langley wall and floor tiles (£/m2)			
'Buchtal' fine grain ceramic wall and floor tiles			
240x115x11 mm - series 1	20.86	194x144x11 mm - series 1	20.23
- series 2	22.90	- series 2	22.21
'Buchtal' rustic glazed wall and floor tiles			
240x115x11 mm - series 1	20.86	194x94x11 mm - series 1	20.38
- series 2	22.34	- series 2	21.81
- series 3	19.67	- series 3	19.31
'Hoganas' frostproof glazed ceramic wall facings			
196x96x8 mm	19.85		
'Hoganas' vitrified paviors			
215x105x18 mm - yellow	22.94	215x105x18 mm - brown	27.51
'Sinzag' glazed ceramic wall facings			
240x115x10 mm - series 1	19.34	240x52x10 mm - series 1	22.04
- series 2	26.57		
- series 3	45.44		
'Sinzag' vitrified ceramic floor tiles; 150x150x11/12 mm			
plain - grey/white speckled	21.46	textured - series 2	22.96
Marley floor tiles (£/m2); 300x300 mm			
'Econoflex' - series 1/2	2.52	anti-static	20.99
- series 4	2.99	'Travertine' - 2.5 mm	4.85
'Marleyflex' - 2.0 mm	3.25	'Vylon'	3.40
- 2.5 mm	3.80		

M SURFACE FINISHES Including overheads and profit at 9.00%	Labour hours	Labour £	Material £	Unit	Total rate £

M40 STONE/CONCRETE/QUARRY/CERAMIC TILING/ MOSAIC

Clay floor quarries; BS 6431; class 1;
Daniel Platt 'Crown' tiles or similar; level
bedding 10 mm thick and jointing in cement
and sand (1:3); butt joints straight both
ways; flush pointing with grout; to cement
and sand base

Work to floors; over 300 mm wide					
150 x 150 x 12.5 mm; red PC £188.11/1000	0.80	8.26	10.62	m2	18.88
150 x 150 x 12.5 mm; brown PC £295.50/1000	0.80	8.26	16.03	m2	24.29
200 x 200 x 19 mm; brown PC £720.74/1000	0.65	6.71	21.69	m2	28.40
Works to floors; in staircase areas or plant rooms					
150 x 150 x 12.5 mm; red	0.90	9.29	10.62	m2	19.91
150 x 150 x 12.5 mm; brown	0.90	9.29	16.03	m2	25.32
200 x 200 x 19 mm; brown	0.75	7.74	21.69	m2	29.43
Work to floors; not exceeding 300 mm wide					
150 x 150 x 12.5 mm; red	0.40	4.13	3.82	m	7.95
150 x 150 x 12.5 mm; brown	0.40	4.13	5.75	m	9.88
200 x 200 x 19 mm; brown	0.33	3.41	8.19	m	11.60
Fair square cutting against flush edges of existing finishings	0.10	1.03	0.48	m	1.51
Raking cutting	0.20	2.06	0.71	m	2.77
Cutting around pipes; not exceeding 0.30 m girth	0.15	1.55	-	nr	1.55
Extra for cutting and fitting into recessed manhole covers 600 x 600 mm; lining up with adjoining work	1.00	10.32	-	nr	10.32
Work to sills; 150 mm wide; rounded edge tiles					
150 x 150 x 12.5 mm; red PC £351.73/1000	0.33	3.41	3.03	m	6.44
150 x 150 x 12.5 mm; brown PC £379.13/1000	0.33	3.41	3.25	m	6.66
Fitted end	0.15	1.55	-	m	1.55
Coved skirtings; 150 mm high; rounded top edge					
150 x 150 x 12.5 mm; red PC £351.73/1000	0.25	2.58	3.03	m	5.61
150 x 150 x 12.5 mm; brown PC £379.13/1000	0.25	2.58	3.25	m	5.83
Ends	0.05	0.52	-	nr	0.52
Angles	0.15	1.55	1.35	nr	2.90

Glazed ceramic wall tiles; BS 6431; fixing
with adhesive; butt joints straight both m
ways; flush pointing with white grout; to
plaster base

Work to walls; over 300 mm wide					
108 x 108 x 4 mm PC £14.27	0.90	12.10	18.11	m2	30.21
152 x 152 x 5.5 mm; white PC £7.95	0.60	8.06	10.88	m2	18.94
152 x 152 x 5.5 mm; light colours PC £9.89	0.60	8.06	13.09	m2	21.15
152 x 152 x 5.5 mm; dark colours PC £12.22	0.60	8.06	15.76	m2	23.82
Extra for RE or REX tile	-	-	-	nr	0.06
Work to walls; in staircase areas or plant rooms					
152 x 152 x 5.5 mm; white	0.67	9.00	10.88	m2	19.88
Work to walls; not exceeding 300 mm wide					
108 x 108 x 4 mm	0.45	6.05	5.67	m	11.72
152 x 152 x 5.5 mm; white	0.30	4.03	3.40	m	7.43
152 x 152 x 5.5 mm; light colours	0.30	4.03	4.09	m	8.12
152 x 152 x 5.5 mm; dark colours	0.30	4.03	4.93	m	8.96
Cutting around pipes; not exceeding 0.30 m girth	0.10	1.34	-	nr	1.34

M SURFACE FINISHES Including overheads and profit at 9.00%	Labour hours	Labour £	Material £	Unit	Total rate £
Work to sills; 150 mm wide; rounded edge **tiles**					
108 x 108 x 4 mm	0.30	4.03	2.86	m	6.89
152 x 152 x 5.5 mm; white	0.25	3.36	1.73	m	5.09
Fitted end	0.10	1.34	-	nr	1.34
198 x 64.5 x 6 mm Langley 'Project Plus' **wall tiles; fixing with adhesive; butt joints** **straight both ways; flush pointing with** **white grout; to plaster base**					
Work to walls					
over 300 mm wide PC £18.45	1.80	24.19	22.89	m2	47.08
not exceeding 300 mm wide	0.70	9.41	6.87	m	16.28
Glazed ceramic floor tiles; level; bedding **10 mm thick and jointing in cement and sand** **(1:3); butt joints straight both ways; flush** **pointing with white grout; to cement and** **sand base**					
Work to floors; over 300 mm wide					
200 x 200 x 8 mm PC £17.37	0.60	8.06	21.03	m2	29.09
Work to floors; not exceeding 300 mm wide					
200 x 200 x 8 mm	0.30	4.03	6.68	m	10.71

M41 TERRAZZO TILING/IN SITU TERRAZZO

	Labour hours	Labour £	Material £	Unit	Total rate £
Terrazzo tiles; cement and white Sicilian **marble aggregate; level; bedding 13 mm thick** **and jointing in cement and sand (1:3); butt** **joints straight both ways; pointing with** **white grout; margins laid in situ; polished;** **to concrete base**					
Work to floors; over 300 mm wide					
305 x 305 x 25 mm	2.30	37.61	17.60	m2	55.21
Work to floors; not exceeding 300 mm wide					
305 x 305 x 25 mm	1.00	16.35	5.28	m	21.63
Add to the above for					
white or coloured cement	-	-	-	m2	2.17
coloured marble aggregate	-	-	-	.m2	2.59
Terrazzo paving; cement and white Sicilian **marble aggregate; polished**					
16 mm work to floors; one coat; level; to **cement and sand base**					
over 300 mm wide	1.80	29.43	7.25	m2	36.68
not exceeding 300 mm wide	0.90	14.72	2.17	m	16.89
Finishing around pipes; not exceeding **0.30 m girth**	0.25	4.09	-	nr	4.09
16 mm work to landings; one coat; level; to **cement and sand base**					
over 300 mm wide	2.40	39.24	7.25	m2	46.49
not exceeding 300 mm wide	1.00	16.35	2.17	m	18.52
Add to the above for					
35 mm cement and sand (1:3) screed	0.80	13.08	4.45	m2	17.53
carborundum grains 0.27 kg/m2	0.50	8.18	0.92	m2	9.10
white or coloured cement	-	-	-	m2	3.11
coloured marble aggregate	-	-	-	m2	2.59
Division strips; set in paving to form **panelling**					
25 x 3 mm ebonite strip	0.20	3.27	0.52	m	3.79
25 x 3 mm brass strip; including anchor wires	0.25	4.09	4.14	m	8.23

M SURFACE FINISHES Including overheads and profit at 9.00%	Labour hours	Labour £	Material £	Unit	Total rate £
M41 TERRAZZO TILING/IN SITU TERRAZZO - cont'd					
Terrazzo paving; cement and white Sicilian marble aggregate; polished - cont'd 16 mm work to treads; one coat; including 16 mm cement and sand (1:3) screed; to concrete base					
225 mm wide	2.50	40.88	2.17	m	43.05
275 mm wide	2.70	44.15	2.38	m	46.53
Return end	0.60	9.81	-	m	9.81
Extra for					
white or coloured cement	-	-	-	m	0.78
38 mm 'Ferodo' nosing	0.25	4.09	4.66	m	8.75
projecting nosing	0.66	10.79	0.62	m	11.41
6 mm skirtings; rounded top edge; coved bottom junction; including 13 mm cement and sand screed; to brickwork or blockwork base					
75 mm wide on face	1.30	21.26	0.47	m	21.73
150 mm wide on face	1.50	24.53	1.04	m	25.57
Ends; fair	0.20	3.27	-	nr	3.27
Angles	0.20	3.27	-	nr	3.27
6 mm outer margin to stairs; to follow profile of and with rounded nosing to treads and risers; fair edge and arris at bottom; to concrete base					
75 mm wide	2.50	40.88	0.83	m	41.71
Angles	0.10	1.64	-	nr	1.64
6 mm riser; one rounded nosing; to concrete base					
150 mm high; plain	2.00	32.70	0.93	m	33.63
150 mm high; undercut	2.00	32.70	0.93	m	33.63
180 mm high; plain	2.10	34.34	1.24	m	35.58
180 mm high; undercut	2.10	34.34	1.24	m	35.58
6 mm wall string to stairs; fair edge and arris at top; coved bottom junction with treads and risers; including 13 mm cement and sand screed; to concrete base					
275 mm (extreme) wide	3.75	61.31	2.38	m	63.69
Ends	0.20	3.27	-	nr	3.27
Angles	0.25	4.09	-	nr	4.09
Ramps	0.33	5.40	-	nr	5.40
Ramped and wreathed corners	0.50	8.18	-	nr	8.18
6 mm outer string to stairs; rounded nosing on top at junction with treads and risers; fair edge and arris at bottom; including 13 mm cement and sand screed; to concrete base					
300 mm (extreme) wide	5.00	81.75	3.62	m	85.37
Ends	0.25	4.09	-	nr	4.09
Angles	0.40	6.54	-	nr	6.54
Ramps	0.50	8.18	-	nr	8.18
Ramped and wreathed corners	0.90	14.72	-	nr	14.72
Terrazzo wall lining; cement and white Sicilian marble aggregate; polished 6 mm work to walls; one coat; to normal dado height; on cement and sand base					
over 300 mm wide	6.00	98.10	5.18	m2	103.28
M42 WOOD BLOCK/COMPOSITION BLOCK/PARQUET FLOORING					
Wood block; Vigers, Stevens & Adams 'Feltwood'; 7.5 mm thick; level; fixing with adhesive; to cement and sand base Work to floors; over 300 mm wide					
Iroko	0.50	5.16	10.35	m2	15.51

M SURFACE FINISHES Including overheads and profit at 9.00% & 5.00%	Labour hours	Labour £	Material £	Unit	Total rate £
Wood blocks 25 mm thick; tongued and grooved **joints; herringbone pattern; level; fixing** **with adhesive; to cement and sand base** Work to floors; over 300 mm wide					
Iroko	-	-	-	m2	**35.97**
Merbau	-	-	-	m2	**35.97**
French Oak	-	-	-	m2	**41.42**
American Oak	-	-	-	m2	**33.79**
Fair square cutting against flush edges of existing finishings	-	-	-	m	**3.38**
Extra for cutting and fitting into recessed duct covers 450 mm wide; lining up with adjoining work	-	-	-	m	**6.54**
Cutting around pipes; not exceeding 0.30 m girth	-	-	-	nr	**0.33**
Extra for cutting and fitting into recessed manhole covers 600 x 600 mm; lining up with adjoining work	-	-	-	nr	**9.81**
Add to wood block flooring over 300 mm wide for					
sanding; one coat sealer; one coat wax polish	-	-	-	m2	**3.27**
sanding; two coats sealer; buffing with steel wool	-	-	-	m2	**2.18**
sanding; three coats polyurethane lacquer; buffing down between coats	-	-	-	m2	**2.18**

ALTERNATIVE FLEXIBLE SHEET MATERIALS (£/m2)

	£		£
Nairn flooring			
'Armourfloor' - 3.2 mm	7.59	'Amourflex' - 2.0 mm	4.95
'Armourflex' - 3.2 mm	6.89		
Marley sheet flooring			
'HD Acoustic'	6.55	'HD Safetread'	9.75
'Format +'	7.95	'Vynatred'	4.55

	Labour hours	Labour £	Material £	Unit	Total rate £
M50 RUBBER/PLASTICS/CORK/LINO/CARPET TILING/ **SHEETING**					
Linoleum sheet; BS 810; Nairn Floors or **similar; level; fixing with adhesive; butt** **joints; to cement and sand base** Work to floors; over 300 mm wide					
3.2 mm; plain	0.40	4.13	8.68	m2	**12.81**
3.2 mm; marbled	0.40	4.13	8.89	m2	**13.02**
Vinyl sheet; Altro 'Safety' range or similar **with welded seams; level; fixing with** **adhesive; to cement and sand base** Work to floors; over 300 mm wide					
2 mm thick; Marine T20	0.60	6.19	10.38	m2	**16.57**
2.5 mm thick; Classic D25	0.70	7.23	13.12	m2	**20.35**
3.5 mm thick; stronghold	0.80	8.26	17.35	m2	**25.61**

M SURFACE FINISHES Including overheads and profit at 9.00% & 5.00%	Labour hours	Labour £	Material £	Unit	Total rate £
M50 RUBBER/PLASTICS/CORK/LINO/CARPET TILING/ SHEETING - cont'd					
Vinyl sheet; heavy duty; Marley 'HD' or similar; level; with welded seams; fixing with adhesive; level; to cement and sand base					
Work to floors; over 300 mm wide					
2 mm thick	0.33	3.41	7.04	m2	**10.45**
2.5 mm thick	0.33	3.41	8.09	m2	**11.50**
2 mm skirtings					
75 mm high	0.10	1.03	0.85	m	**1.88**
100 mm high	0.12	1.24	1.00	m	**2.24**
150 mm high	0.15	1.55	1.14	m	**2.69**
Vinyl sheet; 'Armstrong Rhino Contract'; level; with welded seams; fixing with adhesive; to cement and sand base					
Work to floors; over 300 mm wide					
2.5 mm thick	-	-	-	m2	**11.80**
Vinyl sheet; Marmoleum'; level; with welded seams; fixing with adhesive; to cement and sand base					
Work to floors; over 300 mm wide					
2.5 mm thick	-	-	-	m2	**11.29**
Vinyl tiles; 'Accoflex'; level; fixing with adhesive; butt joints; straight both ways; to cement and sand base					
Work to floors; over 300 mm wide					
300 x 300 x 2.0 mm	-	-	-	m2	**5.90**
Vinyl semi-flexible tiles; Marley 'Marleyflex' or similar; level; fixing with adhesive; butt joints straight both ways; to cement and sand base					
Work to floors; over 300 mm wide					
250 x 250 x 2.0 mm	0.25	2.58	4.33	m2	**6.91**
250 x 250 x 2.5 mm	0.25	2.58	4.96	m2	**7.54**
Vinyl tiles; anti-static; level; fixing with adhesive; butt joints straight both ways; to cement and sand base					
Work to floors; over 300 mm wide					
457 x 457 x 2 mm	0.45	4.65	24.87	m2	**29.52**
Vinyl tiles; 'Polyflex'; level; fixing with adhesive; butt joints; straight both ways;to cement and sand base					
Work to floors; over 300 mm wide					
300 x 300 x 1.5 mm	-	-	-	m2	**6.21**
300 x 300 x 2.0 mm	-	-	-	m2	**6.52**
Vinyl tiles; 'Marley 'HD''; level; fixing with adhesive; butt joints; straight both ways; to cement and sand base					
Work to floors; over 300 mm wide					
300 x 300 x 2.0 mm	-	-	-	m2	**8.21**

M SURFACE FINISHES Including overheads and profit at 9.00%	Labour hours	Labour £	Material £	Unit	Total rate £
Thermoplastic tiles; Marley 'Econoflex' or similar; level; fixing with adhesive; butt joints straight both ways; to cement and sand base					
Work to floors; over 300 mm wide					
250 x 250 x 2.0 mm; (series 2)	0.23	2.37	3.50	m2	5.87
250 x 250 x 2.0 mm; (series 4)	0.23	2.37	4.03	m2	6.40
Linoleum tiles; BS 810; Nairn Floors or similar; level; fixing with adhesive; butt joints straight both ways; to cement and sand base					
Work to floors; over 300 mm wide					
3.2 mm thick (marbled patterns)	0.30	3.10	9.73	m2	12.83
Cork tiles; Wicanders 'Corktile' or similar; level fixing with adhesive; butt joints straight both ways; to cement and sand base					
Work to floors; over 300 mm wide					
300 x 300 x 3.2 mm	0.40	4.13	6.44	m2	10.47
Rubber studded tiles; Altro 'Mondopave' or similar; level; fixing with adhesive; butt joints; straight; to cement and sand base					
Work to floors; over 300 mm wide					
500 x 500 x 2.5 mm; type GS; black	0.60	6.19	14.56	m2	20.75
500 x 500 x 4 mm; type BT; black	0.60	6.19	18.49	m2	24.68
Work to landings; over 300 mm wide					
500 x 500 x 4 mm; type BT; black	0.80	8.26	18.49	m2	26.75
4 mm thick to tread					
275 mm wide	0.50	5.16	5.98	m	11.14
4 mm thick to riser					
185 mm wide	0.60	6.19	4.14	m	10.33
Sundry floor sheeting underlays					
For floor finishings; over 300 mm wide					
building paper to BS 1521; class A;					
75 mm lap	0.06	0.62	0.65	m2	1.27
3.2 mm thick hardboard	0.20	2.06	1.01	m2	3.07
6.0 mm thick plywood	0.30	3.10	3.31	m2	6.41
Stair nosings					
'Ferodo' or equivalent; light duty hard aluminium alloy stair tread nosings; plugged and screwed to concrete					
type SD1 PC £4.72	0.25	2.58	5.67	m	8.25
type SD2 PC £6.52	0.30	3.10	7.73	m	10.83
'Ferodo' or equivalent; heavy duty aluminium alloy stair tread nosings; plugged and screwed to concrete					
type HD1 PC £5.58	0.30	3.10	6.65	m	9.75
type HD2 PC £7.77	0.35	3.61	9.15	m	12.76
Nylon needlepunch tiles; 'Marleyflex' or similar level; fixing with adhesive; to cement and sand base					
Work to floors					
over 300 mm wide	0.22	2.27	7.20	m2	9.47
Heavy duty carpet tiles; 'Heuga 581 Olympic' or similar; PC £14.00/m2; one coat floor sealer cement and sand base					
Work to floors					
over 300 mm wide	0.40	4.13	17.89	m2	22.02

M SURFACE FINISHES Including overheads and profit at 9.00%	Labour hours	Labour £	Material £	Unit	Total rate £
M50 RUBBER/PLASTICS/CORK/LINO/CARPET TILING/ **SHEETING - cont'd**					
Nylon needlepunch carpet; 'Marleyflex' or **similar; fixing; with adhesive; level; to** **cement**					
Work to floors					
over 300 mm wide	0.30	3.10	4.97	m2	8.07
M51 EDGE FIXED CARPETING					
Fitted carpeting; Wilton wool/nylon; 80/20 **velvet pile; heavy domestic plain; PC £30.00/m2**					
Work to floors					
over 300 mm wide	0.50	5.16	36.63	m2	41.79
Work to treads and risers					
over 300 mm wide	1.00	10.32	36.63	m2	46.95
Raking cutting	0.08	0.83	2.52	m2	3.35
Underlay to carpeting; PC £2.75/m2					
Work to floors					
over 300 mm wide	0.08	0.83	3.15	m2	3.98
Sundries					
Carpet gripper fixed to floor; standard edging	0.05	0.52	0.27	m	0.79
M52 DECORATIVE PAPERS/FABRICS					
Lining paper; PC £0.77/roll; **roll; and hanging**					
Plaster walls or columns					
over 300 mm girth	0.20	1.44	0.26	m2	1.70
Plaster ceilings or beams					
over 300 mm girth	0.25	1.80	0.26	m2	2.06
Decorative vinyl wallpaper; PC £4.40/roll; **roll; and hanging**					
Plaster walls or columns					
over 300 mm girth	0.25	1.80	1.13	m2	2.93

BASIC PAINT PRICES (£/5 Litre tin)

	£		£		£
Bituminous	9.27				
Emulsion					
matt	8.88	silk	10.74		
Knotting solution	21.92				
Oil					
gloss	12.56	undercoat	11.97		
Primer/undercoats					
acrylic	10.44	chlorinated			
alkali-resisting	12.42	rubber zinc	15.79	red oxide	16.22
aluminium	14.38	plaster	18.20	wood	11.63
calcium plumbate	20.75	red lead	24.46	zinc phosphate	13.26
Road marking	20.36				

		£		£
Special paints				
aluminium		18.99	fire retardant	
anti-condensation		15.64	- undercoat	23.54
chlorinated rubber			- top coat	31.63
- alkali-resisting		16.41	heat resisting	23.74
- finish		12.63		

M SURFACE FINISHES
Including overheads and profit at 9.00%

Stains and preservatives	£				£
creosote	2.80	Solignum -	'Architectural'		18.40
Cuprinol 'clear'	8.73	-	brown		6.74
linseed oil	13.39	-	cedar		9.84
oil varnish stain	11.27	-	clear		9.53
Sadolin - 'Holdex'	25.13	-	green		9.09
- 'Classic'	21.68	-	hort. green		7.87
- 'Sadovac 35'	10.91	'Multiplus'			10.55
Varnishes					
polyurethene	14.29	yacht			14.49

NOTE: The following prices include for preparing surfaces. Painting woodwork also includes for knotting prior to applying the priming coat and for all stopping of nail holes etc.	Labour hours	Labour £	Material £	Unit	Total rate £

M60 PAINTING/CLEAR FINISHING

One coat primer; PC £11.63/5L; on wood surfaces before fixing
General surfaces

over 300 mm girth	0.15	1.08	0.24	m2	1.32
isolated surfaces not exceeding 300 mm girth	0.06	0.43	0.08	m	0.51
isolated surfaces not exceeding 0.50 m2	0.11	0.79	0.12	nr	0.91

One coat polyurethene sealer; PC £14.29/5L; on wood surfaces before fixing
General surfaces

over 300 mm girth	0.18	1.30	0.25	m2	1.55
isolated surfaces not exceeding 300 mm girth	0.07	0.50	0.08	m	0.58
isolated surfaces not exceeding 0.50 m2	0.13	0.94	0.12	nr	1.06

One coat clear wood preservative; PC £8.73/5L; on wood surfaces before fixing
General surfaces

over 300 mm girth	0.13	0.94	0.21	m2	1.15
isolated surfaces not exceeding 300 mm girth	0.05	0.36	0.07	m	0.43
isolated surfaces not exceeding 0.50 m2	0.10	0.72	0.10	nr	0.82

Two coats emulsion paint; PC £8.88/5L;
Brick or block walls

over 300 mm girth	0.20	1.44	0.39	m2	1.83
Cement render or concrete					
over 300 mm girth	0.18	1.30	0.29	m2	1.59
isolated surfaces not exceeding 300 mm girth	0.08	0.58	0.09	m	0.67
Plaster walls or plaster/plasterboard ceilings					
over 300 mm girth	0.17	1.22	0.29	m2	1.51
over 300 mm girth; in multi-colours	0.27	1.95	0.34	m2	2.29
over 300 mm girth; in staircase areas	0.20	1.44	0.29	m2	1.73
cutting in edges on flush surfaces	0.08	0.58	-	m	0.58
Plaster/plasterboard ceilings					
over 300 mm girth; 3.5 - 5.0 m high	0.19	1.37	0.29	m2	1.66

One mist and two coats emulsion paint
Brick or block walls

over 300 mm girth	0.23	1.66	0.53	m2	2.19
Cement render or concrete					
over 300 mm girth	0.20	1.44	0.43	m2	1.87
isolated surfaces not exceeding 300 mm girth	0.09	0.65	0.12	m	0.77

M SURFACE FINISHES Including overheads and profit at 9.00%	Labour hours	Labour £	Material £	Unit	Total rate £

M60 PAINTING/CLEAR FINISHING - cont'd

One mist and two coats emulsion paint - cont'd
Plaster walls or plaster/plasterboard
ceilings

over 300 mm girth	0.20	1.44	0.39	m2	**1.83**
over 300 mm girth; in multi-colours	0.30	2.16	0.44	m2	**2.60**
over 300 mm girth; in staircase areas	0.23	1.66	0.39	m2	**2.05**
cutting in edges on flush surfaces	0.11	0.79	–	m	**0.79**

Plaster/plasterboard ceilings

over 300 mm girth; 3.5 - 5.0 m high	0.22	1.59	0.39	m2	**1.98**

Textured plastic, 'readymix' finish; PC £0.00/15L
Plasterboard ceilings

over 300 mm girth	0.25	1.80	1.06	m2	**2.86**

Concrete walls or ceilings

over 300 mm girth	0.27	1.95	1.19	m2	**3.14**

Touch up primer; one undercoat and one
finishing coat of gloss oil paint; PC £12.56/5L;
on wood surfaces
General surfaces

over 300 mm girth	0.32	2.31	0.45	m2	**2.76**
isolated surfaces not exceeding 300 mm girth	0.13	0.94	0.14	m	**1.08**
isolated surfaces not exceeding 0.50 m2	0.24	1.73	0.22	nr	**1.95**

Windows and the like

panes over 1.00 m2	0.32	2.31	0.17	m2	**2.48**
panes 0.50 - 1.00 m2	0.37	2.67	0.22	m2	**2.89**
panes 0.10 - 0.50 m2	0.43	3.10	0.28	m2	**3.38**
panes not exceeding 0.10 m2	0.53	3.82	0.34	m2	**4.16**

Touch up primer; two undercoats and one
finishing coat of gloss oil paint; on
wood surfaces
General surfaces

over 300 mm girth	0.45	3.24	0.66	m2	**3.90**
isolated surfaces not exceeding 300 mm girth	0.18	1.30	0.21	m	**1.51**
isolated surfaces not exceeding 0.50 m2	0.33	2.38	0.33	nr	**2.71**

Windows and the like

panes over 1.00 m2	0.45	3.24	0.25	m2	**3.49**
panes 0.50 - 1.00 m2	0.52	3.75	0.33	m2	**4.08**
panes 0.10 - 0.50 m2	0.60	4.32	0.41	m2	**4.73**
panes not exceeding 0.10 m2	0.75	5.40	0.49	m2	**5.89**

Knot; one coat primer; stop; one undercoat
and one finishing coat of gloss oil paint;
on wood surfaces
General surfaces

over 300 mm girth	0.47	3.39	0.74	m2	**4.13**
isolated surfaces not exceeding 300 mm girth	0.19	1.37	0.23	m	**1.60**
isolated surfaces not exceeding 0.50 m2	0.35	2.52	0.37	nr	**2.89**

Windows and the like

panes over 1.00 m2	0.47	3.39	0.27	m2	**3.66**
panes 0.50 - 1.00 m2	0.55	3.96	0.37	m2	**4.33**
panes 0.10 - 0.50 m2	0.65	4.68	0.46	m2	**5.14**
panes not exceeding 0.10 m2	0.80	5.76	0.55	m2	**6.31**

Knot; one coat primer; stop; two undercoats
and one finishing coat of gloss oil paint;
on wood surfaces
General surfaces

over 300 mm girth	0.60	4.32	0.95	m2	**5.27**
isolated surfaces not exceeding 300 mm girth	0.24	1.73	0.30	m	**2.03**
isolated surfaces not exceeding 0.50 m2	0.45	3.24	0.47	nr	**3.71**

M SURFACE FINISHES Including overheads and profit at 9.00%	Labour hours	Labour £	Material £	Unit	Total rate £
Windows and the like					
panes over 1.00 m2	0.60	4.32	0.35	m2	4.67
panes 0.50 - 1.00 m2	0.70	5.04	0.47	m2	5.51
panes 0.10 - 0.50 m2	0.80	5.76	0.59	m2	6.35
panes not exceeding 0.10 m2	1.00	7.20	0.71	m2	7.91
One coat primer; one undercoat and one **finishing coat of gloss oil paint** **Plaster surfaces**					
over 300 mm girth	0.42	3.03	0.82	m2	3.85
One coat primer; two undercoats and one **finishing coat of gloss oil paint** **Plaster surfaces**					
over 300 mm girth	0.55	3.96	1.03	m2	4.99
Touch up primer; one undercoat and one **finishing coat of gloss paint; PC £12.56/5L;** **on iron or steel surfaces** **General surfaces**					
over 300 mm girth	0.32	2.31	0.43	m2	2.74
isolated surfaces not exceeding 300 mm girth	0.13	0.94	0.13	m	1.07
isolated surfaces not exceeding 0.50 m2	0.24	1.73	0.21	nr	1.94
Windows and the like					
panes over 1.00 m2	0.32	2.31	0.16	m2	2.47
panes 0.50 - 1.00 m2	0.37	2.67	0.21	m2	2.88
panes 0.10 - 0.50 m2	0.43	3.10	0.27	m2	3.37
panes not exceeding 0.10 m2	0.53	3.82	0.32	m2	4.14
Structural steelwork					
over 300 mm girth	0.36	2.59	0.43	m2	3.02
Members of roof trusses					
over 300 mm girth	0.48	3.46	0.45	m2	3.91
Ornamental railings and the like; each side **measured overall**					
over 300 mm girth	0.55	3.96	0.48	m2	4.44
Iron or steel radiators					
over 300 mm girth	0.32	2.31	0.45	m2	2.76
Pipes or conduits					
over 300 mm girth	0.48	3.46	0.45	m2	3.91
not exceeding 300 mm girth	0.19	1.37	0.13	m	1.50
Touch up primer; two undercoats and one **finishing coat of gloss oil paint; on** **iron or steel surfaces** **General surfaces**					
over 300 mm girth	0.45	3.24	0.64	m2	3.88
isolated surfaces not exceeding 300 mm girth	0.18	1.30	0.20	m	1.50
isolated surfaces not exceeding 0.50 m2	0.33	2.38	0.32	nr	2.70
Windows and the like					
panes over 1.00 m2	0.45	3.24	0.24	m2	3.48
panes 0.50 - 1.00 m2	0.52	3.75	0.32	m2	4.07
panes 0.10 - 0.50 m2	0.60	4.32	0.40	m2	4.72
panes not exceeding 0.10 m2	0.75	5.40	0.46	m2	5.86
Structural steelwork					
over 300 mm girth	0.51	3.67	0.64	m2	4.31
Members of roof trusses					
over 300 mm girth	0.68	4.90	0.68	m2	5.58
Ornamental railings and the like; each side **measured overall**					
over 300 mm girth	0.77	5.55	0.72	m2	6.27
Iron or steel radiators					
over 300 mm girth	0.45	3.24	0.68	m2	3.92
Pipes or conduits					
over 300 mm girth	0.68	4.90	0.68	m2	5.58
not exceeding 300 mm girth	0.27	1.95	0.20	m	2.15

M SURFACE FINISHES Including overheads and profit at 9.00%	Labour hours	Labour £	Material £	Unit	Total rate £

M60 PAINTING/CLEAR FINISHING - cont'd

**One coat primer; one undercoat and one
finishing coat of gloss oil paint; on
iron or steel surfaces**

General surfaces

over 300 mm girth	0.42	3.03	0.43	m2	3.46
isolated surfaces not exceeding 300 mm girth	0.17	1.22	0.13	m	1.35
isolated surfaces not exceeding 0.50 m2	0.32	2.31	0.21	nr	2.52

Windows and the like

panes over 1.00 m2	0.42	3.03	0.16	m2	3.19
panes 0.50 - 1.00 m2	0.48	3.46	0.21	m2	3.67
panes 0.10 - 0.50 m2	0.56	4.03	0.27	m2	4.30
panes not exceeding 0.10 m2	0.70	5.04	0.32	m2	5.36

Structural steelwork

over 300 mm girth	0.47	3.39	0.43	m2	3.82

Members of roof trusses

over 300 mm girth	0.63	4.54	0.45	m2	4.99

Ornamental railings and the like; each side
measured overall

over 300 mm girth	0.72	5.19	0.48	m2	5.67

Iron or steel radiators

over 300 mm girth	0.42	3.03	0.45	m2	3.48

Pipes or conduits

over 300 mm girth	0.63	4.54	0.45	m2	4.99
not exceeding 300 mm girth	0.25	1.80	0.13	m	1.93

**One coat primer; two undercoats and one
finishing coat of gloss oil paint; on
iron or steel surfaces**

General surfaces

over 300 mm girth	0.55	3.96	0.64	m2	4.60
isolated surfaces not exceeding 300 mm girth	0.22	1.59	0.20	m	1.79
isolated surfaces not exceeding 0.50 m2	0.40	2.88	0.32	nr	3.20

Windows and the like

panes over 1.00 m2	0.55	3.96	0.24	m2	4.20
panes 0.50 - 1.00 m2	0.65	4.68	0.32	m2	5.00
panes 0.10 - 0.50 m2	0.75	5.40	0.40	m2	5.80
panes not exceeding 0.10 m2	0.90	6.48	0.48	m2	6.96

Structural steelwork

over 300 mm girth	0.62	4.47	0.64	m2	5.11

Members of roof trusses

over 300 mm girth	0.82	5.91	0.68	m2	6.59

Ornamental railings and the like; each side
measured overall

over 300 mm girth	0.93	6.70	0.72	m2	7.42

Iron or steel radiators

over 300 mm girth	0.55	3.96	0.68	m2	4.64

Pipes or conduits

over 300 mm girth	0.83	5.98	0.68	m2	6.66
not exceeding 300 mm girth	0.33	2.38	0.20	m	2.58

**Two coats of bituminous paint; PC £9.27/5L;
on iron or steel surfaces**

General surfaces

over 300 mm girth	0.40	2.88	0.32	m2	3.20

Inside of galvanized steel cistern

over 300 mm girth	0.60	4.32	0.32	m2	4.64

**Two coats of boiled linseed oil; PC £13.39/5;
on hardwood surfaces**

General surfaces

over 300 mm girth	0.30	2.16	1.02	m2	3.18
isolated surfaces not exceeding 300 mm girth	0.12	0.86	0.31	m	1.17
isolated surfaces not exceeding 0.50 m2	0.23	1.66	0.51	nr	2.17

M SURFACE FINISHES Including overheads and profit at 9.00% & 5.00%	Labour hours	Labour £	Material £	Unit	Total rate £
Two coats polyurethane; PC £14.29/5L; **on wood surfaces** General surfaces					
over 300 mm girth	0.30	2.16	0.50	m2	2.66
isolated surfaces not exceeding 300 mm girth	0.12	0.86	0.16	m	1.02
isolated surfaces not exceeding 0.50 m2	0.22	1.59	0.25	nr	1.84
Three coats polyurethane; on wood surfaces General surfaces					
over 300 mm girth	0.45	3.24	0.75	m2	3.99
isolated surfaces not exceeding 300 mm girth	0.18	1.30	0.23	m	1.53
isolated surfaces not exceeding 0.50 m2	0.34	2.45	0.37	nr	2.82
One undercoat; PC £23.54/5L; and one **finishing coat; PC £31.63/5L; of 'Albi' clear flame** **retardent surface coating; on wood surfaces** General surfaces					
over 300 mm girth	0.54	3.89	1.19	m2	5.08
isolated surfaces not exceeding 300 mm girth	0.22	1.59	0.38	m	1.97
isolated surfaces not exceeding 0.50 m2	0.42	3.03	0.58	nr	3.61
Two undercoats; PC £23.54/5L; and one **finishing coat; PC £31.63/5L; of 'Albi' clear flame** **retardent surface coating; on wood surfaces** General surfaces					
over 300 mm girth	0.54	3.89	1.83	m2	5.72
isolated surfaces not exceeding 300 mm girth	0.22	1.59	0.58	m	2.17
isolated surfaces not exceeding 0.50 m2	0.42	3.03	0.92	nr	3.95
Seal and wax polish; dull gloss finish **on wood surfaces** General surfaces					
over 300 mm girth	-	-	-	m2	5.31
isolated surfaces not exceeding 300 mm girth	-	-	-	m	2.39
isolated surfaces not exceeding 0.50 m2	-	-	-	nr	3.72
Two coats of 'Sadolins Classic'; PC £21.68/5L; **clear or pigmented; two further coats of** **'Holdex' clear interior silk matt laquer;** **PC £25.13/5L** General surfaces					
over 300 mm girth	0.62	4.47	1.92	m2	6.39
isolated surfaces not exceeding 300 mm girth	0.25	1.80	0.60	m	2.40
isolated surfaces not exceeding 0.50 m2	0.47	3.39	0.96	nr	4.35
Windows and the like					
panes over 1.00 m2	0.62	4.47	0.71	m2	5.18
panes 0.50 - 1.00 m2	0.72	5.19	0.96	m2	6.15
panes 0.10 - 0.50 m2	0.83	5.98	1.21	m2	7.19
panes not exceeding 0.10 m2	1.04	7.49	1.44	m2	8.93
Body in and wax polish; dull gloss finish; **on hardwood surfaces** General surfaces					
over 300 mm girth	-	-	-	m2	7.60
isolated surfaces not exceeding 300 mm girth	-	-	-	m	3.42
isolated surfaces not exceeding 0.50 m2	-	-	-	nr	5.32
Stain; body in and wax polish; dull gloss **finish; on hardwood surface** General surfaces					
over 300 mm girth	-	-	-	m2	9.42
isolated surfaces not exceeding 300 mm girth	-	-	-	m	4.24
isolated surfaces not exceeding 0.50 m2	-	-	-	nr	6.63

M SURFACE FINISHES Including overheads and profit at 9.00% & 5.00%	Labour hours	Labour £	Material £	Unit	Total rate £
M60 PAINTING/CLEAR FINISHING - cont'd					
Seal; two coats of synthetic resin lacquer; **decorative flatted finish; wire down, wax** **and burnish; on wood surfaces**					
General surfaces					
over 300 mm girth	-	-	-	m2	11.97
isolated surfaces not exceeding 300 mm girth	-	-	-	m	5.25
isolated surfaces not exceeding 0.50 m2	-	-	-	nr	8.38
Stain; body in and fully French polish; **full gloss finish; on hardwood surfaces**					
General surfaces					
over 300 mm girth	-	-	-	m2	14.31
isolated surfaces not exceeding 300 mm girth	-	-	-	m	6.45
isolated surfaces not exceeding 0.50 m2	-	-	-	nr	10.02
Stain; fill grain and fully French polish; **full gloss finish; on hardwood surfaces**					
General surfaces					
over 300 mm girth	-	-	-	m2	21.86
isolated surfaces not exceeding 300 mm girth	-	-	-	m	9.84
isolated surfaces not exceeding 0.50 m2	-	-	-	nr	14.57
Stain black; body in and fully French **polish; ebonized finish; on hardwood** **surfaces**					
General surfaces					
over 300 mm girth	-	-	-	m2	27.32
isolated surfaces not exceeding 300 mm girth	-	-	-	m	12.28
isolated surfaces not exceeding 0.50 m2	-	-	-	nr	19.13
M60 PAINTING/CLEAR FINISHING - EXTERNALLY					
Two coats of cement paint, 'Sandtex Matt' **or similar; PC £11.20/5L;**					
Brick or block walls					
over 300 mm girth	0.45	3.24	1.46	m2	4.70
Cement render or concrete walls					
over 300 mm girth	0.40	2.88	0.98	m2	3.86
Roughcast walls					
over 300 mm girth	0.50	3.60	1.95	m2	5.55
One coat sealer and two coats of external **grade emulsion paint, Dulux 'Weathershield'** **or similar**					
Brick or block walls					
over 300 mm girth	0.60	4.32	2.00	m2	6.32
Cement render or concrete walls					
over 300 mm girth	0.50	3.60	1.33	m2	4.93
Concrete soffits					
over 300 mm girth	0.55	3.96	1.33	m2	5.29
One coat sealer (applied by brush) and two **coats of external grade emulsion paint,** **Dulux 'Weathershield' or similar** **(spray applied)**					
Roughcast					
over 300 mm girth	0.60	3.15	2.67	m2	5.82

M SURFACE FINISHES Including overheads and profit at 9.00%	Labour hours	Labour £	Material £	Unit	Total rate £
Touch up primer; one undercoat and one **finishing coat of gloss oil paint; PC £12.56/5L;** **on wood surfaces**					
General surfaces					
over 300 mm girth	0.36	2.59	0.45	m2	3.04
isolated surfaces not exceeding 300 mm girth	0.15	1.08	0.14	m	1.22
isolated surfaces not exceeding 0.50 m2	0.27	1.95	0.22	nr	2.17
Windows and the like					
panes over 1.00 m2	0.36	2.59	0.17	m2	2.76
panes 0.50 - 1.00 m2	0.42	3.03	0.22	m2	3.25
panes 0.10 - 0.50 m2	0.48	3.46	0.28	m2	3.74
panes not exceeding 0.10 m2	0.60	4.32	0.34	m2	4.66
Windows and the like; casements in colours differing from frames					
panes over 1.00 m2	0.40	2.88	0.19	m2	3.07
panes 0.50 - 1.00 m2	0.45	3.24	0.25	m2	3.49
panes 0.10 - 0.50 m2	0.53	3.82	0.31	m2	4.13
panes not exceeding 0.10 m2	0.66	4.76	0.36	m2	5.12
Touch up primer; two undercoats and one **finishing coat of gloss oil paint;** **on wood surfaces**					
General surfaces					
over 300 mm girth	0.50	3.60	0.66	m2	4.26
isolated surfaces not exceeding 300 mm girth	0.20	1.44	0.21	m2	1.65
isolated surfaces not exceeding 0.50 m2	0.38	2.74	0.33	m2	3.07
Windows and the like					
panes over 1.00 m2	0.50	3.60	0.25	m2	3.85
panes 0.50 - 1.00 m2	0.58	4.18	0.33	m2	4.51
panes 0.10 - 0.50 m2	0.67	4.83	0.41	m2	5.24
panes not exceeding 0.10 m2	0.83	5.98	0.49	m2	6.47
Windows and the like; casements in colours differing from frames					
panes over 1.00 m2	0.58	4.18	0.29	m2	4.47
panes 0.50 - 1.00 m2	0.67	4.83	0.37	m2	5.20
panes 0.10 - 0.50 m2	0.77	5.55	0.45	m2	6.00
panes not exceeding 0.10 m2	0.95	6.84	0.53	m2	7.37
Knot; one coat primer; one undercoat and **one finishing coat of gloss oil paint; on** **wood surfaces**					
General surfaces					
over 300 mm girth	0.53	3.82	0.74	m2	4.56
isolated surfaces not exceeding 300 mm girth	0.22	1.59	0.23	m	1.82
isolated surfaces not exceeding 0.50 m2	0.40	2.88	0.37	nr	3.25
Windows and the like					
panes over 1.00 m2	0.53	3.82	0.27	m2	4.09
panes 0.50 - 1.00 m2	0.62	4.47	0.37	m2	4.84
panes 0.10 - 0.50 m2	0.70	5.04	0.46	m2	5.50
panes not exceeding 0.10 m2	0.88	6.34	0.55	m2	6.89
Windows and the like; casements in colours differing from frames					
panes over 1.00 m2	0.58	4.18	0.31	m2	4.49
panes 0.50 - 1.00 m2	0.68	4.90	0.41	m2	5.31
panes 0.10 - 0.50 m2	0.77	5.55	0.67	m2	6.22
panes not exceeding 0.10 m2	0.96	6.92	0.59	m2	7.51
Knot; one coat primer; two undercoats and **one finishing coat of gloss oil paint** **on wood surfaces**					
General surfaces					
over 300 mm girth	0.66	4.76	0.95	m2	5.71
isolated surfaces not exceeding 300 mm girth	0.27	1.95	0.30	m	2.25
isolated surfaces not exceeding 0.50 m2	0.50	3.60	0.47	nr	4.07

M SURFACE FINISHES Including overheads and profit at 9.00%	Labour hours	Labour £	Material £	Unit	Total rate £

M60 PAINTING/CLEAR FINISHING - EXTERNALLY - cont'd

Knot; one coat primer; two undercoats and
one finishing coat of gloss oil paint
on wood surfaces - cont'd

Windows and the like					
panes over 1.00 m2	0.66	4.76	0.35	m2	5.11
panes 0.50 - 1.00 m2	0.77	5.55	0.47	m2	6.02
panes 0.10 - 0.50 m2	0.88	6.34	0.59	m2	6.93
panes not exceeding 0.10 m2	1.10	7.93	0.71	m2	8.64
Windows and the like; casements in colours					
differing from frames					
panes over 1.00 m2	0.76	5.48	0.41	m2	5.89
panes 0.50 - 1.00 m2	0.90	6.48	0.52	m2	7.00
panes 0.10 - 0.50 m2	1.02	7.35	0.81	m2	8.16
panes not exceeding 0.10 m2	1.25	9.01	0.76	m2	9.77

Touch up primer; one undercoat and one
finishing coat of gloss oil paint; PC £12.56/5L;
on iron or steel surfaces

General surfaces					
over 300 mm girth	0.36	2.59	0.43	m2	3.02
isolated surfaces not exceeding 300 mm girth	0.15	1.08	0.13	m	1.21
isolated surfaces not exceeding 0.50 m2	0.27	1.95	0.21	nr	2.16
Windows and the like					
panes over 1.00 m2	0.36	2.59	0.16	m2	2.75
panes 0.50 - 1.00 m2	0.42	3.03	0.21	m2	3.24
panes 0.10 - 0.50 m2	0.48	3.46	0.27	m2	3.73
panes not exceeding 0.10 m2	0.60	4.32	0.32	m2	4.64
Structural steelwork					
over 300 mm girth	0.40	2.88	0.43	m2	3.31
Members of roof trusses					
over 300 mm girth	0.53	3.82	0.45	m2	4.27
Ornamental railings and the like; each side					
measured overall					
over 300 mm girth	0.60	4.32	0.48	m2	4.80
Eaves gutters					
over 300 mm girth	0.65	4.68	0.64	m2	5.32
not exceeding 300 mm girth	0.26	1.87	0.21	m	2.08
Pipes or conduits					
over 300 mm girth	0.53	3.82	0.45	m2	4.27
not exceeding 300 mm girth	0.21	1.51	0.13	m	1.64

Touch up primer; two undercoats and one
finishing coat of gloss oil paint; on
iron or steel surfaces

General surfaces					
over 300 mm girth	0.50	3.60	0.64	m2	4.24
isolated surfaces not exceeding 300 mm girth	0.20	1.44	0.20	m	1.64
isolated surfaces not exceeding 0.50 m2	0.37	2.67	0.32	nr	2.99
Windows and the like					
panes over 1.00 m2	0.50	3.60	0.24	m2	3.84
panes 0.50 - 1.00 m2	0.58	4.18	0.32	m2	4.50
panes 0.10 - 0.50 m2	0.67	4.83	0.40	m2	5.23
panes not exceeding 0.10 m2	0.83	5.98	0.48	m2	6.46
Structural steelwork					
over 300 mm girth	0.57	4.11	0.64	m2	4.75
Members of roof trusses					
over 300 mm girth	0.75	5.40	0.68	m2	6.08
Ornamental railings and the like; each side					
measured overall					
over 300 mm girth	0.85	6.12	0.72	m2	6.84
Eaves gutters					
over 300 mm girth	0.90	6.48	0.95	m2	7.43
not exceeding 300 mm girth	0.36	2.59	0.32	m	2.91
Pipes or conduits					
over 300 mm girth	0.75	5.40	0.68	m2	6.08
not exceeding 300 mm girth	0.30	2.16	0.20	m	2.36

M SURFACE FINISHES Including overheads and profit at 9.00%	Labour hours	Labour £	Material £	Unit	Total rate £
One coat primer; one undercoat and one finishing coat of gloss oil paint; on iron or steel surfaces					
General surfaces					
over 300 mm girth	0.46	3.31	0.43	m2	**3.74**
isolated surfaces not exceeding 300 mm girth	0.19	1.37	0.13	m	**1.50**
isolated surfaces not exceeding 0.50 m2	0.35	2.52	0.21	nr	**2.73**
Windows and the like					
panes over 1.00 m2	0.46	3.31	0.16	m2	**3.47**
panes 0.50 - 1.00 m2	0.53	3.82	0.21	m2	**4.03**
panes 0.10 - 0.50 m2	0.62	4.47	0.27	m2	**4.74**
panes not exceeding 0.10 m2	0.77	5.55	0.32	m2	**5.87**
Structural steelwork					
over 300 mm girth	0.52	3.75	0.43	m2	**4.18**
Members of roof trusses					
over 300 mm girth	0.68	4.90	0.45	m2	**5.35**
Ornamental railings and the like; each side measured overall					
over 300 mm girth	0.78	5.62	0.48	m2	**6.10**
Eaves gutters					
over 300 mm girth	0.82	5.91	0.64	m2	**6.55**
not exceeding 300 mm girth	0.33	2.38	0.21	m	**2.59**
Pipes or conduits					
over 300 mm girth	0.68	4.90	0.45	m2	**5.35**
not exceeding 300 mm girth	0.27	1.95	0.13	m	**2.08**
One coat primer; two undercoats and one finishing coat of gloss oil paint; on iron or steel surfaces					
General surfaces					
over 300 mm girth	0.60	4.32	0.64	m2	**4.96**
isolated surfaces not exceeding 300 mm girth	0.24	1.73	0.20	m	**1.93**
isolated surfaces not exceeding 0.50 m2	0.45	3.24	0.32	nr	**3.56**
Windows and the like					
panes over 1.00 m2	0.60	4.32	0.24	m2	**4.56**
panes 0.50 - 1.00 m2	0.70	5.04	0.32	m2	**5.36**
panes 0.10 - 0.50 m2	0.80	5.76	0.40	m2	**6.16**
panes not exceeding 0.10 m2	1.00	7.20	0.48	m2	**7.68**
Structural steelwork					
over 300 mm girth	0.68	4.90	0.64	m2	**5.54**
Members of roof trusses					
over 300 mm girth	0.90	6.48	0.68	m2	**7.16**
Ornamental railings and the like; each side measured overall					
over 300 mm girth	1.02	7.35	0.72	m2	**8.07**
Eaves gutters					
over 300 mm girth	1.08	7.78	0.95	m2	**8.73**
not exceeding 300 mm girth	0.43	3.10	0.32	m	**3.42**
Pipes or conduits					
over 300 mm girth	0.90	6.48	0.68	m2	**7.16**
not exceeding 300 mm girth	0.36	2.59	0.20	m	**2.79**
Two coats of creosote; PC £2.80/5L; on wood surfaces					
General surfaces					
over 300 mm girth	0.25	1.80	0.13	m2	**1.93**
isolated surfaces not exceeding 300 mm girth	0.10	0.72	0.04	m	**0.76**
Two coats of 'Solignum' wood preservative; PC £9.53/5L; on wood surfaces					
General surfaces					
over 300 mm girth	0.25	1.80	0.46	m2	**2.26**
isolated surfaces not exceeding 300 mm girth	0.10	0.72	0.15	m	**0.87**

M SURFACE FINISHES Including overheads and profit at 9.00% & 5.00%	Labour hours	Labour £	Material £	Unit	Total rate £
M60 PAINTING/CLEAR FINISHING - EXTERNALLY - cont'd					
Three coats of polyurethane; PC £14.29/5L; on wood surfaces					
General surfaces					
over 300 mm girth	0.50	3.60	0.75	m2	**4.35**
isolated surfaces not exceeding 300 mm girth	0.20	1.44	0.23	m	**1.67**
isolated surfaces not exceeding 0.50 m2	0.38	2.74	0.37	nr	**3.11**
Two coats of 'Sadovac 35'primer; PC £10.91/5L; and two coats of 'Classic'; PC £21.68/5L; pigmented; on wood surfaces					
General surfaces					
over 300 mm girth	0.68	4.90	1.82	m2	**6.72**
isolated surfaces not exceeding 300 mm girth	0.27	1.95	0.58	m	**2.53**
Windows and the like					
panes over 1.00 m2	0.68	4.90	0.66	m2	**5.56**
panes 0.50 - 1.00 m2	0.79	5.69	0.91	m2	**6.60**
panes 0.10 - 0.50 m2	0.91	6.56	1.16	m2	**7.72**
panes not exceeding 0.10 m2	1.14	8.21	1.37	m2	**9.58**
Body in with French polish; one coat of lacquer or varnish; on wood surfaces					
General surfaces					
over 300 mm girth	-	-	-	m2	**12.18**
isolated surfaces not exceeding 300 mm girth	-	-	-	m	**5.49**
isolated surfaces not exceeding 0.50 m2	-	-	-	nr	**8.53**

N FURNITURE/EQUIPMENT Including overheads and profit at 9.00%		Labour hours	Labour £	Material £	Unit	Total rate £

N10/11 GENERAL FIXTURES/KITCHEN FITTINGS ETC.

Proprietary items
Closed stove vitreous enamelled finish;
setting in position

571 x 606 x 267 mm	PC £230.00	2.00	23.30	250.70	nr	**274.00**

Closed stove; vitreous enamelled finish
fitted with mild steel barffed boiler;
setting in position

571 x 606 x 267 mm	PC £300.00	2.00	23.30	327.00	nr	**350.30**

Tile surround preslabbed; 406 mm wide
firebrick back; cast iron stool bottom;
black vitreous enamelled fret; assembling,
setting and pointing firebrick back in
fireclay mortar and backing with fine
concrete finished to splay at top, laying
hearth tiles in cement mortar and pointing
in white cement

1372 x 864 x 152 mm	PC £220.05	5.00	58.26	257.84	nr	**316.10**

Fitting components; blockboard
Backs, fronts, sides or divisions;
over 300 mm wide

12 mm thick	PC £21.42	1.30	11.78	23.35	m2	**35.13**
19 mm thick	PC £27.51	1.30	11.78	29.99	m2	**41.77**
25 mm thick	PC £36.10	1.30	11.78	39.35	m2	**51.13**

Shelves or worktops; over 300 mm wide

19 mm thick	PC £28.43	1.30	11.78	30.98	m2	**42.76**
25 mm thick	PC £36.10	1.30	11.78	39.35	m2	**51.13**

Flush doors; lipped on four edges

450 x 750 x 19 mm	PC £15.10	0.35	3.17	16.45	nr	**19.62**
450 x 750 x 25 mm	PC £18.45	0.35	3.17	20.11	nr	**23.28**
600 x 900 x 19 mm	PC £22.47	0.50	4.53	24.49	nr	**29.02**
600 x 900 x 25 mm	PC £27.35	0.50	4.53	29.81	nr	**34.34**

Fitting components; chipboard
Backs, fronts, sides or divisions;
over 300 mm wide

6 mm thick	PC £7.55	1.30	11.78	8.23	m2	**20.01**
9 mm thick	PC £10.24	1.30	11.78	11.16	m2	**22.94**
12 mm thick	PC £12.42	1.30	11.78	13.54	m2	**25.32**
19 mm thick	PC £17.56	1.30	11.78	19.14	m2	**30.92**
25 mm thick	PC £23.74	1.30	11.78	25.87	m2	**37.65**

Shelves or worktops; over 300 mm wide

19 mm thick	PC £18.48	1.30	11.78	20.14	m2	**31.92**
25 mm thick	PC £23.74	1.30	11.78	25.87	m2	**37.65**

Flush doors; lipped on four edges

450 x 750 x 19 mm	PC £12.23	0.35	3.17	13.33	nr	**16.50**
450 x 750 x 25 mm	PC £14.92	0.35	3.17	16.26	nr	**19.43**
600 x 900 x 19 mm	PC £17.95	0.50	4.53	19.56	nr	**24.09**
600 x 900 x 25 mm	PC £21.78	0.50	4.53	23.74	nr	**28.27**

Fitting components; Melamine faced chipboard
Backs, fronts, sides or divisions;
over 300 mm wide

12 mm thick	PC £17.13	1.30	11.78	18.67	m2	**30.45**
19 mm thick	PC £23.08	1.30	11.78	25.16	m2	**36.94**

Shelves or worktops; over 300 mm wide

19 mm thick	PC £24.21	1.30	11.78	26.39	m2	**38.17**

Flush doors; lipped on four edges

450 x 750 x 19 mm	PC £15.21/m2	0.35	3.17	16.58	nr	**19.75**
600 x 900 x 19 mm	PC £22.44/m2	0.50	4.53	24.46	nr	**28.99**

N FURNITURE/EQUIPMENT Including overheads and profit at 9.00%		Labour hours	Labour £	Material £	Unit	Total rate £
N10/11 GENERAL FIXTURES/KITCHEN FITTINGS ETC. - cont'd						
Fitting components; 'Warerite Xcel' standard **colour laminated chipboard type LD2; PC £156.66/10m2**						
Backs, fronts, sides or divisions; over 300 mm wide						
13.2 mm thick	PC £49.21	1.30	11.78	53.64	m2	**65.42**
Shelves or worktops; over 300 mm wide						
13.2 mm thick	PC £49.21	1.30	11.78	53.64	m2	**65.42**
Flush doors; lipped on four edges						
450 x 750 x 13.2 mm	PC £19.89/m2	0.35	3.17	21.68	nr	**24.85**
600 x 900 x 13.2 mm	PC £29.60/m2	0.50	4.53	32.26	nr	**36.79**
Fitting components; plywood						
Backs, fronts, sides or divisions; over 300 mm wide						
6 mm thick	PC £12.25	1.30	11.78	13.35	m2	**25.13**
9 mm thick	PC £16.19	1.30	11.78	17.64	m2	**29.42**
12 mm thick	PC £20.50	1.30	11.78	22.34	m2	**34.12**
19 mm thick	PC £29.71	1.30	11.78	32.38	m2	**44.16**
25 mm thick	PC £39.43	1.30	11.78	42.98	m2	**54.76**
Shelves or worktops; over 300 mm wide						
19 mm thick	PC £30.63	1.30	11.78	33.38	m2	**45.16**
25 mm thick	PC £39.43	1.30	11.78	42.98	m2	**54.76**
Flush doors; lipped on four edges						
450 x 750 x 19 mm	PC £15.58	0.35	3.17	16.98	nr	**20.15**
450 x 750 x 25 mm	PC £19.26	0.35	3.17	20.99	nr	**24.16**
600 x 900 x 19 mm	PC £23.30	0.50	4.53	25.40	nr	**29.93**
600 x 900 x 25 mm	PC £28.68	0.50	4.53	31.26	nr	**35.79**
Fitting components; wrought softwood						
Backs, fronts, sides or divisions; cross-tongued joints; over 300 mm wide						
25 mm thick		1.30	11.78	33.80	m2	**45.58**
Shelves or worktops; cross-tongued joints; over 300 mm wide						
25 mm thick		1.30	11.78	33.80	m2	**45.58**
Bearers						
19 x 38 mm		0.10	0.91	1.89	m	**2.80**
25 x 50 mm		0.10	0.91	2.42	m	**3.33**
50 x 50 mm		0.10	0.91	3.63	m	**4.54**
50 x 75 mm		0.10	0.91	4.90	m	**5.81**
Bearers; framed						
19 x 38 mm		0.13	1.18	3.60	m	**4.78**
25 x 50 mm		0.13	1.18	4.14	m	**5.32**
50 x 50 mm		0.13	1.18	5.33	m	**6.51**
50 x 75 mm		0.13	1.18	6.61	m	**7.79**
Framing to backs, fronts or sides						
19 x 38 mm		0.15	1.36	3.60	m	**4.96**
25 x 50 mm		0.15	1.36	4.14	m	**5.50**
50 x 50 mm		0.15	1.36	5.33	m	**6.69**
50 x 75 mm		0.15	1.36	6.61	m	**7.97**
Flush doors; softwood skeleton or cellular core; plywood facing both sides; lipped on four edges						
450 x 750 x 35 mm		0.35	3.17	18.01	nr	**21.18**
600 x 900 x 35 mm		0.50	4.53	28.51	nr	**33.04**
Add +5% to the above 'Material £ prices' for 'selected' softwood for staining						
Fitting components; selected West African **Mahogany; PC £494.00/m3**						
Bearers						
19 x 38 mm	PC £2.49	0.15	1.36	2.83	m	**4.19**
25 x 50 mm	PC £3.25	0.15	1.36	3.68	m	**5.04**
50 x 50 mm	PC £4.98	0.15	1.36	5.61	m	**6.97**
50 x 75 mm	PC £6.79	0.15	1.36	7.64	m	**9.00**

N FURNITURE/EQUIPMENT Including overheads and profit at 9.00%		Labour hours	Labour £	Material £	Unit	Total rate £
Bearers; framed						
19 x 38 mm	PC £4.79	0.20	1.81	5.35	m	7.16
25 x 50 mm	PC £5.57	0.20	1.81	6.22	m	8.03
50 x 50 mm	PC £7.28	0.20	1.81	8.13	m	9.94
50 x 75 mm	PC £9.09	0.20	1.81	10.16	m	11.97
Framing to backs, fronts or sides						
19 x 38 mm	PC £4.79	0.25	2.26	5.35	m	7.61
25 x 50 mm	PC £5.57	0.25	2.26	6.22	m	8.48
50 x 50 mm	PC £7.28	0.25	2.26	8.13	m	10.39
50 x 75 mm	PC £9.09	0.25	2.26	10.16	m	12.42
Fitting components; Iroko; PC £353.00/m3						
Backs, fronts, sides or divisions;						
cross-tongued joints; over 300 mm wide						
25 mm thick	PC £47.45	1.75	15.85	51.72	m2	67.57
Shelves or worktops; cross-tongued joints;						
over 300 mm wide						
25 mm thick	PC £47.45	1.75	15.85	51.72	m2	67.57
Draining boards; cross-tongued joints;						
over 300 mm wide						
25 mm thick	PC £51.27	1.75	15.85	55.89	m2	71.74
Stopped flutes		-	-	-	m	2.19
Grooves; cross-grain		-	-	-	m	0.52
Bearers						
19 x 38 mm	PC £2.76	0.15	1.36	3.13	m	4.49
25 x 50 mm	PC £3.39	0.15	1.36	3.84	m	5.20
50 x 50 mm	PC £4.93	0.15	1.36	5.56	m	6.92
50 x 75 mm		0.15	1.36	7.36	m	8.72
Bearers; framed						
19 x 38 mm	PC £5.83	0.20	1.81	6.52	m	8.33
25 x 50 mm	PC £6.48	0.20	1.81	7.23	m	9.04
50 x 50 mm	PC £8.02	0.20	1.81	8.96	m	10.77
50 x 75 mm	PC £9.62	0.20	1.81	10.75	m	12.56
Framing to backs, fronts or sides						
19 x 38 mm	PC £5.83	0.25	2.26	6.52	m	8.78
25 x 50 mm	PC £6.48	0.25	2.26	7.23	m	9.49
50 x 50 mm	PC £8.02	0.25	2.26	8.96	m	11.22
50 x 75 mm	PC £9.62	0.25	2.26	10.75	m	13.01
Fitting components; Teak; PC £1400.00/m3						
Backs, fronts, sides or divisions;						
cross-tongued joints; over 300 mm wide						
25 mm thick	PC £118.33	2.00	18.12	128.98	m2	147.10
Shelves or worktops; cross-tongued joints;						
over 300 mm wide						
25 mm thick	PC £118.33	2.00	18.12	128.98	m2	147.10
Draining boards; cross-tongued joints;						
over 300 mm wide						
25 mm thick	PC £121.16	2.00	18.12	132.06	m2	150.18
Stopped flutes		-	-	-	m2	2.19
Grooves; cross-grain		-	-	-	m	0.52
Fitting components; 'Formica' plastic						
laminated plastics coverings fixed with						
adhesive						
1.3 mm thick horizontal grade marbled						
finish over 300 mm wide	PC £15.00	1.60	14.49	24.10	m2	38.59
1.0 mm thick universal backing over						
300 mm wide	PC £1.85	1.20	10.87	5.04	m2	15.91
1.3 mm thick edging strip						
18 mm wide		0.10	0.91	0.75	m	1.66
25 mm wide		0.10	0.91	0.96	m	1.87

N FURNITURE/EQUIPMENT Including overheads and profit at 9.00%		Labour hours	Labour £	Material £	Unit	Total rate £

N10/11 GENERAL FIXTURES/KITCHEN FITTINGS ETC. - cont'd

Fixing kitchen fittings
(Kitchen fittings are largely a matter of
selection and prices vary considerably. PC
supply prices for reasonable quantities for
'Standard' (moderately-priced) kitchen
fiitings have been shown but not extended).

Fixing only to backgrounds requiring
plugging; including any pre-assembly
Wall units

600 x 600 x 300 mm	PC £35.22	1.40	12.68	0.16	nr	12.84
600 x 900 x 300 mm	PC £43.80	1.60	14.49	0.16	nr	14.65
1200 x 600 x 300 mm	PC £63.96	1.85	16.76	0.22	nr	16.98
1200 x 900 x 300 mm	PC £76.65	2.10	19.02	0.22	nr	19.24
Floor units with drawers						
600 x 900 x 500 mm	PC £59.61	1.25	11.32	0.11	nr	11.43
600 x 900 x 600 mm	PC £62.19	1.40	12.68	0.11	nr	12.79
1200 x 900 x 600 mm	PC £110.13	1.70	15.40	0.11	nr	15.51
Laminated plastics worktops to suit last						
500 x 600 mm	PC £11.52	0.40	3.62	-	nr	3.62
600 x 600 mm	PC £12.96	0.45	4.08	-	nr	4.08
1200 x 600 mm	PC £25.92	0.70	6.34	-	nr	6.34
Larder units						
600 x 1950 x 600 mm	PC £100.26	2.45	22.19	0.16	nr	22.35
Sink units (excluding sink top)						
1200 x 900 x 600 mm	PC £106.14	1.80	16.30	0.11	nr	16.41
1500 x 900 x 600	PC £155.04	2.15	19.47	0.16	nr	19.63

6 mm thick rectangular glass mirrors; silver
backed; fixed with chromium plated domed
headed screws; to background requiring
plugging
Mirror with polished edges

356 x 254 mm	PC £2.57	0.80	5.76	5.32	nr	11.08
400 x 300 mm	PC £3.37	0.80	5.76	6.19	nr	11.95
560 x 380 mm	PC £7.10	0.90	6.48	10.25	nr	16.73
640 x 460 mm	PC £8.66	1.00	7.20	11.96	nr	19.16
Mirror with bevelled edges						
356 x 254 mm	PC £5.09	0.80	5.76	8.07	nr	13.83
400 x 300 mm	PC £5.66	0.80	5.76	8.69	nr	14.45
560 x 380 mm	PC £10.06	0.90	6.48	13.48	nr	19.96
640 x 460 mm	PC £12.53	1.00	7.20	16.17	nr	23.37

Matwells
Mild steel matwell; galvanized after
manufacture; comprising 30 x 30 x 3 mm
angle rim; with welded angles and lugs
welded on; to suit mat size

914 x 560 mm	PC £19.79	1.00	10.32	21.57	nr	31.89
1067 x 610 mm	PC £20.56	1.25	12.90	22.42	nr	35.32
1219 x 762 mm	PC £22.81	1.50	15.48	24.87	nr	40.35
Polished aluminium matwell; comprising						
32 x 32 x 5 mm angle rim; with brazed angles						
and lugs brazed on; to suit mat size						
914 x 560 mm	PC £25.34	1.00	10.32	27.62	nr	37.94
1067 x 610 mm	PC £27.00	1.25	12.90	29.43	nr	42.33
1219 x 762 mm	PC £31.14	1.50	15.48	33.94	nr	49.42
Polished brass matwell; comprising						
38 x 38 x 6 mm angle rim; with brazed angles						
and lugs welded on; to suit mat size						
914 x 560 mm	PC £74.88	1.00	10.32	81.62	nr	91.94
1067 x 610 mm	PC £79.85	1.25	12.90	87.03	nr	99.93
1219 x 762 mm	PC £92.02	1.50	15.48	100.30	nr	115.78

N FURNITURE/EQUIPMENT Including overheads and profit at 9.00%		Labour hours	Labour £	Material £	Unit	Total rate £

N13 SANITARY APPLIANCES/FITTINGS

NOTE: Sanitary fittings are largely a matter
of selection and material prices vary
considerably; the PC values given below are
for average quality

Sink; white glazed fireclay; BS 1206; cast iron cantilever brackets						
610 x 455 x 205 mm	PC £62.00	3.00	31.03	72.40	nr	103.43
610 x 455 x 255 mm	PC £69.10	3.00	31.03	80.13	nr	111.16
760 x 455 x 255 mm	PC £104.55	3.00	31.03	118.77	nr	149.80
Sink; stainless steel combined bowl and draining board; chain and self colour plug; to BS 3380						
1050 x 500 mm with bowl						
420 x 350 x 175 mm	PC £78.53	1.75	18.10	85.60	nr	103.70
Sink; stainless steel combined bowl and double draining board; chain and self colour plug to BS 3380						
1550 x 500 mm with bowl						
420 x 350 x 200 mm	PC £97.36	2.00	20.69	106.12	nr	126.81
Lavatory basin; vitreous china; BS 1188; 32 mm chromium plated waste; chain, stay and plug; pair 13 mm chromium plated easy clean pillar taps to BS 1010; painted cantilever brackets; plugged and screwed						
560 x 405 mm; white	PC £45.28	2.30	23.79	49.36	nr	73.15
560 x 405 mm; coloured	PC £52.32	2.30	23.79	57.03	nr	80.82
635 x 455 mm; white	PC £65.87	2.30	23.79	71.80	nr	95.59
635 x 455 mm; coloured	PC £76.60	2.30	23.79	83.49	nr	107.28
Lavatory basin and pedestal; vitreous china; BS 1188; 32 mm chromium plated waste, chain, stay and plug; pair 13 mm chromium plated easy clean pillar taps to BS 1010; pedestal; wall brackets; plugged and screwed						
560 x 405 mm; white	PC £62.31	2.50	25.86	67.92	nr	93.78
560 x 405 mm; coloured	PC £73.40	2.50	25.86	80.01	nr	105.87
635 x 455 mm; white	PC £82.89	2.50	25.86	90.35	nr	116.21
635 x 455 mm; coloured	PC £111.21	2.50	25.86	121.22	nr	147.08
Lavatory basin range; overlap joints; white glazed fireclay; 32 mm chromium plated waste, chain, stay and plug; pair 13 mm chromium plated easy clean pillar taps to BS 1010; painted cast iron cantilever brackets; plugged and screwed						
range of four 560 x 405 mm	PC £228.35	8.80	91.03	248.90	nr	339.93
Add for each additional basin in the range		2.00	20.69	65.30	nr	85.99
Drinking fountain; white glazed fireclay; 19 mm chromium plated waste; self-closing non-concussive tap; regulating valve; plugged and screwed with chromium plated screws	PC £97.25	2.50	25.86	106.00	nr	131.86
Bath; reinforced acrylic rectangular pattern; 40 mm chromium plated overflow chain and plug; 40 mm chromium plated waste; cast brass 'P' trap with plain outlet and overflow connection to BS 1184; pair 20 mm chromium plated easy clean pillar taps to BS 1010						
1700 mm long; white	PC £101.30	3.50	36.20	110.42	nr	146.62
1700 mm long; coloured	PC £101.30	3.50	36.20	110.42	nr	146.62

N FURNITURE/EQUIPMENT Including overheads and profit at 9.00%		Labour hours	Labour £	Material £	Unit	Total rate £

N13 SANITARY APPLIANCES/FITTINGS - cont'd

Bath; enamelled steel; medium gauge
rectangular pattern; 40 mm chromium
plated overflow chain and plug; 40 mm
chromium plated waste; cast brass 'P' trap
with plain outlet and overflow connection
to BS 1184; pair 20 mm chromium plated easy
clean pillar taps to BS 1010

1700 mm long; white	PC £114.40	3.50	36.20	124.70	nr	160.90
1700 mm long; coloured	PC £121.80	3.50	36.20	132.76	nr	168.96

Shower tray; glazed fireclay with outlet
and grated waste; chain and plug; bedding
and pointing in waterproof cement mortar

760 x 760 x 180 mm; white	PC £95.17	3.00	31.03	103.74	nr	134.77
760 x 760 x 180 mm; coloured	PC £134.08	3.00	31.03	146.15	nr	177.18

Shower fitting; riser pipe with mixing
valve and shower rose; chromium plated;
plugging and screwing mixing valve and
pipe bracket

15 mm dia riser pipe; 127 mm dia shower rose	PC £143.38	5.00	51.72	156.29	nr	208.01

WC suite; high level; vitreous china pan;
black plastic seat; 9 litre white vitreous
china cistern and brackets; low pressure
ball valve; galvanized steel flush pipe and
clip; plugged and screwed; mastic joint to
drain

WC suite; white	PC £112.65	3.30	34.14	122.79	nr	156.93
WC suite; coloured	PC £142.55	3.30	34.14	155.38	nr	189.52

WC suite; low level; vitreous china pan;
black plastic seat; 9 litre white vitreous
china cistern and brackets; low pressure
ball valve and plastic flush pipe; plugged
and screwed; mastic joint to drain

WC suite; white	PC £75.01	3.00	31.03	81.76	nr	112.79
WC suite; coloured	PC £100.90	3.00	31.03	109.98	nr	141.01

Slop sink; white glazed fireclay with
hardwood pad; aluminium bucket grating;
vitreous china cistern and porcelain-
enamelled brackets; galvanized steel flush
pipe and clip; plugged and screwed; mastic
joint to drain

slop sink	PC £341.55	3.50	36.20	372.29	nr	408.49

Bowl type wall urinal; white glazed
vitreous china; white vitreous china
automatic flushing cistern and brackets;
38 mm chromium plated waste; chromium
plated flush pipes and spreaders; cistern
brackets and flush pipes plugged and
screwed with chromium plated screws

single; 455 x 380 x 330 mm	PC £105.87	4.00	41.38	115.40	nr	156.78
range of two	PC £175.79	7.50	77.58	191.61	nr	269.19
range of three	PC £235.10	11.00	113.79	256.26	nr	370.05
Add for each additional urinal	PC £69.92	3.20	33.10	76.21	nr	109.31
Add for divisions between	PC £28.93	0.75	7.76	31.53	nr	39.29

N15 SIGNS/NOTICES

**Plain script; in gloss oil paint; on painted
or varnished surfaces**
Capital letters; lower case letters or
numerals

per coat; per 25 mm high		0.10	0.72	-	nr	0.72

Stops

per coat		0.03	0.22	-	nr	0.22

P BUILDING FABRIC SUNDRIES
Including overheads and profit at 9.00%

ALTERNATIVE INSULATION PRICES

	£		£		£		£
Insulation (£/m2)							
Expanded polystyrene							
self-extinguishing grade							
12 mm	0.89	25 mm	1.85	38 mm	2.79	50 mm	3.68
18 mm	1.32						
'Fibreglass'							
'Crown Building Roll'							
60 mm	1.51	80 mm	1.89	100 mm	2.23		
'Crown Wool'							
150 mm	2.99	200 mm	3.91				
'Frametherm' - unfaced							
60 mm	1.19	80 mm	1.58	90 mm	1.77	100 mm	1.92
Sound-deadening quilt type PF; 13 mm - £1.81/m2							

	Labour hours	Labour £	Material £	Unit	Total rate £
P10 SUNDRY INSULATION/PROOFING WORK/ FIRE STOPS					
Building paper and sheets					
Building paper; BS 1521; class A;					
150 mm laps; pinned					
single sided reflective	0.10	0.91	0.98	m2	1.89
double sided reflective	0.10	0.91	1.43	m2	2.34
Mat or quilt insulation					
Glass fibre quilt; Pilkingtons 'Crown					
Wool'; laid over ceiling joists					
60 mm thick PC £1.21	0.10	0.91	1.38	m2	2.29
80 mm thick PC £1.59	0.11	1.00	1.82	m2	2.82
100 mm thick PC £1.89	0.12	1.09	2.16	m2	3.25
150 mm thick PC £2.99	0.13	1.18	3.43	m2	4.61
200 mm thick PC £3.91	0.14	1.27	4.48	m2	5.75
Raking cutting	0.05	0.45	-	m	0.45
Glass fibre building roll; pinned vertically					
to softwood					
60 mm thick PC £1.51	0.15	1.36	1.74	m2	3.10
80 mm thick PC £1.89	0.16	1.45	2.16	m2	3.61
100 mm thick PC £2.23	0.17	1.54	2.55	m2	4.09
Glass fibre flanged building roll; paper					
faces; pinned vertically or to slope between					
timber framing					
60 mm thick PC £1.62	0.18	1.63	1.85	m2	3.48
80 mm thick PC £2.00	0.19	1.72	2.29	m2	4.01
100 mm thick PC £0.00	0.20	1.81	2.67	m2	4.48
Board or slab insulation					
Expanded polystyrene board standard					
grade PC £56.15/m3; fixed with adhesive					
12 mm thick	0.40	3.62	0.99	m2	4.61
25 mm thick	0.42	3.80	1.55	m2	5.35
50 mm thick	0.45	4.08	2.61	m2	6.69
Fire stops					
'Monolux' TRADA firecheck channel;					
intumescent coatings on cut mitres; fixing					
with brass cups and screws					
19 x 44 mm or 19 x 50 mm PC £189.08/36m	0.60	5.43	8.06	m	13.49

P BUILDING FABRIC SUNDRIES Including overheads and profit at 9.00%	Labour hours	Labour £	Material £	Unit	Total rate £
P10 SUNDRY INSULATION/PROOFING WORK/ **FIRE STOPS - cont'd**					
Fire stops - cont'd 'Sealmaster' intumescent fire and smoke seals; pinned into groove in timber type N30; for single leaf half hour door;					
PC £4.55	0.30	2.72	4.96	m	7.68
type N60; for single leaf one hour door					
PC £6.67	0.33	2.99	7.28	m	10.27
type IMN or IMP; for meeting or pivot styles of pair of one hour doors; per style					
PC £6.67	0.33	2.99	7.28	m	10.27
Intumescent plugs in timber; including boring	0.10	0.91	0.23	nr	1.14
Rockwool fire stops; between top of brick/ block wall and concrete soffit					
25 mm deep x 112 mm wide	0.08	0.72	0.79	m	1.51
25 mm deep x 150 mm wide	0.10	0.91	1.06	m	1.97
50 mm deep x 225 mm wide	0.15	1.36	2.67	m	4.03
Fire barriers Rockwool fire barrier between top of suspended ceiling and concrete soffit					
one 50 mm layer x 900 mm wide; half-hour	0.60	5.43	15.28	m	20.71
two 50 mm layers x 900 mm wide; one hour	0.90	8.15	29.42	m	37.57
Dow Chemicals 'Styrofoam 1B'; cold bridging **insulation fixed with adhesive to brick,** **block or concrete base** 25 mm insulation to walls					
over 300 mm wide	0.34	3.51	4.13	m2	7.64
not exceeding 300 mm wide	0.54	5.57	4.27	m2	9.84
25 mm insulation to isolated columns					
over 300 mm wide	0.41	4.23	4.13	m2	8.36
not exceeding 300 mm wide	0.66	6.81	4.27	m2	11.08
25 mm insulation to ceilings					
over 300 mm wide	0.36	3.72	4.13	m2	7.85
not exceeding 300 mm wide	0.58	5.99	4.27	m2	10.26
25 mm insulation to isolated beams					
over 300 mm wide	0.44	4.54	4.13	m2	8.67
not exceeding 300 mm wide	0.70	7.23	4.27	m2	11.50
50 mm insulation to walls					
over 300 mm wide	0.36	3.72	7.07	m2	10.79
not exceeding 300 mm wide	0.58	5.99	7.35	m2	13.34
50 mm insulation to isolated columns					
over 300 mm wide	0.44	4.54	7.07	m2	11.61
not exceeding 300 mm wide	0.70	7.23	7.35	m2	14.58
50 mm insulation to ceilings					
over 300 mm wide	0.39	4.03	7.07	m2	11.10
not exceeding 300 mm wide	0.62	6.40	7.35	m2	13.75
50 mm insulation to isolated beams					
over 300 mm wide	0.47	4.85	7.07	m2	11.92
not exceeding 300 mm wide	0.75	7.74	7.35	m2	15.09
P11 FOAMED/FIBRE/BEAD CAVITY WALL INSULATION					
Cavity wall insulation; injecting 65 mm cavity with					
UF foam	-	-	-	m2	2.73
blown EPS granules	-	-	-	m2	3.27
blown mineral wool	-	-	-	m2	3.49

P BUILDING FABRIC SUNDRIES Including overheads and profit at 9.00%	Labour hours	Labour £	Material £	Unit	Total rate £
P20 UNFRAMED ISOLATED TRIMS/SKIRTINGS/ SUNDRY ITEMS					
Blockboard (Birch faced); 18 mm thick PC £83.30/10m2					
Window boards and the like; rebated;					
hardwood lipped on one edge					
18 x 200 mm	0.25	2.26	2.83	m	5.09
18 x 250 mm	0.28	2.54	3.31	m	5.85
18 x 300 mm	0.31	2.81	3.81	m	6.62
18 x 350 mm	0.34	3.08	4.30	m	7.38
Returned and fitted ends	0.22	1.99	0.28	nr	2.27
Blockboard (Sapele veneered one side); **18 mm thick PC £89.40/10m2**					
Window boards and the like; rebated;					
hardwood lipped on one edge					
18 x 200 mm	0.27	2.45	2.97	m	5.42
18 x 250 mm	0.30	2.72	3.50	m	6.22
18 x 300 mm	0.33	2.99	4.02	m	7.01
18 x 350 mm	0.36	3.26	4.54	m	7.80
Returned and fitted ends	0.22	1.99	0.28	nr	2.27
Blockboard (Afrormosia veneered one side); **18 mm thick PC £101.75/10m2**					
Window boards and the like; rebated;					
hardwood lipped on one edge					
18 x 200 mm	0.27	2.45	3.34	m	5.79
18 x 250 mm	0.30	2.72	3.93	m	6.65
18 x 300 mm	0.33	2.99	4.53	m	7.52
18 x 350 mm	0.36	3.26	5.13	m	8.39
Returned and fitted ends	0.22	1.99	0.31	nr	2.30
Wrought softwood					
Skirtings, picture rails, dado rails and the					
like; splayed or moulded					
19 x 50 mm; splayed	0.10	0.91	1.20	m	2.11
19 x 50 mm; moulded	0.10	0.91	1.30	m	2.21
19 x 75 mm; splayed	0.10	0.91	1.48	m	2.39
19 x 75 mm; moulded	0.10	0.91	1.57	m	2.48
19 x 100 mm; splayed	0.10	0.91	1.77	m	2.68
19 x 100 mm; moulded	0.10	0.91	1.87	m	2.78
19 x 150 mm; moulded	0.12	1.09	2.43	m	3.52
19 x 175 mm; moulded	0.12	1.09	2.71	m	3.80
22 x 100 mm; splayed	0.10	0.91	1.95	m	2.86
25 x 50 mm; moulded	0.10	0.91	1.49	m	2.40
25 x 75 mm; splayed	0.10	0.91	1.77	m	2.68
25 x 100 mm; splayed	0.10	0.91	2.11	m	3.02
25 x 150 mm; splayed	0.12	1.09	2.86	m	3.95
25 x 150 mm; moulded	0.12	1.09	2.97	m	4.06
25 x 175 mm; moulded	0.12	1.09	3.32	m	4.41
25 x 225 mm; moulded	0.14	1.27	4.08	m	5.35
Returned end	0.15	1.36	-	nr	1.36
Mitres	0.10	0.91	-	nr	0.91
Architraves, cover fillets and the like;					
half round; splayed or moulded					
13 x 25 mm; half round	0.12	1.09	0.84	m	1.93
13 x 50 mm; moulded	0.12	1.09	1.13	m	2.22
16 x 32 mm; half round	0.12	1.09	0.97	m	2.06
16 x 38 mm; moulded	0.12	1.09	1.11	m	2.20
16 x 50 mm; moulded	0.12	1.09	1.22	m	2.31
19 x 50 mm; splayed	0.12	1.09	1.20	m	2.29
19 x 63 mm; splayed	0.12	1.09	1.34	m	2.43
19 x 75 mm; splayed	0.12	1.09	1.48	m	2.57

P BUILDING FABRIC SUNDRIES Including overheads and profit at 9.00%	Labour hours	Labour £	Material £	Unit	Total rate £
P20 UNFRAMED ISOLATED TRIMS/SKIRTINGS/ SUNDRY ITEMS - cont'd					
Wrought softwood - cont'd					
Architraves, cover fillets and the like; half round; splayed or moulded					
25 x 44 mm; splayed	0.12	1.09	1.41	m	2.50
25 x 50 mm; moulded	0.12	1.09	1.49	m	2.58
25 x 63 mm; splayed	0.12	1.09	1.57	m	2.66
25 x 75 mm; splayed	0.12	1.09	1.77	m	2.86
32 x 88 mm; moulded	0.12	1.09	2.55	m	3.64
38 x 38 mm; moulded	0.12	1.09	1.50	m	2.59
50 x 50 mm; moulded	0.12	1.09	2.11	m	3.20
Returned end	0.15	1.36	-	nr	1.36
Mitres	0.10	0.91	-	nr	0.91
Stops; screwed on					
16 x 38 mm	0.10	0.91	0.91	m	1.82
16 x 50 mm	0.10	0.91	1.02	m	1.93
19 x 38 mm	0.10	0.91	0.96	m	1.87
25 x 38 mm	0.10	0.91	1.10	m	2.01
25 x 50 mm	0.10	0.91	1.29	m	2.20
Glazing beads and the like					
13 x 16 mm PC £0.53	-	-	0.62	m	0.62
13 x 19 mm PC £0.56	-	-	0.65	m	0.65
13 x 25 mm PC £0.59	-	-	0.69	m	0.69
13 x 25 mm; screwed	0.05	0.45	0.78	m	1.23
13 x 25 mm; fixing with brass cups and screws	0.10	0.91	2.30	m	3.21
16 x 25 mm PC £0.65	-	-	0.76	m	0.76
16 mm; quadrant	0.05	0.45	0.75	m	1.20
19 mm; quadrant or scotia	0.05	0.45	0.82	m	1.27
19 x 36 mm	0.05	0.45	0.92	m	1.37
25 x 38 mm	0.05	0.45	1.04	m	1.49
25 mm; quadrant or scotia	0.05	0.45	0.95	m	1.40
38 mm; scotia	0.05	0.45	1.45	m	1.90
50 mm; scotia	0.05	0.45	2.06	m	2.51
Isolated shelves, worktops, seats and the like					
19 x 150 mm	0.16	1.45	2.34	m	3.79
19 x 200 mm	0.22	1.99	2.87	m	4.86
25 x 150 mm	0.16	1.45	2.87	m	4.32
25 x 200 mm	0.22	1.99	3.56	m	5.55
32 x 150 mm	0.16	1.45	3.45	m	4.90
32 x 200 mm	0.22	1.99	4.39	m	6.38
Isolated shelves, worktops, seats and the like; cross-tongued joints					
19 x 300 mm	0.28	2.54	9.32	m	11.86
19 x 450 mm	0.34	3.08	11.04	m	14.12
19 x 600 mm	0.40	3.62	15.84	m	19.46
25 x 300 mm	0.28	2.54	10.62	m	13.16
25 x 450 mm	0.34	3.08	12.81	m	15.89
25 x 600 mm	0.40	3.62	18.25	m	21.87
32 x 300 mm	0.28	2.54	11.92	m	14.46
32 x 450 mm	0.34	3.08	14.75	m	17.83
32 x 600 mm	0.40	3.62	20.85	m	24.47
Isolated shelves, worktops, seats and the like; slatted with 50 mm wide slats at 75 mm centres					
19 mm thick	1.33	12.05	7.84	m2	19.89
25 mm thick	1.33	12.05	9.21	m2	21.26
32 mm thick	1.33	12.05	10.41	m2	22.46

P BUILDING FABRIC SUNDRIES Including overheads and profit at 9.00%		Labour hours	Labour £	Material £	Unit	Total rate £
Window boards, nosings, bed moulds and **the like; rebated and rounded**						
19 x 75 mm		0.18	1.63	1.64	m	3.27
19 x 150 mm		0.20	1.81	2.51	m	4.32
19 x 225 mm; in one width		0.26	2.36	3.34	m	5.70
19 x 300 mm; cross-tongued joints		0.30	2.72	9.37	m	12.09
25 x 75 mm		0.18	1.63	1.93	m	3.56
25 x 150 mm		0.20	1.81	3.05	m	4.86
25 x 225 mm; in one width		0.26	2.36	4.14	m	6.50
25 x 300 mm; cross-tongued joints		0.30	2.72	10.62	m	13.34
32 x 75 mm		0.18	1.63	2.24	m	3.87
32 x 150 mm		0.20	1.81	3.65	m	5.46
32 x 225 mm; in one width		0.26	2.36	5.06	m	7.42
32 x 300 mm; cross-tongued joints		0.30	2.72	11.93	m	14.65
38 x 75 mm		0.18	1.63	2.50	m	4.13
38 x 150 mm		0.20	1.81	4.18	m	5.99
38 x 225 mm; in one width		0.26	2.36	5.86	m	8.22
38 x 300 mm; cross-tongued joints		0.30	2.72	13.09	m	15.81
Returned and fitted ends		0.15	1.36	-	nr	1.36
Handrails; mopstick						
50 mm dia		0.25	2.26	2.88	m	5.14
Handrails; rounded						
44 x 50 mm		0.25	2.26	2.88	m	5.14
50 x 75 mm		0.27	2.45	3.65	m	6.10
63 x 87 mm		0.30	2.72	5.12	m	7.84
75 x 100 mm		0.35	3.17	5.81	m	8.98
Handrails; moulded						
44 x 50 mm		0.20	1.81	3.15	m	4.96
50 x 75 mm		0.22	1.99	3.93	m	5.92
63 x 87 mm		0.24	2.17	5.40	m	7.57
75 x 100 mm		0.26	2.36	6.08	m	8.44
Add +5% to the above 'Material £ prices' **for 'selected' softwood for staining**						
Selected West African Mahogany; PC £494.00/m3						
Skirtings, picture rails, dado rails **and the like; splayed or moulded**						
19 x 50 mm; splayed	PC £2.04	0.14	1.27	2.39	m	3.66
19 x 50 mm; moulded	PC £2.16	0.14	1.27	2.53	m	3.80
19 x 75 mm; splayed	PC £2.45	0.14	1.27	2.86	m	4.13
19 x 75 mm; moulded	PC £2.56	0.14	1.27	2.99	m	4.26
19 x 100 mm; splayed	PC £2.88	0.14	1.27	3.34	m	4.61
19 x 100 mm; moulded	PC £2.99	0.14	1.27	3.47	m	4.74
19 x 150 mm; moulded	PC £3.81	0.16	1.45	4.41	m	5.86
19 x 175 mm; moulded	PC £4.37	0.16	1.45	5.05	m	6.50
22 x 100 mm; splayed	PC £3.12	0.14	1.27	3.63	m	4.90
25 x 50 mm; moulded	PC £2.43	0.14	1.27	2.84	m	4.11
25 x 75 mm; splayed	PC £2.85	0.14	1.27	3.31	m	4.58
25 x 100 mm; splayed	PC £3.38	0.14	1.27	3.92	m	5.19
25 x 150 mm; splayed	PC £4.63	0.16	1.45	5.36	m	6.81
25 x 150 mm; moulded	PC £4.75	0.16	1.45	5.50	m	6.95
25 x 175 mm; moulded	PC £5.28	0.16	1.45	6.09	m	7.54
25 x 225 mm; moulded	PC £6.35	0.18	1.63	7.33	m	8.96
Returned end		0.22	1.99	-	nr	1.99
Mitres		0.15	1.36	-	nr	1.36
Architraves, cover fillets and the like; **half round; splayed or moulded**						
13 x 25 mm; half round	PC £1.52	0.16	1.45	1.79	m	3.24
13 x 50 mm; moulded	PC £1.91	0.16	1.45	2.24	m	3.69
16 x 32 mm; half round	PC £1.67	0.16	1.45	1.97	m	3.42
16 x 38 mm; moulded	PC £1.87	0.16	1.45	2.19	m	3.64
16 x 50 mm; moulded	PC £2.04	0.16	1.45	2.38	m	3.83
19 x 50 mm; splayed	PC £2.04	0.16	1.45	2.39	m	3.84
19 x 63 mm; splayed	PC £2.26	0.16	1.45	2.63	m	4.08
19 x 75 mm; splayed	PC £2.45	0.16	1.45	2.86	m	4.31

P BUILDING FABRIC SUNDRIES Including overheads and profit at 9.00%		Labour hours	Labour £	Material £	Unit	Total rate £
P20 UNFRAMED ISOLATED TRIMS/SKIRTINGS/ **SUNDRY ITEMS - cont'd**						
Selected West African Mahogany; PC £494.00/m3 - cont'd						
Architraves, cover fillets and the like;						
half round; splayed or moulded						
25 x 44 mm; splayed	PC £2.32	0.16	1.45	2.71	m	4.16
25 x 50 mm; moulded	PC £2.43	0.16	1.45	2.84	m	4.29
25 x 63 mm; splayed	PC £2.58	0.16	1.45	3.01	m	4.46
25 x 75 mm; splayed	PC £2.85	0.16	1.45	3.31	m	4.76
32 x 88 mm; moulded	PC £4.01	0.16	1.45	4.64	m	6.09
38 x 38 mm; moulded	PC £2.47	0.16	1.45	2.88	m	4.33
50 x 50 mm; moulded	PC £3.38	0.16	1.45	3.92	m	5.37
Returned end		0.22	1.99	-	nr	1.99
Mitres		0.15	1.36	-	nr	1.36
Stops; screwed on						
16 x 38 mm	PC £1.62	0.15	1.36	1.89	m	3.25
16 x 50 mm	PC £1.80	0.15	1.36	2.10	m	3.46
19 x 38 mm	PC £1.74	0.15	1.36	2.03	m	3.39
25 x 38 mm	PC £1.93	0.15	1.36	2.25	m	3.61
25 x 50 mm	PC £2.20	0.15	1.36	2.56	m	3.92
Glazing beads and the like						
13 x 16 mm	PC £1.29	-	-	1.47	m	1.47
13 x 19 mm	PC £1.32	-	-	1.52	m	1.52
13 x 25 mm	PC £1.40	-	-	1.60	m	1.60
13 x 25 mm; screwed	PC £1.40	0.08	0.72	1.69	m	2.41
13 x 25 mm; fixing with brass cups						
and screws	PC £1.40	0.15	1.36	3.21	m	4.57
16 x 25 mm	PC £1.45	-	-	1.66	m	1.66
16 mm; quadrant	PC £1.45	0.07	0.63	1.66	m	2.29
19 mm; quadrant or scotia	PC £1.53	0.07	0.63	1.75	m	2.38
19 x 36 mm	PC £1.74	0.07	0.63	1.99	m	2.62
25 x 38 mm	PC £1.93	0.07	0.63	2.22	m	2.85
25 mm; quadrant or scotia	PC £1.76	0.07	0.63	2.01	m	2.64
38 mm; scotia	PC £2.47	0.07	0.63	2.82	m	3.45
50 mm; scotia	PC £3.38	0.07	0.63	3.87	m	4.50
Isolated shelves, worktops, seats						
and the like						
19 x 150 mm	PC £3.92	0.22	1.99	4.49	m	6.48
19 x 200 mm	PC £4.72	0.30	2.72	5.40	m	8.12
25 x 150 mm	PC £4.72	0.22	1.99	5.40	m	7.39
25 x 200 mm	PC £5.74	0.30	2.72	6.57	m	9.29
32 x 150 mm	PC £5.57	0.22	1.99	6.37	m	8.36
32 x 200 mm	PC £6.94	0.30	2.72	7.94	m	10.66
Isolated shelves, worktops, seats and the						
like; cross-tongued joints						
19 x 300 mm	PC £13.04	0.38	3.44	14.92	m	18.36
19 x 450 mm	PC £15.54	0.45	4.08	17.79	m	21.87
19 x 600 mm	PC £22.56	0.55	4.98	25.82	m	30.80
25 x 300 mm	PC £14.86	0.38	3.44	17.00	m	20.44
25 x 450 mm	PC £18.14	0.45	4.08	20.76	m	24.84
25 x 600 mm	PC £26.02	0.55	4.98	29.78	m	34.76
32 x 300 mm	PC £16.67	0.38	3.44	19.07	m	22.51
32 x 450 mm	PC £20.81	0.45	4.08	23.82	m	27.90
32 x 600 mm	PC £29.72	0.55	4.98	34.02	m	39.00
Isolated shelves, worktops, seats and						
the like; slatted with 50 mm wide slats						
at 75 mm centres						
19 mm thick	PC £14.75/m	1.75	15.85	17.49	m2	33.34
25 mm thick	PC £16.63/m	1.75	15.85	19.69	m2	35.54
32 mm thick	PC £18.36/m	1.75	15.85	21.72	m2	37.57

P BUILDING FABRIC SUNDRIES Including overheads and profit at 9.00%		Labour hours	Labour £	Material £	Unit	Total rate £
Window boards, nosings, bed moulds and						
the like; rebated and rounded						
19 x 75 mm	PC £2.73	0.24	2.17	3.25	m	5.42
19 x 150 mm	PC £3.96	0.27	2.45	4.65	m	7.10
19 x 225 mm; in one width	PC £5.19	0.36	3.26	6.06	m	9.32
19 x 300 mm; cross-tongued joints	PC £12.96	0.40	3.62	14.95	m	18.57
25 x 75 mm	PC £3.14	0.24	2.17	3.71	m	5.88
25 x 150 mm	PC £4.79	0.27	2.45	5.60	m	8.05
25 x 225 mm; in one width	PC £6.35	0.36	3.26	7.39	m	10.65
25 x 300 mm; cross-tongued joints	PC £14.70	0.40	3.62	16.94	m	20.56
32 x 75 mm	PC £3.59	0.24	2.17	4.23	m	6.40
32 x 150 mm	PC £5.64	0.27	2.45	6.57	m	9.02
32 x 225 mm; in one width	PC £7.73	0.36	3.26	8.97	m	12.23
32 x 300 mm; cross-tongued joints	PC £16.51	0.40	3.62	19.02	m	22.64
38 x 75 mm	PC £4.60	0.24	2.17	5.39	m	7.56
38 x 150 mm	PC £6.44	0.27	2.45	7.49	m	9.94
38 x 225 mm; in one width	PC £8.90	0.36	3.26	10.30	m	13.56
38 x 300 mm; cross-tongued joints	PC £18.16	0.40	3.62	20.90	m	24.52
Returned and fitted ends		0.23	2.08	-	nr	2.08
Handrails; rounded						
44 x 50 mm	PC £4.59	0.33	2.99	5.25	m	8.24
50 x 75 mm	PC £5.71	0.36	3.26	6.54	m	9.80
63 x 87 mm	PC £7.35	0.40	3.62	8.42	m	12.04
75 x 100 mm	PC £8.91	0.45	4.08	10.20	m	14.28
Handrails; moulded						
44 x 50 mm	PC £4.95	0.33	2.99	5.67	m	8.66
50 x 75 mm	PC £6.08	0.36	3.26	6.95	m	10.21
63 x 87 mm	PC £7.71	0.40	3.62	8.83	m	12.45
75 x 100 mm	PC £9.26	0.45	4.08	10.59	m	14.67
Afrormosia; PC £706.00/m3						
Skirtings, picture rails, dado rails						
and the like; splayed or moulded						
19 x 50 mm; splayed	PC £2.40	0.14	1.27	2.80	m	4.07
19 x 50 mm; moulded	PC £2.52	0.14	1.27	2.94	m	4.21
19 x 75 mm; splayed	PC £2.99	0.14	1.27	3.47	m	4.74
19 x 75 mm; moulded	PC £3.11	0.14	1.27	3.62	m	4.89
19 x 100 mm; splayed	PC £3.57	0.14	1.27	4.14	m	5.41
19 x 100 mm; moulded	PC £3.69	0.14	1.27	4.27	m	5.54
19 x 150 mm; moulded	PC £4.86	0.16	1.45	5.61	m	7.06
19 x 175 mm; moulded	PC £5.61	0.16	1.45	6.47	m	7.92
22 x 100 mm; splayed	PC £3.95	0.14	1.27	4.58	m	5.85
25 x 50 mm; moulded	PC £2.90	0.14	1.27	3.38	m	4.65
25 x 75 mm; splayed	PC £3.53	0.14	1.27	4.10	m	5.37
25 x 100 mm; splayed	PC £4.31	0.14	1.27	4.99	m	6.26
25 x 150 mm; splayed	PC £6.02	0.16	1.45	6.94	m	8.39
25 x 150 mm; moulded	PC £6.14	0.16	1.45	7.09	m	8.54
25 x 175 mm; moulded	PC £6.89	0.16	1.45	7.94	m	9.39
25 x 225 mm; moulded	PC £8.45	0.18	1.63	9.72	m	11.35
Returned end		0.22	1.99	-	nr	1.99
Mitres		0.15	1.36	-	nr	1.36
Architraves, cover fillets and the like;						
half round; splayed or moulded						
13 x 25 mm; half round	PC £1.63	0.16	1.45	1.91	m	3.36
13 x 50 mm; moulded	PC £2.14	0.16	1.45	2.50	m	3.95
16 x 32 mm; half round	PC £1.85	0.16	1.45	2.17	m	3.62
16 x 38 mm; moulded	PC £2.10	0.16	1.45	2.45	m	3.90
16 x 50 mm; moulded	PC £2.33	0.16	1.45	2.72	m	4.17
19 x 50 mm; splayed	PC £2.40	0.16	1.45	2.80	m	4.25
19 x 63 mm; splayed	PC £2.72	0.16	1.45	3.17	m	4.62
19 x 75 mm; splayed	PC £2.99	0.16	1.45	3.47	m	4.92

P BUILDING FABRIC SUNDRIES Including overheads and profit at 9.00%		Labour hours	Labour £	Material £	Unit	Total rate £
P20 UNFRAMED ISOLATED TRIMS/SKIRTINGS/ **SUNDRY ITEMS - cont'd**						
Afrormosia; PC £706.00/m3 - cont'd						
Architraves, cover fillets and the like;						
half round; splayed or moulded						
25 x 44 mm; splayed	PC £2.78	0.16	1.45	3.24	m	4.69
25 x 50 mm; moulded	PC £2.90	0.16	1.45	3.38	m	4.83
25 x 63 mm; splayed	PC £3.17	0.16	1.45	3.69	m	5.14
25 x 75 mm; splayed	PC £3.53	0.16	1.45	4.10	m	5.55
32 x 88 mm; moulded	PC £5.17	0.16	1.45	5.97	m	7.42
38 x 38 mm; moulded	PC £3.02	0.16	1.45	3.50	m	4.95
50 x 50 mm; moulded	PC £4.31	0.16	1.45	4.99	m	6.44
Returned end		0.22	1.99	-	nr	1.99
Mitres		0.15	1.36	-	nr	1.36
Stops; screwed on						
16 x 38 mm	PC £1.87	0.15	1.36	2.17	m	3.53
16 x 50 mm	PC £2.10	0.15	1.36	2.43	m	3.79
19 x 38 mm	PC £2.00	0.15	1.36	2.32	m	3.68
25 x 38 mm	PC £2.27	0.15	1.36	2.63	m	3.99
25 x 50 mm	PC £2.65	0.15	1.36	3.07	m	4.43
Glazing beads and the like						
13 x 16 mm	PC £1.37	-	-	1.57	m	1.57
13 x 19 mm	PC £1.42	-	-	1.62	m	1.62
13 x 25 mm	PC £1.52	-	-	1.74	m	1.74
13 x 25 mm; screwed	PC £1.52	0.08	0.72	1.83	m	2.55
13 x 25 mm; fixing with brass cups						
and screws	PC £1.52	0.15	1.36	3.35	m	4.71
16 x 25 mm	PC £1.61	-	-	1.84	m	1.84
16 mm; quadrant	PC £1.55	0.07	0.63	1.78	m	2.41
19 mm; quadrant or scotia	PC £1.68	0.07	0.63	1.92	m	2.55
19 x 36 mm	PC £2.00	0.07	0.63	2.29	m	2.92
25 x 38 mm	PC £2.53	0.07	0.63	2.89	m	3.52
25 mm; quadrant or scotia	PC £2.01	0.07	0.63	2.30	m	2.93
38 mm; scotia	PC £3.02	0.07	0.63	3.45	m	4.08
50 mm; scotia	PC £4.31	0.07	0.63	4.94	m	5.57
Isolated shelves, worktops, seats						
and the like						
19 x 150 mm	PC £5.01	0.22	1.99	5.74	m	7.73
19 x 200 mm	PC £6.12	0.30	2.72	7.01	m	9.73
25 x 150 mm	PC £6.12	0.22	1.99	7.01	m	9.00
25 x 200 mm	PC £7.59	0.30	2.72	8.68	m	11.40
32 x 150 mm	PC £7.35	0.22	1.99	8.42	m	10.41
32 x 200 mm	PC £9.32	0.30	2.72	10.68	m	13.40
Isolated shelves, worktops, seats and the						
like; cross-tongued joints						
19 x 300 mm	PC £15.38	0.38	3.44	17.61	m	21.05
19 x 450 mm	PC £18.96	0.45	4.08	21.70	m	25.78
19 x 600 mm	PC £27.22	0.55	4.98	31.16	m	36.14
25 x 300 mm	PC £17.97	0.38	3.44	20.57	m	24.01
25 x 450 mm	PC £22.65	0.45	4.08	25.92	m	30.00
25 x 600 mm	PC £32.15	0.55	4.98	36.80	m	41.78
32 x 300 mm	PC £20.55	0.38	3.44	23.53	m	26.97
32 x 450 mm	PC £26.48	0.45	4.08	30.31	m	34.39
32 x 600 mm	PC £37.46	0.55	4.98	42.87	m	47.85
Isolated shelves, worktops, seats and						
the like; slatted with 50 mm wide slats						
at 75 mm centres						
19 mm thick	PC £17.34/m	1.75	15.85	20.52	m2	36.37
25 mm thick	PC £19.91/m	1.75	15.85	23.52	m2	39.37
32 mm thick	PC £22.63/m	1.75	15.85	26.83	m2	42.68
Window boards, nosings, bed moulds and						
the like; rebated and rounded						
19 x 75 mm	PC £3.25	0.24	2.17	3.83	m	6.00
19 x 150 mm	PC £5.02	0.27	2.45	5.87	m	8.32
19 x 225 mm; in one width	PC £6.76	0.36	3.26	7.86	m	11.12

P BUILDING FABRIC SUNDRIES Including overheads and profit at 9.00%		Labour hours	Labour £	Material £	Unit	Total rate £
19 x 300 mm; cross-tongued joints	PC £15.34	0.40	3.62	17.68	m	21.30
25 x 75 mm	PC £3.83	0.24	2.17	4.50	m	6.67
25 x 150 mm	PC £6.21	0.27	2.45	7.22	m	9.67
25 x 225 mm; in one width	PC £8.43	0.36	3.26	9.76	m	13.02
25 x 300 mm; cross-tongued joints	PC £17.81	0.40	3.62	20.50	m	24.12
32 x 75 mm	PC £4.48	0.24	2.17	5.24	m	7.41
32 x 150 mm	PC £7.44	0.27	2.45	8.63	m	11.08
32 x 225 mm; in one width	PC £10.42	0.36	3.26	12.04	m	15.30
32 x 300 mm; cross-tongued joints	PC £20.40	0.40	3.62	23.46	m	27.08
38 x 75 mm	PC £5.02	0.24	2.17	5.87	m	8.04
38 x 150 mm	PC £8.56	0.27	2.45	9.91	m	12.36
38 x 225 mm; in one width	PC £12.07	0.36	3.26	13.93	m	17.19
38 x 300 mm; cross-tongued joints	PC £22.76	0.40	3.62	26.16	m	29.78
Returned and fitted ends		0.23	2.08	-	nr	2.08
Handrails; rounded						
44 x 50 mm	PC £5.51	0.33	2.99	6.31	m	9.30
50 x 75 mm	PC £7.07	0.36	3.26	8.08	m	11.34
63 x 87 mm	PC £10.22	0.40	3.62	11.70	m	15.32
75 x 100 mm	PC £11.68	0.45	4.08	13.37	m	17.45
Handrails; moulded						
44 x 50 mm	PC £5.86	0.33	2.99	6.71	m	9.70
50 x 75 mm	PC £7.42	0.36	3.26	8.49	m	11.75
63 x 87 mm	PC £10.56	0.40	3.62	12.09	m	15.71
75 x 100 mm	PC £12.04	0.45	4.08	13.78	m	17.86

Pin-boards; medium board
Sundeala 'A' pin-board; fixed with adhesive
to backing (measured elsewhere); over 300 mm
wide

		Labour hours	Labour £	Material £	Unit	Total rate £
6.4 mm thick		0.60	5.43	7.87	m2	13.30

Sundries on softwood/hardwood
Extra over fixing with nails for

	Labour hours	Labour £	Material £	Unit	Total rate £
gluing and pinning	0.02	0.18	0.03	m	0.21
masonry nails	0.02	0.18	0.04	m	0.22
steel screws	0.02	0.18	0.06	m	0.24
self-tapping screws	0.02	0.18	0.07	m	0.25
steel screws; gluing	0.04	0.36	0.06	m	0.42
steel screws; sinking; filling heads	0.05	0.45	0.06	m	0.51
steel screws; sinking; pellating over	0.10	0.91	1.69	m	2.60
brass cups and screws	0.15	1.36	0.06	m	1.42
Extra over for					
countersinking	0.02	0.18	-	m	0.18
pellating	0.10	0.91	0.03	m	0.94
Head or nut in softwood					
let in flush	0.05	0.45	-	nr	0.45
Head or nut; in hardwood					
let in flush	0.08	0.72	-	nr	0.72
let in over; pellated	0.20	1.81	-	nr	1.81

P21 IRONMONGERY

Metalwork; mild steel; galvanized
Water bars; groove in timber

	Labour hours	Labour £	Material £	Unit	Total rate £
6 x 30 mm	0.50	4.53	3.30	m	7.83
6 x 40 mm	0.50	4.53	3.98	m	8.51
6 x 50 mm	0.50	4.53	4.67	m	9.20
Dowels; mortice in timber					
8 mm dia x 100 mm long	0.05	0.45	0.05	nr	0.50
10 mm dia x 50 mm long	0.05	0.45	0.13	nr	0.58
Cramps					
25 x 3 x 230 mm girth; one end bent, holed and screwed to softwood; other end fishtailed for building in	0.07	0.63	0.24	nr	0.87

P BUILDING FABRIC SUNDRIES
Including overheads and profit at 9.00%

IRONMONGERY - TYPICAL 'SUPPLY ONLY' PRICES

NOTE: Ironmongery is largely a matter of selection and prices vary considerably;
indicative prices for reasonable quantities of standard quality ironmongery are
given below. The prices for doors include for fixing with and including ordinary
butt hinges, and only the extra cost of fixing hinges are given below.

Bolts (£/each)

barrel	£		£		£		£
152 mm	1.04	203 mm	1.26	254 mm	1.53	305 mm	1.71
straight tower							
102 mm	0.54	152 mm	0.67	203 mm	0.99	254 mm	1.80

Other bolts

	£		£
flush - 203x19 mm brass	4.77	panic - single	35.28
- 203x19 mm SAA	3.92	- double	42.75
garage bolt foot action	6.39	security hinge	2.25
monkey tail - 380 mm	6.93	security mortice	2.34
necked tower - 152 mm	1.08	WC indicator - SAA	8.96
- 203 mm	1.44	- zinc alloy	3.82

Butts and hinges (£/pair)

back flap; steel; medium	£		£		£		£
25x73 mm	0.41	32x82 mm	0.49	38x89 mm	0.58	51x108 mm	0.85
bands and hooks; wrought iron							
305 mm	3.38	406 mm	4.28	457 mm	5.36	609 mm	8.28
butt hinges; brass; washered							
76x51 mm	3.78	102x66 mm	7.70	127x76 mm	9.81		
butt hinges; steel							
extra strong							
76 mm	1.71	102 mm	2.43				
medium							
50 mm	0.23	60 mm	0.23	75 mm	0.27	100 mm	0.54
parliament							
76 mm	2.47	102 mm	4.05				
rising							
76 mm	1.21	102 mm	1.94				
spring; single action							
76 mm	10.13	102 mm	11.97	127 mm	14.13	152 mm	17.28
spring; double action							
76 mm	15.84	102 mm	18.90	127 mm	21.60	152 mm	26.28
strong single flap							
76 mm	0.95	102 mm	1.35				
tee; medium							
230 mm	0.77	300 mm	0.90	375 mm	1.21	450 mm	1.62
washered							
76x49 mm	2.97	101x70 mm	5.36				

Other butts and hinges

	£		£
collinge - 457 mm	26.91	double strap field gate	15.17

Catches (£/each)

magnetic	£		£
cupboard	1.26	door	1.62
roller and ball			
bales	0.77	door	1.21

P BUILDING FABRIC SUNDRIES
Including overheads and profit at 9.00%

	£		£		£

Door closers (£/each)
  floor springs

single action	143.10	double action	160.20		

  overhead door closer; concealed fixing

50.8 kg	71.55	61.6 kg	76.05		

  overhead door closer; surface fixing

45 kg	39.15	67 kg	46.71	91 kg	59.85

  other items

'Perko' closer	4.41	selector	32.58		

Door furniture
  finger plates; 305x76 mm; quality (£/each)

'modest'	2.07	'average'	7.16	'expensive'	13.95

  kicking plates; quality (£/each)
  760x150 mm

'modest'	2.70	'average'	7.52	'expensive'	13.50

  lever latch furniture (£/set)

'modest'	5.18	'average'	13.77	'expensive'	44.91

  lever lock furniture

'modest'	5.36	'average'	11.43	'expensive'	44.19

		£			£

Other door furniture (£/each)

cabin hook and eye	3.24	postal knocker		7.11
centre knob	15.66	pull handles	13.32 -	24.26
door chain	1.67	push pad handle; 150x150 mm		14.17
door viewer	3.55	padlock; close shackle		24.75

Latches (£/each)

cylinder rim night - 'Yale'	10.80	rim	2.70
deadlocking rim night- 'Yale'	21.51	Suffolk	2.84
mortice	3.15		

Locks (£/each)

budget	3.60	rim	7.74
cupboard lock	2.79	'Yale' deadlock	13.14
mortice deadlock	5.27	'Waterloo' mortice	7.11

Shelving; adjustable 'Spur' (£/each)

bracket - 22 cm	1.26	shelf - 1 m x 36 cm		8.05
- 36 cm	2.56	upright - 122 cm		3.74
- 61 cm	4.19	- 240 cm		6.84

Sliding door gear; 'Hillaldam; (£/each - unless otherwise described)
  'Commercial for top hung doors; 365 kg max. weight; 3.6 m max. height

bow handle	4.47	door stop (rubber buffers)	12.92
det. locking bar/padlock	10.22	drop bolt	20.45
door guide - metal	3.55	flush handle	2.51
- timber	4.29	galv. steel top tack	13.00
door hanger - metal	15.50	open side wall bracket	2.35
door hanger - timber	21.01	steel bottom guide	8.30

  'House One' for internal domestic doors

pelmet	8.34	set size 2	11.17

  'Twin fold' for 4 door wardrobes (1.50 m opening)

set size TF60/4	16.99

Window furniture (£/each)

casement fastener; SAA	2.79	metal window locks	14.13
casement stay; SAA	2.61	security mortice bolt	2.34
locking window catch	4.95		

Sundries

hat and coat hooks	1.98	pictograms; stainless steel	2.70
door stops; rubber;to concrete	2.29	numerals; stainless steel	0.45
; to wood	0.36	shelf brackets; japanned	0.18

P BUILDING FABRIC SUNDRIES Including overheads and profit at 9.00%	Labour hours	Labour £	Material £	Unit	Total rate £
P21 IRONMONGERY - cont'd					
Fixing only ironmongery to softwood					
Bolts					
barrel; not exceeding 150 mm long	0.33	2.99	-	nr	2.99
barrel; 150-300 mm long	0.42	3.80	-	nr	3.80
cylindrical mortice; not exceeding					
150 mm long	0.50	4.53	-	nr	4.53
150 - 300 mm long	0.60	5.43	-	nr	5.43
flush; not exceeding 150 mm long	0.50	4.53	-	nr	4.53
flush; 150 - 300 mm long	0.60	5.43	-	nr	5.43
monkey tail; 380 mm long	0.67	6.07	-	nr	6.07
necked; 150 mm long	0.33	2.99	-	nr	2.99
panic; single; locking	2.50	22.64	-	nr	22.64
panic; double; locking	3.50	31.70	-	nr	31.70
WC indicator	0.67	6.07	-	nr	6.07
Butts; extra over for					
rising	0.17	1.54	-	pr	1.54
skew	0.17	1.54	-	pr	1.54
spring; single action	1.30	11.78	-	pr	11.78
spring; double action	1.50	13.59	-	pr	13.59
Catches					
surface mounted	0.17	1.54	-	nr	1.54
mortice	0.33	2.99	-	nr	2.99
Door closers and furniture					
cabin hooks and eyes	0.17	1.54	-	nr	1.54
door selector	0.50	4.53	-	nr	4.53
finger plate	0.17	1.54	-	nr	1.54
floor spring	2.50	22.64	-	nr	22.64
lever furniture	0.25	2.26	-	nr	2.26
handle; not exceeding 150 mm long	0.17	1.54	-	nr	1.54
handle; 150 - 300 mm long	0.25	2.26	-	nr	2.26
handle; flush	0.33	2.99	-	nr	2.99
holder	0.33	2.99	-	nr	2.99
kicking plate	0.33	2.99	-	nr	2.99
letter plate; including perforation	1.33	12.05	-	nr	12.05
overhead door closer; surface fixing	1.25	11.32	-	nr	11.32
overhead door closer; concealed fixing	1.75	15.85	-	nr	15.85
top centre	0.50	4.53	-	nr	4.53
'Perko' door closer	0.67	6.07	-	nr	6.07
rod door closers; 457 mm long	1.00	9.06	-	nr	9.06
Latches					
cylinder rim night latch	0.75	6.79	-	nr	6.79
mortice	0.67	6.07	-	nr	6.07
Norfolk	0.67	6.07	-	nr	6.07
rim	0.50	4.53	-	nr	4.53
Locks					
cupboard	0.42	3.80	-	nr	3.80
mortice	0.83	7.52	-	nr	7.52
mortice budget	0.75	6.79	-	nr	6.79
mortice dead	0.75	6.79	-	nr	6.79
rebated mortice	1.25	11.32	-	nr	11.32
rim	0.50	4.53	-	nr	4.53
rim budget	0.42	3.80	-	nr	3.80
rim dead	0.42	3.80	-	nr	3.80
Sliding door gear for top hung softwood					
timber doors; weight not exceeding 365 kg					
bottom guide; fixed to concrete in groove	0.50	4.53	-	m	4.53
top track	0.25	2.26	-	m	2.26
detachable locking bar and padlock	0.33	2.99	-	nr	2.99
hangers; fixed flush to timber	0.75	6.79	-	nr	6.79
head brackets; bolted to concrete	0.42	3.80	-	nr	3.80

P BUILDING FABRIC SUNDRIES Including overheads and profit at 9.00%	Labour hours	Labour £	Material £	Unit	Total rate £
Window furniture					
casement stay and pin	0.17	1.54	-	nr	1.54
catch; fanlight	0.25	2.26	-	nr	2.26
fastener; cockspur	0.33	2.99	-	nr	2.99
fastener; sash	0.25	2.26	-	nr	2.26
quadrant stay	0.25	2.26	-	nr	2.26
ring catch	0.33	2.99	-	nr	2.99
Sundries					
drawer pull	0.08	0.72	-	nr	0.72
hat and coat hook	0.08	0.72	-	nr	0.72
numerals	0.08	0.72	-	nr	0.72
rubber door stop	0.08	0.72	-	nr	0.72
shelf bracket	0.17	1.54	-	nr	1.54
skirting type door stop	0.17	1.54	-	nr	1.54
Fixing only ironmongery to hardwood					
Bolts					
barrel; not exceeding 150 mm long	0.44	3.99	-	nr	3.99
barrel; 150-300 mm long	0.56	5.07	-	nr	5.07
cylindrical mortice; not exceeding					
150 mm long	0.67	6.07	-	nr	6.07
150 - 300 mm long	0.80	7.25	-	nr	7.25
flush; not exceeding 150 mm long	0.67	6.07	-	nr	6.07
flush; 150 - 300 mm long	0.80	7.25	-	nr	7.25
monkey tail; 380 mm long	0.89	8.06	-	nr	8.06
necked; 150 mm long	0.44	3.99	-	nr	3.99
panic; single; locking	3.33	30.16	-	nr	30.16
panic; double; locking	4.67	42.30	-	nr	42.30
WC indicator	0.89	8.06	-	nr	8.06
Butts; extra over for					
rising	0.23	2.08	-	pr	2.08
skew	0.23	2.08	-	pr	2.08
spring; single action	1.73	15.67	-	pr	15.67
spring; double action	2.00	18.12	-	pr	18.12
Catches					
surface mounted	0.23	2.08	-	nr	2.08
mortice	0.44	3.99	-	nr	3.99
Door closers and furniture					
cabin hooks and eyes	0.23	2.08	-	nr	2.08
door selector	0.67	6.07	-	nr	6.07
finger plate	0.23	2.08	-	nr	2.08
floor spring	3.33	30.16	-	nr	30.16
lever furniture	0.33	2.99	-	nr	2.99
handle; not exceeding 150 mm long	0.23	2.08	-	nr	2.08
handle; 150 - 300 mm long	0.33	2.99	-	nr	2.99
handle; flush	0.44	3.99	-	nr	3.99
holder	0.44	3.99	-	nr	3.99
kicking plate	0.44	3.99	-	nr	3.99
letter plate; including perforation	1.77	16.03	-	nr	16.03
overhead door closer; surface fixing	1.67	15.13	-	nr	15.13
overhead door closer; concealed fixing	2.33	21.10	-	nr	21.10
top centre	0.67	6.07	-	nr	6.07
'Perko' door closer	0.89	8.06	-	nr	8.06
rod door closers; 457 mm long	1.33	12.05	-	nr	12.05
Latches					
cylinder rim night latch	1.00	9.06	-	nr	9.06
mortice	0.89	8.06	-	nr	8.06
Norfolk	0.89	8.06	-	nr	8.06
rim	0.67	6.07	-	nr	6.07
Locks					
cupboard	0.56	5.07	-	nr	5.07
mortice	1.11	10.05	-	nr	10.05
mortice budget	1.00	9.06	-	nr	9.06
mortice dead	1.00	9.06	-	nr	9.06
rebated mortice	1.67	15.13	-	nr	15.13
rim	0.67	6.07	-	nr	6.07
rim budget	0.56	5.07	-	nr	5.07
rim dead	0.56	5.07	-	nr	5.07

P BUILDING FABRIC SUNDRIES Including overheads and profit at 9.00%	Labour hours	Labour £	Material £	Unit	Total rate £
P21 IRONMONGERY - cont'd					
Fixing only ironmongery to hardwood - cont'd					
Sliding door gear for top hung softwood					
timber doors; weight not exceeding 365 kg					
bottom guide; fixed to concrete in groove	0.67	6.07	-	m	6.07
top track	0.33	2.99	-	m	2.99
detachable locking bar and padlock	0.44	3.99	-	nr	3.99
hangers; fixed flush to timber	1.00	9.06	-	nr	9.06
head brackets; bolted to concrete	0.56	5.07	-	nr	5.07
Window furniture					
casement stay and pin	0.23	2.08	-	nr	2.08
catch; fanlight	0.33	2.99	-	nr	2.99
fastener; cockspur	0.44	3.99	-	nr	3.99
fastener; sash	0.33	2.99	-	nr	2.99
quadrant stay	0.33	2.99	-	nr	2.99
ring catch	0.44	3.99	-	nr	3.99
Sundries					
drawer pull	0.11	1.00	-	nr	1.00
hat and coat hook	0.11	1.00	-	nr	1.00
numerals	0.11	1.00	-	nr	1.00
rubber door stop	0.11	1.00	-	nr	1.00
shelf bracket	0.23	2.08	-	nr	2.08
skirting type door stop	0.11	1.00	-	nr	1.00
Sundries					
Rubber door stop plugged and screwed					
to concrete	0.10	0.91	2.53	nr	3.44
P30 TRENCHES/PIPEWAYS/PITS FOR BURIED **ENGINEERING SERVICES**					
Mechanical excavation of trenches to receive					
pipes; grading bottoms; earthwork support;					
filling with excavated material and					
compacting; disposal of surplus soil;					
spreading on site average 50 m					
Pipes not exceeding 200 mm; average depth					
0.50 m deep	0.20	1.35	1.24	m	2.59
0.75 m deep	0.30	2.03	2.07	m	4.10
1.00 m deep	0.60	4.05	3.65	m	7.70
1.25 m deep	0.90	6.08	4.95	m	11.03
1.50 m deep	1.15	7.77	6.48	m	14.25
1.75 m deep	1.40	9.46	8.31	m	17.77
2.00 m deep	1.65	11.15	9.54	m	20.69
Hand excavation of trenches to receive					
pipes; grading bottoms; earthwork support;					
filling with excavated material and					
compacting; disposal of surplus soil;					
spreading on site average 50 m					
Pipes not exceeding 200 mm; average depth					
0.50 m deep	1.00	6.76	-	m	6.76
0.75 m deep	1.50	10.14	-	m	10.14
1.00 m deep	2.20	14.87	0.95	m	15.82
1.25 m deep	3.10	20.95	1.43	m	22.38
1.50 m deep	4.25	28.72	1.72	m	30.44
1.75 m deep	5.60	37.84	2.10	m	39.94
2.00 m deep	6.40	43.25	2.29	m	45.54
Pits for underground stop valves and the					
like; half brick thick walls in common					
bricks in cement mortar (1:3); on in situ					
concrete mix 21.00 N/mm2-20 mm aggregate					
(1:2:4) bed; 100 mm thick; 100 x 100 x 750					
mm deep; internal holes for one small pipe;					
cast iron hinged box cover; bedding in					
cement mortar (1:3)	3.00	34.96	14.94	nr	49.90

P BUILDING FABRIC SUNDRIES Including overheads and profit at 9.00%	Labour hours	Labour £	Material £	Unit	Total rate £

P31 HOLES/CHASES/COVERS/SUPPORTS FOR SERVICES

Builders' work for electrical installations
Cutting away for and making good after
electrician; including cutting or leaving
all holes, notches, mortices, sinkings and
chases, in both the structure and its
coverings, for the following electrical
points
Exposed installation

lighting points	0.30	3.10	-	nr	3.10
socket outlet points	0.50	5.16	-	nr	5.16
fitting outlet points	0.50	5.16	-	nr	5.16
equipment points or control gear points	0.70	7.23	-	nr	7.23
Concealed installation					
lighting points	0.40	4.13	-	nr	4.13
socket outlet points	0.70	7.23	-	nr	7.23
fitting outlet points	0.70	7.23	-	nr	7.23
equipment points or control gear points	1.00	10.32	-	nr	10.32

**Builders' work for other services
installations**
Cutting chases in brickwork
for one pipe; not exceeding 55 mm nominal

size; vertical	0.40	4.66	-	m	4.66
for one pipe; 55 - 110 nominal size; vertical	0.70	8.16	-	m	8.16

Cutting and pinning to brickwork or
blockwork; ends of supports

for pipes not exceeding 55 mm	0.20	2.33	-	m	2.33
for cast iron pipes 55 - 110 mm	0.33	3.85	-	nr	3.85
radiator stays or brackets	0.25	2.91	-	nr	2.91

Cutting holes for pipes or the like;
not exceeding 55 mm nominal size

102 mm brickwork	0.33	2.31	-	nr	2.31
215 mm brickwork	0.55	3.85	-	nr	3.85
327 mm brickwork	0.90	6.30	-	nr	6.30
100 mm blockwork	0.30	2.10	-	nr	2.10
150 mm blockwork	0.40	2.80	-	nr	2.80
215 mm blockwork	0.50	3.50	-	nr	3.50

Cutting holes for pipes or the like;
55 - 110 mm nominal size

102 mm brickwork	0.40	2.80	-	nr	2.80
215 mm brickwork	0.70	4.90	-	nr	4.90
327 mm brickwork	1.10	7.70	-	nr	7.70
100 mm blockwork	0.35	2.45	-	nr	2.45
150 mm blockwork	0.50	3.50	-	nr	3.50
215 mm blockwork	0.60	4.20	-	nr	4.20

Cutting holes for pipes or the like;
over 110 mm nominal size

102 mm brickwork	0.50	3.50	-	nr	3.50
215 mm brickwork	0.85	5.95	-	nr	5.95
327 mm brickwork	1.35	9.45	-	nr	9.45
100 mm blockwork	0.45	5.24	-	nr	5.24
150 mm blockwork	0.60	4.20	-	nr	4.20
215 mm blockwork	0.75	5.25	-	nr	5.25

Add for making good fair face or facings
one side

pipe; not exceeding 55 mm nominal size	0.08	0.93	-	nr	0.93
pipe; 55 - 110 mm nominal size	0.10	1.17	-	nr	1.17
pipe; over 110 mm nominal size	0.12	1.40	-	nr	1.40

Add for fixing sleeve (supply included
elsewhere)

for pipe; small	0.15	1.75	-	nr	1.75
for pipe; large	0.20	2.33	-	nr	2.33
for pipe; extra large	0.30	3.50	-	nr	3.50

P BUILDING FABRIC SUNDRIES Including overheads and profit at 9.00%	Labour hours	Labour £	Material £	Unit	Total rate £
P31 HOLES/CHASES/COVERS/SUPPORTS FOR SERVICES - cont'd					
Builders' work for other services **installations - cont'd**					
Cutting or forming holes for ducts; girth not exceeding 1.00 m					
102 mm brickwork	0.60	4.20	-	nr	4.20
215 mm brickwork	1.00	7.00	-	nr	7.00
327 mm brickwork	1.60	11.20	-	nr	11.20
100 mm blockwork	0.50	3.50	-	nr	3.50
150 mm blockwork	0.70	4.90	-	nr	4.90
215 mm blockwork	0.90	6.30	-	nr	6.30
Cutting or forming holes for ducts; girth 1.00 - 2.00 m					
102 mm brickwork	0.70	4.90	-	nr	4.90
215 mm brickwork	1.20	8.40	-	nr	8.40
327 mm brickwork	1.90	13.30	-	nr	13.30
100 mm blockwork	0.60	4.20	-	nr	4.20
150 mm blockwork	0.80	5.60	-	nr	5.60
215 mm blockwork	1.00	7.00	-	nr	7.00
Cutting or forming holes for ducts; girth 2.00 - 3.00 m					
102 mm brickwork	1.10	7.70	-	nr	7.70
215 mm brickwork	1.90	13.30	-	nr	13.30
327 mm brickwork	3.00	20.99	-	nr	20.99
100 mm blockwork	0.95	6.65	-	nr	6.65
150 mm blockwork	1.30	9.10	-	nr	9.10
215 mm blockwork	1.65	11.55	-	nr	11.55
Cutting or forming holes for ducts; girth 3.00 - 4.00 m					
102 mm brickwork	1.50	10.50	-	nr	10.50
215 mm brickwork	2.50	17.49	-	nr	17.49
327 mm brickwork	4.00	27.99	-	nr	27.99
100 mm blockwork	1.10	7.70	-	nr	7.70
150 mm blockwork	1.50	10.50	-	nr	10.50
215 mm blockwork	1.90	13.30	-	nr	13.30
Mortices in brickwork					
for expansion bolt	0.20	1.40	-	nr	1.40
for 20 mm dia bolt 75 mm deep	0.15	1.05	-	nr	1.05
for 20 mm dia bolt 150 mm deep	0.25	1.75	-	nr	1.75
Mortices in brickwork; grouting with cement mortar (1:1)					
75 x 75 x 200 mm deep	0.30	2.10	0.09	nr	2.19
75 x 75 x 300 mm deep	0.40	2.80	0.13	nr	2.93
Holes in softwood for pipes, bars cables and the like					
12 mm thick	0.04	0.36	-	nr	0.36
25 mm thick	0.06	0.54	-	nr	0.54
50 mm thick	0.10	0.91	-	nr	0.91
100 mm thick	0.15	1.36	-	nr	1.36
Holes in hardwood for pipes, bars, cables and the like					
12 mm thick	0.06	0.54	-	nr	0.54
25 mm thick	0.09	0.82	-	nr	0.82
50 mm thick	0.15	1.36	-	nr	1.36
100 mm thick	0.22	1.99	-	nr	1.99

P BUILDING FABRIC SUNDRIES Including overheads and profit at 9.00%	Labour hours	Labour £	Material £	Unit	Total rate £

Duct covers with frames; cast iron;
Brickhouse Glynwed 'Trucast' or similar;
bedding and pointing frame in cement mortar
(1:3); fixing with 10 mm dia anchor bolts to
concrete at 500 mm centres; including
cutting and pinning anchor bolts
Medium duty; ref 702

	Labour hours	Labour £	Material £	Unit	Total rate £
300 mm wide PC £117.00	2.50	25.81	127.53	m	153.34
Extra for					
ends PC £6.70	-	-	7.31	nr	7.31
right angle frame corner PC £35.17	-	-	38.33	nr	38.33
450 mm wide PC £125.10	2.80	28.90	136.36	m	165.26
Extra for					
ends PC £10.17	-	-	11.08	nr	11.08
right angle frame corner PC £35.17	-	-	38.33	nr	38.33
600 mm wide PC £197.10	3.10	32.00	214.84	m	246.84
Extra for					
ends PC £13.55	-	-	14.77	nr	14.77
right angle frame corner PC £35.17	-	-	38.33	nr	38.33
750 mm wide PC £239.40	3.40	35.10	260.95	m	296.05
Extra for					
ends PC £16.73	-	-	18.24	nr	18.24
right angle frame corner PC £35.17	-	-	38.33	nr	38.33
900 mm wide PC £270.00	3.70	38.19	294.30	m	332.49
Extra for					
ends PC £20.24	-	-	22.06	nr	22.06
right angle frame corner PC £35.17	-	-	38.33	nr	38.33
Heavy duty; ref 704					
300 mm wide PC £203.85	-	-	222.20	m	222.20
Extra for					
ends PC £8.74	-	-	9.53	nr	9.53
right angle frame corner PC £47.66	-	-	51.95	nr	51.95
450 mm wide PC £211.50	3.40	35.10	230.54	m	265.64
Extra for					
ends PC £13.06	-	-	14.24	nr	14.24
right angle frame corner PC £47.66	-	-	51.95	nr	51.95
600 mm wide PC £235.80	3.80	39.22	257.02	m	296.24
Extra for					
ends PC £17.17	-	-	18.71	nr	18.71
right angle frame corner PC £47.66	-	-	51.95	nr	51.95
750 mm wide PC £272.70	4.20	43.35	297.24	m	340.59
Extra for					
ends PC £21.62	-	-	23.57	nr	23.57
right angle frame corner PC £47.66	-	-	51.95	nr	51.95
900 mm wide PC £342.90	4.60	47.48	373.76	m	421.24
Extra for					
ends PC £25.97	-	-	28.30	nr	28.30
right angle frame corner PC £47.66	-	-	51.95	nr	51.95

Q PAVING/PLANTING/FENCING/SITE FURNITURE Including overheads and profit at 9.00%	Labour hours	Labour £	Material £	Unit	Total rate £
Q10 STONE/CONCRETE/BRICK KERBS/EDGINGS/ **CHANNELS**					
Mechanical excavation using a wheeled **hydraulic excavator with a 0.24 m3 bucket**					
Excavating trenches to receive kerb foundation; average size					
300 x 100 mm	0.02	0.14	0.31	m	0.45
450 x 150 mm	0.03	0.20	0.52	m	0.72
600 x 200 mm	0.04	0.27	0.72	m	0.99
Excavating curved trenches to receive kerb foundation; average size					
300 x 100 mm	0.02	0.14	0.41	m	0.55
450 x 150 mm	0.03	0.20	0.62	m	0.82
600 x 200 mm	0.04	0.27	0.83	m	1.10
Hand excavation					
Excavating trenches to receive kerb foundation; average size					
150 x 50 mm	0.03	0.20	-	m	0.20
200 x 75 mm	0.07	0.47	-	m	0.47
250 x 100 mm	0.11	0.74	-	m	0.74
300 x 100 mm	0.14	0.95	-	m	0.95
Excavating curved trenches to receive kerb foundation; average size					
150 x 50 mm	0.04	0.27	-	m	0.27
200 x 75 mm	0.08	0.54	-	m	0.54
250 x 100 mm	0.12	0.81	-	m	0.81
300 x 100 mm	0.15	1.01	-	m	1.01
Plain in situ ready mixed concrete; **7.50 N/mm2 - 40 mm aggregate (1:8);** **PC £42.20/m3; poured on or against** **earth or unblinded hardcore**					
Foundations	1.25	8.75	48.30	m3	57.05
Blinding bed					
not exceeding 150 mm thick	1.85	12.95	48.30	m3	61.25
Plain in situ ready mixed concrete; **11.50 N/mm2 - 40 mm aggregate (1:3:6);** **PC £43.03/m3; poured on or against** **earth or unblinded hardcore**					
Foundations	1.25	8.75	49.25	m3	58.00
Blinding bed					
not exceeding 150 mm thick	1.85	12.95	49.25	m3	62.20
Plain in situ ready mixed concrete; **11.50 N/mm2 - 40 mm aggregate (1:3:6);** **PC £43.03/m3; poured on or against** **earth or unblinded hardcore**					
Foundations	1.25	8.75	52.92	m3	61.67
Blinding bed					
not exceeding 150 mm thick	1.85	12.95	52.92	m3	65.87
Precast concrete kerbs, channels, edgings, **etc.; BS 340; bedded, jointed and pointed in** **cement mortar (1:3); including haunching up** **one side with in situ concrete mix** **11.50 N/mm2 - 40 mm aggregate (1:3:6);** **to concrete base**					
Edging; straight; fig 12					
51 x 152 mm	0.25	2.58	2.15	m	4.73
51 x 203 mm	0.25	2.58	2.43	m	5.01
51 x 254 mm	0.25	2.58	2.55	m	5.13

Q PAVING/PLANTING/FENCING/SITE FURNITURE Including overheads and profit at 9.00%	Labour hours	Labour £	Material £	Unit	Total rate £
Kerb; straight					
127 x 254 mm; fig 7	0.33	3.41	4.17	m	7.58
152 x 305 mm; fig 6	0.33	3.41	5.56	m	8.97
Kerb; curved					
127 x 254 mm; fig 7	0.50	5.16	5.31	m	10.47
152 x 305 mm; fig 6	0.50	5.16	6.71	m	11.87
Channel; 255 x 125 mm; fig 8					
straight	0.33	3.41	4.33	m	7.74
curved	0.50	5.16	5.51	m	10.67
Quadrant; fig 14					
305 x 305 x 152 mm	0.35	3.61	2.93	nr	6.54
305 x 305 x 254 mm	0.35	3.61	2.93	nr	6.54
457 x 457 x 152 mm	0.40	4.13	2.93	nr	7.06
457 x 457 x 254 mm	0.40	4.13	2.93	nr	7.06
Precast concrete drainage channels; Charcon 'Safeticurb' or similar; channels jointed with plastic rings and bedded; jointed and pointed in cement mortar (1:3); including haunching up one side with in situ concrete mix 11.50 N/mm2 - 40 mm aggregate (1:3:6); to concrete base					
Channel; straight; type DBA/3					
248 x 248 mm	0.60	6.19	15.97	m	22.16
End	0.20	2.06	-	nr	2.06
Inspection unit; with cast iron lid					
248 x 248 x 914 mm	0.65	6.71	45.95	nr	52.66
Silt box top; with concrete frame and cast iron lid; set over gully					
500 x 448 x 269 mm; type A	2.00	20.64	84.26	nr	104.90
Q20 HARDCORE/GRANULAR/CEMENT BOUND BASES/ SUB-BASES TO ROADS/PAVINGS					
Mechanical filling with hardcore; PC £6.75/m3 Filling to make up levels over 250 mm thick; depositing; compacting in layers with a 5 tonne roller	0.26	1.76	10.44	m3	12.20
Filling to make up levels not exceeding 250 mm thick; compacting	0.30	2.03	13.91	m3	15.94
Mechanical filling with granular fill; type 1; PC £9.05/t (PC £15.40/m3) Filling to make up levels over 250 mm thick; depositing; compacting in layers with a 5 tonne roller	0.26	1.76	22.63	m3	24.39
Filling to make up levels not exceeding 250 mm thick; compacting	0.30	2.03	29.00	m3	31.03
Mechanical filling with granular fill; type 2; PC £8.75/t (PC £15.00/m3) Filling to make up levels over 250 mm thick; depositing; compacting in layers with a 5 tonne roller	0.26	1.76	22.09	m3	23.85
Filling to make up levels not exceeding 250 mm thick; compacting	0.30	2.03	28.30	m3	30.33
Hand filling with hardcore; PC £6.75/m3 Filling to make up levels over 250 mm thick; depositing; compacting	0.55	3.72	14.59	m3	18.31
Filling to make up levels not exceeding 250 mm thick; compacting	0.66	4.46	18.25	m3	22.71

Q PAVING/PLANTING/FENCING/SITE FURNITURE Including overheads and profit at 9.00%	Labour hours	Labour £	Material £	Unit	Total rate £
Hand filling with coarse ashes Filling to make up levels over 250 mm thick; depositing; compacting in layers with a 2 tonne roller	0.50	3.38	17.95	m3	21.33
Filling to make up levels not exceeding 250 mm thick; compacting	0.60	4.05	22.56	m3	26.61
Hand filling with sand; PC £8.90/t (PC £14.25/m3) Filling to make up levels over 250 mm thick; depositing; compacting in layers with a 2 tonne roller	0.65	4.39	25.79	m3	30.18
Filling to make up levels not exceeding 250 mm thick; compacting	0.77	5.20	32.41	m3	37.61
Surface treatments Compacting					
surfaces of ashes	0.04	0.27	0.11	m2	0.38
filling; blinding with ashes	0.12	0.65	0.59	m2	1.24

Q21 IN SITU CONCRETE ROADS/PAVINGS/BASES

	Labour hours	Labour £	Material £	Unit	Total rate £
Reinforced in situ ready mixed concrete; **normal Portland cement; mix 11.5 N/mm2** **- 20 mm aggregate (1:2:4); PC £46.24/m3** Roads; to hardcore base					
150 - 450 mm thick	1.50	10.50	46.90	m3	57.40
not exceeding 150 mm thick	2.20	15.40	46.90	m3	62.30
Reinforced in situ ready mixed concrete; **normal Portland cement; mix 21.00 N/mm2** Roads; to hardcore base					
150 - 450 mm thick	1.50	10.50	50.40	m3	60.90
not exceeding 150 mm thick	2.20	15.40	50.40	m3	65.80
Reinforced in situ ready mixed concrete; **normal Portland cement; mix 26.00 N/mm2** **- 20 mm aggregate (1:1:5:3); PC £46.24/m3** Roads; to hardcore base					
150 - 450 mm thick	1.50	10.50	52.65	m3	63.15
not exceeding 150 mm thick	2.20	15.40	52.65	m3	68.05
Formwork for in situ concrete Sides of foundations					
not exceeding 250 mm wide	0.40	3.27	1.21	m	4.48
250 - 500 mm wide	0.60	4.91	2.10	m	7.01
500 mm - 1 m wide	0.90	7.36	3.98	m	11.34
Add to above for curved radius 6 m	0.04	0.33	0.15	m	0.48
Steel road forms to in situ concrete Sides of foundations					
150 mm wide	0.20	1.64	0.50	m	2.14
Reinforcement; fabric; BS 4483; lapped; in **roads, footpaths or pavings**					
Ref A142 (2.22 kg/m2) PC £0.77	0.12	0.97	1.00	m2	1.97
Ref A193 (3.02 kg/m2) PC £1.04	0.12	0.97	1.36	m2	2.33
Designed joints in in situ concrete Formed joint; 12.5 mm thick Expandite 'Flexcell' or similar					
not exceeding 150 mm wide	0.30	2.45	1.00	m	3.45
150 - 300 mm wide	0.40	3.27	1.75	m	5.02
300 - 450 mm wide	0.50	4.09	2.71	m	6.80
Formed joint; 25 mm thick Expandite 'Flexcell' or similar					
not exceeding 150 mm wide	0.30	2.45	1.56	m	4.01
150 - 300 mm wide	0.40	3.27	2.77	m	6.04
300 - 450 mm wide	0.50	4.09	4.22	m	8.31
Sealing top 25 mm of joint with rubberized bituminous compound	0.21	1.72	0.78	m	2.50

Q PAVING/PLANTING/FENCING/SITE FURNITURE Including overheads and profit at 5.00%	Labour hours	Labour £	Material £	Unit	Total rate £
Concrete sundries Treating surfaces of unset concrete; grading to cambers; tamping with a 75 mm thick steel shod tamper	0.25	1.75	-	m2	1.75

Q22 COATED MACADAM/ASPHALT ROADS/PAVINGS

In situ finishings
NOTE: The prices for all in situ finishings to roads and footpaths include for work to falls, crossfalls or slopes not exceeding 15 degrees from horizontal; for laying on prepared bases (priced elsewhere) and for rolling with an appropriate roller

Fine graded wearing course; BS 4987:88; clause 2.7.7, tables 34 - 36; 14 mm pre-coated igneous rock chippings; tack coat of bitumen emulsion 19 mm work to roads; one coat					
limestone aggregate	-	-	-	m2	3.22
igneous aggregate	-	-	-	m2	3.27
Close graded bitumen macadam; BS 4987:88; 10 mm graded aggregate to clause 2.7.4 tables 34 - 36; tack coat of bitumen emulsion 30 mm work to roads; one coat					
limestone aggregate	-	-	-	m2	3.35
igneous aggregate	-	-	-	m2	3.47
Bitumen macadam; BS 4987:88; 45 mm thick base course of 20 mm open graded aggregate to clause 2.6.1 tables 5 - 7; 20 mm thick wearing course of 6 mm medium graded aggregate to clause 2.7.6 tables 32 - 33 65 mm work to pavements/footpaths; two coats					
limestone aggregate	-	-	-	m2	6.10
igneous aggregate	-	-	-	m2	6.32
Add to last for 14 mm chippings; sprinkled into wearing course	-	-	-	m2	0.22
Bitumen macadam; BS 4987:88; 50 mm graded aggregate to clause 2.6.2 tables 8 - 10 75 mm work to roads; one coat					
limestone aggregate	-	-	-	m2	5.09
igneous aggregate	-	-	-	m2	6.44
Dense bitumen macadam; BS 4987:88; 50 mm thick base course of 20 mm graded aggregate to clause 2.6.5 tables 15 - 16; 200 pen. binder; 30 mm wearing course of 10 mm graded aggregate to clause 2.7.2 tables 20 - 22 75 mm work to roads; two coats					
limestone aggregate	-	-	-	m2	7.40
igneous aggregate	-	-	-	m2	7.72
Bitumen macadam; BS 4987:88; 50 mm thick base course of 20 mm graded aggregate to clause 2.6.1 tables 5 - 7; 25 mm thick wearing course of 10 mm graded aggregate to clause 2.7.2 tables 20 - 22 75 mm work to roads; two coats					
limestone aggregate	-	-	-	m2	6.22
igneous aggregate	-	-	-	m2	6.44

Q PAVING/PLANTING/FENCING/SITE FURNITURE Including overheads and profit at 9.00%	Labour hours	Labour £	Material £	Unit	Total rate £
Q23 GRAVEL/HOGGIN ROADS/PAVINGS					
Two coat gravel paving; level and to falls; **first layer course clinker aggregate and** **wearing layer fine gravel aggregate**					
50 mm work to paths	0.08	0.83	1.24	m2	**2.07**
63 mm work to paths	0.10	1.03	1.65	m2	**2.68**
Q24 INTERLOCKING BRICK/BLOCK ROADS/PAVINGS					
Concrete interlocking blocks; Marshall's **'Monolok' or similar; to falls or** **crossfalls; bedding 50 mm thick in dry sharp** **sand; filling joints with sharp sand;** **brushed in; vibrated; to earth base** Work to paved areas; over 300 mm wide; interlocking joints					
60 mm thick; grey PC £6.47	0.75	7.74	8.92	m2	**16.66**
80 mm thick; grey PC £7.40	0.80	8.26	10.13	m2	**18.39**
Extra for red stones; forming parking lines	0.05	0.52	0.20	m	**0.72**
Q25 SLAB/BRICK/BLOCK/SETT/COBBLE PAVINGS					
NOTE: Unless otherwise described, prices for pavings do not include for ash or sand beds or bases under.					
Artificial stone paving; Redland Aggregates' **'Texitone' or similar; to falls or** **crossfalls; bedding 25 mm thick in lime** **mortar (1:4) staggered joints; jointing in** **coloured cement mortar (1:3); brushed in; to** **sand base** Work to paved areas; over 300 mm wide					
450 x 600 x 50 mm; grey or coloured PC £2.13/each	0.45	4.65	10.82	m2	**15.47**
600 x 600 x 50 mm; grey or coloured PC £2.47/each	0.42	4.34	9.63	m2	**13.97**
750 x 600 x 50 mm; grey or coloured PC £2.98/each	0.39	4.03	9.40	m2	**13.43**
900 x 750 x 50 mm; grey or coloured PC £3.45/each	0.36	3.72	7.67	m2	**11.39**
Brick paviors; 215 x 103 x 65 mm rough stock **bricks; PC £400.00/1000; to falls or crossfalls;** **bedding 10 mm thick in cement mortar (1:3);** **jointing in cement mortar (1:3); as work** **proceeds; to concrete base** Work to paved areas; over 300 mm wide; straight joints both ways					
bricks laid flat	0.80	9.32	18.85	m2	**28.17**
bricks laid on edge	1.12	13.05	29.69	m2	**42.74**
Work to paved areas; over 300 mm wide; laid to herringbone pattern					
bricks laid flat	1.00	11.65	18.85	m2	**30.50**
bricks laid on edge	1.40	16.31	29.69	m2	**46.00**
Add or deduct for variation of 1.00/1000 in PC of brick paviors					
bricks laid flat	-	-	-	m2	**0.44**
bricks laid on edge	-	-	-	m2	**0.70**

Q PAVING/PLANTING/FENCING/SITE FURNITURE Including overheads and profit at 9.00%	Labour hours	Labour £	Material £	Unit	Total rate £
Cobble paving; 50 - 75 mm PC £40.50/t; to falls or crossfalls; bedding 13 mm thick in cement mortar (1:3); jointing to a height of two thirds of cobbles in dry mortar (1:3); tightly butted, washed and brushed; to concrete base					
Work to paved areas; over 300 mm wide					
regular	4.00	41.29	11.53	m2	**52.82**
laid to pattern	5.00	51.61	11.53	m2	**63.14**
Concrete paving flags; BS 368; to falls or crossfalls; bedding 25 mm thick in lime and sand mortar (1:4); butt joints straight both ways; jointing in cement mortar (1:3); brushed in; to sand base					
Work to paved areas; over 300 mm wide					
450 x 600 x 50 mm; grey PC £1.24/each	0.45	4.65	6.10	m2	**10.75**
450 x 600 x 50 mm; coloured PC £1.83/each	0.45	4.65	8.59	m2	**13.24**
600 x 600 x 50 mm; grey PC £1.45/each	0.42	4.34	5.44	m2	**9.78**
600 x 600 x 50 mm; coloured PC £2.17/each	0.42	4.34	7.72	m2	**12.06**
750 x 600 x 50 mm; grey PC £1.74/each	0.39	4.03	5.28	m2	**9.31**
750 x 600 x 50 mm; coloured PC £2.63/each	0.39	4.03	7.54	m2	**11.57**
900 x 600 x 50 mm; grey PC £2.01/each	0.36	3.72	4.27	m2	**7.99**
900 x 600 x 50 mm; coloured PC £3.10/each	0.36	3.72	6.11	m2	**9.83**
Concrete rectangular blocks; Marshall's 'Keyblok' or similar; to falls or crossfalls; bedding 50 mm thick in dry sharp sand; filling joints with sharp sand; brushed in; vibrated; to earth base					
Work to paved areas; over 300 mm wide; straight joints both ways					
200 x 100 x 65 mm; grey PC £5.95	0.60	6.19	7.98	m2	**14.17**
200 x 100 x 80 mm; grey PC £6.88	0.65	6.71	9.13	m2	**15.84**
Work to paved areas; over 300 mm wide; laid to herringbone pattern					
200 x 100 x 65 mm; coloured PC £6.90	0.75	7.74	9.07	m2	**16.81**
200 x 100 x 80 mm; coloured PC £8.06	0.80	8.26	10.48	m2	**18.74**
Extra for two row boundary edging to herringbone paved areas; 200 mm wide; including a 150 mm high in situ concrete mix 11.5 N/mm2 - 40 mm aggregate (1:3:6) haunching to one side; blocks laid breaking joint					
200 x 100 x 65 mm; coloured	0.25	2.58	1.53	m	**4.11**
200 x 100 x 80 mm; coloured	0.25	2.58	1.62	m	**4.20**
Granite setts; BS 435; 200 x 100 x 100 mm; PC £77.50/t; standard 'C' dressing; tightly butted to falls or crossfalls; bedding 25 mm thick in cement mortar (1:3); filling joints with dry mortar (1:6); washed and brushed; on concrete base					
Work to paved areas; over 300 mm wide					
straight joints	1.60	16.52	24.59	m2	**41.11**
laid to pattern	2.00	20.64	24.59	m2	**45.23**
Two rows of granite setts as boundary edging; 200 mm wide; including a 150 mm high in situ concrete mix 11.5 N/mm2 - 40 mm aggregate (1:3:6) haunching to one side; blocks laid breaking joint	0.70	7.23	6.23	m	**13.46**

Q26 SPECIAL SURFACINGS/PAVINGS FOR SPORT

Sundries					
Painted line on road; one coat					
75 mm wide	0.05	0.36	0.09	m	**0.45**

Q PAVING/PLANTING/FENCING/SITE FURNITURE Including overheads and profit at 9.00%	Labour hours	Labour £	Material £	Unit	Total rate £
Q30 SEEDING/TURFING					
Vegetable soil					
Selected from spoil heaps; grading;					
preparing for turfing or seeding; to general					
surfaces					
average 75 mm thick	0.33	2.23	-	m2	2.23
average 100 mm thick	0.35	2.37	-	m2	2.37
average 125 mm thick	0.38	2.57	-	m2	2.57
average 150 mm thick	0.40	2.70	-	m2	2.70
average 175 mm thick	0.42	2.84	-	m2	2.84
average 200 mm thick	0.44	2.97	-	m2	2.97
Selected from spoil heaps; grading;					
preparing for turfing or seeding; to cutting					
or embankments					
average 75 mm thick	0.37	2.50	-	m2	2.50
average 100 mm thick	0.40	2.70	-	m2	2.70
average 125 mm thick	0.43	2.91	-	m2	2.91
average 150 mm thick	0.45	3.04	-	m2	3.04
average 175 mm thick	0.47	3.18	-	m2	3.18
average 200 mm thick	0.50	3.38	-	m2	3.38
Imported vegetable soil; PC £9.00/m3					
Grading; preparing for turfing or seeding;					
to general surfaces					
average 75 mm thick	0.30	2.03	0.98	m2	3.01
average 100 mm thick	0.32	2.16	1.28	m2	3.44
average 125 mm thick	0.34	2.30	1.86	m2	4.16
average 150 mm thick	0.36	2.43	2.45	m2	4.88
average 175 mm thick	0.38	2.57	2.75	m2	5.32
average 200 mm thick	0.40	2.70	3.04	m2	5.74
Grading; preparing for turfing or seeding;					
to cuttings or embankments					
average 75 mm thick	0.33	2.23	0.98	m2	3.21
average 100 mm thick	0.36	2.43	1.28	m2	3.71
average 125 mm thick	0.38	2.57	1.86	m2	4.43
average 150 mm thick	0.40	2.70	2.45	m2	5.15
average 175 mm thick	0.42	2.84	2.75	m2	5.59
average 200 mm thick	0.44	2.97	3.04	m2	6.01
Fertilizer; PC £0.60/kg					
Fertilizer 0.07 kg/m2; raking in					
general surfaces	0.04	0.27	0.05	m2	0.32
Selected grass seed; PC £2.80/kg					
Grass seed; sowing at a rate of 0.042 kg/m2					
two applications; raking in					
general surfaces	0.07	0.47	0.26	m2	0.73
cuttings or embankments	0.08	0.54	0.26	m2	0.80
Preserved turf from stack on site					
Selected turf					
general surfaces	0.20	1.35	-	m2	1.35
cuttings or embankments; shallow	0.22	1.49	0.14	m2	1.63
cuttings or embankments; steep; pegged	0.30	2.03	0.22	m2	2.25
Imported turf; PC £0.90/m2					
Selected meadow turf					
general surfaces	0.20	1.35	0.98	m2	2.33
cuttings or embankments; shallow	0.22	1.49	1.12	m2	2.61
cuttings or embankments; steep; pegged	0.30	2.03	1.20	m2	3.23

Q PAVING/PLANTING/FENCING/SITE FURNITURE Including overheads and profit at 9.00% & 5.00%	Labour hours	Labour £	Material £	Unit	Total rate £
Q31 PLANTING					
Planting only					
Hedge or shrub plants					
not exceeding 750 mm high	0.25	1.69	-	nr	1.69
750 mm - 1.5 m high	0.60	4.05	-	nr	4.05
Saplings					
not exceeding 3 m high	1.70	11.49	-	nr	11.49

Q40 FENCING

NOTE: The prices for all fencing are to
include for setting posts in position, to a
depth of 0.6 m for fences not exceeding
1.4 m high and of 0.76 m for fences over
1.4 m high.
  The prices allow for excavating post
holes; filling to within 150 mm of ground
level with concrete and all necessary back
filling

	Labour hours	Labour £	Material £	Unit	Total rate £
Strained wire fences; BS 1722 Part 3; 4 mm galvanized mild steel plain wire threaded through posts and strained with eye bolts					
900 mm fencing; three line; concrete posts at 2750 mm centres	-	-	-	m	5.42
Extra for					
end concrete straining post; one strut	-	-	-	nr	26.95
angle concrete straining post; two struts	-	-	-	nr	33.40
1.07 m fencing; five line; concrete posts at 2750 mm centres	-	-	-	m	7.10
Extra for					
end concrete straining post; one strut	-	-	-	nr	47.08
angle concrete straining post; two struts	-	-	-	nr	62.59
1.20 m fencing; six line; concrete posts at 2750 mm centres	-	-	-	m	7.34
Extra for					
end concrete straining post; one strut	-	-	-	nr	47.33
angle concrete straining post; two struts	-	-	-	nr	64.43
1.4 m fencing; seven line; concrete posts at 2750 mm centres	-	-	-	m	7.63
Extra for					
end concrete straining post; one strut	-	-	-	nr	48.91
angle concrete straining post; two struts	-	-	-	nr	65.74
Chainlink fences; BS 1722 Part 1; 3 mm; 50 mm galvanized mild steel mesh; galvanized mild steel tying and line wire; three line wires threaded through posts and strained with eye bolts and winding brackets					
900 mm fencing; galvanized mild steel angle posts at 3 m centres	-	-	-	m	7.73
Extra for					
end steel straining post; one strut	-	-	-	nr	32.59
angle steel straining post; two struts	-	-	-	nr	42.97
900 mm fencing; concrete posts at 3 m centres	-	-	-	m	8.05
Extra for					
end concrete straining post; one strut	-	-	-	nr	32.34
angle concrete straining post; two struts	-	-	-	nr	42.60
1.2 m fencing; galvanized mild steel angle posts at 3 m centres	-	-	-	m	8.94
Extra for					
end steel straining post; one strut	-	-	-	nr	34.82
angle steel straining post; two struts	-	-	-	nr	46.02
1.2 m fencing; concrete posts at 3 m centres	-	-	-	m	9.41
Extra for					
end concrete straining post; one strut	-	-	-	nr	38.08
angle concrete straining post; two struts	-	-	-	nr	46.26

Q PAVING/PLANTING/FENCING/SITE FURNITURE Including overheads and profit at 5.00%	Labour hours	Labour £	Material £	Unit	Total rate £
Q40 FENCING - cont'd					
Chainlink fences; BS 1722 Part 1; 3 mm; **50 mm galvanized mild steel mesh; galvanized** **mild steel tying and line wire; three line** **wires threaded through posts and strained** **with eye bolts and winding brackets - cont'd**					
1.8 m fencing; galvanized mild steel angle posts at 3 m centres	-	-	-	m	12.52
Extra for					
end steel straining post; one strut	-	-	-	nr	52.17
angle steel straining post; two struts	-	-	-	nr	69.95
1.8 m fencing; concrete posts at 3 m centres	-	-	-	m	13.15
Extra for					
end concrete straining post; one strut	-	-	-	nr	54.17
angle concrete straining post; two struts	-	-	-	nr	70.47
Pair of gates and gate posts; gates to match galvanized chain link fencing, with angle framing, braces, etc., complete with hinges, locking bar, lock and bolts; two 100 x 100 mm angle section gate posts; each with one strut					
2.44 x 0.9 m high	-	-	-	nr	341.85
2.44 x 1.2 m high	-	-	-	nr	368.15
2.44 x 1.8 m high	-	-	-	nr	447.04
Chainlink fences; BS 1722 Part 1; 3 mm; **50 mm plastic coated mild steel mesh;** **plastic coated mild steel tying and line** **wire; three line wires threaded through** **posts and strained with eye bolts and** **winding brackets**					
900 mm fencing; galvanized mild steel angle posts at 3 m centres	-	-	-	m	8.52
Extra for					
end steel straining post; one strut	-	-	-	nr	32.59
angle steel straining post; two struts	-	-	-	nr	42.97
900 mm fencing; concrete posts at 3 m centres	-	-	-	m	8.80
Extra for					
end concrete straining post; one strut	-	-	-	nr	32.34
angle concrete straining post; two struts	-	-	-	nr	42.60
1.2 m fencing; galvanized mild steel angle posts at 3 m centres	-	-	-	m	10.14
Extra for					
end steel straining post; one strut	-	-	-	nr	34.90
angle steel straining post; two struts	-	-	-	nr	46.02
1.2 m fencing; concrete posts at 3 m centres	-	-	-	m	10.61
Extra for					
end concrete straining post; one strut	-	-	-	nr	37.03
angle concrete straining post; two struts	-	-	-	nr	46.26
1.8 m fencing; galvanized mild steel angle posts at 3 m centres	-	-	-	m	13.79
Extra for					
end steel straining post; one strut	-	-	-	nr	52.17
angle steel straining post; two struts	-	-	-	nr	69.95
1.8 m fencing; concrete posts at 3 m centres	-	-	-	m	14.42
Extra for					
end concrete straining post; one strut	-	-	-	nr	54.17
angle concrete straining post; two struts	-	-	-	nr	70.47

Q PAVING/PLANTING/FENCING/SITE FURNITURE Including overheads and profit at 5.00%	Labour hours	Labour £	Material £	Unit	Total rate £
Pair of gates and gate posts; gates to match plastic chain link fencing; with angle framing, braces, etc., complete with hinges, locking bar, lock and bolts; two 100 x 100 mm angle section gate posts; each with one strut					
2.44 x 0.9 m high	-	-	-	nr	362.89
2.44 x 1.2 m high	-	-	-	nr	387.08
2.44 x 1.8 m high	-	-	-	nr	470.18
Chain link fences for tennis courts; BS 1722 Part 13; 2.5 mm; 45 mm mesh galvanized mild steel mesh; line and tying wires threaded through 45 x 45 x 5 mm galvanized mild steel angle standards, posts and struts; 60 x 60 x 6 mm angle straining posts and gate posts; straining posts and struts strained with eye bolts and winding brackets Fencing to tennis court 36 x 18 m; including gate 1070 x 1980 mm complete with hinges, locking bar, lock and bolts					
2745 mm fencing; standards at 3 m centres	-	-	-	nr	1754.90
3660 mm fencing; standards at 2.5 m centres	-	-	-	nr	2381.65
Cleft chestnut pale fences; BS 1722 Part 4; pales spaced 51 mm apart; on two lines of galvanized wire; 64 mm dia posts; 76 x 51 mm struts					
900 mm fences; posts at 2.50 m centres	-	-	-	m	6.39
Extra for					
straining post; one strut	-	-	-	nr	11.66
corner straining post; two struts	-	-	-	nr	15.54
1.05 m fences; posts at 2.50 m centres	-	-	-	m	7.02
Extra for					
straining post; one strut	-	-	-	nr	12.85
corner straining post; two struts	-	-	-	nr	17.11
1.20 m fences; posts at 2.25 m centres	-	-	-	m	7.52
Extra for					
straining post; one strut	-	-	-	nr	14.10
corner straining post; two struts	-	-	-	nr	18.80
1.35 m fences; posts at 2.25 m centres	-	-	-	m	7.96
Extra for					
straining post; one strut	-	-	-	nr	15.79
corner straining post; two struts	-	-	-	nr	20.93
Close boarded fencing; BS 1722 Part 5; 76 x 38 mm softwood rails; 89 x 19 mm softwood pales lapped 13 mm; 152 x 25 mm softwood gravel boards; all softwood 'treated'; posts at 3 m centres					
Fences; two rail; concrete posts					
1 m	-	-	-	m	26.70
1.2 m	-	-	-	m	27.58
Fences; three rail; concrete posts					
1.4 m	-	-	-	m	35.85
1.6 m	-	-	-	m	37.23
1.8 m	-	-	-	m	38.67
Fences; two rail; oak posts					
1 m	-	-	-	m	19.30
1.2 m	-	-	-	m	22.06
Fences; three rail; oak posts					
1.4 m	-	-	-	m	24.82
1.6 m	-	-	-	m	28.20
1.8 m	-	-	-	m	31.46

Q PAVING/PLANTING/FENCING/SITE FURNITURE Including overheads and profit at 5.00%	Labour hours	Labour £	Material £	Unit	Total rate £
Q40 FENCING - cont'd					
Precast concrete slab fencing; 305 x 38 x **1753 mm slabs; fitted into twice grooved** **concrete posts at 1830 mm centres** Fences					
1.2 m	-	-	-	m	40.99
1.5 m	-	-	-	m	48.57
1.8 m	-	-	-	m	60.67
Mild steel unclimbable fencing; in rivetted **panels 2440 mm long; 44 x 13 mm flat section** **top and bottom rails; two 44 x 19 mm flat** **section standards, one with foot plate, and** **38 x 13 mm raking stay with foot plate;** **20 mm dia pointed verticals at 120 mm** **centres; two 44 x 19 mm supports 760 mm long** **with ragged ends to bottom rail; the whole** **bolted together; coated with red oxide** **primer; setting standards and stays in** **ground at 2440 mm centres and supports at** **815 mm centres** Fences					
1.67 m	-	-	-	m	58.38
2.13 m	-	-	-	m	65.95
Pair of gates and gate posts, to match mild steel unclimbable fencing; with flat section framing, braces, etc., complete with locking bar, lock, handles, drop bolt, gate stop and holding back catches; two 102 x 102 mm hollow section gate posts with cap and foot plates					
2.44 x 1.67 m	-	-	-	nr	489.11
2.44 x 2.13 m	-	-	-	nr	553.27
4.88 x 1.67 m	-	-	-	nr	972.96
4.88 x 2.13 m	-	-	-	nr	1104.44

R DISPOSAL SYSTEMS Including overheads and profit at 9.00%		Labour hours	Labour £	Material £	Unit	Total rate £
R10 RAINWATER PIPEWORK/GUTTERS						
Aluminium pipes and fittings; BS 2997;						
ears cast on; powder coated finish						
63 mm pipes; plugged and nailed		0.37	3.83	7.26	m	11.09
Extra for						
fittings with one end		0.22	2.28	3.61	nr	5.89
fittings with two ends		0.42	4.34	4.02	nr	8.36
fittings with three ends		0.60	6.21	5.67	nr	11.88
shoe	PC £3.84	0.22	2.28	3.61	nr	5.89
bend	PC £4.22	0.42	4.34	4.02	nr	8.36
single branch	PC £5.50	0.60	6.21	5.67	nr	11.88
offset 229 mm projection	PC £9.73	0.42	4.34	8.92	nr	13.26
offset 305 mm projection	PC £10.85	0.42	4.34	10.14	nr	14.48
access pipe	PC £12.02	-	-	10.62	nr	10.62
connection to clay pipes;						
cement and sand (1:2) joint		0.15	1.55	0.09	nr	1.64
75 mm pipes; plugged and nailed						
	PC £13.46/1.8m	0.40	4.14	8.71	m	12.85
Extra for						
shoe	PC £5.10	0.25	2.59	4.90	nr	7.49
bend	PC £5.33	0.45	4.65	5.16	nr	9.81
single branch	PC £6.62	0.65	6.72	6.99	nr	13.71
offset 229 mm projection	PC £10.75	0.45	4.65	9.75	nr	14.40
offset 305 mm projection	PC £11.90	0.45	4.65	11.00	nr	15.65
access pipe	PC £13.13	-	-	11.41	nr	11.41
connection to clay pipes;						
cement and sand (1:2) joint		0.17	1.76	0.09	nr	1.85
100 mm pipes; plugged and nailed						
	PC £22.35/1.8m	0.45	4.65	14.35	m	19.00
Extra for						
shoe	PC £6.36	0.28	2.90	5.75	nr	8.65
bend	PC £7.23	0.50	5.17	6.69	nr	11.86
single branch	PC £8.63	0.75	7.76	8.71	nr	16.47
offset 229 mm projection	PC £12.10	0.50	5.17	9.81	nr	14.98
offset 305 mm projection	PC £13.45	0.50	5.17	11.28	nr	16.45
access pipe	PC £15.14	-	-	11.53	nr	11.53
connection to clay pipes;						
cement and sand (1:2) joint		0.20	2.07	0.09	nr	2.16
Roof outlets; circular aluminium; with flat						
or domed grate; joint to pipe						
50 mm dia	PC £28.86	0.60	6.21	32.10	nr	38.31
75 mm dia	PC £38.18	0.65	6.72	42.41	nr	49.13
100 mm dia	PC £50.05	0.70	7.24	55.48	nr	62.72
150 mm dia	PC £64.46	0.75	7.76	71.59	nr	79.35
Roof outlets; d-shaped; balcony; with flat						
or domed grate; joint to pipe						
50 mm dia	PC £35.52	0.60	6.21	39.36	nr	45.57
75 mm dia	PC £40.84	0.65	6.72	45.31	nr	52.03
100 mm dia	PC £50.16	0.70	7.24	55.60	nr	62.84
Galvanized wire balloon grating; BS 416						
for pipes or outlets						
50 mm dia	PC £0.94	0.07	0.72	1.02	nr	1.74
63 mm dia	PC £0.94	0.07	0.72	1.02	nr	1.74
75 mm dia	PC £1.04	0.07	0.72	1.13	nr	1.85
100 mm dia	PC £1.15	0.08	0.83	1.25	nr	2.08
Aluminium gutters and fittings; BS 2997;						
powder coated finish						
100 mm half round gutters; on brackets						
screwed to timber	PC £10.10/1.8m	0.35	3.62	7.34	m	10.96
Extra for						
stop end	PC £1.60	0.16	1.66	2.35	nr	4.01
running outlet	PC £3.56	0.33	3.41	3.24	nr	6.65
stop end outlet	PC £3.16	0.16	1.66	3.16	nr	4.82
angle	PC £3.28	0.33	3.41	2.73	nr	6.14

R DISPOSAL SYSTEMS Including overheads and profit at 9.00%		Labour hours	Labour £	Material £	Unit	Total rate £
R10 RAINWATER PIPEWORK/GUTTERS - cont'd						
Aluminium gutters and fittings; BS 2997; powder coated finish - cont'd						
112 mm half round gutters; on brackets						
screwed to timber	PC £10.58/1.8m	0.35	3.62	7.65	m	11.27
Extra for						
stop end	PC £1.68	0.16	1.66	2.45	nr	4.11
running outlet	PC £3.87	0.33	3.41	3.56	nr	6.97
stop end outlet	PC £3.38	0.16	1.66	3.39	nr	5.05
angle	PC £3.70	0.33	3.41	3.14	nr	6.55
125 mm half round gutters; on brackets						
screwed to timber	PC £11.88/1.8m	0.40	4.14	9.25	m	13.39
Extra for						
stop end	PC £2.05	0.18	1.86	3.24	nr	5.10
running outlet	PC £4.19	0.35	3.62	3.82	nr	7.44
stop end outlet	PC £3.59	0.18	1.86	3.89	nr	5.75
angle	PC £4.12	0.35	3.62	4.09	nr	7.71
100 mm ogee gutters; on brackets						
screwed to timber	PC £12.59/1.8m	0.37	3.83	8.83	m	12.66
Extra for						
stop end	PC £1.69	0.17	1.76	2.44	nr	4.20
running outlet	PC £3.88	0.35	3.62	3.38	nr	7.00
stop end outlet	PC £3.04	0.17	1.76	3.04	nr	4.80
angle	PC £3.51	0.35	3.62	2.53	nr	6.15
112 mm ogee gutters; on						
brackets screwed to timber	PC £14.00/1.8m	0.42	4.34	9.74	m	14.08
Extra for						
stop end	PC £1.82	0.17	1.76	2.61	nr	4.37
running outlet	PC £4.24	0.35	3.62	3.66	nr	7.28
stop end outlet	PC £3.63	0.17	1.76	3.36	nr	5.12
angle	PC £4.19	0.35	3.62	3.05	nr	6.67
125 mm ogee gutters; on						
brackets screwed to timber	PC £15.46/1.8m	0.42	4.34	11.45	m	15.79
Extra for						
stop end	PC £1.97	0.19	1.97	3.16	nr	5.13
running outlet	PC £4.63	0.37	3.83	3.98	nr	7.81
stop end outlet	PC £4.12	0.19	1.97	4.15	nr	6.12
angle	PC £4.87	0.37	3.83	4.25	nr	8.08
Cast iron pipes and fittings; BS 460; ears cast on; joints						
50 mm pipes; primed; plugged						
and nailed	PC £15.37/1.8m	0.50	5.17	9.93	m	15.10
Extra for						
fittings with one end		0.30	3.10	7.09	nr	10.19
fittings with two ends		0.55	5.69	3.96	nr	9.65
shoe	PC £7.42	0.30	3.10	7.09	nr	10.19
bend	PC £4.56	0.55	5.69	3.96	nr	9.65
connection to clay pipes;						
cement and sand (1:2) joint		0.13	1.34	0.09	nr	1.43
63 mm pipes; primed; plugged						
and nailed	PC £15.37/1.8m	0.52	5.38	10.01	m	15.39
Extra for						
fittings with one end		0.32	3.31	7.15	nr	10.46
fittings with two ends		0.57	5.90	4.03	nr	9.93
fittings with three ends		0.72	7.45	6.38	nr	13.83
shoe	PC £7.42	0.32	3.31	7.15	nr	10.46
bend	PC £4.56	0.57	5.90	4.03	nr	9.93
single branch	PC £7.03	0.72	7.45	6.38	nr	13.83
offset 229 mm projection	PC £8.08	0.57	5.90	6.87	nr	12.77
offset 305 mm projection	PC £9.47	0.57	5.90	8.04	nr	13.94
connection to clay pipes;						
cement and sand (1:2) joint		0.15	1.55	0.09	nr	1.64

R DISPOSAL SYSTEMS Including overheads and profit at 9.00%		Labour hours	Labour £	Material £	Unit	Total rate £
75 mm pipes; primed, plugged						
and nailed	PC £15.37/1.8m	0.55	5.69	10.10	m	15.79
Extra for						
shoe	PC £7.42	0.35	3.62	7.27	nr	10.89
bend	PC £4.56	0.60	6.21	4.15	nr	10.36
single branch	PC £7.03	0.75	7.76	6.63	nr	14.39
offset 229 mm projection	PC £8.08	0.60	6.21	6.99	nr	13.20
offset 305 mm projection	PC £9.47	0.60	6.21	8.16	nr	14.37
connection to clay pipes;						
cement and sand (1:2) joint		0.17	1.76	0.09	nr	1.85
100 mm pipes; primed, plugged						
and nailed	PC £20.63/1.8m	0.60	6.21	13.60	m	19.81
Extra for						
shoe	PC £9.65	0.40	4.14	9.48	nr	13.62
bend	PC £7.07	0.65	6.72	6.67	nr	13.39
single branch	PC £9.18	0.80	8.28	8.76	nr	17.04
offset 229 mm projection	PC £12.80	0.65	6.72	11.57	nr	18.29
offset 305 mm projection	PC £15.02	0.65	6.72	13.53	nr	20.25
connection to clay pipes;						
cement and sand (1:2) joint		0.20	2.07	0.09	nr	2.16
100 x 75 mm rectangular pipes; primed,						
plugged and nailed	PC £74.01/1.8m	0.60	6.21	47.21	m	53.42
Extra for					nr	
shoe	PC £29.52	0.40	4.14	26.48	nr	30.62
bend	PC £23.97	0.65	6.72	20.43	nr	27.15
offset 229 mm projection	PC £33.34	0.65	6.72	25.80	nr	32.52
offset 305 mm projection	PC £39.44	0.65	6.72	30.84	nr	37.56
connection to clay pipes;						
cement and sand (1:2) joint		0.20	2.07	0.09	nr	2.16
Rainwater head; flat; for pipes						
50 mm dia	PC £5.78	0.55	5.69	6.64	nr	12.33
63 mm dia	PC £5.78	0.57	5.90	6.70	nr	12.60
75 mm dia	PC £5.78	0.60	6.21	6.82	nr	13.03
100 mm dia	PC £13.34	0.65	6.72	15.30	nr	22.02
Rainwater head; rectangular, for pipes						
50 mm dia	PC £12.73	0.55	5.69	14.22	nr	19.91
63 mm dia	PC £12.73	0.57	5.90	14.28	nr	20.18
75 mm dia	PC £12.73	0.60	6.21	14.40	nr	20.61
100 mm dia	PC £25.81	0.65	6.72	28.90	nr	35.62
Roof outlets; cast iron; circular; with						
flat grate; joint to pipe						
50 mm dia	PC £37.04	0.75	7.76	41.02	nr	48.78
75 mm dia	PC £38.84	0.85	8.79	43.11	nr	51.90
100 mm dia	PC £46.60	0.90	9.31	51.72	nr	61.03
Copper wire balloon grating;						
BS 416 for pipes or outlets						
50 mm dia	PC £1.21	0.07	0.72	1.32	nr	2.04
63 mm dia	PC £1.21	0.07	0.72	1.32	nr	2.04
75 mm dia	PC £1.40	0.07	0.72	1.53	nr	2.25
100 mm dia	PC £1.59	0.08	0.83	1.73	nr	2.56
Cast iron gutters and fittings; BS 460						
100 mm half round gutters; primed;						
on brackets; screwed to timber	PC £7.82/1.8m	0.40	4.14	5.98	m	10.12
Extra for						
stop end	PC £0.99	0.17	1.76	1.64	nr	3.40
running outlet	PC £3.03	0.35	3.62	2.78	nr	6.40
angle	PC £3.03	0.35	3.62	2.84	nr	6.46
115 mm half round gutters; primed;						
on brackets; screwed to timber	PC £8.14/1.8m	0.40	4.14	6.24	m	10.38
Extra for						
stop end	PC £1.44	0.17	1.76	2.16	nr	3.92
running outlet	PC £3.35	0.35	3.62	3.10	nr	6.72
angle	PC £3.35	0.35	3.62	3.17	nr	6.79

R DISPOSAL SYSTEMS Including overheads and profit at 9.00%		Labour hours	Labour £	Material £	Unit	Total rate £
R10 RAINWATER PIPEWORK/GUTTERS - cont'd						
Cast iron gutters and fittings; BS 460 - cont'd						
125 mm half round gutters; primed;						
on brackets; screwed to timber	PC £9.52/1.8m	0.45	4.65	7.11	m	11.76
Extra for						
stop end	PC £1.44	0.20	2.07	2.17	nr	4.24
running outlet	PC £3.95	0.40	4.14	3.65	nr	7.79
angle	PC £3.95	0.40	4.14	3.59	nr	7.73
150 mm half round gutters; primed;						
on brackets; screwed to timber	PC £16.29/1.8m	0.50	5.17	11.51	m	16.68
Extra for						
stop end	PC £1.90	0.22	2.28	2.75	nr	5.03
running outlet	PC £5.43	0.45	4.65	4.68	nr	9.33
angle	PC £5.43	0.45	4.65	4.07	nr	8.72
100 mm ogee gutters; primed;						
on brackets; screwed to timber	PC £8.53/1.8m	0.42	4.34	6.30	m	10.64
Extra for						
stop end	PC £0.92	0.18	1.86	1.94	nr	3.80
running outlet	PC £2.70	0.37	3.83	2.38	nr	6.21
angle	PC £2.70	0.37	3.83	2.31	nr	6.14
115 mm ogee gutters; primed;						
on brackets; screwed to timber	PC £9.59/1.8m	0.42	4.34	7.06	m	11.40
Extra for						
stop end	PC £1.32	0.18	1.86	2.41	nr	4.27
running outlet	PC £3.09	0.37	3.83	2.70	nr	6.53
angle	PC £3.09	0.37	3.83	2.55	nr	6.38
125 mm ogee gutters; primed;						
on brackets; screwed to timber	PC £10.05/1.8m	0.47	4.86	7.44	m	12.30
Extra for						
stop end	PC £1.32	0.21	2.17	2.50	nr	4.67
running outlet	PC £3.67	0.42	4.34	3.32	nr	7.66
angle	PC £5.06	0.42	4.34	4.72	nr	9.06
3 mm galvanised heavy pressed steel gutters **and fittings; joggle joints; BS 1091**						
200 x 100 mm (400 mm girth) box gutter;						
screwed to timber		0.65	6.72	13.56	m	20.28
Extra for						
stop end		0.35	3.62	10.65	nr	14.27
running outlet		0.70	7.24	27.45	nr	34.69
stop end outlet		0.35	3.62	35.41	nr	39.03
angle		0.70	7.24	20.53	nr	27.77
381 mm boundary wall gutters;						
screwed to timber		0.65	6.72	12.33	m	19.05
Extra for						
stop end		0.40	4.14	10.08	nr	14.22
running outlet		0.70	7.24	29.77	nr	37.01
stop end outlet		0.35	3.62	37.98	nr	41.60
angle		0.70	7.24	22.52	nr	29.76
457 mm boundary wall gutters;						
screwed to timber		0.75	7.76	13.92	m	21.68
Extra for						
stop end		0.35	3.62	12.21	nr	15.83
running outlet		0.80	8.28	30.63	nr	38.91
stop end outlet		0.40	4.14	39.63	nr	43.77
angle		0.80	8.28	24.73	nr	33.01
uPVC external rainwater pipes and fittings; **BS 4576; slip-in joints**						
50 mm pipes; fixing with pipe or socket						
brackets; plugged and screwed	PC £2.52/2m	0.30	3.10	1.90	m	5.00
Extra for						
fittings with one end		0.20	2.07	0.99	nr	3.06
fittings with two ends		0.30	3.10	1.45	nr	4.55
fittings with three ends		0.40	4.14	1.97	nr	6.11

R DISPOSAL SYSTEMS Including overheads and profit at 9.00%		Labour hours	Labour £	Material £	Unit	Total rate £
shoe	PC £0.74	0.20	2.07	0.99	nr	3.06
bend	PC £1.16	0.30	3.10	1.45	nr	4.55
single branch	PC £1.64	0.40	4.14	1.97	nr	6.11
two bends to form offset 229 mm						
projection	PC £1.72	0.30	3.10	1.60	nr	4.70
connection to clay pipes;						
cement and sand (1:2) joint		0.13	1.34	0.09	nr	1.43
68 mm pipes; fixing with pipe or socket						
brackets; plugged and screwed	PC £2.23/2m	0.33	3.41	1.91	m	5.32
Extra for						
shoe	PC £0.82	0.22	2.28	1.20	nr	3.48
bend	PC £1.43	0.33	3.41	1.87	nr	5.28
single branch	PC £2.50	0.44	4.55	3.04	nr	7.59
two bends to form offset 229 mm						
projection	PC £1.78	0.33	3.41	1.85	nr	5.26
loose drain connector;						
cement and sand (1:2) joint		0.15	1.55	0.09	nr	1.64
110 mm pipes; fixing with pipe or socket						
brackets; plugged and screwed	PC £7.22/3m	0.36	3.72	3.83	m	7.55
Extra for						
shoe	PC £3.05	0.24	2.48	3.65	nr	6.13
bend	PC £4.04	0.36	3.72	4.73	nr	8.45
single branch	PC £5.26	0.48	4.97	6.05	nr	11.02
two bends to form offset 229 mm						
projection	PC £8.08	0.36	3.72	8.03	nr	11.75
loose drain connector;						
cement and sand (1:2) joint		0.35	3.62	3.58	nr	7.20
68.5 mm square pipes; fixing with pipe						
brackets; plugged and screwed	PC £3.77/2.5m	0.33	3.41	2.35	m	5.76
Extra for						
shoe	PC £0.95	0.22	2.28	1.26	nr	3.54
bend	PC £0.97	0.33	3.41	1.28	nr	4.69
single branch	PC £2.47	0.44	4.55	2.91	nr	7.46
two bends to form offset 229 mm						
projection	PC £3.10	0.33	3.41	1.42	nr	4.83
drain connector; square to round;						
cement and sand (1:2) joint		0.20	2.07	2.55	nr	4.62
Rainwater head; rectangular; for pipes						
50 mm dia	PC £4.36	0.45	4.65	5.36	nr	10.01
68 mm dia	PC £3.59	0.47	4.86	4.77	nr	9.63
110 mm dia	PC £8.29	0.55	5.69	10.17	nr	15.86
68.5 mm square	PC £3.52	0.47	4.86	4.28	nr	9.14
uPVC gutters and fittings; BS 4576						
76 mm half round gutters; on						
brackets; screwed to timber	PC £2.02/2m	0.30	3.10	1.58	m	4.68
Extra for						
stop end	PC £0.34	0.13	1.34	0.46	nr	1.80
running outlet	PC £0.95	0.25	2.59	0.86	nr	3.45
stop end outlet	PC £0.95	0.13	1.34	0.96	nr	2.30
angle	PC £0.96	0.25	2.59	1.07	nr	3.66
112 half round gutters; on						
brackets screwed to timber	PC £2.17/2m	0.33	3.41	2.10	m	5.51
Extra for						
stop end	PC £0.59	0.13	1.34	0.83	nr	2.17
running outlet	PC £1.15	0.28	2.90	1.07	nr	3.97
stop end outlet	PC £1.15	0.13	1.34	1.25	nr	2.59
angle	PC £1.29	0.28	2.90	1.59	nr	4.49
152 mm half round gutters; on						
brackets screwed to timber	PC £6.45/2m	0.36	3.72	5.81	m	9.53
Extra for						
stop end	PC £1.44	0.16	1.66	2.12	nr	3.78
running outlet	PC £3.24	0.31	3.21	2.97	nr	6.18
stop end outlet	PC £3.08	0.16	1.66	3.35	nr	5.01
angle	PC £4.22	0.31	3.21	5.13	nr	8.34

R DISPOSAL SYSTEMS Including overheads and profit at 9.00%		Labour hours	Labour £	Material £	Unit	Total rate £
R10 RAINWATER PIPEWORK/GUTTERS - cont'd						
uPVC gutters and fittings; BS 4576 - cont'd 114 mm rectangular gutters; on						
brackets; screwed to timber	PC £2.65/2m	0.33	3.41	2.69	m	6.10
Extra for						
stop end	PC £0.57	0.13	1.34	0.84	nr	2.18
running outlet	PC £1.37	0.28	2.90	1.26	nr	4.16
stop end outlet	PC £1.45	0.13	1.34	1.57	nr	2.91
angle	PC £1.31	0.28	2.90	1.65	nr	4.55

ALTERNATIVE WASTE PIPE AND FITTING PRICES

	£			£			£
ABS waste system (£/each)							
4 m pipe -32 mm	2.04	access plug-32 mm	0.38	cn.to copper-32 mm			0.74
-40 mm	2.50	-40 mm	0.41	-40 mm			0.88
-50 mm	3.00	-50 mm	0.68	sweep bend -32 mm			0.49
pipe brkt-32 mm	0.10	reducer -32 mm	0.34	-40 mm			0.59
-40 mm	0.11	-40 mm	0.38	-50 mm			0.86
-50 mm	0.20	-50 mm	0.49	sweep tee -32 mm			0.70
dl.socket-32 mm	0.31	str.tank cn-32 mm	0.84	-40 mm			0.85
-40 mm	0.36	-40 mm	0.95	-50 mm			1.19
-50 mm	0.57						

		Labour hours	Labour £	Material £	Unit	Total rate £
R11 FOUL DRAINAGE ABOVE GROUND						
Cast iron pipes and fittings; BS 416 50 mm pipes						
primed, eared, plugged and nailed	PC £18.90/1.8m	0.75	7.76	13.24	m	21.00
primed, uneared; fixing with holderbats plugged and screwed; (PC £5.68/nr);	PC £17.76/1.8m	0.85	8.79	15.53	m	24.32
Extra for						
fittings with two ends		0.75	7.76	9.59	nr	17.35
fittings with three ends		1.00	10.34	14.37	nr	24.71
fittings with four ends		1.25	12.93	27.70	nr	40.63
bend; short radius	PC £8.64	0.75	7.76	9.59	nr	17.35
access bend; short radius	PC £21.41	0.75	7.76	23.51	nr	31.27
boss; 38 mm BSP	PC £20.85	0.75	7.76	18.64	nr	26.40
single branch	PC £12.92	1.00	10.34	14.37	nr	24.71
double branch	PC £25.27	1.25	12.93	27.70	nr	40.63
offset 229 mm projection	PC £13.69	0.75	7.76	15.10	nr	22.86
offset 305 mm projection	PC £15.66	0.75	7.76	17.24	nr	25.00
access pipe	PC £18.17	0.75	7.76	15.72	nr	23.48
roof connector; for asphalt	PC £19.10	0.75	7.76	18.67	nr	26.43
roof connector; for roofing felt	PC £37.30	0.80	8.28	40.44	nr	48.72
isolated caulked lead joints		0.65	6.72	1.51	nr	8.23
connection to clay pipes; cement and sand (1:2) joint		0.13	1.34	0.09	nr	1.43
75 mm pipes						
primed, eared, plugged and nailed	PC £18.90/1.8m	0.80	8.28	13.44	m	21.72
primed, uneared; fixing with holderbats plugged and screwed; (PC £5.68/nr);	PC £17.76/1.8m	0.90	9.31	15.69	m	25.00

R DISPOSAL SYSTEMS Including overheads and profit at 9.00%		Labour hours	Labour £	Material £	Unit	Total rate £
Extra for						
bend; short radius	PC £8.64	0.80	8.28	9.83	nr	18.11
access bend; short radius	PC £21.41	0.80	8.28	23.75	nr	32.03
boss; 38 mm BSP	PC £20.85	0.80	8.28	18.88	nr	27.16
single branch	PC £12.92	1.05	10.86	14.71	nr	25.57
double branch	PC £25.27	1.33	13.76	28.18	nr	41.94
offset 229 mm projection	PC £13.69	0.80	8.28	15.34	nr	23.62
offset 305 mm projection	PC £15.66	0.80	8.28	17.48	nr	25.76
access pipe	PC £18.17	0.80	8.28	15.96	nr	24.24
roof connector; for asphalt	PC £19.10	0.80	8.28	18.91	nr	27.19
roof connector; for roofing felt	PC £37.30	0.85	8.79	40.68	nr	49.47
isolated caulked lead joints		0.70	7.24	1.73	nr	8.97
connection to clay pipes;						
cement and sand (1:2) joint		0.17	1.76	0.09	nr	1.85
100 mm pipes						
primed, eared, plugged and						
nailed	PC £25.42/1.8m	0.90	9.31	18.11	m	27.42
primed; uneared; fixing with						
holderbats plugged and						
screwed; (PC £6.45/nr);	PC £24.35/1.8m	1.00	10.34	20.66	m	31.00
Extra for						
W.C. bent connector;						
450 mm long tail	PC £21.02	0.70	7.24	19.70	nr	26.94
'Multikwik' W.C.connector	PC £1.76	0.10	1.03	1.92	nr	2.95
W.C. straight connector;						
300 mm long tail	PC £9.30	0.70	7.24	10.64	nr	17.88
W.C. straight connector;						
450 mm long tail	PC £11.59	0.70	7.24	13.14	nr	20.38
bend; short radius	PC £12.85	0.90	9.31	14.51	nr	23.82
access bend; short radius	PC £26.38	0.90	9.31	29.26	nr	38.57
boss; 38 mm BSP	PC £24.92	0.90	9.31	21.83	nr	31.14
single branch	PC £19.93	1.20	12.41	22.48	nr	34.89
double branch	PC £28.22	1.50	15.52	31.51	nr	47.03
offset 229 mm projection	PC £18.11	0.90	9.31	20.24	nr	29.55
offset 305 mm projection	PC £20.71	0.90	9.31	23.08	nr	32.39
access pipe	PC £19.93	0.90	9.31	16.39	nr	25.70
roof connector; for asphalt	PC £22.88	0.90	9.31	22.26	nr	31.57
roof connector; for roofing felt	PC £42.66	0.95	9.83	46.48	nr	56.31
isolated caulked lead joints		0.80	8.28	2.35	nr	10.63
connection to clay pipes;						
cement and sand (1:2) joint		0.20	2.07	0.09	nr	2.16
150 mm pipes						
primed, eared, plugged and						
nailed	PC £51.52/1.8m	1.15	11.90	35.86	m	47.76
primed; uneared; fixing with						
holderbats plugged and						
screwed; (PC £11.09/nr);	PC £50.19/1.8m	1.25	12.93	40.70	m	53.63
Extra for						
bend; short radius	PC £23.04	1.15	11.90	25.24	nr	37.14
access bend; short radius	PC £33.12	1.15	11.90	36.23	nr	48.13
boss; 38 mm BSP	PC £42.33	1.15	11.90	34.24	nr	46.14
single branch	PC £35.67	1.55	16.03	39.08	nr	55.11
double branch	PC £51.33	1.90	19.65	55.67	nr	75.32
offset 229 mm projection	PC £37.26	1.15	11.90	40.74	nr	52.64
offset 305 mm projection	PC £42.03	1.15	11.90	45.95	nr	57.85
access pipe	PC £32.22	1.15	11.90	23.22	nr	35.12
roof connector; for asphalt	PC £48.78	1.15	11.90	46.74	nr	58.64
roof connector; for roofing felt	PC £71.56	1.20	12.41	77.04	nr	89.45
isolated caulked lead joints	PC £1.14/Kg	1.05	10.86	3.95	nr	14.81
connection to clay pipes;						
cement and sand (1:2) joint	PC £51.82/M3	0.26	2.69	0.12	nr	2.81

R DISPOSAL SYSTEMS Including overheads and profit at 9.00%		Labour hours	Labour £	Material £	Unit	Total rate £
R11 FOUL DRAINAGE ABOVE GROUND - cont'd						
Cast iron 'Timesaver' pipes **and fittings BS 416**						
50 mm pipes						
primed; 2 m lengths; fixing with holderbats						
plugged and screwed;(PC £4.11/nr)						
	PC £16.03/2m	0.55	5.69	13.32	m	**19.01**
Extra for						
fittings with two ends		0.55	5.69	9.13	nr	**14.82**
fittings with three ends		0.75	7.76	15.64	nr	**23.40**
fittings with four ends		0.95	9.83	25.48	nr	**35.31**
bend; short radius	PC £6.16	0.55	5.69	9.13	nr	**14.82**
access bend; short radius	PC £15.20	0.55	5.69	18.98	nr	**24.67**
boss; 38 mm BSP	PC £15.69	0.55	5.69	15.67	nr	**21.36**
single branch	PC £9.28	0.75	7.76	15.64	nr	**23.40**
double branch	PC £15.62	0.95	9.83	25.48	nr	**35.31**
offset 229 mm projection	PC £9.84	0.55	5.69	13.14	nr	**18.83**
offset 305 mm projection	PC £11.21	0.55	5.69	14.63	nr	**20.32**
access pipe	PC £14.83	0.55	5.69	14.73	nr	**20.42**
roof connector; for asphalt	PC £14.80	0.55	5.69	16.44	nr	**22.13**
roof connector; for roofing felt	PC £34.90	0.60	6.21	40.10	nr	**46.31**
isolated 'Timesaver'						
coupling joint	PC £3.49	0.30	3.10	3.81	nr	**6.91**
connection to clay pipes;						
cement and sand (1:2) joint		0.13	1.34	0.09	nr	**1.43**
75 mm pipes						
primed; 3 m lengths; fixing with holderbats						
plugged and screwed;(PC £4.11/nr)						
	PC £23.17/3m	0.55	5.69	11.97	m	**17.66**
primed; 2 m lengths; fixing with holderbats						
plugged and screwed;(PC £4.11/nr)						
	PC £16.03/2m	0.60	6.21	13.52	m	**19.73**
Extra for						
bend; short radius	PC £6.16	0.60	6.21	9.53	nr	**15.74**
access bend; short radius	PC £15.20	0.60	6.21	19.38	nr	**25.59**
boss; 38 mm BSP	PC £15.69	0.60	6.21	16.08	nr	**22.29**
single branch	PC £9.28	0.85	8.79	16.45	nr	**25.24**
double branch	PC £15.62	1.10	11.38	26.71	nr	**38.09**
offset 229 mm projection	PC £9.84	0.60	6.21	13.55	nr	**19.76**
offset 305 mm projection	PC £11.21	0.60	6.21	15.04	nr	**21.25**
access pipe	PC £14.83	0.60	6.21	15.14	nr	**21.35**
roof connector; for asphalt	PC £14.80	0.60	6.21	16.85	nr	**23.06**
roof connector; for roofing felt	PC £34.90	0.65	6.72	40.51	nr	**47.23**
isolated 'Timesaver'						
coupling joint	PC £3.87	0.35	3.62	4.22	nr	**7.84**
connection to clay pipes;						
cement and sand (1:2) joint		0.15	1.55	0.09	nr	**1.64**
100 mm pipes						
primed; 3 m lengths; fixing with holderbats						
plugged and screwed;(PC £4.49/nr)						
	PC £27.97/3m	0.60	6.21	14.42	m	**20.63**
primed; 2 m lengths; fixing with holderbats						
plugged and screwed; (PC £4.49/nr)						
	PC £19.32/2m	0.67	6.93	16.25	m	**23.18**
Extra for						
W.C. bent connector;						
450 mm long tail	PC £20.15	0.60	6.21	25.78	nr	**31.99**
'Multikwik' W.C.connector	PC £1.76	0.10	1.03	1.92	nr	**2.95**
W.C. straight connector;						
300 mm long tail	PC £8.93	0.60	6.21	13.55	nr	**19.76**
bend; short radius		0.67	6.93	13.12	nr	**20.05**
access bend; short radius	PC £18.06	0.67	6.93	23.51	nr	**30.44**
boss; 38 mm BSP	PC £19.68	0.67	6.93	20.63	nr	**27.56**
single branch	PC £13.21	1.00	10.34	22.88	nr	**33.22**
double branch	PC £16.34	1.30	13.45	30.74	nr	**44.19**

R DISPOSAL SYSTEMS Including overheads and profit at 9.00%		Labour hours	Labour £	Material £	Unit	Total rate £
offset 229 mm projection	PC £12.27	0.67	6.93	17.20	nr	24.13
offset 305 mm projection	PC £13.84	0.67	6.93	18.90	nr	25.83
access pipe	PC £15.62	0.67	6.93	16.21	nr	23.14
roof connector; for asphalt	PC £17.51	0.67	6.93	20.38	nr	27.31
roof connector; for roofing felt	PC £42.47	0.72	7.45	49.69	nr	57.14
isolated 'Timesaver' coupling joint	PC £5.05	0.42	4.34	5.50	nr	9.84
transitional clayware socket; cement and sand (1:2) joint		0.40	4.14	12.84	nr	16.98
150 mm pipes primed; 3 m lengths; fixing with holderbats plugged and screwed;(PC £7.72/nr)						
	PC £58.62/3m	0.75	7.76	29.41	m	37.17
primed; 2 m lengths; fixing with holderbats plugged and screwed;(PC £7.72/nr)						
	PC £39.66/2m	0.83	8.59	32.40	m	40.99
Extra for						
bend; short radius	PC £15.26	0.83	8.59	24.17	nr	32.76
access bend; short radius	PC £25.64	0.83	8.59	35.49	nr	44.08
boss; 38 mm BSP	PC £31.77	0.83	8.59	32.66	nr	41.25
single branch	PC £27.18	1.20	12.41	46.44	nr	58.85
double branch	PC £41.72	1.60	16.55	71.12	nr	87.67
offset 229 mm projection	PC £25.22	0.83	8.59	35.03	nr	43.62
offset 305 mm projection	PC £32.39	0.83	8.59	42.84	nr	51.43
roof connector; for asphalt	PC £32.44	0.83	8.59	37.72	nr	46.31
roof connector; for roofing felt	PC £65.03	0.88	9.10	77.55	nr	86.65
isolated 'Timesaver' coupling joint	PC £10.09	0.50	5.17	11.00	nr	16.17
transitional clayware socket; cement and sand (1:2) joint		0.52	5.38	14.61	nr	19.99
Polypropylene (PP) waste pipes and fittings; BS 5254; push fit 'O' - ring joints						
32 mm pipes; fixing with pipe clips; plugged and screwed	PC £1.20/4m	0.22	2.28	0.66	m	2.94
Extra for						
fittings with one end		0.16	1.66	0.42	nr	2.08
fittings with two ends		0.22	2.28	0.60	nr	2.88
fittings with three ends		0.30	3.10	0.87	nr	3.97
access plug	PC £0.38	0.16	1.66	0.42	nr	2.08
double socket	PC £0.41	0.15	1.55	0.44	nr	1.99
male iron to PP coupling	PC £0.71	0.28	2.90	0.78	nr	3.68
sweep bend	PC £0.55	0.22	2.28	0.60	nr	2.88
sweep tee	PC £0.80	0.30	3.10	0.87	nr	3.97
40 mm pipes; fixing with pipe clips; plugged and screwed	PC £1.51/4m	0.27	2.79	0.76	m	3.55
Extra for						
fittings with one end		0.18	1.86	0.45	nr	2.31
fittings with two ends		0.27	2.79	0.65	nr	3.44
fittings with three ends		0.36	3.72	1.01	nr	4.73
access plug	PC £0.41	0.18	1.86	0.45	nr	2.31
double socket	PC £0.44	0.18	1.86	0.48	nr	2.34
reducing set	PC £0.35	0.09	0.93	0.38	nr	1.31
male iron to PP coupling	PC £0.77	0.34	3.52	0.85	nr	4.37
sweep bend	PC £0.60	0.27	2.79	0.65	nr	3.44
sweep tee	PC £0.92	0.36	3.72	1.01	nr	4.73
50 mm pipes; fixing with pipe clips; plugged and screwed	PC £2.18/4m	0.32	3.31	1.16	m	4.47
Extra for						
fittings with one end		0.20	2.07	0.74	nr	2.81
fittings with two ends		0.32	3.31	1.12	nr	4.43
fittings with three ends		0.42	4.34	1.45	nr	5.79
access plug	PC £0.68	0.20	2.07	0.74	nr	2.81
double socket	PC £0.83	0.21	2.17	0.90	nr	3.07
reducing set	PC £0.55	0.10	1.03	0.60	nr	1.63
male iron to PP coupling	PC £0.94	0.40	4.14	1.02	nr	5.16
sweep bend	PC £1.03	0.32	3.31	1.12	nr	4.43
sweep tee	PC £1.33	0.42	4.34	1.45	nr	5.79

R DISPOSAL SYSTEMS Including overheads and profit at 9.00%		Labour hours	Labour £	Material £	Unit	Total rate £
R11 FOUL DRAINAGE ABOVE GROUND - cont'd						
muPVC waste pipes and fittings; BS 5255; solvent welded joints						
32 mm pipes; fixing with pipe clips; plugged and screwed	PC £4.06/4m	0.25	2.59	1.56	m	4.15
Extra for						
fittings with one end		0.17	1.76	0.96	nr	2.72
fittings with two ends		0.25	2.59	0.92	nr	3.51
fittings with three ends		0.33	3.41	1.32	nr	4.73
access plug	PC £0.80	0.17	1.76	0.96	nr	2.72
straight coupling	PC £0.53	0.17	1.76	0.66	nr	2.42
expansion coupling	PC £0.68	0.25	2.59	0.82	nr	3.41
male iron to PVC coupling	PC £0.66	0.31	3.21	0.76	nr	3.97
union coupling	PC £1.61	0.25	2.59	1.84	nr	4.43
sweep bend	PC £0.77	0.25	2.59	0.92	nr	3.51
spigot socket bend	PC £0.77	0.25	2.59	0.92	nr	3.51
sweep tee	PC £1.10	0.33	3.41	1.32	nr	4.73
caulking bush	PC £0.92	0.50	5.17	1.55	nr	6.72
40 mm pipes; fixing with pipe clips; plugged and screwed	PC £4.98/4m	0.30	3.10	1.88	m	4.98
Extra for						
fittings with one end		0.19	1.97	1.11	nr	3.08
fittings with two ends		0.30	3.10	1.03	nr	4.13
fittings with three ends		0.40	4.14	1.61	nr	5.75
fittings with four ends		0.53	5.48	3.90	nr	9.38
access plug	PC £0.95	0.19	1.97	1.11	nr	3.08
straight coupling	PC £0.65	0.20	2.07	0.79	nr	2.86
expansion coupling	PC £0.82	0.30	3.10	0.98	nr	4.08
male iron to PVC coupling	PC £0.77	0.38	3.93	0.89	nr	4.82
union coupling	PC £2.12	0.30	3.10	2.40	nr	5.50
level invert taper	PC £0.66	0.30	3.10	0.80	nr	3.90
sweep bend	PC £0.87	0.30	3.10	1.03	nr	4.13
spigot socket bend	PC £0.86	0.30	3.10	1.02	nr	4.12
sweep tee	PC £1.37	0.40	4.14	1.61	nr	5.75
sweep cross	PC £3.43	0.53	5.48	3.90	nr	9.38
50 mm pipes; fixing with pipe clips; plugged and screwed	PC £7.33/4m	0.35	3.62	2.84	m	6.46
Extra for						
fittings with one end		0.21	2.17	1.76	nr	3.93
fittings with two ends		0.35	3.62	1.45	nr	5.07
fittings with three ends		0.47	4.86	2.73	nr	7.59
fittings with four ends		0.62	6.41	4.47	nr	10.88
access plug	PC £1.54	0.21	2.17	1.76	nr	3.93
straight coupling	PC £0.79	0.23	2.38	0.94	nr	3.32
expansion coupling	PC £1.12	0.35	3.62	1.30	nr	4.92
male iron to PVC coupling	PC £1.12	0.45	4.65	1.26	nr	5.91
union coupling	PC £3.14	0.35	3.62	3.51	nr	7.13
level invert taper	PC £0.88	0.35	3.62	1.04	nr	4.66
sweep bend	PC £1.25	0.35	3.62	1.45	nr	5.07
spigot socket bend	PC £2.03	0.35	3.62	2.30	nr	5.92
sweep tee	PC £2.39	0.47	4.86	2.73	nr	7.59
sweep cross	PC £3.95	0.62	6.41	4.47	nr	10.88
uPVC overflow pipes and fittings; solvent						
19 mm pipes; fixing with pipe clips; plugged and screwed	PC £1.74/4m	0.22	2.28	0.71	m	2.99
Extra for						
splay cut end		0.02	0.21	-	nr	0.21
fittings with one end		0.17	1.76	0.33	nr	2.09
fittings with two ends		0.17	1.76	0.47	nr	2.23
fittings with three ends		0.22	2.28	0.55	nr	2.83
straight connector	PC £0.27	0.17	1.76	0.33	nr	2.09
female iron to PVC coupling	PC £0.51	0.20	2.07	0.58	nr	2.65
bend	PC £0.39	0.17	1.76	0.47	nr	2.23
tee	PC £0.45	0.22	2.28	0.55	nr	2.83
bent tank connector	PC £0.62	0.20	2.07	0.70	nr	2.77

R DISPOSAL SYSTEMS Including overheads and profit at 9.00%		Labour hours	Labour £	Material £	Unit	Total rate £
uPVC pipes and fittings; BS 4514; with **solvent welded joints (unless** **otherwise described)**						
82 mm pipes; fixing with holderbats; plugged						
and screwed (PC £0.97/nr)	PC £10.38/4m	0.40	4.14	4.05	m	8.19
Extra for						
socket plug	PC £1.78	0.20	2.07	2.17	nr	4.24
slip coupling (push-fit)	PC £3.44	0.37	3.83	3.99	nr	7.82
expansion coupling	PC £2.24	0.40	4.14	2.68	nr	6.82
sweep bend	PC £3.14	0.40	4.14	3.66	nr	7.80
boss connector	PC £1.78	0.27	2.79	2.02	nr	4.81
single branch	PC £4.37	0.53	5.48	5.12	nr	10.60
access door	PC £3.80	0.60	6.21	4.25	nr	10.46
connection to clay pipes; caulking ring						
and cement and sand (1:2) joint		0.37	3.83	0.93	nr	4.76
110 mm pipes; fixing with holderbats;						
plugged and screwed (PC £1.00/nr)						
	PC £12.45/4m	0.44	4.55	4.76	m	9.31
Extra for						
socket plug	PC £2.15	0.22	2.28	2.65	nr	4.93
slip coupling (push-fit)	PC £4.30	0.40	4.14	4.99	nr	9.13
expansion coupling	PC £2.26	0.44	4.55	2.76	nr	7.31
WC connector	PC £3.08	0.29	3.00	3.51	nr	6.51
sweep bend	PC £4.30	0.44	4.55	4.99	nr	9.54
WC connecting bend	PC £4.58	0.29	3.00	5.15	nr	8.15
access bend	PC £9.65	0.46	4.76	10.82	nr	15.58
boss connector	PC £1.64	0.29	3.00	2.08	nr	5.08
single branch	PC £5.69	0.58	6.00	6.66	nr	12.66
single branch with access door	PC £11.03	0.60	6.21	12.47	nr	18.68
double branch	PC £13.87	0.73	7.55	15.72	nr	23.27
WC manifold	PC £19.28	0.29	3.00	21.46	nr	24.46
access door	PC £3.80	0.60	6.21	4.25	nr	10.46
access pipe connector	PC £6.79	0.50	5.17	7.70	nr	12.87
connection to clay pipes; caulking ring						
and cement and sand (1:2) joint		0.42	4.34	1.21	nr	5.55
160 mm pipes; fixing with holderbats;						
plugged and screwed (PC £2.50/nr)						
	PC £26.52/4m	0.50	5.17	27.89	m	15.90
Extra for						
socket plug	PC £3.97	0.25	2.59	4.97	nr	7.56
slip coupling (push-fit)	PC £11.02	0.45	4.65	12.65	nr	17.30
expansion coupling	PC £6.82	0.50	5.17	8.07	nr	13.24
sweep bend	PC £9.27	0.50	5.17	10.75	nr	15.92
boss connector	PC £2.24	0.33	3.41	3.08	nr	6.49
single branch	PC £13.06	0.66	6.83	15.20	nr	22.03
double branch	PC £23.12	0.83	8.59	26.49	nr	35.08
access door		0.60	6.21	7.50	nr	13.71
connection to clay pipes; caulking ring						
and cement and sand (1:2) joint		0.50	5.17	2.02	nr	7.19
Weathering apron; for pipe						
82 mm dia	PC £0.88	0.34	3.52	1.08	nr	4.60
110 mm dia	PC £1.04	0.38	3.93	1.28	nr	5.21
160 mm dia	PC £3.12	0.42	4.34	3.72	nr	8.06
Weathering slate; for pipe						
110 mm dia	PC £14.51	0.90	9.31	15.97	nr	25.28
Vent cowl; for pipe						
82 mm dia	PC £0.88	0.33	3.41	1.08	nr	4.49
110 mm dia	PC £0.89	0.33	3.41	1.13	nr	4.54
160 mm dia	PC £2.32	0.33	3.41	2.85	nr	6.26
Roof outlets; circular PVC; with flat or						
domed grate; jointed to pipe						
82 mm dia	PC £10.77	0.55	5.69	11.86	nr	17.55
110 mm dia	PC £10.77	0.60	6.21	11.89	nr	18.10

R DISPOSAL SYSTEMS Including overheads and profit at 9.00%		Labour hours	Labour £	Material £	Unit	Total rate £
R11 FOUL DRAINAGE ABOVE GROUND - cont'd						
Copper, brass and gunmetal ancillaries; **screwed joints to fittings** Brass trap; 'P'; 45 degree outlet; 38 mm seal						
35 mm	PC £10.03	0.48	4.97	10.93	nr	15.90
Brass bath trap; 88.5 degree outlet; shallow seal						
42 mm	PC £7.52	0.54	5.59	8.19	nr	13.78
Copper trap 'P'; two piece						
35 mm with 38 mm seal	PC £5.12	0.48	4.97	5.59	nr	10.56
35 mm with 76 mm seal	PC £5.52	0.48	4.97	6.01	nr	10.98
42 mm with 38 mm seal	PC £7.60	0.54	5.59	8.29	nr	13.88
42 mm with 76 mm seal	PC £7.88	0.54	5.59	8.58	nr	14.17
Copper trap; 'S'; two piece						
35 mm with 38 mm seal	PC £5.52	0.48	4.97	6.01	nr	10.98
35 mm with 76 mm seal	PC £5.75	0.48	4.97	6.26	nr	11.23
42 mm with 38 mm seal	PC £7.91	0.54	5.59	8.62	nr	14.21
42 mm with 76 mm seal	PC £8.18	0.54	5.59	8.91	nr	14.50
Polypropylene ancillaries; screwed joint **to waste fitting** Tubular 'S' trap; bath; shallow seal						
40 mm	PC £5.99	0.55	5.69	6.53	nr	12.22
Trap 'P' two piece; 76 mm seal						
32 mm	PC £1.63	0.38	3.93	1.78	nr	5.71
40 mm	PC £1.88	0.45	4.65	2.05	nr	6.70
Trap 'S' two piece; 76 mm seal						
32 mm	PC £2.08	0.38	3.93	2.26	nr	6.19
40 mm	PC £2.44	0.45	4.65	2.66	nr	7.31
Bottle trap 'P'; 76 mm seal						
32 mm	PC £1.82	0.38	3.93	1.98	nr	5.91
40 mm	PC £2.18	0.45	4.65	2.37	nr	7.02
Bottle trap; 'S'; 76 mm seal						
32 mm	PC £2.20	0.38	3.93	2.39	nr	6.32
40 mm	PC £2.67	0.45	4.65	2.91	nr	7.56

R12 DRAINAGE BELOW GROUND

NOTE: Prices for drain trenches are for
excavation in 'firm' soil and it has been
assumed that earthwork support will only be
required for trenches 1 m or more in depth

Mechanical excavation of trenches to receive
pipes; grading bottoms; earthwork support;
filling with excavated material and
compacting; disposal of surplus soil;
spreading on site average 50 m
Pipes not exceeding 200 mm; average depth

0.50 m deep		0.20	1.35	1.24	m	2.59
0.75 m deep		0.30	2.03	2.07	m	4.10
1.00 m deep		0.60	4.05	3.65	m	7.70
1.25 m deep		0.90	6.08	4.95	m	11.03
1.50 m deep		1.15	7.77	6.48	m	14.25
1.75 m deep		1.40	9.46	8.31	m	17.77
2.00 m deep		1.65	11.15	9.54	m	20.69
2.25 m deep		2.03	13.72	11.75	m	25.47
2.50 m deep		2.40	16.22	13.57	m	29.79
2.75 m deep		2.65	17.91	15.19	m	33.10
3.00 m deep		2.90	19.60	16.81	m	36.41
3.25 m deep		3.13	21.15	17.82	m	38.97
3.50 m deep		3.35	22.64	18.82	m	41.46

R DISPOSAL SYSTEMS Including overheads and profit at 9.00%	Labour hours	Labour £	Material £	Unit	Total rate £
Pipes; 225 mm; average depth					
0.50 m deep	0.20	1.35	1.24	m	2.59
0.75 m deep	0.30	2.03	2.07	m	4.10
1.00 m deep	0.60	4.05	3.65	m	7.70
1.25 m deep	0.90	6.08	4.95	m	11.03
1.50 m deep	1.15	7.77	6.48	m	14.25
1.75 m deep	1.40	9.46	8.31	m	17.77
2.00 m deep	1.65	11.15	9.54	m	20.69
2.25 m deep	2.03	13.72	11.75	m	25.47
2.50 m deep	2.40	16.22	13.57	m	29.79
2.75 m deep	2.65	17.91	15.19	m	33.10
3.00 m deep	2.90	19.60	16.81	m	36.41
3.25 m deep	3.13	21.15	17.82	m	38.97
3.50 m deep	3.35	22.64	18.82	m	41.46
Pipes; 300 mm; average depth					
0.75 m deep	0.34	2.30	2.28	m	4.58
1.00 m deep	0.70	4.73	3.85	m	8.58
1.25 m deep	0.95	6.42	5.16	m	11.58
1.50 m deep	1.25	8.45	6.89	m	15.34
1.75 m deep	1.45	9.80	8.73	m	18.53
2.00 m deep	1.65	11.15	10.16	m	21.31
2.25 m deep	2.03	13.72	12.16	m	25.88
2.50 m deep	2.40	16.22	13.98	m	30.20
2.75 m deep	2.65	17.91	15.60	m	33.51
3.00 m deep	2.90	19.60	17.23	m	36.83
3.25 m deep	3.13	21.15	18.64	m	39.79
3.50 m deep	3.35	22.64	19.23	m	41.87
Pipes; 375 mm; average depth					
0.75 m deep	0.37	2.50	2.69	m	5.19
1.00 m deep	0.75	5.07	4.47	m	9.54
1.25 m deep	1.05	7.10	6.19	m	13.29
1.50 m deep	1.33	8.99	7.93	m	16.92
1.75 m deep	1.55	10.47	9.35	m	19.82
2.00 m deep	1.75	11.83	10.57	m	22.40
2.25 m deep	2.18	14.73	12.99	m	27.72
2.50 m deep	2.60	17.57	15.01	m	32.58
2.75 m deep	2.85	19.26	16.43	m	35.69
3.00 m deep	3.10	20.95	17.85	m	38.80
3.25 m deep	3.35	22.64	19.27	m	41.91
3.50 m deep	3.60	24.33	20.68	m	45.01
Pipes; 450 mm; average depth					
0.75 m deep	0.40	2.70	2.90	m	5.60
1.00 m deep	0.80	5.41	4.68	m	10.09
1.25 m deep	1.15	7.77	6.61	m	14.38
1.50 m deep	1.45	9.80	8.55	m	18.35
1.75 m deep	1.65	11.15	9.97	m	21.12
2.00 m deep	1.90	12.84	11.40	m	24.24
2.25 m deep	2.35	15.88	13.82	m	29.70
2.50 m deep	2.80	18.92	16.05	m	34.97
2.75 m deep	3.05	20.61	17.67	m	38.28
3.00 m deep	3.30	22.30	19.30	m	41.60
3.25 m deep	3.58	24.19	21.13	m	45.32
3.50 m deep	3.85	26.02	22.75	m	48.77
Pipes; 600 mm; average depth					
1.00 m deep	0.85	5.74	5.10	m	10.84
1.25 m deep	1.20	8.11	7.02	m	15.13
1.50 m deep	1.55	10.47	9.59	m	20.06
1.75 m deep	1.80	12.16	11.21	m	23.37
2.00 m deep	2.10	14.19	12.64	m	26.83
2.25 m deep	2.55	17.23	15.48	m	32.71
2.50 m deep	3.00	20.27	18.12	m	38.39
2.75 m deep	3.33	22.50	20.16	m	42.66
3.00 m deep	3.65	24.67	21.99	m	46.66
3.25 m deep	3.93	26.56	23.61	m	50.17
3.50 m deep	4.20	28.38	25.03	m	53.41

R DISPOSAL SYSTEMS	Labour	Labour	Material		Total
Including overheads and profit at 9.00%	hours	£	£	Unit	rate £

R12 DRAINAGE BELOW GROUND - cont'd

Mechanical excavation of trenches to receive pipes; grading bottoms; earthwork support; filling with excavated material and compacting; disposal of surplus soil; spreading on site average 50 m - cont'd

	Labour hours	Labour £	Material £	Unit	Total rate £
Pipes; 900 mm; average depth					
1.25 m deep	1.45	9.80	8.89	m	18.69
1.50 m deep	1.85	12.50	11.66	m	24.16
1.75 m deep	2.15	14.53	13.70	m	28.23
2.00 m deep	2.40	16.22	15.54	m	31.76
2.25 m deep	2.95	19.94	18.79	m	38.73
2.50 m deep	3.50	23.65	21.64	m	45.29
2.75 m deep	3.85	26.02	23.89	m	49.91
3.00 m deep	4.20	28.38	26.13	m	54.51
3.25 m deep	4.55	30.75	28.17	m	58.92
3.50 m deep	4.90	33.11	30.00	m	63.11
Pipes; 1200 mm; average depth					
1.50 m deep	2.10	14.19	13.73	m	27.92
1.75 m deep	2.45	16.56	15.97	m	32.53
2.00 m deep	2.75	18.58	18.24	m	36.82
2.25 m deep	3.34	22.57	21.90	m	44.47
2.50 m deep	4.00	27.03	25.37	m	52.40
2.75 m deep	4.40	29.74	27.82	m	57.56
3.00 m deep	4.80	32.44	30.27	m	62.71
3.25 m deep	5.20	35.14	32.73	m	67.87
3.50 m deep	5.60	37.84	35.18	m	73.02
Extra for breaking up					
brick	1.40	9.46	7.14	m3	16.60
concrete	1.95	13.18	9.79	m3	22.97
reinforced concrete	2.80	18.92	14.28	m3	33.20
concrete 150 mm thick	0.30	2.03	1.64	m2	3.67
tarmacadam 75 mm thick	0.15	1.01	0.86	m2	1.87
tarmacadam and hardcore 150 mm thick	0.20	1.35	1.19	m2	2.54

Hand excavation of trenches to receive pipes; grading bottoms; earthwork support; filling with excavated material and compacting; disposal of surplus soil; spreading on site average 50 m

	Labour hours	Labour £	Material £	Unit	Total rate £
Pipes not exceeding 200 mm; average depth					
0.50 m deep	1.00	6.76	-	m	6.76
0.75 m deep	1.50	10.14	-	m	10.14
1.00 m deep	2.20	14.87	0.95	m	15.82
1.25 m deep	3.10	20.95	1.43	m	22.38
1.50 m deep	4.25	28.72	1.72	m	30.44
1.75 m deep	5.60	37.84	2.10	m	39.94
2.00 m deep	6.40	43.25	2.29	m	45.54
2.25 m deep	8.00	54.06	3.05	m	57.11
2.50 m deep	9.60	64.88	3.62	m	68.50
2.75 m deep	10.55	71.30	4.01	m	75.31
3.00 m deep	11.50	77.72	4.39	m	82.11
3.25 m deep	12.45	84.14	4.77	m	88.91
3.50 m deep	13.40	90.56	5.15	m	95.71
Pipes; 225 mm; average depth					
0.50 m deep	1.00	6.76	-	m	6.76
0.75 m deep	1.50	10.14	-	m	10.14
1.00 m deep	2.20	14.87	0.95	m	15.82
1.25 m deep	3.10	20.95	1.43	m	22.38
1.50 m deep	4.25	28.72	1.72	m	30.44
1.75 m deep	5.60	37.84	2.10	m	39.94
2.00 m deep	6.40	43.25	2.29	m	45.54
2.25 m deep	8.00	54.06	3.05	m	57.11
2.50 m deep	9.60	64.88	3.62	m	68.50
2.75 m deep	10.55	71.30	4.01	m	75.31
3.00 m deep	11.50	77.72	4.39	m	82.11
3.25 m deep	12.45	84.14	4.77	m	88.91
3.50 m deep	13.40	90.56	5.15	m	95.71

R DISPOSAL SYSTEMS Including overheads and profit at 9.00%	Labour hours	Labour £	Material £	Unit	Total rate £
Pipes; 300 mm; average depth					
0.75 m deep	1.75	11.83	-	m	11.83
1.00 m deep	2.55	17.23	0.95	m	18.18
1.25 m deep	3.60	24.33	1.43	m	25.76
1.50 m deep	4.80	32.44	1.72	m	34.16
1.75 m deep	5.60	37.84	2.10	m	39.94
2.00 m deep	6.40	43.25	2.29	m	45.54
2.25 m deep	8.00	54.06	3.05	m	57.11
2.50 m deep	9.60	64.88	3.62	m	68.50
2.75 m deep	10.55	71.30	4.01	m	75.31
3.00 m deep	11.50	77.72	4.39	m	82.11
3.25 m deep	12.45	84.14	4.77	m	88.91
3.50 m deep	13.40	90.56	5.15	m	95.71
Pipes; 375 mm; average depth					
0.75 m deep	1.95	13.18	-	m	13.18
1.00 m deep	2.85	19.26	0.95	m	20.21
1.25 m deep	4.00	27.03	1.43	m	28.46
1.50 m deep	5.33	36.02	1.72	m	37.74
1.75 m deep	6.20	41.90	2.10	m	44.00
2.00 m deep	7.10	47.98	2.29	m	50.27
2.25 m deep	8.90	60.15	3.05	m	63.20
2.50 m deep	10.70	72.31	3.62	m	75.93
2.75 m deep	11.75	79.41	4.01	m	83.42
3.00 m deep	12.80	86.50	4.39	m	90.89
3.25 m deep	13.90	93.94	4.77	m	98.71
3.50 m deep	15.00	101.37	5.15	m	106.52
Pipes; 450 mm; average depth					
0.75 m deep	2.20	14.87	-	m	14.87
1.00 m deep	3.18	21.49	0.95	m	22.44
1.25 m deep	4.47	30.21	1.43	m	31.64
1.50 m deep	5.85	39.53	1.72	m	41.25
1.75 m deep	6.82	46.09	2.10	m	48.19
2.00 m deep	7.80	52.71	2.29	m	55.00
2.25 m deep	9.78	66.09	3.05	m	69.14
2.50 m deep	11.75	79.41	3.62	m	83.03
2.75 m deep	12.93	87.38	4.01	m	91.39
3.00 m deep	14.10	95.29	4.39	m	99.68
3.25 m deep	15.25	103.06	4.77	m	107.83
3.50 m deep	16.40	110.83	5.15	m	115.98
Pipes; 600 mm; average depth					
1.00 m deep	3.50	23.65	0.95	m	24.60
1.25 m deep	5.00	33.79	1.43	m	35.22
1.50 m deep	6.70	45.28	1.72	m	47.00
1.75 m deep	7.75	52.37	2.10	m	54.47
2.00 m deep	8.85	59.81	2.29	m	62.10
2.25 m deep	9.95	67.24	3.05	m	70.29
2.50 m deep	11.10	75.01	3.62	m	78.63
2.75 m deep	13.35	90.22	4.01	m	94.23
3.00 m deep	16.00	108.13	4.39	m	112.52
3.25 m deep	17.33	117.12	4.77	m	121.89
3.50 m deep	18.65	126.04	5.15	m	131.19
Pipes; 900 mm; average depth					
1.25 m deep	6.25	42.24	1.43	m	43.67
1.50 m deep	8.25	55.75	1.72	m	57.47
1.75 m deep	9.60	64.88	2.10	m	66.98
2.00 m deep	10.95	74.00	2.29	m	76.29
2.25 m deep	13.75	92.92	3.05	m	95.97
2.50 m deep	16.55	111.84	3.62	m	115.46
2.75 m deep	18.20	123.00	4.01	m	127.01
3.00 m deep	19.80	133.81	4.39	m	138.20
3.25 m deep	21.45	144.96	4.77	m	149.73
3.50 m deep	23.10	156.11	5.15	m	161.26

R DISPOSAL SYSTEMS Including overheads and profit at 9.00%	Labour hours	Labour £	Material £	Unit	Total rate £
R12 DRAINAGE BELOW GROUND - cont'd					
Hand excavation of trenches to receive **pipes; grading bottoms; earthwork support;** **filling with excavated material and** **compacting; disposal of surplus soil;** **spreading on site average 50 m - cont'd**					
Pipes; 1200 mm; average depth					
1.50 m deep	9.85	66.57	1.72	m	68.29
1.75 m deep	11.45	77.38	2.10	m	79.48
2.00 m deep	13.10	88.53	2.29	m	90.82
2.25 m deep	16.43	111.03	3.05	m	114.08
2.50 m deep	19.75	133.47	3.62	m	137.09
2.75 m deep	21.70	146.65	4.01	m	150.66
3.00 m deep	23.65	159.83	4.39	m	164.22
3.25 m deep	25.58	172.87	4.77	m	177.64
3.50 m deep	27.50	185.85	5.15	m	191.00
Extra for breaking up					
brick	3.00	20.27	3.60	m3	23.87
concrete	4.50	30.41	6.00	m3	36.41
reinforced concrete	6.00	40.55	8.39	m3	48.94
concrete 150 mm thick	0.70	4.73	0.84	m2	5.57
tarmacadam 75 mm thick	0.40	2.70	0.48	m2	3.18
tarmacadam and hardcore 150 mm thick	0.50	3.38	0.60	m2	3.98
Extra for taking up precast concrete paving slabs	0.30	2.03	-	m2	2.03
Sand filling; PC £8.90/t; (PC £14.25/m3)					
Beds; to receive pitch fibre pipes					
600 x 50 mm	0.08	0.54	0.58	m	1.12
700 x 50 mm	0.10	0.68	0.68	m	1.36
800 x 50 mm	0.12	0.81	0.78	m	1.59
Granular (shingle) filling; PC £9.50/t; (PC £17.00/m3)					
Beds; 100 mm thick; to pipes size					
100 mm	0.10	0.68	1.11	m	1.79
150 mm	0.10	0.68	1.30	m	1.98
225 mm	0.12	0.81	1.48	m	2.29
300 mm	0.14	0.95	1.67	m	2.62
375 mm	0.16	1.08	1.85	m	2.93
450 mm	0.18	1.22	2.04	m	3.26
600 mm	0.20	1.35	2.22	m	3.57
Beds; 150 mm thick; to pipes size					
100 mm	0.14	0.95	1.67	m	2.62
150 mm	0.16	1.08	1.85	m	2.93
225 mm	0.18	1.22	2.04	m	3.26
300 mm	0.20	1.35	2.22	m	3.57
375 mm	0.24	1.62	2.78	m	4.40
450 mm	0.26	1.76	2.96	m	4.72
600 mm	0.30	2.03	3.52	m	5.55
Beds and benchings; beds 100 mm thick; to pipes size					
100 mm	0.23	1.55	2.04	m	3.59
150 mm	0.25	1.69	2.04	m	3.73
225 mm	0.30	2.03	2.78	m	4.81
300 mm	0.35	2.37	3.15	m	5.52
375 mm	0.45	3.04	4.26	m	7.30
450 mm	0.52	3.51	4.82	m	8.33
600 mm	0.67	4.53	6.30	m	10.83
Beds and benchings; beds 150 mm thick; to pipes size					
100 mm	0.25	1.69	2.22	m	3.91
150 mm	0.28	1.89	2.41	m	4.30
225 mm	0.35	2.37	3.34	m	5.71
300 mm	0.45	3.04	4.08	m	7.12
375 mm	0.52	3.51	4.82	m	8.33
450 mm	0.62	4.19	5.74	m	9.93
600 mm	0.74	5.00	7.41	m	12.41

R DISPOSAL SYSTEMS Including overheads and profit at 9.00%	Labour hours	Labour £	Material £	Unit	Total rate £
Beds and coverings; 100 mm thick; to pipes size					
100 mm	0.36	2.43	2.78	m	5.21
150 mm	0.45	3.04	3.34	m	6.38
225 mm	0.60	4.05	4.63	m	8.68
300 mm	0.72	4.87	5.56	m	10.43
375 mm	0.87	5.88	6.67	m	12.55
450 mm	1.02	6.89	7.97	m	14.86
600 mm	1.32	8.92	10.19	m	19.11
Beds and coverings; 150 mm thick; to pipes size					
100 mm	0.54	3.65	4.08	m	7.73
150 mm	0.60	4.05	4.63	m	8.68
225 mm	0.78	5.27	5.93	m	11.20
300 mm	0.93	6.28	7.04	m	13.32
375 mm	1.08	7.30	8.34	m	15.64
450 mm	1.29	8.72	10.01	m	18.73
600 mm	1.56	10.54	12.04	m	22.58
In situ ready mixed concrete; normal Portland cement; mix 11.50 N/mm2 - 40 mm aggregate (1:3:6); PC £43.03/m3;					
Beds; 100 mm thick; to pipes size					
100 mm	0.20	1.40	2.35	m	3.75
150 mm	0.20	1.40	2.35	m	3.75
225 mm	0.24	1.68	2.81	m	4.49
300 mm	0.28	1.96	3.28	m	5.24
375 mm	0.32	2.24	3.75	m	5.99
450 mm	0.36	2.52	4.22	m	6.74
600 mm	0.40	2.80	4.69	m	7.49
900 mm	0.48	3.36	5.63	m	8.99
1200 mm	0.64	4.48	7.50	m	11.98
Beds; 150 mm thick; to pipes size					
100 mm	0.28	1.96	3.28	m	5.24
150 mm	0.32	2.24	3.75	m	5.99
225 mm	0.36	2.52	4.22	m	6.74
300 mm	0.40	2.80	4.69	m	7.49
375 mm	0.48	3.36	5.63	m	8.99
450 mm	0.52	3.64	6.10	m	9.74
600 mm	0.60	4.20	7.04	m	11.24
900 mm	0.76	5.32	8.91	m	14.23
1200 mm	0.92	6.44	10.79	m	17.23
Beds and benchings; beds 100 mm thick; to pipes size					
100 mm	0.40	2.80	4.22	m	7.02
150 mm	0.45	3.15	4.69	m	7.84
225 mm	0.54	3.78	5.63	m	9.41
300 mm	0.63	4.41	6.57	m	10.98
375 mm	0.81	5.67	8.44	m	14.11
450 mm	0.95	6.65	9.85	m	16.50
600 mm	1.22	8.54	12.66	m	21.20
900 mm	1.98	13.86	20.64	m	34.50
1200 mm	2.93	20.50	30.49	m	50.99
Beds and benchings; beds 150 mm thick; to pipes size					
100 mm	0.45	3.15	4.69	m	7.84
150 mm	0.50	3.50	5.16	m	8.66
225 mm	0.63	4.41	6.57	m	10.98
300 mm	0.81	5.67	8.44	m	14.11
375 mm	0.95	6.65	9.85	m	16.50
450 mm	1.13	7.91	11.73	m	19.64
600 mm	1.44	10.08	15.01	m	25.09
900 mm	2.30	16.09	23.92	m	40.01
1200 mm	3.24	22.67	33.77	m	56.44

R DISPOSAL SYSTEMS Including overheads and profit at 9.00%	Labour hours	Labour £	Material £	Unit	Total rate £

R12 DRAINAGE BELOW GROUND - cont'd

In situ ready mixed concrete; normal
Portland cement; mix 11.50 N/mm2 - 40 mm
aggregate (1:3:6); PC £43.03/m3; - cont'd
Beds and coverings; 100 mm thick; to pipes
size

100 mm	0.60	4.20	5.63	m	9.83
150 mm	0.70	4.90	6.57	m	11.47
225 mm	1.00	7.00	9.38	m	16.38
300 mm	1.20	8.40	11.26	m	19.66
375 mm	1.45	10.15	13.60	m	23.75
450 mm	1.70	11.90	15.95	m	27.85
600 mm	2.20	15.40	20.64	m	36.04
900 mm	3.35	23.44	31.42	m	54.86
1200 mm	4.60	32.19	43.15	m	75.34

Beds and coverings; 150 mm thick; to pipes
size

100 mm	0.90	6.30	8.44	m	14.74
150 mm	1.00	7.00	9.38	m	16.38
225 mm	1.30	9.10	12.19	m	21.29
300 mm	1.55	10.85	14.54	m	25.39
375 mm	1.80	12.60	16.88	m	29.48
450 mm	2.15	15.05	20.17	m	35.22
600 mm	2.60	18.19	24.39	m	42.58
900 mm	4.25	29.74	39.87	m	69.61
1200 mm	6.00	41.99	56.28	m	98.27

In situ ready mixed concrete; normal
Portland cement; mix 21.00 N/mm2 - 40 mm
aggregate (1:2:4); PC £46.24/m3
Beds; 100 mm thick; to pipes size

100 mm	0.20	1.40	2.52	m	3.92
150 mm	0.20	1.40	2.52	m	3.92
225 mm	0.24	1.68	3.02	m	4.70
300 mm	0.28	1.96	3.53	m	5.49
375 mm	0.32	2.24	4.03	m	6.27
450 mm	0.36	2.52	4.54	m	7.06
600 mm	0.40	2.80	5.04	m	7.84
900 mm	0.48	3.36	6.05	m	9.41
1200 mm	0.64	4.48	8.06	m	12.54

Beds; 150 mm thick; to pipes size

100 mm	0.28	1.96	3.53	m	5.49
150 mm	0.32	2.24	4.03	m	6.27
225 mm	0.36	2.52	4.54	m	7.06
300 mm	0.40	2.80	5.04	m	7.84
375 mm	0.48	3.36	6.05	m	9.41
450 mm	0.52	3.64	6.55	m	10.19
600 mm	0.60	4.20	7.56	m	11.76
900 mm	0.76	5.32	9.58	m	14.90
1200 mm	0.92	6.44	11.59	m	18.03

Beds and benchings; beds 100 mm
thick; to pipes size

100 mm	0.40	2.80	4.54	m	7.34
150 mm	0.45	3.15	5.04	m	8.19
225 mm	0.54	3.78	6.05	m	9.83
300 mm	0.63	4.41	7.06	m	11.47
375 mm	0.81	5.67	9.07	m	14.74
450 mm	0.95	6.65	10.58	m	17.23
600 mm	1.22	8.54	13.61	m	22.15
900 mm	1.98	13.86	22.18	m	36.04
1200 mm	2.93	20.50	32.76	m	53.26

R DISPOSAL SYSTEMS Including overheads and profit at 9.00%	Labour hours	Labour £	Material £	Unit	Total rate £
Beds and benchings; beds 150 mm thick; to pipes size					
100 mm	0.45	3.15	5.04	m	8.19
150 mm	0.50	3.50	5.54	m	9.04
225 mm	0.63	4.41	7.06	m	11.47
300 mm	0.81	5.67	9.07	m	14.74
375 mm	0.95	6.65	10.58	m	17.23
450 mm	1.13	7.91	12.60	m	20.51
600 mm	1.44	10.08	16.13	m	26.21
900 mm	2.30	16.09	25.70	m	41.79
1200 mm	3.24	22.67	36.29	m	58.96
Beds and coverings; 100 mm thick; to pipes size					
100 mm	0.60	4.20	6.05	m	10.25
150 mm	0.70	4.90	7.06	m	11.96
225 mm	1.00	7.00	10.08	m	17.08
300 mm	1.20	8.40	12.10	m	20.50
375 mm	1.45	10.15	14.62	m	24.77
450 mm	1.70	11.90	17.14	m	29.04
600 mm	2.20	15.40	22.18	m	37.58
900 mm	3.35	23.44	33.77	m	57.21
1200 mm	4.60	32.19	46.37	m	78.56
Beds and coverings; 150 mm thick; to pipes size					
100 mm	0.90	6.30	9.07	m	15.37
150 mm	1.00	7.00	10.08	m	17.08
225 mm	1.30	9.10	13.10	m	22.20
300 mm	1.55	10.85	15.62	m	26.47
375 mm	1.80	12.60	18.14	m	30.74
450 mm	2.15	15.05	21.67	m	36.72
600 mm	2.60	18.19	26.21	m	44.40
900 mm	4.25	29.74	42.84	m	72.58
1200 mm	6.00	41.99	60.48	m	102.47

NOTE: The following items unless otherwise
described include for all appropriate
joints/couplings in the running length
The prices for gullies and rainwater shoes,
etc., include for appropriate joints to
pipes and for setting on and surrounding
accessory with site mixed in situ concrete
11.50 N/mm2-40 aggregate (1:3:6)

Cast iron drain pipes and fittings; BS 437;
coated; with caulked lead joints

75 mm pipes

		Labour hours	Labour £	Material £	Unit	Total rate £
laid straight; grey iron	PC £51.40/3m	0.60	4.60	20.01	m	24.61
laid straight; grey iron	PC £32.71/1.8m	0.70	5.37	21.55	m	26.92
in runs not exceeding 3 m long	PC £24.71/.91m	0.95	7.29	32.86	m	40.15
Extra for						
bend; short radius	PC £11.74	0.70	5.37	13.04	nr	18.41
bend; short radius; with						
round access door	PC £21.22	0.70	5.37	23.89	nr	29.26
bend; long radius	PC £17.44 ·	0.70	5.37	18.50	nr	23.87
level invert taper	PC £13.85	0.70	5.37	14.39	nr	19.76
access pipe	PC £42.13	0.70	5.37	46.77	nr	52.14
single branch	PC £22.44	0.95	7.29	23.98	nr	31.27

R DISPOSAL SYSTEMS Including overheads and profit at 9.00%		Labour hours	Labour £	Material £	Unit	Total rate £
R12 DRAINAGE BELOW GROUND - cont'd						
Cast iron drain pipes and fittings; BS 437;						
coated; with caulked lead joints - cont'd						
100 mm pipes						
laid straight; grey iron	PC £54.16/3m	0.70	5.37	21.23	m	26.60
laid straight; grey iron	PC £33.43/1.8m	0.80	6.14	22.34	m	28.48
in runs not exceeding 3 m long	PC £26.97/.91m	1.10	8.44	36.36	m	44.80
in ducts; supported on piers (measured elsewhere)		1.40	10.74	21.51	m	32.25
supported on wall brackets (measured elsewhere)		1.15	8.82	21.51	m	30.33
supported on ceiling hangers (measured elsewhere)		1.40	10.74	21.51	m	32.25
Extra for						
bend; short radius	PC £16.09	0.80	6.14	18.54	nr	24.68
bend; short radius; with round access door	PC £25.25	0.80	6.14	29.00	nr	35.14
bend; long radius	PC £25.07	0.80	6.14	27.70	nr	33.84
bend; long radius; with rectangular access door	PC £45.73	0.80	6.14	51.33	nr	57.47
rest bend	PC £20.40	0.80	6.14	23.46	nr	29.60
level invert taper	PC £16.29	0.80	6.14	17.64	nr	23.78
access pipe	PC £46.08	0.80	6.14	51.74	nr	57.88
single branch	PC £25.45	1.10	8.44	28.12	nr	36.56
single branch; with rectangular access door	PC £58.72	1.10	8.44	66.20	nr	74.64
'Y' branch	PC £57.83	1.10	8.44	65.18	nr	73.62
double branch	PC £69.88	1.35	10.36	79.10	nr	89.46
double branch; with rectangular access door	PC £113.49	1.35	10.36	129.00	nr	139.36
WC connector; 450 mm long	PC £15.52	0.60	4.60	15.15	nr	19.75
WC connector; 600 mm long	PC £24.44	0.60	4.60	23.49	nr	28.09
SW connector	PC £13.67	0.80	6.14	14.09	nr	20.23
150 mm pipes						
laid straight; grey iron	PC £93.05/3m	0.90	6.91	36.44	m	43.35
laid straight; grey iron	PC £51.16/1.8m	1.00	7.67	34.36	m	42.03
in runs not exceeding 3 m long	PC £40.46/.91m	1.35	10.36	55.02	m	65.38
Extra for						
bend; short radius	PC £34.70	1.00	7.67	39.82	nr	47.49
bend; short radius; with round access door	PC £58.98	1.00	7.67	67.61	nr	75.28
bend; long radius	PC £41.10	1.00	7.67	45.23	nr	52.90
bend; long radius; with rectangular access door	PC £84.12	1.00	7.67	94.47	nr	102.14
rest bend	PC £45.65	1.00	7.67	52.37	nr	60.04
level invert taper	PC £25.07	1.00	7.67	26.89	nr	34.56
access pipe	PC £85.40	1.00	7.67	95.94	nr	103.61
single branch	PC £54.86	1.35	10.36	60.93	nr	71.29
single branch; with rectangular access door	PC £115.73	1.35	10.36	130.60	nr	140.96
double branch	PC £108.61	1.70	13.05	122.61	nr	135.66
SW connector	PC £21.16	1.00	7.67	21.51	nr	29.18
225 mm pipes						
laid straight; grey iron	PC £283.33/1.8m	1.60	12.28	269.29	m	281.57
in runs not exceeding 3 m long	PC £163.89/.91m	2.15	16.50	212.16	m	228.66
Extra for						
bend; short radius	PC £132.79	1.60	12.28	158.14	nr	170.42
bend; long radius; with rectangular access door	PC £224.16	1.60	12.28	262.72	nr	275.00
rest bend	PC £159.78	1.60	12.28	189.03	nr	201.31
level invert taper	PC £51.80	1.60	12.28	65.46	nr	77.74
access pipe	PC £179.27	1.60	12.28	211.34	nr	223.62
single branch	PC £172.79	2.15	16.50	207.01	nr	223.51
single branch; with rectangular access door	PC £120.87	2.15	16.50	147.58	nr	164.08
SW connector		1.60	12.28	75.75	nr	88.03

R DISPOSAL SYSTEMS Including overheads and profit at 9.00%		Labour hours	Labour £	Material £	Unit	Total rate £
Accessories in cast iron; with caulked lead **joints to pipes (unless otherwise described)**						
Rainwater shoes; horizontal inlet						
100 mm	PC £47.99	0.75	5.76	55.76	nr	61.52
150 mm	PC £82.50	0.85	6.52	95.53	nr	102.05
Gully fittings; comprising low invert gully trap and round hopper						
75 mm outlet	PC £14.93	1.10	8.44	20.76	nr	29.20
100 mm outlet	PC £25.07	1.20	9.21	32.43	nr	41.64
150 mm outlet	PC £63.60	1.60	12.28	76.58	nr	88.86
Add to above for						
bellmouth 300 mm high; circular plain grating						
75 mm; 175 mm grating	PC £16.17	0.60	4.60	19.18	nr	23.78
100 mm; 200 mm grating	PC £24.36	0.70	5.37	28.70	nr	34.07
150 mm; 250 mm grating	PC £35.01	1.00	7.67	42.10	nr	49.77
bellmouth 300 mm high; circular plain grating as above; one horizontal inlet						
75 x 50 mm	PC £27.06	0.60	4.60	31.22	nr	35.82
100 x 75 mm	PC £26.34	0.70	5.37	31.06	nr	36.43
150 x 100 mm	PC £57.83	1.00	7.67	66.98	nr	74.65
bellmouth 300 mm high; circular plain grating as above; one vertical inlet						
75 x 50 mm	PC £26.11	0.60	4.60	30.18	nr	34.78
100 x 75 mm	PC £30.93	0.70	5.37	36.07	nr	41.44
150 x 100 mm	PC £65.72	1.00	7.67	75.57	nr	83.24
raising piece; 200 mm dia						
75 mm high	PC £12.38	1.20	9.21	18.92	nr	28.13
150 mm high	PC £14.82	1.20	9.21	21.58	nr	30.79
225 mm high	PC £18.84	1.20	9.21	25.97	nr	35.18
300 mm high	PC £22.08	1.20	9.21	29.50	nr	38.71
Yard gully (Deans); trapped; galvanized sediment pan; 267 mm round heavy grating						
100 mm outlet	PC £142.33	3.20	24.56	160.79	nr	185.35
Yard gully (garage); trapless; galvanized sediment pan; 267 mm round heavy grating						
100 mm outlet	PC £121.29	3.00	23.02	137.86	nr	160.88
Yard gully (garage); trapped; with rodding eye; galvanized perforated sediment pan; stopper; round heavy grating						
150 mm outlet; 347 mm grating	PC £532.76	4.00	30.69	589.07	nr	619.76
Grease trap; with internal access; insert galvanized perforated bucket; lid and frame						
450 x 300 x 525 mm deep; 100 mm outlet						
	PC £255.01	3.60	27.62	285.82	nr	313.44
Disconnecting trap; trapped with inspection arm; bridle plate and screw; building in and cutting and fitting brickwork around						
100 mm outlet; 100 mm inlet	PC £113.04	2.50	19.18	129.97	nr	149.15
150 mm outlet; 150 mm inlet	PC £168.50	3.20	24.56	194.23	nr	218.79
Cast iron 'Timesaver' drain pipes and **fittings; BS 437; coated; with mechanical** **coupling joints**						
75 mm pipes						
laid straight	PC £38.78/3m	0.45	3.45	17.59	m	21.04
in runs not exceeding 3 m long	PC £24.20/1m	0.60	4.60	35.04	m	39.64
Extra for						
bend; medium radius	PC £12.09	0.50	3.84	18.96	nr	22.80
single branch	PC £16.80	0.70	5.37	30.38	nr	35.75
isolated 'Timesaver' joint	PC £6.74	0.30	2.30	7.72	nr	10.02

R DISPOSAL SYSTEMS Including overheads and profit at 9.00%		Labour hours	Labour £	Material £	Unit	Total rate £
R12 DRAINAGE BELOW GROUND - cont'd						
Cast iron 'Timesaver' drain pipes and fittings; BS 437; coated; with mechanical coupling joints - cont'd						
100 mm pipes						
laid straight	PC £37.44/3m	0.50	3.84	17.53	m	21.37
in runs not exceeding 3 m long	PC £23.67/1m	0.68	5.22	35.77	m	40.99
Extra for						
bend; medium radius	PC £14.88	0.60	4.60	23.57	nr	28.17
bend; medium radius with access	PC £39.12	0.60	4.60	51.31	nr	55.91
bend; long radius	PC £18.25	0.60	4.60	26.57	nr	31.17
rest bend	PC £17.09	0.60	4.60	26.09	nr	30.69
diminishing pipe	PC £12.67	0.60	4.60	22.32	nr	26.92
single branch	PC £19.77	0.75	5.76	36.56	nr	42.32
single branch; with access	PC £45.57	0.85	6.52	57.41	nr	63.93
double branch	PC £29.76	0.95	7.29	55.82	nr	63.11
double branch; with access	PC £54.98	0.95	7.29	90.18	nr	97.47
isolated 'Timesaver' joint	PC £7.96	0.35	2.69	9.12	nr	11.81
transitional pipe; for WC	PC £11.34	0.50	3.84	19.51	nr	23.35
150 mm pipes						
laid straight	PC £70.52/3m	0.60	4.60	31.02	m	35.62
in runs not exceeding 3 m long	PC £44.04/1m	0.82	6.29	60.92	m	67.21
Extra for						
bend; medium radius	PC £30.05	0.70	5.37	40.87	nr	46.24
bend; medium radius with access	PC £72.64	0.70	5.37	89.62	nr	94.99
bend; long radius	PC £39.12	0.70	5.37	49.64	nr	55.01
diminishing pipe	PC £19.41	0.70	5.37	31.12	nr	36.49
single branch	PC £42.77	0.85	6.52	63.53	nr	70.05
isolated 'Timesaver' joint	PC £9.65	0.42	3.22	11.04	nr	14.26
Accessories in 'Timesaver' cast iron; with mechanical coupling joints						
Rainwater shoes; horizontal inlet						
100 mm	PC £47.99	0.50	3.84	62.10	nr	65.94
150 mm	PC £82.50	0.60	4.60	102.10	nr	106.70
Gully fittings; comprising low invert gully trap and round hopper						
75 mm outlet	PC £12.67	0.90	6.91	23.91	nr	30.82
100 mm outlet	PC £19.77	0.95	7.29	32.99	nr	40.28
150 mm outlet	PC £49.17	1.30	9.98	67.42	nr	77.40
Add to above for						
bellmouth 300 mm high; circular plain grating						
100 mm; 200 mm grating	PC £24.36	0.45	3.45	35.24	nr	38.69
bellmouth 300 mm high; circular plain grating as above; one horizontal inlet						
100 x 100 mm	PC £24.89	0.45	3.45	35.81	nr	39.26
bellmouth 300 mm high; circular plain grating as above; one vertical inlet						
100 x 100 mm	PC £25.81	0.45	3.45	39.50	nr	42.95
Yard gully (Deans); trapped; galvanized sediment pan; 267 mm round heavy grating						
100 mm outlet	PC £151.15	2.90	22.25	168.07	nr	190.32
Yard gully (garage); trapless; galvanized sediment pan; 267 mm round heavy grating						
100 mm outlet	PC £129.25	2.70	20.72	144.19	nr	164.91
Yard gully (garage); trapped; with rodding eye, galvanised perforated sediment pan; stopper; 267 mm round heavy grating						
100 mm outlet; 267 mm grating	PC £287.63	3.00	23.02	317.38	nr	340.40
Grease trap; internal access; galvanized perforated bucket; lid and frame						
20 gal. capacity	PC £620.68	4.00	30.69	542.54	nr	573.23

R DISPOSAL SYSTEMS Including overheads and profit at 9.00%		Labour hours	Labour £	Material £	Unit	Total rate £
Extra strength vitrified clay pipes and **fittings; 'Hepworths' 'SuperSleve'/'HepSleve'** **or similar; plain ends with push-fit** **polypropylene flexible couplings**						
100 mm pipes						
laid straight	PC £1.92	0.25	1.58	2.20	m	3.78
in runs not exceeding 3 m long		0.33	2.09	2.59	m	4.68
Extra for						
bend	PC £1.97	0.20	1.27	3.80	nr	5.07
access bend	PC £12.45	0.20	1.27	15.80	nr	17.07
rest bend	PC £3.26	0.20	1.27	5.28	nr	6.55
access pipe	PC £10.79	0.20	1.27	13.71	nr	14.98
socket adaptor	PC £1.98	0.17	1.08	3.14	nr	4.22
adaptor to flexible pipe	PC £1.92	0.17	1.08	3.07	nr	4.15
saddle	PC £3.93	0.75	4.75	5.57	nr	10.32
single junction	PC £4.15	0.25	1.58	7.17	nr	8.75
single access junction	PC £13.60	0.25	1.58	17.99	nr	19.57
150 mm pipes						
laid straight	PC £4.24	0.30	1.90	4.85	m	6.75
in runs not exceeding 3 m long		0.40	2.53	5.75	m	8.28
Extra for						
bend	PC £4.92	0.24	1.52	9.27	nr	10.79
access bend	PC £20.74	0.24	1.52	27.53	nr	29.05
rest bend	PC £6.32	0.24	1.52	11.03	nr	12.55
taper pipe	PC £4.65	0.24	1.52	7.64	nr	9.16
access pipe	PC £17.75	0.24	1.52	23.67	nr	25.19
socket adaptor	PC £4.75	0.20	1.27	7.55	nr	8.82
adaptor to flexible pipe	PC £3.39	0.20	1.27	6.00	nr	7.27
saddle	PC £6.91	0.90	5.70	10.46	nr	16.16
single junction	PC £7.23	0.30	1.90	13.99	nr	15.89
single access junction	PC £23.14	0.30	1.90	32.39	nr	34.29
Extra strength vitrified clay pipes and **fittings; 'Hepworths' 'HepSeal'; or similar;** **socketted; with push-fit flexible joints**						
100 mm pipes						
laid straight	PC £4.25	0.30	1.90	4.86	m	6.76
in runs not exceeding 3 m long		0.40	2.53	5.33	m	7.86
Extra for						
bend	PC £5.86	0.24	1.52	5.25	nr	6.77
access bend	PC £16.25	0.24	1.52	17.14	nr	18.66
rest bend	PC £7.43	0.24	1.52	7.05	nr	8.57
stopper	PC £2.36	0.15	0.95	2.70	nr	3.65
access pipe	PC £13.47	0.24	1.52	13.47	nr	14.99
socket adaptor	PC £1.98	0.26	1.65	4.16	nr	5.81
saddle	PC £5.57	0.75	4.75	6.38	nr	11.13
single junction	PC £7.79	0.30	1.90	6.97	nr	8.87
single access junction	PC £17.76	0.30	1.90	18.38	nr	20.28
double junction	PC £15.53	0.45	2.85	15.34	nr	18.19
double collar	PC £5.57	0.20	1.27	6.38	nr	7.65
150 mm pipes						
laid straight	PC £5.88	0.35	2.22	6.73	m	8.95
in runs not exceeding 3 m long		0.47	2.98	7.37	m	10.35
Extra for						
bend	PC £10.13	0.28	1.77	9.58	nr	11.35
access bend	PC £25.37	0.28	1.77	27.02	nr	28.79
rest bend	PC £12.32	0.28	1.77	12.08	nr	13.85
stopper	PC £3.37	0.18	1.14	3.86	nr	5.00
taper reducer	PC £15.48	0.28	1.77	15.70	nr	17.47
access pipe	PC £21.12	0.28	1.77	21.48	nr	23.25
socket adaptor	PC £4.75	0.30	1.90	8.70	nr	10.60
saddle	PC £9.20	0.90	5.70	10.53	nr	16.23
single access junction	PC £27.98	0.35	2.22	29.33	nr	31.55
double junction	PC £25.83	0.53	3.36	26.19	nr	29.55
double collar	PC £9.20	0.23	1.46	10.53	nr	11.99

R DISPOSAL SYSTEMS Including overheads and profit at 9.00%		Labour hours	Labour £	Material £	Unit	Total rate £
R12 DRAINAGE BELOW GROUND - cont'd						
Extra strength vitrified clay pipes and **fittings; 'Hepworths' 'HepSeal'; or similar;** **socketted; with push-fit flexible joints - cont'd**						
225 mm pipes						
laid straight	PC £11.51	0.45	2.85	13.18	m	16.03
in runs not exceeding 3 m long		0.60	3.80	14.43	m	18.23
Extra for						
bend	PC £21.18	0.36	2.28	20.29	nr	22.57
access bend	PC £53.49	0.36	2.28	57.27	nr	59.55
rest bend	PC £29.49	0.36	2.28	29.80	nr	32.08
stopper	PC £7.63	0.23	1.46	8.73	nr	10.19
taper reducer	PC £24.48	0.36	2.28	24.06	nr	26.34
access pipe	PC £53.49	0.36	2.28	55.95	nr	58.23
saddle	PC £24.48	1.20	7.60	28.01	nr	35.61
single junction	PC £30.92	0.45	2.85	30.12	nr	32.97
single access junction	PC £63.96	0.45	2.85	67.94	nr	70.79
double junction	PC £61.86	0.68	4.31	64.22	nr	68.53
double collar	PC £20.19	0.30	1.90	23.11	nr	25.01
300 mm pipes						
laid straight	PC £18.03	0.60	3.80	20.63	m	24.43
in runs not exceeding 3 m long		0.80	5.07	22.60	m	27.67
Extra for						
bend	PC £41.76	0.48	3.04	41.61	nr	44.65
access bend	PC £97.14	0.48	3.04	104.99	nr	108.03
rest bend	PC £62.56	0.48	3.04	65.41	nr	68.45
stopper	PC £16.81	0.30	1.90	19.24	nr	21.14
taper reducer	PC £50.08	0.48	3.04	51.12	nr	54.16
access pipe	PC £97.14	0.48	3.04	102.93	nr	105.97
saddle	PC £50.08	1.60	10.13	57.31	nr	67.44
single junction	PC £65.61	0.60	3.80	66.84	nr	70.64
single access junction	PC £116.74	0.60	3.80	125.36	nr	129.16
double junction	PC £131.31	0.90	5.70	139.96	nr	145.66
double collar	PC £32.81	0.40	2.53	37.55	nr	40.08
400 mm pipes						
laid straight	PC £35.03	0.80	5.07	40.09	m	45.16
in runs not exceeding 3 m long		1.06	6.71	43.91	m	50.62
Extra for						
bend	PC £131.79	0.64	4.05	138.80	nr	142.85
single junction	PC £123.48	0.80	5.07	125.28	nr	130.35
450 mm pipes						
laid straight	PC £45.65	1.00	6.33	52.24	m	58.57
in runs not exceeding 3 m long		1.33	8.42	57.23	m	65.65
Extra for						
bend	PC £173.53	0.80	5.07	182.93	nr	188.00
single junction	PC £147.72	1.00	6.33	148.16	nr	154.49
British Standard quality vitrified clay **pipes and fittings; socketted; cement and** **sand (1:2) joints**						
100 mm pipes						
laid straight	PC £2.78	0.40	2.53	3.27	m	5.80
in runs not exceeding 3 m long		0.53	3.36	3.57	m	6.93
Extra for						
bend (short/medium/knuckle)	PC £2.51	0.32	2.03	2.00	nr	4.03
bend (long/rest/elbow)	PC £5.03	0.32	2.03	4.89	nr	6.92
access bend	PC £16.39	0.32	2.03	17.89	nr	19.92
taper	PC £6.60	0.32	2.03	6.59	nr	8.62
access pipe	PC £13.63	0.32	2.03	14.42	nr	16.45
single junction	PC £5.03	0.40	2.53	4.66	nr	7.19
single access junction	PC £18.19	0.40	2.53	19.71	nr	22.24
double junction	PC £10.04	0.60	3.80	10.15	nr	13.95
double collar	PC £3.68	0.27	1.71	4.29	nr	6.00
double access junction	PC £23.22	0.60	3.80	25.24	nr	29.04

R DISPOSAL SYSTEMS Including overheads and profit at 9.00%		Labour hours	Labour £	Material £	Unit	Total rate £
150 mm pipes						
laid straight	PC £4.99	0.45	2.85	5.80	m	8.65
in runs not exceeding 3 m long		0.60	3.80	6.34	m	10.14
Extra for						
bend (short/medium/knuckle)	PC £4.24	0.36	2.28	3.23	nr	5.51
bend (long/rest/elbow)	PC £8.51	0.36	2.28	8.11	nr	10.39
access bend	PC £27.58	0.36	2.28	29.93	nr	32.21
taper	PC £11.11	0.36	2.28	10.91	nr	13.19
access pipe	PC £22.97	0.36	2.28	24.09	nr	26.37
single junction	PC £8.51	0.45	2.85	7.63	nr	10.48
single access junction	PC £30.62	0.45	2.85	32.93	nr	35.78
double junction	PC £17.03	0.68	4.31	16.90	nr	21.21
double collar	PC £6.13	0.30	1.90	7.08	nr	8.98
double access junction	PC £39.17	0.68	4.31	42.23	nr	46.54
225 mm pipes						
laid straight	PC £9.85	0.55	3.48	11.44	m	14.92
in runs not exceeding 3 m long		0.73	4.62	12.52	m	17.14
Extra for						
bend (short/medium/knuckle)	PC £13.25	0.44	2.79	11.95	nr	14.74
bend (long/rest/elbow)	PC £26.54	0.44	2.79	27.17	nr	29.96
access bend	PC £47.01	0.44	2.79	50.59	nr	53.38
taper	PC £24.11	0.44	2.79	24.04	nr	26.83
access pipe	PC £47.01	0.44	2.79	49.46	nr	52.25
single junction	PC £26.54	0.55	3.48	26.21	nr	29.69
single access junction	PC £52.18	0.55	3.48	55.55	nr	59.03
double junction	PC £52.90	0.83	5.26	55.42	nr	60.68
double collar	PC £14.44	0.36	2.28	16.70	nr	18.98
double access junction	PC £78.62	0.83	5.26	84.85	nr	90.11
300 mm pipes						
laid straight	PC £16.06	0.75	4.75	18.55	m	23.30
in runs not exceeding 3 m long		1.00	6.33	20.30	m	26.63
Extra for						
bend (short/medium/knuckle)	PC £26.21	0.60	3.80	24.66	nr	28.46
bend (long/rest/elbow)	PC £52.51	0.60	3.80	54.75	nr	58.55
access bend	PC £93.07	0.60	3.80	101.17	nr	104.97
taper	PC £47.76	0.60	3.80	48.76	nr	52.56
access pipe	PC £93.07	0.60	3.80	99.34	nr	103.14
single junction	PC £52.51	0.75	4.75	53.08	nr	57.83
single access junction	PC £103.38	0.75	4.75	111.31	nr	116.06
double junction	PC £104.93	1.13	7.16	111.42	nr	118.58
double access junction	PC £155.84	1.13	7.16	169.68	nr	176.84
400 mm pipes						
laid straight	PC £30.68	1.00	6.33	35.37	m	41.70
in runs not exceeding 3 m long	PC £30.68	1.33	8.42	38.71	m	47.13
Extra for						
bend (short/medium/knuckle)	PC £94.57	0.80	5.07	97.96	nr	103.03
taper	PC £86.00	0.80	5.07	87.09	nr	92.16
single junction		1.00	6.33	94.70	nr	101.03
450 mm pipes						
laid straight	PC £39.04	1.25	7.92	44.94	m	52.86
in runs not exceeding 3 m long		1.66	10.51	49.20	m	59.71
Extra for						
bend (short/medium/knuckle)	PC £116.49	1.00	6.33	120.18	nr	126.51
taper	PC £105.90	1.00	6.33	106.71	nr	113.04
single junction	PC £116.49	1.25	7.92	115.96	nr	123.88
500 mm pipes						
laid straight	PC £48.86	1.45	9.18	56.26	m	65.44
in runs not exceeding 3 m long		1.95	12.35	61.59	m	73.94
Extra for						
bend (short/medium/knuckle)	PC £138.49	1.15	7.28	142.07	nr	149.35
taper	PC £104.38	1.15	7.28	103.03	nr	110.31
single junction	PC £138.49	1.45	9.18	136.82	nr	146.00

R DISPOSAL SYSTEMS Including overheads and profit at 9.00%		Labour hours	Labour £	Material £	Unit	Total rate £
R12 DRAINAGE BELOW GROUND - cont'd						
Accessories in vitrified clay; set in **concrete; with polypropylene coupling joints** **to pipes**						
Rodding point; with oval aluminium plate						
100 mm	PC £10.85	0.50	3.17	13.90	nr	17.07
Gully fittings; comprising low back trap and square hopper; 150 x 150 mm square gully grid						
100 mm outlet	PC £9.74	0.85	5.38	15.35	nr	20.73
Add to above for						
100 mm back inlet		-	-	-	nr	4.64
100 mm raising pieces	PC £3.71	0.25	1.58	5.11	nr	6.69
Access gully; trapped with rodding eye and integral vertical back inlet; stopper; 150 x 150 mm square gully grid						
100 mm outlet	PC £13.89	0.65	4.12	17.21	nr	21.33
Inspection chamber; comprising base; 300 or 450 mm raising piece; integral alloy cover and frame; 100 mm inlets						
straight through; 2 nr inlets	PC £37.49	2.00	12.67	46.90	nr	59.57
single junction; 3 nr inlets	PC £40.62	2.20	13.93	51.38	nr	65.31
double junction; 4 nr inlets	PC £44.08	2.40	15.20	56.22	nr	71.42
Accessories in propylene; cover set in **concrete; with coupling joints to pipes**						
Inspection chamber; 5 nr 100 mm inlets; cast iron cover and frame						
475 x 585 mm deep	PC £60.26	2.30	14.57	70.20	nr	84.77
475 x 930 mm deep	PC £70.63	2.50	15.83	81.50	nr	97.33
Accessories in vitrified clay; set in **concrete; with cement and sand (1:2) joints** **to pipes**						
Rainwater shoes; with 250 x 100 mm oval access						
100 mm	PC £11.04	0.50	3.17	13.12	nr	16.29
150 mm	PC £18.59	0.60	3.80	21.85	nr	25.65
Gully fittings; comprising low back trap and square hopper; square gully grid						
100 mm outlet; 150 x 150 mm grid	PC £13.25	1.00	6.33	21.90	nr	28.23
150 mm outlet; 225 x 225 mm grid	PC £27.80	1.30	8.23	43.45	nr	51.68
Add to above for						
100 mm back inlet	PC £6.12	-	-	6.67	nr	6.67
100 mm vertical back inlet	PC £6.12	-	-	6.67	nr	6.67
100 mm raising pieces	PC £2.37	0.30	1.90	2.67	nr	4.57
Yard gully (mud); trapped with rodding eye; galvanized square bucket; stopper; square hinged grate and frame						
100 mm outlet; 225 x 225 mm grate	PC £48.83	3.00	19.00	57.33	nr	76.33
150 mm outlet; 300 x 300 mm grate	PC £88.66	4.00	25.33	102.25	nr	127.58
Yard gully (garage); trapped with rodding eye; galvanized perforated round bucket; stopper; round hinged grate and frame						
100 mm outlet; 273 mm grate	PC £59.91	3.00	19.00	70.41	nr	89.41
150 mm outlet; 368 mm grate	PC £103.56	4.00	25.33	118.99	nr	144.32
Road gully; trapped with rodding eye and stopper (grate measured elsewhere)						
300 x 600 x 100 mm outlet	PC £30.03	3.30	20.90	44.36	nr	65.26
300 x 600 x 150 mm outlet	PC £30.03	3.30	20.90	44.36	nr	65.26
400 x 750 x 150 mm outlet	PC £35.84	4.00	25.33	56.72	nr	82.05
450 x 900 x 150 mm outlet	PC £48.49	5.00	31.66	74.02	nr	105.68
Grease trap; with internal access; galvanized perforated bucket; lid and frame						
450 x 300 x 525 mm deep; 100 mm outlet						
	PC £227.13	3.50	22.17	260.29	nr	282.46
600 x 450 x 600 mm deep; 100 mm outlet						
	PC £288.27	4.20	26.60	331.45	nr	358.05

R DISPOSAL SYSTEMS Including overheads and profit at 9.00%		Labour hours	Labour £	Material £	Unit	Total rate £
Interceptor; trapped with inspection arm; lever locking stopper; chain and staple; cement and sand (1:2) joints to pipes; building in, and cutting and fitting brickwork around						
100 mm outlet; 100 mm inlet	PC £38.95	4.00	25.33	52.66	nr	77.99
150 mm outlet; 150 mm inlet	PC £55.23	4.50	28.50	72.42	nr	100.92
225 mm outlet; 225 mm inlet	PC £131.44	5.00	31.66	158.67	nr	190.33
Accessories: grates and covers						
Aluminium alloy gully grids; set in position						
125 x 125 mm	PC £1.43	0.10	0.63	1.56	nr	2.19
150 x 150 mm	PC £1.43	0.10	0.63	1.56	nr	2.19
225 x 225 mm	PC £4.27	0.10	0.63	4.65	nr	5.28
140 mm dia (for 100 mm)	PC £1.43	0.10	0.63	1.56	nr	2.19
197 mm dia (for 150 mm)	PC £2.23	0.10	0.63	2.43	nr	3.06
284 mm dia (for 225 mm)		0.10	0.63	5.21	nr	5.84
Aluminium alloy sealing plates and frames; set in cement and sand (1:3)						
150 x 150 mm	PC £5.55	0.25	1.58	6.14	nr	7.72
225 x 225 mm	PC £10.09	0.25	1.58	11.08	nr	12.66
254 x 150 mm; for access fittings	PC £7.05	0.25	1.58	7.77	nr	9.35
140 mm dia (for 100 mm)	PC £4.49	0.25	1.58	4.98	nr	6.56
190 mm dia (for 150 mm)	PC £6.46	0.25	1.58	7.13	nr	8.71
273 mm dia (for 225 mm)	PC £10.34	0.25	1.58	11.44	nr	13.02
Coated cast iron heavy duty road gratings and frames; BS 497 Tables 6 and 7; bedding and pointing in cement and sand (1:3); one course half brick thick wall in semi- engineering bricks in cement mortar (1:3)						
475 x 475 mm; grade A, ref GA1-450 (131 kg)						
	PC £54.50	2.50	15.83	61.73	nr	77.56
400 x 350 mm; grade A, ref GA2-325 (99 kg)						
	PC £45.86	2.50	15.83	51.89	nr	67.72
500 x 350 mm; grade A, ref GA2-325 (124 kg)						
	PC £65.04	2.50	15.83	73.01	nr	88.84
White vitreous clay floor channels; bedded, **floor channel**						
100 mm half round section						
floor channel	PC £30.19	0.60	3.80	35.00	m	38.80
Extra for						
stop end	PC £18.35	0.40	2.53	20.00	nr	22.53
angle	PC £27.53	0.60	3.80	30.01	nr	33.81
tee piece	PC £32.64	0.80	5.07	35.58	nr	40.65
stop end outlet	PC £23.71	0.50	3.17	25.85	nr	29.02
150 mm half round section						
floor channel	PC £33.50	0.75	4.75	39.01	m	43.76
Extra for						
stop end	PC £20.36	0.50	3.17	22.20	nr	25.37
angle	PC £30.59	0.75	4.75	33.34	nr	38.09
tee piece	PC £21.34	1.00	6.33	39.50	nr	45.83
stop end outlet	PC £31.70	0.60	3.80	34.56	nr	38.36
Channel sump outlet for 150 mm channel						
one inlet	PC £43.84	1.20	7.60	47.78	nr	55.38
two inlets	PC £46.70	1.50	9.50	50.91	nr	60.41
230 mm half round section						
floor channel	PC £53.62	1.00	6.33	62.92	m	69.25
Extra for						
stop end	PC £32.74	0.65	4.12	35.68	nr	39.80
angle	PC £49.07	1.00	6.33	53.48	nr	59.81
tee piece	PC £54.78	1.35	8.55	59.71	nr	68.26
stop end outlet	PC £38.39	0.80	5.07	41.84	nr	46.91

R DISPOSAL SYSTEMS Including overheads and profit at 9.00%		Labour hours	Labour £	Material £	Unit	Total rate £
R12 DRAINAGE BELOW GROUND - cont'd						
White vitreous clay floor channels; bedded, floor channel - cont'd						
100 mm block floor channel	PC £35.00	0.60	3.80	40.28	m	44.08
Extra for						
stop end	PC £21.34	0.40	2.53	23.26	nr	25.79
angle	PC £32.02	0.60	3.80	34.90	nr	38.70
tee piece	PC £32.02	0.80	5.07	34.90	nr	39.97
stop end outlet	PC £26.46	0.50	3.17	28.84	nr	32.01
150 mm block floor channel	PC £38.93	0.75	4.75	45.01	m	49.76
Extra for						
stop end	PC £23.70	0.50	3.17	25.83	nr	29.00
angle	PC £35.57	0.75	4.75	38.77	nr	43.52
tee piece	PC £35.57	1.00	6.33	38.77	nr	45.10
stop end outlet	PC £29.41	0.60	3.80	32.06	nr	35.86
150 mm rebated block floor channel		0.75	4.75	49.92	m	54.67
Extra for						
stop end	PC £26.53	0.50	3.17	28.92	nr	32.09
angle	PC £39.85	0.75	4.75	43.44	nr	48.19
tee piece	PC £39.85	1.00	6.33	43.44	nr	49.77
stop end outlet	PC £32.26	0.60	3.80	35.16	nr	38.96
Accessories; channel gratings and connectors						
Galvanized cast iron medium duty square mesh gratings; bedding and pointing in cement and sand (1:3)						
138 x 13 mm; to suit 100 mm wide channel	PC £39.06	0.70	4.43	42.58	m	47.01
180 x 13 mm; to suit 150 mm wide channel	PC £39.06	0.90	5.70	42.58	m	48.28
Galvanized cast iron medium duty square mesh gratings and frame; galvanized cast iron angle bearers; bedding and pointing in cement and sand (1:3); cutting and pinning lugs to concrete						
148 x 13 mm; to suit 100 mm wide channel	PC £63.33	1.70	10.77	69.03	m	79.80
190 x 13 mm; to suit 150 mm wide channel	PC £63.33	1.90	12.03	69.03	m	81.06
Chromium plated brass domed outlet grating; threaded joint to connector (measured elsewhere)						
50 mm	PC £12.85	0.30	1.90	14.01	nr	15.91
63 mm	PC £16.93	0.30	1.90	18.46	nr	20.36
Cast iron connector; cement and sand (1:2) joint to drain pipe						
75 x 300 mm long; screwed 50 mm	PC £10.65	0.40	2.53	11.69	nr	14.22
75 x 300 mm long; screwed 63 mm	PC £12.44	0.40	2.53	13.65	nr	16.18
Class M tested concrete centrifugally spun pipes and fittings; with flexible joints; BS 5911 Part 1						
300 mm pipes; laid straight	PC £13.06	0.70	4.43	14.94	m	19.37
Extra for						
bend	PC £65.30	0.70	4.43	56.05	nr	60.48
300 x 100 mm single junction	PC £16.50	0.50	3.17	18.89	nr	22.06
450 mm pipes; laid straight	PC £18.82	1.10	6.97	21.54	m	28.51
Extra for						
bend	PC £94.10	1.10	6.97	80.77	nr	87.74
450 x 150 mm single junction	PC £20.00	0.70	4.43	22.89	nr	27.32
600 mm pipes; laid straight	PC £25.47	1.60	10.13	29.15	m	39.28
Extra for						
bend	PC £127.35	1.60	10.13	109.32	nr	119.45
600 x 150 mm single junction	PC £20.00	0.90	5.70	22.89	nr	28.59

R DISPOSAL SYSTEMS Including overheads and profit at 9.00%		Labour hours	Labour £	Material £	Unit	Total rate £
900 mm pipes; laid straight,	PC £51.27	2.80	17.73	58.67	m	76.40
Extra for						
bend	PC £256.35	2.80	17.73	220.04	nr	237.77
900 x 225 mm single junction	PC £38.00	1.10	6.97	43.49	nr	50.46
1200 mm pipes; laid straight	PC £88.20	4.00	25.33	100.94	m	126.27
Extra for						
bend	PC £441.00	4.00	25.33	378.55	nr	403.88
1200 x 300 mm single junction	PC £45.00	1.60	10.13	51.50	nr	61.63

Accessories in precast concrete; top set in with rodding eye and stopper; cement and sand (1:2) joint to pipe
Concrete road gully; BS 556 Part 2; trapped with rodding eye and stopper; cement and sand (1:2) joint to pipe

		Labour hours	Labour £	Material £	Unit	Total rate £
450 mm dia x 1050 mm deep;						
100 or 150 mm outlet		4.75	30.08	38.72	nr	68.80

uPVC pipes and fittings; BS 4660; with lip seal coupling joints
110 mm pipes

		Labour hours	Labour £	Material £	Unit	Total rate £
laid straight	PC £11.46/6m	0.20	1.27	2.52	m	3.79
in runs not exceeding 3 m long	PC £5.73/3m	0.27	1.71	3.04	m	4.75
Extra for						
bend; short radius	PC £5.24	0.16	1.01	5.89	nr	6.90
bend; long radius	PC £8.27	0.16	1.01	8.80	nr	9.81
spigot/socket bend	PC £4.49	0.20	1.27	6.79	nr	8.06
access bend	PC £8.20	0.16	1.01	8.74	nr	9.75
socket plug	PC £2.08	0.05	0.32	2.37	nr	2.69
variable bend	PC £5.73	0.16	1.01	8.11	nr	9.12
inspection pipe	PC £9.94	0.16	1.01	12.93	nr	13.94
adaptor to clay	PC £4.23	0.16	1.01	4.83	nr	5.84
WC connector	PC £3.77	0.20	1.27	6.06	nr	7.33
single junction	PC £7.00	0.20	1.27	7.36	nr	8.63
inspection junction	PC £14.53	0.20	1.27	15.97	nr	17.24
slip coupling	PC £3.12	0.10	0.63	3.58	nr	4.21
160 mm pipes						
laid straight	PC £22.40/6m	0.24	1.52	4.92	m	6.44
in runs not exceeding 3 m long	PC £11.20/3m	0.32	2.03	5.95	m	7.98
Extra for						
bend; short radius	PC £10.04	0.20	1.27	11.29	nr	12.56
spigot/socket bend	PC £9.56	0.25	1.58	14.22	nr	15.80
socket plug	PC £3.61	0.06	0.38	4.13	nr	4.51
inspection pipe	PC £12.91	0.20	1.27	17.85	nr	19.12
adaptor to clay	PC £7.18	0.20	1.27	8.09	nr	9.36
level invert taper	PC £6.12	0.24	1.52	10.08	nr	11.60
single junction	PC £18.44	0.24	1.52	19.83	nr	21.35
inspection junction	PC £23.66	0.24	1.52	25.80	nr	27.32
slip coupling	PC £6.66	0.12	0.76	7.62	nr	8.38

Accessories in uPVC; with lip seal coupling joints to pipes (unless otherwise described)
Access cap assembly

		Labour hours	Labour £	Material £	Unit	Total rate £
110 mm	PC £5.10	0.10	0.63	7.73	nr	8.36
Rodding eye 200 mm; sealed cover	PC £13.08	0.50	3.17	23.48	nr	26.65

Gully fitting; comprising 'P' trap, square hopper 154 x 154 mm grate

		Labour hours	Labour £	Material £	Unit	Total rate £
110 mm outlet	PC £8.87	0.90	5.70	23.87	nr	29.57

Shallow access pipe assembly; 2 nr 110 mm inlets; light duty screw down cover and frame

		Labour hours	Labour £	Material £	Unit	Total rate £
110 x 600 mm deep	PC £32.53	1.00	6.33	40.86	nr	47.19

Shallow branch access assembly; 3 nr 110 mm inlets; light duty screw down cover and frame

		Labour hours	Labour £	Material £	Unit	Total rate £
110 x 600 mm deep	PC £35.95	1.20	7.60	42.66	nr	50.26

R DISPOSAL SYSTEMS Including overheads and profit at 9.00%	Labour hours	Labour £	Material £	Unit	Total rate £

R12 DRAINAGE BELOW GROUND - cont'd

Accessories in uPVC; with lip seal coupling
joints to pipes (unless otherwise described) - cont'd
Inspection chamber; 450 mm dia; 940 mm deep;
heavy duty screw down cover and frame

4 nr 110 mm outlet/inlets PC £69.13	1.80	11.40	122.38	nr	133.78

Kerb to gullies; class B engineering bricks
on edge to three sides in cement mortar
(1:3) rendering in cement mortar (1:3) to
top and two sides and skirting to brickwork
230 mm high; dishing in cement mortar (1:3)
to gully; steel trowelled

230 x 230 mm internally	1.40	8.87	1.49	nr	10.36

Mechanical excavation
Excavating manholes; not exceeding

1 m deep	0.21	1.42	4.35	m3	5.77
2 m deep	0.23	1.55	4.76	m3	6.31
4 m deep	0.27	1.82	5.59	m3	7.41

Hand excavation
Excavating manholes; not exceeding

1 m deep	3.30	22.30	-	m3	22.30
2 m deep	3.90	26.36	-	m3	26.36
4 m deep	5.00	33.79	-	m3	33.79

Earthwork support (average 'risk' prices)
Not exceeding 2 m between opposing faces;
not exceeding

1 m deep	0.15	1.01	0.48	m2	1.49
2 m deep	0.19	1.28	0.57	m2	1.85
4 m deep	0.24	1.62	0.72	m2	2.34

Disposal (mechanical)
Excavated material; depositing on site in
spoil heaps

average 50 m distant	0.07	0.47	1.45	m3	1.92

Removing from site to tip not exceeding
13 km (using lorries)

	-	-	-	m3	8.31

Disposal (hand)
Excavated material; depositing on site in
spoil heaps

average 50 m distant	1.30	8.79	-	m3	8.79

Removing from site to tip not exceeding
13 km (using lorries)

	1.00	6.76	11.92	m3	18.68

Mechanical filling
Excavated material filling to excavations

	0.15	1.01	2.07	m3	3.08

Hand filling
Excavated material filling to excavations

	1.00	6.76	-	m3	6.76

In situ ready mixed concrete; normal
Portland cement; mix 11.50 N/mm2 - 40 mm
aggregate (1:3:6); PC £43.03/m3;
Beds

not exceeding 150 mm thick	3.10	21.69	49.25	m3	70.94
150 - 450 mm thick	2.30	16.09	49.25	m3	65.34
over 450 mm thick	1.90	13.30	49.25	m3	62.55

R DISPOSAL SYSTEMS Including overheads and profit at 9.00%		Labour hours	Labour £	Material £	Unit	Total rate £
Ready mixed in situ concrete; normal Portland cement; mix 21.00 N/mm2-20 mm aggregate (1:2:4); PC £46.24/m3						
Beds						
not exceeding 150 mm thick		3.10	21.69	52.92	m3	74.61
150 - 450 mm thick		2.30	16.09	52.92	m3	69.01
over 450 mm thick		1.90	13.30	52.92	m3	66.22
Site mixed in situ concrete; normal Portland cement; mix 26.00 N/mm2-20 mm aggregate (1:1.5:3); (small quantities); PC £54.75/m3						
Benching in bottoms						
150 - 450 mm average thick		7.00	58.29	59.68	m3	117.97
Reinforced site mixed in situ concrete; normal Portland cement; mix 21.00 N/mm2-20mm aggregate (1:2:4); (small quantities); PC £50.62/m3						
Isolated cover slabs						
not exceeding 150 mm thick		6.00	41.99	55.18	m3	97.17
Reinforcement; fabric to BS 4483; lapped; in beds or suspended slabs						
Ref A98 (1.54 kg/m2)	PC £0.61	0.12	0.97	0.80	m2	1.77
Ref A142 (2.22 kg/m2)	PC £0.77	0.12	0.97	1.00	m2	1.97
Ref A193 (3.02 kg/m2)	PC £1.04	0.12	0.97	1.36	m2	2.33
Formwork to in situ concrete						
Soffits of isolated cover slabs						
horizontal		2.85	23.30	5.45	m2	28.75
Edges of isolated cover slabs						
not exceeding 250 mm high		0.80	6.54	2.00	m	8.54
Precast concrete rectangular access and inspection chambers; 'Brooklyns' chambers or similar; comprising cover frame to receive manhole cover (priced elsewhere) intermediate wall sections and base section with cut outs;bedding;jointing and pointing in cement mortar (1:3) on prepared bed						
Drainage chamber; size 600 x 450 mm internally; depth to invert						
600 mm deep		4.50	28.50	22.56	nr	51.06
900 mm deep		6.00	38.00	28.70	nr	66.70
Drainage chamber; 1200 x 750 mm reducing to 600 x 600 mm; no base unit; depth to invert						
1050 mm deep		7.50	47.50	82.20	nr	129.70
1650 mm deep		9.00	57.00	126.86	nr	183.86
2250 mm deep		11.00	74.34	171.52	nr	245.86
Precast concrete circular manhole rings; BS 5911 Part 1; bedding, jointing and pointing in cement mortar (1:3) on prepared bed						
Chamber or shaft rings; plain						
675 mm	PC £19.98	5.00	31.66	22.18	m	53.84
900 mm	PC £29.21	5.50	34.83	32.45	m	67.28
1050 mm	PC £37.93	6.50	41.16	42.19	m	83.35
1200 mm	PC £50.22	7.50	47.50	55.89	m	103.39
Chamber or shaft rings; reinforced						
1350 mm	PC £77.94	8.50	53.83	86.81	m	140.64
1500 mm	PC £87.03	9.50	60.16	97.43	m	157.59
1800 mm	PC £110.34	12.00	75.99	124.20	m	200.19
2100 mm	PC £197.42	15.00	94.99	220.48	m	315.47
Extra for step irons built in	PC £1.98	0.15	0.95	2.16	nr	3.11

R DISPOSAL SYSTEMS Including overheads and profit at 9.00%	Labour hours	Labour £	Material £	Unit	Total rate £

R12 DRAINAGE BELOW GROUND - cont'd

Precast concrete circular manhole rings;
BS 5911 Part 1; bedding, jointing and
pointing in cement mortar (1:3) on prepared
bed - cont'd

Taper pieces; 675 mm high; from 675 mm to					
900 mm PC £29.07	4.50	28.50	32.04	nr	60.54
1050 mm PC £38.70	5.00	31.66	42.71	nr	74.37
1200 mm PC £51.93	6.00	38.00	57.31	nr	95.31
Taper pieces; 900 mm high; from 675 mm to					
1350 mm PC £91.67	9.50	60.16	101.59	nr	161.75
1500 mm PC £107.24	11.00	69.66	119.18	nr	188.84
1800 mm PC £139.00	14.00	88.66	155.05	nr	243.71
Heavy duty cover slabs; to suit rings					
675 mm PC £15.98	2.75	17.42	17.63	nr	35.05
900 mm PC £26.73	3.00	19.00	29.49	nr	48.49
1050 mm PC £35.01	3.50	22.17	38.69	nr	60.86
1200 mm PC £48.51	4.00	25.33	53.58	nr	78.91
1350 mm PC £59.89	4.50	28.50	66.39	nr	94.89
1500 mm PC £83.25	5.00	31.66	92.29	nr	123.95
1800 mm PC £126.00	6.00	38.00	139.73	nr	177.73
2100 mm PC £261.27	7.00	44.33	287.97	nr	332.30

Common bricks; PC £117.00/1000; in cement mortar (1:3)

Walls to manholes					
one brick thick	2.25	26.22	18.76	m2	44.98
one and a half brick thick	3.20	37.29	28.21	m2	65.50
Projections of footings					
two brick thick	4.35	50.69	37.52	m2	88.21

Class A engineering bricks; PC £340.00/1000
in cement mortar (1:3)

Walls to manholes					
one brick thick	2.50	29.13	49.39	m2	78.52
one and a half brick thick	3.50	40.78	74.15	m2	114.93
Projections of footings					
two brick thick	4.80	55.93	98.78	m2	154.71

Class B engineering bricks; PC £192.00/1000
in cement mortar (1:3)

Walls to manholes					
one brick thick	2.50	29.13	29.06	m2	58.19
one and a half brick thick	3.50	40.78	43.66	m2	84.44
Projections of footings					
two brick thick	4.80	55.93	58.12	m2	114.05

Brickwork sundries

Extra over for fair face; flush smooth pointing					
manhole walls	0.20	2.33	-	m2	2.33
Building ends of pipes into brickwork; making good fair face or rendering					
not exceeding 55 mm nominal size	0.10	1.17	-	nr	1.17
55 - 110 mm nominal size	0.15	1.75	-	nr	1.75
over 110 mm nominal size	0.20	2.33	-	nr	2.33
Step irons; BS 1247; malleable; galvanized; building into joints					
general purpose pattern	0.15	1.75	3.70	nr	5.45

Cement and sand (1:3) in situ finishings;
steel trowelled

13 mm work to manhole walls; one coat; to brickwork base over 300 mm wide	0.70	7.23	0.75	m2	7.98

R DISPOSAL SYSTEMS Including overheads and profit at 9.00%		Labour hours	Labour £	Material £	Unit	Total rate £
Manhole accessories in cast iron						
Petrol trapping bend; coated; 375 x 750 mm;						
building into brickwork						
100 mm	PC £49.37	1.35	10.36	56.16	nr	66.52
150 mm	PC £79.77	1.75	13.43	90.89	nr	104.32
225 mm	PC £147.79	2.85	21.87	167.26	nr	189.13
Cast iron inspection chambers; with bolted						
flat covers; BS 437; bedded in cement mortar						
(1:3); caulked lead joints to pipes						
100 x 100 mm; ref 010; no branches	PC £70.21	1.00	7.67	79.32	nr	86.99
100 x 100 mm; ref 110; one branch	PC £89.38	1.20	9.21	101.61	nr	110.82
100 x 100 mm; ref 111; one branch						
either side	PC £115.67	1.40	10.74	131.44	nr	142.18
100 x 100 mm; ref 212; two branches						
either side	PC £187.48	2.05	15.73	212.06	nr	227.79
100 x 100 mm; ref 313; three branches						
either side	PC £255.18	2.70	20.72	288.20	nr	308.92
150 x 100 mm; ref 110; one branch	PC £121.19	1.60	12.28	138.40	nr	150.68
150 x 100 mm; ref 111; one branch						
either side	PC £155.48	1.80	13.81	176.78	nr	190.59
150 x 100 mm; ref 212; two branches						
either side	PC £223.13	2.55	19.57	252.87	nr	272.44
150 x 100 mm; ref 313; three branches						
either side	PC £302.96	3.30	25.32	342.23	nr	367.55
150 x 150 mm; ref 212; two branches						
either side	PC £310.97	2.75	21.10	351.82	nr	372.92
225 x 100 mm; ref 212; two branches						
either side	PC £446.64	3.30	25.32	499.91	nr	525.23
225 x 100 mm; ref 313; three branches						
either side	PC £563.75	4.20	32.23	629.90	nr	662.13
Cast iron inspection chambers; with bolted						
flat covers; BS 437; bedded in cement mortar						
(1:3); with mechanical coupling joints						
100 x 100 mm; ref 110; one branch	PC £57.23	1.05	8.06	80.41	nr	88.47
100 x 100 mm; ref 111; one branch						
either side	PC £69.63	1.55	11.89	102.61	nr	114.50
150 x 100 mm; ref 110; one branch	PC £81.07	1.25	9.59	108.58	nr	118.17
150 x 100 mm; ref 111; one branch						
either side		1.80	13.81	127.21	nr	141.02
150 x 150 mm; ref 110; one branch	PC £103.79	1.35	10.36	135.18	nr	145.54
150 x 150 mm; ref 111; one branch						
either side	PC £114.49	1.90	14.58	157.35	nr	171.93
Access covers and frames; coated; BS 497						
tables 1-5; bedding frame in cement and sand						
(1:3); cover in grease and sand						
Grade C; light duty; rectangular						
single seal solid top						
450 x 450 mm; ref MC1-45/45						
(31 kg)	PC £19.93	1.50	9.50	22.92	nr	32.42
600 x 450 mm; ref MC1-60/45						
(32 kg)	PC £20.07	1.50	9.50	23.07	nr	32.57
600 x 600 mm; ref MC1-60/60						
(61 kg)	PC £44.57	1.50	9.50	49.77	nr	59.27
Grade C; light duty; rectangular						
single seal recessed						
450 x 450 mm; ref MC1R-45/45						
(43 kg)	PC £32.64	1.50	9.50	36.77	nr	46.27
600 x 450 mm; ref MC1R-60/45						
(43 kg)	PC £40.73	1.50	9.50	45.59	nr	55.09
600 x 600 mm; ref MC1R-60/60						
(53 kg)	PC £56.25	1.50	9.50	62.51	nr	72.01

R DISPOSAL SYSTEMS Including overheads and profit at 9.00%		Labour hours	Labour £	Material £	Unit	Total rate £
R12 DRAINAGE BELOW GROUND - cont'd						
Access covers and frames; coated; BS 497 **tables 1-5; bedding frame in cement and sand** **(1:3); cover in grease and sand - cont'd** Grade C; light duty; rectangular double seal solid top						
450 x 450 mm; ref MC2-45/45						
(51 kg)	PC £30.56	1.50	9.50	34.51	nr	44.01
600 x 450 mm; ref MC2-60/45						
(51 kg)	PC £36.72	1.50	9.50	41.22	nr	50.72
600 x 600 mm; ref MC2-60/60						
(83 kg)	PC £56.45	1.50	9.50	62.72	nr	72.22
Grade C; light duty; rectangular double seal recessed						
450 x 450 mm; ref MC2R-45/45						
56 kg)	PC £50.77	1.50	9.50	56.53	nr	66.03
600 x 450 mm; ref MC2R-60/45						
(71 kg)	PC £70.65	1.50	9.50	78.20	nr	87.70
600 x 600 mm; ref MC2R-60/60						
(75 kg)	PC £98.00	1.50	9.50	108.01	nr	117.51
Grade B; medium duty; circular single seal solid top						
500 mm; ref MB2-50 (106 kg)	PC £56.41	2.00	12.67	62.67	nr	75.34
550 mm; ref MB2-55 (112 kg)	PC £60.31	2.00	12.67	66.93	nr	79.60
600 mm; ref MB2-60 (134 kg)	PC £63.48	2.00	12.67	70.38	nr	83.05
Grade B; medium duty; rectangular single seal solid top						
600 x 450 mm; ref MB2-60/45						
(135 kg)	PC £54.75	2.00	12.67	60.87	nr	73.54
600 x 600 mm; ref MB2-60/60						
(170 kg)	PC £71.20	2.00	12.67	78.80	nr	91.47
Grade B; medium duty; rectangular single seal recessed						
600 x 450 mm; ref MB2R-60/45						
(145 kg)	PC £75.21	2.00	12.67	83.17	nr	95.84
600 x 600 mm; ref MB2R-60/60						
(171 kg)	PC £96.03	2.00	12.67	105.86	nr	118.53
Grade B; 'Chevron'; medium duty; double triangular solid top						
550 mm; ref MB1-55 (125 kg)	PC £54.87	2.00	12.67	61.00	nr	73.67
600 mm; ref MB1-60 (140 kg)	PC £65.62	2.00	12.67	72.72	nr	85.39
Grade A; heavy duty; single triangular solid top						
550 x 455 mm; ref MA-T (196 kg)	PC £78.52	2.50	15.83	86.78	nr	102.61
Grade A; 'Chevron'; heavy duty double triangular solid top						
500 mm; ref MA-50 (164 kg)	PC £91.74	3.00	19.00	101.19	nr	120.19
550 mm; ref MA-55 (176 kg)	PC £89.45	3.00	19.00	98.69	nr	117.69
600 mm; ref MA-60 (230 kg)		3.00	19.00	104.10	nr	123.10
British Standard best quality vitrified clay **channels; bedding and jointing in cement and** **sand (1:2)** Half section straight						
100 mm x 1.00 m long	PC £1.89	0.80	5.07	2.16	nr	7.23
150 mm x 1.00 m long	PC £3.14	1.00	6.33	3.60	nr	9.93
225 mm x 1.00 m long	PC £7.06	1.30	8.23	8.08	nr	16.31
300 mm x 1.00 m long	PC £13.99	1.60	10.13	16.01	nr	26.14
Half section bend						
100 mm	PC £1.93	0.60	3.80	2.21	nr	6.01
150 mm	PC £3.21	0.75	4.75	3.67	nr	8.42
225 mm	PC £10.79	1.00	6.33	12.35	nr	18.68
300 mm	PC £21.42	1.20	7.60	24.51	nr	32.11

R DISPOSAL SYSTEMS Including overheads and profit at 9.00%		Labour hours	Labour £	Material £	Unit	Total rate £
Half section taper straight						
150 mm	PC £8.11	0.70	4.43	9.29	nr	13.72
225 mm	PC £18.09	0.90	5.70	20.70	nr	26.40
300 mm	PC £35.79	1.10	6.97	40.96	nr	47.93
Half section taper bend						
150 mm	PC £12.34	0.90	5.70	14.13	nr	19.83
225 mm	PC £35.53	1.15	7.28	40.67	nr	47.95
300 mm	PC £70.30	1.40	8.87	80.46	nr	89.33
Three quarter section branch bend						
100 mm	PC £4.40	0.50	3.17	5.04	nr	8.21
150 mm	PC £7.41	0.75	4.75	8.48	nr	13.23
225 mm	PC £27.06	1.00	6.33	30.97	nr	37.30
300 mm	PC £53.70	1.33	8.42	61.47	nr	69.89

uPVC channels; with solvent weld or lip seal coupling joints; bedding in cement and sand

		Labour hours	Labour £	Material £	Unit	Total rate £
Half section cut away straight; with coupling either end						
110 mm	PC £13.93	0.30	1.90	15.85	nr	17.75
160 mm	PC £18.69	0.40	2.53	21.26	nr	23.79
Half section cut away long radius bend; with coupling either end						
110 mm	PC £15.81	0.30	1.90	18.28	nr	20.18
160 mm	PC £23.42	0.40	2.53	27.19	nr	29.72
Half section straight channel adaptor; with one coupling						
110 mm	PC £5.79	0.25	1.58	7.85	nr	9.43
160 mm	PC £7.69	0.33	2.09	11.23	nr	13.32
Half section cut away bend						
110 mm	PC £10.37	0.40	2.53	12.29	nr	14.82
Half section bend						
110 mm	PC £2.54	0.33	2.09	3.34	nr	5.43
160 mm	PC £4.93	0.50	3.17	6.63	nr	9.80
Half section channel connector						
110 mm	PC £1.22	0.08	0.51	2.24	nr	2.75
160 mm	PC £2.87	0.10	0.63	5.25	nr	5.88
Half section channel junction						
110 mm	PC £4.13	0.50	3.17	5.16	nr	8.33
160 mm	PC £7.80	0.60	3.80	9.91	nr	13.71
polypropylene slipper bend						
110 mm	PC £5.47	0.40	2.53	6.68	nr	9.21

R13 LAND DRAINAGE

Hand excavation of trenches to receive land
drain pipes; grading bottoms; earthwork
support; filling to within 150 mm of surface
filling to within 150 mm of surface with
gravel rejects; remainder filled with
excavated material and compacting; disposal
of surplus soil; spreading on site
average 50 m

Pipes not exceeding 200 mm; average depth	Labour hours	Labour £	Material £	Unit	Total rate £
0.75 m deep	1.70	11.49	9.21	m	20.70
1.00 m deep	2.25	15.21	12.27	m	27.48
1.25 m deep	3.15	21.29	15.57	m	36.86
1.50 m deep	5.40	36.49	19.63	m	56.12
1.75 m deep	6.40	43.25	23.22	m	66.47
2.00 m deep	7.40	50.01	26.80	m	76.81

R DISPOSAL SYSTEMS Including overheads and profit at 9.00%		Labour hours	Labour £	Material £	Unit	Total rate £
R13 LAND DRAINAGE - cont'd						
Surplus excavated material Removing from site to tip not exceeding 13 km (using lorries)						
machine loaded		-	-	-	m3	7.83
hand loaded		1.00	6.76	11.24	m3	18.00
Clay field drain pipes; BS 1196; unjointed Pipes; laid straight						
75 mm	PC £20.00/100	0.20	1.27	0.75	m	2.02
100 mm	PC £36.00/100	0.25	1.58	1.36	m	2.94
150 mm	PC £75.00/100	0.30	1.90	2.83	m	4.73
Vitrified clay perforated sub-soil pipes; **BS 65; 'Hepworths' 'Hepline' or similar** Pipes; laid straight						
100 mm	PC £2.94	0.22	1.39	3.37	m	4.76
150 mm	PC £5.28	0.27	1.71	6.04	m	7.75
225 mm	PC £9.71	0.36	2.28	11.12	m	13.40

Keep your figures up to date, free of charge

This section, and most of the information in this Price Book, is brought up to date every three months in the *Price Book Update*.

The *Update* is available free to all Price Book purchasers.

To ensure you receive your copy, simply complete the reply card from the centre of the book and return it to us.

S PIPED SUPPLY SYSTEMS
Including overheads and profit at 9.00%

ALTERNATIVE SERVICE PIPE AND FITTING PRICES

	£		£		£		£
Copper pipes to BS 2871 (£/100 m)							
Table X							
6 mm	48.60	10 mm	82.62	67 mm	1198.80	108 mm	2435.40
8 mm	65.88	12 mm	102.60	76 mm	1705.50	133 mm	3007.80
Table Y							
6 mm	65.16	10 mm	112.50	67 mm	2007.00	108 mm	4014.90
8 mm	86.31	12 mm	138.60	76 mm	2258.10		
Table Z							
15 mm	80.91	35 mm	446.40	54 mm	746.10	76 mm	1489.50
22 mm	153.90	42 mm	558.90	67 mm	1055.70	108 mm	2126.70
28 mm	194.40						

		£		£			£
PVC Class E cold water pressure system (£/each)							
(sizes shown as nearest equivalent metric sizes)							
pipe-per m-13 mm	0.77	end cap-13 mm	0.37	elbow-13 mm			0.53
-19 mm	1.07	-19 mm	0.42	-19 mm			0.62
-25 mm	1.39	-25 mm	0.49	-25 mm			0.77
-32 mm	2.24	-32 mm	0.76	-32 mm			1.42
-40 mm	2.95	-40 mm	1.18	-40 mm			1.84
coupling -13 mm	0.38	reducer-32 mm	1.13	tee -13 mm			0.59
-19 mm	0.42	-40 mm	1.34	-19 mm			0.75
-25 mm	0.50	MI.conn-13 mm	0.63	-25 mm			1.09
-32 mm	0.81	-19 mm	0.67	-32 mm			1.56
-40 mm	0.99	-25 mm	1.28	-40 mm			2.31
s.tank.cn.-19 mm	1.51	-32 mm	1.69				
-25 mm	1.69	-40 mm	1.84				

	Labour hours	Labour £	Material £	Unit	Total rate £
S10/S11 HOT AND COLD WATER					
Copper pipes; BS 2871 table X;					
capillary fittings; BS 864					
15 mm pipes; fixing with pipe					
clips; plugged and screwed PC £98.10/100m	0.25	2.59	1.37	m	3.96
Extra for					
made bend	0.15	1.55	-	nr	1.55
fittings with one end	0.11	1.14	0.51	nr	1.65
fittings with two ends	0.17	1.76	0.26	nr	2.02
fittings with three ends	0.25	2.59	0.48	nr	3.07
fittings with four ends	0.35	3.62	3.36	nr	6.98
stop end PC £0.47	0.11	1.14	0.51	nr	1.65
straight union coupling PC £2.09	0.17	1.76	2.27	nr	4.03
copper to iron connector PC £0.76	0.22	2.28	0.83	nr	3.11
elbow PC £0.24	0.17	1.76	0.26	nr	2.02
backplate elbow PC £1.53	0.35	3.62	1.66	nr	5.28
slow bend PC £0.81	0.17	1.76	0.89	nr	2.65
tee; equal PC £0.44	0.25	2.59	0.48	nr	3.07
tee; reducing PC £1.72	0.25	2.59	1.88	nr	4.47
cross PC £3.08	0.35	3.62	3.36	nr	6.98
straight tap connector PC £0.69	0.13	1.34	0.76	nr	2.10
bent tap connector PC £0.84	0.13	1.34	0.92	nr	2.26
straight tank connector; backnut PC £1.48	0.25	2.59	1.62	nr	4.21

S PIPED SUPPLY SYSTEMS Including overheads and profit at 9.00%		Labour hours	Labour £	Material £	Unit	Total rate £
S10/S11 HOT AND COLD WATER - cont'd						
Copper pipes; BS 2871 table X; **capillary fittings; BS 864 - cont'd**						
22 mm pipes; fixing with pipe						
clips; plugged and screwed	PC £193.50/100m	0.26	2.69	2.53	m	5.22
Extra for						
made bend		0.20	2.07	-	nr	2.07
fittings with one end		0.13	1.34	0.85	nr	2.19
fittings with two ends		0.22	2.28	0.53	nr	2.81
fittings with three ends		0.33	3.41	0.98	nr	4.39
fittings with four ends		0.44	4.55	4.33	nr	8.88
stop end	PC £0.78	0.13	1.34	0.85	nr	2.19
reducing coupling	PC £0.59	0.22	2.28	0.64	nr	2.92
straight union coupling	PC £3.22	0.22	2.28	3.51	nr	5.79
copper to iron connector	PC £1.27	0.31	3.21	1.39	nr	4.60
elbow	PC £0.48	0.22	2.28	0.53	nr	2.81
backplate elbow	PC £3.17	0.44	4.55	3.46	nr	8.01
slow bend	PC £1.37	0.22	2.28	1.50	nr	3.78
tee; equal	PC £0.90	0.33	3.41	0.98	nr	4.39
tee; reducing	PC £0.87	0.33	3.41	0.95	nr	4.36
cross	PC £3.98	0.44	4.55	4.33	nr	8.88
straight tap connector	PC £1.06	0.17	1.76	1.16	nr	2.92
bent tap connector	PC £1.61	0.17	1.76	1.75	nr	3.51
straight tank connector; backnut	PC £2.20	0.33	3.41	2.40	nr	5.81
28 mm pipes; fixing with pipe						
clips; plugged and screwed	PC £250.20/100m	0.29	3.00	3.26	m	6.26
Extra for						
made bend		0.25	2.59	-	nr	2.59
fittings with one end		0.15	1.55	1.65	nr	3.20
fittings with two ends		0.28	2.90	0.96	nr	3.86
fittings with three ends		0.41	4.24	2.04	nr	6.28
fittings with four ends		0.56	5.79	6.21	nr	12.00
stop end	PC £1.51	0.15	1.55	1.65	nr	3.20
reducing coupling	PC £1.30	0.28	2.90	1.42	nr	4.32
straight union coupling	PC £4.42	0.28	2.90	4.82	nr	7.72
copper to iron connector	PC £2.00	0.39	4.03	2.18	nr	6.21
elbow	PC £0.88	0.28	2.90	0.96	nr	3.86
slow bend	PC £2.30	0.28	2.90	2.50	nr	5.40
tee; equal	PC £1.88	0.41	4.24	2.04	nr	6.28
tee; reducing	PC £2.02	0.41	4.24	2.20	nr	6.44
cross	PC £5.70	0.56	5.79	6.21	nr	12.00
straight tank connector; backnut	PC £3.03	0.41	4.24	3.30	nr	7.54
35 mm pipes; fixing with pipe						
clips; plugged and screwed	PC £555.30/100m	0.33	3.41	7.02	m	10.43
Extra for						
made bend		0.30	3.10	-	nr	3.10
fittings with one end		0.17	1.76	2.54	nr	4.30
fittings with two ends		0.33	3.41	2.56	nr	5.97
fittings with three ends		0.46	4.76	4.39	nr	9.15
stop end	PC £2.33	0.17	1.76	2.54	nr	4.30
reducing coupling	PC £1.92	0.33	3.41	2.09	nr	5.50
straight union coupling	PC £6.34	0.33	3.41	6.91	nr	10.32
copper to iron connector	PC £3.12	0.44	4.55	3.40	nr	7.95
elbow	PC £2.35	0.33	3.41	2.56	nr	5.97
bend; 91.5 degrees	PC £4.12	0.33	3.41	4.49	nr	7.90
tee; equal	PC £4.03	0.46	4.76	4.39	nr	9.15
tee; reducing	PC £3.82	0.46	4.76	4.17	nr	8.93
pitcher tee; equal or reducing	PC £6.69	0.46	4.76	7.29	nr	12.05
straight tank connector; backnut	PC £4.07	0.46	4.76	4.44	nr	9.20

S PIPED SUPPLY SYSTEMS Including overheads and profit at 9.00%		Labour hours	Labour £	Material £	Unit	Total rate £
42 mm pipes; fixing with pipe						
clips; plugged and screwed	PC £672.30/100m	0.38	3.93	8.70	m	12.63
Extra for						
made bend		0.40	4.14	-	nr	4.14
fittings with one end		0.19	1.97	3.30	nr	5.27
fittings with two ends		0.39	4.03	3.82	nr	7.85
fittings with three ends		0.52	5.38	6.46	nr	11.84
stop end	PC £3.03	0.19	1.97	3.30	nr	5.27
reducing coupling	PC £2.88	0.39	4.03	3.14	nr	7.17
straight union coupling	PC £8.99	0.39	4.03	9.80	nr	13.83
copper to iron connector	PC £3.78	0.50	5.17	4.12	nr	9.29
elbow	PC £3.50	0.39	4.03	3.82	nr	7.85
bend; 91.5 degrees	PC £6.58	0.39	4.03	7.17	nr	11.20
tee; equal	PC £5.93	0.52	5.38	6.46	nr	11.84
tee; reducing	PC £7.01	0.52	5.38	7.65	nr	13.03
pitcher tee; equal or reducing	PC £9.41	0.52	5.38	10.25	nr	15.63
straight tank connector; backnut	PC £5.17	0.52	5.38	5.64	nr	11.02
54 mm pipes; fixing with pipe						
clips; plugged and screwed	PC £870.30/100m	0.45	4.65	11.63	m	16.28
Extra for						
made bend		0.55	5.69	-	nr	5.69
fittings with one end		0.21	2.17	4.88	nr	7.05
fittings with two ends		0.44	4.55	8.86	nr	13.41
fittings with three ends		0.57	5.90	11.90	nr	17.80
stop end	PC £4.48	0.21	2.17	4.88	nr	7.05
reducing coupling	PC £4.23	0.44	4.55	4.61	nr	9.16
straight union coupling	PC £14.62	0.44	4.55	15.93	nr	20.48
copper to iron connector	PC £6.27	0.55	5.69	6.84	nr	12.53
elbow	PC £8.13	0.44	4.55	8.86	nr	13.41
bend; 91.5 degrees	PC £9.83	0.44	4.55	10.71	nr	15.26
tee; equal	PC £10.92	0.57	5.90	11.90	nr	17.80
tee; reducing	PC £11.09	0.57	5.90	12.09	nr	17.99
pitcher tee; equal or reducing	PC £13.55	0.57	5.90	14.76	nr	20.66
straight tank connector; backnut	PC £7.88	0.57	5.90	8.59	nr	14.49
Copper pipes; BS 2871 table X; compression						
fittings; BS 864						
15 mm pipes; fixing with pipe						
clips; plugged and screwed	PC £98.10/100m	0.24	2.48	1.46	m	3.94
Extra for						
made bend		0.15	1.55	-	nr	1.55
fittings with one end		0.10	1.03	0.69	nr	1.72
fittings with two ends		0.15	1.55	0.56	nr	2.11
fittings with three ends		0.22	2.28	0.79	nr	3.07
stop end	PC £0.64	0.10	1.03	0.69	nr	1.72
straight coupling	PC £0.44	0.15	1.55	0.48	nr	2.03
female coupling	PC £0.44	0.20	2.07	0.48	nr	2.55
elbow	PC £0.52	0.15	1.55	0.56	nr	2.11
female wall elbow	PC £1.13	0.30	3.10	1.23	nr	4.33
slow bend	PC £2.10	0.15	1.55	2.29	nr	3.84
bent radiator union; chrome finish	PC £1.58	0.20	2.07	1.72	nr	3.79
tee; equal	PC £0.73	0.22	2.28	0.79	nr	3.07
straight swivel connector	PC £0.90	0.12	1.24	0.98	nr	2.22
bent swivel connector	PC £0.95	0.12	1.24	1.03	nr	2.27
tank coupling; locknut	PC £1.08	0.22	2.28	1.18	nr	3.46
22 mm pipes; fixing with pipe						
clips; plugged and screwed	PC £193.50/100m	0.25	2.59	2.66	m	5.25
Extra for						
made bend		0.20	2.07	-	nr	2.07
fittings with one end		0.12	1.24	0.90	nr	2.14
fittings with two ends		0.20	2.07	0.96	nr	3.03
fittings with three ends		0.30	3.10	1.38	nr	4.48
stop end	PC £0.83	0.12	1.24	0.90	nr	2.14
straight coupling	PC £0.74	0.20	2.07	0.80	nr	2.87
reducing set	PC £0.54	0.06	0.62	0.59	nr	1.21
female coupling	PC £0.67	0.28	2.90	0.72	nr	3.62

S PIPED SUPPLY SYSTEMS Including overheads and profit at 9.00%		Labour hours	Labour £	Material £	Unit	Total rate £
S10/S11 HOT AND COLD WATER - cont'd						
Copper pipes; BS 2871 table X; compression **fittings; BS 864 - cont'd** 22 mm pipes; fixing with pipe clips; plugged and screwed						
elbow	PC £0.88	0.20	2.07	0.96	nr	3.03
female wall elbow	PC £2.59	0.40	4.14	2.82	nr	6.96
slow bend	PC £3.45	0.20	2.07	3.76	nr	5.83
tee; equal	PC £1.27	0.30	3.10	1.38	nr	4.48
tee; reducing	PC £1.81	0.30	3.10	1.97	nr	5.07
straight swivel connector	PC £1.40	0.16	1.66	1.53	nr	3.19
bent swivel connector	PC £2.04	0.16	1.66	2.22	nr	3.88
tank coupling; locknut	PC £0.98	0.30	3.10	1.07	nr	4.17
28 mm pipes; fixing with pipe clips; plugged and screwed	PC £250.20/100m	0.28	2.90	3.61	m	6.51
Extra for						
made bend		0.25	2.59	-	nr	2.59
fittings with one end		0.14	1.45	1.66	nr	3.11
fittings with two ends		0.25	2.59	2.30	nr	4.89
fittings with three ends		0.37	3.83	3.35	nr	7.18
stop end	PC £1.53	0.14	1.45	1.66	nr	3.11
straight coupling	PC £1.71	0.25	2.59	1.86	nr	4.45
reducing set	PC £0.81	0.07	0.72	0.89	nr	1.61
female coupling	PC £1.09	0.35	3.62	1.19	nr	4.81
elbow	PC £2.11	0.25	2.59	2.30	nr	4.89
slow bend	PC £4.63	0.25	2.59	5.04	nr	7.63
tee; equal	PC £3.07	0.37	3.83	3.35	nr	7.18
tee; reducing	PC £3.19	0.37	3.83	3.48	nr	7.31
tank coupling; locknut	PC £1.97	0.37	3.83	2.15	nr	5.98
35 mm pipes; fixing with pipe clips; plugged and screwed	PC £555.30/100m	0.32	3.31	7.60	m	10.91
Extra for						
made bend		0.30	3.10	-	nr	3.10
fittings with one end		0.16	1.66	2.73	nr	4.39
fittings with two ends		0.30	3.10	4.51	nr	7.61
fittings with three ends		0.42	4.34	6.10	nr	10.44
stop end	PC £2.51	0.16	1.66	2.73	nr	4.39
straight coupling	PC £3.19	0.30	3.10	3.48	nr	6.58
reducing set	PC £1.39	0.08	0.83	1.51	nr	2.34
female coupling	PC £2.80	0.40	4.14	3.05	nr	7.19
elbow	PC £4.14	0.30	3.10	4.51	nr	7.61
tee; equal	PC £5.60	0.42	4.34	6.10	nr	10.44
tee; reducing	PC £5.60	0.42	4.34	6.10	nr	10.44
tank coupling; locknut	PC £4.17	0.42	4.34	4.54	nr	8.88
42 mm pipes; fixing with pipe clips; plugged and screwed	PC £672.30/100m	0.37	3.83	9.40	m	13.23
Extra for						
made bend		0.40	4.14	-	nr	4.14
fittings with one end		0.18	1.86	4.49	nr	6.35
fittings with two ends		0.35	3.62	6.36	nr	9.98
fittings with three ends		0.47	4.86	10.16	nr	15.02
stop end	PC £4.12	0.18	1.86	4.49	nr	6.35
straight coupling	PC £4.10	0.35	3.62	4.47	nr	8.09
reducing set	PC £2.18	0.09	0.93	2.38	nr	3.31
female coupling	PC £3.72	0.45	4.65	4.06	nr	8.71
elbow	PC £5.83	0.35	3.62	6.36	nr	9.98
tee; equal	PC £9.32	0.47	4.86	10.16	nr	15.02
tee; reducing	PC £8.73	0.47	4.86	9.51	nr	14.37
54 mm pipes; fixing with pipe clips; plugged and screwed	PC £870.30/100m	0.44	4.55	12.38	m	16.93
Extra for						
made bend		0.55	5.69	-	nr	5.69
fittings with two ends		0.40	4.14	10.44	nr	14.58
fittings with three ends		0.52	5.38	15.90	nr	21.28
straight coupling	PC £6.27	0.40	4.14	6.83	nr	10.97

S PIPED SUPPLY SYSTEMS Including overheads and profit at 9.00%		Labour hours	Labour £	Material £	Unit	Total rate £
reducing set	PC £3.66	0.10	1.03	3.99	nr	5.02
female wall elbow	PC £5.39	0.50	5.17	5.88	nr	11.05
elbow	PC £9.58	0.40	4.14	10.44	nr	14.58
tee; equal	PC £14.59	0.52	5.38	15.90	nr	21.28
tee; reducing	PC £14.81	0.52	5.38	16.15	nr	21.53
Black MDPE pipes; BS 6730; plastic compression fittings						
20 mm pipes; fixing with pipe clips; plugged and screwed	PC £28.70/100m	0.25	2.59	1.09	m	3.68
Extra for						
fittings with two ends		0.20	2.07	1.64	nr	3.71
fittings with three ends		0.30	3.10	2.11	nr	5.21
straight coupling	PC £1.33	0.20	2.07	1.45	nr	3.52
male adaptor	PC £0.83	0.28	2.90	0.91	nr	3.81
elbow	PC £1.51	0.20	2.07	1.64	nr	3.71
tee; equal	PC £1.94	0.30	3.10	2.11	nr	5.21
end cap	PC £1.03	0.12	1.24	1.12	nr	2.36
straight swivel connector	PC £0.91	0.16	1.66	0.99	nr	2.65
bent swivel connector	PC £1.11	0.16	1.66	1.21	nr	2.87
25 mm pipes; fixing with pipe clips; plugged and screwed	PC £35.00/100m	0.28	2.90	1.37	m	4.27
Extra for						
fittings with two ends		0.25	2.59	1.98	nr	4.57
fittings with three ends		0.37	3.83	2.81	nr	6.64
straight coupling	PC £1.64	0.25	2.59	1.79	nr	4.38
reducer	PC £1.62	0.25	2.59	1.77	nr	4.36
male adaptor	PC £0.99	0.35	3.62	1.08	nr	4.70
elbow	PC £1.82	0.25	2.59	1.98	nr	4.57
tee; equal	PC £2.58	0.37	3.83	2.81	nr	6.64
end cap	PC £1.19	0.14	1.45	1.30	nr	2.75
32 mm pipes; fixing with pipe clips; plugged and screwed	PC £57.40/100m	0.32	3.31	1.72	m	5.03
Extra for						
fittings with two ends		0.30	3.10	2.46	nr	5.56
fittings with three ends		0.42	4.34	3.67	nr	8.01
straight coupling	PC £2.30	0.30	3.10	2.50	nr	5.60
reducer	PC £2.14	0.30	3.10	2.33	nr	5.43
male adaptor	PC £1.23	0.40	4.14	1.34	nr	5.48
elbow	PC £2.26	0.30	3.10	2.46	nr	5.56
tee; equal	PC £3.37	0.42	4.34	3.67	nr	8.01
end cap	PC £1.34	0.16	1.66	1.46	nr	3.12
50 mm pipes; fixing with pipe clips; plugged and screwed	PC £147.00/100m	0.35	3.62	2.87	m	6.49
Extra for						
fittings with two ends		0.35	3.62	6.04	nr	9.66
fittings with three ends		0.47	4.86	8.20	nr	13.06
straight coupling	PC £5.15	0.35	3.62	5.62	nr	9.24
reducer	PC £4.75	0.35	3.62	5.18	nr	8.80
male adaptor	PC £3.05	0.45	4.65	3.33	nr	7.98
elbow	PC £5.54	0.35	3.62	6.04	nr	9.66
tee; equal	PC £7.53	0.47	4.86	8.20	nr	13.06
end cap	PC £3.41	0.18	1.86	3.72	nr	5.58
63 mm pipes; fixing with pipe clips; plugged and screwed	PC £213.50/100m	0.40	4.14	3.96	m	8.10
Extra for						
fittings with two ends		0.40	4.14	7.34	nr	11.48
fittings with three ends		0.52	5.38	11.66	nr	17.04
straight coupling	PC £7.84	0.40	4.14	8.55	nr	12.69
reducer	PC £6.73	0.40	4.14	7.34	nr	11.48
male adaptor	PC £4.35	0.50	5.17	4.75	nr	9.92
elbow	PC £6.73	0.40	4.14	7.34	nr	11.48
tee; equal	PC £10.70	0.52	5.38	11.66	nr	17.04
end cap	PC £4.71	0.20	2.07	5.13	nr	7.20

S PIPED SUPPLY SYSTEMS Including overheads and profit at 9.00%		Labour hours	Labour £	Material £	Unit	Total rate £
S10/S11 HOT AND COLD WATER - cont'd						
Stainless steel pipes; BS 4127; stainless **steel capillary fittings**						
15 mm pipes; fixing with pipe						
clips; plugged and screwed	PC £107.35/100m	0.30	3.10	2.30	m	5.40
Extra for						
fittings with two ends		0.20	2.07	3.53	nr	5.60
fittings with three ends		0.30	3.10	5.22	nr	8.32
bend	PC £3.24	0.20	2.07	3.53	nr	5.60
tee; equal	PC £4.79	0.30	3.10	5.22	nr	8.32
tap connector	PC £11.59	0.27	2.79	12.63	nr	15.42
22 mm pipes; fixing with pipe						
clips; plugged and screwed	PC £169.10/100m	0.32	3.31	3.44	m	6.75
Extra for						
fittings with two ends		0.27	2.79	4.20	nr	6.99
fittings with three ends		0.40	4.14	7.13	nr	11.27
reducer	PC £13.40	0.27	2.79	14.61	nr	17.40
bend	PC £3.86	0.27	2.79	4.20	nr	6.99
tee; equal	PC £6.55	0.40	4.14	7.13	nr	11.27
tee; reducing	PC £13.78	0.40	4.14	15.01	nr	19.15
tap connector	PC £21.77	0.38	3.93	23.73	nr	27.66
28 mm pipes; fixing with pipe						
clips; plugged and screwed	PC £247.00/100m	0.35	3.62	4.69	m	8.31
Extra for						
fittings with two ends		0.34	3.52	6.04	nr	9.56
fittings with three ends		0.50	5.17	10.61	nr	15.78
reducer	PC £29.77	0.34	3.52	32.45	nr	35.97
bend	PC £5.54	0.34	3.52	6.04	nr	9.56
tee; equal	PC £9.74	0.50	5.17	10.61	nr	15.78
tee; reducing	PC £17.34	0.50	5.17	18.90	nr	24.07
tap connector	PC £43.80	0.48	4.97	47.74	nr	52.71
Copper, brass and gunmetal ancillaries; **screwed joints to fittings**						
Bibtaps; brass						
chromium plated; capstan head						
15 mm	PC £7.97	0.15	1.55	8.79	nr	10.34
22 mm	PC £10.68	0.20	2.07	11.74	nr	13.81
self closing; 15 mm	PC £10.63	0.15	1.55	11.69	nr	13.24
crutch head with hose union;						
15 mm	PC £4.94	0.15	1.55	5.48	nr	7.03
draincock 15 mm	PC £1.68	0.10	1.03	1.89	nr	2.92
Stopcock; brass/gunmetal						
capillary joints to copper						
15 mm	PC £1.68	0.20	2.07	1.83	nr	3.90
22 mm	PC £3.04	0.27	2.79	3.32	nr	6.11
28 mm	PC £8.67	0.34	3.52	9.45	nr	12.97
compression joints to copper						
15 mm	PC £1.66	0.18	1.86	1.81	nr	3.67
22 mm	PC £2.91	0.24	2.48	3.17	nr	5.65
28 mm	PC £7.57	0.30	3.10	8.26	nr	11.36
compression joints to polyethylene						
13 mm	PC £4.52	0.26	2.69	4.93	nr	7.62
19 mm	PC £6.82	0.33	3.41	7.43	nr	10.84
25 mm	PC £9.85	0.40	4.14	10.74	nr	14.88
Gunmetal 'Fullway' gate valve;						
capillary joints to copper						
15 mm	PC £5.12	0.20	2.07	5.59	nr	7.66
22 mm	PC £6.04	0.27	2.79	6.58	nr	9.37
28 mm	PC £8.25	0.34	3.52	9.00	nr	12.52
35 mm	PC £18.40	0.41	4.24	20.05	nr	24.29
42 mm	PC £21.97	0.47	4.86	23.95	nr	28.81
54 mm	PC £31.86	0.53	5.48	34.72	nr	40.20
Gunmetal stopcock; screwed joints to iron						
15 mm	PC £5.28	0.30	3.10	5.76	nr	8.86
20 mm	PC £8.44	0.40	4.14	9.20	nr	13.34
25 mm	PC £12.27	0.50	5.17	13.37	nr	18.54

S PIPED SUPPLY SYSTEMS Including overheads and profit at 9.00%		Labour hours	Labour £	Material £	Unit	Total rate £
Bronze gate valve; screwed joints to iron						
15 mm	PC £11.17	0.30	3.10	12.18	nr	15.28
20 mm	PC £14.74	0.40	4.14	16.07	nr	20.21
25 mm	PC £19.23	0.50	5.17	20.96	nr	26.13
35 mm	PC £28.38	0.60	6.21	30.93	nr	37.14
42 mm	PC £36.40	0.70	7.24	39.68	nr	46.92
54 mm	PC £53.13	0.80	8.28	57.91	nr	66.19
Chromium plated; pre-setting radiator valve; compression joint; union outlet 15 mm						
	PC £2.85	0.22	2.28	3.11	nr	5.39
Chromium plated; lockshield radiator valve; compression joint; union outlet 15 mm						
	PC £2.85	0.22	2.28	3.11	nr	5.39
Brass ball valves; BS 1212 Part 1; piston type; high pressure; copper float; screwed joint to cistern						
15 mm	PC £2.63	0.25	2.59	4.88	nr	7.47
22 mm	PC £4.82	0.30	3.10	7.82	nr	10.92
25 mm	PC £11.06	0.35	3.62	15.42	nr	19.04
Water tanks/cisterns						
Polyethylene cold water feed and expansion cistern; BS 4213; with covers						
ref PC15; 68 litres	PC £17.32	1.25	12.93	18.88	nr	31.81
ref PC25; 114 litres	PC £22.99	1.45	15.00	25.06	nr	40.06
ref PC40; 182 litres	PC £39.33	1.75	18.10	42.87	nr	60.97
ref PC50; 227 litres	PC £42.12	1.95	20.17	45.90	nr	66.07
GRP cold water storage cistern; with covers						
ref 899.10; 27 litres	PC £19.66	1.10	11.38	21.43	nr	32.81
ref 899.25; 68 litres	PC £31.87	1.25	12.93	34.74	nr	47.67
ref 899.40; 114 litres	PC £37.64	1.45	15.00	41.03	nr	56.03
ref 899.70; 227 litres	PC £75.01	1.95	20.17	81.77	nr	101.94
Storage cylinders/calorifiers						
Copper cylinders; direct; BS 699; grade 3						
ref 1; 350 x 900 mm; 74 litres	PC £39.00	1.50	15.52	42.51	nr	58.03
ref 2; 450 x 750 mm; 98 litres	PC £40.25	1.75	18.10	43.87	nr	61.97
ref 7; 450 x 900 mm; 120 litres	PC £44.61	2.00	20.69	48.63	nr	69.32
ref 8; 450 x 1050 mm; 144 litres	PC £48.75	2.80	28.96	53.14	nr	82.10
ref 9; 450 x 1200 mm; 166 litres	PC £57.98	3.60	37.24	63.20	nr	100.44
Copper cylinders; single feed coil indirect; BS 1566 Part 2; grade 3						
ref 2; 300 x 1500 mm; 96 litres	PC £50.23	2.00	20.69	54.75	nr	75.44
ref 3; 400 x 1050 mm; 114 litres	PC £53.40	2.25	23.27	58.21	nr	81.48
ref 7; 450 x 900 mm; 117 litres	PC £53.00	2.50	25.86	57.77	nr	83.63
ref 8; 450 x 1050 mm; 140 litres	PC £60.42	3.00	31.03	65.86	nr	96.89
ref 9; 450 x 1200 mm; 162 litres	PC £78.74	3.50	36.20	85.82	nr	122.02
Combination copper hot water storage units; coil direct; BS 3198; (hot/cold)						
400 x 900 mm; (65/20 litres)	PC £56.06	2.80	28.96	61.10	nr	90.06
450 x 900 mm; (85/25 litres)	PC £57.98	3.90	40.34	63.20	nr	103.54
450 x 1050 mm; (115/25 litres)	PC £63.75	4.90	50.69	69.48	nr	120.17
450 x 1200 mm; (115/45 litres)	PC £67.36	5.50	56.89	73.43	nr	130.32
Combination copper hot water storage						
450 x 900 mm; (85/25 litres)	PC £76.74	4.40	45.51	83.65	nr	129.16
450 x 1200 mm; (115/45 litres)	PC £83.24	6.00	62.06	90.73	nr	152.79
Galvanized mild steel cylinders; direct; BS 417 Part 2; grade C; welded construction						
ref YM114; 100 litres	PC £81.44	2.50	25.86	88.77	nr	114.63
ref YM141; 123 litres	PC £86.94	3.30	34.14	94.76	nr	128.90
ref YM218; 195 litres	PC £108.96	4.80	49.65	118.77	nr	168.42
Galvanized mild steel cylinders; indirect; BS 1565 Part 2; annular heaters; for vertical fixing						
ref BSG.2m; 136 litres	PC £132.87	4.00	41.38	144.83	nr	186.21
ref BSG.3m; 159 litres	PC £194.18	4.75	49.13	211.66	nr	260.79
ref BSG.3m; 227 litres	PC £198.95	5.50	56.89	216.86	nr	273.75
ref BSG.5m; 364 litres	PC £331.61	6.50	67.24	361.45	nr	428.69
ref BSG.5m; 455 litres	PC £400.21	8.00	82.75	436.23	nr	518.98

S PIPED SUPPLY SYSTEMS Including overheads and profit at 9.00%		Labour hours	Labour £	Material £	Unit	Total rate £
S10/S11 HOT AND COLD WATER - cont'd						
Storage cylinders/calorifiers - cont'd						
Galvanized mild steel cisterns; BS 417						
Part 2; grade A						
ref SCM270;　191 litres	PC £85.62	2.00	20.69	93.33	nr	114.02
ref SCM450/1; 327 litres	PC £107.39	3.00	31.03	117.06	nr	148.09
ref SCM1600; 1227 litres	PC £252.86	6.00	62.06	275.62	nr	337.68
ref SCM2720; 2137 litres	PC £494.54	12.00	124.13	539.05	nr	663.18
ref SCM4540; 3364 litres	PC £701.81	22.00	227.57	764.97	nr	992.54
Galvanized mild steel tanks;						
BS 417 Part 2; grade A;						
ref T25/1;　95 litres	PC £68.20	1.75	18.10	74.34	nr	92.44
ref T30/A; 114 litres	PC £73.08	2.75	28.45	79.66	nr	108.11
ref T40;　155 litres	PC £89.59	4.00	41.38	97.65	nr	139.03
Thermal insulation						
19 mm thick rigid mineral glass fibre						
sectional pipe lagging; plain finish; fixed						
with aluminium bands to steel or copper						
pipework; including working over pipe						
fittings						
around 15/15 mm pipes	PC £1.49	0.07	0.72	1.62	m	2.34
around 20/22 mm pipes	PC £1.57	0.10	1.03	1.71	m	2.74
around 25/28 mm pipes	PC £1.73	0.11	1.14	1.89	m	3.03
around 32/35 mm pipes	PC £1.92	0.12	1.24	2.09	m	3.33
around 40/42 mm pipes	PC £2.04	0.13	1.34	2.22	m	3.56
around 50/54 mm pipes	PC £2.36	0.15	1.55	2.58	m	4.13
19 mm thick rigid mineral glass fibre						
sectional pipe lagging; canvas or class O						
laquered aluminium finish; fixed with						
aluminium bands to steel or copper pipework;						
including working over pipe fittings						
around 15/15 mm pipes	PC £1.87	0.07	0.72	2.03	m	2.75
around 20/22 mm pipes	PC £2.02	0.10	1.03	2.20	m	3.23
around 25/28 mm pipes	PC £2.22	0.11	1.14	2.42	m	3.56
around 32/35 mm pipes	PC £2.42	0.12	1.24	2.64	m	3.88
around 40/42 mm pipes	PC £2.61	0.13	1.34	2.84	m	4.18
around 50/54 mm pipes	PC £3.03	0.15	1.55	3.30	m	4.85
25 mm thick expanded polystyrene lagging						
sets; class O finish; for mild steel						
cisterns to BS 417; complete with fixing						
bands; for cisterns size (ref)						
762 x 584 x 610 mm; (SCM270)	PC £2.88	0.80	8.28	3.14	nr	11.42
1219 x 610 x 610 mm; (SCM450/1)	PC £9.70	0.90	9.31	10.57	nr	19.88
1524 x 1143 x 914 mm; (SCM1600)	PC £21.60	1.10	11.38	23.54	nr	34.92
1829 x 1219 x 1219 mm; (SCM2270)	PC £27.42	1.30	13.45	29.89	nr	43.34
2438 x 1524 x 1219 mm; (SCM4540)	PC £43.67	1.50	15.52	47.60	nr	63.12
50 mm thick glass-fibre filled polyethylene						
insulating jackets for GRP or polyethylene						
cold water cisterns; complete with fixing						
bands; for cisterns size						
445 x 305 x 300 mm;　(18 litres)	PC £1.92	0.40	4.14	2.09	nr	6.23
495 x 368 x 362 mm;　(27 litres)	PC £3.03	0.50	5.17	3.30	nr	8.47
630 x 450 x 420 mm;　(68 litres)	PC £3.03	0.60	6.21	3.30	nr	9.51
665 x 490 x 515 mm;　(91 litres)	PC £4.03	0.70	7.24	4.39	nr	11.63
700 x 540 x 535 mm; (114 litres)	PC £4.03	0.80	8.28	4.39	nr	12.67
955 x 605 x 595 mm; (182 litres)	PC £5.44	0.85	8.79	5.93	nr	14.72
1155 x 640 x 595 mm; (227 litres)	PC £6.20	0.90	9.31	6.75	nr	16.06
80 mm thick glass-fibre filled insulating						
jackets in flame retardant PVC to BS 1763;						
type 1B; segmental type for hot water						
cylinders; complete with fixing bands; for						
cylinders size (ref)						
400 x 900 mm; (2)	PC £5.13	0.33	3.41	5.59	nr	9.00
450 x 750 mm; (5)	PC £5.13	0.33	3.41	5.59	nr	9.00
450 x 900 mm; (7)	PC £5.13	0.33	3.41	5.59	nr	9.00
450 x 1050 mm;(8)	PC £5.76	0.40	4.14	6.28	nr	10.42
500 x 1200 mm;(-)	PC £6.66	0.50	5.17	7.26	nr	12.43

S PIPED SUPPLY SYSTEMS Including overheads and profit at 9.00%	Labour hours	Labour £	Material £	Unit	Total rate £

S13 PRESSURISED WATER

Copper pipes; BS 2871 Part 1 table Y;
annealed; mains pipework; no joints in the
running length; laid in trenches
Pipes

15 mm	PC £190.80/100m	0.10	1.03	2.19	m	3.22
22 mm	PC £332.10/100m	0.11	1.14	3.81	m	4.95
28 mm	PC £490.50/100m	0.12	1.24	5.62	m	6.86
35 mm	PC £767.70/100m	0.13	1.34	8.78	m	10.12
42 mm	PC £927.00/100m	0.15	1.55	10.61	m	12.16
54 mm	PC £1583.10/100m	0.18	1.86	18.12	m	19.98

Blue MDPE pipes; BS 6527; mains
pipework; no joints in the running
length; laid in trenches
Pipes

20 mm	PC £28.70/100m	0.11	1.14	0.32	m	1.46
25 mm	PC £35.00/100m	0.12	1.24	0.40	m	1.64
32 mm	PC £57.40/100m	0.13	1.34	0.66	m	2.00
50 mm	PC £147.00/100m	0.15	1.55	1.68	m	3.23
60 mm	PC £213.50/100m	0.16	1.66	2.45	m	4.11

Steel pipes; BS 1387; heavy weight;
galvanized; mains pipework; screwed joints
in the running length; laid in trenches
Pipes

15 mm	PC £122.81/100m	0.13	1.34	1.40	m	2.74
20 mm	PC £143.43/100m	0.14	1.45	1.64	m	3.09
25 mm	PC £205.27/100m	0.16	1.66	2.35	m	4.01
32 mm	PC £257.17/100m	0.18	1.86	2.94	m	4.80
40 mm	PC £300.03/100m	0.20	2.07	3.43	m	5.50
50 mm	PC £415.72/100m	0.25	2.59	4.76	m	7.35

Ductile iron bitumen coated pipes and
fittings; BS 4772; class K9; Stanton's
'Tyton' water main pipes or similar;
flexible joints

100 mm pipes; laid straight	PC £43.76/5.5m	0.60	4.60	9.47	m	14.07
Extra for						
bend; 45 degrees	PC £16.31	0.60	4.60	21.65	nr	26.25
branch; 45 degrees; socketted	PC £87.13	0.90	6.91	100.78	nr	107.69
tee	PC £27.26	0.90	6.91	35.52	nr	42.43
flanged spigot	PC £12.37	0.60	4.60	15.42	nr	20.02
flanged socket	PC £15.35	0.60	4.60	18.67	nr	23.27
150 mm pipes; laid straight	PC £64.52/5.5m	0.70	5.37	13.83	m	19.20
Extra for						
bend; 45 degrees	PC £28.13	0.70	5.37	34.94	nr	40.31
branch; 45 degrees; socketted	PC £98.53	1.05	8.06	113.82	nr	121.88
tee	PC £42.98	1.05	8.06	53.27	nr	61.33
flanged spigot	PC £21.21	0.70	5.37	25.26	nr	30.63
flanged socket	PC £24.06	0.70	5.37	28.36	nr	33.73
200 mm pipes; laid straight	PC £100.98/5.5m	1.00	7.67	21.52	m	29.19
Extra for						
bend; 45 degrees	PC £55.22	1.00	7.67	65.49	nr	73.16
branch; 45 degrees; socketted	PC £137.95	1.50	11.51	158.31	nr	169.82
tee	PC £83.47	1.50	11.51	98.93	nr	110.44
flanged spigot	PC £33.63	1.00	7.67	39.30	nr	46.97
flanged socket	PC £35.80	1.00	7.67	41.67	nr	49.34

S PIPED SUPPLY SYSTEMS Including overheads and profit at 9.00%		Labour hours	Labour £	Material £	Unit	Total rate £
S32 NATURAL GAS						
Ductile iron bitumen coated pipes and fittings; BS 4772; class K9; Stanton's 'Stanlock' gas main pipes or similar; bolted gland joints						
100 mm pipes; laid straight	PC £48.67/5.5m	0.70	5.37	11.49	m	**16.86**
Extra for						
bend; 45 degrees	PC £17.16	0.70	5.37	29.87	nr	**35.24**
tee	PC £26.43	1.05	8.06	47.40	nr	**55.46**
flanged spigot	PC £12.37	0.70	5.37	20.92	nr	**26.29**
flanged socket	PC £14.92	0.70	5.37	23.70	nr	**29.07**
isolated 'Stanlock' joint	PC £6.83	0.35	2.69	7.44	nr	**10.13**
150 mm pipes; laid straight	PC £73.07/5.5m	0.90	6.91	17.16	m	**24.07**
Extra for						
bend; 45 degrees	PC £29.61	0.90	6.91	48.33	nr	**55.24**
tee	PC £41.88	1.35	10.36	72.40	nr	**82.76**
flanged spigot	PC £21.21	0.90	6.91	33.82	nr	**40.73**
flanged socket	PC £23.39	0.90	6.91	36.20	nr	**43.11**
isolated 'Stanlock' joint	PC £9.82	0.45	3.45	10.70	nr	**14.15**
200 mm pipes; laid straight	PC £106.92/5.5m	1.30	9.98	24.87	m	**34.85**
Extra for						
bend; 45 degrees	PC £53.69	1.30	9.98	79.93	nr	**89.91**
tee	PC £80.39	1.95	14.96	123.30	nr	**138.26**
flanged spigot	PC £33.63	1.30	9.98	50.92	nr	**60.90**
flanged socket	PC £34.77	1.30	9.98	52.17	nr	**62.15**
isolated 'Stanlock' joint	PC £13.09	0.65	4.99	14.27	nr	**19.26**

T MECHANICAL HEATING SYSTEMS ETC. Including overheads and profit at 9.00%	Labour hours	Labour £	Material £	Unit	Total rate £

T10 GAS/OIL FIRED BOILERS

Gas fired domestic boilers; cream or white
enamelled casing; 32 mm BSPT female
flow and return tappings; 102 mm flue socket
13 mm BSPT male draw-off outlet;
electric controls

13.19 kW output PC £241.50	5.00	51.72	263.49	nr	315.21
23.45 kW output PC £313.60	5.50	56.89	342.08	nr	398.97

Smoke flue pipework; light quality 'Duracem'
pipes and fittings; including asbestos
yarn and composition joints in the
running length
75 mm pipes; fixing with wall clips;

plugged and screwed PC £7.89/1.8m	0.35	3.62	5.21	m	8.83
Extra for					
loose sockets PC £2.38	0.40	4.14	3.03	nr	7.17
bend; square and obtuse PC £3.42	0.40	4.14	4.16	nr	8.30
Terminal cone caps; asbestos yarn and					
composition joint PC £13.10	0.40	4.14	14.71	nr	18.85

100 mm pipes; fixing with wall clips;

plugged and screwed PC £10.16/1.8m	0.40	4.14	6.75	m	10.89
Extra for					
loose sockets PC £3.09	0.45	4.65	4.00	nr	8.65
bend; square and obtuse PC £4.28	0.45	4.65	5.30	nr	9.95
Terminal cone caps; asbestos yarn and					
composition joint PC £13.91	0.45	4.65	15.79	nr	20.44

150 mm pipes; fixing with wall clips;

plugged and screwed PC £17.00/1.8m	0.50	5.17	11.27	m	16.44
Extra for					
loose sockets PC £5.30	0.55	5.69	6.82	nr	12.51
bend; square and obtuse PC £6.60	0.55	5.69	8.24	nr	13.93
Terminal cone caps; asbestos yarn and					
composition joint PC £20.99	0.55	5.69	23.92	nr	29.61

Smoke flue pipework; heavy quality 'Duracem'
pipes and fittings; including asbestos
yarn and composition joints in the
running length
125 mm pipes; fixing with wall clips;

plugged and screwed PC £17.47/1.8m	0.45	4.65	11.46	m	16.11
Extra for					
loose sockets PC £5.59	0.50	5.17	6.92	nr	12.09
bend; square and obtuse PC £7.53	0.50	5.17	9.05	nr	14.22
Terminal cone caps; asbestos yarn and					
composition joint PC £15.29	0.50	5.17	17.50	nr	22.67

175 mm pipes; fixing with wall clips;

plugged and screwed PC £32.18/1.8m	0.55	5.69	20.95	m	26.64
Extra for					
loose sockets PC £8.53	0.60	6.21	10.50	nr	16.71
bend; square and obtuse PC £13.79	0.60	6.21	16.24	nr	22.45
Terminal cone caps; asbestos yarn and					
composition joint PC £38.88	0.60	6.21	43.58	nr	49.79

225 mm pipes; fixing with wall clips;

plugged and screwed PC £42.17/1.8m	0.65	6.72	27.39	m	34.11
Extra for					
loose sockets PC £12.00	0.70	7.24	14.57	nr	21.81
bend; square and obtuse PC £23.50	0.70	7.24	27.12	nr	34.36
Terminal cone caps; asbestos yarn and					
composition joint PC £50.24	0.70	7.24	56.26	nr	63.50

T MECHANICAL HEATING SYSTEMS ETC. Including overheads and profit at 9.00%		Labour hours	Labour £	Material £	Unit	Total rate £
T30 MEDIUM TEMPERATURE HOT WATER HEATING						
Steel pipes; BS 1387; black; screwed joints; **malleable iron fittings; BS 143**						
15 mm pipes						
medium weight; fixing with pipe brackets;						
plugged and screwed	PC £68.48/100m	0.30	3.10	1.18	m	4.28
heavy weight; fixing with pipe brackets;						
plugged and screwed	PC £80.38/100m	0.30	3.10	1.32	m	4.42
Extra for						
fittings with one end		0.13	1.34	0.26	nr	1.60
fittings with two ends		0.25	2.59	0.33	nr	2.92
fittings with three ends		0.37	3.83	0.41	nr	4.24
fittings with four ends		0.50	5.17	0.93	nr	6.10
cap	PC £0.24	0.13	1.34	0.26	nr	1.60
socket; equal	PC £0.26	0.25	2.59	0.28	nr	2.87
socket; reducing	PC £0.31	0.25	2.59	0.33	nr	2.92
bend; 90 degree long radius M/F	PC £0.54	0.25	2.59	0.59	nr	3.18
elbow; 90 degree M/F	PC £0.31	0.25	2.59	0.33	nr	2.92
tee; equal	PC £0.37	0.37	3.83	0.41	nr	4.24
cross	PC £0.85	0.50	5.17	0.93	nr	6.10
union	PC £0.82	0.25	2.59	0.89	nr	3.48
isolated screwed joint	PC £0.26	0.30	3.10	0.38	nr	3.48
tank connection; longscrew and						
backnuts; lead washers; joint	PC £1.03	0.37	3.83	1.23	nr	5.06
20 mm pipes						
medium weight; fixing with pipe brackets;						
plugged and screwed	PC £82.95/100m	0.32	3.31	1.39	m	4.70
heavy weight; fixing with pipe brackets;						
plugged and screwed	PC £98.14/100m	0.32	3.31	1.56	m	4.87
Extra for						
fittings with one end		0.16	1.66	0.30	nr	1.96
fittings with two ends		0.33	3.41	0.44	nr	3.85
fittings with three ends		0.47	4.86	0.59	nr	5.45
fittings with four ends		0.66	6.83	1.41	nr	8.24
cap	PC £0.27	0.16	1.66	0.30	nr	1.96
socket; equal	PC £0.31	0.33	3.41	0.33	nr	3.74
socket; reducing	PC £0.37	0.33	3.41	0.41	nr	3.82
bend; 90 degree long radius M/F	PC £0.88	0.33	3.41	0.96	nr	4.37
elbow; 90 degree M/F	PC £0.41	0.33	3.41	0.44	nr	3.85
tee; equal	PC £0.54	0.47	4.86	0.59	nr	5.45
cross	PC £1.29	0.66	6.83	1.41	nr	8.24
union	PC £0.95	0.33	3.41	1.04	nr	4.45
isolated screwed joint	PC £0.31	0.40	4.14	0.44	nr	4.58
tank connection; longscrew and						
backnuts; lead washers; joint	PC £1.20	0.47	4.86	1.41	nr	6.27
25 mm pipes						
medium weight; fixing with pipe brackets;						
plugged and screwed	PC £119.58/100m	0.36	3.72	1.86	m	5.58
heavy weight; fixing with pipe brackets;						
plugged and screwed	PC £143.98/100m	0.36	3.72	2.14	m	5.86
Extra for						
fittings with one end		0.21	2.17	0.37	nr	2.54
fittings with two ends		0.42	4.34	0.74	nr	5.08
fittings with three ends		0.57	5.90	0.85	nr	6.75
fittings with four ends		0.84	8.69	1.78	nr	10.47
cap	PC £0.34	0.21	2.17	0.37	nr	2.54
socket; equal	PC £0.41	0.42	4.34	0.44	nr	4.78
socket; reducing	PC £0.49	0.42	4.34	0.54	nr	4.88
bend; 90 degree long radius M/F	PC £1.26	0.42	4.34	1.37	nr	5.71
elbow; 90 degree M/F	PC £0.68	0.42	4.34	0.74	nr	5.08
tee; equal	PC £0.78	0.57	5.90	0.85	nr	6.75
cross	PC £1.63	0.84	8.69	1.78	nr	10.47
union	PC £1.12	0.42	4.34	1.22	nr	5.56
isolated screwed joint	PC £0.41	0.50	5.17	0.57	nr	5.74
tank connection; longscrew and						
backnuts; lead washers; joint	PC £1.64	0.57	5.90	1.92	nr	7.82

T MECHANICAL HEATING SYSTEMS ETC. Including overheads and profit at 9.00%		Labour hours	Labour £	Material £	Unit	Total rate £
32 mm pipes						
medium weight; fixing with pipe brackets;						
plugged and screwed	PC £149.41/100m	0.42	4.34	2.35	m	6.69
heavy weight; fixing with pipe brackets;						
plugged and screwed	PC £180.35/100m	0.42	4.34	2.71	m	7.05
Extra for						
fittings with one end		0.25	2.59	0.54	nr	3.13
fittings with two ends		0.50	5.17	1.22	nr	6.39
fittings with three ends		0.67	6.93	1.41	nr	8.34
fittings with four ends		1.00	10.34	2.33	nr	12.67
cap	PC £0.49	0.25	2.59	0.54	nr	3.13
socket; equal	PC £0.68	0.50	5.17	0.74	nr	5.91
socket; reducing	PC £0.78	0.50	5.17	0.85	nr	6.02
bend; 90 degree long radius M/F	PC £2.11	0.50	5.17	2.30	nr	7.47
elbow; 90 degree M/F	PC £1.12	0.50	5.17	1.22	nr	6.39
tee; equal	PC £1.29	0.67	6.93	1.41	nr	8.34
cross	PC £2.14	1.00	10.34	2.33	nr	12.67
union	PC £1.90	0.50	5.17	2.08	nr	7.25
isolated screwed joint	PC £0.68	0.60	6.21	0.87	nr	7.08
tank connection; longscrew and						
backnuts; lead washers; joint	PC £1.98	0.67	6.93	2.28	nr	9.21
40 mm pipes						
medium weight; fixing with pipe brackets;						
plugged and screwed	PC £173.70/100m	0.50	5.17	2.82	m	7.99
heavy weight; fixing with pipe brackets;						
plugged and screwed	PC £210.17/100m	0.50	5.17	3.24	m	8.41
Extra for						
fittings with one end		0.29	3.00	0.69	nr	3.69
fittings with two ends		0.58	6.00	1.82	nr	7.82
fittings with three ends		0.78	8.07	1.93	nr	10.00
fittings with four ends		1.16	12.00	3.15	nr	15.15
cap	PC £0.63	0.29	3.00	0.69	nr	3.69
socket; equal	PC £0.92	0.58	6.00	1.00	nr	7.00
socket; reducing	PC £1.02	0.58	6.00	1.11	nr	7.11
bend; 90 degree long radius M/F	PC £2.79	0.58	6.00	3.04	nr	9.04
elbow; 90 degree M/F	PC £1.67	0.58	6.00	1.82	nr	7.82
tee; equal	PC £1.77	0.78	8.07	1.93	nr	10.00
cross	PC £2.89	1.16	12.00	3.15	nr	15.15
union	PC £2.40	0.58	6.00	2.61	nr	8.61
isolated screwed joint	PC £0.92	0.70	7.24	1.15	nr	8.39
tank connection; longscrew and						
backnuts; lead washers; joint	PC £2.73	0.78	8.07	3.11	nr	11.18
50 mm pipes						
medium weight; fixing with pipe brackets;						
plugged and screwed	PC £244.42/100m	0.60	6.21	3.86	m	10.07
heavy weight; fixing with pipe brackets;						
plugged and screwed	PC £292.02/100m	0.60	6.21	4.42	m	10.63
Extra for						
fittings with one end		0.33	3.41	1.30	nr	4.71
fittings with two ends		0.67	6.93	2.33	nr	9.26
fittings with three ends		0.88	9.10	2.78	nr	11.88
fittings with four ends		1.34	13.86	4.89	nr	18.75
cap	PC £1.19	0.33	3.41	1.30	nr	4.71
socket; equal	PC £1.43	0.67	6.93	1.56	nr	8.49
socket; reducing	PC £1.43	0.67	6.93	1.56	nr	8.49
bend; 90 degree long radius M/F	PC £4.76	0.67	6.93	5.19	nr	12.12
elbow; 90 degree M/F	PC £2.14	0.67	6.93	2.33	nr	9.26
tee; equal	PC £2.55	0.88	9.10	2.78	nr	11.88
cross	PC £4.49	1.34	13.86	4.89	nr	18.75
union	PC £3.57	0.67	6.93	3.89	nr	10.82
isolated screwed joint	PC £1.43	0.80	8.28	1.73	nr	10.01
tank connection; longscrew and						
backnuts; lead washers; joint	PC £4.29	0.88	9.10	4.85	nr	13.95

T MECHANICAL HEATING SYSTEMS ETC. Including overheads and profit at 9.00%		Labour hours	Labour £	Material £	Unit	Total rate £
T30 MEDIUM TEMPERATURE HOT WATER HEATING - cont'd						
Steel pipes; BS 1387; galvanized; screwed **joints; galvanized malleable iron fittings;** **BS 143**						
15 mm pipes						
medium weight; fixing with pipe brackets;						
plugged and screwed	PC £105.33/100m	0.30	3.10	1.72	m	4.82
heavy weight; fixing with pipe brackets;						
plugged and screwed	PC £122.81/100m	0.30	3.10	1.92	m	5.02
Extra for						
fittings with one end		0.13	1.34	0.35	nr	1.69
fittings with two ends		0.25	2.59	0.45	nr	3.04
fittings with three ends		0.37	3.83	0.55	nr	4.38
fittings with four ends		0.50	5.17	1.25	nr	6.42
cap	PC £0.32	0.13	1.34	0.35	nr	1.69
socket; equal	PC £0.35	0.25	2.59	0.38	nr	2.97
socket; reducing	PC £0.41	0.25	2.59	0.45	nr	3.04
bend; 90 degree long radius M/F	PC £0.74	0.25	2.59	0.80	nr	3.39
elbow; 90 degree M/F	PC £0.41	0.25	2.59	0.45	nr	3.04
tee; equal	PC £0.51	0.37	3.83	0.55	nr	4.38
cross	PC £1.15	0.50	5.17	1.25	nr	6.42
union	PC £1.10	0.25	2.59	1.20	nr	3.79
isolated screwed joint	PC £0.35	0.30	3.10	0.48	nr	3.58
tank connection; longscrew and						
backnuts; lead washers; joint	PC £1.39	0.37	3.83	1.63	nr	5.46
20 mm pipes						
medium weight; fixing with pipe brackets;						
plugged and screwed	PC £122.25/100m	0.32	3.31	1.96	m	5.27
heavy weight; fixing with pipe brackets;						
plugged and screwed	PC £143.43/100m	0.32	3.31	2.20	m	5.51
Extra for						
fittings with one end		0.16	1.66	0.40	nr	2.06
fittings with two ends		0.33	3.41	0.60	nr	4.01
fittings with three ends		0.47	4.86	0.80	nr	5.66
fittings with four ends		0.66	6.83	1.91	nr	8.74
cap	PC £0.37	0.16	1.66	0.40	nr	2.06
socket; equal	PC £0.41	0.33	3.41	0.45	nr	3.86
socket; reducing	PC £0.51	0.33	3.41	0.55	nr	3.96
bend; 90 degree long radius M/F	PC £1.20	0.33	3.41	1.30	nr	4.71
elbow; 90 degree M/F	PC £0.55	0.33	3.41	0.60	nr	4.01
tee; equal	PC £0.74	0.47	4.86	0.80	nr	5.66
cross	PC £1.75	0.66	6.83	1.91	nr	8.74
union	PC £1.29	0.33	3.41	1.40	nr	4.81
isolated screwed joint	PC £0.41	0.40	4.14	0.56	nr	4.70
tank connection; longscrew and						
backnuts; lead washers; joint	PC £1.62	0.47	4.86	1.87	nr	6.73
25 mm pipes						
medium weight; fixing with pipe brackets;						
plugged and screwed	PC £171.63/100m	0.36	3.72	2.61	m	6.33
heavy weight; fixing with pipe brackets;						
plugged and screwed	PC £205.27/100m	0.36	3.72	2.99	m	6.71
Extra for						
fittings with one end		0.21	2.17	0.50	nr	2.67
fittings with two ends		0.42	4.34	1.00	nr	5.34
fittings with three ends		0.57	5.90	1.15	nr	7.05
fittings with four ends		0.84	8.69	2.41	nr	11.10
cap	PC £0.46	0.21	2.17	0.50	nr	2.67
socket; equal	PC £0.55	0.42	4.34	0.60	nr	4.94
socket; reducing	PC £0.67	0.42	4.34	0.73	nr	5.07
bend; 90 degree long radius M/F	PC £1.70	0.42	4.34	1.86	nr	6.20
elbow; 90 degree M/F	PC £0.92	0.42	4.34	1.00	nr	5.34
tee; equal	PC £1.06	0.57	5.90	1.15	nr	7.05
cross	PC £2.21	0.84	8.69	2.41	nr	11.10
union	PC £1.52	0.42	4.34	1.65	nr	5.99
isolated screwed joint	PC £0.55	0.50	5.17	0.73	nr	5.90
tank connection; longscrew and						
backnuts; lead washers; joint	PC £2.21	0.57	5.90	2.54	nr	8.44

T MECHANICAL HEATING SYSTEMS ETC. Including overheads and profit at 9.00%		Labour hours	Labour £	Material £	Unit	Total rate £
32 mm pipes						
medium weight; fixing with pipe brackets;						
plugged and screwed	PC £214.41/100m	0.42	4.34	3.30	m	7.64
heavy weight; fixing with pipe brackets;						
plugged and screwed	PC £257.17/100m	0.42	4.34	3.78	m	8.12
Extra for						
fittings with three ends		0.25	2.59	0.73	nr	3.32
fittings with two ends		0.50	5.17	1.65	nr	6.82
fittings with three ends		0.67	6.93	1.91	nr	8.84
fittings with four ends		1.00	10.34	3.16	nr	13.50
cap	PC £0.67	0.25	2.59	0.73	nr	3.32
socket; equal	PC £0.92	0.50	5.17	1.00	nr	6.17
socket; reducing	PC £1.06	0.50	5.17	1.15	nr	6.32
bend; 90 degree long radius M/F	PC £2.85	0.50	5.17	3.11	nr	8.28
elbow; 90 degree M/F	PC £1.52	0.50	5.17	1.65	nr	6.82
tee; equal	PC £1.75	0.67	6.93	1.91	nr	8.84
cross	PC £2.90	1.00	10.34	3.16	nr	13.50
union	PC £2.58	0.50	5.17	2.81	nr	7.98
isolated screwed joint	PC £0.92	0.60	6.21	1.13	nr	7.34
tank connection; longscrew and						
backnuts; lead washers; joint	PC £2.68	0.67	6.93	3.04	nr	9.97
40 mm pipes						
medium weight; fixing with pipe brackets;						
plugged and screwed	PC £249.15/100m	0.50	5.17	3.94	m	9.11
heavy weight; fixing with pipe brackets;						
plugged and screwed	PC £300.03/100m	0.50	5.17	4.52	m	9.69
Extra for						
fittings with one end		0.29	3.00	0.93	nr	3.93
fittings with two ends		0.58	6.00	2.46	nr	8.46
fittings with three ends		0.78	8.07	2.61	nr	10.68
fittings with four ends		1.16	12.00	4.26	nr	16.26
cap	PC £0.85	0.29	3.00	0.93	nr	3.93
socket; equal	PC £1.24	0.58	6.00	1.35	nr	7.35
socket; reducing	PC £1.38	0.58	6.00	1.50	nr	7.50
bend; 90 degree long radius M/F	PC £3.77	0.58	6.00	4.11	nr	10.11
elbow; 90 degree M/F	PC £2.25	0.58	6.00	2.46	nr	8.46
tee; equal	PC £2.39	0.78	8.07	2.61	nr	10.68
cross	PC £3.91	1.16	12.00	4.26	nr	16.26
union	PC £3.24	0.58	6.00	3.53	nr	9.53
isolated screwed joint	PC £1.24	0.70	7.24	1.50	nr	8.74
tank connection; longscrew and						
backnuts; lead washers; joint	PC £3.68	0.78	8.07	4.15	nr	12.22
50 mm pipes						
medium weight; fixing with pipe brackets;						
plugged and screwed	PC £349.61/100m	0.60	6.21	5.41	m	11.62
heavy weight; fixing with pipe brackets;						
plugged and screwed	PC £415.72/100m	0.60	6.21	6.18	m	12.39
Extra for						
fittings with one end		0.33	3.41	1.75	nr	5.16
fittings with two ends		0.67	6.93	3.16	nr	10.09
fittings with three ends		0.88	9.10	3.76	nr	12.86
fittings with four ends		1.34	13.86	6.62	nr	20.48
cap	PC £1.61	0.33	3.41	1.75	nr	5.16
socket; equal	PC £1.93	0.67	6.93	2.11	nr	9.04
socket; reducing	PC £1.93	0.67	6.93	2.11	nr	9.04
bend; 90 degree long radius M/F	PC £6.44	0.67	6.93	7.02	nr	13.95
elbow; 90 degree M/F	PC £2.90	0.67	6.93	3.16	nr	10.09
tee; equal	PC £3.45	0.88	9.10	3.76	nr	12.86
cross	PC £6.07	1.34	13.86	6.62	nr	20.48
union	PC £4.83	0.67	6.93	5.26	nr	12.19
isolated screwed joint	PC £1.93	0.80	8.28	2.28	nr	10.56
tank connection; longscrew and						
backnuts; lead washers; joint	PC £5.81	0.88	9.10	6.49	nr	15.59

T MECHANICAL HEATING SYSTEMS ETC. Including overheads and profit at 9.00%		Labour hours	Labour £	Material £	Unit	Total rate £
Radiators; pressed steel panel type, 590 mm high; 3 mm chromium plated air valve; 15 mm chromium plated easy clean straight valve with union; 15 mm chromium plated lockshield valve with union						
1.69 m2 single surface	PC £50.74	2.00	20.69	60.31	nr	81.00
2.12 m2 single surface	PC £58.04	2.25	23.27	68.31	nr	91.58
2.75 m2 single surface	PC £69.21	2.50	25.86	80.48	nr	106.34
3.39 m2 single surgace	PC £85.74	2.75	28.45	98.50	nr	126.95

V ELECTRICAL SYSTEMS Including overheads and profit at 5.00%	Labour hours	Labour £	Material £	Unit	Total rate £
V21 - 22 GENERAL LIGHTING AND LV POWER					

NOTE: The following items indicate
approximate prices for wiring of lighting
and power points complete, including
accessories and socket outlets but excluding
lighting fittings. Consumer control units
are shown separately. For a more detailed
breakdown of these costs and specialist
costs for a complete range of electrical
items reference should be made to
Spon's Mechanical and Electrical Services
Price Book.

	Labour hours	Labour £	Material £	Unit	Total rate £
Consumer control units					
8-way 60 amp SP&N surface mounted insulated consumer control units fitted with miniature circuit breakers including 2 m long 32 mm screwed welded conduit with three runs of 16 mm2 PVC cables ready for final connections by the supply authority	-	-	-	nr	119.00
Extra for current operated ELCB of 30 mA tripping current	-	-	-	nr	53.30
as above but 100 amp metal cased consumer unit and 25 mm2 PVC cables	-	-	-	nr	136.00
Extra for current operated ELCB of 30 mA tripping current	-	-	-	nr	113.00
Final circuits					
Lighting points					
Wired in PVC insulated and PVC sheathed cable in flats and houses; insulated in cavities and roof space; protected where buried by heavy gauge PVC conduit	-	-	-	nr	38.50
As above but in commercial property	-	-	-	nr	46.50
Wired in PVC insulated cable in screwed welded conduit in flats and houses	-	-	-	nr	80.00
As above but in commercial property	-	-	-	nr	97.50
As above but in industrial property	-	-	-	nr	110.00
Wired in MICC cable in flats and houses	-	-	-	nr	67.00
As above but in commercial property	-	-	-	nr	80.50
As above but in industrial property with PVC sheathed cable	-	-	-	nr	93.00
Single 13 amp switched socket outlet points					
Wired in PVC insulated and PVC sheathed cable in flats and houses on a ring main circuit; protected where buried by heavy gauge PVC conduit	-	-	-	nr	42.00
As above but in commercial property	-	-	-	nr	51.00
Wired in PVC insulated cable in screwed welded conduit throughout on a ring main circuit in flats and houses	-	-	-	nr	60.00
As above but in commercial property	-	-	-	nr	70.00
As above but in industrial property	-	-	-	nr	80.50
Wired in MICC cable on a ring main circuit in flats and houses	-	-	-	nr	61.20
As above but in commercial property	-	-	-	nr	71.40
As above but in industrial property with PVC sheathed cable	-	-	-	nr	91.00
Cooker control units					
45 amp circuit including unit wired in PVC insulated and PVC sheathed cable; protected where buried by heavy gauge PVC conduit	-	-	-	nr	82.75
As above but wired in PVC insulated cable in screwed welded conduit	-	-	-	nr	120.00
As above but wired in MICC cable	-	-	-	nr	135.00

W SECURITY SYSTEMS Including overheads and profit at 5.00%	Labour hours	Labour £	Material £	Unit	Total rate £
W20 LIGHTNING PROTECTION					
Flag staff terminal	-	-	-	nr	57.58
Copper strip roof or down conductors					
fixed with bracket or saddle clips					
20 x 3 mm	-	-	-	m	12.75
25 x 3 mm	-	-	-	m	14.22
Aluminium strip roof or down conductors					
fixed with bracket or saddle clips					
20 x 3 mm	-	-	-	m	10.06
25 x 3 mm	-	-	-	m	10.55
Joints in tapes	-	-	-	nr	7.85
Bonding connections to roof and structural					
metalwork	-	-	-	nr	44.15
Testing points	-	-	-	nr	21.58
Earth electrodes					
16 mm driven copper electrodes in 1220 mm					
sectional lengths (2440 mm minimum)	-	-	-	nr	117.72
First 2440 mm driven and tested					
25 x 3 mm copper strip electrode in 457 mm					
deep prepared trench	-	-	-	m	8.53

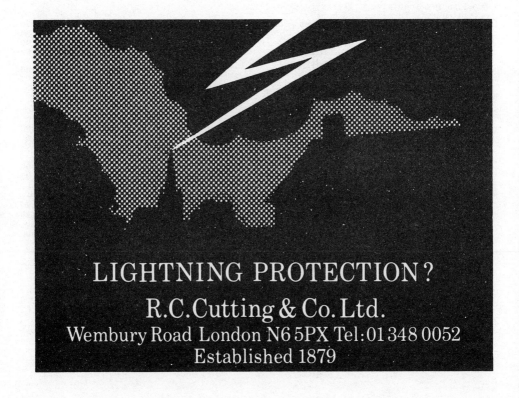

Prices for Measured Work
– Minor Works

INTRODUCTION

The 'Prices for Measured Work - Minor Works' are intended to apply to a small project costing (excluding Preliminaries) about £85 000 in the outer London area.

The format of this section follows that of the 'Major Works' section with minor variations because of the different nature of the work, and reference should be made to the 'Introduction' to that section on page 161.

It has been assumed that reasonable quantities of work are involved, equivalent to quantities for two houses, although clearly this would not apply to all trades and descriptions of work in a project of this value. Where smaller quantities of work are involved it will be necessary to adjust the prices accordingly.

For C Demolition/Alteration/Renovation work, even smaller quantities have been assumed, as can be seen from the stated 'P.C.' of the materials involved.

Where work in an existing building is concerned it has been assumed that the building is vacated and that in all cases there is reasonable access and adequate storage space. Should this not be the case and if any abnormal circumstances have to be taken into account an allowance can be made either by a lump sum addition or by suitably modifying the percentage factor for overheads and profit.

Because of the different nature of the work for which prices are given in this section changes have been made to the percentage additions allowed in 'Major Works'. For those items where detailed breakdowns are given an allowance of 12.5% has been made to labour, plant and material costs for overhead charges and profit. In other cases 5% has been included for the General Contractor's attendance overhead charges and profit.

Labour rates are based upon typical gang costs divided by the number of primary working operatives for the trade concerned; and for general building work include an allowance for trade supervision, but exclude overheads and profit (included in the 'Labour £' column). The 'Labour hours' column gives the total hours allocated to a particular item and the 'Labour £' the consolidated cost of such labour including overhead charges and profit.

'Labour hours' have not always been given for 'spot' items because of the inclusion of Sub-Contractor's labour.

The 'Material Plant £' column includes the cost of removal of debris by skips or lorries and also overhead charges and profit.

No allowance has been made for any Value Added Tax which will probably be payable on the majority of work of this nature.

A PRELIMINARIES/CONTRACT CONDITIONS FOR MINOR WORKS

When pricing 'Preliminaries' all factors affecting the execution of the works
must be considered; some of the more obvious have already been mentioned above.
    As mentioned in 'Preliminaries' in the 'Prices for Measured Work - Major
Works' section (page 183), the current trend is for 'Preliminaries' to be priced
at approximately 12 to 16%, but for alterations and additions work in
particular, care must be exercised in ensuring that all adverse factors are
covered. The reader is advised to identify systematically and separately price
all preliminary items with cost/time implications in order to reflect as
accurately as possible preliminary costs likely to stem from any particular
scheme.
    Where the Standard Form of Contract applies two clauses which will affect the
pricing of Preliminaries should be noted.

    (a) Insurance of the works against Clause 22 Perils
         Clause 22C will apply whereby the Employer and not the Contractor effects
         the insurance.
    (b) Fluctuations
         An allowance for any shortfall in recovery of increased costs under
         whichever clause is contained in the Contract may be covered by the
         inclusion of a lump sum in the Preliminaries or by increasing the prices
         by a suitable percentage

ADDITIONS AND NEW WORKS WITHIN EXISTING BUILDINGS

Depending upon the contract size, either the prices in 'Prices for Measured Work
- Major Works' or those prices in 'Prices for Measured Work - Minor Works' will
best apply.
    It is likely, however, that the excavations for foundations might preclude the
use of mechanical plant, and that it will be necessary to restrict prices to
those applicable to hand excavation.
    If, in any circumstances, less than what might be termed 'normal quantities'
are likely to be involved, it is stressed that actual quotations should be
invited from specialist Sub-contractors for these works.

JOBBING WORK

Jobbing work is outside the scope of this section and no attempt has been made
to include prices for such work.

C DEMOLITION/ALTERATION/RENOVATION	Labour hours	Labour £	Material Plant £	Unit	Total rate £

C10 DEMOLISHING STRUCTURES

Demolition rates vary considerably from one scheme to another; depending on access, type of construction, method of demolition, whether any redundant materials, etc. Therefore it is advisable to obtain specific quotations for each scheme under consideration, but the following rates (excluding scaffolding costs) for simple demolition may be of some assistance for comparison purposes.

	Labour hours	Labour £	Material Plant £	Unit	Total rate £
Demolish to ground level; single-storey brick out-building; timber flat roofs; volume					
50 m3	-	-	-	m3	7.50
200 m3	-	-	-	m3	5.60
500 m3	-	-	-	m3	3.00
Demolish down to ground level; two-storey brick out building; timber joisted suspended floor and timber flat roof; volume					
200 m3	-	-	-	m3	4.20

C20 ALTERATIONS - SPOT ITEMS

Composite 'spot' items

Few exactly similar composite items of alteration work are encountered on different schemes; for this reason it is considered more accurate for the reader to build up the value of such items from individual prices in the following section. However, for estimating purposes, the following 'spot' items have been prepared. Prices do not include for shoring, scaffolding or redecoration.

Form openings

Form opening through 100 mm thick softwood stud partition including framing studwork around, making good boarding and any plaster either side and extending floor finish through opening (new door and frame measured elsewhere)

	Labour hours	Labour £	Material Plant £	Unit	Total rate £
for single door and frame	-	160.00	44.00	nr	204.00
for pair of doors and frame	-	211.00	60.00	nr	271.00

Form opening through internal plastered wall for single door and frame; including cutting structure, quoining or making good jambs, cutting and pinning in suitable precast concrete plate lintel(s), making good plasterwork up to new frame both sides and extending floor finish through opening (new door and frame measured elsewhere)

	Labour hours	Labour £	Material Plant £	Unit	Total rate £
150 mm reinforced concrete wall	-	202.00	55.00	nr	257.00
225 mm reinforced concrete wall	-	262.00	76.00	nr	338.00
half brick thick wall	-	192.00	48.00	nr	240.00
one brick thick wall or two half brick thick skins	-	242.50	73.00	nr	315.50
one and a half brick thick wall	-	297.00	95.00	nr	392.00
two brick thick wall	-	348.00	118.00	nr	466.00
100 mm block wall	-	166.00	56.00	nr	222.00
215 mm block wall	-	204.00	91.00	nr	295.00

C DEMOLITION/ALTERATION/RENOVATION	Labour hours	Labour £	Material Plant £	Unit	Total rate £
C20 ALTERATIONS - SPOT ITEMS					

Form openings - cont'd

Form opening through internal plastered wall
for pair of doors and frame; including cutting
structure, quoining or making good jambs,
cutting and pinning in suitable precast
concrete plate lintel(s), making good
plasterwork up to new frame both sides and
extending floor finish through opening
(new door and frame measured elsewhere)

	Labour hours	Labour £	Material Plant £	Unit	Total rate £
150 mm reinforced concrete wall	-	289.50	80.50	nr	370.00
225 mm reinforced concrete wall	-	363.00	110.00	nr	473.00
half brick thick wall	-	229.50	61.50	nr	291.00
one brick thick wall or two half brick thick skins	-	307.00	91.00	nr	398.00
one and a half brick thick wall	-	380.00	120.00	nr	500.00
two brick thick wall	-	451.50	148.50	nr	600.00
100 mm block wall	-	200.50	67.00	nr	267.50
215 mm block wall	-	253.00	108.50	nr	361.50

Form opening through faced wall 1200 x 1200 mm
(1.44 m2) for new window; including cutting
structure, quoining up jambs, cutting and
pinning in suitable precast concrete boot
lintel with galvanized steel angle bolted on
to support, outer brick soldier course in
facing bricks to match existing (new window
and frame measured elsewhere)

	Labour hours	Labour £	Material Plant £	Unit	Total rate £
one brick thick wall or two half brick thick skins	-	300.50	130.00	nr	430.50
one and a half brick thick wall	-	306.00	122.00	nr	428.00
two brick thick wall	-	362.00	139.00	nr	501.00

Form opening through slated, boarded and
timbered roof; 700 x 1100 mm; for new
rooflight; including cutting structure and
finishings; trimming timbers in rafters and
making good roof coverings (kerb and rooflight
measured separately)

	Labour hours	Labour £	Material Plant £	Unit	Total rate £
	-	199.00	93.00	nr	292.00

Fill openings
Take out door and frame; make good plaster and
skirtings across reveals and head, and leave
as blank opening

	Labour hours	Labour £	Material Plant £	Unit	Total rate £
single door	-	68.50	24.50	nr	93.00
pair of doors	-	80.50	26.50	nr	107.00

Take out door and frame in 100 mm thick
softwood stud partition; block up opening with
timber covered on both sides with boarding or
lining to match existing and extend skirting
both sides

	Labour hours	Labour £	Material Plant £	Unit	Total rate £
single doors	-	100.00	44.50	nr	144.50
pair of doors	-	135.00	55.00	nr	190.00

Take out single door and frame in internal
wall; brick or block up opening; plastering
walls and extend skirting both sides

	Labour hours	Labour £	Material Plant £	Unit	Total rate £
half brick thick	-	120.50	48.00	nr	168.50
one brick thick	-	154.00	76.00	nr	230.00
one and a half brick thick	-	191.00	103.50	nr	294.50
two brick thick	-	231.00	131.50	nr	362.50
100 mm blockwork	-	96.50	43.00	nr	139.50
215 mm blockwork	-	116.50	71.50	nr	188.00

Take out pair of doors and frame in internal
wall; brick or block up opening; plastering
walls and extend skirting both sides

	Labour hours	Labour £	Material Plant £	Unit	Total rate £
half brick thick	-	192.00	82.00	nr	274.00
one brick thick	-	246.50	130.50	nr	377.00
one and a half brick thick	-	303.50	177.50	nr	481.00
two brick thick	-	362.00	225.00	nr	587.00
100 mm blockwork	-	154.50	73.00	nr	227.50
215 mm blockwork	-	185.00	122.00	nr	307.00

C DEMOLITION/ALTERATION/RENOVATION	Labour hours	Labour £	Material Plant £	Unit	Total rate £
Take out 825 x 1406 mm (1.16 m2) sliding sash window and frame in external faced wall; brick up opening with facing bricks on outside to match existing and common bricks on inside; plastered internally					
one brick thick or two half brick thick skins	-	140.00	75.00	nr	215.00
one and a half brick thick	-	150.00	95.00	nr	245.00
two brick thick	-	175.00	115.00	nr	290.00
Take out 825 x 1406 mm (overall) (1.16 m2) curved headed sliding sash window in external stuccoed wall; brick up opening with common bricks; stucco on outside and plaster on inside to match existing					
one brick thick or two half brick thick skins	-	170.00	70.00	nr	240.00
one and a half brick thick	-	200.00	90.00	nr	290.00
two brick thick	-	230.00	115.00	nr	345.00
Take out 825 x 1406 mm (overall) (1.16 m2) curved headed sliding sash window in external masonry faced brick wall; brick up opening with masonry on outside and common bricks on inside; plastered internally					
350 mm wall	-	325.00	320.00	nr	645.00
500 mm wall	-	350.00	350.00	nr	700.00
600 mm wall	-	400.00	375.00	nr	775.00
Other 'spot' items					
Pull down brick chimney to 300 mm below roof level; seal off flues with slates; piece in 'treated' sawn softwood rafters and make good roof coverings over to match existing (scaffolding excluded)					
680 x 680 x 900 mm high above roof	-	130.00	38.00	nr	168.00
Add for each additional 300 mm high	-	19.00	6.00	nr	25.00
680 x 1030 x 900 mm high above roof	-	190.00	55.00	nr	245.00
Add for each additional 300 mm high	-	40.00	12.00	nr	52.00
1030 x 1030 x 900 mm high above roof	-	285.00	80.00	nr	365.00
Add for each additional 300 mm high	-	94.00	26.00	nr	120.00
Carefully take off existing chimney pots and set aside; pull down defective chimney stack to roof level and rebuild using 25% new facing bricks to match existing; provide new lead flashings; parge and core flues, and reset chimney pots including flaunching in cement mortar (scaffolding existing)					
680 x 680 x 900 mm high above roof	-	300.00	90.00	nr	390.00
Add for each additional 300 mm high	-	45.00	14.00	nr	59.00
680 x 1030 x 900 mm high above roof	-	440.00	136.50	nr	576.50
Add for each additional 300 mm high	-	65.00	20.00	nr	85.00
1030 x 1030 x 900 mm high above roof	-	655.00	205.00	nr	860.00
Add for each additional 300 mm high	-	97.00	30.00	nr	127.00
Take out fireplace surround and interior; break up kerb and hearth; block up fireplace opening with half brick thick wall including building in air-brick; plastering wall to finish flush with existing; fixing fibrous plaster ventilator and extending skirting to match existing across front of fireplace					
tiled	-	100.50	44.50	nr	145.00
cast iron; and set aside	-	107.00	27.50	nr	134.50
stone; and set aside	-	173.50	46.00	nr	219.50

C DEMOLITION/ALTERATION/RENOVATION Including overheads and profit at 12.50%	Labour hours	Labour £	Material £	Unit	Total rate £
C20 ALTERATIONS - SPOT ITEMS					
NOTE: All items of removal include for removing debris from site					
Removing/cutting plain/reinforced concrete work					
Break up concrete bed					
100 mm thick	0.54	3.90	2.28	m2	6.18
Break up concrete bed					
150 mm thick	0.80	5.78	4.33	m2	10.11
200 mm thick	1.08	7.80	4.55	m2	12.35
300 mm thick	1.60	11.56	6.68	m2	18.24
Break up reinforced concrete bed					
100 mm thick	0.60	4.33	2.60	m2	6.93
150 mm thick	0.90	6.50	3.84	m2	10.34
200 mm thick	1.20	8.67	5.20	m2	13.87
300 mm thick	1.80	13.00	7.80	m2	20.80
Pull down reinforced concrete column or cut away casing to steel column	12.00	86.67	33.41	m3	120.08
Pull down reinforced concrete beam or cut away casing to steel beam	14.00	101.12	35.89	m3	137.01
Pull down reinforced concrete wall					
100 mm thick	1.20	8.67	3.34	m2	12.01
150 mm thick	1.80	13.00	4.95	m2	17.95
225 mm thick	2.70	19.50	7.43	m2	26.93
300 mm thick	3.60	26.00	10.02	m2	36.02
Pull down reinforced concrete suspended slabs					
100 mm thick	1.00	7.22	3.09	m2	10.31
150 mm thick	1.50	10.83	4.58	m2	15.41
225 mm thick	2.25	16.25	6.88	m2	23.13
300 mm thick	3.00	21.67	9.28	m2	30.95
Cut through reinforced concrete walls to form openings					
150 mm thick	6.00	43.91	8.80	m2	52.71
225 mm thick	8.20	59.94	13.14	m2	73.08
300 mm thick	10.40	75.95	17.73	m2	93.68
Cut through reinforced concrete suspended slabs to form openings					
150 mm thick	5.55	40.68	7.23	m2	47.91
225 mm thick	7.20	52.68	10.89	m2	63.57
300 mm thick	8.80	64.31	14.56	m2	78.87
Break up concrete plinth and make good floor beneath	4.60	33.22	19.74	m3	52.96
Remove precast concrete kerb	0.50	3.61	0.74	m	4.35
Remove precast concrete window sill and set aside for re-use	1.60	11.56	-	m	11.56
Break up concrete hearth	1.80	13.00	1.24	nr	14.24
Removing/cutting brick/blockwork					
Pull down external brick walls; in gauged mortar					
half brick thick	0.70	5.06	1.86	m2	6.92
two half brick thick skins	1.20	8.67	3.96	m2	12.63
one brick thick	1.20	8.67	3.96	m2	12.63
one and a half brick thick	1.70	12.28	6.19	m2	18.47
two brick thick	2.20	15.89	7.92	m2	23.81
Add for plaster, render or pebbledash per side	0.10	0.72	0.37	m2	1.09

C DEMOLITION/ALTERATION/RENOVATION Including overheads and profit at 12.50%	Labour hours	Labour £	Material £	Unit	Total rate £
Pull down external brick walls; in cement mortar					
half brick thick	1.05	7.58	1.86	m2	9.44
two half brick thick skins	1.75	12.64	3.96	m2	16.60
one brick thick	1.75	12.64	3.96	m2	16.60
one and a half brick thick	2.45	17.70	6.19	m2	23.89
two brick thick	3.15	22.75	7.92	m2	30.67
Add for plaster, render or pebbledash per side	0.10	0.72	0.37	m2	1.09
Pull down internal partitions; in gauged mortar					
half brick thick	1.05	7.58	1.86	m2	9.44
one brick thick	1.80	13.00	3.96	m2	16.96
one and a half brick thick	2.55	18.42	6.19	m2	24.61
75 mm blockwork	0.70	5.06	1.36	m2	6.42
90 mm blockwork	0.75	5.42	1.61	m2	7.03
100 mm blockwork	0.80	5.78	1.86	m2	7.64
115 mm blockwork	0.85	6.14	2.10	m2	8.24
125 mm blockwork	0.90	6.50	2.35	m2	8.85
140 mm blockwork	0.95	6.86	2.60	m2	9.46
150 mm blockwork	1.00	7.22	2.85	m2	10.07
190 mm blockwork	1.18	8.52	3.59	m2	12.11
215 mm blockwork	1.30	9.39	3.98	m2	13.37
255 mm blockwork	1.50	10.83	4.70	m2	15.53
Add for plaster per side	0.10	0.72	0.37	m2	1.09
Pull down internal partitions; in cement mortar					
half brick thick	1.60	11.56	1.86	m2	13.42
one brick thick	2.65	19.14	3.96	m2	23.10
one and a half brick thick	3.70	26.72	6.19	m2	32.91
Add for plaster per side	0.10	0.72	0.37	m2	1.09
Break up brick plinths	4.00	28.89	12.38	m3	41.27
Pull down bund walls or piers in cement mortar					
one brick thick	1.40	10.11	3.96	m2	14.07
Pull down walls to roof ventilator housing					
one brick thick	1.60	11.56	3.96	m2	15.52
Pull down brick chimney; to 300 mm below roof level; seal off flues with slates					
680 x 680 x 900 mm high above roof	12.50	90.44	25.51	nr	115.95
Add for each additional 300 mm high	2.50	18.09	5.05	nr	23.14
600 x 1030 x 900 mm high above roof	18.75	135.66	38.30	nr	173.96
Add for each additional 300 mm high	3.74	27.08	7.67	nr	34.75
1030 x 1030 x 900 mm high above roof	28.81	208.45	58.77	nr	267.22
Add for each additional 300 mm high	5.65	40.87	12.12	nr	52.99
Remove fireplace surround and hearth					
interior tiled	1.85	13.36	2.85	nr	16.21
cast iron; and set aside	3.10	22.39	-	nr	22.39
stone; and set aside	8.10	58.50	-	nr	58.50
Remove brick-on-edge coping; prepare walls for raising					
one brick thick	0.45	5.41	0.25	m	5.66
one and a half brick thick	0.60	7.22	0.37	m	7.59
Cut back brick projections flush with adjacent wall					
225 x 112 mm	0.30	3.61	0.12	m	3.73
225 x 225 mm	0.50	6.01	0.25	m	6.26
337 x 225 mm	0.70	8.42	0.37	m	8.79
450 x 225 mm	0.90	10.82	0.50	m	11.32
Cut back brick chimney breast flush with adjacent wall					
half brick thick	1.75	21.05	3.96	m2	25.01
one brick thick	2.35	28.26	6.19	m2	34.45

C DEMOLITION/ALTERATION/RENOVATION Including overheads and profit at 12.50%	Labour hours	Labour £	Material £	Unit	Total rate £
C20 ALTERATIONS - SPOT ITEMS - cont'd					
Removing/cutting brick/blockwork - cont'd					
Cut through brick or block walls or					
partitions; for lintels or beams above					
openings; in gauged mortar					
half brick thick	2.65	31.87	2.10	m2	33.97
one brick thick	4.40	52.92	4.21	m2	57.13
one and a half brick thick	6.15	73.96	6.31	m2	80.27
two brick thick	7.90	95.01	8.42	m2	103.43
75 mm blockwork	1.60	19.24	1.36	m2	20.60
90 mm blockwork	1.80	21.65	1.61	m2	23.26
100 mm blockwork	1.95	23.45	1.86	m2	25.31
115 mm blockwork	2.10	25.26	2.10	m2	27.36
125 mm blockwork	2.20	26.46	2.35	m2	28.81
140 mm blockwork	2.35	28.26	2.60	m2	30.86
150 mm blockwork	2.45	29.46	2.85	m2	32.31
190 mm blockwork	2.73	32.83	3.59	m2	36.42
215 mm blockwork	2.90	34.88	3.96	m2	38.84
255 mm blockwork	3.18	38.24	4.70	m2	42.94
Cut through brick walls or partitions; for					
lintels or beams above openings; in cement					
mortar					
half brick thick	3.80	45.70	2.10	m2	47.80
one brick thick	6.30	75.77	4.21	m2	79.98
one and a half brick thick	8.80	105.83	6.31	m2	112.14
two brick thick	11.30	135.90	8.42	m2	144.32
Cut through brick or block walls or					
partitions; for door or window openings; in					
gauged mortar					
half brick thick	1.35	16.24	2.10	m2	18.34
one brick thick	2.20	26.46	4.21	m2	30.67
one and a half brick thick	3.05	36.68	6.31	m2	42.99
two brick thick	3.95	47.50	8.42	m2	55.92
75 mm blockwork	0.80	9.62	1.36	m2	10.98
90 mm blockwork	0.92	11.06	1.61	m2	12.67
100 mm blockwork	1.00	12.03	1.86	m2	13.89
115 mm blockwork	1.06	12.75	2.10	m2	14.85
125 mm blockwork	1.10	13.23	2.35	m2	15.58
140 mm blockwork	1.16	13.95	2.60	m2	16.55
150 mm blockwork	1.20	14.43	2.85	m2	17.28
190 mm blockwork	1.32	15.87	3.59	m2	19.46
215 mm blockwork	1.45	17.44	3.96	m2	21.40
255 mm blockwork	1.60	19.24	4.70	m2	23.94
Cut through brick or block walls or					
partitions; for door or window openings; in					
cement mortar					
half brick thick	1.90	22.85	2.10	m2	24.95
one brick thick	3.15	37.88	4.21	m2	42.09
one and a half brick thick	4.35	52.31	6.31	m2	58.62
two brick thick	5.65	67.95	8.42	m2	76.37
Quoin up jambs in common bricks; PC £16.09/100;					
in gauged mortar (1:1:6); as the work					
proceeds					
half brick thick or skin of hollow wall	1.00	12.03	3.02	m	15.05
one brick thick	1.50	18.04	6.04	m	24.08
one and a half brick thick	1.95	23.45	9.06	m	32.51
two brick thick	2.40	28.86	12.08	m	40.94
75 mm blockwork	0.63	7.58	3.94	m	11.52
90 mm blockwork	0.67	8.06	4.24	m	12.30
100 mm blockwork	0.70	8.42	4.75	m	13.17
115 mm blockwork	0.76	9.14	5.81	m	14.95
125 mm blockwork	0.80	9.62	6.37	m	15.99
140 mm blockwork	0.86	10.34	6.92	m	17.26
150 mm blockwork	0.90	10.82	7.83	m	18.65
190 mm blockwork	1.01	12.15	9.09	m	21.24
215 mm blockwork	1.08	12.99	11.04	m	24.03
255 mm blockwork	1.19	14.31	12.12	m	26.43

C DEMOLITION/ALTERATION/RENOVATION Including overheads and profit at 12.50%	Labour hours	Labour £	Material £	Unit	Total rate £
Closing at jambs with common brickwork half brick thick					
50 mm cavity; including lead-lined hessian based vertical damp proof course	0.40	4.81	5.56	m	10.37
Quoin up jambs in sand faced facings; PC £45.38/100; in gauged mortar (1:1:6); facing and pointing one side to match existing					
half brick thick or skin of hollow wall	1.25	15.03	3.76	m	18.79
one brick thick	1.50	18.04	7.35	m	25.39
one and a half brick thick	2.30	27.66	10.99	m	38.65
two brick thick	2.80	33.67	14.58	m	48.25
Quoin up jambs in machine made facings; PC £45.38/100; in gauged mortar (1:1:6); facing and pointing one side to match existing					
half brick thick or skin of hollow wall	1.25	15.03	8.41	m	23.44
one brick thick	1.50	18.04	16.64	m	34.68
one and a half brick thick	2.30	27.66	24.92	m	52.58
two brick thick	2.80	33.67	33.15	m	66.82
Quoin up jambs in hand made facings; PC £45.38/100; in gauged mortar (1:1:6); facing and pointing one side to match existing					
half brick thick or skin of hollow wall	1.25	15.03	12.67	m	27.70
one brick thick	1.50	18.04	25.17	m	43.21
one and a half brick thick	2.30	27.66	37.72	m	65.38
two brick thick	2.80	33.67	50.22	m	83.89
Fill existing openings with common brickwork or blockwork in gauged mortar (1:1:6); any cutting and bonding measured separately					
half brick thick PC £16.09/100	1.85	22.25	12.52	m2	34.77
one brick thick	3.05	36.68	25.78	m2	62.46
one and a half brick thick	4.20	50.51	38.74	m2	89.25
two brick thick	5.25	63.14	51.55	m2	114.69
75 mm blockwork PC £6.66	0.92	11.06	8.25	m2	19.31
90 mm blockwork PC £7.20	1.00	12.03	8.94	m2	20.97
100 mm blockwork PC £8.01	1.05	12.63	9.95	m2	22.58
115 mm blockwork PC £9.90	1.13	13.59	12.27	m2	25.86
125 mm blockwork PC £10.80	1.18	14.19	13.39	m2	27.58
140 mm blockwork PC £11.70	1.25	15.03	14.50	m2	29.53
150 mm blockwork PC £13.28	1.30	15.63	16.41	m2	32.04
190 mm blockwork PC £15.30	1.48	17.80	18.99	m2	36.79
215 mm blockwork PC £18.67	1.60	19.24	23.13	m2	42.37
255 mm blockwork PC £20.43	1.78	21.41	25.40	m2	46.81
Cutting and bonding ends to existing;					
half brick thick	0.40	4.81	0.83	m	5.64
one brick thick	0.58	6.98	1.66	m	8.64
one and a half brick thick	0.86	10.34	2.48	m	12.82
two brick thick	1.25	15.03	3.31	m	18.34
75 mm blockwork	0.17	2.04	0.60	m	2.64
90 mm blockwork	0.21	2.53	0.65	m	3.18
100 mm blockwork	0.23	2.77	0.72	m	3.49
115 mm blockwork	0.25	3.01	0.89	m	3.90
125 mm blockwork	0.26	3.13	0.97	m	4.10
140 mm blockwork	0.28	3.37	1.05	m	4.42
150 mm blockwork	0.29	3.49	1.19	m	4.68
190 mm blockwork	0.36	4.33	1.38	m	5.71
215 mm blockwork	0.41	4.93	1.68	m	6.61
255 mm blockwork	0.48	5.77	1.84	m	7.61
half brick thick in facings; to match existing; PC £45.38/100	0.60	7.22	3.67	m2	10.89

C DEMOLITION/ALTERATION/RENOVATION Including overheads and profit at 12.50%	Labour hours	Labour £	Material £	Unit	Total rate £
C20 ALTERATIONS - SPOT ITEMS - cont'd					
Removing/cutting brick/blockwork - cont'd					
Extra over common brickwork for fair face; flush pointing					
walls and the like	0.20	2.41	-	m2	2.41
Extra over common bricks; PC £45.38/100; for facing bricks in Flemish bond; facing and pointing one side					
sand faced facings PC £17.88/100	1.00	12.03	1.51	m2	13.54
machine made facings PC £45.38/100	1.00	12.03	24.71	m2	36.74
hand made facings PC £70.68/100	1.00	12.03	46.06	m2	58.09
ADD or DEDUCT for variation of £10.00/1000 in PC of facing bricks; in flemish bond					
half brick thick	-	-	-	m2	0.73
Fill existing openings in hollow wall with inner skin of common bricks; PC £16.09/100; 50 mm cavity and galvanized steel butter- fly ties; and outer skin of facings; all in gauged mortar (1:1:6); facing and pointing one side					
two half brick thick skins; outer skin sand faced facings	4.60	55.32	25.70	m2	81.02
two half brick thick skins; outer skin machine made facings	4.60	55.32	44.26	m2	99.58
two half brick thick skins; outer skin hand made facings	4.60	55.32	61.34	m2	116.66
Removing stone rubble walling					
Pull down external stone walls in lime mortar					
300 mm thick	1.20	8.67	3.71	m2	12.38
400 mm thick	1.60	11.56	4.95	m2	16.51
600 mm thick	2.40	17.33	7.43	m2	24.76
Pull down stone walls in lime mortar; clean off; set aside for re-use					
300 mm thick	1.80	13.00	1.24	m2	14.24
400 mm thick	2.40	17.33	1.61	m2	18.94
600 mm thick	3.60	26.00	2.47	m2	28.47
Removing metal					
Remove partitions					
corrugated metal partition	0.35	2.53	0.37	m2	2.90
lightweight steel mesh security screen	0.50	3.61	0.62	m2	4.23
solid steel demountable partition	0.75	5.42	0.87	m2	6.29
glazed sheet demountable partition; including removal of glass	1.00	7.22	1.24	m2	8.46
Remove handrails and balustrades					
tubular handrailing and brackets	0.30	2.17	0.12	m	2.29
metal balustrades	0.50	3.61	0.37	m	3.98
Remove					
metal window bars	0.30	2.17	0.12	nr	2.29
plates and bolts located in concrete bases	1.00	7.22	0.12	nr	7.34
steel cat ladder 4 m long	3.00	21.67	1.24	nr	22.91
Remove metal shutter door and track					
6.2 x 4.6 m (12.6 m long track)	12.00	86.67	18.56	nr	105.23
12.4 x 4.6 m (16.4 m long track)	15.00	108.34	37.13	nr	145.47
Removing timber					
Remove roof timbers complete; including rafters, purlins, ceiling joists, plates, etc, (measured flat on plan)	0.33	2.40	1.36	m2	3.76
Remove softwood floor construction					
100 mm deep joists at ground level	0.25	1.81	0.25	m2	2.06
175 mm deep joists at first floor level	0.50	3.61	0.50	m2	4.11
125 mm deep joists at roof level	0.70	5.06	0.37	m2	5.43
Remove individual floor or roof members	0.27	1.97	0.25	m	2.22
Remove infected or decayed floor plates	0.37	2.69	0.25	m	2.94

C DEMOLITION/ALTERATION/RENOVATION Including overheads and profit at 12.50%	Labour hours	Labour £	Material £	Unit	Total rate £
Remove boarding; withdrawing nails					
25 mm thick softwood flooring; at					
ground level	0.37	2.69	0.37	m2	3.06
25 mm thick softwood flooring; at first					
floor level	0.63	4.58	0.37	m2	4.95
25 mm thick softwood roof boarding	0.74	5.38	0.37	m2	5.75
25 mm thick softwood gutter boarding	0.80	5.82	0.37	m2	6.19
22 mm thick chipboard flooring; at first					
floor level	0.37	2.69	0.37	m2	3.06
Remove tilting fillet or roll	0.15	1.09	0.03	m	1.12
Remove fascia or barge boards	0.60	4.37	0.11	m	4.48
Pull down softwood stud partition; including					
finishings both sides etc					
solid	0.45	3.25	1.24	m2	4.49
glazed; including removal of glass	0.60	4.33	1.24	m2	5.57
Remove wall linings; including battening					
behind					
plain sheeting	0.30	2.17	0.50	m2	2.67
matchboarding	0.40	2.89	0.74	m2	3.63
Remove ceiling linings; including battening					
behind					
plain sheeting	0.45	3.25	0.50	m2	3.75
matchboarding	0.60	4.33	0.74	m2	5.07
Remove skirtings, picture rails, dado rails,					
architraves and the like	0.10	0.72	-	m	0.72
Carefully remove skirtings; and set aside					
for re-use in making good	0.25	1.81	-	m	1.81
Remove shelves, window boards and the like	0.33	2.38	0.12	m	2.50
Carefully remove oak dado panelling; and					
stack for subsequent re-use	0.65	9.02	-	m2	9.02
Remove; and set aside or clear away					
single door	0.40	5.55	0.37	nr	5.92
single door and frame or lining	0.80	11.10	0.62	nr	11.72
pair of doors	0.70	9.71	0.74	nr	10.45
pair of doors and frame or lining	1.20	16.65	1.24	nr	17.89
Extra for taking out floor spring box	0.75	10.40	0.25	nr	10.65
casement window and frame	1.20	16.65	0.62	nr	17.27
double hung sash window and frame	1.70	23.58	1.24	nr	24.82
pair of French windows and frame	4.00	55.49	1.86	nr	57.35
Carefully dismantle double hung sash window					
and frame; remove and store for re-use					
elsewhere	2.40	33.29	-	nr	33.29
Remove staircase; including balustrades					
single straight flight	3.50	48.55	12.38	nr	60.93
dogleg flight	5.00	69.36	18.56	m	87.92
Remove handrails and brackets	0.10	1.39	0.37	m	1.76
Remove sloping timber ramp in corridor;					
at change of levels	2.00	27.74	1.86	nr	29.60
Remove bath panels and bearers	0.40	5.55	0.62	nr	6.17
Remove kitchen fittings					
wall units	0.45	6.24	1.67	nr	7.91
floor units	0.30	4.16	2.78	nr	6.94
larder units	0.40	5.55	6.19	nr	11.74
built in cupboards	1.50	20.81	12.38	nr	33.19
Remove pipe casings	0.30	4.16	0.50	m	4.66
Remove ironmongery; in preparation for					
redecoration; and subsequently re-fix;					
including providing any new screws necessary	0.25	3.47	0.12	nr	3.59
Remove, withdrawing nails, etc; making good					
holes					
carpet fixing strip from floors	0.05	0.36	-	m	0.36
curtain track from head of window	0.25	1.81	-	m	1.81
nameplates or numerals from face of door	0.50	3.61	-	nr	3.61
fly screen and frame from window	0.90	6.50	-	nr	6.50
small notice board from walls	0.89	6.43	-	nr	6.43
fire extinguisher and bracket from walls	1.25	9.03	-	nr	9.03

C DEMOLITION/ALTERATION/RENOVATION Including overheads and profit at 12.50%	Labour hours	Labour £	Material £	Unit	Total rate £
C20 ALTERATIONS - SPOT ITEMS - cont'd					
Removing claddings/coverings					
Remove roof coverings					
slates	0.50	3.61	0.25	m2	3.86
slates; set aside for re-use	0.60	4.33	-	m2	4.33
nibbed tiles	0.40	2.89	-	m2	3.14
nibbed tiles; set aside for re-use	0.50	3.61	-	m2	3.61
corrugated asbestos sheeting	0.40	2.89	0.25	m2	3.14
corrugated metal sheeting	0.40	2.89	0.25	m2	3.14
underfelt; and nails	0.05	0.36	0.12	m2	0.48
three layer felt roofing; cleaning base off					
for new coverings	0.25	1.81	0.25	m2	2.06
sheet metal covering	0.50	3.61	0.25	m2	3.86
Remove defective metal flashings					
horizontal	0.20	1.44	0.25	m	1.69
stepped	0.25	1.81	0.12	m	1.93
Turn back bitumen felt and later dress up					
face of new brickwork as skirting; not					
exceeding 150 mm girth	1.00	7.33	0.41	m	7.74
Remove bitumen felt roofing and boarding to					
allow access for work to top of walls or					
beams beneath	0.80	5.86	-	m	5.86
Remove tiling battens; withdraw nails	0.08	0.58	0.12	m2	0.70
Removing waterproofing finishes					
Break up					
asphalt paving	0.60	4.33	-	m2	4.33
asphalt roofing	1.00	7.22	-	m2	7.22
Hack off ashpalt skirtings from walls	0.15	1.08	-	m	1.08
Removing surface finishes					
Remove floor finishings					
carpet and underfelt	0.12	0.87	-	m2	0.87
linoleum sheet flooring	0.10	0.72	-	m2	0.72
carpet gripper	0.03	0.22	-	m2	0.22
Remove floor finishings; prepare screed to					
receive new					
carpet and underfelt	0.60	4.33	0.40	m2	4.73
vinyl or thermoplastic tiles	0.80	5.78	0.40	m2	6.18
Remove woodblock flooring; clean off and set					
aside for re-use	0.75	5.42	-	m2	5.42
Break up flooring					
floor screed	0.45	3.25	0.10	m2	3.35
granolithic flooring and screed	0.60	4.33	0.10	m2	4.43
terrazzo or ceramic floor tiles and screed	1.00	7.22	0.21	m2	7.43
Hack off in situ or tile skirtings from					
walls	0.25	1.81	-	m	1.81
Remove plasterboard wall finishing	0.40	2.89	-	m2	2.89
Hack off wall finishings					
plaster	0.20	1.44	0.62	m2	2.06
cement rendering or pebbledash	0.40	2.89	0.62	m2	3.51
wall tiling and screed	0.50	3.61	0.87	m2	4.48
Remove ceiling finishings					
plasterboard and skim; withdraw nails	0.30	2.17	0.37	m2	2.54
wood lath and plaster; withdraw nails	0.50	3.61	0.62	m2	4.23
suspended ceilings	0.75	5.42	0.62	m2	6.04
plaster moulded cornice; per 25 mm girth	0.15	1.08	0.12	m	1.20
Remove part of plasterboard ceiling to					
facilitate insertion of new steel beam	1.10	7.94	0.87	m	8.81

C DEMOLITION/ALTERATION/RENOVATION Including overheads and profit at 12.50%	Labour hours	Labour £	Material £	Unit	Total rate £
Removing gutterwork/pipework and equipment					
Remove gutterwork and supports					
PVC or asbestos	0.35	2.25	0.12	m	2.37
cast iron	0.40	2.57	0.25	m	2.82
Remove rainwater heads and supports					
PVC or asbestos	0.33	2.12	0.12	nr	2.24
cast iron	0.40	2.57	0.25	nr	2.82
Remove pipework and supports					
PVC or asbestos rainwater stack	0.30	1.93	0.12	m	2.05
cast iron rainwater stack	0.35	2.25	0.25	m	2.50
cast iron jointed soil stack	0.60	3.86	0.37	m	4.23
copper or steel water or gas pipework	0.15	0.97	0.12	m	1.09
cast iron rainwater shoe	0.08	0.51	-	nr	0.51
Remove sanitary fittings and supports; temporarily cap off services; to receive new (measured separately)					
sink or lavatory basin	1.00	8.56	7.43	nr	15.99
bath	2.00	17.11	11.14	nr	28.25
WC suite	1.50	12.83	7.43	nr	20.26
Remove sanitary fittings and supports, complete with associated services, overflows and waste pipes; making good all holes and other works disturbed; bring forward all surfaces ready for redecoration					
sink or lavatory basin	4.00	34.22	11.69	nr	45.91
range of three lavatory basins	8.00	68.45	24.75	nr	93.20
bath	6.00	51.33	15.28	nr	66.61
WC suite	8.00	68.45	12.41	nr	80.86
2 stall urinal	16.00	136.89	24.47	nr	161.36
3 stall urinal	24.00	205.34	44.97	nr	250.31
4 stall urinal	32.00	273.78	65.51	nr	339.29
Remove bathroom toilet fittings; making good works disturbed					
toilet roll holder or soap dispenser	0.30	1.93	-	nr	1.93
towel holder	0.60	3.86	-	nr	3.86
mirror	0.75	4.83	-	nr	4.83
Remove tap	0.10	0.86	-	nr	0.86
Remove the following equipment and ancillaries; cap off services; making good works disturbed (excluding any draining down of system)					
expansion tank 900 x 500 x 900 mm	1.80	15.40	4.95	nr	20.35
hot water cylinder 450 mm dia x 1050 mm high	1.20	10.27	2.10	nr	12.37
cold water tank 1540 x 900 x 900 mm	2.40	20.53	15.22	nr	35.75
cast iron radiator	2.00	17.11	3.96	nr	21.07
gas water heater	4.00	34.22	2.47	nr	36.69
gas fire	2.00	17.11	3.22	nr	20.33
Remove cold water tank and housing on roof; strip out and cap off all associated piping; making good works disturbed and roof finishings					
1540 x 900 x 900 mm	12.00	102.67	18.93	nr	121.60
Remove lagging from pipes					
up to 42 mm dia	0.10	0.64	0.12	m	0.76

C DEMOLITION/ALTERATION/RENOVATION C30 SHORING/TEMPORARY SCREENS ETC.	Labour hour	Labour £	Material Plant £	Unit	Total rate £
Shoring and strutting for large openings					
The requirements for shoring and strutting for the formation of large openings are dependent on a number of factors; for example, the weight of the superimposed structure to be supported, number (if any) of windows above; number of floors and the roof to be strutted, whether raking shores are required, and the depth to a load-bearing surface.					
Prices would best be built up by assessing the use and waste of materials and the labour involved, including getting timber from and returning to yard, cutting away and making good, overheads and profit. This method is considered a more practical way of pricing than endeavouring to price the work on a cubic metre basis of timber used, and has been adopted in preparing the prices of the examples which follow.					
The shoring and strutting for the formation of an opening 6100 mm wide x 2440 mm high through a one brick external wall at ground floor level, timber joisted suspended floor and timber pitched roof would cost:					
Strutting to window openings over proposed new openings	0.60	5.60	1.60	nr	7.20
Plates, struts, braces and hardwood wedges in supports to floors and roof of opening	1.00	9.35	4.90	m	14.25
Other shores and associated work					
Dead shore and needle using die square timber with sole plates, braces, hardwood wedges and steel dogs	20.00	187.00	47.50	nr	234.50
Set of two raking shores using die square timber with 50 mm thick wall piece; hardwood wedges and steel dogs; including forming holes for needles and making good	34.00	317.90	26.10	nr	344.00
Cut holes through one brick wall for die square needle and make good; including facings externally and plaster internally	6.00	56.10	9.65	nr	65.75
Temporary screens and hoardings					
Provide, erect and maintain temporary dust-proof screen; with 50 x 75 mm sawn softwood framing; covered one side with 12 mm plywood					
over 300 mm wide	0.80	7.48	6.57	m2	14.05
Provide, erect and maintain temporary screen; with 50 x 100 mm sawn softwood framing; covered one side with 13 mm insulating board and other side with single layer of polythene sheet					
over 300 mm wide	1.00	9.31	7.44	m2	16.75
Provide, erect and maintain temporary hoardings; with 50 x 100 mm sawn softwood framing; covered one side with 19 mm exterior quality plywood; capped with softwood capping; including three coats of oil paint and afterwards clear away on completion					
over 300 mm wide	1.50	14.03	10.12	m2	24.15

C DEMOLITION/ALTERATION/RENOVATION Including overheads and profit at 12.50%	Labour hours	Labour £	Material £	Unit	Total rate £
C40 REPAIRING/RENOVATING CONCRETE/BRICK/ BLOCK/STONE					
Repairing/renovating plain/reinforced concrete work					
Reinstate plain concrete bed with site-mixed in situ concrete; mix 21.00 N/mm2 - 20 mm aggregate (1:2:4), where opening no longer required					
100 mm thick	0.48	3.51	7.07	m2	10.58
150 mm thick	0.78	5.67	10.60	m2	16.27
Reinstate reinforced concrete bed with site-mixed in situ concrete; mix 21.00 N/mm2 - 20 mm aggregate (1:2:4); including mesh reinforcement; where opening no longer required					
100 mm thick	0.71	5.17	8.76	m2	13.93
150 mm thick	0.98	7.12	12.29	m2	19.41
Reinstate reinforced concrete suspended floor; with site-mixed in situ concrete; mix 26.00 N/mm2 - 20 mm aggregate (1:1:5:3); including mesh reinforcement and formwork; where opening no longer required					
150 mm thick	3.20	23.56	14.40	m2	37.96
225 mm thick	3.75	27.58	20.05	m2	47.63
300 mm thick	4.15	30.47	25.71	m2	56.18
Reinstate 150 x 150 x 150 mm perforation through concrete suspended slab; with site-mixed in situ concrete; mix 21.00 N/mm2 - 20 mm aggregate (1:2:4); including formwork; where opening no longer required	0.92	6.68	0.27	nr	6.95
Clean surfaces of concrete to receive new damp proof membrane	0.15	1.08	-	m2	1.08
Clean out existing minor crack and fill with cement mortar mixed with bonding agent	0.33	2.38	0.44	m	2.82
Cut out existing crack to form 20 x 20 mm groove and fill with fine cement mixed with bonding agent	0.66	4.77	1.82	m	6.59
Make good hole where pipe removed; 150 mm deep					
50 mm dia	0.42	3.03	0.38	nr	3.41
100 mm dia	0.55	3.97	0.56	nr	4.53
150 mm dia	0.70	5.06	0.74	nr	5.80
Add for each additional 25 mm thick up to 300 mm thick					
50 mm dia	0.09	0.65	0.02	nr	0.67
100 mm dia	0.12	0.87	0.02	nr	0.89
150 mm dia	0.15	1.08	0.05	nr	1.13
C40 REPAIRING/RENOVATING CONCRETE/BRICK BLOCK/STONE					
Repairing/renovating brick/blockwork					
Cut out small areas of defective brickwork; replace with new common bricks in gauged mortar (1:1:6)					
half brick thick PC £16.09/100	5.25	63.14	14.62	m2	77.76
one brick thick	9.60	115.45	29.99	m2	145.44
one and a half brick thick	13.60	163.56	43.76	m2	207.32
two brick thick	17.50	210.46	58.52	m2	268.98

C DEMOLITION/ALTERATION/RENOVATION Including overheads and profit at 12.50%		Labour hours	Labour £	Material £	Unit	Total rate £
C40 REPAIRING/RENOVATING CONCRETE/BRICK/ BLOCK/STONE - cont'd						
Repairing/renovating brick/blockwork - cont'd						
Cut out small areas of defective brickwork; replace with new facing brickwork in gauged mortar (1:1:6); half brick thick; facing and pointing one side						
sand faced facings	PC £17.88/100	7.30	87.79	15.90	m2	103.69
machine made facings	PC £45.38/100	7.30	87.79	35.38	m2	123.17
hand made facings	PC £70.68/100	7.30	87.79	53.31	m2	141.10
ADD or DEDUCT for variation of £10.00/1000 in PC of facing bricks; in flemish bond						
half brick thick		-	-	-	m2	0.73
Cut out individual bricks; replace with new common bricks in gauged mortar (1:1:6)						
half brick thick		0.30	3.61	0.28	nr	3.89
Cut out individual bricks; replace with new facing bricks in gauged mortar (1:1:6); half brick thick; facing and pointing one side						
sand faced facings	PC £17.88/100	0.45	5.41	0.30	nr	5.71
machine made facings	PC £45.38/100	0.45	5.41	0.61	nr	6.02
hand made facings	PC £70.68/100	0.45	5.41	0.89	nr	6.30
Cut out staggered cracks and repoint to match existing along brick joints		0.40	4.81	-	m	4.81
Cut out raking crack in brickwork; stitch in new common bricks and repoint to match existing						
half brick thick	PC £16.09/100	3.20	38.48	6.61	m2	45.09
one brick thick		5.85	70.35	13.43	m2	83.78
one and a half brick thick		8.75	105.23	20.28	m2	125.51
Cut out raking crack in brickwork; stitch in new facing bricks; half brick thick; facing and pointing one side to match existing						
sand faced facings	PC £17.88/100	4.80	57.73	7.18	m2	64.91
machine made facings	PC £45.38/100	4.80	57.73	15.95	m2	73.68
hand made facings	PC £70.68/100	4.80	57.73	24.01	m2	81.74
Cut out raking crack in cavity brickwork; stitch in new common bricks; PC £17.88/100 one side; and facing bricks the other side; both skins half brick thick; facing and pointing one side to match existing						
sand faced facings	PC £17.88/100	8.20	98.62	14.30	m2	112.92
machine made facings	PC £45.38/100	8.20	98.62	23.07	m2	121.69
hand made facings	PC £70.68/100	8.20	98.62	31.14	m2	129.76
Make good adjacent work; where intersecting wall removed						
half brick thick		0.30	3.61	0.49	m	4.10
one brick thick		0.40	4.81	0.97	m	5.78
100 mm blockwork		0.25	3.01	0.39	m	3.40
150 mm blockwork		0.29	3.49	0.58	m	4.07
215 mm blockwork		0.35	4.21	0.88	m	5.09
255 mm blockwork		0.39	4.69	0.97	m	5.66
Cut out defective brick soldier arch; renew and repoint to match existing						
sand faced facings	PC £17.88/100	1.95	23.45	3.72	m	27.17
machine made facings	PC £45.38/100	1.95	23.45	8.91	m	32.36
hand made facings	PC £70.68/100	1.95	23.45	13.71	m	37.16
Take down defective parapet wall; 600 mm high; with two courses of tiles and brick coping over; re-build in new facing bricks, tiles and coping stones						
one brick thick		6.66	80.09	62.82	m	142.91
Clean off moss and lichen from walls		0.30	3.61	-	m2	3.61
Clean off lime mortar, sort and stack old bricks for re-use		10.00	120.26	-	1000	120.26

C DEMOLITION/ALTERATION/RENOVATION Including overheads and profit at 12.50%	Labour hours	Labour £	Material £	Unit	Total rate £
Rake out decayed joints of brickwork and point in cement mortar (1:2); to match existing					
walls	0.75	9.02	0.61	m2	9.63
chimney stacks	1.20	14.43	0.61	m2	15.04
Rake out joint in brickwork; re-wedge flashing; and point in cement mortar (1:2) to match existing					
horizontal flashing	0.25	3.01	0.30	m	3.31
stepped flashing	0.37	4.45	0.30	m	4.75
Cut away and renew cement mortar (1:3) angle fillets; 50 mm face width	0.25	3.01	1.58	m	4.59
Cut out ends of joists and plates from walls; make good in common bricks; PC £16.09/100; in cement mortar (1:3)					
175 mm deep joists; 400 mm centres	0.65	7.82	4.40	m	12.22
225 mm deep joists; 400 mm centres	0.80	9.62	5.32	m	14.94
Cut and pin to existing brickwork					
ends of joists	0.40	4.81	-	nr	4.81
Make good hole where small pipe removed					
102 mm brickwork	0.20	1.44	0.05	nr	1.49
215 mm brickwork	0.20	1.44	0.05	nr	1.49
327 mm brickwork	0.20	1.44	0.05	nr	1.49
440 mm brickwork	0.20	1.44	0.05	nr	1.49
100 mm blockwork	0.20	1.44	0.05	nr	1.49
150 mm blockwork	0.20	1.44	0.05	nr	1.49
215 mm blockwork	0.20	1.44	0.05	nr	1.49
255 mm blockwork	0.20	1.44	0.05	nr	1.49
Make good hole and facings one side where small pipe removed					
102 mm brickwork	0.20	2.41	0.37	nr	2.78
215 mm brickwork	0.20	2.41	0.37	nr	2.78
327 mm brickwork	0.20	2.41	0.37	nr	2.78
440 mm brickwork	0.20	2.41	0.37	nr	2.78
Make good hole where large pipe removed					
102 mm brickwork	0.30	2.17	0.11	nr	2.28
215 mm brickwork	0.45	3.25	0.24	nr	3.49
327 mm brickwork	0.60	4.33	0.39	nr	4.72
440 mm brickwork	0.75	5.42	0.53	nr	5.95
100 mm blockwork	0.30	2.17	0.08	nr	2.25
150 mm blockwork	0.35	2.53	0.12	nr	2.65
215 mm blockwork	0.40	2.89	0.16	nr	3.05
255 mm blockwork	0.45	3.25	0.19	nr	3.44
Make good hole and facings one side where large pipe removed					
102 mm brickwork	0.27	3.25	0.66	nr	3.91
215 mm brickwork	0.36	4.33	0.79	nr	5.12
327 mm brickwork	0.45	5.41	0.89	nr	6.30
440 mm brickwork	0.54	6.49	1.16	nr	7.65
Make good hole where extra large pipe removed					
102 mm brickwork	0.40	2.89	0.26	nr	3.15
215 mm brickwork	0.60	4.33	0.53	nr	4.86
327 mm brickwork	0.80	5.78	0.89	nr	6.67
440 mm brickwork	1.00	7.22	1.16	nr	8.38
100 mm blockwork	0.40	2.89	0.24	nr	3.13
150 mm blockwork	0.45	3.25	0.37	nr	3.62
215 mm blockwork	0.50	3.61	0.53	nr	4.14
255 mm blockwork	0.55	3.97	0.63	nr	4.60
Make good hole and facings one side where extra large pipe removed					
102 mm brickwork	0.36	4.33	0.95	nr	5.28
215 mm brickwork	0.48	5.77	1.16	nr	6.93
327 mm brickwork	0.60	7.22	1.47	nr	8.69
440 mm brickwork	0.72	8.66	2.00	nr	10.66

C DEMOLITION/ALTERATION/RENOVATION Including overheads and profit at 12.50%	Labour hours	Labour £	Material £	Unit	Total rate £
C40 REPAIRING/RENOVATING CONCRETE/BRICK/ BLOCK/STONE - cont'd					
Repairing/renovating brick/blockwork - cont'd Cut out small areas of defective uncoursed stonework; rebuild in cement mortar to match existing					
300 mm thick wall	5.60	67.35	11.33	m2	78.68
400 mm thick wall	7.00	84.18	15.33	m2	99.51
600 mm thick wall	10.00	120.26	23.32	m2	143.58
Remove defective capping stones and haunching; replace stones and re-haunch in cement mortar to match existing					
300 mm thick wall	1.35	16.24	3.00	m2	19.24
400 mm thick wall	1.50	18.04	4.00	m2	22.04
600 mm thick wall	1.75	21.05	6.00	m2	27.05
Rake out decayed joints of uncoursed stonework and repoint in cement mortar to match existing walls	1.20	14.43	0.79	m2	15.22
C50 REPAIRING/RENOVATING METAL					
Overhaul and repair metal casement window; adjust and oil ironmongery; bring forward affected parts for redecoration	1.50	10.83	3.34	nr	14.17
C51 REPAIRING/RENOVATING TIMBER					
Cut out infected or decayed structural members; shore up adjacent work; provide and fix new 'treated' sawn softwood members pieced in Floors or flat roofs					
50 x 125 mm	0.40	3.74	2.64	m	6.38
50 x 150 mm	0.44	4.11	3.17	m	7.28
50 x 175 mm	0.48	4.49	3.69	m	8.17
50 x 200 mm	0.52	4.86	4.37	m	9.23
50 x 225 mm	0.56	5.24	5.08	m	10.32
Pitched roofs					
38 x 100 mm	0.36	3.37	1.83	m	5.19
50 x 100 mm	0.45	4.21	2.11	m	6.32
50 x 125 mm	0.50	4.67	2.64	m	7.32
50 x 150 mm	0.55	5.14	3.17	m	8.31
Kerbs, bearers and the like					
50 x 75 mm	0.45	4.21	1.59	m	5.80
50 x 100 mm	0.56	5.24	2.11	m	7.35
75 x 100 mm	0.68	6.36	3.57	m	9.93
Scarfed joint; new to existing; over 450 mm2	1.00	9.35	-	nr	9.35
Scarfed and bolted joint; new existing; including bolt; let in flush; over 450 mm2	1.66	15.52	1.69	nr	17.21
Other repairs Remove or punch in projecting nails; re-fix softwood or hardwood flooring					
loose floor boards	0.15	1.40	-	m2	1.40
floorboards previously set aside	0.80	7.48	0.27	m2	7.75
Remove damaged softwood flooring; provide and fix new 25 mm thick plain edge softwood boarding					
small areas	1.15	10.75	12.99	m2	23.74
individual boards 150 mm wide	0.30	2.80	1.42	m	4.22
Sand down and re-surface existing flooring; prepare body in with shellac and wax polish					
softwood	-	-	-	m2	9.56
hardwood	-	-	-	m2	9.00

C DEMOLITION/ALTERATION/RENOVATION Including overheads and profit at 12.50%	Labour hours	Labour £	Material £	Unit	Total rate £
Fit existing softwood skirting to new frame or architrave					
75 mm high	0.10	0.93	-	m	0.93
150 mm high	0.13	1.22	-	m	1.22
225 mm high	0.16	1.50	-	m	1.50
Piece in a new 25 x 150 mm moulded softwood skirting to match existing where old removed; bring forward for redecoration	0.38	3.17	5.17	m	8.34
Fit short length of salvaged softwood skirting to new reveal; including mitre with existing; fit to new frame; sawn softwood grounds plugged to brickwork; bring forward affected parts for redecoration					
75 mm high	0.34	3.01	0.62	nr	3.63
150 mm high	0.52	4.63	0.70	nr	5.33
225 mm high	0.65	5.79	0.76	nr	6.55
Piece in new 25 x 150 mm moulded softwood skirting to match existing where socket outlet removed; bring forward for redecoration	0.22	1.83	3.06	nr	4.89
Ease and adjust softwood door, oil ironmongery bring forward affected parts for redecoration	0.77	6.97	0.54	nr	7.51
Remove softwood door; ease and adjust; oil ironmongery; re-hang; bring forward affected parts for redecoration	1.20	11.22	0.51	nr	11.73
Remove mortice lock, piece in softwood door, bring forward affected parts for redecoration	1.10	10.09	1.32	nr	11.41
Fix only for salvaged softwood door	1.60	14.96	-	nr	14.96
Remove softwood door; plane 12 mm off bottom edge and re-hang	1.20	11.22	-	nr	11.22
Remove softwood door; alter ironmongery; piece in and rebate door and frame; re-hang on opposite style; bring forward affected parts for redecoration	2.65	24.30	2.17	nr	26.47
Remove softwood panelled door to prepare for fire upgrading; remove ironmongery; replace existing beads with new 25 x 38 mm hardwood beads screwed on; repair where damaged; piece out and re-hang on wider butts; adjust all ironmongery; seal around frame in cement mortar; bring forward affected parts for redecoration (any new glass measured separately)	5.35	49.54	22.07	nr	71.61
Upgrade and face up one side of flush door with 9 mm thick 'Supalux', screwed on	1.25	11.69	22.14	nr	33.83
Upgrade and face up one side of softwood panelled door with 9 mm thick 'Supalux', screwed on; plasterboard infilling in recesses	2.70	25.24	25.71	nr	50.95
Take off existing softwood doorstop; provide and screw on new 25 x 38 mm doorstop; bring forward for redecoration	0.22	2.02	1.76	m	3.78
Cut away defective 75 x 100 mm softwood external door frame; provide and splice in new piece 300 mm long; bed in cement mortar (1:3); point one side; bring forward for redecoration	1.40	12.71	6.86	nr	19.57
Seal roof trap flush with ceiling	0.60	5.61	3.79	nr	9.40
Form opening 762 x 762 mm in existing ceiling for new standard roof trap comprising softwood linings, architraves and 9 mm plywood cover; including all necessary trimming of ceiling joists (making good ceiling plaster measured separately)	2.70	25.24	31.82	nr	57.06

C DEMOLITION/ALTERATION/RENOVATION Including overheads and profit at 12.50%	Labour hours	Labour £	Material £	Unit	Total rate £
C51 REPAIRING/RENOVATING TIMBER - cont'd					
Other repairs - cont'd					
Ease and adjust softwood casement window; oil ironmongery; bring forward affected parts for redecoration	0.52	4.63	0.31	nr	4.94
Remove softwood casement window; ease and adjust; oil ironmongery; re-hang; bring forward affected parts for redecoration	0.77	6.97	0.54	nr	7.51
Remove softwood casement window; repair corner with angle plate; let in flush; oil ironmongery; re-hang; bring forward affected parts for redecoration	2.10	19.44	-	nr	19.44
Splice in or replace top or bottom rail or style of softwood casement window to match existing; bring forward for redecoration (taking off and re-hanging measured separately)	0.92	8.37	8.96	nr	17.33
Renew solid mullion jamb or transom of softwood casement window to match existing; bring forward for redecoration (taking off and re-hanging adjoining casements measured separately)	2.80	26.18	12.71	nr	38.89
Temporary lining 6 mm plywood infill to window whilst casement under repair	0.80	7.48	7.95	nr	15.43
Overhaul softwood double hung sash window; ease, adjust and oil pulley wheels; replace	2.65	24.30	5.82	nr	30.12
Cut away defective part of softwood window sill; provide and splice in new 83 x 65 mm weathered and throated piece 300 mm long; bring forward for redecoration	2.05	18.59	8.21	nr	26.80
Cut off broken stair nosing to tread or landing	0.40	3.74	5.46	nr	9.20
Renew isolated broken softwood tread; including additional framing beneath	1.20	11.22	21.10	nr	32.32
Remove damaged formica; provide and fix new to match existing					
worktop; over 300 mm wide	2.00	18.70	5.24	m2	23.94
edging; 25 mm wide	0.20	1.87	1.16	m	3.03
Renew 6 x 50 mm galvanized water bar	0.50	4.67	1.28	m	5.95
C52 FUNGUS/BEETLE ERADICATION					
Treat timbers with two coats of 'Rentokil' or similar proprietary insecticide and fungicide; PC £12.36/5ltr					
Remove cobwebs, dust and roof insulation; treat by spray application; replace insulation per m3 roof void	-	-	-	m3	6.58
Treat individual timbers by spray application					
boarding	-	-	-	m2	1.46
structural members	-	-	-	m2	2.93
Treat individual timbers by brush application					
boarding	-	-	-	m2	3.66
structural members	-	-	-	m2	4.31
skirtings	-	-	-	m	1.64
Lift necessary floorboards; treat floors by spray application; later re-fix boards	-	-	-	m2	7.63
Treat surfaces of adjoining concrete or brickwork with two coats of 'Rentokil' dry rot fluid by spray application	0.20	1.87	0.80	m2	2.67

C DEMOLITION/ALTERATION/RENOVATION Including overheads and profit at 12.50%	Labour hours	Labour £	Material £	Unit	Total rate £

C53 REPAIRING/RENOVATING GLASS

NOTE: All items of removal include for removing debris from site. The following prices for re-glazing are for re-glazing total windows. Prices for renewing single squares in repairs will be much higher and dependent on the quantity involved.

Cut back putty; clean exposed rebate and re-putty

	Labour hours	Labour £	Material £	Unit	Total rate £
to wood	-	-	-	m	1.40
to metal	-	-	-	m	1.86

Hack out old glass; clean out putty and prepare rebates for new (measured elsewhere)

	Labour hours	Labour £	Material £	Unit	Total rate £
3 mm thick float glass	0.55	4.09	-	m	4.09
4 mm thick float glass	0.63	4.68	-	m	4.68
6 mm thick float glass	0.80	5.95	-	m	5.95

Hack out old glass; clean out putty in rebates; prime for and re-glazing to wood with putty; 0.1 - 0.5 m2 panes

	Labour hours	Labour £	Material £	Unit	Total rate £
3 mm thick clear float glass OQ	3.33	24.76	18.38	m2	43.14
4 mm thick clear float glass	3.55	26.40	20.84	m2	47.24
6 mm thick clear float glass	3.80	28.26	33.79	m2	62.05
4 mm thick patterned glass	3.55	26.40	29.93	m2	56.33
7 mm thick Georgian wired cast glass	3.80	28.26	26.57	m2	54.83
6 mm thick Georgian wired polished plate glass	4.00	29.75	57.94	m2	87.69

C54 REPAIRING/RENOVATING CLADDINGS/COVERINGS

Carefully take off tiles or slates; select and re-fix; including providing 25% new; including nails, sundries, etc. Coverings, sloping

	Labour hours	Labour £	Material £	Unit	Total rate £
asbestos cement 'blue/black' slates; 500 x 250 mm PC £78.60/100	1.10	15.26	5.17	m2	20.43
asbestos cement 'blue/black' slates; 600 x 300 mm PC £108.75/100	1.00	13.87	5.41	m2	19.28
asbestos-free artificial 'blue/black' slates 500 x 250 mm PC £81.75/100	1.10	15.26	5.98	m2	21.24
asbestos-free artificial 'blue/black' slates 600 x 300 mm PC £113.85/100	1.00	13.87	5.62	m2	19.49
natural slates; Welsh blue 510 x 255 mm PC £181.50/100	1.20	16.65	11.27	m2	27.92
natural slates; Welsh blue 600 x 300 mm PC £302.25/100	1.05	14.56	12.30	m2	26.86
second hand natural slates; Welsh blue 510 x 255 mm PC £101.25/100	1.20	16.65	6.72	m2	23.37
second hand natural slates; Welsh blue 600 x 300 mm PC £135.00/100	1.05	14.56	6.14	m2	20.70
clay plain tiles; 'Dreadnought' machine made; 265 x 165 mm PC £38.25/100	1.10	15.26	7.71	m2	22.97
concrete interlocking tiles; Marley 'Ludlow Major' 413 x 330 mm PC £68.25/100	0.70	9.71	2.57	m2	12.28
concrete interlocking tiles; Redland 'Renown' 417 x 330 mm PC £80.10/100	0.70	9.71	2.95	m2	12.66
concrete plain tiles; 267 x 165 mm PC £21.00/100	1.10	15.26	4.52	m2	19.78

C DEMOLITION/ALTERATION/RENOVATION Including overheads and profit at 12.50%	Labour hours	Labour £	Material £	Unit	Total rate £

C54 REPAIRING/RENOVATING CLADDINGS/COVERINGS - cont'd

Carefully take off damaged tiles or slates;
provide and fix new; including nails,
sundries, etc.
Coverings; sloping; areas less than 10 m2

asbestos cement 'blue/black' slates; 500 x 250 mm PC £78.60/100	1.35	18.73	19.10	m2	37.83
asbestos cement 'blue/black' slates; 600 x 300 mm PC £108.75/100	1.25	17.34	18.58	m2	35.92
asbestos-free artificial 'blue/black' slates 500 x 250 mm PC £81.75/100	1.35	18.73	20.06	m2	38.79
asbestos-free artificial 'blue/black' slates 600 x 300 mm PC £113.85/100	1.25	17.34	19.41	m2	36.75
natural slates; Welsh blue 510 x 255 mm	1.45	20.11	5.58	m2	25.69
natural slates; Welsh blue 600 x 300 mm PC £302.25/100	1.30	18.03	49.75	m2	67.78
second hand natural slates; Welsh blue 510 x 255 mm PC £101.25/100	1.45	20.11	24.50	m2	44.61
second hand natural slates; Welsh blue 600 x 300 mm PC £135.00/100	1.30	18.03	21.79	m2	39.82
clay plain tiles; 'Dreadnought' machine made; 265 x 165 mm PC £38.25/100	1.35	18.73	29.27	m2	48.00
concrete interlocking tiles; Marley 'Ludlow Major' 413 x 330 mm PC £68.25/100	0.90	12.48	9.24	m2	21.72
concrete interlocking tiles; Redland 'Renown' 417 x 330 mm PC £80.10/100	0.90	12.48	10.73	m2	23.21
concrete plain tiles; 267 x 165 mm PC £21.00/100	1.35	18.73	16.46	m2	35.19

Individual tiles or slates; isolated; rate
for each

asbestos cement 'blue/black' slates; 500 x 250 mm PC £0.83	0.45	3.35	1.21	nr	4.56
asbestos cement 'blue/black' slates; 600 x 300 mm PC £1.17	0.40	2.97	1.61	nr	4.58
asbestos-free artificial 'blue/black' slates 500 x 250 mm PC £0.88	0.45	3.35	1.28	nr	4.63
asbestos-free artificial 'blue/black' slates 600 x 300 mm PC £1.22	0.40	2.97	1.67	nr	4.64
natural slates; Welsh blue 510 x 255 mm PC £1.94	0.50	3.72	2.58	nr	6.30
natural slates; Welsh blue 600 x 300 mm PC £3.23	0.40	2.97	4.18	nr	7.15
second hand natural slates; Welsh blue 510 x 255 mm PC £1.09	0.50	3.72	1.53	nr	5.25
second hand natural slates; Welsh blue 600 x 300 mm PC £1.44	0.40	2.97	1.94	nr	4.91
clay plain tiles; 'Dreadnought' machine made; 265 x 165 mm PC £0.42	0.20	1.49	0.68	nr	2.17
concrete interlocking tiles; Marley 'Ludlow Major' 413 x 330 mm PC £0.74	0.25	1.86	1.07	nr	2.93
concrete interlocking tiles; Redland 'Renown' 417 x 330 mm PC £0.85	0.25	1.86	1.21	nr	3.07
concrete plain tiles; 267 x 165 mm PC £0.21	0.20	1.49	0.42	nr	1.91

Remove half round ridge or hip tile 300 mm long; provide and fix new	0.75	5.58	2.84	nr	8.42
Make good crack in roofing (any type), with mastic filling	0.40	2.97	1.42	m	4.39
Examine roof battens; renail where loose; and provide and fix 25% new					
19 x 50 mm slating battens at 262 mm centres	0.08	1.11	0.34	m2	1.45
19 x 38 mm tiling battens at 100 mm centres	0.20	2.77	0.77	m2	3.54
Remove roof battens and nails; provide and fix new 'treated' softwood battens throughout					
19 x 50 mm slating battens at 262 mm centres	0.12	1.66	1.42	m2	3.08
19 x 38 mm tiling battens at 100 mm centres	0.25	3.47	2.93	m2	6.40

C DEMOLITION/ALTERATION/RENOVATION Including overheads and profit at 12.50%		Labour hours	Labour £	Material £	Unit	Total rate £
Remove underfelt and nails; provide and fix new						
unreinforced felt	PC £0.74	0.10	1.39	1.11	m2	2.50
reinforced felt	PC £1.70	0.10	1.39	2.45	m2	3.84

C55 REPAIRING/RENOVATING WATERPROOFING FINISHES

	Labour hours	Labour £	Material £	Unit	Total rate £
Cut out crack in asphalt paving; make good to match existing					
13 mm one coat	1.15	8.55	–	m	8.55
20 mm two coats	1.55	11.53	–	m	11.53
30 mm three coats	1.95	14.50	–	m	14.50
Cut out crack in asphalt roof coverings; make good to match existing					
20 mm two coats	1.90	14.13	–	m	14.13

C56 REPAIRING/RENOVATING SURFACE FINISHES

	Labour hours	Labour £	Material £	Unit	Total rate £
Remove defective or damaged wall plaster; re-plaster walls with two coats of gypsum plaster; including dubbing out, joint of new to existing					
small areas	1.60	17.05	3.45	m2	20.50
isolated areas not exceeding 0.50 m2	1.15	12.25	1.91	nr	14.16
Re-plaster walls with two coats of gypsum plaster; including dubbing out, where wall or partition removed; including trimming back existing and joint of new to existing					
150 mm wide	0.63	6.71	0.60	m	7.31
225 mm wide	0.78	8.31	0.83	m	9.14
300 mm wide	0.93	9.91	1.05	m	10.96
Cut back finishings to allow new partition to be built; later re-plaster walls with two coats of gypsum plaster both sides; joints of new to existing	1.20	12.78	1.93	m	14.71
Re-plaster walls with two coats of gypsum plaster; including dubbing out; to face and returns of jamb to new opening; joints of new to existing plaster					
75 mm girth	0.65	6.92	2.17	m	9.09
150 mm girth	0.75	7.99	2.41	m	10.40
225 mm girth	0.85	9.06	2.63	m	11.69
300 mm girth	0.95	10.12	2.83	m	12.95
Re-plaster walls with two coats of gypsum plaster; including bonding agent; to faces and soffits of concrete lintel to new opening; joints of new to existing plaster					
75 mm girth	0.70	7.46	1.92	m	9.38
150 mm girth	0.80	8.52	2.07	m	10.59
225 mm girth	0.90	9.59	2.22	m	11.81
300 mm girth	1.00	10.65	2.34	m	12.99
Remove defective or damaged damp wall plaster; investigate and treat wall; re-plaster with two coats of 'Limelite' renovating plaster; including dubbing out and joint of new to existing					
small areas	1.70	18.11	5.12	m2	23.23
Dubbing out in cement and sand (1:3); average 13 mm thick					
over 300 mm wide	–	–	–	m	8.90
Remove defective or damaged damp wall					
isolated areas not exceeding 0.50 m2	1.22	13.00	2.84	nr	15.84
Remove defective or damaged ceiling plaster; remove laths or cut back boarding; prepare and fix new plasterboard; re-skim with one coat of gypsum; joint of new to existing					
small areas	1.70	18.11	3.11	m2	21.22
isolated areas not exceeding 0.50 m2	1.22	13.00	1.73	nr	14.73

C DEMOLITION/ALTERATION/RENOVATION Including overheads and profit at 12.50%	Labour hours	Labour £	Material £	Unit	Total rate £
C56 REPAIRING/RENOVATING SURFACE FINISHES - cont'd					
Prepare and fix new plasterboard; re-skim with one coat of gypsum where wall or partition removed; including trimming back existing and joint of new to existing					
150 mm wide	0.73	7.78	0.81	m	8.59
225 mm wide	0.88	9.38	0.98	m	10.36
300 mm wide	1.03	10.97	1.14	m	12.11
Cut out; repair existing plaster cracks					
in walls	0.25	2.66	0.27	m	2.93
in ceilings	0.33	3.52	0.18	m	3.70
Make good plaster where items removed or holes left					
small pipe or conduit	0.07	0.75	0.03	nr	0.78
large pipe	0.10	1.07	0.06	nr	1.13
extra large pipe	0.13	1.38	0.09	nr	1.47
light switch	0.08	0.85	0.07	nr	0.92
Make good plasterboard and skim where items removed or holes left					
small pipe or conduit	0.15	1.60	0.30	nr	1.90
large pipe	0.23	2.45	0.45	nr	2.90
extra large pipe	0.30	3.20	0.59	nr	3.79
ceiling rose	0.18	1.92	0.45	nr	2.37
Level and repair floor screed with 'Ardit K10'					
small areas	0.50	5.33	5.25	m2	10.58
isolated areas not exceeding 0.50 m2	0.35	3.73	2.86	nr	6.59
Refix individual loose tiles or blocks					
floor tiles or blocks	0.15	1.60	0.15	nr	1.75
suspended ceiling tiles	0.30	3.20	-	nr	3.20
C57 REPAIRING/RENOVATING GUTTERWORK AND PIPEWORK					
Clean out existing					
rainwater gutters	0.08	0.51	-	m	0.51
rainwater gully	0.20	1.29	-	nr	1.29
rainwater stack; including head, swannecks and shoe (not exceeding 10 m long)	0.75	4.83	-	nr	4.83
Overhaul and re-make leaking joint; including new gutter bolt					
100 mm PVC	0.20	1.71	0.12	nr	1.83
100 mm cast iron	0.80	6.84	0.18	nr	7.02
Remove defective balloon grating; provide and fix in new outlet					
75 mm dia	0.10	1.07	1.86	nr	2.93
100 mm dia	0.10	1.07	2.23	nr	3.30
Overhaul section of rainwater guttering; cut out existing joints; adjust brackets to correct falls; re-make joints; including new gutter bolts					
100 mm PVC	0.25	2.67	0.03	m	2.70
100 mm cast iron	0.90	7.70	0.14	m	7.84
Clean and repair existing crack in lead gutter with wiped soldered joint	0.66	5.65	3.79	m	9.44
Clear blocked wastes without dismantling					
sink	0.50	3.22	-	nr	3.22
WC trap	0.60	3.86	-	nr	3.86
Turn off supplies; dismantle the following fittings; replace washers; re-assemble and test					
15 mm tap	0.25	2.67	0.03	nr	2.70
15 mm ball valve	0.35	3.74	0.06	nr	3.80
Turn off supplies; remove the following fitting; test and replace					
15 mm ball valve	0.50	5.34	5.21	nr	10.55

D GROUNDWORK
Including overheads and profit at 12.50%

Prices are applicable to excavation in firm soil. Multiplying factors for other soils
are as follows:-

	Mechanical	**Hand**
Clay	x 2.00	x 1.20
Compact gravel	x 3.00	x 1.50
Soft chalk	x 4.00	x 2.00
Hard rock	x 5.00	x 6.00
Running sand or silt	x 6.00	x 2.00

	Labour hours	Labour £	Material Plant £	Unit	Total rate £
D20 EXCAVATION AND FILLING					
Site preparation					
Removing trees					
600 mm - 1.50 m girth	22.00	153.45	-	nr	153.45
1.50 - 3.00 m girth	38.50	268.54	-	nr	268.54
over 3.00 m girth	55.00	383.63	-	nr	383.63
Removing tree stumps					
600 mm - 1.50 m girth	1.10	7.67	36.37	nr	44.04
1.50 - 3.00 m girth	1.65	11.51	54.55	nr	66.06
over 3.00 m girth	2.20	15.35	72.73	nr	88.08
Clearing site vegetation	0.04	0.28	-	m2	0.28
Lifting turf for preservation; stacking	0.39	2.72	-	m2	2.72
Excavation using a wheeled hydraulic excavator with a 0.24 m3 bucket					
Excavating topsoil to be preserved					
average 150 mm deep	0.03	0.21	0.76	m2	0.97
Add or deduct for each 25 mm variation in depth	0.01	0.03	0.13	m2	0.16
Excavating to reduce levels; not exceeding					
0.25 m deep	0.11	0.77	2.53	m3	3.30
1 m deep	0.09	0.63	2.03	m3	2.66
2 m deep	0.11	0.77	2.53	m3	3.30
4 m deep	0.14	0.98	3.04	m3	4.02
Excavating basements and the like; starting from reduced level; not exceeding					
1 m deep	0.11	0.77	2.93	m3	3.70
2 m deep	0.09	0.63	2.34	m3	2.97
4 m deep	0.11	0.77	2.93	m3	3.70
6 m deep	0.14	0.98	3.51	m3	4.49
8 m deep	0.16	1.12	4.09	m3	5.21
Excavating pits to receive bases; not exceeding					
0.25 m deep	0.35	2.44	8.78	m3	11.22
1 m deep	0.31	2.16	7.90	m3	10.06
2 m deep	0.35	2.44	8.78	m3	11.22
4 m deep	0.38	2.65	9.65	m3	12.30
6 m deep	0.41	2.86	10.53	m3	13.39
Extra over pit excavation for commencing below ground or reduced level					
1 m below	0.03	0.21	0.88	m3	1.09
2 m below	0.05	0.35	1.17	m3	1.52
3 m below	0.06	0.42	1.46	m3	1.88
4 m below	0.08	0.56	2.05	m3	2.61

D GROUNDWORK Including overheads and profit at 12.50%	Labour hours	Labour £	Material £	Unit	Total rate £
D20 EXCAVATION AND FILLING - cont'd					
Excavation using a wheeled hydraulic **excavator with a 0.24 m3 bucket - cont'd**					
Excavating trenches; not exceeding 0.30 m wide; not exceeding					
0.25 m deep	0.30	2.09	7.61	m3	9.70
1 m deep	0.25	1.74	6.44	m3	8.18
2 m deep	0.30	2.09	7.61	m3	9.70
4 m deep	0.35	2.44	8.78	m3	11.22
6 m deep	0.41	2.86	10.53	m3	13.39
Excavating trenches; over 0.30 m wide; not exceeding					
0.25 m deep	0.28	1.95	7.02	m3	8.97
1 m deep	0.23	1.60	5.85	m3	7.45
2 m deep	0.28	1.95	7.02	m3	8.97
4 m deep	0.31	2.16	7.90	m3	10.06
6 m deep	0.38	2.65	9.65	m3	12.30
Extra over trench excavation for commencing below ground or reduced level					
1 m below	0.03	0.21	0.88	m3	1.09
2 m below	0.05	0.35	1.17	m3	1.52
3 m below	0.06	0.42	1.46	m3	1.88
4 m below	0.08	0.56	2.05	m3	2.61
Extra over trench excavation for curved trench	0.01	0.07	0.29	m3	0.36
Excavating for pile caps and ground beams between piles; not exceeding					
0.25 m deep	0.38	2.65	9.65	m3	12.30
1 m deep	0.35	2.44	8.78	m3	11.22
2 m deep	0.38	2.65	9.65	m3	12.30
Excavating to bench sloping ground to receive filling; not exceeding					
0.25 m deep	0.10	0.70	2.93	m3	3.63
1 m deep	0.08	0.56	2.34	m3	2.90
2 m deep	0.10	0.70	2.93	m3	3.63
Extra over any type of excavation at any depth for excavating					
alongside services	2.25	14.23		m3	14.23
around services crossing excavation	5.20	31.59		m3	31.59
below ground water level	0.11	0.77	2.93	m3	3.70
Extra over excavation for breaking out existing materials					
rock	2.65	18.48	12.54	m3	31.02
concrete	2.25	15.69	9.39	m3	25.08
reinforced concrete	3.20	22.32	13.78	m3	36.10
brickwork; blockwork or stonework	1.60	11.16	6.89	m3	18.05
Extra over excavation for breaking out existing hard pavings					
tarmacadam 75 mm thick	0.17	1.19	0.57	m2	1.76
tarmacadam and hardcore 150 mm thick	0.23	1.60	0.62	m2	2.22
concrete 150 mm thick	0.35	2.44	1.38	m2	3.82
reinforced concrete 150 mm thick	0.52	3.63	2.01	m2	5.64
Working space allowance to excavations					
reduced levels; basements and the like	0.09	0.63	2.34	m2	2.97
pits	0.20	1.40	4.97	m2	6.37
trenches	0.18	1.26	4.68	m2	5.94
pile caps and ground beams between piles	0.21	1.46	5.27	m2	6.73
Extra over excavation for working space for filling in with					
hardcore	0.11	0.77	8.27	m2	9.04
coarse ashes	0.11	0.77	12.32	m2	13.09
sand	0.11	0.77	16.24	m2	17.01
40 - 20 mm gravel	0.11	0.77	19.15	m2	19.92
in situ concrete - 7.50 N/mm2 - 40 mm aggregate (1:8)	0.83	5.99	30.62	m2	36.61

D GROUNDWORK Including overheads and profit at 12.50%	Labour hours	Labour £	Material £	Unit	Total rate £
Hand excavation					
Excavating topsoil to be preserved					
average 150 mm deep	0.28	1.95	-	m2	1.95
Add or deduct for each 25 mm variation					
in depth	0.04	0.28	-	m2	0.28
Excavating to reduce levels; not exceeding					
0.25 m deep	2.00	13.95	-	m3	13.95
1 m deep	2.20	15.35	-	m3	15.35
2 m deep	2.40	16.74	-	m3	16.74
4 m deep	2.65	18.48	-	m3	18.48
Excavating basements and the like; starting					
from reduced level; not exceeding					
1 m deep	2.65	18.48	-	m3	18.48
2 m deep	2.95	20.58	-	m3	20.58
4 m deep	3.85	26.85	-	m3	26.85
6 m deep	4.75	33.13	-	m3	33.13
8 m deep	5.70	39.76	-	m3	39.76
Excavating pits to receive bases; not					
exceeding					
0.25 m deep	3.10	21.62	-	m3	21.62
1 m deep	3.30	23.02	-	m3	23.02
2 m deep	3.85	26.85	-	m3	26.85
4 m deep	4.95	34.53	-	m3	34.53
6 m deep	6.25	43.59	-	m3	43.59
Extra over pit excavation for commencing					
below ground or reduced level					
1 m below	0.55	3.84	-	m3	3.84
2 m below	1.10	7.67	-	m3	7.67
3 m below	1.65	11.51	-	m3	11.51
4 m below	2.20	15.35	-	m3	15.35
Excavating trenches; not exceeding					
0.30 m wide; not exceeding					
0.25 m deep	2.75	19.18	-	m3	19.18
1 m deep	3.05	21.27	-	m3	21.27
2 m deep	3.50	24.41	-	m3	24.41
4 m deep	4.45	31.04	-	m3	31.04
6 m deep	5.65	39.41	-	m3	39.41
Excavating trenches; over 0.30 m wide;					
not exceeding					
0.25 m deep	2.55	17.79	-	m3	17.79
1 m deep	2.75	19.18	-	m3	19.18
2 m deep	3.20	22.32	-	m3	22.32
4 m deep	4.05	28.25	-	m3	28.25
6 m deep	5.15	35.92	-	m3	35.92
Extra over trench excavation for commencing					
below ground or reduced level					
1 m below	0.55	3.84	-	m3	3.84
2 m below	1.10	7.67	-	m3	7.67
3 m below	1.65	11.51	-	m3	11.51
4 m below	2.20	15.35	-	m3	15.35
Extra over trench excavation for curved					
trench	0.55	3.84	-	m3	3.84
Excavating for pile caps and ground beams					
between piles; not exceeding					
0.25 m deep	3.65	25.46	-	m3	25.46
1 m deep	3.95	27.55	-	m3	27.55
2 m deep	4.60	32.09	-	m3	32.09
Excavating to bench sloping ground					
to receive filling; not exceeding					
0.25 m deep	1.65	11.51	-	m3	11.51
1 m deep	2.00	13.95	-	m3	13.95
2 m deep	2.20	15.35	-	m3	15.35
Extra over any type of excavation at any					
depth for excavating					
alongside services	1.10	7.67	-	m3	7.67
around services crossing excavation	2.20	15.35	-	m3	15.35

D GROUNDWORK Including overheads and profit at 12.50%	Labour hours	Labour £	Material £	Unit	Total rate £
D20 EXCAVATION AND FILLING - cont'd					
Hand excavation - cont'd					
Extra over any type of excavation at any depth for excavating					
below ground water level	0.39	2.72	-	m3	2.72
Extra over excavation for breaking out existing materials					
rock	5.50	38.36	7.43	m3	45.79
concrete	4.95	34.53	6.19	m3	40.72
reinforced concrete	6.60	46.04	8.66	m3	54.70
brickwork; blockwork or stonework	3.30	23.02	3.71	m3	26.73
Extra over excavation for breaking out existing hard pavings					
tarmacadam 75 mm thick	0.44	3.07	0.50	m2	3.57
tarmacadam and hardcore 150 mm thick	0.55	3.84	0.62	m2	4.46
concrete 150 mm thick	0.77	5.37	0.87	m2	6.24
reinforced concrete 150 mm thick	0.99	6.91	1.24	m2	8.15
Extra for taking up precast concrete paving slabs	0.33	2.30	-	m2	2.30
Working space allowance to excavations					
reduced levels; basements and the like	2.85	19.88	-	m2	19.88
pits	2.95	20.58	-	m2	20.58
trenches	2.60	18.14	-	m2	18.14
pile caps and ground beams between piles	3.05	21.27	-	m2	21.27
Extra over excavation for working space for filling in with					
hardcore	0.88	6.14	6.75	m2	12.89
coarse ashes	0.79	5.51	10.80	m2	16.31
sand	0.88	6.14	14.72	m2	20.86
40 - 20 mm gravel	0.88	6.14	17.63	m2	23.77
in situ concrete - 7.50 N/mm2					
- 40 mm aggregate (1:8)	1.20	8.67	29.10	m2	37.77
Hand excavation inside existing buildings					
Excavating basements and the like; starting from reduced level; not exceeding					
1 m deep	3.30	23.02	-	m3	23.02
2 m deep	3.70	25.81	-	m3	25.81
4 m deep	4.80	33.48	-	m3	33.48
6 m deep	5.90	41.15	-	m3	41.15
8 m deep	7.15	49.87	-	m3	49.87
Excavating pits to receive bases; not exceeding					
0.25 m deep	3.85	26.85	-	m3	26.85
1 m deep	4.15	28.95	-	m3	28.95
2 m deep	4.80	33.48	-	m3	33.48
4 m deep	6.20	43.25	-	m3	43.25
Extra over pit excavation for commencing below ground or reduced level					
1 m below	0.65	4.53	-	m3	4.53
2 m below	1.35	9.42	-	m3	9.42
3 m below	2.05	14.30	-	m3	14.30
4 m below	2.75	19.18	-	m3	19.18
Excavating trenches; not exceeding 0.30 m wide; not exceeding					
0.25 m deep	3.15	21.97	-	m3	21.97
1 m deep	3.45	24.06	-	m3	24.06
2 m deep	4.00	27.90	-	m3	27.90
4 m deep	5.10	35.57	-	m3	35.57
6 m deep	6.45	44.99	-	m3	44.99
Excavating trenches; over 0.30 m wide; not exceeding					
0.25 m deep	3.15	21.97	-	m3	21.97
1 m deep	3.45	24.06	-	m3	24.06
2 m deep	4.00	27.90	-	m3	27.90
4 m deep	5.10	35.57	-	m3	35.57
6 m deep	6.45	44.99	-	m3	44.99

D GROUNDWORK Including overheads and profit at 12.50%	Labour hours	Labour £	Material £	Unit	Total rate £
Extra over trench excavation for commencing below ground or reduced level					
1 m below	0.65	4.53	-	m3	4.53
2 m below	1.35	9.42	-	m3	9.42
3 m below	2.05	14.30	-	m3	14.30
4 m below	2.75	19.18	-	m3	19.18
Extra over trench excavation for curved trench	0.70	4.88	-	m3	4.88
Extra over any type of excavation at any depth for excavating					
below ground water level	0.07	0.49	1.28	m3	1.77
Extra over excavation for breaking out existing materials					
concrete	6.20	43.25	6.19	m3	49.44
reinforced concrete	8.25	57.54	8.66	m3	66.20
brickwork; blockwork or stonework	4.15	28.95	3.71	m3	32.66
Extra over excavation for breaking out existing hard pavings					
concrete 150 mm thick	0.96	6.70	0.87	m3	7.57
reinforced concrete 150 mm thick	0.96	6.70	0.87	m3	7.57
Working space allowance to excavations					
pits	0.31	2.16	6.63	m2	8.79
trenches	0.06	0.42	1.07	m2	1.49
Earthwork support (average 'risk' prices)					
Not exceeding 2 m between opposing faces; not exceeding					
1 m deep	0.11	0.77	0.34	m2	1.11
2 m deep	0.14	0.98	0.38	m2	1.36
4 m deep	0.18	1.26	0.44	m2	1.70
6 m deep	0.21	1.46	0.53	m2	1.99
2 - 4 m between opposing faces; not exceeding					
1 m deep	0.12	0.84	0.38	m2	1.22
2 m deep	0.15	1.05	0.44	m2	1.49
4 m deep	0.19	1.33	0.53	m2	1.86
6 m deep	0.23	1.60	0.63	m2	2.23
Over 4 m between opposing faces; not exceeding					
1 m deep	0.13	0.91	0.44	m2	1.35
2 m deep	0.17	1.18	0.53	m2	1.71
4 m deep	0.21	1.46	0.63	m2	2.09
6 m deep	0.26	1.82	0.76	m2	2.58
Earthwork support (open boarded)					
Not exceeding 2 m between opposing faces; not exceeding					
1 m deep	0.33	2.30	1.06	m2	3.36
2 m deep	0.42	2.93	1.27	m2	4.20
4 m deep	0.53	3.70	1.59	m2	5.29
6 m deep	0.66	4.60	1.82	m2	6.42
2 - 4 m between opposing faces; not exceeding					
1 m deep	0.37	2.58	1.27	m2	3.85
2 m deep	0.46	3.21	1.61	m2	4.82
4 m deep	0.59	4.12	1.90	m2	6.02
6 m deep	0.73	5.09	2.24	m2	7.33
Over 4 m between opposing faces; not exceeding					
1 m deep	0.42	2.93	1.48	m2	4.41
2 m deep	0.53	3.70	1.86	m2	5.56
4 m deep	0.66	4.60	2.24	m2	6.84
6 m deep	0.84	5.86	2.58	m2	8.44

D GROUNDWORK Including overheads and profit at 12.50%	Labour hours	Labour £	Material £	Unit	Total rate £

D20 EXCAVATION AND FILLING - cont'd

Earthwork support (close boarded)
Not exceeding 2 m between opposing faces;
not exceeding

1 m deep	0.88	6.14	1.90	m2	8.04
2 m deep	1.10	7.67	2.33	m2	10.00
4 m deep	1.40	9.77	2.86	m2	12.63
6 m deep	1.70	11.86	3.39	m2	15.25

2 - 4 m between opposing faces; not exceeding

1 m deep	0.97	6.77	2.12	m2	8.89
2 m deep	1.20	8.37	2.65	m2	11.02
4 m deep	1.40	9.77	3.17	m2	12.94
6 m deep	1.70	11.86	3.70	m2	15.56

Over 4 m between opposing faces;
not exceeding

1 m deep	1.05	7.32	2.33	m2	9.65
2 m deep	1.30	9.07	2.96	m2	12.03
4 m deep	1.55	10.81	3.49	m2	14.30
6 m deep	1.85	12.90	4.02	m2	16.92

Extra over earthwork support for

extending into running silt or the like	0.55	3.84	0.63	m2	4.47
extending below ground water level	0.33	2.30	0.32	m2	2.62
support next to roadways	0.44	3.07	0.53	m3	3.60
curved support	0.03	0.21	0.36	m2	0.57
support left in	0.72	5.02	13.09	m2	18.11

**Earthwork support (average 'risk' prices
- inside existing buildings)**
Not exceeding 2 m between opposing faces;
not exceeding

1 m deep	0.21	1.46	0.53	m2	1.99
2 m deep	0.26	1.81	0.63	m2	2.44
4 m deep	0.33	2.30	0.80	m2	3.10
6 m deep	0.41	2.86	0.91	m2	3.77

2 - 4 m between opposing faces; not exceeding

1 m deep	0.23	1.60	0.63	m2	2.23
2 m deep	0.29	2.02	0.80	m2	2.82
4 m deep	0.37	2.58	0.95	m2	3.53
6 m deep	0.45	3.14	1.12	m2	4.26

Over 4 m between opposing faces;
not exceeding

1 m deep	0.26	1.81	0.74	m2	2.55
2 m deep	0.33	2.30	0.93	m2	3.23
4 m deep	0.41	2.86	1.12	m2	3.98
6 m deep	0.52	3.63	1.33	m2	4.96

Mechanical disposal of excavated materials
Removing from site to tip not exceeding

13 km (using lorries)	-	-	-	m3	9.44

Depositing on site in spoil heaps

average 25 m distant	-	-	-	m3	0.76
average 50 m distant	-	-	-	m3	1.27
average 100 m distant	-	-	-	m3	2.03
average 200 m distant	-	-	-	m3	3.04

Spreading on site

average 25 m distant	0.23	1.60	0.76	m3	2.36
average 50 m distant	0.23	1.60	1.27	m3	2.87
average 100 m distant	0.23	1.60	2.03	m3	3.63
average 200 m distant	0.23	1.60	3.04	m3	4.64

Hand disposal of excavated materials
Removing from site to tip not exceeding

13 km (using lorries)	1.10	7.67	13.54	m3	21.21

D GROUNDWORK Including overheads and profit at 12.50%	Labour hours	Labour £	Material £	Unit	Total rate £
Depositing on site in spoil heaps					
average 25 m distant	1.10	7.67	-	m3	7.67
average 50 m distant	1.45	10.11	-	m3	10.11
average 100 m distant	2.10	14.65	-	m3	14.65
average 200 m distant	3.10	21.62	-	m3	21.62
Spreading on site					
average 25 m distant	1.45	10.11	-	m3	10.11
average 50 m distant	1.75	12.21	-	m3	12.21
average 100 m distant	2.40	16.74	-	m3	16.74
average 200 m distant	3.40	23.72	-	m3	23.72
Hand disposal of excavated materials **inside existing buildings**					
Removing from site to tip not exceeding					
13 km (using lorries)	1.40	9.77	12.76	m3	22.53
Depositing on site in spoil heaps					
average 25 m distant	1.40	9.77	-	m3	9.77
Spreading on site					
average 25 m distant	1.80	12.56	-	m3	12.56
Mechanical filling with excavated material					
Filling to excavations	0.17	1.19	2.03	m3	3.22
Filling to make up levels over					
250 mm thick; depositing; compacting	0.25	1.74	1.52	m3	3.26
Filling to make up levels not exceeding					
250 mm thick; compacting	0.30	2.09	2.03	m3	4.12
Mechanical filling with imported soil					
Filling to make up levels over 250 mm thick;					
depositing; compacting in layers	0.25	1.74	14.18	m3	15.92
Filling to make up levels not exceeding					
250 mm thick; compacting	0.30	2.09	18.23	m3	20.32
Mechanical filling with hardcore; PC £8.00/m3					
Filling to excavations	0.20	1.40	13.78	m3	15.18
Filling to make up levels over					
250 mm thick; depositing; compacting	0.30	2.09	13.28	m3	15.37
Filling to make up levels not exceeding					
250 mm thick; compacting	0.35	2.44	16.93	m3	19.37
Mechanical filling with granular fill type 1; **PC £11.05/t (PC £18.80/m3)**					
Filling to excavations	0.20	1.40	28.97	m3	30.37
Filling to make up levels over					
250 mm thick; depositing; compacting	0.30	2.09	28.46	m3	30.55
Filling to make up levels not exceeding					
250 mm thick; compacting	0.35	2.44	36.37	m3	38.81
Mechanical filling with granular fill type 2; **PC £10.75/t (PC £18.25/m3)**					
Filling to excavations	0.20	1.40	28.20	m3	29.60
Filling to make up levels over					
250 mm thick; depositing; compacting	0.30	2.09	27.69	m3	29.78
Filling to make up levels not exceeding					
250 mm thick; compacting	0.35	2.44	35.38	m3	37.82
Mechanical filling with coarse ashes; PC £12.00/m3					
Filling to make up levels over					
250 mm thick; depositing; compacting	0.28	1.95	19.73	m3	21.68
Filling to make up levels not exceeding					
250 mm thick; compacting	0.32	2.23	25.23	m3	27.46
Mechanical filling with sand; **PC £10.90/t (PC £17.45/m3)**					
Filling to make up levels over					
250 mm thick; depositing; compacting	0.30	2.09	26.56	m3	28.65
Filling to make up levels not exceeding					
250 mm thick; compacting	0.35	2.44	33.94	m3	36.38

D GROUNDWORK Including overheads and profit at 12.50%	Labour hours	Labour £	Material £	Unit	Total rate £
D20 EXCAVATION AND FILLING - cont'd					
Hand filling with excavated material					
Filling to excavations	1.10	7.67	-	m3	7.67
Filling to make up levels over					
250 mm thick; depositing; compacting	1.20	8.37	-	m3	8.37
Filling to make up levels exceeding 250 mm					
thick; multiple handling via spoil heap					
average - 25 m distant	2.65	18.48	-	m3	18.48
Filling to make up levels not exceeding					
250 mm thick; compacting	1.50	10.46	-	m3	10.46
Hand filling with imported soil					
Filling to make up levels over 250 mm thick;					
depositing; compacting in layers	1.20	8.37	12.66	m3	21.03
depositing; compacting in layers	1.50	10.46	16.20	m3	26.66
Hand filling with hardcore; PC £8.00/m3					
Filling to excavations	1.45	10.11	11.25	m3	21.36
Filling to make up levels over					
250 mm thick; depositing; compacting	0.61	4.25	16.82	m3	21.07
Filling to make up levels not exceeding					
250 mm thick; compacting	0.73	5.09	21.08	m3	26.17
Hand filling with granular fill type 1;					
PC £11.05/t (PC £18.80/m3)					
Filling to excavations	1.45	10.11	26.44	m3	36.55
Filling to make up levels over					
250 mm thick; depositing; compacting	0.61	4.25	32.01	m3	36.26
Filling to make up levels not exceeding					
250 mm thick; compacting	0.73	5.09	40.52	m3	45.61
Hand filling with granular fill type 2;					
PC £10.75/t (PC £18.25/m3)					
Filling to excavations	1.45	10.11	25.66	m3	35.77
Filling to make up levels over					
250 mm thick; depositing; compacting	0.61	4.25	31.23	m3	35.48
Filling to make up levels not exceeding					
250 mm thick; compacting	0.73	5.09	39.53	m3	44.62
Hand filling with coarse ashes; PC £12.00/m3					
Filling to make up levels over					
250 mm thick; depositing; compacting	0.55	3.84	23.02	m3	26.86
Filling to make up levels not exceeding					
250 mm thick; compacting	0.66	4.60	29.03	m3	33.63
Hand filling with sand; PC £10.90/t					
(PC £17.45/m3)					
Filling to excavations	0.55	3.84	29.60	m3	33.44
Filling to make up levels over					
250 mm thick; depositing; compacting	0.72	5.02	31.12	m3	36.14
Filling to make up levels not exceeding					
250 mm thick; compacting	0.85	5.93	39.21	m3	45.14
Hand filling with excavated material					
inside existing buildings					
Filling to excavations	1.40	9.77	-	m3	9.77
Hand filling with hardcore; PC £8.00/m3;					
inside existing buildings					
Filling to excavations	1.80	12.56	11.25	m3	23.81
Filling to make up levels over					
250 mm thick; depositing; compacting	0.96	6.70	18.34	m3	25.04
Filling to make up levels not exceeding					
250 mm thick; compacting	0.96	6.70	18.34	m3	25.04
Surface packing to filling					
To vertical or battered faces	0.20	1.40	0.53	m2	1.93

D GROUNDWORK Including overheads and profit at 12.50%	Labour hours	Labour £	Material £	Unit	Total rate £
Surface treatments					
Compacting					
surfaces of ashes	0.04	0.28	-	m2	0.28
filling; blinding with ashes	0.09	0.63	0.90	m2	1.53
filling; blinding with sand	0.09	0.63	1.46	m2	2.09
bottoms of excavations	0.06	0.42	-	m2	0.42
Trimming					
sloping surfaces	0.20	1.40	-	m2	1.40
sloping surfaces; in rock	1.10	7.08	2.47	m2	9.55
Filter membrane; one layer; laid on earth to receive granular material					
'Terram 500'	0.06	0.42	0.45	m2	0.87
'Terram 700'	0.06	0.42	0.45	m2	0.87
'Terram 1000'	0.07	0.49	0.59	m2	1.08
D50 UNDERPINNING					
Mechanical excavation using a wheeled hydraulic off-centre excavator with a 0.24 m3 bucket					
Excavating preliminary trenches; not exceeding					
1 m deep	0.29	2.02	5.91	m3	7.93
2 m deep	0.35	2.44	7.09	m3	9.53
4 m deep	0.40	2.79	8.27	m3	11.06
Extra for breaking up					
concrete 150 mm thick	0.81	5.65	0.87	m2	6.52
Hand excavation					
Excavating preliminary trenches; not exceeding					
1 m deep	3.20	22.32	-	m3	22.32
2 m deep	3.65	25.46	-	m3	25.46
4 m deep	4.70	32.78	-	m3	32.78
Extra for breaking up					
concrete 150 mm thick	0.33	2.30	1.80	m2	4.10
Excavating underpinning pits starting from 1 m below ground level; not exceeding					
0.25 m deep	4.85	33.83	-	m3	33.83
1 m deep	5.30	36.97	-	m3	36.97
2 m deep	6.35	44.29	-	m3	44.29
Excavating underpinning pits starting from 2 m below ground level; not exceeding					
0.25 m deep	5.95	41.50	-	m3	41.50
1 m deep	6.40	44.64	-	m3	44.64
2 m deep	7.45	51.96	-	m3	51.96
Excavating underpinning pits starting from 4 m below ground level; not exceeding					
0.25 m deep	7.05	49.17	-	m3	49.17
1 m deep	7.50	52.31	-	m3	52.31
2 m deep	8.55	59.64	-	m3	59.64
Extra over any type of excavation at any depth for excavating below ground water level	0.39	2.72	-	m3	2.72
Earthwork support (open boarded) in 3 m lengths					
To preliminary trenches; not exceeding 2 m between opposing faces; not exceeding					
1 m deep	0.44	3.07	1.38	m2	4.45
2 m deep	0.55	3.84	1.69	m2	5.53
4 m deep	0.70	4.88	2.12	m2	7.00
To underpinning pits; not exceeding 2 m between opposing faces; not exceeding					
1 m deep	0.48	3.35	1.48	m2	4.83
2 m deep	0.61	4.25	1.90	m2	6.15
4 m deep	0.77	5.37	2.33	m2	7.70

D GROUNDWORK Including overheads and profit at 12.50%	Labour hours	Labour £	Material £	Unit	Total rate £
D50 UNDERPINNING - cont'd					
Earthwork support (closed boarded) **in 3 m lengths**					
To preliminary trenches; not exceeding 2 m between opposing faces; not exceeding					
1 m deep	1.10	7.67	2.33	m2	10.00
2 m deep	1.40	9.77	2.96	m2	12.73
4 m deep	1.70	11.86	3.60	m2	15.46
To underpinning pits; not exceeding 2 m between opposing faces; not exceeding					
1 m deep	1.20	8.37	2.54	m2	10.91
2 m deep	1.50	10.46	3.17	m2	13.63
4 m deep	1.85	12.90	4.02	m2	16.92
Extra over all types of earthwork support for earthwork support left in	0.83	5.79	13.78	m2	19.57
foundations					
150 x 150 mm	0.18	1.26	0.40	m	1.66
150 x 225 mm	0.26	1.81	0.59	m	2.40
150 x 300 mm	0.35	2.44	0.79	m	3.23
300 x 300 mm	0.69	4.81	1.56	m	6.37
Cutting away masonry					
one course high	0.06	0.42	0.12	m	0.54
two courses high	0.15	1.05	0.35	m	1.40
three courses high	0.30	2.09	0.67	m	2.76
four courses high	0.50	3.49	1.11	m	4.60
Preparing underside of existing work to receive new underpinning					
380 mm wide	0.66	4.60	-	m	4.60
600 mm wide	0.88	6.14	-	m	6.14
900 mm wide	1.10	7.67	-	m	7.67
1200 mm wide	1.30	9.07	-	m	9.07
Hand disposal of excavated materials					
Removing from site to tip not exceeding 13 km (using lorries)	1.10	7.67	12.76	m3	20.43
Hand filling with excavated material					
Filling to excavations	1.10	7.67	-	m3	7.67
Surface treatments					
Compacting bottoms of excavations	0.06	0.42	-	m2	0.42
Plain insitu ready mixed concrete; **11.50 N/mm2 - 40 mm aggregate (1:3:6);** **poured against faces of excavation**					
Underpinning					
over 450 mm thick	3.10	22.39	54.61	m3	77.00
150 - 450 mm thick	3.55	25.64	54.61	m3	80.25
not exceeding 150 mm thick	4.25	30.70	54.61	m3	85.31
Extra for working around reinforcement	0.35	2.53	-	m3	2.53
Plain insitu ready mixed concrete; **21.00 N/mm2 - 20 mm aggregate (1:2:4);** **poured against faces of excavation**					
Underpinning					
over 450 mm thick	3.10	22.39	58.63	m3	81.02
150 - 450 mm thick	3.55	25.64	58.63	m3	84.27
not exceeding 150 mm thick	4.25	30.70	58.63	m3	89.33
Extra for working around reinforcement	0.35	2.53	-	m3	2.53

D GROUNDWORK Including overheads and profit at 12.50%		Labour hours	Labour £	Material £	Unit	Total rate £
Sawn formwork						
Sides of foundations in underpinning						
over 1 m high		1.85	15.61	5.97	m2	21.58
not exceeding 250 mm high		0.63	5.32	1.65	m	6.97
250 - 500 mm high		0.98	8.27	3.13	m	11.40
500 mm - 1 m high		1.50	12.66	5.97	m	18.63
Reinforcement						
Reinforcement bars; BS 4449; hot rolled						
plain round mild steel bars; bent						
20 mm	PC £404.50	18.70	156.10	517.65	t	673.75
16 mm	PC £410.50	22.00	183.65	528.53	t	712.18
12 mm	PC £440.50	25.30	211.19	567.76	t	778.95
10 mm	PC £464.50	29.70	247.92	599.90	t	847.82
8 mm	PC £495.50	34.10	284.65	640.31	t	924.96
6 mm	PC £539.50	39.60	330.56	696.08	t	1026.64
8 mm; links or the like	PC £495.50	46.20	385.65	651.69	t	1037.34
6 mm; links or the like	PC £539.50	53.90	449.93	711.24	t	1161.17
Reinforcement bars; BS 4461; cold worked						
deformed square high yield steel bars; bent						
20 mm	PC £408.50	18.70	156.10	522.38	t	678.48
16 mm	PC £414.50	22.00	183.65	533.26	t	716.91
12 mm	PC £444.50	25.30	211.19	572.49	t	783.68
10 mm	PC £468.50	29.70	247.92	604.63	t	852.55
8 mm	PC £499.50	34.10	284.65	645.04	t	929.69
6 mm	PC £543.50	39.60	330.56	700.80	t	1031.36
Common bricks; PC £120.50/1000; **in cement mortar (1:3)**						
Walls in underpinning						
one brick thick		2.75	30.07	20.29	m2	53.36
one and a half brick thick		3.75	45.10	30.51	m2	75.61
two brick thick		4.75	57.12	40.59	m2	97.71
Add or deduct for variation of £10.00/1000 in						
PC of common bricks						
one brick thick		-	-	-	m2	1.42
one and a half brick thick		-	-	-	m2	2.13
two brick thick		-	-	-	m2	2.84
Class A engineering bricks; PC £350.00/1000; **in cement mortar (1:3)**						
Walls in underpinning						
one brick thick		3.00	36.08	60.86	m2	96.94
one and a half brick thick		4.00	48.11	91.36	m2	139.47
two brick thick		5.00	60.13	121.71	m2	181.84
Add or deduct for variation of £10.00/1000 in						
PC of common bricks						
one brick thick		-	-	-	m2	1.42
one and a half brick thick		-	-	-	m2	2.13
two brick thick		-	-	-	m2	2.84
Class B engineering bricks; PC £198.00/1000; **in cement mortar (1:3)**						
Walls in underpinning						
one brick thick		3.00	36.08	31.28	m2	67.36
one and a half brick thick		4.00	48.11	46.99	m2	95.10
two brick thick		5.00	60.13	62.57	m2	122.70
Add or deduct for variation of £10.00/1000 in						
PC of common bricks						
one brick thick		-	-	-	m2	1.42
one and a half brick thick		-	-	-	m2	2.13
two brick thick		-	-	-	m2	2.84

D GROUNDWORK Including overheads and profit at 12.50%	Labour hours	Labour £	Material £	Unit	Total rate £

D50 UNDERPINNING - cont'd

'Pluvex' (hessian based) damp proof course
or similar; PC £3.72/m2; 200 mm laps;
in cement mortar (1:3)
Horizontal

over 225 mm wide	0.29	3.49	4.50	m2	7.99
not exceeding 225 mm wide	0.58	6.98	4.60	m2	11.58

'Hyload' (pitch polymer) damp proof course
or similar; PC £4.43/m2; 150 mm laps;
in cement mortar (1:3)
Horizontal

over 225 mm wide	0.29	3.49	5.36	m2	8.85
not exceeding 225 mm wide	0.58	6.98	5.48	m2	12.46

'Ledkore' grade A (bitumen based lead cored)
damp proof course or similar; PC £12.90/m2;
200 mm laps; in cement mortar (1:3)
Horizontal

over 225 mm wide	0.38	4.57	15.60	m2	20.17
not exceeding 225 mm wide	0.76	9.14	15.96	m2	25.10

Two courses of slates in cement mortar (1:3)
Horizontal

over 225 mm wide	1.70	20.44	24.34	m2	44.78
not exceeding 225 mm wide	2.90	34.88	24.77	m2	59.65

Wedging and pinning
To underside of existing construction with
slates in cement mortar (1:3)

102 mm wall	1.25	15.03	4.60	m	19.63
215 mm wall	1.50	18.04	6.96	m	25.00
317 mm wall	1.70	20.44	9.25	m	29.69

E IN SITU CONCRETE/LARGE PRECAST CONCRETE
Including overheads and profit at 12.50%

BASIC CONCRETE PRICES

	£			£		£
Concrete aggregates (£/tonne)						
40 mm all-in	11.14	40 mm shingle		11.14	10 mm shingle	11.24
20 mm all-in	11.24	20 mm shingle		11.24	sharp sand	10.90

	£		£
Formwork items			
plywood (£/m2)	9.02	timber (£/m3)	188.13

		£
Lightweight aggregates		
'Lytag' (£/m3)		
'fines'	22.40	6-12 mm granular	24.16

	£		£
Portland cement (£/tonne)			
normal - in bags	61.54	normal - in bulk to silos	54.94
sulphate - resisting - plus	9.27	rapid - hardening - plus	3.61

Tying wire for reinforcement - £0.56/kg

MIXED CONCRETE PRICES (£/m3)

The following prices are for ready or site mixed concrete ready for placing including 5% for waste and 12.5% for overheads and profit	Mix 7.50 N/mm2 - 40mm aggre-gate (1:8) £	Mix 11.50 N/mm2 - 40mm aggre-gate (1:3:6) £	Mix 15.00 N/mm2 - 40mm aggre-gate £	Mix 21.00 N/mm2 - 20mm aggre-gate (1:2:4) £	Mix 26.00 N/mm2 - 20mm aggre-gate (1:1.5:3) £	Mix 31.00 N/mm2 - 20mm aggre-gate (1:1:2) £	Mix 40.00 N/mm2 - 20mm aggre-gate £
Ready mixed concrete							
Normal Portland cement	48.50	49.65	50.85	53.30	55.70	58.20	61.75
Sulphate - resistant cement	50.85	52.10	53.30	56.90	59.35	61.75	66.60
Normal Portland cement with water-repellent additive	52.90	52.05	53.25	55.70	58.10	60.60	64.15
Normal Portland cement; air-entrained	49.60	50.75	51.95	54.40	56.80	59.30	62.85
Lightweight concrete using sintered PFA aggregate	-	-	71.50	72.65	75.10	77.50	-
Site mixed concrete							
Normal Portland cement	-	56.50	59.10	62.00	66.00	70.00	75.00
Sulphate - resistant cement	-	58.80	62.00	65.40	71.30	75.00	80.00

	Labour hours	Labour £	Material £	Unit	Total rate £
E10 IN SITU CONCRETE					
Plain in situ ready mixed concrete;					
11.50 N/mm2 - 40 mm aggregate (1:3:6)					
Foundations	1.85	13.36	49.65	m3	**63.01**
Isolated foundations	2.20	15.89	49.65	m3	**65.54**
Beds					
over 450 mm thick	1.45	10.47	49.65	m3	**60.12**
150 - 450 mm thick	1.90	13.72	49.65	m3	**63.37**
not exceeding 150 mm thick	2.75	19.86	49.65	m3	**69.51**
Filling to hollow walls					
not exceeding 150 mm thick	4.90	35.39	49.65	m3	**85.04**

E IN SITU CONCRETE/LARGE PRECAST CONCRETE Including overheads and profit at 12.50%	Labour hours	Labour £	Material £	Unit	Total rate £
E10 IN SITU CONCRETE - cont'd					
Plain in situ ready mixed concrete; **11.50 N/mm2 - 40 mm aggregate (1:3:6);** **poured on or against earth or** **unblinded hardcore**					
Foundations	1.95	14.08	52.13	m3	66.21
Isolated foundations	2.30	16.61	52.13	m3	68.74
Beds					
over 450 mm thick	1.50	10.83	52.13	m3	62.96
150 - 450 mm thick	2.00	14.45	52.13	m3	66.58
not exceeding 150 mm thick	2.90	20.95	52.13	m3	73.08
Plain in situ ready mixed concrete; **21.00 N/mm2 - 20 mm aggregate (1:2:4)**					
Foundations	1.85	13.36	53.30	m3	66.66
Isolated foundations	2.20	15.89	53.30	m3	69.19
Beds					
over 450 mm thick	1.45	10.47	53.30	m3	63.77
150 - 450 mm thick	1.90	13.72	53.30	m3	67.02
not exceeding 150 mm thick	2.75	19.86	53.30	m3	73.16
Filling to hollow walls					
not exceeding 150 mm thick	4.90	35.39	53.30	m3	88.69
Plain in situ ready mixed concrete; **21.00 N/mm2 - 20 mm aggregate (1:2:4);** **poured on or against earth or** **unblinded hardcore**					
Foundations	1.95	14.08	55.97	m3	70.05
Isolated foundations	2.30	16.61	55.97	m3	72.58
Beds					
over 450 mm thick	1.50	10.83	55.97	m3	66.80
150 - 450 mm thick	2.00	14.45	55.97	m3	70.42
not exceeding 150 mm thick	2.90	20.95	55.97	m3	76.92
Reinforced in situ ready mixed concrete; **21.00 N/mm2 - 20 mm aggregate (1:2:4)**					
Foundations	2.30	16.61	53.30	m3	69.91
Ground beams	4.05	29.25	53.30	m3	82.55
Isolated foundations	2.65	19.14	53.30	m3	72.44
Beds					
over 450 mm thick	1.85	13.36	53.30	m3	66.66
150 - 450 mm thick	2.30	16.61	53.30	m3	69.91
not exceeding 150 mm thick	3.15	22.75	53.30	m3	76.05
Slabs					
over 450 mm thick	3.55	25.64	53.30	m3	78.94
150 - 450 mm thick	4.05	29.25	53.30	m3	82.55
not exceeding 150 mm thick	5.05	36.47	53.30	m3	89.77
Coffered or troughed slabs					
over 450 mm thick	4.05	29.25	53.30	m3	82.55
150 - 450 mm thick	4.60	33.22	53.30	m3	86.52
Extra over for laying to slopes					
not exceeding 15 degrees	0.35	2.53	-	m3	2.53
over 15 degrees	0.69	4.98	-	m3	4.98
Walls					
over 450 mm thick	3.75	27.08	53.30	m3	80.38
150 - 450 mm thick	4.25	30.70	53.30	m3	84.00
not exceeding 150 mm thick	5.30	38.28	53.30	m3	91.58
Isolated beams	5.75	41.53	53.30	m3	94.83
Isolated deep beams	6.35	45.86	53.30	m3	99.16
Attached deep beams	5.75	41.53	53.30	m3	94.83
Isolated beam casings	6.35	45.86	53.30	m3	99.16
Isolated deep beam casings	6.90	49.84	53.30	m3	103.14
Attached deep beam casings	6.35	45.86	53.30	m3	99.16

E IN SITU CONCRETE/LARGE PRECAST CONCRETE Including overheads and profit at 12.50%	Labour hours	Labour £	Material £	Unit	Total rate £
Columns	6.90	49.84	53.30	m3	103.14
Column casings	7.60	54.89	53.30	m3	108.19
Staircases	8.65	62.47	53.30	m3	115.77
Upstands	5.50	39.72	53.30	m3	93.02
Reinforced in situ ready mixed concrete; **26.00 N/mm2 - 20 mm aggregate (1:1.5:3)**					
Foundations	2.30	16.61	55.70	m3	72.31
Ground beams	4.05	29.25	55.70	m3	84.95
Isolated foundations	2.65	19.14	55.70	m3	74.84
Beds					
over 450 mm thick	1.85	13.36	55.70	m3	69.06
150 - 450 mm thick	2.30	16.61	55.70	m3	72.31
not exceeding 150 mm thick	3.15	22.75	55.70	m3	78.45
Slabs					
over 450 mm thick	3.55	25.64	55.70	m3	81.34
150 - 450 mm thick	4.05	29.25	55.70	m3	84.95
not exceeding 150 mm thick	5.05	36.47	55.70	m3	92.17
Coffered or troughed slabs					
over 450 mm thick	4.05	29.25	55.70	m3	84.95
150 - 450 mm thick	4.60	33.22	55.70	m3	88.92
Extra over for laying to slopes					
not exceeding 15 degrees	0.35	2.53	-	m3	2.53
over 15 degrees	0.69	4.98	-	m3	4.98
Walls					
over 450 mm thick	3.75	27.08	55.70	m3	82.78
150 - 450 mm thick	4.25	30.70	55.70	m3	86.40
not exceeding 150 mm thick	5.30	38.28	55.70	m3	93.98
Isolated beams	5.75	41.53	55.70	m3	97.23
Isolated deep beams	6.35	45.86	55.70	m3	101.56
Attached deep beams	5.75	41.53	55.70	m3	97.23
Isolated beam casings	6.35	45.86	55.70	m3	101.56
Isolated deep beam casings	6.90	49.84	55.70	m3	105.54
Attached deep beam casings	6.35	45.86	55.70	m3	101.56
Columns	6.90	49.84	55.70	m3	105.54
Column casings	7.60	54.89	55.70	m3	110.59
Staircases	8.65	62.47	55.70	m3	118.17
Upstands	5.50	39.72	55.70	m3	95.42
Reinforced in situ ready mixed concrete; **31.00 N/mm2 - 20 mm aggregate (1:1:2)**					
Beds					
over 450 mm thick	1.85	13.36	58.20	m3	71.56
150 - 450 mm thick	2.30	16.61	58.20	m3	74.81
not exceeding 150 mm thick	3.15	22.75	58.20	m3	80.95
Slabs					
over 450 mm thick	3.55	25.64	58.20	m3	83.84
150 - 450 mm thick	4.05	29.25	58.20	m3	87.45
not exceeding 150 mm thick	5.05	36.47	58.20	m3	94.67
Coffered or troughed slabs					
over 450 mm thick	4.05	29.25	58.20	m3	87.45
150 - 450 mm thick	4.60	33.22	58.20	m3	91.42
Extra over for laying to slopes					
not exceeding 15 degrees	0.35	2.53	-	m3	2.53
over 15 degrees	0.69	4.98	-	m3	4.98
Walls					
over 450 mm thick	3.75	27.08	58.20	m3	85.28
150 - 450 mm thick	4.25	30.70	58.20	m3	88.90
not exceeding 150 mm thick	5.30	38.28	58.20	m3	96.48
Isolated beams	5.75	41.53	58.20	m3	99.73
Isolated deep beams	6.35	45.86	58.20	m3	104.06
Attached deep beams	5.75	41.53	58.20	m3	140.69
Isolated beam casings	6.35	45.86	58.20	m3	104.06
Isolated deep beam casings	6.90	49.84	58.20	m3	108.04
Attached deep beam casings	6.35	45.86	58.20	m3	104.06
Columns	6.90	49.84	58.20	m3	108.04

E IN SITU CONCRETE/LARGE PRECAST CONCRETE Including overheads and profit at 12.50%	Labour hours	Labour £	Material £	Unit	Total rate £
E10 IN SITU CONCRETE - cont'd					
Reinforced in situ ready mixed concrete;					
31.00 N/mm2 - 20 mm aggregate (1:1:2) - cont'd					
Column casings	7.60	54.89	58.20	m3	113.09
Staircases	8.65	62.47	58.20	m3	120.67
Upstands	5.50	39.72	58.20	m3	97.92
Grouting with cement mortar (1:1)					
Stanchion bases					
10 mm thick	1.10	7.94	0.39	nr	8.33
25 mm thick	1.40	10.11	0.92	nr	11.03
Grouting with epoxy resin					
Stanchion bases					
10 mm thick	1.40	10.11	0.30	nr	10.41
25 mm thick	1.65	11.92	0.73	nr	12.65
Filling; plain in situ concrete;					
21.00 N/mm2 - 20 mm aggregate (1:2:4)					
Mortices	0.11	0.79	0.32	nr	1.11
Holes	0.28	2.02	2.30	m3	4.32
Chases					
over 0.01 m2	0.22	1.59	1.28	m3	2.87
not exceeding 0.01 m2	0.17	1.23	0.64	m	1.87
Sheeting to prevent moisture loss					
Building paper; lapped joints					
subsoil grade; horizontal on foundations	0.03	0.22	0.55	m2	0.77
standard grade; horizontal on slabs	0.06	0.43	0.89	m2	1.32
Polyethylene sheeting; lapped joints;					
horizontal on slabs					
65 microns; 0.7 mm thick	0.06	0.43	0.08	m2	0.51
125 microns; 0.13 mm thick	0.06	0.43	0.21	m2	0.64
250 microns; 0.25 mm thick	0.06	0.43	0.41	m2	0.84
'Visqueen' sheeting; lapped joints;					
horizontal on slabs					
1000 Grade; 0.25 mm thick	0.06	0.43	0.35	m2	0.78
1200 Super; 0.30 mm thick	0.07	0.51	0.46	m2	0.97
E20 FORMWORK FOR IN SITU CONCRETE					
Note: Generally all formwork based on four					
uses unless otherwise stated					
Sides of foundations					
over 1 m high	1.60	13.50	4.66	m2	18.16
not exceeding 250 mm high	0.52	4.39	1.43	m	5.82
250 - 500 mm high	0.86	7.26	2.47	m	9.73
500 mm - 1 m high	1.30	10.97	4.66	m	15.63
Sides of foundations; left in					
over 1 m high	1.60	13.50	16.10	m2	29.60
not exceeding 250 mm high	0.52	4.39	4.15	m	8.54
250 - 500 mm high	0.86	7.26	8.13	m	15.39
500 mm - 1 m high	1.30	10.97	16.10	m	27.07
Sides of ground beams and edges of beds					
over 1 m high	1.90	16.03	6.33	m2	22.36
not exceeding 250 mm high	0.58	4.89	1.76	m	6.65
250 - 500 mm high	1.05	8.86	3.24	m	12.10
500 mm - 1 m high	1.45	12.23	6.33	m	18.56
Edges of suspended slabs					
not exceeding 250 mm high	0.86	7.26	2.06	m	9.32
250 - 500 mm high	1.25	10.55	4.09	m	14.64
500 mm - 1 m high	2.00	16.88	8.07	m	24.95

E IN SITU CONCRETE/LARGE PRECAST CONCRETE Including overheads and profit at 12.50%	Labour hours	Labour £	Material £	Unit	Total rate £
Sides of upstands					
over 1 m high	2.30	19.41	8.28	m2	27.69
not exceeding 250 mm high	0.72	6.08	2.15	m	8.23
250 - 500 mm high	1.15	9.70	4.32	m	14.02
500 mm - 1 m high	2.00	16.88	8.28	m	25.16
Steps in top surfaces					
not exceeding 250 mm high	0.58	4.89	2.25	m	7.14
250 - 500 mm high	0.92	7.76	4.71	m	12.47
Steps in soffits					
not exceeding 250 mm high	0.63	5.32	2.25	m	7.57
250 - 500 mm high	1.00	8.44	4.71	m	13.15
Machine bases and plinths					
over 1 m high	1.85	15.61	6.33	m2	21.94
not exceeding 250 mm high	0.58	4.89	1.76	m	6.65
250 - 500 mm high	0.98	8.27	3.24	m	11.51
500 mm - 1 m high	1.45	12.23	6.33	m	18.56
Soffits of slabs; 1.5 - 3 m height to soffit					
not exceeding 200 mm thick	1.95	16.45	6.15	m2	22.60
not exceeding 200 mm thick (5 uses)	1.90	16.03	5.54	m2	21.57
not exceeding 200 mm thick (6 uses)	1.85	15.61	5.11	m2	20.72
200 - 300 mm thick	2.05	17.30	6.71	m2	24.01
300 - 400 mm thick	2.10	17.72	6.88	m2	24.60
400 - 500 mm thick	2.25	18.98	7.16	m2	26.14
500 - 600 mm thick	2.40	20.25	6.15	m2	26.40
Soffits of slabs; not exceeding 200 mm thick					
not exceeding 1.5 m height to soffit	2.05	17.30	7.51	m2	24.81
3 - 4.5 m height to soffit	1.95	16.45	7.34	m2	23.79
4.5 - 6 m height to soffit	2.05	17.30	8.49	m2	25.79
Soffits of landings; 1.5 - 3m height to soffit					
not exceeding 200 mm thick	2.05	17.30	6.42	m2	23.72
200 - 300 mm thick	2.20	18.56	6.98	m2	25.54
300 - 400 mm thick	2.25	18.98	7.16	m2	26.14
Extra over for sloping					
not exceeding 15 degrees	0.23	1.94	-	m2	1.94
over 15 degrees	0.46	3.88	-	m2	3.88
Soffits of coffered or troughed slabs; including 'Cordeck' troughed forms; 300 mm deep; ribs at 600 mm centres and cross ribs at centres of bay; 300 - 400 mm thick					
1.5 - 3 m height to soffit	2.90	24.47	9.73	m2	34.20
3 - 4.5 m height to soffit	3.00	25.31	10.53	m2	35.84
4.5 - 6 m height to soffit	3.10	26.16	11.67	m2	37.83
Soffits of bands and margins to troughed slabs					
horizontal; 300 - 400 mm thick	2.30	19.41	6.89	m2	26.30
Top formwork	1.70	14.34	4.89	m2	19.23
Walls					
vertical	2.30	19.41	7.28	m2	26.69
vertical; interrupted	2.40	20.25	7.60	m2	27.85
vertical; exceeding 3 m high; inside stairwells	2.55	21.52	7.69	m2	29.21
vertical; exceeding 3 m high; inside lift shaft	2.75	23.20	8.28	m2	31.48
battered	3.20	27.00	8.44	m2	35.44
Beams attached to insitu slabs					
square or rectangular; 1.5 - 3 m height to soffit	2.55	21.52	9.15	m2	30.67
square or rectangular; 3 - 4.5 m height to soffit	2.65	22.36	10.31	m2	32.67
square or rectangular; 4.5 - 6 m height to soffit	2.75	23.20	11.48	m2	34.68
Beams attached to walls					
square or rectangular; 1.5 - 3 m height to soffit	2.65	22.36	9.15	m2	31.51

E IN SITU CONCRETE/LARGE PRECAST CONCRETE Including overheads and profit at 12.50%	Labour hours	Labour £	Material £	Unit	Total rate £
E20 FORMWORK FOR IN SITU CONCRETE - cont'd					
Isolated beams					
square or rectangular; 1.5 - 3 m					
height to soffit	2.75	23.20	9.15	m2	32.35
square or rectangular; 3 - 4.5 m					
height to soffit	2.90	24.47	10.31	m2	34.78
square or rectangular; 4.5 - 6 m					
height to soffit	3.00	25.31	11.48	m2	36.79
Beam casings attached to insitu slabs					
square or rectangular; 1.5 - 3 m					
height to soffit	2.65	22.36	9.15	m2	31.51
square or rectangular; 3 - 4.5 m					
height to soffit	2.75	23.20	10.31	m2	33.51
Beam casings attached to walls					
square or rectangular; 1.5 - 3 m					
height to soffit	2.75	23.20	9.15	m2	32.35
Isolated beam casings					
square or rectangular; 1.5 - 3 m					
height to soffit	2.90	24.47	9.15	m2	33.62
square or rectangular; 3 - 4.5 m					
height to soffit	3.00	25.31	10.31	m2	35.62
Extra over for sloping					
not exceeding 15 degrees	2.75	23.20	9.66	m2	32.86
over 15 degrees	2.90	24.47	10.17	m2	34.64
Columns attached to walls					
square or rectangular	2.55	21.52	7.28	m2	28.80
Isolated columns					
square or rectangular	2.65	22.36	7.28	m2	29.64
Column casings attached to walls					
square or rectangular	2.65	22.36	7.28	m2	29.64
Isolated column casings					
square or rectangular	2.75	23.20	7.28	m2	30.48
Extra over for					
Throat	0.06	0.51	0.18	m	0.69
Chamfer					
30 mm wide	0.07	0.59	0.24	m	0.83
60 mm wide	0.08	0.68	0.40	m	1.08
90 mm wide	0.09	0.76	0.82	m	1.58
Rebate or horizontal recess					
12 x 12 mm	0.08	0.68	0.10	m	0.78
25 x 25 mm	0.08	0.68	0.17	m	0.85
25 x 50 mm	0.08	0.68	0.33	m	1.01
50 x 50 mm	0.08	0.68	0.58	m	1.26
Nibs					
50 x 50 mm	0.63	5.32	1.40	m	6.72
100 x 100 mm	0.90	7.59	2.75	m	10.34
100 x 200 mm	1.20	10.13	3.66	m	13.79
Extra over basic formwork for rubbing down, filling and leaving face of concrete smooth					
general surfaces	0.38	3.21	0.11	m2	3.32
edges	0.58	4.89	0.12	m2	5.01
Add to prices for basic formwork for					
curved radius 6 m	27.5%				
curved radius 2 m	50%				
coating with retardant agent	0.02	0.17	0.46	m2	0.63
Wall kickers to both sides					
150 mm high	0.58	4.89	1.43	m	6.32
225 mm high	0.75	6.33	1.88	m	8.21
150 mm high; one side suspended	0.72	6.08	2.33	m	8.41
Wall ends, soffits and steps in walls					
over 1 m wide	2.20	18.56	7.68	m	26.24
not exceeding 250 mm wide	0.69	5.82	2.17	m	7.99
250 - 500 mm wide	1.10	9.28	4.17	m	13.45
500 mm - 1 m wide	1.70	14.34	7.68	m	22.02

E IN SITU CONCRETE/LARGE PRECAST CONCRETE Including overheads and profit at 12.50%	Labour hours	Labour £	Material £	Unit	Total rate £
Openings in walls					
over 1 m wide	2.40	20.25	9.19	m	29.44
not exceeding 250 mm wide	0.75	6.33	2.33	m	8.66
250 - 500 mm wide	1.25	10.55	4.66	m	15.21
500 mm - 1 m wide	1.95	16.45	9.19	m	25.64
Stair flights					
1 m wide; 150 mm waist;					
150 mm undercut risers	5.75	48.52	23.85	m	72.37
2 m wide; 200 mm waist;					
150 mm vertical risers	10.35	87.33	38.13	m	125.46
Mortices; not exceeding 250 mm deep					
not exceeding 500 mm girth	0.17	1.43	1.27	nr	2.70
Holes; not exceeding 250 mm deep					
not exceeding 500 mm girth	0.23	1.94	1.32	nr	3.26
500 mm - 1 m girth	0.29	2.45	2.54	nr	4.99
1 - 2 m girth	0.52	4.39	5.07	nr	9.46
2 - 3 m girth	0.69	5.82	7.61	nr	13.43
Holes; 250 - 500 mm deep					
not exceeding 500 mm girth	0.35	2.95	2.54	nr	5.49
500 mm - 1 m girth	0.44	3.71	5.07	nr	8.78
1 - 2 m girth	0.77	6.50	10.15	nr	16.65
2 - 3 m girth	1.05	8.86	15.22	nr	24.08
Permanent shuttering; left in					
Dufaylite 'Clayboard' shuttering; type KN30; horizontal; under concrete beds; left in					
50 mm thick	0.17	1.43	8.92	m2	10.35
75 mm thick	0.18	1.52	9.55	m2	11.07
100 mm thick	0.20	1.69	9.98	m2	11.67
150 mm thick	0.23	1.94	10.41	m2	12.35
Dufaylite 'Clayboard' shuttering; type KN30; horizontal or vertical (including temporary supports); beneath or to sides of foundations; left in					
50 mm thick; vertical	0.29	2.45	9.58	m2	12.03
400 x 50 mm thick; horizontal	0.10	0.84	3.66	m	4.50
600 x 50 mm thick; horizontal	0.14	1.18	5.48	m	6.66
800 x 50 mm thick; horizontal	0.17	1.43	7.30	m	8.73
75 mm thick; vertical	0.31	2.62	10.23	m2	12.85
400 x 75 mm thick; horizontal	0.11	0.93	3.91	m	4.84
600 x 75 mm thick; horizontal	0.15	1.27	5.87	m	7.14
800 x 75 mm thick; horizontal	0.18	1.52	7.83	m	9.35
100 mm thick; vertical	0.35	2.95	10.67	m2	13.62
400 x 100 mm thick; horizontal	0.13	1.10	4.08	m	5.18
600 x 100 mm thick; horizontal	0.16	1.35	6.13	m	7.48
800 x 100 mm thick; horizontal	0.20	1.69	8.18	m	9.87
Hyrib permanent shuttering and reinforcement ref 2411 to soffits of slabs; left in					
horizontal	1.75	14.77	14.24	m2	29.01

E30 REINFORCEMENT FOR IN SITU CONCRETE

Reinforcement bars; BS 4449; hot rolled
plain round mild steel bars; straight

		Labour hours	Labour £	Material £	Unit	Total rate £
40 mm	PC £395.50	12.10	101.00	495.65	t	596.65
32 mm	PC £382.50	12.70	106.01	484.08	t	590.09
25 mm	PC £375.50	14.90	124.38	479.61	t	603.99
20 mm	PC £373.50	17.10	142.74	481.03	t	623.77
16 mm	PC £374.50	19.80	165.28	486.01	t	651.29
12 mm	PC £394.50	23.10	192.83	513.42	t	706.25
10 mm	PC £403.50	26.40	220.37	527.85	t	748.22
8 mm	PC £414.50	29.70	247.92	544.63	t	792.55
6 mm	PC £453.50	35.20	293.83	594.49	t	888.32

E IN SITU CONCRETE/LARGE PRECAST CONCRETE Including overheads and profit at 12.50%		Labour hours	Labour £	Material £	Unit	Total rate £
E30 REINFORCEMENT FOR IN SITU CONCRETE - cont'd						
Reinforcement bars; BS 4449; hot rolled **plain round mild steel bars; bent**						
40 mm	PC £420.50	12.10	101.00	525.18	t	626.18
32 mm	PC £408.50	14.30	119.37	514.80	t	634.17
25 mm	PC £406.50	16.50	137.73	516.22	t	653.95
20 mm	PC £404.50	18.70	156.10	517.65	t	673.75
16 mm	PC £410.50	22.00	183.65	528.53	t	712.18
12 mm	PC £440.50	25.30	211.19	567.76	t	778.95
10 mm	PC £464.50	29.70	247.92	599.90	t	847.82
8 mm	PC £495.50	34.10	284.65	640.31	t	924.96
6 mm	PC £539.50	39.60	330.56	696.08	t	1026.64
8 mm; links or the like	PC £495.50	46.20	385.65	651.69	t	1037.34
6 mm; links or the like	PC £539.50	53.90	449.93	711.24	t	1161.17
Reinforcement bars; BS 4461; cold worked **deformed square high steel bars; straight**						
40 mm	PC £399.50	11.00	91.82	500.37	t	592.19
32 mm	PC £387.50	12.70	106.01	489.99	t	596.00
25 mm	PC £380.50	14.90	124.38	485.51	t	609.89
20 mm	PC £378.50	17.10	142.74	486.94	t	629.68
16 mm	PC £379.50	19.80	165.28	491.91	t	657.19
12 mm	PC £399.50	23.10	192.83	519.33	t	712.16
10 mm	PC £408.50	26.40	220.37	533.75	t	754.12
8 mm	PC £419.50	29.70	247.92	550.54	t	798.46
6 mm	PC £458.50	35.20	293.83	600.40	t	894.23
Reinforcement bars; BS 4461; cold worked **deformed square high steel bars; bent**						
40 mm	PC £424.50	12.10	101.00	529.90	t	630.90
32 mm	PC £412.50	14.30	119.37	519.52	t	638.89
25 mm	PC £410.50	16.50	137.73	520.95	t	658.68
20 mm	PC £408.50	18.70	156.10	522.38	t	678.48
16 mm	PC £414.50	22.00	183.65	533.26	t	716.91
12 mm	PC £444.50	25.30	211.19	572.49	t	783.68
10 mm	PC £468.50	29.70	247.92	604.63	t	852.55
8 mm	PC £499.50	34.10	284.65	645.04	t	929.69
6 mm	PC £543.50	39.60	330.56	700.80	t	1031.36
Reinforcement fabric; BS 4483; lapped; in **beds or suspended slabs**						
Ref A98 (1.54 kg/m2)	PC £0.65	0.13	1.09	0.88	m2	1.97
Ref A142 (2.22 kg/m2)	PC £0.83	0.13	1.09	1.13	m2	2.22
Ref A193 (3.02 kg/m2)	PC £1.13	0.13	1.09	1.53	m2	2.62
Ref A252 (3.95 kg/m2)	PC £1.47	0.14	1.17	1.98	m2	3.15
Ref A393 (6.16 kg/m2)	PC £2.34	0.17	1.42	3.16	m2	4.58
Ref B196 (3.05 kg/m2)	PC £1.25	0.13	1.09	1.69	m2	2.78
Ref B283 (3.73 kg/m2)	PC £1.48	0.13	1.09	2.00	m2	3.09
Ref B385 (4.53 kg/m2)	PC £1.78	0.14	1.17	2.41	m2	3.58
Ref B503 (5.93 kg/m2)	PC £2.28	0.17	1.42	3.08	m2	4.50
Ref B785 (8.14 kg/m2)	PC £3.12	0.19	1.59	4.21	m2	5.80
Ref B1131 (10.90 kg/m2)	PC £4.21	0.21	1.75	5.68	m2	7.43
Ref C283 (2.61 kg/m2)	PC £1.06	0.13	1.09	1.43	m2	2.52
Ref C385 (3.41 kg/m2)	PC £1.36	0.13	1.09	1.83	m2	2.92
Ref C503 (4.34 kg/m2)	PC £1.67	0.14	1.17	2.25	m2	3.42
Ref C636 (5.55 kg/m2)	PC £2.15	0.15	1.25	2.90	m2	4.15
Ref C785 (6.72 kg/m2)	PC £2.61	0.17	1.42	3.52	m2	4.94
Reinforcement fabric; BS 4483; lapped; in **casings to steel columns or beams**						
Ref D49 (0.77 kg/m2)	PC £0.59	0.28	2.34	0.80	m2	3.14
Ref D98 (1.54 kg/m2)	PC £0.59	0.28	2.34	0.80	m2	3.14

E IN SITU CONCRETE/LARGE PRECAST CONCRETE Including overheads and profit at 12.50%	Labour hours	Labour £	Material £	Unit	Total rate £
E40 DESIGNED JOINTS IN IN SITU CONCRETE					
Expandite 'Flexcell' joint filler; or similar					
Formed joint; 10 mm thick					
not exceeding 150 mm wide	0.33	2.78	1.12	m	3.90
150 - 300 mm wide	0.44	3.71	1.95	m	5.66
300 - 450 mm wide	0.55	4.64	2.99	m	7.63
Formed joint; 12.5 mm thick					
not exceeding 150 mm wide	0.33	2.78	1.18	m	3.96
150 - 300 mm wide	0.44	3.71	2.03	m	5.74
300 - 450 mm wide	0.55	4.64	3.17	m	7.81
Formed joint; 20 mm thick					
not exceeding 150 mm wide	0.33	2.78	1.62	m	4.40
150 - 300 mm wide	0.44	3.71	2.77	m	6.48
300 - 450 mm wide	0.55	4.64	4.39	m	9.03
Formed joint; 25 mm thick					
not exceeding 150 mm wide	0.33	2.78	1.83	m	4.61
150 - 300 mm wide	0.44	3.71	3.25	m	6.96
300 - 450 mm wide	0.55	4.64	4.95	m	9.59
Sealing top of joint with Expandite 'Pliastic' hot poured rubberized bituminous compound					
10 x 25 mm	0.20	1.69	0.41	m	2.10
12.5 x 25 mm	0.21	1.77	0.49	m	2.26
20 x 25 mm	0.22	1.86	0.68	m	2.54
25 x 25 mm	0.23	1.94	0.95	m	2.89
Sealing top of joint with Expandite 'Thioflex 600' cold poured polysulphide rubberized compound					
10 x 25 mm	0.07	0.59	2.76	m	3.35
12.5 x 25 mm	0.08	0.68	3.33	m	4.01
20 x 25 mm	0.09	0.76	4.61	m	5.37
25 x 25 mm	0.10	0.84	6.64	m	7.48
Servicised water stops or similar					
Formed joint; PVC water stop; flat dumbell type; heat welded joints					
100 mm wide PC £41.55/15m	0.24	2.03	3.29	m	5.32
Flat angle PC £2.64	0.30	2.53	3.12	nr	5.65
Vertical angle PC £4.59	0.30	2.53	5.42	nr	7.95
Flat three way intersection PC £5.20	0.41	3.46	6.14	nr	9.60
Vertical three way intersection PC £5.59	0.41	3.46	6.60	nr	10.06
Four way intersection PC £6.52	0.52	4.39	7.71	nr	12.10
170 mm wide PC £58.50/15m	0.28	2.36	4.63	m	6.99
Flat angle PC £2.66	0.33	2.78	3.14	nr	5.92
Vertical angle PC £4.59	0.33	2.78	5.42	nr	8.20
Flat three way intersection PC £5.24	0.44	3.71	6.19	nr	9.90
Vertical three way intersection PC £6.50	0.44	3.71	7.68	nr	11.39
Four way intersection PC £7.11	0.55	4.64	8.40	nr	13.04
210 mm wide PC £69.15/15m	0.31	2.62	5.47	m	8.09
Flat angle PC £4.45	0.35	2.95	5.25	nr	8.20
Vertical angle PC £5.09	0.35	2.95	6.01	nr	8.96
Flat three way intersection PC £6.50	0.46	3.88	7.68	nr	11.56
Vertical three way intersection PC £8.04	0.46	3.88	9.50	nr	13.38
Four way intersection PC £8.15	0.57	4.81	9.63	nr	14.44
250 mm wide PC £101.10/15m	0.33	2.78	8.00	m	10.78
Flat angle PC £5.42	0.37	3.12	6.40	nr	9.52
Vertical angle PC £5.73	0.37	3.12	6.77	nr	9.89
Flat three way intersection PC £7.28	0.48	4.05	8.60	nr	12.65
Vertical three way intersection PC £9.06	0.48	4.05	10.70	nr	14.75
Four way intersection PC £9.38	0.59	4.98	11.08	nr	16.06

E IN SITU CONCRETE/LARGE PRECAST CONCRETE Including overheads and profit at 12.50%		Labour hours	Labour £	Material £	Unit	Total rate £

E40 DESIGNED JOINTS IN IN SITU CONCRETE - cont'd

Serviciced water stops or similar - cont'd
Formed joint; PVC water stop; centre bulb
type; heat welded joints

160 mm wide	PC £56.10/15m	0.28	2.36	4.44	m	6.80
Flat angle	PC £3.50	0.33	2.78	4.14	nr	6.92
Vertical angle	PC £5.75	0.33	2.78	6.80	nr	9.58
Flat three way intersection	PC £7.38	0.44	3.71	8.72	nr	12.43
Vertical three way intersection	PC £9.14	0.44	3.71	10.80	nr	14.51
Four way intersection	PC £8.78	0.55	4.64	10.37	nr	15.01
210 mm wide	PC £80.85/15m	0.31	2.62	6.40	m	9.02
Flat angle	PC £5.20	0.35	2.95	6.14	nr	9.09
Vertical angle	PC £6.84	0.35	2.95	8.08	nr	11.03
Flat three way intersection	PC £6.75	0.46	3.88	7.98	nr	11.86
Vertical three way intersection	PC £10.36	0.46	3.88	12.24	nr	16.12
Four way intersection	PC £10.03	0.57	4.81	11.85	nr	16.66
260 mm wide	PC £114.90/15m	0.33	2.78	9.09	m	11.87
Flat angle	PC £6.79	0.37	3.12	8.02	nr	11.14
Vertical angle	PC £7.45	0.37	3.12	8.80	nr	11.92
Flat three way intersection	PC £10.19	0.48	4.05	12.04	nr	16.09
Vertical three way intersection	PC £12.63	0.48	4.05	14.92	nr	18.97
Four way intersection	PC £12.20	0.59	4.98	14.41	nr	19.39
325 mm wide	PC £193.20/15m	0.36	3.04	15.29	m	18.33
Flat angle	PC £9.83	0.40	3.38	11.61	nr	14.99
Vertical angle	PC £10.15	0.40	3.38	11.99	nr	15.37
Flat three way intersection	PC £14.44	0.51	4.30	17.06	nr	21.36
Vertical three way intersection	PC £14.76	0.51	4.30	17.44	nr	21.74
Four way intersection	PC £16.13	0.62	5.23	19.06	nr	24.29

Formed joint; rubber water stop;
flat dumbell type; sleeved joints

150 mm wide	PC £100.81/9m	0.22	1.86	14.56	m	16.42
Flat angle	PC £28.69	0.22	1.86	33.88	nr	35.74
Vertical angle	PC £28.69	0.22	1.86	33.88	nr	35.74
Flat three way intersection	PC £31.65	0.28	2.36	37.38	nr	39.74
Vertical three way intersection	PC £31.65	0.28	2.36	37.38	nr	39.74
Four way intersection	PC £34.97	0.33	2.78	41.31	nr	44.09
230 mm wide	PC £152.50/9m	0.28	2.36	21.70	m	24.06
Flat angle	PC £34.72	0.24	2.03	41.02	nr	43.05
Vertical angle	PC £34.72	0.24	2.03	41.02	nr	43.05
Flat three way intersection	PC £37.82	0.30	2.53	44.67	nr	47.20
Vertical three way intersection	PC £37.82	0.30	2.53	44.67	nr	47.20
Four way intersection	PC £40.75	0.36	3.04	48.14	nr	51.18

Formed joint; rubber water stop; centre bulb
type; sleeved joints

150 mm wide	PC £116.00/9m	0.22	1.86	16.53	m	18.39
Flat angle	PC £31.43	0.22	1.86	37.13	nr	38.99
Vertical angle	PC £31.43	0.22	1.86	37.13	nr	38.99
Flat three way intersection	PC £34.68	0.28	2.36	40.96	nr	43.32
Vertical three way intersection	PC £34.68	0.28	2.36	40.96	nr	43.32
Four way intersection	PC £37.94	0.33	2.78	44.82	nr	47.60
230 mm wide	PC £173.32/9m	0.28	2.36	24.53	m	26.89
Flat angle	PC £36.92	0.24	2.03	43.62	nr	45.65
Vertical angle	PC £36.92	0.24	2.03	43.62	nr	45.65
Flat three way intersection	PC £38.62	0.30	2.53	45.62	nr	48.15
Vertical three way intersection	PC £38.62	0.30	2.53	45.62	nr	48.15
Four way intersection	PC £44.76	0.36	3.04	52.88	nr	55.92
305 mm wide	PC £285.82/9m	0.33	2.78	40.13	m	42.91
Flat angle	PC £58.01	0.26	2.19	68.52	nr	70.71
Vertical angle	PC £58.01	0.26	2.19	68.52	nr	70.71
Flat three way intersection	PC £70.13	0.33	2.78	82.85	nr	85.63
Vertical three way intersection	PC £70.13	0.33	2.78	82.85	nr	85.63
Four way intersection	PC £87.82	0.40	3.38	103.74	nr	107.12

E IN SITU CONCRETE/LARGE PRECAST CONCRETE Including overheads and profit at 12.50%	Labour hours	Labour £	Material £	Unit	Total rate £
E41 WORKED FINISHES/CUTTING ON IN SITU CONCRETE					
Tamping by mechanical means	0.03	0.22	0.14	m2	0.36
Power floating	0.19	1.37	0.48	m2	1.85
Trowelling	0.36	2.60	-	m2	2.60
Hacking					
by mechanical means	0.36	2.60	0.41	m2	3.01
by hand	0.77	5.56	-	m2	5.56
Wood float finish	0.14	1.01	-	m2	1.01
Tamped finish	0.06	0.43	-	m2	0.43
to falls	0.08	0.58	-	m2	0.58
to crossfalls	0.11	0.79	-	m2	0.79
Spade finish	0.17	1.23	-	m2	1.23
Cutting chases					
not exceeding 50 mm deep; 10 mm wide	0.36	2.60	0.19	m	2.79
not exceeding 50 mm deep; 50 mm wide	0.55	3.97	0.28	m	4.25
not exceeding 50 mm deep; 75 mm wide	0.73	5.27	0.37	m	5.64
50 - 100 mm deep; 75 mm wide	0.99	7.15	0.51	m	7.66
50 - 100 mm deep; 100 mm wide	1.10	7.94	0.56	m	8.50
100 - 150 mm deep; 100 mm wide	1.45	10.47	0.73	m	11.20
100 - 150 mm deep; 150 mm wide	1.75	12.64	0.90	m	13.54
Cutting chases in reinforced concrete					
50 - 100 mm deep; 100 mm wide	1.65	11.92	0.84	m	12.76
100 - 150 mm deep; 100 mm wide	2.20	15.89	1.13	m	17.02
100 - 150 mm deep; 150 mm wide	2.65	19.14	1.35	m	20.49
Cutting rebates					
not exceeding 50 mm deep; 50 mm wide	0.55	3.97	0.28	m	4.25
50 - 100 mm deep; 100 mm wide	1.10	7.94	0.56	m	8.50
Cutting mortices; not exceeding 100 mm deep; making good					
20 mm dia	0.17	1.23	0.07	nr	1.30
50 mm dia	0.19	1.37	0.09	nr	1.46
150 x 150 mm	0.39	2.82	0.21	nr	3.03
300 x 300 mm	0.77	5.56	0.44	nr	6.00
Cutting mortices in reinforced concrete; not exceeding 100 mm deep; making good					
150 x 150 mm	0.61	4.41	0.29	nr	4.70
300 x 300 mm	1.15	8.31	0.61	nr	8.92
Cutting holes; not exceeding 100 mm deep					
50 mm dia	0.39	2.82	0.43	nr	3.25
100 mm dia	0.44	3.18	0.50	nr	3.68
150 x 150 mm	0.50	3.61	0.56	nr	4.17
300 x 300 mm	0.61	4.41	0.68	nr	5.09
Cutting holes; 100 - 200 mm deep					
50 mm dia	0.55	3.97	0.62	nr	4.59
100 mm dia	0.66	4.77	0.74	nr	5.51
150 x 150 mm	0.83	5.99	0.93	nr	6.92
300 x 300 mm	1.05	7.58	1.18	nr	8.76
Cutting holes; 200 - 300 mm deep					
50 mm dia	0.83	5.99	0.93	nr	6.92
100 mm dia	0.99	7.15	1.11	nr	8.26
150 x 150 mm	1.20	8.67	1.36	nr	10.03
300 x 300 mm	1.55	11.19	1.73	nr	12.92
Add for making good fair finish one side					
50 mm dia	0.06	0.43	0.02	nr	0.45
100 mm dia	0.13	0.94	0.02	nr	0.96
150 x 150 mm	0.22	1.59	0.04	nr	1.63
300 x 300 mm	0.44	3.18	0.08	nr	3.26
Add for fixing only sleeve					
50 mm dia	0.11	0.79	-	nr	0.79
100 mm dia	0.24	1.73	-	nr	1.73
150 x 150 mm	0.36	2.60	-	nr	2.60
300 x 300 mm	0.66	4.77	-	nr	4.77

E IN SITU CONCRETE/LARGE PRECAST CONCRETE Including overheads and profit at 12.50%	Labour hours	Labour £	Material £	Unit	Total rate £
E41 WORKED FINISHES/CUTTING ON IN SITU CONCRETE - cont'd					
Cutting holes in reinforced concrete; not exceeding 100 mm deep					
50 mm dia	0.61	4.41	0.68	nr	5.09
100 mm dia	0.66	4.77	0.74	nr	5.51
150 x 150 mm dia	0.77	5.56	0.87	nr	6.43
300 x 300 mm dia	0.94	6.79	1.05	nr	7.84
Cutting holes in reinforced concrete; 100 - 200 mm deep					
50 mm dia	0.83	5.99	0.93	nr	6.92
100 mm dia	0.99	7.15	1.11	nr	8.26
150 x 150 mm dia	1.25	9.03	1.42	nr	10.45
300 x 300 mm dia	1.60	11.56	1.79	nr	13.35
Cutting holes in reinforced concrete; 200 - 300 mm deep					
50 mm dia	1.25	9.03	1.42	nr	10.45
100 mm dia	1.50	10.83	1.67	nr	12.50
150 x 150 mm dia	1.80	13.00	2.04	nr	15.04
300 x 300 mm dia	2.30	16.61	2.60	nr	19.21
E42 ACCESSORIES CAST INTO IN SITU CONCRETE					
Temporary plywood foundation bolt boxes					
75 x 75 x 150 mm	0.50	3.61	0.51	nr	4.12
75 x 75 x 250 mm	0.55	3.97	0.81	nr	4.78
'Expamet' cylindrical expanded steel foundation boxes					
76 mm dia x 152 mm high	0.33	2.38	3.54	nr	5.92
76 mm dia x 305 mm high	0.22	1.59	1.09	nr	2.68
102 mm dia x 457 mm high	0.28	2.02	1.96	nr	3.98
10 mm dia x 100 mm long	0.28	2.02	1.33	nr	3.35
12 mm dia x 120 mm long	0.28	2.02	1.52	nr	3.54
16 mm dia x 160 mm long	0.33	2.38	3.51	nr	5.89
20 mm dia x 200 mm long	0.33	2.38	3.54	nr	5.92
'Abbey' galvanized steel masonry slots; 18 G (1.22 mm)					
3.048 m lengths	0.39	2.82	1.13	m	3.95
76 mm long	0.09	0.65	0.15	nr	0.80
102 mm long	0.09	0.65	0.18	nr	0.83
152 mm long	0.10	0.72	0.23	nr	0.95
229 mm long	0.11	0.79	0.32	nr	1.11
'Unistrut' galvanized steel slotted metal inserts; 2.5 mm thick; end caps and foam filling					
41 x 41 mm; ref P3270	0.44	3.18	4.73	m	7.91
41 x 41 x 75 mm; ref P3249	0.11	0.79	1.79	nr	2.58
41 x 41 x 100 mm; ref P3250	0.11	0.79	1.89	nr	2.68
41 x 41 x 150 mm; ref P3251	0.11	0.79	2.14	nr	2.93
Butterfly type wall ties; casting one end into concrete; other end built into joint of brickwork					
galvanized steel	0.11	0.79	0.08	nr	0.87
stainless steel	0.11	0.79	0.11	nr	0.90
Mild steel fixing cramp; once bent; one end shot fired into concrete; other end fanged and built into joint of brickwork					
200 mm girth	0.17	1.23	0.66	nr	1.89
Sherardized steel floor clips; pinned to surface of concrete					
50 mm wide; standard type	0.09	0.65	0.11	nr	0.76
50 mm wide; direct fix acoustic type	0.11	0.79	0.72	nr	1.51
Hardwood dovetailed fillets					
50 x 50/40 x 1000 mm	0.11	0.79	1.22	nr	2.01
50 x 50/40 x 100 mm	0.09	0.65	0.18	nr	0.83
50 x 50/40 x 200 mm	0.09	0.65	0.24	nr	0.89

E IN SITU CONCRETE/LARGE PRECAST CONCRETE Including overheads and profit at 12.50%	Labour hours	Labour £	Material £	Unit	Total rate £
'Rigifix' galvanized steel plate column guard; 1 m long					
75 mm x 75 mm x 3 mm	0.66	4.77	8.65	nr	13.42
75 mm x 75 mm x 4.5 mm	0.66	4.77	11.55	nr	16.32
'Rigifix' white nylon coated steel plate corner guard; plugged and screwed to concrete with chromium plated domed headed screws					
75 x 75 x 1.5 mm x 1 m long	0.88	6.36	12.72	nr	19.08

E60 PRECAST/COMPOSITE CONCRETE DECKING

Prestressed precast flooring planks; Bison 'Drycast' or similar; cement and sand (1:3) grout between planks and on prepared bearings 100 mm thick suspended slabs; horizontal					
400 mm wide planks	-	-	-	m2	36.45
1200 mm wide planks	-	-	-	m2	33.75
150 mm thick suspended slabs; horizontal					
400 mm wide planks	-	-	-	m2	37.13
1200 mm wide planks	-	-	-	m2	34.43

Prestressed precast concrete beam and block floor; Bison 'Housefloor' or similar; in situ concrete 30 N/mm2 - 10 mm aggregate in filling at wall abutments; cement and sand (1:6) grout brushed in between beams and blocks 155 mm thick suspended slab at ground level; 440 x 215 x 100 mm blocks; horizontal					
beams at 520 mm centres; up to 3.30 m span with a superimposed load of 5 kN/m2	-	-	-	m2	21.60
beams at 295 mm centres; up to 4.35 m span with a superimposed load of 5 kN/m2	-	-	-	m2	24.98

Keep your figures up to date, free of charge

This section, and most of the other information in this Price Book, is brought up to date every three months in the *Price Book Update*.

The *Update* is available free to all Price Book purchasers.

To ensure you receive your copy, simply complete the reply card from the centre of the book and return it to us.

F MASONRY
Including overheads and profit at 12.50%

BASIC MORTAR PRICES

	£		£		£		£
Coloured mortar materials (£/tonne); (excluding cement)							
light	28.69	medium	29.69	dark	27.72	extra dark	30.69
Mortar materials (£/tonne)							
cement	61.54	lime	92.40	sand	10.36	white cement	74.19

Mortar plasticizer - £1.70/Litre

	Labour hours	Labour £	Material £	Unit	Total rate £
F10 BRICK/BLOCK WALLING					
Common bricks; PC £120.50/1000; in cement mortar (1:3)					
Walls					
half brick thick	1.45	17.44	9.75	m2	27.19
one brick thick	2.40	28.86	20.29	m2	49.15
one and a half brick thick	3.25	39.09	30.51	m2	69.60
two brick thick	4.00	48.11	40.59	m2	88.70
Walls; facework one side					
half brick thick	1.60	19.24	9.75	m2	28.99
one brick thick	2.60	31.27	20.29	m2	51.56
one and a half brick thick	3.45	41.49	30.51	m2	72.00
two brick thick	4.20	50.51	40.59	m2	91.10
Walls; facework both sides					
half brick thick	1.70	20.44	9.75	m2	30.19
one brick thick	2.70	32.47	20.29	m2	52.76
one and a half brick thick	3.55	42.69	30.51	m2	73.20
two brick thick	4.30	51.71	40.59	m2	92.30
Walls; built curved mean radius 6 m					
half brick thick	1.90	22.85	10.50	m2	33.35
one brick thick	3.15	37.88	21.78	m2	59.66
Walls; built curved mean radius 1.50 m					
half brick thick	2.40	28.86	11.04	m2	39.90
one brick thick	4.00	48.11	22.86	m2	70.97
Walls; built overhand					
half brick thick	1.80	21.65	9.75	m2	31.40
Walls; building up against concrete including flushing up at back					
half brick thick	1.55	18.64	11.06	m2	29.70
Walls; backing to masonry; cutting and bonding					
one brick thick	2.90	34.88	20.69	m2	55.57
one and a half brick thick	3.85	46.30	31.13	m2	77.43
Honeycomb walls					
half brick thick	1.15	13.83	6.90	m2	20.73
Dwarf support wall					
half brick thick	1.80	21.65	9.75	m2	31.40
one brick thick	2.90	34.88	20.29	m2	55.17
Battering walls					
one and a half brick thick	3.80	45.70	31.13	m2	76.83
two brick thick	4.70	56.52	41.40	m2	97.92
Walls; tapering one side; average					
337 mm thick	4.20	50.51	31.74	m2	82.25
450 mm thick	5.40	64.94	42.21	m2	107.15
Walls; tapering both sides; average					
337 mm thick	4.80	57.73	31.74	m2	89.47
450 mm thick	6.00	72.16	42.21	m2	114.37

F MASONRY Including overheads and profit at 12.50%	Labour hours	Labour £	Material £	Unit	Total rate £
Isolated piers					
one brick thick	3.70	44.50	20.69	m2	65.19
two brick thick	5.75	69.15	41.40	m2	110.55
three brick thick	7.25	87.19	62.09	m2	149.28
Isolated casings to steel columns					
half brick thick	1.85	22.25	9.95	m2	32.20
one brick thick	3.15	37.88	20.69	m2	58.57
Chimney stacks					
one brick thick	3.70	44.50	20.69	m2	65.19
two brick thick	5.75	69.15	41.40	m2	110.55
three brick thick	7.25	87.19	62.09	m2	149.28
Projections; vertical					
225 x 112 mm	0.45	5.41	2.31	m	7.72
225 x 225 mm	0.85	10.22	4.46	m	14.68
337 x 225 mm	1.25	15.03	7.50	m	22.53
440 x 225 mm	1.45	17.44	8.78	m	26.22
Bonding ends to existing					
half brick thick	0.45	5.41	0.72	m	6.13
one brick thick	0.65	7.82	1.44	m	9.26
one and a half brick thick	1.00	12.03	2.16	m	14.19
two brick thick	1.40	16.84	2.89	m	19.73
ADD or DEDUCT to walls for variation of £10.00/1000 in PC of common bricks					
half brick thick	-	-	-	m2	0.73
one brick thick	-	-	-	m2	1.45
one and a half brick thick	-	-	-	m2	2.17
two brick thick	-	-	-	m2	2.84
Extra over walls for sulphate-resisting cement mortar (1:3) in lieu of cement mortar (1:3)					
half brick thick	-	-	-	m2	0.11
one brick thick	-	-	-	m2	0.29
one and a half brick thick	-	-	-	m2	0.44
two brick thick	-	-	-	m2	0.58
Common bricks; PC £120.50/1000; **in gauged mortar (1:1:6)** Walls					
half brick thick	1.45	17.44	9.66	m2	27.10
one brick thick	2.40	28.86	20.05	m2	48.91
one and a half brick thick	3.25	39.09	30.14	m2	69.23
two brick thick	4.00	48.11	40.11	m2	88.22
Walls; facework one side					
half brick thick	1.60	19.24	9.66	m2	28.90
one brick thick	2.60	31.27	20.05	m2	51.32
one and a half brick thick	3.45	41.49	30.14	m2	71.63
two brick thick	4.20	50.51	40.11	m2	90.62
Walls; facework both sides					
half brick thick	1.70	20.44	9.66	m2	30.10
one brick thick	2.70	32.47	20.05	m2	52.52
one and a half brick thick	3.55	42.69	30.14	m2	72.83
two brick thick	4.30	51.71	40.11	m2	91.82
Walls; built curved mean radius 6 m					
half brick thick	1.90	22.85	10.40	m2	33.25
one brick thick	3.15	37.88	21.53	m2	59.41
Walls; built curved mean radius 1.50 m					
half brick thick	2.40	28.86	10.94	m2	39.80
one brick thick	4.00	48.11	22.59	m2	70.70
Walls; built overhand					
half brick thick	1.80	21.65	9.66	m2	31.31
Walls; built up against concrete including flushing up at back					
half brick thick	1.55	18.64	10.88	m2	29.52
Walls; backing to masonry; cutting and bonding					
one brick thick	2.90	34.88	20.45	m2	55.33
one and a half brick thick	3.85	46.30	30.76	m2	77.06

F MASONRY Including overheads and profit at 12.50%	Labour hours	Labour £	Material £	Unit	Total rate £
F10 BRICK/BLOCK WALLING - cont'd					
Common bricks; PC £120.50/1000; **in gauged mortar (1:1:6) - cont'd**					
Honeycomb walls					
half brick thick	1.15	13.83	6.81	m2	20.64
Dwarf support wall					
half brick thick	1.80	21.65	9.66	m2	31.31
one brick thick	2.90	34.88	20.05	m2	54.93
Battering walls					
one and a half brick thick	3.80	45.70	30.76	m2	76.46
two brick thick	4.70	56.52	40.92	m2	97.44
Walls; tapering one side; average					
337 mm thick	4.20	50.51	31.37	m2	81.88
450 mm thick	5.40	64.94	41.73	m2	106.67
Walls; tapering both sides; average					
337 mm thick	4.80	57.73	31.37	m2	89.10
450 mm thick	6.00	72.16	41.73	m2	113.89
Isolated piers					
one brick thick	3.70	44.50	20.45	m2	64.95
two brick thick	5.75	69.15	40.92	m2	110.07
three brick thick	7.25	87.19	61.37	m2	148.56
Isolated casings to steel columns					
half brick thick	1.85	22.25	9.86	m2	32.11
one brick thick	3.15	37.88	20.45	m2	58.33
Chimney stacks					
one brick thick	3.70	44.50	20.45	m2	64.95
two brick thick	5.75	69.15	40.92	m2	110.07
three brick thick	7.25	87.19	61.37	m2	148.56
Projections; vertical					
225 x 112 mm	0.45	5.41	2.29	m	7.70
225 x 225 mm	0.85	10.22	4.42	m	14.64
337 x 225 mm	1.25	15.03	7.44	m	22.47
440 x 225 mm	1.45	17.44	8.70	m	26.14
Bonding ends to existing					
half brick thick	0.45	5.41	0.70	m	6.11
one brick thick	0.65	7.82	1.41	m	9.23
one and a half brick thick	1.00	12.03	2.11	m	14.14
two brick thick	1.40	16.84	2.82	m	19.66
ADD or DEDUCT to walls for variation of £10.00/1000 in PC of common bricks					
half brick thick	-	-	-	m2	0.71
one brick thick	-	-	-	m2	1.42
one and a half brick thick	-	-	-	m2	2.13
two brick thick	-	-	-	m2	2.84
Segmental arches; one ring, 102 mm high on face					
102 mm wide on exposed soffit	2.10	21.49	1.33	m	22.82
215 mm wide on exposed soffit	2.60	25.71	2.86	m	28.57
Segmental arches; two ring, 215 mm high on face					
102 mm wide on exposed soffit	2.55	26.90	1.33	m	28.23
215 mm wide on exposed soffit	3.25	33.52	2.86	m	36.38
Semi-circular arches; one ring, 102 mm high on face					
102 mm wide on exposed soffit	2.45	24.44	1.62	m	26.06
215 mm wide on exposed soffit	2.90	28.24	18.48	m	46.72
Semi-circular arches; two ring, 215 mm high on face					
102 mm wide on exposed soffit	2.45	24.44	1.62	m	26.06
215 mm wide on exposed soffit	2.90	28.24	18.48	m	46.72
Labours on brick fairface					
Fair returns					
half brick wide	0.03	0.36	-	m	0.36
one brick wide	0.06	0.72	-	m	0.72

F MASONRY Including overheads and profit at 12.50%	Labour hours	Labour £	Material £	Unit	Total rate £
Class A engineering bricks; PC £350.00/1000; in cement mortar (1:3)					
Walls					
half brick thick	1.55	18.64	25.43	m2	44.07
one brick thick	2.60	31.27	51.64	m2	82.91
one and a half brick thick	3.45	41.49	77.55	m2	119.04
two brick thick	4.30	51.71	103.28	m2	155.00
Walls; facework one side					
half brick thick	1.70	20.44	26.02	m2	46.46
one brick thick	2.75	33.07	52.82	m2	85.89
one and a half brick thick	3.60	43.29	79.31	m2	122.60
two brick thick	4.50	54.12	105.65	m2	159.77
Walls; facework both sides					
half brick thick	1.80	21.65	26.02	m2	47.67
one brick thick	2.85	34.27	52.82	m2	87.09
one and a half brick thick	3.70	44.50	79.31	m2	123.81
two brick thick	4.60	55.32	105.65	m2	160.97
Walls; built curved mean radius 6 m					
one brick thick	3.45	41.49	55.51	m2	97.00
Walls; backing to masonry; cutting and bonding					
one brick thick	3.10	37.28	54.00	m2	91.28
one and a half brick thick	4.15	49.91	81.09	m2	131.00
Walls; tapering one side; average					
337 mm thick	4.50	54.12	81.08	m2	135.20
450 mm thick	5.75	69.15	108.01	m2	177.16
Walls; tapering both sides; average					
337 mm thick	5.20	62.54	82.86	m2	145.40
450 mm thick	6.55	78.77	110.37	m2	189.14
Isolated piers					
one brick thick	4.00	48.11	52.82	m2	100.93
two brick thick	6.30	75.77	105.65	m2	181.42
three brick thick	7.75	93.20	158.57	m2	251.77
Isolated casings to steel columns					
half brick thick	2.00	24.05	26.61	m2	50.66
one brick thick	3.45	41.49	54.00	m2	95.49
Projections; vertical					
225 x 112 mm	0.50	6.01	6.19	m	12.20
225 x 225 mm	0.90	10.82	11.96	m	22.78
337 x 225 mm	1.40	16.84	20.26	m	37.10
440 x 225 mm	1.55	18.64	23.49	m	42.13
Bonding ends to existing					
half brick thick	0.50	6.01	1.66	m	7.67
one brick thick	0.70	8.42	3.31	m	11.73
one and a half brick thick	1.05	12.63	4.97	m	17.60
two brick thick	1.50	18.04	6.64	m	24.68
ADD or DEDUCT to walls for variation of £10.00/1000 in PC of engineering bricks					
half brick thick	-	-	-	m2	0.71
one brick thick	-	-	-	m2	1.42
one and a half brick thick	-	-	-	m2	2.13
two brick thick	-	-	-	m2	2.84
Class B engineering bricks; PC £198.00/1000; in cement mortar (1:3)					
Walls					
half brick thick	1.55	18.64	14.91	m2	33.55
one brick thick	2.60	31.27	30.60	m2	61.87
one and a half brick thick	3.45	41.49	45.99	m2	87.48
two brick thick	4.30	51.71	61.22	m2	112.93
Walls; facework one side					
half brick thick	1.70	20.44	15.24	m2	35.68
one brick thick	2.75	33.07	31.28	m2	64.35
one and a half brick thick	3.60	43.29	46.99	m2	90.28
two brick thick	4.50	54.12	62.56	m2	116.68

F MASONRY Including overheads and profit at 12.50%	Labour hours	Labour £	Material £	Unit	Total rate £
F10 BRICK/BLOCK WALLING - cont'd					
Class B engineering bricks; PC £198.00/1000; **in cement mortar (1:3) - cont'd**					
Walls; facework both sides					
half brick thick	1.80	21.65	15.24	m2	36.89
one brick thick	2.85	34.27	31.28	m2	65.55
one and a half brick thick	3.70	44.50	46.99	m2	91.49
two brick thick	4.60	55.32	62.56	m2	117.88
Walls; built curved mean radius 6 m					
one brick thick	3.45	41.49	32.89	m2	74.38
Walls; backing to masonry;					
cutting and bonding					
one brick thick	3.10	37.28	31.94	m2	69.22
one and a half brick thick	4.15	49.91	47.99	m2	97.90
Walls; tapering one side; average					
337 mm thick	4.50	54.12	47.99	m2	102.11
450 mm thick	5.75	69.15	63.89	m2	133.04
Walls; tapering both sides; average					
337 mm thick	5.20	62.54	48.99	m2	111.53
450 mm thick	6.55	78.77	65.23	m2	144.00
Isolated piers					
one brick thick	4.00	48.11	31.28	m2	79.38
two brick thick	6.30	75.77	62.56	m2	138.33
three brick thick	7.75	93.20	93.93	m2	187.13
Isolated casings to steel columns					
half brick thick	2.00	24.05	15.58	m2	39.63
one brick thick	3.45	41.49	31.94	m2	73.43
Projections					
225 x 112 mm	0.50	6.01	3.62	m	9.63
225 x 225 mm	0.90	10.82	6.99	m	17.81
337 x 225 mm	1.40	16.84	11.80	m	28.64
440 x 225 mm	1.55	18.64	13.75	m	32.39
Bonding ends to existing					
half brick thick	0.50	6.01	1.02	m	7.04
one brick thick	0.70	8.42	2.06	m	10.48
one and a half brick thick	1.05	12.63	3.08	m	15.71
two brick thick	1.50	18.04	4.12	m	22.16
ADD or DEDUCT to walls for variation of £10.00/1000 in PC of engineering bricks					
half brick thick	-	-	-	m2	0.71
one brick thick	-	-	-	m2	1.42
one and a half brick thick	-	-	-	m2	2.13
two brick thick	-	-	-	m2	2.84
Refractory bricks; PC £546.00/1000; **stretcher bond lining to flue; in fireclay** **cement mortar (1:4); built 50 mm clear of** **flues; one header per m2**					
Walls; vertical; facework one side					
half brick thick	1.95	23.45	38.44	m2	61.89

ALTERNATIVE FACING BRICK PRICES (£/1000)

	£		£
Ibstock facing bricks; 215 x 102.5 x 65 mm			
Aldridge brown blend	311.06	Leicester Anglican Red Rustic	302.82
Cattybrook Gloucester Golden	309.00	Leicester Red Stock	335.78
Himley Dark Brown Rustic	355.35	Roughdales Red Multi Rustic	339.90
Himley Mixed Russet	305.91	Roughdales Trafford Buff Multi	350.20
London Brick Company facing bricks; 215 x 102.5 x 65 mm			
Brecken Grey	117.17	Orton Multi Buff	126.03
Chiltern	124.71	Regency	118.66
Claydon Red Multi	119.93	Sandfaced	124.66
Delph Autumn	117.52	Saxon Gold	121.94

F MASONRY
Including overheads and profit at 12.50%

Edwardian	131.08	Tudor	127.25
Georgian Red Multi	120.56	Victorian	138.54
Heather	122.80	Wansford Multi	125.39
Ironstone	119.64	Windsor	120.13
Milton Buff	118.73		

Redland facing bricks; 215 x 102.5 x 65 mm

Arun	385.22	Southwater class B	303.85
Beare Green restoration red	537.66	Sheppy 'matured' yellow	407.88
Chailey yellow multicoloured	402.73	Stourbridge Sherbourne range	301.79
Cottage mixed multicoloured	303.85	Stourbridge Henley range	288.40
Crowborough multicoloured	362.56	Stourbridge Pennine range	278.10
Dorking	311.06	Stourbridge Stratford range	248.23
Funton yellow London	292.52	Surrey bronze multicoloured	313.12
Hamsey multicoloured	379.04	Tonbridge handmade	495.43
Holbrook Castle range	299.73	Tonbridge handmade(50mm deep)	511.91
Holbrook sandfaced textured	315.18	Tudor (53mm deep)	616.97
Nutbourne sandfaced	277.07	Wealden	385.22
Pevensey red multicoloured	365.65	Wealdmade	430.54
Pluckley multicoloured	383.16		

	Labour hours	Labour £	Material £	Unit	Total rate £
Facing bricks; sand faced; PC £128.75/1000 (unless otherwise stated); in gauged mortar (1:1:6)					
Extra over common bricks; PC £120.50/1000; for facing bricks in					
stretcher bond	0.45	5.41	0.60	m2	6.01
flemish bond with snapped headers	0.55	6.61	3.71	m2	10.32
english bond with snapped headers	0.55	6.61	3.62	m2	10.24
ADD or DEDUCT for variation of £10.00/1000 in PC of facing bricks	-	-	-	m2	0.97
Half brick thick; stretcher bond; facework one side					
walls	1.90	22.85	10.46	m2	33.31
walls; building curved mean radius 6 m	2.75	33.07	11.25	m2	44.32
walls; building curved mean radius 1.50 m	3.45	41.49	11.83	m2	53.32
walls; building overhand	2.30	27.66	10.46	m2	38.12
walls; building up against concrete including flushing up at back	2.00	24.05	11.89	m2	35.94
walls; as formwork; temporary strutting	2.75	33.07	12.79	m2	45.86
walls; panels and aprons; not exceeding 1 m2	2.40	28.86	10.51	m2	39.37
isolated casings to steel columns	2.90	34.88	10.46	m2	45.34
bonding ends to existing	0.75	9.02	1.63	m	10.65
projections; vertical					
225 x 112 mm	0.46	5.53	2.43	m	7.96
337 x 112 mm	0.86	10.34	3.75	m	14.09
440 x 112 mm	1.25	15.03	5.09	m	20.12
Half brick thick; flemish bond with snapped headers; facework one side					
walls	2.20	26.46	11.48	m2	37.94
walls; building curved mean radius 6 m	3.10	37.28	12.06	m2	49.34
walls; building curved mean radius 1.50 m	4.00	48.11	12.62	m2	60.73
walls; building overhand	2.60	31.27	11.48	m2	42.75
walls; building up against concrete including flushing up at back	2.30	27.66	12.91	m2	40.57
walls; as formwork; temporary strutting	3.00	36.08	13.81	m2	49.89
walls; panels and aprons; not exceeding 1 m2	2.70	32.47	11.48	m2	43.95
isolated casings to steel columns	3.15	37.88	10.46	m2	48.34
bonding ends to existing	0.75	9.02	1.63	m	10.65
projections; vertical					
225 x 112 mm	0.58	6.98	2.73	m	9.71
337 x 112 mm	0.98	11.79	4.07	m	15.86
440 x 112 mm	1.40	16.84	5.56	m	22.40

F MASONRY Including overheads and profit at 12.50%	Labour hours	Labour £	Material £	Unit	Total rate £
F10 BRICK/BLOCK WALLING - cont'd					
Facing bricks; sand faced; PC £128.75/1000					
(unless otherwise stated); in gauged					
mortar (1:1:6) - cont'd					
One brick thick; two stretcher skins tied					
together; facework both sides					
walls	3.20	38.48	21.87	m2	60.35
walls; building curved mean radius 6 m	4.50	54.12	23.44	m2	77.56
walls; building curved mean radius 1.50 m	5.50	66.14	26.89	m2	93.03
isolated piers	3.80	45.70	23.72	m2	69.42
bonding ends to existing	0.98	11.79	3.25	m	15.04
One brick thick; flemish bond;					
facework both sides					
walls	3.35	40.29	21.66	m2	61.95
walls; building curved mean radius 6 m	4.60	55.32	23.24	m2	78.56
walls; building curved mean radius 1.50 m	5.75	69.15	26.69	m2	95.84
isolated piers	3.90	46.90	23.52	m2	70.42
bonding ends to existing	0.98	11.79	3.25	m	15.04
projections; vertical					
225 x 225 mm	0.92	11.06	3.22	m	14.28
337 x 225 mm	1.70	20.44	4.70	m	25.14
440 x 225 mm	2.55	30.67	6.29	m	36.96
ADD or DEDUCT for variation of £10.00/1000					
in PC of facing bricks; in stretcher bond					
half brick thick	-	-	-	m2	0.73
one brick thick	-	-	-	m2	1.46
ADD or DEDUCT for variation of £10.00/1000					
in PC of facing bricks; in flemish bond					
half brick thick	-	-	-	m2	0.91
one brick thick	-	-	-	m2	1.82
Extra over facing bricks for					
recessed joints	0.03	0.36	-	m2	0.36
raking out joints and pointing in black					
mortar	0.58	6.98	0.17	m2	7.15
bedding and pointing half brick wall in					
black mortar	-	-	-	m2	0.74
bedding and pointing one brick wall in black					
mortar	-	-	-	m2	1.97
flush plain bands; 225 mm wide stretcher					
bond; horizontal; bricks; PC £150.00/1000	0.29	3.49	0.34	m	3.83
flush quoins; average 320 mm girth; block					
bond vertical; facing bricks; PC £150.00/1000	0.46	5.53	0.32	m	5.85
Flat arches; 215 mm high on face					
102 mm wide exposed soffit	1.32	14.22	1.62	m	15.84
215 mm wide exposed soffit	1.99	21.46	3.24	m	24.70
Flat arches; 215 mm high on face; bullnosed					
specials; PC £47.67/100					
102 mm wide exposed soffit	1.38	14.95	9.07	m	24.02
215 mm wide exposed soffit	2.09	22.66	18.14	m	40.80
Segmental arches; one ring; 215 mm high					
on face					
102 mm wide exposed soffit	2.40	24.74	2.26	m	27.00
215 mm wide exposed soffit	3.60	37.19	4.74	m	41.93
Segmental arches; two ring; 215 mm high					
on face					
102 mm wide exposed soffit	3.10	33.15	2.26	m	35.41
215 mm wide exposed soffit	4.65	49.82	4.74	m	54.56
Segmental arches; 215 mm high on face;					
cut voussoirs; PC £66.17/100					
102 mm wide exposed soffit	2.55	26.54	12.52	m	39.06
215 mm wide exposed soffit	3.75	39.00	25.27	m	64.27
Segmental arches; one and a half ring;					
320 mm high on face; cut voussoirs;					
PC £66.17/100					
102 mm wide exposed soffit	3.45	37.36	23.73	m2	61.09
215 mm wide exposed soffit	5.15	55.83	47.68	m2	103.51

F MASONRY Including overheads and profit at 12.50%	Labour hours	Labour £	Material £	Unit	Total rate £
Semi circular arches; one ring; 215 mm high on face					
102 mm wide exposed soffit	3.00	31.05	2.87	m	33.92
215 mm wide exposed soffit	4.25	44.47	20.98	m	65.45
Semi circular arches; two ring; 215 mm high on face					
102 mm wide exposed soffit	3.70	39.47	2.87	m	42.34
215 mm wide exposed soffit	5.30	57.10	20.98	m	78.08
Semi circular arches; one ring; 215 mm high on face; cut voussoirs PC £66.17/100					
102 mm wide exposed soffit	3.10	32.26	12.82	m	45.08
215 mm wide exposed soffit	4.45	46.88	40.89	m	87.77
Bullseye window 600 mm dia; two rings; 215 mm high on face					
102 mm wide exposed soffit	6.90	78.85	3.41	nr	82.26
215 mm wide exposed soffit	10.35	118.37	6.61	nr	124.98
Bullseye window 1200 mm dia; two rings; 215 mm high on face					
102 mm wide exposed soffit	12.05	136.66	5.20	nr	141.86
215 mm wide exposed soffit	18.15	205.90	10.42	m	216.32
Bullseye window 600 mm dia; one ring; 215 mm high on face; cut voussoirs PC £66.17/100					
102 mm wide exposed soffit	5.75	65.02	29.55	nr	94.57
215 mm wide exposed soffit	8.60	97.32	58.88	nr	156.20
Bullseye window 1200 mm dia; one ring; 215 mm high on face; cut voussoirs					
102 mm wide exposed soffit	10.35	116.22	50.72	nr	166.94
215 mm wide exposed soffit	15.55	174.63	101.45	nr	276.08
ADD or DEDUCT for variation of £10.00/1000 in PC of facing bricks	-	-	-	m	0.30
Sills; horizontal; headers on edge; pointing top and one side; set weathering					
150 x 102 mm	0.81	9.74	2.48	m	12.22
150 x 102 mm; cant headers; PC £61.50/100	0.86	10.34	10.71	m	21.05
Sills; horizontal; headers on flat; pointing top and one side					
150 x 102 mm; bullnosed specials; PC £47.67/100	0.75	9.02	5.96	m	14.98
Coping; horizontal; headers on edge; pointing top and both sides					
215 x 102 mm	0.64	7.70	2.43	m	10.13
260 x 102 mm	1.05	12.63	3.75	m	16.38
215 x 102 mm; double bullnosed specials; PC £48.67/100	0.69	8.30	8.48	m	16.78
260 x 120 mm; single bullnosed specials; PC £47.67/100	1.05	12.63	13.60	m	26.23
ADD or DEDUCT for variation of £10.00/1000 in PC of facing bricks	-	-	-	m	0.30
Facing bricks; machine made facings; PC £339.90/1000 (unless otherwise stated; in gauged mortar (1:1:6) Extra over common bricks; PC £120.50/1000; for facing bricks in					
stretcher bond	0.45	5.41	15.92	m2	21.33
flemish bond with snapped headers	0.55	6.61	19.90	m2	26.51
english bond with snapped headers	0.55	6.61	22.56	m2	29.17
ADD or DEDUCT for variation of £10.00/1000 in PC of facing bricks	-	-	-	m2	0.91

F MASONRY Including overheads and profit at 12.50%	Labour hours	Labour £	Material £	Unit	Total rate £

F10 BRICK/BLOCK WALLING - cont'd

Facing bricks; machine made facings;
PC £339.90/1000 (unless otherwise stated;
in gauged mortar (1:1:6) - cont'd
Half brick thick; stretcher bond; facework
one side

	Labour hours	Labour £	Material £	Unit	Total rate £
walls	1.90	22.85	25.06	m2	47.91
walls; building curved mean radius 6 m	2.75	33.07	27.71	m2	60.78
walls; building curved mean radius 1.50 m	3.45	41.49	29.07	m2	70.56
walls; building overhand	2.30	27.66	25.78	m2	53.44
walls; building up against concrete including flushing up at back	2.00	24.05	27.24	m2	51.27
walls; as formwork; temporary strutting	2.75	33.07	28.11	m2	61.18
walls; panels and aprons; not exceeding 1 m2	2.40	28.86	26.36	m2	55.22
isolated casings to steel columns	2.90	34.88	25.78	m2	60.66
bonding ends to existing	0.75	9.02	1.63	m	10.65
projections; vertical					
225 x 112 mm	0.46	5.53	6.00	m	11.53
337 x 112 mm	0.86	10.34	9.12	m	19.46
440 x 112 mm	1.25	15.03	12.24	m	27.27
Half brick thick; flemish bond with snapped headers; facework one side					
walls	2.20	26.46	27.67	m2	54.13
walls; building curved mean radius 6 m	3.10	37.28	29.82	m2	67.10
walls; building curved mean radius 1.50 m	4.00	48.11	31.18	m2	78.29
walls; building overhand	2.60	31.27	28.46	m2	59.73
walls; building up against concrete including flushing up at back	2.30	27.66	30.79	m2	58.45
walls; as formwork; temporary strutting	3.00	36.08	29.89	m2	65.97
walls; panels and aprons; not exceeding 1 m2	2.70	32.47	28.46	m2	60.93
isolated casings to steel columns	3.15	37.88	25.78	m2	63.66
bonding ends to existing	0.75	9.02	1.63	m	10.65
projections; vertical					
225 x 112 mm	0.58	6.98	6.82	m	13.80
337 x 112 mm	0.98	11.79	9.95	m	21.74
440 x 112 mm	1.40	16.84	13.47	m	30.31
One brick thick; two stretcher skins tied together; facework both sides					
walls	3.20	38.48	51.08	m2	89.56
walls; building curved mean radius 6 m	4.50	54.12	56.16	m2	110.28
walls; building curved mean radius 1.50 m	5.50	66.14	59.09	m2	125.23
isolated piers	3.80	45.70	52.50	m2	98.20
bonding ends to existing	0.98	11.79	3.25	m	15.04
One brick thick; flemish bond; facework both sides					
walls	3.35	40.29	50.86	m2	91.15
walls; building curved mean radius 6 m	4.60	55.32	56.16	m2	111.48
walls; building curved mean radius 1.50 m	5.75	69.15	58.89	m2	128.04
isolated piers	3.90	46.90	52.51	m2	99.41
bonding ends to existing	0.98	11.79	3.25	m	15.04
projections; vertical					
225 x 225 mm	0.92	11.06	11.59	m	22.65
337 x 225 mm	1.70	20.44	19.64	m	40.08
440 x 225 mm	2.55	30.67	22.76	m	53.43
ADD or DEDUCT for variation of £10.00/1000 in PC of facing bricks; in stretcher bond					
half brick thick	-	-	-	m2	0.73
one brick thick	-	-	-	m2	1.46
ADD or DEDUCT for variation of £10.00/1000 in PC of facing bricks; in flemish bond					
half brick thick	-	-	-	m2	0.91
one brick thick	-	-	-	m2	1.82
Extra over facing bricks for					
recessed joints	0.03	0.36	-	m2	0.36
raking out joints and pointing in black mortar	0.58	6.98	0.17	m2	7.15

F MASONRY Including overheads and profit at 12.50%	Labour hours	Labour £	Material £	Unit	Total rate £
bedding and pointing half brick wall in black mortar	-	-	-	m2	0.74
bedding and pointing one brick wall in black mortar	-	-	-	m2	1.97
flush plain bands; 225 mm wide stretcher bond; horizontal; bricks PC £360.00/1000	0.29	3.49	0.51	m	4.00
flush quoins; average 320 mm girth; black bond vertical; bricks PC £360.00/1000	0.46	5.53	0.47	m	6.00
Flat arches; 215 mm high on face					
102 mm wide exposed soffit	1.32	14.22	2.65	m	16.87
215 mm wide exposed soffit	1.99	21.46	5.30	m	26.76
Flat arches; 215 mm high on face; bullnosed specials; PC £115.50/100					
102 mm wide exposed soffit	1.38	14.95	20.55	m	35.50
215 mm wide exposed soffit	2.09	22.66	41.11	m	63.77
Segmental arches; one ring; 215 mm high on face					
102 mm wide exposed soffit	2.40	24.74	3.80	m	28.54
215 mm wide exposed soffit	3.60	37.19	7.82	m	45.01
Segmental arches; two ring; 215 mm high on face					
102 mm wide exposed soffit	3.10	33.15	3.80	m	36.95
215 mm wide exposed soffit	4.65	49.82	7.82	m	57.64
Segmental arches; 215 mm high on face; cut voussoirs; PC £248.60/100					
102 mm wide exposed soffit	2.55	26.54	43.42	m	69.96
215 mm wide exposed soffit	3.75	39.00	87.04	m	126.04
Segmental arches; one and a half ring; 320 mm high on face; cut voussoirs PC £248.60/100					
102 mm wide exposed soffit	3.45	37.36	85.51	m	122.87
215 mm wide exposed soffit	5.15	55.83	171.23	m	227.06
Semi circular arches; one ring; 215 mm high on face					
102 mm wide exposed soffit	3.00	31.05	4.93	m	35.98
215 mm wide exposed soffit	4.25	44.47	25.09	m	69.56
Semi circular arches; two ring; 215 mm high on face					
102 mm wide exposed soffit	3.70	39.47	4.93	m	44.40
215 mm wide exposed soffit	5.30	57.10	25.09	m	82.19
Semi circular arches; one ring; 215 mm high on face; cut voussoirs PC £248.60/100					
102 mm wide exposed soffit	3.10	32.26	43.71	m	75.97
215 mm wide exposed soffit	4.45	46.88	102.66	m	149.54
Bullseye window 600 mm dia; two rings; 215 mm high on face					
102 mm wide exposed soffit	6.90	78.85	51.10	m	129.95
215 mm wide exposed soffit	10.35	118.37	101.99	m	220.36
Bullseye window 1200 mm dia; two rings; 215 mm high on face					
102 mm wide exposed soffit	12.05	136.66	9.32	m	145.98
215 mm wide exposed soffit	18.15	205.90	18.63	m	224.53
Bullseye window 600 mm dia; one ring; 215 mm high on face; cut voussoirs PC £248.60/100					
102 mm wide exposed soffit	5.75	65.02	106.77	m	171.79
215 mm wide exposed soffit	8.60	97.32	213.32	m	310.64
Bullseye window 1200 mm dia; one ring; 215 mm high on face; cut voussoirs PC £248.60/100					
102 mm wide exposed soffit	10.35	116.22	183.93	m	300.15
215 mm wide exposed soffit	15.55	174.63	367.88	m	542.51
ADD or DEDUCT for variation of £10.00/1000 in PC of facing bricks	-	-	-	m2	0.30

F MASONRY Including overheads and profit at 12.50%	Labour hours	Labour £	Material £	Unit	Total rate £

F10 BRICK/BLOCK WALLING - cont'd

Facing bricks; machine made facings;
PC £339.90/1000 (unless otherwise stated;
in gauged mortar (1:1:6) - cont'd

Sills; horizontal; headers on edge; pointing top and one side; set weathering					
150 x 102 mm;	0.81	9.74	6.05	m	15.79
150 x 102 mm; cant headers; PC £115.50/100	0.86	10.34	19.84	m	30.18
Sills; horizontal; headers on flat; pointing top and one side					
150 x 102 mm; bullnosed specials; PC £115.50/100	0.75	9.02	14.22	m	23.24
Coping; horizontal; headers on edge; pointing top and both sides					
215 x 102 mm	0.64	7.70	13.55	m	21.25
260 x 102 mm	1.05	12.63	9.12	m	21.75
215 x 102 mm; double bullnosed specials; PC £115.50/100	0.69	8.30	19.80	m	28.10
260 x 120 mm; single bullnosed specials; PC £115.50/100	1.05	12.63	32.47	m	45.10
ADD or DEDUCT for variation of £10.00/1000 in PC of facing bricks	-	-	-	m	0.30

Facing bricks; hand made; PC £551.05/1000
(unless otherwise stated); in gauged
mortar (1:1:6)

Extra over common bricks; PC £120.50/1000; for facing bricks in					
stretcher bond	0.45	5.41	31.24	m2	36.65
flemish bond with snapped headers	0.55	6.61	39.05	m2	45.66
english bond with snapped headers	0.55	6.61	44.25	m2	50.86
ADD or DEDUCT for variation of £10.00/1000 in PC of facing bricks	-	-	-	m2	0.91
Half brick thick; stretcher bond; facework one side					
walls	1.90	22.85	41.11	m2	63.96
walls; building curved mean radius 6 m	2.75	33.07	44.18	m2	77.25
walls; building curved mean radius 1.50 m	3.45	41.49	46.32	m2	87.81
walls; building overhand	2.30	27.66	41.11	m2	68.77
walls; building up against concrete including flushing up at back	2.00	24.05	42.53	m2	66.58
walls; as formwork; temporary strutting	2.75	33.07	43.43	m2	76.50
walls; panels and aprons; not exceeding 1 m2	2.40	28.86	41.11	m2	69.97
isolated casings to steel columns	2.90	34.88	41.12	m2	76.00
bonding ends to existing	0.75	9.02	1.63	m	10.65
projections; vertical					
225 x 112 mm	0.46	5.53	9.57	m	15.10
337 x 112 mm	0.86	10.34	14.48	m	24.82
440 x 112 mm	1.25	15.03	19.39	m	34.42
Half brick thick; flemish bond with snapped headers; facework one side					
walls	2.20	26.46	45.44	m2	71.90
walls; building curved mean radius 6 m	3.10	37.28	47.59	m2	84.87
walls; building curved mean radius 1.50 m	4.00	48.11	49.73	m2	97.84
walls; building overhand	2.60	31.27	45.44	m2	76.71
walls; building up against concrete including flushing up at back	2.30	27.66	46.87	m2	74.53
walls; as formwork; temporary strutting	3.00	36.08	47.77	m2	83.85
walls; panels and aprons; not exceeding 1 m2	2.70	32.47	45.44	m2	77.91
isolated casings to steel columns	3.15	37.88	41.12	m2	79.00
bonding ends to existing	0.75	9.02	1.63	m	10.65
projections; vertical					
225 x 112 mm	0.58	6.98	10.90	m	17.88
337 x 112 mm	0.98	11.79	15.81	m	27.60
440 x 112 mm	1.40	16.84	21.39	m	38.23

F MASONRY Including overheads and profit at 12.50%	Labour hours	Labour £	Material £	Unit	Total rate £
One brick thick; two stretcher skins tied together; facework both sides					
walls	3.20	38.48	83.52	m2	122.00
walls; building curved mean radius 6 m	4.50	54.12	87.75	m2	141.87
walls; building curved mean radius 1.50 m	5.50	66.14	93.38	m2	159.52
isolated piers	3.80	45.70	83.52	m2	129.22
bonding ends to existing	0.98	11.79	3.25	m	15.04
One brick thick; flemish bond; facework both sides					
walls	3.35	40.29	82.94	m2	123.23
walls; building curved mean radius 6 m	4.60	55.32	89.08	m2	144.40
walls; building curved mean radius 1.50 m	5.75	69.15	93.37	m2	162.52
isolated piers	3.90	46.90	82.94	m2	129.84
bonding ends to existing	0.98	11.79	3.25	m	15.04
projections; vertical					
225 x 225 mm	0.92	11.06	18.48	m	29.54
337 x 225 mm	1.70	20.44	31.38	m	51.82
440 x 225 mm	2.55	30.67	36.29	m	66.96
ADD or DEDUCT for variation of £10.00/1000 in PC of facing bricks; in stretcher bond					
half brick thick	-	-	-	m2	0.73
one brick thick	-	-	-	m2	1.46
ADD or DEDUCT for variation of £10.00/1000 in PC of facing bricks; in flemish bond					
half brick thick	-	-	-	m2	0.91
one brick thick	-	-	-	m2	1.82
Extra over facing bricks for					
recessed joints	0.03	0.36	-	m2	0.36
raking out joints and pointing in black mortar	0.58	6.98	0.17	m2	7.15
bedding and pointing half brick wall in black mortar	-	-	-	m2	0.74
bedding and pointing one brick wall in black mortar	-	-	-	m2	1.97
flush plain bands; 225 mm wide stretcher bond; horizontal; bricks PC £575.00/1000	0.29	3.49	0.41	m	3.90
flush quoins; average 320 mm girth; block bond vertical; bricks PC £575.00/1000	0.46	5.53	0.37	m	5.90
Flat arches; 215 mm high on face					
102 mm wide exposed soffit	1.32	14.22	3.68	m	17.90
215 mm wide exposed soffit	1.99	21.46	7.35	m	28.81
Flat arches; 215 mm high on face; bullnosed specials; PC £115.50/100					
102 mm wide exposed soffit	1.38	14.95	20.55	m	35.50
215 mm wide exposed soffit	2.09	22.66	41.11	m	63.77
Segmental arches; one ring; 215 mm high on face					
102 mm wide exposed soffit	2.40	24.74	5.34	m	30.08
215 mm wide exposed soffit	3.60	37.19	10.90	m	48.09
Segmental arches; two ring; 215 mm high on face					
102 mm wide exposed soffit	3.10	33.15	5.34	m	38.49
215 mm wide exposed soffit	4.65	49.82	10.90	m	60.72
Segmental arches; 215 mm high on face; cut voussoirs; PC £248.60/100					
102 mm wide exposed soffit	2.55	26.54	43.42	m	69.96
215 mm wide exposed soffit	3.75	39.00	87.04	m	126.04
Segmental arches; one and a half ring; 320 mm high on face; cut voussoirs PC £248.60/100					
102 mm wide exposed soffit	3.45	37.36	85.51	m	122.87
215 mm wide exposed soffit	5.15	55.83	171.23	m	227.06
Semi circular arches; one ring; 215 mm high on face					
102 mm wide exposed soffit	3.00	31.05	6.97	m	38.02
215 mm wide exposed soffit	4.25	44.47	29.20	m	73.67

F MASONRY Including overheads and profit at 12.50%	Labour hours	Labour £	Material £	Unit	Total rate £
F10 BRICK/BLOCK WALLING - cont'd					
Facing bricks; hand made; PC £551.05/1000 (unless otherwise stated); in gauged mortar (1:1:6) - cont'd					
Semi circular arches; two ring; 215 mm high on face					
102 mm wide exposed soffit	3.70	39.47	6.97	m	46.44
215 mm wide exposed soffit	5.30	57.10	29.20	m	86.30
Semi circular arches; one ring; 215 mm high on face; cut voussoirs PC £248.60/100					
102 mm wide exposed soffit	3.10	32.26	43.71	m	75.97
215 mm wide exposed soffit	4.45	46.88	102.66	m	149.54
Bullseye window 600 mm dia; two rings; 215 mm high on face					
102 mm wide exposed soffit	6.90	78.85	81.88	m	160.73
215 mm wide exposed soffit	10.35	118.37	163.56	m	281.93
Bullseye window 1200 mm dia; two rings; 215 mm high on face					
102 mm wide exposed soffit	12.05	136.66	13.42	m2	150.08
215 mm wide exposed soffit	18.15	205.90	26.84	m2	232.74
Bullseye window 600 mm dia; one ring; 215 mm high on face; cut voussoirs; PC £248.60/100					
102 mm wide exposed soffit	5.75	65.02	106.70	m	171.72
215 mm wide exposed soffit	8.60	97.32	213.32	m	310.64
Bullseye window 1200 mm dia; one ring; 215 mm high on face; cut voussoirs; PC £248.60/100					
102 mm wide exposed soffit	10.35	116.22	183.93	m	300.15
215 mm wide exposed soffit	15.55	174.63	367.88	m	542.51
ADD or DEDUCT for variation of £10.00/1000 in PC of facing bricks	-	-	-	m	0.30
Sills; horizontal; headers on edge; pointing top and one side; set weathering					
150 x 102 mm;	0.81	9.74	9.62	m	19.36
150 x 102 mm; cant headers; PC £115.50/100	0.86	10.34	19.84	m	30.18
Sills; horizontal; headers on flat; pointing top and one side					
150 x 102 mm; bullnosed specials; PC £115.50/100	0.75	9.02	14.22	m	23.24
Coping; horizontal; headers on edge; pointing top and both sides					
215 x 102 mm	0.64	7.70	17.12	m	24.82
260 x 102 mm	1.05	12.63	14.48	m	27.11
215 x 102 mm; double bullnosed specials; PC £115.50/100	0.69	8.30	19.80	m	28.10
260 x 120 mm; single bullnosed specials; PC £115.50/100	1.05	12.63	32.47	m	45.10
ADD or DEDUCT for variation of £10.00/1000 in PC of facing bricks	-	-	-	m2	0.30
50 mm facing bricks slips; PC £102.30/100; in gauged mortar (1:1:6) built up against concrete including flushing up at back (ties measured elsewhere)					
Walls	2.90	34.88	76.23	m2	111.11
Edges of suspended slabs 200 mm wide	0.86	10.34	15.35	m	25.69
Columns 400 mm wide	1.70	20.44	30.60	m	51.04
Engineering bricks; PC £350.00/1000; and specials at PC £115.50/100; in cement mortar (1:3)					
Steps; all headers-on-edge; edges set with					
215 x 102 mm; horizontal; set weathering	0.81	9.74	19.37	m	29.11
Returned ends pointed	0.23	2.77	2.89	nr	5.66
430 x 102 mm; horizontal; set weathering	1.15	13.83	25.57	m	39.40
Returned ends pointed	0.29	3.49	5.63	nr	9.12

F MASONRY Including overheads and profit at 12.50%	Labour hours	Labour £	Material £	Unit	Total rate £
Labours on brick facework					
Fair cutting					
to curve	0.23	2.77	1.86	m	4.63
Fair returns					
half brick wide	0.06	0.72	-	m	0.72
one brick wide	0.08	0.96	-	m	0.96
one and a half brick wide	0.10	1.20	-	m	1.20
Fair angles formed by cutting					
squint	0.92	11.06	-	m	11.06
birdsmouth	0.81	9.74	-	m	9.74
external; chamfered 25 mm wide	1.15	13.83	-	m	13.83
external; rounded 100 mm radius	1.40	16.84	-	m	16.84
Fair chases					
100 x 50 mm; horizontal	1.70	20.44	-	m	20.44
100 x 50 mm; vertical	1.70	20.44	-	m	20.44
300 x 50 mm; vertical	3.10	37.28	-	m	37.28
Bonding ends to existing					
half brick thick	0.75	9.02	1.63	m	10.65
one brick thick	0.98	11.79	3.25	m	15.04
Centering to brickwork soffits;					
(prices included within arches rates)					
Flat soffits; not exceeding 2 m span					
over 0.3 m wide	2.30	19.41	6.16	m2	25.57
102 mm wide	0.46	3.88	1.10	m	4.98
215 mm wide	0.69	5.82	2.20	m	8.02
Segmental soffits; 1500 mm span 200 mm rise					
102 mm wide	1.55	13.08	2.25	nr	15.33
215 mm wide	2.30	19.41	4.50	nr	23.91
Semicircular soffits; 1500 mm span					
102 mm wide	2.05	17.30	2.67	nr	19.97
215 mm wide	2.75	23.20	5.33	nr	28.53
Bullseye window; 265 mm wide					
600 mm dia	1.70	14.34	3.16	nr	17.50
1200 mm dia	3.45	29.11	5.95	nr	35.06

ALTERNATIVE BLOCK PRICES (£/m2)

	£		£		£		£
Aerated Concrete Durox 'Supablocs'; 630 x 225 mm							
75 mm	4.32	125 mm	8.11	175 mm	11.35	225 mm	14.60
90 mm	5.55	130 mm	8.44	190 mm	11.72	250 mm	16.22
100 mm	6.17	140 mm	9.08	200 mm	12.34	280 mm	18.17
115 mm	7.46	150 mm	9.73	215 mm	13.26		

	£			£			£
ARC Conbloc blocks; 450 x 225 mm							
Cream fair faced							
75 mm solid	5.61	190 mm solid		14.04	190 mm hollow		11.40
100 mm solid	6.71	100 mm hollow		6.53	215 mm hollow		12.78
140 mm solid	9.63	140 mm hollow		8.80			
Fenlite							
90 mm solid	4.38	100 mm solid		4.48	140 mm solid		7.57
Leca thermal							
100 mm Leca 6	6.95	190 mm Leca 6		13.18	125 mm Leca 45		9.37
Standard facing							
100 mm solid	6.13	190 mm solid		12.57	190 mm hollow		10.35
140 mm solid	9.06	140 mm hollow		8.45	215 mm hollow		11.79

	£		£		£		£
Celcon 'Standard' blocks; 450 x 225 mm							
75 mm	4.47	125 mm	7.46	190 mm	11.33	230 mm	13.72
90 mm	5.37	140 mm	8.35	200 mm	11.93	250 mm	14.91
100 mm	5.96	150 mm	8.95	215 mm	12.82	300 mm	17.89

F MASONRY
Including overheads and profit at 12.50%
'Solar' blocks are also available at the same price, in a limited range

Forticrete painting quality blocks; 450 x 225 mm

	£		£		£
100 mm solid	6.41	215 mm solid	13.79	190 mm hollow	10.37
140 mm solid	9.52	100 mm hollow	5.50	215 mm hollow	10.69
190 mm solid	12.72	140 mm hollow	7.63		

Lytag blocks; 450 x 225 mm
3.5 N/mm2 Thermal blocks   Insulating blocks

	£		£		£
75 mm solid	3.42	140 mm solid	6.43	100 mm cellular	7.06
90 mm solid	6.68	190 mm solid	8.71	140 mm cellular	9.09
100 mm solid	4.56				

7.0 N/mm2 Insulating extra strength blocks

100 mm solid	4.72	140 mm solid	6.67	190 mm solid	9.04

10.5 N/mm2 High strength blocks

100 mm solid	5.21	140 mm solid	7.41	190 mm solid	10.02

3.5 N/mm2 and 7.0 N/mm2 Close textured blocks

100 mm solid (3.5)	5.88	140 mm solid (3.5)	8.26	190 mm solid (3.5)	11.16
100 mm solid (7.0)	6.09	140 mm solid (7.0)	8.59	190 mm solid (7.0)	11.60

Tarmac 'Topblocks'; 450 x 225 mm
3.5 N/mm2 'Hemelite' blocks

70/75 mm solid	4.27	140 mm solid	6.59	190 mm solid	8.65
100 mm solid	4.12	150 mm solid	7.00	215 mm solid	10.04

7.0 N/mm2 'Hemelite' blocks

100 mm solid	4.85	190 mm solid	9.35	215 mm solid	10.88
140 mm solid	7.49				

'Toplite' standard blocks

	£		£		£		£
75 mm	4.21	140 mm	7.86	100 mm	5.61	215 mm	12.07
90 mm	5.06	150 mm	8.43	200 mm	11.23		

'Toplite' GTI (thermal) blocks

	£		£		£		£
115 mm	6.63	130 mm	7.50	200 mm	11.54		
125 mm	7.21	150 mm	8.65	215 mm	12.40		

Discounts of 0 - 7.5% available depending on quantity/status

	Labour hours	Labour £	Material £	Unit	Total rate £

F10 BRICK/BLOCK WALLING - cont'd

Lightweight aerated concrete blocks;
Thermalite 'Shield'/'Turbo' blocks or
similar; in gauged mortar (1:2:9)
Walls or partitions or skins of hollow walls

		Labour hours	Labour £	Material £	Unit	Total rate £
75 mm thick	PC £4.57	0.69	8.30	6.18	m2	14.48
90 mm thick	PC £5.49	0.76	9.14	7.38	m2	16.52
100 mm thick	PC £6.10	0.81	9.74	8.24	m2	17.98
115 mm thick	PC £7.39	0.85	10.22	9.92	m2	20.14
125 mm thick	PC £8.02	0.89	10.70	10.83	m2	21.53
130 mm thick	PC £8.34	0.92	11.06	11.24	m2	22.30
140 mm thick	PC £8.54	0.94	11.30	11.52	m2	22.82
150 mm thick	PC £9.16	0.98	11.79	12.29	m2	24.08
190 mm thick	PC £11.59	1.15	13.83	15.56	m2	29.39
200 mm thick	PC £12.20	1.20	14.43	16.31	m2	30.74
215 mm thick	PC £13.11	1.25	15.03	17.57	m2	32.60
255 mm thick	PC £15.54	1.40	16.84	20.76	m2	37.60

F MASONRY Including overheads and profit at 12.50%	Labour hours	Labour £	Material £	Unit	Total rate £
Isolated piers or chimney stacks					
190 mm thick	1.60	19.24	15.56	m2	34.80
215 mm thick	1.70	20.44	17.44	m2	37.88
255 mm thick	1.95	23.45	20.76	m2	44.21
Isolated casings					
75 mm thick	0.81	9.74	6.18	m2	15.92
90 mm thick	0.87	10.46	7.44	m2	17.90
100 mm thick	0.92	11.06	8.24	m2	19.30
115 mm thick	0.97	11.67	9.92	m2	21.59
125 mm thick	1.00	12.03	10.83	m2	22.86
140 mm thick	1.05	12.63	11.52	m2	24.15
Extra over for fair face; flush pointing					
walls; one side	0.11	1.32	-	m2	1.32
walls; both sides	0.20	2.41	-	m2	2.41
Bonding ends to common brickwork					
75 mm blockwork	0.17	2.04	0.50	m	2.54
90 mm blockwork	0.17	2.04	0.59	m	2.63
100 mm blockwork	0.23	2.77	0.65	m	3.42
115 mm blockwork	0.23	2.77	0.78	m	3.55
125 mm blockwork	0.26	3.13	0.88	m	4.01
130 mm blockwork	0.26	3.13	0.92	m	4.05
140 mm blockwork	0.29	3.49	0.94	m	4.43
150 mm blockwork	0.29	3.49	1.00	m	4.49
190 mm blockwork	0.35	4.21	1.28	m	5.49
200 mm blockwork	0.38	4.57	1.35	m	5.92
215 mm blockwork	0.41	4.93	1.43	m	6.36
255 mm blockwork	0.46	5.53	1.72	m	7.25
Lightweight smooth face aerated concrete **blocks; Thermalite 'Smooth Face' blocks or** **similar; in gauged mortar (1:2:9);** **flush pointing one side**					
Walls or partitions or skins of hollow walls					
100 mm thick PC £8.64	1.01	11.71	11.06	m2	22.77
140 mm thick PC £12.07	1.16	13.42	15.50	m2	28.92
150 mm thick PC £13.01	1.21	14.02	16.66	m2	30.68
190 mm thick PC £16.39	1.39	16.04	21.02	m2	37.06
200 mm thick PC £17.24	1.44	16.65	22.07	m2	38.72
215 mm thick PC £18.54	1.56	17.99	23.73	m2	41.72
Isolated piers or chimney stacks					
190 mm thick	1.84	21.46	21.02	m2	42.48
200 mm thick	1.99	23.26	22.07	m2	45.33
215 mm thick	2.21	25.81	23.73	m2	49.54
Isolated casings					
100 mm thick	1.14	13.28	11.06	m2	24.34
140 mm thick	1.26	14.62	15.50	m2	30.12
Extra over for flush pointing					
walls; both sides	0.08	0.96	-	m2	0.96
Bonding ends to common brickwork					
100 mm blockwork	0.29	3.49	0.90	m	4.39
140 mm blockwork	0.32	3.85	1.29	m	5.14
150 mm blockwork	0.35	4.21	1.38	m	5.59
190 mm blockwork	0.40	4.81	1.76	m	6.57
200 mm blockwork	0.44	5.29	1.85	m	7.14
215 mm blockwork	0.46	5.53	1.98	m	7.51
Lightweight aerated high strength concrete **blocks (7 N/mm2); Thermalite 'High Strength'** **blocks or similar; in cement mortar (1:3)**					
Walls or partitions or skins of hollow walls					
100 mm thick PC £7.89	0.86	10.10	10.14	m2	20.24
140 mm thick PC £11.03	0.98	11.50	14.20	m2	25.70
150 mm thick PC £11.82	1.04	12.22	15.19	m2	27.41
190 mm thick PC £15.00	1.22	14.34	19.30	m2	33.64
200 mm thick PC £15.78	1.27	14.94	19.53	m2	34.47
215 mm thick PC £16.96	9.35	73.54	21.79	m2	95.33

F MASONRY Including overheads and profit at 12.50%	Labour hours	Labour £	Material £	Unit	Total rate £

F10 BRICK/BLOCK WALLING - cont'd

Lightweight aerated high strength concrete blocks (7 N/mm2); Thermalite 'High Strength' blocks or similar; in cement mortar (1:3) - cont'd

Isolated piers or chimney stacks					
190 mm thick	1.62	19.15	17.61	m2	36.76
200 mm thick	1.72	20.35	20.27	m2	40.62
215 mm thick	1.98	23.43	21.74	m2	45.17
Isolated casings					
100 mm thick	0.97	11.43	10.14	m2	21.57
140 mm thick	1.11	13.06	14.20	m2	27.26
150 mm thick	1.16	13.66	15.19	m2	28.85
190 mm thick	1.32	15.54	19.30	m2	34.84
200 mm thick	1.47	17.34	20.27	m2	37.61
215 mm thick	1.53	18.02	21.79	m2	39.81
Extra over for fair face; flush pointing					
walls; one side	0.11	1.32	-	m2	1.32
walls; both sides	0.20	2.41	-	m2	2.41
Bonding ends to common brickwork					
100 mm blockwork	0.29	3.49	0.82	m	4.31
140 mm blockwork	0.32	3.85	1.19	m	5.04
150 mm blockwork	0.35	4.21	1.26	m	5.47
190 mm blockwork	0.40	4.81	1.62	m	6.43
200 mm blockwork	0.44	5.29	1.71	m	7.00
215 mm blockwork	0.46	5.53	1.82	m	7.35

Dense aggregate concrete blocks; 'ARC Conbloc' or similar; in gauged mortar (1:2:9)

Walls or partitions or skins of hollow walls					
75 mm thick; solid	PC £4.46	0.81	11.35	5.80	m2	17.15
100 mm thick; solid	PC £4.68	0.92	12.89	6.16	m2	19.05
140 mm thick; hollow	PC £6.85	1.05	14.71	8.98	m2	23.69
140 mm thick; solid	PC £7.37	1.15	16.11	9.63	m2	25.74
190 mm thick; hollow	PC £9.24	1.30	18.21	12.17	m2	30.38
215 mm thick; hollow	PC £9.79	1.45	20.31	12.90	m2	33.21
Isolated piers or chimney stacks					
140 mm thick; hollow		1.45	20.31	8.98	m2	29.29
190 mm thick; hollow		1.85	25.91	12.17	m2	38.08
215 mm thick; hollow		2.05	28.71	12.90	m2	41.61
Isolated casings					
75 mm thick; solid		0.92	12.89	5.80	m2	18.69
100 mm thick; solid		1.05	14.71	6.16	m2	20.87
140 mm thick; solid		1.25	17.51	9.63	m2	27.14
Extra over for fair face; flush pointing					
walls; one side		0.11	1.54	-	m2	1.54
walls; both sides		0.20	2.80	-	m2	2.80
Bonding ends to common brickwork					
75 mm blockwork		0.23	3.22	0.49	m	3.71
100 mm blockwork		0.30	4.20	0.51	m	4.71
140 mm blockwork		0.37	5.18	0.82	m	6.00
190 mm blockwork		0.44	6.16	1.05	m	7.21
215 mm blockwork		0.51	7.14	1.11	m	8.25

Dense aggregate concrete blocks; (7 N/mm2) Forticrete 'Leicester Common' blocks or similar; in cement mortar (1:3)

Walls or partitions or skins of hollow walls					
75 mm thick; solid	PC £4.53	0.81	11.35	5.92	m2	17.27
100 mm thick; hollow	PC £5.00	0.92	12.89	6.60	m2	19.49
100 mm thick; solid	PC £5.82	0.92	12.89	7.62	m2	20.51
140 mm thick; hollow	PC £6.91	1.05	14.71	9.13	m2	23.84
140 mm thick; solid	PC £8.66	1.15	16.11	11.30	m2	27.41
190 mm thick; hollow	PC £9.37	1.30	18.21	12.44	m2	30.65
190 mm thick; solid	PC £11.61	1.45	20.31	15.21	m2	35.52
215 mm thick; hollow	PC £9.71	1.45	20.31	12.91	m2	33.22
215 mm thick; solid	PC £12.51	1.55	21.71	16.38	m2	38.09

F MASONRY Including overheads and profit at 12.50%		Labour hours	Labour £	Material £	Unit	Total rate £
Dwarf support wall						
140 mm thick; solid		1.60	22.41	11.30	m2	33.71
190 mm thick; solid		1.85	25.91	15.21	m2	41.12
215 mm thick; solid		2.05	28.71	16.38	m2	45.09
Isolated piers or chimney stacks						
140 mm thick; hollow		1.45	20.31	9.13	m2	29.44
190 mm thick; hollow		1.85	25.91	12.44	m2	38.35
215 mm thick; hollow		2.05	28.71	12.91	m2	41.62
Isolated casings						
75 mm thick; solid		0.92	12.89	5.92	m2	18.81
100 mm thick; solid		1.05	14.71	7.62	m2	22.33
140 mm thick; solid		1.25	17.51	11.30	m2	28.81
Extra over for fair face; flush pointing						
walls; one side		0.11	1.54	-	m2	1.54
walls; both sides		0.20	2.80	-	m2	2.80
Bonding ends to common brickwork						
75 mm blockwork		0.23	3.22	0.51	m	3.73
100 mm blockwork		0.30	4.20	0.63	m	4.83
140 mm blockwork		0.37	5.18	0.96	m	6.14
190 mm blockwork		0.44	6.16	1.30	m	7.46
215 mm blockwork		0.51	7.14	1.40	m	8.54
Dense aggregate coloured concrete blocks; Forticrete 'Leicester Bathstone'; in coloured gauged mortar (1:1:6); flush pointing one side						
Walls or partitions or skins of hollow walls						
100 mm thick; hollow	PC £11.10	1.05	14.71	14.39	m2	29.10
100 mm thick; solid	PC £13.34	1.05	14.71	17.16	m2	31.87
140 mm thick; hollow	PC £14.70	1.15	16.11	19.09	m2	35.20
140 mm thick; solid	PC £19.72	1.25	17.51	25.30	m2	42.81
190 mm thick; hollow	PC £18.36	1.45	20.31	24.04	m2	44.35
190 mm thick; solid	PC £26.87	1.55	21.71	34.56	m2	56.27
215 mm thick; hollow	PC £19.84	1.55	21.71	25.93	m2	47.64
215 mm thick; solid	PC £28.69	1.65	23.11	36.89	m2	60.00
Isolated piers or chimney stacks						
140 mm thick; solid		1.70	23.81	25.30	m2	49.11
190 mm thick; solid		1.95	27.31	34.56	m2	61.87
215 mm thick; solid		2.20	30.81	37.05	m2	67.86
Extra over for flush pointing						
walls; both sides		0.08	1.12	-	m2	1.12
Extra over blocks for						
100 mm thick lintol blocks; ref D14		0.29	4.06	8.31	m	12.37
140 mm thick lintol blocks; ref H14		0.35	4.90	9.15	m	14.05
140 mm thick quoin blocks; ref H16		0.46	6.44	19.05	m	25.49
140 mm thick cavity closer blocks; ref H17		0.46	6.44	21.36	m	27.80
140 mm thick cill blocks; ref H21		0.35	4.90	13.27	m	18.17
190 mm thick lintol blocks; ref A14		0.46	6.44	10.73	m	17.17
190 mm thick cill blocks; ref A21		0.40	5.60	14.20	m	19.80

F20 NATURAL STONE RUBBLE WALLING

Cotswold Guiting limestone; laid dry	Labour hours	Labour £	Material £	Unit	Total rate £
Uncoursed random rubble walling					
275 mm thick	2.05	28.44	40.61	m2	69.05
350 mm thick	2.40	33.29	51.37	m2	84.66
425 mm thick	2.70	37.45	62.83	m2	100.28
500 mm thick	3.00	41.61	73.58	m2	115.19
Cotswold Guiting limestone; bedded; jointed and pointed in cement - lime mortar (1:2:9)					
Uncoursed random rubble walling; faced and pointed; both sides					
275 mm thick	1.95	27.05	42.46	m2	69.51
350 mm thick	2.05	28.44	53.69	m2	82.13
425 mm thick	2.20	30.52	65.61	m2	96.13
500 mm thick	2.30	31.90	77.29	m2	109.19

F MASONRY Including overheads and profit at 12.50%	Labour hours	Labour £	Material £	Unit	Total rate £
F20 NATURAL STONE RUBBLE WALLING - cont'd					
Cotswold Guiting limestone; **bedded; jointed and pointed in** **cement - lime mortar (1:2:9) - cont'd** Coursed random rubble walling; rough dressed; faced and pointed one side					
114 mm thick	1.70	20.44	52.58	m2	73.02
150 mm thick	2.00	24.05	60.31	m2	84.36
Fair returns on walling					
114 mm wide	0.03	0.36	-	m	0.36
150 mm wide	0.05	0.60	-	m	0.60
275 mm wide	0.08	0.96	-	m	0.96
350 mm wide	0.10	1.20	-	m	1.20
425 mm wide	0.13	1.56	-	m	1.56
500 mm wide	0.15	1.80	-	m	1.80
Fair raking cutting on walling					
114 mm thick	0.23	2.77	7.43	m	10.20
150 mm thick	0.28	3.37	8.53	m	11.90
Level uncoursed rubble walling for damp proof courses and the like					
275 mm wide	0.24	3.33	2.53	m	5.86
350 mm wide	0.25	3.47	2.88	m	6.35
425 mm wide	0.26	3.61	3.22	m	6.83
500 mm wide	0.28	3.88	3.91	m	7.79
Copings formed of rough stones; faced and pointed all round					
275 x 200 mm (average) high	0.59	8.18	8.77	m	16.95
350 x 250 mm (average) high	0.77	10.68	13.29	m	23.97
425 x 300 mm (average) high	0.98	13.59	18.62	m	32.21
500 x 350 mm (average) high	1.20	16.65	25.54	m	42.19
F22 CAST STONE WALLING/DRESSINGS					
Reconstructed limestone walling; 'Bradstone' **100 mm bed weathered Cotswold or** **North Cerney masonry blocks or similar;** **laid to pattern or course recommended;** **bedded, jointed and pointed in approved** **coloured cement - lime mortar (1:2:9)** Walls; facing and pointing one side					
masonry blocks; random uncoursed	1.30	15.63	20.95	m2	36.58
Extra for					
Return ends	0.46	5.53	1.35	m	6.88
Plain 'L' shaped quoins	0.15	1.80	4.63	m	6.43
traditional walling; coursed squared	1.60	19.24	20.24	m2	39.48
squared random rubble	1.60	19.24	20.68	m2	39.92
squared coursed rubble (large module)	1.50	18.04	20.50	m2	38.54
squared coursed rubble (small module)	1.55	18.64	20.84	m2	39.48
squared and pitched rock faced walling; coursed	1.65	19.84	20.24	m2	40.08
rough hewn rockfaced walling; random	1.70	20.44	20.07	m2	40.51
Extra for return ends	0.18	2.16	-	m	2.16
Isolated piers or chimney stacks; facing and pointing one side					
masonry blocks; random uncoursed	1.80	21.65	21.94	m2	43.59
traditional walling; coursed squared	2.25	27.06	21.20	m2	48.26
squared random rubble	2.25	27.06	21.66	m2	48.72
squared coursed rubble (large module)	2.05	24.65	21.48	m2	46.13
squared coursed rubble (small module)	2.20	26.46	21.83	m2	48.29
squared and pitched rock faced walling; coursed	2.35	28.26	21.20	m2	49.46
rough hewn rockfaced walling; random	2.40	28.86	21.02	m2	49.88

F MASONRY Including overheads and profit at 12.50%	Labour hours	Labour £	Material £	Unit	Total rate £
Isolated casings; facing and pointing one side					
masonry blocks; random uncoursed	1.55	18.64	21.94	m2	40.58
traditional walling; coursed squared	1.95	23.45	21.20	m2	44.65
squared random rubble	1.95	23.45	21.66	m2	45.11
squared coursed rubble (large module)	1.80	21.65	21.48	m2	43.13
squared coursed rubble (small module)	1.90	22.85	21.83	m2	44.68
squared and pitched rock faced walling; coursed	2.00	24.05	21.20	m2	45.25
rough hewn rockfaced walling; random	2.05	24.65	21.02	m2	45.67
Fair returns 100 mm wide					
masonry blocks; random uncoursed	0.14	1.68	-	m	1.68
traditional walling; coursed squared	0.17	2.04	-	m	2.04
squared random rubble	0.17	2.04	-	m	2.04
squared coursed rubble (large module)	0.16	1.92	-	m	1.92
squared coursed rubble (small module)	0.16	1.92	-	m	1.92
squared and pitched rock faced walling; coursed	0.17	2.04	-	m	2.04
rough hewn rockfaced walling; random	0.18	2.16	-	m	2.16
Fair raking cutting on masonry blocks 100 mm thick	0.21	2.53	-	m	2.53
Reconstructed limestone dressings; 'Bradstone Architectural' dressings in weathered Cotswold or North Cerney shades or similar; bedded, jointed and pointed in approved coloured cement - lime mortar (1:2:9)					
Copings; twice weathered and throated					
152 x 76 mm; type A	0.38	4.57	9.28	m	13.85
178 x 64 mm; type B	0.38	4.57	9.98	m	14.55
305 x 76 mm; type A	0.46	5.53	17.74	m	23.27
Extra for					
Fair end	-	-	-	nr	3.51
Returned mitred fair end	-	-	-	nr	3.51
Copings; once weathered and throated					
191 x 76 mm	0.38	4.57	10.81	m	15.38
305 x 76 mm	0.46	5.53	17.42	m	22.95
365 x 76 mm	0.46	5.53	18.74	m	24.27
Extra for					
Fair end	-	-	-	nr	3.51
Returned mitred fair end	-	-	-	nr	3.51
Chimney caps; four times weathered and throated; once holed					
553 x 533 x 76 mm	0.46	5.53	23.99	nr	29.52
686 x 686 x 76 mm	0.46	5.53	39.42	nr	44.95
Pier caps; four times weathered and throated					
305 x 305 mm	0.29	3.49	7.96	nr	11.45
381 x 381 mm	0.29	3.49	11.19	nr	14.68
457 x 457 mm	0.35	4.21	15.55	nr	19.76
533 x 533 mm	0.35	4.21	21.59	nr	25.80
Splayed corbels					
457 x 102 x 229 mm	0.17	2.04	10.03	nr	12.07
686 x 102 x 229 mm	0.23	2.77	15.98	nr	18.75
Air bricks					
229 x 142 x 76 mm	0.09	1.08	3.92	nr	5.00
102 x 152 mm lintels; rectangular; reinforced with mild steel bars					
not exceeding 1.22 m long	-	-	-	m	16.29
1.37 - 1.67 m long	-	-	-	m	16.59
1.83 - 1.98 m long	-	-	-	m	18.16
102 x 229 mm lintels; rectangular; reinforced with mild steel bars					
not exceeding 1.67 m long	0.30	3.61	20.33	m	23.94
1.83 - 1.98 m long	0.32	3.85	20.81	m	24.66
2.13 - 2.44 m long	0.35	4.21	20.96	m	25.17
2.59 - 2.90 m long	0.37	4.45	22.85	m	27.30

F MASONRY Including overheads and profit at 12.50%	Labour hours	Labour £	Material £	Unit	Total rate £

F22 CAST STONE WALLING/DRESSINGS - cont'd

Reconstructed limestone dressings;
'Bradstone Architectural' dressings in
weathered Cotswold or North Cerney shades
or similar; bedded, jointed and pointed in
approved coloured cement - lime
mortar (1:2:9) - cont'd

	Labour hours	Labour £	Material £	Unit	Total rate £
197 x 67 mm sills to suit standard softwood windows; stooled at ends					
0.56 - 2.50 m long	0.35	4.21	25.61	m	29.82
Window surround; traditional with label moulding; for single light; sill 146 x 133 mm; jambs 146 x 146 mm; head 146 x 105 mm; including all dowels and anchors					
window size 508 x 1479 mm PC £110.74	1.05	12.63	125.07	nr	137.70
Window surround; traditional with label moulding; three light; for windows 508 x 1219 mm; sill 146 x 133 mm; jambs 146 x 146 mm; head 146 x 103 mm; mullions 146 x 108 mm; including all dowels and anchors					
overall size 1975 x 1479 mm PC £241.19	2.70	32.47	272.16	nr	304.63
Door surround; moulded continuous jambs and head with label moulding; including all dowels and anchors					
door 839 x 1981 mm in 102 x 64 mm frame	1.90	22.85	330.44	nr	353.29

F30 ACCESSORIES/SUNDRY ITEMS FOR BRICK/BLOCK/STONE WALLING

Sundries - brick/block walling
Forming cavities in hollow walls; three wall
ties per m2

	Labour hours	Labour £	Material £	Unit	Total rate £
50 mm cavity; polypropylene ties	0.07	0.84	0.15	m2	0.99
50 mm cavity; galvanized steel butterfly wall ties	0.07	0.84	0.20	m2	1.04
50 mm cavity; galvanized steel twisted wall ties	0.07	0.84	0.27	m2	1.11
50 mm cavity; stainless steel butterfly wall ties	0.07	0.84	0.26	m2	1.10
50 mm cavity; stainless steel twisted wall ties	0.07	0.84	0.68	m2	1.52
75 mm cavity; polypropylene ties	0.07	0.84	0.15	m2	0.99
75 mm cavity; galvanised steel butterfly wall ties	0.07	0.84	0.20	m2	1.04
75 mm cavity; galvanised steel twisted wall ties	0.07	0.84	0.27	m2	1.11
75 mm cavity; stainless steel butterfly wall ties	0.07	0.84	0.26	m2	1.10
75 mm cavity; stainless steel twisted wall ties	0.07	0.84	0.68	m2	1.52
Closing at jambs with common brickwork half brick thick					
50 mm cavity	0.35	4.21	0.97	m	5.18
50 mm cavity; including damp proof course	0.46	5.53	1.53	m	7.06
75 mm cavity	0.35	4.21	1.26	m	5.47
75 mm cavity; including damp proof course	0.46	5.53	1.81	m	7.34
Closing at jambs with blockwork 100 mm thick					
50 mm cavity	0.29	3.49	0.47	m	3.96
50 mm cavity; including damp proof course	0.35	4.21	1.03	m	5.24
75 mm cavity	0.29	3.49	0.66	m	4.15
75 mm cavity; including damp proof course	0.35	4.21	1.22	m	5.43

F MASONRY Including overheads and profit at 12.50%	Labour hours	Labour £	Material £	Unit	Total rate £
Closing at sill with one course common brickwork					
50 mm cavity	0.35	4.21	2.14	m	6.35
50 mm cavity; including damp proof course	0.40	4.81	2.69	m	7.50
75 mm cavity	0.35	4.21	2.14	m	6.35
75 mm cavity; including damp proof course	0.40	4.81	2.69	m	7.50
Closing at top with					
single course of slates	0.17	2.04	2.41	m	4.45
'Westbrick' cavity closer	0.17	2.04	4.14	m	6.18
Cavity wall insulation; 'Dritherm'					
filling 50 mm cavity	0.17	2.04	2.23	m2	4.27
filling 75 mm cavity	0.20	2.41	2.96	m2	5.37
Cavity wall insulation; fixing with insulation retaining ties					
30 mm Celotex RR cavity insulation	0.23	2.77	0.19	m2	2.96
30 mm cavity wall batts	0.23	2.77	2.91	m2	5.68
25 mm Plaschem 'Aerobuild' foil faced polyurethene cavity slabs	0.23	2.77	5.83	m2	8.60
25 mm Styrofoam cavity wall insulation	0.23	2.77	3.01	m2	5.78

ALTERNATIVE DAMP PROOF COURSE PRICES (£/m2)

	£		£		£
Asbestos based					
'Astos'	4.08	'Barchester'	4.08		
Fibre based					
'Challenge'	2.48	'Stormax'	2.50		
Hessian based					
'Callendrite'	3.33	'Nubit'	3.69	'Permaseal'	3.68
Leadcored					
'Astos'	9.92	'Ledumite' grade 2	15.02	'Permalead'	10.80
'Ledkore' grade B	17.54	'Ledumite' grade 3	18.89	'Pluvex'	8.85
'Ledumite' grade 1	10.82	'Nuled'	8.76	'Trindos'	9.85
Pitch polymer					
'Aquagard'	4.58	'Permaflex'	4.58		

Further discounts may be available depending on quantity/status

	Labour hours	Labour £	Material £	Unit	Total rate £
'Pluvex' (hessian based) damp proof course or similar; PC £3.72/m2; 200 mm laps; in gauged mortar (1:1:6)					
Horizontal					
over 225 mm wide	0.29	3.49	4.52	m2	8.01
over 225 mm wide; forming cavity gutters in hollow walls	0.46	5.53	4.52	m2	10.05
not exceeding 225 mm wide	0.58	6.98	4.52	m2	11.50
Vertical					
not exceeding 225 mm wide	0.86	10.34	4.52	m2	14.86
'Pluvex' (fibre based) damp proof course or similar; PC £2.50/m2; 200 mm laps; in gauged mortar (1:1:6)					
Horizontal					
over 225 mm wide	0.29	3.49	3.03	m2	6.52
over 225 mm wide; forming cavity gutters in hollow walls	0.46	5.53	3.02	m2	8.55
not exceeding 225 mm wide	0.58	6.98	3.02	m2	10.00
Vertical					
not exceeding 225 mm wide	0.86	10.34	3.02	m2	13.36

F MASONRY Including overheads and profit at 12.50%	Labour hours	Labour £	Material £	Unit	Total rate £
F30 ACCESSORIES/SUNDRY ITEMS FOR BRICK/BLOCK/ STONE WALLING - cont'd					
'Asbex' (asbestos based) damp proof course or similar; PC £4.08/m2; 200 mm laps; in gauged mortar (1:1:6)					
Horizontal					
over 225 mm wide	0.29	3.49	4.94	m2	8.43
over 225 mm wide; forming cavity gutters					
in hollow walls	0.46	5.53	4.94	m2	10.47
not exceeding 225 mm wide	0.58	6.98	4.94	m2	11.92
Vertical					
not exceeding 225 mm wide	0.86	10.34	4.94	m2	15.28
Polyethelene damp proof course; PC £0.96/m2; 200 mm laps; in gauged mortar (1:1:6)					
Horizontal					
over 225 mm wide	0.29	3.49	1.16	m2	4.65
over 225 mm wide; forming cavity gutters					
in hollow walls	0.46	5.53	1.16	m2	6.69
not exceeding 225 mm wide	0.58	6.98	1.16	m2	8.14
Vertical					
not exceeding 225 mm wide	0.86	10.34	1.16	m2	11.50
'Permabit' bitumen polymer damp proof course or similar; PC £4.58/m2; 150 mm laps; in gauged mortar (1:1:6)					
Horizontal					
over 225 mm wide	0.29	3.49	5.56	m2	9.05
over 225 mm wide; forming cavity gutters					
in hollow walls	0.46	5.53	5.56	m2	11.09
not exceeding 225 mm wide	0.58	6.98	5.56	m2	12.54
Vertical					
not exceeding 225 mm wide	0.86	10.34	5.53	m2	15.87
'Hyload' (pitch polymer) damp proof course or similar; PC £4.43/m2; 150 mm laps; in gauged mortar (1:1:6)					
Horizontal					
over 225 mm wide	0.29	3.49	5.38	m2	8.87
over 225 mm wide; forming cavity gutters					
in hollow walls	0.46	5.53	5.38	m2	10.91
not exceeding 225 mm wide	0.58	6.98	5.38	m2	12.36
Vertical					
not exceeding 225 mm wide	0.86	10.34	5.38	m2	15.72
'Ledkore' grade A (bitumen based lead cored); damp proof course or similar; PC £12.90/m2; 200 mm laps; in gauged mortar (1:1:6)					
Horizontal					
over 225 mm wide	0.38	4.57	15.67	m2	20.24
over 225 mm wide; forming cavity gutters					
in hollow walls	0.61	7.34	15.67	m2	23.01
not exceeding 225 mm wide	0.58	6.98	15.67	m2	22.65
Vertical					
not exceeding 225 mm wide	0.86	10.34	15.60	m2	25.94
Milled lead damp proof course; PC £23.29/m2; BS 1178; 1.80 mm (code 4), 175 mm laps; in cement-lime mortar (1:2:9)					
Horizontal					
over 225 mm wide	2.30	27.66	28.30	m2	55.96
not exceeding 225 mm wide	3.45	41.49	28.30	m2	69.79

F MASONRY Including overheads and profit at 12.50%	Labour hours	Labour £	Material £	Unit	Total rate £
Two courses slates in cement mortar (1:3)					
Horizontal					
over 225 mm wide	1.70	20.44	19.90	m2	40.34
Vertical					
over 225 mm wide	2.60	31.27	20.37	m2	51.64
Silicone injection damp-proofing					
Horizontal; 450 mm centres; make good					
brickwork					
half brick thick	-	-	-	m	10.13
one brick thick	-	-	-	m	11.25
cavity wall comprising two half brick skins	-	-	-	m	14.63
one and a half brick thick	-	-	-	m	16.88
'Synthaprufe' damp proof membrane; PC £28.61/25L;					
three coats brushed on					
Vertical					
not exceeding 150 mm wide	0.09	1.08	0.51	m2	1.59
150 - 225 mm wide	0.11	1.32	0.77	m2	2.09
225 - 300 mm wide	0.15	1.80	1.01	m2	2.81
over 300 mm wide	0.32	3.85	3.38	m2	7.23
'Type X' polypropylene abutment cavity tray					
by Cavity Trays Ltd; built into facing					
brickwork as the work proceeds; complete					
with Code 4 lead flashing					
Intermediate tray with short leads					
(requiring soakers); to suit roof of					
17 - 20 degrees pitch	0.07	0.84	7.67	nr	8.51
21 - 25 degrees pitch	0.07	0.84	10.41	nr	11.25
26 - 45 degrees pitch	0.07	0.84	5.91	nr	6.75
Intermediate tray with long leads (suitable					
only for corrugated roof tiles); to suit					
roof of					
17 - 20 degrees pitch	0.07	0.84	5.44	nr	6.28
21 - 25 degrees pitch	0.07	0.84	5.02	nr	5.86
26 - 45 degrees pitch	0.07	0.84	4.74	nr	5.58
Extra for					
ridge tray	0.07	0.84	4.40	nr	5.24
catchment end tray	0.07	0.84	3.88	nr	4.72
Servicised 'Bitu-thene' self-adhesive cavity					
flashing; type 'CA'; well lapped at joints;					
in gauged mortar (1:1:6)					
Horizontal					
over 225 mm wide	0.98	11.79	0.61	m2	12.40
'Brickforce' galvanised steel joint					
reinforcement					
In walls					
40 mm wide; ref. GBF 40	0.03	0.36	0.19	m	0.55
60 mm wide; ref. GBF 60	0.04	0.48	0.19	m	0.67
100 mm wide; ref. GBF 100	0.05	0.60	0.27	m	0.87
160 mm wide; ref. GBF 160	0.06	0.72	0.37	m	1.09
'Brickforce' stainless steel joint					
reinforcement					
In walls					
40 mm wide; ref. SBF 40	0.03	0.36	1.91	m	2.27
60 mm wide; ref. SBF 60	0.04	0.48	2.09	m	2.57
100 mm wide; ref. SBF 100	0.05	0.60	2.14	m	2.74
160 mm wide; ref. SBF 160	0.06	0.72	2.23	m	2.95

F MASONRY Including overheads and profit at 12.50%	Labour hours	Labour £	Material £	Unit	Total rate £

F30 ACCESSORIES/SUNDRY ITEMS FOR BRICK/BLOCK/STONE WALLING - cont'd

Fillets/pointing/wedging and pinning etc.
Weather or angle fillets in cement
mortar (1:3)

50 mm face width	0.14	1.68	0.08	m	1.76
100 mm face width	0.23	2.77	0.32	m	3.09
Pointing wood frames or sills with mastic					
one side	0.08	0.96	0.46	m	1.42
each side	0.16	1.92	0.93	m	2.85
Pointing wood frames or sills with polysulphide sealant					
one side	0.08	0.96	0.93	m	1.89
each side	0.16	1.92	1.85	m	3.77
Bedding wood plates in cement mortar (1:3)					
100 mm wide	0.07	0.84	0.05	m	0.89
Bedding wood frame in cement mortar (1:3) and point					
one side	0.09	1.08	0.05	m	1.13
each side	0.11	1.32	0.07	m	1.39
one side in mortar; other side in mastic	0.17	2.04	0.51	m	2.55
Wedging and pinning up to underside of existing construction with slates in cement mortar (1:3)					
102 mm wall	0.92	11.06	2.50	m	13.56
215 mm wall	1.15	13.83	4.60	m	18.43
327 mm wall	1.40	16.84	6.70	m	23.54
Raking out joint in brickwork or blockwork for turned-in edge of flashing					
horizontal	0.17	2.04	0.05	m	2.09
stepped	0.23	2.77	0.05	m	2.82
Raking out and enlarging joint in brickwork or blockwork for nib of asphalt					
horizontal	0.23	2.77	-	m	2.77
Cutting grooves in brickwork or blockwork					
for water bars and the like	0.35	4.21	-	m	4.21
for nib of asphalt; horizontal	0.35	4.21	-	m	4.21
Preparing to receive new walls					
top existing 215 mm brick wall	0.23	2.77	-	m	2.77

Sills and tile creasings

Sills; two courses of machine made plain roofing tiles; set weathering; bedded and pointed	0.69	8.30	3.54	m	11.84
Extra over brickwork for two courses of machine made tile creasing; bedded and pointed; projecting 25 mm each side; horizontal					
215 mm wide copings	0.58	6.98	5.35	m	12.33
260 mm wide copings	0.81	9.74	6.87	m	16.61
Galvanised steel coping cramp; built in	0.11	1.32	0.27	nr	1.59

Flue linings etc

Flue linings; True Flue 200 mm refractory concrete square flue linings; rebated joints in refractory mortar (1:2.5)					
linings	0.21	2.53	11.63	m	14.16
offset unit; ref 5u	0.11	1.32	8.71	nr	10.03
offset unit; ref 6u	0.11	1.32	7.83	nr	9.15
off set unit; ref 7u	0.11	1.32	7.31	nr	8.63
pot; ref 8u	0.11	1.32	4.87	nr	6.19
single cap unit; ref 10u	0.29	3.49	19.58	nr	23.07
double cap unit; ref 11u	0.35	4.21	17.55	nr	21.76
U-type lintol with U1 attachment	0.35	4.21	20.69	nr	24.90

F MASONRY Including overheads and profit at 12.50%	Labour hours	Labour £	Material £	Unit	Total rate £
Gas flue linings; True Flue 'Typex 115' **refractory concrete blocks, built in; in** **refractory mortar (1:2.5); cutting brickwork** **or blockwork around**					
recess panel; ref PXY 218	0.11	1.32	3.40	nr	4.72
lintol unit; ref 1Y	0.11	1.32	4.83	nr	6.15
entry and exit unit; ref 2Y	0.11	1.32	6.31	nr	7.63
straight block; ref 3YA	0.11	1.32	3.07	nr	4.39
straight block; ref 3YB	0.11	1.32	3.18	nr	4.50
offset unit; ref 4Y	0.11	1.32	3.81	nr	5.13
reverse rebate unit; ref 5Y	0.11	1.32	3.13	nr	4.45
conversion corbel unit; ref 6Y	0.11	1.32	5.94	nr	7.26
vertical conversion unit; ref 7Y	0.11	1.32	6.27	nr	7.59
back offset unit; ref 8Y	0.11	1.32	9.36	nr	10.68
back offset unit; ref 9Y	0.11	1.32	8.71	nr	10.03
Gas flue linings; True Flue 'Typex 150' **refractory concrete blocks built in; in** **refractory mortar (1:2.5); cutting brickwork** **and or blockwork around**					
recess panel; ref RPX 305	0.11	1.32	4.77	nr	6.09
offset lintol; ref 1X1	0.11	1.32	5.34	nr	6.66
twin channel flue block; ref 2X1	0.11	1.32	4.89	nr	6.21
side offset block; ref 4X1	0.11	1.32	3.89	nr	5.21
back offset block; ref 4X4	0.11	1.32	3.68	nr	5.00
single flue block; ref 5X	0.11	1.32	2.55	nr	3.87
single flue block; ref 5X1	0.11	1.32	3.08	nr	4.40
single flue block; ref 5X2	0.11	1.32	3.13	nr	4.45
single flue block; ref 5X3/5/8	0.11	1.32	3.11	nr	4.43
separating withe; ref 8X	0.11	1.32	1.92	nr	3.24
separating withe; ref 8X4	0.11	1.32	2.82	nr	4.14
end terminal cap unit; ref 9X	0.29	3.49	6.09	nr	9.58
inter terminal cap unit; ref 9X1	0.29	3.49	4.28	nr	7.77
end terminal cap unit; ref 9X6	0.29	3.49	6.32	nr	9.81
inter terminal cap unit; ref 9X7	0.29	3.49	5.15	nr	8.64
single flue cap unit; ref 9X9	0.29	3.49	11.19	nr	14.68
conversion unit; ref 10X1	0.11	1.32	3.53	nr	4.85
angled conversion unit; ref 10X2	0.11	1.32	5.28	nr	6.60
aluminium louvre; ref MLX	0.11	1.32	6.35	nr	7.67
Parging and coring flues with refractory **mortar (1:2.5); sectional area not exceeding** **0.25 m2**					
900 x 900 mm	0.69	8.30	5.72	m	14.02
Ancillaries					
Forming openings; one ring arch over					
225 x 225 mm; one brick facing brickwork making good facings both sides	1.70	20.44	-	nr	20.44
Forming opening through hollow wall; slate **lintel over; sealing 50 cavity with slates** **in cement mortar (1:3); making good fair** **face or facings one side**					
225 x 75 mm	0.23	2.77	1.74	nr	4.51
225 x 150 mm	0.29	3.49	2.55	nr	6.04
225 x 225 mm	0.35	4.21	3.38	nr	7.59
Air bricks; red terracotta; building into **prepared openings**					
215 x 65 mm	0.09	1.08	1.04	nr	2.12
215 x 140 mm	0.09	1.08	1.70	nr	2.78
215 x 215 mm	0.09	1.08	3.62	nr	4.70
Air bricks; cast iron; building into **prepared openings**					
215 x 65 mm	0.09	1.08	3.05	nr	4.13
215 x 140 mm	0.09	1.08	5.54	nr	6.62
215 x 215 mm	0.09	1.08	7.31	nr	8.39

F MASONRY Including overheads and profit at 12.50%		Labour hours	Labour £	Material £	Unit	Total rate £
Bearers; mild steel						
51 x 10 mm flat bar		0.29	3.09	3.32	m	6.41
64 x 13 mm flat bar		0.29	3.09	5.31	m	8.40
50 x 50 x 6 mm angle section		0.40	4.26	2.70	m	6.96
Ends fanged for building in		-	-	-	nr	0.88
Proprietary items						
Ties in walls; 150 mm long butterfly type; building into joints of brickwork or blockwork						
galvanized steel or polypropylene		0.03	0.36	0.08	nr	0.44
stainless steel		0.03	0.36	0.11	nr	0.47
copper		0.03	0.36	0.15	nr	0.51
Ties in walls; 20 x 3 x 150 mm long twisted wall type; building into joints of brickwork or blockwork						
galvanized steel		0.05	0.60	0.11	nr	0.71
stainless steel		0.05	0.60	0.28	nr	0.88
copper		0.05	0.60	0.39	nr	0.99
Anchors in walls; 25 x 3 x 100 mm long; one end dovetailed; other end building into joints of brickwork or blockwork						
galvanized steel		0.07	0.84	0.25	nr	1.09
stainless steel		0.07	0.84	0.40	nr	1.24
copper		0.07	0.84	0.62	nr	1.46
Fixing cramp 25 x 3 x 250 mm long; once bent; fixed to back of frame; other end building into joints of brickwork or blockwork						
galvanized steel		0.07	0.84	0.13	nr	0.97
Chimney pots; red terracotta; plain or cannon-head; setting and flaunching in cement mortar (1:3)						
185 mm dia x 300 mm long	PC £12.75	2.05	24.65	15.23	nr	39.88
185 mm dia x 600 mm long	PC £22.00	2.30	27.66	25.90	nr	53.56
185 mm dia x 900 mm long	PC £29.25	2.55	30.67	34.25	nr	64.92
Galvanized steel lintels; Catnic or similar; built into brickwork or blockwork 'CN7' combined lintel; 143 mm high; for standard cavity walls						
750 mm long	PC £11.20	0.28	2.98	13.23	nr	16.21
900 mm long	PC £13.48	0.33	3.52	15.92	nr	19.44
1200 mm long	PC £17.91	0.39	4.15	21.16	nr	25.31
1500 mm long	PC £23.27	0.44	4.69	27.48	nr	32.17
1800 mm long	PC £28.49	0.50	5.33	33.65	nr	38.98
2100 mm long	PC £33.22	0.55	5.86	39.24	nr	45.10
'CN8' combined lintel; 219 mm high; for standard cavity walls						
2400 mm long	PC £45.11	0.66	7.03	53.29	nr	60.32
2700 mm long	PC £50.89	0.77	8.20	60.11	nr	68.31
3000 mm long	PC £62.06	0.88	9.38	73.31	nr	82.69
3300 mm long	PC £69.08	0.99	10.55	81.60	nr	92.15
3600 mm long	PC £75.66	1.10	11.72	89.37	nr	101.09
3900 mm long	PC £100.75	1.20	12.78	119.01	nr	131.79
4200 mm long	PC £108.46	1.30	13.85	128.12	nr	141.97
'CN92' single lintel; for 75 mm internal walls						
900 mm long	PC £2.00	0.33	3.52	2.37	nr	5.89
1050 mm long	PC £2.36	0.33	3.52	2.79	nr	6.31
1200 mm long	PC £2.65	0.39	4.15	3.13	nr	7.28
'CN102' single lintel; for 100 mm internal walls						
900 mm long	PC £2.47	0.33	3.52	2.91	nr	6.43
1050 mm long	PC £2.88	0.33	3.52	3.40	nr	6.92
1200 mm long	PC £3.14	0.39	4.15	3.72	nr	7.87

F MASONRY Including overheads and profit at 12.50%		Labour hours	Labour £	Material £	Unit	Total rate £

**F31 PRECAST CONCRETE SILLS/LINTELS/COPINGS/
FEATURES**

Mix 21.00 N/mm2 - 20 mm aggregate (1:2:4)
Lintels; plate; prestressed; bedded

100 x 65 x 750 mm	PC £3.00	0.23	2.77	3.90	nr	6.67
100 x 65 x 900 mm	PC £3.20	0.23	2.77	4.13	nr	6.90
100 x 65 x 1050 mm	PC £3.60	0.23	2.77	4.58	nr	7.35
100 x 65 x 1200 mm	PC £4.20	0.23	2.77	5.25	nr	8.02
150 x 65 x 900 mm	PC £4.30	0.29	3.49	5.36	nr	8.85
150 x 65 x 1050 mm	PC £4.70	0.29	3.49	5.81	nr	9.30
150 x 65 x 1200 mm	PC £5.70	0.29	3.49	6.94	nr	10.43
220 x 65 x 900 mm	PC £6.50	0.35	4.21	7.84	nr	12.05
220 x 65 x 1200 mm	PC £8.60	0.35	4.21	10.20	nr	14.41
220 x 65 x 1500 mm	PC £11.00	0.40	4.81	12.90	nr	17.71
265 x 65 x 900 mm	PC £7.50	0.35	4.21	8.96	nr	13.17
265 x 65 x 1200 mm	PC £9.90	0.35	4.21	11.66	nr	15.87
265 x 65 x 1500 mm	PC £12.35	0.40	4.81	14.42	nr	19.23
265 x 65 x 1800 mm	PC £14.83	0.46	5.53	17.21	nr	22.74

Lintels; rectangular; reinforced with mild
steel bars; bedded

100 x 150 x 900 mm	PC £8.00	0.35	4.21	9.53	nr	13.74
100 x 150 x 1050 mm	PC £9.00	0.35	4.21	10.65	nr	14.86
100 x 150 x 1200 mm	PC £10.70	0.35	4.21	12.56	nr	16.77
225 x 150 x 1200 mm	PC £13.50	0.46	5.53	15.71	nr	21.24
225 x 225 x 1800 mm	PC £25.00	0.86	10.34	28.65	nr	38.99

Lintels; boot; reinforced with mild steel
bars; bedded

250 x 225 x 1200 mm	PC £20.00	0.69	8.30	23.03	nr	31.33
275 x 225 x 1800 mm	PC £27.00	1.05	12.63	31.43	nr	44.06

Padstones

300 x 100 x 75 mm	PC £2.00	0.17	2.04	2.78	nr	4.82
225 x 225 x 150 mm	PC £3.00	0.23	2.77	3.90	nr	6.67
450 x 450 x 150 mm	PC £8.00	0.35	4.21	9.53	nr	13.74

Mix 31.00 N/mm2 - 20 mm aggregate (1:1:2)
Copings; once weathered; once throated
bedded
and pointed

152 x 76 mm	PC £9.79	0.38	4.57	11.54	m	16.11
178 x 64 mm	PC £9.79	0.38	4.57	11.54	m	16.11
305 x 76 mm	PC £12.69	0.46	5.53	14.80	m	20.33
Extra for						
fair end	PC £3.57	-	-	4.02	nr	4.02
angles	PC £7.80	-	-	8.77	nr	8.77

Copings; twice weathered; twice throated;
bedded and pointed

152 x 76 mm	PC £9.05	0.38	4.57	10.71	m	15.28
178 x 64 mm	PC £9.05	0.38	4.57	10.71	m	15.28
305 x 76 mm	PC £11.95	0.46	5.53	13.97	m	19.50
Extra for						
fair end	PC £3.57	-	-	4.02	nr	4.02
angles	PC £7.80	-	-	8.77	nr	8.77

G STRUCTURAL/CARCASSING METAL/TIMBER Including overheads and profit at 12.50%	Labour hours	Labour £	Material £	Unit	Total rate £
G10 STRUCTURAL STEEL FRAMING					
Fabricated steelwork; BS 4360; grade 50					
Single beams; universal beams, joists or					
channels PC £415.75	-	-	1060.00	t	1060.00
Single beams; castellated universal beams	-	-	-	t	2574.00
Single beams; rectangular hollow sections	-	-	-	t	1225.00
Single cranked beams; rectangular hollow					
sections	-	-	-	t	1487.50
Built-up beams; universal beams, joists					
or channels and plates PC £415.75	-	-	1533.00	t	1533.00
Latticed beams; angle sections PC £312.15	-	-	1223.00	t	1223.00
Latticed beams; circular hollow sections	-	-	-	t	1376.55
Single columns; universal beams, joists					
or channels PC £415.75	-	-	1000.00	t	1000.00
Single columns; circular hollow sections					
PC £1026.63	-	-	1935.00	t	1935.00
Single columns; rectangular and square					
hollow sections PC £913.75	-	-	1800.00	t	1800.00
Built-up columns; universal beams, joists					
or channels and plates PC £415.75	-	-	1590.00	t	1590.00
Roof trusses; angle sections PC £312.15	-	-	1490.00	t	1490.00
Roof trusses; circular hollow sections					
PC £466.00	-	-	1650.00	t	1650.00
Bolted fittings; other than in connections;					
consisting of cleats, brackets etc.	-	-	-	t	1682.00
Welded fittings; other than in connections;					
consisting of cleats, brackets, etc.	-	-	-	t	1520.00
Erection of fabricated steelwork					
Erection of steelwork on site	-	-	-	t	132.00
Wedging					
stanchion bases	-	-	-	nr	5.25
Holes for other trades; made on site					
16 mm dia; 6 mm thick metal	-	-	-	nr	3.17
16 mm dia; 10 mm thick metal	-	-	-	nr	4.44
16 mm dia; 20 mm thick metal	-	-	-	nr	6.23
Anchorage for stanchion base; including					
plate; holding down bolts; nuts and washers;					
for stanchion size					
152 x 152 mm	-	-	-	nr	9.80
254 x 254 mm	-	-	-	nr	14.65
356 x 406 mm	-	-	-	nr	21.90
Surface treatments off site					
On steelwork; general surfaces					
galvanizing	-	-	-	t	517.50
shot blasting	-	-	-	m2	0.84
grit blast and one coat zinc chromate primer	-	-	-	m2	2.26
touch up primer and one coat of two pack					
epoxy zinc phosphate primer	-	-	-	m2	2.20
G12 ISOLATED STRUCTURAL METAL MEMBERS					
Unfabricated steelwork; BS 4360; grade 43A					
Single beams; joists or channels PC £415.75	-	-	890.00	t	890.00
Erection of steelwork on site	-	-	-	t	136.00
G10 STRUCTURAL STEEL FRAMING					
Unfabricated steelwork; BS 4360; grade 50					
inside existing buildings					
Single beams; joists or channels	-	-	-	t	1035.00
Erection of steelwork within existing					
buildings	-	-	-	t	787.50

G STRUCTURAL/CARCASSING METAL/TIMBER Including overheads and profit at 12.50%	Labour hours	Labour £	Material £	Unit	Total rate £

G12 ISOLATED STRUCTURAL METAL MEMBERS

Naylor open web steel lattice beams; in
single members; one coat red lead primer at
works; raised 3.50 m above ground; ends
built in
Lattice beams (safe distributed loads in
brackets)

	Labour hours	Labour £	Material £	Unit	Total rate £
200 mm deep; to span 5 m (s.d.1. 444 kg/m): ref N200A	1.30	13.85	167.65	nr	181.50
250 mm deep; to span 6 m (s.d.1. 526 kg/m); ref N250B	1.55	16.51	209.72	nr	226.23
300 mm deep; to span 7 m (s.d.1. 466 kg/m); ref N300B	1.75	18.64	246.84	nr	265.48
350 mm deep; to span 8 m (s.d.1. 611 kg/m); ref N350C	2.00	21.31	301.75	nr	323.06
400 mm deep; to span 9 m (s.d.1. 552 kg/m); ref N400C	2.20	23.44	346.56	nr	370.00
450 mm deep; to span 10 m (s.d.1. 728 kg/m); ref N450D	3.30	35.16	426.42	nr	461.58

BASIC TIMBER PRICES

	£		£		£
Hardwood; fair average (£/m3 ex-wharf)					
African Walnut	618.13	Brazil. Mahogany	682.63	Sapele	510.63
Afrormosia	758.95	European Oak	1252.38	Teak	1505.00
Agba	516.00	Iroko	379.48	Utile	602.00
Beech	537.50	Obeche	322.50	W.A.Mahogany	531.05

Softwood
  Carcassing quality (£/m3)

2-4.8 m lengths	212.00	G.S.grade	12.90
4.8-6 m lengths	201.40	S.S.grade	25.80
6-9 m lenths	222.60		

  Joinery quality - £333.25/m3
  'Treatment'(£/m3)
  Pre-treatment of timber by vacuum/pressure
  impregnation, excluding transport costs and
  any subsequent seasoning:-

interior work; min. salt ret. 4 kg/m3	32.25
exterior work; min. salt ret. 5.30 kg/m3	37.63

  Pre-treatment of timber including flame
  proofing

all purposes; min. salt ret. 36 kg/m3	114.00

	Labour hours	Labour £	Material £	Unit	Total rate £

G20 CARPENTRY/TIMBER FRAMING/FIRST FIXING

Sawn softwood; untreated
Floor members

	Labour hours	Labour £	Material £	Unit	Total rate £
38 x 75 mm	0.13	1.22	0.78	m	2.00
38 x 100 mm	0.13	1.22	1.00	m	2.22
38 x 125 mm	0.14	1.31	1.23	m	2.54
38 x 150 mm	0.15	1.40	1.45	m	2.85
50 x 75 mm	0.13	1.22	0.88	m	2.10
50 x 100 mm	0.15	1.40	1.15	m	2.55
50 x 125 mm	0.15	1.40	1.43	m	2.83
50 x 150 mm	0.17	1.59	1.71	m	3.30
50 x 175 mm	0.17	1.59	1.98	m	3.57

G STRUCTURAL/CARCASSING METAL/TIMBER Including overheads and profit at 12.50%	Labour hours	Labour £	Material £	Unit	Total rate £
G20 CARPENTRY/TIMBER FRAMING/FIRST FIXING - cont'd					
Sawn softwood; untreated - cont'd					
Floor members					
50 x 200 mm	0.18	1.68	2.36	m	4.04
50 x 225 mm	0.18	1.68	2.72	m	4.40
50 x 250 mm	0.19	1.78	3.20	m	4.98
75 x 125 mm	0.18	1.68	2.25	m	3.93
75 x 150 mm	0.18	1.68	2.65	m	4.33
75 x 175 mm	0.18	1.68	3.18	m	4.86
75 x 200 mm	0.19	1.78	3.72	m	5.50
75 x 225 mm	0.19	1.78	4.24	m	6.02
75 x 250 mm	0.20	1.87	4.80	m	6.67
100 x 150 mm	0.24	2.24	4.07	m	6.31
100 x 200 mm	0.25	2.34	5.42	m	7.76
100 x 250 mm	0.28	2.62	6.76	m	9.38
100 x 300 mm	0.30	2.80	8.13	m	10.93
Wall or partition members					
25 x 25 mm	0.08	0.75	0.21	m	0.96
25 x 38 mm	0.08	0.75	0.30	m	1.05
25 x 75 mm	0.10	0.93	0.54	m	1.47
38 x 38 mm	0.10	0.93	0.45	m	1.38
38 x 50 mm	0.10	0.93	0.53	m	1.46
38 x 75 mm	0.13	1.22	0.78	m	2.00
38 x 100 mm	0.17	1.59	1.00	m	2.59
50 x 50 mm	0.13	1.22	0.66	m	1.88
50 x 75 mm	0.17	1.59	0.89	m	2.48
50 x 100 mm	0.20	1.87	1.17	m	3.04
50 x 125 mm	0.21	1.96	1.45	m	3.41
75 x 75 mm	0.20	1.87	1.43	m	3.30
75 x 100 mm	0.23	2.15	1.97	m	4.12
100 x 100 mm	0.23	2.15	2.71	m	4.86
Flat roof members					
38 x 75 mm	0.15	1.40	0.80	m	2.20
38 x 100 mm	0.15	1.40	1.00	m	2.40
38 x 125 mm	0.15	1.40	1.23	m	2.63
38 x 150 mm	0.15	1.40	1.45	m	2.85
50 x 100 mm	0.15	1.40	1.15	m	2.55
50 x 125 mm	0.15	1.40	1.43	m	2.83
50 x 150 mm	0.17	1.59	1.71	m	3.30
50 x 175 mm	0.17	1.59	1.98	m	3.57
50 x 200 mm	0.18	1.68	2.36	m	4.04
50 x 225 mm	0.18	1.68	2.72	m	4.40
50 x 250 mm	0.19	1.78	3.20	m	4.98
75 x 150 mm	0.18	1.68	2.64	m	4.32
75 x 175 mm	0.18	1.68	3.17	m	4.85
75 x 200 mm	0.19	1.78	3.72	m	5.50
75 x 225 mm	0.19	1.78	4.24	m	6.02
75 x 250 mm	0.20	1.87	4.80	m	6.67
Pitched roof members					
25 x 100 mm	0.13	1.22	0.66	m	1.88
25 x 125 mm	0.13	1.22	0.80	m	2.02
25 x 150 mm	0.17	1.59	0.96	m	2.55
25 x 150 mm; notching over trussed rafters	0.33	3.09	0.96	m	4.05
25 x 175 mm	0.17	1.59	1.15	m	2.74
25 x 175 mm; notching over trussed rafters	0.33	3.09	1.15	m	4.24
25 x 200 mm	0.20	1.87	1.35	m	3.22
32 x 150 mm; notching over trussed rafters	0.36	3.37	1.36	m	4.73
32 x 175 mm; notching over trussed rafters	0.36	3.37	1.62	m	4.99
32 x 200 mm; notching over trussed rafters	0.36	3.37	1.89	m	5.26
38 x 100 mm	0.17	1.59	1.00	m	2.59
38 x 125 mm	0.17	1.59	1.23	m	2.82
38 x 150 mm	0.17	1.59	1.45	m	3.04
50 x 50 mm	0.13	1.22	0.65	m	1.87
50 x 75 mm	0.17	1.59	0.88	m	2.47

G STRUCTURAL/CARCASSING METAL/TIMBER Including overheads and profit at 12.50%	Labour hours	Labour £	Material £	Unit	Total rate £
50 x 100 mm	0.20	1.87	1.15	m	3.02
50 x 125 mm	0.20	1.87	1.43	m	3.30
50 x 150 mm	0.23	2.15	1.71	m	3.86
50 x 175 mm	0.23	2.15	1.98	m	4.13
50 x 200 mm	0.23	2.15	2.36	m	4.51
50 x 225 mm	0.23	2.15	2.72	m	4.87
75 x 100 mm	0.28	2.62	1.95	m	4.57
75 x 125 mm	0.28	2.62	2.25	m	4.87
75 x 150 mm	0.28	2.62	2.65	m	5.27
100 x 150 mm	0.33	3.09	4.09	m	7.18
100 x 175 mm	0.33	3.09	4.76	m	7.85
100 x 200 mm	0.33	3.09	5.44	m	8.53
100 x 225 mm	0.36	3.37	6.10	m	9.47
100 x 250 mm	0.36	3.37	6.76	m	10.13
Kerbs, bearers and the like					
19 x 100 mm	0.04	0.37	0.53	m	0.90
19 x 125 mm	0.04	0.37	0.66	m	1.03
19 x 150 mm	0.04	0.37	0.78	m	1.15
25 x 75 mm	0.06	0.56	0.54	m	1.10
25 x 100 mm	0.06	0.56	0.66	m	1.22
38 x 75 mm	0.07	0.65	0.78	m	1.43
38 x 100 mm	0.07	0.65	1.00	m	1.65
50 x 75 mm	0.07	0.65	0.88	m	1.53
50 x 100 mm	0.08	0.75	1.15	m	1.90
75 x 100 mm	0.08	0.75	1.94	m	2.69
75 x 125 mm	0.09	0.84	2.24	m	3.08
75 x 150 mm	0.09	0.84	2.64	m	3.48
75 x 150 mm; splayed and rounded	0.11	1.03	4.62	m	5.65
Kerbs, bearers and the like; fixing by bolting					
19 x 100 mm	0.09	0.84	0.53	m	1.37
19 x 125 mm	0.09	0.84	0.66	m	1.50
19 x 150 mm	0.09	0.84	0.78	m	1.62
25 x 75 mm	0.11	1.03	0.54	m	1.57
25 x 100 mm	0.11	1.03	0.66	m	1.69
38 x 75 mm	0.13	1.22	0.78	m	2.00
38 x 100 mm	0.13	1.22	1.00	m	2.22
50 x 75 mm	0.13	1.22	0.88	m	2.10
50 x 100 mm	0.15	1.40	1.15	m	2.55
75 x 100 mm	0.15	1.40	1.95	m	3.35
75 x 125 mm	0.18	1.68	2.25	m	3.93
75 x 150 mm	0.18	1.68	2.64	m	4.32
Herringbone strutting 50 x 50 mm					
to 150 mm deep joists	0.55	5.14	1.52	m	6.66
to 175 mm deep joists	0.55	5.14	1.56	m	6.70
to 200 mm deep joists	0.55	5.14	1.59	m	6.73
to 225 mm deep joists	0.55	5.14	1.62	m	6.76
to 250 mm deep joists	0.55	5.14	1.65	m	6.79
Solid strutting to joists					
50 x 150 mm	0.33	3.09	1.96	m	5.05
50 x 175 mm	0.33	3.09	2.25	m	5.34
50 x 200 mm	0.33	3.09	2.63	m	5.72
50 x 225 mm	0.33	3.09	3.01	m	6.10
50 x 250 mm	0.33	3.09	3.51	m	6.60
Cleats					
225 x 100 x 75 mm	0.22	2.06	0.68	nr	2.74
Sprockets					
50 x 50 x 200 mm	0.17	1.59	0.68	nr	2.27
Extra for stress grading to above timbers					
general structural (GS) grade	-	-	-	m3	14.51
special structural (SS) grade	-	-	-	m3	29.03
Extra for protecting and flameproofing timber with 'Celcure F' protection					
small sections	-	-	-	m3	135.00
large sections	-	-	-	m3	128.25

G STRUCTURAL/CARCASSING METAL/TIMBER Including overheads and profit at 12.50%	Labour hours	Labour £	Material £	Unit	Total rate £
G20 CARPENTRY/TIMBER FRAMING/FIRST FIXING - cont'd					
Sawn softwood; untreated - cont'd					
Wrought faces					
generally	0.33	3.09	-	m2	3.09
50 mm wide	0.04	0.37	-	m	0.37
75 mm wide	0.06	0.56	-	m	0.56
100 mm wide	0.07	0.65	-	m	0.65
Raking cutting					
50 mm thick	0.22	2.06	0.53	m	2.59
75 mm thick	0.28	2.62	0.80	m	3.42
100 mm thick	0.33	3.09	1.06	m	4.15
Curved cutting					
50 mm thick	0.28	2.62	0.74	m	3.36
Scribing					
50 mm thick	0.33	3.09	-	m	3.09
Notching and fitting ends to metal	0.13	1.22	-	nr	1.22
Trimming to openings					
760 x 760 mm; joists 38 x 150 mm	2.40	22.44	0.85	nr	23.29
760 x 760 mm; joists 50 x 175 mm	2.55	23.84	1.48	nr	25.32
1500 x 500 mm; joists 38 x 200 mm	2.75	25.71	1.69	nr	27.40
1500 x 500 mm; joists 50 x 200 mm	2.75	25.71	2.12	nr	27.83
3000 x 1000 mm; joists 50 x 225 mm	4.95	46.28	4.66	nr	50.94
3000 x 1000 mm; joists 75 x 225 mm	5.50	51.42	6.98	nr	58.40
Sawn softwood; 'Tanalised'					
Floor members					
38 x 75 mm	0.13	1.22	0.97	m	2.19
38 x 100 mm	0.13	1.22	1.24	m	2.46
38 x 125 mm	0.14	1.31	1.53	m	2.84
38 x 150 mm	0.15	1.40	1.82	m	3.22
50 x 75 mm	0.13	1.22	1.08	m	2.30
50 x 100 mm	0.15	1.40	1.43	m	2.83
50 x 125 mm	0.15	1.40	1.78	m	3.18
50 x 150 mm	0.17	1.59	2.13	m	3.72
50 x 175 mm	0.17	1.59	2.48	m	4.07
50 x 200 mm	0.18	1.68	2.92	m	4.60
50 x 225 mm	0.18	1.68	3.39	m	5.07
50 x 250 mm	0.19	1.78	4.01	m	5.79
75 x 125 mm	0.18	1.68	2.82	m	4.50
75 x 150 mm	0.18	1.68	3.30	m	4.98
75 x 175 mm	0.18	1.68	3.96	m	5.64
75 x 200 mm	0.19	1.78	4.63	m	6.41
75 x 225 mm	0.19	1.78	5.28	m	7.06
75 x 250 mm	0.20	1.87	5.99	m	7.86
100 x 150 mm	0.24	2.24	5.07	m	7.31
100 x 200 mm	0.25	2.34	6.75	m	9.09
100 x 250 mm	0.28	2.62	8.42	m	11.04
100 x 300 mm	0.30	2.80	10.13	m	12.93
Wall or partition members					
25 x 25 mm	0.08	0.75	0.26	m	1.01
25 x 38 mm	0.08	0.75	0.36	m	1.11
25 x 75 mm	0.10	0.93	0.67	m	1.60
38 x 38 mm	0.10	0.93	0.55	m	1.48
38 x 50 mm	0.10	0.93	0.66	m	1.59
38 x 75 mm	0.13	1.22	0.97	m	2.19
38 x 100 mm	0.17	1.59	1.24	m	2.83
50 x 50 mm	0.13	1.22	0.81	m	2.03
50 x 75 mm	0.17	1.59	1.10	m	2.69
50 x 100 mm	0.20	1.87	1.45	m	3.32
50 x 125 mm	0.21	1.96	1.80	m	3.76
75 x 75 mm	0.20	1.87	1.83	m	3.70
75 x 100 mm	0.23	2.15	2.43	m	4.58
100 x 100 mm	0.23	2.15	3.43	m	5.58

G STRUCTURAL/CARCASSING METAL/TIMBER Including overheads and profit at 12.50%	Labour hours	Labour £	Material £	Unit	Total rate £
Flat roof members					
38 x 75 mm	0.15	1.40	0.97	m	2.37
38 x 100 mm	0.15	1.40	1.24	m	2.64
38 x 125 mm	0.15	1.40	1.53	m	2.93
38 x 150 mm	0.15	1.40	1.82	m	3.22
50 x 100 mm	0.15	1.40	1.43	m	2.83
50 x 125 mm	0.15	1.40	1.78	m	3.18
50 x 150 mm	0.17	1.59	2.13	m	3.72
50 x 175 mm	0.17	1.59	2.48	m	4.07
50 x 200 mm	0.18	1.68	2.92	m	4.60
50 x 225 mm	0.18	1.68	3.39	m	5.07
50 x 250 mm	0.19	1.78	4.01	m	5.79
75 x 150 mm	0.18	1.68	3.30	m	4.98
75 x 175 mm	0.18	1.68	3.95	m	5.63
75 x 200 mm	0.19	1.78	4.63	m	6.41
75 x 225 mm	0.19	1.78	5.28	m	7.06
75 x 250 mm	0.20	1.87	5.99	m	7.86
Pitched roof members					
25 x 100 mm	0.13	1.22	0.80	m	2.02
25 x 125 mm	0.13	1.22	1.00	m	2.22
25 x 150 mm	0.17	1.59	1.19	m	2.78
25 x 150 mm; notching over trussed rafters	0.33	3.09	1.19	m	4.28
25 x 175 mm	0.17	1.59	1.43	m	3.02
25 x 175 mm; notching over trussed rafters	0.33	3.09	1.43	m	4.52
25 x 200 mm	0.20	1.87	1.68	m	3.55
32 x 150 mm; notching over trussed rafters	0.36	3.37	1.68	m	5.05
32 x 175 mm; notching over trussed rafters	0.36	3.37	2.01	m	5.38
32 x 200 mm; notching over trussed rafters	0.36	3.37	2.36	m	5.73
38 x 100 mm	0.17	1.59	1.24	m	2.83
38 x 125 mm	0.17	1.59	1.53	m	3.12
38 x 150 mm	0.17	1.59	1.82	m	3.41
50 x 50 mm	0.13	1.22	0.79	m	2.01
50 x 75 mm	0.17	1.59	1.08	m	2.67
50 x 100 mm	0.20	1.87	1.43	m	3.30
50 x 125 mm	0.20	1.87	1.78	m	3.65
50 x 150 mm	0.23	2.15	2.13	m	4.28
50 x 175 mm	0.23	2.15	2.48	m	4.63
50 x 200 mm	0.23	2.15	2.92	m	5.07
50 x 225 mm	0.23	2.15	3.39	m	5.54
75 x 100 mm	0.28	2.62	2.41	m	5.03
75 x 125 mm	0.28	2.62	2.82	m	5.44
75 x 150 mm	0.28	2.62	3.30	m	5.92
100 x 150 mm	0.33	3.09	5.09	m	8.18
100 x 175 mm	0.33	3.09	5.93	m	9.02
100 x 200 mm	0.33	3.09	6.76	m	9.85
100 x 225 mm	0.36	3.37	7.61	m	10.98
100 x 250 mm	0.36	3.37	8.44	m	11.81
Kerbs, bearers and the like					
19 x 100 mm	0.04	0.37	0.66	m	1.03
19 x 125 mm	0.04	0.37	0.80	m	1.17
19 x 150 mm	0.04	0.37	0.97	m	1.34
25 x 75 mm	0.06	0.56	0.67	m	1.23
25 x 100 mm	0.06	0.56	0.80	m	1.36
38 x 75 mm	0.07	0.65	0.97	m	1.62
38 x 100 mm	0.07	0.65	1.24	m	1.89
50 x 75 mm	0.07	0.65	1.08	m	1.73
50 x 100 mm	0.08	0.75	1.43	m	2.18
75 x 100 mm	0.08	0.75	2.39	m	3.14
75 x 125 mm	0.09	0.84	2.80	Unit	3.64
75 x 150 mm	0.09	0.84	3.29	m	4.13
75 x 150 mm; splayed and rounded	0.11	1.03	5.76	m	6.79

G STRUCTURAL/CARCASSING METAL/TIMBER Including overheads and profit at 12.50%	Labour hours	Labour £	Material £	Unit	Total rate £

G20 CARPENTRY/TIMBER FRAMING/FIRST FIXING - cont'd

Sawn softwood; 'Tanalised' - cont'd
Kerbs, bearers and the like; fixing
by bolting

19 x 100 mm	0.09	0.84	0.66	m	1.50
19 x 125 mm	0.09	0.84	0.80	m	1.64
19 x 150 mm	0.09	0.84	0.97	m	1.81
25 x 75 mm	0.11	1.03	0.68	m	1.71
25 x 100 mm	0.11	1.03	0.80	m	1.83
38 x 75 mm	0.13	1.22	0.97	m	2.19
38 x 100 mm	0.13	1.22	1.24	m	2.46
50 x 75 mm	0.13	1.22	1.08	m	2.30
50 x 100 mm	0.15	1.40	1.43	m	2.83
75 x 100 mm	0.15	1.40	2.41	m	3.81
75 x 125 mm	0.18	1.68	2.82	m	4.50
75 x 150 mm	0.18	1.68	3.29	m	4.97
Herringbone strutting 50 x 50 mm					
to 150 mm deep joists	0.55	5.14	1.84	m	6.98
to 175 mm deep joists	0.55	5.14	1.88	m	7.02
to 200 mm deep joists	0.55	5.14	1.91	m	7.05
to 225 mm deep joists	0.55	5.14	1.96	m	7.10
to 250 mm deep joists	0.55	5.14	2.00	m	7.14
Solid strutting to joists					
50 x 150 mm	0.33	3.09	2.38	m	5.47
50 x 175 mm	0.33	3.09	2.76	m	5.85
50 x 200 mm	0.33	3.09	3.20	m	6.29
50 x 225 mm	0.33	3.09	3.68	m	6.77
50 x 250 mm	0.33	3.09	4.31	m	7.40
Cleats					
225 x 100 x 75 mm	0.22	2.06	0.86	nr	2.92
Sprockets					
50 x 50 x 200 mm	0.17	1.59	0.86	nr	2.45
Extra for stress grading to above timbers					
general structural (GS) grade	-	-	-	m3	14.51
special structural (SS) grade	-	-	-	m3	29.03
Extra for protecting and flameproofing					
timber with 'Celcure F' protection					
small sections	-	-	-	m3	135.00
large sections	-	-	-	m3	128.25
Wrought faces					
generally	0.33	3.09	-	m2	3.09
50 mm wide	0.04	0.37	-	m	0.37
75 mm wide	0.06	0.56	-	m	0.56
100 mm wide	0.07	0.65	-	m	0.65
Raking cutting					
50 mm thick	0.22	2.06	0.67	m	2.73
75 mm thick	0.28	2.62	1.01	m	3.63
100 mm thick	0.33	3.09	1.33	m	4.42
Curved cutting					
50 mm thick	0.28	2.62	0.93	m	3.55
Scribing					
50 mm thick	0.33	3.09	-	m	3.09
Notching and fitting ends to metal	0.13	1.22	-	nr	1.22
Trimming to openings					
760 x 760 mm; joists 38 x 150 mm	2.40	22.44	1.06	nr	23.50
760 x 760 mm; joists 50 x 175 mm	2.55	23.84	1.86	nr	25.70
1500 x 500 mm; joists 38 x 200 mm	2.75	25.71	2.13	nr	27.84
1500 x 500 mm; joists 50 x 200 mm	2.75	25.71	2.66	nr	28.37
3000 x 1000 mm; joists 50 x 225 mm	4.95	46.28	5.85	nr	52.13
3000 x 1000 mm; joists 75 x 225 mm	5.50	51.42	8.78	nr	60.20

G STRUCTURAL/CARCASSING METAL/TIMBER Including overheads and profit at 12.50%	Labour hours	Labour £	Material £	Unit	Total rate £
Trussed rafters, stress graded sawn softwood **pressure impregnated; raised through two** **storeys and fixed in position** 'W' type truss (Fink); 22.5 degree pitch; 450 mm eaves overhang					
5.00 m span	1.75	16.36	20.51	nr	36.87
7.60 m span	1.95	18.23	29.93	nr	48.16
10.00 m span	2.20	20.57	47.12	nr	67.69
'W' type truss (Fink); 30 degree pitch; 450 mm eaves overhang					
5.00 m span	1.75	16.36	21.55	nr	37.91
7.60 m span	1.95	18.23	31.00	nr	49.23
10.00 m span	2.20	20.57	49.31	nr	69.88
'W' type truss (Fink); 45 degree pitch; 450 mm eaves overhang					
4.60 m span	1.75	16.36	23.78	nr	40.14
7.00 m span	1.95	18.23	36.62	nr	54.85
'Mono' type truss; 17.5 degree pitch; 450 mm eaves overhang					
3.30 m span	1.55	14.49	18.31	nr	32.80
5.60 m span	1.75	16.36	28.96	nr	45.32
7.00 m span	2.05	19.16	42.01	nr	61.17
'Mono' type truss; 30 degree pitch; 450 mm eaves overhang					
3.30 m span	1.55	14.49	20.09	nr	34.58
5.60 m span	1.75	16.36	32.61	nr	48.97
7.00 m span	2.05	19.16	46.82	nr	65.98
Attic type truss; 45 degree pitch; 450 mm eaves overhang					
5.00 m span	3.45	32.25	82.90	nr	115.15
7.60 m span	3.65	34.12	63.10	nr	97.22
9.00 m span	3.85	35.99	70.86	nr	106.85
Standard 'Toreboda' glulam timber beams; **Moelven (UK) Ltd.; LB grade whitewood;** **pressure impregnated; phenol resorcinal** **adhesive; clean planed finish; fixed** Laminated roof beams					
56 x 255 mm	0.61	5.70	14.71	m	20.41
66 x 315 mm	0.77	7.20	19.77	m	26.97
90 x 315 mm	0.99	9.26	24.78	m	34.04
90 x 405 mm	1.25	11.69	31.02	m	42.71
115 x 405 mm	1.60	14.96	37.38	m	52.34
115 x 495 mm	2.00	18.70	49.61	m	68.31
115 x 630 mm	2.40	22.44	62.04	m	84.48

ALTERNATIVE FIRST FIXING MATERIAL PRICES

	£			£			£
Chipboard roofing (£/10 m2)							
12 mm	30.53	18 mm		42.03	25 mm		73.80
Non-asbestos boards (£/10 m2) 'Masterboard'							
6 mm	44.30	9 mm		80.30	12 mm		105.80
'Masterclad'; sanded finish							
4.5 mm	35.50	6 mm		46.10	9 mm		72.00
Plywood (£/10 m2) External quality							
12 mm	76.39	15 mm		93.83	25 mm		141.30
Marine quality							
12 mm	67.39	15 mm		85.39	25 mm		136.97

Discounts of 0 - 10% available depending on quantity/status

G STRUCTURAL/CARCASSING METAL/TIMBER Including overheads and profit at 12.50%	Labour hours	Labour £	Material £	Unit	Total rate £
G20 CARPENTRY/TIMBER FRAMING/FIRST FIXING - cont'd					
'Masterboard'; 6 mm thick; PC £44.30/10m2					
Boarding to eaves, verges, fascias and					
the like					
over 300 mm wide	0.77	7.20	5.68	m2	12.88
75 mm wide	0.23	2.15	0.51	m	2.66
150 mm wide	0.26	2.43	0.96	m	3.39
200 mm wide	0.30	2.80	1.24	m	4.04
225 mm wide	0.31	2.90	1.41	m	4.31
250 mm wide	0.32	2.99	1.58	m	4.57
Raking cutting	0.06	0.56	0.27	m	0.83
Plywood; external quality; 18 mm thick; PC £111.49/10m2					
Boarding to eaves, verges, fascias and					
the like					
over 300 mm wide	0.90	8.41	13.81	m2	22.22
75 mm wide	0.28	2.62	1.25	m	3.87
150 mm wide	0.32	2.99	2.35	m	5.34
225 mm wide	0.36	3.37	3.46	m	6.83
Plywood; marine quality; 18 mm thick; PC £98.61/10m2					
Boarding to gutter bottoms or sides;					
butt joints					
over 300 mm wide	1.00	9.35	12.25	m2	21.60
150 mm wide	0.36	3.37	2.08	m	5.45
225 mm wide	0.41	3.83	3.19	m	7.02
300 mm wide	0.46	4.30	4.05	m	8.35
Boarding to eaves, verges, fascias and					
the like					
over 300 mm wide	0.90	8.41	12.25	m2	20.66
75 mm wide	0.28	2.62	1.11	m	3.73
150 mm wide	0.32	2.99	2.08	m	5.07
225 mm wide	0.34	3.18	2.68	m	5.86
Sawn softwood; untreated					
Boarding to gutter bottoms or sides;					
butt joints					
19 mm thick; sloping	1.40	13.09	5.59	m2	18.68
19 mm thick x 75 mm wide	0.39	3.65	0.44	m	4.09
19 mm thick x 150 mm wide	0.44	4.11	0.83	m	4.94
19 mm thick x 225 mm wide	0.50	4.67	1.31	m	5.98
25 mm thick; sloping	1.40	13.09	6.80	m2	19.89
25 mm thick x 75 mm wide	0.39	3.65	0.57	m	4.22
25 mm thick x 150 mm wide	0.44	4.11	1.01	m	5.12
25 mm thick x 225 mm wide	0.50	4.67	1.62	m	6.29
Cesspools with 25 mm thick sides and bottom					
225 x 225 x 150 mm	1.30	12.15	3.22	nr	15.37
300 x 300 x 150 mm	1.55	14.49	4.56	nr	19.05
Firrings					
50 mm wide x 36 mm average depth	0.17	1.59	0.69	m	2.28
50 mm wide x 50 mm average depth	0.17	1.59	0.82	m	2.41
50 mm wide x 75 mm average depth	0.17	1.59	1.23	m	2.82
Bearers					
25 x 50 mm	0.11	1.03	0.41	m	1.44
38 x 50 mm	0.11	1.03	0.57	m	1.60
50 x 50 mm	0.11	1.03	0.69	m	1.72
50 x 75 mm	0.11	1.03	0.92	m	1.95
Angle fillets					
38 x 38 mm	0.11	1.03	0.33	m	1.36
50 x 50 mm	0.11	1.03	0.50	m	1.53
75 x 75 mm	0.13	1.22	0.94	m	2.16

G STRUCTURAL/CARCASSING METAL/TIMBER Including overheads and profit at 12.50%	Labour hours	Labour £	Material £	Unit	Total rate £
Tilting fillets					
19 x 38 mm	0.11	1.03	0.20	m	1.23
25 x 50 mm	0.11	1.03	0.28	m	1.31
38 x 75 mm	0.11	1.03	0.53	m	1.56
50 x 75 mm	0.11	1.03	0.69	m	1.72
75 x 100 mm	0.17	1.59	1.10	m	2.69
Grounds or battens					
13 x 19 mm	0.06	0.56	0.14	m	0.70
13 x 32 mm	0.06	0.56	0.18	m	0.74
25 x 50 mm	0.06	0.56	0.38	m	0.94
Grounds or battens; plugged and screwed					
13 x 19 mm	0.17	1.59	0.20	m	1.79
13 x 32 mm	0.17	1.59	0.24	m	1.83
25 x 50 mm	0.17	1.59	0.44	m	2.03
Open-spaced grounds or battens; at 300 mm centres one way					
25 x 50 mm	0.17	1.59	1.26	m2	2.85
25 x 50 mm; plugged and screwed	0.50	4.67	1.49	m2	6.16
Framework to walls; at 300 mm centres one way and 600 mm centres the other					
25 x 50 mm	0.83	7.76	2.10	m2	9.86
38 x 50 mm	0.83	7.76	2.95	m2	10.71
50 x 50 mm	0.83	7.76	3.60	m2	11.36
50 x 75 mm	0.83	7.76	4.91	m2	12.67
75 x 75 mm	0.83	7.76	8.01	m2	15.77
Framework to walls; at 300 mm centres one way and 600 mm centres the other way; plugged and screwed					
25 x 50 mm	1.40	13.09	2.43	m2	15.52
38 x 50 mm	1.40	13.09	3.29	m2	16.38
50 x 50 mm	1.40	13.09	3.94	m2	17.03
50 x 75 mm	1.40	13.09	5.25	m2	18.34
75 x 75 mm	1.40	13.09	8.35	m2	21.44
Framework to bath panel; at 500 mm centres both ways					
25 x 50 mm	0.99	9.26	2.30	m2	11.56
Framework as bracketing and cradling around steelwork					
25 x 50 mm	1.55	14.49	2.74	m2	17.23
50 x 50 mm	1.65	15.43	4.69	m2	20.12
50 x 75 mm	1.75	16.36	6.38	m2	22.74
Blockings wedged between flanges of steelwork					
50 x 50 x 150 mm	0.13	1.22	0.31	nr	1.53
50 x 75 x 225 mm	0.14	1.31	0.47	nr	1.78
50 x 100 x 300 mm	0.15	1.40	0.63	nr	2.03
Sawn softwood; 'Tanalised'					
Boarding to gutter bottoms or sides; butt joints					
19 mm thick; sloping	1.40	13.09	6.85	m2	19.94
19 mm thick x 75 mm wide	0.39	3.65	0.54	m	4.19
19 mm thick x 150 mm wide	0.44	4.11	1.03	m	5.14
19 mm thick x 225 mm wide	0.50	4.67	1.62	m	6.29
25 mm thick; sloping	1.40	13.09	8.44	m2	21.53
25 mm thick x 75 mm wide	0.39	3.65	0.70	m	4.35
25 mm thick x 150 mm wide	0.44	4.11	1.24	m	5.35
25 mm thick x 225 mm wide	0.50	4.67	2.01	m	6.68
Firrings					
50 mm wide x 36 mm average depth	0.17	1.59	0.82	m	2.41
50 mm wide x 50 mm average depth	0.17	1.59	1.00	m	2.59
50 mm wide x 75 mm average depth	0.17	1.59	1.49	m	3.08
Bearers					
25 x 50 mm	0.11	1.03	0.51	m	1.54
38 x 50 mm	0.11	1.03	0.70	m	1.73
50 x 50 mm	0.11	1.03	0.85	m	1.88
50 x 75 mm	0.11	1.03	1.15	m	2.18
Cesspools with 25 mm thick sides and bottom					
225 x 225 x 150 mm	1.30	12.15	4.00	nr	16.15
300 x 300 x 150 mm	1.55	14.49	5.68	nr	20.17

G STRUCTURAL/CARCASSING METAL/TIMBER Including overheads and profit at 12.50%	Labour hours	Labour £	Material £	Unit	Total rate £
G20 CARPENTRY/TIMBER FRAMING/FIRST FIXING - cont'd					
Sawn softwood; 'Tanalised' - cont'd					
Angle fillets					
38 x 38 mm	0.11	1.03	0.39	m	1.42
50 x 50 mm	0.11	1.03	0.59	m	1.62
75 x 75 mm	0.13	1.22	1.17	m	2.39
Tilting fillets					
19 x 38 mm	0.11	1.03	0.23	m	1.26
25 x 50 mm	0.11	1.03	0.34	m	1.37
38 x 75 mm	0.11	1.03	0.65	m	1.68
50 x 75 mm	0.11	1.03	0.85	m	1.88
75 x 100 mm	0.17	1.59	1.36	m	2.95
Grounds or battens					
13 x 19 mm	0.06	0.56	0.16	m	0.72
13 x 32 mm	0.06	0.56	0.21	m	0.77
25 x 50 mm	0.06	0.56	0.48	m	1.04
Grounds or battens; plugged and screwed					
13 x 19 mm	0.17	1.59	0.20	m	1.79
13 x 32 mm	0.17	1.59	0.28	m	1.87
25 x 50 mm	0.17	1.59	0.54	m	2.13
Open-spaced grounds or battens; at 300 mm centres one way					
25 x 50 mm	0.17	1.59	1.57	m2	3.16
25 x 50 mm; plugged and screwed	0.50	4.67	1.79	m2	6.46
Framework to walls; at 300 mm centres one way and 600 mm centres the other					
25 x 50 mm	0.83	7.76	2.60	m2	10.36
38 x 50 mm	0.83	7.76	3.68	m2	11.44
50 x 50 mm	0.83	7.76	4.48	m2	12.24
50 x 75 mm	0.83	7.76	6.13	m2	13.89
75 x 75 mm	0.83	7.76	10.25	m2	18.01
Framework to walls; at 300 mm centres one way and 600 mm centres the other way; plugged and screwed					
25 x 50 mm	1.40	13.09	2.93	m2	16.02
38 x 50 mm	1.40	13.09	4.02	m2	17.11
50 x 50 mm	1.40	13.09	4.81	m2	17.90
50 x 75 mm	1.40	13.09	6.47	m2	19.56
75 x 75 mm	1.40	13.09	10.58	m2	23.67
Framework to bath panel; at 500 mm centres both ways					
25 x 50 mm	0.99	9.26	2.84	m2	12.10
Framework as bracketing and cradling around steelwork					
25 x 50 mm	1.55	14.49	3.40	m2	17.89
50 x 50 mm	1.65	15.43	5.81	m2	21.24
50 x 75 mm	1.75	16.36	7.95	m2	24.31
Blockings wedged between flanges of steelwork					
50 x 50 x 150 mm	0.13	1.22	0.39	nr	1.61
50 x 75 x 225 mm	0.14	1.31	0.59	nr	1.90
50 x 100 x 300 mm	0.15	1.40	0.78	nr	2.18
Floor fillets set in or on concrete					
38 x 50 mm	0.13	1.22	0.64	m	1.86
50 x 50 mm	0.13	1.22	0.78	m	2.00
Floor fillets fixed to floor clips (priced elsewhere)					
38 x 50 mm	0.11	1.03	0.67	m	1.70
50 x 50 mm	0.11	1.03	0.81	m	1.84
Wrought softwood					
Boarding to gutter bottoms or sides; tongued and grooved joints					
19 mm thick; sloping	1.65	15.43	7.57	m2	23.00
19 mm thick x 75 mm wide	0.44	4.11	0.74	m	4.85
19 mm thick x 150 mm wide	0.50	4.67	1.13	m	5.80

G STRUCTURAL/CARCASSING METAL/TIMBER Including overheads and profit at 12.50%	Labour hours	Labour £	Material £	Unit	Total rate £
19 mm thick x 225 mm wide	0.55	5.14	1.79	m	6.93
25 mm thick; sloping	1.65	15.43	9.81	m2	25.24
25 mm thick x 75 mm wide	0.44	4.11	0.76	m	4.87
25 mm thick x 150 mm wide	0.50	4.67	1.43	m	6.10
25 mm thick x 225 mm wide	0.55	5.14	2.31	m	7.45
Boarding to eaves, verges, fascias and the like					
19 mm thick x over 300 mm wide	1.35	12.62	7.39	m2	20.01
19 mm thick x 150 mm wide; once grooved	0.22	2.06	2.90	m	4.96
25 mm thick x 150 mm wide; once grooved	0.22	2.06	3.54	m	5.60
25 mm thick x 175 mm wide; once grooved	0.24	2.24	3.95	m	6.19
32 mm thick x 225 mm wide; moulded	0.28	2.62	5.95	m	8.57
Mitred angles	0.11	1.03	-	nr	1.03
Rolls					
32 x 44 mm	0.13	1.22	0.65	m	1.87
50 x 50 mm	0.13	1.22	1.04	m	2.26
50 x 75 mm	0.14	1.31	1.63	m	2.94
75 x 75 mm	0.15	1.40	2.41	m	3.81
Wrought softwood; 'Tanalised' Boarding to gutter bottoms or sides; tongued and grooved joints					
19 mm thick; sloping	1.65	15.43	9.32	m2	24.75
19 mm thick x 75 mm wide	0.44	4.11	0.91	m	5.02
19 mm thick x 150 mm wide	0.50	4.67	1.39	m	6.06
19 mm thick x 225 mm wide	0.55	5.14	2.23	m	7.37
25 mm thick; sloping	1.65	15.43	12.08	m2	27.51
25 mm thick x 75 mm wide	0.44	4.11	0.93	m	5.04
25 mm thick x 150 mm wide	0.50	4.67	1.78	m	6.45
25 mm thick x 225 mm wide	0.55	5.14	2.88	m	8.02
Boarding to eaves, verges, fascias and the like					
19 mm thick x over 300 mm wide	1.35	12.62	9.14	m2	21.76
19 mm thick x 150 mm wide; once grooved	0.22	2.06	3.60	m	5.66
25 mm thick x 150 mm wide; once grooved	0.22	2.06	4.41	m	6.47
25 mm thick x 175 mm wide; once grooved	0.24	2.24	4.95	m	7.19
32 mm thick x 225 mm wide; moulded	0.28	2.62	7.41	m	10.03
Mitred angles	0.11	1.03	-	nr	1.03
Rolls					
32 x 44 mm	0.13	1.22	0.80	m	2.02
50 x 50 mm	0.13	1.22	1.29	m	2.51
50 x 75 mm	0.14	1.31	2.02	m	3.33
75 x 75 mm	0.15	1.40	2.99	m	4.39
Labours on softwood boarding Raking cutting					
12 mm thick	0.07	0.65	0.30	m	0.95
19 mm thick	0.09	0.84	0.30	m	1.14
25 mm thick	0.11	1.03	0.42	m	1.45
Boundary cutting					
19 mm thick	0.10	0.93	0.30	m	1.23
Curved cutting					
19 mm thick	0.13	1.22	0.42	m	1.64
Tongued edges and mitred angle					
19 mm thick	0.15	1.40	0.42	m	1.82
Plugging Plugging blockwork					
300 mm centres; both ways	0.13	1.22	0.06	m2	1.28
300 mm centres; one way	0.07	0.65	0.03	m	0.68
isolated	0.03	0.28	0.01	nr	0.29
Plugging brickwork					
300 mm centres; both ways	0.22	2.06	0.06	m2	2.12
300 mm centres; one way	0.11	1.03	0.03	m	1.06
isolated	0.06	0.56	0.01	nr	0.57

G STRUCTURAL/CARCASSING METAL/TIMBER Including overheads and profit at 12.50%	Labour hours	Labour £	Material £	Unit	Total rate £

G20 CARPENTRY/TIMBER FRAMING/FIRST FIXING - cont'd

Plugging - cont'd
Plugging concrete

	Labour hours	Labour £	Material £	Unit	Total rate £
300 mm centres; both ways	0.40	3.74	0.06	m2	3.80
300 mm centres; one way	0.20	1.87	0.03	m	1.90
isolated	0.10	0.93	0.01	nr	0.94

BASIC BOLT PRICES

	£		£		£		£
Black bolts, nuts and washers (£/100)							
Mild steel hex. hdd. bolts/nuts							
M6x50 mm	6.42	M10x50 mm	15.38	M12x100 mm	31.75	M16x140 mm	85.50
M6x80 mm	10.75	M10x80 mm	22.08	M12x140 mm	48.13	M16x180 mm	124.33
M6x100 mm	11.69	M10x100 mm	27.21	M12x180 mm	91.49	M20x80 mm	86.99
M8x50 mm	10.11	M10x140 mm	37.41	M16x50 mm	38.80	M20x100 mm	101.46
M8x80 mm	14.11	M12x50 mm	21.87	M16x80 mm	49.69	M20x140 mm	138.09
M8x100 mm	17.59	M12x80 mm	27.86	M16x100 mm	55.02	M20x180 mm	184.86
Mild steel cup. hdd. bolts/nuts							
M6x50 mm	5.65	M8x75 mm	10.51	M10x100 mm	23.50	M12x100 mm	34.40
M6x75 mm	6.92	M8x100 mm	17.49	M10x150 mm	35.02	M12x150 mm	48.04
M6x100 mm	10.76	M8x150 mm	24.00	M10x200 mm	73.12	M12x200 mm	118.44
M6x150 mm	16.39	M10x50 mm	14.24	M12x50 mm	21.81		
M8x50 mm	8.57	M10x75 mm	17.20	M12x75 mm	27.31		
Mild steel washers; round							
M6	1.47	M10	1.92	M16	4.11	M20	5.09
M8	1.43	M12	3.03				
Mild steel washers; square							
38x38 mm	10.00	50x50 mm	5.81				

	Labour hours	Labour £	Material £	Unit	Total rate £
Metalwork; mild steel; galvanized					
Fix only bolts; 50-200 mm long					
6 mm dia	0.04	0.37	-	nr	0.37
8 mm dia	0.04	0.37	-	nr	0.37
10 mm dia	0.06	0.56	-	nr	0.56
12 mm dia	0.06	0.56	-	nr	0.56
16 mm dia	0.07	0.65	-	nr	0.65
20 mm dia	0.07	0.65	-	nr	0.65
Straps; standard twisted vertical restraint;					
fixing to softwood and brick or blockwork					
30 x 2.5 x 400 mm girth	0.28	2.62	0.73	nr	3.35
30 x 2.5 x 600 mm girth	0.29	2.71	0.93	nr	3.64
30 x 2.5 x 800 mm girth	0.30	2.80	1.23	nr	4.03
30 x 2.5 x 1000 mm girth	0.33	3.09	1.51	nr	4.60
30 x 2.5 x 1200 mm girth	0.34	3.18	1.76	nr	4.94
Timber connectors; round toothed plate;					
for 10 mm or 12 mm dia bolts					
38 mm dia; single sided	0.02	0.19	0.15	nr	0.34
38 mm dia; double sided	0.02	0.19	0.17	nr	0.36
50 mm dia; single sided	0.02	0.19	0.17	nr	0.36
50 mm dia; double sided	0.02	0.19	0.19	nr	0.38
63 mm dia; single sided	0.02	0.19	0.23	nr	0.42
63 mm dia; double sided	0.02	0.19	0.25	nr	0.44
75 mm dia; single sided	0.02	0.19	0.32	nr	0.51
75 mm dia; double sided	0.02	0.19	0.37	nr	0.56
Framing anchor	0.17	1.59	0.48	nr	2.07

G STRUCTURAL/CARCASSING METAL/TIMBER Including overheads and profit at 12.50%		Labour hours	Labour £	Material £	Unit	Total rate £
Joist hangers 1.0 mm thick; for fixing to						
softwood; joint sizes						
50 x 100 mm	PC £0.59	0.13	1.22	0.81	nr	2.03
50 x 125 mm	PC £0.59	0.13	1.22	0.81	nr	2.03
50 x 150 mm	PC £0.59	0.14	1.31	0.81	nr	2.12
50 x 175 mm	PC £0.59	0.14	1.31	0.81	nr	2.12
50 x 200 mm	PC £0.59	0.15	1.40	0.81	nr	2.21
50 x 225 mm	PC £0.59	0.15	1.40	0.81	nr	2.21
50 x 250 mm	PC £0.59	0.17	1.59	0.81	nr	2.40
75 x 150 mm	PC £0.63	0.14	1.31	0.86	nr	2.17
75 x 175 mm	PC £0.63	0.14	1.31	0.86	nr	2.17
75 x 200 mm	PC £0.63	0.15	1.40	0.86	nr	2.26
75 x 225 mm	PC £0.63	0.15	1.40	0.86	nr	2.26
75 x 250 mm	PC £0.63	0.17	1.59	0.86	nr	2.45
100 x 200 mm	PC £0.68	0.17	1.59	0.91	nr	2.50
Joist hangers 2.7 mm thick; for						
building in; joist sizes						
50 x 100 mm	PC £1.10	0.09	0.84	1.36	nr	2.20
50 x 125 mm	PC £1.11	0.09	0.84	1.37	nr	2.21
50 x 150 mm	PC £1.14	0.10	0.93	1.41	nr	2.34
50 x 175 mm	PC £1.16	0.10	0.93	1.43	nr	2.36
50 x 200 mm	PC £1.28	0.11	1.03	1.56	nr	2.59
50 x 225 mm	PC £1.37	0.11	1.03	1.68	nr	2.71
50 x 250 mm	PC £1.78	0.12	1.12	2.16	nr	3.28
75 x 150 mm	PC £1.71	0.10	0.93	2.08	nr	3.01
75 x 175 mm	PC £1.58	0.10	0.93	1.92	nr	2.85
75 x 200 mm	PC £1.73	0.11	1.03	2.10	nr	3.13
75 x 225 mm	PC £1.80	0.11	1.03	2.18	nr	3.21
75 x 250 mm	PC £1.99	0.12	1.12	2.41	nr	3.53
100 x 200 mm	PC £2.00	0.11	1.03	2.42	nr	3.45
Herringbone joist struts; to suit joists at						
400 mm centres	PC £21.05/100	0.33	3.09	1.18	m	4.27
450 mm centres	PC £23.77/100	0.30	2.80	1.18	m	3.98
600 mm centres	PC £26.52/100	0.26	2.43	1.19	m	3.62
Expanding bolts; 'Rawlbolt' projecting type;						
plated; one nut; one washer						
6 mm dia; ref M6 10P		0.09	0.84	0.60	nr	1.44
6 mm dia; ref M6 25P		0.09	0.84	0.67	nr	1.51
6 mm dia; ref M6 60P		0.09	0.84	0.70	nr	1.54
8 mm dia; ref M8 25P		0.09	0.84	0.80	nr	1.64
8 mm dia; ref M8 60P		0.09	0.84	0.85	nr	1.69
10 mm dia; ref M10 15P		0.11	1.03	1.04	nr	2.07
10 mm dia; ref M10 30P		0.11	1.03	1.09	nr	2.12
10 mm dia; ref M10 60P		0.11	1.03	1.13	nr	2.16
12 mm dia; ref M12 15P		0.11	1.03	1.64	nr	2.67
12 mm dia; ref M12 30P		0.11	1.03	1.77	nr	2.80
12 mm dia; ref M12 75P		0.11	1.03	2.20	nr	3.23
16 mm dia; ref M16 35P		0.13	1.22	4.07	nr	5.29
16 mm dia; ref M16 75P		0.13	1.22	4.27	nr	5.49
Expanding bolts; 'Rawlbolt' loose bolt type;						
plated; one bolt; one washer						
6 mm dia; ref M6 10L		0.11	1.03	0.60	nr	1.63
6 mm dia; ref M6 25L		0.11	1.03	0.63	nr	1.66
6 mm dia; ref M6 40L		0.11	1.03	0.64	nr	1.67
8 mm dia; ref M8 25L		0.11	1.03	0.78	nr	1.81
8 mm dia; ref M8 40L		0.11	1.03	0.83	nr	1.86
10 mm dia; ref M10 10L		0.11	1.03	1.01	nr	2.04
10 mm dia; ref M10 25L		0.11	1.03	1.04	nr	2.07
10 mm dia; ref M10 50L		0.11	1.03	1.09	nr	2.12
10 mm dia; ref M10 75L		0.11	1.03	1.13	nr	2.16
12 mm dia; ref M12 10L		0.11	1.03	1.49	nr	2.52
12 mm dia; ref M12 25L		0.11	1.03	1.64	nr	2.67
12 mm dia; ref M12 40L		0.11	1.03	1.72	nr	2.75
12 mm dia; ref M12 60L		0.11	1.03	1.81	nr	2.84
16 mm dia; ref M16 30L		0.13	1.22	4.19	nr	5.41
16 mm dia; ref M16 60L		0.13	1.22	4.19	nr	5.41

G STRUCTURAL/CARCASSING METAL/TIMBER Including overheads and profit at 12.50%	Labour hours	Labour £	Material £	Unit	Total rate £

G20 CARPENTRY/TIMBER FRAMING/FIRST FIXING - cont'd

Metalwork; mild steel; galvanised
Ragbolts; mild steel; one nut; one washer

M10 x 120 mm long	0.11	1.17	1.18	nr	2.35
M12 x 160 mm long	0.14	1.49	1.47	nr	2.96
M20 x 200 mm long	0.17	1.81	3.21	nr	5.02

G32 EDGE SUPPORTED/REINFORCED WOODWOOL SLAB DECKING

Woodwool interlocking reinforced slabs;
Torvale 'Woodcelip' or similar; natural
finish; fixing to timber or steel with
galvanized nails or clips; flat or sloping

50 mm slabs; type 503; max. span 2100 mm					
1800 - 2100 mm lengths PC £11.44	0.55	5.14	13.76	m2	18.90
2400 mm lengths PC £12.07	0.55	5.14	14.51	m2	19.65
2700 - 3000 mm lengths PC £12.39	0.55	5.14	15.07	m2	20.21
75 mm slabs; type 751; max. span 2100 mm					
1800 - 2400 mm lengths PC £16.30	0.61	5.70	19.55	m2	25.25
2700 - 3000 mm lengths PC £16.38	0.61	5.70	19.64	m2	25.34
75 mm slabs; type 752; max. span 2100 mm					
1800 - 2400 mm lengths PC £16.30	0.61	5.70	19.55	m2	25.25
2700 - 3000 mm lengths PC £16.38	0.61	5.70	19.86	m2	25.56
75 mm slabs; type 753; max. span 3600 mm					
2400 mm lengths PC £16.77	0.61	5.70	20.10	m2	25.80
2700 - 3000 mm lengths PC £17.52	0.61	5.70	20.99	m2	26.69
3300 - 3900 mm lengths	0.61	5.70	24.64	m2	30.34
Raking cutting; including additional trim	0.24	2.24	4.43	m	6.67
Holes for pipes and the like	0.13	1.22	-	nr	1.22
100 mm slabs; type 1001; max. span 3600 mm					
3000 mm lengths PC £24.50	0.66	6.17	29.28	m2	35.45
3300 - 3600 mm lengths PC £25.16	0.66	6.17	30.06	m2	36.23
100 mm slabs; type 1002; max. span 3600 mm					
3000 mm lengths PC £25.28	0.66	6.17	30.20	m2	36.37
3300 - 3600 mm lengths PC £25.54	0.66	6.17	30.76	m2	36.93
100 mm slabs; type 1003; max. span 4000 mm					
3000 - 3600 mm lengths PC £23.11	0.66	6.17	27.64	m2	33.81
3900 - 4000 mm lengths PC £23.11	0.66	6.17	27.64	m2	33.81
125 mm slabs; type 1252; max. span 3000 mm					
2400 - 3000 mm lengths PC £25.56	0.66	6.17	30.53	m2	36.70
Extra over slabs for					
pre-screeded deck	-	-	-	m2	1.10
pre-screeded soffit	-	-	-	m2	2.50
pre-screeded deck and soffit	-	-	-	m2	3.17
pre-screeded and proofed deck	-	-	-	m2	2.02
pre-screeded and proofed deck plus pre-screeded soffit	-	-	-	m2	4.72
pre-felted deck (glass fibre)	-	-	-	m2	2.65
pre-felted deck plus pre-screeded soffit	-	-	-	m2	5.06
'Weatherdeck'	-	-	-	m2	1.71

H CLADDING/COVERING Including overheads and profit at 12.50%	Labour hours	Labour £	Material £	Unit	Total rate £

H10 PATENT GLAZING

Patent glazing; aluminium alloy bars 2.44 mm
long at 622 mm centres; fixed to supports
Roof cladding; glazing with 7 mm thick
Georgian wired cast glass

Georgian wired cast glass	-	-	-	m2	94.43
Extra for associated code 4 lead flashings					
top flashing; 210 mm girth	-	-	-	m	33.79
bottom flashing; 240 mm girth	-	-	-	m	43.94
end flashing; 300 mm girth	-	-	-	m	62.51
Wall cladding; glazing with 7 mm thick					
Georgian wired cast glass	-	-	-	m2	101.86
Wall cladding; glazing with 6 mm thick					
plate glass	-	-	-	m2	147.55
Extra for aluminium alloy members					
38 x 38 x 3 mm angle jamb	-	-	-	m	34.26
extruded cill member	-	-	-	m	26.39
extruded channel head and PVC came	-	-	-	m	16.40

H20 RIGID SHEET CLADDING

Eternit 2000 'Glasal' sheet; Eternit TAC Ltd;
flexible neoprene gasket joints; fixing with
stainless steel screws and coloured caps
7.5 mm thick cladding to walls

over 300 mm wide	2.40	33.29	49.14	m2	82.43
not exceeding 300 mm wide	0.80	11.10	23.73	m2	34.83
External angle	0.11	1.53	5.62	m	7.15

7.5 mm thick cladding to eaves; verges
fascias or the like

100 mm wide	0.58	8.05	12.86	m	20.91
150 mm wide	0.63	8.74	15.06	m	23.80
200 mm wide	0.69	9.57	17.24	m	26.81
250 mm wide	0.75	10.40	19.42	m	29.82
300 mm wide	0.81	11.24	22.33	m	33.57

H30 FIBRE CEMENT PROFILED SHEET CLADDING/
COVERING/SIDING

Asbestos-free corrugated sheets; Eternit
'2000' or similar
Roof cladding; sloping not exceeding 50
degrees; fixing to timber purlins with drive
screws

'Profile 3'; natural	PC £6.94	0.23	3.19	11.79	m2	14.98
'Profile 3'; coloured	PC £7.63	0.23	3.19	12.55	m2	15.74
'Profile 6'; natural	PC £6.97	0.29	4.02	11.51	m2	15.53
'Profile 6'; coloured	PC £7.49	0.29	4.02	12.26	m2	16.28
'Profile 6'; natural; insulated; 60 mm glass						
fibre infill; lining panel	PC £6.97	0.52	7.21	27.14	m2	34.35

Roof cladding; sloping not exceeding 50
degrees; fixing to steel purlins with hook
bolts

'Profile 3'; natural	0.29	4.02	12.39	m2	16.41
'Profile 3'; coloured	0.29	4.02	13.13	m2	17.15
'Profile 6'; natural	0.35	4.85	12.03	m2	16.88
'Profile 6'; coloured	0.35	4.85	12.78	m2	17.63
'Profile 6'; natural; insulated; 60 mm glass					
fibre infill; lining panel	0.58	8.05	25.00	m2	33.05

Wall cladding; vertical; fixing to steel
rails with hook bolts

'Profile 3'; natural	0.35	4.85	12.39	m2	17.24
'Profile 3'; coloured	0.35	4.85	13.37	m2	18.22
'Profile 6'; natural	0.40	5.55	12.03	m2	17.58
'Profile 6'; coloured	0.40	5.55	12.78	m2	18.33
'Profile 6'; natural; insulated; 60 mm glass					
fibre infill; lining panel	0.63	8.74	25.00	m2	33.74
Raking cutting	0.17	2.36	1.76	m	4.12
Holes for pipes and the like	0.17	2.36	-	nr	2.36

H CLADDING/COVERING Including overheads and profit at 12.50%	Labour hours	Labour £	Material £	Unit	Total rate £
H30 FIBRE CEMENT PROFILED SHEET CLADDING/ **COVERING/SIDING - cont'd**					
Asbestos-free corrugated sheets; Eternit **'2000' or similar - cont'd** Accessories; to 'Profile 3' cladding; natural					
eaves filler	0.11	1.53	7.68	m	9.21
vertical corrugation closure	0.14	1.94	7.68	m	9.62
apron flashing	0.14	1.94	8.65	m	10.59
underglazing flashing	0.14	1.94	8.31	m	10.25
plain wing or close fitting two-piece adjustable capping to ridge	0.20	2.77	19.28	m	22.05
ventilating two-piece adjustable capping to ridge	0.20	2.77	19.28	m	22.05
Accessories; to 'Profile 3' cladding; coloured					
eaves filler	0.11	1.53	9.20	m	10.73
vertical corrugation closure	0.14	1.94	9.20	m	11.14
apron flashing	0.14	1.94	12.69	m	14.63
underglazing flashing	0.14	1.94	9.98	m	11.92
plain wing or close fitting two-piece adjustable capping to ridge	0.20	2.77	23.03	m	25.80
ventilating two-piece adjustable capping to ridge	0.20	2.77	23.03	m	25.80
Accessories; to 'Profile 6' cladding; natural					
eaves filler	0.11	1.53	6.66	m	8.19
vertical corrugation closure	0.14	1.94	6.66	m	8.60
apron flashing	0.14	1.94	6.87	m	8.81
plain cranked crown to ridge	0.20	2.77	12.68	m	15.45
plain wing, close fitting or north light two-piece adjustable capping to ridge	0.20	2.77	12.19	m	14.96
ventilating two-piece adjustable capping to ridge	0.20	2.77	16.54	m	19.31
Accessories; to 'Profile 6' cladding; coloured					
eaves filler	0.11	1.53	8.01	m	9.54
vertical corrugation closure	0.14	1.94	8.24	m	10.18
apron flashing	0.14	1.94	6.73	m	8.67
plain cranked crown to ridge	0.20	2.77	15.11	m	17.88
plain wing, close fitting or north light two-piece adjustable capping to ridge	0.20	2.77	14.53	m	17.30
ventilating two-piece adjustable capping to ridge	0.20	2.77	19.74	m	22.51
H41 GLASS REINFORCED PLASTICS CLADDING/ **FEATURES**					
Glass fibre translucent sheeting grade AB **class 3** Roof cladding; sloping not exceeding 50 degrees; fixing to timber purlins with drive screws; to suit					
'Profile 3' PC £7.66	0.23	3.19	12.48	m2	15.67
'Profile 6' PC £9.64	0.29	4.02	15.28	m2	19.30
Roof cladding; sloping not exceeding 50 degrees; fixing to steel purlins with hook bolts; to suit					
'Profile 3'	0.29	4.02	13.08	m2	17.10
'Profile 6'	0.35	4.85	15.89	m2	20.74
'Longrib 1000'	0.35	4.85	16.54	m2	21.39

H CLADDING/COVERING
Including overheads and profit at 12.50%

ALTERNATIVE TILE PRICES (£/1000)

	£		£		£
Clay tiles; plain, interlocking and pantile					
Langleys 'Sterreberg' pantiles					
anthracite	1524.00	natural red	1248.00	rustic	1248.00
deep brown	1650.00				
Sandtoft pantiles					
Bold roll 'Roman'	814.00	'Gaelic'	546.70	'County' i'locking	517.00
William Blyth pantiles					
'Barco' bold roll	448.80	'Celtic' (French)	498.30		
Concrete tiles, plain and interlocking					
Marley roof tiles					
'Anglia'	327.00	plain	189.00	'Roman'	455.00
'Ludlow +'	289.00				
Redland roof tiles					
'49'-granule	300.00	'Grovebury'	594.00	'Stonewold Mk 1	806.00
'50 Roman'	534.00				

Discounts of 2.5 - 15% available depending on quantity/status

	Labour hours	Labour £	Material £	Unit	Total rate £
NOTE: The following items of tile roofing unless otherwise described, include for conventional fixing assuming 'normal exposure' with appropriate nails and/or rivets or clips to pressure impregnated softwood battens fixed with galvanized nails; Prices also include for all bedding and pointing at verges; beneath ridge tiles, etc..					
H60 CLAY/CONCRETE ROOF TILING					
Clay interlocking pantiles; Sandtoft Goxhill 'Tudor' red sand faced; PC £806.30/1000; 470 x 285 mm; to 100 mm lap; on 25 x 38 mm battens and type 1F reinforced underlay					
Roof coverings	0.46	6.38	13.79	m2	20.17
Extra over coverings for					
fixing every tile	0.02	0.28	0.32	m2	0.60
eaves course with plastic filler	0.35	4.85	5.01	m	9.86
verges; extra single undercloak course of					
plain tiles	0.35	4.85	2.20	m	7.05
open valleys; cutting both sides	0.21	2.91	4.90	m	7.81
ridge tiles	0.69	9.57	10.90	m	20.47
hip tiles; cutting both sides	0.86	11.93	15.80	m	27.73
Holes for pipes and the like	0.23	3.19	-	nr	3.19
Clay interlocking pantiles; Langley's 'Sterreberg' black glazed; or similar; PC £1650.00/1000; 355 x 240 mm: to 75 mm lap; on 25 x 38 mm battens and type 1F reinforced underlay					
Roof coverings	0.52	7.21	36.75	m2	43.96

H CLADDING/COVERING Including overheads and profit at 12.50%	Labour hours	Labour £	Material £	Unit	Total rate £
H60 CLAY/CONCRETE ROOF TILING - cont'd					
Clay interlocking pantiles; Langley's 'Sterreberg' black glazed; or similar; PC £1650.00/1000; 355 x 240 mm: to 75 mm lap; on 25 x 38 mm battens and type 1F reinforced underlay - cont'd					
Extra over coverings for					
double course at eaves	0.38	5.27	5.72	m	10.99
verges; extra single undercloak course of					
plain tiles	0.35	4.85	7.44	m	12.29
open valleys; cutting both sides	0.21	2.91	14.11	m	17.02
saddleback ridge tiles	0.69	9.57	19.54	m	29.11
saddleback hip tiles; cutting both sides	0.86	11.93	33.65	m	45.58
Holes for pipes and the like	0.23	3.19	-	nr	3.19
Clay pantiles; Sandtoft Goxhill 'Old English' red sand faced; PC £499.40/1000 342 x 241 mm; to 75 mm lap; on 25 x 38 mm battens and type 1F reinforced underlay					
Roof coverings	0.52	7.21	14.59	m2	21.80
Extra over coverings for					
fixing every tile	0.03	0.42	0.53	m2	0.95
other colours	-	-	-	m2	0.92
double course at eaves	0.38	5.27	3.65	m	8.92
verges; extra single undercloak course of					
plain tiles	0.35	4.85	2.20	m	7.05
open valleys; cutting both sides	0.21	2.91	4.27	m	7.18
ridge tiles; tile slips	0.69	9.57	11.70	m	21.27
hip tiles; tile slips; cutting both sides	0.86	11.93	15.97	m	27.90
Holes for pipes and the like	0.23	3.19	-	nr	3.19
Clay pantiles; William Blyth's 'Lincoln' natural; 343 x 280 mm; to 75 mm lap; PC £567.88/1000; on 19 x 38 mm battens and type 1F reinforced underlay					
Roof coverings	0.52	7.21	15.96	m2	23.17
Extra over coverings for					
fixing every tile	0.03	0.42	0.53	m2	0.95
other colours	-	-	-	m2	1.36
double course at eaves	0.38	5.27	3.37	m	8.64
verges; extra single undercloak course of					
plain tiles	0.35	4.85	6.14	m	10.99
open valleys; cutting both sides	0.21	2.91	5.11	m	8.02
ridge tiles; tile slips	0.69	9.57	11.79	m	21.36
hip tiles; tile slips; cutting both sides	0.86	11.93	16.90	m	28.83
Holes for pipes and the like	0.23	3.19	-	nr	3.19
Clay plain tiles; Hinton, Perry and Davenhill 'Dreadnought' smooth red machine- made; PC £267.75/1000; 265 x 165 mm; on 19 x 38 mm battens and type 1F reinforced underlay					
Roof coverings; to 64 mm lap	0.75	10.40	23.25	m2	33.65
Wall coverings; to 38 mm lap	0.92	12.76	20.77	m2	33.53
Extra over coverings for					
25 x 38 mm battens in lieu	-	-	-	m2	0.57
other colours	-	-	-	m2	3.90
ornamental tiles in lieu	-	-	-	m2	8.36
double course at eaves	0.29	4.02	2.00	m	6.02
verges; extra single undercloak course	0.38	5.27	2.95	m	8.22
valley tiles; cutting both sides	0.75	10.40	31.82	m	42.22
bonnet hip tiles; cutting both sides	0.92	12.76	32.96	m	45.72
external vertical angle tiles; supplementary					
nail fixings	0.46	6.38	23.65	m	30.03
half round ridge tiles	0.58	8.05	11.05	m	19.10
Holes for pipes and the like	0.23	3.19	-	nr	3.19

H CLADDING/COVERING Including overheads and profit at 12.50%	Labour hours	Labour £	Material £	Unit	Total rate £
Clay plain tiles; Keymer best hand-made **sand-faced tiles; PC £560.70/1000;** **265 x 165 mm; on 19 x 38 mm battens** **and type 1F reinforced underlay**					
Roof coverings; to 64 mm lap	0.75	10.40	44.02	m2	54.42
Wall coverings; to 38 mm lap	0.92	12.76	39.11	m2	51.87
Extra over coverings for					
25 x 38 mm battens in lieu	-	-	-	m2	0.57
ornamental tiles in lieu	-	-	-	m2	4.25
double course at eaves	0.29	4.02	3.98	m	8.00
verges; extra single undercloak course	0.38	5.27	5.92	m	11.19
valley tiles; cutting both sides	0.75	10.40	23.89	m	34.29
bonnet hip tiles; cutting both sides	0.92	12.76	35.12	m	47.88
external vertical angle tiles; supplementary nail fixings	0.46	6.38	20.72	m	27.10
half round ridge tiles	0.58	8.05	10.51	m	18.56
Holes for pipes and the like	0.23	3.19	-	nr	3.19
Concrete interlocking tiles; Marley 'Bold **Roll' granule finish tiles or similar;** **PC £527.00/1000; 419 x 330 mm;** **to 75 mm lap; on 22 x 38 mm battens** **and type 1F reinforced underlay**					
Roof coverings	0.40	5.55	8.57	m2	14.12
Extra over coverings for					
fixing every tile	0.03	0.42	0.62	m2	1.04
25 x 38 mm battens in lieu	-	-	-	m2	0.08
eaves; eave filler	0.06	0.83	0.82	m	1.65
verges; 150 mm asbestos cement strip undercloak	0.26	3.61	1.79	m	5.40
valley trough tiles; cutting both sides	0.63	8.74	13.18	m	21.92
segmental ridge tiles; tile slips	0.63	8.74	9.98	m	18.72
segmental ridge tiles; tile slips; cutting both sides	0.81	11.24	12.35	m	23.59
dry ridge tiles; segmental including batten sections; unions and filler pieces	0.35	4.85	9.82	m	14.67
segmental monoridge tiles	0.63	8.74	11.42	m	20.16
gas ridge terminal	0.58	8.05	40.37	nr	48.42
Holes for pipes and the like	0.23	3.19	-	nr	3.19
Concrete interlocking tiles; Marley 'Ludlow **Major' granule finish tiles or similar;** **PC £455.00/1000; 413 x 330 mm;** **to 75 mm lap; on 22 x 38 mm battens**					
Roof coverings	0.40	5.55	7.91	m2	13.46
Extra over coverings for					
fixing every tile	0.03	0.42	0.37	m2	0.79
25 x 38 mm battens in lieu	-	-	-	m2	0.09
verges; 150 mm asbestos cement strip undercloak	0.26	3.61	1.79	m	5.40
dry verge system; extruded white pvc	0.17	2.36	5.68	m	8.04
Segmental ridge cap	0.03	0.42	1.74	nr	2.16
valley trough tiles; cutting both sides	0.63	8.74	14.58	m	23.32
segmental ridge tiles	0.58	8.05	5.05	m	13.10
segmental hip tiles; cutting both sides	0.75	10.40	7.20	m	17.60
dry ridge tiles; segmental including batten sections; unions and filler pieces	0.35	4.85	9.82	m	14.67
segmental monoridge tiles	0.58	8.05	10.10	m	18.15
gas ridge terminal	0.58	8.05	40.37	nr	48.42
Holes for pipes and the like	0.23	3.19	-	nr	3.19

H CLADDING/COVERING Including overheads and profit at 12.50%	Labour hours	Labour £	Material £	Unit	Total rate £
H60 CLAY/CONCRETE ROOF TILING - cont'd					
Concrete interlocking tiles; Marley 'Mendip' granule finish double pantiles or similar; PC £527.00/1000; 413 x 330 mm; to 75 mm lap; on 22 x 38 mm battens and type 1F reinforced underlay					
Roof coverings	0.40	5.55	8.79	m2	14.34
Extra over coverings for					
fixing every tile	0.03	0.42	0.37	m2	0.79
25 x 38 mm battens in lieu	-	-	-	m2	0.09
eaves; eave filler	0.03	0.42	0.07	m	0.49
verges; 150 mm asbestos cement strip					
undercloak	0.26	3.61	1.79	m	5.40
dry verge system; extruded white pvc	0.17	2.36	5.68	m	8.04
valley trough tiles; cutting both sides	0.63	8.74	14.92	m	23.66
segmental ridge tiles	0.63	8.74	8.11	m	16.85
segmental hip tiles; cutting both sides	0.81	11.24	10.60	m	21.84
dry ridge tiles; segmental including batten					
sections; unions and filler pieces	0.35	4.85	9.82	m	14.67
segmental monoridge tiles	0.58	8.05	10.10	m	18.15
gas ridge terminal	0.58	8.05	40.37	nr	48.42
Holes for pipes and the like	0.23	3.19	-	nr	3.19
Concrete interlocking tiles; Marley 'Modern' smooth finish tiles or similar; PC £522.00/1000; 413 x 330 mm; to 75 mm lap; on 22 x 38 mm battens and type 1F reinforced underlay					
Roof coverings	0.40	5.55	8.81	m2	14.36
Extra over coverings for					
fixing every tile	0.05	0.69	0.33	m2	1.02
25 x 38 mm battens in lieu	-	-	-	m2	0.09
verges; 150 mm asbestos cement strip					
undercloak	0.32	4.44	2.67	m	7.11
dry verge system, extruded white pvc	0.23	3.19	6.73	m	9.92
'Modern' ridge cap	0.03	0.42	1.74	nr	2.16
valley trough tiles; cutting both sides	0.63	8.74	14.89	m	23.63
'Modern' ridge tiles	0.58	8.05	5.32	m	13.37
'Modern' hip tiles; cutting both sides	0.75	10.40	7.78	m	18.18
dry ridge tiles; 'Modern'; including batten					
sections, unions and filler pieces	0.35	4.85	9.16	m	14.01
monoridge tiles	0.58	8.05	10.10	m	18.15
gas ridge terminal	0.58	8.05	40.37	nr	48.42
Holes for pipes and the like	0.23	3.19	-	nr	3.19
Concrete interlocking tiles; Marley 'Wessex' smooth finish tiles or similar; PC £611.00/1000; 413 x 330 mm; to 75 mm lap; on 22 x 38 mm battens and type 1F reinforced underlay					
Roof coverings	0.40	5.55	9.89	m2	15.44
Extra over coverings for					
fixing every tile	0.05	0.69	0.33	m2	1.02
25 x 38 mm battens in lieu	-	-	-	m2	0.09
verges; 150 mm asbestos cement strip					
undercloak	0.26	3.61	1.79	m	5.40
dry verge system, extruded white pvc	0.17	2.36	5.68	m	8.04
'Modern' ridge cap	0.03	0.42	1.74	nr	2.16
valley trough tiles; cutting both sides	0.63	8.74	15.31	m	24.05
'Modern' ridge tiles	0.63	8.74	6.84	m	15.58
'Modern' hip tiles; cutting both sides	0.81	11.24	9.73	m	20.97
dry ridge tiles; 'Modern'; including batten					
sections, unions and filler pieces	0.35	4.85	10.09	m	14.94
monoridge tiles	0.58	8.05	10.10	m	18.15
gas ridge terminal	0.58	8.05	40.37	nr	48.42
Holes for pipes and the like	0.23	3.19	-	nr	3.19

H CLADDING/COVERING Including overheads and profit at 12.50%	Labour hours	Labour £	Material £	Unit	Total rate £
Concrete interlocking tiles; Redland 'Delta' smooth finish tiles or similar; PC £866.00/1000; 430 x 380 mm; to 75 mm lap; on 22 x 38 mm battens and type 1F reinforced underlay					
Roof coverings	0.40	5.55	10.77	m2	16.32
Extra over coverings for					
fixing every tile	0.03	0.42	0.35	m2	0.77
25 x 38 mm battens in lieu	-	-	-	m2	0.08
eaves; eave filler	0.03	0.42	0.11	m	0.53
verges; extra single undercloak course of plain tiles	0.29	4.02	3.65	m	7.67
dry verge system; extruded white pvc	0.23	3.19	8.72	m	11.91
Ridge end unit	0.03	0.42	2.03	nr	2.45
valley trough tiles; cutting both sides	0.63	8.74	16.69	m	25.43
universal 'Delta' ridge tiles	0.58	8.05	5.48	m	13.53
universal 'Delta' hip tiles; cutting both sides	0.75	10.40	8.98	m	19.38
universal 'Delta' mono-pitch ridge tiles	0.58	8.05	9.40	m	17.45
gas flue terminal; 'Delta' type	0.58	8.05	41.51	nr	49.56
Holes for pipes and the like	0.23	3.19	-	nr	3.19
Concrete interlocking tiles; Redland 'Norfolk' smooth finish pantiles or similar; PC £389.00/1000; 381 x 229 mm; to 75 mm lap; on 22 x 38 mm battens and type 1F reinforced underlay					
Roof coverings	0.52	7.21	9.96	m2	17.17
Extra over coverings for					
fixing every tile	0.06	0.83	0.31	m2	1.14
25 x 38 mm battens in lieu	-	-	-	m2	0.09
eaves; eave filler	0.06	0.83	0.63	m	1.46
verges; extra single undercloak course of plain tiles	0.35	4.85	1.14	m	5.99
valley trough tiles; cutting both sides	0.69	9.57	16.51	m	26.08
segmental ridge tiles	0.69	9.57	8.03	m	17.60
segmental hip tiles; cutting both sides	0.86	11.93	11.36	m	23.29
Holes for pipes and the like	0.23	3.19	-	nr	3.19
Concrete interlocking tiles; Redland 'Regent' granule finish bold roll tiles or similar; PC £594.00/1000; 418 x 332 mm; to 75 mm lap; on 22 x 38 mm battens and type 1F reinforced underlay					
Roof coverings	0.40	5.55	9.37	m2	14.92
Extra over coverings for					
fixing every tile	0.05	0.69	0.52	m2	1.21
25 x 38 mm battens in lieu	-	-	-	m2	0.08
eaves; eave filler	0.06	0.83	0.78	m	1.61
verges; extra single undercloak course of plain tiles	0.29	4.02	2.24	m	6.26
dry verge system; extruded white pvc	0.17	2.36	6.99	m	9.35
Ridge end unit	0.03	0.42	2.03	nr	2.45
cloaked verge system	0.17	2.36	3.91	m	6.27
Blocked end ridge unit	0.03	0.42	3.94	nr	4.36
valley trough tiles; cutting both sides	0.63	8.74	15.86	m	24.60
segmental ridge tiles; tile slips	0.63	8.74	8.03	m	16.77
segmental hip tiles; tile slips; cutting both sides	0.81	11.24	10.70	m	21.94
dry ridge system; segmental ridge tiles; including fixing straps; 'Nuralite' fillets and seals	0.29	4.02	17.69	m	21.71
half round mono-pitch ridge tiles	0.63	8.74	13.66	m	22.40
gas flue terminal; half round type	0.58	8.05	41.51	nr	49.56
Holes for pipes and the like	0.23	3.19	-	nr	3.19

H CLADDING/COVERING Including overheads and profit at 12.50%	Labour hours	Labour £	Material £	Unit	Total rate £
H60 CLAY/CONCRETE ROOF TILING - cont'd					
Concrete interlocking tiles; Redland **'Renown' granule finish tiles or similar;** **PC £534.00/1000; 418 x 330 mm;** **to 75 mm lap; on 22 x 38 mm battens and** **type 1F reinforced underlay**					
Roof coverings	0.40	5.55	8.53	m2	14.08
Extra over coverings for					
fixing every tile	0.03	0.42	0.43	m2	0.85
25 x 38 mm battens in lieu	-	-	-	m2	0.08
verges; extra single undercloak course of					
plain tiles	0.29	4.02	1.14	m	5.16
dry verge system; extruded white pvc	0.17	2.36	6.99	m	9.35
Ridge end unit	0.03	0.42	2.03	nr	2.45
cloaked verge system	0.17	2.36	4.13	m	6.49
Blocked end ridge unit	0.03	0.42	3.94	nr	4.36
valley trough tiles; cutting both sides	0.63	8.74	15.59	m	24.33
segmental ridge tiles	0.58	8.05	5.17	m	13.22
segmental hip tiles; cutting both sides	0.75	10.40	7.57	m	17.97
dry ridge system; segmental ridge tiles; including fixing straps; 'Nuralite' fillets and seals	0.29	4.02	17.69	m	21.71
half round mono-pitch ridge tiles	0.58	8.05	9.01	m	17.06
gas flue terminal; half round type	0.58	8.05	41.51	nr	49.56
Holes for pipes and the like	0.23	3.19	-	nr	3.19
Concrete interlocking tiles; Redland **'Stonewold' smooth finish tiles or similar;** **PC £739.00/1000; 430 x 380 mm;** **to 75 mm lap; on 22 x 38 mm battens and** **type 1F reinforced underlay**					
Roof coverings	0.40	5.55	9.46	m2	15.01
Extra over coverings for					
fixing every tile	0.03	0.42	0.35	m2	0.77
25 x 38 mm battens in lieu	-	-	-	m2	0.08
verges; extra single undercloak course of					
plain tiles	0.35	4.85	3.43	m	8.28
dry verge system; extruded white pvc	0.23	3.19	8.49	m	11.68
Ridge end unit	0.03	0.42	2.03	nr	2.45
valley trough tiles; cutting both sides	0.63	8.74	16.18	m	24.92
universal 'Stonewold' ridge tiles	0.58	8.05	5.48	m	13.53
universal 'Stonewold' hip tiles; cutting both sides	0.75	10.40	8.47	m	18.87
dry ridge system; universal 'Stonewold' ridge tiles; including fixing straps; 'Nuralite' fillets and seals	0.29	4.02	18.22	m	22.24
universal 'Stonewold' mono-pitch ridge tiles	0.58	8.05	9.40	m	17.45
gas flue terminal; 'Stonewold' type	0.58	8.05	41.51	nr	49.56
Holes for pipes and the like	0.23	3.19	-	nr	3.19
Concrete plain tiles; BS 473 and 550 **group A; PC £210.00/1000; 267 x 165 mm;** **on 19 x 38 mm battens and** **type 1F reinforced underlay**					
Roof coverings; to 64 mm lap	0.75	10.40	19.16	m2	29.56
Wall coverings; to 38 mm lap	0.92	12.76	17.15	m2	29.91
Extra over coverings for					
25 x 38 mm battens in lieu	-	-	-	m2	0.57
ornamental tiles in lieu	-	-	-	m2	7.49
double course at eaves	0.29	4.02	1.61	m	5.63
verges; extra single undercloak course	0.38	5.27	2.37	m	7.64
valley tiles; cutting both sides	0.75	10.40	17.75	m	28.15
bonnet hip tiles; cutting both sides	0.92	12.76	18.89	m	31.65
external vertical angle tiles; supplementary nail fixings	0.46	6.38	11.40	m	17.78
segmental ridge tiles	0.58	8.05	7.51	m	15.56
segmental hip tiles; cutting both sides	0.86	11.93	8.17	m	20.10
Holes for pipes and the like	0.23	3.19	-	nr	3.19

H CLADDING/COVERING Including overheads and profit at 12.50%	Labour hours	Labour £	Material £	Unit	Total rate £
Sundries					
Hip irons					
galvanized mild steel; fixing with screws	0.11	1.53	1.89	nr	3.42
Fixing					
lead soakers (supply included elsewhere)	0.09	1.25	-	nr	1.25
Pressure impregnated softwood counter battens; 25 x 50 mm					
450 mm centres	0.08	1.11	0.88	m2	1.99
600 mm centres	0.06	0.83	0.66	m2	1.49
Underlay; BS 747 type 1B; bitumen felt **weighing 14 kg/10 m2;** **PC £14.82/20m2; 75 mm laps**					
To sloping or vertical surfaces	0.03	0.42	0.90	m2	1.32
Underlay; BS 747 type 1F; reinforced bitumen **felt; weighing 22.5 kg/15 m2; PC £17.00/15m2; 75 mm laps** **(prices included within tiling rates)**					
To sloping or vertical surfaces	0.03	0.42	1.37	m2	1.79
H61 FIBRE CEMENT SLATING					
Asbestos-cement slates; Eternit or similar; **to 75 mm lap; on 19 x 50 mm battens and** **type 1F reinforced underlay**					
Coverings; 500 x 250 mm 'blue/black' slates					
roof coverings	0.75	10.40	19.05	m2	29.45
wall coverings	0.98	13.59	19.05	m2	32.64
Coverings; 600 x 300 mm 'blue/black' slates					
roof coverings	0.58	8.05	18.10	m2	26.15
wall coverings	0.75	10.40	18.10	m2	28.50
Extra over slate coverings for					
double course at eaves	0.29	4.02	3.04	m	7.06
verges; extra single undercloak course	0.38	5.27	2.38	m	7.65
open valleys; cutting both sides	0.23	3.19	6.36	m	9.55
valley gutters; cutting both sides	0.63	8.74	21.38	m	30.12
half round ridge tiles	0.58	8.05	15.72	m	23.77
Stop end	0.11	1.53	5.43	nr	6.96
roll top ridge tiles	0.58	8.05	19.67	m	27.72
Stop end	0.11	1.53	8.13	nr	9.66
mono-pitch ridges	0.58	8.05	22.68	m	30.73
Stop end	0.11	1.53	24.64	nr	26.17
duo-pitch ridges	0.58	8.05	19.43	m	27.48
Stop end	0.11	1.53	18.08	nr	19.61
mitred hips; cutting both sides	0.23	3.19	6.36	m	9.55
half round hip tiles; cutting both sides	0.75	10.40	22.09	m	32.49
Holes for pipes and the like	0.23	3.19	-	nr	3.19
Asbestos-free artificial slates; Eternit **'2000' or similar; to 75 mm lap; on** **19 x 50 mm battens and type 1F reinforced** **underlay**					
Coverings; 400 x 200 mm 'blue/black' slates					
roof coverings	0.92	12.76	21.93	m2	34.69
wall coverings	1.20	16.65	21.93	m2	38.58
Coverings; 500 x 250 mm 'blue/black' slates					
roof coverings	0.75	10.40	19.65	m2	30.05
wall coverings	0.98	13.59	19.65	m2	33.24
Coverings; 600 x 300 mm 'blue/black' slates					
roof coverings	0.58	8.05	18.78	m2	26.83
wall coverings	0.75	10.40	18.78	m2	29.18
Coverings; 600 x 300 mm 'brown' or 'rose nuit' slates					
roof coverings	0.58	8.05	18.78	m2	26.83
wall coverings	0.75	10.40	18.78	m2	29.18

H CLADDING/COVERING Including overheads and profit at 12.50%	Labour hours	Labour £	Material £	Unit	Total rate £

H61 FIBRE CEMENT SLATING - cont'd

Asbestos-free artificial slates; Eternit
'2000' or similar; to 75 mm lap; on
19 x 50 mm battens and type 1F reinforced
underlay - cont'd

	Labour hours	Labour £	Material £	Unit	Total rate £
Extra over slate coverings for					
double course at eaves	0.29	4.02	3.17	m	7.19
verges; extra single undercloak course	0.38	5.27	2.48	m	7.75
open valleys; cutting both sides	0.23	3.19	6.66	m	9.85
valley gutters; cutting both sides	0.63	8.74	21.68	m	30.42
half round ridge tiles	0.58	8.05	15.72	m	23.77
Stop end	0.11	1.53	5.43	nr	6.96
roll top ridge tiles	0.69	9.57	19.67	m	29.24
Stop end	0.11	1.53	8.13	nr	9.66
mono-pitch ridges	0.58	8.05	18.56	m	26.61
Stop end	0.11	1.53	24.64	nr	26.17
duo-pitch ridges	0.58	8.05	18.56	m	26.61
Stop end	0.11	1.53	18.08	nr	19.61
mitred hips; cutting both sides	0.23	3.19	6.66	m	9.85
half round hip tiles; cutting both sides	0.75	10.40	22.38	m	32.78
Holes for pipes and the like	0.23	3.19	-	nr	3.19

ALTERNATIVE SLATE PRICES (£/1000)

	£		£		£		£
Natural slates							
Greaves Portmadoc Welsh blue-grey							
Mediums (Class 1)							
305x255 mm	505.00	405x255 mm	750.00	510x255 mm	1210.00	610x305 mm	2216.50
355x255 mm	595.00	460x255 mm	960.00	560x305 mm	1590.00	610x355 mm	2150.00
Strongs (Class 2)							
305x255 mm	490.00	405x255 mm	700.00	510x255 mm	1155.00	610x305 mm	1885.00
355x255 mm	575.00	460x255 mm	900.00	560x305 mm	1540.00	610x355 mm	2050.00

Discounts of 2.5 - 15% available depending on quantity/status

NOTE: The following items of slate roofing unless otherwise described, include for conventional fixing assuming 'normal exposure' with appropriate nails and/or rivets or clips to pressure impregnated softwood battens fixed with galvanized nails; Prices also include for all bedding and pointing at verges; beneath ridge tiles, etc..	Labour hours	Labour £	Material £	Unit	Total rate £

H62 NATURAL SLATING

Natural slates; BS 680 Part 2; Welsh blue;
uniform size; to 75 mm lap; on 25 x 50 mm
battens and type 1F reinforced underlay

	Labour hours	Labour £	Material £	Unit	Total rate £
Coverings; 405 x 255 mm slates					
roof coverings	1.05	14.56	27.66	m2	42.22
wall coverings	1.30	18.03	27.66	m2	45.69
Coverings; 510 x 255 mm slates					
roof coverings	0.86	11.93	32.35	m2	44.28
wall coverings	1.05	14.56	32.35	m2	46.91
Coverings; 610 x 305 mm slates					
roof coverings	0.69	9.57	35.53	m2	45.10
wall coverings	0.86	11.93	35.53	m2	47.46

H CLADDING/COVERING Including overheads and profit at 12.50%	Labour hours	Labour £	Material £	Unit	Total rate £
Extra over coverings for					
double course at eaves	0.35	4.85	6.38	m	11.23
verges; extra single undercloak course	0.48	6.66	5.25	m	11.91
open valleys; cutting both sides	0.25	3.47	13.96	m	17.43
blue/black glazed ware 152 mm half round					
ridge tiles	0.58	8.05	10.20	m	18.25
blue/black glazed ware 125 x 125 mm plain					
angle ridge tiles	0.58	8.05	13.53	m	21.58
mitred hips; cutting both sides	0.25	3.47	13.96	m	17.43
blue/black glazed ware 152 mm half round					
hip tiles; cutting both sides	0.81	11.24	24.16	m	35.40
blue/black glazed ware 125 x 125 mm plain					
angle hip tiles; cutting both sides	0.81	11.24	27.50	m	38.74
Holes for pipes and the like	0.23	3.19	-	nr	3.19
Natural slates; Westmorland green; PC £1265.00/t; random lengths; 457 - 229 mm proportionate widths to 75 mm lap; in diminishing courses; on 25 x 50 mm battens and type 1F underlay					
Roof coverings	1.30	18.03	94.95	m2	112.98
Wall coverings	1.65	22.89	94.95	m2	117.84
Extra over coverings for					
double course at eaves	0.76	10.54	16.26	m	26.80
verges; extra single undercloak course					
slates 152 mm wide	0.86	11.93	12.44	m	24.37
Holes for pipes and the like	0.35	4.85	-	nr	4.85

H63 RECONSTRUCTED STONE SLATING/TILING

	Labour hours	Labour £	Material £	Unit	Total rate £
Reconstructed stone slates; Bradstone 'Cotswold' style or similar; PC £18.36/m2; random lengths 550 - 300 mm; proportional widths; to 80 mm lap; in diminishing courses; on 25 x 50 mm battens and type 1F reinforced underlay					
Roof coverings	1.20	16.65	27.32	m2	43.97
Wall coverings	1.55	21.50	28.85	m2	50.35
Extra over coverings for					
double course at eaves	0.58	8.05	4.45	m	12.50
verges; extra single undercloak course	0.76	10.54	3.62	m	14.16
open valleys; cutting both sides	0.52	7.21	9.29	m	16.50
ridge tile	0.76	10.54	10.90	m	21.44
mitred hips; cutting both sides	0.52	7.21	9.29	m	16.50
hip tile; cutting both sides	1.20	16.65	19.70	m	36.35
Holes for pipes and the like	0.35	4.85	-	nr	4.85
Reconstructed stone slates; Bradstone 'Moordale' style or similar; PC £18.84/m2; random lengths 550 - 450 mm; proportional widths; to 80 mm lap; in diminishing courses; on 25 x 50 mm battens and type 1F reinforced underlay					
Roof coverings	1.10	15.26	27.38	m2	42.64
Wall coverings	1.45	20.11	29.18	m2	49.29
Extra over coverings for					
double course at eaves	0.58	8.05	4.56	m	12.61
verges; extra single undercloak course	0.76	10.54	3.71	m	14.25
ridge tile	0.76	10.54	10.72	m	21.26
mitred hips; cutting both sides	0.52	7.21	9.54	m	16.75
Holes for pipes and the like	0.35	4.85	-	nr	4.85

H CLADDING/COVERING Including overheads and profit at 12.50%	Labour hours	Labour £	Material £	Unit	Total rate £
H64 TIMBER SHINGLING					
Red cedar sawn shingles preservative treated; PC £32.04 per bundle (2.11 m2 cover); uniform length 450 mm; varying widths; to 125 mm lap; on 25 x 100 mm battens and type 1F reinforced underlay					
Roof coverings	1.20	16.65	25.74	m2	42.39
Wall coverings	1.55	21.50	25.74	m2	47.24
Extra over coverings for					
double course at eaves; three rows of battens	0.35	4.85	5.04	m	9.89
verges; extra single undercloak course	0.58	8.05	5.28	m	13.33
open valleys; cutting both sides	0.23	3.19	3.97	m	7.16
selected shingles to form hip capping	1.40	19.42	10.03	m	29.45
Double starter course on last	0.23	3.19	2.18	nr	5.37
Holes for pipes and the like	0.17	2.36	-	nr	2.36
H71 LEAD SHEET COVERINGS/FLASHINGS					
Milled lead; BS 1178; PC £1268.80/t					
1.25 mm (code 3) roof coverings					
flat	3.10	33.10	21.27	m2	54.37
sloping 10 - 50 degrees	3.45	36.83	21.27	m2	58.10
vertical or sloping over 50 degrees	3.80	40.57	21.27	m2	61.84
1.80 mm (code 4) roof coverings					
flat	3.35	35.77	30.62	m2	66.39
sloping 10 - 50 degrees	3.70	39.50	30.62	m2	70.12
vertical or sloping over 50 degrees	4.05	43.24	30.62	m2	73.86
1.80 mm (code 4) dormer coverings					
flat	3.90	41.64	31.34	m2	72.98
sloping 10 - 50 degrees	4.50	48.04	31.34	m2	79.38
vertical or sloping over 50 degrees	4.85	51.78	31.34	m2	83.12
2.24 mm (code 5) roof coverings					
flat	3.55	37.90	38.10	m2	76.00
sloping 10 - 50 degrees	3.90	41.64	38.10	m2	79.74
vertical or sloping over 50 degrees	4.25	45.37	38.10	m2	83.47
2.24 mm (code 5) dormer coverings					
flat	4.25	45.37	39.01	m2	84.38
sloping 10 - 50 degrees	4.70	50.18	39.01	m2	89.19
vertical or sloping over 50 degrees	5.20	55.52	39.01	m2	94.53
2.50 mm (code 6) roof coverings					
flat	3.80	40.57	42.54	m2	83.11
sloping 10 - 50 degrees	4.15	44.31	42.54	m2	86.85
vertical or sloping over 50 degrees	4.50	48.04	42.54	m2	90.58
2.50 mm (code 6) dormer coverings					
flat	4.60	49.11	43.56	m2	92.67
sloping 10 - 50 degrees	4.95	52.85	43.56	m2	96.41
vertical or sloping over 50 degrees	5.40	57.65	43.56	m2	101.21
Dressing over glazing bars and glass	0.38	4.06	-	m	4.06
Soldered dot	1.45	15.48	2.52	nr	18.00
Copper nailing 75 mm spacing	0.23	2.46	0.32	m	2.78
1.80 mm (code 4) lead flashings, etc.					
Flashings; wedging into grooves					
150 mm girth	0.92	9.82	4.63	m	14.45
240 mm girth	1.05	11.21	7.44	m	18.65
Stepped flashings; wedging into grooves					
180 mm girth	1.05	11.21	5.55	m	16.76
270 mm girth	1.15	12.28	8.32	m	20.60
Linings to sloping gutters					
390 mm girth	1.40	14.95	12.08	m	27.03
450 mm girth	1.50	16.01	13.87	m	29.88
750 mm girth	1.85	19.75	23.07	m	42.82
Cappings to hips or ridges					
450 mm girth	1.70	18.15	13.87	m	32.02
600 mm girth	1.85	19.75	18.53	m	38.28

H CLADDING/COVERING Including overheads and profit at 12.50%	Labour hours	Labour £	Material £	Unit	Total rate £
Soakers					
200 x 200 mm	0.17	1.81	1.39	nr	3.20
300 x 300 mm	0.23	2.46	3.12	nr	5.58
Saddle flashings; at intersections of hips					
and ridges; dressing and bossing					
450 x 600 mm	2.05	21.89	9.34	nr	31.23
Slates; with 150 mm high collar					
450 x 450 mm; to suit 50 mm pipe	1.95	20.82	13.08	nr	33.90
450 x 450 mm; to suit 100 mm pipe	2.30	24.56	18.59	nr	43.15
2.24 mm (code 5) lead flashings, etc.					
Flashings; wedging into grooves					
150 mm girth	0.92	9.82	5.77	m	15.59
240 mm girth	1.05	11.21	9.19	m	20.40
Stepped flashings; wedging into grooves					
180 mm girth	1.05	11.21	6.92	m	18.13
270 mm girth	1.15	12.28	10.36	m	22.64
Linings to sloping gutters					
390 mm girth	1.40	14.95	15.00	m	29.95
450 mm girth	1.50	16.01	17.27	m	33.28
750 mm girth	1.85	19.75	28.84	m	48.59
Cappings to hips or ridges					
450 mm girth	1.70	18.15	17.27	m	35.42
600 mm girth	1.85	19.75	23.06	m	42.81
Soakers					
200 x 200 mm	0.17	1.81	1.73	nr	3.54
300 x 300 mm	0.23	2.46	3.82	nr	6.28
Saddle flashings; at intersections of hips					
and ridges; dressing and bossing					
450 x 600 mm	2.05	21.89	12.97	nr	34.86
Slates; with 150 mm high collar					
450 x 450 mm; to suit 50 mm pipe	1.95	20.82	14.79	nr	35.61
450 x 450 mm; to suit 100 mm pipe	2.30	24.56	20.29	nr	44.85

H72 ALUMINIUM SHEET COVERINGS/FLASHINGS

Aluminium roofing; commercial grade; PC £2500.00/t					
0.90 mm roof coverings					
flat	3.45	36.83	7.12	m2	43.95
sloping 10 - 50 degrees	3.80	40.57	7.12	m2	47.69
vertical or sloping over 50 degrees	4.15	44.31	7.12	m2	51.43
0.90 mm dormer coverings					
flat	4.15	44.31	7.28	m2	51.59
sloping 10 - 50 degrees	4.60	49.11	7.28	m2	56.39
vertical or sloping over 50 degrees	5.05	53.92	7.28	m2	61.20
Aluminium nailing; 75 mm spacing	0.23	2.46	0.20	m	2.66
0.90 mm commercial grade aluminium					
flashings, etc.					
Flashings; wedging into grooves					
150 mm girth	0.92	9.82	1.11	m	10.93
240 mm girth	1.05	11.21	1.80	m	13.01
300 mm girth	1.20	12.81	2.23	m	15.04
Stepped flashings; wedging into grooves					
180 mm girth	1.05	11.21	1.33	m	12.54
270 mm girth	1.15	12.28	2.04	m	14.32

H73 COPPER SHEET COVERINGS/FLASHINGS

Copper roofing; BS 2870					
0.56 mm (24 swg) roof coverings					
flat	3.70	39.50	24.44	m2	63.94
sloping 10 - 50 degrees	4.05	43.24	24.44	m2	67.68
vertical or sloping over 50 degrees	4.35	46.44	24.44	m2	70.88
0.56 mm (24 swg) dormer coverings					
flat	4.35	46.44	25.03	m2	71.47
sloping 10 - 50 degrees	4.85	51.78	25.03	m2	76.81
vertical or sloping over 50 degrees	5.30	56.58	25.03	m2	81.61

H CLADDING/COVERING Including overheads and profit at 12.50%	Labour hours	Labour £	Material £	Unit	Total rate £
H73 COPPER SHEET COVERINGS/FLASHINGS - cont'd					
Copper roofing; BS 2870 - cont'd					
0.61 mm (23 swg) roof coverings					
flat PC £4125.00/t	3.70	39.50	28.27	m2	67.77
sloping 10 - 50 degrees	4.05	43.24	28.27	m2	71.51
vertical or sloping over 50 degrees	4.35	46.44	28.27	m2	74.71
0.61 mm (23 swg) dormer coverings					
flat	4.35	46.44	28.93	m2	75.37
sloping 10 - 50 degrees	4.85	51.78	28.93	m2	80.71
vertical or sloping over 50 degrees	5.30	56.58	28.93	m2	85.51
Copper nailing; 75 mm spacing	0.23	2.46	0.32	m	2.78
0.56 mm copper flashings, etc.					
Flashings; wedging into grooves					
150 mm girth	0.92	9.82	3.84	m	13.66
240 mm girth	1.05	11.21	6.16	m	17.37
300 mm girth	1.20	12.81	7.68	m	20.49
Stepped flashings; wedging into grooves					
180 mm girth	1.05	11.21	4.62	m	15.83
270 mm girth	1.15	12.28	6.94	m	19.22
0.61 mm copper flashings, etc.					
Flashings; wedging into grooves					
150 mm girth	0.92	9.82	4.44	m	14.26
240 mm girth	1.05	11.21	7.15	m	18.36
300 mm girth	1.20	12.81	8.88	m	21.69
Stepped flashings; wedging into grooves					
180 mm girth	1.05	11.21	5.36	m	16.57
270 mm girth	1.15	12.28	8.02	m	20.30
H74 ZINC SHEET COVERINGS/FLASHINGS					
Zinc BS 849; PC £2500.00/t					
0.81 mm roof coverings					
flat	3.70	39.50	16.80	m2	56.30
sloping 10 - 50 degrees	4.05	43.24	16.80	m2	60.04
vertical or sloping over 50 degrees	4.35	46.44	16.80	m2	63.24
0.81 mm dormer coverings					
flat	4.35	46.44	17.21	m2	63.65
sloping 10 - 50 degrees	4.85	51.78	17.21	m2	68.99
vertical or sloping over 50 degrees	5.30	56.58	17.21	m2	73.79
0.81 mm zinc flashings, etc.					
Flashings; wedging into grooves					
150 mm girth	0.92	9.82	2.63	m	12.45
240 mm girth	1.05	11.21	4.24	m	15.45
300 mm girth	1.20	12.81	5.29	m	18.10
Stepped flashings; wedging into grooves					
180 mm girth	1.05	11.21	3.19	m	14.40
270 mm girth	1.15	12.28	4.76	m	17.04
H75 STAINLESS STEEL SHEET COVERINGS/ FLASHINGS					
Terne coated stainless steel roofing					
0.38 mm roof coverings					
flat	3.70	39.50	37.65	m2	77.15
sloping 10 - 50 degrees	4.05	43.24	37.65	m2	80.89
vertical or sloping over 50 degrees	4.35	46.44	37.65	m2	84.09
Flashings; wedging into grooves					
150 mm girth	1.15	12.28	7.89	m	20.17
240 mm girth	1.30	13.88	11.43	m	25.31
300 mm girth	1.50	16.01	13.81	m	29.82
Stepped flashings; wedging into grooves					
180 mm girth	1.30	13.88	9.07	m	22.95
270 mm girth	1.45	15.48	12.62	m	28.10

H CLADDING/COVERING Including overheads and profit at 12.50%	Labour hours	Labour £	Material £	Unit	Total rate £

H76 FIBRE BITUMEN THERMOPLASTIC SHEET
COVERINGS/FLASHINGS

Glass fibre reinforced bitumen strip slates;
'Langhome 1000' or similar; PC £19.11/2 m2;
strip pack; 900 x 300 mm mineral finish;
fixed to external plywood boarding
(measured separately)

Roof coverings	0.29	4.02	9.52	m2	13.54
Wall coverings	0.46	6.38	9.52	m2	15.90
Extra over coverings for					
double course at eaves; felt soaker	0.23	3.19	1.90	m	5.09
verges; felt soaker	0.29	4.02	1.80	m	5.82
valley slate; cut to shape; felt soaker					
both sides; cutting both sides	0.52	7.21	4.59	m	11.80
ridge slate; cut to shape	0.35	4.85	2.37	m	7.22
hip slate; cut to shape; cutting both sides	0.52	7.21	4.54	m	11.75
Holes for pipes and the like	0.06	0.83	-	nr	0.83

'Evode Flashband' sealing strips and
flashings; special grey finish
Flashings; wedging at top if required;
pressure bonded; flashband primer before
application; to walls

100 mm girth	0.29	3.10	1.19	m	4.29
150 mm girth	0.38	4.06	1.78	m	5.84
225 mm girth	0.46	4.91	2.69	m	7.60
300 mm girth	0.52	5.55	3.50	m	9.05
450 mm girth	0.69	7.37	6.79	m	14.16

'Nuralite' semi-rigid bitumen membrane
roofing DC12 jointing strip system
PC £27.70/3m2

Roof coverings					
flat	0.58	6.19	15.14	m2	21.33
sloping 10 - 50 degrees	0.63	6.73	15.14	m2	21.87
vertical or sloping over 50 degrees	0.81	8.65	15.14	m2	23.79

'Nuralite' flashings, etc.
Flashings; wedging into grooves

150 mm girth	0.58	6.19	1.92	m	8.11
200 mm girth	0.69	7.37	2.64	m	10.01
250 mm girth	0.69	7.37	3.02	m	10.39
Linings to sloping gutters					
450 mm girth	0.98	10.46	6.55	m	17.01
490 mm girth	1.10	11.74	7.21	m	18.95
Cappings to hips or ridges					
450 mm girth	1.05	11.21	9.67	m	20.88
600 mm girth	1.15	12.28	14.53	m	26.81
Undersoakers					
150 mm girth	0.14	1.49	0.98	nr	2.47
250 mm girth	0.21	2.24	1.03	nr	3.27
Oversoakers					
200 mm girth	0.17	1.81	9.99	nr	11.80
Cavity tray without apron					
450 mm long	0.52	5.55	1.86	nr	7.41
250 mm long	0.29	3.10	1.44	nr	4.54
350 mm long	0.40	4.27	1.73	nr	6.00

J WATERPROOFING Including overheads and profit at 5.00%	Labour hours	Labour £	Material £	Unit	Total rate £
J10 SPECIALIST WATERPROOF RENDERING					
'Sika' waterproof rendering; steel trowelled					
20 mm work to walls; three coat; to concrete base					
over 300 mm wide	-	-	-	m2	35.44
not exceeding 300 mm wide	-	-	-	m2	55.52
25 mm work to walls; three coat; to concrete base					
over 300 mm wide	-	-	-	m2	40.16
not exceeding 300 mm wide	-	-	-	m2	63.79
40 mm work to walls; four coat; to concrete base					
over 300 mm wide	-	-	-	m2	61.43
not exceeding 300 mm wide	-	-	-	m2	94.50
J20 MASTIC ASPHALT TANKING/DAMP PROOF MEMBRANES					
Mastic asphalt to BS 1097					
13 mm one coat coverings to concrete base; flat; subsequently covered					
over 300 mm wide	-	-	-	m2	8.30
225 - 300 mm wide	-	-	-	m2	19.67
150 - 225 mm wide	-	-	-	m2	19.95
not exceeding 150 mm wide	-	-	-	m2	24.61
20 mm two coat coverings to concrete base; flat; subsequently covered					
over 300 mm wide	-	-	-	m2	14.47
225 - 300 mm wide	-	-	-	m2	30.28
150 - 300 mm wide	-	-	-	m2	32.25
not exceeding 150 mm wide	-	-	-	m2	36.18
30 mm three coat coverings to concrete base; flat; subsequently covered					
over 300 mm wide	-	-	-	m2	20.71
225 - 300 mm wide	-	-	-	m2	45.42
150 - 225 mm wide	-	-	-	m2	48.38
not exceeding 150 mm wide	-	-	-	m2	54.28
13 mm two coat coverings to brickwork base; vertical; subsequently covered					
over 300 mm wide	-	-	-	m2	44.90
225 - 300 mm wide	-	-	-	m2	35.80
150 - 225 mm wide	-	-	-	m2	39.63
not exceeding 150 mm wide	-	-	-	m2	47.25
20 mm three coat coverings to brickwork base; vertical; subsequently covered					
over 300 mm wide	-	-	-	m2	61.33
225 - 300 mm wide	-	-	-	m2	53.70
150 - 225 mm wide	-	-	-	m2	59.44
not exceeding 150 mm wide	-	-	-	m2	70.88
Turning 20 mm into groove	-	-	-	m	0.89
Internal angle fillets; subsequently covered	-	-	-	m	5.82
Mastic asphalt to BS 6577					
13 mm one coat coverings to concrete base; flat; subsequently covered					
over 300 mm wide	-	-	-	m2	14.05
225 - 300 mm wide	-	-	-	m2	22.71
150 - 225 mm wide	-	-	-	m2	24.20
not exceeding 150 mm wide	-	-	-	m2	27.20
20 mm two coat coverings to concrete base; flat; subsequently covered					
over 300 mm wide	-	-	-	m2	20.78
225 - 300 mm wide	-	-	-	m2	34.95
150 - 225 mm wide	-	-	-	m2	37.25
not exceeding 150 mm wide	-	-	-	m2	41.84

J WATERPROOFING Including overheads and profit at 5.00%	Labour hours	Labour £	Material £	Unit	Total rate £
30 mm three coat coverings to concrete base; flat; subsequently covered					
over 300 mm wide	-	-	-	m2	30.18
225 - 300 mm wide	-	-	-	m2	52.43
150 - 225 mm wide	-	-	-	m2	55.88
not exceeding 150 mm wide	-	-	-	m2	62.76
13 mm two coat coverings to brickwork base; vertical; subsequently covered					
over 300 mm wide	-	-	-	m2	49.01
225 - 300 mm wide	-	-	-	m2	39.57
150 - 225 mm wide	-	-	-	m2	43.40
not exceeding 150 mm wide	-	-	-	m2	51.06
20 mm three coat coverings to brickwork base; vertical; subsequently covered					
over 300 mm wide	-	-	-	m2	67.64
225 - 300 mm wide	-	-	-	m2	59.36
150 - 225 mm wide	-	-	-	m2	65.10
not exceeding 150 mm wide	-	-	-	m2	76.60
Turning 20 mm into groove	-	-	-	m	0.70
Internal angle fillets; subsequently covered	-	-	-	m	6.51

J21 MASTIC ASPHALT ROOFING/INSULATION/FINISHES

Mastic asphalt to BS 988 20 mm two coat coverings; felt isolating membrane; to concrete (or timber) base; flat or to falls or slopes not exceeding 10 degrees from horizontal					
over 300 mm wide	-	-	-	m2	15.34
225 - 300 mm wide	-	-	-	m2	30.28
150 - 225 mm wide	-	-	-	m2	32.25
not exceeding 150 mm wide	-	-	-	m2	36.21
Add to the above for covering with:					
10 mm limestone chippings in hot bitumen	-	-	-	m2	3.52
coverings with solar reflective paint	-	-	-	m2	3.43
300 x 300 x 8 mm g.r.p. tiles in hot bitumen	-	-	-	m2	51.92
Cutting to line; jointing to old asphalt	-	-	-	m	5.81
13 mm two coat skirtings to brickwork base					
not exceeding 150 mm girth	-	-	-	m	12.91
150 - 225 mm girth	-	-	-	m	14.74
225 - 300 mm girth	-	-	-	m	16.57
13 mm three coat skirtings; expanded metal lathing reinforcement nailed to timber base					
not exceeding 150 mm girth	-	-	-	m	20.29
150 - 225 mm girth	-	-	-	m	24.10
225 - 300 mm girth	-	-	-	m	27.93
13 mm two coat fascias to concrete base					
not exceeding 150 mm girth	-	-	-	m	18.32
150 - 225 mm girth	-	-	-	m	20.17
20 mm two coat linings to channels to concrete base					
not exceeding 150 mm girth	-	-	-	m	31.23
150 - 225 mm girth	-	-	-	m	33.06
225 - 300 mm girth	-	-	-	m	34.89
20 mm two coat lining to cesspools					
250 x 150 x 150 mm deep	-	-	-	nr	42.19
Collars around pipes, standards and like members	-	-	-	nr	14.34

J WATERPROOFING Including overheads and profit at 12.50% & 5.00%	Labour hours	Labour £	Material £	Unit	Total rate £
J21 MASTIC ASPHALT ROOFING/INSULATION/ **FINISHES - cont'd**					
Mastic asphalt to BS 6577					
20 mm two coat coverings; felt isolating					
membrane; to concrete (or timber) base; flat					
or to falls or slopes not exceeding 10					
degrees from horizontal					
over 300 mm wide	-	-	-	m2	16.57
225 - 300 mm wide	-	-	-	m2	34.95
150 - 225 mm wide	-	-	-	m2	37.25
not exceeding 150 mm wide	-	-	-	m2	41.84
Add to the above for covering with:					
10 mm limestone chippings in hot bitumen	-	-	-	m2	3.52
solar reflective paint	-	-	-	m2	3.43
300 x 300 x 8 mm g.r.p. tiles in					
hot bitumen	-	-	-	m2	51.92
Cutting to line; jointing to old asphalt	-	-	-	m	5.81
13 mm two coat skirtings to brickwork base					
not exceeding 150 mm girth	-	-	-	m	14.15
150 - 225 mm girth	-	-	-	m	16.26
225 - 300 mm girth	-	-	-	m	18.35
13 mm three coat skirtings; expanded metal					
lathing reinforcement nailed to timber base					
not exceeding 150 mm girth	-	-	-	m	21.84
150 - 225 mm girth	-	-	-	m	26.03
225 - 300 mm girth	-	-	-	m	30.25
13 mm two coat fascias to concrete base					
not exceeding 150 mm girth	-	-	-	m	20.43
150 - 225 mm girth	-	-	-	m	22.54
20 mm two coat linings to channels to					
concrete base					
not exceeding 150 mm girth	-	-	-	m	34.59
150 - 225 mm girth	-	-	-	m	36.70
225 - 300 mm girth	-	-	-	m	38.78
20 mm two coat lining to cesspools					
250 x 150 x 150 mm deep	-	-	-	nr	42.19
Collars around pipes, standards and like					
members	-	-	-	nr	14.34
Accessories					
Eaves trim; extruded aluminium alloy;					
working asphalt into trim					
'Alutrim'; type A roof edging PC £13.29/2.5m	0.44	6.10	6.36	m	12.46
Angle PC £3.14	0.13	1.80	3.71	nr	5.51
Roof screed ventilator - aluminium alloy					
Extr-aqua-vent; set on screed over and					
including dished sinking; working collar					
around ventilator PC £6.21	1.20	16.65	7.34	nr	23.99
J30 LIQUID APPLIED TANKING/DAMP PROOF **MEMBRANES**					
'Synthaprufe'; blinding with sand;					
horizontal on slabs					
two coats	0.22	1.59	2.13	m2	3.72
three coats	0.31	2.24	3.09	m2	5.33
'Tretolastex 202T'; on vertical surfaces of					
concrete					
two coats	0.22	1.59	2.01	m2	3.60
three coats	0.31	2.24	3.02	m2	5.26
One coat Vandex 'Super' 0.75/m2 slurry; one					
consolidating coat of Vandex 'Premix' 1kg/m2					
slurry; horizontal on beds					
over 225 mm wide	-	-	-	m2	6.00
'Ventrot' hot applied damp proof membrane;					
one coat; horizontal on slabs					
over 225 mm wide	-	-	-	m2	4.24

J WATERPROOFING Including overheads and profit at 12.50% & 5.00%	Labour hours	Labour £	Material £	Unit	Total rate £

J40 FLEXIBLE SHEET TANKING/DAMP PROOF MEMBRANES

'Bituthene' sheeting; lapped joints; horizontal on slabs

standard 500 grade	0.11	0.79	4.99	m2	5.78
1000 grade	0.12	0.87	5.52	m2	6.39
1200 grade	0.13	0.94	8.34	m2	9.28
heavy duty grade	0.14	1.01	6.15	m2	7.16

'Bituthene' sheeting; lapped joints; dressed up vertical face of concrete

1000 grade	0.20	1.44	6.05	m2	7.49

'Kork-pak'; 9.5 mm thick joint filler; set vertically against 'Bituthene' as protection

over 225 mm wide	0.66	4.77	12.12	m2	16.89

'Bituthene' fillet

40 x 40 mm	0.11	0.79	4.44	m	5.23

'Bituthene' reinforcing strip; 300 mm wide

1000 grade	0.11	0.79	1.88	m	2.67

Expandite 'Famflex' waterproof tanking; 150 mm laps

horizontal; over 300 mm wide	0.44	3.18	7.02	m2	10.20
vertical; over 300 mm wide	0.72	5.20	7.02	m2	12.22

J41 BUILT UP FELT ROOF COVERINGS

NOTE: The following items of felt roofing, unless otherwise described, include for conventional lapping, laying and bonding between layers and to base; and laying flat or to falls to cross-falls or to slopes not exceeding 10 degrees - but exclude any insulation etc. (measured separately)

Felt roofing; BS 747; suitable for flat roofs
Three layer coverings type 1B
(two 18 kg/10 m2 and one 25 kg/10 m2)

bitumen fibre based felts	-	-	-	m2	17.87
Extra over top layer type 1B for mineral surfaced layer type 1E	-	-	-	m2	2.06

Three layer coverings type 2B bitumen asbestos based felts

	-	-	-	m2	21.80
Extra over top layer type 2B for mineral surfaced layer type 2E	-	-	-	m2	2.54

Three layer coverings first layer type 3G; subsequent layers type 3B bitumen glass fibre based felt

	-	-	-	m2	16.62
Extra over top layer type 3B for mineral surfaced layer type 3E	-	-	-	m2	2.90

Extra over felt for covering with and bedding in hot bitumen

13 mm granite chippings	-	-	-	m2	7.28
300 x 300 x 8 mm asbestos tiles	-	-	-	m2	55.27
Working into outlet pipes and the like	-	-	-	nr	9.80

Skirtings; three layer; top layer mineral surfaced; dressed over tilting fillet; turned into groove

not exceeding 200 mm girth	-	-	-	m	11.04
200 - 400 mm girth	-	-	-	m	13.66

Coverings to kerbs; three layer

400 - 600 mm girth	-	-	-	m	17.39

J WATERPROOFING Including overheads and profit at 12.50% & 5.00%	Labour hours	Labour £	Material £	Unit	Total rate £

J41 BUILT UP FELT ROOF COVERINGS - cont'd

Felt roofing; BS 747; suitable for flat roofs - cont'd
Linings to gutters; three layer

400 - 600 mm (average) girth	-	-	-	m	22.60
Collars around pipes and the like; three layer mineral surfaced; 150 mm high					
not exceeding 55 mm nominal size	-	-	-	nr	8.70
55 - 110 mm nominal size	-	-	-	nr	10.56

**Felt roofing; BS 747; suitable for pitched
timber roofs; sloping not exceeding
50 degrees**

Two layer coverings; first layer type 2B; second layer 2E; mineral surfaced asbestos based felts	-	-	-	m2	25.17
Three layer coverings; first two layers type 2B; top layer type 2E; mineral surfaced asbestos based felts	-	-	-	m2	35.74

**'Andersons' high performance polyester-based
roofing system**
Two layer coverings; first layer HT 125
underlay; second layer HT 350; fully bonded

to wood; fibre or cork base	0.35	4.85	10.40	m2	15.25
Extra over for					
Top layer mineral surfaced	-	-	-	m2	1.67
13 mm granite chippings	-	-	-	m2	7.28
Third layer of type 3B as underlay for concrete or screeded base	0.17	2.36	1.76	m2	4.12
Working into outlet pipes and the like	0.58	8.05	-	nr	8.05
Welted drip; two layer	0.23	3.19	0.90	m	4.09
Skirtings; two layer; top layer mineral surfaced; dressed over tilting fillet; turned into groove					
not exceeding 200 mm girth	0.17	2.36	2.76	m	5.12
200 - 400 mm girth	0.23	3.19	5.25	m	8.44
Coverings to kerbs; two layer					
400 - 600 mm girth	0.29	4.02	7.20	m	11.22
Linings to gutters; three layer					
400 - 600 mm (average) girth	0.69	9.57	8.93	m	18.50
Collars around pipes and the like; two layer; 150 mm high					
not exceeding 55 mm nominal size	0.38	5.27	0.43	nr	5.70
55 - 110 mm nominal size	0.46	6.38	0.72	nr	7.10

'Ruberglas 120 GP' high performance roofing
Two layer coverings; first and second layers
'Ruberglas 120 GP'; fully bonded to wood;

fibre or cork base	-	-	-	m2	14.75
Extra over for					
Top layer mineral surfaced	-	-	-	m2	1.55
13 mm granite chippings	-	-	-	m2	7.28
Third layer of 'Rubervent 3G' as underlay for concrete or screeded base	-	-	-	m2	6.25
Working into outlet pipes and the like	-	-	-	nr	9.80
Welted drip; two layer	-	-	-	m	5.43
Skirtings; two layer; top layer mineral surfaced; dressed over tilting fillet; turned into groove					
not exceeding 200 mm girth	-	-	-	m	8.00
200 - 400 mm girth	-	-	-	m	10.43

J WATERPROOFING Including overheads and profit at 5.00%	Labour hours	Labour £	Material £	Unit	Total rate £
Coverings to kerbs; two layer					
400 - 600 mm girth	-	-	-	m	17.39
Linings to gutters; three layer					
400 - 600 mm (average) girth	-	-	-	m	25.64
Collars around pipes and the like; two layer; 150 mm high					
not exceeding 55 mm nominal size	-	-	-	nr	8.70
55 - 110 mm nominal size	-	-	-	nr	10.56
'Ruberfort HP 350' high performance roofing					
Two layer coverings; first layer 'Ruberfort HP 180'; second layer 'Ruberfort HP 350'; fully bonded; to wood; fibre or cork base	-	-	-	m2	20.77
Extra over for					
Top layer mineral surfaced	-	-	-	m2	2.66
13 mm granite chippings	-	-	-	m2	7.28
Third layer of 'Rubervent 3G' as underlay for concrete or screeded base	-	-	-	m2	6.25
Working into outlet pipes and the like	-	-	-	nr	9.80
Welted drip; two layer	-	-	-	m	7.44
Skirtings; two layer; top layer mineral surfaced; dressed over tilting fillet; turned into groove					
not exceeding 200 mm girth	-	-	-	m	9.50
200 - 400 mm girth	-	-	-	m	13.61
Coverings to kerbs; two layer					
400 - 600 mm girth	-	-	-	m	21.94
Linings to gutters; three layer					
400 - 600 mm (average) girth	-	-	-	m	33.40
Collars around pipes and the like; two layer; 150 mm high					
not exceeding 55 mm nominal size	-	-	-	nr	8.70
55 - 110 mm nominal size	-	-	-	nr	10.56
'Polybit 350' elastomeric roofing					
Two layer coverings; first layer 'Polybit 180'; second layer 'Polybit 350'; fully bonded to wood; fibre or cork base	-	-	-	m2	22.61
Extra over for					
Top layer mineral surfaced	-	-	-	m2	2.75
13 mm granite chippings	-	-	-	m2	7.28
Third layer of 'Rubervent 3G' as underlay for concrete or screeded base	-	-	-	m2	6.25
Working into outlet pipes and the like	-	-	-	nr	9.75
Welted drip; two layer	-	-	-	m	8.43
Skirtings; two layer; top layer mineral surfaced; dressed over tilting fillet; turned into groove					
not exceeding 200 mm girth	-	-	-	m	9.93
200 - 400 mm girth	-	-	-	m	14.44
Coverings to kerbs; two layer					
400 - 600 mm girth	-	-	-	m	23.39
Linings to gutters; three layer					
400 - 600 mm (average) girth	-	-	-	m	35.80
Collars around pipes and the like; two layer; 150 mm high					
not exceeding 55 mm nominal size	-	-	-	nr	10.56
55 - 110 mm nominal size	-	-	-	nr	14.29
'Hyload 150 E' elastomeric roofing					
Two layer coverings; first layer 'Ruberglas 120 GP'; second layer 'Hyload 150 E' fully bonded to wood; fibre or cork base	-	-	-	m2	21.82
Extra over for					
13 mm granite chippings	-	-	-	m2	7.28
Third layer of 'Rubervent 3G' as underlay for concrete or screeded base	-	-	-	m2	6.25

J WATERPROOFING Including overheads and profit at 5.00%	Labour hours	Labour £	Material £	Unit	Total rate £
J41 BUILT UP FELT ROOF COVERINGS - cont'd					
'Hyload 150 E' elastomeric roofing - cont'd					
Working into outlet pipes and the like	-	-	-	nr	9.75
Welted drip; two layer	-	-	-	m	8.38
Skirtings; two layer; dressed over tilting fillet; turned into groove					
not exceeding 200 mm girth	-	-	-	m	9.41
200 - 400 mm girth	-	-	-	m	13.45
Coverings to kerbs; two layer					
400 - 600 mm girth	-	-	-	m	22.19
Linings to gutters; three layer					
400 - 600 mm (average) girth	-	-	-	m	35.00
Collars around pipes and the like; two layer; 150 mm high					
not exceeding 55 mm nominal size	-	-	-	nr	8.70
55 - 110 mm nominal size	-	-	-	nr	10.56
Felt; 'Paradiene' elastomeric bitumen roofing; perforated crepe paper isolating membrane; first layer 'Paradiene 20'; second layer 'Paradiene 30' pre-finished surface					
Two layer coverings	-	-	-	m2	21.91
Working into outlet pipes and the like	-	-	-	nr	9.14
Welted drip; three layer	-	-	-	m	7.07
Skirtings; three layer; dressed over tilting fillet; turned into groove					
not exceeding 200 mm girth	-	-	-	m	6.47
200 - 400 mm girth	-	-	-	m	11.37
Coverings to kerbs; two layer					
400 - 600 mm girth	-	-	-	m	13.89
Linings to gutters; three layer					
400 - 600 mm (average) girth	-	-	-	m	21.72
Collars around pipes and the like; three layer; 150 mm high					
not exceeding 55 mm nominal size	-	-	-	nr	6.33
55 - 110 mm nominal size	-	-	-	nr	7.03
Metal faced 'Veral' glass cloth reinforced bitumen roofing; first layer 'Veralvent' perforated underlay; second layer 'Veralglas'; third layer 'Veral' natural slate aluminium surfaced					
Three layer coverings	-	-	-	m2	29.17
Working into outlet pipes and the like	-	-	-	nr	9.14
Welted drip; three layer	-	-	-	m	6.93
Skirtings; three layer; dressed over tilting fillet; turned into groove					
not exceeding 200 mm girth	-	-	-	m	12.05
200 - 400 mm girth	-	-	-	m	10.04
Coverings to kerbs; three layer					
400 - 600 mm girth	-	-	-	m	14.58
Linings to gutters; three layer					
400 - 600 mm (average) girth	-	-	-	m	18.95
Collars around pipes and the like; three layer; 150 mm high					
not exceeding 55 mm nominal size	-	-	-	nr	6.33
55 - 110 mm nominal size	-	-	-	nr	7.03

J WATERPROOFING Including overheads and profit at 12.50% & 5.00%	Labour hours	Labour £	Material £	Unit	Total rate £
Accessories					
Eaves trim; extruded aluminium alloy; working felt into trim					
'Alutrim' type F roof edging PC £9.08/2.5m	0.29	4.02	4.37	m	**8.39**
Angle PC £3.14	0.14	1.94	3.71	nr	**5.65**

Roof screed ventilator - aluminium alloy
  'Extr-aqua-vent'; set on screed over and
  including dished sinking; working collar
  around ventilator   PC £4.14
	0.69	9.57	4.89	nr	**14.46**

Insulation board underlays					
Vapour barrier					
reinforced; metal lined	0.03	0.42	4.00	m2	**4.42**
Cork boards; density 112 - 125 kg/m3					
60 mm thick	0.35	4.85	4.96	m2	**9.81**
Foamed glass boards; density 125 - 135 kg/m2					
60 mm thick	0.35	4.85	15.08	m2	**19.93**
Glass fibre boards; density 120 - 130 kg/m2					
60 mm thick	0.35	4.85	9.41	m2	**14.26**
Perlite boards; density 170 - 180 kg/m3					
60 mm thick	0.35	4.85	8.07	m2	**12.92**
Polyurethene boards; density 32 kg/m3					
30 mm thick	0.23	3.19	4.52	m2	**7.71**
35 mm thick	0.23	3.19	5.09	m2	**8.28**
50 mm thick	0.35	4.85	6.61	m2	**11.46**
Wood fibre boards; impregnated; density 220 - 350 kg/m3					
12.7 mm thick	0.23	3.19	2.09	m2	**5.28**
Insulation board overlays					
Dow 'Roofmate SL' extruded polystyrene foam boards					
50 mm thick	0.35	4.85	9.96	m2	**14.81**
75 mm thick	0.35	4.85	14.66	m2	**19.51**
Dow 'Roofmate LG' extruded polystyrene foam boards					
50 mm thick	0.35	4.85	16.93	m2	**21.78**
75 mm thick	0.35	4.85	20.12	m2	**24.97**
100 mm thick	0.35	4.85	23.78	m2	**28.63**
J42 SINGLE LAYER PLASTICS ROOF COVERINGS					
Felt; 'Derbigum' special polyester 4 mm roofing; first layer 'Ventilag' (partial bond) underlay; second layer 'Derbigum SF'; glass reinforced weathering surface					
Two layer coverings	-	-	-	m2	**26.38**
Welted drip; two layer	-	-	-	m	**8.88**
Skirtings; two layer; dressed over tilting fillet; turned into groove					
not exceeding 200 mm girth	-	-	-	m	**13.40**
200 - 400 mm girth	-	-	-	m	**17.14**
Coverings to kerbs; two layer					
400 - 600 mm girth	-	-	-	m	**28.37**
Collars around pipes and the like; two layer; 150 mm high					
not exceeding 55 mm nominal size	-	-	-	nr	**10.56**
55 - 110 mm nominal size	-	-	-	nr	**14.29**

K LININGS/SHEATHING/DRY PARTITIONING	Labour	Labour	Material		Total
Including overheads and profit at 12.50%	hours	£	£	Unit	rate £

ALTERNATIVE SHEET LINING MATERIAL PRICES

	£		£		£			£
Asbestos cement flat sheets (£/10 m2)								
Fully compressed								
4.5 mm	54.81	6 mm	68.05	9 mm		102.95	12 mm	128.15
Semi compressed								
4.5 mm	27.85	6 mm	37.91	9 mm		56.43	12 mm	94.02

Blockboard								
Gaboon faced (£/10 m2)								
16 mm	89.72	18 mm	90.11	22 mm		101.93	25 mm	114.53
			£			£		£
18 mm Decorative faced (£/10 m2)								
Ash		152.85	Mahogany		116.15	Teak		124.00
Beech		130.85	Oak		116.15			
Edgings; self adhesive (£/25 m roll)								
19 mm Mahogany		2.47	19 mm Oak		3.46	19 mm Ash		3.46
25 mm Mahogany		3.63	25 mm Oak		4.12	19 mm Teak		3.46

	£		£		£			£
Chipboard (£/10 m2)								
Standard grade								
3.2 mm	8.35	9 mm	18.85	16 mm	26.68	22 mm		34.97
4 mm	10.55	12 mm	20.47	18 mm	27.83	25 mm		39.69
6 mm	15.54							
Melamine faced								
12 mm	35.59	18 mm	41.89					

Laminboard; Birch faced (£/10 m2)								
16 mm	120.38	18 mm	125.66	22 mm	147.15	25 mm		163.63

Medium density fibreboard (£/10 m2)								
6.5 mm	24.70	12 mm	39.22	17.5 mm	50.49	25 mm		71.45
9 mm	29.81	16 mm	46.92	19 mm	55.28			

	£		£		£			£
Plasterboard (£/100m2)								
Wallboard plank								
9.5 mm	121.41	12.5 mm	145.62	15 mm	204.07	19 mm		254.46
Lath (Thistle baseboard)								
9.5 mm	126.91	12.5 mm	152.56					

	£		£					£
Industrial board								
9.5 mm	238.44	12.5 mm	269.56			Fireline Industrial board		
Fireline board						12.5 mm		306.68
12.5 mm	214.97	15 mm	250.00					

	£				£			£
Plywood (£/10 m2)								
Decorative								
6 mm Afrormosia	45.68							
6 mm Ash	44.50	9 mm Ash		68.13		12 mm Ash		70.44
6 mm Oak	46.46	9 mm Oak		60.00		12 mm Oak		72.39
6 mm Sapele	41.40	9 mm Sapele		55.21		12 mm Sapele		67.35
6 mm Teak	56.00	9 mm Teak		62.70		12 mm Teak		79.75
Prefinished 4 mm random v-grooved decorative								
Afrormosia	66.56	Elm		67.56		Sapele		63.44
Ash	67.88	Oak		71.44		Teak		73.13
Birch	77.50	Knotty Pine		65.69				

Discounts of 0 - 10% available depending on quantity/status

K LININGS/SHEATHING/DRY PARTITIONING Including overheads and profit at 12.50%	Labour hours	Labour £	Material £	Unit	Total rate £
K10 PLASTERBOARD DRY LINING					
Gypsum plasterboard; BS 1230; fixing with nails; joints left open to receive 'Artex'; to softwood base Plain grade tapered edge wallboard					
9.5 mm board to ceilings					
over 300 mm wide	0.31	4.30	1.38	m2	5.68
9.5 mm board to beams					
over 300 mm wide	0.39	5.41	0.87	m	6.28
12.5 mm board to ceilings					
total girth not exceeding 600 mm	0.50	6.94	1.67	m	8.61
total girth 600 - 1200 mm	0.33	4.58	1.92	m2	6.50
12.5 mm board to beams					
total girth not exceeding 600 mm	0.40	5.55	1.20	m	6.75
total girth 600 - 1200 mm	0.53	7.35	2.30	m	9.65
Gypsum plasterboard to BS 1230; fixing with nails; joints filled with joint filler and joint tape to receive direct decoration; to softwood base Plain grade tapered edge wallboard					
9.5 mm board to walls					
wall height 2.40 - 2.70	1.10	15.26	4.88	m	20.14
wall height 2.70 - 3.00	1.25	17.34	5.44	m	22.78
wall height 3.00 - 3.30	1.45	20.11	6.00	m	26.11
wall height 3.30 - 3.60	1.65	22.89	6.57	m	29.46
9.5 mm board to reveals and soffits of openings and recesses					
not exceeding 300 mm wide	0.22	3.05	0.59	m	3.64
width 300 - 600 mm	0.44	6.10	1.16	m	7.26
9.5 mm board to faces of columns					
total girth not exceeding 600 mm	0.55	7.63	1.13	m	8.76
total girth 600 - 1200 mm	1.10	15.26	2.34	m	17.60
total girth 1200 - 1800 mm	1.45	20.11	3.44	m	23.56
9.5 mm board to ceilings					
over 300 mm wide	0.46	6.38	1.81	m2	8.19
9.5 mm board to faces of beams					
total girth not exceeding 600 mm	0.58	8.05	1.16	m	9.21
total girth 600 - 1200 mm	1.15	15.95	2.34	m	18.30
total girth 1200 - 1800 mm	1.50	20.81	3.44	m	24.25
Add for 'Duplex' insulating grade	-	-	-	m2	0.50
12.5 mm board to walls					
wall height 2.40 - 2.70	1.15	15.95	5.74	m	21.69
wall height 2.70 - 3.00	1.30	18.03	6.37	m	24.40
wall height 3.00 - 3.30	1.50	20.81	7.03	m	27.84
wall height 3.30 - 3.60	1.75	24.27	7.69	m	31.96
12.5 mm board to reveals and soffits of openings and recesses					
not exceeding 300 mm wide	0.23	3.19	0.68	m	3.87
width 300 - 600 mm	0.46	6.38	1.37	m	7.75

Keep your figures up to date, free of charge

This section, and most of the other information in this Price Book,
is brought up to date every three months in the *Price Book Update*.

The *Update* is available free to all Price Book purchasers.

To ensure you receive your copy, simply complete the reply card from
the centre of the book and return it to us.

K LININGS/SHEATHING/DRY PARTITIONING Including overheads and profit at 12.50%	Labour hours	Labour £	Material £	Unit	Total rate £
K10 PLASTERBOARD DRY LINING - cont'd					
Gypsum plasterboard to BS 1230; fixing with **nails; joints filled with joint filler and** **joint tape to receive direct decoration; to** **softwood base** **Plain grade tapered edge wallboard - cont'd**					
12.5 mm board to faces of columns					
total girth not exceeding 600 mm	0.57	7.91	1.37	m	9.28
total girth 600 - 1200 mm	1.15	15.95	2.73	m	18.68
total girth 1200 - 1800 mm	1.50	20.81	4.02	m	24.83
12.5 mm board to ceilings					
over 300 mm wide	0.48	6.66	2.14	m2	8.80
12.5 mm board to faces of beams					
total girth not exceeding 600 mm	0.62	8.60	1.37	m	9.97
total girth 600 - 1200 mm	1.25	17.34	2.73	m	20.07
total girth 1200 - 1800 mm	1.60	22.19	4.02	m	26.21
External angle; with joint tape bedded in joint filler; covered with joint finish	0.13	1.80	0.52	m	2.32
Add for 'Duplex' insulating grade	-	-	-	m2	0.50
Tapered edge plank					
19 mm plank to walls					
wall height 2.40 - 2.70	1.20	16.65	9.33	m	25.98
wall height 2.70 - 3.00	1.40	19.42	10.36	m	29.78
wall height 3.00 - 3.30	1.55	21.50	11.40	m	32.90
wall height 3.30 - 3.60	1.85	25.66	11.81	m	37.47
19 mm plank to reveals and soffits of openings and recesses					
not exceeding 300 mm wide	0.24	3.33	1.12	m	4.45
width 300 - 600 mm	0.48	6.66	2.20	m	8.86
19 mm plank to faces of columns					
total girth not exceeding 600 mm	0.59	8.18	2.20	m	10.38
total girth 600 - 1200 mm	1.20	16.65	4.43	m	21.08
total girth 1200 - 1800 mm	1.55	21.50	6.42	m	27.92
19 mm plank to ceilings					
over 300 mm wide	0.51	7.07	3.45	m2	10.53
19 mm plank to faces of beams					
total girth not exceeding 600 mm	0.64	8.88	2.15	m	11.03
total girth 600 - 1200 mm	1.30	18.03	4.43	m	22.46
total girth 1200 - 1800 mm	1.65	22.89	6.42	m	29.31
Thermal board					
25 mm board to walls					
wall height 2.40 - 2.70	1.25	17.34	12.45	m	29.79
wall height 2.70 - 3.00	1.45	20.11	13.79	m	33.90
wall height 3.00 - 3.30	1.60	22.19	15.21	m	37.40
wall height 3.30 - 3.60	1.95	27.05	16.62	m	43.67
25 mm board to reveals and soffits of openings and recesses					
not exceeding 300 mm wide	0.25	3.47	1.46	m	4.93
width 300 - 600 mm	0.51	7.07	2.93	m	10.00
25 mm board to faces of columns					
total girth not exceeding 600 mm	0.62	8.60	2.93	m	11.53
total girth 600 - 1200 mm	1.25	17.34	5.86	m	23.20
total girth 1200 - 1800 mm	1.60	22.19	8.59	m	30.78
25 mm board to ceilings					
over 300 mm wide	0.55	7.63	4.60	m2	12.23
25 mm board to faces of beams					
total girth not exceeding 600 mm	0.66	9.16	2.93	m	12.09
total girth 600 - 1200 mm	1.30	18.03	5.86	m	23.89
total girth 1200 - 1800 mm	1.75	24.27	8.59	m	32.87
50 mm board to walls					
wall height 2.40 - 2.70	1.40	19.42	21.03	m	40.45
wall height 2.70 - 3.00	1.55	21.50	23.39	m	44.89
wall height 3.00 - 3.30	1.70	23.58	25.71	m	49.29
wall height 3.30 - 3.60	2.05	28.44	28.06	m	56.50

K LININGS/SHEATHING/DRY PARTITIONING Including overheads and profit at 12.50%	Labour hours	Labour £	Material £	Unit	Total rate £
50 mm board to reveals and soffits of openings and recesses					
not exceeding 300 mm wide	0.28	3.88	2.46	m	6.34
width 300 - 600 mm	0.55	7.63	4.92	m	12.55
50 mm board to faces of columns					
total girth not exceeding 600 mm	0.66	9.16	4.92	m	14.08
total girth 600 - 1200 mm	1.30	18.03	9.83	m	27.86
total girth 1200 - 1800 mm	1.70	23.58	14.45	m	38.03
50 mm board to ceilings					
over 300 mm wide	0.58	8.05	7.80	m2	15.85
50 mm board to faces of beams					
total girth not exceeding 600 mm	0.69	9.57	4.92	m	14.49
total girth 600 - 1200 mm	1.40	19.42	9.83	m	29.25
total girth 1200 - 1800 mm	1.85	25.66	14.45	m	40.11
White plastic faced gypsum plasterboard to BS 1230; fixing with screws; butt joints; to softwood base **Insulating grade square edge wallboard**					
9.5 mm board to walls					
wall height 2.40 - 2.70	1.10	15.26	9.04	m	24.30
wall height 2.70 - 3.00	1.25	17.34	10.05	m	27.39
wall height 3.00 - 3.30	1.45	20.11	11.05	m	31.16
wall height 3.30 - 3.60	1.65	22.89	12.05	m	34.94
9.5 mm board to reveals and soffits of openings and recesses					
not exceeding 300 mm wide	0.22	3.05	1.05	m	4.10
width 300 - 600 mm	0.44	6.10	2.09	m	8.19
9.5 mm board to faces of columns					
total girth not exceeding 600 mm	0.55	7.63	2.09	m	9.72
total girth 600 - 1200 mm	1.10	15.26	4.15	m	19.41
total girth 1200 - 1800 mm	1.45	20.11	6.11	m	26.22
9.5 mm board to ceilings					
over 300 mm wide	0.46	6.38	3.35	m2	9.73
9.5 mm board to faces of beams					
total girth not exceeding 600 mm	0.58	8.05	2.09	m	10.14
total girth 600 - 1200 mm	1.15	15.95	4.15	m	20.10
total girth 1200 - 1800 mm	1.50	20.81	6.11	m	26.92
12.5 mm board to walls					
wall height 2.40 - 2.70	1.15	15.95	9.05	m	25.00
wall height 2.70 - 3.00	1.30	18.03	10.05	m	28.08
wall height 3.00 - 3.30	1.50	20.81	11.05	m	31.86
wall height 3.30 - 3.60	1.75	24.27	12.05	m	36.33
12.5 mm board to reveals and soffits of openings and recesses					
not exceeding 300 mm wide	0.23	3.19	1.05	m	4.24
width 300 - 600 mm	0.46	6.38	2.09	m	8.47
12.5 mm board to faces of columns					
total girth not exceeding 600 mm	0.57	7.91	2.09	m	10.00
total girth 600 - 1200 mm	1.15	15.95	4.15	m	20.10
total girth 1200 - 1800 mm	1.50	20.81	6.11	m	26.92
12.5 mm board to ceilings					
over 300 mm wide	0.48	6.66	3.71	m2	10.37
12.5 mm board to faces of beams					
total girth not exceeding 600 mm	0.62	8.60	2.32	m	10.92
total girth 600 - 1200 mm	1.25	17.34	4.60	m	21.94
total girth 1200 - 1800 mm	1.60	22.19	6.78	m	28.97
Two layers of gypsum plasterboard to BS 1230; fixing with nails; joints filled					
19 mm two layer board to walls					
wall height 2.40 - 2.70	1.75	24.27	9.22	m	33.49
wall height 2.70 - 3.00	2.00	27.74	10.23	m	37.97
wall height 3.00 - 3.30	2.20	30.52	11.26	m	41.78
wall height 3.30 - 3.60	2.65	36.76	11.66	m	48.42
19 mm two layer board to reveals and soffits of openings and recesses					
not exceeding 300 mm wide	0.33	4.58	1.10	m	5.68
width 300 - 600 mm	0.66	9.16	2.18	m	11.34

K LININGS/SHEATHING/DRY PARTITIONING Including overheads and profit at 12.50%	Labour hours	Labour £	Material £	Unit	Total rate £
K10 PLASTERBOARD DRY LINING - cont'd					
Two layers of gypsum plasterboard to BS 1230; fixing with nails; joints filled - cont'd					
19 mm two layer board to faces of columns					
total girth not exceeding 600 mm	0.88	12.21	2.18	m	14.39
total girth 600 - 1200 mm	1.65	22.89	4.36	m	27.25
total girth 1200 - 1800 mm	2.20	30.52	6.34	m	36.86
19 mm two layer board to ceilings					
over 300 mm wide	0.73	10.13	3.42	m2	13.55
19 mm two layer board to faces of beams					
total girth not exceeding 600 mm	0.94	13.04	2.12	m	15.16
total girth 600 - 1200 mm	1.75	24.27	4.36	m	28.63
total girth 1200 - 1800 mm	2.35	32.30	6.34	m	38.94
25 mm two layer board to walls					
wall height 2.40 - 2.70	1.85	25.66	11.05	m	36.71
wall height 2.70 - 3.00	2.10	29.13	12.27	m	41.40
wall height 3.00 - 3.30	2.40	33.29	13.56	m	46.86
wall height 3.30 - 3.60	2.85	39.53	14.86	m	54.39
25 mm two layer board to reveals and soffits of openings and recesses					
not exceeding 300 mm wide	0.36	4.99	1.31	m	6.30
width 300 - 600 mm	0.72	9.99	2.59	m	12.58
25 mm two layer board to faces of columns					
total girth not exceeding 600 mm	0.95	13.18	2.59	m	15.77
total girth 600 - 1200 mm	1.75	24.27	5.20	m	29.48
total girth 1200 - 1800 mm	2.35	32.60	7.64	m	40.24
25 mm two layer board to ceilings					
over 300 mm wide	0.79	10.96	4.09	m2	15.05
25 mm two layer board to faces of beams					
total girth not exceeding 600 mm	1.00	13.87	2.59	m	16.46
total girth 600 - 1200 mm	1.95	27.05	5.20	m	32.25
total girth 1200 - 1800 mm	2.55	35.37	7.64	m	43.01
Gypsum plasterboard to BS 1230; 3 mm joints; fixed by the 'Thistleboard' system of dry linings; joints; filled with joint filler and joint tape; to receive direct decoration Plain grade tapered edge wallboard					
9.5 mm board to walls					
wall height 2.40 - 2.70	1.15	15.95	4.79	m	20.74
wall height 2.70 - 3.00	1.30	18.03	5.34	m	23.37
wall height 3.00 - 3.30	1.50	20.81	5.89	m	26.70
wall height 3.30 - 3.60	1.75	24.27	6.44	m	30.71
9.5 mm board to reveals and soffits of openings and recesses					
not exceeding 300 mm wide	0.23	3.19	0.56	m	3.75
width 300 - 600 mm	0.46	6.38	1.11	m2	7.49
9.5 mm board to faces of columns					
total girth not exceeding 600 mm	0.57	7.91	1.11	m	9.02
total girth 600 - 1200 mm	1.15	15.95	2.22	m	18.17
total girth 1200 - 1800 mm	1.50	20.81	3.27	m	24.08
Angle; with joint tape bedded in joint filler; covered with joint finish					
internal	0.07	0.97	0.23	m	1.20
external	0.13	1.80	0.52	m	2.32
Vermiculite gypsum cladding; 'Vicuclad' board on and including shaped noggins; fixed with nails and adhesive; joints pointed in adhesive					
25 mm column casings; 2 hour fire protection rating					
over 1 m girth	1.10	15.26	15.05	m2	30.31
150 mm girth	0.33	4.58	4.56	m	9.14
300 mm girth	0.50	6.94	6.83	m	13.77
600 mm girth	0.83	11.51	11.13	m	22.64
900 mm girth	1.05	14.56	14.30	m	28.86

K LININGS/SHEATHING/DRY PARTITIONING Including overheads and profit at 12.50%	Labour hours	Labour £	Material £	Unit	Total rate £
30 mm beam casings; 2 hour fire protection rating					
over 1 m girth	1.20	16.65	18.47	m2	35.11
150 mm girth	0.36	4.99	5.59	m	10.58
300 mm girth	0.55	7.63	8.38	m	16.01
600 mm girth	0.91	12.62	13.70	m	26.32
900 mm girth	1.15	15.95	17.55	m	33.51
55 mm column casings; 4 hour fire protection rating					
over 1 m girth	1.30	18.03	37.90	m2	55.93
150 mm girth	0.40	5.55	11.41	m	16.96
300 mm girth	0.61	8.46	17.12	m	25.58
600 mm girth	0.99	13.73	28.27	m	42.00
900 mm girth	1.25	17.34	36.03	m	53.37
60 mm beam casings; 4 hour fire protection rating					
over 1 m girth	1.45	20.11	39.96	m2	60.08
150 mm girth	0.43	5.96	12.04	m	18.00
300 mm girth	0.66	9.16	18.04	m	27.20
600 mm girth	1.05	14.56	29.82	m	44.38
900 mm girth	1.40	19.42	37.97	m	57.39
Add to the above for					
Plus 3% for work 3.5 - 5 m high					
Plus 6% for work 5 - 6.5 m high					
Plus 12% for work 6.5 - 8 m high					
Plus 18% for work over 8 m high					
Cutting and fitting around steel joists, angles, trunking, ducting, ventilators, pipes, tubes, etc					
over 2 m girth	0.33	4.58	-	m	4.58
not exceeding 0.30 m girth	0.22	3.05	-	nr	3.05
0.30 - 1 m girth	0.28	3.88	-	nr	3.88
1 - 2 m girth	0.39	5.41	-	nr	5.41
K11 RIGID SHEET FLOORING/SHEATHING/LININGS/ CASINGS					
Blockboard (Birch faced)					
12 mm lining to walls					
over 300 mm wide PC £79.93/10m2	0.52	4.86	10.02	m2	14.88
not exceeding 300 mm wide	0.34	3.18	3.29	m	6.47
Raking cutting	0.09	0.84	0.36	m	1.20
Holes for pipes and the like	0.04	0.37	3.60	nr	3.97
18 mm lining to walls					
over 300 mm wide PC £93.71/10m2	0.55	5.14	11.73	m2	16.87
not exceeding 300 mm wide	0.35	3.27	3.84	m	7.11
Raking cutting	0.11	1.03	0.53	m	1.56
Holes for pipes and the like	0.06	0.56	-	nr	0.56
Two-sided 18 mm thick pipe casing; 50 x 50 mm softwood framing; two members plugged to wall					
300 mm girth	1.40	13.09	5.99	m	19.08
450 mm girth	1.50	14.02	7.93	m	21.95
600 mm girth	1.60	14.96	9.87	m	24.83
750 mm girth	1.70	15.89	11.82	m	27.71
Three-sided 18 mm thick pipe casing; 50 x 50 mm softwood framing; two members plugged to wall					
450 mm girth	1.85	17.30	8.64	m	25.94
600 mm girth	2.00	18.70	10.59	m	29.29
750 mm girth	2.10	19.63	12.54	m	32.17
900 mm girth	2.20	20.57	14.47	m	35.04
1050 mm girth	2.30	21.50	16.41	m	37.91
Extra for 400 x 400 mm removable access panel; brass cups and screws; additional framing	1.10	10.28	8.70	nr	18.98

K LININGS/SHEATHING/DRY PARTITIONING Including overheads and profit at 12.50%	Labour hours	Labour £	Material £	Unit	Total rate £
K11 RIGID SHEET FLOORING/SHEATHING/LININGS/ **CASINGS - cont'd**					
Blockboard (Birch faced) - cont'd					
25 mm lining to walls					
over 300 mm wide PC £122.12/10m2	0.59	5.52	15.26	m2	20.78
not exceeding 300 mm wide	0.39	3.65	5.01	m	8.66
Raking cutting	0.13	1.22	0.69	m	1.91
Holes for pipes and the like	0.07	0.65	-	nr	0.65
Chipboard (plain)					
12 mm lining to walls					
over 300 mm wide PC £20.47/10m2	0.42	3.93	2.66	m2	6.59
not exceeding 300 mm wide	0.24	2.24	0.87	m	3.11
Raking cutting	0.07	0.65	0.12	m	0.77
Holes for pipes and the like	0.03	0.28	-	nr	0.28
15 mm lining to walls					
over 300 mm wide PC £26.05/10m2	0.44	4.11	3.35	m2	7.46
not exceeding 300 mm wide	0.26	2.43	1.11	m	3.54
Raking cutting	0.09	0.84	0.15	m	0.99
Holes for pipes and the like	0.04	0.37	-	nr	0.37
Two-sided 15 mm thick pipe casing;					
50 x 50 mm softwood framing; two members					
plugged to wall					
300 mm girth	1.10	10.28	3.47	m	13.75
450 mm girth	1.20	11.22	4.16	m	15.38
600 mm girth	1.30	12.15	4.84	m	16.99
750 mm girth	1.35	12.62	5.53	m	18.15
Three-sided 15 mm thick pipe casing;					
50 x 50 mm softwood framing; two members					
plugged to wall					
450 mm girth	1.65	15.43	4.88	m	20.31
600 mm girth	1.75	16.36	5.56	m	21.92
750 mm girth	1.85	17.30	6.25	m	23.55
900 mm girth	1.95	18.23	6.93	m	25.16
1050 mm girth	2.20	20.57	7.62	m	28.19
Extra for 400 x 400 mm removable access					
panel; brass cups and screws; additional					
framing	1.10	10.28	8.64	nr	18.92
18 mm lining to walls					
over 300 mm wide PC £27.83/10m2	0.46	4.30	3.57	m2	7.87
not exceeding 300 mm wide	0.30	2.80	1.18	m	3.98
Raking cutting	0.11	1.03	0.16	m	1.19
Holes for pipes and the like	0.06	0.56	-	nr	0.56
Chipboard (Melamine faced white matt finish)					
15 mm thick; PC £38.32/10m2; laminated masking strips					
Lining to walls					
over 300 mm wide	1.15	10.75	6.70	m2	17.45
not exceeding 300 mm wide	0.75	7.01	2.92	m	9.93
Raking cutting	0.14	1.31	0.22	m	1.53
Holes for pipes and the like	0.08	0.75	-	nr	0.75
Chipboard boarding and flooring					
Boarding to floors; butt joints					
18 mm thick PC £31.39/10m2	0.33	3.09	3.87	m2	6.96
22 mm thick PC £41.95/10m2	0.36	3.37	5.12	m2	8.49
Boarding to floors; tongued and					
grooved joints					
18 mm thick PC £32.97/10m2	0.35	3.27	4.05	m2	7.32
22 mm thick PC £43.57/10m2	0.39	3.65	5.31	m2	8.96
Boarding to roofs; butt joints					
18 mm thick; pre-felted PC £29.90/10m2	0.36	3.37	3.69	m2	7.06
Raking cutting on 18 mm thick					
chipboard	0.09	0.84	1.77	m	2.61

K LININGS/SHEATHING/DRY PARTITIONING Including overheads and profit at 12.50%		Labour hours	Labour £	Material £	Unit	Total rate £
Durabella 'Westbourne' flooring system or **similar; comprising 19 mm thick tongued and** **grooved chipboard panels secret nailed to** **softwood MK 10X-profiled foam backed battens** **at 600 mm centres; on concrete floor** Flooring tongued and grooved joints						
63 mm thick 36 x 50 mm battens	PC £9.52	0.99	9.26	12.03	m2	21.29
75 mm thick 43 x 50 mm battens	PC £10.23	0.99	9.26	12.91	m2	22.17
Plywood flooring Boarding to floors; tongued and grooved joints						
15 mm thick	PC £111.71/10m2	0.44	4.11	13.36	m2	17.47
18 mm thick	PC £129.78/10m2	0.48	4.49	15.49	m2	19.98
Plywood; external quality; 18 mm thick; PC £111.49/10m2 Boarding to roofs; butt joints						
flat to falls		0.44	4.11	13.33	m2	17.44
sloping		0.47	4.39	13.33	m2	17.72
vertical		0.63	5.89	13.33	m2	19.22
Hardboard to BS 1142 3.2 mm lining to walls						
over 300 mm wide	PC £10.46/10m2	0.33	3.09	1.34	m2	4.43
not exceeding 300 mm wide		0.20	1.87	0.44	m	2.31
Raking cutting		0.02	0.19	0.06	m	0.25
Holes for pipes and the like		0.01	0.09	-	nr	0.09
6.4 mm lining to walls						
over 300 mm wide	PC £20.81/10m2	0.36	3.37	2.63	m2	6.00
not exceeding 300 mm wide		0.23	2.15	0.85	m	3.00
Raking cutting		0.03	0.28	0.12	m	0.40
Holes for pipes and the like		0.02	0.19	-	nr	0.19
Glazed hardboard to BS 1142; PC £47.14/10m2; **on and including 38 x 38 mm wrought** **softwood framing** 3.2 mm thick panel						
to side of bath		2.00	18.70	9.16	nr	27.86
to end of bath		0.77	7.20	3.24	nr	10.44
Insulation board to BS 1142 12.7 mm lining to walls						
over 300 mm wide	PC £18.12/10m2	0.26	2.43	2.37	m2	4.80
not exceeding 300 mm wide		0.15	1.40	0.78	m	2.18
Raking cutting		0.03	0.28	0.10	m	0.38
Holes for pipes and the like		0.01	0.09	-	nr	0.09
19 mm lining to walls						
over 300 mm wide	PC £29.58/10m2	0.29	2.71	3.79	m2	6.50
not exceeding 300 mm wide		0.18	1.68	1.25	m	2.93
Raking cutting		0.06	0.56	0.17	m	0.73
Holes for pipes and the like		0.02	0.19	-	nr	0.19
25 mm lining to walls						
over 300 mm wide	PC £38.28/10m2	0.35	3.27	4.88	m2	8.15
not exceeding 300 mm wide		0.21	1.96	1.62	m	3.58
Raking cutting		0.07	0.65	0.22	m	0.87
Holes for pipes and the like		0.03	0.28	-	nr	0.28
Laminboard (Birch Faced); 18 mm thick PC £125.66/10m2 Lining to walls						
over 300 mm wide		0.58	5.42	15.68	m2	21.10
not exceeding 300 mm wide		0.37	3.46	5.13	m	8.59
Raking cutting		0.12	1.12	0.71	m	1.83
Holes for pipes and the like		0.07	0.65	-	nr	0.65

K LININGS/SHEATHING/DRY PARTITIONING Including overheads and profit at 12.50%	Labour hours	Labour £	Material £	Unit	Total rate £
K11 RIGID SHEET FLOORING/SHEATHING/LININGS/ **CASINGS - cont'd**					
Non-asbestos board; 'Supalux'; natural finish					
6 mm lining to walls					
over 300 mm wide PC £53.80/10m2	0.36	3.37	6.76	m2	10.13
not exceeding 300 mm wide	0.22	2.06	2.21	m	4.27
6 mm lining to ceilings					
over 300 mm wide	0.48	4.49	6.76	m2	11.25
not exceeding 300 mm wide	0.30	2.80	2.21	m	5.01
Raking cutting	0.06	0.56	0.30	m	0.86
Holes for pipes and the like	0.03	0.28	-	nr	0.28
9 mm lining to walls					
over 300 mm wide PC £83.80/10m2	0.40	3.74	10.49	m2	14.23
not exceeding 300 mm wide	0.24	2.24	3.44	m	5.68
9 mm lining to ceilings					
over 300 mm wide	0.50	4.67	10.49	m2	15.16
not exceeding 300 mm wide	0.32	2.99	3.44	m	6.43
Raking cutting	0.07	0.65	0.47	m	1.12
Holes for pipes and the like	0.04	0.37	-	nr	0.37
12 mm lining to walls					
over 300 mm wide PC £114.20/10m2	0.44	4.11	14.26	m2	18.37
not exceeding 300 mm wide	0.26	2.43	4.68	m	7.11
12 mm lining to ceilings					
over 300 mm wide	0.58	5.42	14.26	m2	19.68
not exceeding 300 mm wide	0.35	3.27	4.68	m	7.95
Raking cutting	0.08	0.75	0.64	m	1.39
Holes for pipes and the like	0.06	0.56	-	nr	0.56
Non-asbestos board; 'Supalux'; sanded finish					
6 mm lining to walls					
over 300 mm wide PC £56.50/10m2	0.36	3.37	7.09	m2	10.46
not exceeding 300 mm wide	0.22	2.06	2.32	m	4.38
6 mm lining to ceilings					
over 300 mm wide	0.48	4.49	7.09	m2	11.58
not exceeding 300 mm wide	0.30	2.80	2.32	m	5.12
Raking cutting	0.06	0.56	0.32	m	0.88
Holes for pipes and the like	0.03	0.28	-	nr	0.28
9 mm lining to walls					
over 300 mm wide PC £86.20/10m2	0.40	3.74	10.78	m2	14.52
not exceeding 300 mm wide	0.24	2.24	3.54	m	5.78
9 mm lining to ceilings					
over 300 mm wide	0.50	4.67	10.78	m2	15.45
not exceeding 300 mm wide	0.32	2.99	3.54	m	6.53
Raking cutting	0.07	0.65	0.48	m	1.13
Holes for pipes and the like	0.04	0.37	-	nr	0.37
12 mm lining to walls					
over 300 mm wide PC £114.20/10m2	0.44	4.11	14.26	m2	18.37
not exceeding 300 mm wide	0.26	2.43	4.68	m	7.11
12 mm lining to ceilings					
over 300 mm wide	0.58	5.42	14.26	m2	19.68
not exceeding 300 mm wide	0.35	3.27	4.68	m	7.95
Raking cutting	0.08	0.75	0.64	m	1.39
Holes for pipes and the like	0.06	0.56	-	nr	0.56
Plywood (Russian Birch); internal quality					
4 mm lining to walls					
over 300 mm wide PC £24.02/10m2	0.41	3.83	3.10	m2	6.93
not exceeding 300 mm wide	0.26	2.43	1.02	m	3.45
4 mm lining to ceilings					
over 300 mm wide	0.55	5.14	3.10	m2	8.24
not exceeding 300 mm wide	0.35	3.27	1.02	m	4.29
Raking cutting	0.06	0.56	0.14	m	0.70
Holes for pipes and the like	0.03	0.28	-	nr	0.28
6 mm lining to walls					
over 300 mm wide PC £34.20/10m2	0.44	4.11	4.36	m2	8.47
not exceeding 300 mm wide	0.29	2.71	1.44	m	4.15

K LININGS/SHEATHING/DRY PARTITIONING Including overheads and profit at 12.50%	Labour hours	Labour £	Material £	Unit	Total rate £
6 mm lining to ceilings					
over 300 mm wide	0.58	5.42	4.36	m2	9.78
not exceeding 300 mm wide	0.39	3.65	1.44	m	5.09
Raking cutting	0.06	0.56	0.19	m	0.75
Holes for pipes and the like	0.03	0.28	-	nr	0.28
Two-sided 6 mm thick pipe casings;					
50 x 50 mm softwood framing; two members					
plugged to wall					
300 mm girth	1.30	12.15	3.78	m	15.93
450 mm girth	1.45	13.56	4.62	m	18.18
600 mm girth	1.55	14.49	5.45	m	19.94
750 mm girth	1.65	15.43	6.29	m	21.72
Three-sided 6 mm thick pipe casing;					
50 x 50 mm softwood framing; to members					
plugged to wall					
450 mm girth	1.80	16.83	5.34	m	22.17
600 mm girth	1.95	18.23	6.17	m	24.40
750 mm girth	2.05	19.16	7.01	m	26.17
900 mm girth	2.15	20.10	7.85	m	27.95
1050 mm girth	2.20	20.57	8.68	m	29.25
9 mm lining to walls					
over 300 mm wide PC £47.93/10m2	0.47	4.39	6.06	m2	10.45
not exceeding 300 mm wide	0.31	2.90	1.99	m	4.89
9 mm lining to ceilings					
over 300 mm wide	0.63	5.89	6.06	m2	11.95
not exceeding 300 mm wide	0.41	3.83	1.99	m	5.82
Raking cutting	0.07	0.65	0.27	m	0.92
Holes for pipes and the like	0.04	0.37	-	nr	0.37
12 mm lining to walls					
over 300 mm wide PC £62.78/10m2	0.51	4.77	7.90	m2	12.67
not exceeding 300 mm wide	0.33	3.09	2.59	m	5.68
12 mm lining to ceilings					
over 300 mm wide	0.67	6.26	7.90	m2	14.16
not exceeding 300 mm wide	0.44	4.11	2.59	m	6.70
Raking cutting	0.07	0.65	0.35	m	1.00
Holes for pipes and the like	0.04	0.37	-	nr	0.37
Plywood (Finnish Birch); external quality					
4 mm lining to walls					
over 300 mm wide PC £30.54/10m2	0.41	3.83	3.91	m2	7.74
not exceeding 300 mm wide	0.26	2.43	1.28	m	3.71
4 mm lining to ceilings					
over 300 mm wide	0.55	5.14	3.91	m2	9.05
not exceeding 300 mm wide	0.35	3.27	1.28	m	4.55
Raking cutting	0.06	0.56	0.17	m	0.73
Holes for pipes and the like	0.03	0.28	-	nr	0.28
6 mm lining to walls					
over 300 mm wide PC £45.90/10m2	0.44	4.11	5.81	m2	9.92
not exceeding 300 mm wide	0.29	2.71	1.91	m	4.62
6 mm lining to ceilings					
over 300 mm wide	0.58	5.42	5.81	m2	11.23
not exceeding 300 mm wide	0.39	3.65	1.91	m	5.56
Raking cutting	0.06	0.56	0.26	m	0.82
Holes for pipes and the like	0.03	0.28	-	nr	0.28
Two-sided 6 mm thick pipe casings;					
50 x 50 mm softwood framing; two members					
plugged to wall					
300 mm girth	1.30	12.15	4.21	m	16.36
450 mm girth	1.45	13.56	5.27	m	18.83
600 mm girth	1.55	14.49	6.32	m	20.81
750 mm girth	1.65	15.43	7.38	m	22.81
Three-sided 6 mm thick pipe casing;					
50 x 50 mm softwood framing; to members					
plugged to wall					
450 mm girth	1.80	16.83	5.99	m	22.82
600 mm girth	1.95	18.23	7.04	m	25.27
750 mm girth	2.05	19.16	8.10	m	27.26
900 mm girth	2.15	20.10	9.14	m	29.24
1050 mm girth	2.20	20.57	10.20	m	30.77

K LININGS/SHEATHING/DRY PARTITIONING Including overheads and profit at 12.50%	Labour hours	Labour £	Material £	Unit	Total rate £
K11 RIGID SHEET FLOORING/SHEATHING/LININGS/ CASINGS - cont'd					
Plywood (Finnish Birch); external quality - cont'd					
9 mm lining to walls					
over 300 mm wide PC £59.96/10m2	0.47	4.39	7.55	m2	11.94
not exceeding 300 mm wide	0.31	2.90	2.48	m	5.38
9 mm lining to ceilings					
over 300 mm wide	0.63	5.89	7.55	m2	13.44
not exceeding 300 mm wide	0.41	3.83	2.48	m	6.31
Raking cutting	0.08	0.75	0.34	m	1.09
Holes for pipes and the like	0.06	0.56	-	nr	0.56
12 mm lining to walls					
over 300 mm wide PC £76.39/10m2	0.51	4.77	9.58	m2	14.35
not exceeding 300 mm wide	0.33	3.09	3.16	m	6.25
12 mm lining to ceilings					
over 300 mm wide	0.67	6.26	9.59	m2	15.85
not exceeding 300 mm wide	0.44	4.11	3.16	m	7.27
Raking cutting	0.11	1.03	0.43	m	1.46
Holes for pipes and the like	0.09	0.84	-	nr	0.84
Extra over wall linings fixed with nails; for screwing	0.17	1.59	0.20	m2	1.79
Woodwool unreinforced slabs; Torvale 'Woodcemair' or similar; BS 1105 type SB; natural finish; fixing to timber or steel with galvanized nails or clips; flat or sloping					
50 mm slabs; type 500; max. span 600 mm					
1800 - 2400 mm lengths PC £4.99	0.44	4.11	6.14	m2	10.25
2700 - 3000 mm lengths PC £5.06	0.44	4.11	6.23	m2	10.34
75 mm slabs; type 750; max. span 900 mm					
2100 mm lengths PC £6.46	0.50	4.67	7.92	m2	12.59
2400 - 2700 mm lengths PC £6.55	0.50	4.67	8.07	m2	12.74
3000 mm lengths PC £6.64	0.50	4.67	8.09	m2	12.76
Raking cutting	0.15	1.40	1.68	m	3.08
Holes for pipes and the like	0.13	1.22	-	nr	1.22
100 mm slabs; type 1000; max. span 1200 mm					
3000 - 3600 mm lengths PC £9.45	0.55	5.14	11.41	m2	16.55
Internal quality West African Mahogany veneered plywood; 6 mm thick; PC £40.05/10m2; WAM cover strips					
Lining to walls					
over 300 mm wide	0.77	7.20	6.10	m2	13.30
not exceeding 300 mm wide	0.50	4.67	2.41	m	7.08
K13 RIGID SHEET FINE LININGS/PANELLING					
Formica faced chipboard; 17 mm thick; white matt finish; balancer; PC £180.16/10m2; aluminium cover strips and countersunk screws					
Lining to walls					
over 300 mm wide	2.10	19.63	24.10	m2	43.73
not exceeding 300 mm wide	1.35	12.62	8.54	m	21.16
Lining to isolated columns or the like					
over 300 mm wide	3.15	29.45	26.73	m2	56.18
not exceeding 300 mm wide	1.80	16.83	9.20	m	26.03
K20 TIMBER BOARD FLOORING/SHEATHING/LININGS/ CASINGS					
Sawn softwood; untreated					
Boarding to roofs; 150 mm wide boards; butt joints					
19 mm thick; flat; over 300 mm wide					
PC £62.11/100m	0.50	4.67	5.35	m2	10.02
19 mm thick; flat; not exceeding 300 mm wide	0.33	3.09	1.65	m	4.74

K LININGS/SHEATHING/DRY PARTITIONING Including overheads and profit at 12.50%	Labour hours	Labour £	Material £	Unit	Total rate £
19 mm thick; sloping; over 300 mm wide	0.55	5.14	5.35	m2	10.49
19 mm thick; sloping; not exceeding 300 mm wide	0.36	3.37	1.65	m	5.02
19 mm thick; sloping; laid diagonally; over 300 mm wide	0.69	6.45	5.48	m2	11.93
19 mm thick; sloping; laid diagonally; not exceeding 300 mm wide	0.44	4.11	1.69	m	5.80
25 mm thick; flat; over 300 mm wide; PC £78.95/100m	0.50	4.67	6.72	m2	11.39
25 mm thick; flat; not exceeding 300 mm	0.33	3.09	2.07	m	5.16
25 mm thick; sloping; over 300 mm wide	0.55	5.14	6.72	m2	11.86
25 mm thick; sloping; not exceeding 300 mm wide	0.36	3.37	2.07	m	5.44
Boarding to tops or cheeks of dormers; 150 mm wide boards; butt joints					
19 mm thick; diagonally; over 300 mm wide	0.88	8.23	5.48	m2	13.71
19 mm thick; diagonally; not exceeding 300 mm wide	0.55	5.14	1.65	m	6.79
19 mm thick; diagonally; area not exceeding 1.00 m2; irrespective of width	1.10	10.28	5.59	nr	15.87

Sawn softwood; 'Tanalised'
Boarding to roofs; 150 mm wide boards;
butt joints

19 mm thick; flat; over 300 mm wide PC £77.64/100m	0.50	4.67	6.61	m2	11.28
19 mm thick; flat; not exceeding 300 mm wide	0.33	3.09	2.04	m	5.13
19 mm thick; sloping; over 300 mm wide	0.55	5.14	6.61	m2	11.75
19 mm thick; sloping; not exceeding 300 mm wide	0.36	3.37	2.04	m	5.41
19 mm thick; sloping; laid diagonally; over 300 mm wide	0.69	6.45	6.76	m2	13.21
19 mm thick; sloping; laid diagonally; not exceeding 300 mm wide	0.44	4.11	2.07	m	6.18
25 mm thick; flat; over 300 mm wide; PC £98.69/100m	0.50	4.67	8.32	m2	12.99
25 mm thick; flat; not exceeding 300 mm	0.33	3.09	2.56	m	5.65
25 mm thick; sloping; over 300 mm wide	0.55	5.14	8.32	m2	13.46
25 mm thick; sloping; not exceeding 300 mm wide	0.36	3.37	2.56	m	5.93
Boarding to tops or cheeks of dormers; 150 mm wide boards; butt joints					
19 mm thick; diagonally; over 300 mm wide	0.88	8.23	6.76	m2	14.99
19 mm thick; diagonally; not exceeding 300 mm wide	0.55	5.14	2.07	m	7.21
19 mm thick; diagonally; area not exceeding 1.00 m2; irrespective of width	1.10	10.28	6.76	nr	17.04

Wrought softwood
Boarding to floors; butt joints

19 mm thick x 75 mm wide boards PC £55.87/100m	0.66	6.17	9.57	m2	15.74
19 mm thick x 125 mm wide boards PC £66.47/100m	0.55	5.14	6.82	m2	11.96
22 mm thick x 150 mm wide boards PC £78.51/100m	0.50	4.67	6.68	m2	11.35
25 mm thick x 100 mm wide boards PC £76.92/100m	0.61	5.70	9.76	m2	15.46
25 mm thick x 150 mm wide boards PC £112.18/100m	0.50	4.67	9.41	m2	14.08

25 mm boarding and bearers to floors; butt
joints; in making good where partitions
removed or openings formed (boards running
in direction of partition)

150 mm wide	0.30	2.80	1.66	m	4.46
225 mm wide	0.45	4.21	2.51	m	6.72
300 mm wide	0.60	5.61	3.33	m	8.94

K LININGS/SHEATHING/DRY PARTITIONING Including overheads and profit at 12.50%	Labour hours	Labour £	Material £	Unit	Total rate £
K20 TIMBER BOARD FLOORING/SHEATHING/LININGS/ CASINGS - cont'd					
Wrought softwood - cont'd					
25 mm boarding and bearers to floors; butt joints; in making good where partitions removed or openings formed (boards running at right angles to partition)					
150 mm wide	0.45	4.21	3.49	m	7.70
225 mm wide	0.70	6.54	4.44	m	10.98
300 mm wide	0.90	8.41	5.39	m	13.80
450 mm wide	1.35	12.62	7.14	m	19.76
Boarding to floors; tongued and grooved joints					
19 mm thick x 75 mm wide boards					
PC £55.87/100m	0.77	7.20	10.11	m2	17.31
19 mm thick x 125 mm wide boards					
PC £66.47/100m	0.66	6.17	7.20	m2	13.37
22 mm thick x 150 mm wide boards					
PC £78.51/100m	0.61	5.70	7.07	m2	12.77
25 mm thick x 100 mm wide boards					
PC £76.92/100m	0.72	6.73	10.32	m2	17.05
25 mm thick x 150 mm wide boards					
PC £112.18/100m	0.61	5.70	9.96	m2	15.66
Nosings; tongued to edge of flooring					
19 x 75 mm; once rounded	0.28	2.62	1.68	m	4.30
25 x 75 mm; once rounded	0.28	2.62	2.00	m	4.62
Boarding to internal walls; tongued and grooved and V-jointed					
12 mm thick x 100 mm wide boards					
PC £53.41/100m	0.88	8.23	7.30	m2	15.53
16 mm thick x 100 mm wide boards					
PC £63.00/100m	0.88	8.23	8.53	m2	16.76
19 mm thick x 100 mm wide boards					
PC £74.73/100m	0.88	8.23	10.04	m2	18.27
19 mm thick x 125 mm wide boards					
PC £92.44/100m	0.83	7.76	9.88	m2	17.64
19 mm thick x 125 mm wide boards; chevron pattern	1.30	12.15	10.37	m2	22.52
25 mm thick x 125 mm wide boards					
PC £120.16/100m	0.83	7.76	12.72	m2	20.48
12 mm thick x 100 mm wide					
Knotty Pine boards PC £37.95/100m	0.88	8.23	5.32	m2	13.55
Boarding to internal ceilings; tongued and grooved and V-jointed					
12 mm thick x 100 mm wide boards	1.10	10.28	7.30	m2	17.58
16 mm thick x 100 mm wide boards	1.10	10.28	8.53	m2	18.81
19 mm thick x 100 mm wide boards	1.10	10.28	10.04	m2	20.32
19 mm thick x 125 mm wide boards	1.05	9.82	9.88	m2	19.70
19 mm thick x 125 mm wide boards; chevron pattern	1.55	14.49	10.37	m2	24.86
25 mm thick x 125 mm wide boards	1.05	9.82	12.72	m2	22.54
12 mm thick x 100 mm wide					
Knotty Pine boards	1.10	10.28	5.32	m2	15.60
Boarding to roofs; tongued and grooved joints					
19 mm thick; flat to falls PC £86.49/100m	0.61	5.70	7.75	m2	13.45
19 mm thick; sloping	0.66	6.17	7.75	m2	13.92
19 mm thick; sloping; laid diagonally	0.83	7.76	7.89	m2	15.65
25 mm thick; flat to falls PC £112.18/100m	0.61	5.70	9.96	m2	15.66
25 mm thick; sloping	0.66	6.17	9.96	m2	16.13
Boarding to tops or cheeks of dormers; tongued and grooved joints					
19 mm thick; laid diagonally	1.10	10.28	7.89	m2	18.17

K LININGS/SHEATHING/DRY PARTITIONING Including overheads and profit at 12.50%	Labour hours	Labour £	Material £	Unit	Total rate £
Wrought softwood; 'Tanalised'					
Boarding to roofs; tongued and grooved joints					
19 mm thick; flat to falls PC £108.12/100m	0.61	5.70	9.61	m2	15.31
19 mm thick; sloping	0.66	6.17	9.61	m2	15.78
19 mm thick; sloping; laid diagonally	0.83	7.76	9.78	m2	17.54
25 mm thick; flat to falls PC £140.22/100m	0.61	5.70	12.38	m2	18.08
25 mm thick; sloping	0.66	6.17	12.38	m2	18.55
Boarding to tops or cheeks of dormers; tongued and grooved joints					
19 mm thick; laid diagonally	1.10	10.28	9.78	m2	20.06
Wood strip; 22 mm thick; 'Junckers' pre-treated or similar; tongued and grooved joints; pre-finished boards; level fixing to resilient battens; to cement and sand base					
Strip flooring; over 300 mm wide					
Beech; prime	-	-	-	m2	34.20
Beech; standard	-	-	-	m2	32.31
Beech; sylvia squash	-	-	-	m2	34.20
Oak; quality A	-	-	-	m2	42.82
Wrought hardwood					
Strip flooring to floors; 25 mm thick x 75 mm wide; tongued and grooved joints; secret fixing; surface sanded after laying					
American Oak	-	-	-	m2	38.36
Canadian Maple	-	-	-	m2	30.94
Gurjun	-	-	-	m2	30.94
Iroko	-	-	-	m2	38.36
Raking cutting	-	-	-	m	2.30
Curved cutting	-	-	-	m	3.24
K29 TIMBER FRAMED AND PANELLED PARTITIONS					
Purpose made screen components; wrought softwood					
Panelled partitions					
32 mm thick frame; 9 mm thick					
plywood panels; over 300 mm wide PC £33.08	0.99	9.26	38.15	m2	47.41
38 mm thick frame; 9 mm thick					
plywood panels over 300 mm wide PC £35.28	1.05	9.82	40.68	m2	50.50
50 mm thick frame; 12 mm thick					
plywood panels; over 300 mm wide PC £42.92	1.10	10.28	49.49	m2	59.77
Panelled partitions; mouldings worked on solid both sides					
32 mm thick frame; 9 mm thick					
plywood panels; over 300 mm wide PC £38.22	0.99	9.26	44.08	m2	53.34
38 mm thick frame; 9 mm thick					
plywood panels over 300 mm wide PC £39.99	1.05	9.82	46.11	m2	55.93
50 mm thick frame; 12 mm thick					
plywood panels; over 300 mm wide PC £48.53	1.10	10.28	55.96	m2	66.24
Panelled partitions; mouldings planted on both sides					
32 mm thick frame; 9 mm thick					
plywood panels; over 300 mm wide PC £38.69	0.99	9.26	44.62	m2	53.88
38 mm thick frame; 9 mm thick					
plywood panels over 300 mm wide PC £40.34	1.05	9.82	46.52	m2	56.34
50 mm thick frame; 12 mm thick					
plywood panels; over 300 mm wide PC £48.53	1.10	10.28	55.96	m2	66.24
Panelled partitions; diminishing stiles over 300 mm wide; upper portion open panels for glass; in medium panes					
32 mm thick frame; 9 mm thick					
plywood panels; over 300 mm wide PC £54.96	0.99	9.26	63.37	m2	72.63
38 mm thick frame; 9 mm thick					
plywood panels over 300 mm wide PC £61.45	1.05	9.82	70.86	m2	80.68
50 mm thick frame; 12 mm thick					
plywood panels; over 300 mm wide PC £76.50	1.10	10.28	88.21	m2	98.49

K LININGS/SHEATHING/DRY PARTITIONING Including overheads and profit at 12.50%	Labour hours	Labour £	Material £	Unit	Total rate £

K29 TIMBER FRAMED AND PANELLED PARTITIONS - cont'd

Purpose made screen components; selected
West African Mahogany; PC £531.05/m3
Panelled partitions

32 mm thick frame; 9 mm thick					
plywood panels; over 300 mm wide PC £60.22	1.30	12.15	69.45	m2	**81.60**
38 mm thick frame; 9 mm thick					
plywood panels over 300 mm wide PC £65.22	1.40	13.09	75.21	m2	**88.30**
50 mm thick frame; 12 mm thick					
plywood panels; over 300 mm wide PC £82.11	1.50	14.02	94.69	m2	**108.71**

Panelled partitions; mouldings worked
on solid both sides

32 mm thick frame; 9 mm thick					
plywood panels; over 300 mm wide PC £68.79	1.30	12.15	79.32	m2	**91.47**
38 mm thick frame; 9 mm thick					
plywood panels over 300 mm wide PC £76.85	1.40	13.09	88.62	m2	**101.71**
50 mm thick frame; 12 mm thick					
plywood panels; over 300 mm wide PC £85.27	1.50	14.02	98.33	m2	**112.35**

Panelled partitions; mouldings planted on
both sides

32 mm thick frame; 9 mm thick					
plywood panels; over 300 mm wide PC £86.98	1.30	12.15	100.29	m2	**112.44**
38 mm thick frame; 9 mm thick					
plywood panels over 300 mm wide PC £87.74	1.40	13.09	101.17	m2	**114.26**
50 mm thick frame; 12 mm thick					
plywood panels; over 300 mm wide PC £95.54	1.50	14.02	110.17	m2	**124.19**

Panelled partitions; diminishing stiles
over 300 mm wide; upper portion open panels
for glass; in medium panes

32 mm thick frame; 9 mm thick					
plywood panels over 300 mm wide PC £112.94	1.30	12.15	130.23	m2	**142.38**
38 mm thick frame; 9 mm thick					
plywood panels over 300 mm wide PC £121.35	1.40	13.09	139.92	m2	**153.01**
50 mm thick frame; 12 mm thick					
plywood panels; over 300 mm wide PC £128.79	1.50	14.02	148.52	m2	**162.54**

Purpose made screen components; Afrormosia
PC £758.95/m3
Panelled partitions

32 mm thick frame; 9 mm thick					
plywood panels; over 300 mm wide PC £74.99	1.30	12.15	86.46	m2	**98.61**
38 mm thick frame; 9 mm thick					
plywood panels over 300 mm wide PC £81.32	1.40	13.09	93.77	m2	**106.86**
50 mm thick frame; 12 mm thick					
plywood panels; over 300 mm wide PC £102.49	1.50	14.02	118.18	m2	**132.20**

Panelled partitions; mouldings worked
on solid both sides

32 mm thick frame; 9 mm thick					
plywood panels; over 300 mm wide PC £85.73	1.30	12.15	98.86	m2	**111.01**
38 mm thick frame; 9 mm thick					
plywood panels over 300 mm wide PC £95.94	1.40	13.09	110.64	m2	**123.73**
50 mm thick frame; 12 mm thick					
plywood panels over 300 mm wide PC £106.22	1.50	14.02	122.49	m2	**136.51**

Panelled partitions; mouldings planted on
both sides

32 mm thick frame; 9 mm thick					
plywood panels; over 300 mm wide PC £115.16	1.30	12.15	132.79	m2	**144.94**
38 mm thick frame; 9 mm thick					
plywood panels over 300 mm wide PC £116.07	1.40	13.09	133.84	m2	**146.93**
50 mm thick frame; 12 mm thick					
plywood panels; over 300 mm wide PC £119.16	1.50	14.02	137.40	m2	**151.42**

Panelled partitions; diminishing stiles
over 300 mm wide; upper portion open panels
for glass; in medium panes

32 mm thick frame; 9 mm thick					
plywood panels; over 300 mm wide PC £140.75	1.30	12.15	162.31	m2	**174.46**

K LININGS/SHEATHING/DRY PARTITIONING Including overheads and profit at 12.50%		Labour hours	Labour £	Material £	Unit	Total rate £
38 mm thick frame; 9 mm thick plywood panels over 300 mm wide	PC £151.09	1.40	13.09	174.23	m2	187.32
50 mm thick frame; 12 mm thick plywood panels; over 300 mm wide	PC £160.73	1.50	14.02	185.34	m2	199.36

K31 PLASTERBOARD FIXED PARTITIONS/
INNER WALLS/ LININGS

**'Gyproc' laminated partition; comprising two
skins of gypsum plasterboard bonded to a
centre core of plasterboard square edge
plank 19 mm thick; fixing with nails to
softwood studwork (measured separately);
joints filled with joint filler and joint
tape; to receive direct decoration;
(perimeter studwork measured separately)**
50 mm partition; two outer skins of 12.7 mm
tapered edge wallboard

	Labour hours	Labour £	Material £	Unit	Total rate £
height 2.10 - 2.40 m	2.35	32.60	19.96	m	52.56
height 2.40 - 2.70 m	2.65	36.76	22.53	m	59.29
height 2.70 - 3.00 m	2.95	40.92	25.15	m	66.07
height 3.00 - 3.30 m	3.30	45.78	27.77	m	73.55
height 3.30 - 3.60 m	3.75	52.02	30.39	m	82.41

65 mm partition; two outer skins of 19 mm
tapered edge plank

	Labour hours	Labour £	Material £	Unit	Total rate £
height 2.10 - 2.40 m	2.60	36.07	26.10	m	62.17
height 2.40 - 2.70 m	2.85	39.53	29.42	m	68.95
height 2.70 - 3.00 m	3.30	45.78	32.78	m	78.56
height 3.00 - 3.30 m	3.70	51.32	36.14	m	87.46
height 3.30 - 3.60 m	4.15	57.57	39.50	m	97.07

**Labours and associated additional wrought
softwood studwork**
Floor, wall or ceiling battens

	Labour hours	Labour £	Material £	Unit	Total rate £
25 x 38 mm	0.13	1.80	0.51	m	2.31
Forming openings					
25 x 38 mm framing	0.33	4.58	0.79	m	5.37
Fair ends	0.22	3.05	0.51	m	3.56
Angle	0.33	4.58	0.51	m	5.09
Intersection	0.22	3.05	-	m	3.05

Cutting and fitting around steel joists,
angles, trunking, ducting, ventilators,
pipes, tubes, etc

	Labour hours	Labour £	Material £	Unit	Total rate £
over 2 m girth	0.10	1.39	-	m	1.39
not exceeding 0.30 m girth	0.06	0.83	-	nr	0.83
0.30 - 1 m girth	0.08	1.11	-	nr	1.11
1 - 2 m girth	0.12	1.66	-	nr	1.66

**'Paramount' dry partition comprising
Paramount panels with and including wrought
softwood battens at vertical joints; fixing
with nails to softwood studwork; (perimeter
studwork measured separately)**
Square edge panels; joints filled with
plaster and jute scrim cloth; to receive
plaster (measured elsewhere);
57 mm partition

	Labour hours	Labour £	Material £	Unit	Total rate £
height 2.10 - 2.40 m	1.60	22.19	16.99	m	39.18
height 2.40 - 2.70 m	1.75	24.27	19.13	m	43.40
height 2.70 - 3.00 m	2.00	27.74	21.26	m	49.00
height 3.00 - 3.30 m	2.25	31.21	21.42	m	52.63
height 3.30 - 3.60 m	2.60	36.07	25.61	m	61.68

63 mm partition

	Labour hours	Labour £	Material £	Unit	Total rate £
height 2.10 - 2.40 m	1.75	24.27	19.35	m	43.62
height 2.40 - 2.70 m	1.95	27.05	21.78	m	48.83
height 2.70 - 3.00 m	2.15	29.82	24.19	m	54.01
height 3.00 - 3.30 m	2.40	33.29	26.66	m	59.95
height 3.30 - 3.60 m	2.75	38.15	29.15	m	67.30

K LININGS/SHEATHING/DRY PARTITIONING Including overheads and profit at 12.50%	Labour hours	Labour £	Material £	Unit	Total rate £
K31 PLASTERBOARD FIXED PARTITIONS/ INNER WALLS/ LININGS - cont'd					
'Paramount' dry partition comprising Paramount panels with and including wrought softwood battens at vertical joints; fixing with nails to softwood studwork; (perimeter studwork measured separately) - cont'd Tapered edge panels; joints filled with joint filler and joint tape to receive direct decoration; 57 mm partition					
height 2.10 - 2.40 m	2.25	31.21	18.42	m	49.63
height 2.40 - 2.70 m	2.40	33.29	20.76	m	54.05
height 2.70 - 3.00 m	2.65	36.76	23.09	m	59.85
height 3.00 - 3.30 m	2.90	40.23	25.48	m	65.71
height 3.30 - 3.60 m	3.30	45.78	27.87	m	73.65
63 mm partition					
height 2.10 - 2.40 m	2.40	33.29	20.82	m	54.11
height 2.40 - 2.70 m	2.60	36.07	23.45	m	59.52
height 2.70 - 3.00 m	2.80	38.83	26.08	m	64.91
height 3.00 - 3.30 m	3.10	43.00	28.76	m	71.76
height 3.30 - 3.60 m	3.45	47.86	31.45	m	79.31
Labours and associated additional wrought softwood studwork on 57 mm partition Wall or ceiling battens					
19 x 37 mm	0.13	1.80	0.46	m	2.26
Sole plates; with 19 x 37 x 150 mm long battens spiked on at 600 mm centres					
19 x 57 mm	0.28	3.88	0.85	m	4.73
50 x 57 mm	0.33	4.58	1.35	m	5.93
Forming openings					
37 x 37 mm framing	0.33	4.58	0.46	m	5.04
Fair ends	0.22	3.05	0.46	m	3.51
Angle	0.33	4.58	0.97	m	5.55
Intersection	0.28	3.88	0.66	m	4.54
Cutting and fitting around steel joists, angles, trunking, ducting, ventilators, pipes, tubes, etc					
over 2 m girth	0.11	1.53	-	m	1.53
not exceeding 0.30 m girth	0.07	0.97	-	nr	0.97
0.30 - 1 m girth	0.09	1.25	-	nr	1.25
1 - 2 m girth	0.13	1.80	-	nr	1.80
Plugging; 300 mm centres; one way					
brickwork	0.11	1.53	0.03	m	1.56
concrete	0.20	2.77	0.03	m	2.80
'Gyroc' metal stud partition; comprising 146 mm metal stud frame; with floor channel plugged and screwed to concrete through 38 x 148 mm tanalised softwood sole plate Tapered edge panels; joints filled with joint filler and joint tape to receive direct decoration 171 mm partition; one hour; one layer of 12.5 mm Fireline board each side					
height 2.10 - 2.40 m	4.60	63.81	23.55	m	87.36
height 2.40 - 2.70 m	5.35	74.21	26.06	m	100.27
height 2.70 - 3.00 m	5.95	82.53	28.70	m	111.23
height 3.00 - 3.30 m	6.90	95.71	31.50	m	127.21
height 3.30 - 3.60 m	7.55	104.73	33.70	m	138.43
height 3.60 - 3.90 m	9.00	124.84	36.88	m	161.72
height 3.90 - 4.20 m	9.70	134.55	39.53	m	174.08
Angles	0.22	3.05	1.34	m	4.39
T-junctions	0.22	3.05	1.25	m	4.30
Fair end	0.33	4.58	2.08	m	6.66

K LININGS/SHEATHING/DRY PARTITIONING Including overheads and profit at 12.50%	Labour hours	Labour £	Material £	Unit	Total rate £
Tapered edge panels; joints filled with joint filler and joint tape to receive direct decoration					
196 mm partition; two hour; two layers of 12.5 mm Fireline board both sides					
height 2.10 - 2.40 m	6.60	91.55	37.54	m	129.09
height 2.40 - 2.70 m	7.45	103.34	41.79	m	145.13
height 2.70 - 3.00 m	8.25	114.44	46.19	m	160.63
height 3.00 - 3.30 m	9.65	133.86	50.44	m	184.30
height 3.30 - 3.60 m	10.56	146.48	54.69	m	201.17
height 3.60 - 3.90 m	12.90	178.94	59.61	m	238.55
height 3.90 - 4.20 m	13.90	192.81	64.02	m	256.83
Angles	0.22	3.05	1.34	m	4.39
T-junctions	0.22	3.05	1.25	m	4.30
Fair end	0.39	5.41	2.66	m	8.07

L WINDOWS/DOORS/STAIRS Including overheads and profit at 12.50%	Labour hours	Labour £	Material £	Unit	Total rate £

ALTERNATIVE TIMBER WINDOW PRICES (£/each)

	£		£
Softwood shallow circular bay windows			
3 lightx1200 mm; ref CSB312CV	163.49	4 lightx1350 mm; ref CSB413CV	230.30
3 lightx1350 mm; ref CSB313CV	171.33	4 lightx1500 mm; ref CSB415T	261.03
3 light glass fibre roof unit	72.14	4 light glass fibre roof unit	77.81

	Labour hours	Labour £	Material £	Unit	Total rate £

L10 TIMBER WINDOWS/ROOFLIGHTS/SCREENS/LOUVRES

Standard windows; 'treated' wrought softwood
(refs. refer to Boulton & Paul cat. nos.)
Side hung casement windows without glazing
bars; with 140 mm wide softwood sills;
opening casements and ventilators hung on
rustproof hinges; fitted with aluminized
laquered finish casement stays and
fasteners; knotting and priming by
manufacturer before delivery

		Labour hours	Labour £	Material £	Unit	Total rate £
500 x 750 mm; ref N07V	PC £24.12	0.77	7.20	27.13	nr	34.33
500 x 900 mm; ref N09V	PC £28.12	0.88	8.23	31.63	nr	39.86
600 x 750 mm; ref 107V	PC £30.52	0.88	8.23	34.34	nr	42.57
600 x 750 mm; ref 107C	PC £28.00	0.88	8.23	31.50	nr	39.73
600 x 900 mm; ref 109V	PC £27.86	0.99	9.26	31.34	nr	40.60
600 x 900 mm; ref 109C	PC £33.22	0.88	8.23	37.37	nr	45.60
600 x 1050 mm; ref 110V	PC £32.48	0.88	8.23	36.54	nr	44.77
600 x 1050 mm; ref 110C	PC £34.72	1.10	10.28	39.06	nr	49.34
900 x 900 mm; ref 2N09W	PC £35.95	1.20	11.22	40.44	nr	51.66
900 x 1050 mm; ref 2N10W	PC £36.61	1.25	11.69	41.19	nr	52.88
900 x 1200 mm; ref 2N12W	PC £37.66	1.30	12.15	42.37	nr	54.52
900 x 1350 mm; ref 2N13W	PC £38.57	1.50	14.02	43.39	nr	57.41
900 x 1500 mm; ref 2N15W	PC £39.27	1.55	14.49	44.18	nr	58.67
1200 x 750 mm; ref 207C	PC £37.42	1.25	11.69	42.09	nr	53.78
1200 x 750 mm; ref 207CV	PC £48.93	1.25	11.69	55.05	nr	66.74
1200 x 900 mm; ref 209C	PC £39.52	1.30	12.15	44.45	nr	56.60
1200 x 900 mm; ref 209W	PC £41.72	1.30	12.15	46.94	nr	59.09
1200 x 900 mm; ref 209CV	PC £51.07	1.30	12.15	57.45	nr	69.60
1200 x 1050 mm; ref 210C	PC £41.09	1.50	14.02	46.23	nr	60.25
1200 x 1050 mm; ref 210W	PC £42.59	1.50	14.02	47.92	nr	61.94
1200 x 1050 mm; ref 210T	PC £51.07	1.50	14.02	57.45	nr	71.47
1200 x 1050 mm; ref 210CV	PC £52.78	1.50	14.02	59.38	nr	73.40
1200 x 1200 mm; ref 212C	PC £43.05	1.60	14.96	48.43	nr	63.39
1200 x 1200 mm; ref 212W	PC £43.58	1.60	14.96	49.02	nr	63.98
1200 x 1200 mm; ref 212T	PC £52.82	1.60	14.96	59.42	nr	74.38
1200 x 1200 mm; ref 212CV	PC £54.81	1.60	14.96	61.66	nr	76.62
1200 x 1350 mm; ref 213W	PC £44.59	1.70	15.89	50.16	nr	66.05
1200 x 1350 mm; ref 213CV	PC £57.89	1.70	15.89	65.13	nr	81.02
1200 x 1500 mm; ref 215W	PC £45.85	1.85	17.30	51.58	nr	68.88
1770 x 750 mm; ref 307CC	PC £59.19	1.55	14.49	66.58	nr	81.07
1770 x 900 mm; ref 309CC	PC £62.27	1.85	17.30	70.05	nr	87.35
1770 x 1050 mm; ref 310C	PC £49.77	1.95	18.23	55.99	nr	74.22
1770 x 1050 mm; ref 310T	PC £60.94	1.85	17.30	68.55	nr	85.85
1770 x 1050 mm; ref 310CC	PC £64.96	1.55	14.49	73.08	nr	87.57
1770 x 1050 mm; ref 310WW	PC £71.78	1.55	14.49	80.76	nr	95.25
1770 x 1200 mm; ref 312C	PC £51.49	2.00	18.70	57.92	nr	76.62
1770 x 1200 mm; ref 312T	PC £62.86	2.00	18.70	70.72	nr	89.42
1770 x 1200 mm; ref 312CC	PC £67.69	2.00	18.70	76.15	nr	94.85
1770 x 1200 mm; ref 312WW	PC £73.82	2.00	18.70	83.04	nr	101.74
1770 x 1200 mm; ref 312CVC	PC £79.45	2.00	18.70	89.38	nr	108.08

L WINDOWS/DOORS/STAIRS Including overheads and profit at 12.50%		Labour hours	Labour £	Material £	Unit	Total rate £
1770 x 1350 mm; ref 313CC	PC £72.84	2.10	19.63	81.94	nr	101.57
1770 x 1350 mm; ref 312CC	PC £75.71	2.10	19.63	85.17	nr	104.80
1770 x 1350 mm; ref 313CVC	PC £84.63	2.10	19.63	95.21	nr	114.84
1770 x 1500 mm; ref 315T	PC £66.29	2.20	20.57	74.58	nr	95.15
2340 x 1050 mm; ref 410CWC	PC £91.04	2.15	20.10	102.41	nr	122.51
2340 x 1200 mm; ref 412CWC	PC £94.22	2.25	21.03	106.00	nr	127.03
2340 x 1350 mm; ref 413CWC	PC £99.86	2.40	22.44	112.34	nr	134.78
Top hung casement windows; with 140 mm wide softwood sills; opening casements and ventilators hung on rustproof hinges; fitted with aluminized laquered finish casement stays; knotting and priming by manufacturer before delivery						
600 x 750 mm; ref 107A	PC £30.59	0.88	8.23	34.41	nr	42.64
600 x 900 mm; ref 109A	PC £31.85	0.99	9.26	35.83	nr	45.09
600 x 1050 mm; ref 110A	PC £33.43	1.10	10.28	37.60	nr	47.88
900 x 750 mm; ref 2N07A	PC £38.18	1.15	10.75	42.96	nr	53.71
900 x 900 mm; ref 2N09A	PC £41.51	1.20	11.22	46.70	nr	57.92
900 x 1050 mm; ref 2N10A	PC £43.30	1.25	11.69	48.71	nr	60.40
900 x 1350 mm; ref 2N13AS	PC £50.12	1.50	14.02	56.39	nr	70.41
900 x 1500 mm; ref 2N15AS	PC £51.73	1.55	14.49	58.20	nr	72.69
1200 x 750 mm; ref 207A	PC £43.93	1.25	11.69	49.42	nr	61.11
1200 x 900 mm; ref 209A	PC £46.94	1.30	12.15	52.80	nr	64.95
1200 x 1050 mm; ref 210A	PC £48.76	1.50	14.02	54.85	nr	68.87
1200 x 1050 mm; ref 210AT	PC £65.35	1.50	14.02	73.51	nr	87.53
1200 x 1200 mm; ref 212A	PC £50.44	1.60	14.96	56.74	nr	71.70
1200 x 1200 mm; ref 212AT	PC £68.95	1.60	14.96	77.57	nr	92.53
1200 x 1350 mm; ref 213AS	PC £55.86	1.70	15.89	62.84	nr	78.73
1200 x 1500 mm; ref 215AS	PC £57.61	1.85	17.30	64.81	nr	82.11
1770 x 1050 mm; ref 310A	PC £58.91	1.85	17.30	66.27	nr	83.57
1770 x 1050 mm; ref 310AV	PC £72.24	1.85	17.30	81.27	nr	98.57
1770 x 1200 mm; ref 312A	PC £60.90	2.00	18.70	68.51	nr	87.21
1770 x 1220 mm; ref 312AV	PC £74.31	2.00	18.70	83.59	nr	102.29
2340 x 1200 mm; ref 412A	PC £68.42	2.25	21.03	76.98	nr	98.01
2340 x 1200 mm; ref 412AW	PC £86.63	2.25	21.03	97.45	nr	118.48
2340 x 1500 mm; ref 415AWS	PC £92.23	2.60	24.31	103.75	nr	128.06
High performance top hung reversible windows; with 140 mm wide softwood sills; adjustable ventilators weather stripping; opening sashes and fanlights hung on rustproof hinges; fitted with aluminized laquered espagnolette bolts; knotting and priming by manufacturer before delivery						
600 x 900 mm; ref R0609	PC £76.89	0.99	9.26	86.51	nr	95.77
900 x 900 mm; ref R0909	PC £84.11	1.20	11.22	94.62	nr	105.84
900 x 1050 mm; ref R0910	PC £85.75	1.25	11.69	96.47	nr	108.16
900 x 1200 mm; ref R0912	PC £88.13	1.40	13.09	99.15	nr	112.24
900 x 1500 mm; ref R0915	PC £99.86	1.55	14.49	112.34	nr	126.83
1200 x 1050 mm; ref R1210	PC £93.56	1.50	14.02	105.25	nr	119.27
1200 x 1200 mm; ref R1212	PC £96.04	1.60	14.96	108.05	nr	123.01
1200 x 1500 mm; ref R1215	PC £108.85	1.85	17.30	122.46	nr	139.76
1500 x 1200 mm; ref R1512	PC £107.80	1.85	17.30	121.28	nr	138.58
1800 x 1050 mm; ref R1810	PC £112.60	1.85	17.30	126.67	nr	143.97
1800 x 1200 mm; ref R1812	PC £114.59	2.00	18.70	128.91	nr	147.61
1800 x 1500 mm; ref R1815	PC £123.03	2.10	19.63	138.40	nr	158.03
2400 x 1500 mm; ref R2415	PC £136.50	2.60	24.31	153.56	nr	177.87

L WINDOWS/DOORS/STAIRS Including overheads and profit at 12.50%		Labour hours	Labour £	Material £	Unit	Total rate £

L10 TIMBER WINDOWS/ROOFLIGHTS/SCREENS/LOUVRES - cont'd

Standard windows; 'treated' wrought softwood
(refs. refer to Boulton & Paul cat. nos.) - cont'd
High performance double hung sash windows
with glazing bars; solid frames; 63 x 175 mm
softwood sills; standard flush external
linings; spiral spring balances and sash
catch; knotting and priming by manufacturer
before delivery

635 x 1050 mm; ref DH0610B	PC £110.11	2.20	20.57	123.87	nr	144.44
635 x 1350 mm; ref DH0613B	PC £119.77	2.40	22.44	134.74	nr	157.18
635 x 1650 mm; ref DH0616B	PC £128.38	2.70	25.24	144.43	nr	169.67
860 x 1050 mm; ref DH0810B	PC £120.68	2.55	23.84	135.77	nr	159.61
860 x 1350 mm; ref DH0813B	PC £131.11	2.85	26.64	147.50	nr	174.14
860 x 1650 mm; ref DH0816B	PC £141.54	3.30	30.85	159.23	nr	190.08
1085 x 1050 mm; ref DH1010B	PC £131.81	2.85	26.64	148.29	nr	174.93
1085 x 1350 mm; ref DH1013B	PC £145.15	3.30	30.85	163.29	nr	194.14
1085 x 1650 mm; ref DH1016B	PC £157.33	4.05	37.86	176.99	nr	214.85
1699 x 1050 mm; ref DH1710B	PC £274.56	4.05	37.86	308.88	nr	346.74
1699 x 1350 mm; ref DH1713B	PC £262.92	5.05	47.21	295.79	nr	343.00
1699 x 1650 mm; ref DH1716B	PC £283.01	5.15	48.15	318.39	nr	366.54

Purpose made window casements; 'treated'
wrought softwood
Casements; rebated; moulded

38 mm thick	PC £19.01	-	-	21.39	m2	21.39
50 mm thick	PC £22.58	-	-	25.41	m2	25.41

Casements; rebated; moulded; in medium panes

38 mm thick	PC £27.75	-	-	31.22	m2	31.22
50 mm thick	PC £34.34	-	-	38.63	m2	38.63

Casements; rebated; moulded; with
semi-circular head

38 mm thick	PC £34.68	-	-	39.01	m2	39.01
50 mm thick	PC £42.96	-	-	48.33	m2	48.33

Casements; rebated; moulded; to bullseye
window

38 mm thick x 600 mm dia	PC £62.36/nr	-	-	70.15	m2	70.15
38 mm thick x 900 mm dia	PC £92.51/nr	-	-	104.07	m2	104.07
50 mm thick x 600 mm dia	PC £71.73/nr	-	-	80.70	m2	80.70
50 mm thick x 900 mm dia	PC £106.36/nr	-	-	119.65	m2	119.65

Fitting and hanging casements

square or rectangular		0.55	5.14	-	nr	5.14
semicircular		1.40	13.09	-	nr	13.09
bullseye		2.20	20.57	-	nr	20.57

Purpose made window frames; 'treated'
wrought softwood
Frames; rounded; rebated check grooved

25 x 120 mm	0.15	1.40	6.78	m	8.18
50 x 75 mm	0.15	1.40	7.66	m	9.06
50 x 100 mm	0.18	1.68	9.45	m	11.13
50 x 125 mm	0.18	1.68	11.48	m	13.16
63 x 100 mm	0.18	1.68	11.74	m	13.42
75 x 150 mm	0.20	1.87	19.70	m	21.57
90 x 140 mm	0.20	1.87	25.51	m	27.38

Mullions and transoms; twice rounded,
rebated and check grooved

50 x 75 mm	0.11	1.03	9.31	m	10.34
50 x 100 mm	0.13	1.22	11.08	m	12.30
63 x 100 mm	0.13	1.22	13.37	m	14.59
75 x 150 mm	0.15	1.40	21.33	m	22.73

Sill; sunk weathered, rebated and grooved

75 x 100 mm	0.22	2.06	19.35	m	21.41
75 x 150 mm	0.22	2.06	25.46	m	27.52

Add +5% to the above 'Material £ prices'
for 'selected' softwood for staining

L WINDOWS/DOORS/STAIRS Including overheads and profit at 12.50%		Labour hours	Labour £	Material £	Unit	Total rate £
Purpose made double hung sash windows; **'treated' wrought softwood** Cased frames of 100 x 25 mm grooved inner linings; 114 x 25 mm grooved outer linings; 125 x 38 mm twice rebated head linings; 125 x 32 mm twice rebated grooved pulley stiles; 150 x 13 mm linings; 50 x 19 mm parting slips; 25 x 13 mm parting beads; 25 x 19 mm inside beads; 150 x 75 mm Oak twice sunk weathered throated sill; 50 mm thick rebated and moulded sashes; moulded horns; over 1.25 m2 each; both sashes in medium panes;						
including spiral spring balances	PC £139.76	2.50	23.37	220.01	m2	**243.38**
As above but with cased mullions	PC £147.45	2.75	25.71	228.66	m2	**254.37**
Standard pre-glazed roof windows; 'treated' **Nordic Red Pine and aluminium trimmed** **'Velux' windows, or equivalent; including** **type U flashings and soakers (for tiles and** **pantiles), and sealed double glazing unit;** **trimming opening measured separately**						
550 x 700 mm; ref GGL-9	PC £100.49	5.50	51.42	114.60	nr	**166.02**
550 x 980 mm; ref GGL-6	PC £112.95	5.50	51.42	128.62	nr	**180.04**
700 x 1180 mm; ref GGL-5	PC £134.11	6.60	61.70	152.43	nr	**214.13**
780 x 980 mm; ref GGL-1	PC £129.10	6.05	56.56	146.79	nr	**203.35**
780 x 1400 mm; ref GGL-2	PC £152.01	7.15	66.84	172.56	nr	**239.40**
940 x 1600 mm; ref GGL-3	PC £178.98	7.15	66.84	202.91	nr	**269.75**
1140 x 1180 mm; ref GGL-4	PC £170.97	7.70	71.99	193.89	nr	**265.88**
1340 x 980 mm; ref GGL-7	PC £174.42	7.70	71.99	197.78	nr	**269.77**
1340 x 1400 mm; ref GGL-8	PC £201.79	8.25	77.13	228.57	nr	**305.70**
Standard windows; selected Philippine **Mahogany; preservative stain finish** Side hung casement windows; with 45 x 140 mm hardwood sills; weather stripping; opening sashes on canopy hinges; fitted with fasteners; aluminized lacquered finish ironmongery						
600 x 600 mm; ref SS0606/L	PC £52.78	1.05	9.82	59.38	nr	**69.20**
600 x 900 mm; ref SS0609/L	PC £58.52	1.30	12.15	65.84	nr	**77.99**
600 x 900 mm; ref SV0609/O	PC £78.00	1.05	9.82	87.75	nr	**97.57**
600 x 1050 mm; ref SS0610/L	PC £70.80	1.45	13.56	79.65	nr	**93.21**
600 x 1050 mm; ref SV0610/O	PC £80.32	1.45	13.56	90.36	nr	**103.92**
900 x 900 mm; ref SV0909/O	PC £103.92	1.65	15.43	116.91	nr	**132.34**
900 x 1050 mm; ref SV0910/O	PC £104.64	1.75	16.36	117.72	nr	**134.08**
900 x 1200 mm; ref SV0912/O	PC £107.04	1.85	17.30	120.42	nr	**137.72**
900 x 1350 mm; ref SV0913/O	PC £109.20	2.00	18.70	122.85	nr	**141.55**
900 x 1500 mm; ref SV0915/O	PC £111.44	2.10	19.63	125.37	nr	**145.00**
1200 x 900 mm; ref SS1209/L	PC £109.44	1.85	17.30	123.12	nr	**140.42**
1200 x 900 mm; ref SV1209/O	PC £119.36	1.85	17.30	134.28	nr	**151.58**
1200 x 1050 mm; ref SS1210/L	PC £101.36	2.00	18.70	114.03	nr	**132.73**
1200 x 1050 mm; ref SV1210/O	PC £121.68	2.00	18.70	136.89	nr	**155.59**
1200 x 1200 mm; ref SS1212/L	PC £106.82	2.15	20.10	120.17	nr	**140.27**
1200 x 1200 mm; ref SV1212/O	PC £123.84	2.15	20.10	139.32	nr	**159.42**
1200 x 1350 mm; ref SV1213/O	PC £125.84	2.30	21.50	141.57	nr	**163.07**
1200 x 1500 mm; ref SV1215/O	PC £128.32	2.40	22.44	144.36	nr	**166.80**
1800 x 900 mm; ref SS189D/O	PC £120.40	2.50	23.37	135.45	nr	**158.82**
1800 x 1050 mm; ref SS1810/L	PC £120.40	2.50	23.37	135.45	nr	**158.82**
1800 x 1050 mm; ref SS180D/O	PC £190.24	2.50	23.37	214.02	nr	**237.39**
1800 x 1200 mm; ref SS1812/L	PC £143.92	2.65	24.77	161.91	nr	**186.68**
1800 x 1200mm; ref SS182D/O	PC £175.42	2.65	24.77	197.35	nr	**222.12**
2400 x 1200 mm; ref SS242D/O	PC £194.46	3.10	28.98	218.77	nr	**247.75**

L WINDOWS/DOORS/STAIRS Including overheads and profit at 12.50%		Labour hours	Labour £	Material £	Unit	Total rate £

L10 TIMBER WINDOWS/ROOFLIGHTS/SCREENS/LOUVRES - cont'd

Standard windows; selected Philippine
Mahogany; preservative stain finish - cont'd
Top hung casement windows; with 45 x 140 mm
hardwood sills; weather stripping; opening
sashes on canopy hinges; fitted with
fasteners; aluminized lacquered finish
ironmongery

600 x 900 mm; ref ST0609/O	PC £68.24	1.05	9.82	76.77	nr	86.59
600 x 1050 mm; ref ST0610/O	PC £73.44	1.45	13.56	82.62	nr	96.18
900 x 900 mm; ref ST0909/O	PC £87.04	1.65	15.43	97.92	nr	113.35
900 x 1050 mm; ref ST0910/O	PC £90.00	1.75	16.36	101.25	nr	117.61
900 x 1200 mm; ref ST0912/O	PC £99.44	1.85	17.30	111.87	nr	129.17
900 x 1350 mm; ref ST0913/O	PC £104.96	2.00	18.70	118.08	nr	136.78
900 x 1500 mm; ref ST0915/O	PC £113.52	2.10	19.63	127.71	nr	147.34
1200 x 900 mm; ref ST1209/O	PC £101.36	2.10	19.63	114.03	nr	133.66
1200 x 1050 mm; ref ST1210/O	PC £96.39	2.00	18.70	108.44	nr	127.14
1200 x 1200 mm; ref ST1212/O	PC £104.72	2.15	20.10	117.81	nr	137.91
1200 x 1350 mm; ref ST1213/O	PC £126.88	2.30	21.50	142.74	nr	164.24
1200 x 1500 mm; ref ST1215/O	PC £112.28	2.35	21.97	126.32	nr	148.29
1500 x 1050 mm; ref ST1510/L	PC £134.88	2.30	21.50	151.74	nr	173.24
1500 x 1200 mm; ref ST1512/L	PC £146.88	2.40	22.44	165.24	nr	187.68
1500 x 1350 mm; ref ST1513/L	PC £155.68	2.55	23.84	175.14	nr	198.98
1500 x 1500 mm; ref ST1515/L	PC £166.64	2.75	25.71	187.47	nr	213.18
1800 x 1050 mm; ref ST1810/L	PC £127.68	2.50	23.37	143.64	nr	167.01
1800 x 1200 mm; ref ST1812/L	PC £138.11	2.65	24.77	155.37	nr	180.14
1800 x 1350 mm; ref ST1813/L	PC £145.74	2.75	25.71	163.96	nr	189.67
1800 x 1500 mm; ref ST1815/L	PC £177.68	2.95	27.58	199.89	nr	227.47

Purpose made window casements; selected
West African Mahogany; PC £531.05/m3
Casements; rebated; moulded

38 mm thick	PC £28.39	-	-	31.94	m2	31.94
50 mm thick	PC £37.00	-	-	41.62	m2	41.62

Casements; rebated; moulded; in medium panes

38 mm thick	PC £34.45	-	-	38.75	m2	38.75
50 mm thick	PC £46.09	-	-	51.86	m2	51.86

Casements; rebated; moulded; with
semi-circular head

38 mm thick	PC £43.06	-	-	48.45	m2	48.45
50 mm thick	PC £57.64	-	-	64.84	m2	64.84

Casements; rebated; moulded; to bullseye
window

38 mm thick x 600 mm dia	PC £93.20/nr	-	-	104.85	m2	104.85
38 mm thick x 900 mm dia	PC £139.49/nr	-	-	156.92	m2	156.92
50 mm thick x 600 mm dia	PC £107.17/nr	-	-	120.56	m2	120.56
50 mm thick x 900 mm dia	PC £159.81/nr	-	-	179.79	m2	179.79

Fitting and hanging casements

square or rectangular		0.77	7.20	-	nr	7.20
semicircular		1.85	17.30	-	nr	17.30
bullseye		2.95	27.58	-	nr	27.58

Purpose made window frames; selected West
African Mahogany; PC £531.05/m3
Frames; rounded; rebated check grooved

25 x 120 mm	PC £8.68	0.20	1.87	10.01	m	11.88
50 x 75 mm	PC £9.84	0.20	1.87	11.35	m	13.22
50 x 100 mm	PC £12.13	0.23	2.15	13.98	m	16.13
50 x 125 mm	PC £14.71	0.23	2.15	16.97	m	19.12
63 x 100 mm	PC £15.05	0.23	2.15	17.36	m	19.51
75 x 150 mm	PC £25.28	0.26	2.43	29.14	m	31.57
90 x 140 mm	PC £32.69	0.26	2.43	37.70	m	40.13

L WINDOWS/DOORS/STAIRS Including overheads and profit at 12.50%		Labour hours	Labour £	Material £	Unit	Total rate £
Mullions and transoms; twice rounded, rebated and check grooved						
50 x 75 mm	PC £11.91	0.15	1.40	13.74	m	15.14
50 x 100 mm	PC £14.19	0.18	1.68	16.36	m	18.04
63 x 100 mm	PC £17.12	0.18	1.68	19.74	m	21.42
75 x 150 mm	PC £27.34	0.20	1.87	31.52	m	33.39
Sill; sunk weathered, rebated and grooved						
75 x 100 mm	PC £24.75	0.30	2.80	28.55	m	31.35
75 x 150 mm	PC £32.63	0.30	2.80	37.63	m	40.43
Purpose made double hung sash windows; selected West African Mahogany; PC £531.05/m3 Cased frames of 100 x 25 mm grooved inner linings; 114 x 25 mm grooved outer linings; 125 x 38 mm twice rebated head linings; 125 x 32 mm twice rebated grooved pulley stiles; 150 x 13 mm linings; 50 x 19 mm parting slips; 25 x 13 mm parting beads; 25 x 19 mm inside beads; 150 x 75 mm Oak twice sunk weathered throated sill; 50 mm thick rebated and moulded sashes; moulded horns; over 1.25 m2 each; both sashes in medium panes;						
including spiral spring balances	PC £193.18	3.30	30.85	280.10	m2	310.95
As above but with cased mullions	PC £203.78	3.65	34.12	292.03	m2	326.15
Purpose made window casements; Afrormosia; PC £758.95/m3 Casements; rebated; moulded						
38 mm thick	PC £34.51	-	-	38.82	m2	38.82
50 mm thick	PC £42.40	-	-	47.70	m2	47.70
Casements; rebated; moulded; in medium panes						
38 mm thick	PC £44.73	-	-	50.32	m2	50.32
50 mm thick	PC £53.74	-	-	60.46	m2	60.46
Casements; rebated; moulded; with semi-circular head						
38 mm thick	PC £55.92	-	-	62.91	m2	62.91
50 mm thick	PC £67.20	-	-	75.60	m2	75.60
Casements; rebated; moulded; to bullseye window						
38 mm thick x 600 mm dia	PC £113.36/nr	-	-	127.53	m2	127.53
38 mm thick x 900 mm dia	PC £169.38/nr	-	-	190.55	m2	190.55
50 mm thick x 600 mm dia	PC £130.37/nr	-	-	146.66	m2	146.66
50 mm thick x 900 mm dia	PC £194.94/nr	-	-	219.31	m2	219.31
Fitting and hanging casements						
square or rectangular		0.77	7.20	-	nr	7.20
semicircular		1.85	17.30	-	nr	17.30
bullseye		2.95	27.58	-	nr	27.58
Purpose made window frames; Afrormosia; PC £758.95/m3 Frames; rounded; rebated check grooved						
25 x 120 mm	PC £10.02	0.20	1.87	11.55	m	13.42
50 x 75 mm	PC £11.44	0.20	1.87	13.19	m	15.06
50 x 100 mm	PC £14.25	0.23	2.15	16.44	m	18.59
50 x 125 mm	PC £17.38	0.23	2.15	20.04	m	22.19
63 x 100 mm	PC £17.73	0.23	2.15	20.45	m	22.60
75 x 150 mm	PC £30.07	0.26	2.43	34.68	m	37.11
90 x 140 mm	PC £39.11	0.26	2.43	45.10	m	47.53
Mullions and transoms; twice rounded, rebated and check grooved						
50 x 75 mm	PC £13.51	0.15	1.40	15.58	m	16.98
50 x 100 mm	PC £16.32	0.18	1.68	18.82	m	20.50
63 x 100 mm	PC £19.81	0.18	1.68	22.85	m	24.53
75 x 150 mm	PC £32.14	0.20	1.87	37.05	m	38.92
Sill; sunk weathered, rebated and grooved						
75 x 100 mm	PC £27.94	0.30	2.80	32.21	m	35.01
75 x 150 mm	PC £37.46	0.30	2.80	43.20	m	46.00

L WINDOWS/DOORS/STAIRS Including overheads and profit at 12.50%		Labour hours	Labour £	Material £	Unit	Total rate £
L10 TIMBER WINDOWS/ROOFLIGHTS/SCREENS/LOUVRES - cont'd						
Purpose made double hung sash windows;						
Afrormosia PC £758.95/m3						
Cased frames of 100 x 25 mm grooved inner						
linings; 114 x 25 mm grooved outer linings;						
125 x 38 mm twice rebated head linings; 125						
x 32 mm twice rebated grooved pulley stiles;						
150 x 13 mm linings; 50 x 19 mm parting						
slips; 25 x 13 mm parting beads; 25 x 19 mm						
inside beads; 150 x 75 mm Oak twice sunk						
weathered throated sill; 50 mm thick rebated						
and moulded sashes; moulded horns; over						
1.25 m2 each; both sashes in medium panes;						
including spiral spring balances	PC £216.52	3.30	30.85	280.10	m2	310.95
As above but with cased mullions	PC £228.42	3.65	34.12	292.03	m2	326.15
L11 METAL WINDOWS/ROOFLIGHTS/SCREENS/LOUVRES						
Aluminium fixed and fanlight windows;						
Crittall 'Luminaire' stock sliders or						
similar; white acrylic finish; fixed in						
position; including lugs plugged and screwed						
to brickwork or blockwork; or screwed to						
wooden sub-frame (measured elsewhere)						
Fixed lights; factory glazed with 3, 4 or						
5 mm OQ clear float glass						
600 x 900 mm; ref 6FL9A	PC £47.70	2.40	25.57	54.44	nr	80.01
900 x 1500 mm; ref 9FL15A	PC £67.50	3.45	36.76	76.71	nr	113.47
1200 x 900 mm; ref 12FL9A	PC £72.90	3.45	36.76	82.79	nr	119.55
1200 x 1200 mm; ref 12FL12A	PC £74.70	3.45	36.76	84.81	nr	121.57
1500 x 1200 mm; ref 15FL12A	PC £82.80	3.45	36.76	93.93	nr	130.69
1500 x 1500 mm; ref 15FL15A	PC £168.30	4.40	46.88	190.11	nr	236.99
Fixed lights; site double glazed with						
11 mm clear float glass						
600 x 900 mm; ref 6FL9A	PC £63.90	3.30	35.16	72.66	nr	107.82
900 x 1500 mm; ref 9FL15A	PC £88.20	5.25	55.93	100.00	nr	155.93
1200 x 900 mm; ref 12FL9A	PC £102.60	5.80	61.79	116.20	nr	177.99
1200 x 1200 mm; ref 12FL12A	PC £101.70	6.00	63.92	115.19	nr	179.11
Fixed lights with fanlight over; factory						
glazed with 3, 4 or 5 mm OQ clear float						
glass						
600 x 900 mm; ref 6FV9A	PC £126.90	2.40	25.57	143.54	nr	169.11
600 x 1500 mm; ref 6FV15A	PC £140.40	2.40	25.57	158.73	nr	184.30
900 x 1200 mm; ref 9FV12A	PC £154.80	3.45	36.76	174.93	nr	211.69
900 x 1500 mm; ref 9FV15A	PC £165.60	3.45	36.76	187.08	nr	223.84
Fixed lights with fanlight over; site						
double glazed with 11 mm clear float glass						
600 x 900 mm; ref 6FV9A	PC £152.10	3.30	35.16	171.89	nr	207.05
600 x 1500 mm; ref 6FV15A	PC £181.80	3.90	41.55	205.30	nr	246.85
900 x 1200 mm; ref 9FV12A	PC £200.70	4.75	50.61	226.56	nr	277.17
900 x 1500 mm; ref 9FV15A	PC £216.90	5.80	61.79	244.79	nr	306.58
Vertical slider; factory glazed with						
3; 4 or 5 mm OQ clear float glass						
600 x 900 mm; ref 6VV9A	PC £108.90	2.95	31.43	123.29	nr	154.72
900 x 1100 mm; ref 9VV11A	PC £128.70	2.95	31.43	145.56	nr	176.99
1200 x 1500 mm; ref 12VV15A	PC £156.60	4.15	44.21	176.95	nr	221.16
Vertical slider; factory double glazed						
with 11 mm clear float glass						
600 x 900 mm; ref 6VV9A	PC £177.30	3.75	39.95	200.24	nr	240.19
900 x 1100 mm; ref 9VV11A	PC £208.80	3.75	39.95	235.68	nr	275.63
1200 x 1500 mm; ref 12VV15A	PC £292.50	5.25	55.93	329.84	nr	385.77

L WINDOWS/DOORS/STAIRS Including overheads and profit at 12.50%		Labour hours	Labour £	Material £	Unit	Total rate £
Horizontal slider; factory glazed with						
3, 4 or 5 mm OQ clear float glass						
1200 x 900 mm; ref 12HH9A	PC £139.50	4.15	44.21	157.71	nr	201.92
1200 x 1200 mm; ref 12HH12A	PC £152.10	4.15	44.21	171.89	nr	216.10
1500 x 900 mm; ref 15HH9A	PC £152.10	4.15	44.21	171.89	nr	216.10
1500 x 1300 mm; ref 15HH13A	PC £174.60	4.15	44.21	197.20	nr	241.41
1800 x 1200 mm; ref 18HH12A	PC £184.50	5.80	61.79	208.34	nr	270.13
Horizontal slider; factory double glazed						
with 11 mm clear float glass						
1200 x 900 mm; ref 12HH9A	PC £189.90	5.25	55.93	214.41	nr	270.34
1200 x 1200 mm; ref 12HH12A	PC £208.80	5.25	55.93	235.68	nr	291.61
1500 x 900 mm; ref 15HH9A	PC £207.90	5.25	55.93	234.66	nr	290.59
1500 x 1300 mm; ref 15HH13A	PC £245.70	5.25	55.93	277.19	nr	333.12
1800 x 1200 mm; ref 18HH12A	PC £260.10	6.35	67.65	293.39	nr	361.04
Galvanized steel fixed light; casement and						
fanlight windows; Crittall; 'Homelight'						
range or similar; site glazing measured						
elsewhere; fixed in position; including lugs						
cut and pinned to brickwork or blockwork						
Basic fixed lights; including easy-glaze						
beads						
628 x 292 mm; ref ZNG5	PC £12.38	1.30	13.85	14.25	nr	28.10
628 x 923 mm; ref ZNC5	PC £17.64	1.30	13.85	20.17	nr	34.02
628 x 1513 mm; ref ZNDV5	PC £25.09	1.30	13.85	28.55	nr	42.40
1237 x 292 mm; ref ZNG13	PC £18.90	1.30	13.85	21.59	nr	35.44
1237 x 923 mm; ref ZNC13	PC £24.45	1.95	20.77	27.83	nr	48.60
1237 x 1218 mm; ref ZND13	PC £28.79	1.95	20.77	32.71	nr	53.48
1237 x 1513 mm; ref ZNDV13	PC £31.54	1.95	20.77	35.80	nr	56.57
1846 x 292 mm; ref ZNG14	PC £24.50	1.30	13.85	27.88	nr	41.73
1846 x 923 mm; ref ZNC14	PC £30.20	1.95	20.77	34.29	nr	55.06
1846 x 1513 mm; ref ZNDV14	PC £36.58	2.40	25.57	41.47	nr	67.04
Basic opening lights; including easy-glaze						
beads and weatherstripping						
628 x 292 mm; ref ZNG1	PC £28.11	1.30	13.85	31.94	nr	45.79
1237 x 292 mm; ref ZNG13G	PC £41.72	1.30	13.85	47.25	nr	61.10
1846 x 292 mm; ref ZNG4	PC £71.34	1.30	13.85	80.58	nr	94.43
One-piece composites; including easy-glaze						
beads and weatherstripping						
628 x 923 mm; ref ZNC5F	PC £39.30	1.30	13.85	44.54	nr	58.39
628 x 1513 mm; ref ZNDV5F	PC £47.03	1.30	13.85	53.24	nr	67.09
1237 x 923 mm; ref ZNC2F	PC £77.56	1.95	20.77	87.58	nr	108.35
1237 x 1218 mm; ref ZND2F	PC £91.26	1.95	20.77	102.99	nr	123.76
1237 x 1513 mm; ref ZNDV2V	PC £109.55	1.95	20.77	123.56	nr	144.33
1846 x 923 mm; ref NC4F	PC £119.51	1.95	20.77	134.77	nr	155.54
1846 x 1218 mm; ref ZND10F	PC £115.79	2.40	25.57	130.59	nr	156.16
Reversible windows; including easy-glaze						
beads and weatherstripping						
997 x 923 mm; ref NC13R	PC £104.51	1.70	18.11	117.89	nr	136.00
997 x 1067 mm; ref NCO13R	PC £110.81	1.70	18.11	124.98	nr	143.09
1237 x 923 mm; ref ZNC13R	PC £115.44	2.55	27.17	130.20	nr	157.37
1237 x 1218 mm; ref ZND13R	PC £126.18	2.55	27.17	142.28	nr	169.45
1237 x 1513 mm; ref ZNDV13RS	PC £141.88	2.55	27.17	159.93	nr	187.10
Pressed steel sills; to suit above window						
widths						
628 mm	PC £5.97	0.39	4.15	7.49	nr	11.64
997 mm	PC £8.38	0.50	5.33	10.20	nr	15.53
1237 mm	PC £9.45	0.61	6.50	11.41	nr	17.91
1486 mm	PC £10.97	0.72	7.67	13.12	nr	20.79
1846 mm	PC £12.46	0.83	8.84	14.79	nr	23.63

L WINDOWS/DOORS/STAIRS Including overheads and profit at 12.50%		Labour hours	Labour £	Material £	Unit	Total rate £

L11 METAL WINDOWS/ROOFLIGHTS/SCREENS/LOUVRES - cont'd

Factory finished steel fixed light; casement
and fanlight windows; Crittall polyester
powder coated 'Homelight' range or similar;
site glazing measured elsewhere; fixed in
position; including lugs cut and pinned to
brickwork or blockwork
Basic fixed lights; including easy-glaze
beads

628 x 292 mm; ref ZNG5	PC £15.46	1.30	13.85	17.72	nr	31.57
628 x 923 mm; ref ZNC5	PC £22.04	1.30	13.85	25.12	nr	38.97
628 x 1513 mm; ref ZNDV5	PC £31.36	1.30	13.85	35.60	nr	49.45
1237 x 292 mm; ref ZNG13	PC £23.61	1.30	13.85	26.88	nr	40.73
1237 x 923 mm; ref ZNC13	PC £30.57	1.95	20.77	34.72	nr	55.49
1237 x 1218 mm; ref ZND13	PC £35.99	1.95	20.77	40.81	nr	61.58
1237 x 1513 mm; ref ZNDV13	PC £39.41	1.95	20.77	44.66	nr	65.43
1846 x 292 mm; ref ZNG14	PC £30.62	1.30	13.85	34.77	nr	48.62
1846 x 923 mm; ref ZNC14	PC £37.74	1.95	20.77	42.78	nr	63.55
1846 x 1513 mm; ref ZNDV14	PC £45.72	2.40	25.57	51.76	nr	77.33

Basic opening lights; including easy-glaze
beads and weatherstripping

628 x 292 mm; ref ZNG1	PC £34.18	1.30	13.85	38.78	nr	52.63
1237 x 292 mm; ref ZNG13G	PC £50.83	1.30	13.85	57.51	nr	71.36
1846 x 292 mm; ref ZNG4	PC £86.92	1.30	13.85	98.11	nr	111.96

One-piece composites; including easy-glaze
beads and weatherstripping

628 x 923 mm; ref ZNC5F	PC £48.12	1.30	13.85	54.46	nr	68.31
628 x 1513 mm; ref ZNDV5F	PC £57.79	1.30	13.85	65.34	nr	79.19
1237 x 923 mm; ref ZNC2F	PC £94.19	1.95	20.77	106.28	nr	127.05
1237 x 1218 mm; ref ZND2F	PC £110.93	1.95	20.77	125.12	nr	145.89
1237 x 1513 mm; ref ZNDV2V	PC £132.95	1.95	20.77	149.89	nr	170.66
1846 x 923 mm; ref NC4F	PC £144.80	1.95	20.77	163.22	nr	183.99
1846 x 1218 mm; ref ZND10F	PC £141.02	2.40	25.57	158.97	nr	184.54

Reversible windows; including easy-glaze
beads and weatherstripping

997 x 923 mm; ref NC13R	PC £135.86	1.70	18.11	153.16	nr	171.27
997 x 1067 mm; ref NCO13R	PC £144.05	1.70	18.11	162.38	nr	180.49
1237 x 923 mm; ref ZNC13R	PC £150.08	2.55	27.17	169.16	nr	196.33
1237 x 1218 mm; ref ZND13R	PC £164.03	2.55	27.17	184.85	nr	212.02
1237 x 1513 mm; ref ZNDV13RS	PC £184.44	2.55	27.17	207.82	nr	234.99

Pressed steel sills; to suit above window
widths

628 mm	PC £7.45	0.39	4.15	9.16	nr	13.31
997 mm	PC £10.47	0.50	5.33	12.55	nr	17.88
1237 mm	PC £11.81	0.61	6.50	14.06	nr	20.56
1486 mm	PC £13.72	0.72	7.67	16.21	nr	23.88
1846 mm	PC £15.56	0.83	8.84	18.28	nr	27.12

**L12 PLASTICS WINDOWS/ROOFLIGHTS/SCREENS/
LOUVRES**

uPVC windows to BS 2782; 'Anglian' or
similar; reinforced where appropriate with
aluminium alloy; including standard
ironmongery; cills and glazing; fixed in
position; including lugs plugged and screwed
to brickwork or blockwork
Fixed light; including e.p.d.m. glazing
gaskets and weather seals

600 x 900 mm; single glazed	PC £104.40	3.85	41.02	117.77	nr	158.79
600 x 900 mm; double glazed	PC £117.60	3.85	41.02	132.62	nr	173.64

Casement/fixed light; including e.p.d.m.
glazing gaskets and weather seals

600 x 1200 mm; single glazed	PC £150.00	4.15	44.21	169.07	nr	213.28
600 x 1200 mm; double glazed	PC £163.20	4.15	44.21	183.92	nr	228.13

L WINDOWS/DOORS/STAIRS Including overheads and profit at 12.50%		Labour hours	Labour £	Material £	Unit	Total rate £
1200 x 1200 mm; single glazed	PC £268.80	4.95	52.74	302.72	nr	355.46
1200 x 1200 mm; double glazed	PC £295.20	4.95	52.74	332.42	nr	385.16
1800 x 1200 mm; single glazed	PC £384.00	5.50	58.60	432.32	nr	490.92
1800 x 1200 mm; double glazed	PC £422.40	5.50	58.60	475.52	nr	534.12
'Tilt & Turn' light; including e.p.d.m. glazing gaskets and weather seals						
1200 x 1200 mm; single glazed	PC £216.00	4.95	52.74	243.32	nr	296.06
1200 x 1200 mm; double glazed	PC £241.20	4.95	52.74	271.67	nr	324.41
Cox's 'Skydome' or 'Coxdome'; plugged and screwed to concrete or screwed to timber Rooflight; 'Skydome Mark 3'; single skin; square or rectangular dome						
600 x 600 mm	PC £50.05	1.65	12.27	59.12	nr	71.39
900 x 600 mm	PC £103.50	1.80	13.39	122.27	nr	135.66
900 x 900 mm	PC £104.05	2.00	14.87	122.91	nr	137.78
1200 x 900 mm	PC £125.80	2.15	15.99	148.60	nr	164.59
1200 x 1200 mm	PC £130.60	2.30	17.10	154.27	nr	171.37
1800 x 1200 mm	PC £221.50	2.75	20.45	261.65	nr	282.10
Rooflight; 'Coxdome Mark 5' double skin; square or rectangular dome; extruded aluminium plain splayed upstand with 'hit and miss' ventilators two sides						
600 x 600 mm	PC £222.60	3.30	24.54	255.43	nr	279.97
900 x 600 mm	PC £303.35	3.65	27.14	348.10	nr	375.24
900 x 900 mm	PC £318.90	3.95	29.37	365.94	nr	395.31
1200 x 900 mm	PC £354.15	4.30	31.98	406.38	nr	438.36
1200 x 1200 mm	PC £402.85	4.60	34.21	462.27	nr	496.48
1800 x 1200 mm	PC £554.65	5.50	40.90	636.46	nr	677.36

ALTERNATIVE TIMBER DOOR PRICES (£/each)

	£		£		£
Hardwood doors					
Brazilian Mahogany period doors (£/each); 838 x 1981 x 44 mm					
6 panel	131.89	'Carolina'	117.37	'Kentucky'	134.81
8 panel	114.40	'Gothic'	128.92	'Elizabethan'	125.84
Red Meranti period doors					
6 panel					
762x1981x44 mm	115.56	838x1981x44 mm	115.56	807x2000x44 mm	115.56
8 panel					
762x1981x44 mm	114.40	838x1981x44 mm	114.40	807x2000x44 mm	114.40
Half bow					
762x1981x44 mm	134.31	838x1981x44 mm	134.31	807x2000x44 mm	134.31
'Kentucky'					
762x1981x44 mm	112.53	838x1981x44 mm	112.53	807x2000x44 mm	112.53
Softwood doors					
Casement doors					
2 XG; two panel; beaded					
762x1981x44 mm	32.89	838x1981x44 mm	34.32	807x2000x44 mm	34.32
813x2032x44 mm	34.32	726x2040x44 mm	34.32	826x2040x44 mm	34.32
2 XGG; two panel; beaded					
762x1981x44 mm	30.47	838x1981x44 mm	31.85	807x2000x44 mm	31.85
813x2032x44 mm	31.85	726x2040x44 mm	31.85	826x2040x44 mm	31.85
10; one panel; beaded					
762x1981x44 mm	25.74	838x1981x44 mm	26.79	807x2000x44 mm	26.79
SA; 15 panel; beaded					
762x1981x44 mm	40.70	838x1981x44 mm	42.35	807x2000x44 mm	42.35
22; pair of two panel; beaded					
914x1981x44 mm	67.98	1168x1981x44 mm	67.98		
2 SA; pair of 10 panel; beaded					
914x1981x44 mm	89.27	1168x1981x44 mm	89.27		

L WINDOWS/DOORS/STAIRS
Including overheads and profit at 12.50%

Louvre doors

533x1524x28 mm	13.26	610x1524x28 mm	14.52	686x1981x28 mm	19.69
533x1676x28 mm	14.36	610x1676x28 mm	15.90	762x1981x28 mm	21.84
533x1829x28 mm	15.29	610x1829x28 mm	16.89	- pair of 737x	
533x1981x28 mm	16.28	610x1981x28 mm	18.10	1067x28 mm doors	16.06

Period doors
6 panel

762x1981x44 mm	88.33	838x1981x44 mm	88.33	807x2000x44 mm	88.33

8 panel

762x1981x44 mm	86.45	838x1981x44 mm	86.45	807x2000x44 mm	86.45

Half bow

762x1981x44 mm	85.31	838x1981x44 mm	85.31	807x2000x44 mm	85.31

'Kentucky'

762x1981x44 mm	78.05	838x1981x44 mm	78.05	807x2000x44 mm	78.05

	Labour hours	Labour £	Material £	Unit	Total rate £

L20 TIMBER DOORS/SHUTTERS/HATCHES

Standard matchboarded doors; wrought softwood
Matchboarded, ledged and braced doors; 25 mm
ledges and braces; 19 mm tongued, grooved
and V-jointed; one side vertical boarding

762 x 1981 mm	PC £36.13	1.65	15.43	40.64	nr	56.07
838 x 1981 mm	PC £38.28	1.65	15.43	43.07	nr	58.50

Matchboarded, framed, ledged and braced
doors; 44 mm framing; 25 mm intermediate and
bottom rails; 19 mm tongued, grooved and
V-jointed; one side vertical boarding

762 x 1981 x 44 mm	PC £47.00	2.00	18.70	52.88	nr	71.58
838 x 1981 x 44 mm	PC £48.73	2.00	18.70	54.82	nr	73.52

Standard flush doors; softwood composition
Flush door; internal quality; skeleton or
cellular core; hardboard faced both sides;

457 x 1981 x 35 mm	PC £11.06	1.40	13.09	12.44	nr	25.53
533 x 1981 x 35 mm	PC £11.06	1.40	13.09	12.44	nr	25.53
610 x 1981 x 35 mm	PC £12.43	1.40	13.09	12.44	nr	25.53
686 x 1981 x 35 mm	PC £11.06	1.40	13.09	12.44	nr	25.53
762 x 1981 x 35 mm	PC £11.06	1.40	13.09	12.44	nr	25.53
838 x 1981 x 35 mm	PC £11.55	1.40	13.09	12.99	nr	26.08
526 x 2040 x 40 mm	PC £12.43	1.40	13.09	13.98	nr	27.07
626 x 2040 x 40 mm	PC £12.43	1.40	13.09	13.98	nr	27.07
726 x 2040 x 40 mm	PC £12.43	1.40	13.09	13.98	nr	27.07
826 x 2040 x 40 mm	PC £12.98	1.40	13.09	14.60	nr	27.69

Flush door; internal quality; skeleton or
cellular core; plywood faced both sides;
lipped on two long edges

457 x 1981 x 35 mm	PC £14.36	1.40	13.09	16.15	nr	29.24
533 x 1981 x 35 mm	PC £14.36	1.40	13.09	16.15	nr	29.24
610 x 1981 x 35 mm	PC £14.36	1.40	13.09	16.15	nr	29.24
686 x 1981 x 35 mm	PC £14.36	1.40	13.09	16.15	nr	29.24
762 x 1981 x 35 mm	PC £14.36	1.40	13.09	16.15	nr	29.24
838 x 1981 x 35 mm	PC £15.13	1.40	13.09	17.02	nr	30.11
526 x 2040 x 40 mm	PC £15.13	1.40	13.09	17.02	nr	30.11
626 x 2040 x 40 mm	PC £15.13	1.40	13.09	17.02	nr	30.11
726 x 2040 x 40 mm	PC £15.13	1.40	13.09	17.02	nr	30.11
826 x 2040 x 40 mm	PC £15.68	1.40	13.09	17.63	nr	30.72

Flush door; internal quality; skeleton or
cellular core; Sapele faced both sides;
lipped on all four edges

457 x 1981 x 35 mm	PC £16.17	1.50	14.02	18.19	nr	32.21
533 x 1981 x 35 mm	PC £16.17	1.50	14.02	18.19	nr	32.21
610 x 1981 x 35 mm	PC £16.17	1.50	14.02	18.19	nr	32.21
686 x 1981 x 35 mm	PC £16.17	1.50	14.02	18.19	nr	32.21
762 x 1981 x 35 mm	PC £16.17	1.50	14.02	18.19	nr	32.21
838 x 1981 x 35 mm	PC £16.17	1.50	14.02	18.19	nr	32.21

L WINDOWS/DOORS/STAIRS Including overheads and profit at 12.50%		Labour hours	Labour £	Material £	Unit	Total rate £
Flush door; internal quality; skeleton or cellular core; Teak faced both sides; lipped on all four edges						
457 x 1981 x 35 mm	PC £39.71	1.50	14.02	44.67	nr	58.69
533 x 1981 x 35 mm	PC £34.93	1.50	14.02	39.29	nr	53.31
610 x 1981 x 35 mm	PC £34.93	1.50	14.02	39.29	nr	53.31
686 x 1981 x 35 mm	PC £34.93	1.50	14.02	39.29	nr	53.31
762 x 1981 x 35 mm	PC £41.53	1.50	14.02	46.72	nr	60.74
838 x 1981 x 35 mm	PC £35.97	1.50	14.02	40.47	nr	54.49
526 x 2040 x 40 mm	PC £35.64	1.50	14.02	40.10	nr	54.12
626 x 2040 x 40 mm	PC £35.64	1.50	14.02	40.10	nr	54.12
726 x 2040 x 40 mm	PC £35.64	1.50	14.02	40.10	nr	54.12
826 x 2040 x 40 mm	PC £35.97	1.50	14.02	40.47	nr	54.49
Flush door; half hour fire-resisting (30/30); 'Melador'; laminate faced both sides; hardwood lipped on all edges						
610 x 1981 x 47 mm	PC £196.21	2.50	23.37	220.73	nr	244.10
686 x 1981 x 47 mm	PC £196.21	2.50	23.37	220.73	nr	244.10
762 x 1981 x 47 mm	PC £196.21	2.50	23.37	220.73	nr	244.10
838 x 1981 x 47 mm	PC £196.21	2.50	23.37	220.73	nr	244.10
526 x 2040 x 47 mm	PC £196.21	2.50	23.37	220.73	nr	244.10
626 x 2040 x 47 mm	PC £196.21	2.50	23.37	220.73	nr	244.10
726 x 2040 x 47 mm	PC £196.21	2.50	23.37	220.73	nr	244.10
826 x 2040 x 47 mm	PC £196.21	2.50	23.37	220.73	nr	244.10
Flush door; half-hour fire check (30/20); hardboard faced both sides;						
457 x 1981 x 44 mm	PC £30.01	1.95	18.23	33.76	nr	51.99
533 x 1981 x 44 mm	PC £30.01	1.95	18.23	33.76	nr	51.99
610 x 1981 x 44 mm	PC £30.01	1.95	18.23	33.76	nr	51.99
686 x 1981 x 44 mm	PC £30.01	1.95	18.23	33.76	nr	51.99
762 x 1981 x 44 mm	PC £30.01	1.95	18.23	33.76	nr	51.99
838 x 1981 x 44 mm	PC £31.57	1.95	18.23	35.51	nr	53.74
526 x 2040 x 44 mm	PC £31.17	1.95	18.23	35.06	nr	53.29
626 x 2040 x 44 mm	PC £31.17	1.95	18.23	35.06	nr	53.29
726 x 2040 x 44 mm	PC £31.17	1.95	18.23	35.06	nr	53.29
826 x 2040 x 44 mm	PC £32.74	1.95	18.23	36.84	nr	55.07
Flush door; half-hour fire check (30/20); plywood faced both sides; lipped on all four edges						
457 x 1981 x 44 mm	PC £28.94	1.95	18.23	32.55	nr	50.78
533 x 1981 x 44 mm	PC £28.94	1.95	18.23	32.55	nr	50.78
610 x 1981 x 44 mm	PC £28.94	1.95	18.23	32.55	nr	50.78
686 x 1981 x 44 mm	PC £28.94	1.95	18.23	32.55	nr	50.78
762 x 1981 x 44 mm	PC £28.94	1.95	18.23	32.55	nr	50.78
838 x 1981 x 44 mm	PC £30.42	1.95	18.23	34.22	nr	52.45
526 x 2040 x 44 mm	PC £28.94	1.95	18.23	32.55	nr	50.78
626 x 2040 x 44 mm	PC £28.94	1.95	18.23	32.55	nr	50.78
726 x 2040 x 44 mm	PC £29.73	1.95	18.23	33.44	nr	51.67
826 x 2040 x 44 mm	PC £31.24	1.95	18.23	35.15	nr	53.38
Flush door; half-hour fire check (30/20); Sapele faced both sides; lipped on all four edges						
457 x 1981 x 44 mm	PC £40.91	2.05	19.16	46.03	nr	65.19
533 x 1981 x 44 mm	PC £40.91	2.05	19.16	46.03	nr	65.19
610 x 1981 x 44 mm	PC £40.91	2.05	19.16	46.03	nr	65.19
686 x 1981 x 44 mm	PC £40.91	2.05	19.16	46.03	nr	65.19
762 x 1981 x 44 mm	PC £40.91	2.05	19.16	46.03	nr	65.19
526 x 2040 x 44 mm	PC £41.98	2.05	19.16	47.22	nr	66.38
626 x 2040 x 44 mm	PC £41.98	2.05	19.16	47.22	nr	66.38
726 x 2040 x 44 mm	PC £41.98	2.05	19.16	47.22	nr	66.38
826 x 2040 x 44 mm	PC £43.55	2.05	19.16	49.00	nr	68.16

L WINDOWS/DOORS/STAIRS Including overheads and profit at 12.50%		Labour hours	Labour £	Material £	Unit	Total rate £
L20 TIMBER DOORS/SHUTTERS/HATCHES - cont'd						
Standard flush doors; softwood composition - cont'd						
Flush door; half-hour fire resisting (30/30)						
Sapele faced both sides; lipped on all four						
edges						
457 x 1981 x 44 mm	PC £34.51	2.05	19.16	38.82	nr	57.98
533 x 1981 x 44 mm	PC £35.00	2.05	19.16	39.38	nr	58.54
610 x 1981 x 44 mm	PC £35.42	2.05	19.16	39.85	nr	59.01
686 x 1981 x 44 mm	PC £41.24	2.05	19.16	46.40	nr	65.56
762 x 1981 x 44 mm	PC £42.28	2.05	19.16	47.57	nr	66.73
838 x 1981 x 44 mm	PC £44.12	2.05	19.16	49.64	nr	68.80
526 x 2040 x 44 mm	PC £40.56	2.05	19.16	45.63	nr	64.79
626 x 2040 x 44 mm	PC £41.20	2.05	19.16	46.35	nr	65.51
726 x 2040 x 44 mm	PC £42.76	2.05	19.16	48.11	nr	67.27
826 x 2040 x 44 mm	PC £44.92	2.05	19.16	50.54	nr	69.70
Flush door; one hour fire check (60/45);						
plywood faced both sides; lipped on all						
four edges						
610 x 1981 x 54 mm	PC £152.84	2.20	20.57	171.95	nr	192.52
686 x 1981 x 54 mm	PC £152.84	2.20	20.57	171.95	nr	192.52
762 x 1981 x 54 mm	PC £99.55	2.20	20.57	112.00	nr	132.57
838 x 1981 x 54 mm	PC £106.93	2.20	20.57	120.29	nr	140.86
526 x 2040 x 54 mm	PC £152.84	2.20	20.57	171.95	nr	192.52
626 x 2040 x 54 mm	PC £152.84	2.20	20.57	171.95	nr	192.52
726 x 2040 x 54 mm	PC £99.55	2.20	20.57	112.00	nr	132.57
826 x 2040 x 54 mm	PC £106.93	2.20	20.57	120.29	nr	140.86
Flush door; one hour fire check (60/45);						
Sapele faced both sides; lipped on all						
four edges						
610 x 1981 x 54 mm	PC £159.85	2.30	21.50	179.83	nr	201.33
686 x 1981 x 54 mm	PC £159.85	2.30	21.50	179.83	nr	201.33
762 x 2040 x 54 mm	PC £159.85	2.30	21.50	179.83	nr	201.33
838 x 1981 x 54 mm	PC £159.85	2.30	21.50	179.83	nr	201.33
526 x 2040 x 54 mm	PC £159.85	2.30	21.50	179.83	nr	201.33
626 x 2040 x 54 mm	PC £159.85	2.30	21.50	179.83	nr	201.33
762 x 2040 x 54 mm	PC £159.85	2.30	21.50	179.83	nr	201.33
826 x 2040 x 54 mm	PC £159.85	2.30	21.50	179.83	nr	201.33
Flush door; one hour fire resisting (60/60);						
Sapele faced both sides; lipped on all						
four edges						
610 x 1981 x 54 mm	PC £179.64	2.30	21.50	202.10	nr	223.60
686 x 1981 x 54 mm	PC £179.64	2.30	21.50	202.10	nr	223.60
762 x 1981 x 54 mm	PC £179.64	2.30	21.50	202.10	nr	223.60
838 x 1981 x 54 mm	PC £179.64	2.30	21.50	202.10	nr	223.60
526 x 2040 x 54 mm	PC £179.64	2.30	21.50	202.10	nr	223.60
626 x 2040 x 54 mm	PC £179.64	2.30	21.50	202.10	nr	223.60
726 x 2040 x 54 mm	PC £179.64	2.30	21.50	202.10	nr	223.60
826 x 2040 x 54 mm	PC £179.64	2.30	21.50	202.10	nr	223.60
Flush door; one hour fire-resisting (60/60);						
'Melador'; laminate faced both sides;						
hardwood lipped on all edges						
610 x 1981 x 57 mm	PC £272.03	3.05	28.51	306.04	nr	334.55
686 x 1981 x 57 mm	PC £272.03	3.05	28.51	306.04	nr	334.55
762 x 1981 x 57 mm	PC £272.03	3.05	28.51	306.04	nr	334.55
838 x 1981 x 57 mm	PC £272.03	3.05	28.51	306.04	nr	334.55
526 x 2040 x 57 mm	PC £272.03	3.05	28.51	306.04	nr	334.55
626 x 2040 x 57 mm	PC £272.03	3.05	28.51	306.04	nr	334.55
726 x 2040 x 57 mm	PC £272.03	3.05	28.51	306.04	nr	334.55
826 x 2040 x 57 mm	PC £272.03	3.05	28.51	306.04	nr	334.55
Flush door; external quality; skeleton or						
cellular core; plywood faced both sides;						
lipped on all four edges						
762 x 1981 x 44 mm	PC £27.16	1.65	15.43	30.56	nr	45.99
838 x 1981 x 44 mm	PC £28.39	1.65	15.43	31.93	nr	47.36

L WINDOWS/DOORS/STAIRS Including overheads and profit at 12.50%		Labour hours	Labour £	Material £	Unit	Total rate £
Flush door; external quality with standard glass opening; skeleton or cellular core; plywood faced both sides; lipped on all four edges; including glazing beads						
762 x 1981 x 44 mm	PC £33.77	1.95	18.23	38.00	nr	56.23
838 x 1981 x 44 mm	PC £35.04	1.95	18.23	39.41	nr	57.64
Purpose made panelled doors; wrought softwood Panelled doors; one open panel for glass; including glazing beads						
686 x 1981 x 44 mm	PC £46.09	1.95	18.23	51.86	nr	70.09
762 x 1981 x 44 mm	PC £47.54	1.95	18.23	53.48	nr	71.71
838 x 1981 x 44 mm	PC £48.96	1.95	18.23	55.08	nr	73.31
Panelled doors; two open panel for glass; including glazing beads						
686 x 1981 x 44 mm	PC £63.37	1.95	18.23	71.29	nr	89.52
762 x 1981 x 44 mm	PC £66.76	1.95	18.23	75.11	nr	93.34
838 x 1981 x 44 mm	PC £70.07	1.95	18.23	78.83	nr	97.06
Panelled doors; four 19 mm thick plywood panel; mouldings worked on solid both sides						
686 x 1981 x 44 mm	PC £89.07	1.95	18.23	100.21	nr	118.44
762 x 1981 x 44 mm	PC £92.80	1.95	18.23	104.39	nr	122.62
838 x 1981 x 44 mm	PC £97.47	1.95	18.23	109.65	nr	127.88
Panelled doors; six 19 mm thick panels raised and fielded; mouldings worked on solid both sides						
686 x 1981 x 50 mm	PC £191.15	2.30	21.50	215.05	nr	236.55
762 x 1981 x 50 mm	PC £199.09	2.30	21.50	223.98	nr	245.48
838 x 1981 x 50 mm	PC £209.06	2.30	21.50	235.19	nr	256.69
Rebated edges beaded	-	-	-	-	m	1.12
Rounded edges or heels	-	-	-	-	m	0.57
Weatherboard; fixed to bottom rail		0.28	2.62	2.42	m	5.04
Stopped groove for weatherboard	-	-	-	-	m	0.39
Purpose made panelled doors; selected West African Mahogany; PC £531.05/m3 Panelled doors; one open panel for glass; mouldings worked on the solid one side; 19 x 13 mm beads one side; fixing with brass screws and cups						
686 x 1981 x 50 mm	PC £81.23	2.75	25.71	91.39	nr	117.10
762 x 1981 x 50 mm	PC £83.88	2.75	25.71	94.37	nr	120.08
838 x 1981 x 50 mm	PC £86.52	2.75	25.71	97.33	nr	123.04
686 x 1981 x 63 mm	PC £98.44	3.05	28.51	110.75	nr	139.26
762 x 1981 x 63 mm	PC £101.75	3.05	28.51	114.47	nr	142.98
838 x 1981 x 63 mm	PC £105.01	3.05	28.51	118.13	nr	146.64
Panelled doors; 250 mm wide cross-tongued intermediate rail; two open panels for glass mouldings worked on the solid one side; 19 x 13 mm beads one side; fixing with brass screws and cups						
686 x 1981 x 50 mm	PC £110.94	2.75	25.71	124.81	nr	150.52
762 x 1981 x 50 mm	PC £116.33	2.75	25.71	130.87	nr	156.58
838 x 1981 x 50 mm	PC £121.68	2.75	25.71	136.89	nr	162.60
686 x 1981 x 63 mm	PC £132.85	3.05	28.51	149.45	nr	177.96
762 x 1981 x 63 mm	PC £139.36	3.05	28.51	156.78	nr	185.29
838 x 1981 x 63 mm	PC £145.85	3.05	28.51	164.08	nr	192.59
Panelled doors; four panels; (19 mm for 50 mm thick doors and 25 mm for 63 mm thick doors); mouldings worked on solid both sides						
686 x 1981 x 50 mm	PC £141.50	2.75	25.71	159.18	nr	184.89
762 x 1981 x 50 mm	PC £147.38	2.75	25.71	165.80	nr	191.51
838 x 1981 x 50 mm	PC £154.75	2.75	25.71	174.09	nr	199.80
686 x 1981 x 63 mm	PC £166.54	3.05	28.51	187.36	nr	215.87
762 x 1981 x 63 mm	PC £173.50	3.05	28.51	195.18	nr	223.69
838 x 1981 x 63 mm	PC £182.15	3.05	28.51	204.92	nr	233.43

L WINDOWS/DOORS/STAIRS Including overheads and profit at 12.50%	Labour hours	Labour £	Material £	Unit	Total rate £

L20 TIMBER DOORS/SHUTTERS/HATCHES - cont'd

Purpose made panelled doors; selected West African Mahogany; PC £531.05/m3 - cont'd
Panelled doors; 150 mm wide stiles in one width; 430 mm wide cross-tongued bottom rail; six panels raised and fielded one side (19 mm thick for 50 mm thick doors and 25 mm thick for 63 mm thick doors); mouldings worked on the solid both sides

686 x 1981 x 50 mm PC £297.27	2.75	25.71	334.43	nr	360.14
762 x 1981 x 50 mm PC £309.66	2.75	25.71	348.37	nr	374.08
838 x 1981 x 50 mm PC £325.13	2.75	25.71	365.77	nr	391.48
686 x 1981 x 63 mm PC £364.44	3.05	28.51	409.99	nr	438.50
762 x 1981 x 63 mm PC £379.64	3.05	28.51	427.09	nr	455.60
838 x 1981 x 63 mm PC £398.62	3.05	28.51	448.44	nr	476.95
Rebated edges beaded	-	-	-	m	1.69
Rounded edges or heels	-	-	-	m	0.84
Weatherboard; fixed to bottom rail	0.36	3.37	4.61	m	7.98
Stopped groove for weatherboard	-	-	-	m	0.76

Purpose made panelled doors; Afrormosia; PC £758.95/m3
Panelled doors; one open panel for glass; mouldings worked on the solid one side; 19 x 13 mm beads one side; fixing with brass screws and cups

686 x 1981 x 50 mm PC £94.61	2.75	25.71	106.44	nr	132.15
762 x 1981 x 50 mm PC £97.76	2.75	25.71	109.98	nr	135.69
838 x 1981 x 50 mm PC £100.91	2.75	25.71	113.53	nr	139.24
686 x 1981 x 63 mm PC £115.14	3.05	28.51	129.53	nr	158.04
762 x 1981 x 63 mm PC £119.10	3.05	28.51	133.99	nr	162.50
838 x 1981 x 63 mm PC £123.01	3.05	28.51	138.38	nr	166.89

Panelled doors; 250 mm wide cross-tongued intermediate rail; two open panels for glass mouldings worked on the solid one side; 19 x 13 mm beads one side; fixing with brass screws and cups

686 x 1981 x 50 mm PC £128.59	2.75	25.71	144.67	nr	170.38
762 x 1981 x 50 mm PC £134.97	2.75	25.71	151.85	nr	177.56
838 x 1981 x 50 mm PC £141.30	2.75	25.71	158.97	nr	184.68
686 x 1981 x 63 mm PC £154.89	3.05	28.51	174.25	nr	202.76
762 x 1981 x 63 mm PC £162.62	3.05	28.51	182.95	nr	211.46
838 x 1981 x 63 mm PC £170.34	3.05	28.51	191.63	nr	220.14

Panelled doors; four panels; (19 mm for 50 mm thick doors and 25 mm for 63 mm thick doors); mouldings worked on solid both sides

686 x 1981 x 50 mm PC £163.84	2.75	25.71	184.32	nr	210.03
762 x 1981 x 50 mm PC £170.67	2.75	25.71	192.00	nr	217.71
838 x 1981 x 50 mm PC £179.18	2.75	25.71	201.57	nr	227.28
686 x 1981 x 63 mm PC £194.42	3.05	28.51	218.72	nr	247.23
762 x 1981 x 63 mm PC £202.52	3.05	28.51	227.83	nr	256.34
838 x 1981 x 63 mm PC £212.64	3.05	28.51	239.22	nr	267.73

Panelled doors; 150 mm wide stiles in one width; 430 mm wide cross-tongued bottom rail; six panels raised and fielded one side (19 mm thick for 50 mm thick doors and 25 mm thick for 63 mm thick doors); mouldings worked on the solid both sides

686 x 1981 x 50 mm PC £332.58	2.75	25.71	374.15	nr	399.86
762 x 1981 x 50 mm PC £346.45	2.75	25.71	389.75	nr	415.46
838 x 1981 x 50 mm PC £363.77	2.75	25.71	409.24	nr	434.95
686 x 1981 x 63 mm PC £409.13	3.05	28.51	460.27	nr	488.78

L WINDOWS/DOORS/STAIRS Including overheads and profit at 12.50%		Labour hours	Labour £	Material £	Unit	Total rate £
762 x 1981 x 63 mm	PC £426.18	3.05	28.51	479.45	nr	507.96
838 x 1981 x 63 mm	PC £447.50	3.05	28.51	503.44	nr	531.95
Rebated edges beaded		-	-	-	m	1.69
Rounded edges or heels		-	-	-	m	0.84
Weatherboard; fixed to bottom rail		0.36	3.37	5.21	m	8.58
Stopped groove for weatherboard		-	-	-	m	0.76

Standard joinery sets; wrought softwood
Internal door frame or lining set for 686 x
1981 mm door; all with loose stops unless
rebated; 'finished sizes'

27 x 94 mm lining	PC £15.16	0.88	8.23	17.05	nr	25.28
27 x 107 mm lining	PC £15.93	0.88	8.23	17.92	nr	26.15
35 x 107 mm rebated lining	PC £16.52	0.88	8.23	18.58	nr	26.81
27 x 121 mm lining	PC £17.19	0.88	8.23	19.33	nr	27.56
27 x 121 mm lining with fanlight over						
	PC £24.29	1.05	9.82	27.33	nr	37.15
27 x 133 mm lining	PC £18.52	0.88	8.23	20.83	nr	29.06
35 x 133 mm rebated linings	PC £19.00	0.88	8.23	21.38	nr	29.61
27 x 133 mm lining with fanlight over						
	PC £26.04	1.05	9.82	29.30	nr	39.12
33 x 57 mm frame	PC £12.36	0.88	8.23	13.90	nr	22.13
33 x 57 mm storey height frame	PC £15.19	0.94	8.79	17.09	nr	25.88
33 x 57 mm frame with fanlight over	PC £18.55	1.05	9.82	20.87	nr	30.69
33 x 64 mm frame	PC £13.27	0.88	8.23	14.92	nr	23.15
33 x 64 mm storey height frame	PC £16.17	0.94	8.79	18.19	nr	26.98
33 x 64 mm frame with fanlight over	PC £19.49	1.05	9.82	21.93	nr	31.75
44 x 94 mm frame	PC £20.27	1.00	9.35	22.80	nr	32.15
44 x 94 mm storey height frame	PC £24.12	1.10	10.28	27.13	nr	37.41
44 x 94 mm frame with fanlight over	PC £28.28	1.20	11.22	31.82	nr	43.04
44 x 107 mm frame	PC £23.28	1.00	9.35	26.18	nr	35.53
44 x 107 mm storey height frame	PC £27.27	1.10	10.28	30.67	nr	40.95
44 x 107 mm frame with fanlight over						
	PC £31.22	1.20	11.22	35.12	nr	46.34

Internal door frame or lining set for
762 x 1981 mm door; all with loose stops
unless rebated; 'finished sizes'

27 x 94 mm lining	PC £15.16	0.88	8.23	17.05	nr	25.28
27 x 107 mm lining	PC £15.93	0.88	8.23	17.92	nr	26.15
35 x 107 mm rebated lining	PC £16.52	0.88	8.23	18.58	nr	26.81
27 x 121 mm lining	PC £17.19	0.88	8.23	19.33	nr	27.56
27 x 121 mm lining with fanlight over						
	PC £24.29	1.05	9.82	27.33	nr	37.15
27 x 133 mm lining	PC £18.52	0.88	8.23	20.83	nr	29.06
35 x 133 mm rebated linings	PC £19.00	0.88	8.23	21.38	nr	29.61
27 x 133 mm lining with fanlight over						
	PC £26.04	1.05	9.82	29.30	nr	39.12
33 x 57 mm frame	PC £12.36	0.88	8.23	13.90	nr	22.13
33 x 57 mm storey height frame	PC £15.19	0.94	8.79	17.09	nr	25.88
33 x 57 mm frame with fanlight over	PC £18.55	1.05	9.82	20.87	nr	30.69
33 x 64 mm frame	PC £13.27	0.88	8.23	14.92	nr	23.15
33 x 64 mm storey height frame	PC £16.17	0.94	8.79	18.19	nr	26.98
33 x 64 mm frame with fanlight over	PC £19.49	1.05	9.82	21.93	nr	31.75
44 x 94 mm frame	PC £20.27	1.00	9.35	22.80	nr	32.15
44 x 94 mm storey height frame	PC £24.12	1.10	10.28	27.13	nr	37.41
44 x 94 mm frame with fanlight over	PC £28.28	1.20	11.22	31.82	nr	43.04
44 x 107 mm frame	PC £23.28	1.00	9.35	26.18	nr	35.53
44 x 107 mm storey height frame	PC £27.27	1.10	10.28	30.67	nr	40.95
44 x 107 mm frame with fanlight over						
	PC £31.22	1.20	11.22	35.12	nr	46.34

L WINDOWS/DOORS/STAIRS Including overheads and profit at 12.50%		Labour hours	Labour £	Material £	Unit	Total rate £
L20 TIMBER DOORS/SHUTTERS/HATCHES - cont'd						
Standard joinery sets; wrought softwood - cont'd						
Internal door frame or lining set for						
726 x 2040 mm door; with loose stops						
30 x 94 mm lining	PC £17.61	0.88	8.23	19.81	nr	28.04
30 x 94 mm lining with fanlight over						
	PC £25.20	1.05	9.82	28.35	nr	38.17
30 x 107 mm lining	PC £20.72	0.88	8.23	23.31	nr	31.54
30 x 107 mm lining with fanlight over						
	PC £28.42	1.05	9.82	31.97	nr	41.79
30 x 133 mm lining	PC £23.14	0.88	8.23	26.03	nr	34.26
30 x 133 mm lining with fanlight over						
	PC £30.63	1.05	9.82	34.45	nr	44.27
Internal door frame or lining set for						
826 x 2040 mm door; with loose stops						
30 x 94 mm lining	PC £17.61	0.88	8.23	19.81	nr	28.04
30 x 94 mm lining with fanlight over						
	PC £25.20	1.05	9.82	28.35	nr	38.17
30 x 107 mm lining	PC £20.72	0.88	8.23	23.31	nr	31.54
30 x 107 mm lining with fanlight over						
	PC £28.42	1.05	9.82	31.97	nr	41.79
30 x 133 mm lining	PC £23.14	0.88	8.23	26.03	nr	34.26
30 x 133 mm lining with fanlight over						
	PC £30.63	1.05	9.82	34.45	nr	44.27
Trap door set; 9 mm plywood in 35						
x 107 mm (fin) rebated lining						
762 x 762 mm	PC £19.60	0.88	8.23	22.37	nr	30.60
Serving hatch set; plywood faced doors;						
rebated meeting stiles; hung on nylon						
hinges; in 35 x 140 mm (fin) rebated lining						
648 x 533 mm	PC £37.35	1.45	13.56	42.35	nr	55.91
Purpose made door frames and lining sets;						
wrought softwood						
Jambs and heads; as linings						
32 x 63 mm		0.19	1.78	3.79	m	5.57
32 x 100 mm		0.19	1.78	5.01	m	6.79
32 x 140 mm		0.19	1.78	6.51	m	8.29
Jambs and heads; as frames; rebated, rounded						
and grooved						
38 x 75 mm		0.19	1.78	4.93	m	6.71
38 x 100 mm		0.19	1.78	5.91	m	7.69
38 x 115 mm		0.19	1.78	6.90	m	8.68
38 x 140 mm		0.22	2.06	8.04	m	10.10
50 x 100 mm		0.22	2.06	7.40	m	9.46
50 x 125 mm		0.22	2.06	8.84	m	10.90
63 x 88 mm		0.22	2.06	8.67	m	10.73
63 x 100 mm		0.22	2.06	9.15	m	11.21
63 x 125 mm		0.22	2.06	10.91	m	12.97
75 x 100 mm		0.22	2.06	10.46	m	12.52
75 x 125 mm		0.24	2.24	12.95	m	15.19
75 x 150 mm		0.24	2.24	15.17	m	17.41
100 x 100 mm		0.24	2.24	13.65	m	15.89
100 x 150 mm		0.24	2.24	19.69	m	21.93
Mullions and transoms; in linings						
32 x 63 mm		0.13	1.22	6.16	m	7.38
32 x 100 mm		0.13	1.22	7.37	m	8.59
32 x 140 mm		0.13	1.22	8.88	m	10.10
Mullions and transoms; in frames; twice						
rebated, rounded and grooved						
38 x 75 mm		0.13	1.22	7.42	m	8.64
38 x 100 mm		0.13	1.22	8.41	m	9.63
38 x 115 mm		0.13	1.22	9.25	m	10.47
38 x 140 mm		0.15	1.40	10.39	m	11.79

L WINDOWS/DOORS/STAIRS Including overheads and profit at 12.50%		Labour hours	Labour £	Material £	Unit	Total rate £
50 x 100 mm		0.15	1.40	9.76	m	11.16
50 x 125 mm		0.15	1.40	11.35	m	12.75
63 x 88 mm		0.15	1.40	11.04	m	12.44
63 x 100 mm		0.15	1.40	11.51	m	12.91
75 x 100 mm		0.15	1.40	12.95	m	14.35
Extra for additional labours						
one		0.02	0.19	-	m	0.19
two		0.03	0.28	-	m	0.28
three		0.04	0.37	-	m	0.37
Add +5% to the above 'Material £ prices' for 'selected' softwood for staining						
Purpose made door frames and lining sets; **selected West African Mahogany; PC £531.05/m3**						
Jambs and heads; as linings						
32 x 63 mm	PC £5.26	0.25	2.34	6.06	m	8.40
32 x 100 mm	PC £6.79	0.25	2.34	7.83	m	10.17
32 x 140 mm	PC £8.60	0.25	2.34	9.91	m	12.25
Jambs and heads; as frames; rebated, rounded and grooved						
38 x 75 mm	PC £6.63	0.25	2.34	7.65	m	9.99
38 x 100 mm	PC £8.00	0.25	2.34	9.22	m	11.56
38 x 115 mm	PC £9.17	0.25	2.34	10.57	m	12.91
38 x 140 mm	PC £10.56	0.30	2.80	12.18	m	14.98
50 x 100 mm	PC £10.05	0.30	2.80	11.59	m	14.39
50 x 125 mm	PC £12.06	0.30	2.80	13.91	m	16.71
63 x 88 mm	PC £11.23	0.30	2.80	12.95	m	15.75
63 x 100 mm	PC £12.43	0.30	2.80	14.33	m	17.13
63 x 125 mm	PC £14.88	0.30	2.80	17.16	m	19.96
75 x 100 mm	PC £14.27	0.30	2.80	16.45	m	19.25
75 x 125 mm	PC £17.68	0.33	3.09	20.39	m	23.48
75 x 150 mm	PC £20.73	0.33	3.09	23.91	m	27.00
100 x 100 mm	PC £18.62	0.33	3.09	21.47	m	24.56
100 x 150 mm	PC £26.91	0.33	3.09	31.03	m	34.12
Mullions and transoms; in linings						
32 x 63 mm	PC £8.26	0.18	1.68	9.52	m	11.20
32 x 100 mm	PC £9.80	0.18	1.68	11.29	m	12.97
32 x 140 mm	PC £11.58	0.18	1.68	13.36	m	15.04
Mullions and transoms; in frames; twice rebated, rounded and grooved						
38 x 75 mm	PC £9.79	0.18	1.68	11.28	m	12.96
38 x 100 mm	PC £11.16	0.18	1.68	12.87	m	14.55
38 x 115 mm	PC £12.17	0.18	1.68	14.03	m	15.71
38 x 140 mm	PC £13.56	0.20	1.87	15.64	m	17.51
50 x 100 mm	PC £13.04	0.20	1.87	15.04	m	16.91
50 x 125 mm	PC £15.22	0.20	1.87	17.55	m	19.42
63 x 88 mm	PC £14.22	0.20	1.87	16.41	m	18.28
63 x 100 mm	PC £15.41	0.20	1.87	17.78	m	19.65
75 x 100 mm		0.20	1.87	20.09	m	21.96
Sills; once sunk weathered; once rebated, three times grooved						
63 x 175 mm	PC £28.40	0.36	3.37	32.75	m	36.12
75 x 125 mm	PC £25.81	0.36	3.37	29.77	m	33.14
75 x 150 mm	PC £28.89	0.36	3.37	33.31	m	36.68
Extra for additional labours						
one		0.03	0.28	-	m	0.28
two		0.06	0.56	-	m	0.56
three		0.08	0.75	-	m	0.75
Purpose made door frames and lining sets; **Afrormosia; PC £758.95/m3**						
Jambs and heads; as linings						
32 x 63 mm	PC £6.21	0.25	2.34	7.17	m	9.51
32 x 100 mm	PC £8.15	0.25	2.34	9.39	m	11.73
32 x 140 mm	PC £10.51	0.25	2.34	12.12	m	14.46

L WINDOWS/DOORS/STAIRS Including overheads and profit at 12.50%		Labour hours	Labour £	Material £	Unit	Total rate £
L20 TIMBER DOORS/SHUTTERS/HATCHES - cont'd						
Purpose made door frames and lining sets; **Afrormosia; PC £758.95/m3** - cont'd						
Jambs and heads; as frames; rebated, rounded and grooved						
38 x 75 mm	PC £7.83	0.25	2.34	9.03	m	11.37
38 x 100 mm	PC £9.62	0.25	2.34	11.09	m	13.43
38 x 115 mm	PC £11.14	0.25	2.34	12.85	m	15.19
38 x 140 mm	PC £12.81	0.30	2.80	14.77	m	17.57
50 x 100 mm	PC £10.05	0.30	2.80	14.04	m	16.84
50 x 125 mm	PC £14.73	0.30	2.80	16.99	m	19.79
63 x 88 mm	PC £13.64	0.30	2.80	15.73	m	18.53
63 x 100 mm	PC £15.12	0.30	2.80	17.43	m	20.23
63 x 125 mm	PC £18.24	0.30	2.80	21.04	m	23.84
75 x 100 mm	PC £17.46	0.30	2.80	20.13	m	22.93
75 x 125 mm	PC £21.68	0.33	3.09	25.00	m	28.09
75 x 150 mm	PC £25.53	0.33	3.09	29.44	m	32.53
100 x 100 mm	PC £22.87	0.33	3.09	26.37	m	29.46
100 x 150 mm	PC £33.32	0.33	3.09	38.42	m	41.51
Mullions and transoms; in linings						
32 x 63 mm	PC £9.20	0.18	1.68	10.61	m	12.29
32 x 100 mm	PC £11.14	0.18	1.68	12.85	m	14.53
32 x 140 mm	PC £13.51	0.18	1.68	15.58	m	17.26
Mullions and transoms; in frames; twice rebated, rounded and grooved						
38 x 75 mm	PC £10.99	0.18	1.68	12.67	m	14.35
38 x 100 mm	PC £12.78	0.18	1.68	14.73	m	16.41
38 x 115 mm	PC £14.14	0.18	1.68	16.30	m	17.98
38 x 140 mm	PC £14.14	0.20	1.87	16.30	m	18.17
50 x 100 mm	PC £15.18	0.20	1.87	17.51	m	19.38
50 x 125 mm	PC £17.87	0.20	1.87	20.61	m	22.48
63 x 88 mm	PC £16.67	0.20	1.87	19.23	m	21.10
63 x 100 mm	PC £18.11	0.20	1.87	20.88	m	22.75
75 x 100 mm	PC £20.62	0.20	1.87	23.78	m	25.65
Sills; once sunk weathered; once rebated, three times grooved						
63 x 175 mm	PC £33.09	0.36	3.37	38.17	m	41.54
75 x 125 mm	PC £29.80	0.36	3.37	34.35	m	37.72
75 x 150 mm	PC £33.72	0.36	3.37	38.88	m	42.25
Extra for additional labours						
one		0.03	0.28	-	m	0.28
two		0.06	0.56	-	m	0.56
three		0.08	0.75	-	m	0.75
Door sills; European Oak PC £1252.38/m3						
Sills; once sunk weathered; once rebated, three times grooved						
63 x 175 mm	PC £55.12	0.36	3.37	63.56	m	66.93
75 x 125 mm	PC £49.08	0.36	3.37	56.60	m	59.97
75 x 150 mm	PC £56.07	0.36	3.37	64.66	m	68.03
Extra for additional labours						
one		0.03	0.28	-	m	0.28
two		0.07	0.65	-	m	0.65
three		0.10	0.93	-	m	0.93
Bedding and pointing frames						
Pointing wood frames or sills with mastic						
one side		0.08	0.96	0.46	m	1.42
each side		0.16	1.92	0.93	m	2.85
Pointing wood frames or sills with polysulphide sealant						
one side		0.08	0.96	0.93	m	1.89
each side		0.16	1.92	1.85	m	3.77

L WINDOWS/DOORS/STAIRS Including overheads and profit at 12.50%		Labour hours	Labour £	Material £	Unit	Total rate £
Bedding wood plates in cement mortar (1:3)						
100 mm wide		0.07	0.84	0.05	m	0.89
Bedding wood frame in cement mortar (1:3)						
and point						
one side		0.09	1.08	0.05	m	1.13
each side		0.11	1.32	0.07	m	1.39
one side in mortar; other side in mastic		0.17	2.04	0.51	m	2.55

L21 METAL DOORS/SHUTTERS/HATCHES

Aluminium double glazed sliding patio doors;
Crittall 'Luminaire' or similar; white
acrylic finish; with and including 18 mm
annealed double glazing; fixed in position;
including lugs plugged and screwed to
brickwork or blockwork; or screwed to wooden
sub-frame (measured elsewhere)

Patio doors						
1800 x 2100 mm; ref D18HDL21	PC £486.00	8.25	87.89	549.60	nr	637.49
2400 x 2100 mm; ref D24HDL21	PC £575.55	9.90	105.47	650.34	nr	755.81
3000 x 2100 mm; ref D30HDF21	PC £804.60	11.60	123.58	908.02	nr	1031.60

Galvanized steel 'up and over' type garage
doors; Catnic 'Garador' or similar;
spring counterbalanced; fixed to timber
frame (measured elsewhere)

Garage door						
2135 x 1980 mm; ref MK 3C	PC £114.75	5.50	58.60	130.65	nr	189.25
2135 x 2135 mm; ref MK 3C	PC £126.45	5.80	61.79	143.81	nr	205.60
2400 x 2125 mm; ref MK 3C	PC £149.85	6.60	70.31	170.13	nr	240.44
4270 x 2135 mm; ref 'Carlton'	PC £501.30	9.90	105.47	565.52	nr	670.99

L30 TIMBER STAIRS/WALKWAYS/BALUSTRADES

Standard staircases; wrought softwood
Stairs; 25 mm treads with rounded nosings;
12 mm plywood risers; 32 mm once rounded
strings; bullnose bottom tread; 50 x 75 mm
hardwood handrail; two 32 x 140 mm
balustrade knee rails; 32 x 50 mm stiffeners
and 100 x 100 mm newel posts with hardwood
newel caps on top

straight flight; 838 mm wide; 2688 mm going; 2600 mm rise; with two newel posts		14.90	139.30	324.38	nr	463.68
straight flight; 838 mm wide; 2600 mm rise; with two newel posts and three top treads winding		19.80	185.11	401.64	nr	586.75
dogleg staircase; 838 mm wide; 2600 mm rise; quarter space landing third riser from top; with three newel posts		20.90	195.39	413.98	nr	609.37
as last but with half space landing; one 100 x 200 mm newel post; and two 100 x 100 mm newel posts		22.00	205.67	495.52	nr	701.19

Stairs; 25 mm treads with rounded nosings;
12 mm plywood risers; 32 mm once rounded
strings; with string cappings; bullnose
bottom tread; 50 x 75 mm hardwood handrail;
two 32 x 32 mm balusters per tread and
100 x 100 mm newel post with hardwood newel
caps on top

straight flight; 838 mm wide 2688 going 2600 mm rise with two newel posts		16.50	154.25	318.54	nr	472.79

L WINDOWS/DOORS/STAIRS Including overheads and profit at 12.50%	Labour hours	Labour £	Material £	Unit	Total rate £
L30 TIMBER STAIRS/WALKWAYS/BALUSTRADES - cont'd					
Standard balustrades; wrought softwood Landing balustrade; 50 x 75 mm hardwood handrail; three 32 x 140 mm balustrades knee rails; two 32 x 50 mm stiffeners; one end jointed to newel post; other end built into wall (newel post and mortices both measured separately)					
3 m long	3.30	30.85	102.86	nr	**133.71**
Landing balustrade; 50 x 75 mm hardwood handrail; 32 x 32 mm balusters; one end of handrail jointed to newel post; other end built into wall; balusters housed in at bottom (newel post and mortices both measured separately)					
3 m long	4.95	46.28	83.85	nr	**130.13**
Purpose made staircase components; wrought softwood Board landings; cross-tongued joints; 100 x 50 mm sawn softwood bearers					
25 mm thick	1.10	10.28	46.34	m2	**56.62**
32 mm thick	1.10	10.28	55.09	m2	**65.37**
Treads cross-tongued joints and risers; rounded nosings; tongued, grooved, glued and blocked together; one 175 x 50 mm sawn softwood carriage					
25 mm treads; 19 mm risers	1.65	15.43	58.48	m2	**73.91**
Ends; quadrant	-	-	-	nr	**28.39**
32 mm treads; 25 mm risers	1.65	15.43	68.61	m2	**84.04**
Ends; quadrant	-	-	-	nr	**33.46**
Ends; housed to hardwood	-	-	-	nr	**0.50**
Winders; cross-tongued joints and risers in one width; rounded nosings; tongued, grooved glued and blocked together; one 175 x 50 mm sawn softwood carriage					
25 mm treads; 19 mm risers	2.75	25.71	63.45	m2	**89.16**
32 mm treads; 25 mm risers	2.75	25.71	74.59	m2	**100.30**
Wide ends; housed to hardwood	-	-	-	nr	**1.02**
Narrow ends; housed to hardwood	-	-	-	nr	**0.77**
Closed strings; in one width; 230 mm wide; rounded twice					
32 mm thick	0.55	5.14	10.84	m	**15.98**
38 mm thick	0.55	5.14	13.08	m	**18.22**
50 mm thick	0.55	5.14	17.17	m	**22.31**
Closed strings; cross-tongued joints; 280 mm wide; once rounded; fixing with screws; plugging 450 mm centres					
32 mm thick	0.66	6.17	19.64	m	**25.81**
Extra for short ramp	0.13	1.22	9.81	nr	**11.03**
38 mm thick	0.66	6.17	22.24	m	**28.41**
Extra for short ramp	0.13	1.22	11.01	nr	**12.23**
50 mm thick	0.66	6.17	27.40	m	**33.57**
Ends; fitted	0.11	1.03	0.34	nr	**1.37**
Ends; framed	0.13	1.22	3.27	nr	**4.49**
Extra for tongued heading joint	0.13	1.22	1.69	nr	**2.91**
Extra for short ramp	0.13	1.22	13.69	nr	**14.91**
Closed strings; ramped; crossed tongued joints 280 mm wide; once rounded; fixing with screws; plugging 450 mm centres					
32 mm thick	0.66	6.17	21.60	m	**27.77**
38 mm thick	0.66	6.17	25.67	m	**31.84**
50 mm thick	0.66	6.17	30.14	m	**36.31**
Apron linings; in one width 230 mm wide					
19 mm thick	0.36	3.37	3.80	m	**7.17**
25 mm thick	0.36	3.37	4.65	m	**8.02**

L WINDOWS/DOORS/STAIRS Including overheads and profit at 12.50%	Labour hours	Labour £	Material £	Unit	Total rate £
Handrails; rounded					
44 x 50 mm	0.28	2.62	3.34	m	5.96
50 x 75 mm	0.30	2.80	4.24	m	7.04
63 x 87 mm	0.33	3.09	5.93	m	9.02
75 x 100 mm	0.39	3.65	6.74	m	10.39
Handrails; moulded					
44 x 50 mm	0.28	2.62	3.65	m	6.27
50 x 75 mm	0.30	2.80	4.56	m	7.36
63 x 87 mm	0.33	3.09	6.25	m	9.34
75 x 100 mm	0.39	3.65	7.05	m	10.70
Handrails; rounded; ramped					
44 x 50 mm	0.36	3.37	6.67	m	10.04
50 x 75 mm	0.40	3.74	8.46	m	12.20
63 x 87 mm	0.44	4.11	11.87	m	15.98
75 x 100 mm	0.50	4.67	13.47	m	18.14
Handrails; moulded; ramped					
44 x 50 mm	0.36	3.37	7.31	m	10.68
50 x 75 mm	0.40	3.74	9.12	m	12.86
63 x 87 mm	0.44	4.11	12.50	m	16.61
75 x 100 mm	0.50	4.67	14.11	m	18.78
Add to above for					
grooved once	-	-	-	m	0.30
ends; framed	0.11	1.03	2.60	nr	3.63
ends; framed on rake	0.17	1.59	3.27	nr	4.86
Heading joints on rake; handrail screws					
44 x 50 mm	0.17	1.59	15.95	nr	17.54
50 x 75 mm	0.17	1.59	15.95	nr	17.54
63 x 87 mm	0.17	1.59	20.19	nr	21.78
75 x 100 mm	0.17	1.59	20.19	nr	21.78
Mitres; handrail screws					
44 x 50 mm	0.22	2.06	15.95	nr	18.01
50 x 75 mm	0.22	2.06	15.95	nr	18.01
63 x 87 mm	0.22	2.06	20.19	nr	22.25
75 x 100 mm	0.22	2.06	20.19	nr	22.25
Balusters; stiffeners					
25 x 25 mm	0.09	0.84	1.34	m	2.18
32 x 32 mm	0.09	0.84	1.59	m	2.43
32 x 50 mm	0.09	0.84	1.97	m	2.81
Ends; housed	0.03	0.28	0.63	nr	0.91
Sub rails					
32 x 63 mm	0.36	3.37	2.72	m	6.09
Ends; housed to newel	0.11	1.03	2.60	nr	3.63
Knee rails					
32 x 140 mm	0.44	4.11	4.52	m	8.63
Ends; housed to newel	0.11	1.03	2.60	nr	3.63
Newel posts					
50 x 100 mm; half	0.44	4.11	4.64	m	8.75
75 x 75 mm	0.44	4.11	5.04	m	9.15
100 x 100 mm	0.55	5.14	7.91	m	13.05
Newel caps; splayed on four sides					
62.5 x 125 x 50 mm; half	0.17	1.59	3.05	nr	4.64
100 x 100 x 50 mm	0.17	1.59	3.05	nr	4.64
125 x 125 x 50 mm	0.17	1.59	3.28	nr	4.87

**Purpose made staircase components; selected
West African Mahogany; PC £531.05/m3**
Board landings; cross-tongued joints;
100 x 50 mm sawn softwood bearers

25 mm thick PC £59.73	1.65	15.43	68.35	m2	83.78
32 mm thick PC £72.97	1.65	15.43	83.24	m2	98.67

Treads cross-tongued joints and risers;
rounded nosings; tongued, grooved, glued and
blocked together; one 175 x 50 mm sawn
softwood carriage

25 mm treads; 19 mm risers PC £73.08	2.20	20.57	83.98	m2	104.55
Ends; quadrant PC £42.88	-	-	48.24	nr	48.24

L WINDOWS/DOORS/STAIRS Including overheads and profit at 12.50%		Labour hours	Labour £	Material £	Unit	Total rate £
L30 TIMBER STAIRS/WALKWAYS/BALUSTRADES - cont'd						
Purpose made staircase components; selected						
West African Mahogany; PC £531.05/m3 - cont'd						
Treads cross-tongued joints and risers;						
rounded nosings; tongued, grooved, glued and						
blocked together; one 175 x 50 mm sawn						
softwood carriage						
32 mm treads; 25 mm risers	PC £87.51	2.20	20.57	100.20	m2	120.77
Ends; quadrant	PC £61.25	-	-	68.91	nr	68.91
Ends; housed to hardwood		-	-	-	nr	0.77
Winders; cross-tongued joints and risers in						
one width; rounded nosings; tongued, grooved						
glued and blocked together; one 175 x 50 mm						
sawn softwood carriage						
25 mm treads; 19 mm risers	PC £80.38	3.70	34.59	91.47	m2	126.06
32 mm treads; 25 mm risers	PC £96.27	3.70	34.59	109.35	m2	143.94
Wide ends; housed to hardwood		-	-	-	nr	1.52
Narrow ends; housed to hardwood		-	-	-	nr	1.16
Closed strings; in one width; 230 mm wide;						
rounded twice						
32 mm thick	PC £15.48/m	0.74	6.92	17.49	m2	24.41
38 mm thick	PC £18.54/m	0.74	6.92	20.94	m2	27.86
50 mm thick	PC £24.42/m	0.74	6.92	27.56	m2	34.48
Closed strings; cross-tongued joints; 280 mm						
wide; once rounded; fixing with screws;						
plugging 450 mm centres						
32 mm thick	PC £27.09	0.88	8.23	30.56	m	38.79
Extra for short ramp	PC £13.55	0.20	1.87	15.25	nr	17.12
38 mm thick	PC £30.71	0.88	8.23	34.63	m	42.86
Extra for short ramp	PC £15.37	0.20	1.87	17.29	nr	19.16
50 mm thick	PC £37.94	0.88	8.23	42.76	m	50.99
Ends; fitted	PC £0.82	0.17	1.59	0.92	nr	2.51
Ends; framed	PC £5.06	0.20	1.87	5.70	nr	7.57
Extra for tongued heading joint	PC £2.50	0.20	1.87	2.81	nr	4.68
Extra for short ramp	PC £19.00	0.20	1.87	21.37	nr	23.24
Closed strings; ramped; crossed tongued						
joints 280 mm wide; once rounded; fixing						
with screws; plugging 450 mm centres						
32 mm thick	PC £29.83	0.88	8.23	33.64	m	41.87
38 mm thick	PC £33.81	0.88	8.23	38.11	m	46.34
50 mm thick	PC £41.73	0.88	8.23	47.03	m	55.26
Apron linings; in one width 230 mm wide						
19 mm thick	PC £6.24	0.48	4.49	7.11	m	11.60
25 mm thick	PC £7.53	0.48	4.49	8.55	m	13.04
Handrails; rounded						
44 x 50 mm	PC £5.28	0.36	3.37	6.23	m	9.60
50 x 75 mm	PC £6.56	0.40	3.74	7.76	m	11.50
63 x 87 mm	PC £8.46	0.44	4.11	9.99	m	14.10
75 x 100 mm	PC £10.24	0.50	4.67	12.10	m	16.77
Handrails; moulded						
44 x 50 mm	PC £5.69	0.36	3.37	6.72	m	10.09
50 x 75 mm	PC £6.99	0.40	3.74	8.26	m	12.00
63 x 87 mm	PC £8.87	0.44	4.11	10.48	m	14.59
75 x 100 mm	PC £10.65	0.50	4.67	12.58	m	17.25
Handrails; rounded; ramped						
44 x 50 mm	PC £10.56	0.48	4.49	12.48	m	16.97
50 x 75 mm	PC £13.15	0.53	4.95	15.53	m	20.48
63 x 87 mm	PC £16.94	0.58	5.42	20.01	m	25.43
75 x 100 mm	PC £20.48	0.66	6.17	24.19	m	30.36
Handrails; moulded; ramped						
44 x 50 mm	PC £11.38	0.48	4.49	13.45	m	17.94
50 x 75 mm	PC £13.97	0.53	4.95	16.50	m	21.45
63 x 87 mm	PC £17.73	0.58	5.42	20.95	m	26.37
75 x 100 mm	PC £21.30	0.66	6.17	25.15	m	31.32

L WINDOWS/DOORS/STAIRS Including overheads and profit at 12.50%		Labour hours	Labour £	Material £	Unit	Total rate £
Add to above for						
grooved once		-	-	-	m	0.45
ends; framed	PC £3.84	0.17	1.59	4.32	nr	5.91
ends; framed on rake	PC £4.72	0.24	2.24	5.31	nr	7.55
Heading joints on rake; handrail screws						
44 x 50 mm	PC £17.43	0.24	2.24	19.60	nr	21.84
50 x 75 mm	PC £17.43	0.24	2.24	19.60	nr	21.84
63 x 87 mm	PC £22.33	0.24	2.24	25.12	nr	27.36
75 x 100 mm	PC £22.33	0.24	2.24	25.12	nr	27.36
Mitres; handrail screws						
44 x 50 mm	PC £17.43	0.30	2.80	19.60	nr	22.40
50 x 75 mm	PC £17.43	0.30	2.80	19.60	nr	22.40
63 x 87 mm	PC £22.33	0.30	2.80	25.12	nr	27.92
75 x 100 mm	PC £22.33	0.30	2.80	25.12	nr	27.92
Balusters; stiffeners						
25 x 25 mm	PC £2.57	0.11	1.03	2.90	m	3.93
32 x 32 mm	PC £2.97	0.11	1.03	3.34	m	4.37
32 x 50 mm	PC £3.56	0.11	1.03	4.01	m	5.04
Ends; housed	PC £0.86	0.06	0.56	0.97	nr	1.53
Sub rails						
32 x 63 mm	PC £4.57	0.48	4.49	5.15	m	9.64
Ends; housed to newel	PC £3.47	0.17	1.59	3.90	nr	5.49
Knee rails						
32 x 140 mm	PC £7.07	0.58	5.42	7.96	m	13.38
Ends; housed to newel	PC £3.47	0.17	1.59	3.90	nr	5.49
Newel posts						
50 x 100 mm; half	PC £7.55	0.58	5.42	8.50	m	13.92
75 x 75 mm	PC £8.15	0.58	5.42	9.17	m	14.59
100 x 100 mm	PC £12.52	0.74	6.92	14.09	m	21.01
Newel caps; splayed on four sides						
62.5 x 125 x 50 mm; half	PC £4.84	0.22	2.06	5.45	nr	7.51
100 x 100 x 50 mm	PC £4.84	0.22	2.06	5.45	nr	7.51
125 x 125 x 50 mm	PC £5.22	0.22	2.06	5.88	nr	7.94
Purpose made staircase components;						
Oak; PC £1252.38/m3						
Board landings; cross-tongued joints;						
100 x 50 mm sawn softwood bearers						
25 mm thick	PC £123.33	1.65	15.43	139.89	m2	155.32
32 mm thick	PC £151.58	1.65	15.43	171.68	m2	187.11
Treads cross-tongued joints and risers;						
rounded nosings; tongued, grooved, glued and						
blocked together; one 175 x 50 mm sawn						
softwood carriage						
25 mm treads; 19 mm risers	PC £133.68	2.20	20.57	152.14	m2	172.71
Ends; quadrant	PC £66.86	-	-	75.22	nr	75.22
32 mm treads; 25 mm risers	PC £164.23	2.20	20.57	186.51	m2	207.08
Ends; quadrant	PC £82.13	-	-	92.39	nr	92.39
Ends; housed to hardwood		-	-	-	nr	1.02
Winders; cross-tongued joints and risers in						
one width; rounded nosings; tongued, grooved						
glued and blocked together; one 175 x 50 mm						
sawn softwood carriage						
25 mm treads; 19 mm risers	PC £147.06	3.70	34.59	166.48	m2	201.07
32 mm treads; 25 mm risers	PC £180.65	3.70	34.59	204.27	m2	238.86
Wide ends; housed to hardwood		-	-	-	nr	2.05
Narrow ends; housed to hardwood		-	-	-	nr	1.52
Closed strings; in one width; 230 mm wide;						
rounded twice						
32 mm thick	PC £32.21	0.74	6.92	36.32	m	43.24
38 mm thick	PC £38.34	0.74	6.92	43.21	m	50.13
50 mm thick	PC £50.57	0.74	6.92	56.98	m	63.90

L WINDOWS/DOORS/STAIRS Including overheads and profit at 12.50%		Labour hours	Labour £	Material £	Unit	Total rate £
L30 TIMBER STAIRS/WALKWAYS/BALUSTRADES - cont'd						
Purpose made staircase components;						
Oak; PC £1252.38/m3 - cont'd						
Closed strings; cross-tongued joints; 280 mm						
wide; once rounded; fixing with screws;						
plugging 450 mm centres						
32 mm thick	PC £51.60	0.88	8.23	58.14	m	66.37
Extra for short ramp	PC £25.80	0.20	1.87	29.02	nr	30.89
38 mm thick	PC £59.09	0.88	8.23	66.56	m	74.79
Extra for short ramp	PC £29.55	0.20	1.87	33.25	nr	35.12
50 mm thick	PC £74.21	0.88	8.23	83.57	m	91.80
Ends; fitted	PC £0.96	0.17	1.59	1.08	nr	2.67
Ends; framed	PC £6.52	0.20	1.87	7.34	nr	9.21
Extra for tongued heading joint	PC £3.27	0.20	1.87	3.67	nr	5.54
Extra for short ramp	PC £37.12	0.20	1.87	41.76	nr	43.63
Closed strings; ramped; crossed tongued						
joints 280 mm wide; once rounded; fixing						
with screws; plugging 450 mm centres						
32 mm thick	PC £56.75	0.88	8.23	63.93	m	72.16
38 mm thick	PC £65.01	0.88	8.23	73.22	m	81.45
50 mm thick	PC £81.65	0.88	8.23	91.93	m	100.16
Apron linings; in one width 230 mm wide						
19 mm thick	PC £14.79	0.48	4.49	16.72	m	21.21
25 mm thick	PC £18.57	0.48	4.49	20.98	m	25.47
Handrails; rounded						
44 x 50 mm	PC £10.69	0.36	3.37	12.62	m	15.99
50 x 75 mm	PC £14.48	0.40	3.74	17.10	m	20.84
63 x 87 mm	PC £19.97	0.44	4.11	23.59	m	27.70
75 x 100 mm	PC £25.17	0.50	4.67	29.73	m	34.40
Handrails; moulded						
44 x 50 mm	PC £11.24	0.36	3.37	13.28	m	16.65
50 x 75 mm	PC £15.03	0.40	3.74	17.75	m	21.49
63 x 87 mm	PC £20.52	0.44	4.11	24.24	m	28.35
75 x 100 mm	PC £25.70	0.50	4.67	30.37	m	35.04
Handrails; rounded; ramped						
44 x 50 mm	PC £21.40	0.48	4.49	25.28	m	29.77
50 x 75 mm	PC £28.96	0.53	4.95	34.21	m	39.16
63 x 87 mm	PC £39.96	0.58	5.42	47.20	m	52.62
75 x 100 mm	PC £50.34	0.66	6.17	59.47	m	65.64
Handrails; moulded; ramped						
44 x 50 mm	PC £22.49	0.48	4.49	26.56	m	31.05
50 x 75 mm	PC £30.04	0.53	4.95	35.48	m	40.43
63 x 87 mm	PC £41.04	0.58	5.42	48.48	m	53.90
75 x 100 mm	PC £51.42	0.66	6.17	60.74	m	66.91
Add to above for						
grooved once		-	-	-	m	0.62
ends; framed	PC £5.00	0.17	1.59	5.62	nr	7.21
ends; framed on rake	PC £6.18	0.24	2.24	6.95	nr	9.19
Heading joints on rake; handrail screws						
44 x 50 mm	PC £26.48	0.24	2.24	29.79	nr	32.03
50 x 75 mm	PC £26.48	0.24	2.24	29.79	nr	32.03
63 x 87 mm	PC £32.51	0.24	2.24	36.57	nr	38.81
75 x 100 mm	PC £32.51	0.24	2.24	36.57	nr	38.81
Mitres; handrail screws						
44 x 50 mm	PC £26.48	0.30	2.80	29.79	nr	32.59
50 x 75 mm	PC £26.48	0.30	2.80	29.79	nr	32.59
63 x 87 mm	PC £32.51	0.30	2.80	36.57	nr	39.37
75 x 100 mm	PC £32.51	0.30	2.80	36.57	nr	39.37
Balusters; stiffeners						
25 x 25 mm	PC £4.13	0.11	1.03	4.64	m	5.67
32 x 32 mm	PC £5.28	0.11	1.03	5.94	m	6.97
32 x 50 mm	PC £7.01	0.11	1.03	7.89	m	8.92
Ends; housed	PC £1.13	0.06	0.56	1.27	nr	1.83
Sub rails						
32 x 63 mm	PC £9.00	0.48	4.49	10.12	m	14.61
Ends; housed to newel	PC £4.63	0.17	1.59	5.21	nr	6.80

L WINDOWS/DOORS/STAIRS Including overheads and profit at 12.50%		Labour hours	Labour £	Material £	Unit	Total rate £
Knee rails						
32 x 140 mm	PC £16.22	0.58	5.42	18.25	m	23.67
Ends; housed to newel	PC £4.63	0.17	1.59	5.21	nr	6.80
Newel posts						
50 x 100 mm; half	PC £17.68	0.58	5.42	19.89	m	25.31
75 x 75 mm	PC £19.39	0.58	5.42	21.82	m	27.24
100 x 100 mm	PC £32.11	0.74	6.92	36.12	m	43.04
Newel caps; splayed on four sides						
62.5 x 125 x 50 mm; half	PC £6.99	0.22	2.06	7.86	nr	9.92
100 x 100 x 50 mm	PC £6.99	0.22	2.06	7.86	nr	9.92
125 x 125 x 50 mm	PC £8.12	0.22	2.06	9.13	nr	11.19

L31 METAL STAIRS/WALKWAYS/BALUSTRADES

	Labour hours	Labour £	Material £	Unit	Total rate £
Cat ladders, balustrades and handrails, etc.; mild steel; BS 4360					
Cat ladders; welded construction; 64 x 13 mm bar strings; 19 mm rungs at 250 mm centres; fixing by bolting; 0.46 m wide; 3.05 m high	3.30	35.16	122.12	nr	157.28
Extra for					
ends of strings; bent once; holed once for 10 mm dia bolt	-	-	-	nr	1.61
ends of strings; fanged	-	-	-	nr	0.73
ends of strings; bent in plane	-	-	-	nr	7.61
Balustrades; welded construction; galvanized after manufacture; 1070 mm high; 50 x 50 x 3.2 mm r.h.s. top rail; 38 x 13 mm bottom rail, 50 x 50 x 3.2 mm r.h.s. standards at 1830 mm centres with base plate drilled and bolted to concrete; 13 x 13 mm balusters at 102 mm centres	1.65	17.58	50.63	m	68.21
Balusters; isolated; one end ragged and cemented in; one 76 x 25 x 6 mm flange plate welded on; ground to a smooth finish; countersunk drilled and tap screwed to underside of handrail					
19 x 19 x 914 mm square bar	-	-	-	nr	8.51
Core-rails; joints prepared, welded and ground to a smooth finish; fixing on brackets (measured elsewhere)					
38 x 10 mm flat bar	-	-	-	m	12.15
50 x 8 mm flat bar	-	-	-	m	11.75
Extra for					
ends fanged	-	-	-	nr	0.88
ends scrolled	-	-	-	nr	3.73
ramps in thickness	-	-	-	nr	3.73
wreaths	-	-	-	nr	4.39
Handrails; joints prepared, welded and ground to a smooth finish; fixing on brackets (measured elsewhere)					
38 x 12 mm half oval bar	-	-	-	m	16.20
44 x 13 mm half oval bar	-	-	-	m	17.42
Extra for					
ends fanged	-	-	-	nr	0.88
ends scrolled	-	-	-	nr	3.73
ramps in thickness	-	-	-	nr	3.73
wreaths	-	-	-	nr	4.39
Handrail bracket; comprising 40 x 5 mm plate with mitred and welded angle; one end welded to 100 mm dia x 5 mm backplate; three times holed and plugged and screwed to brickwork; other end scribed and welded to underside of handrail					
140 mm girth	-	-	-	nr	9.43

L WINDOWS/DOORS/STAIRS Including overheads and profit at 12.50% & 5.00%	Labour hours	Labour £	Material £	Unit	Total rate £
L31 METAL STAIRS/WALKWAYS/BALUSTRADES - cont'd					
Holes					
Holes; countersunk; for screws or bolts					
6 mm dia wood screw; 3 mm thick	0.06	0.64	-	nr	0.64
6 mm dia wood screw; 6 mm thick	0.08	0.85	-	nr	0.85
8 mm dia bolt; 6 mm thick	0.08	0.85	-	nr	0.85
10 mm dia bolt; 6 mm thick	0.09	0.96	-	nr	0.96
12 mm dia bolt; 8 mm thick	0.11	1.17	-	nr	1.17
L40 GENERAL GLAZING					
Standard plain glass; BS 952; clear float;					
panes area 0.15 - 4.00 m2					
3 mm thick; glazed with					
putty or bradded beads	-	-	-	m2	14.66
bradded beads and butyl compound	-	-	-	m2	18.23
screwed beads	-	-	-	m2	19.77
screwed beads and butyl compound	-	-	-	m2	22.88
4 mm thick; glazed with					
putty or bradded beads	-	-	-	m2	17.59
bradded beads and butyl compound	-	-	-	m2	21.65
screwed beads	-	-	-	m2	23.22
screwed beads and butyl compound	-	-	-	m2	26.82
5 mm thick; glazed with					
putty or bradded beads	-	-	-	m2	23.27
bradded beads and butyl compound	-	-	-	m2	25.88
screwed beads	-	-	-	m2	27.45
screwed beads and butyl compound	-	-	-	m2	31.05
6 mm thick; glazed with					
putty or bradded beads	-	-	-	m2	24.57
bradded beads and butyl compound	-	-	-	m2	27.18
screwed beads	-	-	-	m2	28.76
screwed beads and butyl compound	-	-	-	m2	32.36
Standard plain glass; BS 952;					
white patterned; panes area 0.15 - 4.00 m2					
4 mm thick; glazed with					
putty or bradded beads	-	-	-	m2	16.68
bradded beads and butyl compound	-	-	-	m2	20.25
screwed beads	-	-	-	m2	21.79
screwed beads and butyl compound	-	-	-	m2	24.91
6 mm thick; glazed with					
putty or bradded beads	-	-	-	m2	24.30
bradded beads and butyl compound	-	-	-	m2	27.95
screwed beads	-	-	-	m2	29.36
screwed beads and butyl compound	-	-	-	m2	32.60
Standard plain glass; BS 952; rough cast;					
panes area 0.15 - 4.00 m2					
6 mm thick; glazed with					
putty or bradded beads	-	-	-	m2	20.13
bradded beads and butyl compound	-	-	-	m2	23.77
screwed beads	-	-	-	m2	25.19
screwed beads and butyl compound	-	-	-	m2	28.43
Standard plain glass; BS 952; Georgian wired					
cast; panes area 0.15 - 4.00 m2					
7 mm thick; glazed with					
putty or bradded beads	-	-	-	m2	21.65
bradded beads and butyl compound	-	-	-	m2	25.29
screwed beads	-	-	-	m2	26.71
screwed beads and butyl compound	-	-	-	m2	29.95
Extra for lining up wired glass	-	-	-	m2	2.23

L WINDOWS/DOORS/STAIRS Including overheads and profit at 5.00%	Labour hours	Labour £	Material £	Unit	Total rate £
Standard plain glass; BS 952; Georgian wired polished; panes area 0.15 - 4.00 m2 6 mm thick; glazed with					
putty or bradded beads	-	-	-	m2	47.10
bradded beads and butyl compound	-	-	-	m2	49.46
screwed beads	-	-	-	m2	50.87
screwed beads and butyl compound	-	-	-	m2	54.11
Extra for lining up wired glass	-	-	-	m2	2.49
Special glass; BS 952; toughened clear float; panes area 0.15 - 4.00 m2 4 mm thick; glazed with					
putty or bradded beads	-	-	-	m2	45.52
bradded beads and butyl compound	-	-	-	m2	47.87
screwed beads	-	-	-	m2	49.29
screwed beads and butyl compound	-	-	-	m2	52.53
5 mm thick; glazed with					
putty or bradded beads	-	-	-	m2	50.48
bradded beads and butyl compound	-	-	-	m2	52.83
screwed beads	-	-	-	m2	54.25
screwed beads and butyl compound	-	-	-	m2	57.49
6 mm thick; glazed with					
putty or bradded beads	-	-	-	m2	54.17
bradded beads and butyl compound	-	-	-	m2	56.52
screwed beads	-	-	-	m2	57.94
screwed beads and butyl compound	-	-	-	m2	61.18
10 mm thick; glazed with					
putty or bradded beads	-	-	-	m2	99.48
bradded beads and butyl compound	-	-	-	m2	101.84
screwed beads	-	-	-	m2	103.25
screwed beads and butyl compound	-	-	-	m2	106.49
Special glass; BS 952; clear laminated safety glass; panes area 0.15 - 4.00 m2 4.4 mm thick; glazed with					
putty or bradded beads	-	-	-	m2	61.09
bradded beads and butyl compound	-	-	-	m2	63.44
screwed beads	-	-	-	m2	64.86
screwed beads and butyl compound	-	-	-	m2	68.10
5.4 mm thick; glazed with					
putty or bradded beads	-	-	-	m2	62.31
bradded beads and butyl compound	-	-	-	m2	64.67
screwed beads	-	-	-	m2	66.08
screwed beads and butyl compound	-	-	-	m2	69.32
6.4 mm thick; glazed with					
putty or bradded beads	-	-	-	m2	65.86
bradded beads and butyl compound	-	-	-	m2	68.21
screwed beads	-	-	-	m2	69.63
screwed beads and butyl compound	-	-	-	m2	72.87
Special glass; BS 952; 'Antisun' solar control float glass infill panels; panes area 0.15 - 4.00 m2 4 mm thick; glazed with					
non-hardening compound to metal	-	-	-	m2	64.37
6 mm thick; glazed with					
non-hardening compound to metal	-	-	-	m2	86.20
10 mm thick; glazed with					
non-hardening compound to metal	-	-	-	m2	171.62
12 mm thick; glazed with					
non-hardening compound to metal	-	-	-	m2	212.49

L WINDOWS/DOORS/STAIRS Including overheads and profit at 12.50% & 5.00%	Labour hours	Labour £	Material £	Unit	Total rate £
L40 GENERAL GLAZING - cont'd					
Special glass; BS 952; 'Pyran' fire-resisting glass; Schott Glass Ltd. 6.5 mm thick rectangular panes; glazed with screwed hardwood beads and Interdens intumescent strip					
300 x 400 mm pane	1.55	21.50	35.71	nr	57.21
400 x 800 mm pane	2.65	36.76	86.60	nr	123.36
500 x 1400 mm pane	4.20	58.26	180.04	nr	238.30
600 x 1800 mm pane	5.30	73.52	270.91	nr	344.43
6.5 mm thick irregular panes; glazed with screwed hardwood beads and Intergens intumescent strip					
300 x 400 mm pane	1.95	27.05	54.28	nr	81.33
400 x 800 mm pane	3.30	45.78	136.11	nr	181.89
500 x 1400 mm pane	5.25	72.82	288.37	nr	361.19
600 x 1800 mm pane	6.60	91.55	438.03	nr	529.58
Glass louvres; BS 952; with long edges ground or smooth 5 mm or 6 mm thick float					
100 mm wide	-	-	-	m	5.81
150 mm wide	-	-	-	m	6.48
7 mm thick Georgian wired cast					
100 mm wide	-	-	-	m	11.72
150 mm wide	-	-	-	m	12.56
6 mm thick Georgian polished wired					
100 mm wide	-	-	-	m	15.85
150 mm wide	-	-	-	m	18.30
Labours on glass/sundries Curved cutting on panes					
4 mm thick	-	-	-	m	1.91
6 mm thick	-	-	-	m	1.91
6 mm thick; wired	-	-	-	m	2.93
Imitation washleather or black velvet strip; as bedding to edge of glass	-	-	-	m	0.79
Intumescent paste to fire doors per side of glass	-	-	-	m	4.50
Drill hole exceeding 6 mm and not exceeding 15 mm dia. through panes					
not exceeding 6 mm thick	-	-	-	nr	1.99
not exceeding 10 mm thick	-	-	-	nr	2.57
not exceeding 12 mm thick	-	-	-	nr	3.20
not exceeding 19 mm thick	-	-	-	nr	3.99
not exceeding 25 mm thick	-	-	-	nr	4.98
Drill hole exceeding 16 mm and not exceeding 38 mm dia. through panes					
not exceeding 6 mm thick	-	-	-	nr	2.83
not exceeding 10 mm thick	-	-	-	nr	3.78
not exceeding 12 mm thick	-	-	-	nr	4.51
not exceeding 19 mm thick	-	-	-	nr	5.66
not exceeding 25 mm thick	-	-	-	nr	7.08
Drill hole over 38 mm dia. through panes					
not exceeding 6 mm thick	-	-	-	nr	5.66
not exceeding 10 mm thick	-	-	-	nr	6.87
not exceeding 12 mm thick	-	-	-	nr	8.13
not exceeding 19 mm thick	-	-	-	nr	10.02
not exceeding 25 mm thick	-	-	-	nr	12.48
Add to the above					
Plus 33% for countersunk holes					
Plus 50% for wired or laminated glass					

L WINDOWS/DOORS/STAIRS Including overheads and profit at 5.00%	Labour hours	Labour £	Material £	Unit	Total rate £
Factory made double hermetically sealed **units; Pilkington's 'Insulight' or similar;** **to wood or metal with butyl non-setting** **compound and screwed or clipped beads**					
Two panes; BS 952; clear float glass; GG; 3 mm or 4 mm thick; 13 mm air space					
2 - 4 m2	-	-	-	m2	59.03
1 - 2 m2	-	-	-	m2	56.95
0.75 - 1 m2	-	-	-	m2	63.41
0.5 - 0.75 m2	-	-	-	m2	72.41
0.35 - 0.5 m2	-	-	-	m2	80.47
0.25 - 0.35 m2	-	-	-	m2	99.62
not exceeding 0.25 m2	-	-	-	m2	99.62
Two panes; BS 952; clear float glass; GG; 5 mm or 6 mm thick; 13 mm air space					
2 - 4 m2	-	-	-	m2	76.02
1 - 2 m2	-	-	-	m2	72.33
0.75 - 1 m2	-	-	-	m2	79.71
0.5 - 0.75 m2	-	-	-	m2	89.39
0.35 - 0.5 m2	-	-	-	m2	99.08
0.25 - 0.35 m2	-	-	-	m2	115.57
not exceeding 0.25 m2	-	-	-	m2	115.57

M SURFACE FINISHES Including overheads and profit at 12.50% & 5.00%	Labour hours	Labour £	Material £	Unit	Total rate £
M10 SAND CEMENT/CONCRETE/GRANOLITHIC SCREEDS/ FLOORING					
Cement and sand (1:3); steel trowelled					
Work to floors; one coat; level and to falls					
not exceeding 15 degrees from horizontal;					
to concrete base; over 300 mm wide					
32 mm	0.44	6.10	1.88	m2	7.98
40 mm	0.46	6.38	2.26	m2	8.64
48 mm	0.50	6.94	2.67	m2	9.61
50 mm	0.51	7.07	2.74	m2	9.81
60 mm	0.54	7.49	3.15	m2	10.64
65 mm	0.56	7.77	3.41	m2	11.18
70 mm	0.58	8.05	3.68	m2	11.73
75 mm	0.61	8.46	3.94	m2	12.40
Finishing around pipes; not exceeding					
0.30 m girth	0.01	0.11	-	nr	0.11
Add to the above for work					
to falls and crossfalls and to slopes					
not exceeding 15 degrees from horizontal	0.03	0.32	-	m2	0.32
to slopes over 15 degrees from horizontal	0.11	1.17	-	m2	1.17
two coats of surface hardener brushed on	0.04	0.43	1.26	m2	1.69
water-repellent additive incorporated in					
the mix	-	-	-	m2	0.83
oil-repellent additive incorporated in the					
mix	-	-	-	m2	2.61
Cement and sand (1:3) beds and backings					
Work to floors; one coat; level; to concrete					
base; screeded; over 300 mm wide					
25 mm thick	-	-	-	m2	5.91
50 mm thick	-	-	-	m2	7.97
75 mm thick	-	-	-	m2	10.34
100 mm thick	-	-	-	m2	12.70
Work to floors; one coat; level; to concrete					
base; steel trowelled; over 300 mm wide					
25 mm thick	-	-	-	m2	5.91
50 mm thick	-	-	-	m2	7.97
75 mm thick	-	-	-	m2	10.34
100 mm thick	-	-	-	m2	12.70
Granolithic paving; cement and granite					
chippings 5 mm down (1:2.5); steel trowelled					
Work to floors; one coat; level; laid on					
concrete while green; over 300 mm wide					
20 mm	0.33	5.57	3.93	m2	9.50
25 mm	0.39	6.58	4.74	m2	11.32
Work to floors; two coat; laid on hacked					
concrete with slurry; over 300 mm wide					
38 mm	0.66	11.14	6.32	m2	17.46
50 mm	0.77	12.99	8.10	m2	21.09
75 mm	0.99	16.71	11.93	m2	28.64
Work to landings; one coat; level; laid					
on concrete while green; over 300 mm wide					
20 mm	0.50	8.44	3.93	m2	12.37
25 mm	0.50	8.44	4.74	m2	13.18
Work to landings; two coat; laid on hacked					
concrete with slurry; over 300 mm wide					
38 mm	0.83	14.01	6.32	m2	20.33
50 mm	0.94	15.86	8.10	m2	23.96
75 mm	1.10	18.56	11.93	m2	30.49
Finishing around pipes; not exceeding					
0.30 m girth	0.06	1.01	-	nr	1.01

M SURFACE FINISHES Including overheads and profit at 12.50% & 5.00%	Labour hours	Labour £	Material £	Unit	Total rate £
Add to the above over 300 mm wide for					
1.35 kg/m2 carborundum grains trowelled in	0.07	1.18	1.14	m2	2.32
two coats of surface hardener brushed on	0.02	0.34	1.53	m2	1.87
liquid hardening additive incorporated					
in the mix	-	-	-	m2	2.38
oil-repellent additive incorporated in					
the mix	-	-	-	m2	2.61
25 mm work to treads; one coat; to concrete					
base					
225 mm wide	0.99	16.71	5.17	m	21.88
275 mm wide	0.99	16.71	5.26	m	21.97
Return end	0.20	3.38	-	nr	3.38
Extra for 38 mm 'Ferodo' nosing	0.22	3.71	7.15	m	10.86
13 mm skirtings; rounded top edge and coved					
bottom junction; to brickwork or blockwork					
base					
75 mm wide on face	0.61	10.29	0.74	m	11.03
150 mm wide on face	0.83	14.01	1.12	m	15.13
Ends; fair	0.06	1.01	-	nr	1.01
Angles	0.08	1.35	-	nr	1.35
13 mm outer margin to stairs; to follow					
profile of and with rounded nosing to treads					
and risers; fair edge and arris at bottom;					
to concrete base					
75 mm wide	0.99	16.71	1.49	m	18.20
Angles	0.08	1.35	-	nr	1.35
13 mm wall string to stairs; fair edge and					
arris on top; coved bottom junction with					
treads and risers; to brickwork or blockwork					
base					
275 mm (extreme) wide	0.88	14.85	1.49	m	16.34
Ends	0.06	1.01	-	nr	1.01
Angles	0.08	1.35	-	nr	1.35
Ramps	0.09	1.52	-	nr	1.52
Ramped and wreathed corners	0.11	1.86	-	nr	1.86
13 mm outer string to stairs; rounded nosing					
on top at junction with treads and risers;					
fair edge and arris at bottom; to concrete					
base					
300 mm (extreme) wide	0.88	14.85	2.48	m	17.33
Ends	0.06	1.01	-	nr	1.01
Angles	-	-	-	nr	1.18
Ramps	0.09	1.52	-	nr	1.52
Ramped and wreathed corners	0.11	1.86	-	nr	1.86
19 mm skirtings; rounded top edge and coved					
bottom junction; to brickwork or blockwork					
base					
75 mm wide on face	0.61	10.29	1.24	m	11.53
150 mm wide on face	0.83	14.01	1.86	m	15.87
Ends; fair	0.06	1.01	-	nr	1.01
Angles	0.08	1.35	-	nr	1.35
19 mm riser; one rounded nosing; to concrete					
base					
150 mm high; plain	0.99	16.71	4.20	m	20.91
150 mm high; undercut	0.99	16.71	4.20	m	20.91
180 mm high; plain	0.99	16.71	4.27	m	20.98
180 mm high; undercut	0.99	16.71	4.27	m	20.98

M11 MASTIC ASPHALT FLOORING

Mastic asphalt paving to BS 1076; black 20 mm one coat coverings; felt isolating membrane; to concrete base; flat					
over 300 mm wide	-	-	-	m2	14.68
225 - 300 mm wide	-	-	-	m2	27.84
150 - 225 mm wide	-	-	-	m2	28.24
not exceeding 150 mm wide	-	-	-	m2	34.80

M SURFACE FINISHES Including overheads and profit at 5.00%	Labour hours	Labour £	Material £	Unit	Total rate £
M11 MASTIC ASPHALT FLOORING - cont'd					
Mastic asphalt paving to BS 1076; black - cont'd					
25 mm one coat coverings; felt isolating					
membrane; to concrete base; flat					
over 300 mm wide	-	-	-	m2	17.48
225 - 300 mm wide	-	-	-	m2	33.16
150 - 225 mm wide	-	-	-	m2	33.61
not exceeding 150 mm wide	-	-	-	m2	41.46
20 mm three coat skirtings to brickwork base					
not exceeding 150 mm girth	-	-	-	m	16.45
150 - 225 mm girth	-	-	-	m	19.20
225 - 300 mm girth	-	-	-	m	21.92
Mastic asphalt paving; acid-resisting; black					
20 mm one coat coverings; felt isolating					
membrane; to concrete base; flat					
over 300 mm wide	-	-	-	m2	17.31
225 - 300 mm wide	-	-	-	m2	30.49
150 - 225 mm wide	-	-	-	m2	30.87
not exceeding 150 mm wide	-	-	-	m2	37.46
25 mm one coat coverings; felt isolating					
membrane; to concrete base; flat					
over 300 mm wide	-	-	-	m2	20.11
225 - 300 mm wide	-	-	-	m2	38.17
150 - 225 mm wide	-	-	-	m2	38.67
not exceeding 150 mm wide	-	-	-	m2	47.71
20 mm three coat skirtings to brickwork base					
not exceeding 150 mm girth	-	-	-	m	19.41
150 - 225 mm girth	-	-	-	m	22.64
225 - 300 mm girth	-	-	-	m	25.86
Mastic asphalt paving to BS 6577; black					
20 mm one coat coverings; felt isolating					
membrane; to concrete base; flat					
over 300 mm wide	-	-	-	m2	23.39
225 - 300 mm wide	-	-	-	m2	37.81
150 - 225 mm wide	-	-	-	m2	40.29
not exceeding 150 mm wide	-	-	-	m2	45.27
25 mm one coat coverings; felt isolating					
membrane; to concrete base; flat					
over 300 mm wide	-	-	-	m2	28.34
225 - 300 mm wide	-	-	-	m2	45.83
150 - 225 mm wide	-	-	-	m2	48.84
not exceeding 150 mm wide	-	-	-	m2	54.86
20 mm three coat skirtings to brickwork base					
not exceeding 150 mm girth	-	-	-	m	18.41
150 - 225 mm girth	-	-	-	m	21.77
225 - 300 mm girth	-	-	-	m	25.12
Mastic asphalt paving to BS 1451; red					
15 mm one coat coverings; felt isolating					
membrane; to concrete base; flat					
over 300 mm wide	-	-	-	m2	16.28
225 - 300 mm wide	-	-	-	m2	30.87
150 - 225 mm wide	-	-	-	m2	31.30
not exceeding 150 mm wide	-	-	-	m2	38.62
20 mm three coat skirtings to brickwork base					
not exceeding 150 mm girth	-	-	-	m	16.85
150 - 225 mm girth	-	-	-	m	19.70

M SURFACE FINISHES	Labour	Labour	Material		Total
Including overheads and profit at 12.50% & 5.00%	hours	£	£	Unit	rate £

M12 TROWELLED BITUMEN/RESIN/RUBBER LATEX FLOORING

Latex cement floor screeds; steel trowelled
Work to floors; level; to concrete base;
over 300 mm wide

3 mm thick; one coat	0.19	2.02	1.80	m2	3.82
5 mm thick; two coats	0.24	2.56	2.81	m2	5.37

Isocrete K screeds; steel trowelled
Work to floors; level; to concrete base;
over 300 mm wide

35 mm thick; plus polymer bonder coat	-	-	-	m2	9.08
40 mm thick	-	-	-	m2	9.60
45 mm thick	-	-	-	m2	10.13
50 mm thick	-	-	-	m2	10.71

Work to floors; to falls or cross-falls; to
concrete base; over 300 mm wide

55 mm (average) thick	-	-	-	m2	6.36
60 mm (average) thick	-	-	-	m2	7.24
65 mm (average) thick	-	-	-	m2	7.77
75 mm (average) thick	-	-	-	m2	8.82
90 mm (average) thick	-	-	-	m2	10.50

'Synthanite' floor screeds; steel trowelled
Work to floors; one coat; level; paper felt
underlay; to concrete base; over 300 mm wide

25 mm thick	0.55	5.86	12.82	m2	18.68
50 mm thick	0.66	7.03	20.35	m2	27.38
75 mm thick	0.77	8.20	24.48	m2	32.68

Bituminous lightweight insulating roof screeds
'Bit-Ag' or similar roof screed; to falls or
cross-falls; bitumen felt vapour barrier;
over 300 mm wide

75 mm (average) thick	-	-	-	m2	18.35
100 mm (average) thick	-	-	-	m2	21.41

BASIC PLASTER PRICES

	£		£
Plaster prices (£/tonne)			
BS 1191 Part 1; class A			
CB stucco	59.20		
BS 1191 Part 1; class B			
'Thistle' board	66.42	'Thistle' finish	67.89
BS 1191 Part 1; class C			
'Limelite' - backing	143.43	'Limelite' - renovating	191.25
- finishing	115.38		
Pre-mixed lightweight; BS 1191 Part 2			
'Carlite' - bonding	111.36	'Carlite' - finishing	88.88
- browning	109.01	- metal lathing	112.58
- browning HSB	119.18		
Projection 102.94			
Tyrolean 'Cullamix'	179.52		

Plastering sand £10.36/tonne

Sundries			
ceramic tile - adhesive (£/5L)	6.74	plasterboard nails (£/25kg)	41.17
- grout (£/kg)	0.74	scrim (£/100 m roll)	3.68
impact adhesive (£/5L)	17.67		

M SURFACE FINISHES Including overheads and profit at 12.50% & 5.00%	Labour hours	Labour £	Material £	Unit	Total rate £
	Labour hours	Labour £	Material £	Unit	Total rate £

M20 PLASTERED/RENDERED/ROUGHCAST COATINGS

Prepare and brush down; two coats of 'Unibond' or similar bonding agent; PC £14.88/5
Brick or block walls

over 300 mm wide	0.15	1.60	0.51	m2	2.11
Concrete walls or ceilings					
over 300 mm wide	0.12	1.28	0.40	m2	1.68

Cement and sand (1:3) beds and backings
10 mm work to walls; one coat; to brickwork or blockwork base

over 300 mm wide	-	-	-	m2	6.97
not exceeding 300 mm wide	-	-	-	m	4.19

12 mm work to walls; one coat; to brickwork or blockwork base

over 300 mm wide	-	-	-	m2	7.62
not exceeding 300 mm wide	-	-	-	m	4.61

13 mm work to walls; one coat; to brickwork or blockwork base; screeded

over 300 mm wide	-	-	-	m2	8.86
not exceeding 300 mm wide	-	-	-	m	5.02

15 mm work to walls; one coat; to brickwork or blockwork base

over 300 mm wide	-	-	-	m2	8.98
not exceeding 300 mm wide	-	-	-	m	5.37

Cement and sand (1:3); steel trowelled
13 mm work to walls; two coats; to brickwork or blockwork base

over 300 mm wide	-	-	-	m2	9.98
not exceeding 300 mm wide	-	-	-	m	5.91

16 mm work to walls; two coats; to brickwork or blockwork base

over 300 mm wide	-	-	-	m2	10.87
not exceeding 300 mm wide	-	-	-	m	6.50

19 mm work to walls; two coats; to brickwork or blockwork base

over 300 mm wide	-	-	-	m2	11.81
not exceeding 300 mm wide	-	-	-	m	7.09

Add to the above over 300 mm wide for first coat in water-repellent cement

	-	-	-	m2	2.07
finishing coat in coloured cement	-	-	-	m2	5.02

Cement-lime-sand (1:2:9); steel trowelled
19 mm work to walls; two coats; to brickwork or blockwork base

over 300 mm wide	-	-	-	m2	11.22
not exceeding 300 mm wide	-	-	-	m	6.73

Cement-lime-sand (1:1:6); steel trowelled
19 mm work to ceilings; three coats; to metal lathing base

over 300 mm wide	-	-	-	m2	15.95
not exceeding 300 mm wide	-	-	-	m	9.45

M SURFACE FINISHES Including overheads and profit at 5.00%	Labour hours	Labour £	Material £	Unit	Total rate £
Plaster; first and finishing coats of 'Carlite' pre-mixed lightweight plaster; steel trowelled					
13 mm work to walls; two coats; to brickwork or blockwork base (or 10 mm work to concrete base)					
over 300 mm wide	-	-	-	m2	7.68
over 300 mm wide; in staircase areas or plant rooms	-	-	-	m2	9.45
not exceeding 300 mm wide	-	-	-	m	4.73
13 mm work to isolated piers or columns; two					
coats over 300 mm wide	-	-	-	m2	12.40
not exceeding 300 mm wide	-	-	-	m	7.38
10 mm work to ceilings; two coats; to concrete base					
over 300 mm wide	-	-	-	m2	7.97
over 300 mm wide; 3.5 - 5 m high	-	-	-	m2	8.15
over 300 mm wide; in staircase areas or plant rooms	-	-	-	m2	9.45
not exceeding 300 mm wide	-	-	-	m	4.96
10 mm work to isolated beams; two coats; to concrete base					
over 300 mm wide	-	-	-	m2	11.81
over 300 mm wide; 3.5 - 5 m high	-	-	-	m2	12.70
not exceeding 300 mm wide	-	-	-	m	7.09
Plaster; one coat 'Snowplast' plaster; steel trowelled					
13 mm work to walls; one coat; to brickwork or blockwork base					
over 300 mm wide	-	-	-	m2	8.86
over 300 mm wide; in staircase areas or plant rooms	-	-	-	m2	10.04
not exceeding 300 mm wide	-	-	-	m	5.08
13 mm work to isolated columns; one coat					
over 300 mm wide	-	-	-	m2	12.40
not exceeding 300 mm wide	-	-	-	m	7.32
Plaster; first coat of cement and sand (1:3); finishing coat of 'Thistle' class B plaster; steel trowelled					
13 mm work to walls; two coats; to brickwork or blockwork base					
over 300 mm wide	-	-	-	m2	10.34
over 300 mm wide; in staircase areas or plant rooms	-	-	-	m2	11.81
not exceeding 300 mm wide	-	-	-	m	6.26
13 mm work to isolated columns; two coats					
over 300 mm wide	-	-	-	m2	13.88
not exceeding 300 mm wide	-	-	-	m	8.50
Plaster; first coat of cement-lime-sand (1:1:6); finishing coat of 'Sirapite' class B plaster; steel trowelled					
13 mm work to walls; two coats; to brickwork or blockwork base					
over 300 mm wide	-	-	-	m2	10.34
over 300 mm wide; in staircase areas or plant rooms	-	-	-	m2	11.93
not exceeding 300 mm wide	-	-	-	m	6.20
13 mm work to isolated columns; two coats					
over 300 mm wide	-	-	-	m2	14.18
not exceeding 300 mm wide	-	-	-	m	8.50

M SURFACE FINISHES Including overheads and profit at 5.00%	Labour hours	Labour £	Material £	Unit	Total rate £
M20 PLASTERED/RENDERED/ROUGHCAST COATINGS - cont'd					
Plaster; first coat of 'Limelite' renovating					
plaster; finishing coat of 'Limelite'					
finishing plaster; steel trowelled					
13 mm work to walls; two coats; to brickwork					
or blockwork base					
over 300 mm wide	-	-	-	m2	10.63
over 300 mm wide; in staricase areas or					
compartments under 4 m2	-	-	-	m2	12.76
not exceeding 300 mm wide	-	-	-	m	6.38
Dubbing out existing walls with undercoat					
plaster; average 6 mm thick					
over 300 mm wide	-	-	-	m2	4.26
not exceeding 300 mm wide	-	-	-	m	2.54
Dubbing out existing walls with undercoat					
plaster; average 12 mm thick					
over 300 mm wide	-	-	-	m2	5.76
not exceeding 300 mm wide	-	-	-	m	3.44
Plaster; one coat 'Thistle' projection					
plaster; steel trowelled					
13 mm work to walls; one coat; to brickwork					
or blockwork base					
over 300 mm wide	-	-	-	m2	11.22
over 300 mm wide; in staircase areas or					
plant rooms	-	-	-	m2	12.70
not exceeding 300 mm wide	-	-	-	m	6.79
6 mm work to isolated columns; one coat					
over 300 mm wide	-	-	-	m2	12.11
not exceeding 300 mm wide	-	-	-	m	7.09
Plaster; first, second and finishing coats					
of 'Carlite' pre-mixed lightweight plaster;					
steel trowelled					
13 mm work to ceilings; three coats to metal					
lathing base					
over 300 mm wide	-	-	-	m2	12.11
over 300 mm wide; in staircase areas or					
plant rooms	-	-	-	m2	13.58
not exceeding 300 mm wide	-	-	-	m	7.32
13 mm work to swept soffit of metal lathing					
arch former					
not exceeding 300 mm wide	-	-	-	m	18.31
300 - 400 mm wide	-	-	-	m	19.49
13 mm work to vertical face of metal					
lathing arch former					
not exceeding 0.5 m2 per side	-	-	-	nr	35.73
0.5 m2 - 1 m2 per side	-	-	-	nr	43.12
Tyrolean decorative rendering; 13 mm first					
coat of cement-lime-sand (1:1:6); finishing					
three coats of 'Cullamix' applied with					
approved hand operated machine external					
To walls; four coats; to brickwork or					
blockwork base					
over 300 mm wide	-	-	-	m2	25.40
not exceeding 300 mm wide	-	-	-	m	15.36
'Mineralite' decorative rendering; first					
coat of cement-lime-sand (1:0.5:4.5);					
finishing coat of 'Mineralite'; applied by					
Specialist Subcontractor					
17 mm work to walls; to brickwork or					
blockwork base					
over 300 mm wide	-	-	-	m2	42.45
not exceeding 300 mm wide	-	-	-	m	21.23
form 9 x 6 mm expansion joint and point					
in one part polysulphide mastic					m	7.31

M SURFACE FINISHES Including overheads and profit at 12.50% & 5.00%	Labour hours	Labour £	Material £	Unit	Total rate £
Plaster; one coat 'Thistle' board finish; steel trowelled **(prices included within plasterboard rates** 3 mm work to walls or ceilings; one coat; to plasterboard base					
over 300 mm wide	-	-	-	m2	5.32
over 300 mm wide; in staircase areas or plant rooms	-	-	-	m2	6.79
not exceeding 300 mm wide	-	-	-	m	3.25
Plaster; one coat 'Thistle' board finish; **steel trowelled 3 mm work to walls or** **ceilings; one coat on and including gypsum** **plasterboard; BS 1230; fixing with nails;** **3 mm joints filled with plaster and jute** **scrim cloth; to softwood base; plain grade** **baseboard or lath with rounded edges** 9.5 mm board to walls					
over 300 mm wide	0.55	7.63	2.80	m2	10.43
not exceeding 300 mm wide	0.28	3.88	0.89	m	4.77
9.5 mm board to walls; in staircase areas or plant rooms					
over 300 mm wide	0.66	9.16	2.80	m2	11.96
not exceeding 300 mm wide	0.33	4.58	0.89	m	5.47
9.5 mm board to isolated columns					
over 300 mm wide	0.73	10.13	2.82	m2	12.95
not exceeding 300 mm wide	0.36	4.99	0.89	m	5.88
9.5 mm board to ceilings					
over 300 mm wide	0.62	8.60	2.80	m2	11.40
over 300 mm wide; 3.5 - 5 m high	0.73	10.13	2.80	m2	12.93
not exceeding 300 mm wide	0.31	4.30	0.89	m	5.19
9.5 mm board to ceilings; in staircase areas or plant rooms					
over 300 mm wide	0.73	10.13	2.80	m2	12.93
not exceeding 300 mm wide	0.36	4.99	0.89	m	5.88
9.5 mm board to isolated beams					
over 300 mm wide	0.77	10.68	2.82	m2	13.50
not exceeding 300 mm wide	0.39	5.41	0.86	m	6.27
Add for 'Duplex' insulating grade	-	-	-	m2	0.50
12.5 mm board to walls					
over 300 mm wide	0.62	8.60	3.33	m2	11.93
not exceeding 300 mm wide	0.31	4.30	1.05	m	5.35
12.5 mm board to walls; in staircase areas or plant rooms					
over 300 mm wide	0.73	10.13	3.33	m2	13.46
not exceeding 300 mm wide	0.36	4.99	1.05	m	6.04
12.5 mm board to isolated columns					
over 300 mm wide	0.77	10.68	3.35	m2	14.03
not exceeding 300 mm wide	0.39	5.41	1.06	m	6.47
12.5 mm board to ceilings					
over 300 mm wide	0.66	9.16	3.33	m2	12.49
over 300 mm wide; 3.5 - 5 m high	0.77	10.68	3.33	m2	14.01
not exceeding 300 mm wide	0.33	4.58	1.05	m	5.63
12.5 mm board to ceilings; in staircase areas or plant rooms					
over 300 mm wide	0.77	10.68	3.33	m2	14.01
not exceeding 300 mm wide	0.39	5.41	1.05	m	6.46
12.5 mm board to isolated beams					
over 300 mm wide	0.84	11.65	3.35	m2	15.00
not exceeding 300 mm wide	0.42	5.83	1.06	m	6.89
Add for 'Duplex' insulating grade	-	-	-	m2	0.50

M SURFACE FINISHES Including overheads and profit at 12.50% & 5.00%	Labour hours	Labour £	Material £	Unit	Total rate £
M20 PLASTERED/RENDERED/ROUGHCAST COATINGS - cont'd					
Accessories					
'Expamet' render beads; white PVC nosings;					
to brickwork or blockwork base					
external stop bead; ref 1222	0.07	0.97	1.75	m	2.72
'Expamet' plaster beads; to brickwork or					
blockwork base					
angle bead; ref 550	0.08	1.11	0.48	m	1.59
architrave bead; ref 579	0.10	1.39	0.75	m	2.14
stop bead; ref 563	0.07	0.97	0.57	m	1.54
M22 SPRAYED MINERAL FIBRE COATINGS					
Prepare and apply by spray Mandolite P20					
fire protection on structural steel/metalwork					
16 mm (one hour) fire protection					
to walls and columns	-	-	-	m2	8.44
to ceilings and beams	-	-	-	m2	9.28
to isolated metalwork	-	-	-	m2	21.09
22 mm (one and a half hour) fire protection					
to walls and columns	-	-	-	m2	10.69
to ceilings and beams	-	-	-	m2	11.76
to isolated metalwork	-	-	-	m2	24.58
28 mm (two hour) fire protection					
to walls and columns	-	-	-	m2	11.81
to ceilings and beams	-	-	-	m2	12.99
to isolated metalwork	-	-	-	m2	25.99
52 mm (four hour) fire protection					
to walls and columns	-	-	-	m2	24.19
to ceilings and beams	-	-	-	m2	26.61
to isolated metalwork	-	-	-	m2	48.38
M30 METAL MESH LATHING/ANCHORED REINFORCEMENT					
FOR PLASTERED COATING					
Accessories					
Pre-formed galvanised expanded steel					
arch-frames; 'Truline' or similar; semi-					
circular; to suit walls up to 230 mm thick					
375 mm radius; for 800 mm opening;					
ref SC 750 PC £15.65	0.28	3.88	17.61	nr	21.49
425 mm radius; for 850 mm opening;					
ref SC 850 PC £16.61	0.28	3.88	18.69	nr	22.57
450 mm radius; for 900 mm opening;					
ref SC 900 PC £19.28	0.28	3.88	21.69	nr	25.57
600 mm radius; for 1200 mm opening;					
ref SC 1200 PC £23.99	0.28	3.88	26.99	nr	30.87
'Newlath' damp-free lathing or similar;					
PC £48.88/15; shot-fired to brick or block					
or concrete base					
Lining to walls					
over 300 mm wide	0.25	3.47	5.73	m2	9.19
not exceeding 300 mm wide	0.15	2.08	1.89	m	3.98
Lathing; Expamet 'BB' expanded metal					
lathing or similar; BS 1369; 50 mm laps					
6 mm mesh linings to ceilings; fixing with					
staples; to softwood base; over 300 mm wide					
ref BB263; 0.500 mm thick PC £2.33	0.33	4.58	2.76	m2	7.34
ref BB264; 0.675 mm thick PC £2.73	0.33	4.58	3.24	m2	7.81
6 mm mesh linings to ceilings; fixing with					
wire; to steelwork; over 300 mm wide					
ref BB263; 0.500 mm thick	0.35	4.85	2.84	m2	7.69
ref BB264; 0.675 mm thick	0.35	4.85	3.32	m2	8.17

M SURFACE FINISHES Including overheads and profit at 12.50% & 5.00%	Labour hours	Labour £	Material £	Unit	Total rate £
6 mm mesh linings to ceilings; fixings with wire; to steelwork; not exceeding 300 mm wide					
ref BB263; 0.500 mm thick	0.22	3.05	0.89	m	3.94
ref BB264; 0.675 mm thick	0.22	3.05	1.04	m	4.09
Raking cutting	0.22	3.05	0.90	m	3.95
Cutting and fitting around pipes; not exceeding 0.30 m girth	0.22	3.05	1.05	nr	4.10
Lathing; Expamet 'Riblath' or 'Spraylath' **stiffened expanded metal lathing or similar;** **50 mm laps** 10 mm mesh lining to walls; fixing with nails; to softwood base; over 300 mm wide					
Riblath ref 269; 0.30 mm thick PC £2.91	0.28	3.88	3.45	m2	7.33
Riblath ref 271; 0.50 mm thick PC £3.38	0.28	3.88	4.00	m2	7.88
Spraylath ref 273; 0.50 mm thick	0.28	3.88	4.89	m2	8.77
10 mm mesh lining to walls; fixing with nails; to softwood base; not exceeding 300 mm wide					
Riblath ref 269; 0.30 mm thick	0.17	2.36	1.09	m	3.45
Riblath ref 271; 0.50 mm thick	0.17	2.36	1.26	m	3.62
Spraylath ref 273; 0.50 mm thick	0.17	2.36	1.53	m	3.89
10 mm mesh lining to walls; fixing to brick or blockwork; over 300 mm wide					
Red-rib ref 274; 0.50 mm thick PC £3.73	0.22	3.05	6.89	m2	9.94
Stainless steel Riblath ref 267; 0.30 mm thick PC £8.19	0.22	3.05	12.15	m2	15.20
10 mm mesh lining to ceilings; fixing with wire; to steelwork; over 300 mm wide					
Riblath ref 269; 0.30 mm thick	0.35	4.85	3.56	m2	8.41
Riblath ref 271; 0.50 mm thick	0.35	4.85	4.11	m2	8.96
Spraylath ref 273; 0.50 mm thick	0.35	4.85	5.00	m2	9.85
Raking cutting	0.17	2.36	0.70	m	3.06
Cutting and fitting around pipes; not exceeding 0.30 m girth	0.06	0.83	-	nr	0.83
M31 FIBROUS PLASTER					
Fibrous plaster; fixing with screws; **plugging; countersinking; stopping; filling** **and pointing joints with plaster** 16 mm plain slab coverings to ceilings					
over 300 mm wide	-	-	-	m2	101.64
not exceeding 300 mm wide	-	-	-	m	32.76
Coves; not exceeding 150 mm girth					
per 25 mm girth	-	-	-	m	4.75
Coves; 150 - 300 mm girth					
per 25 mm girth	-	-	-	m	5.85
Cornices					
per 25 mm girth	-	-	-	m	5.85
Cornice enrichments					
per 25 mm girth; depending on degree of enrichments	-	-	-	m	7.17
Fibrous plaster; fixing with plaster wadding **filling and pointing joints with plaster; to** **steel base** 16 mm plain slab coverings to ceilings					
over 300 mm wide	-	-	-	m2	101.64
not exceeding 300 mm wide	-	-	-	m	32.76
16 mm plain casings to stanchions					
per 25 mm girth	-	-	-	m	2.85
16 mm plain casings to beams					
per 25 mm girth	-	-	-	m	2.85

M SURFACE FINISHES	Labour	Labour	Material		Total
Including overheads and profit at 12.50%	hours	£	£	Unit	rate £

M31 FIBROUS PLASTER - cont'd

Gyproc cove; fixing with adhesive; filling and pointing joints with plaster
Cove

	Labour hours	Labour £	Material £	Unit	Total rate £
125 mm girth	0.22	2.34	1.05	m	3.39
Angles	0.04	0.43	0.31	nr	0.74

ALTERNATIVE TILE MATERIALS

	£		£
Dennis Ruabon clay floor quarries (£/1000)			
Heather brown			
150x72x12.5 mm	171.81	194x94x12.5 mm	227.13
150x150x12.5 mm	226.21	194x194x12.5 mm	438.61
Red			
150x150x12.5 mm; square	159.74	150x150x12.5 mm; hexagonal	252.20
Rustic			
150x150x12.5 mm	235.17		
Daniel Platt heavy duty floor tiles (£/m2)			
'Ferrolite' flat; 150x150x9 mm			
black	11.25	steel	11.00
cream	11.00	red	9.45
'Ferrolite' flat; 150x150x12 mm			
black	14.93	chocolate	14.39
cream	13.19	red	13.59
'Ferrundum' anti-slip; 150x150x12 mm			
black	23.64	red	22.47
chocolate	23.13		
Langley wall and floor tiles (£/m2)			
'Buchtal' fine grain ceramic wall and floor tiles			
240x115x11 mm - series 1	26.99	194x144x11 mm - series 1	26.20
- series 2	29.27	- series 2	28.41
'Buchtal' rustic glazed wall and floor tiles			
240x115x11 mm - series 1	26.99	194x94x11 mm - series 1	26.31
- series 2	28.64	- series 2	27.88
- series 3	25.67	- series 3	25.09
'Hoganas' frostproof glazed ceramic wall facings			
196x96x8 mm	23.92		
'Hoganas' vitrified paviors			
215x105x18 mm - yellow	27.64	215x105x18 mm - brown	38.34
'Sinzag' glazed ceramic wall facings			
240x115x10 mm - series 1	25.11	240x52x10 mm - series 1	28.41
- series 2	33.26		
- series 3	54.05		
'Sinzag' vitrified ceramic floor tiles; 150x150x11/12 mm			
plain - grey/white speckled	27.52	textured - series 2	29.20
Marley floor tiles (£/m2); 300x300 mm			
'Econoflex' - series 1/2	2.65	anti-static	22.04
- series 4	3.14	'Travertine' - 2.5 mm	5.09
'Marleyflex' - 2.0 mm	3.41	'Vylon'	3.57
- 2.5 mm	3.99		

M SURFACE FINISHES Including overheads and profit at 12.50%	Labour hours	Labour £	Material £	Unit	Total rate £

M40 STONE/CONCRETE/QUARRY/CERAMIC TILING/ MOSAIC

Clay floor quarries; BS 6431; class 1;
Daniel Platt 'Crown' tiles or similar; level
bedding 10 mm thick and jointing in cement
and sand (1:3); butt joints straight both
ways; flush pointing with grout; to cement
and sand base

Work to floors; over 300 mm wide					
150 x 150 x 12.5 mm; red PC £225.74/1000	0.88	9.38	13.03	m2	22.41
150 x 150 x 12.5 mm; brown PC £354.60/1000	0.88	9.38	19.72	m2	29.10
200 x 200 x 19 mm; brown PC £864.89/1000	0.72	7.67	26.75	m2	34.42
Works to floors; in staircase areas or plant rooms					
150 x 150 x 12.5 mm; red	0.99	10.55	13.03	m2	23.58
150 x 150 x 12.5 mm; brown	0.99	10.55	19.72	m2	30.27
200 x 200 x 19 mm; brown	0.83	8.84	26.75	m2	35.59
Work to floors; not exceeding 300 mm wide					
150 x 150 x 12.5 mm; red	0.44	4.69	4.68	m	9.37
150 x 150 x 12.5 mm; brown	0.44	4.69	7.07	m	11.76
200 x 200 x 19 mm; brown	0.36	3.84	10.10	m	13.94
Fair square cutting against flush edges of existing finishings	0.11	1.17	0.60	m	1.77
Raking cutting	0.22	2.34	0.88	m	3.22
Cutting around pipes; not exceeding 0.30 m girth	0.17	1.81	-	nr	1.81
Extra for cutting and fitting into recessed manhole covers 600 x 600 mm; lining up with adjoining work	1.10	11.72	-	nr	11.72
Work to sills; 150 mm wide; rounded edge tiles					
150 x 150 x 12.5 mm; red PC £422.07/1000	0.36	3.84	3.73	m	7.57
150 x 150 x 12.5 mm; brown PC £454.95/1000	0.36	3.84	4.00	m	7.84
Fitted end	0.17	1.81	-	m	1.81
Coved skirtings; 150 mm high; rounded top edge					
150 x 150 x 12.5 mm; red PC £422.07/1000	0.28	2.98	3.73	m	6.71
150 x 150 x 12.5 mm; brown PC £454.95/1000	0.28	2.98	4.00	m	6.98
Ends	0.06	0.64	-	nr	0.64
Angles	0.17	1.81	1.67	nr	3.48

Glazed ceramic wall tiles; BS 6431; fixing
with adhesive; butt joints straight both m
ways; flush pointing with white grout; to
plaster base

Work to walls; over 300 mm wide					
108 x 108 x 4 mm PC £15.70	0.99	13.73	20.75	m2	34.48
152 x 152 x 5.5 mm; white PC £9.94	0.66	9.16	13.96	m2	23.12
152 x 152 x 5.5 mm; light colours PC £13.85	0.66	9.16	18.57	m2	27.73
152 x 152 x 5.5 mm; dark colours PC £17.48	0.66	9.16	22.86	m2	32.02
Extra for RE or REX tile	-	-	-	nr	0.08
Work to walls; in staircase areas or plant rooms					
152 x 152 x 5.5 mm; white	0.74	10.26	13.96	m2	24.22
Work to walls; not exceeding 300 mm wide					
108 x 108 x 4 mm	0.50	6.94	6.49	m	13.43
152 x 152 x 5.5 mm; white	0.33	4.58	4.36	m	8.94
152 x 152 x 5.5 mm; light colours	0.33	4.58	5.81	m	10.39
152 x 152 x 5.5 mm; dark colours	0.33	4.58	7.15	m	11.73
Cutting around pipes; not exceeding 0.30 m girth	0.11	1.53	-	nr	1.53

M SURFACE FINISHES Including overheads and profit at 12.50% & 5.00%	Labour hours	Labour £	Material £	Unit	Total rate £
M40 STONE/CONCRETE/QUARRY/CERAMIC TILING/ **MOSAIC - cont'd**					
Glazed ceramic wall tiles; BS 6431; fixing **with adhesive; butt joints straight both m** **ways; flush pointing with white grout; to** **plaster base - cont'd**					
Work to sills; 150 mm wide; rounded edge tiles					
108 x 108 x 4 mm	0.33	4.58	3.30	m	7.88
152 x 152 x 5.5 mm; white	0.28	3.88	2.22	m	6.10
Fitted end	0.11	1.53	-	nr	1.53
198 x 64.5 x 6 mm Langley 'Project Plus' **wall tiles; fixing with adhesive; butt joints** **straight both ways; flush pointing with** **white grout; to plaster base**					
Work to walls					
over 300 mm wide PC £20.30	2.00	27.74	26.18	m2	53.92
not exceeding 300 mm wide	0.77	10.68	7.85	m	18.53
Glazed ceramic floor tiles; level; bedding **10 mm thick and jointing in cement and sand** **(1:3); butt joints straight both ways; flush** **pointing with white grout; to cement and** **sand base**					
Work to floors; over 300 mm wide					
200 x 200 x 8 mm PC £26.06	0.66	9.16	32.07	m2	41.23
Work to floors; not exceeding 300 mm wide					
200 x 200 x 8 mm	0.33	4.58	10.16	m	14.74
M42 WOOD BLOCK/COMPOSITION BLOCK/PARQUET **FLOORING**					
Wood block; Vigers, Stevens & Adams **'Feltwood'; 7.5 mm thick; level; fixing with** **adhesive; to cement and sand base**					
Work to floors; over 300 mm wide					
Iroko	0.55	5.86	11.34	m2	17.20
Wood blocks 25 mm thick; tongued and grooved **joints; herringbone pattern; level; fixing** **with adhesive; to cement and sand base**					
Work to floors; over 300 mm wide					
Iroko	-	-	-	m2	42.69
Merbau	-	-	-	m2	42.69
French Oak	-	-	-	m2	49.16
American Oak	-	-	-	m2	40.11
Fair square cutting against flush edges of existing finishings	-	-	-	m	4.01
Extra for cutting and fitting into recessed duct covers 450 mm wide; lining up with adjoining work	-	-	-	m	7.76
Cutting around pipes; not exceeding 0.30 m girth	-	-	-	nr	0.39
Extra for cutting and fitting into recessed manhole covers 600 x 600 mm; lining up with adjoining work	-	-	-	nr	11.64
Add to wood block flooring over 300 mm wide for					
sanding; one coat sealer; one coat wax polish	-	-	-	m2	3.88
sanding; two coats sealer; buffing with steel wool	-	-	-	m2	2.59
sanding; three coats polyurethane lacquer; buffing down between coats	-	-	-	m2	2.59

M SURFACE FINISHES
Including overheads and profit at 12.50% & 5.00%

ALTERNATIVE FLEXIBLE SHEET MATERIALS (£/m2)

	£		£
Nairn flooring			
'Armourfloor' - 3.2 mm	7.99	'Amourflex' - 2.0 mm	5.21
'Armourflex' - 3.2 mm	7.25		
Marley sheet flooring			
'HD Acoustic'	6.88	'HD Safetread'	10.24
'Format +'	8.35	'Vynatred'	4.78

	Labour hours	Labour £	Material £	Unit	Total rate £
M50 RUBBER/PLASTICS/CORK/LINO/CARPET TILING/ SHEETING					
Linoleum sheet; BS 810; Nairn Floors or similar; level; fixing with adhesive; butt joints; to cement and sand base					
Work to floors; over 300 mm wide					
3.2 mm; plain	0.44	4.69	9.44	m2	**14.13**
3.2 mm; marbled	0.44	4.69	9.66	m2	**14.35**
Vinyl sheet; Altro 'Safety' range or similar with welded seams; level; fixing with adhesive; to cement and sand base					
Work to floors; over 300 mm wide					
2 mm thick; Marine T20	0.66	7.03	10.97	m2	**18.00**
2.5 mm thick; Classic D25	0.77	8.20	14.45	m2	**22.65**
3.5 mm thick; stronghold	0.88	9.38	19.32	m2	**28.70**
Vinyl sheet; heavy duty; Marley 'HD' or similar; level; with welded seams; fixing with adhesive; level; to cement and sand base					
Work to floors; over 300 mm wide					
2 mm thick	0.36	3.84	8.83	m2	**12.67**
2.5 mm thick	0.36	3.84	8.80	m2	**12.64**
2 mm skirtings					
75 mm high	0.11	1.17	0.92	m	**2.09**
100 mm high	0.13	1.38	1.09	m	**2.47**
150 mm high	0.17	1.81	1.24	m	**3.05**
Vinyl sheet; 'Armstrong Rhino Contract'; level; with welded seams; fixing with adhesive; to cement and sand base					
Work to floors; over 300 mm wide					
2.5 mm thick	-	-	-	m2	**12.83**
Vinyl sheet; Marmoleum'; level; with welded seams; fixing with adhesive; to cement and sand base					
Work to floors; over 300 mm wide					
2.5 mm thick	-	-	-	m2	**12.26**
Vinyl tiles; 'Accoflex'; level; fixing with adhesive; butt joints; straight both ways; to cement and sand base					
Work to floors; over 300 mm wide					
300 x 300 x 2.0 mm	-	-	-	m2	**6.41**

M SURFACE FINISHES Including overheads and profit at 12.50% & 5.00%	Labour hours	Labour £	Material £	Unit	Total rate £
M50 RUBBER/PLASTICS/CORK/LINO/CARPET TILING/ SHEETING - cont'd					
Vinyl semi-flexible tiles; Marley 'Marleyflex' or similar; level; fixing with adhesive; butt joints straight both ways; to cement and sand base					
Work to floors; over 300 mm wide					
250 x 250 x 2.0 mm	0.28	2.98	4.71	m2	7.69
250 x 250 x 2.5 mm	0.28	2.98	5.39	m2	8.37
Vinyl tiles; anti-static; level; fixing with adhesive; butt joints straight both ways; to cement and sand base					
Work to floors; over 300 mm wide					
457 x 457 x 2 mm	0.50	5.33	27.03	m2	32.36
Vinyl tiles; 'Polyflex'; level; fixing with adhesive; butt joints; straight both ways;to cement and sand base					
Work to floors; over 300 mm wide					
300 x 300 x 1.5 mm	-	-	-	m2	6.75
300 x 300 x 2.0 mm	-	-	-	m2	7.09
Vinyl tiles; 'Marley 'HD''; level; fixing with adhesive; butt joints; straight both ways; to cement and sand base					
Work to floors; over 300 mm wide					
300 x 300 x 2.0 mm	-	-	-	m2	8.92
Thermoplastic tiles; Marley 'Econoflex' or similar; level; fixing with adhesive; butt joints straight both ways; to cement and sand base					
Work to floors; over 300 mm wide					
250 x 250 x 2.0 mm; (series 2)	0.25	2.66	3.80	m2	6.46
250 x 250 x 2.0 mm; (series 4)	0.25	2.66	4.39	m2	7.05
Linoleum tiles; BS 810; Nairn Floors or similar; level; fixing with adhesive; butt joints straight both ways; to cement and sand base					
Work to floors; over 300 mm wide					
3.2 mm thick (marbled patterns)	0.33	3.52	10.57	m2	14.09
Cork tiles; Wicanders 'Corktile' or similar; level fixing with adhesive; butt joints straight both ways; to cement and sand base					
Work to floors; over 300 mm wide					
300 x 300 x 3.2 mm	0.44	4.69	7.00	m2	11.69
Rubber studded tiles; Altro 'Mondopave' or similar; level; fixing with adhesive; butt joints; straight; to cement and sand base					
Work to floors; over 300 mm wide					
500 x 500 x 2.5 mm; type GS; black	0.66	7.03	17.22	m2	24.25
500 x 500 x 4 mm; type BT; black	0.66	7.03	21.86	m2	28.89
Work to landings; over 300 mm wide					
500 x 500 x 4 mm; type BT; black	0.88	9.38	21.86	m2	31.24
4 mm thick to tread					
275 mm wide	0.55	5.86	7.08	m	12.94
4 mm thick to riser					
185 mm wide	0.66	7.03	4.89	m	11.92

M SURFACE FINISHES Including overheads and profit at 12.50%		Labour hours	Labour £	Material £	Unit	Total rate £
Sundry floor sheeting underlays For floor finishings; over 300 mm wide building paper to BS 1521; class A;						
75 mm lap		0.07	0.75	0.74	m2	1.49
3.2 mm thick hardboard		0.22	2.34	1.18	m2	3.52
6.0 mm thick plywood		0.33	3.52	3.85	m2	7.37
Stair nosings 'Ferodo' or equivalent; light duty hard aluminium alloy stair tread nosings; plugged and screwed to concrete						
type SD1	PC £5.24	0.28	2.98	6.48	m	9.46
type SD2	PC £7.24	0.33	3.52	8.85	m	12.37
'Ferodo' or equivalent; heavy duty aluminium alloy stair tread nosings; plugged and screwed to concrete						
type HD1	PC £6.20	0.33	3.52	7.62	m	11.14
type HD2	PC £8.63	0.39	4.15	10.49	m	14.64
Nylon needlepunch tiles; 'Marleyflex' or similar level; fixing with adhesive; to cement and sand base Work to floors						
over 300 mm wide		0.24	2.56	7.82	m2	10.38
Heavy duty carpet tiles; 'Heuga 581 Olympic' or similar; PC £15.40/m2; one coat floor sealer cement and sand base Work to floors						
over 300 mm wide		0.44	4.69	20.39	m2	25.08
Nylon needlepunch carpet; 'Marleyflex' or similar; fixing; with adhesive; level; to cement Work to floors						
over 300 mm wide		0.33	3.52	5.40	m2	8.92
M51 EDGE FIXED CARPETING						
Fitted carpeting; Wilton wool/nylon; 80/20 velvet pile; heavy domestic plain; PC £35.00/m2 Work to floors						
over 300 mm wide		0.55	5.86	44.06	m2	49.92
Work to treads and risers						
over 300 mm wide		1.10	11.72	44.06	m2	55.78
Raking cutting		0.09	0.96	3.04	m2	4.00
Underlay to carpeting; PC £3.00/m2 Work to floors						
over 300 mm wide		0.09	0.96	3.54	m2	4.50
Sundries Carpet gripper fixed to floor; standard edging		0.06	0.64	0.31	m	0.95
M52 DECORATIVE PAPERS/FABRICS						
Lining paper; PC £0.84/roll; roll; and hanging Plaster walls or columns						
over 300 mm girth		0.21	1.56	0.29	m2	1.85
Plaster ceilings or beams						
over 300 mm girth		0.26	1.93	0.29	m2	2.22
Decorative vinyl wallpaper; PC £4.80/roll; roll; and hanging Plaster walls or columns						
over 300 mm girth		0.26	1.93	1.27	m2	3.20

M SURFACE FINISHES Including overheads and profit at 12.50%	Labour hours	Labour £	Material £	Unit	Total rate £
M60 PREPARATION OF EXISTING SURFACES **- INTERNALLY**					
NOTE: The prices for preparation given hereunder assume that existing surfaces are in 'fair condition' and should be increased for badly dilapidated surfaces.					
Wash down walls; cut out and make good cracks					
emulsion painted surfaces; including					
bringing forward bare patches	0.10	0.74	-	m2	0.74
gloss painted surfaces	0.08	0.59	-	m2	0.59
Wash down ceilings; cut out and make good **cracks**					
distempered surfaces	0.12	0.89	-	m2	0.89
emulsion painted surfaces; including					
bringing forward bare patches	0.14	1.04	-	m2	1.04
gloss painted surfaces	0.12	0.89	-	m2	0.89
Wash down plaster moulded cornices; cut out **and make good cracks**					
distempered surfaces	0.18	1.34	-	m2	1.34
emulsion painted surfaces; including					
bringing forward bare patches	0.21	1.56	-	m2	1.56
Wash and rub down iron or steel surfaces; **bringing forward**					
General surfaces					
over 300 mm girth	0.16	1.19	-	m2	1.19
isolated surfaces not exceeding 300 mm girth	0.07	0.52	-	m	0.52
isolated surfaces not exceeding 0.50 m2	0.12	0.89	-	nr	0.89
Windows and the like					
panes over 1.00 m2	0.16	1.19	-	m2	1.19
panes 0.50 - 1.00 m2	0.18	1.34	-	m2	1.34
panes 0.10 - 0.50 m2	0.21	1.56	-	m2	1.56
panes not exceeding 0.10 m2	0.27	2.01	-	m2	2.01
Wash and rub down wood surfaces; prime bare **patches; bringing forward**					
General surfaces					
over 300 mm girth	0.25	1.86	-	m2	1.86
isolated surfaces not exceeding 300 mm girth	0.10	0.74	-	m	0.74
isolated surfaces not exceeding 0.50 m2	0.19	1.41	-	nr	1.41
Windows and the like					
panes over 1.00 m2	0.25	1.86	-	m2	1.86
panes 0.50 - 1.00 m2	0.29	2.16	-	m2	2.16
panes 0.10 - 0.50 m2	0.33	2.45	-	m2	2.45
panes not exceeding 0.10 m2	0.42	3.12	-	m2	3.12
Wash down and remove paint with chemical **stripper from iron, steel or wood surfaces**					
General surfaces					
over 300 mm girth	0.75	5.58	-	m2	5.58
isolated surfaces not exceeding 300 mm girth	0.33	2.45	-	m	2.45
isolated surfaces not exceeding 0.50 m2	0.56	4.16	-	nr	4.16
Windows and the like					
panes over 1.00 m2	0.96	7.14	-	m2	7.14
panes 0.50 - 1.00 m2	1.10	8.18	-	m2	8.18
panes 0.10 - 0.50 m2	1.28	9.52	-	m2	9.52
panes not exceeding 0.10 m2	1.60	11.90	-	m2	11.90

M SURFACE FINISHES Including overheads and profit at 12.50%	Labour hours	Labour £	Material £	Unit	Total rate £
Burn off and rub down to remove paint from **iron, steel or wood surfaces**					
General surfaces					
over 300 mm girth	0.93	6.92	0.18	m2	7.10
isolated surfaces not exceeding 300 mm girth	0.42	3.12	0.07	m	3.19
isolated surfaces not exceeding 0.50 m2	0.70	5.21	0.07	nr	5.28
Windows and the like					
panes over 1.00 m2	1.20	8.92	-	m2	8.92
panes 0.50 - 1.00 m2	1.38	10.26	-	m2	10.26
panes 0.10 - 0.50 m2	1.60	11.90	-	m2	11.90
panes not exceeding 0.10 m2	2.00	14.87	-	m2	14.87

BASIC PAINT PRICES (£/5 Litre tin)

	£		£		£
Bituminous	10.08				
Emulsion					
matt	9.65	silk	11.68		
Knotting solution	23.83				
Oil					
gloss	13.65	undercoat	13.01		
Primer/undercoats					
acrylic	11.35	chlorinated			
alkali-resisting	13.50	rubber zinc	17.16	red oxide	17.63
aluminium	15.63	plaster	19.79	wood	12.64
calcium plumbate	22.55	red lead	26.59	zinc phosphate	14.41
Road marking	22.13				

	£		£
Special paints			
aluminium	20.64	fire retardant	
anti-condensation	17.00	- undercoat	26.15
chlorinated rubber		- top coat	35.15
- alkali-resisting	17.84	heat resisting	25.80
- finish	13.73		
Stains and preservatives			
creosote	3.15	Solignum - 'Architectural'	20.60
Cuprinol 'clear'	9.77	- brown	7.55
linseed oil	14.60	- cedar	11.01
oil varnish stain	12.25	- clear	10.67
Sadolin - 'Holdex'	26.61	- green	10.18
- 'Classic'	22.95	- hort. green	8.81
- 'Sadovac 35'	11.55	'Multiplus'	11.81
Varnishes			
polyurethene	15.54	yacht	15.75

NOTE: The following prices include for preparing surfaces. Painting woodwork also includes for knotting prior to applying the priming coat and for all stopping of nail holes etc.	Labour hours	Labour £	Material £	Unit	Total rate £
M60 PAINTING/CLEAR FINISHING					
One coat primer; PC £12.64/5L; on **wood surfaces before fixing**					
General surfaces					
over 300 mm girth	0.16	1.19	0.27	m2	1.46
isolated surfaces not exceeding 300 mm girth	0.06	0.45	0.08	m	0.53
isolated surfaces not exceeding 0.50 m2	0.12	0.89	0.13	nr	1.02

M SURFACE FINISHES Including overheads and profit at 12.50%	Labour hours	Labour £	Material £	Unit	Total rate £
M60 PAINTING/CLEAR FINISHING - cont'd					
One coat polyurethene sealer; PC £15.54/5L; on wood surfaces before fixing					
General surfaces					
over 300 mm girth	0.19	1.41	0.28	m2	**1.69**
isolated surfaces not exceeding 300 mm girth	0.07	0.52	0.09	m	**0.61**
isolated surfaces not exceeding 0.50 m2	0.14	1.04	0.14	nr	**1.18**
One coat clear wood preservative; PC £9.77/5L; on wood surfaces before fixing					
General surfaces					
over 300 mm girth	0.14	1.04	0.24	m2	**1.28**
isolated surfaces not exceeding 300 mm girth	0.05	0.37	0.08	m	**0.45**
isolated surfaces not exceeding 0.50 m2	0.11	0.82	0.12	nr	**0.94**
Two coats emulsion paint; PC £9.65/5L;					
Brick or block walls					
over 300 mm girth	0.21	1.56	0.43	m2	**1.99**
Cement render or concrete					
over 300 mm girth	0.19	1.41	0.33	m2	**1.74**
isolated surfaces not exceeding 300 mm girth	0.08	0.59	0.10	m	**0.69**
Plaster walls or plaster/plasterboard ceilings					
over 300 mm girth	0.18	1.34	0.33	m2	**1.67**
over 300 mm girth; in multi-colours	0.28	2.08	0.38	m2	**2.46**
over 300 mm girth; in staircase areas	0.21	1.56	0.33	m2	**1.89**
cutting in edges on flush surfaces	0.08	0.59	-	m	**0.59**
Plaster/plasterboard ceilings					
over 300 mm girth; 3.5 - 5.0 m high	0.20	1.49	0.33	m2	**1.82**
One mist and two coats emulsion paint					
Brick or block walls					
over 300 mm girth	0.24	1.78	0.60	m2	**2.38**
Cement render or concrete					
over 300 mm girth	0.21	1.56	0.48	m2	**2.04**
isolated surfaces not exceeding 300 mm girth	0.09	0.67	0.13	m	**0.80**
Plaster walls or plaster/plasterboard ceilings					
over 300 mm girth	0.21	1.56	0.43	m2	**1.99**
over 300 mm girth; in multi-colours	0.32	2.38	0.49	m2	**2.87**
over 300 mm girth; in staircase areas	0.24	1.78	0.43	m2	**2.21**
cutting in edges on flush surfaces	0.12	0.89	-	m	**0.89**
Plaster/plasterboard ceilings					
over 300 mm girth; 3.5 - 5.0 m high	0.23	1.71	0.43	m2	**2.14**
Textured plastic, 'readymix' finish; PC £0.00/15L					
Plasterboard ceilings					
over 300 mm girth	0.26	1.93	1.22	m2	**3.15**
Concrete walls or ceilings					
over 300 mm girth	0.28	2.08	1.37	m2	**3.45**
Touch up primer; one undercoat and one finishing coat of gloss oil paint; PC £13.65/5L; on wood surfaces					
General surfaces					
over 300 mm girth	0.34	2.53	0.50	m2	**3.03**
isolated surfaces not exceeding 300 mm girth	0.14	1.04	0.16	m	**1.20**
isolated surfaces not exceeding 0.50 m2	0.25	1.86	0.25	nr	**2.11**
Windows and the like					
panes over 1.00 m2	0.34	2.53	0.19	m2	**2.72**
panes 0.50 - 1.00 m2	0.39	2.90	0.25	m2	**3.15**
panes 0.10 - 0.50 m2	0.45	3.35	0.31	m2	**3.66**
panes not exceeding 0.10 m2	0.56	4.16	0.38	m2	**4.54**

M SURFACE FINISHES Including overheads and profit at 12.50%	Labour hours	Labour £	Material £	Unit	Total rate £
Touch up primer; two undercoats and one **finishing coat of gloss oil paint; on** **wood surfaces**					
General surfaces					
over 300 mm girth	0.47	3.50	0.74	m2	4.24
isolated surfaces not exceeding 300 mm girth	0.19	1.41	0.23	m	1.64
isolated surfaces not exceeding 0.50 m2	0.35	2.60	0.37	nr	2.97
Windows and the like					
panes over 1.00 m2	0.47	3.50	0.28	m2	3.78
panes 0.50 - 1.00 m2	0.55	4.09	0.37	m2	4.46
panes 0.10 - 0.50 m2	0.63	4.68	0.46	m2	5.14
panes not exceeding 0.10 m2	0.79	5.87	0.55	m2	6.42
Knot; one coat primer; stop; one undercoat **and one finishing coat of gloss oil paint;** **on wood surfaces**					
General surfaces					
over 300 mm girth	0.49	3.64	0.82	m2	4.46
isolated surfaces not exceeding 300 mm girth	0.20	1.49	0.26	m	1.75
isolated surfaces not exceeding 0.50 m2	0.37	2.75	0.41	nr	3.16
Windows and the like					
panes over 1.00 m2	0.49	3.64	0.31	m2	3.95
panes 0.50 - 1.00 m2	0.58	4.31	0.41	m2	4.72
panes 0.10 - 0.50 m2	0.68	5.06	0.51	m2	5.57
panes not exceeding 0.10 m2	0.84	6.25	0.61	m2	6.86
Knot; one coat primer; stop; two undercoats **and one finishing coat of gloss oil paint;** **on wood surfaces**					
General surfaces					
over 300 mm girth	0.63	4.68	1.06	m2	5.74
isolated surfaces not exceeding 300 mm girth	0.25	1.86	0.33	m	2.19
isolated surfaces not exceeding 0.50 m2	0.47	3.50	0.53	nr	4.03
Windows and the like					
panes over 1.00 m2	0.63	4.68	0.39	m2	5.07
panes 0.50 - 1.00 m2	0.73	5.43	0.53	m2	5.96
panes 0.10 - 0.50 m2	0.84	6.25	0.66	m2	6.91
panes not exceeding 0.10 m2	1.05	7.81	0.79	m2	8.60
One coat primer; one undercoat and one **finishing coat of gloss oil paint**					
Plaster surfaces					
over 300 mm girth	0.44	3.27	0.93	m2	4.20
One coat primer; two undercoats and one **finishing coat of gloss oil paint**					
Plaster surfaces					
over 300 mm girth	0.58	4.31	1.16	m2	5.47
Touch up primer; one undercoat and one **finishing coat of gloss paint; PC £13.65/5L;** **on iron or steel surfaces**					
General surfaces					
over 300 mm girth	0.34	2.53	0.48	m2	3.01
isolated surfaces not exceeding 300 mm girth	0.14	1.04	0.15	m	1.19
isolated surfaces not exceeding 0.50 m2	0.25	1.86	0.24	nr	2.10
Windows and the like					
panes over 1.00 m2	0.34	2.53	0.18	m2	2.71
panes 0.50 - 1.00 m2	0.39	2.90	0.24	m2	3.14
panes 0.10 - 0.50 m2	0.45	3.35	0.30	m2	3.65
panes not exceeding 0.10 m2	0.56	4.16	0.36	m2	4.52
Structural steelwork					
over 300 mm girth	0.38	2.83	0.48	m2	3.31
Members of roof trusses					
over 300 mm girth	0.50	3.72	0.51	m2	4.23

M SURFACE FINISHES Including overheads and profit at 12.50%	Labour hours	Labour £	Material £	Unit	Total rate £
M60 PAINTING/CLEAR FINISHING - cont'd					
Touch up primer; one undercoat and one **finishing coat of gloss paint; PC £13.65/5L;** **on iron or steel surfaces - cont'd**					
Ornamental railings and the like; each side measured overall					
over 300 mm girth	0.58	4.31	0.54	m2	4.85
Iron or steel radiators					
over 300 mm girth	0.34	2.53	0.51	m2	3.04
Pipes or conduits					
over 300 mm girth	0.50	3.72	0.51	m2	4.23
not exceeding 300 mm girth	0.20	1.49	0.15	m	1.64
Touch up primer; two undercoats and one **finishing coat of gloss oil paint; on** **iron or steel surfaces**					
General surfaces					
over 300 mm girth	0.47	3.50	0.71	m2	4.21
isolated surfaces not exceeding 300 mm girth	0.19	1.41	0.22	m	1.63
isolated surfaces not exceeding 0.50 m2	0.35	2.60	0.36	nr	2.96
Windows and the like					
panes over 1.00 m2	0.47	3.50	0.27	m2	3.77
panes 0.50 - 1.00 m2	0.55	4.09	0.36	m2	4.45
panes 0.10 - 0.50 m2	0.63	4.68	0.45	m2	5.13
panes not exceeding 0.10 m2	0.79	5.87	0.52	m2	6.39
Structural steelwork					
over 300 mm girth	0.54	4.02	0.71	m2	4.73
Members of roof trusses					
over 300 mm girth	0.71	5.28	0.76	m2	6.04
Ornamental railings and the like; each side measured overall					
over 300 mm girth	0.81	6.02	0.80	m2	6.82
Iron or steel radiators					
over 300 mm girth	0.47	3.50	0.76	m2	4.26
Pipes or conduits					
over 300 mm girth	0.71	5.28	0.76	m2	6.04
not exceeding 300 mm girth	0.28	2.08	0.22	m	2.30
One coat primer; one undercoat and one **finishing coat of gloss oil paint; on** **iron or steel surfaces**					
General surfaces					
over 300 mm girth	0.44	3.27	0.48	m2	3.75
isolated surfaces not exceeding 300 mm girth	0.18	1.34	0.15	m	1.49
isolated surfaces not exceeding 0.50 m2	0.34	2.53	0.24	nr	2.77
Windows and the like					
panes over 1.00 m2	0.44	3.27	0.18	m2	3.45
panes 0.50 - 1.00 m2	0.50	3.72	0.24	m2	3.96
panes 0.10 - 0.50 m2	0.59	4.39	0.30	m2	4.69
panes not exceeding 0.10 m2	0.73	5.43	0.36	m2	5.79
Structural steelwork					
over 300 mm girth	0.49	3.64	0.48	m2	4.12
Members of roof trusses					
over 300 mm girth	0.66	4.91	0.51	m2	5.42
Ornamental railings and the like; each side measured overall					
over 300 mm girth	0.76	5.65	0.54	m2	6.19
Iron or steel radiators					
over 300 mm girth	0.44	3.27	0.51	m2	3.78
Pipes or conduits					
over 300 mm girth	0.66	4.91	0.51	m2	5.42
not exceeding 300 mm girth	0.26	1.93	0.15	m	2.08

M SURFACE FINISHES Including overheads and profit at 12.50%	Labour hours	Labour £	Material £	Unit	Total rate £
One coat primer; two undercoats and one **finishing coat of gloss oil paint; on** **iron or steel surfaces**					
General surfaces					
over 300 mm girth	0.58	4.31	0.71	m2	5.02
isolated surfaces not exceeding 300 mm girth	0.23	1.71	0.22	m	1.93
isolated surfaces not exceeding 0.50 m2	0.42	3.12	0.36	nr	3.48
Windows and the like					
panes over 1.00 m2	0.58	4.31	0.27	m2	4.58
panes 0.50 - 1.00 m2	0.68	5.06	0.36	m2	5.42
panes 0.10 - 0.50 m2	0.79	5.87	0.45	m2	6.32
panes not exceeding 0.10 m2	0.94	6.99	0.54	m2	7.53
Structural steelwork					
over 300 mm girth	0.65	4.83	0.71	m2	5.54
Members of roof trusses					
over 300 mm girth	0.86	6.40	0.76	m2	7.16
Ornamental railings and the like; each side measured overall					
over 300 mm girth	0.98	7.29	0.80	m2	8.09
Iron or steel radiators					
over 300 mm girth	0.58	4.31	0.76	m2	5.07
Pipes or conduits					
over 300 mm girth	0.87	6.47	0.76	m2	7.23
not exceeding 300 mm girth	0.35	2.60	0.22	m	2.82
Two coats of bituminous paint; PC £10.08/5L; **on iron or steel surfaces**					
General surfaces					
over 300 mm girth	0.42	3.12	0.36	m2	3.48
Inside of galvanized steel cistern					
over 300 mm girth	0.63	4.68	0.36	m2	5.04
Two coats of boiled linseed oil; PC £14.60/5; **on hardwood surfaces**					
General surfaces					
over 300 mm girth	0.32	2.38	1.15	m2	3.53
isolated surfaces not exceeding 300 mm girth	0.13	0.97	0.35	m	1.32
isolated surfaces not exceeding 0.50 m2	0.24	1.78	0.58	nr	2.36
Two coats polyurethane; PC £15.54/5L; **on wood surfaces**					
General surfaces					
over 300 mm girth	0.32	2.38	0.56	m2	2.94
isolated surfaces not exceeding 300 mm girth	0.13	0.97	0.17	m	1.14
isolated surfaces not exceeding 0.50 m2	0.23	1.71	0.28	nr	1.99
Three coats polyurethane; on wood surfaces					
General surfaces					
over 300 mm girth	0.47	3.50	0.84	m2	4.34
isolated surfaces not exceeding 300 mm girth	0.19	1.41	0.26	m	1.67
isolated surfaces not exceeding 0.50 m2	0.36	2.68	0.42	nr	3.10
One undercoat; PC £26.15/5L; and one **finishing coat; PC £35.15/5L; of 'Albi' clear flame** **retardent surface coating; on wood surfaces**					
General surfaces					
over 300 mm girth	0.57	4.24	1.37	m2	5.61
isolated surfaces not exceeding 300 mm girth	0.23	1.71	0.43	m	2.14
isolated surfaces not exceeding 0.50 m2	0.44	3.27	0.67	nr	3.94
Two undercoats; PC £26.15/5L; and one **finishing coat; PC £35.15/5L; of 'Albi' clear flame** **retardent surface coating; on wood surfaces**					
General surfaces					
over 300 mm girth	0.57	4.24	2.10	m2	6.34
isolated surfaces not exceeding 300 mm girth	0.23	1.71	0.67	m	2.38
isolated surfaces not exceeding 0.50 m2	0.44	3.27	1.05	nr	4.32

M SURFACE FINISHES Including overheads and profit at 12.50%	Labour hours	Labour £	Material £	Unit	Total rate £
M60 PAINTING/CLEAR FINISHING - cont'd					
Seal and wax polish; dull gloss finish on wood surfaces General surfaces					
over 300 mm girth	-	-	-	m2	5.62
isolated surfaces not exceeding 300 mm girth	-	-	-	m	2.54
isolated surfaces not exceeding 0.50 m2	-	-	-	nr	3.94
Two coats of 'Sadolins Classic'; PC £22.95/5L; clear or pigmented; two further coats of 'Holdex' clear interior silk matt laquer; PC £26.61/5L General surfaces					
over 300 mm girth	0.65	4.83	2.09	m2	6.92
isolated surfaces not exceeding 300 mm girth	0.26	1.93	0.66	m	2.59
isolated surfaces not exceeding 0.50 m2	0.49	3.64	1.05	nr	4.69
Windows and the like					
panes over 1.00 m2	0.65	4.83	0.77	m2	5.60
panes 0.50 - 1.00 m2	0.76	5.65	1.05	m2	6.70
panes 0.10 - 0.50 m2	0.87	6.47	1.32	m2	7.79
panes not exceeding 0.10 m2	1.10	8.18	1.57	m2	9.75
Body in and wax polish; dull gloss finish; on hardwood surfaces General surfaces					
over 300 mm girth	-	-	-	m2	8.05
isolated surfaces not exceeding 300 mm girth	-	-	-	m	3.63
isolated surfaces not exceeding 0.50 m2	-	-	-	nr	5.63
Stain; body in and wax polish; dull gloss finish; on hardwood surface General surfaces					
over 300 mm girth	-	-	-	m2	9.98
isolated surfaces not exceeding 300 mm girth	-	-	-	m	4.49
isolated surfaces not exceeding 0.50 m2	-	-	-	nr	7.02
Seal; two coats of synthetic resin lacquer; decorative flatted finish; wire down, wax and burnish; on wood surfaces General surfaces					
over 300 mm girth	-	-	-	m2	12.68
isolated surfaces not exceeding 300 mm girth	-	-	-	m	5.56
isolated surfaces not exceeding 0.50 m2	-	-	-	nr	8.88
Stain; body in and fully French polish; full gloss finish; on hardwood surfaces General surfaces					
over 300 mm girth	-	-	-	m2	15.16
isolated surfaces not exceeding 300 mm girth	-	-	-	m	6.84
isolated surfaces not exceeding 0.50 m2	-	-	-	nr	10.62
Stain; fill grain and fully French polish; full gloss finish; on hardwood surfaces General surfaces					
over 300 mm girth	-	-	-	m2	23.15
isolated surfaces not exceeding 300 mm girth	-	-	-	m	10.42
isolated surfaces not exceeding 0.50 m2	-	-	-	nr	15.44
Stain black; body in and fully French polish; ebonized finish; on hardwood surfaces General surfaces					
over 300 mm girth	-	-	-	m2	28.94
isolated surfaces not exceeding 300 mm girth	-	-	-	m	13.01
isolated surfaces not exceeding 0.50 m2	-	-	-	nr	20.26

M SURFACE FINISHES Including overheads and profit at 12.50%	Labour hours	Labour £	Material £	Unit	Total rate £
M60 PREPARATION OF EXISTING SURFACES - EXTERNALLY					
Wash and rub down iron or steel surfaces; bringing forward					
General surfaces					
over 300 mm girth	0.20	1.49	-	m2	1.49
isolated surfaces not exceeding 300 mm girth	0.08	0.59	-	m	0.59
isolated surfaces not exceeding 0.50 m2	0.15	1.12	-	nr	1.12
Windows and the like					
panes over 1.00 m2	0.20	1.49	-	m2	1.49
panes 0.50 - 1.00 m2	0.23	1.71	-	m2	1.71
panes 0.10 - 0.50 m2	0.27	2.01	-	m2	2.01
panes not exceeding 0.10 m2	0.33	2.45	-	m2	2.45
Wash and rub down wood surfaces; prime bare patches; bringing forward					
General surfaces					
over 300 mm girth	0.33	2.45	-	m2	2.45
isolated surfaces not exceeding 300 mm girth	0.13	0.97	-	m	0.97
isolated surfaces not exceeding 0.50 m2	0.25	1.86	-	nr	1.86
Windows and the like					
panes over 1.00 m2	0.33	2.45	-	m2	2.45
panes 0.50 - 1.00 m2	0.38	2.83	-	m2	2.83
panes 0.10 - 0.50 m2	0.44	3.27	-	m2	3.27
panes not exceeding 0.10 m2	0.55	4.09	-	m2	4.09
Wash down and remove paint with chemical stripper from iron, steel or wood surfaces					
General surfaces					
over 300 mm girth	1.00	7.44	-	m2	7.44
isolated surfaces not exceeding 300 mm girth	0.45	3.35	-	m	3.35
isolated surfaces not exceeding 0.50 m2	0.75	5.58	-	nr	5.58
Windows and the like					
panes over 1.00 m2	1.28	9.52	-	m2	9.52
panes 0.50 - 1.00 m2	1.47	10.93	-	m2	10.93
panes 0.10 - 0.50 m2	1.70	12.64	-	m2	12.64
panes not exceeding 0.10 m2	2.14	15.91	-	m2	15.91
Burn off and rub down to remove paint from iron, steel or wood surfaces					
General surfaces					
over 300 mm girth	1.25	9.30	0.22	m2	9.52
isolated surfaces not exceeding 300 mm girth	0.56	4.16	0.03	m	4.19
isolated surfaces not exceeding 0.50 m2	0.94	6.99	0.08	nr	7.07
Windows and the like					
panes over 1.00 m2	1.60	11.90	-	m2	11.90
panes 0.50 - 1.00 m2	1.84	13.68	-	m2	13.68
panes 0.10 - 0.50 m2	2.13	15.84	-	m2	15.84
panes not exceeding 0.10 m2	2.67	19.85	-	m2	19.85
M60 PAINTING/CLEAR FINISHING - EXTERNALLY					
Two coats of cement paint, 'Sandtex Matt' or similar; PC £12.60/5L;					
Brick or block walls					
over 300 mm girth	0.47	3.50	1.70	m2	5.20
Cement render or concrete walls					
over 300 mm girth	0.42	3.12	1.13	m2	4.25
Roughcast walls					
over 300 mm girth	0.52	3.87	2.27	m2	6.14

M SURFACE FINISHES Including overheads and profit at 12.50%	Labour hours	Labour £	Material £	Unit	Total rate £
M60 PAINTING/CLEAR FINISHING - EXTERNALLY - cont'd					
One coat sealer and two coats of external **grade emulsion paint, Dulux 'Weathershield'** **or similar**					
Brick or block walls					
over 300 mm girth	0.63	4.68	2.29	m2	**6.97**
Cement render or concrete walls					
over 300 mm girth	0.52	3.87	1.53	m2	**5.40**
Concrete soffits					
over 300 mm girth	0.58	4.31	1.53	m2	**5.84**
One coat sealer (applied by brush) and two **coats of external grade emulsion paint,** **Dulux 'Weathershield' or similar** **(spray applied)**					
Roughcast					
over 300 mm girth	0.63	3.43	3.06	m2	**6.49**
Touch up primer; one undercoat and one **finishing coat of gloss oil paint; PC £13.65/5L;** **on wood surfaces**					
General surfaces					
over 300 mm girth	0.38	2.83	0.50	m2	**3.33**
isolated surfaces not exceeding 300 mm girth	0.16	1.19	0.16	m	**1.35**
isolated surfaces not exceeding 0.50 m2	0.28	2.08	0.25	nr	**2.33**
Windows and the like					
panes over 1.00 m2	0.38	2.83	0.19	m2	**3.02**
panes 0.50 - 1.00 m2	0.44	3.27	0.25	m2	**3.52**
panes 0.10 - 0.50 m2	0.50	3.72	0.31	m2	**4.03**
panes not exceeding 0.10 m2	0.63	4.68	0.38	m2	**5.06**
Windows and the like; casements in colours differing from frames					
panes over 1.00 m2	0.42	3.12	0.22	m2	**3.34**
panes 0.50 - 1.00 m2	0.47	3.50	0.28	m2	**3.78**
panes 0.10 - 0.50 m2	0.56	4.16	0.34	m2	**4.50**
panes not exceeding 0.10 m2	0.69	5.13	0.41	m2	**5.54**
Touch up primer; two undercoats and one **finishing coat of gloss oil paint;** **on wood surfaces**					
General surfaces					
over 300 mm girth	0.52	3.87	0.74	m2	**4.61**
isolated surfaces not exceeding 300 mm girth	0.21	1.56	0.23	m2	**1.79**
isolated surfaces not exceeding 0.50 m2	0.40	2.97	0.37	m2	**3.34**
Windows and the like					
panes over 1.00 m2	0.52	3.87	0.28	m2	**4.15**
panes 0.50 - 1.00 m2	0.61	4.54	0.37	m2	**4.91**
panes 0.10 - 0.50 m2	0.70	5.21	0.46	m2	**5.67**
panes not exceeding 0.10 m2	0.87	6.47	0.55	m2	**7.02**
Windows and the like; casements in colours differing from frames					
panes over 1.00 m2	0.61	4.54	0.32	m2	**4.86**
panes 0.50 - 1.00 m2	0.70	5.21	0.41	m2	**5.62**
panes 0.10 - 0.50 m2	0.81	6.02	0.51	m2	**6.53**
panes not exceeding 0.10 m2	1.00	7.44	0.60	m2	**8.04**
Knot; one coat primer; one undercoat and **one finishing coat of gloss oil paint; on** **wood surfaces**					
General surfaces					
over 300 mm girth	0.56	4.16	0.82	m2	**4.98**
isolated surfaces not exceeding 300 mm girth	0.23	1.71	0.26	m	**1.97**
isolated surfaces not exceeding 0.50 m2	0.42	3.12	0.41	nr	**3.53**

M SURFACE FINISHES Including overheads and profit at 12.50%	Labour hours	Labour £	Material £	Unit	Total rate £
Windows and the like					
panes over 1.00 m2	0.56	4.16	0.31	m2	**4.47**
panes 0.50 - 1.00 m2	0.65	4.83	0.41	m2	**5.24**
panes 0.10 - 0.50 m2	0.73	5.43	0.51	m2	**5.94**
panes not exceeding 0.10 m2	0.92	6.84	0.61	m2	**7.45**
Windows and the like; casements in colours differing from frames					
panes over 1.00 m2	0.61	4.54	0.35	m2	**4.89**
panes 0.50 - 1.00 m2	0.71	5.28	0.45	m2	**5.73**
panes 0.10 - 0.50 m2	0.81	6.02	0.75	m2	**6.77**
panes not exceeding 0.10 m2	1.00	7.44	0.66	m2	**8.10**
Knot; one coat primer; two undercoats and one finishing coat of gloss oil paint on wood surfaces					
General surfaces					
over 300 mm girth	0.69	5.13	1.06	m2	**6.19**
isolated surfaces not exceeding 300 mm girth	0.28	2.08	0.33	m	**2.41**
isolated surfaces not exceeding 0.50 m2	0.52	3.87	0.53	nr	**4.40**
Windows and the like					
panes over 1.00 m2	0.69	5.13	0.39	m2	**5.52**
panes 0.50 - 1.00 m2	0.81	6.02	0.53	m2	**6.55**
panes 0.10 - 0.50 m2	0.92	6.84	0.66	m2	**7.50**
panes not exceeding 0.10 m2	1.15	8.55	0.79	m2	**9.34**
Windows and the like; casements in colours differing from frames					
panes over 1.00 m2	0.80	5.95	0.45	m2	**6.40**
panes 0.50 - 1.00 m2	0.94	6.99	0.59	m2	**7.58**
panes 0.10 - 0.50 m2	1.05	7.81	0.91	m2	**8.72**
panes not exceeding 0.10 m2	1.30	9.67	0.85	m2	**10.52**
Touch up primer; one undercoat and one finishing coat of gloss oil paint; PC £13.65/5L; on iron or steel surfaces					
General surfaces					
over 300 mm girth	0.38	2.83	0.48	m2	**3.31**
isolated surfaces not exceeding 300 mm girth	0.16	1.19	0.15	m	**1.34**
isolated surfaces not exceeding 0.50 m2	0.28	2.08	0.24	nr	**2.32**
Windows and the like					
panes over 1.00 m2	0.38	2.83	0.18	m2	**3.01**
panes 0.50 - 1.00 m2	0.44	3.27	0.24	m2	**3.51**
panes 0.10 - 0.50 m2	0.50	3.72	0.30	m2	**4.02**
panes not exceeding 0.10 m2	0.63	4.68	0.36	m2	**5.04**
Structural steelwork					
over 300 mm girth	0.42	3.12	0.48	m2	**3.60**
Members of roof trusses					
over 300 mm girth	0.56	4.16	0.51	m2	**4.67**
Ornamental railings and the like; each side measured overall					
over 300 mm girth	0.63	4.68	0.54	m2	**5.22**
Eaves gutters					
over 300 mm girth	0.68	5.06	0.72	m2	**5.78**
not exceeding 300 mm girth	0.27	2.01	0.24	m	**2.25**
Pipes or conduits					
over 300 mm girth	0.56	4.16	0.51	m2	**4.67**
not exceeding 300 mm girth	0.22	1.64	0.15	m	**1.79**
Touch up primer; two undercoats and one finishing coat of gloss oil paint; on iron or steel surfaces					
General surfaces					
over 300 mm girth	0.52	3.87	0.71	m2	**4.58**
isolated surfaces not exceeding 300 mm girth	0.21	1.56	0.22	m	**1.78**
isolated surfaces not exceeding 0.50 m2	0.39	2.90	0.36	nr	**3.26**

M SURFACE FINISHES Including overheads and profit at 12.50%	Labour hours	Labour £	Material £	Unit	Total rate £

M60 PAINTING/CLEAR FINISHING - EXTERNALLY - cont'd

Touch up primer; two undercoats and one
finishing coat of gloss oil paint; on
iron or steel surfaces - cont'd

Windows and the like					
panes over 1.00 m2	0.52	3.87	0.27	m2	4.14
panes 0.50 - 1.00 m2	0.61	4.54	0.36	m2	4.90
panes 0.10 - 0.50 m2	0.70	5.21	0.45	m2	5.66
panes not exceeding 0.10 m2	0.87	6.47	0.54	m2	7.01
Structural steelwork					
over 300 mm girth	0.60	4.46	0.71	m2	5.17
Members of roof trusses					
over 300 mm girth	0.79	5.87	0.76	m2	6.63
Ornamental railings and the like; each side measured overall					
over 300 mm girth	0.89	6.62	0.80	m2	7.42
Eaves gutters					
over 300 mm girth	0.94	6.99	1.07	m2	8.06
not exceeding 300 mm girth	0.38	2.83	0.36	m	3.19
Pipes or conduits					
over 300 mm girth	0.79	5.87	0.76	m2	6.63
not exceeding 300 mm girth	0.32	2.38	0.22	m	2.60

One coat primer; one undercoat and one
finishing coat of gloss oil paint; on
iron or steel surfaces

General surfaces					
over 300 mm girth	0.48	3.57	0.48	m2	4.05
isolated surfaces not exceeding 300 mm girth	0.20	1.49	0.15	m	1.64
isolated surfaces not exceeding 0.50 m2	0.37	2.75	0.24	nr	2.99
Windows and the like					
panes over 1.00 m2	0.48	3.57	0.18	m2	3.75
panes 0.50 - 1.00 m2	0.56	4.16	0.24	m2	4.40
panes 0.10 - 0.50 m2	0.65	4.83	0.30	m2	5.13
panes not exceeding 0.10 m2	0.81	6.02	0.36	m2	6.38
Structural steelwork					
over 300 mm girth	0.55	4.09	0.48	m2	4.57
Members of roof trusses					
over 300 mm girth	0.71	5.28	0.51	m2	5.79
Ornamental railings and the like; each side measured overall					
over 300 mm girth	0.82	6.10	0.54	m2	6.64
Eaves gutters					
over 300 mm girth	0.86	6.40	0.72	m2	7.12
not exceeding 300 mm girth	0.35	2.60	0.24	m	2.84
Pipes or conduits					
over 300 mm girth	0.71	5.28	0.51	m2	5.79
not exceeding 300 mm girth	0.28	2.08	0.15	m	2.23

One coat primer; two undercoats and one
finishing coat of gloss oil paint; on
iron or steel surfaces

General surfaces					
over 300 mm girth	0.63	4.68	0.71	m2	5.39
isolated surfaces not exceeding 300 mm girth	0.25	1.86	0.22	m	2.08
isolated surfaces not exceeding 0.50 m2	0.47	3.50	0.36	nr	3.86
Windows and the like					
panes over 1.00 m2	0.63	4.68	0.27	m2	4.95
panes 0.50 - 1.00 m2	0.73	5.43	0.36	m2	5.79
panes 0.10 - 0.50 m2	0.84	6.25	0.45	m2	6.70
panes not exceeding 0.10 m2	1.05	7.81	0.54	m2	8.35
Structural steelwork					
over 300 mm girth	0.71	5.28	0.71	m2	5.99
Members of roof trusses					
over 300 mm girth	0.94	6.99	0.76	m2	7.75

M SURFACE FINISHES Including overheads and profit at 12.50%	Labour hours	Labour £	Material £	Unit	Total rate £
Ornamental railings and the like; each side measured overall					
over 300 mm girth	1.05	7.81	0.80	m2	8.61
Eaves gutters					
over 300 mm girth	1.15	8.55	1.07	m2	9.62
not exceeding 300 mm girth	0.45	3.35	0.36	m	3.71
Pipes or conduits					
over 300 mm girth	0.94	6.99	0.76	m2	7.75
not exceeding 300 mm girth	0.38	2.83	0.22	m	3.05
Two coats of creosote; PC £3.15/5L; on wood surfaces					
General surfaces					
over 300 mm girth	0.26	1.93	0.16	m2	2.09
isolated surfaces not exceeding 300 mm girth	0.11	0.82	0.05	m	0.87
Two coats of 'Solignum' wood preservative; PC £10.67/5L; on wood surfaces					
General surfaces					
over 300 mm girth	0.26	1.93	0.53	m2	2.46
isolated surfaces not exceeding 300 mm girth	0.11	0.82	0.17	m	0.99
Three coats of polyurethane; PC £15.54/5L; on wood surfaces					
General surfaces					
over 300 mm girth	0.52	3.87	0.84	m2	4.71
isolated surfaces not exceeding 300 mm girth	0.21	1.56	0.26	m	1.82
isolated surfaces not exceeding 0.50 m2	0.40	2.97	0.42	nr	3.39
Two coats of 'Sadovac 35'primer; PC £11.55/5L; and two coats of 'Classic'; PC £22.95/5L; pigmented; on wood surfaces					
General surfaces					
over 300 mm girth	0.71	5.28	1.99	m2	7.27
isolated surfaces not exceeding 300 mm girth	0.28	2.08	0.63	m	2.71
Windows and the like					
panes over 1.00 m2	0.71	5.28	0.72	m2	6.00
panes 0.50 - 1.00 m2	0.83	6.17	1.00	m2	7.17
panes 0.10 - 0.50 m2	0.96	7.14	1.27	m2	8.41
panes not exceeding 0.10 m2	1.20	8.92	1.50	m2	10.42
Body in with French polish; one coat of lacquer or varnish; on wood surfaces					
General surfaces					
over 300 mm girth	-	-	-	m2	12.90
isolated surfaces not exceeding 300 mm girth	-	-	-	m	5.81
isolated surfaces not exceeding 0.50 m2	-	-	-	nr	9.03

N FURNITURE/EQUIPMENT Including overheads and profit at 12.50%		Labour hours	Labour £	Material £	Unit	Total rate £
N10/11 GENERAL FIXTURES/KITCHEN FITTINGS ETC.						
Proprietary items						
Closed stove vitreous enamelled finish;						
setting in position						
571 x 606 x 267 mm	PC £253.00	2.30	27.66	284.63	nr	312.29
Closed stove; vitreous enamelled finish						
fitted with mild steel barffed boiler;						
setting in position						
571 x 606 x 267 mm	PC £330.00	2.30	27.66	371.25	nr	398.91
Tile surround preslabbed; 406 mm wide						
firebrick back; cast iron stool bottom;						
black vitreous enamelled fret; assembling,						
setting and pointing firebrick back in						
fireclay mortar and backing with fine						
concrete finished to splay at top, laying						
hearth tiles in cement mortar and pointing						
in white cement						
1372 x 864 x 152 mm	PC £242.06	5.75	69.15	292.73	nr	361.88
Fitting components; blockboard						
Backs, fronts, sides or divisions;						
over 300 mm wide						
12 mm thick	PC £24.64	1.45	13.56	27.72	m2	41.28
19 mm thick	PC £31.64	1.45	13.56	35.59	m2	49.15
25 mm thick	PC £41.52	1.45	13.56	46.71	m2	60.27
Shelves or worktops; over 300 mm wide						
19 mm thick	PC £32.69	1.45	13.56	36.78	m2	50.34
25 mm thick	PC £41.52	1.45	13.56	46.71	m2	60.27
Flush doors; lipped on four edges						
450 x 750 x 19 mm	PC £17.36	0.39	3.65	19.53	nr	23.18
450 x 750 x 25 mm	PC £21.22	0.39	3.65	23.88	nr	27.53
600 x 900 x 19 mm	PC £25.84	0.55	5.14	29.07	nr	34.21
600 x 900 x 25 mm	PC £31.46	0.55	5.14	35.39	nr	40.53
Fitting components; chipboard						
Backs, fronts, sides or divisions;						
over 300 mm wide						
6 mm thick	PC £8.68	1.45	13.56	9.77	m2	23.33
9 mm thick	PC £11.78	1.45	13.56	13.25	m2	26.81
12 mm thick	PC £14.29	1.45	13.56	16.07	m2	29.63
19 mm thick	PC £20.19	1.45	13.56	22.71	m2	36.27
25 mm thick	PC £27.30	1.45	13.56	30.71	m2	44.27
Shelves or worktops; over 300 mm wide						
19 mm thick	PC £21.25	1.45	13.56	23.91	m2	37.47
25 mm thick	PC £27.30	1.45	13.56	30.71	m2	44.27
Flush doors; lipped on four edges						
450 x 750 x 19 mm	PC £14.06	0.39	3.65	15.82	nr	19.47
450 x 750 x 25 mm	PC £17.16	0.39	3.65	19.30	nr	22.95
600 x 900 x 19 mm	PC £20.64	0.55	5.14	23.22	nr	28.36
600 x 900 x 25 mm	PC £25.05	0.55	5.14	28.18	nr	33.32
Fitting components; Melamine faced chipboard						
Backs, fronts, sides or divisions;						
over 300 mm wide						
12 mm thick	PC £19.70	1.45	13.56	22.16	m2	35.72
19 mm thick	PC £26.54	1.45	13.56	29.86	m2	43.42
Shelves or worktops; over 300 mm wide						
19 mm thick	PC £27.84	1.45	13.56	31.32	m2	44.88
Flush doors; lipped on four edges						
450 x 750 x 19 mm	PC £17.49/m2	0.39	3.65	19.67	nr	23.32
600 x 900 x 19 mm	PC £25.81/m2	0.55	5.14	29.03	nr	34.17
Fitting components; 'Warerite Xcel' standard						
colour laminated chipboard type LD2; PC £180.16/10m2						
Backs, fronts, sides or divisions;						
over 300 mm wide						
13.2 mm thick	PC £56.59	1.45	13.56	63.67	m2	77.23

N FURNITURE/EQUIPMENT Including overheads and profit at 12.50%		Labour hours	Labour £	Material £	Unit	Total rate £
Shelves or worktops; over 300 mm wide						
13.2 mm thick	PC £56.59	1.45	13.56	63.67	m2	77.23
Flush doors; lipped on four edges						
450 x 750 x 13.2 mm	PC £22.87/m2	0.39	3.65	25.73	nr	29.38
600 x 900 x 13.2 mm	PC £34.04/m2	0.55	5.14	38.30	nr	43.44
Fitting components; plywood						
Backs, fronts, sides or divisions;						
over 300 mm wide						
6 mm thick	PC £14.08	1.45	13.56	15.85	m2	29.41
9 mm thick	PC £18.62	1.45	13.56	20.94	m2	34.50
12 mm thick	PC £23.57	1.45	13.56	26.52	m2	40.08
19 mm thick	PC £34.17	1.45	13.56	38.44	m2	52.00
25 mm thick	PC £45.35	1.45	13.56	51.02	m2	64.58
Shelves or worktops; over 300 mm wide						
19 mm thick	PC £35.22	1.45	13.56	39.63	m2	53.19
25 mm thick	PC £45.35	1.45	13.56	51.02	m2	64.58
Flush doors; lipped on four edges						
450 x 750 x 19 mm	PC £17.91	0.39	3.65	20.15	nr	23.80
450 x 750 x 25 mm	PC £22.15	0.39	3.65	24.92	nr	28.57
600 x 900 x 19 mm	PC £26.80	0.55	5.14	30.15	nr	35.29
600 x 900 x 25 mm	PC £32.98	0.55	5.14	37.10	nr	42.24
Fitting components; wrought softwood						
Backs, fronts, sides or divisions;						
cross-tongued joints; over 300 mm wide						
25 mm thick		1.45	13.56	40.12	m2	53.68
Shelves or worktops; cross-tongued joints;						
over 300 mm wide						
25 mm thick		1.45	13.56	40.12	m2	53.68
Bearers						
19 x 38 mm		0.11	1.03	2.25	m	3.28
25 x 50 mm		0.11	1.03	2.88	m	3.91
50 x 50 mm		0.11	1.03	4.30	m	5.33
50 x 75 mm		0.11	1.03	5.82	m	6.85
Bearers; framed						
19 x 38 mm		0.14	1.31	4.27	m	5.58
25 x 50 mm		0.14	1.31	4.92	m	6.23
50 x 50 mm		0.14	1.31	6.34	m	7.65
50 x 75 mm		0.14	1.31	7.84	m	9.15
Framing to backs, fronts or sides						
19 x 38 mm		0.17	1.59	4.27	m	5.86
25 x 50 mm		0.17	1.59	4.92	m	6.51
50 x 50 mm		0.17	1.59	6.34	m	7.93
50 x 75 mm		0.17	1.59	7.84	m	9.43
Flush doors; softwood skeleton or cellular						
core; plywood facing both sides; lipped on						
four edges						
450 x 750 x 35 mm		0.39	3.65	21.37	nr	25.02
600 x 900 x 35 mm		0.55	5.14	33.84	nr	38.98
Add +5% to the above 'Material £ prices'						
for 'selected' softwood for staining						
Fitting components; selected West African						
Mahogany; PC £531.05/m3						
Bearers						
19 x 38 mm	PC £2.86	0.17	1.59	3.36	m	4.95
25 x 50 mm	PC £3.73	0.17	1.59	4.37	m	5.96
50 x 50 mm	PC £5.72	0.17	1.59	6.66	m	8.25
50 x 75 mm	PC £7.81	0.17	1.59	9.07	m	10.66
Bearers; framed						
19 x 38 mm	PC £5.51	0.22	2.06	6.36	m	8.42
25 x 50 mm	PC £6.40	0.22	2.06	7.38	m	9.44
50 x 50 mm	PC £8.37	0.22	2.06	9.65	m	11.71
50 x 75 mm	PC £10.46	0.22	2.06	12.06	m	14.12

N FURNITURE/EQUIPMENT Including overheads and profit at 12.50%		Labour hours	Labour £	Material £	Unit	Total rate £

N10/11 GENERAL FIXTURES/KITCHEN FITTINGS ETC. - cont'd

Fitting components; selected West African Mahogany; PC £531.05/m3 - cont'd
Framing to backs, fronts or sides

19 x 38 mm	PC £5.51	0.28	2.62	6.36	m	8.98
25 x 50 mm	PC £6.40	0.28	2.62	7.38	m	10.00
50 x 50 mm	PC £8.37	0.28	2.62	9.65	m	12.27
50 x 75 mm	PC £10.46	0.28	2.62	12.06	m	14.68

Fitting components; Iroko; PC £379.48/m3
Backs, fronts, sides or divisions;
cross-tongued joints; over 300 mm wide

25 mm thick	PC £54.57	1.95	18.23	61.39	m2	79.62

Shelves or worktops; cross-tongued joints;
over 300 mm wide

25 mm thick	PC £54.57	1.95	18.23	61.39	m2	79.62

Draining boards; cross-tongued joints;
over 300 mm wide

25 mm thick	PC £58.97	1.95	18.23	66.34	m2	84.57
Stopped flutes		-	-	-	m	2.60
Grooves; cross-grain		-	-	-	m	0.62

Bearers

19 x 38 mm	PC £3.17	0.17	1.59	3.72	m	5.31
25 x 50 mm	PC £3.90	0.17	1.59	4.57	m	6.16
50 x 50 mm	PC £5.67	0.17	1.59	6.60	m	8.19
50 x 75 mm		0.17	1.59	8.74	m	10.33

Bearers; framed

19 x 38 mm	PC £6.70	0.22	2.06	7.73	m	9.79
25 x 50 mm	PC £7.45	0.22	2.06	8.59	m	10.65
50 x 50 mm	PC £9.22	0.22	2.06	10.63	m	12.69
50 x 75 mm	PC £11.06	0.22	2.06	12.76	m	14.82

Framing to backs, fronts or sides

19 x 38 mm	PC £6.70	0.28	2.62	7.73	m	10.35
25 x 50 mm	PC £7.45	0.28	2.62	8.59	m	11.21
50 x 50 mm	PC £9.22	0.28	2.62	10.63	m	13.25
50 x 75 mm	PC £11.06	0.28	2.62	12.76	m	15.38

Fitting components; Teak; PC £1505.00/m3
Backs, fronts, sides or divisions;
cross-tongued joints; over 300 mm wide

25 mm thick	PC £136.08	2.20	20.57	153.09	m2	173.66

Shelves or worktops; cross-tongued joints;
over 300 mm wide

25 mm thick	PC £136.08	2.20	20.57	153.09	m2	173.66

Draining boards; cross-tongued joints;
over 300 mm wide

25 mm thick	PC £139.34	2.20	20.57	156.75	m2	177.32
Stopped flutes		-	-	-	m2	2.60
Grooves; cross-grain		-	-	-	m	0.62

Fitting components; 'Formica' plastic laminated plastics coverings fixed with adhesive
1.3 mm thick horizontal grade marbled
finish over 300 mm wide

	PC £18.00	1.75	16.36	29.63	m2	45.99

1.0 mm thick universal backing over
300 mm wide

	PC £2.22	1.30	12.15	6.02	m2	18.17

1.3 mm thick edging strip

18 mm wide		0.11	1.03	0.92	m	1.95
25 mm wide		0.11	1.03	1.19	m	2.22

Fixing kitchen fittings
(Kitchen fittings are largely a matter of
selection and prices vary considerably. PC
supply prices for reasonable quantities for
'Standard' (moderately-priced) kitchen
fiitings have been shown but not extended).

N FURNITURE/EQUIPMENT Including overheads and profit at 12.50%		Labour hours	Labour £	Material £	Unit	Total rate £
Fixing only to backgrounds requiring **plugging; including any pre-assembly** Wall units						
600 x 600 x 300 mm	PC £41.09	1.55	14.49	0.19	nr	14.68
600 x 900 x 300 mm	PC £51.10	1.75	16.36	0.19	nr	16.55
1200 x 600 x 300 mm	PC £74.62	2.05	19.16	0.26	nr	19.42
1200 x 900 x 300 mm	PC £89.43	2.30	21.50	0.26	nr	21.76
Floor units with drawers						
600 x 900 x 500 mm	PC £69.55	1.40	13.09	0.13	nr	13.22
600 x 900 x 600 mm	PC £72.56	1.55	14.49	0.13	nr	14.62
1200 x 900 x 600 mm	PC £128.49	1.85	17.30	0.13	nr	17.43
Laminated plastics worktops to suit last						
500 x 600 mm	PC £13.44	0.44	4.11	-	nr	4.11
600 x 600 mm	PC £15.12	0.50	4.67	-	nr	4.67
1200 x 600 mm	PC £30.24	0.77	7.20	-	nr	7.20
Larder units						
600 x 1950 x 600 mm	PC £116.97	2.70	25.24	0.19	nr	25.43
Sink units (excluding sink top)						
1200 x 900 x 600 mm	PC £123.83	2.00	18.70	0.13	nr	18.83
1500 x 900 x 600	PC £180.88	2.35	21.97	0.19	nr	22.16
6 mm thick rectangular glass mirrors; silver **backed; fixed with chromium plated domed** **headed screws; to background requiring** **plugging** Mirror with polished edges						
356 x 254 mm	PC £2.71	0.88	6.54	5.78	nr	12.32
400 x 300 mm	PC £3.55	0.88	6.54	6.73	nr	13.27
560 x 380 mm	PC £7.47	0.99	7.36	11.14	nr	18.50
640 x 460 mm	PC £9.12	1.10	8.18	12.99	nr	21.17
Mirror with bevelled edges						
356 x 254 mm	PC £5.36	0.88	6.54	8.76	nr	15.30
400 x 300 mm	PC £5.96	0.88	6.54	9.44	nr	15.98
560 x 380 mm	PC £10.59	0.99	7.36	14.65	nr	22.01
640 x 460 mm	PC £13.19	1.10	8.18	17.57	nr	25.75
Matwells Mild steel matwell; galvanized after manufacture; comprising 30 x 30 x 3 mm angle rim; with welded angles and lugs welded on; to suit mat size						
914 x 560 mm	PC £21.77	1.10	11.72	24.49	nr	36.21
1067 x 610 mm	PC £22.62	1.40	14.92	25.45	nr	40.37
1219 x 762 mm	PC £25.10	1.65	17.58	28.23	nr	45.81
Polished aluminium matwell; comprising 32 x 32 x 5 mm angle rim; with brazed angles and lugs brazed on; to suit mat size						
914 x 560 mm	PC £28.15	1.10	11.72	31.67	nr	43.39
1067 x 610 mm	PC £30.00	1.40	14.92	33.75	nr	48.67
1219 x 762 mm	PC £34.60	1.65	17.58	38.93	nr	56.51
Polished brass matwell; comprising 38 x 38 x 6 mm angle rim; with brazed angles and lugs welded on; to suit mat size						
914 x 560 mm	PC £83.20	1.10	11.72	93.60	nr	105.32
1067 x 610 mm	PC £88.72	1.40	14.92	99.81	nr	114.73
1219 x 762 mm	PC £102.24	1.65	17.58	115.02	nr	132.60

Keep your figures up to date, free of charge

This section, and most of the other information in this Price Book, is brought up to date every three months in the *Price Book Update*.

The *Update* is available free to all Price Book purchasers.

To ensure you receive your copy, simply complete the reply card from the centre of the book and return it to us.

N FURNITURE/EQUIPMENT Including overheads and profit at 12.50%		Labour hours	Labour £	Material £	Unit	Total rate £

N13 SANITARY APPLIANCES/FITTINGS

NOTE: Sanitary fittings are largely a matter
of selection and material prices vary
considerably; the PC values given below are
for average quality
Sink; white glazed fireclay; BS 1206;
cast iron cantilever brackets

610 x 455 x 205 mm	PC £88.60	3.60	38.43	105.17	nr	143.60
610 x 455 x 255 mm	PC £106.80	3.60	38.43	125.64	nr	164.07
760 x 455 x 255 mm	PC £149.35	3.60	38.43	173.51	nr	211.94

Sink; stainless steel combined bowl and
draining board; chain and self colour plug;
to BS 3380
1050 x 500 mm with bowl

420 x 350 x 175 mm	PC £107.82	2.10	22.42	121.30	nr	143.72

Sink; stainless steel combined bowl and
double draining board; chain and self
colour plug to BS 3380
1550 x 500 mm with bowl

420 x 350 x 200 mm	PC £133.68	2.40	25.62	150.39	nr	176.01

Lavatory basin; vitreous china; BS 1188;
32 mm chromium plated waste; chain,
stay and plug; pair 13 mm chromium plated
easy clean pillar taps to BS 1010; painted
cantilever brackets; plugged and screwed

560 x 405 mm; white	PC £61.85	2.75	29.36	69.58	nr	98.94
560 x 405 mm; coloured	PC £71.09	2.75	29.36	79.98	nr	109.34
635 x 455 mm; white	PC £88.87	2.75	29.36	99.98	nr	129.34
635 x 455 mm; coloured	PC £102.95	2.75	29.36	115.82	nr	145.18

Lavatory basin and pedestal; vitreous
china; BS 1188; 32 mm chromium plated
waste, chain, stay and plug; pair 13 mm
chromium plated easy clean pillar taps to
BS 1010; pedestal; wall brackets;
plugged and screwed

560 x 405 mm; white	PC £84.19	3.00	32.03	94.71	nr	126.74
560 x 405 mm; coloured	PC £98.75	3.00	32.03	111.09	nr	143.12
635 x 455 mm; white	PC £97.67	3.00	32.03	109.88	nr	141.91
635 x 455 mm; coloured	PC £130.62	3.00	32.03	146.95	nr	178.98

Lavatory basin range; overlap joints;
white glazed fireclay; 32 mm chromium
plated waste, chain, stay and plug; pair
13 mm chromium plated easy clean pillar
taps to BS 1010; painted cast iron
cantilever brackets; plugged and screwed

range of four 560 x 405 mm	PC £314.00	10.56	112.74	353.25	nr	465.99
Add for each additional basin in						
the range		2.40	25.62	91.25	nr	116.87

Drinking fountain; white glazed fireclay;
19 mm chromium plated waste; self-closing
non-concussive tap; regulating valve;
plugged and screwed with chromium

plated screws	PC £138.90	3.00	32.03	156.26	nr	188.29

Bath; reinforced acrylic rectangular
pattern; 40 mm chromium plated overflow
chain and plug; 40 mm chromium plated
waste; cast brass 'P' trap with plain outlet
and overflow connection to BS 1184; pair
20 mm chromium plated easy clean pillar taps
to BS 1010

1700 mm long; white	PC £144.70	4.20	44.84	162.79	nr	207.63
1700 mm long; coloured	PC £144.70	4.20	44.84	162.79	nr	207.63

N FURNITURE/EQUIPMENT Including overheads and profit at 12.50%		Labour hours	Labour £	Material £	Unit	Total rate £
Bath; enamelled steel; medium gauge rectangular pattern; 40 mm chromium plated overflow chain and plug; 40 mm chromium plated waste; cast brass 'P' trap with plain outlet and overflow connection to BS 1184; pair 20 mm chromium plated easy clean pillar taps to BS 1010						
1700 mm long; white	PC £158.60	4.20	44.84	178.43	nr	223.27
1700 mm long; coloured	PC £169.20	4.20	44.84	190.35	nr	235.19
Shower tray; glazed fireclay with outlet and grated waste; chain and plug; bedding and pointing in waterproof cement mortar						
760 x 760 x 180 mm; white	PC £125.44	3.60	38.43	141.12	nr	179.55
760 x 760 x 180 mm; coloured	PC £176.52	3.60	38.43	198.59	nr	237.02
Shower fitting; riser pipe with mixing valve and shower rose; chromium plated; plugging and screwing mixing valve and pipe bracket						
15 mm dia riser pipe; 127 mm dia shower rose	PC £200.73	6.00	64.06	225.82	nr	289.88
WC suite; high level; vitreous china pan; black plastic seat; 9 litre white vitreous china cistern and brackets; low pressure ball valve; galvanized steel flush pipe and clip; plugged and screwed; mastic joint to drain						
WC suite; white	PC £150.17	3.95	42.17	168.94	nr	211.11
WC suite; coloured	PC £186.48	3.95	42.17	209.79	nr	251.96
WC suite; low level; vitreous china pan; black plastic seat; 9 litre white vitreous china cistern and brackets; low pressure ball valve and plastic flush pipe; plugged and screwed; mastic joint to drain						
WC suite; white	PC £99.61	3.60	38.43	112.06	nr	150.49
WC suite; coloured	PC £130.64	3.60	38.43	146.97	nr	185.40
Slop sink; white glazed fireclay with hardwood pad; aluminium bucket grating; vitreous china cistern and porcelain- enamelled brackets; galvanized steel flush pipe and clip; plugged and screwed; mastic joint to drain						
slop sink	PC £487.90	4.20	44.84	548.89	nr	593.73
Bowl type wall urinal; white glazed vitreous china; white vitreous china automatic flushing cistern and brackets; 38 mm chromium plated waste; chromium plated flush pipes and spreaders; cistern brackets and flush pipes plugged and screwed with chromium plated screws						
single; 455 x 380 x 330 mm	PC £142.60	4.80	51.25	160.43	nr	211.68
range of two	PC £236.41	9.00	96.09	265.96	nr	362.05
range of three	PC £315.91	13.20	140.93	355.40	nr	496.33
Add for each additional urinal	PC £93.80	3.85	41.10	105.53	nr	146.63
Add for divisions between		0.90	9.61	42.76	nr	52.37

N15 SIGNS/NOTICES

Plain script; in gloss oil paint; on painted or varnished surfaces Capital letters; lower case letters or numerals	Labour hours	Labour £	Material £	Unit	Total rate £
per coat; per 25 mm high	0.11	0.82	-	nr	0.82
Stops					
per coat	0.03	0.22	-	nr	0.22

P BUILDING FABRIC SUNDRIES
Including overheads and profit at 12.50%

ALTERNATIVE INSULATION PRICES

Insulation (£/m2)
  Expanded polystyrene
  self-extinguishing grade

12 mm	0.94	25 mm	1.97	38 mm	2.98	50 mm	3.92
18 mm	1.41						

  'Fibreglass'
  'Crown Building Roll'

60 mm	2.08	80 mm	2.60	100 mm	3.06	

  'Crown Wool'

150 mm	4.49	200 mm	5.87	

  'Frametherm' - unfaced

60 mm	1.63	80 mm	2.18	90 mm	2.43	100 mm	2.64

  Sound-deadening quilt type PF; 13 mm - £2.72/m2

	Labour hours	Labour £	Material £	Unit	Total rate £
P10 SUNDRY INSULATION/PROOFING WORK/ FIRE STOPS					
Building paper and sheets					
Building paper; BS 1521; class A; 150 mm laps; pinned					
single sided reflective	0.11	1.03	1.19	m2	**2.22**
double sided reflective	0.11	1.03	1.73	m2	**2.76**
Mat or quilt insulation					
Glass fibre quilt; Pilkingtons 'Crown Wool'; laid over ceiling joists					
60 mm thick PC £1.81	0.11	1.03	2.14	m2	**3.17**
80 mm thick PC £2.39	0.12	1.12	2.82	m2	**3.94**
100 mm thick PC £2.84	0.13	1.22	3.35	m2	**4.57**
150 mm thick PC £4.49	0.14	1.31	5.30	m2	**6.61**
200 mm thick PC £5.87	0.15	1.40	6.93	m2	**8.33**
Raking cutting	0.06	0.56	-	m	**0.56**
Glass fibre building roll; pinned vertically to softwood					
60 mm thick PC £2.08	0.17	1.59	2.45	m2	**4.04**
80 mm thick PC £2.60	0.18	1.68	3.07	m2	**4.75**
100 mm thick PC £3.06	0.19	1.78	3.61	m2	**5.39**
Glass fibre flanged building roll; paper faces; pinned vertically or to slope between timber framing					
60 mm thick PC £2.23	0.20	1.87	2.63	m2	**4.50**
80 mm thick PC £2.76	0.21	1.96	3.26	m2	**5.22**
100 mm thick PC £3.21	0.22	2.06	3.79	m2	**5.85**
Board or slab insulation					
Expanded polystyrene board standard grade PC £59.89/m3; fixed with adhesive					
12 mm thick	0.44	4.11	1.40	m2	**5.51**
25 mm thick	0.46	4.30	2.33	m2	**6.63**
50 mm thick	0.50	4.67	4.08	m2	**8.75**
Fire stops					
'Monolux' TRADA firecheck channel; intumescent coatings on cut mitres; fixing with brass cups and screws					
19 x 44 mm or 19 x 50 mm PC £236.37/36m	0.66	6.17	10.25	m	**16.42**

P BUILDING FABRIC SUNDRIES Including overheads and profit at 12.50%	Labour hours	Labour £	Material £	Unit	Total rate £
'Sealmaster' intumescent fire and smoke seals; pinned into groove in timber					
type N30; for single leaf half hour door;					
PC £5.08	0.33	3.09	5.72	m	8.81
type N60; for single leaf one hour door					
PC £7.46	0.36	3.37	8.40	m	11.77
type IMN or IMP; for meeting or pivot styles of pair of one hour doors; per style					
PC £7.46	0.36	3.37	8.40	m	11.77
Intumescent plugs in timber; including boring	0.11	1.03	0.27	nr	1.30
Rockwool fire stops; between top of brick/ block wall and concrete soffit					
25 mm deep x 112 mm wide	0.09	0.84	0.90	m	1.74
25 mm deep x 150 mm wide	0.11	1.03	1.20	m	2.23
50 mm deep x 225 mm wide	0.17	1.59	3.03	m	4.62
Fire barriers					
Rockwool fire barrier between top of suspended ceiling and concrete soffit					
one 50 mm layer x 900 mm wide; half-hour	0.66	6.17	17.34	m	23.51
two 50 mm layers x 900 mm wide; one hour	1.00	9.35	33.39	m	42.74
Dow Chemicals 'Styrofoam 1B'; cold bridging insulation fixed with adhesive to brick, block or concrete base					
25 mm insulation to walls					
over 300 mm wide	0.37	3.94	5.35	m2	9.29
not exceeding 300 mm wide	0.59	6.29	5.53	m2	11.82
25 mm insulation to isolated columns					
over 300 mm wide	0.45	4.79	5.35	m2	10.14
not exceeding 300 mm wide	0.73	7.78	5.53	m2	13.31
25 mm insulation to ceilings					
over 300 mm wide	0.40	4.26	5.35	m2	9.61
not exceeding 300 mm wide	0.64	6.82	5.53	m2	12.35
25 mm insulation to isolated beams					
over 300 mm wide	0.48	5.11	5.35	m2	10.46
not exceeding 300 mm wide	0.77	8.20	5.53	m2	13.73
50 mm insulation to walls					
over 300 mm wide	0.40	4.26	9.29	m2	13.55
not exceeding 300 mm wide	0.64	6.82	9.67	m2	16.49
50 mm insulation to isolated columns					
over 300 mm wide	0.48	5.11	9.29	m2	14.40
not exceeding 300 mm wide	0.77	8.20	9.67	m2	17.87
50 mm insulation to ceilings					
over 300 mm wide	0.43	4.58	9.29	m2	13.87
not exceeding 300 mm wide	0.68	7.24	9.67	m2	16.91
50 mm insulation to isolated beams					
over 300 mm wide	0.52	5.54	9.29	m2	14.83
not exceeding 300 mm wide	0.83	8.84	9.67	m2	18.51
P11 FOAMED/FIBRE/BEAD CAVITY WALL INSULATION					
Cavity wall insulation; injecting 65 mm cavity with					
UF foam	-	-	-	m2	3.09
blown EPS granules	-	-	-	m2	3.38
blown mineral wool	-	-	-	m2	3.60
P20 UNFRAMED ISOLATED TRIMS/SKIRTINGS/ SUNDRY ITEMS					
Blockboard (Birch faced); 18 mm thick PC £93.71/10m2					
Window boards and the like; rebated; hardwood lipped on one edge					
18 x 200 mm	0.28	2.62	3.32	m	5.94
18 x 250 mm	0.31	2.90	3.88	m	6.78
18 x 300 mm	0.34	3.18	4.45	m	7.63
18 x 350 mm	0.37	3.46	5.01	m	8.47
Returned and fitted ends	0.24	2.24	0.34	nr	2.58

P BUILDING FABRIC SUNDRIES Including overheads and profit at 12.50%	Labour hours	Labour £	Material £	Unit	Total rate £
P20 UNFRAMED ISOLATED TRIMS/SKIRTINGS/ SUNDRY ITEMS - cont'd					
Blockboard (Sapele veneered one side); **18 mm thick PC £100.58/10m2** Window boards and the like; rebated; hardwood lipped on one edge					
18 x 200 mm	0.30	2.80	3.48	m	6.28
18 x 250 mm	0.33	3.09	4.08	m	7.17
18 x 300 mm	0.36	3.37	4.69	m	8.06
18 x 350 mm	0.40	3.74	5.30	m	9.04
Returned and fitted ends	0.24	2.24	0.34	nr	2.58
Blockboard (Afrormosia veneered one side); **18 mm thick PC £114.47/10m2** Window boards and the like; rebated; hardwood lipped on one edge					
18 x 200 mm	0.30	2.80	3.90	m	6.70
18 x 250 mm	0.33	3.09	4.60	m	7.69
18 x 300 mm	0.36	3.37	5.29	m	8.66
18 x 350 mm	0.40	3.74	5.98	m	9.72
Returned and fitted ends	0.24	2.24	0.37	nr	2.61
Wrought softwood Skirtings, picture rails, dado rails and the like; splayed or moulded					
19 x 50 mm; splayed	0.11	1.03	1.42	m	2.45
19 x 50 mm; moulded	0.11	1.03	1.54	m	2.57
19 x 75 mm; splayed	0.11	1.03	1.77	m	2.80
19 x 75 mm; moulded	0.11	1.03	1.86	m	2.89
19 x 100 mm; splayed	0.11	1.03	2.10	m	3.13
19 x 100 mm; moulded	0.11	1.03	2.21	m	3.24
19 x 150 mm; moulded	0.13	1.22	2.90	m	4.12
19 x 175 mm; moulded	0.13	1.22	3.22	m	4.44
22 x 100 mm; splayed	0.11	1.03	2.32	m	3.35
25 x 50 mm; moulded	0.11	1.03	1.78	m	2.81
25 x 75 mm; splayed	0.11	1.03	2.10	m	3.13
25 x 100 mm; splayed	0.11	1.03	2.51	m	3.54
25 x 150 mm; splayed	0.13	1.22	3.40	m	4.62
25 x 150 mm; moulded	0.13	1.22	3.52	m	4.74
25 x 175 mm; moulded	0.13	1.22	3.94	m	5.16
25 x 225 mm; moulded	0.15	1.40	4.84	m	6.24
Returned end	0.17	1.59	-	nr	1.59
Mitres	0.11	1.03	-	nr	1.03
Architraves, cover fillets and the like; half round; splayed or moulded					
13 x 25 mm; half round	0.13	1.22	1.01	m	2.23
13 x 50 mm; moulded	0.13	1.22	1.34	m	2.56
16 x 32 mm; half round	0.13	1.22	1.15	m	2.37
16 x 38 mm; moulded	0.13	1.22	1.32	m	2.54
16 x 50 mm; moulded	0.13	1.22	1.46	m	2.68
19 x 50 mm; splayed	0.13	1.22	1.42	m	2.64
19 x 63 mm; splayed	0.13	1.22	1.59	m	2.81
19 x 75 mm; splayed	0.13	1.22	1.77	m	2.99
25 x 44 mm; splayed	0.13	1.22	1.67	m	2.89
25 x 50 mm; moulded	0.13	1.22	1.78	m	3.00
25 x 63 mm; splayed	0.13	1.22	1.86	m	3.08
25 x 75 mm; splayed	0.13	1.22	2.10	m	3.32
32 x 88 mm; moulded	0.13	1.22	3.02	m	4.24
38 x 38 mm; moulded	0.13	1.22	1.79	m	3.01
50 x 50 mm; moulded	0.13	1.22	2.51	m	3.73
Returned end	0.17	1.59	-	nr	1.59
Mitres	0.11	1.03	-	nr	1.03

P BUILDING FABRIC SUNDRIES Including overheads and profit at 12.50%		Labour hours	Labour £	Material £	Unit	Total rate £
Stops; screwed on						
16 x 38 mm		0.11	1.03	1.07	m	2.10
16 x 50 mm		0.11	1.03	1.21	m	2.24
19 x 38 mm		0.11	1.03	1.14	m	2.17
25 x 38 mm		0.11	1.03	1.31	m	2.34
25 x 50 mm		0.11	1.03	1.53	m	2.56
Glazing beads and the like						
13 x 16 mm	PC £0.61	-	-	0.74	m	0.74
13 x 19 mm	PC £0.64	-	-	0.77	m	0.77
13 x 25 mm	PC £0.68	-	-	0.82	m	0.82
13 x 25 mm; screwed		0.06	0.56	0.93	m	1.49
13 x 25 mm; fixing with brass cups and screws		0.11	1.03	2.78	m	3.81
16 x 25 mm	PC £0.74	-	-	0.91	m	0.91
16 mm; quadrant		0.06	0.56	0.89	m	1.45
19 mm; quadrant or scotia		0.06	0.56	0.98	m	1.54
19 x 36 mm		0.06	0.56	1.10	m	1.66
25 x 38 mm		0.06	0.56	1.24	m	1.80
25 mm; quadrant or scotia		0.06	0.56	1.13	m	1.69
38 mm; scotia		0.06	0.56	1.73	m	2.29
50 mm; scotia		0.06	0.56	2.44	m	3.00
Isolated shelves, worktops, seats and the like						
19 x 150 mm		0.18	1.68	2.78	m	4.46
19 x 200 mm		0.24	2.24	3.41	m	5.65
25 x 150 mm		0.18	1.68	3.41	m	5.09
25 x 200 mm		0.24	2.24	4.22	m	6.46
32 x 150 mm		0.18	1.68	4.09	m	5.77
32 x 200 mm		0.24	2.24	5.21	m	7.45
Isolated shelves, worktops, seats and the like; cross-tongued joints						
19 x 300 mm		0.31	2.90	11.07	m	13.97
19 x 450 mm		0.37	3.46	13.09	m	16.55
19 x 600 mm		0.44	4.11	18.81	m	22.92
25 x 300 mm		0.31	2.90	12.61	m	15.51
25 x 450 mm		0.37	3.46	15.20	m	18.66
25 x 600 mm		0.44	4.11	21.66	m	25.77
32 x 300 mm		0.31	2.90	14.15	m	17.05
32 x 450 mm		0.37	3.46	17.51	m	20.97
32 x 600 mm		0.44	4.11	24.76	m	28.87
Isolated shelves, worktops, seats and the like; slatted with 50 mm wide slats at 75 mm centres						
19 mm thick		1.45	13.56	9.31	m2	22.87
25 mm thick		1.45	13.56	10.94	m2	24.50
32 mm thick		1.45	13.56	12.37	m2	25.93
Window boards, nosings, bed moulds and the like; rebated and rounded						
19 x 75 mm		0.20	1.87	1.96	m	3.83
19 x 150 mm		0.22	2.06	2.99	m	5.05
19 x 225 mm; in one width		0.29	2.71	3.97	m	6.68
19 x 300 mm; cross-tongued joints		0.33	3.09	11.13	m	14.22
25 x 75 mm		0.20	1.87	2.29	m	4.16
25 x 150 mm		0.22	2.06	3.62	m	5.68
25 x 225 mm; in one width		0.29	2.71	4.92	m	7.63
25 x 300 mm; cross-tongued joints		0.33	3.09	12.61	m	15.70
32 x 75 mm		0.20	1.87	2.66	m	4.53
32 x 150 mm		0.22	2.06	4.33	m	6.39
32 x 225 mm; in one width		0.29	2.71	6.00	m	8.71
32 x 300 mm; cross-tongued joints		0.33	3.09	14.16	m	17.25
38 x 75 mm		0.20	1.87	2.98	m	4.85
38 x 150 mm		0.22	2.06	4.96	m	7.02
38 x 225 mm; in one width		0.29	2.71	6.95	m	9.66
38 x 300 mm; cross-tongued joints		0.33	3.09	15.55	m	18.64
Returned and fitted ends		0.17	1.59	-	nr	1.59
Handrails; mopstick						
50 mm dia		0.28	2.62	3.42	m	6.04

P BUILDING FABRIC SUNDRIES Including overheads and profit at 12.50%		Labour hours	Labour £	Material £	Unit	Total rate £
P20 UNFRAMED ISOLATED TRIMS/SKIRTINGS/ **SUNDRY ITEMS - cont'd**						
Wrought softwood - cont'd						
Handrails; rounded						
44 x 50 mm		0.28	2.62	3.42	m	6.04
50 x 75 mm		0.30	2.80	4.34	m	7.14
63 x 87 mm		0.33	3.09	6.08	m	9.17
75 x 100 mm		0.39	3.65	6.90	m	10.55
Handrails; moulded						
44 x 50 mm		0.22	2.06	3.74	m	5.80
50 x 75 mm		0.24	2.24	4.67	m	6.91
63 x 87 mm		0.26	2.43	6.41	m	8.84
75 x 100 mm		0.29	2.71	7.22	m	9.93
Add +5% to the above 'Material £ prices' for 'selected' softwood for staining						
Selected West African Mahogany; PC £531.05/m3						
Skirtings, picture rails, dado rails						
and the like; splayed or moulded						
19 x 50 mm; splayed	PC £2.35	0.15	1.40	2.84	m	4.24
19 x 50 mm; moulded	PC £2.49	0.15	1.40	3.00	m	4.40
19 x 75 mm; splayed	PC £2.82	0.15	1.40	3.39	m	4.79
19 x 75 mm; moulded	PC £2.95	0.15	1.40	3.55	m	4.95
19 x 100 mm; splayed	PC £3.31	0.15	1.40	3.98	m	5.38
19 x 100 mm; moulded	PC £3.44	0.15	1.40	4.12	m	5.52
19 x 150 mm; moulded	PC £4.38	0.18	1.68	5.24	m	6.92
19 x 175 mm; moulded	PC £5.02	0.18	1.68	5.99	m	7.67
22 x 100 mm; splayed	PC £3.59	0.15	1.40	4.30	m	5.70
25 x 50 mm; moulded	PC £2.80	0.15	1.40	3.37	m	4.77
25 x 75 mm; splayed	PC £3.28	0.15	1.40	3.93	m	5.33
25 x 100 mm; splayed	PC £3.88	0.15	1.40	4.65	m	6.05
25 x 150 mm; splayed	PC £5.33	0.18	1.68	6.36	m	8.04
25 x 150 mm; moulded	PC £5.47	0.18	1.68	6.52	m	8.20
25 x 175 mm; moulded	PC £6.07	0.18	1.68	7.24	m	8.92
25 x 225 mm; moulded	PC £7.31	0.20	1.87	8.70	m	10.57
Returned end		0.24	2.24	-	nr	2.24
Mitres		0.17	1.59	-	nr	1.59
Architraves, cover fillets and the like;						
half round; splayed or moulded						
13 x 25 mm; half round	PC £1.74	0.18	1.68	2.13	m	3.81
13 x 50 mm; moulded	PC £2.19	0.18	1.68	2.65	m	4.33
16 x 32 mm; half round	PC £1.93	0.18	1.68	2.34	m	4.02
16 x 38 mm; moulded	PC £2.15	0.18	1.68	2.61	m	4.29
16 x 50 mm; moulded	PC £2.34	0.18	1.68	2.83	m	4.51
19 x 50 mm; splayed	PC £2.35	0.18	1.68	2.84	m	4.52
19 x 63 mm; splayed	PC £2.60	0.18	1.68	3.13	m	4.81
19 x 75 mm; splayed	PC £2.82	0.18	1.68	3.39	m	5.07
25 x 44 mm; splayed	PC £2.67	0.18	1.68	3.21	m	4.89
25 x 50 mm; moulded	PC £2.80	0.18	1.68	3.37	m	5.05
25 x 63 mm; splayed	PC £2.97	0.18	1.68	3.57	m	5.25
25 x 75 mm; splayed	PC £3.28	0.18	1.68	3.93	m	5.61
32 x 88 mm; moulded	PC £4.61	0.18	1.68	5.51	m	7.19
38 x 38 mm; moulded	PC £2.84	0.18	1.68	3.42	m	5.10
50 x 50 mm; moulded	PC £3.88	0.18	1.68	4.65	m	6.33
Returned end		0.24	2.24	-	nr	2.24
Mitres		0.17	1.59	-	nr	1.59
Stops; screwed on						
16 x 38 mm	PC £1.86	0.17	1.59	2.24	m	3.83
16 x 50 mm	PC £2.07	0.17	1.59	2.49	m	4.08
19 x 38 mm	PC £2.00	0.17	1.59	2.41	m	4.00
25 x 38 mm	PC £2.22	0.17	1.59	2.67	m	4.26
25 x 50 mm	PC £2.53	0.17	1.59	3.04	m	4.63

P BUILDING FABRIC SUNDRIES Including overheads and profit at 12.50%		Labour hours	Labour £	Material £	Unit	Total rate £
Glazing beads and the like						
13 x 16 mm	PC £1.48	-	-	1.74	m	1.74
13 x 19 mm	PC £1.52	-	-	1.80	m	1.80
13 x 25 mm	PC £1.61	-	-	1.90	m	1.90
13 x 25 mm; screwed	PC £1.61	0.09	0.84	2.01	m	2.85
13 x 25 mm; fixing with brass cups and screws	PC £1.61	0.17	1.59	3.85	m	5.44
16 x 25 mm	PC £1.67	-	-	1.97	m	1.97
16 mm; quadrant	PC £1.67	0.08	0.75	1.97	m	2.72
19 mm; quadrant or scotia	PC £1.76	0.08	0.75	2.08	m	2.83
19 x 36 mm	PC £2.00	0.08	0.75	2.36	m	3.11
25 x 38 mm	PC £2.22	0.08	0.75	2.63	m	3.38
25 mm; quadrant or scotia	PC £2.02	0.08	0.75	2.39	m	3.14
38 mm; scotia	PC £2.84	0.08	0.75	3.35	m	4.10
50 mm; scotia	PC £3.88	0.08	0.75	4.58	m	5.33
Isolated shelves, worktops, seats and the like						
19 x 150 mm	PC £4.51	0.24	2.24	5.33	m	7.57
19 x 200 mm	PC £5.43	0.33	3.09	6.41	m	9.50
25 x 150 mm	PC £5.43	0.24	2.24	6.41	m	8.65
25 x 200 mm	PC £6.60	0.33	3.09	7.79	m	10.88
32 x 150 mm	PC £6.40	0.24	2.24	7.56	m	9.80
32 x 200 mm	PC £7.98	0.33	3.09	9.43	m	12.52
Isolated shelves, worktops, seats and the like; cross-tongued joints						
19 x 300 mm	PC £15.00	0.42	3.93	17.72	m	21.65
19 x 450 mm	PC £17.87	0.50	4.67	21.11	m	25.78
19 x 600 mm	PC £25.95	0.61	5.70	30.65	m	36.35
25 x 300 mm	PC £17.08	0.42	3.93	20.18	m	24.11
25 x 450 mm	PC £20.86	0.50	4.67	24.64	m	29.31
25 x 600 mm	PC £29.92	0.61	5.70	35.35	m	41.05
32 x 300 mm	PC £19.17	0.42	3.93	22.65	m	26.58
32 x 450 mm	PC £23.94	0.50	4.67	28.28	m	32.95
32 x 600 mm	PC £34.18	0.61	5.70	40.38	m	46.08
Isolated shelves, worktops, seats and the like; slatted with 50 mm wide slats at 75 mm centres						
19 mm thick	PC £16.97/m	1.95	18.23	20.76	m2	38.99
25 mm thick	PC £19.13/m	1.95	18.23	23.37	m2	41.60
32 mm thick	PC £21.12/m	1.95	18.23	25.77	m2	44.00
Window boards, nosings, bed moulds and the like; rebated and rounded						
19 x 75 mm	PC £3.14	0.26	2.43	3.86	m	6.29
19 x 150 mm	PC £4.55	0.30	2.80	5.53	m	8.33
19 x 225 mm; in one width	PC £5.97	0.40	3.74	7.20	m	10.94
19 x 300 mm; cross-tongued joints	PC £14.90	0.44	4.11	17.76	m	21.87
25 x 75 mm	PC £3.61	0.26	2.43	4.40	m	6.83
25 x 150 mm	PC £5.51	0.30	2.80	6.66	m	9.46
25 x 225 mm; in one width	PC £7.31	0.40	3.74	8.78	m	12.52
25 x 300 mm; cross-tongued joints	PC £16.90	0.44	4.11	20.12	m	24.23
32 x 75 mm	PC £4.13	0.26	2.43	5.02	m	7.45
32 x 150 mm	PC £6.49	0.30	2.80	7.81	m	10.61
32 x 225 mm; in one width	PC £8.89	0.40	3.74	10.64	m	14.38
32 x 300 mm; cross-tongued joints	PC £18.99	0.44	4.11	22.58	m	26.69
38 x 75 mm	PC £4.55	0.26	2.43	5.53	m	7.96
38 x 150 mm	PC £7.40	0.30	2.80	8.89	m	11.69
38 x 225 mm; in one width	PC £10.23	0.40	3.74	12.23	m	15.97
38 x 300 mm; cross-tongued joints	PC £20.88	0.44	4.11	24.81	m	28.92
Returned and fitted ends		0.25	2.34	-	nr	2.34
Handrails; rounded						
44 x 50 mm	PC £5.28	0.36	3.37	6.23	m	9.60
50 x 75 mm	PC £6.56	0.40	3.74	7.76	m	11.50
63 x 87 mm	PC £8.46	0.44	4.11	9.99	m	14.10
75 x 100 mm	PC £10.24	0.50	4.67	12.10	m	16.77

P BUILDING FABRIC SUNDRIES Including overheads and profit at 12.50%		Labour hours	Labour £	Material £	Unit	Total rate £
P20 UNFRAMED ISOLATED TRIMS/SKIRTINGS/ **SUNDRY ITEMS - cont'd**						
Selected West African Mahogany; PC £531.05/m3 - cont'd						
Handrails; moulded						
44 x 50 mm	PC £5.69	0.36	3.37	6.72	m	10.09
50 x 75 mm	PC £6.99	0.40	3.74	8.26	m	12.00
63 x 87 mm	PC £8.87	0.44	4.11	10.48	m	14.59
75 x 100 mm	PC £10.65	0.50	4.67	12.58	m	17.25
Afrormosia; PC £758.95/m3						
Skirtings, picture rails, dado rails						
and the like; splayed or moulded						
19 x 50 mm; splayed	PC £2.76	0.15	1.40	3.32	m	4.72
19 x 50 mm; moulded	PC £2.89	0.15	1.40	3.48	m	4.88
19 x 75 mm; splayed	PC £3.44	0.15	1.40	4.12	m	5.52
19 x 75 mm; moulded	PC £3.57	0.15	1.40	4.29	m	5.69
19 x 100 mm; splayed	PC £4.11	0.15	1.40	4.92	m	6.32
19 x 100 mm; moulded	PC £4.24	0.15	1.40	5.08	m	6.48
19 x 150 mm; moulded	PC £5.58	0.18	1.68	6.66	m	8.34
19 x 175 mm; moulded	PC £6.45	0.18	1.68	7.68	m	9.36
22 x 100 mm; splayed	PC £4.54	0.15	1.40	5.43	m	6.83
25 x 50 mm; moulded	PC £3.34	0.15	1.40	4.01	m	5.41
25 x 75 mm; splayed	PC £4.06	0.15	1.40	4.86	m	6.26
25 x 100 mm; splayed	PC £4.96	0.15	1.40	5.92	m	7.32
25 x 150 mm; splayed	PC £6.93	0.18	1.68	8.25	m	9.93
25 x 150 mm; moulded	PC £7.06	0.18	1.68	8.40	m	10.08
25 x 175 mm; moulded	PC £7.93	0.18	1.68	9.43	m	11.11
25 x 225 mm; moulded	PC £9.71	0.20	1.87	11.54	m	13.41
Returned end		0.24	2.24	-	nr	2.24
Mitres		0.17	1.59	-	nr	1.59
Architraves, cover fillets and the like;						
half round; splayed or moulded						
13 x 25 mm; half round	PC £1.87	0.18	1.68	2.27	m	3.95
13 x 50 mm; moulded	PC £2.46	0.18	1.68	2.96	m	4.64
16 x 32 mm; half round	PC £2.13	0.18	1.68	2.58	m	4.26
16 x 38 mm; moulded	PC £2.41	0.18	1.68	2.92	m	4.60
16 x 50 mm; moulded	PC £2.68	0.18	1.68	3.23	m	4.91
19 x 50 mm; splayed	PC £2.76	0.18	1.68	3.32	m	5.00
19 x 63 mm; splayed	PC £3.13	0.18	1.68	3.76	m	5.44
19 x 75 mm; splayed	PC £3.44	0.18	1.68	4.12	m	5.80
25 x 44 mm; splayed	PC £3.20	0.18	1.68	3.85	m	5.53
25 x 50 mm; moulded	PC £3.34	0.18	1.68	4.01	m	5.69
25 x 63 mm; splayed	PC £3.65	0.18	1.68	4.37	m	6.05
25 x 75 mm; splayed	PC £4.06	0.18	1.68	4.86	m	6.54
32 x 88 mm; moulded	PC £5.95	0.18	1.68	7.09	m	8.77
38 x 38 mm; moulded	PC £3.47	0.18	1.68	4.16	m	5.84
50 x 50 mm; moulded	PC £4.96	0.18	1.68	5.92	m	7.60
Returned end		0.24	2.24	-	nr	2.24
Mitres		0.17	1.59	-	nr	1.59
Stops; screwed on						
16 x 38 mm	PC £2.15	0.17	1.59	2.58	m	4.17
16 x 50 mm	PC £2.41	0.17	1.59	2.90	m	4.49
19 x 38 mm	PC £2.30	0.17	1.59	2.75	m	4.34
25 x 38 mm	PC £2.61	0.17	1.59	3.12	m	4.71
25 x 50 mm	PC £3.05	0.17	1.59	3.65	m	5.24
Glazing beads and the like						
13 x 16 mm	PC £1.57	-	-	1.86	m	1.86
13 x 19 mm	PC £1.63	-	-	1.92	m	1.92
13 x 25 mm	PC £1.74	-	-	2.06	m	2.06
13 x 25 mm; screwed	PC £1.74	0.09	0.84	2.17	m	3.01
13 x 25 mm; fixing with brass cups						
and screws	PC £1.74	0.17	1.59	4.02	m	5.61
16 x 25 mm	PC £1.85	-	-	2.18	m	2.18
16 mm; quadrant	PC £1.79	0.08	0.75	2.11	m	2.86
19 mm; quadrant or scotia	PC £1.94	0.08	0.75	2.29	m	3.04

P BUILDING FABRIC SUNDRIES Including overheads and profit at 12.50%		Labour hours	Labour £	Material £	Unit	Total rate £
19 x 36 mm	PC £2.30	0.08	0.75	2.71	m	3.46
25 x 38 mm	PC £2.90	0.08	0.75	3.44	m	4.19
25 mm; quadrant or scotia	PC £2.31	0.08	0.75	2.73	m	3.48
38 mm; scotia	PC £3.47	0.08	0.75	4.09	m	4.84
50 mm; scotia	PC £4.96	0.08	0.75	5.86	m	6.61
Isolated shelves, worktops, seats and the like						
19 x 150 mm	PC £5.77	0.24	2.24	6.81	m	9.05
19 x 200 mm	PC £7.04	0.33	3.09	8.32	m	11.41
25 x 150 mm	PC £7.04	0.24	2.24	8.32	m	10.56
25 x 200 mm	PC £8.72	0.33	3.09	10.31	m	13.40
32 x 150 mm	PC £8.46	0.24	2.24	9.99	m	12.23
32 x 200 mm	PC £10.72	0.33	3.09	12.67	m	15.76
Isolated shelves, worktops, seats and the like; cross-tongued joints						
19 x 300 mm	PC £17.69	0.42	3.93	20.89	m	24.82
19 x 450 mm	PC £21.81	0.50	4.67	25.76	m	30.43
19 x 600 mm	PC £31.31	0.61	5.70	36.99	m	42.69
25 x 300 mm	PC £20.67	0.42	3.93	24.41	m	28.34
25 x 450 mm	PC £26.05	0.50	4.67	30.77	m	35.44
25 x 600 mm	PC £36.98	0.61	5.70	43.68	m	49.38
32 x 300 mm	PC £23.64	0.42	3.93	27.92	m	31.85
32 x 450 mm	PC £30.46	0.50	4.67	35.97	m	40.64
32 x 600 mm	PC £43.08	0.61	5.70	50.89	m	56.59
Isolated shelves, worktops, seats and the like; slatted with 50 mm wide slats at 75 mm centres						
19 mm thick	PC £19.95/m	1.95	18.23	24.37	m2	42.60
25 mm thick	PC £22.89/m	1.95	18.23	27.93	m2	46.16
32 mm thick	PC £26.02/m	1.95	18.23	31.85	m2	50.08
Window boards, nosings, bed moulds and the like; rebated and rounded						
19 x 75 mm	PC £3.73	0.26	2.43	4.56	m	6.99
19 x 150 mm	PC £5.78	0.30	2.80	6.97	m	9.77
19 x 225 mm; in one width	PC £7.78	0.40	3.74	9.33	m	13.07
19 x 300 mm; cross-tongued joints	PC £17.64	0.44	4.11	20.98	m	25.09
25 x 75 mm	PC £4.40	0.26	2.43	5.35	m	7.78
25 x 150 mm	PC £7.14	0.30	2.80	8.58	m	11.38
25 x 225 mm; in one width	PC £9.69	0.40	3.74	11.59	m	15.33
25 x 300 mm; cross-tongued joints	PC £20.48	0.44	4.11	24.33	m	28.44
32 x 75 mm	PC £5.15	0.26	2.43	6.23	m	8.66
32 x 150 mm	PC £8.55	0.30	2.80	10.25	m	13.05
32 x 225 mm; in one width	PC £11.98	0.40	3.74	14.30	m	18.04
32 x 300 mm; cross-tongued joints	PC £23.46	0.44	4.11	27.85	m	31.96
38 x 75 mm	PC £5.78	0.26	2.43	6.97	m	9.40
38 x 150 mm	PC £9.84	0.30	2.80	11.77	m	14.57
38 x 225 mm; in one width	PC £13.88	0.40	3.74	16.54	m	20.28
38 x 300 mm; cross-tongued joints	PC £26.17	0.44	4.11	31.06	m	35.17
Returned and fitted ends		0.25	2.34	-	nr	2.34
Handrails; rounded						
44 x 50 mm	PC £6.34	0.36	3.37	7.49	m	10.86
50 x 75 mm	PC £8.13	0.40	3.74	9.60	m	13.34
63 x 87 mm	PC £11.75	0.44	4.11	13.89	m	18.00
75 x 100 mm	PC £13.44	0.50	4.67	15.87	m	20.54
Handrails; moulded						
44 x 50 mm	PC £6.74	0.36	3.37	7.97	m	11.34
50 x 75 mm	PC £8.53	0.40	3.74	10.08	m	13.82
63 x 87 mm	PC £12.15	0.44	4.11	14.35	m	18.46
75 x 100 mm	PC £13.85	0.50	4.67	16.36	m	21.03

Pin-boards; medium board
Sundeala 'A' pin-board; fixed with adhesive to backing (measured elsewhere); over 300 mm wide

6.4 mm thick		0.66	6.17	9.11	m2	15.28

P BUILDING FABRIC SUNDRIES Including overheads and profit at 12.50%	Labour hours	Labour £	Material £	Unit	Total rate £
P20 UNFRAMED ISOLATED TRIMS/SKIRTINGS/ **SUNDRY ITEMS - cont'd**					
Sundries on softwood/hardwood					
Extra over fixing with nails for					
gluing and pinning	0.02	0.19	0.03	m	0.22
masonry nails	0.02	0.19	0.05	m	0.24
steel screws	0.02	0.19	0.07	m	0.26
self-tapping screws	0.02	0.19	0.09	m	0.28
steel screws; gluing	0.04	0.37	0.07	m	0.44
steel screws; sinking; filling heads	0.06	0.56	0.07	m	0.63
steel screws; sinking; pellating over	0.11	1.03	2.06	m	3.09
brass cups and screws	0.17	1.59	0.07	m	1.66
Extra over for					
countersinking	0.02	0.19	-	m	0.19
pellating	0.11	1.03	0.03	m	1.06
Head or nut in softwood					
let in flush	0.06	0.56	-	nr	0.56
Head or nut; in hardwood					
let in flush	0.09	0.84	-	nr	0.84
let in over; pellated	0.22	2.06	-	nr	2.06
P21 IRONMONGERY					
Metalwork; mild steel; galvanized					
Water bars; groove in timber					
6 x 30 mm	0.55	5.14	3.81	m	8.95
6 x 40 mm	0.55	5.14	4.60	m	9.74
6 x 50 mm	0.55	5.14	5.39	m	10.53
Dowels; mortice in timber					
8 mm dia x 100 mm long	0.06	0.56	0.05	nr	0.61
10 mm dia x 50 mm long	0.06	0.56	0.15	nr	0.71
Cramps					
25 x 3 x 230 mm girth; one end bent, holed					
and screwed to softwood; other end					
fishtailed for building in	0.08	0.75	0.27	nr	1.02
Fixing only ironmongery to softwood					
Bolts					
barrel; not exceeding 150 mm long	0.36	3.37	-	nr	3.37
barrel; 150-300 mm long	0.46	4.30	-	nr	4.30
cylindrical mortice; not exceeding					
150 mm long	0.55	5.14	-	nr	5.14
150 - 300 mm long	0.66	6.17	-	nr	6.17
flush; not exceeding 150 mm long	0.55	5.14	-	nr	5.14
flush; 150 - 300 mm long	0.66	6.17	-	nr	6.17
monkey tail; 380 mm long	0.74	6.92	-	nr	6.92
necked; 150 mm long	0.36	3.37	-	nr	3.37
panic; single; locking	2.75	25.71	-	nr	25.71
panic; double; locking	3.85	35.99	-	nr	35.99
WC indicator	0.74	6.92	-	nr	6.92
Butts; extra over for					
rising	0.19	1.78	-	pr	1.78
skew	0.19	1.78	-	pr	1.78
spring; single action	1.45	13.56	-	pr	13.56
spring; double action	1.65	15.43	-	pr	15.43
Catches					
surface mounted	0.19	1.78	-	nr	1.78
mortice	0.36	3.37	-	nr	3.37
Door closers and furniture					
cabin hooks and eyes	0.19	1.78	-	nr	1.78
door selector	0.55	5.14	-	nr	5.14
finger plate	0.19	1.78	-	nr	1.78
floor spring	2.75	25.71	-	nr	25.71
lever furniture	0.28	2.62	-	nr	2.62
handle; not exceeding 150 mm long	0.19	1.78	-	nr	1.78

P BUILDING FABRIC SUNDRIES Including overheads and profit at 12.50%	Labour hours	Labour £	Material £	Unit	Total rate £
handle; 150 - 300 mm long	0.28	2.62	-	nr	2.62
handle; flush	0.36	3.37	-	nr	3.37
holder	0.36	3.37	-	nr	3.37
kicking plate	0.36	3.37	-	nr	3.37
letter plate; including perforation	1.45	13.56	-	nr	13.56
overhead door closer; surface fixing	1.40	13.09	-	nr	13.09
overhead door closer; concealed fixing	1.95	18.23	-	nr	18.23
top centre	0.55	5.14	-	nr	5.14
'Perko' door closer	0.74	6.92	-	nr	6.92
rod door closers; 457 mm long	1.10	10.28	-	nr	10.28
Latches					
cylinder rim night latch	0.83	7.76	-	nr	7.76
mortice	0.74	6.92	-	nr	6.92
Norfolk	0.74	6.92	-	nr	6.92
rim	0.55	5.14	-	nr	5.14
Locks					
cupboard	0.46	4.30	-	nr	4.30
mortice	0.91	8.51	-	nr	8.51
mortice budget	0.83	7.76	-	nr	7.76
mortice dead	0.83	7.76	-	nr	7.76
rebated mortice	1.40	13.09	-	nr	13.09
rim	0.55	5.14	-	nr	5.14
rim budget	0.46	4.30	-	nr	4.30
rim dead	0.46	4.30	-	nr	4.30
Sliding door gear for top hung softwood **timber doors; weight not exceeding 365 kg**					
bottom guide; fixed to concrete in groove	0.55	5.14	-	m	5.14
top track	0.28	2.62	-	m	2.62
detachable locking bar and padlock	0.36	3.37	-	nr	3.37
hangers; fixed flush to timber	0.83	7.76	-	nr	7.76
head brackets; bolted to concrete	0.46	4.30	-	nr	4.30
Window furniture					
casement stay and pin	0.19	1.78	-	nr	1.78
catch; fanlight	0.28	2.62	-	nr	2.62
fastener; cockspur	0.36	3.37	-	nr	3.37
fastener; sash	0.28	2.62	-	nr	2.62
quadrant stay	0.28	2.62	-	nr	2.62
ring catch	0.36	3.37	-	nr	3.37
Sundries					
drawer pull	0.09	0.84	-	nr	0.84
hat and coat hook	0.09	0.84	-	nr	0.84
numerals	0.09	0.84	-	nr	0.84
rubber door stop	0.09	0.84	-	nr	0.84
shelf bracket	0.19	1.78	-	nr	1.78
skirting type door stop	0.19	1.78	-	nr	1.78
Fixing only ironmongery to hardwood					
Bolts					
barrel; not exceeding 150 mm long	0.48	4.49	-	nr	4.49
barrel; 150-300 mm long	0.62	5.80	-	nr	5.80
cylindrical mortice; not exceeding					
150 mm long	0.74	6.92	-	nr	6.92
150 - 300 mm long	0.88	8.23	-	nr	8.23
flush; not exceeding 150 mm long	0.74	6.92	-	nr	6.92
flush; 150 - 300 mm long	0.88	8.23	-	nr	8.23
monkey tail; 380 mm long	0.98	9.16	-	nr	9.16
necked; 150 mm long	0.48	4.49	-	nr	4.49
panic; single; locking	3.65	34.12	-	nr	34.12
panic; double; locking	5.15	48.15	-	nr	48.15
WC indicator	0.98	9.16	-	nr	9.16
Butts; extra over for					
rising	0.25	2.34	-	pr	2.34
skew	0.25	2.34	-	pr	2.34
spring; single action	1.90	17.76	-	pr	17.76
spring; double action	2.20	20.57	-	pr	20.57
Catches					
surface mounted	0.25	2.34	-	nr	2.34
mortice	0.48	4.49	-	nr	4.49

P BUILDING FABRIC SUNDRIES Including overheads and profit at 12.50%	Labour hours	Labour £	Material £	Unit	Total rate £
P21 IRONMONGERY - cont'd					
Fixing only ironmongery to hardwood - cont'd					
Door closers and furniture					
cabin hooks and eyes	0.25	2.34	-	nr	2.34
door selector	0.74	6.92	-	nr	6.92
finger plate	0.25	2.34	-	nr	2.34
floor spring	3.65	34.12	-	nr	34.12
lever furniture	0.36	3.37	-	nr	3.37
handle; not exceeding 150 mm long	0.25	2.34	-	nr	2.34
handle; 150 - 300 mm long	0.36	3.37	-	nr	3.37
handle; flush	0.48	4.49	-	nr	4.49
holder	0.48	4.49	-	nr	4.49
kicking plate	0.48	4.49	-	nr	4.49
letter plate; including perforation	1.95	18.23	-	nr	18.23
overhead door closer; surface fixing	1.85	17.30	-	nr	17.30
overhead door closer; concealed fixing	2.55	23.84	-	nr	23.84
top centre	0.74	6.92	-	nr	6.92
'Perko' door closer	0.98	9.16	-	nr	9.16
rod door closers; 457 mm long	1.45	13.56	-	nr	13.56
Latches					
cylinder rim night latch	1.10	10.28	-	nr	10.28
mortice	0.98	9.16	-	nr	9.16
Norfolk	0.98	9.16	-	nr	9.16
rim	0.74	6.92	-	nr	6.92
Locks					
cupboard	0.62	5.80	-	nr	5.80
mortice	1.20	11.22	-	nr	11.22
mortice budget	1.10	10.28	-	nr	10.28
mortice dead	1.10	10.28	-	nr	10.28
rebated mortice	1.85	17.30	-	nr	17.30
rim	0.74	6.92	-	nr	6.92
rim budget	0.62	5.80	-	nr	5.80
rim dead	0.62	5.80	-	nr	5.80
Sliding door gear for top hung softwood timber doors; weight not exceeding 365 kg					
bottom guide; fixed to concrete in groove	0.74	6.92	-	m	6.92
top track	0.36	3.37	-	m	3.37
detachable locking bar and padlock	0.48	4.49	-	nr	4.49
hangers; fixed flush to timber	1.10	10.28	-	nr	10.28
head brackets; bolted to concrete	0.62	5.80	-	nr	5.80
Window furniture					
casement stay and pin	0.25	2.34	-	nr	2.34
catch; fanlight	0.36	3.37	-	nr	3.37
fastener; cockspur	0.48	4.49	-	nr	4.49
fastener; sash	0.36	3.37	-	nr	3.37
quadrant stay	0.36	3.37	-	nr	3.37
ring catch	0.48	4.49	-	nr	4.49
Sundries					
drawer pull	0.12	1.12	-	nr	1.12
hat and coat hook	0.12	1.12	-	nr	1.12
numerals	0.12	1.12	-	nr	1.12
rubber door stop	0.12	1.12	-	nr	1.12
shelf bracket	0.25	2.34	-	nr	2.34
skirting type door stop	0.12	1.12	-	nr	1.12
Sundries					
Rubber door stop plugged and screwed to concrete	0.11	1.03	2.90	nr	3.93

P BUILDING FABRIC SUNDRIES Including overheads and profit at 12.50%	Labour hours	Labour £	Material £	Unit	Total rate £

P30 TRENCHES/PIPEWAYS/PITS FOR BURIED ENGINEERING SERVICES

Mechanical excavation of trenches to receive pipes; grading bottoms; earthwork support; filling with excavated material and compacting; disposal of surplus soil; spreading on site average 50 m
Pipes not exceeding 200 mm; average depth

0.50 m deep	0.23	1.60	1.28	m	2.88
0.75 m deep	0.35	2.44	2.14	m	4.58
1.00 m deep	0.69	4.81	3.84	m	8.65
1.25 m deep	1.05	7.32	5.22	m	12.54
1.50 m deep	1.30	9.07	6.82	m	15.89
1.75 m deep	1.60	11.16	8.74	m	19.90
2.00 m deep	1.90	13.25	10.02	m	23.27

Hand excavation of trenches to receive pipes; grading bottoms; earthwork support; filling with excavated material and compacting; disposal of surplus soil; spreading on site average 50 m
Pipes not exceeding 200 mm; average depth

0.50 m deep	1.10	7.67	-	m	7.67
0.75 m deep	1.65	11.51	-	m	11.51
1.00 m deep	2.40	16.74	1.06	m	17.80
1.25 m deep	3.40	23.72	1.59	m	25.31
1.50 m deep	4.70	32.78	1.90	m	34.68
1.75 m deep	6.15	42.90	2.33	m	45.23
2.00 m deep	7.05	49.17	2.54	m	51.71

Pits for underground stop valves and the like; half brick thick walls in common bricks in cement mortar (1:3); on in situ concrete mix 21.00 N/mm2-20 mm aggregate (1:2:4) bed; 100 mm thick; 100 x 100 x 750 mm deep; internal holes for one small pipe; cast iron hinged box cover; bedding in cement mortar (1:3)

	3.45	41.49	16.67	nr	58.16

P31 HOLES/CHASES/COVERS/SUPPORTS FOR SERVICES

Builders' work for electrical installations
Cutting away for and making good after electrician; including cutting or leaving all holes, notches, mortices, sinkings and chases, in both the structure and its coverings, for the following electrical points
Exposed installation

lighting points	0.35	3.73	-	nr	3.73
socket outlet points	0.58	6.18	-	nr	6.18
fitting outlet points	0.58	6.18	-	nr	6.18
equipment points or control gear points	0.81	8.63	-	nr	8.63
Concealed installation					
lighting points	0.46	4.90	-	nr	4.90
socket outlet points	0.81	8.63	-	nr	8.63
fitting outlet points	0.81	8.63	-	nr	8.63
equipment points or control gear points	1.15	12.25	-	nr	12.25

P BUILDING FABRIC SUNDRIES Including overheads and profit at 12.50%	Labour hours	Labour £	Material £	Unit	Total rate £

P31 HOLES/CHASES/COVERS/SUPPORTS FOR SERVICES - cont'd

Builders' work for other services installations

	Labour hours	Labour £	Material £	Unit	Total rate £
Cutting chases in brickwork					
for one pipe; not exceeding 55 mm nominal size; vertical	0.46	5.53	-	m	5.53
for one pipe; 55 - 110 nominal size; vertical	0.81	9.74	-	m	9.74
Cutting and pinning to brickwork or blockwork; ends of supports					
for pipes not exceeding 55 mm	0.23	2.77	-	m	2.77
for cast iron pipes 55 - 110 mm	0.38	4.57	-	nr	4.57
radiator stays or brackets	0.29	3.49	-	nr	3.49
Cutting holes for pipes or the like; not exceeding 55 mm nominal size					
102 mm brickwork	0.38	2.74	-	nr	2.74
215 mm brickwork	0.63	4.55	-	nr	4.55
327 mm brickwork	1.05	7.58	-	nr	7.58
100 mm blockwork	0.35	2.53	-	nr	2.53
150 mm blockwork	0.46	3.32	-	nr	3.32
215 mm blockwork	0.58	4.19	-	nr	4.19
Cutting holes for pipes or the like; 55 - 110 mm nominal size					
102 mm brickwork	0.46	3.32	-	nr	3.32
215 mm brickwork	0.81	5.85	-	nr	5.85
327 mm brickwork	1.25	9.03	-	nr	9.03
100 mm blockwork	0.40	2.89	-	nr	2.89
150 mm blockwork	0.58	4.19	-	nr	4.19
215 mm blockwork	0.69	4.98	-	nr	4.98
Cutting holes for pipes or the like; over 110 mm nominal size					
102 mm brickwork	0.58	4.19	-	nr	4.19
215 mm brickwork	0.98	7.08	-	nr	7.08
327 mm brickwork	1.55	11.19	-	nr	11.19
100 mm blockwork	0.52	6.25	-	nr	6.25
150 mm blockwork	0.69	4.98	-	nr	4.98
215 mm blockwork	0.86	6.21	-	nr	6.21
Add for making good fair face or facings one side					
pipe; not exceeding 55 mm nominal size	0.09	1.08	-	nr	1.08
pipe; 55 - 110 mm nominal size	0.11	1.32	-	nr	1.32
pipe; over 110 mm nominal size	0.14	1.68	-	nr	1.68
Add for fixing sleeve (supply included elsewhere)					
for pipe; small	0.17	2.04	-	nr	2.04
for pipe; large	0.23	2.77	-	nr	2.77
for pipe; extra large	0.35	4.21	-	nr	4.21
Cutting or forming holes for ducts; girth not exceeding 1.00 m					
102 mm brickwork	0.69	4.98	-	nr	4.98
215 mm brickwork	1.15	8.31	-	nr	8.31
327 mm brickwork	1.85	13.36	-	nr	13.36
100 mm blockwork	0.58	4.19	-	nr	4.19
150 mm blockwork	0.81	5.85	-	nr	5.85
215 mm blockwork	1.05	7.58	-	nr	7.58
Cutting or forming holes for ducts; girth 1.00 - 2.00 m					
102 mm brickwork	0.81	5.85	-	nr	5.85
215 mm brickwork	1.40	10.11	-	nr	10.11
327 mm brickwork	2.20	15.89	-	nr	15.89
100 mm blockwork	0.69	4.98	-	nr	4.98
150 mm blockwork	0.92	6.64	-	nr	6.64
215 mm blockwork	1.15	8.31	-	nr	8.31

P BUILDING FABRIC SUNDRIES Including overheads and profit at 12.50%		Labour hours	Labour £	Material £	Unit	Total rate £
Cutting or forming holes for ducts; **girth 2.00 - 3.00 m**						
102 mm brickwork		1.25	9.03	-	nr	9.03
215 mm brickwork		2.20	15.89	-	nr	15.89
327 mm brickwork		3.45	24.92	-	nr	24.92
100 mm blockwork		1.10	7.94	-	nr	7.94
150 mm blockwork		1.50	10.83	-	nr	10.83
215 mm blockwork		1.90	13.72	-	nr	13.72
Cutting or forming holes for ducts; **girth 3.00 - 4.00 m**						
102 mm brickwork		1.70	12.28	-	nr	12.28
215 mm brickwork		2.90	20.95	-	nr	20.95
327 mm brickwork		4.60	33.22	-	nr	33.22
100 mm blockwork		1.25	9.03	-	nr	9.03
150 mm blockwork		1.70	12.28	-	nr	12.28
215 mm blockwork		2.20	15.89	-	nr	15.89
Mortices in brickwork						
for expansion bolt		0.23	1.66	-	nr	1.66
for 20 mm dia bolt 75 mm deep		0.17	1.23	-	nr	1.23
for 20 mm dia bolt 150 mm deep		0.29	2.09	-	nr	2.09
Mortices in brickwork; grouting with cement **mortar (1:1)**						
75 x 75 x 200 mm deep		0.35	2.53	0.10	nr	2.63
75 x 75 x 300 mm deep		0.46	3.32	0.15	nr	3.47
Holes in softwood for pipes, bars cables **and the like**						
12 mm thick		0.04	0.37	-	nr	0.37
25 mm thick		0.07	0.65	-	nr	0.65
50 mm thick		0.11	1.03	-	nr	1.03
100 mm thick		0.17	1.59	-	nr	1.59
Holes in hardwood for pipes, bars, cables **and the like**						
12 mm thick		0.07	0.65	-	nr	0.65
25 mm thick		0.10	0.93	-	nr	0.93
50 mm thick		0.17	1.59	-	nr	1.59
100 mm thick		0.24	2.24	-	nr	2.24
Duct covers with frames; cast iron; **Brickhouse Glynwed 'Trucast' or similar;** **bedding and pointing frame in cement mortar** **(1:3); fixing with 10 mm dia anchor bolts to** **concrete at 500 mm centres; including** **cutting and pinning anchor bolts** **Medium duty; ref 702**						
300 mm wide	PC £130.00	2.75	29.30	146.25	m	175.55
Extra for						
ends	PC £3.61	-	-	4.07	nr	4.07
right angle frame corner	PC £42.20	-	-	47.48	nr	47.48
450 mm wide	PC £139.00	3.10	33.03	156.38	m	189.41
Extra for						
ends	PC £12.20	-	-	13.72	nr	13.72
right angle frame corner	PC £42.20	-	-	47.48	nr	47.48
600 mm wide	PC £219.00	3.40	36.22	246.38	m	282.60
Extra for						
ends	PC £16.26	-	-	18.29	nr	18.29
right angle frame corner	PC £42.20	-	-	47.48	nr	47.48
750 mm wide	PC £266.00	3.75	39.95	299.25	m	339.20
Extra for						
ends	PC £20.08	-	-	22.59	nr	22.59
right angle frame corner	PC £42.20	-	-	47.48	nr	47.48
900 mm wide	PC £300.00	4.05	43.15	337.50	m	380.65
Extra for						
ends	PC £24.29	-	-	27.32	nr	27.32
right angle frame corner	PC £42.20	-	-	47.48	nr	47.48

P BUILDING FABRIC SUNDRIES Including overheads and profit at 12.50%		Labour hours	Labour £	Material £	Unit	Total rate £
P31 HOLES/CHASES/COVERS/SUPPORTS FOR SERVICES - cont'd						
Duct covers with frames; cast iron;						
Brickhouse Glynwed 'Trucast' or similar;						
bedding and pointing frame in cement mortar						
(1:3); fixing with 10 mm dia anchor bolts to						
concrete at 500 mm centres; including						
cutting and pinning anchor bolts - cont'd						
Heavy duty; ref 704						
300 mm wide	PC £226.50	-	-	254.81	m	254.81
Extra for						
ends	PC £10.49	-	-	11.80	nr	11.80
right angle frame corner	PC £57.19	-	-	64.34	nr	64.34
450 mm wide	PC £235.00	3.75	39.95	264.38	m	304.33
Extra for						
ends	PC £15.68	-	-	17.64	nr	17.64
right angle frame corner	PC £57.19	-	-	64.34	nr	64.34
600 mm wide	PC £262.00	4.20	44.75	294.75	m	339.50
Extra for						
ends	PC £20.60	-	-	23.18	nr	23.18
right angle frame corner	PC £57.19	-	-	64.34	nr	64.34
750 mm wide	PC £303.00	4.60	49.01	340.88	m	389.89
Extra for						
ends	PC £25.94	-	-	29.19	nr	29.19
right angle frame corner	PC £57.19	-	-	64.34	nr	64.34
900 mm wide	PC £381.00	5.05	53.80	428.63	m	482.43
Extra for						
ends	PC £31.16	-	-	35.06	nr	35.06
right angle frame corner	PC £57.19	-	-	64.34	nr	64.34

Q PAVING/PLANTING/FENCING/SITE FURNITURE Including overheads and profit at 12.50%	Labour hours	Labour £	Material £	Unit	Total rate £

Q10 STONE/CONCRETE/BRICK KERBS/EDGINGS/ CHANNELS

Mechanical excavation using a wheeled hydraulic excavator with a 0.24 m3 bucket
Excavating trenches to receive kerb foundation; average size

300 x 100 mm	0.02	0.14	0.32	m	0.46
450 x 150 mm	0.03	0.21	0.53	m	0.74
600 x 200 mm	0.05	0.35	0.75	m	1.10

Excavating curved trenches to receive kerb foundation; average size

300 x 100 mm	0.02	0.14	0.43	m	0.57
450 x 150 mm	0.03	0.21	0.64	m	0.85
600 x 200 mm	0.05	0.35	0.86	m	1.21

Hand excavation
Excavating trenches to receive kerb foundation; average size

150 x 50 mm	0.03	0.21	-	m	0.21
200 x 75 mm	0.08	0.56	-	m	0.56
250 x 100 mm	0.12	0.84	-	m	0.84
300 x 100 mm	0.15	1.05	-	m	1.05

Excavating curved trenches to receive kerb foundation; average size

150 x 50 mm	0.04	0.28	-	m	0.28
200 x 75 mm	0.09	0.63	-	m	0.63
250 x 100 mm	0.13	0.91	-	m	0.91
300 x 100 mm	0.17	1.19	-	m	1.19

Plain in situ ready mixed concrete; 7.50 N/mm2 - 40 mm aggregate (1:8); PC £43.11/m3; poured on or against earth or unblinded hardcore

Foundations	1.45	10.47	50.92	m3	61.39
Blinding bed					
not exceeding 150 mm thick	2.15	15.53	50.92	m3	66.45

Plain in situ ready mixed concrete; 11.50 N/mm2 - 40 mm aggregate (1:3:6); PC £44.13/m3; poured on or against earth or unblinded hardcore

Foundations	1.45	10.47	52.13	m3	62.60
Blinding bed					
not exceeding 150 mm thick	2.15	15.53	52.13	m3	67.66

Plain in situ ready mixed concrete; 11.50 N/mm2 - 40 mm aggregate (1:3:6); PC £44.13/m3; poured on or against earth or unblinded hardcore

Foundations	1.45	10.47	55.97	m3	66.44
Blinding bed					
not exceeding 150 mm thick	2.15	15.53	55.97	m3	71.50

Precast concrete kerbs, channels, edgings, etc.; BS 340; bedded, jointed and pointed in cement mortar (1:3); including haunching up one side with in situ concrete mix 11.50 N/mm2 - 40 mm aggregate (1:3:6); to concrete base
Edging; straight; fig 12

51 x 152 mm	0.29	3.09	2.59	m	5.68
51 x 203 mm	0.29	3.09	2.94	m	6.03
51 x 254 mm	0.29	3.09	3.10	m	6.19

Q PAVING/PLANTING/FENCING/SITE FURNITURE Including overheads and profit at 12.50%	Labour hours	Labour £	Material £	Unit	Total rate £
Q10 STONE/CONCRETE/BRICK KERBS/EDGINGS/ CHANNELS - cont'd					
Precast concrete kerbs, channels, edgings, etc.; BS 340; bedded, jointed and pointed in cement mortar (1:3); including haunching up one side with in situ concrete mix 11.50 N/mm2 - 40 mm aggregate (1:3:6); to concrete base - cont'd					
Kerb; straight					
127 x 254 mm; fig 7	0.38	4.05	5.05	m	9.10
152 x 305 mm; fig 6	0.38	4.05	6.79	m	10.84
Kerb; curved					
127 x 254 mm; fig 7	0.58	6.18	6.47	m	12.65
152 x 305 mm; fig 6	0.58	6.18	8.20	m	14.38
Channel; 255 x 125 mm; fig 8					
straight	0.38	4.05	5.25	m	9.30
curved	0.58	6.18	6.71	m	12.89
Quadrant; fig 14					
305 x 305 x 152 mm	0.40	4.26	3.58	nr	7.84
305 x 305 x 254 mm	0.40	4.26	3.58	nr	7.84
457 x 457 x 152 mm	0.46	4.90	3.58	nr	8.48
457 x 457 x 254 mm	0.46	4.90	3.58	nr	8.48
Precast concrete drainage channels; Charcon 'Safeticurb' or similar; channels jointed with plastic rings and bedded; jointed and pointed in cement mortar (1:3); including haunching up one side with in situ concrete mix 11.50 N/mm2 - 40 mm aggregate (1:3:6); to concrete base					
Channel; straight; type DBA/3					
248 x 248 mm	0.69	7.35	19.44	m	26.79
End	0.23	2.45	-	nr	2.45
Inspection unit; with cast iron lid					
248 x 248 x 914 mm	0.75	7.99	56.56	nr	64.55
Silt box top; with concrete frame and cast iron lid; set over gully					
500 x 448 x 269 mm; type A	2.30	24.50	103.90	nr	128.40
Q20 HARDCORE/GRANULAR/CEMENT BOUND BASES/ SUB-BASES TO ROADS/PAVINGS					
Mechanical filling with hardcore; PC £8.00/m3 Filling to make up levels over 250 mm thick; depositing; compacting in layers with a					
5 tonne roller	0.30	2.09	12.53	m3	14.62
Filling to make up levels not exceeding 250 mm thick; compacting	0.35	2.44	16.61	m3	19.05
Mechanical filling with granular fill; type 1; PC £11.05/t (PC £18.80/m3) Filling to make up levels over 250 mm thick; depositing; compacting in layers with a					
5 tonne roller	0.30	2.09	28.14	m3	30.23
Filling to make up levels not exceeding 250 mm thick; compacting	0.35	2.44	36.05	m3	38.49
Mechanical filling with granular fill; type 2; PC £10.75/t (PC £18.25/m3) Filling to make up levels over 250 mm thick; depositing; compacting in layers with a					
5 tonne roller	0.30	2.09	27.37	m3	29.46
Filling to make up levels not exceeding 250 mm thick; compacting	0.35	2.44	35.06	m3	37.50

Q PAVING/PLANTING/FENCING/SITE FURNITURE Including overheads and profit at 12.50%	Labour hours	Labour £	Material £	Unit	Total rate £
Hand filling with hardcore; PC £8.00/m3					
Filling to make up levels over					
250 mm thick; depositing; compacting	0.61	4.25	16.82	m3	**21.07**
Filling to make up levels not exceeding					
250 mm thick; compacting	0.73	5.09	21.08	m3	**26.17**
Hand filling with coarse ashes; PC £12.00/m3					
Filling to make up levels over 250 mm thick;					
depositing; compacting in layers with a					
2 tonne roller	0.55	3.84	23.02	m3	**26.86**
Filling to make up levels not exceeding					
250 mm thick; compacting	0.66	4.60	29.03	m3	**33.63**
Hand filling with sand; PC £10.90/t (PC £17.45/m3)					
Filling to make up levels over 250 mm thick;					
depositing; compacting in layers with a					
2 tonne roller	0.72	5.02	31.12	m3	**36.14**
Filling to make up levels not exceeding					
250 mm thick; compacting	0.85	5.93	39.21	m3	**45.14**
Surface treatments					
Compacting					
surfaces of ashes	0.05	0.35	0.12	m2	**0.47**
filling; blinding with ashes	0.09	0.63	0.93	m2	**1.56**
Q21 IN SITU CONCRETE ROADS/PAVINGS/BASES					
Reinforced in situ ready mixed concrete;					
normal Portland cement; mix 11.5 N/mm2					
- 20 mm aggregate (1:2:4); PC £47.38/m3					
Roads; to hardcore base					
150 - 450 mm thick	1.70	12.28	49.65	m3	**61.93**
not exceeding 150 mm thick	2.55	18.42	49.65	m3	**68.07**
Reinforced in situ ready mixed concrete;					
normal Portland cement; mix 21.00 N/mm2					
Roads; to hardcore base					
150 - 450 mm thick	1.70	12.28	53.30	m3	**65.58**
not exceeding 150 mm thick	2.55	18.42	53.30	m3	**71.72**
Reinforced in situ ready mixed concrete;					
normal Portland cement; mix 26.00 N/mm2					
- 20 mm aggregate (1:1:5:3); PC £47.38/m3					
Roads; to hardcore base					
150 - 450 mm thick	1.70	12.28	55.70	m3	**67.98**
not exceeding 150 mm thick	2.55	18.42	55.70	m3	**74.12**
Formwork for in situ concrete					
Sides of foundations					
not exceeding 250 mm wide	0.46	3.88	1.43	m	**5.31**
250 - 500 mm wide	0.69	5.82	2.46	m	**8.28**
500 mm - 1 m wide	1.05	8.86	4.64	m	**13.50**
Add to above for curved radius 6 m	0.05	0.42	0.18	m	**0.60**
Steel road forms to in situ concrete					
Sides of foundations					
150 mm wide	0.22	1.86	0.52	m	**2.38**

Reinforcement; fabric; BS 4483; lapped; in **roads, footpaths or pavings**		Labour hours	Labour £	Material £	Unit	Total rate £
Ref A142 (2.22 kg/m2)	PC £0.83	0.13	1.09	1.13	m2	**2.22**
Ref A193 (3.02 kg/m2)	PC £1.13	0.13	1.09	1.53	m2	**2.62**

Q PAVING/PLANTING/FENCING/SITE FURNITURE Including overheads and profit at 12.50%	Labour hours	Labour £	Material £	Unit	Total rate £
Q21 IN SITU CONCRETE ROADS/PAVINGS/BASES - cont'd					
Designed joints in in situ concrete Formed joint; 12.5 mm thick Expandite 'Flexcell' or similar					
not exceeding 150 mm wide	0.33	2.78	1.17	m	3.95
150 - 300 mm wide	0.44	3.71	2.03	m	5.74
300 - 450 mm wide	0.55	4.64	3.17	m	7.81
Formed joint; 25 mm thick Expandite 'Flexcell' or similar					
not exceeding 150 mm wide	0.33	2.78	1.83	m	4.61
150 - 300 mm wide	0.44	3.71	3.25	m	6.96
300 - 450 mm wide	0.55	4.64	4.94	m	9.58
Sealing top 25 mm of joint with rubberized bituminous compound	0.23	1.94	0.85	m	2.79
Concrete sundries Treating surfaces of unset concrete; grading to cambers; tamping with a 75 mm thick steel shod tamper	0.28	2.02	-	m2	2.02
Q22 COATED MACADAM/ASPHALT ROADS/PAVINGS					
Fine graded wearing course; BS 4987:88; clause 2.7.7, tables 34 - 36; 14 mm pre- coated igneous rock chippings; tack coat of bitumen emulsion 19 mm work to roads; one coat					
limestone aggregate	-	-	-	m2	8.94
igneous aggregate	-	-	-	m2	9.00
Close graded bitumen macadam; BS 4987:88; 10 mm graded aggregate to clause 2.7.4 tables 34 - 36; tack coat of bitumen emulsion 30 mm work to roads; one coat					
limestone aggregate	-	-	-	m2	9.26
igneous aggregate	-	-	-	m2	9.37
Bitumen macadam; BS 4987:88; 45 mm thick base course of 20 mm open graded aggregate to clause 2.6.1 tables 5 - 7; 20 mm thick wearing course of 6 mm medium graded aggregate to clause 2.7.6 tables 32 - 33 65 mm work to pavements/footpaths; two coats					
limestone aggregate	-	-	-	m2	11.25
igneous aggregate	-	-	-	m2	11.48
Add to last for 14 mm chippings; sprinkled into wearing course	-	-	-	m2	0.45
Bitumen macadam; BS 4987:88; 50 mm graded aggregate to clause 2.6.2 tables 8 - 10 75 mm work to roads; one coat					
limestone aggregate	-	-	-	m2	11.08
igneous aggregate	-	-	-	m2	11.60
Dense bitumen macadam; BS 4987:88; 50 mm thick base course of 20 mm graded aggregate to clause 2.6.5 tables 15 - 16; 200 pen. binder; 30 mm wearing course of 10 mm graded aggregate to clause 2.7.2 tables 20 - 22 75 mm work to roads; two coats					
limestone aggregate	-	-	-	m2	12.30
igneous aggregate	-	-	-	m2	12.61

Q PAVING/PLANTING/FENCING/SITE FURNITURE Including overheads and profit at 12.50%	Labour hours	Labour £	Material £	Unit	Total rate £
Bitumen macadam; BS 4987:88; 50 mm thick **base course of 20 mm graded aggregate to** **clause 2.6.1 tables 5 - 7; 25 mm thick** **wearing course of 10 mm graded aggregate to** **clause 2.7.2 tables 20 - 22** 75 mm work to roads; two coats					
limestone aggregate	-	-	-	m2	11.37
igneous aggregate	-	-	-	m2	11.60

Q24 INTERLOCKING BRICK/BLOCK ROADS/PAVINGS

Concrete interlocking blocks; Marshall's
'Monolok' or similar; to falls or
crossfalls; bedding 50 mm thick in dry sharp
sand; filling joints with sharp sand;
brushed in; vibrated; to earth base
Work to paved areas; over 300 mm wide;
interlocking joints

60 mm thick; grey	PC £7.76	0.86	9.16	11.08	m2	20.24
80 mm thick; grey	PC £8.88	0.92	9.80	12.59	m2	22.39
Extra for red stones; forming parking lines		0.06	0.64	0.26	m	0.90

Q25 SLAB/BRICK/BLOCK/SETT/COBBLE PAVINGS

NOTE: Unless otherwise described, prices for
pavings do not include for ash or sand beds
or bases under.

Artificial stone paving; Redland Aggregates'
'Texitone' or similar; to falls or
crossfalls; bedding 25 mm thick in lime
mortar (1:4) staggered joints; jointing in
coloured cement mortar (1:3); brushed in; to
sand base
Work to paved areas; over 300 mm wide
450 x 600 x 50 mm;

grey or coloured	PC £2.56/each	0.52	5.54	13.26	m2	18.80
600 x 600 x 50 mm; grey or coloured	PC £2.96/each	0.48	5.11	11.78	m2	16.89
750 x 600 x 50 mm; grey or coloured	PC £3.58/each	0.45	4.79	11.50	m2	16.29
900 x 750 x 50 mm; grey or coloured	PC £4.14/each	0.41	4.37	9.36	m2	13.73

Brick paviors; 215 x 103 x 65 mm rough stock
bricks; PC £410.00/1000; to falls or crossfalls;
bedding 10 mm thick in cement mortar (1:3);
jointing in cement mortar (1:3); as work
proceeds; to concrete base
Work to paved areas; over 300 mm wide;
straight joints both ways

bricks laid flat	0.92	11.06	20.14	m2	31.20
bricks laid on edge	1.30	15.63	31.65	m2	47.28

Work to paved areas; over 300 mm wide; laid
to herringbone pattern

bricks laid flat	1.15	13.83	20.14	m2	33.97
bricks laid on edge	1.60	19.24	31.65	m2	50.89

Add or deduct for variation of   1.00/1000
in PC of brick paviors
Work to paved areas; over 300 mm wide; laid
to herringbone pattern

bricks laid flat	-	-	-	m2	0.45
bricks laid on edge	-	-	-	m2	0.72

Q PAVING/PLANTING/FENCING/SITE FURNITURE Including overheads and profit at 12.50%	Labour hours	Labour £	Material £	Unit	Total rate £

Q25 SLAB/BRICK/BLOCK/SETT/COBBLE PAVINGS - cont'd

Cobble paving; 50 - 75 mm PC £56.00/t; to
falls or crossfalls; bedding 13 mm thick in
cement mortar (1:3); jointing to a height of
two thirds of cobbles in dry mortar (1:3);
tightly butted, washed and brushed; to
concrete base
Work to paved areas; over 300 mm wide

regular	4.60	49.01	15.72	m2	64.73
laid to pattern	5.75	61.26	15.72	m2	76.98

Concrete paving flags; BS 368; to falls or
crossfalls; bedding 25 mm thick in lime and
sand mortar (1:4); butt joints straight both
ways; jointing in cement mortar (1:3);
brushed in; to sand base
Work to paved areas; over 300 mm wide

450 x 600 x 50 mm; grey PC £1.49/each	0.52	5.54	7.50	m2	13.04
450 x 600 x 50 mm; coloured PC £2.20/each	0.52	5.54	10.60	m2	16.14
600 x 600 x 50 mm; grey PC £1.74/each	0.48	5.11	6.69	m2	11.80
600 x 600 x 50 mm; coloured PC £2.60/each	0.48	5.11	9.51	m2	14.62
750 x 600 x 50 mm; grey PC £2.09/each	0.45	4.79	6.49	m2	11.28
750 x 600 x 50 mm; coloured PC £3.16/each	0.45	4.79	9.30	m2	14.09
900 x 600 x 50 mm; grey PC £2.41/each	0.41	4.37	5.24	m2	9.61
900 x 600 x 50 mm; coloured PC £3.72/each	0.41	4.37	7.53	m2	11.90

Concrete rectangular blocks; Marshall's
'Keyblok' or similar; to falls or
crossfalls; bedding 50 mm thick in dry sharp
sand; filling joints with sharp sand;
brushed in; vibrated; to earth base
Work to paved areas; over 300 mm wide;
straight joints both ways

200 x 100 x 65 mm; grey PC £7.14	0.69	7.35	9.91	m2	17.26
200 x 100 x 80 mm; grey PC £8.26	0.75	7.99	11.34	m2	19.33

Work to paved areas; over 300 mm wide; laid
to herringbone pattern

200 x 100 x 65 mm; coloured PC £8.28	0.86	9.16	11.25	m2	20.41
200 x 100 x 80 mm; coloured PC £9.67	0.92	9.80	13.02	m2	22.82

Extra for two row boundary edging to
herringbone paved areas; 200 mm wide;
including a 150 mm high in situ concrete
mix 11.5 N/mm2 - 40 mm aggregate (1:3:6)
haunching to one side; blocks laid
breaking joint

200 x 100 x 65 mm; coloured	0.29	3.09	1.83	m	4.92
200 x 100 x 80 mm; coloured	0.29	3.09	1.94	m	5.03

Granite setts; BS 435; 200 x 100 x 100 mm;
PC £96.00/t; standard 'G' dressing; tightly
butted to falls or crossfalls; bedding
25 mm thick in cement mortar (1:3); filling
joints with dry mortar (1:6); washed and
brushed; on concrete base
Work to paved areas; over 300 mm wide

straight joints	1.85	19.71	31.18	m2	50.89
laid to pattern	2.30	24.50	31.18	m2	55.68

Two rows of granite setts as boundary
edging; 200 mm wide; including a 150 mm high
in situ concrete mix 11.5 N/mm2 - 40 mm
aggregate (1:3:6) haunching to one side;

blocks laid breaking joint	0.81	8.63	7.61	m	16.24

Q26 SPECIAL SURFACINGS/PAVINGS FOR SPORT

Sundries
Painted line on road; one coat

75 mm wide	0.06	0.45	0.10	m	0.55

Q PAVING/PLANTING/FENCING/SITE FURNITURE Including overheads and profit at 12.50%	Labour hours	Labour £	Material £	Unit	Total rate £
Q30 SEEDING/TURFING					
Vegetable soil					
Selected from spoil heaps; grading; preparing for turfing or seeding; to general surfaces					
average 75 mm thick	0.36	2.51	-	m2	2.51
average 100 mm thick	0.39	2.72	-	m2	2.72
average 125 mm thick	0.42	2.93	-	m2	2.93
average 150 mm thick	0.44	3.07	-	m2	3.07
average 175 mm thick	0.46	3.21	-	m2	3.21
average 200 mm thick	0.48	3.35	-	m2	3.35
Selected from spoil heaps; grading; preparing for turfing or seeding; to cutting or embankments					
average 75 mm thick	0.41	2.86	-	m2	2.86
average 100 mm thick	0.44	3.07	-	m2	3.07
average 125 mm thick	0.47	3.28	-	m2	3.28
average 150 mm thick	0.50	3.49	-	m2	3.49
average 175 mm thick	0.52	3.63	-	m2	3.63
average 200 mm thick	0.55	3.84	-	m2	3.84
Imported vegetable soil; PC £11.00/m3					
Grading; preparing for turfing or seeding; to general surfaces					
average 75 mm thick	0.33	2.30	1.24	m2	3.54
average 100 mm thick	0.35	2.44	1.61	m2	4.05
average 125 mm thick	0.37	2.58	2.35	m2	4.93
average 150 mm thick	0.40	2.79	3.09	m2	5.88
average 175 mm thick	0.42	2.93	3.47	m2	6.40
average 200 mm thick	0.44	3.07	3.84	m2	6.91
Grading; preparing for turfing or seeding; to cuttings or embankments					
average 75 mm thick	0.36	2.51	1.24	m2	3.75
average 100 mm thick	0.40	2.79	1.61	m2	4.40
average 125 mm thick	0.42	2.93	2.35	m2	5.28
average 150 mm thick	0.44	3.07	3.09	m2	6.16
average 175 mm thick	0.46	3.21	3.47	m2	6.68
average 200 mm thick	0.48	3.35	3.84	m2	7.19
Fertilizer; PC £0.65/kg					
Fertilizer 0.07 kg/m2; raking in					
general surfaces	0.04	0.28	0.05	m2	0.33
Selected grass seed; PC £3.10/kg					
Grass seed; sowing at a rate of 0.042 kg/m2 two applications; raking in					
general surfaces	0.08	0.56	0.29	m2	0.85
cuttings or embankments	0.09	0.63	0.29	m2	0.92
Preserved turf from stack on site					
Selected turf					
general surfaces	0.22	1.53	-	m2	1.53
cuttings or embankments; shallow	0.24	1.67	0.16	m2	1.83
cuttings or embankments; steep; pegged	0.33	2.30	0.24	m2	2.54
Imported turf; PC £1.00/m2					
Selected meadow turf					
general surfaces	0.22	1.53	1.13	m2	2.66
cuttings or embankments; shallow	0.24	1.67	1.29	m2	2.96
cuttings or embankments; steep; pegged	0.33	2.30	1.37	m2	3.67

Q PAVING/PLANTING/FENCING/SITE FURNITURE Including overheads and profit at 12.50% & 5.00%	Labour hours	Labour £	Material £	Unit	Total rate £
Q31 PLANTING					
Planting only					
Hedge or shrub plants					
not exceeding 750 mm high	0.28	1.95	-	nr	1.95
750 mm - 1.5 m high	0.66	4.60	-	nr	4.60
Saplings					
not exceeding 3 m high	1.85	12.90	-	nr	12.90

Q40 FENCING

NOTE: The prices for all fencing are to
include for setting posts in position, to a
depth of 0.6 m for fences not exceeding
1.4 m high and of 0.76 m for fences over
1.4 m high.
  The prices allow for excavating post
holes; filling to within 150 mm of ground
level with concrete and all necessary back
filling

Strained wire fences; BS 1722 Part 3; 4 mm **galvanized mild steel plain wire threaded** **through posts and strained with eye bolts**					
900 mm fencing; three line; concrete posts					
at 2750 mm centres	-	-	-	m	6.26
Extra for					
end concrete straining post; one strut	-	-	-	nr	27.87
angle concrete straining post; two struts	-	-	-	nr	34.61
1.07 m fencing; five line; concrete posts at					
2750 mm centres	-	-	-	m	7.57
Extra for					
end concrete straining post; one strut	-	-	-	nr	48.12
angle concrete straining post; two struts	-	-	-	nr	64.06
1.20 m fencing; six line; concrete posts at					
2750 mm centres	-	-	-	m	7.63
Extra for					
end concrete straining post; one strut	-	-	-	nr	48.91
angle concrete straining post; two struts	-	-	-	nr	65.22
1.4 m fencing; seven line; concrete posts at					
2750 mm centres	-	-	-	m	7.85
Extra for					
end concrete straining post; one strut	-	-	-	nr	49.44
angle concrete straining post; two struts	-	-	-	nr	66.27
Chainlink fences; BS 1722 Part 1; 3 mm; **50 mm galvanized mild steel mesh; galvanized** **mild steel tying and line wire; three line** **wires threaded through posts and strained** **with eye bolts and winding brackets**					
900 mm fencing; galvanized mild steel angle					
posts at 3 m centres	-	-	-	m	8.05
Extra for					
end steel straining post; one strut	-	-	-	nr	33.24
angle steel straining post; two struts	-	-	-	nr	44.18
900 mm fencing; concrete posts at 3 m centres	-	-	-	m	8.31
Extra for					
end concrete straining post; one strut	-	-	-	nr	33.13
angle concrete straining post; two struts	-	-	-	nr	43.92
1.2 m fencing; galvanized mild steel angle					
posts at 3 m centres	-	-	-	m	9.41
Extra for					
end steel straining post; one strut	-	-	-	nr	35.34
angle steel straining post; two struts	-	-	-	nr	47.54

Q PAVING/PLANTING/FENCING/SITE FURNITURE Including overheads and profit at 5.00%	Labour hours	Labour £	Material £	Unit	Total rate £
1.2 m fencing; concrete posts at 3 m centres	-	-	-	m	9.68
Extra for					
end concrete straining post; one strut	-	-	-	nr	38.81
angle concrete straining post; two struts	-	-	-	nr	47.23
1.8 m fencing; galvanized mild steel angle posts at 3 m centres	-	-	-	m	13.15
Extra for					
end steel straining post; one strut	-	-	-	nr	54.17
angle steel straining post; two struts	-	-	-	nr	71.42
1.8 m fencing; concrete posts at 3 m centres	-	-	-	m	13.62
Extra for					
end concrete straining post; one strut	-	-	-	nr	55.49
angle concrete straining post; two struts	-	-	-	nr	72.05
Pair of gates and gate posts; gates to match galvanized chain link fencing, with angle framing, braces, etc., complete with hinges, locking bar, lock and bolts; two 100 x 100 mm angle section gate posts; each with one strut					
2.44 x 0.9 m high	-	-	-	nr	376.57
2.44 x 1.2 m high	-	-	-	nr	404.97
2.44 x 1.8 m high	-	-	-	nr	492.28
Chainlink fences; BS 1722 Part 1; 3 mm; 50 mm plastic coated mild steel mesh; plastic coated mild steel tying and line wire; three line wires threaded through posts and strained with eye bolts and winding brackets					
900 mm fencing; galvanized mild steel angle posts at 3 m centres	-	-	-	m	8.84
Extra for					
end steel straining post; one strut	-	-	-	nr	33.24
angle steel straining post; two struts	-	-	-	nr	44.18
900 mm fencing; concrete posts at 3 m centres	-	-	-	m	9.10
Extra for					
end concrete straining post; one strut	-	-	-	nr	33.13
angle concrete straining post; two struts	-	-	-	nr	43.92
1.2 m fencing; galvanized mild steel angle posts at 3 m centres	-	-	-	m	10.60
Extra for					
end steel straining post; one strut	-	-	-	nr	35.34
angle steel straining post; two struts	-	-	-	nr	47.54
1.2 m fencing; concrete posts at 3 m centres	-	-	-	m	10.88
Extra for					
end concrete straining post; one strut	-	-	-	nr	38.81
angle concrete straining post; two struts	-	-	-	nr	47.23
1.8 m fencing; galvanized mild steel angle posts at 3 m centres	-	-	-	m	14.42
Extra for					
end steel straining post; one strut	-	-	-	nr	54.17
angle steel straining post; two struts	-	-	-	nr	71.42
1.8 m fencing; concrete posts at 3 m centres	-	-	-	m	14.89
Extra for					
end concrete straining post; one strut	-	-	-	nr	55.49
angle concrete straining post; two struts	-	-	-	nr	71.53
Pair of gates and gate posts; gates to match plastic chain link fencing; with angle framing, braces, etc., complete with hinges, locking bar, lock and bolts; two 100 x 100 mm angle section gate posts; each with one strut					
2.44 x 0.9 m high	-	-	-	nr	399.71
2.44 x 1.2 m high	-	-	-	nr	426.01
2.44 x 1.8 m high	-	-	-	nr	517.52

Q PAVING/PLANTING/FENCING/SITE FURNITURE Including overheads and profit at 5.00%	Labour hours	Labour £	Material £	Unit	Total rate £
Q40 FENCING - cont'd					
Chain link fences for tennis courts; BS 1722 Part 13; 2.5 mm; 45 mm mesh galvanized mild steel mesh; line and tying wires threaded through 45 x 45 x 5 mm galvanized mild steel angle standards, posts and struts; 60 x 60 x 6 mm angle straining posts and gate posts; straining posts and struts strained with eye bolts and winding brackets Fencing to tennis court 36 x 18 m; including gate 1070 x 1980 mm complete with hinges, locking bar, lock and bolts					
2745 mm fencing; standards at 3 m centres	-	-	-	nr	1811.25
3660 mm fencing; standards at 2.5 m centres	-	-	-	nr	2458.13
Cleft chestnut pale fences; BS 1722 Part 4; pales spaced 51 mm apart; on two lines of galvanized wire; 64 mm dia posts; 76 x 51 mm struts					
900 mm fences; posts at 2.50 m centres	-	-	-	m	6.60
Extra for					
straining post; one strut	-	-	-	nr	12.03
corner straining post; two struts	-	-	-	nr	16.04
1.05 m fences; posts at 2.50 m centres	-	-	-	m	7.25
Extra for					
straining post; one strut	-	-	-	nr	13.26
corner straining post; two struts	-	-	-	nr	17.66
1.20 m fences; posts at 2.25 m centres	-	-	-	m	7.76
Extra for					
straining post; one strut	-	-	-	nr	14.55
corner straining post; two struts	-	-	-	nr	19.41
1.35 m fences; posts at 2.25 m centres	-	-	-	m	8.22
Extra for					
straining post; one strut	-	-	-	nr	16.30
corner straining post; two struts	-	-	-	nr	21.61
Close boarded fencing; BS 1722 Part 5; 76 x 38 mm softwood rails; 89 x 19 mm softwood pales lapped 13 mm; 152 x 25 mm softwood gravel boards; all softwood 'treated'; posts at 3 m centres Fences; two rail; concrete posts					
1 m	-	-	-	m	27.56
1.2 m	-	-	-	m	28.46
Fences; three rail; concrete posts					
1.4 m	-	-	-	m	37.00
1.6 m	-	-	-	m	38.42
1.8 m	-	-	-	m	39.91
Fences; two rail; oak posts					
1 m	-	-	-	m	19.92
1.2 m	-	-	-	m	22.77
Fences; three rail; oak posts					
1.4 m	-	-	-	m	25.62
1.6 m	-	-	-	m	29.11
1.8 m	-	-	-	m	32.47
Precast concrete slab fencing; 305 x 38 x 1753 mm slabs; fitted into twice grooved concrete posts at 1830 mm centres Fences					
1.2 m	-	-	-	m	42.31
1.5 m	-	-	-	m	50.13
1.8 m	-	-	-	m	62.62

Q PAVING/PLANTING/FENCING/SITE FURNITURE Including overheads and profit at 5.00%	Labour hours	Labour £	Material £	Unit	Total rate £
Mild steel unclimbable fencing; in rivetted panels 2440 mm long; 44 x 13 mm flat section top and bottom rails; two 44 x 19 mm flat section standards, one with foot plate, and 38 x 13 mm raking stay with foot plate; 20 mm dia pointed verticals at 120 mm centres; two 44 x 19 mm supports 760 mm long with ragged ends to bottom rail; the whole bolted together; coated with red oxide primer; setting standards and stays in ground at 2440 mm centres and supports at 815 mm centres					
Fences					
1.67 m	-	-	-	m	59.96
2.13 m	-	-	-	m	68.90
Pair of gates and gate posts, to match mild steel unclimbable fencing; with flat section framing, braces, etc., complete with locking bar, lock, handles, drop bolt, gate stop and holding back catches; two 102 x 102 mm hollow section gate posts with cap and foot plates					
2.44 x 1.67 m	-	-	-	nr	538.56
2.44 x 2.13 m	-	-	-	nr	609.04
4.88 x 1.67 m	-	-	-	nr	1070.81
4.88 x 2.13 m	-	-	-	nr	1214.92

R DISPOSAL SYSTEMS Including overheads and profit at 12.50%		Labour hours	Labour £	Material £	Unit	Total rate £
R10 RAINWATER PIPEWORK/GUTTERS						
Aluminium pipes and fittings; BS 2997; **ears cast on; powder coated finish**						
63 mm pipes; plugged and nailed		0.44	4.70	9.07	m	13.77
Extra for						
fittings with one end		0.26	2.78	4.48	nr	7.26
fittings with two ends		0.50	5.34	4.99	nr	10.33
fittings with three ends		0.72	7.69	7.02	nr	14.71
shoe	PC £4.67	0.26	2.78	4.48	nr	7.26
bend	PC £5.13	0.50	5.34	4.99	nr	10.33
single branch	PC £6.67	0.72	7.69	7.02	nr	14.71
offset 229 mm projection	PC £11.82	0.50	5.34	11.14	nr	16.48
offset 305 mm projection	PC £13.18	0.50	5.34	12.67	nr	18.01
access pipe	PC £14.59	-	-	13.26	nr	13.26
connection to clay pipes; cement and sand (1:2) joint		0.18	1.92	0.10	nr	2.02
75 mm pipes; plugged and nailed						
	PC £16.35/1.8m	0.48	5.12	10.86	m	15.98
Extra for						
shoe	PC £6.19	0.30	3.20	6.09	nr	9.29
bend	PC £6.47	0.54	5.77	6.40	nr	12.17
single branch	PC £8.04	0.78	8.33	8.63	nr	16.96
offset 229 mm projection	PC £13.06	0.54	5.77	12.16	nr	17.93
offset 305 mm projection	PC £14.45	0.54	5.77	13.73	nr	19.50
access pipe	PC £15.95	-	-	14.24	nr	14.24
connection to clay pipes; cement and sand (1:2) joint		0.20	2.14	0.10	nr	2.24
100 mm pipes; plugged and nailed						
	PC £27.14/1.8m	0.54	5.77	17.93	m	23.70
Extra for						
shoe	PC £7.73	0.60	6.41	7.12	nr	13.53
bend	PC £8.78	0.34	3.63	8.31	nr	11.94
single branch	PC £10.48	-	-	10.75	nr	10.75
offset 229 mm projection	PC £14.70	0.90	9.61	12.21	nr	21.82
offset 305 mm projection	PC £16.33	0.60	6.41	14.05	nr	20.46
access pipe	PC £18.39	-	-	14.36	nr	14.36
connection to clay pipes; cement and sand (1:2) joint		0.24	2.56	0.10	nr	2.66
Roof outlets; circular aluminium; with flat or domed grate; joint to pipe						
50 mm dia	PC £32.07	0.72	7.69	36.81	nr	44.50
75 mm dia	PC £42.42	0.78	8.33	48.63	nr	56.96
100 mm dia	PC £55.61	0.84	8.97	63.63	nr	72.60
150 mm dia	PC £71.62	0.90	9.61	82.10	nr	91.71
Roof outlets; d-shaped; balcony; with flat or domed grate; joint to pipe						
50 mm dia	PC £39.47	0.72	7.69	45.14	nr	52.83
75 mm dia	PC £45.38	0.78	8.33	51.96	nr	60.29
100 mm dia	PC £55.73	0.84	8.97	63.76	nr	72.73
Galvanized wire balloon grating; BS 416 for pipes or outlets						
50 mm dia	PC £1.10	0.08	0.85	1.24	nr	2.09
63 mm dia	PC £1.10	0.08	0.85	1.24	nr	2.09
75 mm dia	PC £1.22	0.08	0.85	1.37	nr	2.22
100 mm dia	PC £1.35	0.10	1.07	1.52	nr	2.59
Aluminium gutters and fittings; BS 2997; **powder coated finish**						
100 mm half round gutters; on brackets						
screwed to timber	PC £12.27/1.8m	0.42	4.48	9.18	m	13.66
Extra for						
stop end	PC £1.95	0.19	2.03	2.94	nr	4.97
running outlet	PC £4.32	0.40	4.27	4.06	nr	8.33
stop end outlet	PC £3.83	0.19	2.03	3.96	nr	5.99
angle	PC £3.98	0.40	4.27	3.42	nr	7.69

R DISPOSAL SYSTEMS Including overheads and profit at 12.50%		Labour hours	Labour £	Material £	Unit	Total rate £
112 mm half round gutters; on brackets						
screwed to timber	PC £12.85/1.8m	0.42	4.48	9.58	m	14.06
Extra for						
stop end	PC £2.04	0.19	2.03	3.06	nr	5.09
running outlet	PC £4.70	0.40	4.27	4.45	nr	8.72
stop end outlet	PC £4.11	0.19	2.03	4.24	nr	6.27
angle	PC £4.50	0.40	4.27	3.93	nr	8.20
125 mm half round gutters; on brackets						
screwed to timber	PC £14.42/1.8m	0.48	5.12	11.59	m	16.71
Extra for						
stop end	PC £2.49	0.22	2.35	4.05	nr	6.40
running outlet	PC £5.09	0.42	4.48	4.77	nr	9.25
stop end outlet	PC £4.36	0.22	2.35	4.86	nr	7.21
angle	PC £5.00	0.42	4.48	5.12	nr	9.60
100 mm ogee gutters; on brackets						
screwed to timber	PC £15.29/1.8m	0.44	4.70	11.07	m	15.77
Extra for						
stop end	PC £2.05	0.20	2.14	3.05	nr	5.19
running outlet	PC £4.71	0.42	4.48	4.22	nr	8.70
stop end outlet	PC £3.94	0.20	2.14	3.80	nr	5.94
angle	PC £4.27	0.42	4.48	3.16	nr	7.64
112 mm ogee gutters; on						
brackets screwed to timber	PC £17.00/1.8m	0.50	5.34	12.20	m	17.54
Extra for						
stop end	PC £2.21	0.20	2.14	3.26	nr	5.40
running outlet	PC £5.14	0.42	4.48	4.57	nr	9.05
stop end outlet	PC £4.40	0.20	2.14	4.20	nr	6.34
angle	PC £5.09	0.42	4.48	3.81	nr	8.29
125 mm ogee gutters; on						
brackets screwed to timber	PC £18.78/1.8m	0.50	5.34	14.35	m	19.69
Extra for						
stop end	PC £2.40	0.23	2.46	3.94	nr	6.40
running outlet	PC £5.62	0.44	4.70	4.97	nr	9.67
stop end outlet	PC £5.00	0.23	2.46	5.18	nr	7.64
angle	PC £5.92	0.44	4.70	5.32	nr	10.02
Cast iron pipes and fittings; BS 460; ears cast on; joints						
50 mm pipes; primed; plugged						
and nailed	PC £18.08/1.8m	0.60	6.41	12.06	m	18.47
Extra for						
fittings with one end		0.36	3.84	8.58	nr	12.42
fittings with two ends		0.66	7.05	4.79	nr	11.84
shoe	PC £8.73	0.36	3.84	8.58	nr	12.42
bend	PC £5.36	0.66	7.05	4.79	nr	11.84
connection to clay pipes; cement and sand (1:2) joint		0.16	1.71	0.10	nr	1.81
63 mm pipes; primed; plugged						
and nailed	PC £18.08/1.8m	0.62	6.62	12.14	m	18.76
Extra for						
fittings with one end		0.38	4.06	8.65	nr	12.71
fittings with two ends		0.68	7.26	4.86	nr	12.12
fittings with three ends		0.86	9.18	7.70	nr	16.88
shoe	PC £8.73	0.38	4.06	8.65	nr	12.71
bend	PC £5.36	0.68	7.26	4.86	nr	12.12
single branch	PC £8.27	0.86	9.18	7.70	nr	16.88
offset 229 mm projection	PC £9.51	0.68	7.26	8.31	nr	15.57
offset 305 mm projection	PC £11.14	0.68	7.26	9.73	nr	16.99
connection to clay pipes; cement and sand (1:2) joint		0.18	1.92	0.10	nr	2.02

R DISPOSAL SYSTEMS Including overheads and profit at 12.50%		Labour hours	Labour £	Material £	Unit	Total rate £
R10 RAINWATER PIPEWORK/GUTTERS - cont'd						
Cast iron pipes and fittings; BS 460; **ears cast on; joints - cont'd** 75 mm pipes; primed, plugged						
and nailed	PC £18.08/1.8m	0.66	7.05	12.25	m	**19.30**
Extra for						
shoe	PC £8.73	0.42	4.48	8.79	nr	**13.27**
bend	PC £5.36	0.72	7.69	5.00	nr	**12.69**
single branch	PC £8.27	0.90	9.61	7.98	nr	**17.59**
offset 229 mm projection	PC £9.51	0.72	7.69	8.45	nr	**16.14**
offset 305 mm projection	PC £11.14	0.72	7.69	9.88	nr	**17.57**
connection to clay pipes; cement and sand (1:2) joint		0.20	2.14	0.10	nr	**2.24**
100 mm pipes; primed, plugged						
and nailed	PC £24.27/1.8m	0.72	7.69	16.49	m	**24.18**
Extra for						
shoe	PC £11.35	0.48	5.12	11.45	nr	**16.57**
bend	PC £8.32	0.78	8.33	8.05	nr	**16.38**
single branch	PC £10.80	0.96	10.25	10.54	nr	**20.79**
offset 229 mm projection	PC £15.06	0.78	8.33	13.99	nr	**22.32**
offset 305 mm projection	PC £17.67	0.78	8.33	16.38	nr	**24.71**
connection to clay pipes; cement and sand (1:2) joint		0.24	2.56	0.10	nr	**2.66**
100 x 75 mm rectangular pipes; primed, plugged and nailed	PC £87.07/1.8m	0.72	7.69	57.29	m	**64.98**
Extra for					nr	
shoe	PC £34.73	0.48	5.12	32.10	nr	**37.22**
bend	PC £28.20	0.78	8.33	24.76	nr	**33.09**
offset 229 mm projection	PC £39.22	0.78	8.33	31.28	nr	**39.61**
offset 305 mm projection	PC £46.40	0.78	8.33	37.40	nr	**45.73**
connection to clay pipes; cement and sand (1:2) joint		0.24	2.56	0.10	nr	**2.66**
Rainwater head; flat; for pipes						
50 mm dia	PC £6.80	0.66	7.05	8.04	nr	**15.09**
63 mm dia	PC £6.80	0.68	7.26	8.11	nr	**15.37**
75 mm dia	PC £6.80	0.72	7.69	8.25	nr	**15.94**
100 mm dia	PC £15.69	0.78	8.33	18.52	nr	**26.85**
Rainwater head; rectangular, for pipes						
50 mm dia	PC £14.98	0.66	7.05	17.24	nr	**24.29**
63 mm dia	PC £14.98	0.68	7.26	17.31	nr	**24.57**
75 mm dia	PC £14.98	0.72	7.69	17.45	nr	**25.14**
100 mm dia	PC £30.37	0.78	8.33	35.04	nr	**43.37**
Roof outlets; cast iron; circular; with flat grate; joint to pipe						
50 mm dia		0.90	9.61	49.76	nr	**59.37**
75 mm dia	PC £45.69	1.00	10.68	52.29	nr	**62.97**
100 mm dia	PC £54.82	1.10	11.74	62.74	nr	**74.48**
Copper wire balloon grating; BS 416 for pipes or outlets						
50 mm dia	PC £1.42	0.08	0.85	1.60	nr	**2.45**
63 mm dia	PC £1.42	0.08	0.85	1.60	nr	**2.45**
75 mm dia	PC £1.65	0.08	0.85	1.86	nr	**2.71**
100 mm dia		0.10	1.07	2.10	nr	**3.17**
Cast iron gutters and fittings; BS 460 100 mm half round gutters; primed;						
on brackets; screwed to timber	PC £9.20/1.8m	0.48	5.12	7.25	m	**12.37**
Extra for						
stop end	PC £1.16	0.20	2.14	1.99	nr	**4.13**
running outlet	PC £3.57	0.42	4.48	3.37	nr	**7.85**
angle	PC £3.57	0.42	4.48	3.45	nr	**7.93**
115 mm half round gutters; primed;						
on brackets; screwed to timber	PC £9.58/1.8m	0.48	5.12	7.57	m	**12.69**
Extra for						
stop end	PC £1.69	0.20	2.14	2.62	nr	**4.76**
running outlet	PC £3.94	0.42	4.48	3.76	nr	**8.24**
angle	PC £3.94	0.42	4.48	3.85	nr	**8.33**

R DISPOSAL SYSTEMS Including overheads and profit at 12.50%	Labour hours	Labour £	Material £	Unit	Total rate £
125 mm half round gutters; primed;					
on brackets; screwed to timber PC £11.20/1.8m	0.54	5.77	8.63	m	14.40
Extra for					
stop end PC £1.69	0.24	2.56	2.63	nr	5.19
running outlet PC £4.65	0.48	5.12	4.43	nr	9.55
angle PC £4.65	0.48	5.12	4.35	nr	9.47
150 mm half round gutters; primed;					
on brackets; screwed to timber PC £19.16/1.8m	0.60	6.41	13.98	m	20.39
Extra for					
stop end PC £2.24	0.26	2.78	3.33	nr	6.11
running outlet PC £6.39	0.54	5.77	5.68	nr	11.45
angle PC £6.39	0.54	5.77	4.93	nr	10.70
100 mm ogee gutters; primed;					
on brackets; screwed to timber PC £10.04/1.8m	0.50	5.34	7.65	m	12.99
Extra for					
stop end PC £1.08	0.22	2.35	2.35	nr	4.70
running outlet PC £3.18	0.44	4.70	2.88	nr	7.58
angle PC £3.18	0.44	4.70	2.79	nr	7.49
115 mm ogee gutters; primed;					
on brackets; screwed to timber PC £11.28/1.8m	0.50	5.34	8.57	m	13.91
Extra for					
stop end PC £1.55	0.22	2.35	2.92	nr	5.27
running outlet PC £3.63	0.44	4.70	3.27	nr	7.97
angle PC £3.63	0.44	4.70	3.09	nr	7.79
125 mm ogee gutters; primed;					
on brackets; screwed to timber PC £11.82/1.8m	0.56	5.98	9.04	m	15.02
Extra for					
stop end PC £1.55	0.25	2.67	3.03	nr	5.70
running outlet PC £4.32	0.50	5.34	4.02	nr	9.36
angle PC £5.95	0.50	5.34	5.72	nr	11.06
3 mm galvanised heavy pressed steel gutters **and fittings; joggle joints; BS 1091**					
200 x 100 mm (400 mm girth) box gutter;					
screwed to timber	0.78	8.33	15.26	m	23.59
Extra for					
stop end	0.42	4.48	12.00	nr	16.48
running outlet	0.84	8.97	30.89	nr	39.86
stop end outlet	0.42	4.48	39.82	nr	44.30
angle	0.84	8.97	23.20	nr	32.17
381 mm boundary wall gutters;					
screwed to timber	0.78	8.33	13.88	m	22.21
Extra for					
stop end	0.48	5.12	11.36	nr	16.48
running outlet	0.84	8.97	33.52	nr	42.49
stop end outlet	0.42	4.48	42.72	nr	47.20
angle	0.84	8.97	25.44	nr	34.41
457 mm boundary wall gutters;					
screwed to timber	0.90	9.61	15.68	m	25.29
Extra for					
stop end	0.42	4.48	13.75	nr	18.23
running outlet	0.96	10.25	34.49	nr	44.74
stop end outlet	0.48	5.12	44.58	nr	49.70
angle	0.96	10.25	27.95	nr	38.20

R DISPOSAL SYSTEMS Including overheads and profit at 12.50%		Labour hours	Labour £	Material £	Unit	Total rate £
R10 RAINWATER PIPEWORK/GUTTERS - cont'd						
uPVC external rainwater pipes and fittings;						
BS 4576; slip-in joints						
50 mm pipes; fixing with pipe or socket						
brackets; plugged and screwed	PC £2.94/2m	0.36	3.84	2.29	m	6.13
Extra for						
fittings with one end		0.24	2.56	1.19	nr	3.75
fittings with two ends		0.36	3.84	1.74	nr	5.58
fittings with three ends		0.48	5.12	2.37	nr	7.49
shoe	PC £0.87	0.24	2.56	1.19	nr	3.75
bend	PC £1.36	0.36	3.84	1.74	nr	5.58
single branch	PC £1.91	0.48	5.12	2.37	nr	7.49
two bends to form offset 229 mm						
projection	PC £2.02	0.36	3.84	1.92	nr	5.76
connection to clay pipes;						
cement and sand (1:2) joint		0.16	1.71	0.10	nr	1.81
68 mm pipes; fixing with pipe or socket						
brackets; plugged and screwed	PC £2.60/2m	0.40	4.27	2.29	m	6.56
Extra for						
shoe	PC £0.95	0.26	2.78	1.45	nr	4.23
bend	PC £1.67	0.40	4.27	2.26	nr	6.53
single branch	PC £2.92	0.53	5.66	3.66	nr	9.32
two bends to form offset 229 mm						
projection	PC £2.08	0.40	4.27	2.22	nr	6.49
loose drain connector;						
cement and sand (1:2) joint		0.18	1.92	0.10	nr	2.02
110 mm pipes; fixing with pipe or socket						
brackets; plugged and screwed	PC £8.42/3m	0.43	4.59	4.62	m	9.21
Extra for						
shoe	PC £3.56	0.29	3.10	4.40	nr	7.50
bend	PC £4.71	0.43	4.59	5.69	nr	10.28
single branch	PC £6.13	0.58	6.19	7.29	nr	13.48
two bends to form offset 229 mm						
projection	PC £9.42	0.43	4.59	9.66	nr	14.25
loose drain connector;						
cement and sand (1:2) joint		0.42	4.48	4.31	nr	8.79
68.5 mm square pipes; fixing with pipe						
brackets; plugged and screwed	PC £4.40/2.5m	0.40	4.27	2.83	m	7.10
Extra for						
shoe	PC £1.11	0.26	2.78	1.51	nr	4.29
bend	PC £1.13	0.40	4.27	1.54	nr	5.81
single branch	PC £2.88	0.53	5.66	3.50	nr	9.16
two bends to form offset 229 mm						
projection	PC £3.62	0.40	4.27	1.71	nr	5.98
drain connector; square to round;						
cement and sand (1:2) joint		0.24	2.56	3.07	nr	5.63
Rainwater head; rectangular; for pipes						
50 mm dia	PC £5.09	0.54	5.77	6.46	nr	12.23
68 mm dia	PC £4.19	0.56	5.98	5.74	nr	11.72
110 mm dia	PC £9.67	0.66	7.05	12.25	nr	19.30
68.5 mm square	PC £4.11	0.56	5.98	5.16	nr	11.14
uPVC gutters and fittings; BS 4576						
76 mm half round gutters; on						
brackets; screwed to timber	PC £2.36/2m	0.36	3.84	1.91	m	5.75
Extra for						
stop end	PC £0.39	0.16	1.71	0.56	nr	2.27
running outlet	PC £1.11	0.30	3.20	1.03	nr	4.23
stop end outlet	PC £1.11	0.16	1.71	1.15	nr	2.86
angle	PC £1.12	0.30	3.20	1.28	nr	4.48
112 half round gutters; on						
brackets screwed to timber	PC £2.53/2m	0.40	4.27	2.53	m	6.80
Extra for						
stop end	PC £0.69	0.16	1.71	1.00	nr	2.71
running outlet	PC £1.34	0.34	3.63	1.28	nr	4.91
stop end outlet	PC £1.34	0.16	1.71	1.51	nr	3.22
angle	PC £1.51	0.34	3.63	1.91	nr	5.54

R DISPOSAL SYSTEMS Including overheads and profit at 12.50%		Labour hours	Labour £	Material £	Unit	Total rate £
152 mm half round gutters; on brackets screwed to timber	PC £7.53/2m	0.43	4.59	7.00	m	11.59
Extra for						
stop end	PC £1.68	0.19	2.03	2.55	nr	4.58
running outlet	PC £3.78	0.37	3.95	3.58	nr	7.53
stop end outlet	PC £3.60	0.19	2.03	4.03	nr	6.06
angle	PC £4.93	0.37	3.95	6.18	nr	10.13
114 mm rectangular gutters; on brackets; screwed to timber	PC £3.09/2m	0.40	4.27	3.24	m	7.51
Extra for						
stop end	PC £0.67	0.16	1.71	1.02	nr	2.73
running outlet	PC £1.60	0.34	3.63	1.52	nr	5.15
stop end outlet	PC £1.69	0.16	1.71	1.89	nr	3.60
angle	PC £1.53	0.34	3.63	1.98	nr	5.61

ALTERNATIVE WASTE PIPE AND FITTING PRICES

	£		£		£
ABS waste system (£/each)					
4 m pipe -32 mm	2.38	access plug-32 mm	0.45	cn.to copper-32 mm	0.86
-40 mm	2.91	-40 mm	0.48	-40 mm	1.02
-50 mm	3.50	-50 mm	0.79	sweep bend -32 mm	0.57
pipe brkt-32 mm	0.12	reducer -32 mm	0.40	-40 mm	0.69
-40 mm	0.13	-40 mm	0.45	-50 mm	1.01
-50 mm	0.23	-50 mm	0.57	sweep tee -32 mm	0.81
dl.socket-32 mm	0.36	str.tank cn-32 mm	0.98	-40 mm	0.99
-40 mm	0.42	-40 mm	1.11	-50 mm	1.39
-50 mm	0.67				

£ R11 FOUL DRAINAGE ABOVE GROUND		Labour hours	Labour £	Material £	Unit	Total rates
Cast iron pipes and fittings; BS 416						
50 mm pipes						
primed, eared, plugged and nailed	PC £22.23/1.8m	0.90	9.61	16.04	m	25.65
primed, uneared; fixing with holderbats plugged and screwed; (PC £6.68/nr);	PC £20.89/1.8m	1.00	10.68	18.79	m	29.47
Extra for						
fittings with two ends		0.90	9.61	11.53	nr	21.14
fittings with three ends		1.20	12.81	17.27	nr	30.08
fittings with four ends		1.50	16.01	33.40	nr	49.41
bend; short radius	PC £10.16	0.90	9.61	11.53	nr	21.14
access bend; short radius	PC £25.19	0.90	9.61	28.43	nr	38.04
boss; 38 mm BSP	PC £24.53	0.90	9.61	22.52	nr	32.13
single branch	PC £15.20	1.20	12.81	17.27	nr	30.08
double branch	PC £29.73	1.50	16.01	33.40	nr	49.41
offset 229 mm projection	PC £16.11	0.90	9.61	18.22	nr	27.83
offset 305 mm projection	PC £18.42	0.90	9.61	20.82	nr	30.43
access pipe	PC £21.38	0.90	9.61	18.98	nr	28.59
roof connector; for asphalt	PC £22.47	0.90	9.61	22.55	nr	32.16
roof connector; for roofing felt	PC £43.88	0.96	10.25	48.99	nr	59.24
isolated caulked lead joints		0.78	8.33	1.73	nr	10.06
connection to clay pipes; cement and sand (1:2) joint		0.16	1.71	0.10	nr	1.81
75 mm pipes						
primed, eared, plugged and nailed	PC £22.23/1.8m	0.96	10.25	16.28	m	26.53

R DISPOSAL SYSTEMS Including overheads and profit at 12.50%		Labour hours	Labour £	Material £	Unit	Total rate £
R11 FOUL DRAINAGE ABOVE GROUND - cont'd						
Cast iron pipes and fittings; BS 416 - cont'd						
75 mm pipes						
primed, uneared; fixing with						
holderbats plugged and						
screwed; (PC £6.68/nr);	PC £20.89/1.8m	1.10	11.74	18.96	m	30.70
Extra for						
bend; short radius	PC £10.16	0.96	10.25	11.80	nr	22.05
access bend; short radius	PC £25.19	0.96	10.25	28.71	nr	38.96
boss; 38 mm BSP	PC £24.53	0.96	10.25	22.80	nr	33.05
single branch	PC £15.20	1.25	13.35	17.66	nr	31.01
double branch	PC £29.73	1.60	17.08	33.95	nr	51.03
offset 229 mm projection	PC £16.11	0.96	10.25	18.49	nr	28.74
offset 305 mm projection	PC £18.42	0.96	10.25	21.09	nr	31.34
access pipe	PC £21.38	0.96	10.25	19.25	nr	29.50
roof connector; for asphalt	PC £22.47	0.96	10.25	22.83	nr	33.08
roof connector; for roofing felt	PC £43.88	1.00	10.68	49.27	nr	59.95
isolated caulked lead joints		0.84	8.97	1.98	nr	10.95
connection to clay pipes;						
cement and sand (1:2) joint		0.20	2.14	0.10	nr	2.24
100 mm pipes						
primed, eared, plugged and						
nailed	PC £29.90/1.8m	1.10	11.74	21.93	m	33.67
primed; uneared; fixing with						
holderbats plugged and						
screwed; (PC £7.59/nr);	PC £28.65/1.8m	1.20	12.81	24.99	m	37.80
Extra for						
W.C. bent connector;						
450 mm long tail	PC £24.73	0.84	8.97	23.75	nr	32.72
'Multikwik' W.C.connector	PC £2.07	0.12	1.28	2.33	nr	3.61
W.C. straight connector;						
300 mm long tail	PC £10.94	0.84	8.97	12.74	nr	21.71
W.C. straight connector;						
450 mm long tail	PC £13.64	0.84	8.97	15.78	nr	24.75
bend; short radius	PC £15.12	1.10	11.74	17.45	nr	29.19
access bend; short radius	PC £31.04	1.10	11.74	35.36	nr	47.10
boss; 38 mm BSP	PC £29.32	1.10	11.74	26.33	nr	38.07
single branch	PC £23.45	1.45	15.48	27.04	nr	42.52
double branch	PC £33.20	1.80	19.22	37.90	nr	57.12
offset 229 mm projection	PC £21.30	1.10	11.74	24.40	nr	36.14
offset 305 mm projection	PC £24.36	1.10	11.74	27.84	nr	39.58
access pipe	PC £23.45	1.10	11.74	19.73	nr	31.47
roof connector; for asphalt	PC £26.92	1.10	11.74	26.85	nr	38.59
roof connector; for roofing felt	PC £50.19	1.15	12.28	56.26	nr	68.54
isolated caulked lead joints		0.96	10.25	2.69	nr	12.94
connection to clay pipes;						
cement and sand (1:2) joint		0.24	2.56	0.10	nr	2.66
150 mm pipes						
primed, eared, plugged and						
nailed	PC £60.61/1.8m	1.40	14.95	43.46	m	58.41
primed; uneared; fixing with						
holderbats plugged and						
screwed; (PC £13.05/nr);	PC £59.05/1.8m	1.50	16.01	49.25	m	65.26
Extra for						
bend; short radius	PC £27.10	1.40	14.95	30.35	nr	45.30
access bend; short radius	PC £38.96	1.40	14.95	43.69	nr	58.64
boss; 38 mm BSP	PC £49.80	1.40	14.95	41.27	nr	56.22
single branch	PC £41.96	1.85	19.75	46.99	nr	66.74
double branch	PC £60.39	2.30	24.56	66.99	nr	91.55
offset 229 mm projection	PC £43.83	1.40	14.95	49.17	nr	64.12
offset 305 mm projection	PC £49.45	1.40	14.95	55.49	nr	70.44
access pipe	PC £37.91	1.40	14.95	27.89	nr	42.84
roof connector; for asphalt	PC £57.39	1.40	14.95	56.45	nr	71.40
roof connector; for roofing felt	PC £84.19	1.45	15.48	93.24	nr	108.72
isolated caulked lead joints		1.25	13.35	4.52	nr	17.87
connection to clay pipes;						
cement and sand (1:2) joint		0.31	3.31	0.14	nr	3.45

R DISPOSAL SYSTEMS Including overheads and profit at 12.50%		Labour hours	Labour £	Material £	Unit	Total rate £
Cut into existing soil stack; provide and insert new single branch						
50 mm	PC £18.24	3.00	32.03	46.36	nr	78.39
50 mm; with access door	PC £36.26	3.00	32.03	66.64	nr	98.67
75 mm	PC £18.24	3.00	32.03	47.46	nr	79.49
75 mm; with access door	PC £36.26	3.00	32.03	67.74	nr	99.77
100 mm	PC £28.14	3.85	41.10	67.36	nr	108.46
100 mm; with access door	PC £47.28	3.85	41.10	88.89	nr	129.99
150 mm	PC £50.35	4.70	50.18	123.60	nr	173.78
150 mm; with access door	PC £66.53	4.70	50.18	141.80	nr	191.98

Cast iron 'Timesaver' pipes
and fittings BS 416
50 mm pipes
  primed; 2 m lengths; fixing with holderbats
  plugged and screwed;(PC £4.84/nr)						

	PC £18.86/2m	0.66	7.05	16.17	m	23.22
Extra for						
fittings with two ends		0.66	7.05	11.08	nr	18.13
fittings with three ends		0.90	9.61	18.99	nr	28.60
fittings with four ends		1.15	12.28	30.94	nr	43.22
bend; short radius	PC £7.25	0.66	7.05	11.08	nr	18.13
access bend; short radius	PC £17.88	0.66	7.05	23.04	nr	30.09
boss; 38 mm BSP	PC £18.46	0.66	7.05	19.03	nr	26.08
single branch	PC £10.92	0.90	9.61	18.99	nr	28.60
double branch	PC £18.38	1.15	12.28	30.94	nr	43.22
offset 229 mm projection	PC £11.58	0.66	7.05	15.95	nr	23.00
offset 305 mm projection	PC £13.19	0.66	7.05	17.77	nr	24.82
access pipe	PC £17.45	0.66	7.05	17.89	nr	24.94
roof connector; for asphalt	PC £17.41	0.66	7.05	19.97	nr	27.02
roof connector; for roofing felt	PC £41.06	0.72	7.69	48.69	nr	56.38
isolated 'Timesaver' coupling joint	PC £4.11	0.36	3.84	4.62	nr	8.46
connection to clay pipes; cement and sand (1:2) joint	PC £59.23/M3	0.16	1.71	0.10	nr	1.81

75 mm pipes
  primed; 3 m lengths; fixing with holderbats
  plugged and screwed;(PC £4.84/nr)						

	PC £27.26/3m	0.66	7.05	14.54	m	21.59
primed; 2 m lengths; fixing with holderbats plugged and screwed;(PC £4.84/nr)						
	PC £18.86/2m	0.72	7.69	16.42	m	24.11
Extra for						
bend; short radius	PC £7.25	0.72	7.69	11.58	nr	19.27
access bend; short radius	PC £17.88	0.72	7.69	23.54	nr	31.23
boss; 38 mm BSP	PC £18.46	0.72	7.69	19.52	nr	27.21
single branch	PC £10.92	1.00	10.68	19.98	nr	30.66
double branch	PC £18.38	1.30	13.88	32.43	nr	46.31
offset 229 mm projection	PC £11.58	0.72	7.69	16.45	nr	24.14
offset 305 mm projection	PC £13.19	0.72	7.69	18.26	nr	25.95
access pipe	PC £17.45	0.72	7.69	18.38	nr	26.07
roof connector; for asphalt	PC £17.41	0.72	7.69	20.46	nr	28.15
roof connector; for roofing felt	PC £41.06	0.78	8.33	49.19	nr	57.52
isolated 'Timesaver' coupling joint	PC £4.55	0.42	4.48	5.12	nr	9.60
connection to clay pipes; cement and sand (1:2) joint		0.18	1.92	0.10	nr	2.02

100 mm pipes
  primed; 3 m lengths; fixing with holderbats
  plugged and screwed;(PC £5.28/nr)						

	PC £32.91/3m	0.72	7.69	17.51	m	25.20

R DISPOSAL SYSTEMS Including overheads and profit at 12.50%		Labour hours	Labour £	Material £	Unit	Total rate £
R11 FOUL DRAINAGE ABOVE GROUND - cont'd						
Cast iron 'Timesaver' pipes and fittings BS 416 - cont'd 100 mm pipes primed; 2 m lengths; fixing with holderbats plugged and screwed; (PC £5.28/nr)						
	PC £22.73/2m	0.80	8.54	19.74	m	28.28
Extra for W.C. bent connector;						
450 mm long tail	PC £23.70	0.72	7.69	31.30	nr	38.99
'Multikwik' W.C.connector	PC £2.07	0.12	1.28	2.33	nr	3.61
W.C. straight connector;						
300 mm long tail	PC £10.50	0.72	7.69	16.45	nr	24.14
bend; short radius	PC £10.04	0.80	8.54	15.93	nr	24.47
access bend; short radius	PC £21.25	0.80	8.54	28.54	nr	37.08
boss; 38 mm BSP	PC £23.15	0.80	8.54	25.05	nr	33.59
single branch	PC £15.54	1.20	12.81	27.78	nr	40.59
double branch	PC £19.22	1.55	16.55	37.32	nr	53.87
offset 229 mm projection	PC £14.44	0.80	8.54	20.88	nr	29.42
offset 305 mm projection	PC £16.28	0.80	8.54	22.95	nr	31.49
access pipe	PC £18.38	0.80	8.54	19.69	nr	28.23
roof connector; for asphalt	PC £20.60	0.80	8.54	24.74	nr	33.28
roof connector; for roofing felt	PC £49.97	0.86	9.18	60.34	nr	69.52
isolated 'Timesaver' coupling joint	PC £5.94	0.50	5.34	6.68	nr	12.02
transitional clayware socket; cement and sand (1:2) joint		0.48	5.12	15.59	nr	20.71
150 mm pipes primed; 3 m lengths; fixing with holderbats plugged and screwed;(PC £9.08/nr)						
	PC £68.96/3m	0.90	9.61	35.71	m	45.32
primed; 2 m lengths; fixing with holderbats plugged and screwed;(PC £9.08/nr)						
	PC £46.66/2m	1.00	10.68	39.35	m	50.03
Extra for						
bend; short radius	PC £17.95	1.00	10.68	29.35	nr	40.03
access bend; short radius	PC £30.17	1.00	10.68	43.10	nr	53.78
boss; 38 mm BSP	PC £37.38	1.00	10.68	39.66	nr	50.34
single branch	PC £31.98	1.45	15.48	56.39	nr	71.87
double branch		1.90	20.28	86.35	nr	106.63
offset 229 mm projection	PC £29.67	1.00	10.68	42.53	nr	53.21
offset 305 mm projection	PC £38.10	1.00	10.68	52.02	nr	62.70
roof connector; for asphalt	PC £38.17	1.00	10.68	45.80	nr	56.48
roof connector; for roofing felt	PC £76.50	1.05	11.21	94.17	nr	105.38
isolated 'Timesaver' coupling joint	PC £11.87	0.60	6.41	13.35	nr	19.76
transitional clayware socket; cement and sand (1:2) joint		0.62	6.62	17.74	nr	24.36
Polypropylene (PP) waste pipes and fittings; BS 5254; push fit 'O' - ring joints 32 mm pipes; fixing with pipe clips; plugged and screwed	PC £1.40/4m	0.26	2.78	0.80	m	3.58
Extra for						
fittings with one end		0.19	2.03	0.50	nr	2.53
fittings with two ends		0.26	2.78	0.72	nr	3.50
fittings with three ends		0.36	3.84	1.05	nr	4.89
access plug	PC £0.45	0.19	2.03	0.50	nr	2.53
double socket	PC £0.48	0.18	1.92	0.54	nr	2.46
male iron to PP coupling	PC £0.83	0.34	3.63	0.93	nr	4.56
sweep bend	PC £0.64	0.26	2.78	0.72	nr	3.50
sweep tee	PC £0.93	0.36	3.84	1.05	nr	4.89

R DISPOSAL SYSTEMS Including overheads and profit at 12.50%		Labour hours	Labour £	Material £	Unit	Total rate £
40 mm pipes; fixing with pipe clips; plugged and screwed	PC £1.76/4m	0.32	3.42	0.92	m	4.34
Extra for						
fittings with one end		0.22	2.35	0.54	nr	2.89
fittings with two ends		0.32	3.42	0.79	nr	4.21
fittings with three ends		0.43	4.59	1.21	nr	5.80
access plug	PC £0.48	0.22	2.35	0.54	nr	2.89
double socket	PC £0.52	0.22	2.35	0.58	nr	2.93
reducing set	PC £0.41	0.11	1.17	0.46	nr	1.63
male iron to PP coupling		0.41	4.38	1.02	nr	5.40
sweep bend	PC £0.70	0.32	3.42	0.79	nr	4.21
sweep tee	PC £1.08	0.43	4.59	1.21	nr	5.80
50 mm pipes; fixing with pipe clips; plugged and screwed	PC £2.55/4m	0.38	4.06	1.39	m	5.45
Extra for						
fittings with one end		0.24	2.56	0.89	nr	3.45
fittings with two ends		0.38	4.06	1.35	nr	5.41
fittings with three ends		0.50	5.34	1.75	nr	7.09
access plug	PC £0.79	0.24	2.56	0.89	nr	3.45
double socket	PC £0.97	0.25	2.67	1.09	nr	3.76
reducing set	PC £0.64	0.12	1.28	0.72	nr	2.00
male iron to PP coupling	PC £1.09	0.48	5.12	1.23	nr	6.35
sweep bend	PC £1.20	0.38	4.06	1.35	nr	5.41
sweep tee	PC £1.55	0.50	5.34	1.75	nr	7.09
muPVC waste pipes and fittings; BS 5255; solvent welded joints						
32 mm pipes; fixing with pipe clips; plugged and screwed	PC £4.74/4m	0.30	3.20	1.89	m	5.09
Extra for						
fittings with one end		0.20	2.14	1.15	nr	3.29
fittings with two ends		0.30	3.20	1.11	nr	4.31
fittings with three ends		0.40	4.27	1.60	nr	5.87
access plug	PC £0.94	0.20	2.14	1.15	nr	3.29
straight coupling	PC £0.62	0.20	2.14	0.79	nr	2.93
expansion coupling	PC £0.79	0.30	3.20	0.99	nr	4.19
male iron to PVC coupling	PC £0.77	0.37	3.95	0.92	nr	4.87
union coupling	PC £1.88	0.30	3.20	2.22	nr	5.42
sweep bend	PC £0.90	0.30	3.20	1.11	nr	4.31
spigot socket bend	PC £0.90	0.30	3.20	1.11	nr	4.31
sweep tee	PC £1.29	0.40	4.27	1.60	nr	5.87
caulking bush	PC £1.07	0.60	6.41	1.84	nr	8.25
40 mm pipes; fixing with pipe clips; plugged and screwed	PC £5.81/4m	0.36	3.84	2.27	m	6.11
Extra for						
fittings with one end		0.23	2.46	1.34	nr	3.80
fittings with two ends		0.36	3.84	1.24	nr	5.08
fittings with three ends		0.48	5.12	1.94	nr	7.06
fittings with four ends		0.64	6.83	4.69	nr	11.52
access plug	PC £1.11	0.23	2.46	1.34	nr	3.80
straight coupling	PC £0.76	0.24	2.56	0.95	nr	3.51
expansion coupling	PC £0.96	0.36	3.84	1.18	nr	5.02
male iron to PVC coupling	PC £0.90	0.46	4.91	1.07	nr	5.98
union coupling	PC £2.48	0.36	3.84	2.89	nr	6.73
level invert taper	PC £0.77	0.36	3.84	0.96	nr	4.80
sweep bend	PC £1.02	0.36	3.84	1.24	nr	5.08
spigot socket bend	PC £1.00	0.36	3.84	1.22	nr	5.06
sweep tee	PC £1.60	0.48	5.12	1.94	nr	7.06
sweep cross	PC £4.00	0.64	6.83	4.69	nr	11.52

R DISPOSAL SYSTEMS Including overheads and profit at 12.50%		Labour hours	Labour £	Material £	Unit	Total rate £
R11 FOUL DRAINAGE ABOVE GROUND - cont'd						
muPVC waste pipes and fittings; BS 5255; solvent welded joints - cont'd 50 mm pipes; fixing with pipe clips; plugged and screwed	PC £8.55/4m	0.42	4.48	3.44	m	7.92
Extra for						
fittings with one end		0.25	2.67	2.46	nr	5.13
fittings with two ends		0.42	4.48	1.74	nr	6.22
fittings with three ends		0.56	5.98	3.29	nr	9.27
fittings with four ends		0.74	7.90	5.38	nr	13.28
access plug	PC £2.10	0.25	2.67	2.46	nr	5.13
straight coupling	PC £0.92	0.28	2.99	1.14	nr	4.13
expansion coupling	PC £1.30	0.42	4.48	1.56	nr	6.04
male iron to PVC coupling	PC £1.30	0.54	5.77	1.52	nr	7.29
union coupling	PC £3.67	0.42	4.48	4.22	nr	8.70
level invert taper	PC £1.02	0.42	4.48	1.25	nr	5.73
sweep bend	PC £1.46	0.42	4.48	1.74	nr	6.22
spigot socket bend	PC £2.37	0.42	4.48	2.77	nr	7.25
sweep tee	PC £2.79	0.56	5.98	3.29	nr	9.27
sweep cross	PC £4.61	0.74	7.90	5.38	nr	13.28
uPVC overflow pipes and fittings; solvent 19 mm pipes; fixing with pipe clips; plugged and screwed	PC £2.03/4m	0.26	2.78	0.87	m	3.65
Extra for						
splay cut end		0.02	0.21	-	nr	0.21
fittings with one end		0.20	2.14	0.40	nr	2.54
fittings with two ends		0.20	2.14	0.56	nr	2.70
fittings with three ends		0.26	2.78	0.66	nr	3.44
straight connector	PC £0.32	0.20	2.14	0.40	nr	2.54
female iron to PVC coupling	PC £0.59	0.24	2.56	0.70	nr	3.26
bend	PC £0.46	0.20	2.14	0.56	nr	2.70
tee	PC £0.52	0.26	2.78	0.66	nr	3.44
bent tank connector	PC £0.72	0.24	2.56	0.84	nr	3.40
uPVC pipes and fittings; BS 4514; with solvent welded joints (unless otherwise described) 82 mm pipes; fixing with holderbats; plugged and screwed (PC £1.13/nr)	PC £12.11/4m	0.48	5.12	4.87	m	9.99
Extra for						
socket plug	PC £2.07	0.24	2.56	2.62	nr	5.18
slip coupling (push-fit)	PC £4.02	0.44	4.70	4.80	nr	9.50
expansion coupling	PC £2.61	0.48	5.12	3.22	nr	8.34
sweep bend	PC £3.67	0.48	5.12	4.41	nr	9.53
boss connector	PC £1.91	0.32	3.42	2.43	nr	5.85
single branch	PC £5.10	0.64	6.83	6.17	nr	13.00
access door	PC £4.43	0.72	7.69	5.12	nr	12.81
connection to clay pipes; caulking ring and cement and sand (1:2) joint		0.44	4.70	1.12	nr	5.82
110 mm pipes; fixing with holderbats; plugged and screwed (PC £1.17/nr)	PC £14.53/4m	0.53	5.66	5.73	m	11.39
Extra for						
socket plug	PC £2.51	0.26	2.78	3.19	nr	5.97
slip coupling (push-fit)	PC £5.02	0.48	5.12	6.01	nr	11.13
expansion coupling	PC £2.63	0.53	5.66	3.32	nr	8.98
WC connector	PC £3.59	0.35	3.74	4.22	nr	7.96
sweep bend	PC £5.02	0.53	5.66	6.01	nr	11.67
WC connecting bend	PC £5.35	0.35	3.74	6.20	nr	9.94
access bend	PC £11.26	0.55	5.87	13.02	nr	18.89
boss connector	PC £1.91	0.35	3.74	2.51	nr	6.25
single branch	PC £6.64	0.70	7.47	8.02	nr	15.49
single branch with access door	PC £12.87	0.72	7.69	15.02	nr	22.71
double branch	PC £16.18	0.88	9.40	18.93	nr	28.33
WC manifold	PC £22.49	0.35	3.74	25.84	nr	29.58

R DISPOSAL SYSTEMS Including overheads and profit at 12.50%		Labour hours	Labour £	Material £	Unit	Total rate £
access door	PC £4.43	0.72	7.69	5.12	nr	12.81
access pipe connector	PC £7.92	0.60	6.41	9.27	nr	15.68
connection to clay pipes; caulking ring						
and cement and sand (1:2) joint		0.50	5.34	1.46	nr	6.80
160 mm pipes; fixing with holderbats;						
plugged and screwed (PC £2.92/nr) PC £30.94/4m		0.60	6.41	33.58	m	19.32
Extra for						
socket plug	PC £4.63	0.30	3.20	5.99	nr	9.19
slip coupling (push-fit)	PC £12.85	0.54	5.77	15.23	nr	21.00
expansion coupling	PC £7.95	0.60	6.41	9.72	nr	16.13
sweep bend	PC £10.82	0.60	6.41	12.94	nr	19.35
boss connector	PC £2.61	0.40	4.27	3.71	nr	7.98
single branch	PC £15.24	0.79	8.43	18.31	nr	26.74
double branch	PC £26.98	1.00	10.68	31.89	nr	42.57
access door	PC £7.91	0.72	7.69	9.03	nr	16.72
connection to clay pipes; caulking ring						
and cement and sand (1:2) joint		0.60	6.41	2.43	nr	8.84
Weathering apron; for pipe						
82 mm dia	PC £1.03	0.41	4.38	1.30	nr	5.68
110 mm dia	PC £1.21	0.46	4.91	1.54	nr	6.45
160 mm dia	PC £3.64	0.50	5.34	4.48	nr	9.82
Weathering slate; for pipe						
110 mm dia	PC £16.93	1.10	11.74	19.23	nr	30.97
Vent cowl; for pipe						
82 mm dia	PC £1.03	0.40	4.27	1.30	nr	5.57
110 mm dia	PC £1.04	0.40	4.27	1.35	nr	5.62
160 mm dia	PC £2.71	0.40	4.27	3.44	nr	7.71
Roof outlets; circular PVC; with flat or						
domed grate; jointed to pipe						
82 mm dia	PC £12.57	0.66	7.05	14.28	nr	21.33
110 mm dia	PC £12.57	0.72	7.69	14.32	nr	22.01
Cut into existing soil stack; provide and						
insert new single branch						
82 mm	PC £6.19	1.00	10.68	7.39	nr	18.07
110 mm	PC £7.97	1.10	11.74	9.51	nr	21.25
110 mm; with access door	PC £15.62	1.10	11.74	18.12	nr	29.86
160 mm	PC £18.50	1.20	12.81	21.98	nr	34.79
Copper, brass and gunmetal ancillaries;						
screwed joints to fittings						
Brass trap; 'P'; 45 degree outlet;						
38 mm seal						
35 mm	PC £12.04	0.58	6.19	13.54	nr	19.73
Brass bath trap; 88.5 degree outlet;						
shallow seal						
42 mm	PC £9.02	0.65	6.94	10.15	nr	17.09
Copper trap 'P'; two piece						
35 mm with 38 mm seal	PC £6.15	0.58	6.19	6.92	nr	13.11
35 mm with 76 mm seal	PC £6.62	0.58	6.19	7.45	nr	13.64
42 mm with 38 mm seal	PC £9.12	0.65	6.94	10.26	nr	17.20
42 mm with 76 mm seal	PC £9.45	0.65	6.94	10.63	nr	17.57
Copper trap; 'S'; two piece						
35 mm with 38 mm seal	PC £6.62	0.58	6.19	7.45	nr	13.64
35 mm with 76 mm seal	PC £6.90	0.58	6.19	7.76	nr	13.95
42 mm with 38 mm seal	PC £9.49	0.65	6.94	10.68	nr	17.62
42 mm with 76 mm seal	PC £9.81	0.65	6.94	11.04	nr	17.98
Polypropylene ancillaries; screwed joint						
to waste fitting						
Tubular 'S' trap; bath; shallow seal						
40 mm	PC £6.99	0.66	7.05	7.86	nr	14.91
Trap 'P' two piece; 76 mm seal						
32 mm	PC £1.90	0.46	4.91	2.14	nr	7.05
40 mm	PC £2.20	0.54	5.77	2.47	nr	8.24
Trap 'S' two piece; 76 mm seal						
32 mm	PC £2.42	0.46	4.91	2.72	nr	7.63
40 mm	PC £2.84	0.54	5.77	3.20	nr	8.97

R DISPOSAL SYSTEMS Including overheads and profit at 12.50%		Labour hours	Labour £	Material £	Unit	Total rate £
R11 FOUL DRAINAGE ABOVE GROUND - cont'd						
Polypropylene ancillaries; screwed joint to waste fitting - cont'd						
Bottle trap 'P'; 76 mm seal						
32 mm	PC £2.12	0.46	4.91	2.39	nr	7.30
40 mm	PC £2.54	0.54	5.77	2.86	nr	8.63
Bottle trap; 'S'; 76 mm seal						
32 mm	PC £2.56	0.46	4.91	2.88	nr	7.79
40 mm	PC £3.12	0.54	5.77	3.50	nr	9.27
R12 DRAINAGE BELOW GROUND						
Mechanical excavation of trenches to receive pipes; grading bottoms; earthwork support; filling with excavated material and compacting; disposal of surplus soil; spreading on site average 50 m						
Pipes not exceeding 200 mm; average depth						
0.50 m deep		0.23	1.60	1.28	m	2.88
0.75 m deep		0.35	2.44	2.14	m	4.58
1.00 m deep		0.69	4.81	3.84	m	8.65
1.25 m deep		1.05	7.32	5.22	m	12.54
1.50 m deep		1.30	9.07	6.82	m	15.89
1.75 m deep		1.60	11.16	8.74	m	19.90
2.00 m deep		1.90	13.25	10.02	m	23.27
2.25 m deep		2.35	16.39	12.36	m	28.75
2.50 m deep		2.75	19.18	14.28	m	33.46
2.75 m deep		3.05	21.27	15.99	m	37.26
3.00 m deep		3.35	23.37	17.69	m	41.06
3.25 m deep		3.60	25.11	18.76	m	43.87
3.50 m deep		3.85	26.85	19.82	m	46.67
Pipes; 225 mm; average depth						
0.50 m deep		0.23	1.60	1.28	m	2.88
0.75 m deep		0.35	2.44	2.14	m	4.58
1.00 m deep		0.69	4.81	3.84	m	8.65
1.25 m deep		1.05	7.32	5.22	m	12.54
1.50 m deep		1.30	9.07	6.82	m	15.89
1.75 m deep		1.60	11.16	8.74	m	19.90
2.00 m deep		1.90	13.25	10.02	m	23.27
2.25 m deep		2.35	16.39	12.36	m	28.75
2.50 m deep		2.75	19.18	14.28	m	33.46
2.75 m deep		3.05	21.27	15.99	m	37.26
3.00 m deep		3.35	23.37	17.69	m	41.06
3.25 m deep		3.60	25.11	18.76	m	43.87
3.50 m deep		3.85	26.85	19.82	m	46.67
Pipes; 300 mm; average depth						
0.75 m deep		0.39	2.72	2.35	m	5.07
1.00 m deep		0.81	5.65	4.05	m	9.70
1.25 m deep		1.10	7.67	5.43	m	13.10
1.50 m deep		1.45	10.11	7.25	m	17.36
1.75 m deep		1.65	11.51	9.17	m	20.68
2.00 m deep		1.90	13.25	10.66	m	23.91
2.25 m deep		2.35	16.39	12.79	m	29.18
2.50 m deep		2.75	19.18	14.71	m	33.89
2.75 m deep		3.05	21.27	16.41	m	37.68
3.00 m deep		3.35	23.37	18.12	m	41.49
3.25 m deep		3.60	25.11	19.61	m	44.72
3.50 m deep		3.85	26.85	20.25	m	47.10
Extra for breaking up						
brick		1.60	11.16	7.37	m3	18.53
concrete		2.25	15.69	10.10	m3	25.79
reinforced concrete		3.20	22.32	14.74	m3	37.06
concrete 150 mm thick		0.35	2.44	1.69	m2	4.13
tarmacadam 75 mm thick		0.17	1.19	0.89	m2	2.08
tarmacadam and hardcore 150 mm thick		0.23	1.60	1.23	m2	2.83

R DISPOSAL SYSTEMS Including overheads and profit at 12.50%	Labour hours	Labour £	Material £	Unit	Total rate £
Hand excavation of trenches to receive **pipes; grading bottoms; earthwork support;** **filling with excavated material and** **compacting; disposal of surplus soil;** **spreading on site average 50 m**					
Pipes not exceeding 200 mm; average depth					
0.50 m deep	1.10	7.67	-	m	7.67
0.75 m deep	1.65	11.51	-	m	11.51
1.00 m deep	2.40	16.74	1.06	m	17.80
1.25 m deep	3.40	23.72	1.59	m	25.31
1.50 m deep	4.70	32.78	1.90	m	34.68
1.75 m deep	6.15	42.90	2.33	m	45.23
2.00 m deep	7.05	49.17	2.54	m	51.71
2.25 m deep	8.80	61.38	3.39	m	64.77
2.50 m deep	10.56	73.66	4.02	m	77.68
2.75 m deep	11.60	80.91	4.44	m	85.35
3.00 m deep	12.70	88.58	4.87	m	93.45
3.25 m deep	13.70	95.56	5.29	m	100.85
3.50 m deep	14.70	102.53	5.71	m	108.24
Pipes; 225 mm; average depth					
0.50 m deep	1.10	7.67	-	m	7.67
0.75 m deep	1.65	11.51	-	m	11.51
1.00 m deep	2.40	16.74	1.06	m	17.80
1.25 m deep	3.40	23.72	1.59	m	25.31
1.50 m deep	4.70	32.78	1.90	m	34.68
1.75 m deep	6.15	42.90	2.33	m	45.23
2.00 m deep	7.05	49.17	2.54	m	51.71
2.25 m deep	8.80	61.38	3.39	m	64.77
2.50 m deep	10.56	73.66	4.02	m	77.68
2.75 m deep	11.60	80.91	4.44	m	85.35
3.00 m deep	12.70	88.58	4.87	m	93.45
3.25 m deep	13.70	95.56	5.29	m	100.85
3.50 m deep	14.70	102.53	5.71	m	108.24
Pipes; 300 mm; average depth					
0.75 m deep	1.95	13.60	-	m	13.60
1.00 m deep	2.80	19.53	1.06	m	20.59
1.25 m deep	3.95	27.55	1.59	m	29.14
1.50 m deep	5.30	36.97	1.90	m	38.87
1.75 m deep	6.15	42.90	2.33	m	45.23
2.00 m deep	7.05	49.17	2.54	m	51.71
2.25 m deep	8.80	61.38	3.39	m	64.77
2.50 m deep	10.56	73.66	4.02	m	77.68
2.75 m deep	11.60	80.91	4.44	m	85.35
3.00 m deep	12.70	88.58	4.87	m	93.45
3.25 m deep	13.70	95.56	5.29	m	100.85
3.50 m deep	14.70	102.53	5.71	m	108.24
Extra for breaking up					
brick	3.30	23.02	3.71	m3	26.73
concrete	4.95	34.53	6.19	m3	40.72
reinforced concrete	6.60	46.04	8.66	m3	54.70
concrete 150 mm thick	0.77	5.37	0.87	m2	6.24
tarmacadam 75 mm thick	0.44	3.07	0.50	m2	3.57
tarmacadam and hardcore 150 mm thick	0.55	3.84	0.62	m2	4.46
Extra for taking up precast concrete paving slabs	0.33	2.30	-	m2	2.30
Sand filling; PC £10.90/t; (PC £17.45/m3)					
Beds; to receive pitch fibre pipes					
600 x 50 mm	0.09	0.63	0.74	m	1.37
700 x 50 mm	0.11	0.77	0.86	m	1.63
800 x 50 mm	0.13	0.91	0.98	m	1.89
Granular (shingle) filling; PC £11.00/t; (PC £20.00/m3)					
Beds; 100 mm thick; to pipes size					
100 mm	0.11	0.77	1.35	m	2.12
150 mm	0.11	0.77	1.58	m	2.35
225 mm	0.13	0.91	1.80	m	2.71
300 mm	0.15	1.05	2.03	m	3.08

R DISPOSAL SYSTEMS Including overheads and profit at 12.50%	Labour hours	Labour £	Material £	Unit	Total rate £
R12 DRAINAGE BELOW GROUND - cont'd					
Granular (shingle) filling; PC £11.00/t; (PC £20.00/m3) - cont'd					
Beds; 150 mm thick; to pipes size					
100 mm	0.15	1.05	2.03	m	**3.08**
150 mm	0.18	1.26	2.25	m	**3.51**
225 mm	0.20	1.40	2.47	m	**3.87**
300 mm	0.22	1.53	2.70	m	**4.23**
Beds and benchings; beds 100 mm thick; to pipes size					
100 mm	0.25	1.74	2.47	m	**4.21**
150 mm	0.28	1.95	2.47	m	**4.42**
225 mm	0.33	2.30	3.38	m	**5.68**
300 mm	0.39	2.72	3.83	m	**6.55**
Beds and benchings; beds 150 mm thick; to pipes size					
100 mm	0.28	1.95	2.70	m	**4.65**
150 mm	0.31	2.16	2.93	m	**5.09**
225 mm	0.39	2.72	4.05	m	**6.77**
300 mm	0.50	3.49	4.95	m	**8.44**
Beds and coverings; 100 mm thick; to pipes size					
100 mm	0.40	2.79	3.38	m	**6.17**
150 mm	0.50	3.49	4.05	m	**7.54**
225 mm	0.66	4.60	5.63	m	**10.23**
300 mm	0.79	5.51	6.75	m	**12.26**
Beds and coverings; 150 mm thick; to pipes size					
100 mm	0.59	4.12	4.95	m	**9.07**
150 mm	0.66	4.60	5.63	m	**10.23**
225 mm	0.86	6.00	7.20	m	**13.20**
300 mm	1.00	6.98	8.55	m	**15.53**
In situ ready mixed concrete; normal **Portland cement; mix 11.50 N/mm2 - 40 mm** **aggregate (1:3:6); PC £44.13/m3;**					
Beds; 100 mm thick; to pipes size					
100 mm	0.23	1.66	2.48	m	**4.14**
150 mm	0.23	1.66	2.48	m	**4.14**
225 mm	0.28	2.02	2.98	m	**5.00**
300 mm	0.32	2.31	3.48	m	**5.79**
Beds; 150 mm thick; to pipes size					
100 mm	0.32	2.31	3.48	m	**5.79**
150 mm	0.37	2.67	3.97	m	**6.64**
225 mm	0.41	2.96	4.47	m	**7.43**
300 mm	0.46	3.32	4.96	m	**8.28**
Beds and benchings; beds 100 mm thick; to pipes size					
100 mm	0.46	3.32	4.47	m	**7.79**
150 mm	0.52	3.76	4.96	m	**8.72**
225 mm	0.62	4.48	5.96	m	**10.44**
300 mm	0.72	5.20	6.95	m	**12.15**
Beds and benchings; beds 150 mm thick; to pipes size					
100 mm	0.52	3.76	4.96	m	**8.72**
150 mm	0.58	4.19	5.46	m	**9.65**
225 mm	0.72	5.20	6.95	m	**12.15**
300 mm	0.93	6.72	8.94	m	**15.66**
Beds and coverings; 100 mm thick; to pipes size					
100 mm	0.69	4.98	5.96	m	**10.94**
150 mm	0.81	5.85	6.95	m	**12.80**
225 mm	1.15	8.31	9.93	m	**18.24**
300 mm	1.40	10.11	11.92	m	**22.03**

R DISPOSAL SYSTEMS Including overheads and profit at 12.50%	Labour hours	Labour £	Material £	Unit	Total rate £
Beds and coverings; 150 mm thick; to pipes size					
100 mm	1.05	7.58	8.94	m	16.52
150 mm	1.15	8.31	9.93	m	18.24
225 mm	1.50	10.83	12.91	m	23.74
300 mm	1.80	13.00	15.39	m	28.39
In situ ready mixed concrete; normal Portland cement; mix 21.00 N/mm2 - 40 mm aggregate (1:2:4); PC £47.38/m3					
Beds; 100 mm thick; to pipes size					
100 mm	0.23	1.66	2.67	m	4.33
150 mm	0.23	1.66	2.67	m	4.33
225 mm	0.28	2.02	3.20	m	5.22
300 mm	0.32	2.31	3.73	m	6.04
Beds; 150 mm thick; to pipes size					
100 mm	0.32	2.31	3.73	m	6.04
150 mm	0.37	2.67	4.26	m	6.93
225 mm	0.41	2.96	4.80	m	7.76
300 mm	0.46	3.32	5.33	m	8.65
Beds and benchings; beds 100 mm thick; to pipes size					
100 mm	0.46	3.32	4.80	m	8.12
150 mm	0.52	3.76	5.33	m	9.09
225 mm	0.62	4.48	6.40	m	10.88
300 mm	0.72	5.20	7.46	m	12.66
Beds and benchings; beds 150 mm thick; to pipes size					
100 mm	0.52	3.76	5.33	m	9.09
150 mm	0.58	4.19	5.86	m	10.05
225 mm	0.72	5.20	7.46	m	12.66
300 mm	0.93	6.72	9.59	m	16.31
Beds and coverings; 100 mm thick; to pipes size					
100 mm	0.69	4.98	6.40	m	11.38
150 mm	0.81	5.85	7.46	m	13.31
225 mm	1.15	8.31	10.66	m	18.97
300 mm	1.40	10.11	12.79	m	22.90
Beds and coverings; 150 mm thick; to pipes size					
100 mm	1.05	7.58	9.59	m	17.17
150 mm	1.15	8.31	10.66	m	18.97
225 mm	1.50	10.83	13.86	m	24.69
300 mm	1.80	13.00	16.52	m	29.52

NOTE: The following items unless otherwise described include for all appropriate joints/couplings in the running length
The prices for gullies and rainwater shoes, etc., include for appropriate joints to pipes and for setting on and surrounding accessory with site mixed in situ concrete 11.50 N/mm2-40 aggregate (1:3:6)

Cast iron drain pipes and fittings; BS 437; coated; with caulked lead joints						
75 mm pipes						
laid straight; grey iron	PC £60.18/3m	0.69	5.46	24.13	m	29.59
laid straight; grey iron	PC £38.29/1.8m	0.81	6.42	25.98	m	32.40
in runs not exceeding 3 m long		1.10	8.71	39.59	m	48.30
Extra for						
bend; short radius		0.81	6.42	15.65	nr	22.07
bend; short radius; with round access door		0.81	6.42	28.76	nr	35.18
bend; long radius		0.81	6.42	22.26	nr	28.68
level invert taper		0.81	6.42	17.29	nr	23.71
access pipe		0.81	6.42	56.40	nr	62.82
single branch		1.10	8.71	28.81	nr	37.52

R DISPOSAL SYSTEMS Including overheads and profit at 12.50%		Labour hours	Labour £	Material £	Unit	Total rate £
R12 DRAINAGE BELOW GROUND - cont'd						
Cast iron drain pipes and fittings; BS 437; coated; with caulked lead joints - cont'd						
100 mm pipes						
laid straight; grey iron	PC £63.41/3m	0.81	6.42	25.61	m	32.03
laid straight; grey iron	PC £39.13/1.8m	0.92	7.29	26.92	m	34.21
in runs not exceeding 3 m long	PC £31.57/.91m	1.25	9.90	43.78	m	53.68
in ducts; supported on piers (measured elsewhere)		1.60	12.67	25.96	m	38.63
supported on wall brackets (measured elsewhere)		1.30	10.30	25.96	m	36.26
supported on ceiling hangers (measured elsewhere)		1.60	12.67	25.96	m	38.63
Extra for						
bend; short radius	PC £18.84	0.92	7.29	22.25	nr	29.54
bend; short radius; with round access door	PC £29.56	0.92	7.29	34.90	nr	42.19
bend; long radius	PC £29.35	0.92	7.29	33.32	nr	40.61
bend; long radius; with rectangular access door	PC £53.53	0.92	7.29	61.89	nr	69.18
rest bend	PC £23.88	0.92	7.29	28.20	nr	35.49
level invert taper	PC £19.07	0.92	7.29	21.17	nr	28.46
access pipe	PC £53.95	0.92	7.29	62.38	nr	69.67
single branch	PC £29.80	1.25	9.90	33.77	nr	43.67
single branch; with rectangular access door	PC £68.75	1.25	9.90	79.78	nr	89.68
'Y' branch	PC £67.70	1.25	9.90	78.54	nr	88.44
double branch	PC £81.82	1.55	12.28	95.29	nr	107.57
double branch; with rectangular access door	PC £132.86	1.55	12.28	155.59	nr	167.87
WC connector; 450 mm long	PC £18.17	0.69	5.46	18.15	nr	23.61
WC connector; 600 mm long	PC £28.61	0.69	5.46	28.24	nr	33.70
SW connector	PC £16.01	0.92	7.29	16.95	nr	24.24
150 mm pipes						
laid straight; grey iron	PC £108.94/3m	1.05	8.32	43.95	m	52.27
laid straight; grey iron	PC £59.89/1.8m	1.15	9.11	41.38	m	50.49
in runs not exceeding 3 m long	PC £47.36/.91m	1.55	12.28	66.21	m	78.49
Extra for						
bend; short radius	PC £40.62	1.15	9.11	47.87	nr	56.98
bend; short radius; with round access door	PC £69.05	1.15	9.11	81.46	nr	90.57
bend; long radius	PC £48.12	1.15	9.11	54.42	nr	63.53
bend; long radius; with rectangular access door	PC £98.48	1.15	9.11	113.91	nr	123.02
rest bend	PC £53.45	1.15	9.11	63.03	nr	72.14
level invert taper	PC £29.35	1.15	9.11	32.24	nr	41.35
access pipe	PC £99.98	1.15	9.11	115.69	nr	124.80
single branch	PC £64.22	1.55	12.28	73.25	nr	85.53
single branch; with rectangular access door	PC £135.49	1.55	12.28	157.45	nr	169.73
double branch	PC £127.15	1.95	15.44	147.67	nr	163.11
SW connector	PC £24.77	1.15	9.11	25.86	nr	34.97
225 mm pipes						
laid straight; grey iron	PC £331.70/1.8m	1.85	14.65	325.17	m	339.82
in runs not exceeding 3 m long	PC £191.87/.91m	2.45	19.40	255.94	m	275.34
Extra for						
bend; short radius	PC £155.46	1.85	14.65	190.71	nr	205.36
bend; long radius; with rectangular access door	PC £262.43	1.85	14.65	317.06	nr	331.71
rest bend	PC £187.06	1.85	14.65	228.03	nr	242.68
level invert taper	PC £60.65	1.85	14.65	78.71	nr	93.36
access pipe	PC £209.88	1.85	14.65	254.99	nr	269.64
single branch	PC £202.30	2.45	19.40	249.57	nr	268.97
single branch; with rectangular access door	PC £141.50	2.45	19.40	177.77	nr	197.17
SW connector		1.85	14.65	91.34	nr	105.99

R DISPOSAL SYSTEMS Including overheads and profit at 12.50%		Labour hours	Labour £	Material £	Unit	Total rate £
Accessories in cast iron; with caulked lead **joints to pipes (unless otherwise described)**						
Rainwater shoes; horizontal inlet						
100 mm	PC £56.25	0.86	6.81	67.26	nr	74.07
150 mm	PC £96.65	0.98	7.76	115.19	nr	122.95
Gully fittings; comprising low invert						
gully trap and round hopper						
75 mm outlet	PC £17.48	1.25	9.90	24.86	nr	34.76
100 mm outlet	PC £29.35	1.40	11.09	38.92	nr	50.01
150 mm outlet	PC £74.46	1.85	14.65	92.15	nr	106.80
Add to above for						
bellmouth 300 mm high; circular						
plain grating						
75 mm; 175 mm grating	PC £18.92	0.69	5.46	23.08	nr	28.54
100 mm; 200 mm grating	PC £28.52	0.81	6.42	34.55	nr	40.97
150 mm; 250 mm grating	PC £40.98	1.15	9.11	50.63	nr	59.74
bellmouth 300 mm high; circular						
plain grating as above;						
one horizontal inlet						
75 x 50 mm	PC £31.68	0.69	5.46	37.62	nr	43.08
100 x 75 mm	PC £30.84	0.81	6.42	37.39	nr	43.81
150 x 100 mm	PC £67.70	1.15	9.11	80.69	nr	89.80
bellmouth 300 mm high; circular						
plain grating as above;						
one vertical inlet						
75 x 50 mm	PC £30.56	0.69	5.46	36.37	nr	41.83
100 x 75 mm	PC £36.21	0.81	6.42	43.43	nr	49.85
150 x 100 mm	PC £76.93	1.15	9.11	91.07	nr	100.18
raising piece; 200 mm dia						
75 mm high	PC £14.56	1.40	11.09	22.60	nr	33.69
150 mm high	PC £17.43	1.40	11.09	25.83	nr	36.92
225 mm high	PC £22.17	1.40	11.09	31.16	nr	42.25
300 mm high		1.40	11.09	35.45	nr	46.54
Yard gully (Deans); trapped; galvanized						
sediment pan; 267 mm round heavy grating						
100 mm outlet	PC £166.68	3.70	29.30	194.06	nr	223.36
Yard gully (garage); trapless; galvanized						
sediment pan; 267 mm round heavy grating						
100 mm outlet	PC £142.69	3.45	27.32	167.07	nr	194.39
Yard gully (garage); trapped; with rodding						
eye; galvanized perforated sediment pan;						
stopper; round heavy grating						
150 mm outlet; 347 mm grating	PC £623.72	4.60	36.43	711.35	nr	747.78
Grease trap; with internal access; insert						
galvanized perforated bucket; lid and frame						
450 x 300 x 525 mm deep; 100 mm outlet						
	PC £300.01	4.15	32.87	346.63	nr	379.50
Disconnecting trap; trapped with inspection						
arm; bridle plate and screw; building in and						
cutting and fitting brickwork around						
100 mm outlet; 100 mm inlet	PC £134.56	2.90	22.97	159.21	nr	182.18
150 mm outlet; 150 mm inlet	PC £200.27	3.70	29.30	237.54	nr	266.84
Cast iron 'Timesaver' drain pipes and						
fittings; BS 437; coated; with mechanical						
coupling joints						
75 mm pipes						
laid straight	PC £45.62/3m	0.52	4.12	21.36	m	25.48
in runs not exceeding 3 m long	PC £28.47/1m	0.69	5.46	42.55	m	48.01
Extra for						
bend; medium radius	PC £14.22	0.58	4.59	23.03	nr	27.62
single branch	PC £19.77	0.81	6.42	36.89	nr	43.31
isolated 'Timesaver' joint	PC £7.93	0.35	2.77	9.37	nr	12.14

R DISPOSAL SYSTEMS Including overheads and profit at 12.50%		Labour hours	Labour £	Material £	Unit	Total rate £
R12 DRAINAGE BELOW GROUND - cont'd						
Cast iron 'Timesaver' drain pipes and fittings; BS 437; coated; with mechanical coupling joints - cont'd						
100 mm pipes						
laid straight	PC £44.05/3m	0.58	4.59	21.28	m	25.87
in runs not exceeding 3 m long	PC £27.85/1m	0.78	6.18	43.44	m	49.62
Extra for						
bend; medium radius	PC £17.51	0.69	5.46	28.63	nr	34.09
bend; medium radius with access	PC £46.02	0.69	5.46	62.30	nr	67.76
bend; long radius	PC £21.47	0.69	5.46	32.26	nr	37.72
rest bend	PC £20.10	0.69	5.46	31.68	nr	37.14
diminishing pipe	PC £14.90	0.69	5.46	27.10	nr	32.56
single branch	PC £23.26	0.86	6.81	44.40	nr	51.21
single branch; with access	PC £53.61	0.98	7.76	69.70	nr	77.46
double branch	PC £35.01	1.10	8.71	67.78	nr	76.49
double branch; with access	PC £64.68	1.10	8.71	109.50	nr	118.21
isolated 'Timesaver' joint	PC £9.37	0.40	3.17	11.07	nr	14.24
transitional pipe; for WC	PC £13.34	0.58	4.59	23.70	nr	28.29
150 mm pipes						
laid straight	PC £82.97/3m	0.69	5.46	37.66	m	43.12
in runs not exceeding 3 m long	PC £51.81/1m	0.94	7.44	73.97	m	81.41
Extra for						
bend; medium radius	PC £35.35	0.81	6.42	49.63	nr	56.05
bend; medium radius with access	PC £85.46	0.81	6.42	108.82	nr	115.24
bend; long radius	PC £46.02	0.81	6.42	60.27	nr	66.69
diminishing pipe	PC £22.84	0.81	6.42	37.79	nr	44.21
single branch	PC £50.32	0.98	7.76	77.13	nr	84.89
isolated 'Timesaver' joint	PC £11.35	0.48	3.80	13.41	nr	17.21
Accessories in 'Timesaver' cast iron; with mechanical coupling joints						
Rainwater shoes; horizontal inlet						
100 mm	PC £56.25	0.58	4.59	75.11	nr	79.70
150 mm	PC £96.65	0.69	5.46	123.43	nr	128.89
Gully fittings; comprising low invert gully trap and round hopper						
75 mm outlet	PC £14.90	1.05	8.32	28.90	nr	37.22
100 mm outlet	PC £23.26	1.10	8.71	39.92	nr	48.63
150 mm outlet	PC £57.85	1.50	11.88	81.70	nr	93.58
Add to above for						
bellmouth 300 mm high; circular plain grating						
100 mm; 200 mm grating	PC £28.52	0.52	4.12	42.63	nr	46.75
bellmouth 300 mm high; circular plain grating as above; one horizontal inlet						
100 x 100 mm	PC £29.27	0.52	4.12	43.47	nr	47.59
bellmouth 300 mm high; circular plain grating as above; one vertical inlet						
100 x 100 mm	PC £30.36	0.52	4.12	47.95	nr	52.07
Yard gully (Deans); trapped; galvanized sediment pan; 267 mm round heavy grating						
100 mm outlet	PC £177.83	3.35	26.53	203.91	nr	230.44
Yard gully (garage); trapless; galvanized sediment pan; 267 mm round heavy grating						
100 mm outlet	PC £152.06	3.10	24.55	174.92	nr	199.47
Yard gully (garage); trapped; with rodding eye, galvanised perforated sediment pan; stopper; 267 mm round heavy grating						
100 mm outlet; 267 mm grating	PC £338.23	3.45	27.32	385.01	nr	412.33
Grease trap; internal access; galvanized perforated bucket; lid and frame						
20 gal. capacity	PC £776.29	4.60	36.43	719.35	nr	755.78

R DISPOSAL SYSTEMS Including overheads and profit at 12.50%		Labour hours	Labour £	Material £	Unit	Total rate £
Extra strength vitrified clay pipes and fittings; 'Hepworths' 'SuperSleve'/'HepSleve' or similar; plain ends with push-fit polypropylene flexible couplings						
100 mm pipes						
laid straight	PC £2.11	0.29	1.90	2.50	m	4.40
in runs not exceeding 3 m long		0.38	2.48	2.94	m	5.42
Extra for						
bend	PC £2.17	0.23	1.50	4.32	nr	5.82
access bend	PC £13.70	0.23	1.50	17.94	nr	19.44
rest bend	PC £3.59	0.23	1.50	5.99	nr	7.49
access pipe	PC £11.87	0.23	1.50	15.56	nr	17.06
socket adaptor	PC £2.18	0.20	1.31	3.56	nr	4.87
adaptor to flexible pipe	PC £2.11	0.20	1.31	3.49	nr	4.80
saddle	PC £4.32	0.86	5.62	6.32	nr	11.94
single junction	PC £4.57	0.29	1.90	8.14	nr	10.04
single access junction	PC £14.96	0.29	1.90	20.42	nr	22.32
150 mm pipes						
laid straight	PC £4.66	0.35	2.29	5.51	m	7.80
in runs not exceeding 3 m long		0.46	3.01	6.53	m	9.54
Extra for						
bend	PC £5.41	0.28	1.83	10.53	nr	12.36
access bend	PC £22.81	0.28	1.83	31.26	nr	33.09
rest bend	PC £6.95	0.28	1.83	12.52	nr	14.35
taper pipe	PC £5.12	0.28	1.83	8.67	nr	10.50
access pipe	PC £19.53	0.28	1.83	26.88	nr	28.71
socket adaptor	PC £5.23	0.23	1.50	8.57	nr	10.07
adaptor to flexible pipe	PC £3.73	0.23	1.50	6.81	nr	8.31
saddle	PC £7.60	1.05	6.86	11.87	nr	18.73
single junction	PC £7.95	0.35	2.29	15.87	nr	18.16
single access junction	PC £25.45	0.35	2.29	36.78	nr	39.07
Extra strength vitrified clay pipes and fittings; 'Hepworths' 'HepSeal'; or similar; socketted; with push-fit flexible joints						
100 mm pipes						
laid straight	PC £4.68	0.35	2.29	5.52	m	7.81
in runs not exceeding 3 m long		0.46	3.01	6.05	m	9.06
Extra for						
bend	PC £6.45	0.28	1.83	5.96	nr	7.79
access bend	PC £17.88	0.28	1.83	19.45	nr	21.28
rest bend	PC £8.12	0.28	1.83	8.00	nr	9.83
stopper	PC £2.60	0.17	1.11	3.07	nr	4.18
access pipe	PC £14.82	0.28	1.83	15.30	nr	17.13
socket adaptor	PC £2.18	0.30	1.96	4.73	nr	6.69
saddle	PC £6.13	0.86	5.62	7.24	nr	12.86
single junction	PC £8.57	0.35	2.29	7.91	nr	10.20
single access junction	PC £19.54	0.35	2.29	20.86	nr	23.15
double junction	PC £17.08	0.52	3.40	17.42	nr	20.82
double collar	PC £6.13	0.23	1.50	7.24	nr	8.74
150 mm pipes						
laid straight	PC £6.47	0.40	2.61	7.64	m	10.25
in runs not exceeding 3 m long		0.54	3.53	8.37	m	11.90
Extra for						
bend	PC £11.14	0.32	2.09	10.87	nr	12.96
access bend	PC £27.91	0.32	2.09	30.67	nr	32.76
rest bend	PC £13.55	0.32	2.09	13.72	nr	15.81
stopper	PC £3.71	0.21	1.37	4.38	nr	5.75
taper reducer	PC £17.03	0.32	2.09	17.82	nr	19.91
access pipe	PC £23.23	0.32	2.09	24.38	nr	26.47
socket adaptor	PC £5.23	0.35	2.29	9.87	nr	12.16
saddle	PC £10.12	1.05	6.86	11.96	nr	18.82
single access junction	PC £30.78	0.40	2.61	33.30	nr	35.91
double junction	PC £28.41	0.61	3.99	29.74	nr	33.73
double collar	PC £10.12	0.26	1.70	11.96	nr	13.66

R DISPOSAL SYSTEMS Including overheads and profit at 12.50%		Labour hours	Labour £	Material £	Unit	Total rate £
R12 DRAINAGE BELOW GROUND - cont'd						
Extra strength vitrified clay pipes and **fittings; 'Hepworths' 'HepSeal'; or similar;** **socketted; with push-fit flexible joints - cont'd**						
225 mm pipes						
laid straight	PC £12.66	0.52	3.40	14.95	m	18.35
in runs not exceeding 3 m long		0.69	4.51	16.38	m	20.89
Extra for						
bend	PC £23.30	0.41	2.68	23.03	nr	25.71
access bend	PC £58.84	0.41	2.68	65.01	nr	67.69
rest bend	PC £32.44	0.41	2.68	33.83	nr	36.51
stopper	PC £8.39	0.26	1.70	9.91	nr	11.61
taper reducer	PC £26.93	0.41	2.68	27.33	nr	30.01
access pipe	PC £58.84	0.41	2.68	63.52	nr	66.20
saddle	PC £26.93	1.40	9.15	31.81	nr	40.96
single junction	PC £34.01	0.52	3.40	34.20	nr	37.60
single access junction	PC £70.36	0.52	3.40	77.12	nr	80.52
double junction	PC £68.05	0.78	5.10	72.91	nr	78.01
double collar	PC £22.21	0.35	2.29	26.23	nr	28.52
300 mm pipes						
laid straight	PC £19.83	0.69	4.51	23.43	m	27.94
in runs not exceeding 3 m long		0.92	6.01	25.65	m	31.66
Extra for						
bend	PC £45.94	0.55	3.59	47.23	nr	50.82
access bend	PC £106.85	0.55	3.59	119.20	nr	122.79
rest bend	PC £68.82	0.55	3.59	74.26	nr	77.85
stopper	PC £18.49	0.35	2.29	21.84	nr	24.13
taper reducer	PC £55.09	0.55	3.59	58.05	nr	61.64
access pipe	PC £106.85	0.55	3.59	116.85	nr	120.44
saddle	PC £55.09	1.85	12.09	65.07	nr	77.16
single junction	PC £72.17	0.69	4.51	75.88	nr	80.39
single access junction	PC £128.41	0.69	4.51	142.31	nr	146.82
double junction	PC £144.44	1.05	6.86	158.91	nr	165.77
double collar	PC £36.09	0.46	3.01	42.63	nr	45.64
British Standard quality vitrified clay **pipes and fittings; socketted; cement and** **sand (1:2) joints**						
100 mm pipes						
laid straight	PC £3.06	0.46	3.01	3.71	m	6.72
in runs not exceeding 3 m long		0.61	3.99	4.06	m	8.05
Extra for						
bend (short/medium/knuckle)	PC £2.76	0.37	2.42	2.28	nr	4.70
bend (long/rest/elbow)	PC £5.53	0.37	2.42	5.55	nr	7.97
access bend	PC £18.03	0.37	2.42	20.32	nr	22.74
taper	PC £7.26	0.37	2.42	7.48	nr	9.90
access pipe	PC £14.99	0.37	2.42	16.37	nr	18.79
single junction	PC £5.53	0.46	3.01	5.30	nr	8.31
single access junction	PC £20.01	0.46	3.01	22.39	nr	25.40
double junction	PC £11.04	0.69	4.51	11.55	nr	16.06
double collar	PC £4.05	0.31	2.03	4.88	nr	6.91
double access junction	PC £25.54	0.69	4.51	28.67	nr	33.18
150 mm pipes						
laid straight	PC £5.49	0.52	3.40	6.58	m	9.98
in runs not exceeding 3 m long		0.69	4.51	7.20	m	11.71
Extra for						
bend (short/medium/knuckle)	PC £4.66	0.41	2.68	3.66	nr	6.34
bend (long/rest/elbow)	PC £9.36	0.41	2.68	9.22	nr	11.90
access bend	PC £30.34	0.41	2.68	33.99	nr	36.67
taper	PC £12.22	0.41	2.68	12.40	nr	15.08
access pipe	PC £25.27	0.41	2.68	27.35	nr	30.03

R DISPOSAL SYSTEMS Including overheads and profit at 12.50%		Labour hours	Labour £	Material £	Unit	Total rate £
single junction	PC £9.36	0.52	3.40	8.67	nr	12.07
single access junction	PC £33.68	0.52	3.40	37.39	nr	40.79
double junction	PC £18.73	0.78	5.10	19.19	nr	24.29
double collar	PC £6.74	0.35	2.29	8.04	nr	10.33
double access junction	PC £43.09	0.78	5.10	47.96	nr	53.06
225 mm pipes						
laid straight	PC £10.84	0.63	4.12	13.00	m	17.12
in runs not exceeding 3 m long		0.84	5.49	14.22	m	19.71
Extra for						
bend (short/medium/knuckle)	PC £14.58	0.51	3.33	13.58	nr	16.91
bend (long/rest/elbow)	PC £29.19	0.51	3.33	30.85	nr	34.18
access bend	PC £51.71	0.51	3.33	57.44	nr	60.77
taper	PC £26.52	0.51	3.33	27.31	nr	30.64
access pipe	PC £51.71	0.51	3.33	56.17	nr	59.50
single junction	PC £29.19	0.63	4.12	29.76	nr	33.88
single access junction	PC £57.40	0.63	4.12	63.08	nr	67.20
double junction	PC £58.19	0.95	6.21	62.94	nr	69.15
double collar	PC £15.88	0.41	2.68	18.96	nr	21.64
double access junction	PC £86.48	0.95	6.21	96.36	nr	102.57
300 mm pipes						
laid straight	PC £17.67	0.86	5.62	21.07	m	26.69
in runs not exceeding 3 m long		1.15	7.52	23.06	m	30.58
Extra for						
bend (short/medium/knuckle)	PC £28.83	0.69	4.51	28.00	nr	32.51
bend (long/rest/elbow)	PC £57.76	0.69	4.51	62.17	nr	66.68
access bend	PC £102.38	0.69	4.51	114.87	nr	119.38
taper	PC £52.54	0.69	4.51	55.38	nr	59.89
access pipe	PC £102.38	0.69	4.51	112.79	nr	117.30
single junction	PC £57.76	0.86	5.62	60.28	nr	65.90
single access junction	PC £113.72	0.86	5.62	126.38	nr	132.00
double junction	PC £115.42	1.30	8.50	126.51	nr	135.01
double access junction	PC £171.42	1.30	8.50	192.66	nr	201.16

Accessories in vitrified clay; set in concrete; with polypropylene coupling joints to pipes						
Rodding point; with oval aluminium plate						
100 mm	PC £11.94	0.58	3.79	15.81	nr	19.60
Gully fittings; comprising low back trap and square hopper; 150 x 150 mm square gully grid						
100 mm outlet	PC £10.71	0.98	6.41	17.52	nr	23.93
Add to above for						
100 mm back inlet		-	-	-	nr	5.27
100 mm raising pieces		0.29	1.90	5.80	nr	7.70
Access gully; trapped with rodding eye and integral vertical back inlet; stopper; 150 x 150 mm square gully grid						
100 mm outlet	PC £15.28	0.75	4.90	19.58	nr	24.48
Inspection chamber; comprising base; 300 or 450 mm raising piece; integral alloy cover and frame; 100 mm inlets						
straight through; 2 nr inlets	PC £41.24	2.30	15.03	53.39	nr	68.42
single junction; 3 nr inlets	PC £44.69	2.55	16.67	58.48	nr	75.15
double junction; 4 nr inlets	PC £48.49	2.75	17.97	63.97	nr	81.94

Accessories in propylene; cover set in concrete; with coupling joints to pipes						
Inspection chamber; 5 nr 100 mm inlets; cast iron cover and frame						
475 x 585 mm deep	PC £66.29	2.65	17.32	79.86	nr	97.18
475 x 930 mm deep	PC £77.69	2.90	18.96	92.69	nr	111.65

R DISPOSAL SYSTEMS Including overheads and profit at 12.50%		Labour hours	Labour £	Material £	Unit	Total rate £

R12 DRAINAGE BELOW GROUND - cont'd

**Accessories in vitrified clay; set in
concrete; with cement and sand (1:2) joints
to pipes**
Rainwater shoes; with 250 x 100 mm oval
access

100 mm	PC £12.14	0.58	3.79	14.94	nr	18.73
150 mm	PC £20.45	0.69	4.51	24.87	nr	29.38

Gully fittings; comprising low back trap and
square hopper; square gully grid

100 mm outlet; 150 x 150 mm grid	PC £15.60	1.15	7.52	24.96	nr	32.48
150 mm outlet; 225 x 225 mm grid	PC £32.70	1.50	9.80	49.44	nr	59.24
Add to above for						
100 mm back inlet	PC £6.73	-	-	7.57	nr	7.57
100 mm vertical back inlet	PC £6.73	-	-	7.57	nr	7.57
100 mm raising pieces	PC £2.61	0.35	2.29	3.03	nr	5.32

Yard gully (mud); trapped with rodding eye;
galvanized square bucket; stopper; square
hinged grate and frame

100 mm outlet; 225 x 225 mm grate	PC £53.72	3.45	22.55	65.23	nr	87.78
150 mm outlet; 300 x 300 mm grate	PC £97.52	4.60	30.07	116.28	nr	146.35

Yard gully (garage); trapped with rodding
eye; galvanized perforated round bucket;
stopper; round hinged grate and frame

100 mm outlet; 273 mm grate	PC £65.90	3.45	22.55	80.11	nr	102.66
150 mm outlet; 368 mm grate	PC £113.91	4.60	30.07	135.31	nr	165.38

Road gully; trapped with rodding eye and
stopper (grate measured elsewhere)

300 x 600 x 100 mm outlet	PC £33.03	3.80	24.84	50.77	nr	75.61
300 x 600 x 150 mm outlet	PC £33.03	3.80	24.84	50.77	nr	75.61
400 x 750 x 150 mm outlet	PC £39.42	4.60	30.07	65.01	nr	95.08
450 x 900 x 150 mm outlet	PC £53.34	5.75	37.58	84.78	nr	122.36

Grease trap; with internal access;
galvanized perforated bucket; lid and frame

450 x 300 x 525 mm deep; 100 mm outlet						
	PC £249.84	4.05	26.47	295.96	nr	322.43
600 x 450 x 600 mm deep; 100 mm outlet						
	PC £317.10	4.85	31.70	376.91	nr	408.61

Interceptor; trapped with inspection arm;
lever locking stopper; chain and staple;
cement and sand (1:2) joints to pipes;
building in, and cutting and fitting
brickwork around

100 mm outlet; 100 mm inlet	PC £42.85	4.60	30.07	60.15	nr	90.22
150 mm outlet; 150 mm inlet	PC £60.75	5.20	33.99	82.65	nr	116.64
225 mm outlet; 225 mm inlet	PC £144.58	5.75	37.58	180.68	nr	218.26

Accessories: grates and covers
Aluminium alloy gully grids; set in position

125 x 125 mm	PC £1.57	0.11	0.72	1.77	nr	2.49
150 x 150 mm	PC £1.57	0.11	0.72	1.77	nr	2.49
225 x 225 mm	PC £4.70	0.11	0.72	5.28	nr	6.00
140 mm dia (for 100 mm)	PC £1.57	0.11	0.72	1.77	nr	2.49
197 mm dia (for 150 mm)	PC £2.45	0.11	0.72	2.76	nr	3.48
284 mm dia (for 225 mm)	PC £5.26	0.11	0.72	5.92	nr	6.64

Aluminium alloy sealing plates and frames;
set in cement and sand (1:3)

150 x 150 mm	PC £6.11	0.29	1.90	6.97	nr	8.87
225 x 225 mm	PC £11.10	0.29	1.90	12.59	nr	14.49
254 x 150 mm; for access fittings	PC £7.76	0.29	1.90	8.82	nr	10.72
140 mm dia (for 100 mm)	PC £4.94	0.29	1.90	5.66	nr	7.56
190 mm dia (for 150 mm)	PC £7.11	0.29	1.90	8.09	nr	9.99
273 mm dia (for 225 mm)	PC £11.37	0.29	1.90	13.00	nr	14.90

R DISPOSAL SYSTEMS Including overheads and profit at 12.50%		Labour hours	Labour £	Material £	Unit	Total rate £
Coated cast iron heavy duty road gratings and frames; BS 497 Tables 6 and 7; bedding and pointing in cement and sand (1:3); one course half brick thick wall in semi-engineering bricks in cement mortar (1:3)						
475 x 475 mm; grade A, ref GA1-450 (131 kg)						
	PC £62.87	2.90	18.96	73.26	nr	92.22
400 x 350 mm; grade A, ref GA2-325 (99 kg)						
	PC £52.90	2.90	18.96	61.60	nr	80.56
500 x 350 mm; grade A, ref GA2-325 (124 kg)						
	PC £75.02	2.90	18.96	86.71	nr	105.67
White vitreous clay floor channels; bedded, floor channel						
100 mm half round section						
floor channel	PC £42.77	0.69	4.51	51.05	m	55.56
Extra for						
stop end	PC £25.99	0.46	3.01	29.24	nr	32.25
angle	PC £39.00	0.69	4.51	43.87	nr	48.38
tee piece	PC £46.24	0.92	6.01	52.02	nr	58.03
stop end outlet	PC £33.59	0.58	3.79	37.79	nr	41.58
150 mm half round section						
floor channel	PC £47.46	0.86	5.62	56.85	m	62.47
Extra for						
stop end	PC £28.85	0.58	3.79	32.46	nr	36.25
angle	PC £43.33	0.86	5.62	48.75	nr	54.37
tee piece	PC £30.23	1.15	7.52	57.76	nr	65.28
stop end outlet	PC £44.91	0.69	4.51	50.53	nr	55.04
Channel sump outlet for 150 mm channel						
one inlet	PC £62.10	1.40	9.15	69.86	nr	79.01
two inlets	PC £66.16	1.70	11.11	74.43	nr	85.54
230 mm half round section						
floor channel	PC £75.96	1.15	7.52	91.58	m	99.10
Extra for						
stop end	PC £46.38	0.75	4.90	52.17	nr	57.07
angle	PC £69.51	1.15	7.52	78.20	nr	85.72
tee piece	PC £77.61	1.55	10.13	87.31	nr	97.44
stop end outlet	PC £54.38	0.92	6.01	61.18	nr	67.19
100 mm block floor channel	PC £49.58	0.69	4.51	58.83	m	63.34
Extra for						
stop end	PC £30.23	0.46	3.01	34.00	nr	37.01
angle	PC £45.36	0.69	4.51	51.03	nr	55.54
tee piece	PC £45.36	0.92	6.01	51.03	nr	57.04
stop end outlet	PC £37.49	0.58	3.79	42.17	nr	45.96
150 mm block floor channel	PC £55.16	0.86	5.62	65.68	m	71.30
Extra for						
stop end	PC £33.58	0.58	3.79	37.77	nr	41.56
angle	PC £50.39	0.86	5.62	56.69	nr	62.31
tee piece	PC £50.39	1.15	7.52	56.69	nr	64.21
stop end outlet	PC £41.67	0.69	4.51	46.88	nr	51.39
150 mm rebated block						
floor channel		0.86	5.62	72.87	m	78.49
Extra for						
stop end	PC £37.59	0.58	3.79	42.29	nr	46.08
angle	PC £56.46	0.86	5.62	63.51	nr	69.13
tee piece	PC £56.46	1.15	7.52	63.51	nr	71.03
stop end outlet	PC £45.70	0.69	4.51	51.41	nr	55.92

R DISPOSAL SYSTEMS Including overheads and profit at 12.50%		Labour hours	Labour £	Material £	Unit	Total rate £
R12 DRAINAGE BELOW GROUND - cont'd						
Accessories; channel gratings and connectors						
Galvanized cast iron medium duty square mesh						
gratings; bedding and pointing in cement and						
sand (1:3)						
138 x 13 mm; to suit 100 mm wide						
channel	PC £47.74	0.81	5.29	53.71	m	59.00
180 x 13 mm; to suit 150 mm wide						
channel	PC £47.74	1.05	6.86	53.71	m	60.57
Galvanized cast iron medium duty square mesh						
gratings and frame; galvanized cast iron						
angle bearers; bedding and pointing in						
cement and sand (1:3); cutting and pinning						
lugs to concrete						
148 x 13 mm; to suit 100 mm wide						
channel	PC £77.41	1.95	12.75	87.08	m	99.83
190 x 13 mm; to suit 150 mm wide						
channel	PC £77.41	2.20	14.38	87.08	m	101.46
Chromium plated brass domed outlet grating;						
threaded joint to connector (measured						
elsewhere)						
50 mm	PC £15.05	0.35	2.29	16.93	nr	19.22
63 mm	PC £19.82	0.35	2.29	22.30	nr	24.59
Cast iron connector; cement and sand (1:2)						
joint to drain pipe						
75 x 300 mm long; screwed 50 mm	PC £12.47	0.46	3.01	14.13	nr	17.14
75 x 300 mm long; screwed 63 mm	PC £14.57	0.46	3.01	16.49	nr	19.50
Accessories in precast concrete; top set in						
with rodding eye and stopper; cement and						
sand (1:2) joint to pipe						
Concrete road gully; BS 556 Part 2; trapped						
with rodding eye and stopper; cement and						
sand (1:2) joint to pipe						
450 mm dia x 1050 mm deep;						
100 or 150 mm outlet		5.45	35.62	45.09	nr	80.71
uPVC pipes and fittings; BS 4660; with lip						
seal coupling joints						
110 mm pipes						
laid straight	PC £13.37/6m	0.23	1.50	3.02	m	4.52
in runs not exceeding 3 m long	PC £6.69/3m	0.31	2.03	3.65	m	5.68
Extra for						
bend; short radius	PC £6.11	0.18	1.18	7.09	nr	8.27
bend; long radius	PC £9.65	0.18	1.18	10.61	nr	11.79
spigot/socket bend	PC £5.24	0.23	1.50	8.18	nr	9.68
access bend	PC £9.56	0.18	1.18	10.53	nr	11.71
socket plug	PC £2.42	0.06	0.39	2.86	nr	3.25
variable bend	PC £6.69	0.18	1.18	9.77	nr	10.95
inspection pipe	PC £11.60	0.18	1.18	15.57	nr	16.75
adaptor to clay	PC £4.94	0.18	1.18	5.80	nr	6.98
WC connector	PC £4.40	0.23	1.50	7.30	nr	8.80
single junction	PC £8.17	0.23	1.50	8.87	nr	10.37
inspection junction	PC £16.95	0.23	1.50	19.22	nr	20.72
slip coupling	PC £3.64	0.11	0.72	4.30	nr	5.02
160 mm pipes						
laid straight	PC £26.13/6m	0.28	1.83	5.92	m	7.75
in runs not exceeding 3 m long	PC £13.06/3m	0.37	2.42	7.16	m	9.58
Extra for						
bend; short radius	PC £11.72	0.23	1.50	13.60	nr	15.10
spigot/socket bend	PC £11.15	0.29	1.90	17.13	nr	19.03
socket plug	PC £4.21	0.07	0.46	4.97	nr	5.43
inspection pipe	PC £15.06	0.23	1.50	21.50	nr	23.00
adaptor to clay	PC £8.37	0.23	1.50	9.74	nr	11.24
level invert taper	PC £7.14	0.28	1.83	12.13	nr	13.96

R DISPOSAL SYSTEMS Including overheads and profit at 12.50%		Labour hours	Labour £	Material £	Unit	Total rate £
single junction	PC £21.51	0.28	1.83	23.87	nr	25.70
inspection junction	PC £27.60	0.28	1.83	31.06	nr	32.89
slip coupling	PC £7.77	0.14	0.92	9.18	nr	10.10
Accessories in uPVC; with lip seal coupling **joints to pipes (unless otherwise described)** Access cap assembly						
110 mm	PC £5.95	0.11	0.72	9.30	nr	10.02
Rodding eye 200 mm; sealed cover	PC £15.26	0.58	3.79	28.18	nr	31.97
Gully fitting; comprising 'P' trap, square hopper 154 x 154 mm grate						
110 mm outlet	PC £10.35	1.05	6.86	28.64	nr	35.50
Shallow access pipe assembly; 2 nr 110 mm inlets; light duty screw down cover and frame						
110 x 600 mm deep	PC £37.95	1.15	7.52	49.14	nr	56.66
Shallow branch access assembly; 3 nr 110 mm inlets; light duty screw down cover and frame						
110 x 600 mm deep	PC £41.94	1.40	9.15	51.31	nr	60.46
Inspection chamber; 450 mm dia; 940 mm deep; heavy duty screw down cover and frame						
4 nr 110 mm outlet/inlets	PC £80.65	2.05	13.40	147.23	nr	160.63
Kerb to gullies; class B engineering bricks on edge to three sides in cement mortar (1:3) rendering in cement mortar (1:3) to top and two sides and skirting to brickwork 230 mm high; dishing in cement mortar (1:3) to gully; steel trowelled						
230 x 230 mm internally		1.60	10.46	1.64	nr	12.10
Mechanical excavation Excavating manholes; not exceeding						
1 m deep		0.24	1.67	4.49	m3	6.16
2 m deep		0.26	1.81	4.92	m3	6.73
4 m deep		0.31	2.16	5.77	m3	7.93
Hand excavation Excavating manholes; not exceeding						
1 m deep		3.65	25.46	-	m3	25.46
2 m deep		4.30	29.99	-	m3	29.99
4 m deep		5.50	38.36	-	m3	38.36
Earthwork support (average 'risk' prices) Not exceeding 2 m between opposing faces; not exceeding						
1 m deep		0.17	1.19	0.53	m2	1.72
2 m deep		0.21	1.46	0.63	m2	2.09
4 m deep		0.26	1.81	0.80	m2	2.61
Disposal (mechanical) Excavated material; depositing on site in spoil heaps						
average 50 m distant		0.08	0.56	1.50	m3	2.06
Removing from site to tip not exceeding 13 km (using lorries)		-	-	-	m3	9.44
Disposal (hand) Excavated material; depositing on site in spoil heaps						
average 50 m distant		1.45	10.11	-	m3	10.11
Removing from site to tip not exceeding 13 km (using lorries)		1.10	7.67	13.54	m3	21.21

R DISPOSAL SYSTEMS Including overheads and profit at 12.50%	Labour hours	Labour £	Material £	Unit	Total rate £
R12 DRAINAGE BELOW GROUND - cont'd					
Mechanical filling					
Excavated material filling to excavations	0.17	1.19	2.14	m3	3.33
Hand filling					
Excavated material filling to excavations	1.10	7.67	-	m3	7.67
In situ ready mixed concrete; normal **Portland cement; mix 11.50 N/mm2 - 40 mm** **aggregate (1:3:6); PC £44.13/m3;** Beds					
not exceeding 150 mm thick	3.55	25.64	52.13	m3	77.77
150 - 450 mm thick	2.65	19.14	52.13	m3	71.27
over 450 mm thick	2.20	15.89	52.13	m3	68.02
Ready mixed in situ concrete; normal **Portland cement; mix 21.00 N/mm2-20 mm** **aggregate (1:2:4); PC £47.38/m3** Beds					
not exceeding 150 mm thick	3.55	25.64	55.97	m3	81.61
150 - 450 mm thick	2.65	19.14	55.97	m3	75.11
over 450 mm thick	2.20	15.89	55.97	m3	71.86
Site mixed in situ concrete; normal Portland **cement; mix 26.00 N/mm2-20 mm aggregate** **(1:1.5:3); (small quantities); PC £60.87/m3** Benching in bottoms					
150 - 450 mm average thick	8.05	69.19	68.48	m3	137.67
Reinforced site mixed in situ concrete; **normal Portland cement; mix 21.00 N/mm2-20mm** **aggregate (1:2:4); (small quantities); PC £57.11/m3** Isolated cover slabs					
not exceeding 150 mm thick	6.90	49.84	64.25	m3	114.09
Reinforcement; fabric to BS 4483; lapped; in **beds or suspended slabs**					
Ref A98 (1.54 kg/m2) PC £0.65	0.13	1.09	0.88	m2	1.97
Ref A142 (2.22 kg/m2) PC £0.83	0.13	1.09	1.13	m2	2.22
Ref A193 (3.02 kg/m2) PC £1.13	0.13	1.09	1.53	m2	2.62
Formwork to in situ concrete Soffits of isolated cover slabs					
horizontal	3.30	27.84	6.14	m2	33.98
Edges of isolated cover slabs					
not exceeding 250 mm high	0.92	7.76	2.26	m	10.02
Precast concrete rectangular access and **inspection chambers; 'Brooklyns' chambers** **or similar; comprising cover frame to** **receive manhole cover (priced elsewhere)** **intermediate wall sections and base section** **with cut outs;bedding;jointing and pointing** **in cement mortar (1:3) on prepared bed** Drainage chamber; size 600 x 450 mm internally; depth to invert					
600 mm deep	5.20	33.99	25.94	nr	59.93
900 mm deep	6.90	45.10	32.99	nr	78.09
Drainage chamber; 1200 x 750 mm reducing to 600 x 600 mm; no base unit; depth to invert					
1050 mm deep	8.65	56.54	94.36	nr	150.90
1650 mm deep	10.35	67.65	145.63	nr	213.28
2250 mm deep	12.70	88.58	196.90	nr	285.48

R DISPOSAL SYSTEMS Including overheads and profit at 12.50%	Labour hours	Labour £	Material £	Unit	Total rate £
Common bricks; PC £120.50/1000; in cement mortar (1:3)					
Walls to manholes					
one brick thick	2.60	31.27	20.29	m2	51.56
one and a half brick thick	3.70	44.50	30.51	m2	75.01
Projections of footings					
two brick thick	5.00	60.13	40.59	m2	100.72
Class A engineering bricks; PC £350.00/1000 **in cement mortar (1:3)**					
Walls to manholes					
one brick thick	2.90	34.88	52.82	m2	87.70
one and a half brick thick	4.05	48.71	79.31	m2	128.02
Projections of footings					
two brick thick	5.50	66.14	105.65	m2	171.79
Class B engineering bricks; PC £198.00/1000 **in cement mortar (1:3)**					
Walls to manholes					
one brick thick	2.90	34.88	31.28	m2	66.16
one and a half brick thick	4.05	48.71	46.99	m2	95.70
Projections of footings					
two brick thick	5.50	66.14	62.56	m2	128.70
Brickwork sundries					
Extra over for fair face; flush smooth pointing					
manhole walls	0.23	2.77	-	m2	2.77
Building ends of pipes into brickwork; making good fair face or rendering					
not exceeding 55 mm nominal size	0.11	1.32	-	nr	1.32
55 - 110 mm nominal size	0.17	2.04	-	nr	2.04
over 110 mm nominal size	0.23	2.77	-	nr	2.77
Step irons; BS 1247; malleable; galvanized; building into joints					
general purpose pattern	0.17	2.04	4.42	nr	6.46
Cement and sand (1:3) in situ finishings; **steel trowelled**					
13 mm work to manhole walls; one coat; to					
brickwork base over 300 mm wide	0.77	8.20	0.89	m2	9.09
Manhole accessories in cast iron					
Petrol trapping bend; coated; 375 x 750 mm; building into brickwork					
100 mm PC £57.80	1.55	12.28	67.72	nr	80.00
150 mm PC £93.38	2.00	15.84	109.58	nr	125.42
225 mm PC £173.03	3.30	26.14	201.73	nr	227.87
Cast iron inspection chambers; with bolted **flat covers; BS 437; bedded in cement mortar** **(1:3); caulked lead joints to pipes**					
100 x 100 mm; ref 010; no branches PC £82.20	1.15	9.11	95.69	nr	104.80
100 x 100 mm; ref 110; one branch PC £104.64	1.40	11.09	122.55	nr	133.64
100 x 100 mm; ref 111; one branch					
either side PC £135.42	1.60	12.67	158.52	nr	171.19
100 x 100 mm; ref 212; two branches					
either side PC £219.49	2.35	18.61	255.79	nr	274.40
100 x 100 mm; ref 313; three branches					
either side PC £298.75	3.10	24.55	347.65	nr	372.20
150 x 100 mm; ref 110; one branch PC £141.88	1.85	14.65	166.89	nr	181.54
150 x 100 mm; ref 111; one branch					
either side PC £182.03	2.05	16.24	213.21	nr	229.45
150 x 100 mm; ref 212; two branches					
either side PC £261.23	2.95	23.36	305.00	nr	328.36
150 x 100 mm; ref 313; three branches					
either side PC £354.68	3.80	30.10	412.83	nr	442.93

R DISPOSAL SYSTEMS Including overheads and profit at 12.50%		Labour hours	Labour £	Material £	Unit	Total rate £
R12 DRAINAGE BELOW GROUND - cont'd						
Cast iron inspection chambers; with bolted **flat covers; BS 437; bedded in cement mortar** **(1:3); caulked lead joints to pipes - cont'd**						
150 x 150 mm; ref 212; two branches either side	PC £364.07	3.15	24.95	424.36	nr	449.31
225 x 100 mm; ref 212; two branches either side	PC £522.90	3.80	30.10	603.35	nr	633.45
225 x 100 mm; ref 313; three branches either side	PC £660.00	4.85	38.41	760.28	nr	798.69
Cast iron inspection chambers; with bolted **flat covers; BS 437; bedded in cement mortar** **(1:3); with mechanical coupling joints**						
100 x 100 mm; ref 110; one branch	PC £67.09	1.20	9.50	97.35	nr	106.85
100 x 100 mm; ref 111; one branch either side	PC £81.92	1.80	14.26	124.57	nr	138.83
150 x 100 mm; ref 110; one branch	PC £95.38	1.45	11.48	131.82	nr	143.30
150 x 100 mm; ref 111; one branch either side		2.05	16.24	154.44	nr	170.68
150 x 150 mm; ref 110; one branch	PC £122.11	1.55	12.28	164.12	nr	176.40
150 x 150 mm; ref 111; one branch either side	PC £134.69	2.20	17.42	191.04	nr	208.46
Access covers and frames; coated; BS 497 **tables 1-5; bedding frame in cement and sand** **(1:3); cover in grease and sand**						
Grade C; light duty; rectangular single seal solid top						
450 x 450 mm; ref MC1-45/45 (31 kg)	PC £22.99	1.70	11.11	27.28	nr	38.39
600 x 450 mm; ref MC1-60/45 (32 kg)	PC £23.15	1.70	11.11	27.46	nr	38.57
600 x 600 mm; ref MC1-60/60 (61 kg)	PC £51.41	1.70	11.11	59.25	nr	70.36
Grade C; light duty; rectangular single seal recessed						
450 x 450 mm; ref MC1R-45/45 (43 kg)	PC £37.65	1.70	11.11	43.77	nr	54.88
600 x 450 mm; ref MC1R-60/45 (43 kg)	PC £46.98	1.70	11.11	54.27	nr	65.38
600 x 600 mm; ref MC1R-60/60 (53 kg)	PC £64.89	1.70	11.11	74.42	nr	85.53
Grade C; light duty; rectangular double seal solid top						
450 x 450 mm; ref MC2-45/45 (51 kg)	PC £35.25	1.70	11.11	41.08	nr	52.19
600 x 450 mm; ref MC2-60/45 (51 kg)	PC £42.36	1.70	11.11	49.07	nr	60.18
600 x 600 mm; ref MC2-60/60 (83 kg)	PC £65.11	1.70	11.11	74.67	nr	85.78
Grade C; light duty; rectangular double seal recessed						
450 x 450 mm; ref MC2R-45/45 56 kg)	PC £58.57	1.70	11.11	67.30	nr	78.41
600 x 450 mm; ref MC2R-60/45 (71 kg)	PC £81.49	1.70	11.11	93.10	nr	104.21
600 x 600 mm; ref MC2R-60/60 (75 kg)	PC £113.04	1.70	11.11	128.59	nr	139.70
Grade B; medium duty; circular single seal solid top						
500 mm; ref MB2-50 (106 kg)	PC £65.06	2.30	15.03	74.61	nr	89.64
550 mm; ref MB2-55 (112 kg)	PC £69.56	2.30	15.03	79.68	nr	94.71
600 mm; ref MB2-60 (134 kg)	PC £73.22	2.30	15.03	83.79	nr	98.82

R DISPOSAL SYSTEMS Including overheads and profit at 12.50%		Labour hours	Labour £	Material £	Unit	Total rate £
Grade B; medium duty; rectangular single seal solid top						
600 x 450 mm; ref MB2-60/45 (135 kg)	PC £63.15	2.30	15.03	72.46	nr	87.49
600 x 600 mm; ref MB2-60/60 (170 kg)	PC £82.13	2.30	15.03	93.81	nr	108.84
Grade B; medium duty; rectangular single seal recessed						
600 x 450 mm; ref MB2R-60/45 (145 kg)	PC £86.75	2.30	15.03	99.01	nr	114.04
600 x 600 mm; ref MB2R-60/60 (171 kg)	PC £110.77	2.30	15.03	126.03	nr	141.06
Grade B; 'Chevron'; medium duty; double triangular solid top						
550 mm; ref MB1-55 (125 kg)	PC £63.29	2.30	15.03	72.62	nr	87.65
600 mm; ref MB1-60 (140 kg)	PC £75.69	2.30	15.03	86.57	nr	101.60
Grade A; heavy duty; single triangular solid top						
550 x 455 mm; ref MA-T (196 kg)	PC £90.57	2.90	18.96	103.31	nr	122.27
Grade A; 'Chevron'; heavy duty double triangular solid top						
500 mm; ref MA-50 (164 kg)	PC £105.82	3.45	22.55	120.47	nr	143.02
550 mm; ref MA-55 (176 kg)	PC £103.18	3.45	22.55	117.50	nr	140.05
600 mm; ref MA-60 (230 kg)	PC £108.90	3.45	22.55	123.93	nr	146.48
British Standard best quality vitrified clay **channels; bedding and jointing in cement and** **sand (1:2)**						
Half section straight						
100 mm x 1.00 m long	PC £2.08	0.92	6.01	2.45	nr	8.46
150 mm x 1.00 m long	PC £3.45	1.15	7.52	4.08	nr	11.60
225 mm x 1.00 m long	PC £7.77	1.50	9.80	9.18	nr	18.98
300 mm x 1.00 m long	PC £15.39	1.85	12.09	18.18	nr	30.27
Half section bend						
100 mm	PC £2.12	0.69	4.51	2.51	nr	7.02
150 mm	PC £3.53	0.86	5.62	4.17	nr	9.79
225 mm	PC £11.87	1.15	7.52	14.02	nr	21.54
300 mm	PC £23.56	1.40	9.15	27.83	nr	36.98
Half section taper straight						
150 mm	PC £8.92	0.81	5.29	10.54	nr	15.83
225 mm	PC £19.90	1.05	6.86	23.50	nr	30.36
300 mm	PC £39.37	1.25	8.17	46.51	nr	54.68
Half section taper bend						
150 mm	PC £13.57	1.05	6.86	16.04	nr	22.90
225 mm	PC £39.08	1.30	8.50	46.16	nr	54.66
300 mm	PC £77.33	1.60	10.46	91.35	nr	101.81
Three quarter section branch bend						
100 mm	PC £4.84	0.58	3.79	5.72	nr	9.51
150 mm	PC £8.15	0.86	5.62	9.63	nr	15.25
225 mm	PC £29.77	1.15	7.52	35.16	nr	42.68
300 mm	PC £59.07	1.55	10.13	69.77	nr	79.90
uPVC channels; with solvent weld or lip seal **coupling joints; bedding in cement and sand** Half section cut away straight; with coupling either end						
110 mm	PC £16.25	0.35	2.29	19.09	nr	21.38
160 mm	PC £21.81	0.46	3.01	25.60	nr	28.61
Half section cut away long radius bend; with coupling either end						
110 mm	PC £18.45	0.35	2.29	22.01	nr	24.30
160 mm	PC £27.32	0.46	3.01	32.74	nr	35.75
Half section straight channel adaptor; with one coupling						
110 mm	PC £6.76	0.29	1.90	9.45	nr	11.35
160 mm	PC £8.97	0.38	2.48	13.52	nr	16.00

R DISPOSAL SYSTEMS Including overheads and profit at 12.50%		Labour hours	Labour £	Material £	Unit	Total rate £
R12 DRAINAGE BELOW GROUND - cont'd						
uPVC channels; with solvent weld or lip seal coupling joints; bedding in cement and sand - cont'd						
Half section cut away bend						
110 mm	PC £12.10	0.46	3.01	14.79	nr	**17.80**
Half section bend						
110 mm	PC £2.97	0.38	2.48	4.02	nr	**6.50**
160 mm	PC £5.75	0.58	3.79	7.98	nr	**11.77**
Half section channel connector						
110 mm	PC £1.43	0.09	0.59	2.70	nr	**3.29**
160 mm	PC £3.35	0.11	0.72	6.32	nr	**7.04**
Half section channel junction						
110 mm	PC £4.82	0.58	3.79	6.20	nr	**9.99**
160 mm	PC £9.10	0.69	4.51	11.93	nr	**16.44**
polypropylene slipper bend						
110 mm	PC £6.38	0.46	3.01	8.05	nr	**11.06**
R13 LAND DRAINAGE						
Hand excavation of trenches to receive land drain pipes; grading bottoms; earthwork support; filling to within 150 mm of surface filling to within 150 mm of surface with gravel rejects; remainder filled with excavated material and compacting; disposal of surplus soil; spreading on site average 50 m						
Pipes not exceeding 200 mm; average depth						
0.75 m deep		1.85	12.90	11.38	m	**24.28**
1.00 m deep		2.50	17.44	15.17	m	**32.61**
1.25 m deep		3.45	24.06	19.22	m	**43.28**
1.50 m deep		5.95	41.50	24.24	m	**65.74**
1.75 m deep		7.05	49.17	28.66	m	**77.83**
2.00 m deep		8.15	56.85	33.11	m	**89.96**
Surplus excavated material						
Removing from site to tip not exceeding 13 km (using lorries)						
machine loaded		-	-	-	m3	**8.89**
hand loaded		1.10	7.67	12.76	m3	**20.43**
Clay field drain pipes; BS 1196; unjointed						
Pipes; laid straight						
75 mm	PC £22.00/100	0.23	1.50	0.86	m	**2.36**
100 mm	PC £39.60/100	0.29	1.90	1.55	m	**3.45**
150 mm	PC £82.50/100	0.35	2.29	3.22	m	**5.51**
Vitrified clay perforated sub-soil pipes; **BS 65; 'Hepworths' 'Hepline' or similar**						
Pipes; laid straight						
100 mm	PC £3.23	0.25	1.63	3.82	m	**5.45**
150 mm	PC £5.81	0.31	2.03	6.86	m	**8.89**
225 mm	PC £10.68	0.41	2.68	12.61	m	**15.29**

S PIPED SUPPLY SYSTEMS Including overheads and profit at 12.50%	Labour hours	Labour £	Material £	Unit	Total rate £

ALTERNATIVE SERVICE PIPE AND FITTING PRICES

	£		£		£		£
Copper pipes to BS 2871 (£/100 m)							
Table X							
6 mm	54.00	10 mm	91.80	67 mm	1332.00	108 mm	2706.00
8 mm	73.20	12 mm	114.00	76 mm	1895.00	133 mm	3342.00
Table Y							
6 mm	72.40	10 mm	125.00	67 mm	2230.00	108 mm	4461.00
8 mm	95.90	12 mm	154.00	76 mm	2509.00		
Table Z							
15 mm	89.90	35 mm	496.00	54 mm	829.00	76 mm	1655.00
22 mm	171.00	42 mm	621.00	67 mm	1173.00	108 mm	2363.00
28 mm	216.00						

		£			£			£
PVC Class E cold water pressure system (£/each)								
(sizes shown as nearest equivalent metric sizes)								
pipe-per m	-13 mm	0.86	end cap	-13 mm	0.41	elbow	-13 mm	0.59
	-19 mm	1.19		-19 mm	0.47		-19 mm	0.69
	-25 mm	1.54		-25 mm	0.55		-25 mm	0.86
	-32 mm	2.49		-32 mm	0.84		-32 mm	1.58
	-40 mm	3.28		-40 mm	1.31		-40 mm	2.05
coupling	-13 mm	0.42	reducer	-32 mm	1.25	tee	-13 mm	0.66
	-19 mm	0.47		-40 mm	1.49		-19 mm	0.83
	-25 mm	0.56	MI.conn	-13 mm	0.70		-25 mm	1.21
	-32 mm	0.90		-19 mm	0.75		-32 mm	1.73
	-40 mm	1.10		-25 mm	1.42		-40 mm	2.57
s.tank.cn.	-19 mm	1.68		-32 mm	1.88			
	-25 mm	1.88		-40 mm	2.05			

	Labour hours	Labour £	Material £	Unit	Total rate £
S12 HOT AND COLD WATER (SMALL SCALE)					
Copper pipes; BS 2871 table X;					
capillary fittings; BS 864					
15 mm pipes; fixing with pipe					
clips; plugged and screwed PC £109.00/100m	0.30	3.20	1.60	m	4.80
Extra for					
made bend	0.18	1.92	-	nr	1.92
fittings with one end	0.13	1.39	0.63	nr	2.02
fittings with two ends	0.20	2.14	0.32	nr	2.46
fittings with three ends	0.30	3.20	0.60	nr	3.80
fittings with four ends	0.42	4.48	4.16	nr	8.64
stop end PC £0.56	0.13	1.39	0.63	nr	2.02
straight union coupling PC £2.50	0.20	2.14	2.82	nr	4.96
copper to iron connector PC £0.92	0.26	2.78	1.03	nr	3.81
elbow PC £0.29	0.20	2.14	0.32	nr	2.46
backplate elbow PC £1.83	0.42	4.48	2.06	nr	6.54
slow bend PC £0.97	0.20	2.14	1.10	nr	3.24
tee; equal PC £0.53	0.30	3.20	0.60	nr	3.80
tee; reducing PC £2.07	0.30	3.20	2.32	nr	5.52
cross PC £3.70	0.42	4.48	4.16	nr	8.64
straight tap connector PC £0.83	0.16	1.71	0.94	nr	2.65
bent tap connector PC £1.01	0.16	1.71	1.13	nr	2.84
straight tank connector; backnut PC £1.78	0.30	3.20	2.00	nr	5.20

S PIPED SUPPLY SYSTEMS Including overheads and profit at 12.50%		Labour hours	Labour £	Material £	Unit	Total rate £
S12 HOT AND COLD WATER (SMALL SCALE) - cont'd						
Copper pipes; BS 2871 table X;						
capillary fittings; BS 864 - cont'd						
22 mm pipes; fixing with pipe						
clips; plugged and screwed	PC £215.00/100m	0.31	3.31	2.93	m	6.24
Extra for						
made bend		0.24	2.56	-	nr	2.56
fittings with one end		0.16	1.71	1.05	nr	2.76
fittings with two ends		0.26	2.78	0.65	nr	3.43
fittings with three ends		0.40	4.27	1.21	nr	5.48
fittings with four ends		0.53	5.66	5.37	nr	11.03
stop end	PC £0.93	0.16	1.71	1.05	nr	2.76
reducing coupling	PC £0.71	0.26	2.78	0.79	nr	3.57
straight union coupling	PC £3.86	0.26	2.78	4.35	nr	7.13
copper to iron connector	PC £1.53	0.37	3.95	1.72	nr	5.67
elbow	PC £0.58	0.26	2.78	0.65	nr	3.43
backplate elbow	PC £3.81	0.53	5.66	4.28	nr	9.94
slow bend	PC £1.65	0.26	2.78	1.85	nr	4.63
tee; equal	PC £1.08	0.40	4.27	1.21	nr	5.48
tee; reducing	PC £1.04	0.40	4.27	1.17	nr	5.44
cross	PC £4.77	0.53	5.66	5.37	nr	11.03
straight tap connector	PC £1.28	0.20	2.14	1.44	nr	3.58
bent tap connector	PC £1.93	0.20	2.14	2.17	nr	4.31
straight tank connector; backnut	PC £2.64	0.40	4.27	2.97	nr	7.24
28 mm pipes; fixing with pipe						
clips; plugged and screwed	PC £278.00/100m	0.35	3.74	3.78	m	7.52
Extra for						
made bend		0.30	3.20	-	nr	3.20
fittings with one end		0.18	1.92	2.04	nr	3.96
fittings with two ends		0.34	3.63	1.19	nr	4.82
fittings with three ends		0.49	5.23	2.53	nr	7.76
fittings with four ends		0.67	7.15	7.69	nr	14.84
stop end	PC £1.81	0.18	1.92	2.04	nr	3.96
reducing coupling	PC £1.56	0.34	3.63	1.76	nr	5.39
straight union coupling	PC £5.31	0.34	3.63	5.97	nr	9.60
copper to iron connector	PC £2.40	0.47	5.02	2.70	nr	7.72
elbow	PC £1.06	0.34	3.63	1.19	nr	4.82
slow bend	PC £2.76	0.34	3.63	3.10	nr	6.73
tee; equal	PC £2.25	0.49	5.23	2.53	nr	7.76
tee; reducing	PC £2.42	0.49	5.23	2.72	nr	7.95
cross	PC £6.84	0.67	7.15	7.69	nr	14.84
straight tank connector; backnut	PC £3.64	0.49	5.23	4.09	nr	9.32
35 mm pipes; fixing with pipe						
clips; plugged and screwed	PC £617.00/100m	0.40	4.27	8.12	m	12.39
Extra for						
made bend		0.36	3.84	-	nr	3.84
fittings with one end		0.20	2.14	3.15	nr	5.29
fittings with two ends		0.40	4.27	3.17	nr	7.44
fittings with three ends		0.55	5.87	5.43	nr	11.30
stop end	PC £2.80	0.20	2.14	3.15	nr	5.29
reducing coupling	PC £2.30	0.40	4.27	2.59	nr	6.86
straight union coupling	PC £7.60	0.40	4.27	8.55	nr	12.82
copper to iron connector	PC £3.75	0.53	5.66	4.21	nr	9.87
elbow	PC £2.81	0.40	4.27	3.17	nr	7.44
bend; 91.5 degrees	PC £4.95	0.40	4.27	5.57	nr	9.84
tee; equal	PC £4.83	0.55	5.87	5.43	nr	11.30
tee; reducing	PC £4.59	0.55	5.87	5.16	nr	11.03
pitcher tee; equal or reducing	PC £8.03	0.55	5.87	9.03	nr	14.90
straight tank connector; backnut	PC £4.89	0.55	5.87	5.50	nr	11.37
42 mm pipes; fixing with pipe						
clips; plugged and screwed	PC £747.00/100m	0.46	4.91	10.07	m	14.98
Extra for						
made bend		0.48	5.12	-	nr	5.12
fittings with one end		0.23	2.46	4.09	nr	6.55
fittings with two ends		0.47	5.02	4.73	nr	9.75
fittings with three ends		0.62	6.62	8.00	nr	14.62
stop end	PC £3.64	0.23	2.46	4.09	nr	6.55

S PIPED SUPPLY SYSTEMS Including overheads and profit at 12.50%		Labour hours	Labour £	Material £	Unit	Total rate £
reducing coupling	PC £3.46	0.47	5.02	3.89	nr	8.91
straight union coupling	PC £10.79	0.47	5.02	12.13	nr	17.15
copper to iron connector	PC £4.54	0.60	6.41	5.10	nr	11.51
elbow	PC £4.20	0.47	5.02	4.73	nr	9.75
bend; 91.5 degrees	PC £7.90	0.47	5.02	8.88	nr	13.90
tee; equal	PC £7.11	0.62	6.62	8.00	nr	14.62
tee; reducing	PC £8.42	0.62	6.62	9.47	nr	16.09
pitcher tee; equal or reducing	PC £11.29	0.62	6.62	12.70	nr	19.32
straight tank connector; backnut	PC £6.21	0.62	6.62	6.98	nr	13.60
54 mm pipes; fixing with pipe clips; plugged and screwed	PC £967.00/100m	0.54	5.77	13.48	m	19.25
Extra for						
made bend		0.66	7.05	-	nr	7.05
fittings with one end		0.25	2.67	6.05	nr	8.72
fittings with two ends		0.53	5.66	10.97	nr	16.63
fittings with three ends		0.68	7.26	14.74	nr	22.00
stop end	PC £5.38	0.25	2.67	6.05	nr	8.72
reducing coupling	PC £5.07	0.53	5.66	5.71	nr	11.37
straight union coupling	PC £17.54	0.53	5.66	19.73	nr	25.39
copper to iron connector	PC £7.53	0.66	7.05	8.47	nr	15.52
elbow	PC £9.75	0.53	5.66	10.97	nr	16.63
bend; 91.5 degrees	PC £11.79	0.53	5.66	13.27	nr	18.93
tee; equal	PC £13.10	0.68	7.26	14.74	nr	22.00
tee; reducing	PC £13.31	0.68	7.26	14.97	nr	22.23
pitcher tee; equal or reducing	PC £16.25	0.68	7.26	18.29	nr	25.55
straight tank connector; backnut	PC £9.46	0.68	7.26	10.64	nr	17.90
Cut into existing copper services; provide and insert new capillary jointed tee; equal or reducing						
15 mm	PC £0.53	0.50	5.34	0.60	nr	5.94
22 mm	PC £1.04	0.58	6.19	1.17	nr	7.36
28 mm	PC £2.42	0.66	7.05	2.72	nr	9.77
35 mm	PC £4.59	0.74	7.90	5.16	nr	13.06
42 mm	PC £8.42	0.80	8.54	9.47	nr	18.01
54 mm	PC £13.31	0.86	9.18	14.97	nr	24.15
Copper pipes; BS 2871 table X; compression fittings; BS 864						
15 mm pipes; fixing with pipe clips; plugged and screwed	PC £109.00/100m	0.29	3.10	1.71	m	4.81
Extra for						
made bend		0.18	1.92	-	nr	1.92
fittings with one end		0.12	1.28	0.86	nr	2.14
fittings with two ends		0.18	1.92	0.70	nr	2.62
fittings with three ends		0.26	2.78	0.98	nr	3.76
stop end	PC £0.76	0.12	1.28	0.86	nr	2.14
straight coupling	PC £0.53	0.18	1.92	0.60	nr	2.52
female coupling	PC £0.53	0.24	2.56	0.60	nr	3.16
elbow	PC £0.62	0.18	1.92	0.70	nr	2.62
female wall elbow	PC £1.35	0.36	3.84	1.52	nr	5.36
slow bend	PC £2.52	0.18	1.92	2.84	nr	4.76
bent radiator union; chrome finish	PC £1.89	0.24	2.56	2.13	nr	4.69
tee; equal	PC £0.87	0.26	2.78	0.98	nr	3.76
straight swivel connector	PC £1.08	0.14	1.49	1.21	nr	2.70
bent swivel connector	PC £1.13	0.14	1.49	1.28	nr	2.77
tank coupling; locknut	PC £1.29	0.26	2.78	1.46	nr	4.24
22 mm pipes; fixing with pipe clips; plugged and screwed	PC £215.00/100m	0.30	3.20	3.09	m	6.29
Extra for						
made bend		0.24	2.56	-	nr	2.56
fittings with one end		0.14	1.49	1.12	nr	2.61
fittings with two ends		0.24	2.56	1.19	nr	3.75
fittings with three ends		0.36	3.84	1.71	nr	5.55
stop end	PC £0.99	0.14	1.49	1.12	nr	2.61
straight coupling	PC £0.88	0.24	2.56	0.99	nr	3.55
reducing set	PC £0.65	0.07	0.75	0.73	nr	1.48

S PIPED SUPPLY SYSTEMS Including overheads and profit at 12.50%		Labour hours	Labour £	Material £	Unit	Total rate £
S12 HOT AND COLD WATER (SMALL SCALE) - cont'd						
Copper pipes; BS 2871 table X; compression						
fittings; BS 864 - cont'd						
22 mm pipes; fixing with pipe						
clips; plugged and screwed						
female coupling	PC £0.80	0.34	3.63	0.90	nr	4.53
elbow	PC £1.06	0.24	2.56	1.19	nr	3.75
female wall elbow	PC £3.11	0.48	5.12	3.50	nr	8.62
slow bend	PC £4.14	0.24	2.56	4.66	nr	7.22
tee; equal	PC £1.52	0.36	3.84	1.71	nr	5.55
tee; reducing	PC £2.17	0.36	3.84	2.44	nr	6.28
straight swivel connector	PC £1.68	0.19	2.03	1.89	nr	3.92
bent swivel connector	PC £2.44	0.19	2.03	2.75	nr	4.78
tank coupling; locknut	PC £1.18	0.36	3.84	1.32	nr	5.16
28 mm pipes; fixing with pipe						
clips; plugged and screwed	PC £278.00/100m	0.34	3.63	4.21	m	7.84
Extra for						
made bend		0.30	3.20	-	nr	3.20
fittings with one end		0.17	1.81	2.06	nr	3.87
fittings with two ends		0.30	3.20	2.85	nr	6.05
fittings with three ends		0.44	4.70	4.15	nr	8.85
stop end	PC £1.83	0.17	1.81	2.06	nr	3.87
straight coupling	PC £2.05	0.30	3.20	2.31	nr	5.51
reducing set	PC £0.97	0.08	0.85	1.10	nr	1.95
female coupling	PC £1.31	0.42	4.48	1.47	nr	5.95
elbow	PC £2.54	0.30	3.20	2.85	nr	6.05
slow bend	PC £5.55	0.30	3.20	6.25	nr	9.45
tee; equal	PC £3.69	0.44	4.70	4.15	nr	8.85
tee; reducing	PC £3.83	0.44	4.70	4.31	nr	9.01
tank coupling; locknut	PC £2.37	0.44	4.70	2.66	nr	7.36
35 mm pipes; fixing with pipe						
clips; plugged and screwed	PC £617.00/100m	0.38	4.06	8.83	m	12.89
Extra for						
made bend		0.36	3.84	-	nr	3.84
fittings with one end		0.19	2.03	3.38	nr	5.41
fittings with two ends		0.36	3.84	5.58	nr	9.42
fittings with three ends		0.50	5.34	7.56	nr	12.90
stop end	PC £3.01	0.19	2.03	3.38	nr	5.41
straight coupling	PC £3.83	0.36	3.84	4.31	nr	8.15
reducing set	PC £1.66	0.10	1.07	1.87	nr	2.94
female coupling	PC £3.36	0.48	5.12	3.78	nr	8.90
elbow	PC £4.96	0.36	3.84	5.58	nr	9.42
tee; equal	PC £6.72	0.50	5.34	7.56	nr	12.90
tee; reducing	PC £6.72	0.50	5.34	7.56	nr	12.90
tank coupling; locknut	PC £5.00	0.50	5.34	5.62	nr	10.96
42 mm pipes; fixing with pipe						
clips; plugged and screwed	PC £747.00/100m	0.44	4.70	10.93	m	15.63
Extra for						
made bend		0.48	5.12	-	nr	5.12
fittings with one end		0.22	2.35	5.56	nr	7.91
fittings with two ends		0.42	4.48	7.87	nr	12.35
fittings with three ends		0.56	5.98	12.59	nr	18.57
stop end	PC £4.94	0.22	2.35	5.56	nr	7.91
straight coupling	PC £4.92	0.42	4.48	5.54	nr	10.02
reducing set	PC £2.62	0.11	1.17	2.95	nr	4.12
female coupling	PC £4.47	0.54	5.77	5.03	nr	10.80
elbow	PC £7.00	0.42	4.48	7.87	nr	12.35
tee; equal	PC £11.19	0.56	5.98	12.59	nr	18.57
tee; reducing	PC £10.47	0.56	5.98	11.78	nr	17.76

S PIPED SUPPLY SYSTEMS Including overheads and profit at 12.50%		Labour hours	Labour £	Material £	Unit	Total rate £
54 mm pipes; fixing with pipe clips; plugged and screwed	PC £967.00/100m	0.53	5.66	14.41	m	20.07
Extra for						
made bend		0.66	7.05	-	nr	7.05
fittings with two ends		0.48	5.12	12.93	nr	18.05
fittings with three ends		0.62	6.62	19.69	nr	26.31
straight coupling	PC £7.52	0.48	5.12	8.46	nr	13.58
reducing set	PC £4.39	0.12	1.28	4.94	nr	6.22
female wall elbow	PC £6.47	0.60	6.41	7.28	nr	13.69
elbow	PC £11.49	0.48	5.12	12.93	nr	18.05
tee; equal	PC £17.51	0.62	6.62	19.69	nr	26.31
tee; reducing	PC £17.77	0.62	6.62	20.00	nr	26.62
Black MDPE pipes; BS 6730; plastic **compression fittings**						
20 mm pipes; fixing with pipe clips; plugged and screwed	PC £32.80/100m	0.30	3.20	1.32	m	4.52
Extra for						
fittings with two ends		0.24	2.56	1.94	nr	4.50
fittings with three ends		0.36	3.84	2.49	nr	6.33
straight coupling	PC £1.52	0.24	2.56	1.71	nr	4.27
male adaptor	PC £0.95	0.34	3.63	1.07	nr	4.70
elbow	PC £1.72	0.24	2.56	1.94	nr	4.50
tee; equal	PC £2.22	0.36	3.84	2.49	nr	6.33
end cap	PC £1.18	0.14	1.49	1.32	nr	2.81
straight swivel connector	PC £1.04	0.19	2.03	1.17	nr	3.20
bent swivel connector	PC £1.26	0.19	2.03	1.42	nr	3.45
25 mm pipes; fixing with pipe clips; plugged and screwed	PC £40.00/100m	0.34	3.63	1.64	m	5.27
Extra for						
fittings with two ends		0.30	3.20	2.34	nr	5.54
fittings with three ends		0.44	4.70	3.31	nr	8.01
straight coupling	PC £1.87	0.30	3.20	2.11	nr	5.31
reducer	PC £1.86	0.30	3.20	2.09	nr	5.29
male adaptor	PC £1.13	0.42	4.48	1.27	nr	5.75
elbow	PC £2.08	0.30	3.20	2.34	nr	5.54
tee; equal	PC £2.94	0.44	4.70	3.31	nr	8.01
end cap	PC £1.36	0.17	1.81	1.53	nr	3.34
32 mm pipes; fixing with pipe clips; plugged and screwed	PC £65.60/100m	0.38	4.06	2.05	m	6.11
Extra for						
fittings with two ends		0.36	3.84	2.91	nr	6.75
fittings with three ends		0.50	5.34	4.33	nr	9.67
straight coupling	PC £2.62	0.36	3.84	2.95	nr	6.79
reducer	PC £2.45	0.36	3.84	2.75	nr	6.59
male adaptor	PC £1.40	0.48	5.12	1.58	nr	6.70
elbow	PC £2.58	0.36	3.84	2.91	nr	6.75
tee; equal	PC £3.85	0.50	5.34	4.33	nr	9.67
end cap	PC £1.54	0.19	2.03	1.73	nr	3.76
50 mm pipes; fixing with pipe clips; plugged and screwed	PC £168.00/100m	0.42	4.48	3.41	m	7.89
Extra for						
fittings with two ends		0.42	4.48	7.13	nr	11.61
fittings with three ends		0.56	5.98	9.68	nr	15.66
straight coupling	PC £5.89	0.42	4.48	6.62	nr	11.10
reducer	PC £5.43	0.42	4.48	6.11	nr	10.59
male adaptor	PC £3.49	0.54	5.77	3.92	nr	9.69
elbow	PC £6.34	0.42	4.48	7.13	nr	11.61
tee; equal	PC £8.60	0.56	5.98	9.68	nr	15.66
end cap	PC £3.90	0.22	2.35	4.38	nr	6.73

S PIPED SUPPLY SYSTEMS Including overheads and profit at 12.50%		Labour hours	Labour £	Material £	Unit	Total rate £
S12 HOT AND COLD WATER (SMALL SCALE) - cont'd						
Black MDPE pipes; BS 6730; plastic						
compression fittings - cont'd						
63 mm pipes; fixing with pipe						
clips; plugged and screwed	PC £244.00/100m	0.48	5.12	4.69	m	9.81
Extra for						
fittings with two ends		0.48	5.12	8.66	nr	13.78
fittings with three ends		0.62	6.62	13.75	nr	20.37
straight coupling	PC £8.96	0.48	5.12	10.08	nr	15.20
reducer	PC £7.70	0.48	5.12	8.66	nr	13.78
male adaptor	PC £4.98	0.60	6.41	5.60	nr	12.01
elbow	PC £7.70	0.48	5.12	8.66	nr	13.78
tee; equal	PC £12.22	0.62	6.62	13.75	nr	20.37
end cap	PC £5.38	0.24	2.56	6.06	nr	8.62
Stainless steel pipes; BS 4127; stainless						
steel capillary fittings						
15 mm pipes; fixing with pipe						
clips; plugged and screwed	PC £124.30/100m	0.36	3.84	2.76	m	6.60
Extra for						
fittings with two ends		0.24	2.56	4.22	nr	6.78
fittings with three ends		0.36	3.84	6.24	nr	10.08
bend	PC £3.75	0.24	2.56	4.22	nr	6.78
tee; equal	PC £5.54	0.36	3.84	6.24	nr	10.08
tap connector	PC £13.42	0.32	3.42	15.10	nr	18.52
22 mm pipes; fixing with pipe						
clips; plugged and screwed	PC £195.80/100m	0.38	4.06	4.14	m	8.20
Extra for						
fittings with two ends		0.32	3.42	5.02	nr	8.44
fittings with three ends		0.48	5.12	8.53	nr	13.65
reducer	PC £15.52	0.32	3.42	17.46	nr	20.88
bend	PC £4.47	0.32	3.42	5.02	nr	8.44
tee; equal	PC £7.58	0.48	5.12	8.53	nr	13.65
tee; reducing	PC £15.95	0.48	5.12	17.94	nr	23.06
tap connector	PC £25.21	0.46	4.91	28.36	nr	33.27
28 mm pipes; fixing with pipe						
clips; plugged and screwed	PC £286.00/100m	0.42	4.48	5.62	m	10.10
Extra for						
fittings with two ends		0.41	4.38	7.21	nr	11.59
fittings with three ends		0.60	6.41	12.68	nr	19.09
reducer	PC £34.47	0.41	4.38	38.78	nr	43.16
bend	PC £6.41	0.41	4.38	7.21	nr	11.59
tee; equal	PC £11.28	0.60	6.41	12.68	nr	19.09
tee; reducing	PC £20.08	0.60	6.41	22.58	nr	28.99
tap connector	PC £50.71	0.58	6.19	57.05	nr	63.24
Copper, brass and gunmetal ancillaries;						
screwed joints to fittings						
Bibtaps; brass						
chromium plated; capstan head						
15 mm	PC £9.29	0.18	1.92	10.56	nr	12.48
22 mm	PC £12.43	0.24	2.56	14.12	nr	16.68
self closing; 15 mm	PC £12.38	0.18	1.92	14.05	nr	15.97
crutch head with hose union;						
15 mm	PC £5.75	0.18	1.92	6.59	nr	8.51
draincock 15 mm	PC £2.02	0.12	1.28	2.34	nr	3.62
Stopcock; brass/gunmetal						
capillary joints to copper						
15 mm	PC £2.02	0.24	2.56	2.27	nr	4.83
22 mm	PC £3.65	0.32	3.42	4.11	nr	7.53
28 mm	PC £10.41	0.41	4.38	11.71	nr	16.09
compression joints to copper						
15 mm	PC £1.99	0.22	2.35	2.24	nr	4.59
22 mm	PC £3.49	0.29	3.10	3.93	nr	7.03
28 mm	PC £9.09	0.36	3.84	10.22	nr	14.06

S PIPED SUPPLY SYSTEMS Including overheads and profit at 12.50%		Labour hours	Labour £	Material £	Unit	Total rate £
compression joints to polyethylene						
13 mm	PC £5.43	0.30	3.20	6.10	nr	9.30
19 mm	PC £8.18	0.40	4.27	9.20	nr	13.47
25 mm	PC £11.82	0.48	5.12	13.30	nr	18.42
Gunmetal 'Fullway' gate valve; capillary joints to copper						
15 mm	PC £6.15	0.24	2.56	6.92	nr	9.48
22 mm	PC £7.25	0.32	3.42	8.16	nr	11.58
28 mm	PC £9.90	0.41	4.38	11.14	nr	15.52
35 mm	PC £22.08	0.49	5.23	24.83	nr	30.06
42 mm	PC £26.37	0.56	5.98	29.66	nr	35.64
54 mm	PC £38.23	0.64	6.83	43.01	nr	49.84
Gunmetal stopcock; screwed joints to iron						
15 mm	PC £5.76	0.36	3.84	6.48	nr	10.32
20 mm	PC £9.20	0.48	5.12	10.35	nr	15.47
25 mm	PC £13.38	0.60	6.41	15.05	nr	21.46
Bronze gate valve; screwed joints to iron						
15 mm	PC £12.29	0.36	3.84	13.82	nr	17.66
20 mm	PC £16.21	0.48	5.12	18.24	nr	23.36
25 mm	PC £21.15	0.60	6.41	23.80	nr	30.21
35 mm	PC £31.22	0.72	7.69	35.12	nr	42.81
42 mm	PC £40.04	0.84	8.97	45.05	nr	54.02
54 mm	PC £58.44	0.96	10.25	65.75	nr	76.00
Chromium plated; pre-setting radiator valve; compression joint; union outlet 15 mm						
	PC £3.42	0.26	2.78	3.85	nr	6.63
Chromium plated; lockshield radiator valve; compression joint; union outlet 15 mm						
	PC £3.42	0.26	2.78	3.85	nr	6.63
Brass ball valves; BS 1212 Part 1; piston type; high pressure; copper float; screwed joint to cistern						
15 mm	PC £2.89	0.30	3.20	5.55	nr	8.75
22 mm	PC £5.47	0.36	3.84	9.15	nr	12.99
25 mm	PC £12.17	0.42	4.48	17.51	nr	21.99
Water tanks/cisterns						
Polyethylene cold water feed and expansion cistern; BS 4213; with covers						
ref PC15; 68 litres	PC £19.49	1.50	16.01	21.92	nr	37.93
ref PC25; 114 litres	PC £25.87	1.75	18.68	29.10	nr	47.78
ref PC40; 182 litres	PC £44.24	2.10	22.42	49.77	nr	72.19
ref PC50; 227 litres	PC £47.37	2.35	25.09	53.30	nr	78.39
GRP cold water storage cistern; with covers						
ref 899.10; 27 litres	PC £22.94	1.30	13.88	25.81	nr	39.69
ref 899.25; 68 litres	PC £37.19	1.50	16.01	41.83	nr	57.84
ref 899.40; 114 litres	PC £43.92	1.75	18.68	49.40	nr	68.08
ref 899.70; 227 litres	PC £87.52	2.35	25.09	98.46	nr	123.55
Storage cylinders/calorifiers						
Copper cylinders; direct; BS 699; grade 3						
ref 1; 350 x 900 mm; 74 litres	PC £41.05	1.80	19.22	46.18	nr	65.40
ref 2; 450 x 750 mm; 98 litres	PC £42.37	2.10	22.42	47.67	nr	70.09
ref 7; 450 x 900 mm; 120 litres	PC £46.96	2.40	25.62	52.83	nr	78.45
ref 8; 450 x 1050 mm; 144 litres	PC £51.32	3.35	35.77	57.74	nr	93.51
ref 9; 450 x 1200 mm; 166 litres	PC £61.03	4.30	45.91	68.66	nr	114.57
Copper cylinders; single feed coil indirect; BS 1566 Part 2; grade 3						
ref 2; 300 x 1500 mm; 96 litres	PC £52.87	2.40	25.62	59.48	nr	85.10
ref 3; 400 x 1050 mm; 114 litres	PC £56.21	2.70	28.83	63.24	nr	92.07
ref 7; 450 x 900 mm; 117 litres	PC £55.79	3.00	32.03	62.76	nr	94.79
ref 8; 450 x 1050 mm; 140 litres	PC £63.60	3.60	38.43	71.55	nr	109.98
ref 9; 450 x 1200 mm; 162 litres	PC £82.88	4.20	44.84	93.24	nr	138.08
Combination copper hot water storage units; coil direct; BS 3198; (hot/cold)						
400 x 900 mm; (65/20 litres)	PC £59.01	3.35	35.77	66.39	nr	102.16
450 x 900 mm; (85/25 litres)	PC £61.03	4.70	50.18	68.66	nr	118.84
450 x 1050 mm; (115/25 litres)	PC £67.10	5.90	62.99	75.49	nr	138.48
450 x 1200 mm; (115/45 litres)	PC £70.91	6.60	70.46	79.77	nr	150.23

S PIPED SUPPLY SYSTEMS Including overheads and profit at 12.50%		Labour hours	Labour £	Material £	Unit	Total rate £
S12 HOT AND COLD WATER (SMALL SCALE) - cont'd						
Storage cylinders/calorifiers - cont'd						
Combination copper hot water storage						
450 x 900 mm; (85/25 litres)	PC £80.78	5.30	56.58	90.88	nr	147.46
450 x 1200 mm; (115/45 litres)	PC £87.62	7.20	76.87	98.57	nr	175.44
Thermal insulation						
19 mm thick rigid mineral glass fibre						
sectional pipe lagging; plain finish; fixed						
with aluminium bands to steel or copper						
pipework; including working over pipe						
fittings						
around 15/15 mm pipes	PC £1.74	0.08	0.85	1.95	m	2.80
around 20/22 mm pipes	PC £1.83	0.12	1.28	2.06	m	3.34
around 25/28 mm pipes	PC £2.02	0.13	1.39	2.28	m	3.67
around 32/35 mm pipes	PC £2.24	0.14	1.49	2.52	m	4.01
around 40/42 mm pipes	PC £2.38	0.16	1.71	2.68	m	4.39
around 50/54 mm pipes	PC £2.76	0.18	1.92	3.10	m	5.02
19 mm thick rigid mineral glass fibre						
sectional pipe lagging; canvas or class O						
laquered aluminium finish; fixed with						
aluminium bands to steel or copper pipework;						
including working over pipe fittings						
around 15/15 mm pipes	PC £2.18	0.08	0.85	2.45	m	3.30
around 20/22 mm pipes	PC £2.36	0.12	1.28	2.65	m	3.93
around 25/28 mm pipes	PC £2.59	0.13	1.39	2.91	m	4.30
around 32/35 mm pipes	PC £2.82	0.14	1.49	3.17	m	4.66
around 40/42 mm pipes	PC £3.04	0.16	1.71	3.43	m	5.14
around 50/54 mm pipes	PC £3.54	0.18	1.92	3.98	m	5.90
25 mm thick expanded polystyrene lagging						
sets; class O finish; for mild steel						
cisterns to BS 417; complete with fixing						
bands; for cisterns size (ref)						
762 x 584 x 610 mm; (SCM270)	PC £3.36	0.96	10.25	3.78	nr	14.03
1219 x 610 x 610 mm; (SCM450/1)	PC £11.31	1.10	11.74	12.73	nr	24.47
1524 x 1143 x 914 mm; (SCM1600)	PC £25.20	1.30	13.88	28.35	nr	42.23
1829 x 1219 x 1219 mm; (SCM2270)	PC £31.99	1.55	16.55	35.99	nr	52.54
2438 x 1524 x 1219 mm; (SCM4540)	PC £50.95	1.80	19.22	57.31	nr	76.53
50 mm thick glass-fibre filled polyethylene						
insulating jackets for GRP or polyethylene						
cold water cisterns; complete with fixing						
bands; for cisterns size						
445 x 305 x 300 mm; (18 litres)	PC £2.18	0.48	5.12	2.45	nr	7.57
495 x 368 x 362 mm; (27 litres)	PC £3.43	0.60	6.41	3.86	nr	10.27
630 x 450 x 420 mm; (68 litres)	PC £3.43	0.72	7.69	3.86	nr	11.55
665 x 490 x 515 mm; (91 litres)	PC £4.56	0.84	8.97	5.14	nr	14.11
700 x 540 x 535 mm; (114 litres)	PC £4.56	0.96	10.25	5.14	nr	15.39
955 x 605 x 595 mm; (182 litres)	PC £6.16	1.00	10.68	6.93	nr	17.61
1155 x 640 x 595 mm; (227 litres)	PC £7.02	1.10	11.74	7.90	nr	19.64
80 mm thick glass-fibre filled insulating						
jackets in flame retardant PVC to BS 1763;						
type 1B; segmental type for hot water						
cylinders; complete with fixing bands; for						
cylinders size (ref)						
400 x 900 mm; (2)	PC £5.98	0.40	4.27	6.73	nr	11.00
450 x 750 mm; (5)	PC £5.98	0.40	4.27	6.73	nr	11.00
450 x 900 mm; (7)	PC £5.98	0.40	4.27	6.73	nr	11.00
450 x 1050 mm;(8)	PC £6.72	0.48	5.12	7.56	nr	12.68
500 x 1200 mm;(-)	PC £7.77	0.60	6.41	8.74	nr	15.15

S PIPED SUPPLY SYSTEMS Including overheads and profit at 12.50%	Labour hours	Labour £	Material £	Unit	Total rate £

S13 PRESSURISED WATER

Copper pipes; BS 2871 Part 1 table Y;
annealed; mains pipework; no joints in the
running length; laid in trenches
Pipes

15 mm	PC £212.00/100m	0.12	1.28	2.51	m	3.79
22 mm	PC £369.00/100m	0.13	1.39	4.35	m	5.74
28 mm	PC £545.00/100m	0.14	1.49	6.44	m	7.93
35 mm	PC £853.00/100m	0.16	1.71	10.08	m	11.79
42 mm	PC £1030.00/100m	0.18	1.92	12.17	m	14.09
54 mm	PC £1759.00/100m	0.22	2.35	20.78	m	23.13

Blue MDPE pipes; BS 6527; mains
pipework; no joints in the running
length; laid in trenches
Pipes

20 mm	PC £32.80/100m	0.13	1.39	0.39	m	1.78
25 mm	PC £40.00/100m	0.14	1.49	0.47	m	1.96
32 mm	PC £65.60/100m	0.16	1.71	0.77	m	2.48
50 mm	PC £168.00/100m	0.18	1.92	1.98	m	3.90
60 mm	PC £244.00/100m	0.19	2.03	2.88	m	4.91

Steel pipes; BS 1387; heavy weight;
galvanized; mains pipework; screwed joints
in the running length; laid in trenches
Pipes

15 mm	PC £134.96/100m	0.16	1.71	1.60	m	3.31
20 mm	PC £157.61/100m	0.17	1.81	1.86	m	3.67
25 mm	PC £225.57/100m	0.19	2.03	2.66	m	4.69
32 mm	PC £282.60/100m	0.22	2.35	3.34	m	5.69
40 mm	PC £329.70/100m	0.24	2.56	3.89	m	6.45
50 mm	PC £456.84/100m	0.30	3.20	5.40	m	8.60

Ductile iron bitumen coated pipes and
fittings; BS 4772; class K9; Stanton's
'Tyton' water main pipes or similar;
flexible joints

100 mm pipes; laid straight	PC £51.48/5.5m	0.69	5.46	11.50	m	16.96
Extra for						
bend; 45 degrees	PC £19.19	0.69	5.46	26.29	nr	31.75
branch; 45 degrees; socketted	PC £102.50	1.05	8.32	122.37	nr	130.69
tee	PC £32.07	1.05	8.32	43.13	nr	51.45
flanged spigot	PC £14.55	0.69	5.46	18.72	nr	24.18
flanged socket	PC £18.06	0.69	5.46	22.67	nr	28.13
150 mm pipes; laid straight	PC £75.90/5.5m	0.81	6.42	16.79	m	23.21
Extra for						
bend; 45 degrees	PC £33.09	0.81	6.42	42.42	nr	48.84
branch; 45 degrees; socketted	PC £115.92	1.20	9.50	138.21	nr	147.71
tee	PC £50.57	1.20	9.50	64.69	nr	74.19
flanged spigot	PC £24.95	0.81	6.42	30.67	nr	37.09
flanged socket	PC £28.30	0.81	6.42	34.44	nr	40.86
200 mm pipes; laid straight	PC £118.80/5.5m	1.15	9.11	26.12	m	35.23
Extra for						
bend; 45 degrees	PC £64.96	1.15	9.11	79.52	nr	88.63
branch; 45 degrees; socketted	PC £162.29	1.70	13.46	192.23	nr	205.69
tee	PC £98.20	1.70	13.46	120.13	nr	133.59
flanged spigot	PC £39.56	1.15	9.11	47.72	nr	56.83
flanged socket	PC £42.12	1.15	9.11	50.60	nr	59.71

S PIPED SUPPLY SYSTEMS Including overheads and profit at 12.50%		Labour hours	Labour £	Material £	Unit	Total rate £
S32 NATURAL GAS						
Ductile iron bitumen coated pipes and fittings; BS 4772; class K9; Stanton's 'Stanlock' gas main pipes or similar; bolted gland joints						
100 mm pipes; laid straight	PC £57.26/5.5m	0.81	6.42	13.95	m	**20.37**
Extra for						
bend; 45 degrees	PC £20.19	0.81	6.42	36.26	nr	**42.68**
tee	PC £31.09	1.20	9.50	57.56	nr	**67.06**
flanged spigot	PC £14.55	0.81	6.42	25.40	nr	**31.82**
flanged socket	PC £17.55	0.81	6.42	28.78	nr	**35.20**
isolated 'Stanlock' joint	PC £8.03	0.40	3.17	9.03	nr	**12.20**
150 mm pipes; laid straight	PC £85.97/5.5m	1.05	8.32	20.84	m	**29.16**
Extra for						
bend; 45 degrees	PC £34.84	1.05	8.32	58.69	nr	**67.01**
tee	PC £49.27	1.55	12.28	87.91	nr	**100.19**
flanged spigot	PC £24.95	1.05	8.32	41.06	nr	**49.38**
flanged socket	PC £27.52	1.05	8.32	43.95	nr	**52.27**
isolated 'Stanlock' joint	PC £11.55	0.52	4.12	12.99	nr	**17.11**
200 mm pipes; laid straight	PC £125.79/5.5m	1.50	11.88	30.19	m	**42.07**
Extra for						
bend; 45 degrees	PC £63.17	1.50	11.88	97.05	nr	**108.93**
tee	PC £94.58	2.25	17.82	149.72	nr	**167.54**
flanged spigot	PC £39.56	1.50	11.88	61.83	nr	**73.71**
flanged socket	PC £40.91	1.50	11.88	63.35	nr	**75.23**
isolated 'Stanlock' joint	PC £15.40	0.75	5.94	17.33	nr	**23.27**

T MECHANICAL HEATING SYSTEMS ETC. Including overheads and profit at 12.50%	Labour hours	Labour £	Material £	Unit	Total rate £

T10 GAS/OIL FIRED BOILERS

Gas fired domestic boilers; cream or white
enamelled casing; 32 mm BSPT female
flow and return tappings; 102 mm flue socket
13 mm BSPT male draw-off outlet;
electric controls

13.19 kW output PC £265.65	6.00	64.06	299.16	nr	363.22
23.45 kW output PC £344.96	6.60	70.46	388.38	nr	458.84

Smoke flue pipework; light quality 'Duracem'
pipes and fittings; including asbestos
yarn and composition joints in the
running length

75 mm pipes; fixing with wall clips;					
plugged and screwed PC £9.56/1.8m	0.42	4.48	6.48	m	10.96
Extra for					
loose sockets PC £2.89	0.48	5.12	3.75	nr	8.87
bend; square and obtuse PC £4.15	0.48	5.12	5.17	nr	10.29
Terminal cone caps; asbestos yarn and					
composition joint PC £15.88	0.48	5.12	18.36	nr	23.48
100 mm pipes; fixing with wall clips;					
plugged and screwed PC £12.32/1.8m	0.48	5.12	8.41	m	13.53
Extra for					
loose sockets PC £3.75	0.54	5.77	4.94	nr	10.71
bend; square and obtuse PC £5.19	0.54	5.77	6.56	nr	12.33
Terminal cone caps; asbestos yarn and					
composition joint PC £16.86	0.54	5.77	19.69	nr	25.46
150 mm pipes; fixing with wall clips;					
plugged and screwed PC £20.61/1.8m	0.60	6.41	14.05	m	20.46
Extra for					
loose sockets PC £6.42	0.66	7.05	8.42	nr	15.47
bend; square and obtuse PC £8.00	0.66	7.05	10.20	nr	17.25
Terminal cone caps; asbestos yarn and					
composition joint PC £25.44	0.66	7.05	29.82	nr	36.87

Smoke flue pipework; heavy quality 'Duracem'
pipes and fittings; including asbestos
yarn and composition joints in the
running length

125 mm pipes; fixing with wall clips;					
plugged and screwed PC £21.18/1.8m	0.54	5.77	14.29	m	20.06
Extra for					
loose sockets PC £6.77	0.60	6.41	8.58	nr	14.99
bend; square and obtuse PC £9.13	0.60	6.41	11.23	nr	17.64
Terminal cone caps; asbestos yarn and					
composition joint PC £18.53	0.60	6.41	21.81	nr	28.22
175 mm pipes; fixing with wall clips;					
plugged and screwed PC £39.01/1.8m	0.66	7.05	26.13	m	33.18
Extra for					
loose sockets PC £10.34	0.72	7.69	13.01	nr	20.70
bend; square and obtuse PC £16.72	0.72	7.69	20.19	nr	27.88
Terminal cone caps; asbestos yarn and					
composition joint PC £47.13	0.72	7.69	54.40	nr	62.09
225 mm pipes; fixing with wall clips;					
plugged and screwed PC £51.12/1.8m	0.78	8.33	34.18	m	42.51
Extra for					
loose sockets PC £14.54	0.84	8.97	18.08	nr	27.05
bend; square and obtuse PC £28.49	0.84	8.97	33.77	nr	42.74
Terminal cone caps; asbestos yarn and					
composition joint PC £60.90	0.84	8.97	70.23	nr	79.20

T MECHANICAL HEATING SYSTEMS ETC. Including overheads and profit at 12.50%		Labour hours	Labour £	Material £	Unit	Total rate £
T30 MEDIUM TEMPERATURE HOT WATER HEATING						
Steel pipes; BS 1387; black; screwed joints; malleable iron fittings; BS 143						
15 mm pipes						
medium weight; fixing with pipe brackets;						
plugged and screwed	PC £75.25/100m	0.36	3.84	1.39	m	5.23
heavy weight; fixing with pipe brackets;						
plugged and screwed	PC £88.33/100m	0.36	3.84	1.53	m	5.37
Extra for						
fittings with one end		0.16	1.71	0.32	nr	2.03
fittings with two ends		0.30	3.20	0.41	nr	3.61
fittings with three ends		0.44	4.70	0.50	nr	5.20
fittings with four ends		0.60	6.41	1.13	nr	7.54
cap	PC £0.28	0.16	1.71	0.32	nr	2.03
socket; equal	PC £0.30	0.30	3.20	0.34	nr	3.54
socket; reducing	PC £0.36	0.30	3.20	0.41	nr	3.61
bend; 90 degree long radius M/F	PC £0.64	0.30	3.20	0.72	nr	3.92
elbow; 90 degree M/F	PC £0.36	0.30	3.20	0.41	nr	3.61
tee; equal	PC £0.44	0.44	4.70	0.50	nr	5.20
cross	PC £1.00	0.60	6.41	1.13	nr	7.54
union	PC £0.96	0.30	3.20	1.08	nr	4.28
isolated screwed joint	PC £0.30	0.36	3.84	0.46	nr	4.30
tank connection; longscrew and						
backnuts; lead washers; joint	PC £1.35	0.44	4.70	1.65	nr	6.35
20 mm pipes						
medium weight; fixing with pipe brackets;						
plugged and screwed	PC £91.15/100m	0.38	4.06	1.62	m	5.68
heavy weight; fixing with pipe brackets;						
plugged and screwed	PC £107.85/100m	0.38	4.06	1.81	m	5.87
Extra for						
fittings with one end		0.19	2.03	0.36	nr	2.39
fittings with two ends		0.40	4.27	0.54	nr	4.81
fittings with three ends		0.56	5.98	0.72	nr	6.70
fittings with four ends		0.79	8.43	1.71	nr	10.14
cap	PC £0.32	0.19	2.03	0.36	nr	2.39
socket; equal	PC £0.36	0.40	4.27	0.41	nr	4.68
socket; reducing	PC £0.44	0.40	4.27	0.50	nr	4.77
bend; 90 degree long radius M/F	PC £1.04	0.40	4.27	1.17	nr	5.44
elbow; 90 degree M/F	PC £0.48	0.40	4.27	0.54	nr	4.81
tee; equal	PC £0.64	0.56	5.98	0.72	nr	6.70
cross	PC £1.52	0.79	8.43	1.71	nr	10.14
union	PC £1.12	0.40	4.27	1.26	nr	5.53
isolated screwed joint	PC £0.36	0.48	5.12	0.53	nr	5.65
tank connection; longscrew and						
backnuts; lead washers; joint	PC £1.57	0.56	5.98	1.89	nr	7.87
25 mm pipes						
medium weight; fixing with pipe brackets;						
plugged and screwed	PC £131.41/100m	0.43	4.59	2.16	m	6.75
heavy weight; fixing with pipe brackets;						
plugged and screwed	PC £158.22/100m	0.43	4.59	2.48	m	7.07
Extra for						
fittings with one end		0.25	2.67	0.45	nr	3.12
fittings with two ends		0.50	5.34	0.90	nr	6.24
fittings with three ends		0.68	7.26	1.04	nr	8.30
fittings with four ends		1.00	10.68	2.16	nr	12.84
cap	PC £0.40	0.25	2.67	0.45	nr	3.12
socket; equal	PC £0.48	0.50	5.34	0.54	nr	5.88
socket; reducing	PC £0.58	0.50	5.34	0.65	nr	5.99
bend; 90 degree long radius M/F	PC £1.48	0.50	5.34	1.67	nr	7.01
elbow; 90 degree M/F	PC £0.80	0.50	5.34	0.90	nr	6.24
tee; equal	PC £0.92	0.68	7.26	1.04	nr	8.30
cross	PC £1.92	1.00	10.68	2.16	nr	12.84
union	PC £1.32	0.50	5.34	1.49	nr	6.83
isolated screwed joint	PC £0.48	0.60	6.41	0.69	nr	7.10
tank connection; longscrew and						
backnuts; lead washers; joint	PC £2.15	0.68	7.26	2.57	nr	9.83

T MECHANICAL HEATING SYSTEMS ETC. Including overheads and profit at 12.50%		Labour hours	Labour £	Material £	Unit	Total rate £
32 mm pipes						
medium weight; fixing with pipe brackets;						
plugged and screwed	PC £164.19/100m	0.50	5.34	2.72	m	8.06
heavy weight; fixing with pipe brackets;						
plugged and screwed	PC £198.19/100m	0.50	5.34	3.13	m	8.47
Extra for						
fittings with one end		0.30	3.20	0.65	nr	3.85
fittings with two ends		0.60	6.41	1.49	nr	7.90
fittings with three ends		0.80	8.54	1.71	nr	10.25
fittings with four ends		1.20	12.81	2.84	nr	15.65
cap	PC £0.58	0.30	3.20	0.65	nr	3.85
socket; equal	PC £0.80	0.60	6.41	0.90	nr	7.31
socket; reducing	PC £0.92	0.60	6.41	1.04	nr	7.45
bend; 90 degree long radius M/F	PC £2.48	0.60	6.41	2.79	nr	9.20
elbow; 90 degree M/F	PC £1.32	0.60	6.41	1.49	nr	7.90
tee; equal	PC £1.52	0.80	8.54	1.71	nr	10.25
cross	PC £2.52	1.20	12.81	2.84	nr	15.65
union	PC £2.24	0.60	6.41	2.52	nr	8.93
isolated screwed joint	PC £0.80	0.72	7.69	1.05	nr	8.74
tank connection; longscrew and						
backnuts; lead washers; joint	PC £2.58	0.80	8.54	3.05	nr	11.59
40 mm pipes						
medium weight; fixing with pipe brackets;						
plugged and screwed	PC £190.88/100m	0.60	6.41	3.27	m	9.68
heavy weight; fixing with pipe brackets;						
plugged and screwed	PC £230.96/100m	0.60	6.41	3.75	m	10.16
Extra for						
fittings with one end		0.35	3.74	0.83	nr	4.57
fittings with two ends		0.70	7.47	2.21	nr	9.68
fittings with three ends		0.94	10.04	2.34	nr	12.38
fittings with four ends		1.40	14.95	3.83	nr	18.78
cap	PC £0.74	0.35	3.74	0.83	nr	4.57
socket; equal	PC £1.08	0.70	7.47	1.22	nr	8.69
socket; reducing	PC £1.20	0.70	7.47	1.35	nr	8.82
bend; 90 degree long radius M/F	PC £3.28	0.70	7.47	3.69	nr	11.16
elbow; 90 degree M/F	PC £1.96	0.70	7.47	2.21	nr	9.68
tee; equal	PC £2.08	0.94	10.04	2.34	nr	12.38
cross	PC £3.40	1.40	14.95	3.83	nr	18.78
union	PC £2.82	0.70	7.47	3.17	nr	10.64
isolated screwed joint	PC £1.08	0.84	8.97	1.39	nr	10.36
tank connection; longscrew and						
backnuts; lead washers; joint	PC £3.56	0.94	10.04	4.18	nr	14.22
50 mm pipes						
medium weight; fixing with pipe brackets;						
plugged and screwed	PC £268.59/100m	0.72	7.69	4.47	m	12.16
heavy weight; fixing with pipe brackets;						
plugged and screwed	PC £320.90/100m	0.72	7.69	5.10	m	12.79
Extra for						
fittings with one end		0.40	4.27	1.58	nr	5.85
fittings with two ends		0.80	8.54	2.84	nr	11.38
fittings with three ends		1.05	11.21	3.38	nr	14.59
fittings with four ends		1.60	17.08	5.94	nr	23.02
cap	PC £1.40	0.40	4.27	1.58	nr	5.85
socket; equal	PC £1.68	0.80	8.54	1.89	nr	10.43
socket; reducing	PC £1.68	0.80	8.54	1.89	nr	10.43
bend; 90 degree long radius M/F	PC £5.60	0.80	8.54	6.30	nr	14.84
elbow; 90 degree M/F	PC £2.52	0.80	8.54	2.84	nr	11.38
tee; equal	PC £3.00	1.05	11.21	3.38	nr	14.59
cross	PC £5.28	1.60	17.08	5.94	nr	23.02
union	PC £4.20	0.80	8.54	4.73	nr	13.27
isolated screwed joint	PC £1.68	0.96	10.25	2.09	nr	12.34
tank connection; longscrew and						
backnuts; lead washers; joint	PC £5.59	1.05	11.21	6.49	nr	17.70

T MECHANICAL HEATING SYSTEMS ETC. Including overheads and profit at 12.50%		Labour hours	Labour £	Material £	Unit	Total rate £
T30 MEDIUM TEMPERATURE HOT WATER HEATING - cont'd						
Steel pipes; BS 1387; galvanized; screwed joints; galvanized malleable iron fittings; BS 143						
15 mm pipes						
medium weight; fixing with pipe brackets;						
plugged and screwed	PC £115.75/100m	0.36	3.84	2.01	m	5.85
heavy weight; fixing with pipe brackets;						
plugged and screwed	PC £134.96/100m	0.36	3.84	2.24	m	6.08
Extra for						
fittings with one end		0.16	1.71	0.43	nr	2.14
fittings with two ends		0.30	3.20	0.56	nr	3.76
fittings with three ends		0.44	4.70	0.68	nr	5.38
fittings with four ends		0.60	6.41	1.55	nr	7.96
cap	PC £0.39	0.16	1.71	0.43	nr	2.14
socket; equal	PC £0.41	0.30	3.20	0.46	nr	3.66
socket; reducing	PC £0.50	0.30	3.20	0.56	nr	3.76
bend; 90 degree long radius M/F	PC £0.88	0.30	3.20	0.99	nr	4.19
elbow; 90 degree M/F	PC £0.50	0.30	3.20	0.56	nr	3.76
tee; equal	PC £0.61	0.44	4.70	0.68	nr	5.38
cross	PC £1.38	0.60	6.41	1.55	nr	7.96
union	PC £1.32	0.30	3.20	1.49	nr	4.69
isolated screwed joint	PC £0.41	0.36	3.84	0.59	nr	4.43
tank connection; longscrew and						
backnuts; lead washers; joint	PC £1.68	0.44	4.70	2.01	nr	6.71
20 mm pipes						
medium weight; fixing with pipe brackets;						
plugged and screwed	PC £134.34/100m	0.38	4.06	2.29	m	6.35
heavy weight; fixing with pipe brackets;						
plugged and screwed	PC £157.61/100m	0.38	4.06	2.56	m	6.62
Extra for						
fittings with one end		0.19	2.03	0.50	nr	2.53
fittings with two ends		0.40	4.27	0.74	nr	5.01
fittings with three ends		0.56	5.98	0.99	nr	6.97
fittings with four ends		0.79	8.43	2.35	nr	10.78
cap	PC £0.44	0.19	2.03	0.50	nr	2.53
socket; equal	PC £0.50	0.40	4.27	0.56	nr	4.83
socket; reducing	PC £0.61	0.40	4.27	0.68	nr	4.95
bend; 90 degree long radius M/F	PC £1.43	0.40	4.27	1.61	nr	5.88
elbow; 90 degree M/F	PC £0.66	0.40	4.27	0.74	nr	5.01
tee; equal	PC £0.88	0.56	5.98	0.99	nr	6.97
cross	PC £2.09	0.79	8.43	2.35	nr	10.78
union	PC £1.54	0.40	4.27	1.73	nr	6.00
isolated screwed joint	PC £0.50	0.48	5.12	0.68	nr	5.80
tank connection; longscrew and						
backnuts; lead washers; joint	PC £1.94	0.56	5.98	2.31	nr	8.29
25 mm pipes						
medium weight; fixing with pipe brackets;						
plugged and screwed	PC £188.60/100m	0.43	4.59	3.02	m	7.61
heavy weight; fixing with pipe brackets;						
plugged and screwed	PC £225.57/100m	0.43	4.59	3.45	m	8.04
Extra for						
fittings with one end		0.25	2.67	0.62	nr	3.29
fittings with two ends		0.50	5.34	1.24	nr	6.58
fittings with three ends		0.68	7.26	1.42	nr	8.68
fittings with four ends		1.00	10.68	2.97	nr	13.65
cap	PC £0.55	0.25	2.67	0.62	nr	3.29
socket; equal	PC £0.66	0.50	5.34	0.74	nr	6.08
socket; reducing	PC £0.80	0.50	5.34	0.90	nr	6.24
bend; 90 degree long radius M/F	PC £2.04	0.50	5.34	2.29	nr	7.63
elbow; 90 degree M/F	PC £1.10	0.50	5.34	1.24	nr	6.58
tee; equal	PC £1.27	0.68	7.26	1.42	nr	8.68
cross	PC £2.64	1.00	10.68	2.97	nr	13.65
union	PC £1.82	0.50	5.34	2.04	nr	7.38
isolated screwed joint	PC £0.66	0.60	6.41	0.89	nr	7.30
tank connection; longscrew and						
backnuts; lead washers; joint	PC £2.66	0.68	7.26	3.14	nr	10.40

T MECHANICAL HEATING SYSTEMS ETC. Including overheads and profit at 12.50%		Labour hours	Labour £	Material £	Unit	Total rate £
32 mm pipes						
medium weight; fixing with pipe brackets;						
plugged and screwed	PC £235.62/100m	0.50	5.34	3.83	m	9.17
heavy weight; fixing with pipe brackets;						
plugged and screwed	PC £282.60/100m	0.50	5.34	4.38	m	9.72
Extra for						
fittings with three ends		0.30	3.20	0.90	nr	4.10
fittings with two ends		0.60	6.41	2.04	nr	8.45
fittings with three ends		0.80	8.54	2.35	nr	10.89
fittings with four ends		1.20	12.81	3.90	nr	16.71
cap	PC £0.80	0.30	3.20	0.90	nr	4.10
socket; equal	PC £1.10	0.60	6.41	1.24	nr	7.65
socket; reducing	PC £1.27	0.60	6.41	1.42	nr	7.83
bend; 90 degree long radius M/F	PC £3.41	0.60	6.41	3.84	nr	10.25
elbow; 90 degree M/F	PC £1.82	0.60	6.41	2.04	nr	8.45
tee; equal	PC £2.09	0.80	8.54	2.35	nr	10.89
cross	PC £3.47	1.20	12.81	3.90	nr	16.71
union	PC £3.08	0.60	6.41	3.47	nr	9.88
isolated screwed joint	PC £1.10	0.72	7.69	1.39	nr	9.08
tank connection; longscrew and						
backnuts; lead washers; joint	PC £3.21	0.80	8.54	3.76	nr	12.30
40 mm pipes						
medium weight; fixing with pipe brackets;						
plugged and screwed	PC £273.79/100m	0.60	6.41	4.59	m	11.00
heavy weight; fixing with pipe brackets;						
plugged and screwed	PC £329.70/100m	0.60	6.41	5.24	m	11.65
Extra for						
fittings with one end		0.35	3.74	1.14	nr	4.88
fittings with two ends		0.70	7.47	3.03	nr	10.50
fittings with three ends		0.94	10.04	3.22	nr	13.26
fittings with four ends		1.40	14.95	5.26	nr	20.21
cap	PC £1.02	0.35	3.74	1.14	nr	4.88
socket; equal	PC £1.49	0.70	7.47	1.67	nr	9.14
socket; reducing	PC £1.65	0.70	7.47	1.86	nr	9.33
bend; 90 degree long radius M/F	PC £4.51	0.70	7.47	5.07	nr	12.54
elbow; 90 degree M/F	PC £2.70	0.70	7.47	3.03	nr	10.50
tee; equal	PC £2.86	0.94	10.04	3.22	nr	13.26
cross	PC £4.68	1.40	14.95	5.26	nr	20.21
union	PC £3.88	0.70	7.47	4.36	nr	11.83
isolated screwed joint	PC £1.49	0.84	8.97	1.85	nr	10.82
tank connection; longscrew and						
backnuts; lead washers; joint	PC £4.41	0.94	10.04	5.13	nr	15.17
50 mm pipes						
medium weight; fixing with pipe brackets;						
plugged and screwed	PC £384.19/100m	0.72	7.69	6.29	m	13.98
heavy weight; fixing with pipe brackets;						
plugged and screwed	PC £456.84/100m	0.72	7.69	7.15	m	14.84
Extra for						
fittings with one end		0.40	4.27	1.81	nr	6.08
fittings with two ends		0.80	8.54	3.90	nr	12.44
fittings with three ends		1.05	11.21	4.64	nr	15.85
fittings with four ends		1.60	17.08	8.17	nr	25.25
cap	PC £1.61	0.40	4.27	1.81	nr	6.08
socket; equal	PC £2.31	0.80	8.54	2.60	nr	11.14
socket; reducing	PC £2.31	0.80	8.54	2.60	nr	11.14
bend; 90 degree long radius M/F	PC £7.70	0.80	8.54	8.66	nr	17.20
elbow; 90 degree M/F	PC £3.47	0.80	8.54	3.90	nr	12.44
tee; equal	PC £4.13	1.05	11.21	4.64	nr	15.85
cross	PC £7.26	1.60	17.08	8.17	nr	25.25
union	PC £5.78	0.80	8.54	6.50	nr	15.04
isolated screwed joint	PC £2.31	0.96	10.25	2.80	nr	13.05
tank connection; longscrew and						
backnuts; lead washers; joint	PC £6.96	1.05	11.21	8.03	nr	19.24

T MECHANICAL HEATING SYSTEMS ETC. Including overheads and profit at 12.50%		Labour hours	Labour £	Material £	Unit	Total rate £
Radiators; pressed steel panel type, 590 mm high; 3 mm chromium plated air valve; 15 mm chromium plated easy clean straight valve with union; 15 mm chromium plated lockshield valve with union						
1.69 m2 single surface	PC £56.45	2.40	25.62	69.70	nr	95.32
2.12 m2 single surface	PC £64.48	2.70	28.83	78.79	nr	107.62
2.75 m2 single surface	PC £76.77	3.00	32.03	92.61	nr	124.64
3.39 m2 single surgace	PC £94.95	3.30	35.23	113.07	nr	148.30

V ELECTRICAL SYSTEMS Including overheads and profit at 5.00%	Labour hours	Labour £	Material £	Unit	Total rate £

V21 - 22 GENERAL LIGHTING AND LV POWER

NOTE: The following items indicate
approximate prices for wiring of lighting
and power points complete, including
accessories and socket outlets but excluding
lighting fittings. Consumer control units
are shown separately. For a more detailed
breakdown of these costs and specialist
costs for a complete range of electrical
items reference should be made to
Spon's Mechanical and Electrical Services
Price Book.

	Labour hours	Labour £	Material £	Unit	Total rate £
Consumer control units					
8-way 60 amp SP&N surface mounted insulated consumer control units fitted with miniature circuit breakers including 2 m long 32 mm screwed welded conduit with three runs of 16 mm2 PVC cables ready for final connections by the supply authority	-	-	-	nr	136.00
Extra for current operated ELCB of 30 mA tripping current	-	-	-	nr	61.00
as above but 100 amp metal cased consumer unit and 25 mm2 PVC cables	-	-	-	nr	155.25
Extra for current operated ELCB of 30 mA tripping current	-	-	-	nr	130.00
Final circuits					
Lighting points					
Wired in PVC insulated and PVC sheathed cable in flats and houses; insulated in cavities and roof space; protected where buried by heavy gauge PVC conduit	-	-	-	nr	44.00
As above but in commercial property	-	-	-	nr	53.00
Wired in PVC insulated cable in screwed welded conduit in flats and houses	-	-	-	nr	90.50
As above but in commercial property	-	-	-	nr	111.25
As above but in industrial property	-	-	-	nr	127.00
Wired in MICC cable in flats and houses	-	-	-	nr	76.33
As above but in commercial property	-	-	-	nr	92.00
As above but in industrial property with PVC sheathed cable	-	-	-	nr	106.00
Single 13 amp switched socket outlet points					
Wired in PVC insulated and PVC sheathed cable in flats and houses on a ring main circuit; protected where buried by heavy gauge PVC conduit	-	-	-	nr	48.00
As above but in commercial property	-	-	-	nr	58.25
Wired in PVC insulated cable in screwed welded conduit throughout on a ring main circuit in flats and houses	-	-	-	nr	68.50
As above but in commercial property	-	-	-	nr	80.00
As above but in industrial property	-	-	-	nr	92.00
Wired in MICC cable on a ring main circuit in flats and houses	-	-	-	nr	70.00
As above but in commercial property⁻	-	-	-	nr	81.50
As above but in industrial property with PVC sheathed cable	-	-	-	nr	103.50
Cooker control units					
45 amp circuit including unit wired in PVC insulated and PVC sheathed cable; protected where buried by heavy gauge PVC conduit	-	-	-	nr	94.50
As above but wired in PVC insulated cable in screwed welded conduit	-	-	-	nr	137.00
As above but wired in MICC cable	-	-	-	nr	154.00

W SECURITY SYSTEMS Including overheads and profit at 5.00%	Labour hours	Labour £	Material £	Unit	Total rate £
W20 LIGHTNING PROTECTION					
Flag staff terminal	-	-	-	nr	**62.74**
Copper strip roof or down conductors fixed with bracket or saddle clips					
20 x 3 mm	-	-	-	m	**13.89**
25 x 3 mm	-	-	-	m	**15.50**
Aluminium strip roof or down conductors fixed with bracket or saddle clips					
20 x 3 mm	-	-	-	m	**10.95**
25 x 3 mm	-	-	-	m	**11.49**
Joints in tapes	-	-	-	nr	**8.55**
Bonding connections to roof and structural metalwork	-	-	-	nr	**48.09**
Testing points	-	-	-	nr	**23.51**
Earth electrodes					
16 mm driven copper electrodes in 1220 mm sectional lengths (2440 mm minimum)	-	-	-	nr	**128.25**
First 2440 mm driven and tested					
25 x 3 mm copper strip electrode in 457 mm deep prepared trench	-	-	-	m	**9.30**

PART IV

Approximate Estimating

This part of the book contains the following sections:

Keep your figures up to date, free of charge

This section, and most of the other information in this Price Book, is brought up to date every three months in the *Price Book Update*.

The *Update* is available free to all Price Book purchasers.

To ensure you receive your copy, simply complete the reply card from the centre of the book and return it to us.

BCIS

Building Cost Information Service

Publications available from BCIS,
85/87 Clarence Street, Kingston upon Thames, KT1 1RB.
(Telephone 01-546 7554)

Prices include postage and packing within UK.
PLEASE MAKE CHEQUES PAYABLE TO RICS.

BCIS SUBSCRIPTION SERVICE*

1989/90 annual subscription rates
(from April 1989).

Chartered surveyors:	Main subscription £200.00
	Supplementary subscription £95.00
Non chartered surveyors:	Main subscription £250.00
	Supplementary subscription £100.00

BCIS stationery

Detailed form of cost analysis:	£3.00 per set of 4
Concise form of cost analysis:	£2.00 per set of 4
A4 divider cards:	£5.00 per set
A5 divider cards:	£2.50 per set
Ring binder:	£5.50 each (incl. VAT)

STANDARD FORM OF COST ANALYSIS

Principles, Instructions and Definitions:	£3.00 each

BCIS QUARTERLY REVIEW OF BUILDING PRICES*

Annual subscription — 4 issues:	£150.00 per annum
Supplementary subscriptions:	£50.00 per annum
Single issues:	£50.00 each

BCIS BUILDING COST TRENDS WALLCHART

1989 edition *(available May 1989)*:	£5.00 each

GUIDE TO HOUSE REBUILDING COSTS FOR INSURANCE VALUATION*

1989 edition:	£20.00 each

DAYWORK RATES*

Annual updating service	£100.00 per annum

*Further details available on request

Building Costs and Tender Prices Index

The tables which follow show the changes in building costs and tender prices since 1976.

To avoid confusion it is essential that the term 'building costs' and 'tender price' are clearly defined and understood.

'Building costs' are the costs actually incurred by the builder in the course of his business the major ones being those for labour and materials.

'Tender price' is the price for which a builder offers to erect a building. This includes 'building costs' but also takes into account market considerations such as the availability of labour and materials and the prevailing economic situation. This means that in 'boom' periods when there is a surfeit of building work to be done 'tender prices' may increase at a greater rate than 'building costs' whilst in a period when work is scarce 'tender prices' may actually fall when building costs are rising.

Building costs

This table reflects the fluctuations since 1976 in wages and materials costs to the builder. In compiling the table the proportion of labour to material has been assumed to be 40:60. The wages element has been assessed from a contract wages sheet revalued for each variation in labour costs whilst the changes in the cost of materials have been based on the indices prepared by the Department of Trade and Industry. No allowance has been made for changes in productivity, plus rates or hours worked which may occur in particular conditions and localities.

1976 = 100

Year	First quarter	Second quarter	Third quarter	Fourth quarter	Annual average
1976	93	97	104	107	100
1977	109	112	116	117	114
1978	118	120	127	129	124
1979	131	135	149	153	142
1980	157	161	180	181	170
1981	182	185	195	199	190
1982	203	206	214	216	210
1983	217	219	227	229	223
1984	230	232	239	241	236
1985	243	245	252	254	249
1986	256	258	266	267	262
1987	270	272	281	282	276
1988	284	286	299	302	293
1989	305 (P)	308 (F)	322 (F)		

Note: P = Provisional   F = Forecast

Tender prices

This table reflects the changes in tender prices since 1976. It indicates the level of pricing contained in the lowest competitive fluctuating tenders for new work in the outer London area (over £300 000 in value) compared with a common base.

1976 = 100

Year	First quarter	Second quarter	Third quarter	Fourth quarter	Annual average
1976	97	98	102	103	100
1977	105	105	109	110	107
1978	113	116	126	139	124
1979	142	146	160	167	154
1980	179	200	192	188	190
1981	199	193	190	195	194
1982	191	188	195	195	192
1983	198	200	198	200	199
1984	205	206	214	215	210
1985	215	219	219	220	218
1986	221	226	234	234	229
1987	242	249	265	279	258
1988	289	299	321	328	309
1989	341 (P)	352 (F)	370 (F)		

Regional variations

As well as being aware of inflationary trends when preparing an estimate, it is also important to establish the appropriate price level. Regional variations in price levels can be significant. The boom in London and the South East throughout 1987 and 1988, that has resulted in Tender price increases approaching 20% per annum in both those years, has begun to ripple out to many of the regions. Although workload in the South East continues to increase, it is possible that in 1989 Tender price increases in some regions could be greater than in the South East as a period of catching up occurs. Location adjustment factors to assist with the preparation of initial estimates are shown in the following table. For ease of reference the table shows both the forecast third quarter 1989 tender price indices for each region of the country and the percentage adjustment to the Prices for Measured Work. In addition, a band of percentage adjustments is shown to reflect the range of price levels within each region.

Region	Forecast third quarter 1989 tender price index	Percentage adjustment to Major Works section	Range of percentage adjustments
East Anglia	326	-10.9%	- 8% to -15%
East Midlands	278	-24%	-21% to -28%
Inner London	389	+ 6.3%	+ 2% to +12%
Northern	289	-21%	-16% to -25%
Northern Ireland	211	-42.3%	-38% to -46%
North West	296	-19.1%	-15% to -23%
Outer London	370	+ 1.1%	- 5% to + 6%
Scotland	296	-19.1%	-16% to -26%
South East	344	- 6%	-13% to + 3%
South West	296	-19.1%	-15% to -22%
Wales	289	-21%	-17% to -28%
West Midlands	289	-21%	-15% to -23%
Yorkshire and Humberside	281	-23.2%	-18% to -27%

Special further adjustment to the above percentages may be necessary when considering city centre or very isolated locations.

Note: P = Provisional   F = Forecast

The following example illustrates the adjustment of an estimate prepared using Spon's A&B 1990, to a price level that reflects the forecast Outer London market conditions for competitive tenders in the third quarter 1989:

		£
A	Value of items priced using Spon's A&B 1990 i.e. Tender price index 366	1 230 000
B	Adjustment to increase value of A to forecast price level for third quarter 1989 i.e. Tender price index 370 $\frac{(370 - 366)}{366} \times 100 = 1.09\%$ add 1.09% say	13 400
		---------
		1 243 400
C	Value of items priced using competitive quotations that reflect the market conditions in the third quarter 1989	650 000
		---------
		1 893 400
D	Allowance for preliminaries +14% say	265 100

E	Total value of estimate at third quarter 1989 price levels	£ 2 158 500

Alternatively, for a similar estimate in East Anglia:

£

A	Value of items priced using Spon's A&B 1990 i.e. Tender price index 366	1 230 000
B	Adjustment to reduce value of A to forecast price level for third quarter 1989 for East Anglia (from Regional variation table) ie Tender price index 326 $\frac{(366 - 326)}{326} \times 100 = 12.27\%$ deduct 12.27% say	151 000
		---------
		1 079 000
C	Value of items priced using competitive quotations that reflect the market conditions in the third quarter 1989	650 000
		---------
		1 729 000
D	Allowance for preliminaries +14% say	242 000

E	Total value of estimate at third quarter 1989 price levels	£ 1 971 000

Building Prices per Square Metre

Prices given under this heading are average prices, on a 'fluctuating basis', for typical buildings during the third quarter 1989 with a tender price level index of 366 (1976 = 100). Unless otherwise stated, prices do not allow for external works, other than those adjacent to the building, furniture, loose or special equipment and are, of course, exclusive of fees for professional services.

Prices are based upon the total floor area of all storeys, measured between external walls and without deduction for internal walls, columns, stairwells, liftwells and the like.

As in previous editions it is emphasized that the prices must be treated with reserve in that they represent the average of prices from our records and cannot provide more than a rough guide to the probable cost of a building.

In many instances normal commercial pressures together with a limited range of available specifications ensure that a single rate is sufficient to indicate the prevailing average price. However, where such restrictions do not apply a range has been given; this is not to suggest that figures outside this range will not be encountered, but simply that the calibre of such a type of building can itself vary significantly.

For assistance with the compilation of a closer estimate, or of a 'Cost Plan' the reader is directed to the 'Approximate Estimates' sections.

As elsewhere in this edition, prices do not include Value Added Tax, which should be applied at the current rate to all non-domestic building.

Utilities, civil engineering facilities	Square metre excluding VAT £
Multi-storey car parks	
split level	230 to 255
flat slab	245 to 300
warped	275 to 330
Underground car parks	
partially underground under buildings	360 to 435
completely underground under buildings	435 to 525
completely underground with landscaped roof	475 to 570
Surface car parking	40 to 55
Railway stations	1200 to 1900
Bus and coach stations	570 to 680
Bus garages	625 to 685
Petrol stations	750 to 1000
Garage showrooms	450 to 750
Garages, domestic	275 to 450
Airport passenger terminal buildings (excluding aprons)	
national standard	1100 to 1300
international standard	1400 to 2000
Airport facility buildings	
large hangars	1100 to 1350
workshops and small hangers	700 to 800
TV, radio and video studios	800 to 1300
Telephone exchanges	600 to 900
Telephone engineering centres	475 to 625
Branch Post Offices	700 to 950

	Square metre excluding VAT £

Industrial facilities (CI/SfB 2)

Postal Delivery Offices/Sorting Offices	550 to 750
Mortuaries	1200 to 1500
Sub-stations	800 to 950
Agricultural storage buildings	280 to 440
Factories	
for letting (incoming services only)	240 to 325
for letting (including lighting, power and heating)	315 to 430
nursery units (including lighting, power and heating)	395 to 570
workshops	450 to 570
maintenance/motor transport workshops	450 to 750
owner occupation-for light industrial use	400 to 570
owner occupation-for heavy industrial use	750 to 850
Factory/office buildings - high technology production	
for letting (shell and core only)	435 to 570
for letting (ground floor shell, first floor offices)	700 to 900
for owner occupation (controlled environment, fully finished)	900 to 1200
Laboratory workshops and offices	825 to 1025
High technology laboratory workshop centres, air conditioned	1900 to 2400
Warehouses	
low bay (6 - 8 m high) for letting (no heating)	240 to 290
low bay for owner occupation (including heating)	340 to 420
high bay (9 - 18 m high) for owner occupation (including heating)	450 to 570
Cold stores, refrigerated stores	450 to 570

Administrative, commercial protective service facilities (CI/SfB 3)

Embassies	1300 to 1900
County Courts	1150 to 1350
Magistrates Courts	900 to 1100
Civic offices	
non air conditioned	825 to 950
fully air conditioned	1100 to 1300
Probation/Registrar Offices	625 to 750
Offices for letting	
low rise, non air conditioned	600 to 850
low rise, air conditioned	750 to 950
medium rise, non air conditioned	700 to 950
medium rise, air conditioned	850 to 1150
high rise, non air conditioned	900 to 1200
high rise, air conditioned	1150 to 1600
Offices for owner occupation	
low rise, non air conditioned	750 to 950
low rise, air conditioned	900 to 1200
medium rise, non air conditioned	900 to 1100
medium rise, air conditioned	1150 to 1450
high rise, air conditioned	1500 to 1850
Offices, prestige	
medium rise	1500 to 1850
high rise	1850 to 2500
Large trading floors in medium rise offices	2000 to 2350
Two storey ancillary office accommodation to warehouses/factories	625 to 700
Fitting out offices	
basic fitting out including carpets, decorations, partitions and services	200 to 250
good quality fitting out including carpets, decorations, partitions, ceilings, furniture and services	400 to 485

	Square metre excluding VAT £
high quality fitting out including raised floors and carpets, decorations, partitions, ceilings, furniture, air conditioning and electrical services	700 to 870
Office refurbishment including air conditioning	
good quality	675 to 875
high quality	1200 to 1600
Banks	
local	1025 to 1200
city centre	1650 to 2200
Building Society Branch Offices	900 to 1150
Shop shells	
small	435 to 550
large including department stores and supermarkets	400 to 450
Fitting out shell for small shop (including shop fittings)	
simple store	450 to 515
fashion store	825 to 1025
Fitting out shell for department store or supermarket	
excluding shop fittings	500 to 650
including shop fittings	750 to 1025
Retail Warehouses	
shell	285 to 410
fitting out	200 to 230
Shopping centres	
Malls including fitting out	
comfort cooled	1500 to 1700
air-conditioned	1700 to 1900
Retail area shells, capped off services	450 to 550
Landlord's back-up areas, management offices, plant rooms	
non air-conditioned	625 to 700
*Ambulance stations	500 to 750
Ambulance controls centre	750 to 1100
Fire stations	775 to 1050
Police stations	775 to 1125

Health and welfare facilities (CI/SfB 4)

*District general hospitals	970 to 1250
Hospice	950 to 1150
Private hospitals	1350 to 1700
Hospital laboratories	1100 to 1400
Ward blocks	800 to 1000
Geriatric units	850 to 1100
Hospital teaching centres	650 to 925
*Health centres	625 to 825
*Day centres	700 to 950
Group practice surgeries	625 to 750
*Homes for the physically handicapped	
houses	725 to 975
*Homes for the mentally handicapped	650 to 875
Geriatric day hospital	650 to 850
Nursing homes, convalescent homes	650 to 950
*Children's homes	525 to 800
*Homes for the aged	600 to 800
*Observation and assessment units	600 to 900

Recreational facilities (CI/SfB 5)

Public houses	700 to 975
Kitchen blocks (including fitting out)	1300 to 1500
Dining blocks and canteens in shop and factory	700 to 1000
Restaurants	850 to 1100

	Square metre excluding VAT £

Recreational facilities (CI/SfB 5) - cont'd

Community centres	575 to 825
General purpose halls	600 to 825
Visitors' centres	850 to 1100
Youth clubs	650 to 750
Arts and drama centres	875 to 1025
Theatres, including seating and stage equipment	
large-over 500 seats	1400 to 1650
workshop - less than 500 seats	975 to 1250
Concert halls, including seating and stage equipment	1600 to 2700
Cinema	
shell	425 to 480
fitting out including all equipment, air-conditioned	875 to 1025
Exhibition centres	1025 to 1400
Swimming pools	
international standard	1150 to 1375
local authority standard	1025 to 1300
school standard	850 to 975
leisure pools, including wave making equipment	1300 to 1700
Ice rinks	950 to 1075
Rifle ranges	650 to 825
Leisure centre	
dry	575 to 850
wet	675 to 1025
Sports halls including changing	575 to 800
School gymnasiums	575 to 700
Squash courts	575 to 775
Indoor bowls halls	500 to 675
Sports pavilions	
Social and changing	575 to 800
changing only	675 to 750
changing and public toilets	850 to 975
Grandstands	
simple stands	500 to 575
first class stands with ancillary accommodation	900 to 1100
Clubhouses	675 to 900

Religious facilities (CI/SfB 6)

Temples, mosques, synagogues	1000 to 1250
Churches	725 to 1050
Crematoria	1075 to 1250

Educational, scientific, information facilities (CI/SfB 7)

Nursery Schools	725 to 1050
*Primary/junior school	600 to 775
*Secondary/middle schools	550 to 700
*Extensions to schools	
classrooms	650 to 725
residential	725 to 800
laboratories	900 to 975
Sixth form colleges	675 to 875
*Special schools	600 to 850
*Polytechnics	
Students Union buildings	650 to 750
arts buildings	575 to 675
scientific laboratories	775 to 975
*Training colleges	775 to 925
Management training centres	850 to 1025

	Square metre excluding VAT £
*Universities	
arts buildings	800 to 925
science buildings	900 to 1150
College/University Libraries	675 to 925
Laboratories and offices, low level servicing	800 to 1025
Laboratories (specialist, controlled environment)	1600 to 1800
Computer buildings	900 to 1700
Museums	
national, including full air conditioning and standby generator	2400 to 2850
regional, including full air conditioning	1100 to 1850
local, air conditioned	875 to 1250
conversion of existing warehouse to regional standard	875 to 1300
conversion of existing warehouse to local standard	750 to 1100
Libraries	
city centre	900 to 1050
branch	675 to 900

Residential facilities (CI/SfB 8)

*Local authority and housing association schemes	
Bungalows	
semi-detached	525 to 625
terraced	450 to 550
Two storey housing	
detached	500 to 600
semi-detached	450 to 550
terraced	400 to 500
Three storey housing	
semi-detached	425 to 525
terraced	375 to 475
low rise flats excluding lifts	475 to 650
medium rise flats including lifts	525 to 750
Sheltered housing with wardens' accommodation	525 to 650
Terraced blocks of garages	375 to 425
Private developments	
Single detached houses	625 to 825
Houses - two or three storey	425 to 575
Flats	
standard	525 to 675
luxury	850 to 1075
Warehouse conversions to apartments	625 to 950
Hotels	
5 star city centre hotel	1500 to 2000
4 star city/provincial centre hotel	1200 to 1800
3 star city/provincial hotel	950 to 1500
3/2 star, provincial hotel	750 to 1250
3/2 star, provincial hotel bedroom extension	700 to 1000
*Students' residences	600 to 750
Hostels	575 to 800

Common facilities, other facilities (CI/SfB 9)

Conference centres	1250 to 1700
Public conveniences	1150 to 1850

* Refer also to 'Cost Limits and Allowances' in following section

BUILDING & CIVIL ENGINEERING CONTRACTORS

EASTBOURNE
16-20 South Street
Telephone 21300

LONDON
83 Coborn Road E3 2DB
Telephone 01 980 0013

HASTINGS
5 Warrior Square
Telephone 423888

BRIGHTON
160 Eastern Road
Telephone 698281

MILTON KEYNES
Bleak Hall
Telephone 679222

SPECIALIST COMPANIES

TIMBER FRAME & TRUSS MANUFACTURERS
Llewellyn Homes Ltd · Eastbourne 35271
Bleak Hall · Milton Keynes 679222

SHOPFITTERS
G Bainbridge & Son Ltd · Eastbourne 37222

PLANT HIRE & SALES
Bleak Hall · Milton Keynes 679222
Courtlands Road · Eastbourne 35276

Approximate Estimates

(incorporating Comparative Prices)

Estimating by means of priced approximate quantities is always more accurate than by using overall prices per square metre. Prices given in this section, which is arranged in elemental order, are derived from 'Prices for Measured Work - Major Works' section, but also include for all incidental items and labours which are normally measured separately in Bills of Quantities. As in other sections, they have been established with a tender price index level of 366 (1976 = 100). They do not include for preliminaries, details of which are given in Part II and which may amount to approximately 14% of the value of the measured work or fees for professional services.

This year, the **Approximate Estimates** section has been re-drafted to include a considerable expansion of items and a consolidation of items from the previous **Approximate Estimates and Comparative Prices** sections.

The Comparative prices section previously consisted of schedules of commonly used forms of construction and types and qualities of materials, but provided no guidance as to the types of buildings in which certain materials were used.

The authors have therefore taken this opportunity to include not only these items but have also extracted an extensive range of structural and specification alternatives from their Cost Information library to indicate typical specifications and ranges of cost for a number of building types.

Price ranges have been completed for those building types for commonly-used specifications and where they vary indicate for different building types, the effect of different structural or specifications standards/ranges for each building type.

Whilst every effort has been made to ensure the accuracy of these figures, they have been prepared for approximate estimate purposes and on no account should be used for the preparation of tenders.

Unless otherwise described, units denoted as m2 refer to appropriate area unit (rather than gross floor) areas.

As elsewhere in this edition, prices do not include Value Added Tax, which should be applied at the current rate to all non-domestic building.

Item nr.	SPECIFICATIONS	Unit	RESIDENTIAL £ range £
1.0	**SUBSTRUCTURE**	**ground floor plan area** (unless otherwise described)	

comprising:-

Trench fill foundations
Strip foundations
Strip or base foundations
Raft foundations
Piled foundations
Other foundations/Extras
Basements

Trench fill foundations
Foundations; concrete bed and thickening under partitions; for two storey residential

Item nr.	SPECIFICATIONS	Unit	RESIDENTIAL £ range £
1.0.1	1 m deep	m2	75.00 - 85.00
	Extra for		
1.0.2	each additional 0.25 m deep	m2	12.00 - 14.00
1.0.3	each additional storey	m2	10.00 - 12.00

Foundations; hollow ground floor; timber and boarding for two storey residential

Item nr.	SPECIFICATIONS	Unit	RESIDENTIAL £ range £
1.0.4	1 m deep	m2	90.00 -110.00

Strip foundations
Foundations; brickwork; concrete bed and thickening under partitions; for two storey residential

Item nr.	SPECIFICATIONS	Unit	RESIDENTIAL £ range £
1.0.5	1 m deep; 265 mm cavity brickwork	m2	85.00 - 95.00
	Extra for		
1.0.6	additional 0.25 m deep	m2	12.00 - 14.00
1.0.7	each additional storey	m2	10.00 - 12.00
1.0.8	1 m deep; 322.5 mm solid brickwork	m2	90.00 -100.00
	Extra for		
1.0.9	additional 0.25 m deep	m2	13.00 - 15.00

Foundations; brickwork; hollow ground floor; timber and boarding; for two storey residential

Item nr.	SPECIFICATIONS	Unit	RESIDENTIAL £ range £
1.0.10	1 m deep; 265 mm cavity brickwork	m2	100.00 -120.00
	Extra for		
1.0.11	each additional 0.25 m deep	m2	12.00 - 14.00

Strip or base foundations
Foundations in good ground; reinforced concrete bed; for one storey commercial development

Item nr.	SPECIFICATIONS	Unit	RESIDENTIAL £ range £
1.0.12	shallow foundations	m2	- -
1.0.13	deep foundations	m2	- -

Foundations in good ground; reinforced concrete bed; for two storey commercial development

Item nr.	SPECIFICATIONS	Unit	RESIDENTIAL £ range £
1.0.14	shallow foundations	m2	- -
1.0.15	deep foundations	m2	- -
	Extra for		
1.0.16	each additional storey	m2	- -

Raft foundations
Raft on poor ground for commercial development

Item nr.	SPECIFICATIONS	Unit	RESIDENTIAL £ range £
1.0.17	one storey	m2	- -
1.0.18	two storey	m2	- -
	Extra for		
1.0.19	each additional storey	m2	- -

INDUSTRIAL £ range £	RETAILING £ range £	LEISURE £ range £	OFFICES £ range £	HOTELS £ range £	Item nr.
- -	- -	- -	- -	- -	1.0.1
- -	- -	- -	- -	- -	1.0.2
- -	- -	- -	- -	- -	1.0.3
- -	- -	- -	- -	- -	1.0.4
- -	- -	- -	- -	- -	1.0.5
- -	- -	- -	- -	- -	1.0.6
- -	- -	- -	- -	- -	1.0.7
- -	- -	- -	- -	- -	1.0.8
- -	- -	- -	- -	- -	1.0.9
- -	- -	- -	- -	- -	1.0.10
- -	- -	- -	- -	- -	1.0.11
35.00 - 50.00	40.00 - 65.00	40.00 - 60.00	- -	- -	1.0.12
60.00 - 80.00	65.00 - 90.00	60.00 - 80.00	- -	- -	1.0.13
60.00 - 75.00	70.00 - 90.00	65.00 - 85.00	70.00 - 85.00	70.00 - 85.00	1.0.14
80.00 -110.00	90.00 -125.00	85.00 -120.00	85.00 -120.00	85.00 -120.00	1.0.15
17.00 - 20.00	15.00 - 18.00	15.00 - 18.00	15.00 - 18.00	15.00 - 18.00	1.0.16
75.00 -120.00	90.00 -115.00	100.00 -140.00	- -	- -	1.0.17
110.00 -150.00	125.00 -175.00	120.00 -170.00	120.00 -170.00	120.00 -170.00	1.0.18
17.00 - 20.00	15.00 - 18.00	15.00 - 18.00	15.00 - 18.00	15.00 - 18.00	1.0.19

Item nr.	SPECIFICATIONS	Unit	RESIDENTIAL £ range £
1.0	SUBSTRUCTURE - cont'd ground floor plan area (unless otherwise described)		
	Piled foundations		
	Foundations in poor ground; reinforced concrete slab and ground beams; for two storey residential		
1.0.20	short bore piled	m2	125.00 -160.00
1.0.21	fully piled	m2	150.00 -190.00
	Foundations in poor ground; hollow ground floor; timber and boarding; for two storey residential		
1.0.22	short bore piled	m2	145.00 -180.00
	Foundations in poor ground; reinforced concrete slab; for one storey commercial development		
1.0.23	short bore piles to columns only	m2	- -
1.0.24	short bore piles	m2	- -
1.0.25	fully piled	m2	- -
	Foundations in poor ground with reinforced concrete slab and ground beams; for two storey commercial development		
1.0.26	short bore piles	m2	- -
1.0.27	fully piled	m2	- -
	Extra for		
1.0.28	each additional storey	m2	- -
	Foundations in bad ground; inner city redevelopment; reinforced concrete slab and ground beams; for two storey commercial development		
1.0.29	fully piled	m2	- -
	Other foundations/alternative slabs/extras		
	Cantilevered foundations in good ground; reinforced		
1.0.30	concrete slab for two storey commercial development	m2	- -
	Underpinning foundations of existing buildings		
1.0.31	abutting site	m	- -
	Extra to substructure rates for		
1.0.32	watertight pool construction	m2	- -
1.0.33	ice pad	m2	- -
	Reinforced concrete bed including excavation and hardcore under		
1.0.34	150 mm thick	m2	25.00 - 30.00
1.0.35	200 mm thick	m2	30.00 - 36.00
1.0.36	300 mm thick	m2	- -
	Hollow ground floor with timber and boarding, including excavation, concrete and hardcore under		
1.0.37	300 mm deep	m2	32.50 - 40.00
	Extra for		
1.0.38	sound reducing quilt in screed	m2	3.00 - 4.00
1.0.39	50 mm insulation under slab and at edges	m2	3.00 - 5.00
1.0.40	75 mm insulation under slab and at edges	m2	- -
	suspended precast concrete slabs in lieu of insitu		
1.0.41	slab	m2	5.00 - 7.50
	Basement basement floor/wall area		
	Reinforced concrete basement floors		
1.0.42	non-waterproofed	m2	- -
1.0.43	waterproofed	m2	- -
	Reinforced concrete basement walls		
1.0.44	non-waterproofed	m2	- -
1.0.45	waterproofed	m2	- -
1.0.46	sheet piled	m2	- -
1.0.47	diaphragm walling	m2	- -
	Extra for		
1.0.48	each additional storey		- -

INDUSTRIAL £ range £	RETAILING £ range £	LEISURE £ range £	OFFICES £ range £	HOTELS £ range £	Item nr.
- -	- -	- -	- -	- -	1.0.20
- -	- -	- -	- -	- -	1.0.21
- -	- -	- -	- -	- -	1.0.22
70.00 -110.00	80.00 -110.00	70.00 -110.00	- -	- -	1.0.23
95.00 -125.00	100.00 -130.00	100.00 -130.00	100.00 -130.00	100.00 -130.00	1.0.24
125.00 -170.00	150.00 -200.00	150.00 -200.00	150.00 -200.00	150.00 -200.00	1.0.25
- -	110.00 -140.00	110.00 -140.00	110.00 -140.00	110.00 -140.00	1.0.26
- -	160.00 -210.00	160.00 -210.00	160.00 -210.00	160.00 -210.00	1.0.27
17.00 - 20.00	10.00 - 12.00	10.00 - 12.00	10.00 - 12.00	10.00 - 12.00	1.0.28
- -	200.00 -260.00	- -	200.00 -350.00	200.00 -300.00	1.0.29
- -	250.00 -320.00	250.00 -320.00	250.00 -320.00	250.00 -320.00	1.0.30
- -	600.00 -800.00	600.00 -800.00	600.00 -800.00	600.00 -800.00	1.0.31
- -	- -	100.00 -140.00	- -	- -	1.0.32
- -	- -	110.00 -165.00	- -	- -	1.0.33
30.00 - 35.00	28.00 - 35.00	28.00 - 33.00	28.00 - 33.00	28.00 - 33.00	1.0.34
35.00 - 45.00	33.00 - 42.00	33.00 - 40.00	33.00 - 40.00	33.00 - 40.00	1.0.35
45.00 - 60.00	40.00 - 55.00	40.00 - 50.00	40.00 - 50.00	40.00 - 50.00	1.0.36
- -	- -	- -	- -	- -	1.0.37
3.00 - 5.00	3.00 - 5.00	3.00 - 5.00	3.00 - 5.00	3.00 - 5.00	1.0.38
5.00 - 7.00	5.00 - 7.00	5.00 - 7.00	5.00 - 7.00	5.00 - 7.00	1.0.39
6.00 - 9.00	6.00 - 9.00	6.00 - 9.00	6.00 - 9.00	6.00 - 9.00	1.0.40
12.00 - 15.00	10.00 - 12.00	10.00 - 12.00	10.00 - 12.00	10.00 - 12.00	1.0.41
- -	45.00 - 55.00	45.00 - 55.00	45.00 - 55.00	45.00 - 55.00	1.0.42
- -	55.00 - 70.00	55.00 - 70.00	55.00 - 70.00	55.00 - 70.00	1.0.43
- -	130.00 -160.00	130.00 -160.00	130.00 -160.00	130.00 -160.00	1.0.44
- -	150.00 -190.00	150.00 -190.00	150.00 -190.00	150.00 -190.00	1.0.45
- -	275.00 -325.00	275.00 -325.00	275.00 -325.00	275.00 -325.00	1.0.46
- -	300.00 -350.00	300.00 -350.00	300.00 -350.00	300.00 -350.00	1.0.47
- -	+ 20%	+ 20%	+ 20%	+ 20%	1.0.48

Item nr.	SPECIFICATIONS	Unit	RESIDENTIAL £ range £

2.0   **SUPERSTRUCTURE**

2.1   **FRAME TO ROOF**   **roof plan area**
                                          (unless otherwise described)

comprising:-

Reinforced concrete frame
Precast concrete frame
Steel frame
Other frames/extras

Item nr.	SPECIFICATIONS	Unit	RESIDENTIAL	£ range £
	Reinforced concrete frame			
2.1.1	Columns	m2	-	-
2.1.2	Columns and beams; 7.2 x 7.2 m grid	m2	-	-
	Extra for			
2.1.3	spans 7.5 - 15 m	m2	-	-
	Precast concrete frame			
	Portal frame			
2.1.4	industrial	m2	-	-
2.1.5	industrial; to small units	m2	-	-
2.1.6	retailing	m2	-	-
	Steel frame			
	Columns			
2.1.7	unprotected	m2	-	-
2.1.8	fully protected	m2	-	-
	Portal frame			
2.1.9	unprotected	m2	-	-
2.1.10	unprotected; to small units	m2	-	-
2.1.11	protected columns only	m2	-	-
2.1.12	fully protected	m2	-	-
2.1.13	protected; to small units	m2	-	-
	Columns and beams; to flat roofs			
2.1.14	unprotected	m2	-	-
2.1.15	unprotected; to small units	m2	-	-
2.1.16	unprotected; large span	m2	-	-
2.1.17	unprotected; very large span	m2	-	-
2.1.18	protected columns only	m2	-	-
2.1.19	fully protected	m2	-	-
2.1.20	fully protected; insulation	m2	-	-
	Other frames/extras			
	Space deck on steel frame			
2.1.21	unprotected	m2	-	-
2.1.22	Exposed steel frame for tent/mast structures	m2	-	-
	Columns and beams to 18 m high bay warehouse			
2.1.23	unprotected	m2	-	-
	Columns and beams to mansard			
2.1.24	protected	m2	-	-
	Feature columns and beams to glazed atruim roof			
2.1.25	unprotected	m2	-	-

INDUSTRIAL £ range £	RETAILING £ range £	LEISURE £ range £	OFFICES £ range £	HOTELS £ range £	Item nr.
25.00 - 35.00	28.00 - 40.00	28.00 - 40.00	28.00 - 35.00	28.00 - 35.00	2.1.1
45.00 - 70.00	50.00 - 75.00	50.00 - 70.00	50.00 - 70.00	50.00 - 75.00	2.1.2
15.00 - 40.00	20.00 - 45.00	15.00 - 40.00	15.00 - 40.00	15.00 - 40.00	2.1.3
45.00 - 60.00	- -	- -	- -	- -	2.1.4
60.00 - 80.00	- -	- -	- -	- -	2.1.5
- -	50.00 - 70.00	- -	- -	- -	2.1.6
25.00 - 35.00	30.00 - 45.00	30.00 - 40.00	30.00 - 40.00	- -	2.1.7
30.00 - 40.00	35.00 - 50.00	35.00 - 45.00	35.00 - 45.00	- -	2.1.8
35.00 - 50.00	40.00 - 60.00	- -	- -	- -	2.1.9
45.00 - 65.00	- -	- -	- -	- -	2.1.10
40.00 - 60.00	48.00 - 72.00	45.00 - 60.00	- -	- -	2.1.11
55.00 - 70.00	60.00 - 85.00	55.00 - 72.00	- -	- -	2.1.12
75.00 - 90.00	- -	- -	- -	- -	2.1.13
40.00 - 50.00	45.00 - 55.00	45.00 - 55.00	50.00 - 60.00	50.00 - 60.00	2.1.14
70.00 - 80.00	- -	- -	- -	- -	2.1.15
70.00 - 85.00	75.00 - 90.00	78.00 -100.00	- -	- -	2.1.16
75.00 - 90.00	80.00 -100.00	85.00 -105.00	- -	- -	2.1.17
60.00 - 70.00	55.00 - 75.00	52.00 - 70.00	- -	- -	2.1.18
60.00 - 75.00	60.00 - 80.00	65.00 - 90.00	60.00 - 85.00	60.00 - 90.00	2.1.19
- -	65.00 - 90.00	75.00 -100.00	70.00 - 95.00	70.00 -100.00	2.1.20
80.00 -150.00	85.00 -150.00	85.00 -150.00	85.00 -150.00	- -	2.1.21
130.00 -200.00	130.00 -200.00	130.00 -200.00	- -	- -	2.1.22
85.00 -115.00	- -	- -	- -	- -	2.1.23
- -	70.00 -100.00	70.00 -100.00	70.00 -100.00	70.00 -100.00	2.1.24
- -	- -	- -	80.00 -120.00	- -	2.1.25

Item nr.	SPECIFICATIONS	Unit	RESIDENTIAL £ range £

2.2   **FRAME & UPPER FLOORS (COMBINED)**   **upper floor area**
                              (unless otherwise described)

comprising:-

Softwood floors; no frame
Softwood floors; steel frame
Reinforced concrete floors; no frame
Reinforced concrete floors and frame
Reinforced concrete floors; steel frame
Precast concrete floors; no frame
Precast concrete floors; reinforced concrete frame
Precast concrete floors and frame
Precast concrete floors; steel frame
Other floor and frame constructions/extras

Softwood floors; no frame
Joisted floor; supported on layers; 22 mm chipboard t & g
flooring; herring bone strutting; no coverings or finishes

Item nr.	Description	Unit	£ range £
2.2.1	150 x 50 mm joists	m2	20.00 - 23.00
2.2.2	175 x 50 mm joists	m2	22.00 - 25.00
2.2.3	200 x 50 mm joists	m2	23.50 - 26.50
2.2.4	225 x 50 mm joists	m2	25.00 - 28.00
2.2.5	250 x 50 mm joists	m2	27.00 - 29.00
2.2.6	275 x 50 mm joists	m2	29.00 - 32.00
	Joisted floor; average depth; plasterboard; skim;		
2.2.7	emulsion; vinyl flooring and painted softwood skirtings	m2	50.00 - 60.00

Softwood construction; steel frame
Joisted floor; average depth; plasterboard; skim;

| 2.2.8 | emulsion; vinyl flooring and painted softwood skirtings | m2 | - | - |

Reinforced concrete floors; no frame
Suspended slab; no coverings or finishes

2.2.9	3.65 m span; 3.00 KN/m2 loading?	m2	38.00 - 40.00
2.2.10	4.25 m span; 3.00 KN/m2 loading?	m2	44.00 - 47.00
2.2.11	2.75 m span; 8.00 KN/m2 loading?	m2	-	-
2.2.12	3.35 m span; 8.00 KN/m2 loading?	m2	-	-
2.2.13	4.25 m span; 8.00 KN/m2 loading?	m2	-	-
	Suspended slab; plaster; emulsion; vinyl flooring and painted softwood skirtings		
2.2.14	150 mm thick	m2	70.00 - 80.00
2.2.15	225 mm thick	m2	90.00 -100.00

Reinforced concrete floors and frame
Suspended slab; average depth; plaster; emulsion; vinyl
flooring and painted softwood skirtings

2.2.16	up to six storeys	m2	-	-
2.2.17	seven to twelve storeys	m2	-	-
2.2.18	thirteen to eighteen storeys	m2	-	-
	Extra for		
2.2.19	section 20 fire regulations	m2	-	-
	Wide span suspended slab		
2.2.20	up to six storeys	m2	-	-

Reinforced concrete floors; steel frame
Suspended slab; average depth; 'Holorib' permanent steel
shuttering; protected steel frame; plaster; emulsion;
vinyl flooring and painted softwood skirtings

2.2.21	up to six storeys	m2	-	-
	Extra for		
2.2.22	spans 7.5 to 15 m	m2	-	-
2.2.23	seven to twelve storeys	m2	-	-
	Extra for		
2.2.24	section 20 fire regulations	m2	-	-

INDUSTRIAL £ range £	RETAILING £ range £	LEISURE £ range £	OFFICES £ range £	HOTELS £ range £	Item nr.
- -	- -	- -	- -	- -	2.2.1
- -	- -	- -	- -	- -	2.2.2
- -	- -	- -	- -	- -	2.2.3
- -	- -	- -	- -	- -	2.2.4
- -	- -	- -	- -	- -	2.2.5
- -	- -	- -	- -	- -	2.2.6
- -	- -	- -	- -	- -	2.2.7
65.00 - 70.00	75.00 -100.00	70.00 - 95.00	70.00 - 95.00	70.00 - 95.00	2.2.8
- -	- -	38.00 - 44.00	38.00 - 44.00	38.00 - 44.00	2.2.9
- -	- -	44.00 - 47.00	44.00 - 52.00	44.00 - 52.00	2.2.10
38.00 - 44.00	38.00 - 44.00	38.00 - 44.00	38.00 - 44.00	38.00 - 44.00	2.2.11
42.00 - 50.00	42.00 - 50.00	42.00 - 50.00	42.00 - 50.00	42.00 - 50.00	2.2.12
54.00 - 60.00	54.00 - 60.00	54.00 - 60.00	54.00 - 62.00	54.00 - 62.00	2.2.13
75.00 - 95.00	70.00 - 95.00	70.00 - 90.00	70.00 - 85.00	70.00 - 95.00	2.2.14
100.00 -120.00	90.00 -115.00	90.00 -110.00	90.00 -105.00	90.00 -115.00	2.2.15
120.00 -150.00	100.00 -135.00	110.00 -140.00	100.00 -125.00	100.00 -130.00	2.2.16
- -	- -	- -	135.00 -200.00	- -	2.2.17
- -	- -	- -	180.00 -250.00	- -	2.2.18
- -	- -	- -	8.00 - 10.00	- -	2.2.19
130.00 -160.00	110.00 -145.00	120.00 -150.00	110.00 -135.00	110.00 -140.00	2.2.20
150.00 -180.00	130.00 -170.00	140.00 -170.00	130.00 -160.00	130.00 -165.00	2.2.21
15.00 - 40.00	20.00 - 45.00	15.00 - 40.00	15.00 - 40.00	15.00 - 40.00	2.2.22
- -	- -	- -	170.00 -240.00	- -	2.2.23
- -	- -	- -	15.00 - 18.00	- -	2.2.24

Item nr.	SPECIFICATIONS	Unit	RESIDENTIAL £ range £

2.2   **FRAME & UPPER FLOORS (COMBINED)** - cont'd   **upper floor area**
                                                  (unless otherwise described)

Reinforced concrete floors; steel frame - cont'd
Suspended slab; average depth; protected steel frame;
plaster; emulsion; vinyl flooring and painted softwood
skirtings

2.2.25	up to six storeys	m2	- -
2.2.26	seven to twelve storeys	m2	- -
	Extra for		
2.2.27	section 20 fire regulations	m2	- -

Precast concrete floors; no frame
Suspended slab; 75 mm screed; no coverings or finishes

2.2.28	3 m span; 5.00 KN/m2 loading	m2	22.00 - 30.00
2.2.29	6 m span; 5.00 KN/m2 loading	m2	23.00 - 31.00
2.2.30	7.5 m span; 5.00 KN/m2 loading	m2	24.00 - 32.00
2.2.31	3 m span; 8.50 KN/m2 loading	m2	- -
2.2.32	6 m span; 8.50 KN/m2 loading	m2	- -
2.2.33	7.5 m span; 8.50 KN/m2 loading	m2	- -
2.2.34	3 m span; 12.50 KN/m2 loading	m2	- -
2.2.35	6 m span; 12.50 KN/m2 loading	m2	- -

Suspended slab; average depth; plaster; emulsion; vinyl

2.2.36	flooring and painted softwood skirtings	m2	55.00 - 70.00

Precast concrete floors; reinforced concrete frame
Suspended slab; average depth; plaster; emulsion; vinyl

2.2.37	flooring and painted softwood skirtings	m2	- -

Precast concrete floors and frame
Suspended slab; average depth; plaster; emulsion; vinyl

2.2.38	flooring and painted softwood skirtings	m2	- -

Precast concrete floors; steel frame
Suspended slabs; average depth; unprotected steel frame;
plaster; emulsion; vinyl flooring; and painted
softwood skirtings

2.2.39	up to three storeys	m2	- -

Suspended slabs; average depth; protected steel frame;
plaster; emulsion; vinyl flooring and painted
softwood skirtings

2.2.40	up to six storeys	m2	- -
2.2.41	seven to twelve storeys	m2	- -

Other floor and frame construction/extras

2.2.42	Reinforced concrete cantilevered balcony	nr	1300 -1800
2.2.43	Reinforced concrete cantilevered walkways	m2	- -
2.2.44	Reinforced concrete walkways and supporting frame	m2	- -
	Reinforced concrete core with steel umbrella frame		
2.2.45	twelve to twenty four storeys	m2	- -
	Extra for		
2.2.46	wrought formwork	m2	3.00 - 4.00
2.2.47	sound reducing quilt in screed	m2	3.00 - 4.00
2.2.48	insulation to avoid cold bridging	m2	3.00 - 5.00

INDUSTRIAL £ range £	RETAILING £ range £	LEISURE £ range £	OFFICES £ range £	HOTELS £ range £	Item nr.
155.00 -190.00	135.00 -180.00	140.00 -180.00	135.00 -170.00	135.00 -170.00	2.2.25
-	-	-	180.00 -250.00	-	2.2.26
-	-	-	15.00 - 18.00	-	2.2.27
-	28.00 - 33.00	28.00 - 33.00	28.00 - 33.00	28.00 - 33.00	2.2.28
-	29.00 - 35.00	29.00 - 35.00	29.00 - 34.00	29.00 - 35.00	2.2.29
-	30.00 - 40.00	30.00 - 40.00	30.00 - 35.00	30.00 - 40.00	2.2.30
30.00 - 35.00	-	-	-	-	2.2.31
32.50 - 37.50	-	-	-	-	2.2.32
35.00 - 40.00	-	-	-	-	2.2.33
32.00 - 37.00	-	-	-	-	2.2.34
37.50 - 42.50	-	-	-	-	2.2.35
65.00 - 80.00	60.00 - 90.00	60.00 - 85.00	60.00 - 80.00	60.00 - 90.00	2.2.36
85.00 -110.00	80.00 -115.00	80.00 -110.00	80.00 -105.00	80.00 -115.00	2.2.37
85.00 -140.00	80.00 -150.00	80.00 -140.00	80.00 -140.00	80.00 -150.00	2.2.38
95.00 -150.00	90.00 -140.00	90.00 -130.00	80.00 -130.00	90.00 -140.00	2.2.39
140.00 -170.00	120.00 -160.00	130.00 -160.00	120.00 -150.00	120.00 -155.00	2.2.40
-	-	-	160.00 -225.00	-	2.2.41
-	-	-	-	-	2.2.42
-	100.00 -120.00	-	-	-	2.2.43
-	110.00 -135.00	-	-	-	2.2.44
-	-	-	225.00 -300.00	-	2.2.45
3.00 - 4.00	3.00 - 6.00	3.00 - 6.00	3.00 - 6.00	3.00 - 8.00	2.2.46
3.00 - 5.00	3.00 - 5.00	3.00 - 5.00	3.00 - 5.00	3.00 - 5.00	2.2.47
5.00 - 7.00	5.00 - 7.00	5.00 - 7.00	5.00 - 7.00	5.00 - 7.00	2.2.48

Item nr.	SPECIFICATIONS	Unit	RESIDENTIAL £ range £

2.3   **ROOF**   **roof plan area**
                                  (unless otherwise described)

comprising:-

Softwood flat roofs
Softwood trussed pitched roofs
Steel trussed pitched roofs
Concrete flat roofs
Flatroof decking and finishes
Roof claddings
Rooflights/patent glazing and glazed roofs
Comparative over/underlays
Comparative tiling and slating finishes/perimeter treatments
Comparative cladding finishes/perimeter treatments
Comparative waterproofing finishes/perimeter treatments

Softwood Flat roofs
Structure only comprising roof joists; 100 x 50 mm wall
plates; herring-bone strutting; 50 mm woodwool slabs;
no coverings or finishes

Item nr.	Description	Unit	£ range £
2.3.1	150 x 50 mm joists	m2	27.00 - 30.00
2.3.2	175 x 50 mm joists	m2	28.50 - 32.00
2.3.3	200 x 50 mm joists	m2	30.00 - 33.00
2.3.4	225 x 50 mm joists	m2	31.50 - 35.00
2.3.5	250 x 50 mm joists	m2	33.00 - 37.00
2.3.6	275 x 50 mm joists	m2	34.50 - 38.50

Structure only comprising roof joists; 100 x 50 mm wall
plates; herring-bone strutting; 25 mm softwood boarding;
no coverings or finishes

Item nr.	Description	Unit	£ range £
2.3.7	150 x 50 mm joists	m2	29.00 - 32.00
2.3.8	175 x 50 mm joists	m2	30.50 - 34.00
2.3.9	200 x 50 mm joists	m2	32.00 - 35.00
2.3.10	225 x 50 mm joists	m2	33.50 - 37.00
2.3.11	250 x 50 mm joists	m2	35.00 - 39.00
2.3.12	275 x 50 mm joists	m2	36.50 - 40.50

Roof joists; average depth; 25 mm softwood boarding;
PVC rainwater goods; plasterboard; skim and emulsion

Item nr.	Description	Unit	£ range £
2.3.13	three layer felt and chippings	m2	62.00 - 80.00
2.3.14	two coat asphalt and chippings	m2	60.00 - 90.00

Softwood trussed pitched roofs
Structure only comprising 75 x 50 mm Fink roof
trusses @ 600 mm centres (measured on plan)

Item nr.	Description	Unit	£ range £
2.3.15	22.5° pitch	m2	12.00 - 15.00

Structure only comprising 100 x 38 mm Fink roof
trusses @ 600 mm centres (measured on plan)

Item nr.	Description	Unit	£ range £
2.3.16	30° pitch	m2	14.00 - 17.00
2.3.17	35° pitch	m2	14.50 - 17.50
2.3.18	40° pitch	m2	16.00 - 20.00

Structure only comprising 100 x 50 mm Fink roof
trusses @ 375 mm centres (measured on plan)

Item nr.	Description	Unit	£ range £
2.3.19	30° pitch	m2	24.00 - 27.00
2.3.20	35° pitch	m2	24.50 - 27.50
2.3.21	40° pitch	m2	26.00 - 30.00
	Extra for		
2.3.22	forming dormers	nr	350.00 -500.00

Structure only for Mansard roof comprising
100 x 50 mm roof trusses @ 600 mm centres

Item nr.	Description	Unit	£ range £
2.3.23	70° pitch	m2	18.00 - 24.00

Fink roof trusses; narrow span; 100 mm insulation; PVC
rainwater goods; plasterboard; skim and emulsion

Item nr.	Description	Unit	£ range £
2.3.24	concrete interlocking tile coverings	m2	65.00 - 80.00
2.3.25	composition slate coverings	m2	80.00 - 95.00
2.3.26	plain clay tile coverings	m2	95.00 -110.00
2.3.27	natural slate coverings	m2	100.00 -115.00
2.3.28	reconstructed stone coverings	m2	105.00 -120.00
	Extra for		
2.3.29	150 mm insulation in lieu of 100 mm	m2	2.25 - 3.25

INDUSTRIAL £ range £		RETAILING £ range £	LEISURE £ range £		OFFICES £ range £	HOTELS £ range £	Item nr.
-	-	27.00 - 30.00	-	-	27.00 - 30.00	27.00 - 30.00	2.3.1
-	-	28.50 - 32.00	-	-	28.50 - 32.00	28.50 - 32.00	2.3.2
-	-	30.00 - 33.00	-	-	30.00 - 33.00	30.00 - 33.00	2.3.3
-	-	31.50 - 35.00	-	-	31.50 - 35.00	31.50 - 35.00	2.3.4
-	-	33.00 - 37.00	-	-	33.00 - 37.00	33.00 - 37.00	2.3.5
-	-	34.50 - 38.50	-	-	34.50 - 38.50	34.50 - 38.50	2.3.6
-	-	29.00 - 32.00	-	-	29.00 - 32.00	29.00 - 32.00	2.3.7
-	-	30.50 - 34.00	-	-	30.50 - 34.00	30.50 - 34.00	2.3.8
-	-	32.00 - 35.00	-	-	32.00 - 35.00	32.00 - 35.00	2.3.9
-	-	33.50 - 37.00	-	-	33.50 - 37.00	33.50 - 37.00	2.3.10
-	-	35.00 - 39.00	-	-	35.00 - 39.00	35.00 - 39.00	2.3.11
-	-	36.50 - 40.50	-	-	36.50 - 40.50	36.50 - 40.50	2.3.12
-	-	62.00 - 80.00	-	-	62.00 - 80.00	62.00 - 80.00	2.3.13
-	-	60.00 - 90.00	-	-	60.00 - 90.00	60.00 - 90.00	2.3.14
-	-	13.00 - 16.00	-	-	13.00 - 16.00	13.00 - 17.00	2.3.15
-	-	15.00 - 18.00	-	-	15.00 - 18.00	15.00 - 19.00	2.3.16
-	-	15.50 - 18.50	-	-	15.50 - 18.50	15.50 - 19.50	2.3.17
-	-	17.00 - 21.00	-	-	17.00 - 21.00	17.00 - 22.00	2.3.18
-	-	25.00 - 28.00	-	-	25.00 - 28.00	25.00 - 29.00	2.3.19
-	-	25.50 - 28.50	-	-	25.50 - 28.50	25.50 - 29.50	2.3.20
-	-	27.00 - 31.00	-	-	27.00 - 31.00	27.00 - 32.00	2.3.21
-	-	350.00 -500.00	-	-	350.00 -500.00	350.00 -500.00	2.3.22
-	-	19.00 - 25.00	-	-	19.00 - 25.00	19.00 - 26.00	2.3.23
-	-	70.00 - 95.00	-	-	70.00 - 85.00	75.00 - 90.00	2.3.24
-	-	85.00 -110.00	-	-	85.00 -100.00	90.00 -105.00	2.3.25
-	-	100.00 -125.00	-	-	100.00 -115.00	105.00 -120.00	2.3.26
-	-	105.00 -135.00	-	-	105.00 -120.00	110.00 -125.00	2.3.27
-	-	110.00 -140.00	-	-	110.00 -125.00	115.00 -130.00	2.3.28
-	-	2.50 - 3.50	-	-	2.50 - 3.50	2.50 - 3.50	2.3.29

Item nr.	SPECIFICATIONS	Unit	RESIDENTIAL £ range £

2.3   ROOF - cont'd   **roof plan area**
                                      (unless otherwise described)

Softwood trussed pitched roofs - cont'd
Monopitch roof trusses; 100 mm insulation; PVC
rainwater goods; plasterboard; skim and emulsion

Item nr.	SPECIFICATIONS	Unit	RESIDENTIAL £ range £
2.3.30	concrete interlocking tile coverings	m2	67.00 - 85.00
2.3.31	composition slate coverings	m2	82.00 -100.00
2.3.32	plain clay tile coverings	m2	97.00 -115.00
2.3.33	natural slate coverings	m2	102.00 -120.00
2.3.34	reconstructed stone coverings	m2	107.00 -125.00

Dormer roof tresses; 100 mm insulation; PVC rainwater
goods; plasterboard; skim and emulsion

Item nr.	SPECIFICATIONS	Unit	RESIDENTIAL £ range £
2.3.35	concrete interlocking tile coverings	m2	90.00 -110.00
2.3.36	composition slate coverings	m2	105.00 -125.00
2.3.37	plain clay tile coverings	m2	120.00 -140.00
2.3.38	natural slate coverings	m2	125.00 -145.00
2.3.39	reconstructed stone coverings	m2	130.00 -150.00
	Extra for		
2.3.40	end of terrace/semi-detached configeration	m2	24.00 - 27.00
2.3.41	hipped roof configeration	m2	25.00 - 30.00

Steel trussed pitched roofs
Fink roof trusses; wide span; 100 mm insulation

Item nr.	SPECIFICATIONS	Unit	RESIDENTIAL £ range £
2.3.42	concrete interlocking tile coverings	m2	- -
2.3.43	composition slate coverings	m2	- -
2.3.44	clay plain tile coverings	m2	- -
2.3.45	natural slate coverings	m2	- -
2.3.46	reconstructed stone coverings	m2	- -
	Extra for		
2.3.47	150 mm insulation in lieu of 100 mm	m2	- -

Steel roof trusses and beams; thermal and accoustic
insulation

Item nr.	SPECIFICATIONS	Unit	RESIDENTIAL £ range £
2.3.48	aluminium profiled composite cladding	m2	- -
2.3.49	copper roofing on boarding	m2	- -

Steel roof and glulam beams; thermal and accoustic
insulation

Item nr.	SPECIFICATIONS	Unit	RESIDENTIAL £ range £
2.3.50	aluminium profiled composite cladding	m2	- -

Concrete flat roofs
Structure only comprising reinforced concrete
suspended slab; no coverings or finishes

Item nr.	SPECIFICATIONS	Unit	RESIDENTIAL £ range £
2.3.51	3.65 m span; 3.00 KN/m2 loading	m2	38.00- 40.00
2.3.52	4.25 m span; 3.00 KN/m2 loading	m2	44.00 - 47.00
2.3.53	2.75 m span; 8.00 KN/m2 loading	m2	- -
2.3.54	3.65 m span; 8.00 KN/m2 loading	m2	- -
2.3.55	4.25 m span; 8.00 KN/m2 loading	m2	- -

Precast concrete suspended slab;average depth;
100 mm insulation; PVC rainwater goods; plaster and
emulsion

Item nr.	SPECIFICATIONS	Unit	RESIDENTIAL £ range £
2.3.56	two coat asphalt coverings and chippings	m2	- -
2.3.57	polyester roofing	m2	- -

Reinforced concrete or waffle suspended slabs; average
depth; 100 mm insulation; PVC rainwater goods; plaster
and emulsion

Item nr.	SPECIFICATIONS	Unit	RESIDENTIAL £ range £
2.3.58	two coat asphalt coverings and chippings	m2	- -
2.3.59	two coat asphalt coverings and paving slabs	m2	- -
	Extra for		
2.3.60	150 mm insulation in lieu of 100 mm	m2	- -

Reinforced concrete suspended slabs; average depth;
150 mm insulation

Item nr.	SPECIFICATIONS	Unit	RESIDENTIAL £ range £
2.3.61	two coat asphalt coverings and 80 mm gravel ballast	m2	- -

INDUSTRIAL £ range £	RETAILING £ range £	LEISURE £ range £	OFFICES £ range £	HOTELS £ range £	Item nr.
-	72.00 -100.00	-	72.00 - 95.00	77.00 - 95.00	2.3.30
-	87.00 -115.00	-	87.00 -105.00	92.00 -110.00	2.3.31
-	102.00 -130.00	-	102.00 -120.00	107.00 -125.00	2.3.32
-	107.00 -140.00	-	107.00 -125.00	112.00 -130.00	2.3.33
-	112.00 -145.00	-	112.00 -130.00	117.00 -135.00	2.3.34
-	95.00 -130.00	-	95.00 -120.00	100.00 -125.00	2.3.35
-	110.00 -145.00	-	110.00 -135.00	115.00 -140.00	2.3.36
-	125.00 -160.00	-	125.00 -150.00	130.00 -155.00	2.3.37
-	130.00 -165.00	-	130.00 -155.00	135.00 -160.00	2.3.38
-	135.00 -170.00	-	135.00 -160.00	140.00 -165.00	2.3.39
-	-	-	-	-	2.3.40
-	-	-	-	-	2.3.41
-	100.00 -125.00	95.00 -120.00	100.00 -115.00	105.00 -120.00	2.3.42
-	115.00 -140.00	110.00 -130.00	115.00 -130.00	120.00 -135.00	2.3.43
-	130.00 -155.00	125.00 -150.00	130.00 -145.00	135.00 -150.00	2.3.44
-	135.00 -165.00	130.00 -160.00	135.00 -150.00	140.00 -155.00	2.3.45
-	140.00 -170.00	135.00 -165.00	140.00 -155.00	145.00 -160.00	2.3.46
-	-	-	-	-	2.3.47
150.00 -175.00	140.00 -165.00	135.00 -160.00	140.00 -155.00	-	2.3.48
-	150.00 -185.00	155.00 -180.00	160.00 -175.00	-	2.3.49
-	150.00 -200.00	145.00 -195.00	150.00 -190.00	-	2.3.50
-	-	38.00 - 44.00	38.00 - 44.00	38.00 - 44.00	2.3.51
-	-	44.00 - 47.00	44.00 - 52.00	44.00 - 52.00	2.3.52
38.00 - 44.00	38.00 - 44.00	38.00 - 44.00	38.00 - 44.00	38.00 - 44.00	2.3.53
42.00 - 50.00	42.00 - 50.00	42.00 - 50.00	42.00 - 50.00	42.00 - 50.00	2.3.54
54.00 - 60.00	54.00 - 60.00	54.00 - 60.00	54.00 - 60.00	54.00 - 60.00	2.3.55
85.00 -110.00	80.00 -120.00	80.00 -115.00	80.00 -110.00	80.00 -120.00	2.3.56
87.50 -100.00	82.50 -110.00	82.50 -105.00	82.50 -100.00	82.50 -110.00	2.3.57
90.00 -115.00	85.00 -125.00	85.00 -120.00	85.00 -115.00	85.00 -125.00	2.3.58
120.00 -145.00	115.00 -155.00	115.00 -150.00	115.00 -145.00	115.00 -160.00	2.3.59
5.00 - 8.00	5.00 - 8.00	5.00 - 8.00	5.00 - 8.00	5.00 - 8.00	2.3.60
105.00 -130.00	100.00 -140.00	100.00 -135.00	100.00 -130.00	100.00 -140.00	2.3.61

Item nr.	SPECIFICATIONS	Unit	RESIDENTIAL £ range £
2.3	**ROOF - cont'd** roof plan area (unless otherwise described)		
	Flat roof decking and finishes		
	Woodwool roof decking		
	50 mm thick; two coat asphalt coverings to BS 988		
2.3.62	and chippings	m2	- -
	Galvanised steel roof decking; 100 mm insulation; three layer felt roofing and chippings		
2.3.63	0.7 mm thick; 2.38 m span	m2	- -
2.3.64	0.7 mm thick; 2.96 m span	m2	- -
2.3.65	0.7 mm thick; 3.74 m span	m2	- -
2.3.66	0.7 mm thick; 5.13 m span	m2	- -
	Aluminium roof decking; 100 mm insulation; three layer felt roofing and chippings		
2.3.67	0.9 mm thick; 1.79 m span	m2	- -
2.3.68	0.9 mm thick; 2.34 m span	m2	- -
	Metal decking; 100 mm insulation; on wood/steel open lattice beams		
2.3.69	three layer felt roofing and chippings	m2	- -
	three layer high performance felt roofing		
2.3.70	and chippings	m2	- -
2.3.71	two coats asphalt coverings and chippings	m2	- -
	Extra for		
2.3.72	150 mm insulation in lieu of 100 mm	m2	- -
	Metal decking to mansard; excluding frame; 50 mm insulation		
	two coat asphalt coverings and chippings on decking;		
2.3.73	natural slate covering to mansard faces	m2	- -
	Roof claddings		
	Non-asbestos profiled cladding		
2.3.74	'profile 3'; natural	m2	- -
2.3.75	'profile 3'; coloured	m2	- -
2.3.76	'profile 6'; natural	m2	- -
2.3.77	'profile 6'; coloured	m2	- -
2.3.78	'profile 6'; natural; insulated; inner lining panel	m2	- -
	Non-asbestos profiled cladding on steel purlins		
2.3.79	insulated	m2	- -
2.3.80	insulated; with 10% transluscent sheets	m2	- -
2.3.81	insulated; plasterboard inner lining on metal tees	m2	- -
	Asbestos cement profiled cladding on steel purlins		
2.3.82	insulated	m2	- -
2.3.83	insulated; with 10% transluscent sheets	m2	- -
2.3.84	insulated; plasterboard inner lining on metal tees	m2	- -
2.3.85	insulated; steel sheet liner on metal tees	m2	- -
	PVF2 coated galvanised steel profiled cladding		
2.3.86	0.72 mm thick; 'profile 20B'	m2	- -
2.3.87	0.72 mm thick; 'profile TOP 40'	m2	- -
2.3.88	0.72 mm thick; 'profile 45'	m2	- -
	Extra for		
	80 mm insulation and 0.4 mm thick coated inner lining		
2.3.89	sheet	m2	- -
	PVF2 coated galvanised steel profiled cladding on steel purlins		
2.3.90	insulated	m2	- -
2.3.91	insulated; plasterboard inner lining on metal tees	m2	- -
	insulated; plasterboard inner lining on metal tees;		
2.3.92	with 1% fire vents	m2	- -
	insulated; plasterboard inner lining on metal tees;		
2.3.93	with 2.5% fire vents	m2	- -
2.3.94	insulated; coloured inner lining panel	m2	- -
	insulated; coloured inner lining panel; with 1%		
2.3.95	fire vents	m2	- -
	insulated; coloured inner lining panel; with		
2.3.96	2.5% fire vents	m2	- -
2.3.97	insulated; sandwich panel	m2	- -

INDUSTRIAL £ range £	RETAILING £ range £	LEISURE £ range £	OFFICES £ range £	HOTELS £ range £	Item nr.
35.00 - 50.00	37.50 - 47.50	40.00 - 50.00	35.00 - 45.00	40.00 - 50.00	2.3.62
40.00 - 50.00	40.00 - 50.00	40.00 - 50.00	40.00 - 50.00	40.00 - 50.00	2.3.63
41.00 - 51.00	41.00 - 51.00	41.00 - 51.00	41.00 - 51.00	41.00 - 51.00	2.3.64
42.00 - 52.00	42.00 - 52.00	42.00 - 52.00	42.00 - 52.00	42.00 - 52.00	2.3.65
43.00 - 53.00	43.00 - 53.00	43.00 - 53.00	43.00 - 53.00	43.00 - 53.00	2.3.66
48.00 - 58.00	48.00 - 58.00	48.00 - 58.00	48.00 - 58.00	48.00 - 58.00	2.3.67
50.00 - 60.00	50.00 - 60.00	50.00 - 60.00	50.00 - 60.00	50.00 - 60.00	2.3.68
72.00 - 87.50	77.00 - 92.50	77.00 - 92.50	77.00 - 92.50	77.00 - 95.00	2.3.69
75.00 - 92.50	80.00 - 92.50	80.00 - 97.50	80.00 - 97.50	80.00 -100.00	2.3.70
70.00 - 95.00	75.00 -100.00	75.00 -100.00	75.00 -100.00	80.00 -105.00	2.3.71
5.00 - 8.00	5.00 - 8.00	5.00 - 8.00	5.00 - 8.00	5.00 - 8.00	2.3.72
-	130.00 -215.00	120.00 -200.00	120.00 -200.00	130.00 -230.00	2.3.73
13.00 - 15.00	-	-	-	-	2.3.74
14.00 - 16.00	-	-	-	-	2.3.75
14.00 - 16.00	-	-	-	-	2.3.76
15.00 - 17.00	-	-	-	-	2.3.77
28.00 - 33.00	-	-	-	-	2.3.78
26.00 - 30.00	-	-	-	-	2.3.79
29.00 - 33.00	-	-	-	-	2.3.80
40.00 - 48.00	-	-	-	-	2.3.81
25.00 - 29.00	-	-	-	-	2.3.82
28.00 - 32.00	-	-	-	-	2.3.83
39.00 - 47.00	-	-	-	-	2.3.84
38.00 - 46.00	-	-	-	-	2.3.85
17.50 - 22.00	22.50 - 30.00	22.50 - 30.00	-	-	2.3.86
17.00 - 21.00	22.00 - 29.00	22.00 - 29.00	-	-	2.3.87
19.00 - 24.00	24.00 - 32.50	24.00 - 32.50	-	-	2.3.88
11.00 - 12.00	11.00 - 12.00	11.00 - 12.00	-	-	2.3.89
30.00 - 40.00	35.00 - 45.00	35.00 - 45.00	-	-	2.3.90
44.00 - 58.00	50.00 - 62.00	50.00 - 62.00	-	-	2.3.91
55.00 - 70.00	60.00 - 75.00	60.00 - 75.00	-	-	2.3.92
65.00 - 80.00	70.00 - 85.00	70.00 - 85.00	-	-	2.3.93
44.00 - 57.00	50.00 - 65.00	50.00 - 65.00	-	-	2.3.94
55.00 - 68.00	60.00 - 73.00	60.00 - 73.00	-	-	2.3.95
65.00 - 80.00	70.00 - 85.00	70.00 - 85.00	-	-	2.3.96
100.00 -150.00	120.00 -170.00	120.00 -170.00	-	-	2.3.97

Item nr.	SPECIFICATIONS	Unit	RESIDENTIAL £ range £

2.3 **ROOF - cont'd**   **roof plan area**
                                        (unless otherwise described)

Item nr.	SPECIFICATIONS	Unit	£ range £	
	Roof claddings - cont'd			
	Pre-painted 'Rigidal' aluminium profiled cladding			
2.3.98	0.7 mm thick; type WA6	m2	-	-
2.3.99	0.9 mm thick; type A7	m2	-	-
	PVF2 coated aluminium profiled cladding on steel purlins			
2.3.100	insulated; plasterboard inner lining on metal tees	m2	-	-
2.3.101	insulated; coloured inner lining panel	m2	-	-
	Rooflights/patent glazing and glazed roofs			
	Rooflights			
2.3.102	standard pvc	m2	-	-
2.3.103	feature/ventilating	m2	-	-
	Patent glazing; including flashings			
	standard aluminium georgian wired			
2.3.104	single glazed	m2	-	-
2.3.105	double glazed	m2	-	-
	purpose made polyester powder coated aluminium;			
2.3.106	double glazed low emissivity glass	m2	-	-
2.3.107	feature; to covered walkways	m2	-	-
	Glazed roofing on framing; to covered walkways			
2.3.108	feature; single glazed	m2	-	-
2.3.109	feature; double glazed barrel vault	m2	-	-
2.3.110	feature; very expensive	m2	-	-
	Comparative over/underlays			
	Roofing felt; unreinforced			
2.3.111	sloping (measured on face)	m2	1.20 -	1.50
	Roofing felt; reinforced			
2.3.112	sloping (measured on face)	m2	1.50 -	1.80
	sloping (measured on plan)			
2.3.113	20° pitch	m2	1.80 -	2.20
2.3.114	30° pitch	m2	2.00 -	2.40
2.3.115	35° pitch	m2	2.40 -	2.60
2.3.116	40° pitch	m2	2.50 -	2.70
2.3.117	Building paper	m2	2.00 -	2.50
2.3.118	Vapour barrier	m2	3.00 -	4.25
	Insulation quilt; laid over ceiling joists			
2.3.119	60 mm thick	m2	2.30 -	2.60
2.3.120	80 mm thick	m2	2.80 -	3.10
2.3.121	100 mm thick	m2	3.25 -	3.50
2.3.122	150 mm thick	m2	4.60 -	5.00
2.3.123	200 mm thick	m2	5.75 -	6.25
	Wood fibre insulation boards; impregnated; density 220 - 350 kg/m3			
2.3.124	12.7 mm thick	m2	-	-
	Polystyrene insulation boards; fixed vertically with adhesive			
2.3.125	12 mm thick	m2	-	-
2.3.126	25 mm thick	m2	-	-
2.3.127	50 mm thick	m2	-	-
2.3.128	Ballast	m2	-	-
	Polyurethene insulation boards; density 32 kg/m3			
2.3.129	30 mm thick	m2	-	-
2.3.130	35 mm thick	m2	-	-
2.3.131	50 mm thick	m2	-	-
	Cork insulation boards; density 112 - 125 kg/m3			
2.3.132	60 mm thick	m2	-	-
	Glass fibre insulation boards; density 120 - 130 kg/m2			
2.3.133	60 mm thick	m2	-	-
	Extruded polystyrene foam boards			
2.3.134	50 mm thick	m2	-	-
2.3.135	50 mm thick; with cement topping	m2	-	-
2.3.136	75 mm thick	m2	-	-
	Perlite insulation board; density 170 - 180 kg/m3			
2.3.137	60 mm thick	m2	-	-
	Foam glass insulation board; density 125 - 135 kg/m2			
2.3.138	60 mm thick	m2	-	-

INDUSTRIAL £ range £	RETAILING £ range £	LEISURE £ range £	OFFICES £ range £	HOTELS £ range £	Item nr.
22.50 - 27.50	27.50 - 35.00	27.50 - 35.00	-	-	-	-	-	-	2.3.98
27.50 - 32.50	32.50 - 40.00	32.50 - 40.00	-	-	-	-	-	-	2.3.99
48.00 - 58.00	53.00 - 65.00	53.00 - 65.00	-	-	-	-	-	-	2.3.100
50.00 - 60.00	57.00 - 68.00	57.00 - 68.00	-	-	-	-	-	-	2.3.101
100.00 -180.00	100.00 -180.00	100.00 -180.00	100.00 -180.00	100.00 -180.00	2.3.102
-	-	180.00 -320.00	180.00 -320.00	180.00 -320.00	180.00 -320.00	2.3.103
120.00 -170.00	130.00 -180.00	130.00 -180.00	130.00 -180.00	140.00 -200.00	2.3.104
150.00 -190.00	160.00 -200.00	160.00 -200.00	160.00 -200.00	170.00 -220.00	2.3.105
180.00 -210.00	190.00 -220.00	190.00 -220.00	190.00 -220.00	200.00 -250.00	2.3.106
-	-	200.00 -350.00	200.00 -350.00	200.00 -350.00	200.00 -375.00	2.3.107
-	-	275.00 -400.00	275.00 -400.00	275.00 -400.00	300.00 -450.00	2.3.108
-	-	400.00 -600.00	400.00 -600.00	400.00 -600.00	400.00 -650.00	2.3.109
-	-	600.00 -750.00	600.00 -750.00	600.00 -750.00	600.00 -800.00	2.3.110
-	-	1.20 - 1.50	-	-	1.20 - 1.50	1.20 - 1.50	2.3.111
-	-	1.50 - 1.80	-	-	1.50 - 1.80	1.50 - 1.80	2.3.112
-	-	1.80 - 2.20	-	-	1.80 - 2.20	1.80 - 2.20	2.3.113
-	-	2.00 - 2.40	-	-	2.00 - 2.40	2.00 - 2.40	2.3.114
-	-	2.40 - 2.60	-	-	2.40 - 2.60	2.40 - 2.60	2.3.115
-	-	2.50 - 2.70	-	-	2.50 - 2.70	2.50 - 2.70	2.3.116
-	-	2.00 - 2.50	-	-	2.00 - 2.50	2.00 - 2.50	2.3.117
-	-	3.00 - 4.25	-	-	3.00 - 4.25	3.00 - 4.25	2.3.118
-	-	2.30 - 2.60	-	-	2.30 - 2.60	2.30 - 2.60	2.3.119
-	-	2.80 - 3.10	-	-	2.80 - 3.10	2.80 - 3.10	2.3.120
-	-	3.25 - 3.50	-	-	3.25 - 3.50	3.25 - 3.50	2.3.121
-	-	4.60 - 5.00	-	-	4.60 - 5.00	4.60 - 5.00	2.3.122
-	-	5.75 - 6.25	-	-	5.75 - 6.25	5.75 - 6.25	2.3.123
-	-	4.50 - 6.00	-	-	-	-	-	-	2.3.124
5.00 - 6.00	5.00 - 6.00	5.00 - 6.00	5.00 - 6.00	5.00 - 6.00	2.3.125
6.00 - 7.00	6.00 - 7.00	6.00 - 7.00	6.00 - 7.00	6.00 - 7.00	2.3.126
7.00 - 8.00	7.00 - 8.00	7.00 - 8.00	7.00 - 8.00	7.00 - 8.00	2.3.127
5.00 - 8.00	5.00 - 8.00	5.00 - 8.00	5.00 - 8.00	5.00 - 8.00	2.3.128
7.00 - 8.00	7.00 - 8.00	7.00 - 8.00	7.00 - 8.00	7.00 - 8.00	2.3.129
8.00 - 9.00	8.00 - 9.00	8.00 - 9.00	8.00 - 9.00	8.00 - 9.00	2.3.130
10.00 - 11.00	10.00 - 11.00	10.00 - 11.00	10.00 - 11.00	10.00 - 11.00	2.3.131
9.00 - 12.00	9.00 - 12.00	9.00 - 12.00	9.00 - 12.00	9.00 - 12.00	2.3.132
12.00 - 14.00	12.00 - 14.00	12.00 - 14.00	12.00 - 14.00	12.00 - 14.00	2.3.133
12.00 - 14.00	12.00 - 14.00	12.00 - 14.00	12.00 - 14.00	12.00 - 14.00	2.3.134
17.50 - 20.00	17.50 - 20.00	17.50 - 20.00	17.50 - 20.00	17.50 - 20.00	2.3.135
15.00 - 18.00	15.00 - 18.00	15.00 - 18.00	15.00 - 18.00	15.00 - 18.00	2.3.136
12.00 - 14.00	12.00 - 14.00	12.00 - 14.00	12.00 - 14.00	12.00 - 14.00	2.3.137
16.50 - 19.00	16.50 - 19.00	16.50 - 19.00	16.50 - 19.00	16.50 - 19.00	2.3.138

Item nr.	SPECIFICATIONS	Unit	RESIDENTIAL £ range £
2.3	**ROOF - cont'd** roof plan area (unless otherwise described)		
	Comparative over/underlays - cont'd Screeds to receive roof coverings		
2.3.139	50 mm cement and sand screed	m2	7.00 - 8.00
2.3.140	60 mm (av.) 'Isocrete K' screed; density 500 kg	m2	7.00 - 8.00
2.3.141	75 mm lightweight bituminous screed and vapour barrier	m2	16.00 - 18.00
2.3.142	100 mm lightweight bituminous screed and vapour barrier	m2	18.00 - 20.00
	50 mm Woodwool slabs; unreinforced		
2.3.143	sloping (measured on face)	m2	9.00 - 11.00
	sloping (measured on plan)		
2.3.144	20° pitch	m2	10.00 - 12.00
2.3.145	30° pitch	m2	12.50 - 14.50
2.3.146	35° pitch	m2	13.70 - 15.70
2.3.147	40° pitch	m2	14.20 - 16.20
2.3.148	50 mm Woodwool slabs; unreinforced; on and including steel purlins @ 600 mm centres	m2	15.00 - 18.00
	19 mm 'Tanalised' softwood boarding		
2.3.149	sloping (measured on face)	m2	10.00 - 12.00
	sloping (measured on plan)		
2.3.150	20° pitch	m2	11.00 - 13.00
2.3.151	30° pitch	m2	13.50 - 15.50
2.3.152	35° pitch	m2	15.00 - 17.00
2.3.153	40° pitch	m2	15.50 - 17.50
	25 mm 'Tanalised' softwood boarding		
2.3.154	sloping (measured on face)	m2	11.50 - 13.50
	sloping (measured on plan)		
2.3.155	20° pitch	m2	12.75 - 14.75
2.3.156	30° pitch	m2	15.50 - 17.50
2.3.157	35° pitch	m2	17.30 - 19.30
2.3.158	40° pitch	m2	18.00 - 20.00
	18 mm External quality plywood boarding		
2.3.159	sloping (measured on face)	m2	15.00 - 18.00
	sloping (measured on plan)		
2.3.160	20° pitch	m2	16.50 - 19.50
2.3.161	30° pitch	m2	20.40 - 23.50
2.3.162	35° pitch	m2	23.00 - 26.00
2.3.163	40° pitch	m2	23.50 - 27.00
	Comparative tiling and slating finishes/perimeter treatments (including underfelt, battening, eaves courses and ridges) Concrete troughed interlocking tiles; 413 x 300 mm; 75 mm lap		
2.3.164	sloping (measured on face)	m2	14.50 - 17.50
	sloping (measured on plan)		
2.3.165	30° pitch	m2	19.00 - 22.00
2.3.166	35° pitch	m2	21.00 - 24.00
2.3.167	40° pitch	m2	22.00 - 25.00
	Concrete interlocking slates; 430 x 330 mm; 75 mm lap		
2.3.168	sloping (measured on face)	m2	14.90 - 17.90
	sloping (measured on plan)		
2.3.169	30° pitch	m2	19.50 - 22.50
2.3.170	35° pitch	m2	21.60 - 24.60
2.3.171	40° pitch	m2	22.30 - 25.30
	Glass fibre reinforced bitumen slates; 900 x 300 mm; fixed to boarding (measured elsewhere)		
2.3.172	sloping (measured on face)	m2	15.00 - 18.00
	sloping (measured on plan)		
2.3.173	30° pitch	m2	19.00 - 22.00
2.3.174	35° pitch	m2	21.30 - 24.30
2.3.175	40° pitch	m2	22.00 - 25.00

INDUSTRIAL £ range £	RETAILING £ range £	LEISURE £ range £	OFFICES £ range £	HOTELS £ range £	Item nr.
-	7.00 - 8.00	-	7.00 - 8.00	7.00 - 8.00	2.3.139
-	7.00 - 8.00	-	7.00 - 8.00	7.00 - 8.00	2.3.140
-	16.00 - 18.00	-	16.00 - 18.00	16.00 - 18.00	2.3.141
-	18.00 - 20.00	-	18.00 - 20.00	18.00 - 20.00	2.3.142
9.00 - 11.00	-	-	9.00 - 11.00	-	2.3.143
10.00 - 12.00	-	-	10.00 - 12.00	-	2.3.144
12.50 - 14.50	-	-	12.50 - 14.50	-	2.3.145
13.70 - 15.70	-	-	13.70 - 15.70	-	2.3.146
14.20 - 16.20	-	-	14.20 - 16.20	-	2.3.147
15.00 - 18.00	-	-	15.00 - 18.00	-	2.3.148
-	10.00 - 12.00	-	10.00 - 12.00	10.00 - 12.00	2.3.149
-	11.00 - 13.00	-	11.00 - 13.00	11.00 - 13.00	2.3.150
-	13.50 - 15.50	-	13.50 - 15.50	13.50 - 15.50	2.3.151
-	15.00 - 17.00	-	15.00 - 17.00	15.00 - 17.00	2.3.152
-	15.50 - 17.50	-	15.50 - 17.50	15.50 - 17.50	2.3.153
-	11.50 - 13.50	-	11.50 - 13.50	11.50 - 13.50	2.3.154
-	12.75 - 14.75	-	12.75 - 14.75	12.75 - 14.75	2.3.155
-	15.50 - 17.50	-	15.50 - 17.50	15.50 - 17.50	2.3.156
-	17.30 - 19.30	-	17.30 - 19.30	17.30 - 19.30	2.3.157
-	18.00 - 20.00	-	18.00 - 20.00	18.00 - 20.00	2.3.158
-	15.00- 18.00	-	15.00 - 18.00	15.00 - 18.00	2.3.159
-	16.50 - 19.50	-	16.50 - 19.50	16.50 - 19.50	2.3.160
-	20.40 - 23.50	-	20.40 - 23.50	20.40 - 23.50	2.3.161
-	23.00 - 26.00	-	23.00 - 26.00	23.00 - 26.00	2.3.162
-	23.50 - 27.00	-	23.50 - 27.00	23.50 - 27.00	2.3.163
-	14.50 - 17.50	-	14.50 - 17.50	14.50 - 17.50	2.3.164
-	19.00 - 22.00	-	19.00 - 22.00	19.00 - 22.00	2.3.165
-	21.00 - 24.00	-	21.00 - 24.00	21.00 - 24.00	2.3.166
-	22.00 - 25.00	-	22.00 - 25.00	22.00 - 25.00	2.3.167
-	14.90 - 17.90	-	14.90 - 17.90	14.90 - 17.90	2.3.168
-	19.50 - 22.50	-	19.50 - 22.50	19.50 - 22.50	2.3.169
-	21.60 - 24.60	-	21.60 - 24.60	21.60 - 24.60	2.3.170
-	22.30 - 25.30	-	22.30 - 25.30	22.30 - 25.30	2.3.171
-	15.00 - 18.00	-	15.00 - 18.00	15.00 - 18.00	2.3.172
-	19.00 - 22.00	-	19.00 - 22.00	19.00 - 22.00	2.3.173
-	21.30 - 24.30	-	21.30 - 24.30	21.30 - 24.30	2.3.174
-	22.00 - 25.00	-	22.00 - 25.00	22.00 - 25.00	2.3.175

Item nr.	SPECIFICATIONS	Unit	RESIDENTIAL £ range £

2.3   **ROOF - cont'd**   **roof plan area**
                                        (unless otherwise described)

Comparative tiling and slating finishes/perimeter treatments - cont'd

Concrete bold roll interlocking tiles; 418 x 332 mm; 75 mm lap

Item nr.	SPECIFICATIONS	Unit	£ range £
2.3.176	sloping (measured on face)	m2	14.50 - 17.50
	sloping (measured on plan)		
2.3.177	30° pitch	m2	19.00 - 22.00
2.3.178	35° pitch	m2	21.00 - 24.00
2.3.179	40° pitch	m2	21.50 - 24.50
	Tudor clay pantiles; 470 x 285 mm; 100 mm lap		
2.3.180	sloping (measured on face)	m2	22.00 - 27.00
	sloping (measured on plan)		
2.3.181	30° pitch	m2	28.50 - 33.50
2.3.182	35° pitch	m2	31.25 - 36.25
2.3.183	40° pitch	m2	32.00 - 37.00
	Natural red pantiles; 337 x 241 mm; 76 mm head and 38 mm side laps		
2.3.184	sloping (measured on face)	m2	24.00 - 29.00
	sloping (measured on plan)		
2.3.185	30° pitch	m2	31.00 - 36.00
2.3.186	35° pitch	m2	34.00 - 39.00
2.3.187	40° pitch	m2	35.00 - 40.00
	Blue asbestos cement slates; 600 x 300 mm; 75 mm lap		
2.3.188	sloping (measured on face)	m2	24.50 - 29.50
	sloping (measured on plan)		
2.3.189	30° pitch	m2	32.00 - 37.00
2.3.190	35° pitch	m2	35.00 - 40.00
2.3.191	40° pitch	m2	36.00 - 41.00
	Blue composition (non-asbestos) slates; 600 x 300 mm; 75 mm lap		
2.3.192	sloping (measured on face)	m2	25.00 - 30.00
2.3.193	sloping to mansard (measured on face)	m2	35.00 - 40.00
	sloping (measured on plan)		
2.3.194	30° pitch	m2	33.00 - 38.00
2.3.195	35° pitch	m2	36.00 - 41.00
2.3.196	40° pitch	m2	37.00 - 42.00
2.3.197	vertical to mansard; including 18 mm blockboard (measured on face)	m2	50.00 - 60.00
	Concrete plain tiles; 267 x 165 mm; 64 mm lap		
2.3.198	sloping (measured on face)	m2	30.00 - 35.00
	sloping (measured on plan)		
2.3.199	30° pitch	m2	39.00 - 44.00
2.3.200	35° pitch	m2	43.00 - 48.00
2.3.201	40° pitch	m2	45.00 - 50.00
	Machine made clay plain tiles; 267 x 165 mm; 64 mm lap		
2.3.202	sloping (measured on face)	m2	34.00 - 39.00
	sloping (measured on plan)		
2.3.203	30° pitch	m2	45.00 - 50.00
2.3.204	35° pitch	m2	49.00 - 54.00
2.3.205	40° pitch	m2	51.00 - 56.00
	Black 'Sterreberg' glazed interlocking pantiles; 355 x 240 mm; 76 mm head and 38 mm side laps		
2.3.206	sloping (measured on face)	m2	38.50 - 46.00
	sloping (measured on plan)		
2.3.207	30° pitch	m2	50.00 - 57.50
2.3.208	35° pitch	m2	55.00 - 62.50
2.3.209	40° pitch	m2	56.50 - 64.00
	Red cedar sawn shingles; 450 mm long; 125 mm lap		
2.3.210	sloping (measured on face)	m2	41.00 - 49.00
	sloping (measured on plan)		
2.3.211	30° pitch	m2	54.00 - 62.00
2.3.212	35° pitch	m2	59.00 - 67.00
2.3.213	40° pitch	m2	61.00 - 69.00

INDUSTRIAL £ range £	RETAILING £ range £	LEISURE £ range £	OFFICES £ range £	HOTELS £ range £	Item nr.
- -	14.50 - 17.50	- -	14.50 - 17.50	14.50 - 17.50	2.3.176
- -	19.00 - 22.00	- -	19.00 - 22.00	19.00 - 22.00	2.3.177
- -	21.00 - 24.00	- -	21.00 - 24.00	21.00 - 24.00	2.3.178
- -	21.50 - 24.50	- -	21.50 - 24.50	21.50 - 24.50	2.3.179
- -	22.00 - 27.00	- -	22.00 - 27.00	22.00 - 27.00	2.3.180
- -	28.50 - 33.50	- -	28.50 - 33.50	28.50 - 33.50	2.3.181
- -	31.25 - 36.25	- -	31.25 - 36.25	31.25 - 36.25	2.3.182
- -	32.00 - 37.00	- -	32.00 - 37.00	32.00 - 37.00	2.3.183
- -	24.00 - 29.00	- -	24.00 - 29.00	24.00 - 29.00	2.3.184
- -	31.00 - 36.00	- -	31.00 - 36.00	31.00 - 36.00	2.3.185
- -	34.00 - 39.00	- -	34.00 - 39.00	34.00 - 39.00	2.3.186
- -	35.00 - 40.00	- -	35.00 - 40.00	35.00 - 40.00	2.3.187
- -	24.50 - 29.50	- -	24.50 - 29.50	24.50 - 29.50	2.3.188
- -	32.00 - 37.00	- -	32.00 - 37.00	32.00 - 37.00	2.3.189
- -	35.00 - 40.00	- -	35.00 - 40.00	35.00 - 40.00	2.3.190
- -	36.00 - 41.00	- -	36.00 - 41.00	36.00 - 41.00	2.3.191
- -	25.00 - 30.00	- -	25.00 - 30.00	25.00 - 30.00	2.3.192
- -	35.00 - 40.00	- -	35.00 - 40.00	35.00 - 40.00	2.3.193
- -	33.00 - 38.00	- -	33.00 - 38.00	33.00 - 38.00	2.3.194
- -	36.00 - 41.00	- -	36.00 - 41.00	36.00 - 41.00	2.3.195
- -	37.00 - 42.00	- -	37.00 - 42.00	37.00 - 42.00	2.3.196
- -	50.00 - 60.00	- -	50.00 - 60.00	50.00 - 60.00	2.3.197
- -	30.00 - 35.00	- -	30.00 - 35.00	30.00 - 35.00	2.3.198
- -	39.00 - 44.00	- -	39.00 - 44.00	39.00 - 44.00	2.3.199
- -	43.00 - 48.00	- -	43.00 - 48.00	43.00 - 48.00	2.3.200
- -	45.00 - 50.00	- -	45.00 - 50.00	45.00 - 50.00	2.3.201
- -	34.00 - 39.00	- -	34.00 - 39.00	34.00 - 39.00	2.3.202
- -	45.00 - 50.00	- -	45.00 - 50.00	45.00 - 50.00	2.3.203
- -	49.00 - 54.00	- -	49.00 - 54.00	49.00 - 54.00	2.3.204
- -	51.00 - 56.00	- -	51.00 - 56.00	51.00 - 56.00	2.3.205
- -	38.50 - 46.00	- -	38.50 - 46.00	38.50 - 46.00	2.3.206
- -	50.00 - 57.50	- -	50.00 - 57.50	50.00 - 57.50	2.3.207
- -	55.00 - 62.50	- -	55.00 - 62.50	55.00 - 62.50	2.3.208
- -	56.50 - 64.00	- -	56.50 - 64.00	56.50 - 64.00	2.3.209
- -	41.00 - 49.00	- -	41.00 - 49.00	41.00 - 49.00	2.3.210
- -	54.00 - 62.00	- -	54.00 - 62.00	54.00 - 62.00	2.3.211
- -	59.00 - 67.00	- -	59.00 - 67.00	59.00 - 67.00	2.3.212
- -	61.00 - 69.00	- -	61.00 - 69.00	61.00 - 69.00	2.3.213

Item nr.	SPECIFICATIONS	Unit	RESIDENTIAL £ range £
2.3	ROOF - cont'd	roof plan area (unless otherwise described)	
	Comparative tiling and slating finishes/perimeter treatments - cont'd		
	Welsh natural slates; 510 x 255 mm; 76 mm lap		
2.3.214	sloping (measured on face)	m2	42.00 - 49.00
	sloping (measured on plan)		
2.3.215	30° pitch	m2	55.00 - 64.00
2.3.216	35° pitch	m2	61.00 - 68.00
2.3.217	40° pitch	m2	63.00 - 70.00
	Welsh slates; 610 x 305 mm; 76 mm lap		
2.3.218	sloping (measured on face)	m2	43.00 - 50.00
	sloping (measured on plan)		
2.3.219	30° pitch	m2	56.00 - 63.00
2.3.220	35° pitch	m2	62.00 - 69.00
2.3.221	40° pitch	m2	64.00 - 71.00
	Reconstructed stone slates; random lengths; 80 mm lap		
2.3.222	sloping (measured on face)	m2	43.00 - 50.00
	sloping (measured on plan)		
2.3.223	30° pitch	m2	57.00 - 64.00
2.3.224	35° pitch	m2	62.50 - 70.00
2.3.225	40° pitch	m2	64.50 - 72.00
	Handmade sandfaced plain tiles; 267 x 165 mm; 64 mm lap		
2.3.226	sloping (measured on face)	m2	53.00 - 60.00
	sloping (measured on plan)		
2.3.227	30° pitch	m2	70.00 - 77.00
2.3.228	35° pitch	m2	78.00 - 85.00
2.3.229	40° pitch	m2	81.00 - 88.00
	Westmorland green slates; random sizes; 76 mm lap		
2.3.230	sloping (measured on face)	m2	100.00 -110.00
	sloping (measured on plan)		
2.3.231	30° pitch	m2	135.00 -145.00
2.3.232	35° pitch	m2	150.00 -160.00
2.3.233	40° pitch	m2	155.00 -165.00
	Verges to sloping roofs; 250 x 25 mm **perimeter length** painted softwood bargeboard		
2.3.234	6 mm 'Masterboard' soffit lining 150 mm wide	m	14.00 - 16.00
2.3.235	19 x 150 mm painted softwood soffit	m	16.00 - 18.00
	Eaves to sloping roofs; 200 x 25 mm painted softwood fascia; 6 mm 'Masterboard' soffit lining 225 mm wide		
2.3.236	100 mm PVC gutter	m	19.00 - 25.00
2.3.237	150 mm PVC gutter	m	24.00 - 30.00
2.3.238	100 mm cast iron gutter; decorated	m	30.00 - 35.00
2.3.239	150 mm cast iron gutter; decorated	m	36.00 - 42.00
	Eaves to sloping roofs; 200 x 25 mm painted softwood fascia; 19 x 225 mm painted softwood soffit		
2.3.240	100 mm PVC gutter	m	22.00 - 28.00
2.3.241	150 mm PVC gutter	m	27.00 - 33.00
2.3.242	100 mm cast iron gutter; decorated	m	32.50 - 37.50
2.3.243	150 mm cast iron gutter; decorated	m	39.00 - 45.00
	Rainwater pipes; fixed to backgrounds; including offsets and shoe		
2.3.244	75 mm PVC	m	8.00 - 10.00
2.3.245	100 mm PVC	m	10.00 - 12.00
2.3.246	75 mm cast iron; decorated	m	23.00 - 27.00
2.3.247	100 mm cast iron; decorated	m	28.00 - 32.00
	Ridges		
2.3.248	concrete half round tiles	m	13.00 - 16.00
2.3.249	machine-made clay half round tiles	m	15.00 - 18.00
2.3.250	hand-made clay half round tiles	m	16.00 - 25.00
	Hips; including mitreing roof tiles		
2.3.251	concrete half round tiles	m	17.00 - 22.00
2.3.252	machine-made clay half round tiles	m	19.00 - 24.00
2.3.253	hand-made clay half round tiles	m	20.00 - 30.00
2.3.254	hand-made clay bonnet hip tiles	m	35.00 - 45.00

INDUSTRIAL £ range £	RETAILING £ range £	LEISURE £ range £	OFFICES £ range £	HOTELS £ range £	Item nr.
-	42.00 - 49.00	-	42.00 - 49.00	42.00 - 49.00	2.3.214
-	55.00 - 64.00	-	55.00 - 64.00	55.00 - 64.00	2.3.215
-	61.00 - 68.00	-	61.00 - 68.00	61.00 - 68.00	2.3.216
-	63.00 - 70.00	-	63.00 - 70.00	63.00 - 70.00	2.3.217
-	43.00 - 50.00	-	43.00 - 50.00	43.00 - 50.00	2.3.218
-	56.00 - 63.00	-	56.00 - 63.00	56.00 - 63.00	2.3.219
-	62.00 - 69.00	-	62.00 - 69.00	62.00 - 69.00	2.3.220
-	64.00 - 71.00	-	64.00 - 71.00	64.00 - 71.00	2.3.221
-	43.00 - 50.00	-	43.00 - 50.00	43.00 - 50.00	2.3.222
-	57.00 - 64.00	-	57.00 - 64.00	57.00 - 64.00	2.3.223
-	62.50 - 70.00	-	62.50 - 70.00	62.50 - 70.00	2.3.224
-	64.50 - 72.00	-	64.50 - 72.00	64.50 - 72.00	2.3.225
-	53.00 - 60.00	-	53.00 - 60.00	53.00 - 60.00	2.3.226
-	70.00 - 77.00	-	70.00 - 77.00	70.00 - 77.00	2.3.227
-	78.00 - 85.00	-	78.00 - 85.00	78.00 - 85.00	2.3.228
-	81.00 - 88.00	-	81.00 - 88.00	81.00 - 88.00	2.3.229
-	100.00 -110.00	-	100.00 -110.00	100.00 -110.00	2.3.230
-	135.00 -145.00	-	135.00 -145.00	135.00 -145.00	2.3.231
-	150.00 -160.00	-	150.00 -160.00	150.00 -160.00	2.3.232
-	155.00 -165.00	-	155.00 -165.00	155.00 -165.00	2.3.233
-	14.00 - 16.00	-	14.00 - 16.00	14.00 - 16.00	2.3.234
-	16.00 - 18.00	-	16.00 - 18.00	16.00 - 18.00	2.3.235
-	19.00 - 25.00	-	19.00 - 25.00	19.00 - 25.00	2.3.236
-	24.00 - 30.00	-	24.00 - 30.00	24.00 - 30.00	2.3.237
-	30.00 - 35.00	-	30.00 - 35.00	30.00 - 35.00	2.3.238
-	36.00 - 42.00	-	36.00 - 42.00	36.00 - 42.00	2.3.239
-	22.00 - 28.00	-	22.00 - 28.00	22.00 - 28.00	2.3.240
-	27.00 - 33.00	-	27.00 - 33.00	27.00 - 33.00	2.3.241
-	32.50 - 37.50	-	32.50 - 37.50	32.50 - 37.50	2.3.242
-	39.00 - 45.00	-	39.00 - 45.00	39.00 - 45.00	2.3.243
-	8.00 - 10.00	-	8.00 - 10.00	8.00 - 10.00	2.3.244
-	10.00 - 12.00	-	10.00 - 12.00	10.00 - 12.00	2.3.245
-	23.00 - 27.00	-	23.00 - 27.00	23.00 - 27.00	2.3.246
-	28.00 - 32.00	-	28.00 - 32.00	28.00 - 32.00	2.3.247
-	13.00 - 16.00	-	13.00 - 16.00	13.00 - 16.00	2.3.248
-	15.00 - 18.00	-	15.00 - 18.00	15.00 - 18.00	2.3.249
-	16.00 - 25.00	-	16.00 - 25.00	16.00 - 25.00	2.3.250
-	17.00 - 22.00	-	17.00 - 22.00	17.00 - 22.00	2.3.251
-	19.00 - 24.00	-	19.00 - 24.00	19.00 - 24.00	2.3.252
-	20.00 - 30.00	-	20.00 - 30.00	20.00 - 30.00	2.3.253
-	35.00 - 45.00	-	35.00 - 45.00	35.00 - 45.00	2.3.254

Item nr.	SPECIFICATIONS	Unit	RESIDENTIAL £ range £
2.3	ROOF - cont'd	roof plan area (unless otherwise described)	
	Comparative cladding finishes **(including underfelt, labours etc.)**		
	0.91 mm Aluminium roofing; commercial grade		
2.3.255	flat	m2	- -
	0.91 mm Aluminium roofing; commercial grade; fixed to boarding (included)		
2.3.256	sloping (measured on face)	m2	- -
	sloping (measured on plan)		
2.3.257	20° pitch	m2	- -
2.3.258	30° pitch	m2	- -
2.3.259	35° pitch	m2	- -
2.3.260	40° pitch	m2	- -
	0.81 mm Zinc roofing		
2.3.261	flat	m2	- -
	0.81 mm Zinc roofing; fixed to boarding (included)		
2.3.262	sloping (measured on face)	m2	- -
	sloping (measured on plan)		
2.3.263	20° pitch	m2	- -
2.3.264	30° pitch	m2	- -
2.3.265	35° pitch	m2	- -
2.3.266	40° pitch	m2	- -
	Copper roofing		
2.3.267	0.56 mm thick; flat	m2	- -
2.3.268	0.61 mm thick; flat	m2	- -
	Copper roofing; fixed to boarding (included)		
2.3.269	0.56 mm thick; sloping (measured on face)	m2	- -
	0.56 mm thick; sloping (measured on plan)		
2.3.270	20° pitch	m2	- -
2.3.271	30° pitch	m2	- -
2.3.272	35° pitch	m2	- -
2.3.273	40° pitch	m2	- -
2.3.274	0.61 mm thick; sloping (measured on face)	m2	- -
	0.61 mm thick; sloping (measured on plan)		
2.3.275	20° pitch	m2	- -
2.3.276	30° pitch	m2	- -
2.3.277	35° pitch	m2	- -
2.3.278	40° pitch	m2	- -
	Lead roofing		
2.3.279	code 4 sheeting; flat	m2	- -
2.3.280	code 5 sheeting; flat	m2	- -
2.3.281	code 6 sheeting; flat	m2	- -
	Lead roofing; fixed to boarding (included)		
2.3.282	code 4 sheeting; sloping (measured on face)	m2	- -
	code 4 sheeting; sloping (measured on plan)		
2.3.283	20° pitch	m2	- -
2.3.284	30° pitch	m2	- -
2.3.285	35° pitch	m2	- -
2.3.286	40° pitch	m2	- -
2.3.287	code 6 sheeting; sloping (measured on face)	m2	- -
	code 6 sheeting; sloping (measured on plan)		
2.3.288	20° pitch	m2	- -
2.3.289	30° pitch	m2	- -
2.3.290	35° pitch	m2	- -
2.3.291	40° pitch	m2	- -
	code 6 sheeting; vertical to mansard; including		
2.3.292	insulation (measured on face)	m2	- -
	Comparative waterproof finishes/perimeter treatments Liquid applied coatings		
2.3.293	solar reflective paint	m2	- -
2.3.294	spray applied bitumen	m2	- -
2.3.295	spray applied co-polymer	m2	- -
2.3.296	spray applied polyurethene	m2	- -

INDUSTRIAL £ range £	RETAILING £ range £	LEISURE £ range £	OFFICES £ range £	HOTELS £ range £	Item nr.
-	-	40.00 - 45.00	40.00 - 45.00	40.00 - 45.00	40.00 - 45.00	2.3.255
-	-	43.50 - 48.50	43.50 - 48.50	43.50 - 48.50	43.50 - 48.50	2.3.256
-	-	48.00 - 54.00	48.00 - 54.00	48.00 - 54.00	48.00 - 54.00	2.3.257
-	-	59.00 - 64.00	59.00 - 64.00	59.00 - 64.00	59.00 - 64.00	2.3.258
-	-	65.00 - 70.00	65.00 - 70.00	65.00 - 70.00	65.00 - 70.00	2.3.259
-	-	67.50 - 72.50	67.50 - 72.50	67.50 - 72.50	67.50 - 72.50	2.3.260
-	-	50.00 - 55.00	50.00 - 55.00	50.00 - 55.00	50.00 - 55.00	2.3.261
-	-	54.00 - 60.00	54.00 - 60.00	54.00 - 60.00	54.00 - 60.00	2.3.262
-	-	59.00 - 65.00	59.00 - 65.00	59.00 - 65.00	59.00 - 65.00	2.3.263
-	-	72.00 - 78.00	72.00 - 78.00	72.00 - 78.00	72.00 - 78.00	2.3.264
-	-	80.00 - 86.00	80.00 - 86.00	80.00 - 86.00	80.00 - 86.00	2.3.265
-	-	84.00 - 90.00	84.00 - 90.00	84.00 - 90.00	84.00 - 90.00	2.3.266
-	-	57.00 - 63.00	57.00 - 63.00	57.00 - 63.00	57.00 - 63.00	2.3.267
-	-	60.00 - 66.00	60.00 - 66.00	60.00 - 66.00	60.00 - 66.00	2.3.268
-	-	60.00 - 66.00	60.00 - 66.00	60.00 - 66.00	60.00 - 60.00	2.3.269
-	-	66.00 - 72.00	66.00 - 72.00	66.00 - 72.00	66.00 - 72.00	2.3.270
-	-	80.00 - 86.00	80.00 - 86.00	80.00 - 86.00	80.00 - 86.00	2.3.271
-	-	90.00 - 96.00	90.00 - 96.00	90.00 - 96.00	90.00 - 96.00	2.3.272
-	-	93.00 - 99.00	93.00 - 99.00	93.00 - 99.00	93.00 - 99.00	2.3.273
-	-	64.00 - 70.00	64.00 - 70.00	64.00 - 70.00	64.00 - 70.00	2.3.274
-	-	70.00 - 76.00	70.00 - 76.00	70.00 - 76.00	70.00 - 76.00	2.3.275
-	-	86.00 - 92.00	86.00 - 92.00	86.00 - 92.00	86.00 - 92.00	2.3.276
-	-	96.00 -102.00	96.00 -102.00	96.00 -102.00	96.00 -102.00	2.3.277
-	-	99.00 -105.00	99.00 -105.00	99.00 -105.00	99.00 -105.00	2.3.278
-	-	62.00 - 68.00	62.00 - 68.00	62.00 - 68.00	62.00 - 68.00	2.3.279
-	-	72.00 - 78.00	72.00 - 78.00	72.00 - 78.00	72.00 - 78.00	2.3.280
-	-	78.00 - 85.00	78.00 - 85.00	78.00 - 85.00	78.00 - 85.00	2.3.281
-	-	65.00 - 71.00	65.00 - 71.00	65.00 - 71.00	65.00 - 71.00	2.3.282
-	-	71.00 - 77.00	71.00 - 77.00	71.00 - 77.00	71.00 - 77.00	2.3.283
-	-	87.00 - 93.00	87.00 - 93.00	87.00 - 93.00	87.00 - 93.00	2.3.284
-	-	97.50 -104.00	97.50 -104.00	97.50 -104.00	97.50 -104.00	2.3.285
-	-	100.00 -107.00	100.00 -107.00	100.00 -107.00	100.00 -107.00	2.3.286
-	-	82.00 - 90.00	82.00 - 90.00	82.00 - 90.00	82.00 - 90.00	2.3.287
-	-	90.00 - 98.00	90.00 - 98.00	90.00 - 98.00	90.00 - 98.00	2.3.288
-	-	110.00 -118.00	110.00 -118.00	110.00 -118.00	110.00 -118.00	2.3.289
-	-	122.00 -130.00	122.00 -130.00	122.00 -130.00	122.00 -130.00	2.3.290
-	-	127.00 -135.00	127.00 -135.00	127.00 -135.00	127.00 -135.00	2.3.291
-	-	110.00 -160.00	110.00 -160.00	110.00 -160.00	110.00 -160.00	2.3.292
-	-	1.50 - 2.50	1.50 - 2.50	1.50 - 2.50	1.50 - 2.50	2.3.293
-	-	5.00 - 8.00	5.00 - 8.00	5.00 - 8.00	5.00 - 8.00	2.3.294
-	-	6.00 - 9.00	6.00 - 9.00	6.00 - 9.00	6.00 - 9.00	2.3.295
-	-	11.00 - 13.00	11.00 - 13.00	11.00 - 13.00	11.00 - 13.00	2.3.296

Item nr.	SPECIFICATIONS	Unit	RESIDENTIAL £ range £

2.3   **ROOF** - cont'd   **roof plan area**
                                          (unless otherwise described)

Comparative waterproof finishes/perimeter treatments - cont'd
20 mm Two coat asphalt roofing; laid flat; on felt
underlay

2.3.297	to BS 988	m2	-	-
2.3.298	to BS 6577	m2	-	-
	Extra for			
2.3.299	solar reflective paint	m2	-	-
2.3.300	limestone chipping finish	m2	-	-
2.3.301	grp tiles in hot bitumen	m2	-	-

20 mm Two coat reinforced asphaltic compound; laid
flat; on felt underlay

2.3.302	to BS 6577	m2	-	-

Built-up bitumen felt roofing; laid flat

2.3.303	three layer glass fibre roofing	m2	-	-
2.3.304	three layer asbestos based roofing	m2	-	-
	Extra for			
2.3.305	granite chipping finish	m2	-	-

Built-up self-finished asbestos based bitumen felt
roofing; laid sloping

2.3.306	two layer roofing (measured on face)	m2	17.00 -	20.00
	two layer roofing (measured on plan)			
2.3.307	35° pitch	m2	26.00 -	29.00
2.3.308	40° pitch	m2	27.00 -	30.00
2.3.309	three layer roofing (measured on face)	m2	24.00 -	27.00
	three layer roofing (measured on plan)			
2.3.310	20° pitch	m2	36.00 -	39.00
2.3.311	30° pitch	m2	37.00 -	40.00

Elastomeric single ply roofing; laid flat

2.3.312	EPDM membrane; laid loose	m2	-	-
2.3.313	Bytyl rubber membrane; laid loose	m2	-	-
	Extra for			
2.3.314	ballast	m2	-	-

Thermoplastic single ply roofing; laid flat
    PVC membrane

2.3.315	laid loose	m2	-	-
2.3.316	mechanically fixed	m2	-	-
2.3.317	fully adhered	m2	-	-
2.3.318	CPE membrane; laid loose	m2	-	-
2.3.319	CSPG membrane; fully adhered	m2	-	-
2.3.320	PIB membrane; laid loose	m2	-	-
	Extra for			
2.3.321	ballast	m2	-	-

High performance built-up felt roofing; laid flat
    two layer self-finished 'Paradiene' elastomeric

2.3.322	bitumen roofing	m2	-	-
	three layer 'Ruberglas 120 GP' felt roofing;			
2.3.323	granite chipping finish	m2	-	-
	'Andersons' three layer self-finished polyester			
2.3.324	based bitumen felt roofing	m2	-	-
	three layer polyester based modified bitumen felt			
2.3.325	roofing	m2	-	-
2.3.326	two layer 'Derbigum' self-finished polyester roofing	m2	-	-
	'Anderson' three layer polyester based bitumen			
2.3.327	felt roofing; granite chipping finish	m2	-	-
	three layer metal faced 'Veral' glass cloth reinforced			
2.3.328	bitumen roofing	m2	-	-
	three layer 'Ruberfort HP 350' felt roofing; granite			
2.3.329	chipping finish	m2	-	-
	three layer 'Hyload 150E' elastomeric roofing; granite			
2.3.330	chipping finish	m2	-	-
	three layer 'Polybit 350' elastomeric roofing; granite			
2.3.331	chipping finish	m2	-	-

INDUSTRIAL £ range £	RETAILING £ range £	LEISURE £ range £	OFFICES £ range £	HOTELS £ range £	Item nr.
-	12.00 - 15.00	12.00 - 15.00	12.00 - 15.00	12.00 - 15.00	2.3.297
-	16.00 - 20.00	16.00 - 20.00	16.00 - 20.00	16.00 - 20.00	2.3.298
-	2.00 - 2.50	2.00 - 2.50	2.00 - 2.50	2.00 - 2.50	2.3.299
-	2.00 - 4.00	2.00 - 4.00	2.00 - 4.00	2.00 - 4.00	2.3.300
-	32.00 - 35.00	32.00 - 35.00	32.00 - 35.00	32.00 - 35.00	2.3.301
-	16.00 - 20.00	16.00 - 20.00	16.00 - 20.00	16.00 - 20.00	2.3.302
-	13.00 - 17.00	13.00 - 17.00	13.00 - 17.00	13.00 - 17.00	2.3.303
-	16.00 - 20.00	16.00 - 20.00	16.00 - 20.00	16.00 - 20.00	2.3.304
-	2.00 - 4.00	2.00 - 4.00	2.00 - 4.00	2.00 - 4.00	2.3.305
-	-	-	-	-	2.3.306
-	-	-	-	-	2.3.307
-	-	-	-	-	2.3.308
-	-	-	-	-	2.3.309
-	-	-	-	-	2.3.310
-	-	-	-	-	2.3.311
16.00 - 19.00	16.00 - 19.00	16.00 - 19.00	16.00 - 19.00	16.00 - 19.00	2.3.312
18.00 - 21.00	18.00 - 21.00	18.00 - 21.00	18.00 - 21.00	18.00 - 21.00	2.3.313
5.00 - 8.00	5.00 - 8.00	5.00 - 8.00	5.00 - 8.00	5.00 - 8.00	2.3.314
16.00 - 19.00	16.00 - 19.00	16.00 - 19.00	16.00 - 19.00	16.00 - 19.00	2.3.315
20.00 - 23.00	20.00 - 23.00	20.00 - 23.00	20.00 - 23.00	20.00 - 23.00	2.3.316
22.00 - 25.00	22.00 - 25.00	22.00 - 25.00	22.00 - 25.00	22.00 - 25.00	2.3.317
18.50 - 22.00	18.50 - 22.00	18.50 - 22.00	18.50 - 22.00	18.50 - 22.00	2.3.318
18.50 - 22.00	18.50 - 22.00	18.50 - 22.00	18.50 - 22.00	18.50 - 22.00	2.3.319
20.50 - 25.00	20.50 - 25.00	20.50 - 25.00	20.50 - 25.00	20.50 - 25.00	2.3.320
5.00 - 8.00	5.00 - 8.00	5.00 - 8.00	5.00 - 8.00	5.00 - 8.00	2.3.321
19.50 - 21.50	19.50 - 21.50	19.50 - 21.50	19.50 - 21.50	19.50 - 21.50	2.3.322
21.75 - 24.00	21.75 - 24.00	21.75 - 24.00	21.75 - 24.00	21.75 - 24.00	2.3.323
22.00 - 24.50	22.00 - 24.50	22.00 - 24.50	22.00 - 24.50	22.00 - 24.50	2.3.324
-	-	-	-	-	2.3.325
24.00 - 26.50	24.00 - 26.50	24.00 - 26.50	24.00 - 26.50	24.00 - 26.50	2.3.326
25.25 - 27.75	25.25 - 27.75	25.25 - 27.75	25.25 - 27.75	25.25 - 27.75	2.3.327
26.70 - 29.40	26.70 - 29.40	26.70 - 29.40	26.70 - 29.40	26.70 - 29.40	2.3.328
27.80 - 30.60	27.80 - 30.60	27.80 - 30.60	27.80 - 30.60	27.80 - 30.60	2.3.329
28.80 - 31.60	28.80 - 31.60	28.80 - 31.60	28.80 - 31.60	28.80 - 31.60	2.3.330
29.50 - 32.50	29.50 - 32.50	29.50 - 32.50	29.50 - 32.50	29.50 - 32.50	2.3.331

Item nr.	SPECIFICATIONS	Unit	RESIDENTIAL £ range £

2.3 **ROOF** - cont'd   **roof plan area**
(unless otherwise described)

Comparative waterproof finishes/perimeter treatments - cont'd
Torch on roofing; laid flat

2.3.332	three layer polyester-based modified bitumen roofing	m2	- -
2.3.333	two layer polymeric isotropic roofing	m2	- -
	Extra for		
2.3.334	granite chipping finish	m2	- -

Edges to flat felt roofs; softwood splayed fillet;
280 x 25 mm painted softwood fascia; no gutter

2.3.335	aluminium edge trim	m	26.00 - 27.50

Edges to flat roofs; code 4 lead drip dressed into
gutter; 230 x 25 mm painted softwood fascia

2.3.336	100 mm PVC gutter	m	25.00 - 32.00
2.3.337	150 mm PVC gutter	m	30.00 - 37.00
2.3.338	100 mm cast iron gutter; decorated	m	37.50 - 45.00
2.3.339	150 mm cast iron gutter; decorated	m	45.00 - 55.00

2.4 **STAIRS**   **storey**
(unless otherwise described)

comprising:-

Timber construction
Reinforced concrete construction
Metal construction
Comparative finishes/balustrading

Timber construction
Softwood staircase; softwood balustrades and hardwood
handrail; plasterboard; skim and emulsion to soffit

2.4.1	2.6 m rise; standard; straight flight	nr	500.00 -750.00
2.4.2	2.6 m rise; standard; top three treads winding	nr	600.00 -825.00
2.4.3	2.6 m rise; standard; dogleg	nr	700.00 -900.00

Oak staircase; balusters and handrails; plasterboard;
skim and emulsion to soffit

2.4.4	2.6 m rise; purpose-made; dogleg	nr	- -
	Plus or minus for		
2.4.5	each 300 mm variation in storey height	nr	- -

Reinforced concrete construction
Escape staircase; granolithic finish; mild steel
balustrades and handrails

2.4.6	3 m rise; dogleg	nr	2500 -3000
	Plus or minus for		
2.4.7	each 300 mm variation in storey height	nr	250.00 -300.00

Staircase; terrazzo finish; mild steel balustrades
and handrails; plastered soffit; balustrades and
staircase soffit decorated

2.4.8	3 m rise; dogleg	nr	- -
	Plus or minus for		
2.4.9	each 300 mm variation in storey height	nr	- -

Staircase; terrazzo finish; stainless steel
balustrades and handrails; plastered and decorated
soffit

2.4.10	3 m rise; dogleg	nr	- -
	Plus or minus for		
2.4.11	each 300 mm variation in storey height	nr	- -

Staircase; high quality finishes; stainless steel
and glass balustrades; plastered and decorated soffit

2.4.12	3 m rise; dogleg	nr	- -
	Plus or minus for		
2.4.13	each 300 mm variation in storey height	nr	- -

INDUSTRIAL £ range £	RETAILING £ range £	LEISURE £ range £	OFFICES £ range £	HOTELS £ range £	Item nr.
-	-	20.00 - 23.00	20.00 - 23.00	20.00 - 23.00	20.00 - 23.00	2.3.332
-	-	20.00 - 23.00	20.00 - 23.00	20.00 - 23.00	20.00 - 23.00	2.3.333
-	-	3.00 - 5.00	3.00 - 5.00	3.00 - 5.00	3.00 - 5.00	2.3.334
-	-	26.00 - 27.50	26.00 - 27.50	26.00 - 27.50	26.00 - 27.50	2.3.335
-	-	25.00 - 32.00	-	-	25.00 - 32.00	25.00 - 32.00	2.3.336
-	-	30.00 - 37.00	-	-	30.00 - 37.00	30.00 - 37.00	2.3.337
-	-	37.50 - 45.00	-	-	37.50 - 45.00	37.50 - 45.00	2.3.338
-	-	45.00 - 55.00	-	-	45.00 - 55.00	45.00 - 55.00	2.3.339

INDUSTRIAL £ range £	RETAILING £ range £	LEISURE £ range £	OFFICES £ range £	HOTELS £ range £	Item nr.
-	-	-	-	-	-	-	-	-	-	2.4.1
-	-	-	-	-	-	-	-	-	-	2.4.2
-	-	-	-	-	-	-	-	-	-	2.4.3
-	-	-	-	-	-	-	-	4500 - 6000	2.4.4
-	-	-	-	-	-	-	-	700.00 - 800.00	2.4.5
3500 - 4500	2400 - 4500	3250 - 4000	3250 - 4000	2500 - 5000	2.4.6
350.00 - 450.00	240.00 - 450.00	325.00 - 400.00	325.00 - 400.00	250.00 - 500.00	2.4.7
5000 - 6000	4000 - 6500	4750 - 6000	4750 - 6500	4000 - 7000	2.4.8
500.00 - 600.00	400.00 - 650.00	475.00 - 600.00	475.00 - 650.00	400.00 - 700.00	2.4.9
-	-	5000 - 8000	5500 - 7500	5500 - 8000	5000 - 8500	2.4.10
-	-	500.00 - 800.00	550.00 - 750.00	550.00 - 800.00	500.00 - 850.00	2.4.11
10000 - 13500	11000 - 15000	10000 - 13500	9500 - 13000	12000 - 16000	2.4.12
1000 - 1350	1100 - 1500	1000 - 1350	950 - 1300	1200 - 1600	2.4.13

Item nr.	SPECIFICATIONS	Unit	RESIDENTIAL £ range £	
2.4	**STAIRS** - cont'd	**Storey** (unless otherwise described)		
	Metal construction			
	Steel access/fire ladder			
2.4.14	3 m high	nr	-	-
2.4.15	4 m high; epoxide finished	nr	-	-
	Light duty metal staircase; galvanised finish; perforated treads; no risers; balustrades and handrails; decorated			
2.4.16	3 m rise; spiral; 1548 mm diameter	nr	-	-
	Plus or minus for			
2.4.17	each 300 mm variation in storey height	nr	-	-
2.4.18	3 m rise; spiral; 1936 mm diameter	nr	-	-
	Plus or minus for			
2.4.19	each 300 mm variation in storey height	nr	-	-
2.4.20	3 m rise; spiral; 2072 mm diameter	nr	-	-
	Plus or minus for			
2.4.21	each 300 mm variation in storey height	nr	-	-
2.4.22	3 m rise; straight; 760 mm wide	nr	-	-
	Plus or minus for			
2.4.23	each 300 mm variation in storey height	nr	-	-
2.4.24	3 m rise; straight; 900 mm wide	nr	-	-
	Plus or minus for			
2.4.25	each 300 mm variation in storey height	nr	-	-
2.4.26	3 m; straight; 1070 mm wide	nr	-	-
	Plus or minus for			
2.4.27	each 300 mm variation in storey height	nr	-	-
	Heavy duty cast iron staircase; perforated treads; no risers; balustrades and handrails; decorated			
2.4.28	3 m rise; spiral; 1548 mm diameter	nr	-	-
	Plus or minus for			
2.4.29	each 300 mm variation in storey height	nr	-	-
2.4.30	3 m rise; straight	nr	-	-
	Plus or minus for			
2.4.31	each 300 mm variation in storey height	nr	-	-
	Feature metal staircase; galvanised finish; perforated treads; no risers; decorated			
2.4.32	3 m rise; spiral; balustrades and handrails	nr	-	-
	Plus or minus for			
2.4.33	each 300 mm variation in storey height	nr	-	-
2.4.34	3 m rise; dogleg; hardwood balustrades and handrails	nr	-	-
2.4.35	3 m rise; dogleg; stainless steel balustrades and handrails	nr	-	-
	Feature metal staircase; galvanised finish; concrete treads; balustrades and handrails; decorated			
2.4.36	3 m rise; dogleg	nr	-	-
	Feature metal staircase to water chute; steel springers; stainless steel treads in-filled with tiling; landings every one metre rise			
2.4.37	9 m rise; spiral	nr	-	-
	Galvanised steel catwalk; nylon coated balustrading			
2.4.38	450 mm wide	m		
	Comparative finishes/balustrading			
	Finishes to treads and risers			
2.4.39	PVC floor tiles including screeds	nr	-	-
2.4.40	granolithic	nr	-	-
2.4.41	heavy duty carpet	nr	-	-
2.4.42	terrazzo	nr	-	-
	Wall handrails			
2.4.43	PVC covered mild steel rail on brackets	nr	-	-
2.4.44	hardwood handrail on brackets	nr	-	-
2.4.45	stainless steel handrail on brackets	nr	-	-

INDUSTRIAL £ range £	RETAILING £ range £	LEISURE £ range £	OFFICES £ range £	HOTELS £ range £	Item nr.
350.00 -550.00	- -	350.00 -550.00	- -	- -	2.4.14
- -	- -	500.00 -850.00	- -	- -	2.4.15
1700 -2100	- -	- -	- -	- -	2.4.16
170.00 -210.00	- -	- -	- -	- -	2.4.17
2000 -2400	- -	- -	- -	- -	2.4.18
200.00 -240.00	- -	- -	- -	- -	2.4.19
21000 -2500	- -	- -	- -	- -	2.4.20
210.00 -250.00	- -	- -	- -	- -	2.4.21
2000 -2500	- -	- -	- -	- -	2.4.22
200.00 -250.00	- -	- -	- -	- -	2.4.23
2100 -2600	- -	- -	- -	- -	2.4.24
21.00 -260.00	- -	- -	- -	- -	2.4.25
2200 -2700	- -	- -	- -	- -	2.4.26
220.00 -270.00	- -	- -	- -	- -	2.4.27
2750 -3500	- -	- -	- -	- -	2.4.28
275.00 -350.00	- -	- -	- -	- -	2.4.29
3000 -4000	- -	- -	- -	- -	2.4.30
300.00 -400.00	- -	- -	- -	- -	2.4.31
- -	- -	3750 -4250	- -	- -	2.4.32
- -	- -	375.00 -425.00	- -	- -	2.4.33
- -	- -	4000 -5500	- -	- -	2.4.34
- -	- -	4500 -7500	- -	- -	2.4.35
- -	- -	7000 -9500	- -	- -	2.4.36
- -	- -	22500 -27500	- -	- -	2.4.37
200.00 -240.00	- -	210.00 -250.00	- -	- -	2.4.38
- -	600.00 -800.00	600.00 -800.00	600.00 -800.00	600.00 -800.00	2.4.39
- -	900.00 -1000	900.00 -1000	900.00 -1000	900.00 -1000	2.4.40
- -	1200 -1500	1200 -1500	1200 -1500	1200 -1500	2.4.41
- -	2500 -3000	2500 -3000	2500 -3000	2500 -3000	2.4.42
- -	200.00 -300.00	200.00 -300.00	200.00 -300.00	200.00 -300.00	2.4.43
- -	600.00 -1000	600.00 -1000	600.00 -1000	600.00 -1000	2.4.44
- -	2500 -3000	2500 -3000	2500 -3000	2500 -3000	2.4.45

Item nr.	SPECIFICATIONS	Unit	RESIDENTIAL £ range £
2.4	**STAIRS** - cont'd	**Storey**	
	(unless otherwise described)		
	Comparative finishes/balustrading - cont'd		
	Balustrading and handrails		
2.4.46	mild steel balustrades and PVC covered handrails	nr	- -
2.4.47	mild steel balustrades and hardwood handrails	nr	- -
2.4.48	stainless steel balustrades and handrails	nr	- -
2.4.49	stainless steel and glass balustrades	nr	- -
2.5	**EXTERNAL WALLS**	**external wall area**	
	(unless otherwise described)		

comprising:-

Timber framed walling
Brick/block walling
Reinforced concrete walling
Panelled walling
Wall claddings
Curtain/glazed walling
Comparative external finishes

	Timber framed walling		
	Structure only comprising softwood studs at 400 x 600 mm centres; head and sole plates		
2.5.1	125 x 50 mm	m2	12.00 - 15.00
	125 x 50 mm; one layer of double sided building		
2.5.2	paper	m2	14.00 - 18.00
	Softwood stud wall; vapour barrier and plasterboard inner lining; decorated		
2.5.3	PVC weatherboard outer lining	m2	50.00 - 65.00
2.5.4	tile hanging on battens outer lining	m2	60.00 - 72.00
	Brick/block walling		
	Autoclaved aerated light weight block walls		
2.5.5	100 mm thick	m2	17.00 - 20.00
2.5.6	140 mm thick	m2	21.00 - 24.00
2.5.7	190 mm thick	m2	27.00 - 31.00
	Dense aggregate block walls		
2.5.8	100 mm thick	m2	18.00 - 21.00
2.5.9	140 mm thick	m2	23.00 - 26.00
	Coloured dense aggregate masonry block walls		
2.5.10	100 mm thick; hollow	m2	30.00 - 33.00
2.5.11	100 mm thick; solid	m2	33.00 - 36.00
2.5.12	140 mm thick; hollow	m2	34.00 - 40.00
2.5.13	140 mm thick; solid	m2	45.00 - 50.00
	Common brick solid walls; bricks PC £117.00/1000		
2.5.14	half brick thick	m2	23.00 - 27.00
2.5.15	one brick thick	m2	45.00 - 50.00
2.5.16	one and a half brick thick	m2	64.00 - 70.00
	Add or deduct for each variation of £10.00/1000 in PC value		
2.5.17	half brick thick	m2	0.70 - 1.00
2.5.18	one brick thick	m2	1.40 - 1.70
2.5.19	one and a half brick thick	m2	2.10 - 2.40
	Extra for		
2.5.20	fair face one side	m2	1.50 - 2.00
	Engineering brick walls; class B; bricks PC £192.00/1000		
2.5.21	half brick thick	m2	30.00 - 34.00
2.5.22	one brick thick	m2	57.00 - 63.00
	Facing brick walls; sand faced facings; bricks PC £125.00/1000		
2.5.23	half brick thick; pointed one side	m2	30.00 - 34.00
2.5.24	one brick thick; pointed both sides	m2	56.00 - 62.00

INDUSTRIAL £ range £	RETAILING £ range £	LEISURE £ range £	OFFICES £ range £	HOTELS £ range £	Item nr.
-	-	600 -750	600 -750	600 -750	600 -750	2.4.46
-	-	1100 -1500	1100 -1500	1100 -1500	1100 -1500	2.4.47
-	-	4500 -5500	4500 -5500	4500 -5500	4500 -5500	2.4.48
-	-	6000 -10000	6000 -10000	6000 -10000	6000 -10000	2.4.49
-	-	-	-	-	-	-	-	-	-	2.5.1
-	-	-	-	-	-	-	-	-	-	2.5.2
-	-	-	-	-	-	-	-	-	-	2.5.3
-	-	-	-	-	-	-	-	-	-	2.5.4
16.00 - 19.00	16.00 - 19.00	16.00 - 19.00	16.00 - 19.00	18.00 - 21.00	2.5.5
20.00 - 23.00	20.00 - 23.00	20.00 - 23.00	20.00 - 23.00	22.00 - 25.00	2.5.6
26.00 - 31.00	26.00 - 31.00	26.00 - 31.00	26.00 - 31.00	28.00 - 32.00	2.5.7
17.00 - 20.00	17.00 - 20.00	17.00 - 20.00	17.00 - 20.00	19.00 - 22.00	2.5.8
22.00 - 25.00	22.00 - 25.00	22.00 - 25.00	22.00 - 25.00	24.00 - 27.00	2.5.9
28.50 - 31.50	28.50 - 31.50	28.50 - 31.50	28.50 - 31.50	31.00 - 34.00	2.5.10
31.00 - 34.00	31.00 - 34.00	31.00 - 34.00	31.00 - 34.00	34.00 - 37.00	2.5.11
35.00 - 38.00	35.00 - 38.00	35.00 - 38.00	35.00 - 38.00	38.00 - 42.00	2.5.12
43.00 - 48.00	43.00 - 48.00	43.00 - 48.00	43.00 - 48.00	46.00 - 52.00	2.5.13
22.00 - 26.00	22.00 - 26.00	22.00 - 26.00	22.00 - 26.00	24.00 - 28.00	2.5.14
43.00 - 48.00	43.00 - 48.00	43.00 - 48.00	43.00 - 48.00	47.00 - 52.00	2.5.15
62.00 - 68.00	62.00 - 68.00	62.00 - 68.00	62.00 - 68.00	66.00 - 72.00	2.5.16
0.70 - 1.00	0.70 - 1.00	0.70 - 1.00	0.70 - 1.00	0.70 - 1.00	2.5.17
1.40 - 1.70	1.40 - 1.70	1.40 - 1.70	1.40 - 1.70	1.40 - 1.70	2.5.18
2.10 - 2.40	2.10 - 2.40	2.10 - 2.40	2.10 - 2.40	2.10 - 2.40	2.5.19
1.50 - 2.00	1.50 - 2.00	1.50 - 2.00	1.50 - 2.00	1.50 - 2.00	2.5.20
29.00 - 33.00	29.00 - 33.00	29.00 - 33.00	29.00 - 33.00	31.00 - 35.00	2.5.21
55.00 - 61.00	55.00 - 61.00	55.00 - 61.00	55.00 - 61.00	60.00 - 65.00	2.5.22
29.00 - 33.00	29.00 - 33.00	29.00 - 33.00	29.00 - 33.00	31.00 - 35.00	2.5.23
54.00 - 60.00	54.00 - 60.00	54.00 - 60.00	54.00 - 60.00	58.00 - 64.00	2.5.24

Item nr.	SPECIFICATIONS	Unit	RESIDENTIAL £ range £
2.5	**EXTERNAL WALLS** - cont'd **external wall area** (unless otherwise described)		
	Brick/block walling - cont'd Facing brick walls; machine-made facings; bricks PC £320.00/1000		
2.5.25	half brick thick; pointed one side	m2	45.00 - 50.00
2.5.26	half brick thick; built against concrete	m2	47.50 - 53.00
2.5.27	one brick thick; pointed both sides	m2	85.00 - 95.00
	Facing bricks solid walls; hand-made facings; bricks PC £520.00/1000		
2.5.28	half brick thick; pointed one side	m2	60.00 - 66.00
2.5.29	one brick thick; pointed both sides	m2	115.00 -125.00
	Add or deduct for each variation of £10.00/1000 in PC value		
2.5.30	half brick thick	m2	0.70 - 1.00
2.5.31	one brick thick	m2	1.40 - 1.70
	Composite solid walls; facing brick on outside; bricks PC £320.00/1000 and common brick on inside; bricks PC £117.00/1000		
2.5.32	one brick thick; pointed one side	m2	70.00 - 77.50
	Extra for		
2.5.33	weather pointing as a separate operation	m2	4.00 - 6.00
2.5.34	one and a half brick thick; pointed one side	m2	90.00 -100.00
	Composite cavity wall; block outer skin; 50 mm insulation; lightweight block inner skin		
2.5.35	outer block rendered	m2	50.00 - 62.50
	Extra for		
2.5.36	heavyweight block inner skin	m2	1.00 - 2.00
2.5.37	fair face one side	m2	0.50 - 2.00
2.5.38	75 mm cavity insulation	m2	1.00 - 2.00
2.5.39	100 mm cavity insulation	m2	2.00 - 3.00
2.5.40	plaster and emulsion	m2	10.00 - 12.00
	outer block rendered; no insulation; inner skin		
2.5.41	insulating	m2	50.00 - 65.00
2.5.42	outer block roughcast	m2	55.00 - 70.00
2.5.43	coloured masonry outer block	m2	60.00 - 75.00
	Composite cavity wall; facing brick outer skin; 50 mm insulation; plasterboard on stud inner skin; emulsion		
2.5.44	sand-faced; PC £125.00/1000	m2	58.00 - 68.00
2.5.45	machine-made facings; PC £320.00/1000	m2	72.50 - 82.50
2.5.46	hand-made facings; PC £520.00/1000	m2	86.00 - 96.00
	Composite cavity wall; facing brick outer skin; lightweight block inner skin; plaster and emulsion		
2.5.47	sand-faced facings; PC £125.00/1000	m2	53.00 - 65.00
2.5.48	machine-made facings; PC £320.00/1000	m2	68.50 - 80.00
2.5.49	hand-made facings; PC £520.00/1000	m2	80.00 - 95.00
	Add or deduct for		
2.5.50	each variation of £10.00/1000 in PC value	m2	0.70 - 1.00
	Extra for		
2.5.51	heavyweight block inner skin	m2	1.00 - 2.00
2.5.52	insulating block inner skin	m2	2.00 - 5.00
2.5.53	30 mm cavity wall slab	m2	2.50 - 5.50
2.5.54	50 mm cavity insulation	m2	3.00 - 3.50
2.5.55	75 mm cavity insulation	m2	4.00 - 4.50
2.5.56	100 mm cavity insulation	m2	5.00 - 5.50
2.5.57	weather-pointing as a separate operation	m2	4.00 - 6.00
2.5.58	purpose made feature course to windows	m2	5.00 - 10.00
	Composite cavity wall; facing brick outer skin; 50 mm insulation; common brick inner skin; fair face on inside		
2.5.59	sand-faced facings; PC £125.00/1000	m2	56.00 - 70.00
2.5.60	machine-made facings; PC £320.00/1000	m2	74.00 - 85.00
2.5.61	hand-made facings; PC £520.00/1000	m2	88.00 -100.00
	Composite cavity wall; facing brick outer skin; 50 mm insulation; common brick inner skin; plaster and emulsion		
2.5.62	sand-faced facings; PC £125.00/1000	m2	64.00 - 78.00
2.5.63	machine-made facings; PC £320.00/1000	m2	82.00 - 92.50
2.5.64	hand-made facings; PC £520.00/1000	m2	96.00 -110.00
	Composite cavity wall; coloured masonry block; outer and		
2.5.65	inner skins; fair faced both sides	m2	90.00 -120.00

INDUSTRIAL £ range £	RETAILING £ range £	LEISURE £ range £	OFFICES £ range £	HOTELS £ range £	Item nr.
43.00 - 48.00	45.00 - 50.00	45.00 - 50.00	45.00 - 50.00	47.00 - 52.00	2.5.25
45.50 - 51.00	47.50 - 53.00	47.50 - 53.00	47.50 - 53.00	50.00 - 55.00	2.5.26
82.00 - 92.00	85.00 - 95.00	85.00 - 95.00	85.00 - 95.00	89.00 -100.00	2.5.27
58.00 - 64.00	60.00 - 66.00	60.00 - 66.00	60.00 - 66.00	63.00 - 70.00	2.5.28
110.00 -120.00	115.00 -125.00	115.00 -125.00	115.00 -125.00	120.00 -130.00	2.5.29
0.70 - 1.00	0.70 - 1.00	0.70 - 1.00	0.70 - 1.00	0.70 - 1.00	2.5.30
1.40 - 1.70	1.40 - 1.70	1.40 - 1.70	1.40 - 1.70	1.40 - 1.70	2.5.31
68.00 - 75.00	70.00 - 77.50	70.00 - 77.50	70.00 - 77.50	73.50 - 80.00	2.5.32
4.00 - 6.00	4.00 - 6.00	4.00 - 6.00	4.00 - 6.00	4.00 - 6.00	2.5.33
86.00 - 95.00	90.00 -100.00	90.00 -100.00	90.00 -100.00	95.00 -105.00	2.5.34
48.00 - 60.00	50.00 - 62.50	50.00 - 62.50	50.00 - 62.50	52.50 - 65.00	2.5.35
1.00 - 2.00	1.00 - 2.00	1.00 - 2.00	1.00 - 2.00	1.00 - 2.00	2.5.36
0.50 - 2.00	0.50 - 2.00	0.50 - 2.00	0.50 - 2.00	0.50 - 2.00	2.5.37
1.00 - 2.00	1.00 - 2.00	1.00 - 2.00	1.00 - 2.00	1.00 - 2.00	2.5.38
2.00 - 3.00	2.00 - 3.00	2.00 - 3.00	2.00 - 3.00	2.00 - 3.00	2.5.39
9.50 - 11.50	10.00 - 12.00	10.00 - 12.00	10.00 - 12.00	10.50 - 12.50	2.5.40
48.00 - 63.00	50.00 - 65.00	50.00 - 65.00	50.00 - 65.00	52.00 - 67.00	2.5.41
53.00 - 68.00	55.00 - 70.00	55.00 - 70.00	55.00 - 70.00	57.00 - 72.00	2.5.42
58.00 - 72.00	60.00 - 75.00	60.00 - 75.00	60.00 - 75.00	63.00 - 78.00	2.5.43
56.00 - 66.00	58.00 - 68.00	58.00 - 68.00	58.00 - 68.00	60.00 - 70.00	2.5.44
70.00 - 80.00	72.50 - 82.50	72.50 - 82.50	72.50 - 82.50	75.00 - 85.00	2.5.45
82.00 - 92.00	86.00 - 96.00	86.00 - 96.00	86.00 - 96.00	90.00 -100.00	2.5.46
51.00 - 63.00	53.00 - 65.00	53.00 - 65.00	53.00 - 65.00	56.00 - 68.00	2.5.47
65.00 - 77.00	68.50 - 80.00	68.50 - 80.00	68.50 - 80.00	72.50 - 84.00	2.5.48
77.00 - 95.00	80.00 - 95.00	80.00 - 95.00	80.00 - 95.00	84.00 -100.00	2.5.49
0.70 - 1.00	0.70 - 1.00	0.70 - 1.00	0.70 - 1.00	0.70 - 1.00	2.5.50
1.00 - 2.00	1.00 - 2.00	1.00 - 2.00	1.00 - 2.00	1.00 - 2.00	2.5.51
2.00 - 5.00	2.00 - 5.00	2.00 - 5.00	2.00 - 5.00	2.00 - 5.00	2.5.52
2.50 - 5.50	2.50 - 5.50	2.50 - 5.50	2.50 - 5.50	2.50 - 5.50	2.5.53
3.00 - 3.50	3.00 - 3.50	3.00 - 3.50	3.00 - 3.50	3.00 - 3.50	2.5.54
4.00 - 4.50	4.00 - 4.50	4.00 - 4.50	4.00 - 4.50	4.00 - 4.50	2.5.55
5.00 - 5.50	5.00 - 5.50	5.00 - 5.50	5.00 - 5.50	5.00 - 5.50	2.5.56
4.00 - 6.00	4.00 - 6.00	4.00 - 6.00	4.00 - 6.00	4.00 - 6.00	2.5.57
5.00 - 10.00	5.00 - 10.00	5.00 - 10.00	5.00 - 10.00	5.00 - 10.00	2.5.58
54.00 - 68.00	56.00 - 70.00	56.00 - 70.00	56.00 - 70.00	58.00 - 72.00	2.5.59
71.00 - 82.00	74.00 - 85.00	74.00 - 85.00	74.00 - 85.00	77.00 - 88.00	2.5.60
84.00 - 95.00	88.00 -100.00	88.00 -100.00	88.00 -100.00	92.00 -105.00	2.5.61
62.00 - 75.00	64.00 - 78.00	64.00 - 78.00	64.00 - 78.00	67.00 - 82.00	2.5.62
80.00 - 90.00	82.00 - 92.50	82.00 - 92.50	82.00 - 92.50	86.00 - 96.00	2.5.63
91.00 -105.00	96.00 -110.00	96.00 -110.00	96.00 -110.00	100.00 -115.00	2.5.64
86.00 -115.00	90.00 -120.00	90.00 -120.00	90.00 -120.00	95.00 -125.00	2.5.65

Item nr.	SPECIFICATIONS	Unit	RESIDENTIAL £ range £
2.5	**EXTERNAL WALLS** - cont'd	**external wall area** (unless otherwise described)	
	Reinforced concrete walling Insitu reinforced concrete 25.5 N/mm2; 13 kg/m2 reinforcement; formwork both sides		
2.5.66	150 mm thick	m2	80.00 - 90.00
2.5.67	225 mm thick	m2	90.00 -100.00
	Panelled walling Precast concrete panels; including insulation; lining and fixings		
2.5.68	standard panels	m2	- -
2.5.69	standard panels; exposed aggregate finish	m2	- -
2.5.70	brick clad panels	m2	- -
2.5.71	reconstructed stone faced panels	m2	- -
2.5.72	natural stone faced panels	m2	- -
2.5.73	marble or granite faced panels	m2	- -
	GRP/laminate panels; including battens; back-up walls; plaster and emulsion		
2.5.74	melamined finished solid laminate panels	m2	- -
2.5.75	GRP single skin panels	m2	- -
2.5.76	GRP double skin panels	m2	- -
2.5.77	GRP insulated sandwich panels	m2	- -
	Wall claddings Non-asbestos profiled cladding		
2.5.78	'profile 3'; natural	m2	- -
2.5.79	'profile 3'; coloured	m2	- -
2.5.80	'profile 6'; natural	m2	- -
2.5.81	'profile 6'; coloured	m2	- -
2.5.82	insulated; inner lining of plasterboard	m2	- -
2.5.83	'profile 6'; natural; insulated; inner lining panel	m2	- -
2.5.84	insulated; with 2.8 m high block inner skin; emulsion	m2	- -
2.5.85	insulated; with 2.8 m high block inner skin; plasterboard lining on metal tees; emulsion	m2	- -
	Asbestos cement profiled cladding on steel rails		
2.5.86	insulated; with 2.8 m high block inner skin; emulsion	m2	- -
2.5.87	insulated; with 2.8 m high block inner skin; plasterboard lining on metal tees; emulsion	m2	- -
	PVF2 coated galvanised steel profiled cladding		
2.5.88	0.60 mm thick; 'profile 20B'; corrugations vertical	m2	- -
2.5.89	0.60 mm thick; 'profile 30'; corrugations vertical	m2	- -
2.5.90	0.60 mm thick; 'profile TOP 40'; corrugations vertical	m2	- -
2.5.91	0.60 mm thick; 'profile 60B'; corrugations vertical	m2	- -
2.5.92	0.60 mm thick; 'profile 30'; corrugations horizontal	m2	- -
2.5.93	0.60 mm thick; 'profile 60B' corrugations horizontal	m2	- -
	Extra for		
2.5.94	80 mm insulation and 0.4 mm thick coated inner lining sheet	m2	
	PVF2 coated galvanised steel profiled cladding on steel rails		
2.5.95	insulated	m2	- -
2.5.96	2.8 m high insulating block inner skin; emulsion	m2	- -
2.5.97	2.8 m high insulating block inner skin; plasterboard lining on metal tees; emulsion	m2	- -
2.5.98	insulated; coloured inner lining panel	m2	- -
2.5.99	insulated; full-height insulating block inner skin; plaster and emulsion	m2	- -
2.5.100	insulated; metal sandwich panel system	m2	- -
	PVF2 coated aluminium profiled cladding on steel rails		
2.5.101	insulated	m2	- -
2.5.102	insulated; plasterboard lining on metal tees; emulsion	m2	- -
2.5.103	insulated; coloured inner lining panel	m2	- -
2.5.104	insulated; full-height insulating block inner skin; plaster and emulsion	m2	- -
	Extra for		
2.5.105	heavyweight block inner skin	m2	- -

INDUSTRIAL £ range £	RETAILING £ range £	LEISURE £ range £	OFFICES £ range £	HOTELS £ range £	Item nr.
75.00 - 85.00	80.00 - 90.00	80.00 - 90.00	80.00 - 90.00	80.00 - 90.00	2.5.66
85.00 - 90.00	90.00 -100.00	90.00 -100.00	90.00 -100.00	90.00 -100.00	2.5.67
80.00 -130.00	90.00 -130.00	90.00 -130.00	90.00 -130.00	90.00 -140.00	2.5.68
130.00 -180.00	140.00 -180.00	140.00 -180.00	140.00 -180.00	140.00 -190.00	2.5.69
170.00 -210.00	180.00 -210.00	180.00 -210.00	180.00 -210.00	180.00 -220.00	2.5.70
-	280.00 -350.00	280.00 -350.00	280.00 -350.00	280.00 -375.00	2.5.71
-	400.00 -500.00	400.00 -500.00	400.00 -500.00	400.00 -525.00	2.5.72
-	450.00 -550.00	450.00 -550.00	450.00 -550.00	450.00 -600.00	2.5.73
110.00 -140.00	120.00 -140.00	120.00 -140.00	-	-	2.5.74
115.00 -150.00	125.00 -150.00	125.00 -150.00	100.00 -120.00	-	2.5.75
160.00 -200.00	170.00 -200.00	170.00 -200.00	150.00 -190.00	-	2.5.76
-	200.00 -250.00	200.00 -250.00	190.00 -240.00	190.00 -240.00	2.5.77
15.00 - 17.00	-	-	-	-	2.5.78
16.00 - 18.00	-	-	-	-	2.5.79
16.00 - 18.00	-	-	-	-	2.5.80
17.00 - 19.00	-	-	-	-	2.5.81
30.00 - 36.00	-	-	-	-	2.5.82
30.00 - 36.00	-	-	-	-	2.5.83
26.00 - 28.00	-	-	-	-	2.5.84
34.00 - 40.00	-	-	-	-	2.5.85
25.00 - 27.00	-	-	-	-	2.5.86
33.00 - 39.00	-	-	-	-	2.5.87
19.00 - 24.00	24.00 - 31.00	24.00 - 31.00	-	-	2.5.88
19.00 - 24.00	24.00 - 31.00	24.00 - 31.00	-	-	2.5.89
18.00 - 23.00	23.00 - 30.00	23.00 - 30.00	-	-	2.5.90
22.00 - 27.00	27.00 - 34.00	27.00 - 34.00	-	-	2.5.91
20.00 - 25.00	25.00 - 32.00	25.00 - 32.00	-	-	2.5.92
23.00 - 28.00	28.00 - 35.00	28.00 - 35.00	-	-	2.5.93
11.00 - 12.00	11.00 - 12.00	11.00 - 12.00	-	-	2.5.94
32.00 - 42.00	-	-	-	-	2.5.95
40.00 - 50.00	-	-	-	-	2.5.96
48.00 - 58.00	-	-	-	-	2.5.97
45.00 - 57.50	53.00 - 68.00	53.00 - 68.00	-	-	2.5.98
65.00 - 85.00	70.00 - 90.00	70.00 - 90.00	70.00 - 95.00	-	2.5.99
120.00 -180.00	140.00 -190.00	130.00 -180.00	150.00 -180.00	-	2.5.100
36.00 - 46.00	-	-	-	-	2.5.101
50.00 - 60.00	-	-	-	-	2.5.102
52.00 - 65.00	60.00 - 72.00	60.00 - 72.00	-	-	2.5.103
70.00 - 90.00	75.00 - 95.00	75.00 - 95.00	75.00 -100.00	-	2.5.104
1.00 - 2.00	1.00 - 2.00	1.00 - 2.00	1.00 - 2.00	-	2.5.105

Item nr.	SPECIFICATIONS	Unit	RESIDENTIAL £ range £
2.5	**EXTERNAL WALLS** - cont'd **external wall area** (unless otherwise described)		
	Wall claddings - cont'd		
	PVF2 coated aluminium profiled cladding on steel rails - cont'd		
2.5.106	insulated; aluminium sandwich panel system	m2	- -
	insulated; aluminium sandwich panel system; on framing;		
2.5.107	on block inner skin; fair face one side	m2	- -
	Other cladding systems		
	vitreous enamelled insulated steel sandwich panel system; with non-asbestos fibre insulating board on		
2.5.108	inner face	m2	- -
	Formalux sandwich panel system; with coloured lining		
2.5.109	tray; on steel cladding rails	m2	- -
2.5.110	aluminium over-cladding system rain screen	m2	- -
	natural stone cladding on full-height insulating		
2.5.111	block inner skin; plaster and emulsion	m2	- -
	Curtain/glazed walling		
	Single glazed anodised aluminium curtain walling		
	economical; including part-height block back-up wall;		
2.5.112	plaster and emulsion	m2	- -
	Extra over single 6 mm float glass for		
2.5.113	double glazing unit with two 6 mm float glass skins	m2	- -
	double glazing unit with one 6 mm 'Antisun' skin		
2.5.114	and one 6 mm float glass skin	m2	- -
	economical; including part-height block back-up wall;		
2.5.115	plaster and emulsion	m2	- -
2.5.116	economical; including infill panels	m2	- -
	Extra for		
2.5.117	50 mm insulation	m2	- -
2.5.118	polyester powder coating in lieu of anodised finish	m2	- -
2.5.119	bronze anodising in lieu of anodised finish	m2	- -
2.5.120	good quality	m2	- -
2.5.121	good quality; 35% opening lights	m2	- -
2.5.122	high quality	m2	- -
2.5.123	high quality; fire rated	m2	- -
	Suspended panelled glazing system (e.g. Planar)		
2.5.124	10 mm toughened float glass panels	m2	- -
	Extra for		
2.5.125	solar control glass	m2	- -
	supporting structure of aluminium; stainless steel		
2.5.126	or glass fins	m2	- -
2.5.127	High quality structural glazing to entrance elevation	m2	- -
	Comparative external finishes		
	Comparative concrete wall finishes		
2.5.128	wrought formwork one side including rubbing down	m2	- -
2.5.129	shotblasting to expose aggregate	m2	- -
2.5.130	bush hammering to expose aggregate	m2	- -
	Comparative insitu finishes		
2.5.131	two coats cement paint	m2	- -
2.5.132	cement and sand plain face rendering	m2	- -
2.5.133	three-coat Tyrolean rendering; including backing	m2	- -
2.5.134	'Mineralite' decorative rendering; including backing	m2	- -
	Comparative claddings		
	25 mm tongued and grooved 'tanalised' softwood		
2.5.135	boarding; including battens	m2	- -
	25 mm tongued and grooved Western Red Cedar boarding		
2.5.136	including battens	m2	- -
2.5.137	machine-made tiles; including battens	m2	- -
2.5.138	best hand-made sand-faced tiles; including battens	m2	- -
	20 x 20 mm mosaic glass or ceramic; in common colours;		
2.5.139	fixed on prepared surface	m2	- -

INDUSTRIAL £ range £	RETAILING £ range £	LEISURE £ range £	OFFICES £ range £	HOTELS £ range £	Item nr.
100.00 -130.00	130.00 -170.00	130.00 -170.00	140.00 -180.00	-	2.5.106
-	180.00 -275.00	-	-	-	2.5.107
120.00 -150.00	-	-	-	-	2.5.108
150.00 -180.00	160.00 -190.00	160.00 -190.00	160.00 -190.00	-	2.5.109
190.00 -210.00	-	-	-	-	2.5.110
-	-	-	325.00 -475.00	350.00 -500.00	2.5.111
200.00 -270.00	210.00 -280.00	210.00 -280.00	210.00 -280.00	220.00 -300.00	2.5.112
60.00 - 70.00	60.00 - 70.00	60.00 - 70.00	60.00 - 70.00	60.00 - 70.00	2.5.113
100.00 -110.00	100.00 -110.00	100.00 -110.00	100.00 -110.00	100.00 -110.00	2.5.114
225.00 -280.00	240.00 -300.00	240.00 -300.00	240.00 -300.00	250.00 -325.00	2.5.115
230.00 -330.00	250.00 -350.00	250.00 -350.00	250.00 -350.00	275.00 -400.00	2.5.116
11.00 - 13.00	11.00 - 13.00	11.00 - 13.00	11.00 - 13.00	11.00 - 13.00	2.5.117
5.00 - 13.00	5.00 - 13.00	5.00 - 13.00	5.00 - 13.00	5.00 - 13.00	2.5.118
12.00 - 20.00	12.00 - 20.00	12.00 - 20.00	12.00 - 20.00	12.00 - 20.00	2.5.119
340.00 -430.00	350.00 -450.00	350.00 -460.00	360.00 -460.00	350.00 -470.00	2.5.120
410.00 -490.00	420.00 -510.00	420.00 -510.00	420.00 -520.00	420.00 -530.00	2.5.121
450.00 -600.00	480.00 -660.00	460.00 -630.00	480.00 -640.00	460.00 -650.00	2.5.122
-	-	-	800.00 -1200	-	2.5.123
-	275.00 -350.00	275.00 -350.00	275.00 -350.00	-	2.5.124
-	50.00 -150.00	50.00 -150.00	50.00 -150.00	-	2.5.125
-	30.00 -115.00	30.00 -115.00	30.00 -115.00	-	2.5.126
-	450.00 -700.00	450.00 -700.00	450.00 -700.00	-	2.5.127
-	2.50 - 4.00	2.50 - 4.00	2.50 - 4.00	2.50 - 4.00	2.5.128
-	4.50 - 6.00	4.50 - 6.00	4.50 - 6.00	4.50 - 6.00	2.5.129
-	9.00 - 12.00	9.00 - 12.00	9.00 - 12.00	9.00 - 12.00	2.5.130
-	5.00 - 7.00	5.00 - 7.00	5.00 - 7.00	5.00 - 7.00	2.5.131
-	10.00 - 14.00	10.00 - 14.00	10.00 - 14.00	10.00 - 14.00	2.5.132
-	24.00 - 28.00	24.00 - 28.00	24.00 - 28.00	24.00 - 28.00	2.5.133
-	40.00 - 48.00	40.00 - 48.00	40.00 - 48.00	40.00 - 48.00	2.5.134
-	22.00 - 26.00	22.00 - 26.00	22.00 - 26.00	22.00 - 26.00	2.5.135
-	28.00 - 33.00	28.00 - 33.00	28.00 - 33.00	28.00 - 33.00	2.5.136
-	33.00 - 36.00	33.00 - 36.00	33.00 - 36.00	33.00 - 36.00	2.5.137
-	38.00 - 42.00	38.00 - 42.00	38.00 - 42.00	38.00 - 42.00	2.5.138
-	70.00 - 80.00	70.00 - 80.00	70.00 - 80.00	70.00 - 80.00	2.5.139

Item nr.	SPECIFICATIONS	Unit	RESIDENTIAL £ range £
2.5	**EXTERNAL WALLS** - cont'd	external wall area (unless otherwise described)	
	Comparative external finishes		
	75 mm Portland stone facing slabs and fixing; including		
2.5.140	clamps	m2	- -
	75 mm Ancaster stone facing slabs and fixing; including		
2.5.141	clamps	m2	- -
	Comparative curtain wall finishes;		
	extra over aluminium mill finish for		
2.5.142	natural anodising	m2	- -
2.5.143	polyester powder coating	m2	- -
2.5.144	bronze anodising	m2	- -
2.6	**WINDOWS AND EXTERNAL DOORS**	window and external door area (unless otherwise described)	
	comprising:-		
	Softwood windows and external doors		
	Steel windows and external doors		
	Steel roller shutters		
	Hardwood windows and external doors		
	UPVC windows and external doors		
	Aluminium windows, entrance screens and doors		
	Stainless steel entrance screens and doors		
	Shop fronts, shutters and grilles		
	Softwood windows and external doors		
	Standard windows; painted		
2.6.1	single glazed	m2	140.00 -180.00
2.6.2	double glazed	m2	175.00 -215.00
	Purpose-made windows; painted		
2.6.3	single glazed	m2	180.00 -220.00
2.6.4	double glazed	m2	225.00 -275.00
	Standard external softwood doors and hardwood frames;		
	doors painted; including ironmongery		
2.6.5	two panelled door; plywood panels	nr	230.00 -270.00
2.6.6	solid flush door	nr	260.00 -300.00
2.6.7	two panelled door; glazed panels	nr	500.00 -550.00
	heavy-duty solid flush door		
2.6.8	single leaf	nr	- -
2.6.9	double leaf	nr	- -
	Extra for		
2.6.10	emergency fire exit door	nr	- -
	Steel windows and external doors		
	Standard windows		
2.6.11	single glazed; galvanised; painted	m2	150.00 -190.00
2.6.12	single glazed; powder-coated	m2	155.00 -195.00
2.6.13	double glazed; galvanised; painted	m2	190.00 -230.00
2.6.14	double glazed; powder coated	m2	195.00 -235.00
	Purpose-made windows		
2.6.15	double glazed; powder coated	m2	- -
	Steel roller shutters		
	Shutters; galvanised		
2.6.16	manual	m2	- -
2.6.17	electric	m2	- -
2.6.18	manual; insulated	m2	- -
2.6.19	electric; insulated	m2	- -
2.6.20	electric; insulated; fire-resistant	m2	- -
	Hardwood windows and external doors		
	Standard windows; stained or UPVC coated		
2.6.21	single glazed	m2	200.00 -270.00
2.6.22	double glazed	m2	250.00 -320.00

INDUSTRIAL £ range £	RETAILING £ range £	LEISURE £ range £	OFFICES £ range £	HOTELS £ range £	Item nr.
- -	- -	- -	240.00 -300.00	240.00 -350.00	2.5.140
- -	- -	- -	270.00 - 330.00	- -	2.5.141
- -	15.00 - 20.00	15.00 - 20.00	15.00 - 20.00	15.00 - 20.00	2.5.142
- -	20.00 - 33.00	20.00 - 33.00	20.00 - 33.00	20.00 - 33.00	2.5.143
- -	27.00 - 40.00	27.00 - 40.00	27.00 - 40.00	27.00 - 40.00	2.5.144
130.00 -170.00	130.00 -170.00	140.00 -180.00	140.00 -180.00	140.00 -190.00	2.6.1
170.00 -210.00	170.00 -210.00	180.00 -220.00	180.00 -220.00	180.00 -230.00	2.6.2
180.00 -220.00	180.00 -210.00	190.00 -230.00	185.00 -225.00	185.00 -240.00	2.6.3
220.00 -260.00	220.00 -260.00	220.00 -260.00	220.00 -255.00	230.00 -270.00	2.6.4
-	-	-	-	-	2.6.5
350.00 -600.00	270.00 -310.00	270.00 -310.00	270.00 -310.00	280.00 -320.00	2.6.6
-	-	-	-	-	2.6.7
400.00 -600.00	450.00 -550.00	450.00 -550.00	-	-	2.6.8
700.00 -1000	850.00 -950.00	850.00 -950.00	-	-	2.6.9
- -	150.00 -225.00	150.00 -225.00	-	-	2.6.10
140.00 -180.00	140.00 -180.00	150.00 -190.00	150.00 -190.00	150.00 -200.00	2.6.11
145.00 -185.00	145.00 -185.00	155.00 -195.00	155.00 -195.00	155.00 -210.00	2.6.12
175.00 -220.00	175.00 -220.00	185.00 -230.00	185.00 -230.00	190.00 -240.00	2.6.13
180.00 -225.00	180.00 -225.00	190.00 -235.00	190.00 -235.00	195.00 -250.00	2.6.14
- -	250.00 -310.00	250.00 -310.00	250.00 -310.00	270.00 -330.00	2.6.15
150.00 -190.00	140.00 -180.00	140.00 -180.00	-	-	2.6.16
180.00 -250.00	170.00 -240.00	170.00 -240.00	-	-	2.6.17
240.00 -290.00	230.00 -280.00	230.00 -280.00	-	-	2.6.18
275.00 -350.00	260.00 -340.00	260.00 -340.00	-	-	2.6.19
670.00 -800.00	650.00 -800.00	650.00 -800.00	-	-	2.6.20
250.00 -300.00	250.00 -300.00	250.00 -300.00	250.00 -300.00	260.00 -325.00	2.6.21
320.00 -380.00	320.00 -380.00	320.00 -380.00	320.00 -380.00	320.00 -400.00	2.6.22

Item nr.	SPECIFICATIONS	Unit	RESIDENTIAL £ range £
2.6	**WINDOWS AND EXTERNAL DOORS** **Window and external door area** - cont'd (unless otherwise described)		
	Hardwood windows and external doors - cont'd Purpose-made windows; stained or UPVC coated		
2.6.23	single glazed	m2	235.00 -305.00
2.6.24	double glazed	m2	285.00 -360.00
	Upvc windows and external doors Purpose-made windows		
2.6.25	double glazed	m2	400.00 -500.00
	Extra for		
2.6.26	tinted glass	m2	- -
	Aluminium windows, entrance screens and doors Standard windows; anodised finish		
2.6.27	single glazed; horizontal sliding sash	m2	- -
2.6.28	single glazed; vertical sliding sash	m2	- -
2.6.29	single glazed; casement; in hardwood sub-frame	m2	210.00 -285.00
2.6.30	double glazed; vertical sliding sash	m2	- -
2.6.31	double glazed; casement; in hardwood sub-frame	m2	250.00 -340.00
	Purpose-made windows		
2.6.32	single glazed	m2	- -
2.6.33	double glazed	m2	- -
2.6.34	double glazed; feature; with precast concrete surrounds	m2	- -
	Purpose-made entrance screens and doors		
2.6.35	double glazed	m2	- -
	Purpose-made revolving door		
2.6.36	2000 mm dia.; double glazed	nr	- -
	Stainless steel entrance screens and doors Purpose-made screen; double glazed		
2.6.37	with manual doors	m2	- -
2.6.38	with automatic doors	m2	- -
	Purpose-made revolving door		
2.6.39	2000 mm dia; double glazed	nr	- -
	Shop fronts, shutters and grilles shop front length		
2.6.40	Temporary timber shop fronts	m	- -
2.6.41	Grilles or shutters	m	- -
2.6.42	Fire shutters; power-operated	m	- -
	Shop front flat facade; glass in aluminium framing; manual		
2.6.43	centre doors only	m	- -
	flat facade; glass in aluminium framing; automatic		
2.6.44	centre doors only	m	- -
	hardwood and glass; including high enclosed window		
2.6.45	beds	m	- -
	high quality; marble or granite plasters and stair		
2.6.46	risers; window beds and backings; illuminated signs	m	- -

INDUSTRIAL £ range £	RETAILING £ range £	LEISURE £ range £	OFFICES £ range £	HOTELS £ range £	Item nr.
300.00 -360.00	300.00 -360.00	300.00 -360.00	300.00 -360.00	300.00 -380.00	2.6.23
360.00 -430.00	360.00 -430.00	360.00 -430.00	360.00 -430.00	360.00 -440.00	2.6.24
375.00 -450.00	375.00 -450.00	375.00 -450.00	375.00 -450.00	375.00 -450.00	2.6.25
19.00 - 24.00	19.00 - 24.00	19.00 - 24.00	19.00 - 24.00	19.00 - 24.00	2.6.26
180.00 -215.00	180.00 -215.00	180.00 -215.00	180.00 -215.00	180.00 -225.00	2.6.27
275.00 -325.00	275.00 -325.00	275.00 -325.00	275.00 -325.00	275.00 -340.00	2.6.28
-	-	-	-	-	2.6.29
300.00 -360.00	350.00 -430.00	350.00 -430.00	350.00 -430.00	350.00 -450.00	2.6.30
-	-	-	-	-	2.6.31
180.00 -250.00	180.00 -250.00	180.00 -250.00	180.00 -250.00	180.00 -250.00	2.6.32
400.00 -480.00	400.00 -480.00	400.00 -480.00	400.00 -480.00	400.00 -500.00	2.6.33
-	-	1000 -1500	-	-	2.6.34
-	500.00 -800.00	500.00 -800.00	500.00 -800.00	-	2.6.35
-	17500 -22500	17500 -22500	17500 -22500	17500 -22500	2.6.36
-	800.00 -1200	-	-	-	2.6.37
-	1000 -1400	-	-	-	2.6.38
-	24000 -30000	24000 -30000	24000 -30000	24000 -30000	2.6.39
-	35.00 - 45.00	-	-	-	2.6.40
-	400.00 -800.00	-	-	-	2.6.41
-	700.00 -1000	-	-	-	2.6.42
-	800.00 -1800	-	-	-	2.6.43
-	1000 -2200	-	-	-	2.6.44
-	2900 -3400	-	-	-	2.6.45
-	3300 -4400	-	-	-	2.6.46

Keep your figures up to date, free of charge

This section, and most of the other information in this Price Book, is brought up to date every three months in the *Price Book Update.*

The *Update* is available free to all Price Book purchasers.

To ensure you receive your copy, simply complete the reply card from the centre of the book and return it to us.

Item nr.	SPECIFICATIONS	Unit	RESIDENTIAL £ range £
2.7	**INTERNAL WALLS, PARTITIONS AND DOORS** internal wall area (unless otherwise described)		
	comprising:-		
	Timber or metal stud partitions and doors		
	Brick/block partitions and doors		
	Reinforced concrete walls		
	Solid partitioning and doors		
	Glazed partitioning and doors		
	Special partitioning and doors		
	WC/Changing cubicles		
	Comparative doors/door linings/frames		
	Perimeter treatments		
	Timber or metal stud partitions and doors Structure only comprising softwood studs at 400 x 600 mm centres; head and sole plates		
2.7.1	100 x 38 mm	m2	10.00 - 12.00
	Softwood stud and plasterboard partitions		
2.7.2	57 mm Paramount dry partition	m2	18.25 - 20.00
2.7.3	65 mm Paramount dry partition	m2	20.00 - 22.00
2.7.4	50 mm laminated partition	m2	19.50 - 21.50
2.7.5	65 mm laminated partition	m2	23.00 - 25.00
2.7.6	65 mm laminated partition; emulsioned both sides	m2	27.00 - 30.00
2.7.7	100 mm partition; taped joints; emulsioned both sides	m2	30.00 - 35.00
2.7.8	100 mm partition; skim and emulsioned both sides	m2	35.00 - 40.00
2.7.9	150 mm partition as party wall; skim and emulsioned both sides	m2	42.00 - 50.00
	Metal stud and plasterboard partitions		
2.7.10	170 mm partition; one hour; taped joints; emulsioned both sides	m2	- -
2.7.11	200 mm partition; two hour; taped joints; emulsioned both sides	m2	- -
2.7.12	325 mm two layer partition; cavity insulation	m2	- -
2.7.13	Extra for curved work		- -
	Metal stud and plasterboard partitions; emulsioned both sides; softwood doors and frames; painted		
2.7.14	170 mm partition	m2	- -
2.7.15	200 mm partition; insulated	m2	- -
	Stud or plasterboard partitions; softwood doors and frames; painted		
2.7.16	partition; plastered and emulsioned both sides	m2	60.00 - 75.00
	Extra for		
2.7.17	vinyl paper in lieu of emulsion	m2	4.00 - 6.00
2.7.18	hardwood doors and frames in lieu of softwood	m2	17.00 - 20.00
2.7.19	partition; plastered and vinyled both sides	m2	70.00 - 85.00
	Stud or plasterboard partitions; hardwood doors and frames		
2.7.20	partition; plastered and emulsioned both sides	m2	77.00 - 95.00
2.7.21	partition; plastered and vinyled both sides	m2	87.00 -105.00
	Brick/block partitions and doors Autoclaved aerated/lightweight block partitions		
2.7.22	75 mm thick	m2	13.50 - 15.00
2.7.23	100 mm thick	m2	17.00 - 20.00
2.7.24	130 mm thick; insulating	m2	21.00 - 23.00
2.7.25	150 mm thick	m2	22.50 - 24.50
2.7.26	190 mm thick	m2	27.00 - 31.00
	Extra for		
2.7.27	fair face both sides	m2	2.00 - 5.00
2.7.28	curved work		+10% to +20%
2.7.29	average thickness; fair face both sides	m2	23.00 - 28.00
2.7.30	average thickness; fair face and emulsioned both sides	m2	27.00 - 33.00
2.7.31	average thickness; plastered and emulsioned both sides	m2	41.00 - 47.00
	Concrete block partitions		
2.7.32	to retail units	m2	- -

INDUSTRIAL £ range £	RETAILING £ range £	LEISURE £ range £	OFFICES £ range £	HOTELS £ range £	Item nr.
-	-	-	-	-	2.7.1
-	-	-	-	-	2.7.2
-	-	-	-	-	2.7.3
-	-	-	-	-	2.7.4
-	-	-	-	-	2.7.5
-	-	-	-	-	2.7.6
-	-	-	-	-	2.7.7
-	-	-	-	-	2.7.8
-	-	-	-	-	2.7.9
36.00 - 43.00	38.00 - 45.00	38.00 - 45.00	38.00 - 45.00	39.00 - 46.00	2.7.10
50.00 - 55.00	52.00 - 57.00	52.00 - 57.00	52.00 - 57.00	53.00 - 58.00	2.7.11
85.00 -115.00	90.00 -120.00	90.00 -120.00	90.00 -120.00	90.00 -120.00	2.7.12
+50%	+50%	+50%	+50%	+50%	2.7.13
48.00 - 63.00	50.00 - 65.00	50.00 - 65.00	50.00 - 65.00	50.00 - 67.50	2.7.14
63.00 - 78.00	65.00 - 80.00	65.00 - 80.00	65.00 - 80.00	65.00 - 85.00	2.7.15
58.00 - 73.00	60.00 - 75.00	60.00 - 75.00	60.00 - 75.00	60.00 - 80.00	2.7.16
-	4.00 - 6.00	4.00 - 6.00	4.00 - 6.00	4.00 - 10.00	2.7.17
17.00 - 20.00	17.00 - 20.00	17.00 - 20.00	17.00 - 20.00	17.00 - 25.00	2.7.18
68.00 - 82.50	70.00 - 85.00	70.00 - 85.00	70.00 - 85.00	70.00 - 90.00	2.7.19
75.00 - 90.00	77.00 - 95.00	77.00 - 95.00	77.00 - 95.00	77.00 -100.00	2.7.20
85.00 -100.00	87.00 -105.00	87.00 -105.00	87.00 -105.00	87.00 -120.00	2.7.21
12.00 - 14.50	13.50 - 15.00	13.50 - 15.00	13.50 - 15.00	14.50 - 16.00	2.7.22
16.00 - 19.00	17.00 - 20.00	17.00 - 20.00	17.00 - 20.00	18.00 - 21.00	2.7.23
20.00 - 22.00	21.00 - 23.00	21.00 - 23.00	21.00 - 23.00	22.00 - 24.00	2.7.24
21.50 - 23.50	22.50 - 24.50	22.50 - 24.50	22.50 - 24.50	23.50 - 24.50	2.7.25
26.00 - 30.00	27.00 - 31.00	27.00 - 31.00	27.00 - 31.00	28.00 - 32.00	2.7.26
2.00 - 5.00	2.00 - 5.00	2.00 - 5.00	2.00 - 5.00	2.00 - 5.00	2.7.27
+10% to +20%	+10% to +20%	+10% to +20%	+10% to +20%	+10% to +20%	2.7.28
22.00 - 27.00	23.00 - 28.00	23.00 - 28.00	23.00 - 28.00	24.00 - 29.00	2.7.29
26.00 - 32.00	27.00 - 33.00	27.00 - 33.00	27.00 - 33.00	28.00 - 34.00	2.7.30
40.00 - 46.00	41.00 - 47.00	41.00 - 47.00	41.00 - 47.00	42.00 - 50.00	2.7.31
-	35.00 - 40.00	-	-	-	2.7.32

Item nr.	SPECIFICATIONS	Unit	RESIDENTIAL £ range £
2.7	**INTERNAL WALLS, PARTITIONS AND DOORS** internal wall area - cont'd (unless otherwise described)		
	Brick/block partitions and doors - cont'd		
	Dense aggregate block partitions		
2.7.33	average thickness; fair face both sides	m2	26.00 - 31.00
2.7.34	average thickness; fair face and emulsioned both sides	m2	30.00 - 36.00
2.7.35	average thickness; plastered and emulsioned both sides	m2	44.00 - 50.00
	Coloured dense aggregate masonry block partition		
2.7.36	fair face both sides	m2	- -
	Common brick partitions; bricks PC £117.00/1000		
2.7.37	half brick thick	m2	23.00 - 27.00
2.7.38	half brick thick; fair face both sides	m2	26.00 - 31.00
2.7.39	half brick thick; fair face and emulsioned both sides	m2	30.00 - 36.00
2.7.40	half brick thick; plastered and emulsioned both sides	m2	43.00 - 55.00
2.7.41	one brick thick	m2	45.00 - 50.00
2.7.42	one brick thick; fair face both sides	m2	48.00 - 54.00
2.7.43	one brick thick; fair face and emulsioned both sides	m2	52.00 - 59.00
2.7.44	one brick thick; plastered and emulsioned both sides	m2	65.00 - 75.00
	Block partitions; softwood doors and frames; painted		
2.7.45	partition	m2	38.00 - 50.00
2.7.46	partition; fair face both sides	m2	40.00 - 52.00
2.7.47	partition; fair face and emulsioned both sides	m2	44.00 - 57.00
2.7.48	partition; plastered and emulsioned both sides	m2	58.00 - 73.00
	Block partitions; hardwood doors and frames		
2.7.49	partition	m2	55.00 - 70.00
2.7.50	partition; plastered and emulsioned both sides	m2	75.00 - 95.00
	Reinforced concrete walls		
	Walls		
2.7.51	150 mm thick	m2	80.00 - 90.00
2.7.52	150 mm thick; plastered and emulsioned both sides	m2	100.00 -115.00
	Solid partitioning and doors		
	Patent partitioning; softwood doors		
2.7.53	frame and sheet	m2	- -
2.7.54	frame and panel	m2	- -
2.7.55	panel to panel	m2	- -
2.7.56	economical	m2	- -
	Patent partitioning; hardwood doors		
2.7.57	economical	m2	- -
	Demountable partitioning; hardwood doors		
2.7.58	medium quality; vinyl-faced	m2	- -
2.7.59	high quality; vinyl-faced	m2	- -
	Glazed partitioning and doors		
	Aluminium internal patent glazing		
2.7.60	single glazed	m2	- -
2.7.61	double glazed	m2	- -
	Demountable steel partitioning and doors		
2.7.62	medium quality	m2	- -
2.7.63	high quality	m2	- -
	Demountable aluminium/steel partitioning and doors		
2.7.64	high quality	m2	- -
2.7.65	high quality; sliding	m2	- -
	Stainless steel glazed manual doors and screens		
2.7.66	high quality; to inner lobby of malls	m2	- -
	Special partitioning and doors		
	Demountable fire partitions		
2.7.67	enamelled steel; half hour	m2	- -
2.7.68	stainless steel; half hour	m2	- -
	Soundproof partitions; hardwood doors		
2.7.69	luxury veneered	m2	- -
	Folding screens		
2.7.70	gym divider; electronically operated	m2	- -
2.7.71	bar divider	m2	- -

INDUSTRIAL £ range £	RETAILING £ range £	LEISURE £ range £	OFFICES £ range £	HOTELS £ range £	Item nr.
25.00 - 30.00	26.00 - 31.00	26.00 - 31.00	26.00 - 31.00	27.00 - 32.00	2.7.33
29.00 - 35.00	30.00 - 36.00	30.00 - 36.00	30.00 - 36.00	31.00 - 37.00	2.7.34
43.00 - 48.00	44.00 - 50.00	44.00 - 50.00	44.00 - 50.00	45.00 - 52.50	2.7.35
-	-	50.00 - 60.00	-	-	2.7.36
22.00 - 26.00	23.00 - 27.00	23.00 - 27.00	23.00 - 27.00	24.00 - 28.00	2.7.37
25.00 - 30.00	26.00 - 31.00	26.00 - 31.00	26.00 - 31.00	27.00 - 32.00	2.7.38
29.00 - 35.00	30.00 - 36.00	30.00 - 36.00	30.00 - 36.00	31.00 - 37.00	2.7.39
42.00 - 53.00	43.00 - 55.00	43.00 - 55.00	43.00 - 55.00	45.00 - 60.00	2.7.40
43.00 - 48.00	45.00 - 50.00	45.00 - 50.00	45.00 - 50.00	47.00 - 52.00	2.7.41
46.00 - 52.00	48.00 - 54.00	48.00 - 54.00	48.00 - 54.00	50.00 - 56.00	2.7.42
50.00 - 57.00	52.00 - 59.00	52.00 - 59.00	52.00 - 59.00	54.00 - 61.00	2.7.43
63.00 - 72.00	65.00 - 75.00	65.00 - 75.00	65.00 - 75.00	67.50 - 80.00	2.7.44
36.00 - 48.00	38.00 - 50.00	38.00 - 50.00	38.00 - 50.00	40.00 - 50.00	2.7.45
38.00 - 50.00	40.00 - 52.00	40.00 - 52.00	40.00 - 52.00	42.00 - 52.00	2.7.46
42.00 - 54.00	44.00 - 57.00	44.00 - 57.00	44.00 - 57.00	46.00 - 56.00	2.7.47
56.00 - 70.00	58.00 - 73.00	58.00 - 73.00	58.00 - 73.00	58.00 - 78.00	2.7.48
53.00 - 67.00	55.00 - 70.00	55.00 - 70.00	55.00 - 70.00	59.50 - 72.50	2.7.49
73.00 - 92.00	75.00 - 95.00	75.00 - 95.00	75.00 - 95.00	77.00 -100.00	2.7.50
75.00 - 85.00	80.00 - 90.00	80.00 - 90.00	80.00 - 90.00	80.00 - 90.00	2.7.51
95.00 -110.00	100.00 -115.00	100.00 -115.00	100.00 -115.00	100.00 -120.00	2.7.52
50.00 - 95.00	55.00 -100.00	55.00 -100.00	55.00 -100.00	55.00 -120.00	2.7.53
42.50 - 78.00	45.00 - 80.00	45.00 - 80.00	45.00 - 80.00	45.00 -100.00	2.7.54
62.50 -130.00	66.00 -135.00	66.00 -135.00	66.00 -135.00	66.00 -150.00	2.7.55
65.00 - 85.00	70.00 - 90.00	70.00 - 90.00	70.00 - 90.00	70.00 -110.00	2.7.56
70.00 - 90.00	75.00 - 95.00	75.00 - 95.00	75.00 - 95.00	75.00 -100.00	2.7.57
95.00 -120.00	100.00 -125.00	100.00 -125.00	100.00 -125.00	100.00 -130.00	2.7.58
120.00 -170.00	125.00 -175.00	125.00 -175.00	125.00 -175.00	130.00 -190.00	2.7.59
70.00 -100.00	-	-	-	-	2.7.60
120.00 -150.00	-	-	-	-	2.7.61
130.00 -160.00	-	-	135.00 -170.00	-	2.7.62
160.00 -200.00	-	-	180.00 -210.00	-	2.7.63
-	-	200.00 -350.00	-	275.00 -350.00	2.7.64
-	-	475.00 -575.00	-	-	2.7.65
-	275.00 -750.00	-	-	-	2.7.66
300.00 -450.00	350.00 -480.00	350.00 -480.00	350.00 -480.00	-	2.7.67
-	600.00 -750.00	600.00 -750.00	600.00 -750.00	-	2.7.68
150.00 -200.00	160.00 -225.00	160.00 -225.00	160.00 -225.00	175.00 -250.00	2.7.69
-	-	110.00 -120.00	-	-	2.7.70
-	-	300.00 -330.00	-	-	2.7.71

Item nr.	SPECIFICATIONS	Unit	RESIDENTIAL £ range £

2.7 **INTERNAL WALLS, PARTITION AND DOORS**   **internal wall area**
      - cont'd   (unless otherwise described)

Special partitioning and doors - cont'd

2.7.72	Squash court glass back wall and door	m2	- -
	WC/Changing cubicles	**each**	
2.7.73	WC cubicles	nr	- -
	Changing cubicles		
2.7.74	aluminium	nr	- -
2.7.75	aluminium; textured glass and bench seating	nr	- -

Comparative doors/door linings/frames
Standard softwood doors; excluding ironmongery; linings
and frames
   40 mm flush; hollow core; painted

2.7.76	726 x 2040 mm	nr	44.00 - 50.00
2.7.77	826 x 2040 mm	nr	47.00 - 53.00

   40 mm flush; hollow core; plywood faced; painted

2.7.78	726 x 2040 mm	nr	47.50 - 52.50
2.7.79	826 x 2040 mm	nr	50.50 - 56.00

   40 mm flush; hollow core; Sapele veneered hard
   board faced

2.7.80	726 x 2040 mm	nr	55.00 - 65.00
2.7.81	826 x 2040 mm	nr	58.00 - 68.00

   40 mm flush; hollow core; Teak veneered hard
   board faced

2.7.82	726 x 2040 mm	nr	75.00 - 85.00
2.7.83	826 x 2040 mm	nr	78.00 - 88.00

Standard softwood fire-doors; excluding ironmongery;
linings and frames
   44 mm flush; half hour fire check; plywood faced;
   painted

2.7.84	726 x 2040 mm	nr	65.00 - 75.00
2.7.85	826 x 2040 mm	nr	70.00 - 80.00

   44 mm flush; half hour fire check; Sapele veneered
   hardboard faced

2.7.86	726 x 2040 mm	nr	82.50 - 92.50
2.7.87	826 x 2040 mm	nr	87.50 - 97.50

   54 mm flush; one hour fire check; Sapele veneered
   hardboard faced

2.7.88	726 x 2040 mm	nr	195.00 -220.00
2.7.89	826 x 2040 mm	nr	200.00 -225.00

   54 mm flush; one hour fire resisting; Sapele veneered
   hardboard faced

2.7.90	726 x 2040 mm	nr	215.00 -240.00
2.7.91	826 x 2040 mm	nr	220.00 -250.00

Purpose-made softwood doors; excluding ironmongery;
linings and frames
   44 mm four panel door; painted

2.7.92	726 x 2040 mm	nr	125.00 -135.00
2.7.93	826 x 2040 mm	nr	130.00 -140.00

Purpose-made Mahogany doors; excluding ironmongery;
linings and frames
   50 mm four panel door; wax polished

2.7.94	726 x 2040 mm	nr	215.00 -240.00
2.7.95	826 x 2040 mm	nr	225.00 -250.00

Purpose-made softwood door frames/linings; painted
including grounds
   32 x 100 mm lining

2.7.96	726 x 2040 mm opening	nr	67.50 - 75.00
2.7.97	826 x 2040 mm opening	nr	69.00 - 76.50

   32 x 140 mm lining

2.7.98	726 x 2040 mm opening	nr	74.00 - 85.00
2.7.99	826 x 2040 mm opening	nr	76.00 - 87.00

   32 x 250 mm Cross-tongued lining

2.7.100	726 x 2040 mm opening	nr	98.50 -110.00
2.7.101	826 x 2040 mm opening	nr	100.00 -112.50

INDUSTRIAL £ range £	RETAILING £ range £	LEISURE £ range £	OFFICES £ range £	HOTELS £ range £	Item nr.
- -	- -	160.00 -200.00	- -	- -	2.7.72
250.00 -375.00	255.00 -375.00	255.00 -375.00	250.00 -550.00	310.00 -550.00	2.7.73
- -	- -	300.00 - 550.00	- -	- -	2.7.74
- -	- -	500.00 - 650.00	- -	- -	2.7.75
44.00 - 50.00	44.00 - 50.00	44.00 - 50.00	44.00 - 50.00	44.00 - 50.00	2.7.76
47.00 - 53.00	47.00 - 53.00	47.00 - 53.00	47.00 - 53.00	47.00 - 53.00	2.7.77
47.50 - 52.50	47.50 - 52.50	47.50 - 52.50	47.50 - 52.50	47.50 - 52.50	2.7.78
50.50 - 56.00	50.50 - 56.00	50.50 - 56.00	50.50 - 56.00	50.50 - 56.00	2.7.79
55.00 - 65.00	55.00 - 65.00	55.00 - 65.00	55.00 - 65.00	55.00 - 65.00	2.7.80
58.00 - 68.00	58.00 - 68.00	58.00 - 68.00	58.00 - 68.00	58.00 - 68.00	2.7.81
75.00 - 85.00	75.00 - 85.00	75.00 - 85.00	75.00 - 85.00	75.00 - 85.00	2.7.82
78.00 - 88.00	78.00 - 88.00	78.00 - 88.00	78.00 - 88.00	78.00 - 88.00	2.7.83
65.00 - 75.00	65.00 - 75.00	65.00 - 75.00	65.00 - 75.00	65.00 - 75.00	2.7.84
70.00 - 80.00	70.00 - 80.00	70.00 - 80.00	70.00 - 80.00	70.00 - 80.00	2.7.85
82.50 - 92.50	82.50 - 92.50	82.50 - 92.50	82.50 - 92.50	82.50 - 92.50	2.7.86
87.50 - 97.50	87.50 - 97.50	87.50 - 97.50	87.50 - 97.50	87.50 - 97.50	2.7.87
195.00 -220.00	195.00 -220.00	195.00 -220.00	195.00 -220.00	195.00 -220.00	2.7.88
200.00 -225.00	200.00 -225.00	200.00 -225.00	200.00 -225.00	200.00 -225.00	2.7.89
215.00 -240.00	215.00 -240.00	215.00 -240.00	215.00 -240.00	215.00 -240.00	2.7.90
220.00 -250.00	220.00 -250.00	220.00 -250.00	220.00 -250.00	220.00 -250.00	2.7.91
125.00 -135.00	125.00 -135.00	125.00 -135.00	125.00 -135.00	125.00 -135.00	2.7.92
130.00 -140.00	130.00 -140.00	130.00 -140.00	130.00 -140.00	130.00 -140.00	2.7.93
215.00 -240.00	215.00 -240.00	215.00 -240.00	215.00 -240.00	215.00 -240.00	2.7.94
225.00 -250.00	225.00 -250.00	225.00 -250.00	225.00 -250.00	225.00 -250.00	2.7.95
67.50 - 75.00	67.50 - 75.00	67.50 - 75.00	67.50 - 75.00	67.50 - 75.00	2.7.96
69.00 - 76.50	69.00 - 76.50	69.00 - 76.50	69.00 - 76.50	69.00 - 76.50	2.7.97
74.00 - 85.00	74.00 - 85.00	74.00 - 85.00	74.00 - 85.00	74.00 - 85.00	2.7.98
76.00 - 87.00	76.00 - 87.00	76.00 - 87.00	76.00 - 87.00	76.00 - 87.00	2.7.99
98.50 -110.00	98.50 -110.00	98.50 -110.00	98.50 -110.00	98.50 -110.00	2.7.100
100.00 -112.50	100.00 -112.50	100.00 -112.50	100.00 -112.50	100.00 -112.50	2.7.101

Item nr.	SPECIFICATIONS	Unit	RESIDENTIAL £ range £
2.7	**INTERNAL WALLS, PARTITION AND DOORS** **internal wall area** - cont'd (unless otherwise described)		
	Comparative doors/door linings/frames - cont'd		
	32 x 375 mm Cross-tongued lining		
2.7.102	726 x 2040 mm opening	nr	130.00 -140.00
2.7.103	826 x 2040 mm opening	nr	132.50 -142.50
	75 x 100 mm rebated frame		
2.7.104	726 x 2040 mm opening	nr	87.50 - 92.50
2.7.105	826 x 2040 mm opening	nr	92.50 - 97.50
	Purpose-made Mahogany door frames/linings; wax polished including grounds		
	32 x 100 mm lining		
2.7.106	726 x 2040 mm opening	nr	90.00 - 97.50
2.7.107	826 x 2040 mm opening	nr	92.50 -100.00
	32 X 140 mm lining		
2.7.108	726 x 2040 mm opening	nr	100.00 -110.00
2.7.109	826 x 2040 mm opening	nr	102.50 -112.50
	32 x 250 mm cross-tongued lining		
2.7.110	726 x 2040 mm opening	nr	112.50 -122.50
2.7.111	826 x 2040 mm opening	nr	115.00 -125.00
	32 x 375 mm cross-tongued lining		
2.7.112	726 x 2040 mm opening	nr	170.00 -185.00
2.7.113	826 x 2040 mm opening	nr	175.00 -190.00
	75 x 100 mm rebated frame		
2.7.114	726 x 2040 mm opening	nr	120.00 -130.00
2.7.115	826 x 2040 mm opening	nr	122.50 -132.50
	Standard softwood doors and frames;including ironmongery and painting		
2.7.116	flush; hollow core	nr	160.00 -200.00
2.7.117	flush; hollow core; hardwood faced	nr	170.00 -220.00
	flush; solid core		
2.7.118	single leaf	nr	190.00 -250.00
2.7.119	double leaf	nr	280.00 -375.00
2.7.120	flush; solid core; hardwood faced	nr	200.00 -260.00
2.7.121	four panel door	nr	280.00 -350.00
	Purpose-made softwood doors and hardwood frames; including ironmongery; painting and polishing		
	flush; solid core; heavy duty		
2.7.122	single leaf	nr	- -
2.7.123	double leaf	nr	- -
	flush; solid core; heavy duty; plastic laminate faced		
2.7.124	single leaf	nr	- -
2.7.125	double leaf	nr	- -
	Purpose-made softwood fire doors and hardwood frames; including ironmongery; painting and polishing		
	flush; one hour fire resisting		
2.7.126	single leaf	nr	- -
2.7.127	double leaf	nr	- -
	flush; one hour fire resisting; plastic laminate faced		
2.7.128	single leaf	nr	- -
2.7.129	double leaf	nr	- -
	Purpose-made softwood doors and pressed steel frames		
2.7.130	flush; half hour fire check; plastic laminate faced	nr	- -
	Purpose-made Mahogany doors and frames; including ironmongery and polishing		
2.7.131	four panel door	nr	- -
	Perimeter treatments		
	Precast concrete lintels; in block walls		
2.7.132	75 mm wide	m	7.00 - 10.00
2.7.133	100 mm wide	m	8.00 - 11.00
	Precast concrete lintels; in brick walls		
2.7.134	half brick thick	m	8.00 - 11.00
2.7.135	one brick thick	m	12.00 - 15.00
	Purpose-made Softwood architraves; painted; including grounds		
	25 x 50 mm; to both sides of openings		
2.7.136	726 x 2040 mm opening	nr	65.00 - 70.00
2.7.137	826 x 2040 mm opening	nr	66.50 - 71.50

INDUSTRIAL £ range £	RETAILING £ range £	LEISURE £ range £	OFFICES £ range £	HOTELS £ range £	Item nr.
130.00 -140.00	130.00 -140.00	130.00 -140.00	130.00 -140.00	130.00 -140.00	2.7.102
132.50 -142.50	132.50 -142.50	132.50 -142.50	132.50 -142.50	132.50 -142.50	2.7.103
87.50 - 92.50	87.50 - 92.50	87.50 - 92.50	87.50 - 92.50	87.50 - 92.50	2.7.104
92.50 - 97.50	92.50 - 97.50	92.50 - 97.50	92.50 - 97.50	92.50 - 97.50	2.7.105
90.00 - 97.50	90.00 - 97.50	90.00 - 97.50	90.00 - 97.50	90.00 - 97.50	2.7.106
92.50 -100.00	92.50 -100.00	92.50 -100.00	92.50 -100.00	92.50 -100.00	2.7.107
100.00 -110.00	100.00 -110.00	100.00 -110.00	100.00 -110.00	100.00 -110.00	2.7.108
102.50 -112.50	102.50 -112.50	102.50 -112.50	102.50 -112.50	102.50 -112.50	2.7.109
112.50 -122.50	112.50 -122.50	112.50 -122.50	112.50 -122.50	112.50 -122.50	2.7.110
115.00 -125.00	115.00 -125.00	115.00 -125.00	115.00 -125.00	115.00 -125.00	2.7.111
170.00 -185.00	170.00 -185.00	170.00 -185.00	170.00 -185.00	170.00 -185.00	2.7.112
175.00 -190.00	175.00 -190.00	175.00 -190.00	175.00 -190.00	175.00 -190.00	2.7.113
120.00 -130.00	120.00 -130.00	120.00 -130.00	120.00 -130.00	120.00 -130.00	2.7.114
122.50 -132.50	122.00 -132.50	122.50 -132.50	122.50 -132.50	122.50 -132.50	2.7.115
160.00 -200.00	160.00 -200.00	160.00 -200.00	160.00 -200.00	160.00 -200.00	2.7.116
170.00 -220.00	170.00 -220.00	170.00 -220.00	170.00 -220.00	170.00 -220.00	2.7.117
190.00 -250.00	190.00 -250.00	190.00 -250.00	190.00 -250.00	190.00 -250.00	2.7.118
280.00 -375.00	280.00 -375.00	280.00 -375.00	280.00 -375.00	280.00 -375.00	2.7.119
200.00 -260.00	200.00 -260.00	200.00 -260.00	200.00 -260.00	200.00 -260.00	2.7.120
280.00 -350.00	280.00 -350.00	280.00 -350.00	280.00 -350.00	280.00 -350.00	2.7.121
480.00 -550.00	480.00 -550.00	480.00 -550.00	480.00 -550.00	480.00 -550.00	2.7.122
650.00 -825.00	650.00 -825.00	650.00 -825.00	650.00 -825.00	650.00 -825.00	2.7.123
580.00 -650.00	580.00 -650.00	580.00 -650.00	580.00 -650.00	580.00 -650.00	2.7.124
800.00 -900.00	800.00 -900.00	800.00 -900.00	800.00 -900.00	800.00 -900.00	2.7.125
630.00 -700.00	630.00 -700.00	630.00 -700.00	630.00 -700.00	630.00 -700.00	2.7.126
800.00 -975.00	800.00 -975.00	800.00 -975.00	800.00 -975.00	800.00 -975.00	2.7.127
780.00 -850.00	780.00 -850.00	780.00 -850.00	780.00 -850.00	780.00 -850.00	2.7.128
1000 -1080	1000 -1080	1000 -1080	1000 -1080	1000 -1080	2.7.129
725.00 -875.00	725.00 -875.00	725.00 -875.00	725.00 -875.00	725.00 -875.00	2.7.130
600.00 -700.00	600.00 -700.00	600.00 -700.00	600.00 -700.00	600.00 -700.00	2.7.131
7.00 - 10.00	7.00 - 10.00	7.00 - 10.00	7.00 - 10.00	7.00 - 10.00	2.7.132
8.00 - 11.00	8.00 - 11.00	8.00 - 11.00	8.00 - 11.00	8.00 - 11.00	2.7.133
8.00 - 11.00	8.00 - 11.00	8.00 - 11.00	8.00 - 11.00	8.00 - 11.00	2.7.134
12.00 - 15.00	12.00 - 15.00	12.00 - 15.00	12.00 - 15.00	12.00 - 15.00	2.7.135
65.00 - 70.00	65.00 - 70.00	65.00 - 70.00	65.00 - 70.00	65.00 - 70.00	2.7.136
66.50 - 71.50	66.50 - 71.50	66.50 - 71.50	66.50 - 71.50	66.50 - 71.50	2.7.137

Item nr.	SPECIFICATIONS	Unit	RESIDENTIAL £ range £
2.7	**INTERNAL WALLS, PARTITION AND DOORS** - cont'd (unless otherwise described)	**each**	
	Perimeter treatments - cont'd Purpose-made Mahogany architraves; wax polished; including grounds 25 x 50 mm; to both sides of openings		
2.7.138	726 x 2040 mm opening	nr	105.50 -115.50
2.7.139	826 x 2040 mm opening	nr	107.00 -117.00
3.1	**WALL FINISHES**	**wall finish area**	
	(unless otherwise described)		
	comprising:-		
	Sheet/board finishes **In situ wall finishes** **Rigid tile/panel finishes**		
	Sheet/board finishes Dry plasterboard lining; taped joints; for direct decoration		
3.1.1	9.5 mm Gyproc Wallboard	m2	7.50 - 10.50
	Extra for		
3.1.2	insulating grade	m2	0.50 - 0.60
3.1.3	insulating grade; plastic faced	m2	1.60 - 1.80
3.1.4	12.5 mm Gyproc Wallboard (half-hour fire-resisting)	m2	8.00 - 11.00
	Extra for		
3.1.5	insulating grade	m2	0.50 - 0.60
3.1.6	insulating grade; plastic faced	m2	1.40 - 1.60
	two layers of 12.5 mm Gyproc Wallboard (one hour		
3.1.7	fire-resisting)	m2	14.00 - 18.00
3.1.8	9 mm Supalux (half-hour fire-resisting)	m2	14.00 - 18.00
	Dry plasterboard lining; taped joints; for direct decoration; fixed to wall on dabs		
3.1.9	9.5 mm Gyproc Wallboard	m2	8.00 - 11.00
	Dry plasterboard lining; taped joints; for direct decoration; including metal tees		
3.1.10	9.5 mm Gyproc Wallboard	m2	- -
3.1.11	12.5 mm Gyproc Wallboard	m2	- -
	Dry lining/sheet panelling; including battens; plugged to wall		
3.1.12	6.4 mm hardboard	m2	9.00 - 10.00
3.1.13	9.5 mm Gyproc Wallboard	m2	12.50 - 16.50
3.1.14	6 mm birch faced plywood	m2	14.00 - 16.00
3.1.15	6 mm WAM plywood	m2	17.50 - 20.00
3.1.16	15 mm chipboard	m2	12.50 - 14.00
3.1.17	15 mm melamine faced chipboard	m2	20.00 - 22.50
3.1.18	13.2 mm 'Formica' faced chipboard	m2	30.00 - 45.00
	Timber boarding/panelling; on and including battens; plugged to wall		
3.1.19	12 mm softwood boarding	m2	20.00 - 25.00
3.1.20	25 mm softwood boarding	m2	25.00 - 27.50
3.1.21	hardwood panelling; t, g & v-jointed	m2	45.00 - 90.00
	In situ wall finishes Extra over common brickwork for		
3.1.22	fair face and pointing both sides	m2	3.00 - 4.00
	Comparative finishes		
3.1.23	one mist and two coats emulsion paint	m2	2.00 - 3.00
3.1.24	multi-coloured gloss paint	m2	3.75 - 5.00
3.1.25	two coats of lightweight plaster	m2	8.00 - 10.00
3.1.26	9.5 mm Gyproc Wallboard and skim coat	m2	10.00 - 12.00
3.1.27	12.5 mm Gyproc Wallboard and skim coat	m2	11.00 - 13.00
3.1.28	two coats of 'Thistle' plaster	m2	10.00 - 12.50
3.1.29	plaster and emulsion	m2	10.00 - 15.00

INDUSTRIAL £ range £	RETAILING £ range £	LEISURE £ range £	OFFICES £ range £	HOTELS £ range £	Item nr.
105.50 -115.50	105.50 -115.50	105.50 -115.50	105.50 -115.50	105.50 -115.50	2.7.138
107.00 -117.00	107.00 -117.00	107.00 -117.00	107.00 -117.00	107.00 -117.00	2.7.139
7.00 - 10.00	7.50 - 10.50	7.50 - 10.50	7.50 - 10.50	8.00 - 12.00	3.1.1
0.50 - 0.60	0.50 - 0.60	0.50 - 0.60	0.50 - 0.60	0.50 - 0.60	3.1.2
1.60 - 1.80	1.60 - 1.80	1.60 - 1.80	1.60 - 1.80	1.60 - 1.80	3.1.3
7.50 - 10.50	8.00 - 11.00	8.00 - 11.00	8.00 - 11.00	8.50 - 12.50	3.1.4
0.50 - 0.60	0.50 - 0.60	0.50 - 0.60	0.50 - 0.60	0.50 - 0.60	3.1.5
1.40 - 1.60	1.40 - 1.60	1.40 - 1.60	1.40 - 1.60	1.40 - 1.60	3.1.6
13.50 - 17.50	14.00 - 18.00	14.00 - 18.00	14.00 - 18.00	15.00 - 20.00	3.1.7
13.50 - 17.50	14.00 - 18.00	14.00 - 18.00	14.00 - 18.00	15.00 - 20.00	3.1.8
7.50 - 10.50	8.00 - 11.00	8.00 - 11.00	8.00 - 11.00	8.50 - 11.50	3.1.9
17.00 - 20.00	- -	- -	- -	- -	3.1.10
18.00 - 21.00	- -	- -	- -	- -	3.1.11
8.50 - 9.50	9.00 - 10.00	9.00 - 10.00	9.00 - 10.00	9.50 - 11.00	3.1.12
12.00 - 16.00	12.50 - 16.50	12.50 - 16.50	12.50 - 16.50	13.00 - 17.50	3.1.13
13.50 - 15.50	14.00 - 16.00	14.00 - 16.00	14.00 - 16.00	14.50 - 17.50	3.1.14
17.00 - 19.50	17.50 - 20.00	17.50 - 20.00	17.50 - 20.00	18.00 - 21.00	3.1.15
12.00 - 13.50	12.50 - 14.00	12.50 - 14.00	12.50 - 14.00	13.00 - 15.00	3.1.16
19.50 - 22.00	20.00 - 22.50	20.00 - 22.50	20.00 - 22.50	21.00 - 25.00	3.1.17
29.00 - 43.00	30.00 - 45.00	30.00 - 45.00	30.00 - 45.00	30.00 - 50.00	3.1.18
19.00 - 24.00	20.00 - 25.00	20.00 - 25.00	20.00 - 25.00	21.00 - 26.00	3.1.19
24.00 - 26.50	25.00 - 27.50	25.00 - 27.50	25.00 - 27.50	26.00 - 29.00	3.1.20
43.00 - 87.00	45.00 - 90.00	45.00 - 90.00	45.00 - 90.00	47.00 - 94.00	3.1.21
3.00 - 4.00	3.00 - 4.00	3.00 - 4.00	3.00 - 4.00	3.00 - 4.00	3.1.22
2.00 - 3.00	2.00 - 3.00	2.00 - 3.00	2.00 - 3.00	2.00 - 3.00	3.1.23
3.75 - 5.00	3.75 - 5.00	3.75 - 5.00	3.75 - 5.00	3.75 - 5.00	3.1.24
7.50 - 9.50	8.00 - 10.00	8.00 - 10.00	8.00 - 10.00	8.50 - 11.00	3.1.25
9.50 - 11.50	10.00 - 12.00	10.00 - 12.00	10.00 - 12.00	10.50 - 12.50	3.1.26
10.50 - 12.50	11.00 - 13.00	11.00 - 13.00	11.00 - 13.00	11.50 - 13.50	3.1.27
10.00 - 12.50	10.00 - 12.50	10.00 - 12.50	10.00 - 12.50	10.50 - 13.00	3.1.28
9.50 - 14.00	10.00 - 15.00	10.00 - 15.00	10.00 - 15.00	11.00 - 17.50	3.1.29

Item nr.	SPECIFICATIONS	Unit	RESIDENTIAL £ range £
3.1	**WALL FINISHES** - cont'd	**wall finish area**	
	(unless otherwise described)		
	In situ wall finishes - cont'd		
	Extra for		
3.1.30	gloss paint in lieu of emulsion	m2	1.75 - 2.00
3.1.31	two coat render and emulsion	m2	17.50 - 22.50
3.1.32	plaster and vinyl	m2	14.00 - 20.00
3.1.33	plaster and fabric	m2	14.00 - 30.00
3.1.34	squash court plaster 'including markings'	m2	- -
3.1.35	6 mm terrazzo wall lining; including backing	m2	- -
3.1.36	glass reinforced gypsum	m2	- -
	Rigid tile/panel finishes		
	Ceramic wall tiles; including backing		
3.1.37	economical	m2	30.00 - 50.00
3.1.38	medium quality	m2	40.00 - 60.00
	high quality; to toilet blocks, kitchens and first aid		
3.1.39	rooms	m2	- -
	high quality; to changing areas, toilets, showers and		
3.1.40	fitness areas	m2	- -
	Porcelain mosaic tiling; including backing		
3.1.41	to swimming pool lining; walls and floors	m2	- -
	'Roman Travertine' marble wall linings; polished		
3.1.42	19 mm thick	m2	- -
3.1.43	40 mm thick	m2	- -
3.1.44	Metal mirror cladding panels	m2	- -
3.2	**FLOOR FINISHES**	**floor finish area**	
	(unless otherwise described)		
	comprising:-		
	Sheet/board flooring		
	Insitu screed and floor finishes		
	Rigid tile/slab finishes		
	Parquet/Wood block finishes		
	Flexible tiling/sheet finishes		
	Carpet tiles/Carpetting		
	Access floors and finishes		
	Perimeter treatments and sundries		
	Sheet/board flooring		
	Chipboard flooring; t & g joints		
3.2.1	18 mm thick	m2	7.50 - 9.00
3.2.2	22 mm thick	m2	9.00 - 11.00
	Wrought softwood flooring		
3.2.3	25 mm thick; butt joints	m2	15.00 - 17.00
3.2.4	25 mm thick; butt joints; cleaned off and polished	m2	18.00 - 21.00
3.2.5	25 mm thick; t & g joints	m2	17.00 - 20.00
3.2.6	25 mm thick; t & g joints; cleaned off and polished	m2	20.00 - 24.00
	Wrought softwood t & g strip flooring; 25 mm thick;		
3.2.7	polished; including fillets	m2	24.00 - 30.00
	Wrought hardwood t & g strip flooring; 25 mm thick;		
	polished		
3.2.8	Maple	m2	- -
3.2.9	Gurjun	m2	- -
3.2.10	Iroko	m2	- -
3.2.11	American Oak	m2	- -
	Wrought hardwood t & g strip flooring; 25 mm thick;		
	polished; including fillets		
3.2.12	Maple	m2	- -
3.2.13	Gurjun	m2	- -
3.2.14	Iroko	m2	- -
3.2.15	American Oak	m2	- -
	Wrought hardwood t & g strip flooring; 25 mm thick;		
	polished; including rubber pads		
3.2.16	Maple	m2	- -

INDUSTRIAL £ range £		RETAILING £ range £		LEISURE £ range £		OFFICES £ range £		HOTELS £ range £		Item nr.
1.75	2.00	1.75	2.00	1.75	2.00	1.75	2.00	1.75	2.00	3.1.30
17.00	22.00	17.50	22.50	17.60	22.50	17.50	22.50	18.00	24.00	3.1.31
-	-	14.00	20.00	14.00	20.00	14.00	20.00	15.00	21.00	3.1.32
-	-	14.00	30.00	14.00	30.00	14.00	30.00	15.00	32.00	3.1.33
-	-	-	-	20.00	25.00	-	-	-	-	3.1.34
-	-	140.00	175.00	140.00	175.00	140.00	175.00	150.00	190.00	3.1.35
-	-	130.00	190.00	130.00	190.00	130.00	190.00	140.00	200.00	3.1.36
27.50	45.00	30.00	50.00	30.00	50.00	30.00	50.00	30.00	50.00	3.1.37
35.00	55.00	40.00	60.00	40.00	60.00	40.00	60.00	40.00	65.00	3.1.38
-	-	55.00	70.00	55.00	70.00	55.00	70.00	55.00	75.00	3.1.39
-	-	-	-	70.00	85.00	-	-	-	-	3.1.40
-	-	-	-	45.00	60.00	-	-	-	-	3.1.41
-	-	220.00	300.00	220.00	300.00	220.00	300.00	220.00	300.00	3.1.42
-	-	280.00	350.00	280.00	350.00	280.00	350.00	280.00	350.00	3.1.43
-	-	240.00	390.00	220.00	380.00	-	-	250.00	400.00	3.1.44
-	-	-	-	-	-	-	-	-	-	3.2.1
-	-	-	-	-	-	-	-	-	-	3.2.2
-	-	-	-	-	-	-	-	-	-	3.2.3
-	-	-	-	-	-	-	-	-	-	3.2.4
-	-	-	-	-	-	-	-	-	-	3.2.5
-	-	-	-	-	-	-	-	-	-	3.2.6
-	-	-	-	-	-	-	-	-	-	3.2.7
-	-	-	-	-	-	32.50	37.50	33.00	40.00	3.2.8
-	-	-	-	32.50	37.50	32.50	37.50	33.00	40.00	3.2.9
-	-	-	-	-	-	40.00	45.00	-	-	3.2.10
-	-	-	-	-	-	40.00	45.00	-	-	3.2.11
-	-	-	-	-	-	37.00	42.00	37.50	45.00	3.2.12
-	-	-	-	37.00	42.00	37.00	42.00	37.50	45.00	3.2.13
-	-	-	-	-	-	44.00	50.00	-	-	3.2.14
-	-	-	-	-	-	44.00	50.00	-	-	3.2.15
-	-	-	-	45.00	50.00	-	-	-	-	3.2.16

Item nr.	SPECIFICATIONS	Unit	RESIDENTIAL £ range £
3.2	**FLOOR FINISHES** - cont'd **floor finish area** (unless otherwise described)		
	In situ screed and floor finishes		
	Extra over concrete floor for		
3.2.17	power floating	m2	- -
3.2.18	power floating; surface hardener	m2	- -
	Latex cement screeds		
3.2.19	3 mm thick; one coat	m2	- -
3.2.20	5 mm thick; two coat	m2	- -
3.2.21	Rubber latex non-slip solution and epoxy sealant	m2	- -
	Cement and sand (1:3) screeds		
3.2.22	25 mm thick	m2	5.50 - 8.00
3.2.23	50 mm thick	m2	7.50 - 10.00
3.2.24	75 mm thick	m2	9.50 - 12.50
	Cement and sand (1:3) paving		
3.2.25	paving	m2	7.00 - 9.00
3.2.26	32 mm thick; surface hardener	m2	9.00 - 13.00
3.2.27	Screed only (for subsequent finish)	m2	11.00 - 16.00
3.2.28	Screed only (for subsequent finish); allowance for skirtings	m2	13.00 - 19.00
	Mastic asphalt paving		
3.2.29	20 mm thick; BS 1076; black	m2	- -
3.2.30	15 mm thick; BS 1451; red	m2	- -
	Granolithic		
3.2.31	20 mm thick	m2	- -
3.2.32	25 mm thick	m2	- -
3.2.33	25 mm thick; including screed	m2	- -
3.2.34	38 mm thick; including screed	m2	- -
	Synthanite; on and including building paper		
3.2.35	25 mm thick	m2	- -
3.2.36	50 mm thick	m2	- -
3.2.37	75 mm thick	m2	- -
	Acrylic polymer floor finish		
3.2.38	10 mm thick	m2	- -
	Epoxy floor finish		
3.2.39	1.5 - 2 mm thick	m2	- -
3.2.40	5 - 6 mm thick	m2	- -
	Polyester resin floor finish		
3.2.41	5 - 9 mm thick	m2	- -
	Terrazzo paving; divided into squares with ebonite strip; polished		
3.2.42	16 mm thick	m2	- -
3.2.43	16 mm thick; including screed	m2	- -
	Rigid Tile/slab finishes		
	Quarry tile flooring		
3.2.44	150 x 150 x 12.5 mm thick; red	m2	- -
3.2.45	150 x 150 x 12.5 mm thick; brown	m2	- -
3.2.46	200 x 200 x 19 mm thick; brown	m2	- -
3.2.47	average tiling	m2	- -
3.2.48	tiling; including screed	m2	- -
3.2.49	tiling; including screed and allowance for skirtings	m2	- -
	Brick paving		
3.2.50	paving	m2	- -
3.2.51	paving; including screed	m2	- -
	Glazed ceramic tile flooring		
3.2.52	100 x 100 x 9 mm thick; red	m2	- -
3.2.53	150 x 150 x 12 mm thick; red	m2	- -
3.2.54	100 x 100 x 9 mm thick; black	m2	- -
3.2.55	150 x 150 x 12 mm thick; black	m2	- -
3.2.56	150 x 150 x 12 mm thick; antislip	m2	- -
3.2.57	fully vitrified	m2	- -
3.2.58	fully vitrified; including screed	m2	- -
3.2.59	fully vitrified; including screed and allowance for skirtings	m2	- -
3.2.60	high quality; to service areas; kitchen and toilet blocks; including screed	m2	- -
3.2.61	high quality; to foyer; fitness and bar areas; including screed	m2	- -

INDUSTRIAL £ range £	RETAILING £ range £	LEISURE £ range £	OFFICES £ range £	HOTELS £ range £	Item nr.
4.00 - 8.00	3.00 - 7.50	-	-	-	3.2.17
8.00 - 11.00	6.00 - 9.50	-	-	-	3.2.18
4.00 - 5.00	-	-	4.00 - 4.50	-	3.2.19
5.50 - 7.00	-	-	5.50 - 6.50	-	3.2.20
7.00 - 15.00	-	-	-	-	3.2.21
5.00 - 7.50	5.50 - 8.00	5.50 - 8.00	5.50 - 8.00	5.50 - 8.00	3.2.22
7.50 - 9.50	7.50 - 10.00	7.50 - 10.00	7.50 - 10.00	7.50 - 10.00	3.2.23
9.00 - 12.00	9.50 - 12.50	9.50 - 12.50	9.50 - 12.50	9.50 - 12.50	3.2.24
6.50 - 8.50	7.00 - 9.00	7.00 - 9.00	7.00 - 9.00	7.00 - 9.00	3.2.25
8.00 - 12.00	9.00 - 13.00	9.00 - 13.00	9.00 - 13.00	9.00 - 13.00	3.2.26
10.00 - 15.00	11.00 - 16.00	11.00 - 16.00	11.00 - 16.00	11.00 - 16.00	3.2.27
12.00 - 18.00	13.00 - 19.00	13.00 - 19.00	13.00 - 19.00	13.00 - 20.00	3.2.28
15.00 - 17.50	16.00 - 18.00	16.00 - 18.00	-	-	3.2.29
10.50 - 12.50	11.00 - 13.00	11.00 - 13.00	-	-	3.2.30
9.00 - 13.00	9.50 - 13.50	9.50 - 13.50	9.50 - 13.50	9.50 - 13.50	3.2.31
10.50 - 15.00	11.00 - 16.00	11.00 - 16.00	11.00 - 16.00	11.00 - 16.00	3.2.32
16.00 - 21.00	17.00 - 22.00	17.00 - 22.00	17.00 - 22.00	17.00 - 22.00	3.2.33
23.00 - 29.00	24.00 - 30.00	24.00 - 30.00	24.00 - 30.00	24.00 - 30.00	3.2.34
17.50 - 21.00	18.00 - 22.00	18.00 - 22.00	18.00 - 22.00	-	3.2.35
25.00 - 28.50	26.00 - 30.00	26.00 - 30.00	26.00 - 30.00	-	3.2.36
30.00 - 35.00	32.00 - 36.00	32.00 - 36.00	32.00 - 36.00	-	3.2.37
18.00 - 22.00	19.00 - 23.00	19.00 - 23.00	19.00 - 23.00	-	3.2.38
18.00 - 23.00	19.00 - 24.00	19.00 - 24.00	19.00 - 24.00	-	3.2.39
36.00 - 41.00	37.50 - 42.50	37.50 - 42.50	37.50 - 42.50	-	3.2.40
41.00 - 46.00	42.50 - 47.50	42.50 - 47.50	42.50 - 47.50	-	3.2.41
-	40.00 - 45.00	40.00 - 45.00	40.00 - 45.00	40.00 - 50.00	3.2.42
-	60.00 - 70.00	60.00 - 70.00	60.00 - 70.00	60.00 - 80.00	3.2.43
19.00 - 22.00	20.00 - 23.00	20.00 - 23.00	20.00 - 23.00	21.00 - 24.00	3.2.44
25.00 - 29.00	26.00 - 30.00	26.00 - 30.00	26.00 - 30.00	27.00 - 31.00	3.2.45
29.00 - 35.00	30.00 - 36.00	30.00 - 36.00	30.00 - 36.00	31.00 - 37.00	3.2.46
19.00 - 35.00	20.00 - 36.00	20.00 - 36.00	20.00 - 36.00	21.00 - 37.00	3.2.47
27.00 - 47.00	28.00 - 48.00	28.00 - 48.00	28.00 - 48.00	30.00 - 50.00	3.2.48
38.00 - 58.00	40.00 - 60.00	40.00 - 60.00	40.00 - 60.00	42.00 - 63.00	3.2.49
-	33.00 - 48.00	33.00 - 48.00	33.00 - 48.00	33.00 - 50.00	3.2.50
-	42.00 - 60.00	42.00 - 60.00	42.00 - 60.00	42.00 - 60.00	3.2.51
-	30.00 - 34.00	30.00 - 34.00	30.00 - 34.00	31.00 - 35.00	3.2.52
-	26.00 - 30.00	26.00 - 30.00	26.00 - 30.00	27.00 - 31.00	3.2.53
-	33.00 - 36.00	33.00 - 36.00	33.00 - 36.00	34.00 - 37.00	3.2.54
-	27.00 - 33.00	27.00 - 33.00	27.00 - 33.00	28.00 - 34.00	3.2.55
-	33.00 - 35.00	33.00 - 35.00	33.00 - 35.00	34.00 - 36.00	3.2.56
-	35.00 - 50.00	35.00 - 50.00	35.00 - 50.00	36.00 - 52.00	3.2.57
-	42.00 - 63.00	42.00 - 63.00	42.00 - 63.00	44.00 - 65.00	3.2.58
-	48.00 - 75.00	48.00 - 75.00	48.00 - 75.00	48.00 - 80.00	3.2.59
-	-	66.00 - 77.50	-	-	3.2.60
-	-	70.00- 80.00	-	-	3.2.61

Item nr.	SPECIFICATIONS	Unit	RESIDENTIAL £ range £
3.2	**FLOOR FINISHES** - cont'd **floor finish area** (unless otherwise described)		
	Rigid Tile/slab finishes - cont'd		
	Glazed ceramic tile flooring - cont'd		
	high quality; to pool surround, bottoms, steps and		
3.2.62	changing room; including screed	m2	- -
	Porcelain mosaic paving; including screed		
3.2.63	to swimming pool lining, walls and floors	m2	- -
	Extra for		
	non-slip finish to pool lining, beach and changing		
3.2.64	areas and showers	m2	- -
	Terrazzo tile flooring		
3.2.65	28 mm thick white Sicilian marble aggregate tiling	m2	- -
3.2.66	tiling; including screed	m2	- -
	York stone		
3.2.67	50 mm thick paving	m2	- -
3.2.68	paving; including screed	m2	- -
	Slate		
3.2.69	200 x 400 x 10 mm blue-grey	m2	- -
3.2.70	Otta riven	m2	- -
3.2.71	Otta honed (polished)	m2	- -
	Portland stone		
3.2.72	50 mm thick paving	m2	- -
	Fine sanded 'Roman Travertine' marble		
3.2.73	20 mm thick paving	m2	- -
3.2.74	paving; including screed	m2	- -
3.2.75	paving; including screed and allowance for skirtings	m2	- -
	Granite		
3.2.76	20 mm thick paving	m2	- -
	Parquet/wood block finishes		
	Parquet flooring; polished		
3.2.77	8 mm Gurjun 'Feltwood'	m2	20.00 - 24.00
	Wrought hardwood block floorings; 25 mm thick; polished tongued and grooved joints; herringbone pattern		
3.2.78	Merbau	m2	- -
3.2.79	Iroko	m2	- -
3.2.80	Iroko; including screed	m2	- -
3.2.81	Oak	m2	- -
3.2.82	Oak; including screed	m2	- -
	Composition block flooring		
3.2.83	174 x 57 mm blocks	m2	- -
	Flexible tiling/sheet finishes		
	Thermoplastic tile flooring		
3.2.84	2 mm thick (series 2)	m2	6.50 - 7.50
3.2.85	2 mm thick (series 4)	m2	7.00 - 8.00
3.2.86	2 mm thick; including screed	m2	15.00 - 17.00
	Cork tile flooring		
3.2.87	3.2 mm thick	m2	11.00 - 13.00
3.2.88	3.2 mm thick; including screed	m2	19.00 - 22.00
3.2.89	6.3 mm thick	m2	- -
3.2.90	6.3 mm thick; including screed	m2	- -
	Vinyl tile flooring		
3.2.91	2 mm thick; semi-flexible tiles	m2	8.00 - 10.00
3.2.92	2 mm thick; fully flexible tiles	m2	7.00 - 9.00
3.2.93	2.5 mm thick; semi-flexible tiles	m2	8.50 - 11.00
3.2.94	tiling; including screed	m2	17.50 - 22.00
3.2.95	tiling; including screed and allowance for skirtings	m2	20.00 - 26.00
3.2.96	tiling; antistatic	m2	- -
3.2.97	tiling; antistatic; including screed	m2	- -
	Vinyl sheet flooring; heavy duty		
3.2.98	2 mm thick	m2	- -
3.2.99	2.5 mm thick	m2	- -
3.2.100	3 mm thick; needle felt backed	m2	- -
3.2.101	3 mm thick; foam backed	m2	- -
3.2.102	sheeting; including screed and allowance for skirtings	m2	- -

INDUSTRIAL £ range £	RETAILING £ range £	LEISURE £ range £	OFFICES £ range £	HOTELS £ range £	Item nr.
- -	- -	72.50 - 85.00	- -	- -	3.2.62
- -	- -	45.00 - 60.00	- -	- -	3.2.63
- -	- -	1.00 - 2.00	- -	- -	3.2.64
- -	57.50 - 70.00	57.50 - 70.00	57.50 - 70.00	60.00 - 75.00	3.2.65
- -	77.50 - 90.00	77.50 - 90.00	77.50 - 90.00	80.00 - 95.00	3.2.66
- -	85.00 -110.00	85.00 -110.00	85.00 -110.00	90.00 -120.00	3.2.67
- -	93.00 -122.00	93.00 -122.00	93.00 -122.00	100.00 -135.00	3.2.68
- -	105.00 -115.00	105.00 -115.00	105.00 -115.00	- -	3.2.69
- -	115.00 -125.00	115.00 -125.00	115.00 -125.00	- -	3.2.70
- -	125.00 -138.00	125.00 -138.00	125.00 -138.00	- -	3.2.71
- -	150.00 -180.00	150.00 - 180.00	150.00 -180.00	- -	3.2.72
- -	160.00 -190.00	160.00 -190.00	160.00 -190.00	175.00 -200.00	3.2.73
- -	180.00 -220.00	180.00 -220.00	180.00 -220.00	190.00 -230.00	3.2.74
- -	240.00 -300.00	240.00 -300.00	240.00 -300.00	250.00 -320.00	3.2.75
- -	200.00 -340.00	200.00 -340.00	200.00 -340.00	210.00 -375.00	3.2.76
- -	- -	- -	- -	- -	3.2.77
- -	42.00 - 46.00	42.00 - 46.00	42.00 - 46.00	43.00 - 49.00	3.2.78
- -	42.00 - 46.00	42.00 - 46.00	42.00 - 46.00	43.00 - 49.00	3.2.79
- -	50.00 - 60.00	50.00 - 60.00	50.00 - 60.00	52.00 - 62.00	3.2.80
- -	40.00 - 55.00	40.00 - 55.00	40.00 - 55.00	41.00 - 57.50	3.2.81
- -	48.00 - 70.00	48.00 - 70.00	48.00 - 70.00	50.00 - 72.50	3.2.82
- -	50.00 - 55.00	50.00 - 55.00	50.00 - 55.00	52.00 - 57.50	3.2.83
- -	- -	- -	- -	- -	3.2.84
- -	- -	- -	- -	- -	3.2.85
- -	- -	- -	- -	- -	3.2.86
- -	- -	- -	- -	- -	3.2.87
- -	- -	- -	- -	- -	3.2.88
- -	15.00 - 17.00	15.00 - 17.00	15.00 - 17.00	15.00 - 18.00	3.2.89
- -	23.00 - 27.50	23.00 - 27.50	23.00 - 27.50	24.00 - 30.00	3.2.90
- -	8.00 - 10.00	8.00 - 10.00	8.00 - 10.00	8.00 - 10.00	3.2.91
- -	7.00 - 9.00	7.00 - 9.00	7.00 - 9.00	7.00 - 9.00	3.2.92
- -	8.50 - 11.00	8.50 - 11.00	8.50 - 11.00	8.50 - 11.00	3.2.93
- -	17.50 - 22.00	17.50 - 22.00	17.50 - 22.00	17.50 - 22.00	3.2.94
- -	20.00 - 26.00	20.00 - 26.00	20.00 - 26.00	20.00 - 26.00	3.2.95
- -	32.00 - 37.00	32.00 - 37.00	32.00 - 37.00	32.00 - 37.00	3.2.96
- -	40.00 - 50.00	40.00 - 50.00	40.00 - 50.00	40.00 - 50.00	3.2.97
- -	12.00 - 14.00	12.00 - 14.00	12.00 - 14.00	12.00 - 14.00	3.2.98
- -	13.00 - 15.00	13.00 - 15.00	13.00 - 15.00	13.00 - 15.00	3.2.99
- -	9.00 - 11.00	9.00 - 11.00	9.00 - 11.00	9.00 - 11.00	3.2.100
- -	12.00 - 15.00	12.00 - 15.00	12.00 - 15.00	12.00 - 15.00	3.2.101
- -	24.00 - 27.00	24.00 - 27.00	24.00 - 27.00	24.00 - 27.00	3.2.102

Item nr.	SPECIFICATIONS	Unit	RESIDENTIAL £ range £
3.2	**FLOOR FINISHES** - cont'd floor finish area (unless otherwise described)		
	Flexible tiling/sheet finishes - cont'd 'Altro' safety flooring		
3.2.103	2 mm thick; Marine T20	m2	- -
3.2.104	2.5 mm thick; Classic D25	m2	- -
3.2.105	3.5 mm thick; stronghold	m2	- -
3.2.106	flooring	m2	- -
3.2.107	flooring; including screed	m2	- -
	Linoleum tile flooring		
3.2.108	3.2 mm thick; coloured	m2	- -
3.2.109	3.2 mm thick; coloured; including screed	m2	- -
	Linoleum sheet flooring		
3.2.110	3.2 mm thick; coloured	m2	- -
3.2.111	3.2 mm thick; marbled; including screed	m2	- -
	Rubber tile flooring; smooth; ribbed or studded tiles		
3.2.112	2.5 mm thick	m2	- -
3.2.113	5 mm thick	m2	- -
3.2.114	5 mm thick; including screed	m2	- -
	Carpet tiles/Carpetting		
3.2.115	Underlay	m2	4.00 - 5.00
	Carpet tiles		
3.2.116	nylon needlepunch (stick down)	m2	10.00 - 12.00
3.2.117	80% animal hair; 20% wool cord	m2	- -
3.2.118	100% wool	m2	- -
3.2.119	80% wool; 20% nylon antistatic	m2	- -
	economical; including screed and allowance for		
3.2.120	skirtings	m2	- -
3.2.121	good quality	m2	- -
3.2.122	good quality; including screed	m2	- -
	good quality; including screed and allowance for		
3.2.123	skirtings	m2	- -
	Carpet; including underlay		
3.2.124	nylon needlepunch	m2	14.00 - 17.00
3.2.125	100% acrylic; light duty	m2	- -
3.2.126	80% animal hair; 20% wool cord	m2	- -
3.2.127	open-weave matting poolside carpet (no underlay)	m2	- -
3.2.128	80% wool; 20% acrylic; light duty	m2	- -
3.2.129	100% acrylic; heavy duty	m2	- -
3.2.130	cord	m2	- -
3.2.131	100% wool	m2	- -
3.2.132	good quality; including screed	m2	- -
	good quality (grade 5); including screed and allowance		
3.2.133	for skirtings	m2	- -
3.2.134	80% wool; 20% acrylic; heavy duty	m2	- -
3.2.135	Wilton/pile carpet	m2	- -
3.2.136	high quality; including screed	m2	- -
	high quality; including screed and allowance for		
3.2.137	skirtings	m2	- -
	Access floors and finishes Shallow void block and battened floors		
3.2.138	chipboard on softwood battens; partial access	m2	- -
3.2.139	chipboard on softwood cradles (for uneven floors)	m2	- -
	chipboard on softwood battens and cross battens;		
3.2.140	full access	m2	- -
	fibre and particle board; on lightweight concrete		
3.2.141	pedestal blocks	m2	- -
	Shallow void block and battened floors; including carpet-tile finish		
	fibre and particle board; on lightweight concrete		
3.2.142	pedestal blocks	m2	- -

INDUSTRIAL £ range £	RETAILING £ range £	LEISURE £ range £	OFFICES £ range £	HOTELS £ range £	Item nr.
- -	18.00 - 22.00	18.00 - 22.00	18.00 - 22.00	- -	3.2.103
- -	22.00 - 26.00	22.00 - 26.00	22.00 - 26.00	- -	3.2.104
- -	27.50 - 32.50	27.50 - 32.50	27.50 - 32.50	- -	3.2.105
- -	18.00 - 32.50	18.00 - 32.50	18.00 - 32.50	- -	3.2.106
- -	26.00 - 45.00	26.00 - 45.00	26.00 - 45.00	- -	3.2.107
- -	14.00 - 16.00	14.00 - 16.00	14.00 - 16.00	- -	3.2.108
- -	22.00 - 30.00	22.00 - 30.00	22.00 - 30.00	- -	3.2.109
- -	14.00 - 16.00	14.00 - 16.00	14.00 - 16.00	- -	3.2.110
- -	22.00 - 30.00	22.00 - 30.00	22.00 - 30.00	- -	3.2.111
- -	22.00 - 26.00	22.00 - 26.00	22.00 - 26.00	- -	3.2.112
- -	26.00 - 30.00	26.00 - 30.00	26.00 - 30.00	- -	3.2.113
- -	34.00 - 45.00	34.00 - 45.00	34.00 - 45.00	- -	3.2.114
- -	4.00 - 5.00	4.00 - 5.00	4.00 - 5.00	- -	3.2.115
- -	10.00 - 12.00	10.00 - 12.00	10.00 - 12.00	- -	3.2.116
- -	18.00 - 20.00	18.00 - 20.00	18.00 - 20.00	- -	3.2.117
- -	25.00 - 30.00	25.00 - 30.00	25.00 - 30.00	- -	3.2.118
- -	26.00 - 36.00	26.00 - 36.00	26.00 - 36.00	- -	3.2.119
- -	27.50 - 30.00	27.50 - 30.00	27.50 - 30.00	- -	3.2.120
- -	25.00 - 36.00	25.00 - 36.00	25.00 - 36.00	25.00 - 40.00	3.2.121
- -	33.00 - 45.00	33.00 - 45.00	33.00 - 45.00	33.00 - 50.00	3.2.122
- -	37.00 - 50.00	37.00 - 50.00	37.00 - 50.00	37.50 - 55.00	3.2.123
- -	14.00 - 17.00	14.00 - 17.00	14.00 - 17.00	- -	3.2.124
- -	17.00 - 20.00	17.00 - 20.00	17.00 - 20.00	- -	3.2.125
- -	21.00 - 26.00	21.00 - 26.00	21.00 - 26.00	- -	3.2.126
- -	- -	17.50 - 22.50	- -	- -	3.2.127
- -	26.00 - 34.00	26.00 - 34.00	26.00 - 34.00	- -	3.2.128
- -	27.00 - 32.00	27.00 - 32.00	27.00 - 32.00	- -	3.2.129
- -	- -	32.00 - 36.00	- -	- -	3.2.130
- -	32.00 - 42.00	32.00 - 42.00	32.00 - 42.00	32.00 - 45.00	3.2.131
- -	40.00 - 55.00	40.00 - 55.00	40.00 - 55.00	42.00 - 60.00	3.2.132
- -	42.50 - 57.50	42.50 - 57.50	42.50 - 57.50	45.00 - 62.00	3.2.133
- -	43.00 - 50.00	43.00 - 50.00	43.00 - 50.00	45.00 - 52.50	3.2.134
- -	50.00 - 55.00	50.00 - 55.00	50.00 - 55.00	52.50 - 57.50	3.2.135
- -	58.00 - 65.00	58.00 - 65.00	58.00 - 65.00	60.00 - 67.50	3.2.136
- -	60.00 - 67.50	60.00 - 67.50	60.00 - 67.50	60.00 - 70.00	3.2.137
- -	- -	- -	18.00 - 22.00	- -	3.2.138
- -	- -	- -	21.50 - 25.00	- -	3.2.139
- -	- -	- -	26.00 - 29.00	- -	3.2.140
- -	- -	- -	25.20 - 30.00	- -	3.2.141
- -	- -	- -	37.50 - 42.50	- -	3.2.142

Item nr.	SPECIFICATIONS	Unit	RESIDENTIAL £ range £

3.2   **FLOOR FINISHES - cont'd**   **floor finish area**
                                        (unless otherwise described)

Access floors and finishes - cont'd
Access floors; excluding finish
   600 x 600 mm chipboard panels; faced both sides with
   galvanised steel sheet; on adjustable steel/aluminium
   pedestals; cavity height 100 - 300 mm high

3.2.143	light grade duty	m2	- -
3.2.144	medium grade duty	m2	- -
3.2.145	heavy grade duty	m2	- -
3.2.146	extra heavy grade duty	m2	- -

   600 x 600 mm chipboard panels; faced both sides
   with galvanised steel sheet; on adjustable steel/
   aluminium pedestals; cavity height 300 - 600 mm high

3.2.147	medium grade duty	m2	- -
3.2.148	heavy grade duty	m2	- -
3.2.149	extra heavy grade duty	m2	- -

Access floor with medium quality carpetting

3.2.150	'Durabella' suspended floors	m2	- -
3.2.151	'Buroplan' partial access raised floor	m2	- -
3.2.152	'Pedestal' partial access raised floor	m2	- -
3.2.153	Modular floor; 100% access raised floor	m2	- -

Access floor with high quality carpetting

3.2.154	computer loading; 100% access raised floor	m2	- -

Common floor coverings bonded to access floor panels

3.2.155	heavy-duty fully flexible vinyl to BS 3261, type A	m2	- -
3.2.156	fibre-bonded carpet	m2	- -
3.2.157	high-pressure laminate to BS 3794 class D	m2	- -
3.2.158	anti-static grade fibre-bonded carpet	m2	- -
3.2.159	anti-static grade sheet PVC to BS 3261	m2	- -
3.2.160	low loop tufted carpet	m2	- -

Perimeter treatments and sundries
Comparative skirtings

3.2.161	25 x 75 mm softwood skirting; painted; including grounds	m	8.00 - 9.00
3.2.162	25 x 100 mm Mahogany skirting; polished; including grounds	m	11.00 - 12.00
3.2.163	12.5 x 150 mm Quarry tile skirting; including backing	m	10.50 - 12.50
3.2.164	13 x 75 mm granolithic skirting; including backing	m	15.00 - 18.00
3.2.165	6 x 75 mm terrazzo; including backing	m	26.00 - 30.00

Entrance matting in aluminium-framed

3.2.166	matwell	m2	- -

3.3   **CEILING FINISHES**   **ceiling finish area**
                                        (unless otherwise described)

comprising:-

In situ/board finishes
Suspended and integrated ceilings

In situ/board finishes
Decoration only to soffits

3.3.1	to exposed steelwork	m2	- -
3.3.2	to concrete soffits	m2	2.00 - 3.00
3.3.3	one mist and two coats emulsion paint; to plaster/ plasterboard	m2	2.00 - 3.00

Plaster to soffits

3.3.4	lightweight plaster	m2	8.50 - 10.50
3.3.5	plaster and emulsion	m2	10.50 - 15.50
	Extra for		
3.3.6	gloss paint in lieu of emulsion	m2	1.75 - 2.00

INDUSTRIAL £ range £	RETAILING £ range £	LEISURE £ range £	OFFICES £ range £	HOTELS £ range £	Item nr.
-	-	-	-	-	-	-	-	38.00 - 48.00	-	-	3.2.143
-	-	-	-	-	-	-	-	40.00 - 47.50	-	-	3.2.144
-	-	-	-	-	-	-	-	47.50 - 60.00	-	-	3.2.145
-	-	-	-	-	-	-	-	54.00 - 60.00	-	-	3.2.146
-	-	-	-	-	-	45.00 - 50.00	-	-	3.2.147
-	-	-	-	-	-	50.00 - 60.00	-	-	3.2.148
-	-	-	-	-	-	55.00 - 60.00	-	-	3.2.149
-	-	-	-	-	-	47.50 - 52.50	-	-	3.2.150
-	-	-	-	-	-	54.00 - 60.00	-	-	3.2.151
-	-	-	-	-	-	60.00 - 75.00	-	-	3.2.152
-	-	-	-	-	-	80.00 -100.00	-	-	3.2.153
-	-	-	-	-	-	90.00 -125.00	-	-	3.2.154
-	-	-	-	-	-	6.00 - 17.00	-	-	3.2.155
-	-	-	-	-	-	7.00 - 12.00	-	-	3.2.156
-	-	-	-	-	-	7.00 - 19.00	-	-	3.2.157
-	-	-	-	-	-	8.50 - 13.00	-	-	3.2.158
-	-	-	-	-	-	11.00 - 17.00	-	-	3.2.159
-	-	-	-	-	-	13.00 - 19.00	-	-	3.2.160
7.50 - 8.50	8.00 - 9.00	8.00 - 9.00	8.00 - 9.00	8.50 - 10.00	3.2.161
10.50 - 11.50	11.00 - 12.00	11.00 - 12.00	11.00 - 12.00	11.50 - 12.50	3.2.162
10.00 - 12.00	10.50 - 12.50	10.50 - 12.50	10.50 - 12.50	11.00 - 13.00	3.2.163
14.50 - 17.50	15.00 - 18.00	15.00 - 18.00	15.00 - 18.00	15.50 - 19.00	3.2.164
-	-	26.00 - 30.00	26.00 - 30.00	26.00 - 30.00	27.00 - 32.00	3.2.165
-	-	230.00 -320.00	220.00 -295.00	225.00 -300.00	225.00 -300.00	3.2.166
3.00 - 4.00	2.00 - 3.00	-	-	-	-	-	-	3.3.1
2.00 - 3.00	2.00 - 3.00	2.00 - 3.00	2.00 - 3.00	2.00 - 3.00	3.3.2
2.00 - 3.00	2.00 - 3.00	2.00 - 3.00	2.00 - 3.00	2.00 - 3.00	3.3.3
8.00 - 10.00	8.00 - 10.00	8.00 - 10.00	8.00 - 10.00	9.00 - 11.50	3.3.4
10.00 - 15.00	10.50 - 15.50	10.50 - 15.50	10.50 - 15.50	11.50 - 18.00	3.3.5
1.75 - 2.00	1.75 - 2.00	1.75 - 2.00	1.75 - 2.00	1.75 - 2.00	3.3.6

Item nr.	SPECIFICATIONS	Unit	RESIDENTIAL £ range £
3.3	**CEILING FINISHES** - cont'd **ceiling finish area** (unless otherwise described)		
	In situ/board finishes - cont'd **Plasterboard to soffits**		
3.3.7	9.5 mm Gyproc lath and skim coat	m2	11.00 - 13.00
3.3.8	9.5 mm Gyproc insulating lath and skim coat	m2	11.50 - 13.50
3.3.9	plasterboard, skim and emulsion	m2	13.00 - 16.00
	Extra for		
3.3.10	gloss paint in lieu of emulsion	m2	1.75 - 2.00
3.3.11	plasterboard and Artex	m2	9.00 - 11.00
3.3.12	plasterboard, Artex and emulsion	m2	11.00 - 14.00
3.3.13	plaster and emulsion; including metal lathing	m2	18.00 - 25.00
	Other board finishes; with fire-resisting properties; excluding decoration		
3.3.14	12.5 mm Gyproc Fireline; half hour	m2	- -
3.3.15	6 mm Supalux; half hour	m2	- -
3.3.16	two layers of 12.5 mm Gyproc Wallboard; half hour	m2	- -
3.3.17	two layers of 12.5 mm Gyproc Fireline; one hour	m2	- -
3.3.18	9 mm Supalux; one hour; on fillets	m2	- -
	Specialist plasters; to soffits		
3.3.19	sprayed accoustic plaster; self-finished	m2	- -
3.3.20	rendering; 'Tyrolean' finish	m2	- -
	Other ceiling finishes		
3.3.21	50 mm wood wool slabs as permanent lining	m2	- -
3.3.22	12 mm Pine tongued and grooved boarding	m2	14.00 - 16.50
3.3.23	16 mm Softwood tongued and grooved boardings	m2	17.50 - 20.00
	Suspended and integrated ceilings **Suspended ceiling**		
3.3.24	economical; exposed grid	m2	- -
3.3.25	jointless; plasterboard	m2	- -
3.3.26	semi-concealed grid	m2	- -
3.3.27	medium quality; 'Minatone'; concealed grid	m2	- -
3.3.28	high quality; 'Travertone'; concealed grid	m2	- -
	Other suspended ceilings		
3.3.29	metal linear strip; 'Dampa'/'Luxalon'	m2	- -
3.3.30	metal tray	m2	- -
3.3.31	egg-crate	m2	- -
3.3.32	open grid; 'Formalux'/'Dimension'	m2	- -
	Integrated ceilings		
3.3.33	coffered; with steel services	m2	- -
3.4	**DECORATIONS** **surface area** (unless otherwise described)		
	comprising:-		
	Comparative wall and ceiling finishes **Comparative steel/metalwork finishes** **Comparative woodwork finishes**		
	Comparative wall and ceiling finishes **Emulsion**		
3.4.1	two coats	m2	1.75 - 2.25
3.4.2	one mist and two coats	m2	2.00 - 3.00
	Artex plastic compound		
3.4.3	one coat; textured	m2	2.75 - 3.75
3.4.4	Wall paper	m2	4.00 - 6.00
3.4.5	Hessian wall coverings	m2	- -
	Gloss		
3.4.6	primer and two coats	m2	3.85 - 4.75
3.4.7	primer and three coats	m2	5.00 - 6.00
	Comparative steel/metalwork finishes **Primer**		
3.4.8	only	m2	- -
3.4.9	grit blast and one coat zinc chromate primer touch up primer and one coat of two pack epoxy	m2	- -

INDUSTRIAL £ range £	RETAILING £ range £	LEISURE £ range £	OFFICES £ range £	HOTELS £ range £	Item nr.
10.50 - 12.50	11.00 - 13.00	11.00 - 13.00	11.00 - 13.00	11.50 - 13.50	3.3.7
11.00 - 13.00	11.50 - 13.50	11.50 - 13.50	11.50 - 13.50	12.00 - 14.00	3.3.8
12.50 - 15.50	13.00 - 16.00	13.00 - 16.00	13.00 - 16.00	13.50 - 16.50	3.3.9
1.75 - 2.00	1.75 - 2.00	1.75 - 2.00	1.75 - 2.00	1.75 - 2.00	3.3.10
8.50 - 10.50	9.00 - 11.00	9.00 - 11.00	9.00 - 11.00	9.50 - 11.50	3.3.11
10.50 - 13.50	11.00 - 14.00	11.00 - 14.00	11.00 - 14.00	11.50 - 14.50	3.3.12
17.50 - 24.00	18.00 - 25.00	18.00 - 25.00	18.00 - 25.00	18.50 - 26.00	3.3.13
9.00 - 11.00	9.25 - 11.25	9.25 - 11.25	9.25 - 11.25	10.00 - 12.00	3.3.14
11.00 - 12.00	11.50 - 12.50	11.50 - 12.50	11.50 - 12.50	12.00 - 13.00	3.3.15
12.50 - 14.50	13.00 - 15.00	13.00 - 15.00	13.00 - 15.00	13.50 - 15.50	3.3.16
14.50 - 17.00	15.00 - 17.50	15.00 - 17.50	15.00 - 17.50	15.50 - 18.00	3.3.17
16.50 - 19.50	17.00 - 20.00	17.00 - 20.00	17.00 - 20.00	17.50 - 21.00	3.3.18
22.50 - 30.00	-	-	-	-	3.3.19
-	-	23.50 - 33.00	-	-	3.3.20
-	-	-	10.50 - 12.50	-	3.3.21
-	14.00 - 16.50	14.00 - 16.50	14.00 - 16.50	14.50 - 17.50	3.3.22
-	17.50 - 20.00	17.50 - 20.00	17.50 - 20.00	17.50 - 20.00	3.3.23
18.00 - 23.00	20.00 - 25.00	20.00 - 25.00	20.00 - 25.00	20.00 - 28.00	3.3.24
22.00 - 28.00	24.00 - 30.00	24.00 - 30.00	24.00 - 30.00	24.00 - 33.00	3.3.25
24.00 - 30.00	26.00 - 32.50	26.00 - 32.50	26.00 - 32.50	26.00 - 35.00	3.3.26
26.00 - 35.00	28.00 - 37.50	28.00 - 37.50	28.00 - 37.50	28.00 - 40.00	3.3.27
30.00 - 37.50	32.50 - 40.00	32.50 - 40.00	32.50 - 40.00	32.50 - 44.00	3.3.28
38.00 - 48.00	40.00 - 50.00	40.00 - 50.00	40.00 - 50.00	-	3.3.29
32.00 - 44.00	33.00 - 45.00	33.00 - 45.00	33.00 - 45.00	-	3.3.30
38.00 - 77.50	40.00 - 80.00	40.00 - 80.00	40.00 - 80.00	-	3.3.31
67.50 - 82.50	70.00 - 85.00	70.00 - 85.00	70.00 - 85.00	-	3.3.32
80.00 -125.00	80.00 -125.00	80.00 -125.00	80.00 -125.00	-	3.3.33
1.75 - 2.25	1.75 - 2.25	1.75 - 2.25	1.75 - 2.25	1.75 - 2.25	3.4.1
2.00 - 3.00	2.00 - 3.00	2.00 - 3.00	2.00 - 3.00	2.00 - 3.00	3.4.2
2.75 - 3.75	2.75 - 3.75	2.75 - 3.75	2.75 - 3.75	2.75 - 3.75	3.4.3
4.00 - 6.00	4.00 - 6.00	4.00 - 6.00	4.00 - 6.00	4.00 - 10.00	3.4.4
-	9.00 - 12.00	9.00 - 12.00	9.00 - 12.00	10.00 - 15.00	3.4.5
3.85 - 4.75	3.85 - 4.75	3.85 - 4.75	3.85 - 4.75	3.85 - 4.75	3.4.6
5.00 - 6.00	5.00 - 6.00	5.00 - 6.00	5.00 - 6.00	5.00 - 6.00	3.4.7
0.75 - 1.25	-	-	-	-	3.4.8
1.50 - 2.25	-	-	-	-	3.4.9

Item nr.	SPECIFICATIONS	Unit	RESIDENTIAL £ range £

3.4   **DECORATIONS** - cont'd   **surface area**
                                        (unless otherwise described)

Comparative steel/metalwork finishes - cont'd

Item nr.	SPECIFICATIONS	Unit	RESIDENTIAL £ range £	
3.4.9	grit blast and one coat zinc chromate primer touch up primer and one coat of two pack epoxy	m2	-	-
3.4.10	zinc phosphate primer	m2	-	-
	Gloss			
3.4.11	three coats	m2	5.00 -	6.00
	Sprayed mineral fibre			
3.4.12	one hour	m2	-	-
3.4.13	two hour	m2	-	-
	Sprayed vermiculate cement			
3.4.14	one hour	m2	-	-
3.4.15	two hour	m2	-	-
	Intumescent coating with decorative top seal			
3.4.16	half hour	m2	-	-
3.4.17	one hour	m2	-	-
	Comparative woodwork finishes			
	Primer			
3.4.18	only	m2	1.30 -	1.50
	Gloss			
3.4.19	two coats; touch up primer	m2	2.75 -	3.30
3.4.20	three coats; touch up primer	m2	3.90 -	4.75
3.4.21	primer and two coat	m2	4.15 -	7.00
3.4.22	primer and three coat	m2	5.25 -	6.30
	Polyurethene laquer			
3.4.23	two coats	m2	2.70 -	3.25
3.4.24	three coats	m2	4.00 -	4.80
	Flame-retardent paint			
3.4.25	three coats	m2	5.75 -	6.90
	Wax polish			
3.4.26	seal	m2	7.60 -	9.15
3.4.27	stain and body-in	m2	9.45 -	11.35
	French polish			
3.4.28	stain and body-in	m2	14.30 -	17.00

4.1   **FITTINGS AND FURNISHINGS**   **gross internal area**
                                         or individual units

comprising:-

Residential fittings
Comparative fittings/sundries
Industrial/Office furniture, fittings and equipment
Retail furniture, fittings and equipment
Leisure furniture, fittings and equipment
Hotel furniture, fittings and equipment

Residential fittings
Kitchen fittings for residential units

Item nr.	SPECIFICATIONS	Unit	RESIDENTIAL £ range £
4.1.1	one person flat/bed-sit	nr	450 -900
4.1.2	two person flat/house	nr	600 -1500
4.1.3	three person flat/house	nr	700 -2100
4.1.4	four person house	nr	800 -2700
4.1.5	five person house	nr	1000 -5000

Comparative fittings/sundries
Individual kitchen fittings

Item nr.	SPECIFICATIONS	Unit	RESIDENTIAL £ range £
4.1.6	600 x 600 x 300 mm wall unit	nr	55.00 - 65.00
4.1.7	1200 x 900 x 300 mm wall unit	nr	90.00 -100.00
4.1.8	500 x 900 x 600 mm floor unit	nr	90.00 -100.00
4.1.9	600 x 500 x 195 mm store cupboard	nr	130.00 -140.00
4.1.10	1200 x 900 x 600 mm sink unit (excluding top)	nr	135.00 -145.00
	Comparative wrot softwood shelving		
4.1.11	25 x 225 mm; including black japanned brackets	m	9.00 - 10.00
4.1.12	25 mm thick slatted shelving; including bearers	m2	35.00 - 40.00
4.1.13	25 mm thick cross-tongued shelving; including bearers	m2	45.00 - 50.00

INDUSTRIAL £ range £	RETAILING £ range £	LEISURE £ range £	OFFICES £ range £	HOTELS £ range £	Item nr.
1.50 - 2.25	- -	- -	- -	- -	3.4.9
2.00 - 2.50	- -	- -	- -	- -	3.4.10
- -	- -	- -	- -	- -	3.4.11
9.00 - 13.00	- -	- -	- -	- -	3.4.12
14.00 - 17.00	- -	- -	- -	- -	3.4.13
10.00 - 14.00	- -	- -	- -	- -	3.4.14
12.00 - 17.00	- -	- -	- -	- -	3.4.15
15.50 - 17.00	15.50 - 17.00	15.50 - 17.00	15.50 - 17.00	- -	3.4.16
25.00 - 30.00	25.00 - 30.00	25.00 - 30.00	25.00 - 30.00	- -	3.4.17
1.30 - 1.50	1.30 - 1.50	1.30 - 1.50	1.30 - 1.50	1.30 - 1.50	3.4.18
2.75 - 3.30	2.75 - 3.30	2.75 - 3.30	2.75 - 3.30	2.75 - 3.30	3.4.19
3.90 - 4.75	3.90 - 4.75	3.90 - 4.75	3.90 - 4.75	3.90 - 4.75	3.4.20
4.15 - 7.00	4.15 - 7.00	4.15 - 7.00	4.15 - 7.00	4.15 - 7.00	3.4.21
5.25 - 6.30	5.25 - 6.30	5.25 - 6.30	5.25 - 6.30	5.25 - 6.30	3.4.22
2.70 - 3.25	2.70 - 3.25	2.70 - 3.25	2.70 - 3.25	2.70 - 3.25	3.4.23
4.00 - 4.80	4.00 - 4.80	4.00 - 4.80	4.00 - 4.80	4.00 - 4.80	3.4.24
5.75 - 6.90	5.75 - 6.90	5.75 - 6.90	5.75 - 6.90	5.75 - 6.90	3.4.25
7.60 - 9.15	7.60 - 9.15	7.60 - 9.15	7.60 - 9.15	7.60 - 9.15	3.4.26
9.45 - 11.35	9.45 - 11.35	9.45 - 11.35	9.45 - 11.35	9.45 - 11.35	3.4.27
14.30 - 17.00	14.30 - 17.00	14.30 - 17.00	14.30 - 17.00	14.30 - 17.00	3.4.28
- -	- -	- -	- -	- -	4.1.1
- -	- -	- -	- -	- -	4.1.2
- -	- -	- -	- -	- -	4.1.3
- -	- -	- -	- -	- -	4.1.4
- -	- -	- -	- -	- -	4.1.5
- -	- -	- -	- -	- -	4.1.6
- -	- -	- -	- -	- -	4.1.7
- -	- -	- -	- -	- -	4.1.8
- -	- -	- -	- -	- -	4.1.9
- -	- -	- -	- -	- -	4.1.10
9.00 - 10.00	9.00 - 10.00	9.00 - 10.00	9.00 - 10.00	9.00 - 10.00	4.1.11
35.00 - 40.00	35.00 - 40.00	35.00 - 40.00	35.00 - 40.00	35.00 - 40.00	4.1.12
45.00 - 50.00	45.00 - 50.00	45.00 - 50.00	45.00 - 50.00	45.00 - 50.00	4.1.13

Item nr.	SPECIFICATIONS	Unit	RESIDENTIAL £ range £
4.1	**FITTINGS AND FURNISHINGS** - cont'd gross internal area or individual areas (unless otherwise described) **Industrial/Office furniture, fittings and equipment** Reception desk, shelves and cupboards for general areas		
4.1.14	economical	m2	- -
4.1.15	medium quality	m2	- -
4.1.16	high quality	m2	- -
	Extra for		
4.1.17	high quality finishes to reception areas	m2	- -
4.1.18	full kitchen equipment (one cover/20m2)	m2	- -
	Furniture and fittings to general office areas		
4.1.19	economical	m2	- -
4.1.20	medium quality	m2	- -
4.1.21	high quality	m2	- -
	Retail fitting out, furniture, fittings and equipment Mall furniture etc.		
4.1.22	minimal provision	m2	- -
4.1.23	good provision	m2	- -
4.1.24	internal planting	m2	- -
4.1.25	glazed metal balustrades to voids	m	- -
4.1.26	feature pond and fountain	nr	- -
4.1.27	Fitting out a retail warehouse	m2	- -
	Fitting out shell for small shop (including shop fittings)		
4.1.28	simple store	m2	- -
4.1.29	fashion store	m2	- -
	Fitting out shell for department store or supermarket		
4.1.30	excluding shop fittings	m2	- -
4.1.31	including shop fittings	m2	- -
	Special fittings/equipment refrigerated installation for cold stores; display		
4.1.32	fittings in food stores	m2	- -
	food court furniture; fittings and special finishes		
4.1.33	(excluding catering display units)	m2	- -
4.1.34	bakery ovens	nr	- -
4.1.35	refuse compactors	nr	- -
	Leisure furniture, fittings and equipment General fittings		
4.1.36	internal planting	m2	- -
	signs, notice-boards, shelving, fixed seating,		
4.1.37	curtains and blinds	m2	- -
4.1.38	electric hand-dryers, incinerators, mirrors	m2	- -
	Specific fittings		
4.1.39	lockers, coin return locks	nr	- -
4.1.40	kitchen units; excluding equipment	nr	- -
4.1.41	folding sun bed	nr	- -
4.1.42	security grille	nr	- -
4.1.43	entrance balustrading and control turnstile	nr	- -
4.1.44	sports nets, screens etc. in a medium sized sports hall	nr	- -
	reception counter, fittings and reception counter		
4.1.45	screen	nr	- -
4.1.46	bar and fittings	nr	- -
4.1.47	telescopic seatings	nr	- -
	Swimming pool fittings		
4.1.48	metal balustrades to pool areas	m	- -
4.1.49	skimmer grilles to pool edge	nr	- -
	stainless steel pool access ladder		
4.1.50	1700 mm high	nr	- -
4.1.51	2500 mm high	nr	- -
	Leisure pool fittings		
4.1.52	stainless steel lighting post	nr	- -
4.1.53	water cannons	nr	- -
4.1.54	fountain or water sculpture	nr	- -
4.1.55	loudspeaker tower	nr	- -
	grp water chute; 65 - 80 m long; steel supports,		
4.1.56	spiral stairs and balustrading	nr	- -

INDUSTRIAL £ range £	RETAILING £ range £	LEISURE £ range £	OFFICES £ range £	HOTELS £ range £	Item nr.
3.00 - 6.00	- -	- -	4.80 - 9.60	- -	4.1.14
5.00 - 10.00	- -	- -	7.50 - 12.50	- -	4.1.15
8.00 - 15.00	- -	- -	11.00 - 18.00	- -	4.1.16
- -	- -	- -	5.00 - 7.00	- -	4.1.17
7.00 - 9.00	- -	- -	8.00 - 10.00	- -	4.1.18
- -	- -	- -	6.00 - 8.00	- -	4.1.19
- -	- -	- -	8.00 - 12.00	- -	4.1.20
- -	- -	- -	14.00 - 20.00	- -	4.1.21
- -	7.00 - 14.00	- -	- -	- -	4.1.22
- -	30.00 - 36.00	- -	- -	- -	4.1.23
- -	16.00 - 20.00	- -	- -	- -	4.1.24
- -	525.00 -675.00	- -	- -	- -	4.1.25
- -	30000 -45000	- -	- -	- -	4.1.26
- -	190.00 -215.00	- -	- -	- -	4.1.27
- -	475.00 -525.00	- -	- -	- -	4.1.28
- -	700 -1000	- -	- -	- -	4.1.29
- -	500.00 - 650.00	- -	- -	- -	4.1.30
- -	700 -1000	- -	- -	- -	4.1.31
- -	20.00 - 60.00	- -	- -	- -	4.1.32
- -	625.00 -675.00	- -	- -	- -	4.1.33
- -	8000 -12000	- -	- -	- -	4.1.34
- -	8500 -15500	- -	- -	- -	4.1.35
- -	- -	13.00 - 15.00	- -	- -	4.1.36
- -	- -	8.75 - 10.00	- -	- -	4.1.37
- -	- -	2.15 - 3.35	- -	- -	4.1.38
- -	- -	100.00 -140.00	- -	- -	4.1.39
- -	- -	1650 -2200	- -	- -	4.1.40
- -	- -	3850 -4400	- -	- -	4.1.41
- -	- -	5500 -7700	- -	- -	4.1.42
- -	- -	10000 -11000	- -	- -	4.1.43
- -	- -	14000 -16000	- -	- -	4.1.44
- -	- -	16500 -27500	- -	- -	4.1.45
- -	- -	27500 -33000	- -	- -	4.1.46
- -	- -	38500 -44000	- -	- -	4.1.47
- -	- -	275.00 -440.00	- -	- -	4.1.48
- -	- -	220.00 -270.00	- -	- -	4.1.49
- -	- -	800.00 -950.00	- -	- -	4.1.50
- -	- -	1100 -1250	- -	- -	4.1.51
- -	- -	110.00 -140.00	- -	- -	4.1.52
- -	- -	750 -1000	- -	- -	4.1.53
- -	- -	3000 -15000	- -	- -	4.1.54
- -	- -	4000 -4500	- -	- -	4.1.55
- -	- -	100000 -125000	- -	- -	4.1.56

Item nr.	SPECIFICATIONS	Unit	RESIDENTIAL £ range £
4.1	**FITTINGS AND FURNISHINGS** - cont'd	**gross internal area or individual units** (unless otherwise described)	
	Hotel furniture, fittings and equipment		
	General fittings		
4.1.57	signs	m2	- -
4.1.58	housekeeper and bolt	m2	- -
	Fittings for specific areas		
4.1.59	bedrooms (including bathrooms)	bedroom	- -
4.1.60	bedroom (operating supplies)	bedroom	- -
4.1.61	bedroom corridors	m2	- -
4.1.62	public areas	m2	- -
4.1.63	restaurant (operating supplies)	m2	- -
4.1.64	kitchen (including special equipment)	m2	- -
4.1.65	laundry (including special equipment)	nr	- -
5.1	**SANITARY AND DISPOSAL INSTALLATIONS**	**gross internal area** (unless otherwise described)	

comprising:-

Comparative sanitary fittings/sundries
Sanitary and disposal installations

Item nr.	SPECIFICATIONS	Unit	RESIDENTIAL £ range £
	Comparative sanitary fittings/sundries		
	Individual sanitary appliances (including fittings)		
	lavatory basins		
5.1.1	white	nr	150.00 -170.00
5.1.2	coloured	nr	180.00 -200.00
	low level WC's; on ground floor		
5.1.3	white	nr	130.00 -150.00
5.1.4	coloured	nr	160.00 -180.00
	low level WC's; one of a range; on upper floors		
5.1.5	white	nr	250.00 -275.00
5.1.6	coloured	nr	280.00 -310.00
	Extra for		
5.1.7	single storey PVC stack and connection to drain	nr	125.00 -150.00
5.1.8	bowl-type wall urinal; fireclay; white	nr	- -
	shower tray; including fittings		
5.1.9	white	nr	350.00 -400.00
5.1.10	coloured	nr	385.00 -425.00
5.1.11	sink; white; fireclay	nr	- -
	sink; stainless steel		
5.1.12	single drainer	nr	170.00 -200.00
5.1.13	double drainer	nr	200.00 -230.00
	bath		
5.1.14	acrylic; white	nr	250.00 -300.00
5.1.15	acrylic; coloured	nr	250.00 -300.00
5.1.16	enamelled steel; white	nr	275.00 -325.00
5.1.17	enamelled steel; coloured	nr	300.00 -350.00
	Soil waste stacks; 3.15 m storey height; branch and connection to drain		
5.1.18	100 mm PVC	nr	225.00 -250.00
	Extra for		
5.1.19	additional floors	nr	110.00 -125.00
5.1.20	100 mm cast iron; decorated	nr	- -
	Extra for		
5.1.21	additional floors	nr	- -
	Sanitary and disposal installations		
	Residential units		
	range including WC, wash handbasin, bath and kitchen		
5.1.22	sink	nr	1100 -1900
	range including WC, wash handbasin, bidet, bath		
5.1.23	and kitchen sink	nr	1400 -2250
	range including two WC's, two wash handbasins, bath		
5.1.24	and kitchen sink	nr	1600 -2700

INDUSTRIAL £ range £		RETAILING £ range £		LEISURE £ range £		OFFICES £ range £		HOTELS £ range £		Item nr.
-	-	-	-	-	-	-	-	1.80	- 3.00	4.1.57
-	-	-	-	-	-	-	-	6.00	- 10.00	4.1.58
-	-	-	-	-	-	-	-	4000	-8000	4.1.59
-	-	-	-	-	-	-	-	200.00	-300.00	4.1.60
-	-	-	-	-	-	-	-	60.00	-120.00	4.1.61
-	-	-	-	-	-	-	-	500	-1000	4.1.62
-	-	-	-	-	-	-	-	70.00	- 90.00	4.1.63
-	-	-	-	-	-	-	-	700.00	-800.00	4.1.64
-	-	-	-	-	-	-	-	87500	-117500	4.1.65
130.00	-160.00	150.00	-200.00	150.00	-200.00	150.00	-200.00	150.00	-250.00	5.1.1
-	-	-	-	180.00	-240.00	180.00	-240.00	180.00	-240.00	5.1.2
125.00	-150.00	130.00	-150.00	130.00	-150.00	130.00	-150.00	130.00	-175.00	5.1.3
-	-	-	-	160.00	-200.00	160.00	-200.00	160.00	-225.00	5.1.4
240.00	-275.00	250.00	-275.00	250.00	-275.00	250.00	-280.00	250.00	-300.00	5.1.5
-	-	-	-	280.00	-310.00	280.00	-325.00	280.00	-350.00	5.1.6
110.00	-130.00	125.00	-150.00	125.00	-150.00	125.00	-150.00	125.00	-150.00	5.1.7
175.00	-200.00	185.00	-225.00	185.00	-225.00	185.00	-225.00	-	-	5.1.8
325.00	-375.00	350.00	-400.00	350.00	-400.00	350.00	-400.00	350.00	-500.00	5.1.9
-	-	-	-	350.00	-425.00	350.00	-450.00	350.00	-650.00	5.1.10
160.00	-200.00	160.00	-280.00	-	-	160.00	-280.00	-	-	5.1.11
-	-	-	-	-	-	-	-	-	-	5.1.12
-	-	-	-	-	-	-	-	-	-	5.1.13
-	-	-	-	-	-	-	-	250.00	-350.00	5.1.14
-	-	-	-	-	-	-	-	250.00	-350.00	5.1.15
-	-	-	-	-	-	-	-	275.00	-375.00	5.1.16
-	-	-	-	-	-	-	-	300.00	-400.00	5.1.17
225.00	-250.00	225.00	-250.00	225.00	-250.00	225.00	-250.00	225.00	-250.00	5.1.18
110.00	-125.00	110.00	-125.00	110.00	-125.00	110.00	-125.00	110.00	-125.00	5.1.19
450.00	-475.00	450.00	-475.00	450.00	-475.00	450.00	-475.00	450.00	-475.00	5.1.20
225.00	-250.00	225.00	-250.00	225.00	-250.00	225.00	-250.00	225.00	-250.00	5.1.21
-	-	-	-	-	-	-	-	-	-	5.1.22
-	-	-	-	-	-	-	-	-	-	5.1.23
-	-	-	-	-	-	-	-	-	-	5.1.24

Item nr.	SPECIFICATIONS	Unit	RESIDENTIAL £ range £
5.1	**SANITARY AND DISPOSAL INSTALLATIONS** **- cont'd**	**gross internal area** (unless otherwise described)	
	Sanitary and disposal installations - cont'd **Residential units - cont'd** range including two WC's, two wash handbasins, bidet		
5.1.25	bath and kithen sink	nr	19.00 - 29.00
	Extra for		
5.1.26	rainwater pipe per storey	nr	50.00 - 60.00
5.1.27	soil pipe per storey	nr	110.00 -125.00
5.1.28	shower over bath	nr	275.00 -400.00
	Industrial buildings warehouse		
5.1.29	minimum provision	m2	- -
5.1.30	high provision	m2	- -
	production unit		
5.1.31	minimum provision	m2	- -
5.1.32	minimum provision; area less than 1000 m2	m2	- -
5.1.33	high provision	m2	- -
	Retailing outlets		
5.1.34	to superstore	m2	- -
	to shopping centre malls; public conveniences; branch		
5.1.35	connections to shop shells	m2	- -
5.1.36	fitting out public conveniences in shopping mall	block	- -
5.1.37	**Leisure buildings**	m2	- -
	Offices and industrial office buildings		
5.1.38	speculative; low rise; area less than 1000 m2	m2	- -
5.1.39	speculative; low rise	m2	- -
5.1.40	speculative; medium rise; area less than 1000 m2	m2	- -
5.1.41	speculative; medium rise	m2	- -
5.1.42	speculative; high rise	m2	- -
5.1.43	owner-occupied; low rise; area less than 1000 m2	m2	- -
5.1.44	owner-occupied; low rise	m2	- -
5.1.45	owner-occupied; medium rise; area less than 1000 m2	m2	- -
5.1.46	owner-occupied; medium rise	m2	- -
5.1.47	owner-occupied; high rise	m2	- -
	Hotels WC, bath, shower, basin to each bedroom; sanitary		
5.1.48	accommodation to public areas	m2	- -
5.2	**WATER INSTALLATIONS**	**gross internal area**	
	Hot and cold water installations		
5.2.1	Complete installations	m2	12.00 - 20.00
	To mall public conveniences; branch connections to		
5.2.2	shop shells	m2	- -
5.3	**HEATING, AIR-CONDITIONING AND** **VENTILATING INSTALLATIONS**	**gross internal area serviced** (unless otherwise described)	

comprising:-

Solid fuel radiator heating
Gas or oil-fired radiator heating
Gas or oil-fired convector heating
Electric and under floor heating
Hot air systems
Ventilation systems
Heating and ventilation systems
Comfort cooling systems
Full air-conditioning systems

INDUSTRIAL £ range £	RETAILING £ range £	LEISURE £ range £	OFFICES £ range £	HOTELS £ range £	Item nr.
-	-	-	-	-	5.1.25
-	-	-	-	-	5.1.26
-	-	-	-	-	5.1.27
-	-	-	-	-	5.1.28
7.00 - 10.00	-	-	-	-	5.1.29
10.00 - 15.00	-	-	-	-	5.1.30
10.00 - 15.00	-	-	-	-	5.1.31
12.50 - 20.00	-	-	-	-	5.1.32
11.00 - 18.00	-	-	-	-	5.1.33
-	2.50 - 6.00	-	-	-	5.1.34
-	5.00 - 8.00	-	-	-	5.1.35
-	3750 -6000	-	-	-	5.1.36
-	-	8.75 - 11.00	-	-	5.1.37
3.50 - 9.50	-	-	3.50 - 9.50	-	5.1.38
7.00 - 12.00	-	-	7.00 - 12.00	-	5.1.39
-	-	-	7.50 - 12.50	-	5.1.40
-	-	-	10.00 - 15.00	-	5.1.41
-	-	-	10.00 - 15.00	-	5.1.42
-	-	-	6.00 - 12.00	-	5.1.43
6.00 - 12.00	-	-	10.00 - 15.00	-	5.1.44
10.00 - 15.00	-	-	10.00 - 15.00	-	5.1.45
-	-	-	12.00 - 17.00	-	5.1.46
-	-	-	14.00 - 19.00	-	5.1.47
-	-	-	-	17.50 - 42.50	5.1.48
9.00 - 15.00	-	16.00 - 23.00	9.50 - 13.00	25.00 - 30.00	5.2.1
-	3.50 - 6.00	-	-	-	5.2.2

Keep your figures up to date, free of charge

This section, and most of the other information in this Price Book, is brought up to date every three months in the *Price Book Update*.

The *Update* is available free to all Price Book purchasers.

To ensure you receive your copy, simply complete the reply card from the centre of the book and return it to us.

Item nr.	SPECIFICATIONS	Unit	RESIDENTIAL £ range £
5.3	**HEATING, AIR-CONDITIONING AND VENTILATING INSTALLATIONS** - cont'd	**gross internal area serviced** (unless otherwise described)	
	Solid fuel radiator heating		
5.3.1	Chimney stack, hearth and surround to independent residential unit	nr	1500 -1800
	Chimney, hot water service and central heating for		
5.3.2	two radiators	nr	2250 -2600
5.3.3	three radiators	nr	2600 -2900
5.3.4	four radiators	nr	2900 -3200
5.3.5	five radiators	nr	3200 -3500
5.3.6	six radiators	nr	3500 -3800
5.3.7	seven radiators	nr	3800 -4400
	Gas or oil-fired radiator heating		
	Gas-fired hot water service and central heating for		
5.3.8	three radiators	nr	1700 -2350
5.3.9	four radiators	nr	2300 -2550
5.3.10	five radiators	nr	2550 -2700
5.3.11	six radiators	nr	2700 -2900
5.3.12	seven radiators	nr	2900 -3400
	Oil-fired hot water service, tank and central heating for		
5.3.13	seven radiators	nr	2700 -3600
5.3.14	ten radiators	nr	3000 -4000
5.3.15	LPHW radiator system	m2	23.00 - 35.00
5.3.16	speculative; area less than 1000 m2	m2	- -
5.3.17	speculative	m2	- -
5.3.18	owner-occupied; area less than 1000 m2	m2	- -
5.3.19	owner-occupied	m2	- -
5.3.20	LPHW radiant panel system	m2	- -
5.3.21	speculative; less than 1000 m2	m2	- -
5.3.22	speculative	m2	- -
5.3.23	LPHW fin tube heating	m2	- -
	Gas or oil-fired convector heating		
5.3.24	LPHW convector system	m2	- -
5.3.25	speculative; area less than 1000 m2	m2	- -
5.3.26	speculative	m2	- -
5.3.27	owner-occupied; area less than 1000 m2	m2	- -
5.3.28	owner-occupied	m2	- -
	LPHW sill-line connector system		
5.3.29	owner-occupied	m2	- -
	Electric and under floor heating		
5.3.30	Panel heaters	m2	- -
5.3.31	Skirting heaters	m2	- -
5.3.32	Storage heaters	m2	- -
5.3.33	Under floor heating in changing areas	m2	- -
	Hot air systems		
	Hot water service and ducted hot air heating to		
5.3.34	three rooms	nr	1500 -1940
5.3.35	five rooms	nr	2050 -2300
5.3.36	Elvaco warm air heating	m2	- -
5.3.37	Gas-fired hot air space heating	m2	- -
5.3.38	Warm air curtains; 1.6 m long electrically-operated	nr	- -
5.3.39	Hot water-operated; including supply pipework	nr	- -
	Ventilation systems		
	Local ventilation to		
5.3.40	WC's	nr	175.00 -220.00
5.3.41	toilet areas	m2	- -
5.3.42	bathroom and toilet areas	m2	- -
5.3.43	Air extract system	m2	- -
5.3.44	Air supply and extract system	m2	- -
5.3.45	Service yard vehicle extract system	m2	- -

INDUSTRIAL £ range £	RETAILING £ range £	LEISURE £ range £	OFFICES £ range £	HOTELS £ range £	Item nr.
- -	- -	- -	- -	- -	5.3.1
- -	- -	- -	- -	- -	5.3.2
- -	- -	- -	- -	- -	5.3.3
- -	- -	- -	- -	- -	5.3.4
- -	- -	- -	- -	- -	5.3.5
- -	- -	- -	- -	- -	5.3.6
- -	- -	- -	- -	- -	5.3.7
- -	- -	- -	- -	-	5.3.8
- -	- -	- -	- -	-	5.3.9
- -	- -	- -	- -	-	5.3.10
- -	- -	- -	- -	-	5.3.11
- -	- -	- -	- -	-	5.3.12
- -	- -	- -	- -	-	5.3.13
- -	- -	- -	- -	-	5.3.14
33.00 - 46.00	33.00 - 60.00	42.50 - 54.00	- -	50.00 - 70.00	5.3.15
- -	- -	- -	41.00 - 51.00	- -	5.3.16
- -	- -	- -	44.00 - 60.00	- -	5.3.17
- -	- -	- -	45.00 - 58.50	- -	5.3.18
- -	- -	- -	47.00 - 66.00	- -	5.3.19
- -	40.00 - 60.00	50.00 - 60.00	- -	- -	5.3.20
44.00 - 52.00	- -	- -	- -	- -	5.3.21
46.00 - 54.00	- -	- -	- -	- -	5.3.22
58.00 - 85.00	- -	- -	- -	- -	5.3.23
35.00 - 46.00	33.00 - 63.00	45.00 - 56.00	- -	57.50 - 73.50	5.3.24
- -	- -	- -	45.00 - 54.00	- -	5.3.25
- -	- -	- -	47.00 - 63.00	- -	5.3.26
- -	- -	- -	50.00 - 60.00	- -	5.3.27
- -	- -	- -	55.00 - 70.00	- -	5.3.28
- -	- -	- -	75.00 -100.00	- -	5.3.29
- -	- -	- -	10.00 - 13.00	- -	5.3.30
- -	- -	- -	15.00 - 20.00	- -	5.3.31
- -	- -	- -	18.00 - 23.00	- -	5.3.32
- -	- -	37.50 - 47.50	- -	- -	5.3.33
- -	- -	- -	- -	- -	5.3.34
- -	- -	- -	- -	- -	5.3.35
- -	- -	- -	54.00 - 67.50	- -	5.3.36
20.00 - 40.00	- -	- -	- -	- -	5.3.37
- -	1650 -2200	- -	- -	- -	5.3.38
- -	3300 -3800	- -	- -	- -	5.3.39
- -	- -	- -	- -	- -	5.3.40
- -	- -	- -	3.00 - 7.50	- -	5.3.41
- -	- -	- -	- -	17.50 - 22.50	5.3.42
- -	27.50 - 37.50	27.50 - 37.50	27.50 - 37.50	30.00 - 40.00	5.3.43
- -	- -	- -	40.00 - 60.00	- -	5.3.44
- -	27.50 - 37.50	- -	- -	- -	5.3.45

Item nr.	SPECIFICATIONS	Unit	RESIDENTIAL £ range £

5.3 · **HEATING, AIR-CONDITIONING AND VENTILATING INSTALLATIONS - cont'd** · **gross internal area serviced** (unless otherwise described)

Heating and ventilation systems

Item nr.	SPECIFICATIONS	Unit	£	range	£
5.3.46	Space heating and ventilation - economical	m2	-		-
5.3.47	Heating and ventilation	m2	-		-
5.3.48	Warm air heating and ventilation	m2	-		-
5.3.49	Hot air heating and ventilation to shopping malls; including automatic remote vents in rooflights	m2	-		-
	Extra for				
5.3.50	comfort cooling	m2	-		-
5.3.51	full air-conditioning	m2	-		-
	Comfort cooling systems				
5.3.52	Fan coil/induction systems	m2	-		-
5.3.53	speculative; area less than 1000 m2	m2	-		-
5.3.54	speculative	m2	-		-
5.3.55	owner-occupied; area less than 1000 m2	m2	-		-
5.3.56	owner-occupied	m2	-		-
5.3.57	VAV system	m2	-		-
5.3.58	speculative; area less than 1000 m2	m2	-		-
5.3.59	speculative	m2	-		-
5.3.60	owner-occupied; area less than 1000 m2	m2	-		-
5.3.61	owner-occupied	m2	-		-
	Full air-conditioning				
	Stand-alone air-conditioning unit systems				
5.3.62	air supply and extract	m2	-		-
5.3.63	air supply and extract; including heat re-claim	m2	-		-
	Extra for				
5.3.64	automatic control installation	m2	-		-
5.3.65	Full air-conditioning with dust and humidity control	m2	-		-
5.3.66	Fan coil/induction systems	m2	-		-
5.3.67	speculative; area less than 1000 m2	m2	-		-
5.3.68	speculative	m2	-		-
5.3.69	owner-occupied; area less than 1000 m2	m2	-		-
5.3.70	owner-occupied	m2	-		-
5.3.71	VAV system	m2	-		-
5.3.72	speculative; area less than 1000 m2	m2	-		-
5.3.73	speculative	m2	-		-
5.3.74	owner-occupied; area less than 1000 m2	m2	-		-
5.3.75	owner-occupied	m2	-		-

5.4 · **ELECTRICAL INSTALLATIONS** · **gross internal area serviced** (unless otherwise described)

comprising:-

Mains and sub-mains switchgear and distribution
Lighting installation
Lighting and power installations
Comparative fittings/rates per point

Item nr.	SPECIFICATIONS	Unit	£	range	£
	Mains and sub-mains switchgear and distribution				
5.4.1	Mains intake only	m2	-		-
5.4.2	Mains switchgear only	m2	2.00	-	3.00
	Mains and sub-mains distribution				
5.4.3	to floors only	m2	-		-
5.4.4	to floors; including small power and supplies to equipment	m2	-		-
5.4.5	to floors; including lighting and power to landlords areas and supplies to equipment	m2	-		-
5.4.6	to floors; including power, communication and supplies to equipment	m2	-		-
5.4.7	to shop units; including fire alarms and telephone distribution	m2	-		-

INDUSTRIAL £ range £	RETAILING £ range £	LEISURE £ range £	OFFICES £ range £	HOTELS £ range £	Item nr.
17.50 - 38.50	18.25 - 33.60	-	-	-	5.3.46
45.00 - 55.00	-	-	-	-	5.3.47
-	80.00 -105.00	80.00 -105.00	80.00 -105.00	90.00 -125.00	5.3.48
-	75.00 - 95.00	-	-	-	5.3.49
-	60.00 - 90.00	-	-	-	5.3.50
-	70.00 -130.00	-	-	-	5.3.51
-	-	-	-	125.00 -165.00	5.3.52
-	-	-	125.00 -145.00	125.00 -165.00	5.3.53
-	-	-	130.00 -155.00	-	5.3.54
-	-	-	135.00 -165.00	-	5.3.55
-	-	-	140.00 -175.00	-	5.3.56
-	125.00 -180.00	-	-	150.00 -220.00	5.3.57
-	-	-	140.00 -200.00	-	5.3.58
-	-	-	145.00 -205.00	-	5.3.59
-	-	-	150.00 -210.00	-	5.3.60
-	-	-	155.00 -220.00	-	5.3.61
-	-	105.00 -140.00	-	-	5.3.62
-	-	115.00 -150.00	-	-	5.3.63
-	-	10.00 - 30.00	-	-	5.3.64
130.00 -200.00	-	-	-	-	5.3.65
-	140.00 -205.00	-	-	145.00 -220.00	5.3.66
-	-	-	130.00 -165.00	-	5.3.67
-	-	-	135.00 -185.00	-	5.3.68
-	-	-	140.00 -205.00	-	5.3.69
-	-	-	145.00 -220.00	-	5.3.70
-	-	-	-	160.00 -250.00	5.3.71
-	-	-	155.00 -205.00	-	5.3.72
-	-	-	160.00 -220.00	-	5.3.73
-	-	-	165.00 -235.00	-	5.3.74
-	-	-	170.00 -250.00	-	5.3.75
1.30 - 2.50	-	-	-	-	5.4.1
3.25 - 6.50	3.00 - 7.00	-	-	-	5.4.2
3.85 - 7.50	-	15.00 - 27.50	11.00 - 20.00	-	5.4.3
-	-	11.00 - 13.00	-	-	5.4.4
7.50 - 16.50	-	-	27.50 - 45.00	-	5.4.5
-	-	-	-	45.00 - 65.00	5.4.6
-	4.50 - 11.00	-	-	-	5.4.7

Item nr.	SPECIFICATIONS	Unit	RESIDENTIAL £ range £

5.4 **ELECTRICAL INSTALLATIONS** **gross internal area serviced**
- cont'd **(unless otherwise described)**

Lighting installation
Lighting to

Item nr.	SPECIFICATIONS	Unit	RESIDENTIAL £ range £
5.4.8	warehouse area	m2	- -
5.4.9	production area	m2	- -
5.4.10	General lighting; including luminaries	m2	- -
5.4.11	Emergency lighting	m2	- -
5.4.12	standby generators only	m2	- -
5.4.13	Underwater lighting	m2	- -

Lighting and power installations
Lighting and power to residential units

Item nr.	SPECIFICATIONS	Unit	RESIDENTIAL £ range £
5.4.14	one person flat/bed-sit	nr	750 -1150
5.4.15	two person flat/house	nr	900 -1650
5.4.16	three person flat/house	nr	1050 -2000
5.4.17	four person house	nr	1200 -2450
5.4.18	five/six person house	nr	1500 -3000
	Extra for		
5.4.19	intercom	nr	300.00 - 350.00

Lighting and power to industrial buildings

Item nr.	SPECIFICATIONS	Unit	RESIDENTIAL £ range £
5.4.20	warehouse area	m2	- -
5.4.21	production area	m2	- -
5.4.22	production area; high provision	m2	- -
5.4.23	office area	m2	- -
5.4.24	office area; high provision	m2	- -

Lighting and power to retail outlets

Item nr.	SPECIFICATIONS	Unit	RESIDENTIAL £ range £
5.4.25	shopping mall and landlords' areas	m2	- -

Lighting and power to offices

Item nr.	SPECIFICATIONS	Unit	RESIDENTIAL £ range £
5.4.26	speculative office areas; average standard	m2	- -
5.4.27	speculative office areas; high standard	m2	- -
5.4.28	owner-occupied office areas; average standard	m2	- -
5.4.29	owner-occupied office areas; high standard	m2	- -

Comparative fittings/rates per point

Item nr.	SPECIFICATIONS	Unit	RESIDENTIAL £ range £
5.4.30	Consumer control unit	nr	120.00 -130.00

Fittings; excluding lamps or light fittings

Item nr.	SPECIFICATIONS	Unit	RESIDENTIAL £ range £
5.4.31	lighting point; PVC cables	nr	38.00 - 42.00
5.4.32	lighting point; PVC cables in screwed conduits	nr	80.00 - 90.00
5.4.33	lighting point; MICC cables	nr	67.00 - 75.00

switch socket outlet; PVC cables

Item nr.	SPECIFICATIONS	Unit	RESIDENTIAL £ range £
5.4.34	single	nr	42.00 - 45.00
5.4.35	double	nr	50.00 - 55.00

switch socket outlet; PVC cables in screwed conduit

Item nr.	SPECIFICATIONS	Unit	RESIDENTIAL £ range £
5.4.36	single	nr	60.00 - 65.00
5.4.37	double	nr	68.00 - 75.00

switch socket outlet; MICC cables

Item nr.	SPECIFICATIONS	Unit	RESIDENTIAL £ range £
5.4.38	single	nr	61.00 - 66.00
5.4.39	double	nr	70.00 - 77.50
5.4.40	Immersion heater point (excluding heater)	nr	60.00 - 70.00
5.4.41	Cooker point; including control unit	nr	85.00 -135.00

5.5 **GAS INSTALLATION**

Item nr.	SPECIFICATIONS	Unit	RESIDENTIAL £ range £
5.5.1	Connection charge	nr	100.00 -150.00

Supply to heaters within shopping mall and capped off

Item nr.	SPECIFICATIONS	Unit	RESIDENTIAL £ range £
5.5.2	supply to shop shells	m2	- -

INDUSTRIAL £ range £	RETAILING £ range £	LEISURE £ range £	OFFICES £ range £	HOTELS £ range £	Item nr.
16.00 - 30.00	- -	- -	- -	- -	5.4.8
20.00 - 33.00	- -	- -	- -	- -	5.4.9
- -	- -	20.00 - 32.50	- -	17.00 - 27.00	5.4.10
- -	- -	7.00 - 11.00	- -	3.00 - 5.50	5.4.11
2.00 - 8.00	2.00 - 8.00	2.00 - 8.00	2.00 - 8.00	2.00 - 8.00	5.4.12
- -	- -	5.00 - 10.00	- -	- -	5.4.13
- -	- -	- -	- -	- -	5.4.14
- -	- -	- -	- -	- -	5.4.15
- -	- -	- -	- -	- -	5.4.16
- -	- -	- -	- -	- -	5.4.17
- -	- -	- -	- -	- -	5.4.18
- -	- -	- -	- -	- -	5.4.19
30.00 - 50.00	- -	- -	- -	- -	5.4.20
35.00 - 55.00	- -	- -	- -	- -	5.4.21
50.00 - 65.00	- -	- -	- -	- -	5.4.22
75.00 - 90.00	- -	- -	- -	- -	5.4.23
100.00 -115.00	- -	- -	- -	- -	5.4.24
- -	50.00 - 80.00	- -	- -	- -	5.4.25
- -	- -	- -	65.00 - 95.00	- -	5.4.26
- -	- -	- -	80.00 -100.00	- -	5.4.27
- -	- -	- -	90.00 -115.00	- -	5.4.28
- -	- -	- -	105.00 -130.00	- -	5.4.29
- -	- -	- -	- -	- -	5.4.30
93.00 -100.00	46.50 - 50.00	46.50 - 50.00	46.50 - 50.00	46.50 - 50.00	5.4.31
110.00 -120.00	97.50 -107.50	97.50 -107.50	97.50 -107.50	97.50 -107.50	5.4.32
95.00 -105.00	80.00 - 90.00	80.00 - 90.00	80.00 - 90.00	80.00 - 90.00	5.4.33
91.00 - 97.50	51.00 - 56.00	51.00 - 56.00	51.00 - 56.00	51.00 - 56.00	5.4.34
100.00 -107.50	60.00 - 65.00	60.00 - 65.00	60.00 - 65.00	60.00 - 65.00	5.4.35
80.00 - 87.50	70.00 - 75.00	70.00 - 75.00	70.00 - 75.00	70.00 - 75.00	5.4.36
90.00 - 97.50	80.00 - 85.00	80.00 - 85.00	80.00 - 85.00	80.00 - 85.00	5.4.37
82.00 - 88.00	71.00 - 76.00	71.00 - 76.00	71.00 - 76.00	71.00 - 76.00	5.4.38
90.00 -110.00	80.00 - 87.50	80.00 - 87.50	80.00 - 87.50	80.00 - 87.50	5.4.39
- -	- -	- -	- -	- -	5.4.40
- -	- -	- -	- -	- -	5.4.41
- -	- -	- -	- -	- -	5.5.1
- -	2.75 - 4.00	- -	- -	- -	5.5.2

Item nr.	SPECIFICATIONS	Unit	RESIDENTIAL £ range £

5.6   **LIFT AND CONVEYOR INSTALLATIONS**   lift or escalator
                                              (unless otherwise described)
           comprising:-

           Passenger lifts
           Escalators
           Goods lifts
           Dock levellers

           Electro-hydraulic passenger lifts
           eight to twelve person lifts

Item nr.	SPECIFICATIONS	Unit	RESIDENTIAL	£ range £
5.6.1	8 person; 0.3 m/sec ; 2 - 3 levels	nr	-	-
5.6.2	8 person; 0.63 m/sec ; 3 - 5 levels	nr	-	-
5.6.3	8 person; 1.5 m/sec ; 3 levels	nr	-	-
5.6.4	8 person; 1.6 m/sec ; 6 - 9 levels	nr	-	-
5.6.5	10 person; 0.63 m/sec ; 2 - 4 levels	nr	-	-
5.6.6	10 person; 1.0 m/sec ; 2 levels	nr	-	-
5.6.7	10 person; 1.0 m/sec ; 3 - 6 levels	nr	-	-
5.6.8	10 person; 1.0 m/sec ; 3 - 6 levels	nr	-	-
5.6.9	10 person; 1.6 m/sec ; 4 levels	nr	-	-
5.6.10	10 person; 1.6 m/sec ; 10 levels	nr	-	-
5.6.11	10 person; 1.6 m/sec ; 13 levels	nr	-	-
5.6.12	10 person; 2.5 m/sec ; 18 levels	nr	-	-
5.6.13	12 person; 1.0 m/sec ; 2 - 3 levels	nr	-	-
5.6.14	12 person; 1.0 m/sec ; 3 - 6 levels	nr	-	-
	Thirteen person lifts			
5.6.15	13 person; 0.3 m/sec ; 2 - 3 levels	nr	-	-
5.6.16	13 person; 0.63 m/sec ; 4 levels	nr	-	-
5.6.17	13 person; 1.0 m/sec ; 2 levels	nr	-	-
5.6.18	13 person; 1.0 m/sec ; 2 - 3 levels	nr	-	-
5.6.19	13 person; 1.0 m/sec ; 3 - 6 levels	nr	-	-
5.6.20	13 person; 1.6 m/sec ; 7 - 11 levels	nr	-	-
5.6.21	13 person; 2.5 m/sec ; 7 - 11 levels	nr	-	-
5.6.22	13 person; 2.5 m/sec ; 12 - 15 levels	nr	-	-
	Sixteen person lifts			
5.6.23	16 person; 1.0 m/sec ; 2 levels	nr	-	-
5.6.24	16 person; 1.0 m/sec ; 3 levels	nr	-	-
5.6.25	16 person; 1.0 m/sec ; 3 - 6 levels	nr	-	-
5.6.26	16 person; 1.6 m/sec ; 3 - 6 levels	nr	-	-
5.6.27	16 person; 1.6 m/sec ; 7 - 11 levels	nr	-	-
5.6.28	16 person; 2.5 m/sec ; 7 - 11 levels	nr	-	-
5.6.29	16 person; 2.5 m/sec ; 12 - 15 levels	nr	-	-
5.6.30	16 person; 3.5 m/sec ; 12 - 15 levels	nr	-	-
	Twenty one person lifts			
5.6.31	21 person/bed; 0.4 m/sec ; 3 levels	nr	-	-
5.6.32	21 person; 0.6 m/sec ; 2 levels	nr	-	-
5.6.33	21 person; 1.0 m/sec ; 4 levels	nr	-	-
5.6.34	21 person; 1.6 m/sec ; 4 levels	nr	-	-
5.6.35	21 person; 1.6 m/sec ; 7 - 11 levels	nr	-	-
5.6.36	21 person; 1.6 m/sec ; 10 levels	nr	-	-
5.6.37	21 person; 1.6 m/sec ; 13 levels	nr	-	-
5.6.38	21 person; 2.5 m.sec ; 7 - 11 levels	nr	-	-
5.6.39	21 person; 2.5 m/sec ; 12 - 15 levels	nr	-	-
5.6.40	21 person; 2.5 m/sec ; 18 levels	nr	-	-
5.6.41	21 person; 3.5 m/sec ; 12 - 15 levels	nr	-	-
	Extra for			
5.6.42	enhanced finish to car	nr	-	-
5.6.43	glass backed observation car	nr	-	-
	Ten person wall climber lifts			
5.6.44	10 person; 0.5 m/sec ; 2 levels	nr	-	-
	Escalators			
	30° Escalator; 0.5 m/sec; enamelled steel glass balustrades			
5.6.45	3.5 m rise; 600 mm step width	nr	-	-
5.6.46	3.5 m rise; 800 mm step width	nr	-	-
5.6.47	3.5 m rise; 1000 mm step width	nr	-	-
	Extra for			
5.6.48	enhanced finish	nr	-	-

INDUSTRIAL £ range £		RETAILING £ range £		LEISURE £ range £		OFFICES £ range £		HOTELS £ range £		Item nr.
-	-	-	-	19500	-27500	18500	-26500	-	-	5.6.1
-	-	-	-	-	-	33000	-50000	-	-	5.6.2
45000	-58000	-	-	-	-	-	-	-	-	5.6.3
-	-	-	-	-	-	60000	-72000	-	-	5.6.4
-	-	-	-	-	-	40000	-52000	-	-	5.6.5
-	-	-	-	-	-	34000	-36000	-	-	5.6.6
-	-	-	-	-	-	41500	-53000	-	-	5.6.7
-	-	-	-	-	-	45000	-57500	-	-	5.6.8
-	-	-	-	-	-	48000	-60000	-	-	5.6.9
-	-	-	-	-	-	66000	-78000	-	-	5.6.10
-	-	-	-	-	-	74000	-92000	-	-	5.6.11
-	-	-	-	-	-	125000	-145000	-	-	5.6.12
22500	-45000	-	-	-	-	-	-	-	-	5.6.13
-	-	-	-	-	-	-	-	52500	-64000	5.6.14
-	-	-	-	-	-	22000	-30000	-	-	5.6.15
-	-	-	-	-	-	45000	-57000	-	-	5.6.16
-	-	54000	-60000	-	-	-	-	-	-	5.6.17
45000	-59000	-	-	-	-	-	-	-	-	5.6.18
-	-	-	-	-	-	46000	-60000	53000	-67000	5.6.19
-	-	-	-	-	-	60000	-75000	64000	-80000	5.6.20
-	-	-	-	-	-	95000	-110000	-	-	5.6.21
-	-	-	-	-	-	110000	-127500	-	-	5.6.22
-	-	52000	-65000	-	-	-	-	-	-	5.6.23
-	-	60000	-70000	-	-	-	-	-	-	5.6.24
-	-	-	-	-	-	53000	-67000	-	-	5.6.25
-	-	-	-	-	-	55000	-70000	-	-	5.6.26
-	-	-	-	-	-	77000	-88000	-	-	5.6.27
-	-	-	-	-	-	100000	-115000	-	-	5.6.28
-	-	-	-	-	-	115000	-135000	-	-	5.6.29
-	-	-	-	-	-	120000	-140000	-	-	5.6.30
-	-	-	-	-	-	37500	-42500	-	-	5.6.31
-	-	-	-	-	-	55000	-66000	-	-	5.6.32
-	-	-	-	-	-	66000	-77000	-	-	5.6.33
-	-	-	-	-	-	58000	-66000	-	-	5.6.34
-	-	-	-	-	-	73000	-90000	-	-	5.6.35
-	-	-	-	-	-	76000	-93000	-	-	5.6.36
-	-	-	-	-	-	84000	-100000	-	-	5.6.37
-	-	-	-	-	-	110000	-130000	-	-	5.6.38
-	-	-	-	-	-	125000	-144000	-	-	5.6.39
-	-	-	-	-	-	140000	-160000	-	-	5.6.40
-	-	-	-	-	-	132000	-152000	-	-	5.6.41
-	-	-	-	-	-	-	-	7000	-8000	5.6.42
-	-	-	-	-	-	-	-	8000	-16000	5.6.43
-	-	120000	-160000	-	-	120000	-160000	120000	-160000	5.6.44
-	-	-	-	-	-	42000	-54000	-	-	5.6.45
-	-	-	-	-	-	48000	-60000	-	-	5.6.46
-	-	60000	-80000	55000	-75000	55000	-67000	57500	-75000	5.6.47
-	-	18000	-24000	12000	-18000	12000	-18000	18000	-25000	5.6.48

Item nr.	SPECIFICATIONS	Unit	RESIDENTIAL £ range £

5.6 **LIFT AND CONVEYOR INSTALLATIONS** -cont'd   **lift or escalator**
(unless otherwise described)

Escalators - cont'd
30° Escalator; 0.5 m/sec; enamelled steel glass
balustrades - cont'd

5.6.49	4.4 m rise; 800 mm step width	nr	-	-
5.6.50	4.4 m rise; 1000 mm step width	nr	-	-
5.6.51	5.2 m rise; 800 mm step width	nr	-	-
5.6.52	5.2 m rise; 1000 mm step width	nr	-	-
5.6.53	6.0 m rise; 800 mm step width	nr	-	-
5.6.54	6.0 m rise; 1000 mm step width	nr	-	-
	Extras (per escalator)			
5.6.55	under step lighting	nr	-	-
5.6.56	under handrail lighting	nr	-	-
5.6.57	stainless steel balustrades	nr	-	-
5.6.58	mirror glass cladding to sides and soffits	nr	-	-
5.6.59	heavy duty chairs	nr	-	-

Goods lifts

5.6.60	Hoist	nr	-	-
	Kitchen service hoist			
5.6.61	50 kg; 2 levels	nr	-	-
	Electric heavy duty goods lifts			
5.6.62	500 kg; 2 levels	nr	-	-
5.6.63	500 kg; 2 - 3 levels	nr	-	-
5.6.64	500 kg; 5 levels	nr	-	-
5.6.65	1000 kg; 2 levels	nr	-	-
5.6.66	1000 kg; 2 - 5 levels	nr	-	-
5.6.67	1000 kg; 5 levels	nr	-	-
5.6.68	1500 kg; 3 levels	nr	-	-
5.6.69	1500 kg; 4 levels	nr	-	-
5.6.70	1500 kg; 7 levels	nr	-	-
5.6.71	2000 kg; 2 levels	nr	-	-
5.6.72	2000 kg; 3 levels	nr	-	-
5.6.73	3000 kg; 2 levels	nr	-	-
5.6.74	3000 kg; 3 levels	nr	-	-
	Oil hydraulic heavy duty goods lifts			
5.6.75	500 kg; 3 levels	nr	-	-
5.6.76	1000 kg; 3 levels	nr	-	-
5.6.77	2000 kg; 3 levels	nr	-	-
5.6.78	3500 kg; 4 levels	nr	-	-

Dock levellers

5.6.79	Dock levellers	nr	-	-
5.6.80	Dock leveller and canopy	nr	-	-

5.7 **PROTECTIVE, COMMUNICATION AND**   **gross internal area served**
SPECIAL INSTALLATIONS   (unless otherwise described)

comprising:-

Fire fighting/protective installations
Security/communication installations
Special installations

Fire fighting/protective installations
Fire alarms/appliances

5.7.1	loose fire-fighting equipment	m2	-	-
5.7.2	smoke detectors, alarms and controls	m2	-	-
5.7.3	hosereels, dry risers and extinguishers	m2	-	-
	Sprinkler installations			
	landlords areas; supply to shop shells; including			
5.7.4	fire alarms, appliances etc.	m2	-	-
	single level sprinkler systems, alarms and smoke			

INDUSTRIAL £ range £	RETAILING £ range £	LEISURE £ range £	OFFICES £ range £	HOTELS £ range £	Item nr.
- -	- -	- -	50000 -63000	- -	5.6.49
- -	- -	- -	57000 -69000	- -	5.6.50
- -	- -	- -	54000 -66000	- -	5.6.51
- -	- -	- -	60000 -72000	- -	5.6.52
- -	- -	- -	57000 -69000	- -	5.6.53
- -	- -	- -	63000 -75000	- -	5.6.54
- -	- -	- -	300 -360	- -	5.6.55
- -	- -	- -	1100 -2200	- -	5.6.56
- -	- -	- -	1200 -2400	- -	5.6.57
- -	- -	- -	8800 -11000	- -	5.6.58
- -	- -	- -	11000 -14500	- -	5.6.59
3750 -15000	- -	- -	- -	- -	5.6.60
- -	- -	5000 -5500	- -	- -	5.6.61
- -	38000 -43000	- -	37500 -42500	37500 -42500	5.6.62
16000 -23000	- -	- -	- -	- -	5.6.63
- -	- -	- -	45000 -56000	- -	5.6.64
- -	42000 -50000	- -	40000 -48000	40000 -48000	5.6.65
21000 -30000	- -	- -	- -	- -	5.6.66
- -	- -	- -	- -	48000 -56000	5.6.67
45000 -54000	- -	- -	50000 -60000	- -	5.6.68
- -	- -	- -	53000 -64000	- -	5.6.69
- -	- -	- -	55000 -70000	- -	5.6.70
- -	- -	- -	46000 -56000	- -	5.6.71
- -	- -	- -	57000 -68000	- -	5.6.72
- -	- -	- -	54000 -64000	- -	5.6.73
- -	- -	- -	66000 -80000	- -	5.6.74
50000 -60000	- -	- -	- -	- -	5.6.75
54000 -63000	- -	- -	- -	- -	5.6.76
62000 -72000	- -	- -	- -	- -	5.6.77
55000 -66000	- -	- -	- -	- -	5.6.78
5000 -13000	5000 -13000	- -	- -	5000 -13000	5.6.79
7000 -18000	7000 -18000	- -	- -	- -	5.6.80
- -	- -	0.20 - 0.25	- -	- -	5.7.1
2.50 - 5.00	2.50 - 5.00	5.00 - 6.00	2.50 - 5.00	5.50 - 10.00	5.7.2
4.00 - 9.00	4.50 - 9.00	- -	4.50 - 9.00	4.30 - 8.60	5.7.3
- -	7.50 - 13.00	- -	- -	- -	5.7.4

Item nr.	SPECIFICATIONS	Unit	RESIDENTIAL £ range £

5.7   **PROTECTIVE, COMMUNICATION AND   gross internal area served**
           SPECIAL INSTALLATIONS - cont'd (unless otherwise described)

Fire fighting/protective installations - cont'd
Sprinkler installations - cont'd

5.7.5	detectors; low hazard	m2	- -
	Extra for		
5.7.6	ordinary hazard	m2	- -
	single level sprinkler systems; alarms and smoke		
5.7.7	detectors; ordinary hazard	m2	- -
	double level sprinkler systems; alarms and smoke		
5.7.8	detectors; high hazard	m2	- -
	Smoke vents		
5.7.9	automatic smoke vents over glazed shopping mall	m2	- -
5.7.10	smoke control ventilation to atria	m2	- -
5.7.11	Lightning protection	m2	- -

Security/communication installations

5.7.12	Clock installation	m2	- -
5.7.13	Security alarm system	m2	- -
5.7.14	Telephone system	m2	- -
5.7.15	Public address, television aerial and clocks	m2	- -
5.7.16	Closed-circuit television	m2	- -
5.7.17	Public address system	m2	- -
5.7.18	Closed-circuit television and public address system	m2	- -

Special installations
Window cleaning equipment

5.7.19	twin track	m	- -
5.7.20	manual trolley/cradle	nr	- -
5.7.21	automatic trolley/cradle	nr	- -
5.7.22	Refrigeration installation for ice rinks	m2 ice	- -
5.7.23	Pool water treatment installation	m2 pool	- -
5.7.24	Laundry chute	nr	- -
5.7.25	Sauna	nr	- -
5.7.26	Jacuzzi installation	nr	- -
5.7.27	Wave machine; four chamber wave generation equipment	nr	- -
	Swimming pool; size 1300 x 6.00 m		
	Extra over cost including structure; finishings		
5.7.28	ventilation; heating and filtration	m2	- -

5.8   **BUILDERS' WORK IN CONNECTION   gross internal area**
           WITH SERVICES

General builders work to

5.8.1	mains supplies; lighting and power to landlords areas	m2	- -
5.8.2	central heating and electrical installation	m2	- -
5.8.3	space heating and electrical installation	m2	- -
5.8.4	central heating; electrical and lift installations	m2	- -
	space heating; electrical and ventilation		
5.8.5	installations	m2	- -
5.8.6	air-conditioning	m2	- -
5.8.7	air-conditioning and electrical installation	m2	- -
5.8.8	air-conditioning; electrical and lift installations	m2	- -
	General builders work; including allowance for plant rooms; to		
5.8.9	central heating and electrical installations	m2	- -
5.8.10	central heating, electrical and lift installations	m2	- -
5.8.11	air-conditioning	m2	- -
5.8.12	air-conditioning and electrical installation	m2	- -
5.8.13	air-conditioning; electrical and lift installations	m2	- -

INDUSTRIAL £ range £	RETAILING £ range £	LEISURE £ range £	OFFICES £ range £	HOTELS £ range £	Item nr.
9.00 - 14.00	10.00 - 14.00	-	-	-	5.7.5
4.00 - 5.00	4.00 - 5.00	-	-	-	5.7.6
13.00 - 18.00	14.00 - 18.00	-	12.00 - 17.00	10.50 - 16.00	5.7.7
20.00 - 25.00	22.00 - 26.00	-	20.00 - 25.00	-	5.7.8
-	14.00 - 22.00	-	-	-	5.7.9
-	-	-	42.50 - 50.00	42.50 - 52.50	5.7.10
0.50 - 0.60	1.00 - 2.00	0.50 - 0.60	0.75 - 1.25	0.75 - 1.25	5.7.11
-	-	0.10 - 1.00	-	-	5.7.12
-	-	1.50 - 2.00	1.50 - 2.00	-	5.7.13
-	-	0.80 - 1.60	0.80 - 1.60	-	5.7.14
2.00 - 3.00	-	2.20 - 3.30	-	2.00 - 4.25	5.7.15
-	3.00 - 3.50	3.00 - 3.50	-	-	5.7.16
-	8.00 - 9.00	8.00 - 9.00	-	-	5.7.17
-	17.50 - 35.00	11.00 - 13.00	-	-	5.7.18
-	110.00 -125.00	-	100.00 -115.00	107.50 -120.00	5.7.19
-	7200 -8200	-	7300 -8250	7000 -8000	5.7.20
-	16000 -20000	-	17500 -20000	16000 -20000	5.7.21
-	-	375.00 -450.00	-	-	5.7.22
-	-	400.00 -425.00	-	-	5.7.23
-	-	10000 -12000	-	10000 -12000	5.7.24
-	-	-	-	6000 -12000	5.7.25
-	-	12000 -15000	-	12000 -15000	5.7.26
-	-	37500 -52500	-	-	5.7.27
-	-	-	-	750.00 -850.00	5.7.28
1.40 - 4.20	-	-	-	-	5.8.1
3.00 - 10.00	5.00 - 11.00	5.00 - 11.00	7.50 - 11.00	7.50 - 11.00	5.8.2
4.00 - 11.00	7.00 - 12.00	7.00 - 12.00	9.00 - 12.00	9.00 - 12.00	5.8.3
4.50 - 12.00	7.50 - 12.50	7.50 - 12.50	10.00 - 13.00	10.00 - 13.00	5.8.4
7.00 - 15.00	10.00 - 16.00	10.00 - 16.00	14.00 - 18.00	14.00 - 18.00	5.8.5
11.00 - 17.00	13.00 - 18.00	13.00 - 18.00	16.00 - 20.00	16.00 - 20.00	5.8.6
12.50 - 18.00	15.00 - 20.00	15.00 - 20.00	18.00 - 22.00	18.00 - 12.00	5.8.7
14.00 - 20.00	17.00 - 22.00	17.00 - 22.00	20.00 - 24.00	20.00 - 24.00	5.8.8
-	20.00 - 25.00	20.00 - 25.00	25.00 - 30.00	25.00 - 30.00	5.8.9
-	25.00 - 30.00	25.00 - 30.00	30.00 - 35.00	30.00 - 35.00	5.8.10
-	35.00 - 40.00	35.00 - 40.00	45.00 - 50.00	45.00 - 50.00	5.8.11
-	45.00 - 50.00	45.00 - 50.00	55.00 - 60.00	55.00 - 60.00	5.8.12
-	55.00 - 60.00	55.00 - 60.00	65.00 - 70.00	65.00 - 70.00	5.8.13

Item nr.	SPECIFICATIONS	Unit	RESIDENTIAL £ range £
6.1	**SITE WORK** surface area (unless otherwise described)		
	comprising:-		
	Preparatory excavation and sub-bases **Seeded and planted areas** **Paved areas** **Car parking alternatives** **Roads and barriers** **Fencing and screen walls etc.**		
	Preparatory excavation and sub-bases Excavating		
6.1.1	top soil; average 150 mm deep	m2	1.00 - 1.75
	top soil; average 225 mm deep; preserving in spoil heaps		
6.1.2	by machine	m2	2.00 - 2.50
6.1.3	by hand	m2	5.40 - 7.00
	to reduce levels; not exceeding 225 mm deep		
6.1.4	by machine	m2	2.50 - 2.75
6.1.5	by hand	m2	7.00 - 8.00
6.1.6	to form new foundation levels and contours	m2	2.50 - 7.50
	Filling; imported top soil		
6.1.7	150 mm thick; spread and levelled for planting	m2	5.00 - 6.00
	Comparative sub-bases/beds		
6.1.8	50 mm sand	m2	1.90 - 2.20
6.1.9	75 mm ashes	m2	2.00 - 2.25
6.1.10	75 mm sand	m2	2.80 - 3.20
6.1.11	50 mm (1:8) blinding concrete	m2	3.00 - 4.00
6.1.12	100 mm granular fill	m2	3.10 - 3.50
6.1.13	150 mm granular fill	m2	4.65 - 5.00
6.1.14	75 mm (1:3:6) blinding concrete	m2	5.00 - 5.50
	Seeded and planted areas Plant supply, planting, maintenance and 12 months guarantee		
6.1.15	seeded areas	m2	3.00 - 6.00
6.1.16	turfed areas	m2	4.00 - 8.00
	Planted areas (per m2 of surface area)		
6.1.17	herbaceous plants	m2	3.50 - 4.50
6.1.18	climbing plants	m2	4.50 - 7.50
6.1.19	general planting	m2	10.00 - 20.00
6.1.20	woodland	m2	- -
6.1.21	shrubbed planting	m2	- -
6.1.22	dense planting	m2	- -
6.1.23	shrubbed area including allowance for small trees	m2	- -
	Trees		
6.1.24	advanced nursery stock trees (12 - 20 cm girth)	tree	- -
	semi-mature trees; 5 - 8 m high		
6.1.25	coniferous	tree	- -
6.1.26	deciduous	tree	- -
	Sports facilities		
6.1.27	Putting green	hole	- -
6.1.28	Par 3 golf course	hole	- -
	Paved areas Gravel paving rolled to falls and cambers		
6.1.29	50 mm thick	m2	2.00 - 3.00
6.1.30	paving on sub-base; including excavation	m2	7.50 - 9.00
	Cold bitumen emulsion paving; in three layers		
6.1.31	25 mm thick	m2	4.00 - 5.00
	Tarmacadam paving; two layers; limestone or igneous chipping finish		
6.1.32	65 mm thick	m2	7.00 - 10.00
6.1.33	paving on sub-base; including excavation	m2	17.00 - 23.00

INDUSTRIAL £ range £	RETAILING £ range £	LEISURE £ range £	OFFICES £ range £	HOTELS £ range £	Item nr.
1.00 - 1.75	1.00 - 1.75	1.00 - 1.75	1.00 - 1.75	1.00 - 1.75	6.1.1
2.00 - 2.50	2.00 - 2.50	2.00 - 2.50	2.00 - 2.50	2.00 - 2.50	6.1.2
5.40 - 7.00	5.40 - 7.00	5.40 - 7.00	5.40 - 7.00	5.40 - 7.00	6.1.3
2.50 - 2.75	2.50 - 2.75	2.50 - 2.75	2.50 - 2.75	2.50 - 2.75	6.1.4
7.00 - 8.00	7.00 - 8.00	7.00 - 8.00	7.00 - 8.00	7.00 - 8.00	6.1.5
2.50 - 7.50	2.50 - 7.50	2.50 - 7.50	2.50 - 7.50	2.50 - 7.50	6.1.6
5.00 - 6.00	5.00 - 6.00	5.00 - 6.00	5.00 - 6.00	5.00 - 6.00	6.1.7
1.90 - 2.20	1.90 - 2.20	1.90 - 2.20	1.90 - 2.20	1.90 - 2.20	6.1.8
2.00 - 2.25	2.00 - 2.25	2.00 - 2.25	2.00 - 2.25	2.00 - 2.25	6.1.9
2.80 - 3.20	2.80 - 3.20	2.80 - 3.20	2.80 - 3.20	2.80 - 3.20	6.1.10
3.00 - 4.00	3.00 - 4.00	3.00 - 4.00	3.00 - 4.00	3.00 - 4.00	6.1.11
3.10 - 3.50	3.10 - 3.50	3.10 - 3.50	3.10 - 3.50	3.10 - 3.50	6.1.12
4.65 - 5.00	4.65 - 5.00	4.65 - 5.00	4.65 - 5.00	4.65 - 5.00	6.1.13
5.00 - 5.50	5.00 - 5.50	5.00 - 5.50	5.00 - 5.50	5.00 - 5.50	6.1.14
3.00 - 6.00	3.00 - 6.00	3.00 - 6.00	3.00 - 6.00	3.00 - 6.00	6.1.15
4.00 - 8.00	4.00 - 8.00	4.00 - 8.00	4.00 - 8.00	4.00 - 8.00	6.1.16
3.50 - 4.50	3.50 - 4.50	3.50 - 4.50	3.50 - 4.50	3.50 - 4.50	6.1.17
4.50 - 7.50	4.50 - 7.50	4.50 - 7.50	4.50 - 7.50	4.50 - 7.50	6.1.18
10.00 - 20.00	10.00 - 20.00	10.00 - 20.00	10.00 - 20.00	10.00 - 20.00	6.1.19
15.00 - 30.00	- -	- -	- -	- -	6.1.20
20.00 - 55.00	20.00 - 40.00	20.00 - 40.00	20.00 - 40.00	20.00 - 40.00	6.1.21
- -	- -	- -	25.00 - 50.00	- -	6.1.22
30.00 - 80.00	30.00 - 70.00	30.00 - 70.00	30.00 - 70.00	30.00 - 70.00	6.1.23
125.00 -150.00	125.00 -150.00	125.00 -150.00	125.00 -150.00	125.00 -150.00	6.1.24
400 -1000	400 -1000	400 -1000	400 -1000	400 -1000	6.1.25
600 -1500	600 -1500	600 -1500	600 -1500	600 -1500	6.1.26
- -	- -	1500 -2000	- -	- -	6.1.27
- -	- -	10000 -20000	- -	- -	6.1.28
2.00 - 3.00	2.00 - 3.00	2.00 - 3.00	2.00 - 3.00	2.00 - 3.00	6.1.29
7.50 - 9.00	7.50 - 9.00	7.50 - 9.00	7.50 - 9.00	7.50 - 9.00	6.1.30
4.00 - 5.00	4.00 - 5.00	4.00 - 5.00	4.00 - 5.00	4.00 - 5.00	6.1.31
7.00 - 10.00	7.00 - 10.00	7.00 - 10.00	7.00 - 10.00	7.00 - 10.00	6.1.32
17.00 - 23.00	17.00 - 23.00	17.00 - 23.00	17.00 - 23.00	17.00 - 23.00	6.1.33

Item nr.	SPECIFICATIONS	Unit	RESIDENTIAL £ range £

6.1   SITE WORK - cont'd   **surface area**
                                                     (unless otherwise described)

Paved areas -cont'd
Precast concrete paving slabs

Item nr.	SPECIFICATIONS	Unit	RESIDENTIAL £ range £
6.1.34	50 mm thick	m2	8.00 - 13.00
6.1.35	50 mm thick 'Texitone' slabs	m2	12.00 - 16.00
6.1.36	slabs on sub-base; including excavation	m2	20.00 - 28.00
	Precast concrete block paviours		
6.1.37	65 mm 'Keyblok' grey rectangular blocks	m2	15.50 - 18.50
	'Monolok' grey interlocking blocks		
6.1.38	60 mm thick	m2	17.00 - 19.00
6.1.39	80 mm thick	m2	19.00 - 21.00
6.1.40	paviours on sub-base; including excavation	m2	30.00 - 36.00
	Brick paviours		
	229 x 114 x 38 mm paving bricks		
6.1.41	laid flat	m2	29.00 - 35.00
6.1.42	laid to herring-bone pattern	m2	45.00 - 52.50
6.1.43	paviours on sub-base; including excavation	m2	42.00 - 67.50
	Granite setts		
6.1.44	200 x 100 x 100 mm setts	m2	43.00 - 48.00
6.1.45	setts on sub-base; including excavation	m2	55.00 - 65.00
	York stone slab paving		
6.1.46	paving on sub-base; including excavation	m2	55.00 - 60.00
	Cobblestone paving		
6.1.47	50 - 75 mm	m2	55.00 - 65.00
6.1.48	cobblestones on sub-base; including excavation	m2	65.00 - 80.00

Car Parking alternatives
Surface level parking; including lighting and drainage

Item nr.	SPECIFICATIONS	Unit	RESIDENTIAL £ range £
6.1.49	tarmacadam on sub-base	car	875 -975
6.1.50	concrete interlocking blocks	car	1400 -1700
6.1.51	at ground level with deck or building over	car	- -
	Garages etc		
6.1.52	single car park	nr	600 -1000
6.1.53	single; traditional construction; in a block	nr	2000 -2800
6.1.54	single; traditional construction; pitched roof	nr	4500 -5800
6.1.55	double; traditional construction; pitched roof	nr	6200 -8000
	Multi-storey parking; including lighting and drainage on roof of two-storey shopping centre; including		
6.1.56	ramping and strengthening structure	car	- -
6.1.57	split level/parking ramp	car	- -
6.1.58	multi-storey flat slab	car	- -
6.1.59	multi-storey warped slab	car	- -
	Extra for		
6.1.60	feature cladding and pitched roof	car	- -
	Underground parking; including lighting and drainage partially underground; natural ventilation; no		
6.1.61	sprinklers	car	- -
	completely underground; mechanical ventilation and		
6.1.62	sprinklers	car	- -
	completely underground; mechanical ventilation;		
6.1.63	sprinklers and landscaped roof	car	- -

Roads and barriers
Tarmacadam road surfacing; two layers

Item nr.	SPECIFICATIONS	Unit	RESIDENTIAL £ range £
6.1.64	75 mm thick	m2	8.00 - 10.00
6.1.65	tarmacadam road on sub-base; including excavation	m2	20.00 - 27.50
	Tarmacadam roads; two layers; 75 mm thick; including excavation; 225 mm hardcore bed; kerbs and channels both sides on concrete foundations		
6.1.66	4.9 m wide road	m2	120.00 -130.00
6.1.67	6.1 m wide road	m2	140.00 -160.00
6.1.68	7.32 m wide road	m2	160.00 -175.00
	Concrete road surfacing		
	insitu reinforced concrete on sub-base; including		
6.1.69	excavation	m2	27.50 - 35.00

INDUSTRIAL £ range £	RETAILING £ range £	LEISURE £ range £	OFFICES £ range £	HOTELS £ range £	Item nr.
8.00 - 13.00	8.00 - 13.00	8.00 - 15.00	8.00 - 12.00	8.00 - 15.00	6.1.34
12.00 - 16.00	12.00 - 16.00	12.00 - 18.00	12.00 - 17.00	12.00 - 18.00	6.1.35
20.00 - 28.00	20.00 - 28.00	20.00 - 33.00	20.00 - 30.00	22.00 - 33.00	6.1.36
15.50 - 18.50	15.50 - 18.50	15.50 - 18.50	15.50 - 18.50	15.50 - 18.50	6.1.37
17.00 - 19.00	17.00 - 19.00	17.00 - 19.00	17.00 - 19.00	17.00 - 19.00	6.1.38
19.00 - 21.00	19.00 - 21.00	19.00 - 21.00	19.00 - 21.00	19.00 - 21.00	6.1.39
30.00 - 36.00	30.00 - 36.00	30.00 - 36.00	30.00 - 36.00	30.00 - 36.00	6.1.40
29.00 - 35.00	29.0 - 35.00	29.00 - 35.00	29.00 - 35.00	29.00 - 35.00	6.1.41
45.00 - 52.50	45.00 - 52.50	45.00 - 52.50	45.00 - 52.50	45.00 - 52.50	6.1.42
42.00 - 67.50	42.00 - 67.50	42.00 - 67.50	42.00 - 67.50	47.00 - 67.50	6.1.43
43.00 - 48.00	43.00 - 48.00	43.00 - 48.00	43.00 - 48.00	43.00 - 48.00	6.1.44
55.00 - 65.00	55.00 - 65.00	55.00 - 65.00	55.00 - 65.00	55.00 - 65.00	6.1.45
55.00 - 60.00	55.00 - 60.00	55.00 - 60.00	55.00 - 60.00	55.00 - 60.00	6.1.46
55.00 - 65.00	55.00 - 65.00	55.00 - 65.00	55.00 - 65.00	55.00 - 65.00	6.1.47
65.00 - 80.00	65.00 - 80.00	65.00 - 80.00	65.00 - 80.00	65.00 - 80.00	6.1.48
875 -975	875 -975	875 -975	875 -975	875 -975	6.1.49
1400 -1700	1400 -1700	1400 -1700	1400 -1700	1400 -1700	6.1.50
- -	4750 -5750	4750 -5750	4750 -5750	4750 -5750	6.1.51
- -	- -	- -	- -	- -	6.1.52
- -	- -	- -	- -	- -	6.1.53
- -	- -	- -	- -	- -	6.1.54
- -	- -	- -	- -	- -	6.1.55
- -	5250 -6250	5250 -6250	5250 -6250	5250 -6250	6.1.56
- -	5400 -6400	5400 -6400	5400 -6400	5400 -6400	6.1.57
- -	6500 -8000	6500 -8000	6500 -8000	6500 -8000	6.1.58
- -	7250 -8750	7250 -8750	7250 -8750	7250 -8750	6.1.59
- -	900 -1500	900 -1500	900 -1500	900 -1500	6.1.60
- -	9250 -11000	9250 -11000	9250 -11000	9250 -11000	6.1.61
- -	11000 -13750	11000 -13750	11000 -13750	11000 -13750	6.1.62
- -	13000 -15500	13000 -15500	13000 -15500	13000 -15500	6.1.63
7.50 - 9.50	8.00 - 10.00	8.00 - 10.00	8.00 - 10.00	8.00 - 10.00	6.1.64
19.00 - 26.00	20.00 - 27.50	20.00 - 27.50	20.00 - 27.50	20.00 - 27.50	6.1.65
115.00 -125.00	120.00 -130.00	120.00 -130.00	120.00 -130.00	120.00 -130.00	6.1.66
135.00 -145.00	140.00 -160.00	140.00 -160.00	140.00 -160.00	140.00 -160.00	6.1.67
155.00 -170.00	160.00 -175.00	160.00 -175.00	160.00 -175.00	160.00 -175.00	6.1.68
35.00 - 45.00	27.50 - 35.00	27.50 - 35.00	27.50 - 35.00	27.50 - 35.00	6.1.69

Item nr.	SPECIFICATIONS	Unit	RESIDENTIAL £ range £

6.1   SITE WORK - cont'd   **surface area**
                              (unless otherwise described)

Roads and barriers - cont'd
Reinforced concrete roads; 21 N/mm2 concrete; 150 mm
thick; mesh reinforcements and expansion joints;
including excavation; 150 mm granular filling;
blinding; kerbs and channels both sides on concrete
foundations

Item nr.	SPECIFICATIONS	Unit	£ range £
6.1.70	4.9 m wide road	m	140.00 -150.00
6.1.71	6.1 m wide road	m	160.00 -180.00
6.1.72	7.32 m wide road	m	180.00 -210.00
6.1.73	'Grasscrete' precast concrete fire paths	m2	30.00 - 33.00
6.1.74	Precast concrete planks and topping; as car park ramps	m2	- -
6.1.75	Two lane road; including drainage and lighting	m2	- -
6.1.76	Two lane; short span bridge	m2	- -
6.1.77	Roundabout on existing dual carriageway; including perimeter road, drainage and lighting, signs and disruption while under construction	nr	- -
	Guard rails and parking bollards etc.		
6.1.78	open metal post and rail fencing 1 m high	m	- -
6.1.79	galvanised steel post and rail fencing 2 m high	m	- -
6.1.80	steel guard rails and vehicle barriers	m	- -
	parking bollards		
6.1.81	precast concrete	nr	87.50 -105.00
6.1.82	steel	nr	145.00 -190.00
6.1.83	cast iron	nr	175.00 -225.00
6.1.84	vehicle control barrier; manual pole	nr	- -
6.1.85	precast concrete lamp-post	nr	- -
6.1.86	galvanised steel cycle stand	nr	- -
6.1.87	galvanised steel flag staff	nr	- -

Fencing and screen walls etc
Chain link fencing; plastic coated

Item nr.	SPECIFICATIONS	Unit	£ range £
6.1.88	1.22 m high	m	12.50 - 14.50
6.1.89	1.8 m high	m2	17.50 - 20.00
	Timber fencing		
6.1.90	1.2 m high chestnut pale fencing	m	17.00 - 19.00
6.1.91	1.8 m high close-boarded fencing	m2	40.00 - 50.00
	Screen walls; one brick thick; including foundations etc		
6.1.92	1.8 m high facing brick screen wall	m2	200.00 -250.00
6.1.93	1.8 m high coloured masonry block boundary wall	m2	220.00 -275.00
	Tennis courts, complete with tennis post, net, netting surround; proprietary 'non-attention' or coated macadam surface		
6.1.94	one court	nr	12000 -15000
6.1.95	two courts side by side	nr	22000 -28000
6.1.96	squash courts, independent building, including shower	nr	60000 -70000
	Demolish existing buildings		
6.1.97	of brick construction	m3	4.20 - 7.50

6.2   DRAINAGE   **gross internal area**
                              (unless otherwise described)

comprising:-

Overall £/m2 allowances
Comparative pipework
Comparative manholes

Overall £/m2 allowances

Item nr.	SPECIFICATIONS	Unit	£ range £
6.2.1	Site drainage (per m2 of paved areas)	m2	7.00 - 13.00
6.2.2	Building drainage (per m2 of gross floor area)	m2	7.00 - 15.00
6.2.3	Drainage work beyond the boundary of the site and final connection	nr	2000 -10000

INDUSTRIAL £ range £	RETAILING £ range £	LEISURE £ range £	OFFICES £ range £	HOTELS £ range £	Item nr.
135.00 -145.00	140.00 -150.00	140.00 -150.00	140.00 -150.00	140.00 -150.00	6.1.70
155.00 -175.00	160.00 -180.00	160.00 -180.00	160.00 -180.00	160.00 -180.00	6.1.71
175.00 -200.00	180.00 -210.00	180.00 -210.00	180.00 -210.00	180.00 -210.00	6.1.72
29.00 - 32.00	30.00 - 33.00	30.00 - 33.00	30.00 - 33.00	30.00 - 33.00	6.1.73
110.00 -135.00	115.00 -140.00	115.00 -140.00	115.00 -140.00	115.00 -140.00	6.1.74
- -	300.00 -450.00	300.00 -450.00	300.00 -450.00	300.00 -450.00	6.1.75
- -	600.00 -750.00	600.00 -750.00	600.00 -750.00	- -	6.1.76
- -	300000 -450000	- -	- -	- -	6.1.77
110.00 -130.00	110.00 -130.00	110.00 -130.00	110.00 -130.00	- -	6.1.78
120.00 -160.00	120.00 -160.00	120.00 -160.00	120.00 -160.00	- -	6.1.79
40.00 - 60.00	40.00 - 60.00	40.00 - 60.00	40.00 - 60.00	- -	6.1.80
87.50 -105.00	87.50 -105.00	87.50 -105.00	87.50 -105.00	87.50 -105.00	6.1.81
145.00 -190.00	145.00 -190.00	145.00 -190.00	145.00 -190.00	145.00 -190.00	6.1.82
175.00 -225.00	175.00 -225.00	175.00 -225.00	175.00 -225.00	175.00 -225.00	6.1.83
- -	775.00 -825.00	775.00 -825.00	775.00 -825.00	775.00 -825.00	6.1.84
- -	87.50 -105.00	87.50 -105.00	87.50 -105.00	87.50 -105.00	6.1.85
- -	32.50 - 42.50	32.50 - 42.50	32.50 - 42.50	32.50 - 42.50	6.1.86
- -	880 -1100	880 -1100	800 -1100	800 -1100	6.1.87
12.50 - 14.50	12.50 - 14.50	12.50 - 14.50	12.50 - 14.50	12.50 - 14.50	6.1.88
17.50 - 20.00	17.50 - 20.00	17.50 - 20.00	17.50 - 20.00	17.50 - 20.00	6.1.89
17.00 - 19.00	17.00 - 19.00	17.00 - 19.00	17.00 - 19.00	17.00 - 19.00	6.1.90
40.00 - 50.00	40.00 - 50.00	40.00 - 50.00	40.00 - 50.00	40.00 - 50.00	6.1.91
200.00 -250.00	200.00 -250.00	200.00 -250.00	200.00 -250.00	200.00 -250.00	6.1.92
220.00 -275.00	220.00 -275.00	220.00 -275.00	220.00 -275.00	220.00 -275.00	6.1.93
12000 -15000	12000 -15000	12000 -15000	12000 -15000	12000 -15000	6.1.94
22000 -28000	22000 -28000	22000 -28000	22000 -28000	22000 -28000	6.1.95
60000 -70000	60000 -70000	60000 -70000	60000 -70000	60000 -70000	6.1.96
4.20 - 7.50	4.20 - 7.50	4.20 - 7.50	4.20 - 7.50	4.20 - 7.50	6.1.97
11.00 - 17.00	9.00 - 15.00	7.00 - 13.00	8.00 - 12.50	9.50 - 18.50	6.2.1
8.00 - 14.00	10.00 - 15.00	11.00 - 16.00	7.00 - 10.00	7.00 - 15.00	6.2.2
2000 -10000	2000 -10000	2000 -10000	2000 -10000	2000 -10000	6.2.3

Item nr.	SPECIFICATIONS	Unit	RESIDENTIAL £ range £

6.2   **DRAINAGE** - cont'd   **gross internal area**
                                      (unless otherwise described)

Comparative pipework
Vitrified clay pipes; flexible couplings; including
150 mm granular bed and benching; in trench 1 m deep

6.2.4	100 mm 'Supersleve' pipe	m	17.00 - 19.50
6.2.5	150 mm 'Supersleve' pipe	m	20.50 - 23.50
6.2.6	100 mm 'Hepseal' pipe	m	20.00 - 23.00
6.2.7	150 mm 'Hepseal' pipe	m	23.00 - 26.00

Vitrified clay pipes; cement and sand joints; including
150 mm concrete bed and benching in trench 1 m deep

6.2.8	100 mm pipe	m	19.00 - 21.75
6.2.9	150 mm pipe	m	22.75 - 25.75

UPVC pipes; including 150 mm granular bed and benching;
in trench 1 m deep

6.2.10	100 mm pipe	m	23.50 - 27.50
6.2.11	150 mm pipe	m	26.50 - 31.50

Cast iron pipes; caulked lead joints; including
150 mm concrete bed; in trench 1 m deep

6.2.12	100 mm pipe	m	45.00 - 52.50
6.2.13	150 mm pipe	m	65.00 - 75.00
	Extra for		
	each additional 250 mm in trench not exceeding		
6.2.14	2 m deep	m	3.25 - 4.00
	each additional 250 mm in trench over 2 m and		
6.2.15	not exceeding 4 m deep	m	3.50 - 4.50

Comparative manholes
Precast concrete ring manhole; including excavation;
base, cover, channels and benching bottoms

6.2.16	1.2 m dia x 1500 m deep	nr	1150 -1250

Brick manhole; including excavation; concrete base,
cover, channels and benching bottoms

6.2.17	686 x 457 x 600 mm deep	nr	235.00 -275.00
	Extra for		
	each additional 300 mm		
6.2.18	up to 2 m deep internally	nr	75.00 - 85.00
6.2.19	980 x 686 x 600 mm deep	nr	375.00 -425.00
	Extra for		
	each additional 300 mm		
6.2.20	up to 2 m deep internally	nr	95.00 -110.00
6.2.21	1370 x 800 x 1500 mm deep	nr	850.00 -950.00
	Extra for		
	685 x 685 mm shaft for each additional 1300 mm		
6.2.22	up to 4 m deep internally	nr	100.00 -120.00
6.2.23	100 mm three-quarter section branch bends	nr	8.20 - 10.00
6.2.24	150 mm three-quarter section branch bends	nr	13.25 - 15.00

6.3   **EXTERNAL SERVICES**   **gross internal area**

Incoming mains services (electricity, gas, water,

6.3.1	BT etc)	m2	2.00 - 5.00
6.3.2	External lighting (per m2 of lighted area)	m2	2.00 - 3.00

INDUSTRIAL £ range £	RETAILING £ range £	LEISURE £ range £	OFFICES £ range £	HOTELS £ range £	Item nr.
17.00 - 19.50	17.00 - 19.50	17.00 - 19.50	17.00 - 19.50	17.00 - 19.50	6.2.4
20.50 - 23.50	20.50 - 23.50	20.50 - 23.50	20.50 - 23.50	20.50 - 23.50	6.2.5
20.00 - 23.00	20.00 - 23.00	20.00 - 23.00	20.00 - 23.00	20.00 - 23.00	6.2.6
23.00 - 26.00	23.00 - 26.00	23.00 - 26.00	23.00 - 26.00	23.00 - 26.00	6.2.7
19.00 - 21.75	19.00 - 21.75	19.00 - 21.75	19.00 - 21.75	19.00 - 21.75	6.2.8
22.75 - 25.75	22.75 - 25.75	22.75 - 25.75	22.75 - 25.75	22.75 - 25.75	6.2.9
23.50 - 27.50	23.50 - 27.50	23.50 - 27.50	23.50 - 27.50	23.50 - 27.50	6.2.10
26.50 - 31.50	26.50 - 31.50	26.50 - 31.50	26.50 - 31.50	26.50 - 31.50	6.2.11
45.00 - 52.50	45.00 - 52.50	45.00 - 52.50	45.00 - 52.50	45.00 - 52.50	6.2.12
65.00 - 75.00	65.00 - 75.00	65.00 - 75.00	65.00 - 75.00	65.00 - 75.00	6.2.13
3.25 - 4.00	3.25 - 4.00	3.25 - 4.00	3.25 - 4.00	3.25 - 4.00	6.2.14
3.50 - 4.50	3.50 - 4.50	3.50 - 4.50	3.50 - 4.50	3.50 - 4.50	6.2.15
1150 -1250	1150 -1250	1150 -1250	1150 -1250	1150 -1250	6.2.16
235.00 -275.00	235.00 -275.00	235.00 -275.00	235.00 -275.00	235.00 -275.00	6.2.17
75.00 - 85.00	75.00 - 85.00	75.00 - 85.00	75.00 - 85.00	75.00 - 85.00	6.2.18
375.00 -425.00	375.00 -425.00	375.00 -425.00	375.00 -425.00	375.00 -425.00	6.2.19
95.00 -110.00	95.00 -110.00	95.00 -110.00	95.00 -110.00	95.00 -110.00	6.2.20
850.00 -950.00	850.00 -950.00	850.00 -950.00	850.00 -950.00	850.00 -950.00	6.2.21
100.00 -120.00	100.00 -120.00	100.00 -120.00	100.00 -120.00	100.00 -120.00	6.2.22
8.20 - 10.00	8.20 - 10.00	8.20 - 10.00	8.20 - 10.00	8.20 - 10.00	6.2.23
13.25 - 15.00	13.25 - 15.00	13.25 - 15.00	13.25 - 15.00	13.25 - 15.00	6.2.24
4.50 - 7.50	10.00 - 17.50	17.50 - 22.50	5.00 - 10.00	10.00 - 17.50	6.3.1
2.00 - 3.00	3.50 - 12.50	6.00 - 10.00	2.00 - 3.00	3.50 - 12.50	6.3.2

ARCHITECTS' AND BUILDERS' TITLES

SHOPPING CENTRE DESIGN
N K Scott

An assessment of the various approaches to the design of shopping centres in modern urban environments, this book discusses important international examples, and stresses the need for creative and integrated schemes. The growth in the last 25 years of shopping as a major leisure activity can be linked to the emergence of efficient, attractive central retail developments. Designers of these high profile structures must reconcile the differing interests of clients, planners, consumers, environmentalists and local politicians. *Shopping Centre Design* details the processes by which town centre retail developments are created in modern urban environments. Throughout, the book stresses the need for creative and integrated schemes and highlights some of the technical details.

Contents: The early years: 1960 to 1970. How it all started. The middle years: 1970 to 1980. Reappraisal: counting the cost. The last decade: the debate about 'Style'. Current design trends. Planning and the environment lobby. Shopping out of town. The architect's role. Refurbishment. Development competitions. Fees. Index.

1989 Hardback 0747 60045 7 £35.00 250pp Van Nostrand Reinhold International

THE WAY WE BUILD NOW
Form, scale and technique
A Orton

This book documents current practice in the structural and constructional design of buildings. The first part presents an overview of materials and structural forms taking the point of view of the designer, architect and engineer. The second part is an extensive examination of over 70 case studies. They have been carefully selected and tightly structured to present a summary of established modern methods of building construction. It contains copious ready-reference charts of design information, numerous photographs and meticulous axonometric drawings. The book is international in scope. Dual units are used throughout (SI and Imperial) and nearly half the case studies are taken from the USA. Cases are also drawn from Canada, Europe, Africa, Malaysia, Hong Kong as well as 25 from the UK.

Contents: Acknowledgements. Introduction. Building materials. Building structures. Building physics. Fire safety. Houses and community buildings. Office buildings. Public buildings. Residential buildings. Highly serviced buildings. Sport and storage buildings. Open structure buildings. High rise buildings. Appendix. References. Bibliography.

1987 Paperback 0747 60011 2 £32.50 544pp Van Nostrand Reinhold International

These books may be obtained from your usual bookshop. In case of difficulty please write to the address below or telephone the Order Department on 0264 332424.

 **E & F N SPON**
11 New Fetter Lane, London EC4P 4EE

Costs Limits and Allowances

Information given under this heading is based upon the cost targets currently in force for buildings financed out of public funds, i.e., hospitals, schools, universities and public authority housing. The information enables the cost limit for a scheme to be calculated and is not intended to be a substitute for estimates prepared from drawings and specifications.

The cost limits are generally set as target costs based upon the user accommodation, i.e., in the case of universities they are given per metre of usable floor area. However ad-hoc additions can be agreed with the relevant Authority in exceptional circumstances.

The documents setting out cost targets are almost invariably complex and cover a range of differing circumstances. They should be studied carefully before being applied to any scheme; this study should preferably be undertaken in consultation with a Chartered Quantity Surveyor.

The cost limits for Public Authority housing generally known as the Housing Cost Yardstick have been replaced by a system of new procedures. The cost criteria prepared by the Department of the Enviroment that have superseded the Housing Cost Yardstick are intended as indicators and not cost limits.

HOSPITAL BUILDINGS

The information and tables which follow are contained in the Concise 4 database (issued to Regional Health Authorities) version 5.0 and Capricode (obtainable from HMSO) by kind permission of the Controller of Her Majesty's Stationery office. The tables, which should be read in conjunction with the Building notes (as issued to May 1989) concerned, provide the departmental cost limit related to the functional content of a project.

The cost allowance figures relate either to a QSSD index level of 277 - effective on 16th August 1988 or a PIPSH index level of 157 - effective on 14th November 1988. Both index levels will be revised to respond to a +5% movement in the indices and Authorities will be informed when such revisions take place.

The cost allowance for Teaching Hospitals have not been included in this section as they are partly funded by the UGC.

The reader should check that the levels of the following cost allowances are current when used for submitting new schemes.

The works cost compromises the departmental cost for a scheme plus the on-costs.

DEPARTMENTAL COSTS

The Departmental Cost section provides tables from which a capital sum is calculated, based on (or derived from) the functional content, representing the total departmental cost. Where cost allowances are not appropriate, the costs for individual departments should be justified on the basis of similar accommodation in other schemes.

In interpolating for functional units which fall between two of the figures given in the tables the cost should be rounded to the nearest:-

- £1 000 for departments costing over £50 000
- £500 for departments costing £2 000 - £50 000
- £100 for departments costing £1 000 - £20 000
- £10 for departments costing less than £1 000

ON COSTS

These are capital costs arising from interaction of the building and its site, (for example the cost of communications between departments, external works, and services, additional energy saving measures, auxiliary building and abnormals).

COMMUNICATIONS

Space and/or main engineering services between departments for movement of people supplies and services (including lifts, stairs and main ducts or shafts).

EXTERNAL WORKS

Building and engineering works external to the outer wall of the building but forming
an integral part of the scheme, and main engineering supply services internal to the
building but outside the confines of individual departments.

ABNORMALS

Exceptional or ad-hoc factors, usually arising from site difficulties and constraints
which increase capital costs, (for example, demolitions, adverse soil conditions,
poor bearing capacity of ground works or alterations to existing buildings).

EXTENSION OF AN EXISTING HOSPITAL

Where it is intended to add to the departmental or service provision of an extant hospital
by attaching to and/or extending the existing structure, the Works cost should be
calculated as set out above for new separate hospital buildings.

UPGRADING OF EXISTING ACCOMMODATION

The Works cost for upgrading work should be estimated in each case. To establish that
the cost is economic, it should be compared to a notional cost limit calculated on the
cost of a functionally equivalent new building discounted for the life of the upgraded
building and abated by the value of sound elements to be re-used in the existing
building.

HOSPITAL BUILDINGS

REF. NO.	SERVICE	DEPARTMENT	ACCOMMODATION
01.01.01	**In-Patients Services**	General Acute Wards	Ward
02			Essential Complementary Accommodation: Orthopaedic Equipment Store
03			Seminar Room
04			Senior Nurse's Office
05			Clinical Teacher's Office
06			Relative's Room
07			Relative's WC
08			Relative's Shower/Wash
09			Shared Accommodation (for single ward)
10			Optional Accommodation: Cook Chill Trolley Holding Room
01.02.01		Children	Nursing Section: Shared Ancillary Accommodation
02			Independent Ancillary Accommodation
03			Day Unit
04			Essential Complementary Accommodation: Staff Locker Room (with central changing)
05			Staff Changing (with local changing)
06			Seminar Room
07			Nursing Officer's Office
08			Office/Interview Room
09			Parent's Bedroom
10			Classroom
11			Store
12			Teacher's Base
13			Adolescent Day Room
01.03.01		Elderly	Nursing Section Type A (acute)
02			Type B (rehabilitation)
03			Essential Complementary Accommodation: Seminar Room
04			Doctor's Office
05			Nursing Officer's Office
06			Staff Changing (with local changing)
07			Day Hospital
08			Administrative Centre
01.04.01		Maternity	Delivery Suite
02			Ward Accommodation
03			Neo-Natal Unit

HOSPITAL BUILDINGS

FUNCTIONAL SIZE		BUILDING NOTE NO.	INDEX	COST GUIDE £	REMARKS
1 Bed	)	4	QSSD	17 342	
1 Store	)			1 454	
1 Room	)			10 708	
1 Office	)			7 114	
1 Office	)	4	QSSD	7 114	
1 Room	)			5 315	
1 WC	)			3 501	
1 Shower	)			3 859	
1 Ward	)			13 659	
1 Room	)	4	QSSD	17 641	
20 Beds	)			404 966	See HN(83)26.
20 Beds	)	23	QSSD	416 040	(i) See HN(83)26.
8 Beds	)			129 675	(ii) For Children's OPD see 04.08.01
1 Room	)			5 266	)
1 Room	)			8 052	)
1 Room	)			7 795	)
1 Office	)			6 266	)
1 Office	)			6 748	)
1 Bedroom	)	23	QSSD	7 543	) See HN(83)26.
10 Places	)			26 645	)
20 Places	)			31 977	)
1 Store	)			7 147	)
1 Base	)			5 468	)
1 Room	)			11 775	)
24 Beds	)	37	QSSD	495 193	(i) See HN(80)21. (ii) Costs include, as additional accommodation, an allowance for a relatives'/hairdressing room.
24 Beds	)	37	QSSD	514 223	(i) See HN(80)21. (ii) Costs include, as additional accommodation, an allowance for a relatives'/hairdressing room.
1 Room	)			11 753	)
1 Office	)	37	QSSD	7 274	)See HN(80)21.
1 Office	)			6 972	)
1 Room	)			10 548	)
25 Places	)	37	QSSD	493 993	)See HN(80)21.
40 Places	)			572 116	)
200 000 Pop.	)	37	QSSD	69 987	See HN(80)21.
1 Suite	)	21	QSSD	696 163	
1 Ward	)	21	QSSD	398 729	
6 Cots	)			235 360	
8 Cots	)			245 224	
10 Cots	)	21	QSSD	273 041	
12 Cots	)			300 542	
20 Cots	)			396 335	

HOSPITAL BUILDINGS

REF. NO.	SERVICE	DEPARTMENT	ACCOMMODATION
01.04.04	**In-Patients'**	Maternity	Essential Complementary Accommodation:
	Service	(continued)	Doctor's Office/Overnight Stay
05	**(continued)**		Dayroom
06			Seminar/Staff Room
07			Consulting/Examination Room
08			Milk Kitchen Store
09			Cleaner's Room
10			Parents' Bedroom
11			Parents' Shower
12			Parents' WC
13			Office
14			Incubator Transport Bay
15			Optional Accommodation:
			Staff Cloaks
16			Interview/Examination Room
17			Emergency Obstetric Store
18			Milk Bank
01.05.01		Intensive	
		Therapy	
02.01.01	**Main**		
	Operating	Operating	Operating Suites
	Facilities		
02			TSSU (for departments with no nearby CSSD)
03			Sterile Store (where no TSSU provided)
03.01.01	**Diagnostic and**	Radio-	Department
	treatment	diagnostic	
	Facilities		
02			Essential Complementary Accommodation:
			Seminar Room
03			workshop
04			Optional Accommodation:
			Darkroom
05			Lavage Room
06			Special Procedures Changing Room
07			Anaesthetic Room
08			Barium Preparation Room
09			Radiologist's Office
10			Typist's Room
11			Cooling
03.02.01		Radiosotopes	
		(sub-regional)	

HOSPITAL BUILDINGS

FUNCTIONAL SIZE		BUILDING NOTE NO.	INDEX	COST GUIDE £	REMARKS
1 Office	)			6 221	
1 Room	)			7 171	
1 Room	)			7 033	
1 Room	)			9 142	
1 Store	)			7 229	
1 Room	)	21	QSSD	4 640	
1 Bedroom	)			5 315	
1 Shower	)			3 859	
1 WC	)			3 501	
1 Office	)			5 266	
1 Bay	)			2 097	
1 Room	)			8 088	
1 Room	)			11 235	
1 Store	)	21	QSSD	2 323	
1 Store	)			4 538	
8 Beds	)	27	QSSD	228 334	Piped medical gases included.
1 Theatres	)			411 201	(i) BN 26 not applicable to 1,2 or
2	)			602 411	3 theatres.
3	)			787 469	(ii) Plant rooms excluded.
4	)	26	QSSD	977 511	(iii) Cooling, piped medical gases
5	)			1 169 516	and improved ventilation included.
6	)			1 372 324	
7	)			1 569 116	
8	)			1 768 498	
4 Theatres	)			152 261	Cooling and emergency sterilizer
5	)			162 350	included
6	)	26	QSSD	172 435	
7	)			183 875	
8	)			195 360	
4 Theatres	)			30 980	
5	)			35 273	
6	)	26	QSSD	39 603	
7	)			43 896	
8	)			48 181	
4 R/D Rooms	)			542 485	See HN(85)1.
5	)			623 546	
6	)			692 622	
7	)			781 860	
8	)	6	QSSD	854 642	
9	)			977 433	
10	)			1 088 308	
11	)			1 199 629	
12	)			1 271 818	
1 Room	)	6	QSSD	22 517	)See HN(85)1.
1 Room	)			9 892	)
1 Room	)			5 438	)
1 Room	)			10 523	)
1 Room	)			4 676	)
1 Room	)	6	QSSD	15 703	)See HN(85)1.
1 Room	)			7 457	)
1 Office	)			8 205	)
1 Room	)			4 920	)
1 Unit	)			10 526	)
1 Department	)		QSSD	135 686	See Design Guide issued October 1973.

HOSPITAL BUILDINGS

REF. NO.	SERVICE	DEPARTMENT	ACCOMMODATION
03.03.01	**Diagnostic and treatment Facilities (continued)**	Pathology	Area Laboratory
02			Public Health Services Laboratory (Supplementary)
03			Reference Laboratory (supplementary): Cytology
04			Trace Elements
05			Toxicology
06			Immunology
07			Chromosones
08			Automation
09			Neuropathology
03.04.01		Pharmacy	In-Patient Dispensing
02			Sterile Suite (small)
03			Sterile Suite (large)
04			Out-Patient Dispensing (attached)
05			Out-Patient Dispensing (remote)
06			Drug Information (hospital level)
07			Drug Information (district level)
08			Essential Complementary Accommodation: Seminar Room
09			Flammable Store
10			Medical Gas Cylinder Store
11			Computer Facilities
12			In-process Control Laboratory
13			Cytotoxic Preparation Room
14			Optional Accommodation: General Office/Reprographic Service

HOSPITAL BUILDINGS

FUNCTIONAL SIZE		BUILDING NOTE NO.	INDEX	COST GUIDE £	REMARKS
39 LSUs	)	15	QSSD	1 392 097	See Annex II to HN(80)21 for 'Howie' revisions to Building Note
6.5 LSUs	)	15	QSSD	176 781	See Annex II to HN(80)21 for 'Howie' revisions to Building Note
1 LSUs 2	))	15	QSSD	20 398 41 218	(i) See Annex II to HN(80)21 for 'Howie' revisions to Building Note (ii) 1 LSU serves up to 500 000 pop 2 LSUs serve up to 1 000 000 pop.
0.5 LSUs 1	))	15	QSSD	18 213 36 858	(i) See Annex II to HN(80)21 for 'Howie' revisions to Building Note (ii) 0.5 LSU serves up to 500 000 pop 1 LSU serves up to 1 000 000 pop
0.5 LSUs 1 LSUs	))	15	QSSD	18 213 36 858	(i) See Annex II to HN(80)21 for 'Howie' revisions to Building Note (ii) 0.5 LSU serves up to 500 000 pop 1 LSU serves up to 1 000 000 pop
1.5 LSUs 3 LSUs	))	15	QSSD	35 647 70 857	(i) See Annex II to HN(80)21 for 'Howie' revisions to Building Note. (ii) 1.5 LSUs serve up to 500 000 pop 3 LSUs serve up to 1 000 000 pop.
1 LSUs	)	15	QSSD	26 492	(i) See Annex II to HN(80)21 for 'Howie' revisions to Building Note. (ii) 1 LSU serves up to 1 000 000 pop.
2 LSUs 4	))	15	QSSD	64 851 130 137	(i) See Annex II to HN(80)21 for 'Howie' revisions to Building Note (ii) 2 LSUs serve up to 500 000 pop 4 LSUs serve up to 1 000 000 pop.
2 LSUs 3	))	15	QSSD	52 647 79 577	(i) See Annex II to HN(80)21 for 'Howie revisions to Building Note. (ii) 2 LSUs serve up to 500 000 pop 3 LSUs serve up to 1 000 000 pop.
450 Beds 600 601 1 200	))))	29	QSSD	359 326 359 326 429 570 429 570	
1 Suite	)		QSSD	74 664	
1 Suite	)	29	QSSD	141 307	
1 Unit	)	29	QSSD	24 045	
1 Suite	)	29	QSSD	38 131	
1 Unit	)	29	QSSD	14 442	
1 Unit	)	29	QSSD	55 288	
8 Persons 20 1 Store 1 Store 1 Room 1 Laboratory 1 Room	)))))))	29	QSSD	10 537 18 955 5 092 12 190 12 781 27 454 12 112	
1 Office	)	29	QSSD	11 512	

HOSPITAL BUILDINGS

REF. NO.	SERVICE	DEPARTMENT	ACCOMMODATION
03.04.15	**Diagnostic and treatment**	Pharmacy (continued)	Sterile Manufacturing
16	**Facilities (continued)**		Non-sterile Manufacturing
17			Assembly
18			Quality Control
19			Purchase and Distribution
20			Radio-Pharmaceuticals Preparation (closed procedure)
21			Radio-Pharmaceuticals Preparation (open procedure)
22			Essential Complementary Accommodation: Still Room
23			Porters' Base
24			Goods Reception/Loading Bay
25			Flammable/Hazardous Liquid Filling Area
26			Refuse Collection Store
27			Laboratory Workshop
28			Microbiological Testing Suite
29			Optional Accommodation: Tablet Packing
30			Engineer's Office
03.05.01		Mortuary and Post-Mortem	
03.06.01		Medical Photography	
03.07.01		Anaesthetic Services	
03.08.01		Rehabilitation	Department (excluding Hydrotherapy)
02			Hydrotherapy Section
04.01.01	**Out-Patient Services**	Out-Patient	Consulting Suites
04.02.01		Adult Acute Day Patients	Nursing Section
02			Treatment Suite
03			Treatment Room and Treatment Bathroom
04.03.01		Dental	Department

HOSPITAL BUILDINGS

FUNCTIONAL SIZE		BUILDING NOTE NO.	INDEX	COST GUIDE £	REMARKS
1 Unit	)	29	QSSD	734 848	
1 Unit	)	29	QSSD	304 833	
1 Unit	)	29	QSSD	160 807	
1 Unit	)	29	QSSD	182 505	
1 Unit	)	29	QSSD	301 914	
1 Area	)	29	QSSD	75 146	
1 Area	)	29	QSSD	110 290	
1 Room	)			83 416	
1 Base	)			4 922	
1 Bay	)			14 038	
1 Area	)	29	QSSD	16 307	
1 Store	)			19 266	
1 Workshop	)			12 461	
1 Suite	)			71 535	
1 Line	)	29	QSSD	58 192	
1 Office	)			4 858	
300 Bed	)			124 223	See Annex II to HN(80)21 for
500	)	20	QSSD	145 544	'Howie' revisions to Building
800	)			176 729	Note.
1 Department	)	19	QSSD	50 816	
1 Department	)		QSSD	71 563	See Design Guide issued Oct. 1971
150 000 Pop	)			551 100	(i) See Design Guide issued August
200 000	)		QSSD	627 103	1974 replacing Building Notes Nos
250 000	)			703 104	8 and 9. (ii) Allowance for the Hydro-therapy Pool is to be ADDED if required.
1 Pool	)		QSSD	113 138	See Design Guide issued Aug. 1974 replacing Building Notes Nos 8 & 9
36 Dr Sessions	)			286 723	Piped medical gases included.
72	)	12	QSSD	474 772	
108	)			664 850	
144	)			852 900	
1 Section	)	38	QSSD	281 764	See HN(81)33.
1 Suite	)	38	QSSD	123 218	See HN(81)33.
1 Suite	)	38	QSSD	30 387	(i) See (81)33. (ii) Alternative to Treatment Suite.

FUNCTIONAL SIZE		BUILDING NOTE NO.	INDEX	COST GUIDE £	Sessions:	Surgeries:
1 Surgery	)			53 708	up to 9	1
2	)			74 037	10 to 18	2
3	)	28	QSSD	166 621	19 to 27	3
4	)			190 964	28 to 36	4
5	)			214 692	37 to 45	5

HOSPITAL BUILDINGS

REF. NO.	SERVICE	DEPARTMENT	ACCOMMODATION
04.03.02	**Out-Patient Services (continued**	Dental (continued)	Orthodontic Supplementary
04.04.01		Opthalmic	Clinic
02			Essential Complementary Accommodation: Staff Locker Room (with central changing)
03			Optional Accommodation: Dispensing Optician
04.05.01		Dermatology (Supplementary to main OPD)	
04.06.01		Ear, Nose and Throat	
04.07.01		Special Treatment	Clinic
04.08.01		Children	Out-Patient
02			Comprehensive Assessment: Next to OPD
03			Independent
04.09.01		Accident and Emergency	Department
04.09.02			Essential Complementary Accommodation: Staff Changing (Male)
03			Staff Changing (Female)
04			Recovery (One Bed)
05			Nursing Officer's Office
06			Optional Accommodation: Major Treatment Room
04.10.01		Neuro-physiology	Clinic
04.11.01		Maternity	Clinic
02			Essential Complementary Accommodation: Ultra-Sound Room
03			Snack Bar and Store
04			Optional Accommodation: Treatment Room
05			Staff Cloaks
05.01.01	**Psychiatric Patients' Services**	Acute Mental Illness	Ward (annex to Day Hospital)

HOSPITAL BUILDINGS

FUNCTIONAL SIZE		BUILDING NOTE NO.	INDEX	COST GUIDE £	REMARKS
					Sessions: Surgeries:
1 Surgery	)			65 788	up to 9 1
2	)	28	QSSD	84 274	10 to 18 2
3	)			108 060	19 to 27 3
3 Consultants	)			233 630	See HN(81)33.
4	)	39	QSSD	258 798	
5	)			297 648	
6	)			335 353	
1 Room	)	39	QSSD	5 022	See HN(81)33.
1 Optician	)	39	QSSD	11 756	See HN(81)33.
1 Department	)		QSSD	29 573	(i) See HN(76)193 and Design Guide issued October 1976. (ii) Department serves a population of 250 00 to 30 000.
150 000 Pop	)			245 001	See Design Guide issued February
200 000	)		QSSD	261 831	1974.
250 000	)			279 199	
300 000 Pop	)		QSSD	216 337	See Design Guide issued September 1974.
1 Department	)	23	QSSD	130 763	(i) See HN(83)26. (ii) Department comprises of 3 C/E rooms.
1 Unit	)	23	QSSD	111 991	)See HN(83)26.
1 Unit	)			166 688	)
10 000 Attndncs	)			364 042	
20 000	)			414 358	
30 000	)	22	QSSD	545 518	
50 000	)			641 281	
70 000	)			736 825	
1 Room	)			7 427	
1 Room	)			12 978	
1 Room	)	22	QSSD	8 887	
1 Room	)			7 113	
1 Room	)	22	QSSD	35 321	
9 Suite	)		QSSD	489 282	
1 Clinic	)	21	QSSD	314 534	
1 Room	)			10 667	
1 Bar	)	21	QSSD	9 853	
1 Room	)	21	QSSD	10 548	
1 Room	)			8 088	
10 Beds	)			185 438	
15	)	35	QSSD	235 541	
20	)			302 178	
25	)			358 956	

HOSPITAL BUILDINGS

REF. NO.	SERVICE	DEPARTMENT	ACCOMMODATION
05.01.02	**Psychiatric Patients' Services (continued)**	Acute Mental Illness (continued)	Ward (independent from Day Hospital)
03			Ward (elderly)
04			Essential Complementary Accommodation (Wards):
			Special Bedroom
05			En Suite WC/Wash
06			Day Hospital
07			Essential Complementary Accommodation (Day Hospital):
			Treatment/Clean Utility
08			Staff Dining/Rest Room (small)
09			Staff Dining/Rest Room (large)
10			Optional Accommodation (Day Hospital):
			Patients' Utility Room
11			Cleaner's Room
12			Disposal Room
13			Office/Interview Room
14			Day Hospital for the Elderly
15			Optional Accommodation (Day Hospital for the Elderly):
			Treatment Bathroom
16			Assisted Shower
17			Out-Patients' Suite
18			ECT Suite
20			Administrative Centre (small)
21			Administrative Centre (large)
05.02.01		Mental Handicap Hospital Units	Adult Residential Unit
02			Children's Residential Unit
03			Adult Day Care Unit
05.03.01		Mental Handicap Community Units	Normally Handicapped

HOSPITAL BUILDINGS

FUNCTIONAL SIZE		BUILDING NOTE NO.	INDEX	COST GUIDE £	REMARKS
10 Beds	)			240 296	
15	)	35	QSSD	322 277	
20	)			418 997	
25	)			491 680	
10 Beds	)			245 793	
15	)	35	QSSD	336 512	
20	)			416 517	
25	)			487 494	
1 Bedroom	)	35	QSSD	9 861	
1 Suite	)			7 657	
10 Places	)			139 831	Community location
20	)	35	QSSD	211 000	
30	)			298 307	
40	)			348 111	
1 Room	)			10 046	
1 Room	)	35	QSSD	7 967	
1 Room	)			12 058	
1 Room	)			7 872	
1 Room	)	35	QSSD	6 765	
1 Room	)			4 061	
1 Room	)			7 157	
10 Places	)			166 997	Community location
20	)	35	QSSD	235 607	
30	)			356 393	
40	)			441 110	
1 Bathroom	)	35	QSSD	13 682	
1 Shower	)			8 205	
3 C/E Rooms	)	35	QSSD	87 105	Only required when main OPD not
6	)			165 417	used.
1 Suite	)	35	QSSD	78 841	
1 Centre	)	35	QSSD	61 352	
1 Centre	)	35	QSSD	109 614	
12 Places	)		QSSD	147 782	See HN(77)58, Annex III to HN(80)21
24	)			251 796	and Design Bulletins Nos 1 and 2.
8 Places	)			121 002	See HN(77)58, Annex III to HN(80)21
16	)		QSSD	205 623	and Design Bulletins Nos 1 and 2.
24	)			285 354	
60 Places	)		QSSD	313 329	See HN(77)58, Annex III to HN(80)2
115	)			459 413	and Design Bulletins Nos 1 and 2.
8 Places	)			159 462	See HN(80)21 Annex III for on-cost
12	)			212 261	details.
16	)		PIPSH	265 055	
20	)			308 579	
24	)			352 101	

HOSPITAL BUILDINGS

REF. NO.	SERVICE	DEPARTMENT	ACCOMMODATION
05.03.02	Psychiatric Patients' Services (continued)	Mental Handicap Community Unit (continued)	Heavily Handicapped
06 01.01	Teaching Facilities	Nurses' Training School	Students' Accommodation
02			Pupils' Accommodation
06.02.01		Post-Graduate Medical Centre	
06.03.01		Education Centre	
07.01.01	Administration Services	Administration	General
02			Main Entrance
03			Medical Records
04			Group Accommodation
08.01.01	Staff Facilities	Catering (Dining)	Dining Rooms
02			Servery
03			Essential Complementary Accomodation: Furniture Store
08.02.01		Residential Accommodation for Staff	

HOSPITAL BUILDINGS

FUNCTIONAL SIZE	BUILDING NOTE NO.	INDEX	COST GUIDE £	REMARKS
8 Places)			167 488	See HN(80)21 Annex III for on-cost
12)			223 141	details.
16)		PIPSH	278 793	
20)			324 455	
24)			370 117	
100 Students)			151 655	
200)	14	QSSD	208 539	
300)			265 732	
400)			323 126	
20 Pupils)			64 727	
40)	14	QSSD	78 009	
60)			91 280	
80)			104 030	
1 Centre)		QSSD	219 938	See Design Guide issued Oct 1968
1 Centre)		QSSD	65 652	(i) See DS 71/73. (ii) Serves 800 beds. (iii) Provides facilities for staff other than doctors or nurses.
9 Points)			287 773	300 beds = 9 points
12)	18	QSSD	338 339	450 beds = 12 points
16)			401 340	600 beds = 16 points
20)			464 343	800 beds = 20 points
9 Points)			59 549	(i) 300 beds = 9 points
12)	18	QSSD	64 954	450 beds = 12 points
16)			66 704	600 beds = 16 points
20)			73 350	800 beds = 20 points (ii) Central Telephone Installation excluded: see Department 09.02.01
9 points)			137 215	300 beds = 9 points
12)	18	QSSD	168 571	450 beds = 12 points
16)			206 157	600 beds = 16 points
20)			243 743	800 beds = 20 points
19 points)			154 635	Points on same scale as Department
30)	18	QSSD	211 578	07.01.01 (General Administration).
40)			263 690	
300 Meals)	10	QSSD	194 354	See HN(85)21.
600)			300 908	
300 Meals)	10	QSSD	131 290	See HN(85)21.
600)			170 928	
1 Store)	10	QSSD	7 861	See HN(85)21.
100 Points)			123 725	(i) For each additional point
150)			174 061	after the first 500, allow £746.
200)	24	PIPSH	223 218	(ii) Points scale per person:
250)			268 643	Scale A: 6.5 points
500)			517 401	Scale B: 8.5 points
				Scale C: 12.5 points
				Scale D: 19.5 points
				Scale E: 24 points
				Scale F: 22 points
				Scale G: 28 points.

HOSPITAL BUILDINGS

REF. NO.	SERVICE	DEPARTMENT	ACCOMMODATION
08.03.01	Staff Facilities (continued)	Occupational Health Centre for Staff	
08.04.01		Staff Changing and Storage of Uniforms (excluding medical staff)	Manual System
02			Semi-Automatic System
03			Fully Automatic System
04			Essential Complementary Accommodation: Sewing Room
08.04.05			Optional Accommodation: Additional Storage for Uniforms (Manual system)
06			(Automatic system)
07			Addition for Storage on Hangers (Manual System)
09.01.01	Service Facilities	CSSD	
09.02.01		Telephone Services	Operator's Suite and Equipment Room
02			Telephone Equipment
09.03.01		Catering (Kithcens)	Central Kitchens
02			Optional Accommodation: Facilities for Future Expansion of 600-meal Kitchen
03			Bulk Store

HOSPITAL BUILDINGS

FUNCTIONAL SIZE		BUILDING NOTE NO.	INDEX	COST GUIDE £	REMARKS
1 Centre	)		QSSD	67 355	(i) See Design Guide issued April 1973. (ii)Centre serves 1000 beds.
300 Places	)			138 129	See HN(83)26.
600	)	41	QSSD	204 592	
900	)			278 011	
1 200	)			356 266	
300 Places	)			145 226	See HN(83)26.
600	)	41	QSSD	221 007	
900	)			302 778	
1 200	)			385 781	
300 Places	)			126 420	See HN(83)26.
600	)	41	QSSD	173 543	
900	)			227 206	
1 200	)			286 274	
1 Seamstress	)	41	QSSD	8 906	See HN(83)26.
2	)			13 706	
300 Places	)	41	QSSD	1 884	)(i) For example, with central
300 Places	)			9 158	)issue of uniforms where staff use)local changing facilites.)(ii) See HN(83)26.
300 Places	)	41	QSSD	4 094	See HN(83)26.
1 Department	)	13	QSSD	240 223	
1 Cabinets	)			36 877	
2	)	48	QSSD	45 542	
3	)			53 605	
200 Extensions	)				
300	)				
400	)				
500	)				
600	)				
700	)				
800	)				
900	)				
1 000	)				
1 100	)				
1 200	)				
1 300	)				
1 400	)				
1 500	)				
1 600	)				
600 Meals	)	10	QSSD	954 381	See HN(85)21.
1 200	)			1 170 400	
1 Facility	)	10	QSSD	28 404	See HN(85)21.
600 Meals	)	10	QSSD	85 729	See HN(85)21.
1 200	)			107 315	

HOSPITAL BUILDINGS

REF. NO.	SERVICE	DEPARTMENT	ACCOMMODATION
09.03.04	**Service Facilities (continued)**	Catering (Kithcens) (continued)	Optional Accommodation: Storeman's Office
09.04.01		Laundries	Laundry
02			Infants' Napkin Section
03			Special Personal Clothing
04			Dry Cleaning Facility
09.04.05			Storage at Sending Hospital
10.01.01	**Hospital Engineering and Works Services**	Boiler Houses and Fuel Storage	
10.02.01		Works	Main Department
02			Lock-up Store and Workshop
11.01.01	**Community Health Services**	Health Centre	Primary Care Services: General Medical Practitioner Services

HOSPITAL BUILDINGS

FUNCTIONAL SIZE		BUILDING NOTE NO.	INDEX	COST GUIDE £	REMARKS
1 Office	)	10	QSSD	4 537	See HN(85)21.
55 000 Art/wk	)			1 115 703	See HN(77)65.
120 000	)	25	QSSD	2 104 158	
180 000	)			2 988 838	
250 000	)			3 899 099	
1 Section	)	25	QSSD	49 359	(i) See HN(77)65. (ii) Cost allowance applies to over 10 000 articles/week
3 500 Art/wk	)			64 286	See HN(77)65
5 000	)	25	QSSD	85 568	
7 000	)			118 401	
600 Kilos/wk	)			62 300	See HN(77)65.
950	)	25	QSSD	77 430	
1 450	)			105 520	
2 400	)			151 547	
300 Bed	)			23 060	See HN(77)65.
400	)			32 262	
600	)	25	QSSD	43 977	
800	)			49 627	
1 000	)			64 231	
1 200	)			72 773	
1 760 K/watts	)			366 593	(i) See HN(81)33. (ii) Cost Allowances are for oil-fired installations: Coal-fired installations should be taken as approximately 50% increase on these figures.
4 400	)			586 941	
6 150	)	16	QSSD	711 635	
8 800	)			868 054	
11 700	)			994 485	
20 500	)			1 283 205	
300 Bed	)			154 386	(i) Compound, garages and a separate lock-up store are included (ii) Beds are total in group served by the Department.
600	)			154 386	
601	)			202 052	
800	)			202 052	
801	)			249 716	
1 200	)	34	QSSD	249 716	
1 201	)			299 811	
2 000	)			299 811	
2 001	)			349 397	
3 000	)			349 397	
3 001	)			389 689	
4 000	)			389 689	
300 Bed	)			4 775	(i) To serve hospital or group remote from the Works Department. (ii) Beds are total in group served by the Department.
600	)			4 775	
601	)			7 615	
800	)			7 615	
801	)			9 615	
1 200	)	34	QSSD	9 615	
1 201	)			11 556	
2 000	)			11 556	
2 001	)			13 141	
3 000	)			13 141	
3 001	)			13 991	
4 000	)			13 991	
3 GMPs	)			58 818	)These allowances do not attract
6	)	36	QSSD	86 468	)the full on-costs of HNPN6. See
9	)			135 132	)HN(76)54 for on-cost norms.
12	)			168 194	)

HOSPITAL BUILDINGS

REF. NO.	SERVICE	DEPARTMENT	ACCOMMODATION
11.01.02	Community Health Services (continued)	Health Centre (continued)	Community Health Services, Health Visiting, District Nursing and Midwifery and Welfare Foods
03			Shared Facilities: Related to General Medical Practitioners
04			Related to Community Health Services
05			Other Services: Chiropody
06			Speech Therapy and/or Child Health Assessment
07			Social Services
11.01.08			Dental Services: School and Priority Dental Services
09			General Dental Services
10			Hospital Consultant Services
11			Pharmaceutical Services
11.02.01		Health Authority Clinic	
11.03.01		Ambulance Station	
20.01.01	Teaching Hospital	Administration	Medical Records: Reading Space for Students
02			Additional Storage Space for Records
03			Changing Accommodation

HOSPITAL BUILDINGS FUNCTIONAL SIZE		BUILDING NOTE NO.	INDEX	COST GUIDE £	REMARKS
7 500 Pop	)			63 652	)These allowances do not attract
15 000	)	36	QSSD	87 590	)the full on-costs of HBNP6. See
22 500	)			112 246	)HN(76)54 for on-cost norms.
30 000	)			140 051	)
3 GMPs	)			71 012	)
6	)			103 792	)
9	)			138 522	)These allowances do not attract
12	)	36	QSSD	171 657	)the full on-costs of HBPN6. See
7 500 Pop	)			86 020	)HN(76)54 for on-cost norms.
15 000	)			110 462	)
22 500	)			130 514	)
30 000	)			157 267	)
1 Chiropodist	)			8 676	)
2	)			17 404	)
1 Therapist	)			13 670	)These allowances do not attract
2	)	36	QSSD	20 343	)the full on-costs of HBPN6. See
1 Soc. Worker	)			5 130	)HN(76)54 for on-cost norms.
2	)			5 662	)
3	)			11 238	)
1 Dentist	)			47 400	)
2	)			60 954	)
3	)			77 308	)These allowances do not attract
4	)	36	QSSD	91 427	)the full on-costs of HBPN6. See
1 Dentist	)			31 254	)HN(76)54 for on-costs norms.
2	)			45 195	)
3	)			73 073	)
4	)			87 585	)
9 Dr Sessions	)	36	QSSD	9 185	)These allowances do not attract
18	)			18 501	)the full on-costs of HBPN6. See
					)HN(76)54 for on-cost norms.
15 000 Pop	)	36	QSSD	30 035	)(i) Cost allowances relate to
15 001	)			39 874	)populations of up to 15 000.
					)(ii)These allowances do not
					)attract the full on-costs of
					)HBPN6. See HN(76)54 for on-cost
					)norms.
7 000 Pop	)			121 636	See LABN3.
10 000	)			121 636	
10 001	)		QSSD	136 470	
20 000	)			136 470	
20 001	)			154 270	
30 000	)			154 270	
					Draft new guidance under way. Contact A. Maun at DHHS.
1 Space	)		QSSD		(i) See DS 65/74. (ii) UGC funds £1 866 (QSSD base 100) excluding VAT.
1 Space	)		QSSD		(i) See ds 65/74. (ii) UGC funds £8 049 (QSSD base 100) excluding VAT.
1 Space	)		QSSD		(i) See DS 65/74. (ii) UGC funds £79 (QSSD base 100) excluding VAT.

HOSPITAL BUILDINGS

REF. NO.	SERVICE	DEPARTMENT	ACCOMMODATION
20.02.01	**Teaching Hospital (continued)**	Adult Nursing Units (general acute nursing sections, including geriatric assessment and rehabilitation)	Seminar Rooms
02			Clinical Investigation Area
03			Urine Testing Room
04			Student's Room
20.03.01		Psychiatric	Interview Room
02			Seminar/Demonstration Room
20.04.01		Geriatric Units (applicable to units not covered by Adult Nursing Units)	Seminar/Demonstration Room
20.05.01		Special Units: Intensive Therapy and Coronary Care	Seminar Room
20.06.01		Maternity	Bed Areas: Seminar Room
02			Clinical Investigation Area
03			Urine Test Rooms
04			Ante-Natal Clinic: Seminar Room
05			Labour Suite: Students' Common Room
06			Operating Theatre Suite: CCTV to one Seminar Room
07			Student Changing Accommodation
08			Special Care Baby Units: Seminar Room
20.07.01		Children	Bed Areas: Seminar Room
02			Clinical Investigation Area
03			Urine Test Room
04			Students' Room
05			Out-Patients: Seminar Room

HOSPITAL BUILDINGS

FUNCTIONAL SIZE		BUILDING NOTE NO.	INDEX	COST GUIDE £	REMARKS
2 Rooms	)		QSSD	29 038	(i) See DS 65/74. (ii) Funded by the NHS.
1 Area	)		QSSD	26 210	(i) See DS 65/74. (ii) Funded by the NHS.
1 Room	)		QSSD	3 579	(i) See DS 65/74. (ii) Funded by the NHS.
1 Room	)		QSSD	6 623	(i) See DS 65/74. (ii) Funded by the NHS.
6 Rooms	)		QSSD	28 487	(i) See DS 65/74. (ii) Funded by the NHS.
1 Room	)		QSSD	21 889	(i) See DS 65/74. (ii) Funded by the NHS.
1 Room	)		QSSD	21 889	(i) See DS 65/74. (ii) Funded by the NHS.
1 Room	)		QSSD	16 415	(i) See DS 65/74. (ii) Funded by the NHS.
1 Room	)			21 889	)
1 Area	)		QSSD	26 210	)(i) See DS 65/74. (ii) Funded by
2 Rooms	)			7 158	)the NHS.
1 Room	)		QSSD	16 415	(i) See DS 65/74. (ii) Funded by the NHS.
1 Room	)		QSSD		(i) See DS 65/74. (ii) UGC funds £3 112 (QSSD base 100) excluding VAT.
1 Unit	)		QSSD		(i) See DS 65/74. (ii) UGC funds £101 (QSSD base 100) excluding VAT
1 Room	)		QSSD		(i) See DS 65/74. (ii) UGC funds £786 (QSSD base 100) excluding VAT
1 Room	)		QSSD	16 415	(i) See DS 65/74. (ii) Funded by the NHS.
1 Room	)		QSSD	16 415	(i) See DS 65/74. (ii) Funded by the NHS.
2 Areas	)		QSSD	52 417	(i) See DS 65/74. (ii) Funded by the NHS.
1 Room	)		QSSD	3 579	(i) See DS 65/74. (ii) Funded by the NHS.
1 Room	)		QSSD	6 623	(i) See DS 65/74. (ii) Funded by the NHS.
1 Room	)		QSSD	21 889	(i) See DS 65/74. (ii) Funded by the NHS.

HOSPITAL BUILDINGS

REF. NO.	SERVICE	DEPARTMENT	ACCOMMODATION
20.08.01	**Teaching Hospital (continued)**	Operating Suites	Demonstration/Seminar Room
02			Student Changing Accommodation
20.09.01		Pathology	Additional Laboratory Space
20.10.01		Mortuary and Post Mortem Room	Observation Gallery
02			Student Changing Accommodation
20.11.01		Rehabilitation	Seminar Room
20.12.01		Out-Patients	Consulting Suites: Increase in Doctor Sessions
02			Seminar Room
03			Adult Acute Day Patients Treatment Suite: Student Changing Accommodation
20.13.01		Accident and Emergency	Examination Rooms
02			Seminar Room
03			Students' Rest Room
04			Operating Theatre: Student Changing Accommodation

HOSPITAL BUILDINGS

FUNCTIONAL SIZE		BUILDING NOTE NO.	INDEX	COST GUIDE £	REMARKS
1 Room	)		QSSD		(i) See DS 65/74. (ii) UGC funds £7 680 (QSSD base 100) excluding VAT.
1 Room	)		QSSD		(i) See DS 65/74. (ii) UGC funds £786 (QSSD base 100) excluding VAT
1 Space	)		QSSD	14 698	(i) See DS 65/74. (ii) Funded by the NHS.
1 Gallery	)		QSSD		(i) See DS 65/74. (ii) UGC funds £122 (QSSD base 100) excluding VAT
1 Room	)		QSSD		(i) See DS 65.74. (ii) UGC funds £786 QSSD base 100 excluding VAT
1 Room	)		QSSD	16 415	(i) See DS 65/74. (ii) Funded by the NHS.
Dr Sessions	)	12	QSSD		(i) See DS 65/74. (ii) Increase the number of doctor sessions by 10 percent. Cost allowances are on the same scale as those provided for in Department 04.01.01. (iii) Funded by the NHS.
1 Room	)		QSSD	21 889	(i) See DS 65/74. (ii) Funded by the NHS.
1 Room	)		QSSD		(i) See DS 65/74. (ii) UGC funds £786 (QSSD base 100) excluding VAT
2 Rooms	)		QSSD	13 049	(i) See DS 65/74. (ii) Funded by the NHS.
1 Room	)		QSSD	16 415	(i) See DS 65/74. (ii) Funded by the NHS.
1 Room	)		QSSD		(i) See DS 65/74. (ii) UGC funds £1 794 (QSSD base 100) excluding VAT
1 Room	)		QSSD		(i) See DS 65/74. (ii) UGC funds £786 (QSSD base 100) excluding VAT

Keep your figures up to date, free of charge

This section, and most of the other information in this Price Book, is brought up to date every three months in the *Price Book Update*.

The *Update* is available free to all Price Book purchasers.

To ensure you receive your copy, simply complete the reply card from the centre of the book and return it to us.

UNIVERSITY BUILDINGS

The Universities Funding Council's document University Building Projects - Notes on
Control and Guidance 1986, contains notes to assist universities in investment appraisal
studies, preparing briefs and establishing expenditure limits for projects.
   The notes contain areas derived from the Council's 'Planning Norms for University
Buildings', and costs that are an expression of the standards and facilities considered
appropriate by the Council.
   The following notional unit areas taken from the notes are intended as a guide for
planning purposes.

ACADEMIC AREAS		Academic Staff: Student ratio	Usable Area/FTE			Inclusion for academic staff m2
			Ug m2	PgC m2	PgR m2	
Departmental areas including centrally timetabled seminar rooms for:						
1 + 2	Preclinical medicine and dentistry	1:8	14.1	14.1	20.4	3.8
3	Clinical medicine	1:6	6.5	6.5	22.2	5.2
4	Clinical dentistry	1:6	10.5	13.0	16.2	4.2
5	Studies allied to medicine	(as for Physical sciences)				
6	Biological sciences	1:9	9.2	12.2	19.1	3.4
	Experimental Psychology	1:11	8.2	10.5	19.5	3.0
7	Agriculture & Forestry	(as for Biological sciences with special additions)				
	Veterinary science	(as for Biological sciences with special additions)				
8	Physical sciences	1:8	9.8	12.3	18.4	3.8
9	Mathematics	1:11	3.6	3.6	5.0	1.3
10	Computer studies	1:11	7.3	10.2	11.2	2.2
11	Engineering and technology	1:9	9.8	16.6	17.9	3.2
12	Architecture, Building and Planning	1:8	9.8	9.8	9.4	2.3
13	Geography	1.14	5.4	6.8	7.1	1.0
	Economics	1.12	2.4	2.4	4.9	1.2
14	Politics, Law and other social studies	1:12	2.4	2.4	4.9	1.2
15	Business Management	1:11	3.3	3.3	5.0	1.3
	Accountancy	1.12	2.4	2.4	4.9	1.2
17	Languages	1.10	3.5	3.5	5.2	1.4
18	Humanities	1.11	2.6	2.6	5.2	1.3
	Archaeology	1.10	5.5	6.8	7.2	1.4
19	Art and Design	1.10	9.1	9.1	8.8	1.9
	Music, Drama	1:8	7.9	7.9	9.0	2.2
20	Education	1.11	5.0	5.0	4.8	1.3

Non departmental academic areas:
Lecture Theatres, Provision for Ug and PgC only   0.5
Library
   Basic provision   1.25
   Addition where necessary for law students   0.8
   Expansion provision   0.2
   Special collection   Ad hoc
   Reserve book stores   50 m2 + 3.5 m2/1000 Vols

NON-ACADEMIC AREAS

Administration (including maintenance)	up to 3000 FTE	0.8
	over 3000 FTE	0.5
Social, Dining (excluding Kitchens), Health	up to 3000 FTE	1.4
	over 3000 FTE	1.3
Social, Health (clinical medical only)		1.3
Sports Facilities (buildings)		
Indoor	up to 3000 FTE	0.5
	300 to 6000 FTE	0.1
	over 6000 FTE	0.3
Outdoor	up to 3000 FTE	0.2
	over 3000 FTE	0.1

UNIVERSITY BUILDINGS

The following cost guidance taken from the notes is intended to assist in establishing expenditure limits for minor works projects (less than £1 m). The rates are expressed as the cost per metre of usable floor area at 1st quarter 1987 tender price levels and include for the normal balance area.

The Expenditure Limit for major projects (ie over £1 m) will be related to the facilities that are agreed to be necessary and may well differ from a limit based on the following rates.

Subject Group	Basic Rate (1st quarter 1987 tender price levels)	Additions for specialist facilities	
		Specialist area as % of total usable area	Rate (1st quarter 1987 tender price levels)
	£/usable m2	%	£/usable m2
1 + 2 Preclinical medicine and dentistry	585	80	210
3 Clinical medicine	585	55	130
4 Clinical dentistry	590	65	170
5 Studies allied to			
medicine - Nursing	535	50	50
- Pharmacy	580	80	175
6 Biological sciences	580	80	175
Experimental psychology	535	60	75
8 Physical sciences			
Physics	555	75	55
Geology	575	70	170
Chemistry	575	70	170
9 Mathematics	535	45	5
10 Computer studies	535	60	10
11 Engineering and technology			
Equipment dominated	460	To be agreed ad hoc	
Person orientated	535	To be agreed ad hoc	
12 Architecture, Planning	535	60	20
13 Geography	535	50	20
14 Politics, Law and other social studies	535	--	--
15 Business management	535	15	5
17 Languages	535	15	25
18 Humanities	535	--	--
Archaeology	535	45	20
19 Creative Arts			
Music	535	50	90
Drama	535	35	30
Art and Design	535	60	20
20 Education	535	50	50

NOTE: Rates for specialist facilities to be applied to total usable area

Type of accommodation	Total rate (1st quarter 1987 tender price levels) £/usable m2
Lecture theatres	765
Main libraries	650
Social, Dining, Unions	580
Maintenance Workshops	415
Internal Sports (exluding swimming pools)	505

UNIVERSITY BUILDINGS

A separate document exists for the appraisal of and establishing expenditure limits for medical projects - the University Building Projects Notes on Control and Guidance for Medical Projects 1986. These notes deal with the provision of university accommodation for medecine and dentistry and should be read in conjunction with Notes on Control and Guidance of University Building Projects (NOCAG) 1986. These notes only apply to medical projects in which either the NHS has no share or for joint UFC/NHS projects which are managed by the university. When the Health Authority manages the work and controls expenditure, then these projects are subject to NHS building control procedures.

The following notional unit areas, taken from Annex C, are included in the Council's planning norms and notional limits on which the cost limits are based.

The figures, based on staff: student ratios of 1:8 for preclinical and 1:6 for clinical medicine and dentistry, are:-

Total academic usable area per FTE student

	UG m2	PGC m2	PGR m2	m2	
Preclinical M & D	14.1	14.1	20.4	3.8	) included in total
Clinical - M	6.5	6.5	22.2	5.2	) usable areas for
Clinical - D	10.5	13.0	16.1	4.2	) academic staff use

CATEGORY

General teaching: Usable area m2
    small groups (up to six)   6.5 for 6 places
    larger groups   1.2 per place
    class, seminar rooms (informal seating)   1.9 per place
    class, seminar rooms (tables and chairs)   2.3 per place

Note: 0.35m2 per FTE UG and PGC student is included in
notional unit area for general teaching.

Lecture theatres:
    lecture theatres (close seating)   1.0 per place
    lecture theatres (clinical demonstration):
    additional area for patients in one clinical
    theatre per school:
        demonstration   18.5
        waiting, preparation   18.5

                                                             37.0

Note: Norm = 1 place per 2 FTE (UG and PGC) students.

Offices:
    Professors (including small group space)   18.5 per room
    Tutorial staff (including small group space)   13.5 per room
    Non tutorial staff (e.g. researchers, secretaries)   7.0 per person
    Chief technician   9.3 per person

Laboratory places (teaching):
    (except Anatomy dissecting room)
    preparation + stores   5.0 per place
                                                             2.5 per place
                                                             ---
                                                             7.5

    Anatomy department:
        dissecting room   2.5 per place
        preparation + stores   1.5 per place
        museum   0.5 per place

                                                             4.7

UNIVERSITY BUILDINGS

	Usable area m2

Clinical dental:
    preparation + stores

                                        5.0 per place
                                        2.5 per place

                                        7.5

Laboratory places (individual research):
  Medical and all preclinical academic staff:
    research space   11.0 per place
    ancillary   5.5 per place

                                        16.5

  Medical and all preclinical research fellows,
  postgraduates and NHS staff:
    research space   10.0 per place
    ancillary   6.0 per place

                                        16.0

  Dental academic staff:
    research space   7.3 per place
    ancillary   3.7 per place

                                        11.0

  Dental research fellows, postgraduates,
  and NHS staff:
    research space   7.3 per place
    ancillary   3.7 per place

                                        11.0

AVA facilities:
  Photographic and TV studios including ancillary   0.3 per U/G
  areas (to be planned in association with   student
  hospital Medical Illustration Department if on
    hospital site).

Animal houses (other than in departmental ancillary
                areas):   Assessed ad hoc

Libraries, central administration including) for   -   See non-medical
maintenance, social and indoor sports   ) medical   space standards
facilities.   ) students   and notional unit
areas

Dining and kitchens (if not provided with NHS   -   See non-medical
                facilities)   notional unit
areas

OTHER UNIVERSITY AREAS IN TEACHING HOSPITALS

Teaching and Research areas in Hospital Departments (Standard Teaching Additions)

The following basic areas were agreed in 1974 between the UFC and Health Departments (DHSS and SHHD) as being required for clinical teaching in a main teaching hospital in addition to:

a)   teaching, research, and social facilities provided by the university for medical school staff and students, including postgraduate students taking a university degree or diploma, and

b)   facilities provided by the NHS for the postgraduate education and training of doctors or dentists and the training of nurses, physiotherapists and laboratory technicians.

UNIVERSITY BUILDINGS

The basic areas listed below are subject to adjustment where specialised teaching is
carried out in hospitals peripheral to the main teaching hospital(s).

Department	Net area m2	
	UFC-funded	NHS-funded
Administration:		
medical records	63.5	-
student changing	0.4 per FTE student	-
Acute nursing units (incl. Geriatric Assessment and Rehabilitation)	-	105.5 per 60.72 beds
Psychiatric Units	-	91.0
Geriatric Units	-	37.0
Intensive Care/Therapy Units	-	28.0
Maternity Department	18.5 + (4.0 per theatre)	142.0
Childrens Department	-	158.5
Main Operating Department	37.0 + (4.0 per theatre)	-
Pathology	-	20.5

Mortuary/Postmortem Room:
gallery   1.0 (per annual student intake)

                                      2

student changing   4.0 per table   -

Rehabilitation Department (Physiotherapy, Occupational Therapy)	-	28.0
Outpatients Department	4.0 per theatre	37.0 per 100 doctor sessions
Accident and Emergency Department	11.0 + (4.0 per theatre)	50.0

For details see DHSS CIS notes and Health Departments/UFC document entitled 'Teaching
Hospital Space Requirements'.

OTHER AREAS

Dining and Kitchens   -   assessed on the basis of 1 meal per FTE student at NHS space
                          and cost rates.

On-call residence   -   where needed on hospital sites for teaching purposes in
                          addition to the students' normal place of residence, these
                          should normally be provided for numbers not exceeding one-
                          third of the annual intake of clinical medical students at
                          each medical school and in accordance with NHS nurses Scale
                          A residences excluding common room facilities.

Car parking   -   surface parking for medical school academic and technician
                          staff on a hospital site need not exceed 1 place to every 3
                          FTE students.

UNIVERSITY BUILDINGS

UNIVERSITY ACCOMMODATION - DEPARTMENTAL COSTS

The following cost guidance, taken from Annex D, are at 1st quarter 1986 tender price levels and should be applied to the departmental area or functional unit for university accommodation in each category, whether the space is to be used exclusively by the university or provided as part of a joint facility to be shared with NHS users, and aggregated to form the university's departmental cost limit.

The figures should be discussed with the Health Authority and included in their schedules of area and cost for the whole project at appropriate stages in the design of the project.

Departmental rates are not given for preclinical academic accommodation and if a case should arise where such accommodation is to be provided under Health Authority control using HBPN no. 6 procedures, reference should be made to the Universities Funding Council.

Rates for University Academic and Non-Academic Areas:

Academic areas	£/m2 Departmental area
Basic rate for:	
Clinical medicine	385
Clinical dentistry	

Additional rate for specialist facilities expressed as a percentage of the specialist usable area to be applied to the total <u>departmental area</u>:

*40% (clinical medicine)		155
*50% (clinical dentistry)		190
75% (clinical medicine)	) whole usable area	245
	) with specialist area	
75% (clinical dentistry)	) facilities	265

*These percentages represent a full range of specialist facilities for a complete department included in the notional unit areas given in Annex C.

Non-academic areas

Libraries, central administration, social space and other non specialised areas	385
Lecture theatres	610

Other areas

Animal houses, AVA, photographic/medical illustration, etc.	ad hoc
Teaching and Research areas in Hospital departments (Standard Addtions)	NHS rates
Dining Rooms and Kitchens	NHS rates
On-call residence places	NHS rates

For further information on Rates for other university areas, cost apportionment, on-costs and professional fees, refer to Annex D of Notes on Control and Guidance for Medical Projects 1986.

EDUCATIONAL BUILDINGS (Procedures and Cost Guidance)

In January 1986 the Department of Education and Science introduced new procedures for the approval of educational building projects.

In its Administrative Memorandum 1/86 the Department currently requires detailed individual approval only for projects costing £2 millions and over. Minor works i.e., costing less than £200,000, require no approval except where the authority requires such or where statutory notices are involved. Projects of £200,000, up to £2 millions require approval which will be given automatically upon the submission of particulars of the scheme and performance data relating to it.

Capital works and repair projects supported by grant from the Department will continue to require individual approval.

Data relating to the performance of educational building (area and cost standards) will be published periodically and tender price data will continue to be issued quarterly. The current forecast cost of £482 per square metre (Q2 1989) is based upon a public sector tender price index of 317 (1975 = 100) and relates to a mixture of types of accommodation normally found in <u>complete</u> new schools and further education establishments.

LOCAL AUTHORITY HOUSING AND HOUSING ASSSOCIATION SCHEMES

The cost limit and allowances arrangements for Local Authority and Housing Association schemes were previously very similar, with the Department of Enviroment admissable cost limits (ACL's) corresponding closely with the Housing Corporation's total indicative costs (TIC's) system.

However, 1989 has seen a divergence of procedures brought about by the Housing Corporation introducing a new system based on total cost indicator (TCI) tables.

Radically altering the previous arrangements, the new TCI's, which come into effect from 1st April 1989, represent not just development costs, but estimates of final costs at practical completion including professional fees, interest charges etc. There also now exist some divergence between groups and regions applicable to the two schemes.

LOCAL AUTHORITY HOUSING SCHEMES

In 1988 the Secretary of State for the Environment abolished the housing project control system as as set out in DOE Circular 23/82, and substituted arrangements for assessing admissable costs for subsidy on local authority capital expenditure on housing. These arrangements, set out in DOE Circular 5/88 do not cover:-

   (a) slum clearance (see DOE Circulars 129/74 and 42/75);
   (b) improvement for sale and homesteading (see DOE Circular 20/80);
   (c) local authority funded housing association schemes (see DOE Circular 14/83);
   (d) grants and loans to the private sector. (For grants see DOE Circular 21/80 and the
       Department's letter of 5 February 1987 about revenue subsidies to private
       landlords);
   (e) environmental works in General Improvement Areas and Housing Action Areas

The new system is based on a comparison of average actual resource costs with the appropriate figure from a table of admissible cost limits (ACLs). An ACL figure should represent a reasonable average cost for a dwelling of a specified size in a specified area; for 1989/90 the same figures will apply to both new build and renovation schemes, although this may not be the case in other years. As regards dwellings built to meet special housing needs, a further specified amount (only one of two options) will be added to the appropriate figure in the ACL table and it is the sum of these 2 amounts that will be the ACL for comparison with resource costs. The first step in calculating the amount of new admissible costs on housing capital expenditure is to determine the lower of the average actual resources costs per dwelling and the appropriate ACL figure. The intention is that an ACL should cover:-

      for new build: the current market value of the land, construction costs
      and fees
      for renovation: the pre-improvement value of the dwelling, works costs
      and fees

The latest ACLs and revised cost groups for 1989/90 have been published in DOE Circular 10/89, which replace those tables and annexes previously contained within DOE Circular 5/88.

For the benefit of readers, these revisions have been reproduced here in full.

DOE Circular 10/89 - Capital expenditures by Local Authorities under part II of the Housing act 1985: Admissible costs for subsidy 1989/90

   1. Since 1 April 1988 local authorities have been calculating new admissible costs for
      subsidy on housing capital expenditure by means of the admissable costs limits
      (ACLs) system. In consequence of paragraph 26 of DOE Circular 5/88 the distribution
      of local authorities into various cost groups has been reviewed and the local
      authority associations and local authorities have been consulted upon the outcome
      of the review. This circular now publishes the Secretary of State's decisions as
      regards the table of ACLs and the revised cost groups for the year 1989/90.
      Annexes A and B to this circular therefore replace Annexes A and B to DOE Circular
      5/88 as from 1 April 1989. Expressions used in this circular which are used in the
      Housing Subsidies and Accounting Manual 1981 have, unless the contrary intention
      appears, the meaning they bear in the manual (e.g."year" means "financial year").

LOCAL AUTHORITY HOUSING SCHEMES

Thresholds

2. The system of thresholds prescribing expenditure on renovation that is excluded
    from the averaging and comparison with ACLs is set out in paragraph 17 of DOE
    Circular 5/88. As local authorities will be aware, this has been reviewed. A
    proposal to dispense with the contract threshold was included in the list of
    amendments to the General Determination of Reckonable Expenditure (Appendix A to
    the Manual) on which the Department is currently consulting the local authority
    associations. Without any prejudice to those consultations, authorities are
    advised to proceed for the time being on the assumption that the proposed new rules
    on thresholds are to apply from 1 April 1989. Consultations close on 31 March.
    Responses will then be considered and a decision issued as soon as possible.

Building for Sale

3. The reference to the cost of dwellings built expressly for outright sale in Annex C
    to DOE Circular 5/88 is inconsistent with the provisions of the Manual. Even if
    the local authority puts up the working capital, the costs would be recouped if the
    dwelling was sold without being tenanted. Paragraph 1 of Annex C is therefore
    replaced with the following:

    "The cost of dwellings built expressly for the outright sale even if met by the
    authority should not normally be included in the comparison between resource
    costs and ACLs. It may transpire however that after completion a local
    authority wishes or is obliged, for example under a buy-in guarantee provision,
    to transfer them into its rented stock, or to dispose of them on shared
    ownership terms. In such cases the Department's consent is required to an
    appropriate amount of admissible costs being added to the authority's total.
    In considering whether to give such consent the Department may have regard to
    the level of expenditure it considers reasonable in such cirumstances."

Provision for Review

4. The Department hopes that in future years it will be able to complete the annual
    review sooner than proved possible in this first year of the system, but it may not
    be able to do that by November. It will aim to provide authorities with more of
    the evidence on which the annual review is based; this has been a point of which
    the Department was pressed this year. Paragraph 26 of Circular 5/88 is therefore
    cancelled from 1 April 1989 and replaced with the following:

    "The distribution of local authorities in ACL cost groups will be reviewed
    annually in consultation with the local authority associations. Details of the
    sources and data used by the Department in its review will be announced when
    proposals for the changes in cost groups for the ensuing year are issued for
    comment, together with the methodology for updating the ACLs. The table of
    ACLs and the cost groups will be published as early as possible before the year
    to which they relate."

Statistics

5. Local authorities are reminded of the need to complete forms LH 1-6 and the
    advantages of their doing so (see paragraph 25 of DOE Circular 5/88). It was
    proposed in the consultation letter of 22 December, on the Review of ACLs for
    1989/90, that the revised General Determination of Reckonable Expenditure would
    render inadmissable any costs that have not been reported on these LH forms. The
    Department will give further thought to this proposal.

New Financial Regime

6. Authorities with a negative entitlement to subsidy for 1989/90 will know from
    paragraphs 15(i) and 25 of the consultation paper "New Financial Regime for Local
    Authority Housing in England and Wales" issued on 27 July 1988 and paragraph 4 of
    the Note issued on 20 October 1988 that they should assess costs of new tenders
    against ACLs because this will be taken into account in their base amount for HRA
    subsidy on 1 April 1990.

For further information on how to implement these new procedures the reader is referred
to DOE Circulars 5/88 and 10/89 and also advised to check that the ACLs and other cost
allowances provided here are valid at the time they are being used.

LOCAL AUTHORITY HOUSING SCHEMES

ANNEX A - NEW BUILD AND REHABILITATION

ADMISSIBLE COST LIMITS 1988/1989 - £ per dwelling

General needs accommodation

Dwelling floor area m2	A1 £	A £	B £	C £	D £	E £	F £	G £
25	62 400	54 700	45 700	38 200	32 000	26 600	22 300	18 600
35	73 200	64 200	53 700	44 800	37 400	31 300	26 100	21 800
45	83 600	73 300	61 300	51 200	42 700	35 700	29 800	24 900
55	93 900	82 400	68 900	57 500	48 100	40 100	33 500	28 000
65	104 400	91 600	76 500	63 900	53 400	44 600	37 200	31 100
75	114 300	100 300	83 900	70 000	58 400	48 800	40 800	34 100
85	124 400	109 100	91 100	76 100	63 600	53 100	44 400	37 100
95	133 500	117 100	97 900	81 800	68 300	57 000	47 600	39 800
105	143 100	125 500	104 800	87 600	73 200	61 100	51 100	42 700
115	152 300	133 600	111 500	93 200	77 800	65 000	54 400	45 400

Adjustments

Add for special needs (1 only of the following)

	A1	A	B	C	D	E	F	G
Old Person or Disabled Wheelchair	14 600	12 800	10 700	9 000	7 500	6 200	5 200	4 400
Old Person Warden Supervised or Frail Elderly	31 400	27 500	23 000	19 200	16 000	13 400	11 200	9 300

LOCAL AUTHORITY HOUSING SCHEMES

ANNEX B - ADMISSIBLE COST LIMITS GROUPS

Group A1 comprising:

        Camden
        Kensington & Chelsea
        Westminster

Group A comprising:

        The following **London Boroughs**

Barnet	Hounslow
Brent	Islington
Ealing	Kingston-upon-Thames
Enfield	Lambeth
Hackney	Richmond-upon-Thames
Hammersmith & Fulham	Southwark
Haringey	Tower Hamlets
Harrow	Wandsworth

        City of London

        The following Districts in the Counties of:

Essex	**Surrey**
Epping Forest	Elmbridge

Group B comprising:

        The following **London Boroughs**

Bexley	Merton
Bromley	Newham
Croydon	Redbridge
Greenwich	Sutton
Hillingdon	Waltham Forest
Lewisham	

        The following Districts in the Counties of:

Berkshire	**Buckinghamshire**
Bracknell Forest	Chiltern
Slough	South Buckinghamshire
Windsor & Maidenhead	

Hertfordshire	**Surrey**
Broxbourne	Epsom & Ewell
East Hertfordshire	Mole Valley
Hertsmere	Reigate & Banstead
St. Albans	Runnymede
Three Rivers	Spelthorne
Watford	Tandridge
Welwyn Hatfield	Woking

LOCAL AUTHORITY HOUSING SCHEMES

ANNEX B - ADMISSIBLE COST LIMITS GROUPS - cont'd

Group C comprising:

The following **London Boroughs**

Barking & Dagenham
Havering

The following Districts in the Counties of:

Bedfordshire Luton	**Hertfordshire** all Districts not listed above
Berkshire Reading Wokingham	**Isles of Scilly**
Buckinghamshire Wycombe	**Kent** Dartford Gravesham Sevenoaks Tunbridge Wells
Cambridgeshire Cambridge	**Oxfordshire** Oxford South Oxfordshire
East Sussex Brighton Hove	**Surrey** all Districts not listed above
Essex Basildon Brentwood Castle Point Chelmsford Harlow Rochford Southend-on-Sea Thurrock Uttlesford	**West Sussex** Chichester Crawley Horsham Mid. Sussex
Hampshire Basingstoke & Deane Hart Rushmore	

LOCAL AUTHORITY HOUSING SCHEMES

ANNEX B - ADMISSIBLE COST LIMITS GROUPS - cont'd

Group D comprising:

The following Districts in the Counties of:

Avon
Bath

Bedfordshire
all Districts not listed above

Berkshire
all Districts not listed above

Buckinghamshire
Aylesbury Vale

Cambridgeshire
Huntingdon
South Cambridgeshire

Dorset
Bournemouth
Christchurch
East Dorset
Poole

East Sussex
all Districts not listed above

Essex
all Districts not listed above

Hampshire
all Districts not listed above

Kent
Ashford
Gillingham
Maidstone
Rochester upon Medway
Shepway
Swale
Tonbridge & Malling

Norfolk
Broadland
Norwich

Oxfordshire
West Oxfordshire
Vale of White Horse

Suffolk
Babergh
Forest Heath
Ipswich
Mid-Suffolk
St. Edmundsbury
Suffolk Coastal

West Sussex
all Districts not listed above

Wiltshire
Kennet
North Wiltshire
Salisbury
Thamesdown

Group E comprising:

The following Districts in the Counties of:

Avon
all Districts not listed above

Buckinghamshire
all Districts not listed above

Cambridgeshire
East Cambridgeshire
Peterborough

Isle of Wight
Medina
South Wight

Kent
all Districts not listed above

Norfolk
all Districts not listed above

LOCAL AUTHORITY HOUSING SCHEMES

ANNEX B - ADMISSIBLE COST LIMITS GROUPS - cont'd

Group E (continued) comprising:

Devon
East Devon
Exeter
Mid-Devon
North Devon
Plymouth
South Hams
Teignbridge
Torbay

Dorset
all Districts not listed above

Gloucestershire
Cheltenham
Cotswold
Gloucester
Stroud
Tewkesbury

Hereford & Worcester
Bromsgrove

Northamptonshire
Northampton
South Northamptonshire

Oxfordshire
all Districts not listed above

Somerset
all Districts

Suffolk
all Districts not listed above

Warwickshire
Stratford-on-Avon
Warwick

Wiltshire
all Districts not listed above

The following District in the former Metropolitan County of:

West Midlands
Birmingham
Solihull

Group F comprising:

The following Districts in the former Metropolitan Counties of:

Greater Manchester
Bury
Manchester
Salford
Stockport
Tameside
Trafford

Merseyside
all Districts

South Yorkshire
Sheffield

Tyne & Wear
Newcastle upon Tyne

West Midlands
all Districts not listed above

West Yorkshire
Bradford
Leeds

LOCAL AUTHORITY HOUSING SCHEMES

ANNEX B - ADMISSIBLE COST LIMITS GROUPS - cont'd

Group F (continued) comprising:

The following Districts in the Counties of:

Cambridgeshire
all Districts not listed above

Cheshire
Chester
Congleton
Ellesmere Port & Neston
Halton
Macclesfield
Vale Royal

Cleveland
Hartlepool

Cornwall
all Districts

Cumbria
all Districts

Derbyshire
High Peak
Derbyshire Dales

Devon
all Districts not listed above

Durham
Darlington
Durham
Easington
Sedgefield
Teesdale

Gloucestershire
all Districts not listed above

Hereford & Worcester
all Districts not listed above

Humberside
Beverley
East Yorkshire
Hull

Lancashire
Blackpool
Burnley
Chorley
Fylde
Lancaster
Pendle
Preston
Ribble Valley
South Ribble
West Lancashire
Wyre

Leicestershire
all Districts

The following Districts in the Counties of:

Lincolnshire
Boston
Lincoln
South Holland
South Kesteven

Northamptonshire
all Districts not listed above

Northumberland
Alnwick
Berwick-upon-Tweed
Castle Morpeth
Tynedale

Nottinghamshire
Nottingham
Rushcliffe

Shropshire
Bridgnorth
Shrewsbury & Atcham
South Shropshire
The Wrekin

Staffordshire
East Staffordshire
Lichfield
South Staffordshire
Tamworth

LOCAL AUTHORITY HOUSING SCHEMES

ANNEX B - ADMISSIBLE COST LIMITS GROUPS - cont'd

Group F (continued) comprising:

North Yorkshire	**Warwickshire**
Craven	all Districts not listed above
Harrowgate	
Ryedale	
Scarborough	
Selby	
York	

Group G comprising:

The following Districts in the former Metropolitan Counties of:

Greater Manchester	**Tyne & Wear**
all Districts not listed above	all Districts not listed above
South Yorkshire	**West Yorkshire**
all Districts not listed above	all Districts not listed above

The following Districts in the Counties of:

Cheshire	**Lincolnshire**
all Districts not listed above	all Districts not listed above
Cleveland	**Northumberland**
all Districts not listed above	all Districts not listed above
Derbyshire	**North Yorkshire**
all Districts not listed above	all Districts not listed above
Durham	**Nottinghamshire**
all Districts not listed above	all Districts not listed above
Humberside	**Shropshire**
all Districts not listed above	all Districts not listed above
Lancashire	**Staffordshire**
all Districts not listed above	all Districts not listed above

HOUSING ASSOCIATION SCHEMES

For the benefit of readers, the new TCI system instigated by Housing Corporation Circular HC 11/89 has been reproduced here in full.

Housing Corporation Circular HC 11/89 - Total cost indicators for housing association schemes new build and rehabilitation for rent and related issues from April 1989.

Total cost indicators for housing association schemes 1989/90: new build and rehabilitation for rent, and related issues.

Summary

Advises association of:

    (i)   the total cost indicators, tranche percentages and standard on-costs for new build and rehabilitation for rent (tariff and non-tariff) schemes approved under the 1988 Act effective from 1st April 1989;

    (ii)   the total indicative costs for schemes approved under the 1985 Act effective from 1st April 1989.

This Circular supersedes Circular HC 37/88 for tariff and non-tariff schemes approved under the 1988 Act and amends it for schemes approved under the 1985 Act.

1. **Introduction**

    1.1   The new system of cost controls will have the two-fold purpose of securing value for money and determining the maximum amount of grant appropriate to a particular scheme or programme of schemes. The nucleus of the system is a series of Total Cost Indicator (TCI) tables which set out representative costs for different types of project in different parts of the country.

    1.2   TCI differ from Total Indicative Costs (TIC) in the following ways:

        i)   the TCI allow for costs other than direct development costs including interest on development period loans, associations' development administration costs (formerly A & D allowances) and other sundry costs, as well as the professional fees included in TIC;

        ii)   they represent estimates of final costs at practical completion.

    1.3   The TCI, in combination with the grant percentage applicable to a particular project or programme of projects, will ensure that public funding is confined to a proportion of reasonable costs for schemes of a particular type in a particular area. The basic TCI table is for new build, 2 to 3 storey, general needs dwellings, of specified floor areas and design occupancy. Standard multipliers are then applied to provide figures for different storey heights, rehabilitation, special needs (e.g. sheltered schemes of various kinds), schemes not involving acquisition of a site or a property, and other variables. The TCI will apply equally to schemes funded by the Housing Corporation and by individual local authorities.

    1.4   This Circular sets out the TCI tranche percentages and standard on-costs for schemes approved under the 1988 Act:

        i)   the TCI are described in paragraph 2 and set out in Appendix 1;

        ii)   the tranche percentages are defined and set out in paragraph 3;

        iii) the standard on-costs are described in paragraph 4 and set out in Appendix 4.

    The adjustments to the TIC published in Circular HC 37/88 which will apply to schemes funded under transitional HAG arrangements (see Circular HC 46/88) are set out in paragraph 5. The TCI and TIC apply to the same cost groups as defined in Appendix 2.

HOUSING ASSOCIATION SCHEMES

1.5   This Circular forms part of a wider group of documents related to a new system that will be published before 1st April 1989, including the following:

   i)   Grant Rates Circular;

   ii)   parallel circulars describing other aspects of the new system;

   iii)   the Procedure Guides for tariff and non-tariff housing for rent, including the design and contracting requirements;

   iv)   Good Practice Guide;

   v)   Capital Grant Determinations.

2.   Total cost indicators

2.1   The TCI cover self-contained and shared accommodation designed to general needs, sheltered, frail elderly and wheelchair design criteria. The figures in Appendix 1 supersede the cost criteria published in Circular HC 37/88 and are for use from 1st April 1989 for schemes approved under the 1988 Act. They cover land/property purchase, works and on-costs, all as defined in Appendix 3.

2.2   TCI are set out in Appendix 1 as follows:

   (i)   A table of basic costs per dwelling (see Table 1) related to the floor area of the dwelling (in self-contained dwellings) or per person (in shared accommodation). Floor area for this purpose is defined in Appendix 5. The figures apply to both fixed price and variation of price building contracts. Separate figures are shown for each cost group.

   (ii)   Key multipliers in Table 2. Separate multipliers are shown for acquisition and works schemes, off the shelf new build schemes and for the purchase of existing satisfactory dwellings, works only, re-improvements etc. It should be noted that:

       -     only one key multiplier should be used for each dwelling type;

       -     the key multiplier for new build, acquisition and works is 1.00; i.e. the table of basic cost is for this type;

       -     separate multipliers are shown for each cost group.

   (iii) Supplementary multipliers in Table 3. These apply to all cost groups. They apply both to new build and rehabilitation, unless stated otherwise. They can be applied sequentially. Separate multipliers are shown for acquisition and works schemes and others, e.g. works only and re-improvement etc.

       Note: Only one supplementary multiplier from sections (a), (b) and (d) of Table 3 may be used for each dwelling type.

2.3   Further explanatory notes and definitions are set out in Appendix 5.

3.   Tranche percentages

3.1   These are proportions of Housing Association Grant payable at various stages of a scheme approved under the 1988 Act.

3.2   The following tranche percentages apply as appropriate from 1st April 1989. The tranches are payable at three key stages, at the time of:

   (i)   exchange of purchase contracts;

   (ii)   the main contract works starting on site;

   (iii) practical completion of the scheme or phase of the building contact.

HOUSING ASSOCIATION SCHEMES

   3.3   The tranches, set out in order of payment, are as follows:

                                    Scheme type

Cost Group	New build			Rehabilitation			Works only and Re-improvement	
	(i)	(ii)	(iii)	(i)	(ii)	(iii)	(ii)	(iii)
A & B	30%	50%	20%	50%	30%	20%	80%	20%
C,D & E	20%	60%	20%	40%	40%	20%	80%	20%
F and G	10%	70%	20%	30%	50%	20%	80%	20%

4.   **Standard on-costs**

   TCI, as noted above, will cover all the costs of a scheme, a full list of which is set
   out at Appendix 3. It will therefore be necessary in future to increase the estimate
   of acquisition and works costs to allow for all the standard on-costs. The approach
   to be adopted in identifying the total eligible capital costs for a scheme will be to
   add to the sum of acquisition and works costs, as defined in Appendix 3, the relevant
   standard percentage, as set out in Appendix 4, of those acquisition and works costs.
   The appropriate on-cost percentage for a particular scheme is determined on the basis
   of predominant dwelling type, and will vary according to TCI cost group and the
   general purpose of the scheme. It should be noted that professional fees will in
   future be treated as standard on-costs, and it will not be necessary (as hitherto) to
   provide estimates of actual fees.

5.   **Total Indicative Costs**

   5.1   TIC will apply to transitional schemes and will continue to apply to those
         shared housing schemes proceeding under the 1985 Act from 1st April 1989. (See
         Circular HC 46/88 on Transitional Arrangements for Housing Schemes).

   5.2   Following review the Secretary of State for the Environment has decided to
         increase the level of TIC in Circular HC 37/88 for new build and rehabilitation
         schemes by 7% in all TIC groups as from 1st April 1989.

   5.3   **The updating addition** to the acquisition will be calculated on form S6 using the
         updating percentages listed in the following table which supersedes the table
         given in paragraph 1.5 of Circular HC 37/88.

Year of acquisition (completion date of acquisition)	Cost criteria group			
	A & B %	C & D %	E %	F & G %
1989 (from 1st Apr)	Nil	Nil	Nil	Nil
1989 (up to 31st Mar)	Nil	5	5	5
1988 (calendar year)	15	25	30	25
1987 "	35	45	55	40
1986 "	50	55	60	45
1985 "	65	70	65	50
1984 "	90	90	75	60
1983 "	110	105	80	65
1982 "	125	120	90	75
1981 "	125	120	90	75
1980 "	185	170	110	95
1979 "	280	250	145	130
1978 and earlier	345	305	165	150

6.   **Further reviews of TCI and TIC**

   TCI and TIC will be reviewed annually with six monthly up-dates in between.

HOUSING ASSOCIATION SCHEMES

APPENDIX 1 - TABLE 1 - TOTAL COST INDICATORS

TOTAL DWELLING COSTS 1989/90 - £ PER DWELLING

Dwelling Dwelling floor area (m2)	Probable occupancy (persons)	Group						
		A	B	C	D	E	F	G
Up to 30	1	59 000	51 000	44 100	38 100	32 900	28 500	24 600
Exceeding 30 and not exceeding 40	1 & 2	70 000	60 500	52 300	45 200	39 100	33 800	29 200
Exceeding 40 and not exceeding 50	2 & 3	80 800	69 800	60 300	52 200	45 100	39 000	33 700
Exceeding 50 and not exceeding 60	2 & 3	91 300	78 900	68 200	59 000	51 000	44 100	38 100
Exceeding 60 and not exceeding 70	3 & 4	101 600	87 800	75 900	65 600	56 700	49 000	42 400
Exceeding 70 and not exceeding 80	3,4 & 5	111 000	95 900	82 900	71 700	62 000	53 600	46 300
Exceeding 80 and not exceeding 90	4,5 & 6	120 100	103 800	89 700	77 600	67 000	58 000	50 100
Exceeding 90 and not exceeding 100	5 & 6	128 900	111 400	96 300	83 300	72 000	62 200	53 800
Exceeding 100 and not exceeding 110	6 & 7	137 100	118 500	102 400	88 500	76 500	66 200	57 200
Exceeding 110 and not exceeding 120	6 & over	144 700	125 100	108 200	93 500	80 800	69 900	60 400

APPENDIX 1 - TABLE 2 - KEY MULTIPLIERS By cost group

Note - Only one of the following to be used.

		Group						
		A	B	C	D	E	F	G
a)	**New build**							
	i) Acquisition & works	1.00	1.00	1.00	1.00	1.00	1.00	1.00
	ii) Off-the-shelf	0.85	0.86	0.87	0.87	0.88	0.89	0.90
	iii) Works only	0.75	0.77	0.78	0.80	0.82	0.83	0.85
b)	**Rehabilitation**							
	i) Acquisition & works	1.15	1.12	1.08	1.05	1.02	0.98	0.95
	ii) Existing satisfactory dwellings	0.95	0.93	0.92	0.90	0.88	0.87	0.85
	iii) Vacant works only and re-improvements	0.60	0.60	0.60	0.60	0.60	0.60	0.60
	iv) Tenanted works only and re-improvements	0.65	0.65	0.65	0.65	0.65	0.65	0.65

HOUSING ASSOCIATION SCHEMES

APPENDIX 1 - TABLE 3 - SUPPLEMENTARY MULTIPLIERS All cost groups.

Note   a) New build and rehabilitation, unless stated otherwise.

       b) Can be applied sequentially.

       c) Only one supplementary multiplier from sections (a),(b) and (d) may be used for each dwelling type.

		Acquisition and works and off-the-shelf	Works only and re-improvements
a)	**Storey height**		
	i) New build		
	- 4,5 & 6 storey	1.05	1.05
	- 1 and 7+ storeys	1.10	1.15
	ii) Rehabilitation		
	- New lifts provided	1.10	1.15
b)	**Sheltered**		
	i) Category 1	1.15	1.20
	plus common room	1.20	1.30
	plus common room and		
	heated circulation area	1.30	1.45
	ii) Category 2	1.35	1.50
	iii) Frail elderly	1.60	1.85
c)	**Wheelchair**	1.20	1.30
d)	**Shared**		
	Bedspaces in unit		
	i) 2 and 3	1.10	1.15
	ii) 4 to 6	0.95	0.95
	iii) 7 to 10	0.90	0.85
	iv) 11 to 50	0.85	0.80
	v) Over 50	0.80	0.70
e)	**Single family house.** Rehabilitation only	1.05	1.05
f)	**Special Projects** Dwellings attracting SPPA under the 1985 Act.	1.05	1.05

HOUSING ASSOCIATION SCHEMES

APPENDIX 2 - GROUPINGS

The cost TCI shall apply to schemes in groups defined as follows:

Group A comprising:

        The following **London Boroughs**

Barnet	Islington
Brent	Kensington & Chelsea
Camden	Kingston-upon-Thames
Ealing	Lambeth
Enfield	Richmond-upon-Thames
Hackney	Southwark
Hammersmith & Fulham	Tower Hamlets
Haringey	Wandsworth
Hounslow	City of Westminster

        City of London

        The following Districts in the Counties of:

Essex	**Surrey**
Epping Forest	Elmbridge

Group B comprising:

        The following **London Boroughs**

Bexley	Lewisham
Bromley	Merton
Croydon	Newham
Harrow	Redbridge
Hillingdon	Sutton
Greenwich	Waltham Forest

        The following Districts in the Counties of:

Berkshire	**Buckinghamshire**
Bracknell	Chiltern
Slough	South Buckinghamshire
Windsor & Maidenhead	

Hertfordshire	**Surrey**
Broxbourne	Epsom & Ewell
East Hertfordshire	Mole Valley
Hertsmere	Reigate & Banstead
St. Albans	Runneymede
Three Rivers	Spelthorne
Watford	Tandridge
Welwyn Hatfield	Woking

Group C comprising:

        The following **London Boroughs**

Barking & Dagenham
Havering

        The following Districts in the Counties of:

Bedfordshire	**Hampshire**
Luton	Basingstoke & Deane
	Hart
	Rushmoor

Berkshire	
Reading	
Wokingham	

HOUSING ASSOCIATION SCHEMES

APPENDIX 2 - GROUPINGS - cont'd

Group C (continued) comprising:

Buckinghamshire Wycombe	**Hertfordshire** all Districts not listed above
Cambridgeshire Cambridge	**Isles of Scilly**
East Sussex Brighton Hove	**Kent** Dartford Sevenoaks Tunbridge Wells
Essex Basildon Brentwood Chelmsford Harlow Southend-on-Sea Uttlesford	**Oxfordshire** Oxford South Oxfordshire **Surrey** all Districts not listed above
	West Sussex Crawley Horsham Mid. Sussex

Group D comprising:

The following Districts in the Counties of:

Avon Bath	**Kent** Ashford Gillingham Gravesham
Bedfordshire all Districts not listed above	Maidstone Rochester-upon-Medway Shepway Swale
Berkshire all Districts not listed above	Tonbridge & Malling
Buckinghamshire Aylesbury Vale	**Oxfordshire** West Oxfordshire Vale of White Horse
Cambridgeshire Huntingdon South Cambridgeshire	**Suffolk** Babergh Forest Heath Ipswich Mid. Suffolk
Dorset Bournemouth Christchurch Poole Wimborne	St.Edmondsbury Suffolk Coastal
East Sussex all Districts not listed above	**West Sussex** all Districts not listed above
Essex all districts not listed above	

HOUSING ASSOCIATION SCHEMES

APPENDIX 2 - GROUPINGS - cont'd

Group D (continued) comprising:

Hampshire all Districts not listed above	**Wiltshire** Kennet North Wiltshire Salisbury Thamesdown

Group E comprising:

The following Districts in the Counties of:

Avon all Districts not listed above	**Kent** all Districts not listed above
Buckinghamshire all Districts not listed above	**Norfolk** all Districts
Cambridgeshire East Cambridgeshire Peterborough	**Northamptonshire** Northampton South Northamptonshire
Devon East Devon Exeter Mid-Devon North Devon Plymouth South Hams Teignbridge Torbay	**Oxfordshire** all Districts not listed above **Somerset** all Districts **Suffolk** all Districts not listed above
Dorset all Districts not listed above	**Warwickshire** Stratford on Avon Warwick
Gloucestershire Cheltenham Cotswold Gloucester Tewkesbury	**West Midlands** Birmingham Solihull
Hereford & Worcester Bromsgrove	**Wiltshire** all Districts not listed above

The County of:
Isle of Wight

HOUSING ASSOCIATION SCHEMES

APPENDIX 2 - GROUPINGS - cont'd

Group F comprising:

The following Districts in the Counties of:

Cambridgeshire
all Districts not listed above

Cheshire
Chester
Congleton
Ellesmere Port & Neston
Halton
Macclesfield
Vale Royal

Cleveland
Hartlepool

Cornwall
all Districts

Cumbria
all Districts

Derbyshire
High Peak
Derbyshire Dales

Devon
all Districts not listed above

Durham
Darlington
Durham
Easington
Sedgefield
Teesdale

Gloucestershire
all Districts not listed above

Greater Manchester
Bury
Manchester
Salford
Stockport
Tameside
Trafford

Hereford & Worcester
all Districts not listed above

Humberside
Beverley
East Yorkshire
Kingston upon Hull

Lancashire
Blackpool
Burnley
Chorley
Fylde
Lancaster
Pendle
Preston
Ribble Valley
South Ribble
West Lancashire
Wyre

Leicestershire
all Districts

Lincolnshire
Boston
Lincoln
South Holland
South Kesteven

Merseyside
all Districts

South Yorkshire
Sheffield

Tyne & Wear
Newcastle upon Tyne

West Midlands
all Districts not listed above

West Yorkshire
Bradford
Leeds

HOUSING ASSOCIATION SCHEMES

APPENDIX 2 - GROUPINGS - cont'd

Group F (continued) comprising:

The following Districts in the Counties of:

Northamptonshire
all Districts not listed above

Nottinghamshire
Nottingham
Rushcliffe

Northumberland
Alnwick
Berwick upon Tweed
Castle Morpeth
Tynedale

Shropshire
Bridgnorth
Shrewsbury & Atcham
South Shropshire
The Wrekin

North Yorkshire
Craven
Harrogate
Ryedale
Scarborough
Selby
York

Staffordshire
East Staffordshire
Lichfield
South Staffordshire
Tamworth

Warwickshire
all Districts not listed above

Group G comprising:

The following Districts in the Counties of:

Cheshire
all Districts not listed above

Lincolnshire
all Districts not listed above

Cleveland
all Districts not listed above

Northumberland
all Districts not listed above

Derbyshire
all Districts not listed above

North Yorkshire
all Districts not listed above

Durham
all Districts not listed above

Nottinghamshire
all Districts not listed above

Greater Manchester
all Districts not listed above

South Yorkshire
all Districts not listed above

Humberside
all Districts not listed above

Shropshire
all Districts not listed above

Lancashire
all Districts not listed above

Staffordshire
all Districts not listed above

Tyne & Wear
all Districts not listed above

West Yorkshire
all Districts not listed above

HOUSING ASSOCIATION SCHEMES

APPENDIX 3 - TCI DEFINITION

The TCI include the following:

1. **ACQUISITION**

    (i)   Purchase price of land/property

2. **WORKS**

    (i)   Main works contract costs (including where applicable, adjustments for
           additional claims and fluctuations)

    (ii)   Major Site Development Works (where applicable), including piling, soil
           stabilisation, road and sewer construction, major demolition, statutory
           agreements and associated bonds

    (iii) Major Pre Works (Rehabilitation), where applicable

    (iv)   VAT on the above, where applicable

3. **ON-COSTS**

    (i)   Legal fees, disbursements and expenses

    (ii)   Stamp duty

    (iii) Interest charges on development period loans

    (iv)   Building society or other valuation and administration fees

    (v)   Fees for building control and planning permission

    (vi)   In house or external consultants' fees, disbursements and expenses

    (vii) Insurance premiums (except contract insurance included in works cost)

    (viii) Contract performance bond premiums

    (ix)   Borrowing administration charges

    (x)   Associations' development administration costs (formerly A & D allowances,
           excluding Co-operative promotional allowance)

    (xi)   Furniture

    (xii) Home loss and disturbance payments; void rates

    (xiii) Preliminary minor site development works (New build) and pre-works
           (Rehabilitation)

    (xiv) VAT on the above, where applicable

APPENDIX 4 - SCHEDULE OF STANDARD ON-COSTS

1. One standard on-cost shall be determined for each scheme. It is determined by:
   a)   the TCI cost group;
   b)   the general purpose of the scheme;
   and is applied to the sum of the acquisition and works costs of a scheme.

2. The standard on-cost percentage will be the total of:
   a)   the appropriate key on-cost (only one will be appropriate) plus
   b)   the total of one or more of the appropriate supplementary on-costs.

   Note The appropriate on-cost will be that which applies to:
             a)   for key on-costs - the predominant dwelling type determined by the
                   largest number of persons in total;
             b)   for supplementary on-costs - over 50% of the persons in the scheme.

HOUSING ASSOCIATION SCHEMES

APPENDIX 4 - SCHEDULE OF STANDARD ON-COSTS - cont'd

Standard on-cost percentage	A	B	C	Group D	E	F	G
a) key on-costs (all +)							
i) New build	20	21	22	22	23	24	25
ii) Rehabilitation and re-improvement	18	19	20	20	21	22	23
iii) Off-the shelf and existing satisfactory dwellings	13	13	14	14	14	15	15
b) Supplementary on-costs							
i) Tenanted rehabilitation (acquisition and works)	+22	+21	+19	+18	+17	+15	+14
ii) Vacant works only and re-improvements	- 2	- 2	- 2	- 2	- 2	- 2	- 2
iii) Tenanted works only and re-improvements	+11	+10	+10	+ 9	+ 8	+ 8	+ 7
iv) Shared	+ 2	+ 2	+ 3	+ 3	+ 3	+ 4	+ 4
v) Sheltered with communal accommodation	+ 1	+ 1	+ 1	+ 1	+ 1	+ 1	+ 1
vi) Frail elderly	+ 2	+ 2	+ 2	+ 2	+ 2	+ 2	+ 2
vii) Special projects attracting SPPA under 1985 Act	+ 2	+ 2	+ 3	+ 3	+ 3	+ 4	+ 4

APPENDIX 5 - EXPLANATORY NOTES AND DEFINITIONS

1. **Works costs**

   Works costs are all-inclusive and thus include all dwelling and site works and car accommodation.

2. **Unit Size and Number of Occupants**

   a) **Generally**

   The unit size in square metres relates to the total floor area of the resultant unit. (For a definition of floor area see below). The probable occupancy figure is shown as a guide. The number of occupants is related to the number of single and double bedrooms provided. The TCI for a unit where the total floor area exceeds 120 m2 will be appropriate cost for a unit of 110 - 120 m2 plus, for each additional; 10 m2 or part thereof, the difference between the cost of a unit of 100 - 110 m2 and a unit of 110 - 120 m2.

   b) **Shared accommodation**

   The term shared housing is used to describe accommodation for two or more persons with shared facilities, e.g. bathroom, communal living room or dining room. The number of occupants/bedspaces in a shared unit of accommodation is exclusive of those in self-contained staff accommodation or other self-contained accommodation. There may be a number of shared units within the same scheme.

HOUSING ASSOCIATION SCHEMES

APPENDIX 5 - EXPLANATORY NOTES AND DEFINITIONS cont'd

c)   **Total floor area**

Self-contained dwellings

Area is measured to the internal faces of the main containing walls on each floor of the dwelling and includes the space, on plan, taken up by private staircases, partitions, internal walls (but not 'party' or similar walls), chimney breasts, flues and heating appliances.

It includes the area of tenant's storage space, except where a non-habitable basement or attic is used exclusively as storage space.

It excludes:

(i)   any space where the height to the ceiling is less than 1.5 m (e.g. areas in rooms with sloping ceilings, external dustbin enclosures):

(ii)   any porch, covered way, etc. open to the air;

(iii) any garage except that, where a garage is integral with or adjoining the dwelling, the excess over 12 m2 qualifies as storage space and as part of the area of the dwelling;

(iv)   all balconies (private, escape and access) and decks:

(v)   all public access space (e.g. communal entrances, staircases, corridors);

(vi)   all space for communal facilities or services;

(vii) all space for purposes other than housing (e.g. commercial premises).

Shared accommodation

Area is measured to the internal faces of the main containing walls on each floor of the unit and includes all space for shared facilities or services and the space, on the plan, taken up by staircases, partitions, internal walls, chimney breasts, flues and heating appliances.

It includes the area of the resident's storage space, except where a non-habitable basement or attic is used exclusively as storage space.

It excludes:

(i)   any space where the height of the ceiling is less than 1.5 m (e.g. areas in rooms with sloping ceilings, external dustbin enclosures);

(ii)   any porch, covered way, etc. open to the air;

(iv)   all balconies (private, escape and access) and decks;

(iii) any garage except that, where a garage is integral with or adjoining the dwelling, the excess over 12 m2 qualifies as storage space and as part of the area of the dwelling;

(v)   all space for purposes other than housing (e.g. commercial premises).

The appropriate floor area should be calculated by taking the total floor area of the shared unit (see above) excluding self-contained staff accommodation or other self-contained accommodation, and dividing by the total number of bedspaces in the shared unit, excluding those in self-contained accommodation.

HOUSING ASSOCIATION SCHEMES

APPENDIX 5 - **EXPLANATORY NOTES AND DEFINITIONS** cont'd

3. **Storey Heights**

   Storey heights relate to the number of storeys in the block or building in which
   the dwelling occurs, regardless of the storey height of the dwelling itself. In
   the case of mixed use blocks the number of storeys shall include those not used for
   housing. If a block is of two or more different storey heights and a dwelling is
   in more than one part, the TCI shall be related to the number of storeys where the
   greater part of the dwelling is situated.

4. **Single Family House Multiplier**

   This multiplier applies to houses or bungalows improved to form single dwellings.
   It does not apply to flats or maisonettes, unless a number of flats or maisonettes
   are converted to form a single house or bungalow.

5. **Accommodation for Extended Families**

   The TCI for self-contained dwellings for extended families of eight or more persons
   shall be calculated on the basis set out for shared accommodation, but with each
   bedroom counting as one bedspace.

   Dwellings of this type should incorporate particular design features relevant to
   the needs of extended families, for example, two living rooms, two kitchens or
   dining rooms, extra utility space, or special washing or sanitary facilities. The
   Housing Corporation will expect associations to identify the needs met through this
   provision at application for project approval stage.

6. **Wheelchair accommodation**

   The TCI multiplier allows for the higher standards set out in the Design and
   Contracting Requirements, but not any fixed additional equipment to meet the
   particular needs of identified disabled persons being housed. These are met by
   100% grants for Adaptions for the Physically Handicapped (see Procedure Guides).

Keep your figures up to date, free of charge

This section, and most of the other information in this Price Book,
is brought up to date every three months in the *Price Book Update*.

The *Update* is available free to all Price Book purchasers.

To ensure you receive your copy, simply complete the reply card from
the centre of the book and return it to us.

ESSENTIAL BOOKS ON LAW FROM E & F N SPON

AVOIDING CLAIMS
A Practical guide to limiting liability in the construction industry

M Coombes Davies

Avoiding Claims explores the pre-contractual, contractual, and post-contractual liabilities of those involved in the construction industry, from architects, engineers and quantity surveyors to contractors and sub-contractors. It presents an important yet frequently confusing subject by way of a question-and-answer format. Flow diagrams explain procedure in arbitration and in the courts, and the book features 'cautionary tales' — the cases where others got it wrong. Readers' questions are answered simply, directly and concisely, and the book is applicable not only for everyday use in the construction industry but also as a reference guide for problem solving.

1989 Hardback 0 419 14620 2 £27.50 224pp E & F N Spon

THE PRESENTATION AND SETTLEMENT OF CONTRACTORS' CLAIMS

G Trickey

Contract disputes, often involving large sums of money, regularly occur in the building industry. Many books have been written on the legal side of such disputes, but nothing currently available provides detailed practical advice for the professional in the construction industry to the extent offered in this book. Claims under both the 1983 and 1980 editions of the Joint Contracts Tribunal Standard Form of Building Contract are covered. The book considers the legal interpretation of the clauses and includes practical examples of how a claim entitlement should be calculated. In addition to the traditional grounds of claims, the book covers extensions of time, nominations, determination of the contractor's employment and fluctuations. A final section considers the control of claims. Throughout the book the emphasis is on practical applications, with many worked examples of a fictional building contract.

1983 Hardback 0 419 11430 0 £34.50 400pp E & F N Spon

These books may be obtained from your usual bookshop. In case of difficulty please write to the address below or telephone the Order Department on 0264 332424.

 **E & F N SPON**
11 New Fetter Lane, London EC4P 4EE

Property Insurance

The problem of adequately covering by insurance the loss and damage caused to buildings by fire and other perils has been highlighted in recent years by the increasing rate of inflation.

There are a number of schemes available to the building owner wishing to insure his property against the usual risk. Traditionally the insured value must be sufficient to cover the actual cost of reinstating the building. This means that in addition to assessing the current value an estimate has also to be made of the increases likely to occur during the period of the policy and of rebuilding which, for a moderate size building, could amount to a total of three years. Obviously such an estimate is difficult to make with any degree of accuracy, if it is too low the insured may be penalized under the terms of the policy and if too high will result in the payment of unnecessary premiums. There are variations on the traditional method of insuring which aim to reduce the effects of over estimating and details of these are available from the appropriate offices. For the convenience of readers who may wish to make use of the information contained in this publication in calculating insurance cover required the following may be of interest.

1 PRESENT COST
The current rebuilding costs may be ascertained in a number of ways:

(a) Where the actual building cost is known this may be updated by reference to tender indices (page 724);
(b) By reference to average published prices per square metre of floor area (page 727). In this case it is important to understand clearly the method of measurement used to calculate the total floor area on which the rates have been based which were current at November 1989;
(c) By professional valuation;
(d) By comparison with the known cost of another similar building

Whichever of these methods is adopted regard must be paid to any special conditions that may apply, i.e., a confined site, complexity of design or any demolition and site clearance that may be required.

2 ALLOWANCE FOR INFLATION
The 'Present Cost' when established will usually, under the conditions of the policy, be the rebuilding cost on the first day of the policy period. To this must be added a sum to cover future increases. For this purpose, using the historical indices on pages 723 and 724, as a base and taking account of the likely change in building costs and tender climate the following annual average indices are predicted for the future.

	Cost index	Tender index
1987	276	259
1988	293	309
1989	315	360
1990	339	397
1991	364	428
1992	390	459
1993	417	489

3 FEES
To the total of 1 and 2 above must be added an allowance for fees.

4 VALUE ADDED TAX
To the total of 1 to 3 above must be added Value Added Tax. Previously relief
may have been given to total reconstructions following a fire damage etc.
However revisions to previous VAT legislation contained in the 1989 Finance Bill
require that such work now attracts VAT, and the limit of insurance cover should
be raised to allow for this.

5 EXAMPLE
An assessment for insurance cover is required in mid 1989 for a
property which cost £200.000.00 when completed in mid 1976.

		£
Present cost		
Known cost at mid 1976		200 000.00
Predicted tender index mid 1989 = 360		
Tender index mid 1976 = 100		
Increases in tender index = 260%		
applied to known cost =		520 000.00

Present cost (excluding any allowance for demolition)		720 000.00

		£
Allowance for inflation		
Present cost at day one of policy		720 000.00
Allow for changes in tender levels during 12 months currency		
of policy		
Predicted tender index at mid 1990 = 397		
Predicted tender index at mid 1989 = 360		
Increase in tender index = 10%		
applied to present cost =	say	72 000.00
		----------
		792 000.00

Assuming that total damage is suffered on the last day of the
currency of the policy and that planning and documentation would
require a period of twelve months before re-building and could
commence then a further similar allowance must be made
Predicted tender index at mid 1991 = 428
Predicted tender index at mid 1990 = 397
Increase in tender index = 7.81%

applied to adjusted present cost =	say	62 000.00
		----------
		854 000.00

Assuming that total reinstatement would take two years allowance
must be made for the increases in costs which would directly or
indirectly be met under a building contract.
Predicted cost index at mid 1993 = 489
Predicted cost index at mid 1991 = 428
Increases in cost index = 14.2%. This is the total increase
at the end of two years and the amount applicable to the contract

cost incurred over this period might be about half, say 7.19%	say	61 000.00

Estimated cost of reinstatement		915 000.00

SUMMARY	£
Estimated cost of reinstatement	915 000.00
Add professional fees at say 16%	146 000.00

	1061 000.00
Add for Value Added Tax (currently plus 15%)	159 000.00

Total insurance cover required, say	£ 1220 000.00

PART V

Tables and Memoranda

This part of the book contains the following sections:

ARCHITECTS' AND BUILDERS' TITLES FROM E & F N SPON

BUILDING FAILURES
Diagnosis and avoidance
W H Ransom

In recent years building failures and the resulting lawsuits and awards for damages have frequently been in the news. The biggest headlines may have been reserved for structural failures and complete collapses, but we should not forget the less newsworthy failures such as leaky roofs, damp walls, dropped foundations and rotted timber. Their cost in simple cash terms is considerable, but perhaps even more important is the personal discomfort they cause, and the waste of resources and effort involved in repeating a job which should have been done right the first time. Nearly all such failures are caused by faulty design or execution rather than by faulty materials. The mistakes which produce them result not from lack of knowledge but from a failure to apply it properly. This book gives practical guidance on the prevention of failure by describing the nature and cause of the most common defects in buildings, and then showing how they should be avoided in design and construction.

1987 Hardback 0 419 14260 6 £18.00 174pp Paperback 0 419 14270 3 £8.95 E & F N Spon

HAZARDOUS BUILDING MATERIALS
S R Curwell and C G March

Compiled from contributions by a distinguished group of experts in the fields of medicine, toxicology, occupational health and environmental science, this unique book breaks new ground in providing a detailed review of building materials known, or suspected, to have dangerous effects. An essential guide for architects, specifiers and many others involved in the design and construction process, it also evaluates suitable substitutes for dangerous materials, assessing them for cost, safety and performance.

Contents: Foreword. Introduction. Hazards to health from building materials. Asbestos and other natural materials. Man-made mineral fibres. Metals. Lead in building materials. Plastics and toxic chemicals. **Data Sheets.** Roofing slates. Troughed sheeting. Eaves soffit. Verge-pitched roof. Rainwater pipes and gutters. Roof flashing. Pipe Sleeve flashing. Roof insulation. Roofing felt. Flat roof coverings – asphalt and bitumen. Flat roof promenade tiles. Cavity wall insulation. Timber framed wall insulation. Sealants to door and window frames. Glazing putty. Patent glazing. Leaded lights. Ceiling and wall linings. Floorboarding. Fire doors. Floor tile and sheet. Textured paint coatings. Undercoat and finishing paint. Priming paint. Varnish and wood stain. Timber preservatives. Cold water supply pipework. Fittings for copper pipe. Pipe insulation. Cold water storage tanks. Hot and cold water tank insulation. Central heating boiler insulation. Flue pipes. **Hazardous Materials in Existing Buildings.** Appendix. Index.

1986 Paperback 0 419 13740 8 £8.95 150pp E & F N Spon

These books may be obtained from your usual bookshop. In case of difficulty please write to the address below or telephone the Order Department on 0264 332424.

 **E & F N SPON**
11 New Fetter Lane, London EC4P 4EE

Tables and Memoranda

CONVERSION TABLES

	Unit	Conversion factors	

Length

Millimetre	mm	1 in = 25.4 mm	1 mm = 0.0394 in
Centimetre	cm	1 in = 2.54 cm	1 cm = 0.3937 in
Metre	m	1 ft = 0.3048 m	1 m = 3.2808 ft
Kilometre	km	1 yd = 0.0144 m	or 1.0936 yd
		1 mile= 1.6093 km	1 km = 0.6214 mile

Note: 1 cm = 10 mm   1 ft = 12 in
       1 m   = 100 mm   1 yd = 3 ft
       1 km = 1 000 m   1 mile= 1 760 yd

Area

Square Millimetre	mm2	1 in2 = 645.2 mm2	1 mm2= 0.0016 in2
Square Centimetre	cm2	1 in2 = 6.4516 cm2	1 cm2= 1.1550 in2
Square Metre	m2	1 ft2 = 0.0929 m2	1 m = 10.764 ft2
		1 yd2 = 0.8361 m2	1 m2 = 1.1960 yd2
Square Kilometre	km2	1 mile2= 2.590 km2	1 km2= 0.3861 mile2

Note: 1 cm2 = 100 mm2   1 ft2 = 144 in2
       1 m2   = 10 000 cm2   1 yd2 = 9 ft2
       1 km2 = 100 hectares   1 acre= 4 840 yd2
                               1 mile2= 640 acres

Volume

Cubic Centimetre	cm3	1 cm3 = 0.0610 in3	1 in3 = 16.387 cm3
Cubic Decimetre	dm3	1 dm3 = 0.0353 ft3	1 ft3 = 28.329 dm3
Cubic metre	m3	1 m3 = 35.3147 ft3	1 ft3 = 0.0283 m3
		1 m3 = 1.3080 yd3	1 yd3 = 0.7646 m3
Litre	l	1 l = 1.76 pint	1 pint= 0.5683 l
		= 2.113 US pt	= 0.4733 US l

.ote: 1 dm3 = 1 000 cm3   1 ft3 = 1 728 in3
       1 m3 = 1 000 dm3   1 yd3 = 27 ft3
       1 l   = 1 dm3   1 pint= 20 fl oz
       1 hl = 100 l   1 gal = 8 pints

Mass

Milligram	mg	1 mg = 0.0154 grain	1 grain= 64.935 mg
Gram	g	1 g = 0.0353 oz	1 oz = 28.35 g
Kilogram	kg	1 kg = 2.2046 lb	1 lb = 0.4536 kg
Tonne	t	1 t = 0.9842 ton	1 ton = 1.016 t

CONVERSION TABLES

	Unit	Conversion factors

Note: 1 g = 1000 mg   1 oz   = 437.5 grains
        1 kg= 1000 g   1 lb   = 16 oz
        1 t = 1000 kg   1 stone= 14 lb
                               1 cwt = 112 lb
                               1 ton = 20 cwt

Force

Newton	N	1 lb f = 4.448 N	1 kg f = 9.807 N
Kilonewton	kN	1 lb f = 0.004448 kN	1 ton f= 9.964 kN
Meganewton	mN	100 ton f= 0.9964 mN	

Pressure and stress

Kilonewton per   1 lb f/in2 = 6.895 kN/m2
square metre   kN/m2   1 bar   = 100 kN/m2
Meganewton per   1 ton f/ft2= 107.3 kN/m2 = 0.1073 mN/m2
square metre   mN/m2   1 kg f/cm2 = 98.07 kN/m2
                                1 lb f/ft2 = 0.04788 kN/m2

Coefficient of consolidation
(Cv) or swelling

Square metre per year m2/year   1 cm2/s   = 3 154 m2/year
                                  1 ft2/year = 0.0929 m2/year

Coefficent of permeability

Metre per second   m/s   1 cm/s   = 0.01 m/s
Metre per year   m/year   1 ft/year = 0.3048 m/year
                                = 0.9651 x (10)8 m/s

Temperature

Degree celcius $^\circ$C $^\circ C = \dfrac{5}{9} (^\circ F - 32^\circ)$ $^\circ F = \dfrac{9}{5} (^\circ C + 32^\circ)$

FORMULAE

Two dimensional figures

Figure   **Area**

Square   (side) sq

Rectangle   Length x breadth

FORMULAE

Figure	Area
Triangle	0.5 x base x height or N(s(s - a)(s-b)(s-c)) where s = 0.5 x the sum of the three sides and a, b and c are the lengths of the three sides. or a2 = b2 + c2 - 2 x bc x COS A where A is the angle opposite side a
Hexagon	2.6 x (side) sq
Octagon	4.83 x (side) sq
Trapezoid	height x 0.5(base + top)
Circle	3.142 x radius sq or 0.7854 x diameter sq (circumference = 2 x 3.142 x radius or 3.142 x diameter)
Sector of a circle	0.5 x length of arc x radius
Segment of a circle	area of sector - area of triangle
Ellipse	3.142 x AB (where A = 0.5 x height and B = 0.5 x length)
Bellmouth	$\dfrac{3 \times radius\ sq}{14}$

Three dimensional figures

Figure	Volume	Surface Area
Prism	Area of base x height	circumference of base x height
Cube	(side) cubed	6 x (side) sq
Cylinder	3.142 x radius sq x height	2 x 3.142 x radius x (height - radius)
Sphere	$\dfrac{4 \times 3.142 \times radius\ cubed}{3}$	4 x 3.142 x radius sq
Segment of a sphere	$\left(\dfrac{3.142 \times h}{6}\right) \times [3 \times r2 + h2]$	2 x 3.142 x r x h
Pyramid	$\dfrac{1}{3}$ of area of base x height	0.5 x circumference of base x slant height
Cone	$\dfrac{1}{3} \times 3.142 \times radius\ sq \times h$	3.142 x radius x slant height

FORMULAE

Three dimensional figures

Figure	Volume	Surface area
Frustrum of a pyramid	0.33 x height [A + B + sq root (AB)] where A is the area of the large end and B is the area of the small end	0.5 mean circumference x slant height
Frustrum of a cone	(0.33 x 3.142 x height (R sq + r sq + R x r)) where R is the radius of the large end and r is the radius of the small end	(3.142 x slant height x (R + r))

Other formulae

Formulae	Description
Pythagoras theorem	A2 = B2 + C2 where A is the hypotenuse of a right-angled triangle and B and C are the two adjacent sides
Simpsons Rule	Volume = $\dfrac{x(y^\wedge + yn) + 2(y1 + y3 + y5) + 4(y2 + y4)}{3}$
	trench must be split into even sections e.g. 1-7, the areas at intermediate even cross-sections (No 2, 4, 6 etc.) are each multiplied by 4 and the areas at intermediate uneven cross-sections (No 3,5 etc.) are each multiplied by 2 and the end cross-sections taken once only. The sum of these areas is multiplied by 0.33 of the distance between the cross-sections to give the total volume.
Trapezoidal Rule	(0.16 x [Total length of trench] x [area of first section x 4 times area of middle section + area of last section])

Note: Both Simpsons and Trapezoidal Rule are useful in calculating the volume of an irregular trench accurately.

DESIGN LOADINGS FOR BUILDINGS

Definitions

Dead load: The load due to the weight of all walls, permanent partitions, floors, roofs and finishes, including services and all other permanent construction.

Imposed load: The load assumed to be produced by the intended occupancy or use, including the weight of moveable partitions, distributed, concentrated, impact, inertia and snow loads, but excluding wind loads.

Distributed load: The uniformly distributed static loads per square metre of plan area which provide for the effects of normal use. Where no values are given for concentrated load it may be assumed that the tabulated distributed load is adequate for design purposes.

Note: The general recommendations are not applicable to certain a typical usages particularly where mechanical stacking, plant of machinery are to be installed and in these cases the designer should determine the loads from a knowledge of the equipment and processes likely to be employed.

The additional imposed load to provide for partitions, where their positions are not shown on the plans, on beams and floors, where these are capable of effective lateral distributional of the load, is a uniformly distributed load per square metre of not less than one-third of the weight per metre run by the partitions **but not less than 1kN/m2.**

DESIGN LOADINGS FOR BUILDINGS

Floor area usage	Distributed load kN/m2	Concentrated load kN/300 mm2
Industrial occupancy class (workshops, factories)		
Founderies	20.0	-
Cold storage	5.0 for each metre of storage height with a minimum of 15.0	9.0
Paper storage, for printing plants	4.0 for each metre of storage height	9.0
Storage, other than types listed seperately	2.4 for each metre of storage height	7.0
Type storage and other areas in printing plants	12.5	9.0
Boiler rooms, motor rooms, fan rooms and the like, including the weight of machinery	7.5	4.5
Factories, workshops and similar buildings	5.0	4.5
Corridors, hallways, foot-bridges, etc. subject to loads greater than for crowds, such as wheeled vehicles, trolleys and the like	5.0	4.5
Corridors, hallways, stairs, landings, footbridges, etc.	4.0	4.5
Machinery halls, circulation spaces therein	4.0	4.5
Laboratories (including equipment), kitchens, laundries	3.0	4.5
Workrooms, light without storage	2.5	1.8
Toilet rooms	2.0	-
Cat walks	-	1.0 at 1m centres
Institutional and educational occupancy class (prisons, hospitals, schools, colleges)		
Dense mobile stacking (books) on mobile trolleys	4.8 for each metre of stack height but with a minimum of 9.6	7.0
Stack rooms (books)	2.4 for each metre of stack height but with a minimum of 6.5	7.0

DESIGN LOADINGS FOR BUILDINGS

Floor area usage	Distributed load kN/m2	Concentrated load kN/300 mm2
Stationery stores	4.0 for each metre of storage height	9.0
Boiler rooms, motor rooms, fan rooms and the like, including the weight of machinery	7.5	4.5
Corridors, hallways, etc. subject to loads greater than from crowds, such as wheeled vehicles, trolleys and the like	5.0	4.5
Drill rooms and drill halls	5.0	9.0
Assembly areas without fixed seating, stages gymnasia	5.0	3.6
Bars	5.0	-
Projection rooms	5.0	-
Corridors, hallways, aisles, stairs, landings, foot-bridges, etc.	4.0	4.5
Reading rooms with book storage, e.g. libraries	4.0	4.5
Assembly areas with fixed seating	4.0	-
Laboratories (including equipment), kitchens, laundries	3.0	4.5
Classrooms, chapels	3.0	2.7
Reading rooms without book storage	2.5	4.5
Areas for equipment	2.0	1.8
X-ray rooms, operating rooms, utility rooms	2.0	4.5
Dining rooms, lounges, billiard rooms	2.0	2.7
Dressing rooms, hospital bedrooms and wards	2.0	1.8
Toilet rooms	2.0	-
Bedrooms, dormitories	1.5	1.8

DESIGN LOADINGS FOR BUILDINGS

Floor area usage	Distributed load kN/m2	Concentrated load kN/300 mm2

Institutional and educational occupancy class (prisons, hospitals, schools, colleges - cont'd

Balconies	same as rooms to which they give access but with a minimum of 4.0	1.5 per metre run concentrated at the outer edge
Fly galleries	4.5 kN per metre run distributed uniformly over the width	-
Cat walks	-	1.0 at 1 m centres

Offices occupancy class (offices, banks)

Stationery stores	4.0 for each metre of storage height	9.0
Boiler rooms, motor rooms, fan rooms and the like, including the weight of machinery	7.5	4.5
Corridors, hallways, etc. subject to loads greater than from crowds, such as wheeled vehicles, trolleys and the like	5.0	4.5
File rooms, filing and storage space	5.0	4.5
Corridors, hallways, stairs, landings, footbridges, etc.	4.0	4.5
Offices with fixed computers or similar equipment	3.5	4.5
Laboratories (including equipment), kitchens, laundries	3.0	-
Banking halls	3.0	4.5
Offices for general use	2.5	2.7
Toilet rooms	2.0	-
Balconies	Same as rooms to which they give access but with a minimum of 4.0	1.5 per metre run concentrated at the outer edge
Cat walks	-	1.0 at 1m centre

DESIGN LOADINGS FOR BUILDINGS

Floor area usage	Distributed load kN/m2	Concentrated load kN/300 mm2
Public assembly occupancy class (halls, auditoria, restaurants, museums, libraries, non-residential clubs, theatres, broadcasting studios, grandstands		
Dense mobile stacking (books) on mobile trucks	4.8 for each metre of stack height but with a minimum of 9.6	7.0
Stack rooms (books)	2.4 for each metre of stack height but with a minimum of 6.5	7.0
Boiler rooms, motor rooms, fan rooms and the like, including the weight of machinery	7.5	4.5
Stages	7.5	4.5
Corridors, hallways, etc. subject to loads greater than from crowds, such as wheeled vehicles, trolleys and the like. Corridors, stairs, and passageways in grandstands	5.0	4.5
Drill rooms and drill halls	5.0	9.0
Assembly areas without fixed seating: dance halls, gymnasia, grandstands	5.0	3.6
Projection rooms, bars	5.0	-
Museum floors and art galleries for exhibition purposes	4.0	4.5
Corridors, hallways, stairs, landings, footbridges, etc.	4.0	4.5
Reading rooms with book storage, e.g. libraries	4.0	4.5
Assembly areas with fixed seating	4.0	-
Kitchens, laundries	3.0	4.5
Chapels, churches	3.0	2.7
Reading rooms without book storage	2.5	4.5
Grids	2.5	-
Areas for equipment	2.0	1.8
Dining rooms, lounges, billiard rooms	2.0	2.7

DESIGN LOADINGS FOR BUILDINGS

Floor area usage	Distributed load kN/m2	Concentrated loads kN/300 mm2
Public assembly occupancy class (halls, auditoria, restaurants, museums, libraries, non-residential clubs, theatres, broadcasting studios, grandstands - cont'd		
Dressing rooms	2.0	1.8
Toilet rooms	2.0	-
Balconies	Same as rooms to which they give access but with a minimum of 4.0	1.5 per metre run concentrated at the outer edge
Fly galleries	4.5 kN per metre run distributed uniformly over the width	
Cat walks	-	1.0 at 1 m centres
Residential occupancy class		
Self contained dwelling units		
All	1.5	1.4
Apartment houses, boarding houses, lodging houses, guest houses, hostels, residential clubs and communal areas in blocks of flats		
Boiler rooms, motor rooms, fan rooms and the like including the weight of machinery	7.5	4.5
Communal kitchens, laundries	3.0	4.5
Dining rooms, lounges, billiard rooms	2.0	2.7
Toilet rooms	2.0	-
Bedrooms, dormitories	1.5	1.8
Corridors, hallways, stairs, landings, footbridges, etc.	3.0	4.5
Balconies	Same as rooms to which they give access but with a minimum of 3.0	1.5 per metre run concentrated at the outer edge
Cat walks	-	1.0 at 1 m centres

DESIGN LOADINGS FOR BUILDINGS

Floor usage area	Distributed load kN/m2	Concentrated load kN/300 mm2
Hotels and Motels		
Boiler rooms, motor rooms, fan rooms and the like, including the weight of machinery	7.5	4.5
Assembly areas without fixed seating, dance halls	5.0	3.6
Bars	5.0	-
Assembly areas with fixed seating	4.0	-
Corridors, hallways, stairs, landings, footbridges, etc.	4.0	4.5
Kitchens, laundries	3.0	4.5
Dining rooms, lounges, billiard rooms	2.0	2.7
Bedrooms	2.0	1.8
Toilet rooms	2.0	-
Balconies	Same as rooms to which they give access but with a minimum of 4.0	1.5 per metre run concentrated at the outer edge
Cat Walks	-	1.0 at 1 m centres
Retail occupancy class (shops, departmental stores, supermarkets)		
Cold storage	5.0 for each metre of storage height with a minimum of 15.0	9.0
Stationery stores	4.0 for each metre of storage height	9.0
Storage, other than types separately	2.4 for each metre of storage height	7.0
Boiler rooms, motor rooms, fan rooms and the like, including the weight of machinery	7.5	4.5
Corridors, hallways, etc. subject to loads greater than from crowds, such as wheeled vehicles, trolleys and the like	5.0	4.5

DESIGN LOADINGS FOR BUILDINGS

Floor area usage	Distribution load kN/m2	Concentrated load kN/300 mm2
Retail occupancy class (shops, departmental stores, supermarkets) - cont'd		
Corridors, hallways, stairs, landings, footbridges, etc.	4.0	4.5
Shop floors for the display and sale of merchandise	4.0	3.6
Kitchens, laundries	3.0	4.5
Toilet rooms	2.0	-
Balconies	Same as rooms to which they give access but with a minimum of 4.0	1.5 per metre run concentrated at the outer edge
Cat walks	-	1.0 at 1 m centres
Storage occupancy class (warehouses)		
Cold storage	5.0 for each metre of storage height with a mimimum of 15.0	9.0
Dense mobile stacking (books) on mobile trucks	4.8 for each metre of storage height with a minimum of 15.0	7.0
Paper storage, for printing plants	4.0 for each metre of storage height	9.0
Stationery stores	4.0 for each metre of storage height	9.0
Storage, other than types listed separately, warehouses	2.4 for each metre of storage height	7.0
Motor rooms, fan rooms and the like, including the weight of machinery	7.5	4.5
Corridors, hallways, footbridges, etc. subject to loads greater than for crowds, such as wheeled vehicles, trolleys and the like	5.0	4.5
Cat walks	-	1.0 at 1 m centres

DESIGN LOADINGS FOR BUILDINGS

Floor area usage	Distribution load kN/m2	Concentrated load kN/300 mm2
Vehicular occupancy class (garages, car parks, vehicle access ramps)		
Motor rooms, fan rooms and the like, including the weight of machinery	7.5	4.5
Driveways and vehicle ramps, other than in garages for the parking only of passenger vehicles and light vans not exceeding 2500 kg gross mass	5.0	9.0
Repair workshops for all types of vehicles, parking for vehicles exceeding 2500 kg gross mass including driveways and ramps	5.0	9.0
Footpaths, terraces and plazas leading from ground level with no obstruction to vehicular traffic, pavement lights	5.0	9.0
Corridors, hallways, stairs, landings, footbridges, etc. subject to crowd loading	4.0	4.5
Footpaths, terraces and plazas leading from ground level but restricted to pedestrian traffic only	4.0	4.5
Car parking only, for passenger vehicles and light vans not exceeding 2500 kg gross mass including garages, driveways and ramps.	2.5	9.0
Cat walks	-	1.0 at 1 m centres

PLANNING PARAMETERS

Functional unit areas

As a 'rule of thumb' guide to establish a cost per functional unit, or as a check on economy of design in terms of floor area, the following indicative functional unit areas have been derived from historical data:-

Car parking	- surface	19 m2/car
	- multi-storey	25 m2/car
	- basement	27 m2/car
Concert Halls		8 m2/seat
Halls of residence	- college/polytechnic	25-35 m2/bedroom
	- university	30-50 m2/bedroom
Hospitals	- district general	65-85 m2/bed
	- teaching	120 + m2/bed
	- private	m2/bed
Hotels	- economy	40 m2/bedroom
	- five star	60 + m2/bedroom

Housing

		Gross floor area
Private developer:	1 Bedroom Flat	45 - 40 m2
	2 Bedroom Flat	55 - 65 m2
	2 Bedroom House	55 - 65 m2
	3 Bedroom House	70 - 90 m2
	4 Bedroom House	90 - 100 m2

Offices	- high density open plan	10 m2/person to
	- low density cellular	20 m2/person
Schools	- nursery	3 - 5 m2/child
	- secondary	6 - 8 m2/child
	- boarding	10 -12 m2/child
Theatres	- small, local	3 m2/seat to
	- large, prestige	7 m2/seat

PLANNING PARAMETERS

Recommended sizes of various sports facilities

Archery (Clout)	:	7.3 m firing area Range 109.728 (Women) 146.304 (Men) 182.88 (Normal range)
Baseball	:	Overall 60 m x 70 m
Basketball	:	14.0 m x 26.0 m
Camogie	:	91 - 110 m x 54 - 68 m
Discus and Hammer	:	Safety cage 2.74 m square Landing area 45 arc (65^ safety) 70 m radius
Football, American	:	Pitch 109.8 m x 48.8 m Overall 118.94 m x 57.94 m
Football, Association	:	NPFA rules Senior pitches 96 - 100 m x 60 - 64 m Junior pitches 90 m x 46 - 55 m International 100 - 110 m x 64 - 75 m
Football, Australian	:	Overall 135 - 185 m x 110 - 155 m
Football, Canadian	:	Overall 145.74 m x 59.47 m
Football, Gaelic	:	128 - 146.4 m x 76.8 - 91.50 m
Football, Rugby League	:	111 - 122 m x 68.0 m
Football, Rugby Union	:	144.00 m max x 69.0 m
Handball	:	91 - 110 m x 55 - 65 m
Hockey	:	91.5 m x 54.9 m
Hurling	:	137 m x 82 m
Javelin	:	Runway 36.5 m x 4.270 m Landing area 80 - 95 m long 48 m wide
Jump, High	:	Running area 38.8 m x 19 m Landing area 5 m x 4 m
Jump, Long	:	Runway 45 m x 1.220 m Landing area 9 m x 2.750 m
Jump, Triple	:	Runway 45 m x 1.22 m Landing area 7.3 m x 2.75 m
Korfball	:	90 m x 40 m
Lacrosse	:	(Mens) 100 m x 55 m (Womens) 110 m x 73 m
Netball		15.250 m x 30.480 m

PLANNING PARAMETERS

Recommended sizes of various sports facilities - cont'd

Pole Vault	:	Runway 45 m x 1.220 m Landing area 5 m x 5 m
Polo	:	275 m x 183 m
Rounders	:	Overall 19 m x 17 m
400m Running Track	:	115.61 m bend length x 2 84.39 m straight length x 2 Overall 176.91 m long x 92.52 m wide
Shot Putt	:	Base 2.135 m dia Landing area 65^ arc 25 m radius from base
Shinty	:	128 - 183 m x 64 - 91.5 m
Tennis	:	Court 23.77 m x 10.97 m Overall minimum 36.27 m x 18.29 m
Tug-of-war	:	46 m x 5 m

SOUND INSULATION

Sound reduction requirements as Building Regulations (E1/2/3)

The Building Regulations on airborne and impact sound (E1/2/3) state simply that both airborne and impact sound must be reasonably reduced in floors and walls. No minimum reduction is given but the following tables give example sound reductions for various types of constructions.

Sound reductions of typical walls

	Average sound reduction (dB)
13 mm Fibreboard	20
16 mm Plasterboard	25
6 mm Float glass	30
16 mm Plasterboard, plastered both sides	35
75 mm Plastered concrete blockwork (100 mm)	44
110 mm half brick wall, half brick thick, plastered both sides	43
220 mm Brick wall one brick thick, plastered both sides	48
Timber stud partitioning with plastered metal lathing both sides	35
Cupboards used as partitions	30
Cavity block wall, plastered both sides	42
75 mm Breeze block cavity wall, plastered both sides	50

SOUND INSULATION   **Average sound reductions (dB)**

100 mm Breeze block cavity wall, plastered both   55
sides including 50 mm air-gap and plasterboard
suspended ceiling

As above with 150 mm Breeze blocks   65

19 mm T & G boarding on timber joists including   32
plasterboard ceiling and plaster skim coat

As above including metal lash/and plaster ceiling   37

As above with solid sound proofing material between   55
joists approx 98 kg per sq metre

As above with floating floor of T & G boarding on   75
batten and soundproofing quilt

Impact noise is particularly difficult to reduce satisfactorily.
The following are the most efficient methods of reducing such sound.

1) Carpet on underlay of rubber or felt.
2) Pugging between joists (e.g. Slag Wool) and
3) A good suspended ceiling system.

THERMAL INSULATION

Thermal properties of various building elements

Thickness	Material	(m2k/W) R	(W/m2K) U - Value
-	Internal and external surface resistance	0.18	-
	Air-gap cavity	0.18	-
103 mm	Brick skin	0.12	-
	Dense concrete block		
100 mm	Arc conbloc	0.09	11.11
140 mm	Arc conbloc	0.13	7.69
190 mm	Arc conbloc	0.18	5.56
	Lightweight aggregate block		
100 mm	Celcon standard	0.59	1.69
125 mm	Celcon standard	0.74	1.35
150 mm	Celcon standard	0.88	1.14
200 mm	Celcon standard	1.18	0.85
	Lightweight aggregate thermal block		
125 mm	Celcon solar	1.14	0.88
150 mm	Celcon solar	1.36	0.74
200 mm	Celcon solar	1.82	0.55
	Insulating board		
25 mm	Drithern	0.69	1.45
50 mm	Drithern	1.39	0.72
75 mm	Drithern	2.08	0.48
13 mm	Lightweight plaster 'Carlite'	0.07	14.29
13 mm	Dense plaster 'Thistle'	0.02	50.00

THERMAL INSULATION

Thermal properties of various building elements - cont'd

Thickness	Material	(m2K/W) R	(W/m2k) U - Value
	Plasterboard		
9.5 mm	British gypsum	0.06	16.67
12.7 mm	British gypsum	0.08	12.50
40 mm	Screed	0.10	10.00
150 mm	Reinforced concrete	0.12	8.33
100 mm	Dow roofmate insultation	3.57	0.28

RESISTANCE TO THE PASSAGE OF HEAT

Provisions meeting the requirement set out in the Building Regulations (L2/3):-

a) **Dwellings**   **Minimum U - Value**

Roof	0.35
Exposed wall	0.60
Exposed floor	0.60

b) **Residential, Offices, Shops and Assembly Buildings**

Roof	0.06
Exposed wall	0.60
Exposed floor	0.60

c) **Industrial Storage and Other Buildings**

Roof	0.70
Exposed wall	0.70
Exposed floor	0.70

TYPICAL CONSTRUCTIONS MEETING THERMAL REQUIREMENTS

External wall, masonry construction:-

Concrete blockwork   **U - Value**

200 mm lightweight concrete block, 25 mm air-gap, 10 mm
plasterboard   0.68

200 mm lightweight concrete block, 20 mm EPS slab, 10 mm
plasterboard   0.54

200 mm lightweight concrete block, 25 mm air-gap, 25 mm
EPS slab, 10 mm plasterboard   0.46

Brick/Cavity/Brick

105 mm brickwork, 50 mm UF foam, 105 mm brickwork, 3 mm
lightweight plaster   0.55

TYPICAL CONSTRUCTIONS MEETING THERMAL REQUIREMENTS

	U - Value
105 mm brickwork, 50 mm cavity, 125 mm Thermalite block, 3 mm lightweight plaster	0.59
105 mm brickwork, 50 mm cavity, 130 mm Thermalite block, 3 mm lightweight plaster	0.57
105 mm brickwork, 50 mm cavity, 130 mm Thermalite block, 3 mm dense plaster	0.59
105 mm brickwork, 50 mm cavity, 100 mm Thermalite block, foilbacked plasterboard	0.55
105 mm brickwork, 50 mm cavity, 115 mm Thermalite block, 9.5 mm plasterboard	0.58
105 mm brickwork, 50 mm cavity, 115 mm Thermalite block, foilbacked plasterboard	0.52
105 mm brickwork, 50 mm cavity, 125 mm Theramlite block, 9.5 mm plasterboard	0.55
105 mm brickwork, 50 mm cavity, 100 mm Thermalite block, 25 mm insulating plasterboard	0.53
105 mm brickwork, 50 mm cavity, 125 mm Thermalite block, 25 mm insulating plasterboard	0.47
105 mm brickwork, 25 mm cavity, 25 mm insulation, 100 mm Thermalite block, lightweight plaster	0.47
105 mm brickwork, 25 mm cavity, 25 mm insulation, 115 mm Thermalite block, lightweight plaster	0.44

Block/Cavity/Block

Render, 100 mm 'SHIELD' block, 50 mm cavity, 100 mm Thermalite block, lightweight plaster	0.50
Render, 100 mm 'SHIELD' block, 50 mm cavity, 115 mm Thermalite block, lightweight plaster	0.47
Render, 100 mm 'SHIELD' block, 50 mm cavity, 125 mm Thermalite block, lightweight plaster	0.45

Tile hanging

10 mm tile on battens and felt, 150 mm Thermalite block, lightweight plaster	0.57
25 mm insulating plasterboard	0.46
10 mm tile on battens and felt, 190 mm Thermalite block, lightweight plaster	0.47
25 mm insulating plasterboard	0.40
10 mm tile on battens and felt, 200 mm Thermalite block, lightweight plaster	0.45

TYPICAL CONSTRUCTIONS MEETING THERMAL REQUIREMENTS

	U - Value
External walls, masonry construction:- - cont'd	
25 mm insulated plasterboard	0.38
10 mm tile on battens, breather paper, 25 mm air-gap, 50 mm glass fibre quilts, 10 mm plasterboard	0.56
10 mm tile on battens, breather paper, 25 mm air-gap, 75 mm glass fibre quilts, 10 mm plasterboard	0.41
10 mm tile on battens, breather paper, 25 mm air-gap, 100 mm glass fibre quilts, 10 mm plasterboard	0.33

Pitched roofs

Slate or concrete tiles, felt, airspace, 'Rockwool' flexible slabs laid between rafters, plasterboard

Slab	40 mm thick	0.62
	50 mm thick	0.52
	60 mm thick	0.45
	75 mm thick	0.38
	100 mm thick	0.29

Concrete tiles, sarking felt, rollbatts between joists, plasterboard

Insulation	100 mm thick	0.31
	120 mm thick	0.26
	140 mm thick	0.23
	160 mm thick	0.21

Steel frame 'Rockwool' insulation sandwiched between steel exterior profiled sheeting and interior sheet lining

Insulation	60 mm thick	0.53
	80 mm thick	0.41
	100 mm thick	0.34

Steel frame, steel profiled sheeting, 'Rockwool' insulation over purlins and plasterboard lining

Insulation	60 mm thick	0.51
	80 mm thick	0.38
	100 mm thick	0.32
	120 mm thick	0.27
	140 mm thick	0.24
	160 mm thick	0.21

Flat roofs

Asphalt, 'Rockwool' roof slabs, 25 mm timber boarding, timber joists and 9.5 mm plasterboard

Insulation	30 mm thick	0.68
	40 mm thick	0.57
	50 mm thick	0.49
	60 mm thick	0.44
	70 mm thick	0.39
	80 mm thick	0.35
	90 mm thick	0.32
	100 mm thick	0.29

TYPICAL CONSTRUCTIONS MEETING THERMAL REQUIREMENTS

U - Value

Asphalt, 'Rockwool' roof slabs on 150 mm dense
concrete deck and screed with 16 mm plaster finish

	Insulation	40 mm thick	0.68
		50 mm thick	0.57
		60 mm thick	0.49
		70 mm thick	0.43
		80 mm thick	0.39
		90 mm thick	0.35
		100 mm thick	0.32

Asphalt, 'Rockwool' roof slabs on 150 mm dense
concrete deck and screed with suspended
plasterboard ceiling

	Insulation	40 mm thick	0.60
		50 mm thick	0.52
		60 mm thick	0.45
		70 mm thick	0.40
		80 mm thick	0.36
		90 mm thick	0.33
		100 mm thick	0.30

Steel frame, asphalt on insulation slabs on troughed
steel decking

	Insulation	50 mm thick	0.59
		60 mm thick	0.51
		70 mm thick	0.45
		80 mm thick	0.39
		90 mm thick	0.35
		100 mm thick	0.33

Steel frame, asphalt on insulation slabs on troughed
steel decking including suspended plasterboard
ceiling

	Insulation	40 mm thick	0.67
		50 mm thick	0.57
		60 mm thick	0.49
		70 mm thick	0.43
		80 mm thick	0.38
		90 mm thick	0.34
		100 mm thick	0.32

WEIGHTS OF VARIOUS MATERIALS

AGGREGATES	lbs/ft3	lbs/yd3	kg/m3
Ashes	50	1350	800
Cement (Portland)	90	2430	1441
Chalk	140	3780	2240
Chippings (stone)	110	2970	1762
Clinker (furnace)	50	1350	800
(concrete)	90	2430	1441
Ballast or stone	140	3780	2241
Pumice	70	1890	1121
Gravel	110	2970	1762
Lime:			
Chalk (lump)	44	1188	704
Ground	60	1620	961
Quick	55	1485	880
Sand:			
Dry	100	2700	1601
Wet	88	2376	1281
Water	62	1674	933
Shale	125	3371	2000
Whinstone	125	3371	2000
Broken stone	107	2881	1709
Pitch	72	1944	1152

METALS			
Aluminium	162	4374	2559
Brass	525	14175	8129
Bronze	524	14148	8113
Gunmetal	528	14256	8475
Iron:			
Cast	450	12150	7207
Wrought	480	12960	7687
Lead	708	19116	11260
Tin	465	12555	7448
Zinc	466	12582	7464

STONE AND BRICKWORK			
Blockwork:			
Aerated	41	1095	650
Dense concrete	112	3034	1800
Lightweight concrete	75	2023	1200
Pumice concrete	67	1820	1080
Brickwork:			
Common Fletton	125	3375	1822
Glazed brick	130	3510	2080
Staffordshire Blue	135	3645	2162
Red Engineering	140	3780	2240
Concrete	115	3105	1841
Stone:			
Artificial	140	3780	2242
Bath	140	3780	2242
Blue Pennant	168	4536	2682
Cragleith	145	3915	2322
Darley Dale	148	3996	2370
Forest of Dean	149	4023	2386
Granite	166	4482	2642
Marble	170	4590	2742
Portland	135	3645	2170
Slate	180	4860	2882
York	150	4050	2402
Terra-cotta	132	3564	2116

WEIGHTS OF VARIOUS MATERIALS

WOOD	lbs/ft3	lbs/yd3	kg/m3
Blockboard	31 - 44	843 - 1180	500 - 700
Cork Bark	5	135	80
Hardboard:			
Standard	59 - 62	1584 - 1686	940 - 1000
Tempered	59 - 66	1584 - 1787	940 - 1060
Wood chipboard:			
Type I	41 - 47	1096 - 1264	650 - 750
Type II	42 - 50	1146 - 1349	680 - 800
Type III	41 - 50	1096 - 1349	650 - 800
Type II/III	42 - 50	1146 - 1349	680 - 800
Laminboard	31 - 44	843 - 1180	500 - 700
Timber:			
Ash	50	1350	800
Baltic spruce	30	810	480
Beech	51	1377	816
Birch	45	1215	720
Box	60	1620	961
Cedar	30	810	480
Chestnut	40	1080	640
Ebony	76	2052	1217
Elm	39	1053	624
Greenheart	60	1620	961
Jarrah	51	1377	816
Maple	47	1269	752
Mahogany:			
Honduras	36	972	576
Spanish	66	1782	1057
Oak:			
English	53	1431	848
American	45	1215	720
Austrian & Turkish	44	1188	704
Pine:			
Pitchpine	50	1350	800
Red Deal	36	972	576
Yellow Deal	33	891	528
Spruce	31	837	496
Sycamore	38	1026	530
Teak:			
African	60	1620	961
Indian	41	1107	656
Moulmein	46	1242	736
Walnut:			
English	41	1107	496
Black	45	1215	720

EXCAVATION AND EARTHWORK

Transport capacities

Type of vehicle	Capacity of vehicle	
	cu yards (solid)	cu metres (solid)
Standard wheelbarrow	0.10	0.08
2 ton truck (2.03 t)	1.50	1.15
3 ton truck (3.05 t)	2.25	1.72
4 ton truck (4.06 t)	2.90	2.22
5 ton truck (5.08 t)	3.50	2.68
6 ton truck (6.10 t)	4.50	3.44
2 cu yard dumper (1.53 m3)	1.50	1.15
3 cu yard dumper (2.29 m3)	2.25	1.72
6 cu yard dumper (4.59 m3)	4.50	3.44
10 cu yard dumper (7.65 m3)	7.50	5.73

Planking and strutting

Maximum depth of excavation in various soils without the use of earthwork support

Ground conditions	Feet (ft)	Metres (m)
Compact soil	12	3.66
Drained loam	6	1.83
Dry sand	1	0.30
Gravelly earth	2	0.61
Ordinary earth	3	0.91
Stiff clay	10	3.05

It is important to note that the above table should only be used as a guide. Each case must be taken on its merits and, as the limited distances, given above, are approached, careful watch must be kept for the slightest signs of caving in.

Baulkage of soils after excavation

	Approximate bulk of 1 m3 after excavation
Vegetable soil and loam	25 - 30%
Soft clay	30 - 40%
Stiff clay	10 - 15%
Gravel	20 - 25%
Sand	40 - 50%
Chalk	40 - 50%
Rock, weathered	30 - 40%
Rock, unweathered	50 - 60%

CONCRETE WORK

Approximate average weights of materials

Materials	Percentage of voids (%)	Weight per m3 (kg)
Sand	39	1660
Gravel 10 - 20 mm	45	1440
Gravel 35 - 75 mm	42	1555
Crushed stone	50	1330
Crushed granite		
(over 15 mm)	50	1345
(n.e. 15 mm)	47	1440
'All-in' ballast	32	1800

CONCRETE WORK

Common mixes for various types of work per m3

Recommended mix	Class of work suitable for:-	Cement (kg)	Sand (kg) (kg)	Course Aggregate	No. of 50 kg bags of cement per m3 of combined aggregate
1:3:6	Roughest type of mass concrete such as footings, road haunching over 300 mm thick.	208	905	1509	4.00
1:2.5:5	Mass concrete of better class than 1:3:6 such as bases for machinery, walls below ground etc.	249	881	1474	5.00
1:2:4	Most ordinary uses of concrete, such as mass walls above ground, road slabs etc. and general reinforced concrete work.	304	889	1431	6.00
1:1.5:3	Watertight floors, pavements and walls, tanks, pits, steps, paths, surface of 2 course roads, reinforced concrete where extra strength is required.	371	801	1336	7.50
1:1:2	Work of thin section such as fence posts and small precast work.	511	720	1206	10.50

Bar reinforcement

Cross-sectional area and mass

Nominal sizes (m)		Cross-sectional area (mm2)	Mass per metre run (kg)
6*	28.3	0.222	
8	50.3	0.395	
10	78.5	0.616	
12	113.1	0.888	
16	201.1	1.579	
20	314.2	2.466	
25	490.9	3.854	
32	804.2	6.313	
40	1256.6	9.864	
50*	1963.5	15.413	

*Where a bar larger than 40 mm is to be used the recommended size is 50 mm. Where a bar smaller than 8 mm is to be used the recommended size is 6 mm.

CONCRETE WORK

Fabric reinforcement

Preferred range of designated fabric types and stock sheet sizes

Fabric reference	Longitudinal wires			Cross wires			Mass
	Nominal wire size (mm)	Pitch (mm)	Area (mm2/m)	Nominal wire size (mm)	Pitch (mm)	Area (mm2/m)	(kg/m2/m)
Square mesh							
A393	10	200	393	10	200	393	6.16
A252	8	200	252	8	200	252	3.95
A193	7	200	193	7	200	193	3.02
A142	6	200	142	6	200	142	2.22
A98	5	200	98	5	200	98	1.54
Structural mesh							
B1131	12	100	1131	8	200	252	10.90
B785	10	100	785	8	200	252	8.14
B503	8	100	503	8	200	252	5.93
B385	7	100	385	7	200	193	4.53
B283	6	100	283	7	200	193	3.73
B196	5	100	196	7	200	193	3.05
Long mesh							
C785	10	100	785	6	400	70.8	6.72
C636	9	100	636	6	400	70.8	5.55
C503	8	100	503	5	400	49.0	4.34
C385	7	100	385	5	400	49.0	3.41
C283	6	100	283	5	400	49.0	2.61
Wrapping mesh							
D98	5	200	98	5	200	98	1.54
D49	2.5	100	49	2.5	100	49	0.77

	Longitudinal wires	Cross wires	Sheet area
Stock sheet size	Length 4.8 m	Width 2.4 m	11.52 m2

CONCRETE WORK

Average weight kg/m3 of steelwork reinforcement in concrete for various building elements

Substructure	kg/m3 concrete
Pile caps	110 - 150
Tie beams	130 - 170
Ground beams	230 - 330
Bases	90 - 130
Footings	70 - 110
Retaining walls	110 - 150

Superstructure	
Slabs - one way	75 - 125
Slabs - two way	65 - 135
Plate slab	95 - 135
Cant slab	90 - 130
Ribbed floors	80 - 120
Columns	200 - 300
Beams	250 - 350
Stairs	130 - 170
Walls - normal	30 - 70
Walls - wind	50 - 90

Note: For exposed elements add the following % :

Walls 50%, Beams 100%, Columns 15%

BRICKWORK AND BLOCKWORK

Number of bricks required for various types of work

Description	Brick size	
	215 x 103.5 x 50 mm	215 x 103.5 x 65 mm
Half brick thick		
Stretcher bond	72	58
English bond	108	86
English garden wall bond	90	72
Flemish bond	96	79
Flemish garden wall bond	83	66
One brick thick and cavity wall of two half brick skins		
Stretcher bond	144	116

BRICKWORK AND BLOCKWORK

Quantities of bricks and mortar required per m2 of walling

Description	Unit	No of bricks required	Mortar required (cubic metres)		
			No frogs	Single frogs	Double frogs
Standard bricks					
Brick size 215 x 102.5 x 50 mm					
half brick wall (103 mm)	m2	72	0.022	0.027	0.032
2 x half brick cavity wall (270 mm)	m2	144	0.044	0.054	0.064
one brick wall (215 mm)	m2	144	0.052	0.064	0.076
one and a half brick wall (322 mm)	m2	216	0.073	0.091	0.108
Mass brickwork	m3	576	0.347	0.413	0.480
Brick size 215 x 102.5 x 65 mm					
half brick wall (103 mm)	m2	58	0.019	0.022	0.026
2 x half brick cavity wall (270 mm)	m2	116	0.038	0.045	0.055
one brick wall (215 mm)	m2	116	0.046	0.055	0.064
one and a half brick wall (322 mm)	m2	174	0.063	0.074	0.088
Mass brickwork	m3	464	0.307	0.360	0.413
Metric modular bricks			**Perforated**		
Brick size 200 x 100 x 75 mm					
90 mm thick	m2	67	0.016	0.019	
190 mm thick	m2	133	0.042	0.048	
290 mm thick	m2	200	0.068	0.078	
Brick size 200 x 100 x 100 mm					
90 mm thick	m2	50	0.013	0.016	
190 mm thick	m2	100	0.036	0.041	
290 mm thick	m2	150	0.059	0.067	
Brick size 300 x 100 x 75 mm					
90 mm thick	m2	33	-	0.015	
Brick size 300 x 100 x 100 mm					
90 mm thick	m2	44	0.015	0.018	

Note: Assuming 10 mm deep joints.

BRICKWORK AND BLOCKWORK

Concrete blockwork

Standard available block sizes

Block	Length x height		
	Co-ordinating size	Work size	Thicknesses (work size)
A	400 x 100 400 x 200	390 x 90 440 x 190	(75, 90, 100, (140 & 190 mm
	450 x 225	440 x 215	(75, 90, 100 (140, 190, & 215 mm
B	400 x 100 400 x 200	390 x 90 390 x 190	(75, 90, 100 (140 & 190 mm
	450 x 200 450 x 225 450 x 300 600 x 200 600 x 225	440 x 190 440 x 215 440 x 290 590 x 190 590 x 215	((75, 90, 100 (140, 190, & 215 mm ((
C	400 x 200 450 x 200 450 x 225 450 x 300 600 x 200 600 x 225	390 x 190 440 x 190 440 x 215 440 x 290 590 x 190 590 x 215	(((60 & 75 mm (((

Mortar required per m2 blockwork (9.88 blocks/m2)

Wall thickness	75	90	100	125	140	190	215
Mortar m3/m2	0.005	0.006	0.007	0.008	0.009	0.013	0.014

ROOFING

Total roof loadings for various types of tiles/slates

	Roof load (slope) kg/m2		
	Slate/Tile	Roofing underlay and battens	Total dead load kg/m2
Asbestos cement slate (600 x 300 mm)	21.50	3.14	24.64
Clay tile			
interlocking	67.00	5.50	72.50
plain	43.50	2.87	46.37
Concrete tile			
interlocking	47.20	2.69	49.89
plain	78.20	5.50	83.70
Natural slate (18" x 10")	35.40	3.40	38.80

ROOFING

Total roof loadings for various types of tiles/slates - cont'd

Roof load (plan) kg/m2

	Dead load	Imposed load	Total roof load kg/m2
Asbestos cement slate (600 x 300 mm)	28.45	76.50	104.95
Clay tile			
interlocking	53.54	76.50	130.04
plain	83.71	76.50	160.21
Concrete tile			
interlocking	57.60	76.50	134.10
plain	96.64	76.50	173.14
Natural slate (18" x 10")	44.80	76.50	121.30

Tiling data

Product	Size of slates/tiles	Lap (mm)	Gauge of battens	No. slates per m2	Battens (m/m2)	Weight as laid (kg/m2)
CEMENT SLATES						
Eternit T.A.C. slates (Duracem)						
	600 x 300 mm	100	250	13.4	4.00	19.50
		90	255	13.1	3.92	19.20
		80	260	12.9	3.85	19.00
		70	265	12.7	3.77	18.60
	600 x 350 mm	100	250	11.5	4.00	19.50
		90	255	11.2	3.92	19.20
	500 x 250 mm	100	200	20.0	5.00	20.00
		90	205	19.5	4.88	19.50
		80	210	19.1	4.76	19.00
		70	215	18.6	4.65	18.60
	400 x 200 mm	90	155	32.3	6.45	20.80
		80	160	31.3	6.25	20.20
		70	165	30.3	6.06	19.60
CONCRETE TILES/SLATES						
Redland Roofing						
Stonewold slate	430 x 380 mm	75	355	8.2	2.82	51.20
Double Roman tile	418 x 330 mm	75	355	8.2	2.91	45.50
Grovebury pantile	418 x 332 mm	75	343	9.7	2.91	47.90
Norfolk pantile	381 x 227 mm	75	306	16.3	3.26	44.01
		100	281	17.8	3.56	48.06
Renown interlocking tile	418 x 330 mm	75	343	9.7	2.91	46.40
'49' tile	381 x 227 mm	75	306	16.3	3.26	44.80
		100	281	17.8	3.56	48.95
Plain, vertical tiling	265 x 165 mm	35	115	52.7	8.70	62.20

ROOFING

Product	Size of slate	Lap (mm)	Gauge of battens	No. slates per m2	Battens (m/m2)	Weight as laid (kg/m2)

Marley Roofing

Product	Size of slate	Lap (mm)	Gauge of battens	No. slates per m2	Battens (m/m2)	Weight as laid (kg/m2)
Bold roll tile	419 x 330 mm	75	344	9.7	2.90	47.00
		100	-	10.5	3.20	51.00
Modern roof tile	413 x 330 mm	75	338	10.2	3.00	54.00
		100	-	11.0	3.20	58.00
Ludlow major	413 x 330 mm	75	338	10.2	3.00	45.00
		100	-	11.0	3.20	49.00
Ludlow plus	380 x 230 mm	75	305	16.1	3.30	47.00
		100	-	17.5	3.60	51.00
Mendip tile	413 x 330 mm	75	338	10.2	3.00	47.00
		100	-	11.0	3.20	51.00
Wessex	413 x 330 mm	75	338	10.2	3.00	54.00
		100	-	11.0	3.20	58.00
Plain tile	265 x 165 mm	65	100	60.0	10.00	76.00
		75	95	64.0	10.50	81.00
		85	90	68.0	11.30	86.00
Plain vertical tiles (feature)	265 x 165 mm	35	110	53.0	8.70	67.00
		34	115	56.0	9.10	71.00

CLAY TILES

Product	Size of slate	Lap (mm)	Gauge of battens	No. slates per m2	Battens (m/m2)	Weight as laid (kg/m2)
Crossley Ltd Old English pantile	342 x 241 mm	75	267	18.5	3.76	47.00
French pattern pantiles	342 x 241 mm	75	267	18.8	3.76	48.00
Bold roll Roman tiles	342 x 266 mm	75	267	17.8	3.76	50.00

Slate nails, quantity per kilogram

	Type			
	Plain wire	Galvanised wire	Copper nail	Zinc nail
28.5 mm long	325	305	325	415
34.4 mm long	286	256	254	292
50.8 mm long	242	224	194	200

Metal sheet coverings

Thicknesses and weights of sheet metal coverings

Lead to BS 1178

BS Code No	3 4	5 6	7	8		
Colour Code	Green	Blue	Red	Black	White	Orange
Thickness (mm)	1.25	1.80	2.24	2.50	3.15	3.55
kg/m2	14.18	20.41	25.40	28.36	35.72	40.26

ROOFING

Metal sheet coverings

Thicknesses and weights of sheet metal coverings

Copper to BS 2870

Thickness (mm)	0.60	0.70
Bay width		
Roll (mm)	500	650
Seam (mm)	525	600
Standard width to form bay	600	750
Normal length of sheet	1.80	1.80

kg/m2

Zinc to BS 849

Zinc Gauge (Nr)	9	10	11	12	13	14	15	16
Thickness (mm)	0.43	0.48	0.56	0.64	0.71	0.79	0.91	1.04
kg/m2	3.1	3.2	3.8	4.3	4.8	5.3	6.2	7.0

Aluminium to BS 4868

Thickness (mm)	0.5	0.6	0.7	0.8	0.9	1.0	1.2
kg/m2	12.8	15.4	17.9	20.5	23.0	25.6	30.7

Bitumen felt roofing

Masses per unit area of components and lengths of rolls

Type of felt	Nominal mass per unit area (kg/10m)	Nominal mass per unit area of fibre base (g/m2)	Nominal length of roll (m)
Class 1			
1B fine granule surfaced bitumen	14	220	10 or 20
	18	330	10 or 20
	25	470	10
1E mineral surfaced bitumen	38	470	10
1F reinforced bitumen	15	160 (fibre)	15
		110 (hessian)	
1F reinforced bitumen, aluminium faced	13	160 (fibre)	15
		110 (hessian)	

ROOFING

Type of felt	Nominal mass per unit area (kg/10m)	Nominal mass per unit area of fibre base (g/m2)	Nominal length of roll (m)
Class 2			
2B fine granule surfaced bitumen asbestos	18	500	10 or 20
2E mineral surfaced bitumen asbestos	38	600	10
Class 3			
3B fine granule surfaced bitumen glass fibre	18	60	20
3E mineral surfaced bitumen glass fibre	28	60	10
3E venting base layer bitumen glass fibre	32	60*	10
3H venting base layer bitumen glass fibre	17	60*	20

* Excluding effect of perforations

WOODWORK

Conversion tables (for timber only)

Inches	Millimetres	Feet	Metres
1	25	1	0.300
2	50	2	0.600
3	75	3	0.900
4	100	4	1.200
5	125	5	1.500
6	150	6	1.800
7	175	7	2.100
8	200	8	2.400
9	225	9	2.700
10	250	10	3.000
11	275	11	3.300
12	300	12	3.600
13	325	13	3.900
14	350	14	4.200
15	375	15	4.500
16	400	16	4.800
17	425	17	5.100
18	450	18	5.400
19	475	19	5.700
20	500	20	6.000
21	525	21	6.300
22	550	22	6.600
23	575	23	6.900
24	600	24	7.200

WOODWORK

Planed softwood

The finished end section size of planed timber is usually 3/16" less than the original size from which it is produced. This however varies slightly dependant upon availability of material and origin of the species used.

Standards (timber) to cubic metres and cubic metres to standards (timber)

Cubic metres	Cubic metres standards	Standards
4.672	1	0.214
9.344	2	0.428
14.017	3	0.642
18.689	4	0.856
23.361	5	1.070
28.033	6	1.284
32.706	7	1.498
37.378	8	1.712
42.050	9	1.926
46.722	10	2.140
93.445	20	4.281
140.167	30	6.421
186.890	40	8.561
233.612	50	10.702
280.335	60	12.842
327.057	70	14.982
373.779	80	17.122
420.502	90	19.263
467.224	100	21.403

Standards (timber) to cubic metres and cubic metres to standards (timber)

1 cu metre = 35.3148 cu ft = 0.21403 std

1 cu ft = 0.028317 cu metres

1 std = 4.67227 cu metres

WOODWORK

Basic sizes of sawn softwood available (cross sectional areas)

Thickness (mm)   Width (mm)

Thickness	75	100	125	150	175	200	225	250	300
16	X	X	X	X					
19	X	X	X	X					
22	X	X	X	X					
25	X	X	X	X	X	X	X	X	X
32	X	X	X	X	X	X	X	X	X
36	X	X	X	X					
38	X	X	X	X	X	X	X		
44	X	X	X	X	X	X	X	X	X
47*	X	X	X	X	X	X	X	X	X
50	X	X	X	X	X	X	X	X	X
63	X	X	X	X	X	X	X		
75		X	X	X	X	X	X	X	X
100		X		X		X		X	X
150				X		X			X
200						X			
250								X	
300									X

* This range of widths for 47 mm thickness will usually be found to be available in construction quality only.

Note: The smaller sizes below 100 mm thick and 250 mm width are normally but not exclusively of European origin. Sizes beyond this are usually of North and South American origin.

Basic lengths of sawn softwood available (metres)

1.80	2.10	3.00	4.20	5.10	6.00	7.20
	2.40	3.30	4.50	5.40	6.30	
	2.70	3.60	4.80	5.70	6.60	
		3.90	6.90			

Note: Lengths of 6.00 m and over will generally only be available from North American species and may have to be recut from larger sizes.

WOODWORK

Reductions from basic size to finished size by planing of two opposed faces - cont'd

Purpose	Reductions from basic sizes for timber			
	15 - 35 mm	36 - 100 mm	101 - 150 mm	over 150 mm
a) constructional timber	3 mm	3 mm	5 mm	6 mm
b) Matching interlocking boards	4 mm	4 mm	6 mm	6 mm
c) Wood trim not specified in BS 584	5 mm	7 mm	7 mm	9 mm
d) Joinery and cabinet work	7 mm	9 mm	11 mm	13 mm

Note: The reduction of width or depth is overall the extreme size and is exclusive of any reduction of the face by the machining of a tongue or lap joints.

Maximum spans for various roof trusses

Maximum permissible spans for rafters for Fink trussed rafters

Basic size (mm)	Actual size (mm)	Pitch (degrees)								
		15 (m)	17.5 (m)	20 (m)	22.5 (m)	25 (m)	27.5 (m)	30 (m)	32.5 (m)	35 (m)
38 x 75	35 x 72	6.03	6.16	6.29	6.41	6.51	6.60	6.70	6.80	6.90
38 x 100	35 x 97	7.48	7.67	7.83	7.97	8.10	8.22	8.34	8.47	8.61
38 x 125	35 x 120	8.80	9.00	9.20	9.37	9.54	9.68	9.82	9.98	10.16
44 x 75	41 x 72	6.45	6.59	6.71	6.83	6.93	7.03	7.14	7.24	7.35
44 x 100	41 x 97	8.05	8.23	8.40	8.55	8.68	8.81	8.93	9.09	9.22
44 x 125	41 x 120	9.38	9.60	9.81	9.99	10.15	10.31	10.45	10.64	10.81
50 x 75	47 x 72	6.87	7.01	7.13	7.25	7.35	7.45	7.53	7.67	7.78
50 x 100	47 x 97	8.62	8.80	8.97	9.12	9.25	9.38	9.50	9.66	9.80
50 x 125	47 x 120	10.01	10.24	10.44	10.62	10.77	10.94	11.00	11.00	11.00

Sizes of internal and external doorsets

Description	Internal		External	
	Size (mm)	Permissible deviation	Size (mm)	Permissible deviation
Co-ordinating dimension: height of door leaf height sets	2100		2100	
Co-ordinating dimension: height of ceiling height set	2300		2300	
	2350		2350	
	2400		2400	
	2700		2700	
	3000		3000	

WOODWORK

Description	Internal		External	
	Size (mm)	Permissible deviation	Size (mm)	Permissible deviation
Co-ordinating dimension: width of all door sets	600 S		900 S	
	700 S		1000 S	
S = Single leaf set	800 S&D		1200 D	
D = Double leaf set	900 S&D		1500 D	
	1000 S&D		1800 D	
	1200 D		2100 D	
	1500 D			
	1800 D			
	2100 D			
Work size: height of door leaf height set	2090	± 2.0	2095	± 2.0
Work size: height of ceiling height set	2285	)	2295	)
	2335	)	2345	)
	2385	) ± 2.0	2395	) ± 2.0
	2685	)	2695	)
	2985	)	2995	)
Work size: width of all door sets	590 S	)	895 S	)
	690 S	)	995 S	)
S = Single leaf set	790 S&D	)	1195 D	) ± 2.0
D = Double leaf set	890 S&D	)	1495 D	)
	990 S&D	) ± 2.0	1795 D	)
	1190 D	)	2095 D	)
	1490 D	)		
	1790 D	)		
	2090 D	)		
Width of door leaf in single leaf sets	526 F	)	806 F&P	)
F = Flush leaf	626 F	)	906 F&P	) ± 1.5
P = Panel leaf	726 F&P	) ± 1.5		
	826 F&P	)		
	926 F&P	)		
Width of door leaf in double leaf sets	362 F	)	552 F&P	)
F = Flush leaf	412 F	)	702 F&P	) ± 1.5
P = Panel leaf	426 F	)	852 F&P	)
	562 F&P	) ± 1.5	1002 F&P	)
	712 F&P	)		
	826 F&P	)		
	1012 F&P	)		
Door leaf height for all door sets	2040	± 1.5	1994	± 1.5

STRUCTURAL STEELWORK

Tables showing the mass and surface area per metre run for various steel members

Size (mm)	(kg/m)	Surface area per m2
Universal beams		
914 x 3419	388	3.404
	343	3.382
914 x 305	289	2.988
	253	2.967
	224	2.948
	201	2.932
838 x 292	226	2.791
	194	2.767
	176	2.754
762 x 267	197	2.530
	173	2.512
	147	2.493
686 x 254	170	2.333
	152	2.320
	140	2.310
	125	2.298
610 x 305	238	2.421
	179	2.381
	149	2.361
610 x 229	140	2.088
	125	2.075
	113	2.064
	101	2.053
533 x 210	122	1.872
	109	1.860
	101	1.853
	92	1.844
	82	1.833
457 x 191	98	1.650
	89	1.641
	82	1.633
	74	1.625
	67	1.617
457 x 152	82	1.493
	74	1.484
	67	1.474
	60	1.487
	52	1.476
406 x 178	74	1.493
	67	1.484
	60	1.476
	54	1.468
406 x 140	46	1.332
	39	1.320

STRUCTURAL STEELWORK

Size (mm)	(kg/m)	Surface area per m2
Universal beams		
356 x 171	67	1.371
	57	1.358
	51	1.351
	45	1.343
356 x 127	39	1.169
	33	1.160
305 x 165	54	1.245
	46	1.235
	40	1.227
305 x 127	48	1.079
	42	1.069
	37	1.062
305 x 102	33	1.006
	28	0.997
	25	0.988
254 x 146	43	1.069
	37	1.060
	31	1.050
254 x 102	28	0.900
	25	0.893
	22	0.887
203 x 133	30	0.912
	25	0.904
Universal columns		
356 x 406	634	2.525
	551	2.475
	467	2.425
	393	2.379
	340	2.346
	287	2.132
	235	2.279
356 x 368	202	2.187
	177	2.170
	153	2.154
	129	2.137
305 x 305	283	1.938
	240	1.905
	198	1.872
	158	1.839
	137	1.822
	118	1.806
	97	1.789
254 x 254	167	1.576
	132	1.543
	107	1.519
	89	1.502
	73	1.485

STRUCTURAL STEELWORK

Tables showing the mass and surface area per metre run for various steel members - cont'd

size (mm)	kg/m	Surface area per m2
203 x 203	86	1.236
	71	1.218
	60	1.204
	52	1.194
	46	1.187
152 x 152	37	0.912
	30	0.900
	23	0.889

Joists

size (mm)	kg/m	Surface area per m2
254 x 203	81.85	1.213
254 x 114	37.20	0.898
203 x 152	52.09	0.931
152 x 127	37.20	0.735
127 x 114	29.76	0.644
127 x 114	26.79	0.649
114 x 114	26.79	0.617
102 x 102	23.07	0.547
89 x 89	19.35	0.475
76 x 76	12.65	0.410

**Circular hollow
sections - outside dia (mm)**

outside dia (mm)	kg/m	Surface area per m2
21.3	1.43	0.067
26.9	1.43	0.085
33.7	1.87	0.106
	2.41	0.106
	2.93	0.106
42.4	2.55	0.133
	3.09	0.133
	3.79	0.133
48.3	3.56	0.152
	4.37	0.152
	5.34	0.152

STRUCTURAL STEELWORK

Outside diameter (mm)	kg/m	Surface area per m2
60.3	4.51	0.189
	5.55	0.189
	6.82	0.189
76.1	5.75	0.239
	7.11	0.239
	8.77	0.239
88.9	6.76	0.279
	8.38	0.279
	10.30	0.279
114.3	9.83	0.359
	13.50	0.359
	16.80	0.395
139.7	16.60	0.439
	20.70	0.439
	26.00	0.439
	32.00	0.439
168.3	20.10	0.529
	25.20	0.529
	31.60	0.529
	39.00	0.529
193.7	25.10	0.609
	29.10	0.609
	36.60	0.609
	45.30	0.609
	55.90	0.609
	70.10	0.609
219.1	33.10	0.688
	41.60	0.688
	51.60	0.688
	63.70	0.688
	80.10	0.688
	98.20	0.688
273.0	41.40	0.858
	52.30	0.858
	64.90	0.858
	80.30	0.858
	101.00	0.858
	125.00	0.858
	153.00	0.858
323.9	62.30	1.020
	77.40	1.020
	96.00	1.020
	121.00	1.020
	150.00	1.020
	184.00	1.020
406.4	97.80	1.280
	121.00	1.280
	154.00	1.280
	191.00	1.280
	235.00	1.280
	295.00	1.280

STRUCTURAL STEELWORK

Tables showing the mass and surface area per metre run for various steel
members - cont'd

Outside diameter (mm)	kg/m	Surface area per m2
457.0	110.0	1.440
	137.0	1.440
	174.0	1.440
	216.0	1.440
	266.0	1.440
	335.0	1.440
	411.0	1.440

**Square hollow
sections - size (mm)**

20 x 20	1.12	0.076
	1.39	0.074
30 x 30	2.21	0.114
	2.65	0.113
40 x 40	3.03	0.154
	3.66	0.153
	4.46	0.151
50 x 50	4.66	0.193
	5.72	0.191
	6.97	0.189
60 x 60	5.67	0.233
	6.97	0.231
	8.54	0.229
70 x 70	7.46	0.272
	10.10	0.269
80 x 80	8.59	0.312
	11.70	0.309
	14.40	0.306
90 x 90	9.72	0.352
	13.30	0.349
	16.40	0.346
100 x 100	12.00	0.391
	14.80	0.389
	18.40	0.386
	22.90	0.383
	27.90	0.379
120 x 120	18.00	0.469
	22.30	0.466
	27.90	0.463
	34.20	0.459

STRUCTURAL STEELWORK

Size (mm)	kg/m	Surface area per m2
150 x 150	22.70	0.589
	28.30	0.586
	35.40	0.583
	43.60	0.579
	53.40	0.573
	66.40	0.566
180 x 180	34.20	0.706
	43.00	0.703
	53.00	0.699
	65.20	0.693
	81.40	0.686
200 x 200	38.20	0.786
	48.00	0.783
	59.30	0.779
	73.00	0.773
	91.50	0.766
250 x 250	48.10	0.986
	60.50	0.983
	75.00	0.979
	92.60	0.973
	117.00	0.966
300 x 300	90.70	1.180
	112.00	1.170
	142.00	1.170
350 x 350	106.00	1.380
	132.00	1.370
	167.00	1.370
400 x 400	122.00	1.580
	152.00	1.570

**Rectangular
hollow sections**

Size (mm)	kg/m	Surface area per m2
50 x 30	3.03	0.154
	3.66	0.153
60 x 40	4.66	0.193
	5.72	0.191
80 x 40	5.67	0.232
	6.97	0.231
90 x 50	7.46	0.272
	10.10	0.269
100 x 50	7.18	0.293
	8.86	0.291
	10.90	0.289
100 x 60	8.59	0.312
	11.70	0.309
	14.40	0.306

STRUCTURAL STEELWORK

Tables showing the mass and surface area per metre run for various steel members - cont'd

Size (mm)	kg/m	Surface area per m2
120 x 60	9.72	0.352
	13.30	0.349
	16.40	0.346
120 x 80	14.80	0.389
	18.40	0.386
	22.90	0.383
	27.90	0.379
150 x 100	18.70	0.489
	23.30	0.486
	29.10	0.483
	35.70	0.479
160 x 80	18.00	0.469
	22.30	0.466
	27.90	0.463
	34.20	0.459
200 x 100	22.70	0.589
	28.30	0.586
	35.40	0.583
	43.60	0.579
	53.40	0.573
250 x 150	38.20	0.785
	48.00	0.783
	59.30	0.779
	73.00	0.773
	91.50	0.766
300 x 200	48.10	0.986
	60.50	0.983
	75.00	0.979
	92.60	0.973
	117.00	0.966
400 x 200	90.70	1.180
	112.00	1.170
	142.00	1.170
450 x 250	106.00	1.380
	132.00	1.370
	167.00	1.370

STRUCTURAL STEELWORK

Size (mm)		kg/m	Surface area per m2
Channels			
432 x 102		65.54	1.217
381 x 102		55.10	1.118
305 x 102		46.18	0.966
305 x 89		41.69	0.920
254 x 89		35.74	0.820
254 x 76		28.29	0.774
229 x 89		32.76	0.770
229 x 76		26.06	0.725
203 x 89		29.78	0.720
203 x 76		23.82	0.675
178 x 89		26.81	0.671
178 x 76		20.84	0.625
152 x 89		23.84	0.621
152 x 76		17.88	0.575
127 x 64		14.90	0.476

Angles - sum of leg lengths	Thickness (mm)		
50	3	1.11	0.10
	4	1.45	0.10
	5	1.77	0.10
80	4	2.42	0.16
	5	2.97	0.16
	6	3.52	0.16
90	4	2.74	0.18
	5	3.38	0.18
	6	4.00	0.18
100	5	3.77	0.20
	6	4.47	0.20
	8	5.82	0.20
115	5	4.35	0.23
	6	5.16	0.23
	8	6.75	0.23
120	5	4.57	0.24
	6	5.42	0.24
	8	7.09	0.24
	10	8.69	0.24
125	6	5.65	0.25
	8	7.39	0.25

STRUCTURAL STEELWORK

Tables showing the mass and surface area per metre run for various steel members - cont'd

Sum of leg lengths	Thickness (mm)	kg/m	Surface area per m2
200	8	12.20	0.40
	10	15.00	0.40
	12	17.80	0.40
	15	21.90	0.40
225	10	17.00	0.45
	12	20.20	0.45
	15	24.80	0.45
240	8	14.70	0.48
	10	18.20	0.48
	12	21.60	0.48
	15	26.60	0.48
300	10	23.00	0.60
	12	27.30	0.60
	15	33.80	0.60
	18	40.10	0.60
350	12	32.00	0.70
	15	39.60	0.70
	18	47.10	0.70
400	16	48.50	0.80
	18	54.20	0.80
	20	59.90	0.80
	24	71.10	0.80

PLUMBING AND MECHANICAL INSTALLATIONS

Dimensions and weights of tubes

Outside diameter (mm)	Internal dia (mm)	Weight per m (kg)	Internal dia (mm)	Weight per m (kg)	Internal dia (mm)	Weight per m (kg)
Copper to BS 2871 Part 1						
	Table X		Table Y		Table Z	
6	4.80	0.0911	4.40	0.1170	5.00	0.0774
8	6.80	0.1246	6.40	0.1617	7.00	0.1054
10	8.80	0.1580	8.40	0.2064	9.00	0.1334
12	10.80	0.1914	10.40	0.2511	11.00	0.1612
15	13.60	0.2796	13.00	0.3923	14.00	0.2031
18	16.40	0.3852	16.00	0.4760	16.80	0.2918
22	20.22	0.5308	19.62	0.6974	20.82	0.3589
28	26.22	0.6814	25.62	0.8985	26.82	0.4594
35	32.63	1.1334	32.03	1.4085	33.63	0.6701
42	39.63	1.3675	39.03	1.6996	40.43	0.9216
54	51.63	1.7691	50.03	2.9052	52.23	1.3343
76.1	73.22	3.1287	72.22	4.1437	73.82	2.5131
108	105.12	4.4666	103.12	7.3745	105.72	3.5834
133	130.38	5.5151	-	-	130.38	5.5151
159	155.38	8.7795	-	-	156.38	6.6056

PLUMBING AND MECHANICAL INSTALLATIONS

Dimensions and weights of tubes - cont'd

Nominal size	Outside diameter max	min	Wall thickness	Weight	Weight screwed and socketed
(mm)	(mm)	(mm)	(mm)	(kg/m)	kg/m

Steel pipes to BS 1387

Light gauge

6	10.1	9.7	1.80	0.361	0.364
8	13.6	13.2	1.80	0.517	0.521
10	17.1	16.7	1.80	0.674	0.680
15	21.4	21.0	2.00	0.952	0.961
20	26.9	26.4	2.35	1.410	1.420
25	33.8	33.2	2.65	2.010	2.030
32	42.5	41.9	2.65	2.580	2.610
40	48.4	47.8	2.90	3.250	3.290
50	60.2	59.6	2.90	4.110	4.180
65	76.0	75.2	3.25	5.800	5.920
80	88.7	87.9	3.25	6.810	6.980
100	113.9	113.0	3.65	9.890	10.200

Medium gauge

6	10.4	9.8	2.00	0.407	0.410
8	13.9	13.3	2.35	0.650	0.654
10	17.4	16.8	2.35	0.852	0.858
15	21.7	21.1	2.65	1.220	1.230
20	27.2	26.6	2.65	1.580	1.590
25	34.2	33.4	3.25	2.440	2.460
32	42.9	42.1	3.25	3.140	3.170
40	48.8	48.0	3.25	3.610	3.650
50	60.8	59.8	3.65	5.100	5.170
65	76.6	75.4	3.65	6.510	6.630
80	89.5	88.1	4.05	8.470	8.640

PLUMBING AND MECHANICAL INSTALLATIONS

Nominal size	Outside diameter max	min	Wall thickness	Weight	Weight screwed and socketed
(mm)	(mm)	(mm)	(mm)	(kg/m)	(kg/m)
100	114.9	113.3	4.50	12.100	12.400
125	140.6	138.7	4.85	16.200	16.700
150	166.1	164.1	4.85	19.200	19.800

Heavy gauge

6	10.4	9.8	2.65	0.493	0.496
8	13.9	13.3	2.90	0.769	0.773
10	17.4	16.8	2.90	1.020	1.030
15	21.7	21.1	3.25	1.450	1.460
20	27.2	26.6	3.25	1.900	1.910
25	34.2	33.4	4.05	2.970	2.990
32	42.9	42.1	4.05	3.840	3.870
40	48.8	48.0	4.05	4.430	4.470
50	60.8	59.8	4.50	6.170	6.240
65	76.6	75.4	4.50	7.900	8.020
80	89.5	88.1	4.85	10.100	10.300
100	114.9	113.3	5.40	14.400	14.700
125	140.6	138.7	5.40	17.800	18.300
150	166.1	164.1	5.40	21.200	21.800

Stainless steel pipes to BS 4127 Part 2

8	8.045	7.940	0.60	0.1120	
10	10.045	9.940	0.60	0.1419	
12	12.045	11.940	0.60	0.1718	
15	15.045	14.940	0.60	0.2174	
18	18.045	17.940	0.70	0.3046	
22	22.055	21.950	0.70	0.3748	
28	28.055	27.950	0.80	0.5469	
35	35.070	34.965	1.00	0.8342	

PLUMBING AND MECHANICAL INSTALLATIONS

Maximum distances between pipe supports

Pipe material	BS nominal pipe size inch	mm	Pipes fitted vertically support distances in metres	Pipes fitted horizontally onto low gradients support distances in metres
Copper	0.50	15.0	1.90	1.3
	0.75	22.0	2.50	1.9
	1.00	28.0	2.50	1.9
	1.25	35.0	2.80	2.5
	1.50	42.0	2.80	2.5
	2.00	54.0	3.90	2.5
	2.50	67.0	3.90	2.8
	3.00	76.1	3.90	2.8
	4.00	108.0	3.90	2.8
	5.00	133.0	3.90	2.8
	6.00	159.0	3.90	2.8
muPVC	1.25	32.0	1.20	0.5
	1.50	40.0	1.20	0.5
	2.00	50.0	1.20	0.6
Polypropylene	1.25	32.0	1.20	0.5
	1.50	40.0	1.20	0.5
uPVC	-	82.4	1.20	0.5
	-	110.0	1.80	0.9
	-	160.0	1.80	1.2

Litres of water storage required per person in various types of building

Type of building	Storage per person (litres)
Houses and flats	90
Hostels	90
Hotels	135
Nurse's home and medical quarters	115
Offices with canteens	45
Offices without canteens	35
Restaurants, per meal served	7
Boarding schools	90
Day schools	30

PLUMBING AND MECHANICAL INSTALLATIONS

Cold water plumbing - thickness of insulation required against frost

Bore of tube		Pipework within buildings declared thermal conductivity (W/m degrees C)		
		Up to 0.040	0.041 to 0.055	0.056 to 0.070
(mm)	(in)	Minimum thickness of insulation (mm)		
15		32	50	75
20		32	50	75
25		32	50	75
32		32	50	75
40		32	50	75
50		25	32	50
65		25	32	50
80		25	32	50
100		19	25	38

Cisterns

Capacities and dimensions of galvanised mild steel cisterns from BS 417

Capacity (litres)	BS type	Dimensions (mm)		
		length	width	depth
18	SCM 45	457	305	305
36	SCM 70	610	305	371
54	SCM 90	610	406	371
68	SCM 110	610	432	432
86	SCM 135	610	457	482
114	SCM 180	686	508	508
159	SCM 230	736	559	559
191	SCM 270	762	584	610
227	SCM 320	914	610	584
264	SCM 360	914	660	610
327	SCM 450/1	1220	610	610
336	SCM 450/2	965	686	686
423	SCM 570	965	762	787
491	SCM 680	1090	864	736
709	SCM 910	1170	889	889

Capacities of cold water polypropylene storage cisterns from BS 4213

Capacity (litres)	BS type	Maximum height (mm)
18	PC 4	310
36	PC 8	380
68	PC 15	430
91	PC 20	510
114	PC 25	530
182	PC 40	610
227	PC 50	660
273	PC 60	660
318	PC 70	660
455	PC 100	760

HEATING AND HOT WATER INSTALLATIONS

Storage capacity and recommended power of hot water storage boilers

Type of building	Storage at 65^C (litres per person)	Boiler power to 65^C (kW per person)
Flats and dwellings		
(a) Low rent properties	25	0.5
(b) Medium rent properties	30	0.7
(c) High rent properties	45	1.2
Nurses homes	45	0.9
Hostels	30	0.7
Hotels		
(a) Top quality - upmarket	45	1.2
(b) Average quality - low market	35	0.9
Colleges and schools		
(a) Live-in accommodation	25	0.7
(b) Public comprehensive	5	0.1
Factories	5	0.1
Hospitals		
(a) General	30	1.5
(b) Infectious	45	1.5
(c) Infirmaries	25	0.6
(d) Infirmaries (inc. laundry facilities)	30	0.9
(e) Maternity	30	2.1
(f) Mental	25	0.7
Offices	5	0.1
Sports pavilions	35	0.3

Thickness of thermal insulation for heating installations

Size of tube (mm)	Up to 0.025	Declared thermal conductivity		
		0.026 to 0.040	0.041 to 0.055	0.056 to 0.070
		Minimum thickness of insulation		
LTHW Systems				
15	25	25	38	38
20	25	32	38	38
25	25	38	38	38
32	32	38	38	50
40	32	38	38	50
50	38	38	50	50
65	38	50	50	50
80	38	50	50	50
100	38	50	50	63
125	38	50	50	63
150	50	50	63	63
200	50	50	63	63
250	50	63	63	63
300	50	63	63	63
Flat surfaces	50	63	63	63

HEATING AND HOT WATER INSTALLATIONS

Size of tube (mm)	Up to 0.025	0.026 to 0.040	0.041 to 0.055	0.056 to 0.070
MTHW Systems and condensate				
		Declared thermal conductivity		
15	25	38	38	38
20	32	38	38	50
25	38	38	38	50
32	38	50	50	50
40	38	50	50	50
50	38	50	50	50
65	38	50	50	50
80	50	50	50	63
100	50	63	63	63
125	50	63	63	63
150	50	63	63	63
200	50	63	63	63
250	50	63	63	75
300	63	63	63	75
Flat surfaces	63	63	63	75
HTHW Systems and steam				
15	38	50	50	50
20	38	50	50	50
25	38	50	50	50
32	50	50	50	63
40	50	50	50	63
50	50	50	75	75
65	50	63	75	75
80	50	63	75	75
100	63	63	75	100
125	63	63	100	100
150	63	63	100	100
200	63	63	100	100
250	63	75	100	100
300	63	75	100	100
Flat surfaces	63	75	100	100

HEATING AND HOT WATER INSTALLATIONS

Capacities and dimensions of copper indirect cylinders (coil type) from BS 1566

Capacity (litres)	BS Type	External diameter (mm)		External height over dome (mm)
96	0	300	1600	
72	1	350		900
96	2	400		900
114	3	400	1050	
84	4	450		675
95	5	450		750
106	6	450		825
117	7	450		900
140	8	450	1050	
162	9	450	1200	
206	9 E	450	1500	
190	10	500	1200	
245	11	500	1500	
280	12	600	1200	
360	13	600	1500	
440	14	600	1800	

		Dimensions (mm) Internal diameter		Height
109	BSG 1M	457		762
136	BSG 2M	457		914
159	BSG 3M	457	1067	
227	BSG 4M	508	1270	
273	BSG 5M	508	1473	
364	BSG 6M	610	1372	
455	BSG 7M	610	1753	
123	BSG 8M	457		838

Comparison of energy costs (January 1987)

Energy form	Calorific value per unit of supply			Average price	
	Average price	Gross (MJ)	Nett (MJ)	Gross (£/GJ)	Nett (£/GJ)
Electricity - direct	3.66p/kwh	3.6	3.6	10.17	10.17
- off peak	1.70p/kwh	3.6	3.6	4.72	4.72
Natural gas	36.2p/therm	105.5	95.2	3.43	3.80
Fuel oil - 35 seconds	15.63p/l	38.1	35.7	4.10	4.38
- 200 seconds	14.11p/l	40.4	38.2	3.49	3.69
- 950 seconds	12.04p/l	40.6	38.4	2.97	3.14
- 3500 seconds	11.03p/l	41.1	38.8	2.68	2.84
Propane	£226.00/tonne	50000	46300	4.52	4.88
Butane	£190.00/tonne	49300	45800	3.85	4.04
Coal - singles	£66.60/tonne	28400	27200	2.35	2.45
- smalls	£61.60/tonne	28100	26900	2.19	2.29
Industrial coke	£95.80/tonne	27900	27500	3.43	3.48

VENTILATION AND AIR-CONDITIONING

Typical fresh air supply factors in typical situations

Building type	Litres of fresh air per second per person	Litres of fresh air per second per m2 floor area
General offices	5 - 8	1.3
Board rooms	18 - 25	6.0
Private offices	5 - 12	1.2 - 2.0
Dept. stores	5 - 8	3.0
Factories	20 - 30	0.8
Garages	-	8.0
Bars	12 - 18	-
Dance halls	8 - 12	-
Hotel rooms	8 - 12	1.7
Schools	14	-
Assembly halls	14	-
Drawing offices	16	-

Note: As a global figure for fresh air allow per 1000 m2 1.2 m3/second.

Typical air-changes per hour in typical situations

Building type	Air changes per hour
Residences	1 - 2
Churches	1 - 2
Storage buildings	1 - 2
Libraries	3 - 4
Book stacks	1 - 2
Banks	5 - 6
Offices	4 - 6
Assembly halls	5 - 10
Laboratories	4 - 6
Internal bathrooms	5 - 6
Laboratories - internal	6 - 8
Restaurants/cafes	10 - 15
Canteens	8 - 12
Small kitchens	20 - 40
Large kitchens	10 - 20
Boiler houses	15 - 30

GLAZING

Float and polished plate glass

Nominal thickness (mm)	Tolerance on thickness (mm)	(kg/m2)	Approximate weight (mm)	Normal maximum size
3	+ 0.2		7.5	2140 x 1220
4	+ 0.2		10.0	2760 x 1220
5	+ 0.2		12.5	3180 x 2100
6	+ 0.2		15.0	4600 x 3180
10	+ 0.3		25.0)	
12	+ 0.3		30.0)	6000 x 3300
15	+ 0.5		37.5	3050 x 3000
19	+ 1.0		47.5)	
25	+ 1.0		63.5)	3000 x 2900

Clear sheet glass

Nominal thickness (mm)	Tolerance on thickness (mm)	(kg/m2)	Approximate weight (mm)	Normal maximum size
2 *	+ 0.2		5.0	1920 x 1220
3	+ 0.3		7.5	2130 x 1320
4	+ 0.3		10.0	2760 x 1220
5 *	+ 0.3		12.5)	
6 *	+ 0.3		15.0)	2130 x 2400

Cast glass

Nominal thickness (mm)	Tolerance on thickness (mm)	(kg/m2)	Approximate weight (mm)	Normal maximum size
3	+ 0.4 - 0.2		6.0)	
4	+ 0.5		7.5)	2140 x 1280
5	+ 0.5		9.5	2140 x 1320
6	+ 0.5		11.5)	
10	+ 0.8		21.5)	3700 x 1280

Wired glass

(Cast wired glass)

Nominal thickness (mm)	Tolerance on thickness (mm)	(kg/m2)	Approximate weight (mm)	Normal maximum size
6	+ 0.3 - 0.7		-))	3700 x 1840
7	+ 0.7		-)	

(Polished wire glass)

Nominal thickness (mm)	Tolerance on thickness (mm)	(kg/m2)	Approximate weight (mm)	Normal maximum size
6	+ 1.0		-	330 x 1830

* The 5 mm and 6 mm thickness are known as 'thick drawn sheet'. Although 2 mm sheet glass is available it is not recommended for general glazing purposes.

DRAINAGE

Width required for trenches for various diameters of pipes

Pipe diameter (mm)	Trench n.e. 1.5 m deep	Trench over 1.5 m deep
n.e. 100 mm	450 mm	600 mm
100 - 150 mm	500 mm	650 mm
150 - 225 mm	600 mm	750 mm
225 - 300 mm	650 mm	800 mm
300 - 400 mm	750 mm	900 mm
400 - 450 mm	900 mm	1050 mm
450 - 600 mm	1100 mm	1300 mm

Weights and dimensions of typically sized uPVC pipes

Nominal size	Mean outside diameter (mm) min	max	Wall thickness	Weight kg per metre
Standard pipes				
82.4	82.4	82.7	3.2	1.2
110.0	110.0	110.4	3.2	1.6
160.0	160.0	160.6	4.1	3.0
200.0	200.0	200.6	4.9	4.6
250.0	250.0	250.7	6.1	7.2

Perforated pipes

Heavy grade as above

Thin wall

82.4	82.4	82.7	1.7	-
110.0	110.0	110.4	2.2	-
160.0	160.0	160.6	3.2	-

Vitrified clay pipes

Product	Nominal diameter (mm)	Effective pipe length (mm)	Limits of bore load per metre length min	max	Crushing strength (kN/m)	Weight kg/pipe (/m)
Supersleve	100	1600	96	105	35.00	15.63 (9.77)
Hepsleve	150	1600	146	158	22.00 (normal)	36.50 (22.81)
Hepseal	150	1500	146	158	22.00	37.04 (24.69)
	225	1750	221	235	28.00	95.24 (54.42)
	300	2500	295	313	34.00	196.08 (78.43)
	400	2500	394	414	44.00	357.14 (142.86)
	450	2500	444	464	44.00	500.00 (200.00)
	500	2500	494	514	48.00	555.56 (222.22)
	600	3000	591	615	70.00	847.46 (282.47)
	700	3000	689	719	81.00	1111.11 (370.37)
	800	3000	788	822	86.00	1351.35 (450.35)
	1000	3000	985	1027	120.00	2000.00 (666.67)

DRAINAGE

Vitrified clay pipes - cont'd

Product	Nominal diameter	Effective pipe length	Limits of bore load per metre length		Crushing strength	Weight kg/pipe
	(mm)	(mm)	min	max	(kN/m)	(/m)
Hepline	100	1250	95	107	22.00	15.15 (12.12)
	150	1500	145	160	22.00	32.79 (21.86)
	225	1850	219	239	28.00	74.07 (40.04)
	300	1850	292	317	34.00	105.26 (56.90)
Hepduct (Conduit)	90	1500	-	-	28.00	12.05 (8.03)
	100	1600	-	-	28.00	14.29 (8.93)
	125	1250	-	-	22.00	21.28 (17.02)
	150	1250	-	-	22.00	28.57 (22.86)
	225	1850	-	-	28.00	64.52 (34.88)
	300	1850	-	-	34.00	111.11 (60.06)

USEFUL ADDRESSES FOR FURTHER INFORMATION

ACOUSTICAL INVESTIGATION &
RESEARCH ORGANISATION LTD
Duxon's Turn, Maylands Avenue,
Hemel Hempstead, Herts HP2 4SB
Hemel Hempstead (0442) 47146/7

AGGREGATE CONCRETE BLOCK
ASSOCIATION
60 Charles Street, Leicester LE1 1FB
Leicester (0533) 536161

ALUMINIUM EXTRUDERS ASSOCIATION
Broadway House, Calthorpe Road,
Five Ways, Birmingham B15 1TN
021-455 0311
Telex BIRCOM G 338024 ALFED

ALUMINIUM FEDERATION LTD
Broadway House, Calthorpe Road,
Five Ways, Birmingham B15 1TN
021-455 0311
Telex BIRCOM G 338024 ALFED

ALUMINIUM WINDOW ASSOCIATION
Ivy Bank House, Ivy Bank Park,
Entry Hill, Bath BA2 5NF
Bath (0225) 835811

AMERICAN PLYWOOD ASSOCIATION
101 Wigmore Street, London W1H 9AB
01-629 3437/8
Telex 296009 USAGOF G

ARBORICULTURAL ADVISORY &
INFORMATION SERVICE
Forest Research Station,
Alice Holt Lodge, Farnham
Surrey GU10 4LH
Bentley (0420) 22255

ARBORICULTURAL ASSOCIATION
Ampfield House, Ampfield,
Romsey, Hants SO5 9PA
Braishfield (0794) 68717

ARCHITECTURAL ADVISORY SERVICE
(APPLIED & ANODIC METAL FINISHES)
Unit 5, Royal London Estate,
29 North Acton Road, Willesden,
London NW10 6PD
01-965 4677/0833

ARCHITECTURAL ALUMINIUM ASSOCIATION
193 Forest Road, Tunbridge Wells
Kent TN2 5JA
Tunbridge Wells (0892) 30630

ARCHITECTURAL ASSOCIATION
34/36 Bedford Square, London
WC1B 3ES
01-636 0974

ARMS (ASSOCIATION OF ROOFING
MATERIALS SUPPLIERS)
c/o Allan Harris & Sons Ltd, Station Rd,
St Georges, Weston-Super-Mare, Avon
BS22 0XN
Weston-Super-Mare (0934) 511166
Alloa (0259) 721010
Darwen (0254) 771722
London 01-485 1791

ASBESTOS INFORMATION CENTRE LTD
St Andrews House, 22-28 High
Epsom, Surrey KT18 8AH
Epsom (037 27) 42055
Telex 21120 Ref 2526

ASBESTOS REMOVAL CONTRACTORS
ASSOCIATION
45 Sheen Lane, London SW14 8AB
01-876 4415/6
Telex 927298

BRITISH AGGREGATE CONSTRUCTION
MATERIALS INDUSTRIES LTD
156 Buckingham Palace Road, London
SW1W 9TR
01-730 8194

BRITISH AIR CONDITIONING APPROVALS
BOARD
30 Millbank, London SW1P 4RD
01-834 8827

BRITISH AIRPORTS AUTHORITY
Head Office, Gatwick Airport,
Gatwick, West Sussex RH6 0HZ
Gatwick (0293) 517755
Telex 877995 BAA HQG

BRITISH ANODISING ASSOCIATION
Broadway House, Calthorpe Road,
Five Ways, Birmingham B15 1TN
021-455 0311
Telex BIRCOM-G-338024 ALFED

BRITISH ARCHITECTURAL LIBRARY
RIBA, 66 Portland Place, London
W1N 4AD
01-580 5533

BRITISH ASSOCIATION OF LANDSCAPE
INDUSTRIES
9 Henry Street, Keighley, West Yorkshire
BD21 3DR
Keighley (0535) 606 139

BRITISH BATH MANUFACTURERS
ASSOCIATION
Fleming House, 134 Renfrew Street,
Glasgow G3 6TG
041-332 0826
Telex 779433

BRITISH BLIND AND SHUTTER
ASSOCIATION
5 Greenfield Crescent, Edgbaston,
Birmingham B15 3BE
021-454 2177
Telex 336006

BRITISH BOARD OF AGREEMENT
PO Box 195, Bucknalls Lane, Garston,
Herts, Watford WD2 7NG
Garston (0923) 670844
Telex 946240 Ref 19006505

USEFUL ADDRESSES FOR FURTHER INFORMATION

**BRITISH CARPET MANUFACTURERS
ASSOCIATION**
Fourth Floor, Royalty House, 72 Dean St,
London W1V 5HB
01-734 9853

**BRITISH CERAMIC RESEARCH
ASSOCIATION LTD**
Queens Road, Penkhull, Stoke-on-Trent
ST4 7LQ
Stoke-on-Trent (0782) 45431
Telex 36228 DCRA G

BRITISH CERAMIC TILE COUNCIL
Federation House, Station Rd,
Stoke-on-Trent ST4 2RU
Stoke-on-Trent (0782) 45147

**BRITISH CLAYWARE LAND DRAIN
INDUSTRY**
Federation House, Station Rd,
Stoke-on-Trent ST4 2TJ
Stoke-on-Trent (0782) 416256
Telex 367446

**BRITISH COMBUSTION EQUIPMENT
MANUFACTURERS ASSOCIATION**
The Fernery, Market Place, Midhurst,
West Sussex GU29 9DP
Midhurst (073 081) 2782

**BRITISH CONCRETE MASONRY
ASSOCIATION**
St John's Works, Bedford MK42 0DL
Bedford (0234) 63171/9

**BRITISH CONCRETE PUMPING ASSOCIATION
LTD**
1 Highline Cottage, Lodge Hill, Newtown,
Nr Southampton, Hants
Fareham (0329) 232220

**BRITISH CONSTRUCTIONAL STEELWORK
ASSOCIATION LTD**
35 Old Queen St, London SW1H 9HZ
01-222 2254
Telex 27523

**BRITISH CONTRACT FURNISHING
ASSOCIATION**
PO Box 384, London N12 8HF
01-445 8694
Telex 8951182 GECOMS GBCFA

**BRITISH CUBICLE MANUFACTURERS
ASSOCIATION**
17 Bridge Street, Evesham, Worcs
WR11 4SQ
Evesham (0386) 6560

BRITISH DECORATORS ASSOCIATION
6 Haywra Street, Harrogate, North Yorkshire
HG1 5BL
Harrogate (0423) 67292/3

BRITISH EFFLUENT & WATER ASSOCIATION
51 Castle Street, High Wycombe, Bucks
HP13 6RN
High Wycombe (0494) 444544

**BRITISH ELECTRICAL & ALLIED
MANUFACTURERS ASSOCIATION**
Leicester House, 8 Leicester St, London
WC2H 7BN
01-437 0678

**BRITISH ELECTRICAL SYSTEMS
ASSOCIATION (BESA)**
Granville Chambers, 2 Radford St,
Stone, Staffs ST15 8DA
Stone (0785) 812426

**BRITISH FIRE PROTECTION SYSTEMS
ASSOCIATION LTD**
48a Eden Street, Kingston-upon-Thames,
Surrey KT1 1EE
Kingston-upon-Thames 01-549 5855

BRITISH FIRE SERVICES ASSOCIATION
86 London Road, Leicester LE2 0QR
Leicester (0533) 542879

BRITISH FLAT ROOFING COUNCIL
PO Box 125, Haywards Heath, West Sussex
RH16 3TJ
Haywards Heath (0444) 416681/2

**BRITISH FLOOR COVERING
MANUFACTURERS ASSOCIATION**
125 Queens Road, Brighton BN1 3YW
Brighton (0273) 29271
Telex 87595

**BRITISH FLUE & CHIMNEY
MANUFACTURERS ASSOCIATION**
Nicholson House, High St, Maidenhead,
Berks SL6 1LF
Maidenhead (0628) 34667

**BRITISH FURNITURE MANUFACTURERS
FEDERATED ASSOCIATIONS**
30 Harcourt Street, London W1H 2AA
01-724 0854
Telex 269592 EXFURN G

**BRITISH GLASS INDUSTRY RESEARCH
ASSOCIATION**
Northumberland Road, Sheffield S10 2UA
Sheffield (0742) 686201

**BRITISH GYPSUM ARCHITECTS ADVISORY
SERVICE**
Westfield, 360 Singlewell Rd, Gravesend,
Kent DA11 7RZ
Gravesend (0474) 534251
Telex 96439

**BRITISH INDEPENDENT STEEL
PRODUCERS ASSOCIATION**
5 Cromwell Road, London SW7 2HX
01-581 0231

**BRITISH INDUSTRIAL FASTENERS
FEDERATION**
Queen's House, Queen's Rd, Coventry
CV1 3EG
Coventry (0203) 22325
Telex 311650

USEFUL ADDRESSES FOR FURTHER INFORMARTION

BRITISH INSTITUTE OF INTERIOR DESIGN
1c Devonshire Avenue, Beeston, Notts
NG9 1BS
Nottingham (0602) 221255

**BRITISH KITCHEN FURNITURE
MANUFACTURERS**
c/o Building Employers Confederation,
82 New Cavendish St, London W1M 8AD
01-580 5588
Telex 265763

**BRITISH LAMINATED PLASTICS
FABRICATORS ASSOCIATION**
5 Belgrave Square, London SW1X 8PH
01-235 9483
Telex 895 1528 PLAFED G

**BRITISH LEAD MANUFACTURERS
ASSOCIATION**
68 High St, Weybridge, Surrey KT13 8BL
Weybridge (0932) 56621

BRITISH LIBRARY LENDING DIVISION
Boston Spa, Wetherby, West Yorks LS23 7BQ
Boston Spa (0937) 843434
Telex 557381

**BRITISH LIBRARY, SCIENCE REFERENCE
LIBRARY**
25 Southampton Buildings, Chancery Lane,
London WC2A 1AW
01-405 8721, ex 3344/5
Telex 266959

**BRITISH LOCK MANUFACTURERS
ASSOCIATION**
5 Greenfield Crescent, Edgbaston
Birmingham B15 3BE
021-454 2177
Telex 336006

**BRITISH MALLEABLE TUBE FITTINGS
ASSOCIATION**
105 Meadow View Road, Catford, London
SE6 3NJ
01-698 8856

**BRITISH NON-FERROUS METALS
FEDERATION**
Crest House, 7 Highfield Road, Edgbaston,
Birmingham B15 3ED
021-454 7766
Telex 339161

**BRITISH NON-FERROUS METALS
TECHNICAL CENTRE**
Grove Laboratories, Detchworth Rd,
Wantage, Oxon OX12 9BJ
Wantage (023 57) 2992
Telex 837166

BRITISH PLASTICS FEDERATION
5 Belgrave Square, London SW1X 8PH
01-235 9483
Information 01-235 9888
Telex 895 1528

**BRITISH PRECAST CONCRETE
FEDERATION**
60 Charles St, Leicester LE1 1FB
Leicester (0533) 536161

**BRITISH PUMP MANUFACTURERS
ASSOCIATION**
3 Pannells Court, Chertsey St, Guildford,
Surrey GU1 4EU
Guildford (0483) 37997/8

**BRITISH READY MIXED CONCRETE
ASSOCIATION**
Shepperton House, Green Lane,
Shepperton, Middlesex TW17 8DN
Walton-on-Thames (0932) 243232

BRITISH REFRIGERATION ASSOCIATION
Nicholson House, High St, Maidenhead,
Berks SL6 1LF
Maidenhead (0628) 3466745

**BRITISH REINFORCEMENT
MANUFACTURERS ASSOCIATION (BRMA)**
15 Tooks Court, London EC4 1LA
01-831 7581
Telex 23485 WHAM G

BRITISH ROAD FEDERATION
Cowdray House, 6 Portugal St, London
WC2A 2HG
01-242 1285

**BRITISH RUBBER MANUFACTURERS
ASSOCIATION LTD**
90/91 Tottenham Court Road, London
W1P 0BR
01-580 2794
Telex 8813271 GECOMS G

BRITISH STANDARDS INSTITUTION
2 Park Street, London W1A 2BS
01-629 9000
Enquiries: Milton Keynes (0908) 320066

BRITISH STEEL CORPORATION
Swinden Laboratories, Moorgate,
Rotherham S60 3AR
Rotherham (0709) 60166
Telex 547279

**BRITISH VALVE MANUFACTURERS
ASSOCIATION LTD**
3 Pannells Court, Chertsey Street,
Guildford, Surrey GU1 4EU
Guildford (0483) 37379

**BRITISH WELDED STEEL TUBE
MANUFACTURERS ASSOCIATION**
38 Pamela Road, Northfield, Birmingham
B31 2QG
021-475 3583

**BRITISH WOOD PRESERVING
ASSOCIATION**
Premier House, 150 Southampton Row,
London WC1B 5AL
01-837 8217

USEFUL ADDRESSES FOR FURTHER INFORMATION

BRITISH WOODWORKING FEDERATION
82 New Cavendish St, London W1M 8AD
01-580 5588
Telex 265763

BUILDING CENTRE: BRISTOL
The Building Centre, Stonebridge House,
Colston Avenue, The Centre, Bristol
BS1 4TW
Management Bristol: (0272) 22953
Information Bristol: (0272) 277002

BUILDING CENTRE: COVENTRY
Coventry Building Information Centre,
Dept of Architecture & Planning, Tower
Block, Council Offices, Much Park St,
Coventry CV1 5RT
Coventry (0203) 25555, ext 2512
Telex 31469

BUILDING CENTRE: DURHAM
Northern Counties Building
Information Centre, c/o NFBTE
Green Lane, Durham DH1 3JI
Durham (0385) 62611
Telex 537544

BUILDING CENTRE: LONDON
The Building Centre, 26 Store St,
London WC1E 7BT
Administration: 01-637 1022
Telex 261507 Ref 3324
Bookshop: 01-637 3151
Information Service: (0344) 884999

BUILDING CENTRE: MANCHESTER
The Building Centre, 113-115 Portland
Street, Manchester M1 6FB
061-236 6933/9802

BUILDING CENTRE: PETERBOROUGH
Building Materials Information Service,
22 Broadway, Peterborough PE1 1RU
Peterborough (0733) 314239

BUILDING CENTRE: SCOTLAND
The Building Centre Scotland 1971 Ltd,
Macdata Unit, 47 High Street, Paisley
PA1 2AL
041-840 1199

BUILDING CENTRE: STOKE-ON-TRENT
The Building Information Centre,
Stoke-on-Trent Cauldon College,
The Concourse, Stoke Rd, Shelton,
Stoke-on-Trent ST4 2DG
Stoke-on-Trent (0782) 29561

BUILDING EMPLOYERS CONFEDERATION
82 New Cavendish St, London W1M 8AD
01-580 5588

BUILDING MAINTENANCE INFORMATION
85/87 Clarence Street, Kingston-upon-Thames,
Surrey KT1 1RB
Kingston-upon-Thames 01-546 7555

**BUILDING RESEARCH ESTABLISHMENT:
SCOTLAND**
Kelvin Rd, East Kilbride, Glasgow G75 0RZ
East Kilbride (035 52) 33001

**BUILDING SERVICES RESEARCH
AND INFORMATION ASSOCIATION**
Old Bracknell Lane West, Bracknell,
Berks RG12 4AH
Bracknell (0344) 426511
Telex 848288

**CALCIUM SILICATE BRICK
ASSOCIATION**
24 Fearnley Rd, Welwyn Garden City,
Herts AL8 0HW
Welwyn Garden (070 73) 24538

**CATERING EQUIPMENT MANUFACTURERS
ASSOCIATION (CEMA)**
14 Pall Mall, London SW1Y 5LZ
01-930 0461
Telex 24282

CAVITY FOAM BUREAU
9-11 The Hayes, Cardiff CF1 1NU
Cardiff (0222) 388621
Telex 497629

CEMENT ADMIXTURES ASSOCIATION
2a High St, Hythe, Southampton SO4 6YW
Southampton (0703) 842765

CEMENT AND CONCRETE ASSOCIATION
Wexham Springs, Slough, Berks SL3 6PL
Fulmer (028 16) 2727
Telex 848352

CEMENT MAKERS' FEDERATION
Terminal House, 52 Grosvenor Gardens,
London SW1W 0AH
01-730 2148
Telex 261700 CEMFED

**CHARTERED INSTITUTE OF
ARBITRATORS**
75 Cannon Street, London EC4N 5BH
01-236 8761
Telex 893466 CIARB G

CHARTERED INSTITUTE OF BUILDING
Englemere, Kings Ride, Ascot,
Berks SL5 8BJ
Ascot (0990) 23355

**CHARTERED INSTITUTION OF
BUILDING SERVICES**
Delta House, 222 Balham High Road,
London SW12 9BS
01-675 5211

**CHIPBOARD PROMOTION
ASSOCIATION LTD**
50 Station Road, Marlow, Bucks SL7 1NN
Marlow (062 84) 3022

**CLAY PIPE DEVELOPMENT
ASSOCIATION**
Drayton House, 30 Gordon Street,
London WC1H 0AN
01-388 0025/6

USEFUL ADDRESSES FOR FURTHER INFORMATION

CLAY ROOFING TILE COUNCIL
Federation House, Station Road,
Stoke-on-Trent ST4 2TJ
Stoke-on-Trent (0782) 416256

COLD ROLLED SECTIONS ASSOCIATIONS
Centre City Tower, 7 Hill Street,
Birmingham, B5 4UU
021-643 5494
Telex 339420 ROBSON G

**COMMITTEE OF ASSOCIATIONS OF
SPECIALIST ENGINEERING CONTRACTORS**
ESCA House, 34 Palace Court, Bayswater,
London W2 4JG
01-229 2488
Telex 27929

**CONCRETE BRICK MANUFACTURERS
ASSOCIATION**
c/o British Precast Concrete Federation,
60 Charles St, Leicester LE1 1FP
Leicester (0533) 536161

**CONCRETE PIPE ASSOCIATION
OF GREAT BRITAIN**
60 Charles Street, Leicester LE1 1FB
Leicester (0533) 536161

CONCRETE SOCIETY
Devon House, 12-15 Dartmouth
Street, London SW1H 9BL
01-222 1822

CONFEDERATION OF BRITISH INDUSTRY
Centre Point, 103 New Oxford Street,
London WC1A 1DU
01-379 7400
Telex 21352

**CONSTRADO (CONSTRUCTIONAL
STEEL RESEARCH AND DEVELOPMENT
ORGANISATION)**
NLA Tower, 12 Addiscombe Road,
Croydon CR9 3JH
Croydon 01-688 2688, 01-686 0366

CONTRACT FLOORING ASSOCIATION
23 Chippenham Mews, London W9 2AN
01-286 4499

**CONTRACTORS MECHANICAL PLANT
ENGINEERS**
20 Knave Wood Rd, Kemsing, Sevenoaks,
Kent TN15 6RH
Otford (09592) 2628

**COPPER CYLINDER AND BOILER
MANUFACTURERS ASSOCIATION**
56 Oxford St, Manchester M1 6EU
061-236 0384

COPPER DEVELOPMENT ASSOCIATION
Orchard House, Mutton Lane,
Potters Bar, Herts EN6 3AP
Potters Bar (0707) 50711
Telex 27711

**COPPER TUBE FITTINGS
MANUFACTURERS ASSOCIATION**
7 Highfield Road, Birmingham B15 3ED
021-454 7766
Telex 339161

**COUNCIL OF BRITISH CERAMIC
SANITARY WARE MANUFACTURERS**
Federation House, Station Rd,
Stoke-on-Trent ST4 2RT
Stoke-on-Trent (0782) 48675

CP/M USERS GROUP (UK)
72 Mill Road, Hawley, Dartford,
Kent DA2 7RZ
Dartford (0322) 22669

**DECORATIVE LIGHTING
ASSOCIATION LTD**
Bishops Castle, Shropshire SY9 5LE
Clun (058 84) 658

**DOOR & SHUTTER MANUFACTURERS
ASSOCIATION**
5 Greenfield Crescent, Edgbaston,
Birmingham B15 3BE
021-454 2177
Telex 336006

**DRAUGHT PROOFING ADVISORY
ASSOCIATION**
PO Box 12, Haslemere, Surrey GU27 3AN
Haslemere (0428) 54011

**DRY LINING AND PARTITION
ASSOCIATION**
82 New Cavendish Street, London W1M 8AD
01-580 5588
Telex 265763

DRY STONE WALLING ASSOCIATION
YFC Office, National Agricultural
Centre, Kenilworth, Warwickshire CV8 2LG
Coventry (0203) 56131

**DUCTILE IRON PRODUCERS
ASSOCIATION**
8th Floor, Bridge House, 121 Smallbrook,
Queensway, Birmingham B5 4JP
021-643 3377

**ELECTRIC CABLE MAKERS
CONFEDERATION**
56 Palace Road, East Molesey,
Surrey KT8 9DW
East Molesey 01-941 4079
Telex 24893

**ELECTRICAL CONTRACTORS
ASSOCIATION (ECA)**
Esca House, 34 Palace Court, Bayswater,
London W2 4HY
01-229 1266
Telex 27929

USEFUL ADDRESSES FOR FURTHER INFORMATION

ELECTRICAL CONTRACTORS
ASSOCIATION OF SCOTLAND
23 Heriot Row, Edinburgh EH3 6EW
031-225 7221

ELECTRICAL INSTALLATION EQUIPMENT
MANUFACTURERS ASSOCIATION
Leicester House, 8 Leicester St,
London WC2H 7BN
01-437 0678
Telex 263536

ENERGY EFFICIENCY OFFICE
Room 1312, Department of Energy,
Thames House South, Milbank,
London SW1P 4QJ
01-211 7156

ENERGY SYSTEMS TRADE ASSOCIATION
LTD (ESTA)
PO Box 16, Stroud, Glos GL5 5EB
Amberley (045 387) 3568

EUROBUILD
26 Rue la Perouse, 75116 Paris
(010 33) 7201020
Telex 611975 F

EXTERNAL WALL INSULATION
ASSOCIATION
PO Box 12, Haslemere, Surrey GU27 3AN
Haslemere (0428) 54011
Telex 858819 PRADMNG

FABRIC CARE RESEARCH ASSOCIATION
Forest House Laboratories,
Knaresborough Road, Harrogate,
North Yorks HG2 7LZ
Harrogate (0423) 882301

FARM BUILDINGS ASSOCIATION
National Agricultural Centre, Stoneleigh,
Kenilworth, Warwickshire CV8 2LG
Coventry (0203) 22345

FEDERATION OF BUILDING & CIVIL
ENGINEERING CONTRACTORS (NI) LTD
143 Malone Rd, Belfast BT9 6SU
Belfast (0232) 661711

FEDERATION OF CIVIL ENGINEERING
CONTRACTORS
Cowdray House, 6 Portugal Street,
London WC2A 2HH
01-404 4020

FEDERATION OF RESIN FORMULATORS
AND APPLICATORS LTD (FERFA)
16 Courtmoor Avenue, Fleet,
Aldershot, Hampshire
GU13 9UF
Fleet (025 144) 6936

FEDERATION OF MANUFACTURERS OF
CONSTRUCTION EQUIPMENT & CRANES
7-15 Lansdowne Rd, Croydon, Surrey CR9 2P1
Croydon 01-688 4422
Telex 917857 BINDER G

FEDERATION OF MASTER BUILDERS
33 John Street, London WC1N 2BB
01-242 7583

FEDERATION OF PILING SPECIALIST
Dickens House, 15 Tooks Court, London
EC4A 1LA
01-831 7581
Telex 23485

FEDERATION OF WIRE ROPE
MANUFACTURERS OF GREAT BRITIAN
PO Box 121, The Fountain Precinct,
1 Balm Green, Sheffield S1 3AF
Sheffield (0742) 751234
Telex 54170

FLAT ROOFING CONTRACTORS ADVISORY
BOARD
Maxwelton House, 41/43 Boltro Road,
Haywards Heath, West Sussex RH16 1BJ
Haywards Heath (0444) 451835/6

FENCING CONTRACTORS ASSOCIATION
St Johns House, 23 St Johns Rd,
Watford WD1 1PY
Watford (0923) 27236

FIBRE BONDED CARPET MANUFACTURERS
ASSOCIATION
3 Manchester Rd, Bury, Lancs BL9 0DR
Bury 061-764 1114

FIBRE CEMENT MANUFACTURERS
ASSOCIATION LTD (FCMA)
PO Box 92, Elmswell, Bury St, Edmunds,
Suffolk IP30 9HS
Elmswell (0359) 40963

FINNISH PLYWOOD INTERNATIONAL
PO Box 99, Welwyn Garden City,
Herts AL6 0HS
Bulls Green (043 879) 746

FLAT GLASS MANUFACTURERS
ASSOCIATION
Prescot Road, St Helens, Merseyside
WA10 3TT
St Helens (0744) 28882

FLEXIBLE ROOFING ASSOCIATION
125 Queen's Road, Brighton,
East Sussex BN1 3YW
Brighton (0273) 33322

GLASS AND GLAZING FEDERATION
6 Mount Row, London W1Y 6DY
01-409 0545

GLASS MANUFACTURERS FEDERATION
19 Portland Place, London W1N 4BH
01-580 6952
Telex 27470

GLASSFIBRE REINFORCEMENT
CEMENT ASSOCIATION
5 Upper Bar, Newport, Shropshire TF10 7EH
Newport (0952) 811397

USEFUL ADDRESSES FOR FURTHER INFORMATION

GLAZED AND FLOOR TILE HOME TRADE ASSOCIATION
Federation House, Station Road,
Stoke-on-Trent ST4 2RU
Stoke-on-Trent (0782) 45147

GYPSUM PRODUCTS DEVELOPMENT ASSOCIATION
360 Singlewell Rd, Gravesend, Kent DA11 7RZ
Gravesend (0474) 332314
Telex 96439

HEAT PUMP AND AIR CONDITIONING BUREAU
30 Millbank, London SW1P 4RD
01-834 8827/8

HEAT PUMP MANUFACTURERS ASSOCIATION
Nicholson House, High St, Maidenhead,
Berks SL6 1LF
Maidenhead (0628) 34667

HEATING AND VENTILATING CONTRACTORS ASSOCIATION
ESCA House, 34 Palace Court, Bayswater,
London W2 4JG
01-229 2488
Telex 27929 ESCA G

HEATING, VENTILATING AND AIR CONDITIONING MANUFACTURERS ASSOCIATION
Nicholson House, High St, Maidenhead,
Berks SL6 1LF
Maidenhead (0628) 34667

HOUSING CORPORATION
149 Tottenham Court Road, London W1P 0BN
01-387 9466

INDUSTRIAL BUILDING BUREAU
33 Upper St, London N1 0PN
01-359 9877

INSTITUTE OF ACOUSTICS
25 Chambers St, Edinburgh EH1 1HU
031-225 2143

INSTITUTE OF ASPHALT TECHNOLOGY
Unit 18, Central Trading Estate, Staines,
Middlesex TW18 4XE
Staines (0784) 65387

INSTITUTE OF CONCRETE TECHNOLOGY
PO Box 52, Slough SL3 6PL
Postal enquiries only

INSTITUTE OF DOMESTIC HEATING AND ENVIRONMENTAL ENGINEERS
37a High Road, Benfleet, Essex SS7 3BR
Benfleet (037 45) 54266

INSTITUTE OF PLUMBING
64 Station Lane, Hornchurch, Essex RM12 6NB
Hornchurch (040 24) 72791

INSTITUTE OF SHEET METAL ENGINEERING
Queensway House, 2 Queensway, Redhill,
Surrey RH1 1QS
Redhill (0737) 68611

INSTITUTE OF WASTES MANAGEMENT
3 Albion Place, Northampton NN1 1UD
Northampton (0604) 20426

INSTITUTE OF WATER POLLUTION CONTROL
Ledstone House, 53 London Road,
Maidstone, Kent ME16 8JH
Maidstone (0622) 62034

INSTITUTION OF BRITISH ENGINEERS
Regency House, 3 Marlborough Place,
Brighton, East Sussex BN1 1UB
Brighton (0273) 601399

INSTITUTION OF CIVIL ENGINEERS
1-7 Great George St, London SW1P 3AA
01-222 7722

INSTITUTION OF ELECTRICAL AND ELECTRONICS INCORPORATED ENGINEERS
Savoy Hill House, Savoy Hill,
London WC2R 0BS
01-836 3357

INSTITUTION OF ELECTRICAL ENGINEERS
Savoy Place, London WC2R 0BL
01-240 1871
Telex 261176 IEELDN G

INSTITUTION OF STRUCTURAL ENGINEERS
11 Upper Belgrave Street, London
SW1X 8BH
01-235 4535

JOINT CONTRACTS TRIBUNAL
66 Portland Place, London W1N 4AD
01-580 5533, 01-580 5588

LEAD DEVELOPMENT ASSOCIATION
34 Berkeley Square, London W1X 6AJ
01-499 8422
Telex 281286

NATIONAL HOUSE-BUILDING COUNCIL
58 Portland Place, London W1N 4BU
01-637 1248/9

NATIONAL PAVING AND KERB ASSOCIATION
60 Charles St, Leicester LE1 1FB
Leicester (0533) 536161

USEFUL ADDRESSES FOR FURTHER INFORMATION

PARTITIONING INDUSTRY ASSOCIATION
1 Landsale Avenue, Solihull,
West Midlands B92 0PP
021-705 9270

PATENT GLAZING CONFERENCE
13 Upper High Street, Epsom,
Surrey KT17 4QY
Epsom (037 27) 2919

PIPELINE INDUSTRIES GUILD
17 Grosvenor Crescent, London SW1X 7ES
01-235 7938

**PITCH FIBRE PIPE ASSOCIATION
OF GREAT BRITAIN**
c/o Croda Hydrocarbons Ltd, Pipes
Division, PO Box 16, Weeland Rd,
Knottingley, West Yorkshire WF11 8DZ
Knottingley (0977) 87161
Telex 8814171

PLASTERERS CRAFT GUILD
56 Burton Rd, Kingston-upon-Thames,
Surrey KT2 5TF
Kingston-upon-Thames 01-546 1470

**PLASTIC PIPE MANUFACTURERS
SOCIETY**
89 Cornwall Street, Birmingham B3 3BY
021-236 1866

PLASTICS AND RUBBER INSTITUTE
11 Hobart Place, London SW1W 0HL
01-245 9555
Telex 912881 PRI

**PLASTICS BATH MANUFACTURERS
ASSOCIATION**
12th Floor, Fleming House, 134 Renfrew St,
Glasgow G3 6TG
041-332 0826

**PLASTICS TANKS AND CISTERNS
MANUFACTURERS ASSOCIATION**
8 Belmain Close, Grange Road, Ealing,
London W5 5BY
01-579 6081

POST-TENSIONING ASSOCIATION
CCL Systems Ltd, Cabca House, 296 Elwell
Road, Surbiton, Surrey, KT6 7AH
Surbiton 01-390 1122

PRECAST CONCRETE FRAME ASSOCIATION
60 Charles St, Leicester LE1 1FB
Leicester (0533) 536161

**PREFABRICATED BUILDING MANUFACTURERS
ASSOCIATION OF GREAT BRITAIN**
Westgate House, Chalk Lane, Epsom,
Surrey KT18 7AJ
Epsom (037 27) 40044

PRESTRESSED CONCRETE ASSOCIATION
60 Charles St, Leicester LE1 1FB
Leicester (0533) 536161

**PROPERTY SERVICES AGENCY
DEPARTMENT OF THE ENVIRONMENT**
Whitgift Centre, Wellesley Road,
Croydon CR9 3LY
Croydon 01-686 8710
Architectural Services: 01-686 8710
Civil Accommodation/Estate Surveying
Services: 01-928 7999
Defence Services 1 (Army, Overseas):
01-397 5266
Defence Services 2 (Air/Navy):
01-686 8710
Diplomatic/Post Offices Services:
01-686 5622
Engineering/QS Services: 01 686 3499

**ROYAL INSTITUTE OF BRITISH
ARCHITECTS**
66 Portland Place, London W1N 4AD
01-580 5533

**ROYAL INSTITUTION OF CHARTERED
SURVEYORS**
12 Great George Street, London SW1P 3AD
01-222 7000
Telex 915443 RICS G

SAND AND GRAVEL ASSOCIATION
3rd Floor, 32-36 Fleet Lane,
London EC4M 4YA
01-236 0224

**SCOTTISH BUILDING EMPLOYERS
FEDERATION**
13 Woodside Crescent, Glasgow G3 7UP
041-332 7144
Telex 779657
Aberdeen (0224) 643838
Edinburgh 031-226 4907
Inverness (0463) 237626

SCOTTISH DEVELOPMENT AGENCY
120 Bothwell Street, Glasgow G2 7JP
041-248 2700
Telex 777600

**SCOTTISH PRECAST CONCRETE
MANUFACTURERS ASSOCIATION**
9 Princes Street, Falkirk FK1 1LS
Falkirk (0324) 22088

**SCOTTISH SPECIAL HOUSING
ASSOCIATION**
15/21 Palmerston Place, Edinburgh
EH12 5AJ
031-225 1281

**SCOTTISH WIREWORK MANUFACTURERS
ASSOCIATION**
36 Renfield Street, Glasgow, G2 1BD
041-248 6161

USEFUL ADDRESSES FOR FURTHER INFORMATION

SOCIETY OF CHAIN LINK FENCING
MANUFACTURERS
PO Box 121, The Fountain Precinct,
1 Balm Green, Sheffield S1 3AF
Sheffield (0742) 751234
Telex 54170

SOCIETY OF GLASS TECHNOLOGY
20 Hallam Gate Road, Broom Hill,
Sheffield S10 5BT
Sheffield (0742) 663168

SPONS A & B EDITORS
Davis Langdon Computer Systems
11 Portland Square
Bristol BS2 8ST
Team: David Wood, Sally Higgins
Margaret Jones, Marlene McAllister
Bristol (0272) 232283

STAINLESS STEEL ADVISORY CENTRE
Shepcote Lane, PO Box 161,
Sheffield S9 1TR
Sheffield (0742) 440060, 441224
Telex 547025

STAINLESS STEEL FABRICATORS
ASSOCIATION OF GREAT BRITAIN
14 Knoll Road, Dorking, Surrey RH4 3EW
Dorking (0306) 884079

STEEL CASTINGS RESEARCH & TRADE
ASSOCIATION
East Bank Road, Sheffield S2 3PT
Sheffield (0742) 28647

STEEL LINTEL MANUFACTURERS
ASSOCIATION
c/o PO Box 10, British Steel Corporation,
Newport, Gwent NP9 0XN
Newport (0633) 272281

STEEL WINDOW ASSOCIATION
26 Store Street, London WC1E 7JR
01-637 3571/2

STONE FEDERATION
82 New Cavendish Street, London W1M 8AD
01-580 5588
Telex 265763

STRUCTURAL INSULATION ASSOCIATION
45 Sheen Lane, London SW14 8AB
01-876 4415/6
Telex 927298 ALLEN G

SUSPENDED ACCESS EQUIPMENT
MANUFACTURERS ASSOCIATION
82 New Cavendish Street, London W1M 8AD
01-580 5588
Telex 265763

SUSPENDED CEILINGS ASSOCIATION
29 High Street, Hemel Hempstead,
Herts HP1 3AA
Hemel Hempstead (0442) 40313

SWEDISH FINNISH TIMBER COUNCIL
21 Carolgate, Retford, Notts DN22 6BZ
Retford (0777) 706616/7

SWIMMING POOL AND ALLIED TRADES
ASSOCIATION (SPATA)
Faraday House, 17 Essendene Road,
Caterham, Surrey CR3 5PB
Caterham (0883) 40110

TAR INDUSTRIES SERVICES
Mill Lane, Wingerworth, Chesterfield,
Derbyshire S42 6NG
Chesterfield (0246) 76823
Telex 547061

THERMAL INSULATION CONTRACTORS
ASSOCIATION
Kensway House, 388 High Rd, Ilford,
Essex IG1 1TL
Ilford 01-514 2120

THERMAL INSULATION MANUFACTURERS
AND SUPPLIERS ASSOCIATION
45 Sheen Lane, London, SW14 8AB
01-876 4415
Telex 927298 ALLEN

TIMBER TRADE FEDERATION OF THE
UNITED KINGDOM
Clareville House, 26-27 Oxenden Street,
London SW1Y 4EL
01-839 1891
Telex 8954628

TOWN AND COUNTRY PLANNING
ASSOCIATION
17 Carlton House Terrace, London SW1 5AS
01-930 8903

TRADA
Stocking Lane, Hughenden Valley, High
Wycombe, Bucks HP14 4ND
Naphill (024 024) 3091
Telex 83292

UNITED KINGDOM WOOD WOOL
ASSOCIATION
Gordon House, Oakleigh Road South,
New Southgate, London N11 1HL
01-368 1266
Telex 21252

VERMICULITE INFORMATION SERVICES
COUNCIL LTD
Mark House, The Square, Lightwater,
Surrey GU18 5SS
Bagshot (0276) 71617

VITREOUS ENAMEL DEVELOPMENT
New House, High Street, Ticehurst,
Wadhurst, East Sussex TN5 7AL
Ticehurst (0580) 200152

USEFUL ADDRESSES FOR FURTHER INFORMATION

**WALLCOVERING MANUFACTURERS
ASSOCIATION**
Alembic House, 93 Albert Embankment,
London SE1 7TY
01-582 1185

WELDING INSTITUTE
Abingdon Hall, Abingdon,
Cambridge CB1 6AL
Cambridge (0223) 891162
3AE
Fax (0223) 892588
Telex 81183

**WESTERN WOOD PRODUCTS ASSOCIATION
(USA)**
6AJ
69 Wigmore St, London W1H 9LG
01-486 7488/9

**WIRE GOODS MANUFACTURERS
ASSOCIATION**
Kensington House, 136 Suffolk St,
Queensway, Birmingham B1 1LL
021-643 4488
Telex 338876 GRBHAMG

**WOOD WOOL SLAB MANUFACTURERS
ASSOCIATION**
10 Great George St, London SW1P

01-222 5315

ZINC DEVELOPMENT ASSOCIATION
34 Berkeley Square, London W1X

01-499 6636
Telex 261286